Foundations of Elastoplasticity: Subloading Surface Model

Koichi Hashiguchi

Foundations of Elastoplasticity: Subloading Surface Model

Springer

Koichi Hashiguchi
MSC Software Ltd.
Tokyo
Japan

ISBN 978-3-319-84021-5 ISBN 978-3-319-48821-9 (eBook)
DOI 10.1007/978-3-319-48821-9

Softcover reprint of the hardcover 3rd edition 2017

Printed on acid-free paper

This Springer imprint is published by Springer Nature
The registered company is Springer International Publishing AG
The registered company address is: Gewerbestrasse 11, 6330 Cham, Switzerland

The original version of the book was revised:
For detailed information please see Erratum.
The erratum to the book is available
at 10.1007/978-3-319-48821-9_22

Preface

The elastoplasticity has highly developed responding to the rapid advance of industries during the last half century. In the conventional plasticity, the interior of the yield surface was assumed to be an elastic domain so that the plastic strain rate due to the rate of stress inside the yield surface is not predicted, resulting in the following limitations and the difficulties: (1) The tangent modulus changes discontinuously at the yield point; (2) the yield judgment whether the stress reaches the yield surface is required with the determination of the offset (permanent, i.e., plastic strain) value at yielding which is accompanied with an arbitrariness (although 0.2 % plastic strain is often used); and (3) the operation to pullback the stress to the yield surface is required when it goes out from the yield surface by large loading increments in numerical calculation. Then, the fierce fight has started aiming at formulating the rigorous plasticity model capable of predicting cyclic loading behavior of solids and structures under the machine vibrations, the earthquakes, etc., at the middle of the last century. The fight would have come to the end by the creation of the *subloading surface model* by which the above-mentioned limitations and the difficulties are resolved, describing the cyclic loading behavior rigorously with the high efficiency in numerical calculation.

All the elastoplastic models other than the subloading surface model are the *cyclic kinematic hardening models*, e.g. the *multi surface model*, the *two (so-called bounding) surface model* and the *superposed-kinematic hardening (so-called Chaboche) model* in which a small yield surface enclosing the elastic domain translates as the plastic strain rate develops. The models inheriting the yield surface enclosing a purely-elastic domain are required to incorporate smaller and smaller yield surfaces one after another endlessly depending on the stress level and the stress amplitude, like an infinity mirror or a nest of boxes, so that they fall into the endless pit without any substantial resolution of the problem.

The basic features of the subloading surface model are itemized as follows:

(1) It is based on the quite natural concept that the plastic deformation is developed as the stress approaches the yield surface and thus it possesses the high generality and the capability of describing accurately irreversible deformations.

(2) It fulfills the smoothness condition, describing always continuous variation of tangent stiffness modulus and thus depicting the smooth elastic-plastic transition.
(3) It possesses the automatic controlling functions to attract the stress to the normal-yield surface, so that the stress is pulled-back to the yield surface when it goes over the surface in numerical calculations. In addition, the normal-isotropic hardening surface varies so as to involve always the kinematic hardening variable, i.e. the back stress in the stagnation phenomenon of isotropic hardening observed after the re-yielding in the stress reversal event.
(4) It is capable of describing the finite deformation and rotation under an infinitesimal elastic deformation.

By virtue of these rigorous physical backgrounds, the subloading surface model is endowed with the following rigorous descriptions distinguishable from the other elastoplastic constitutive models.

1. Plastic strain rate is predicted even for any low stress level. Then, cyclic loading behavior is predicted accurately even for infinitesimal loading amplitude. *The other models, e.g. the multi, the two and the superposed kinematic hardening models are incapable of describing the cyclic loading behavior appropriately.*
2. Plastic deformation of various solids unlimited to metals, e.g. soils, etc. can be described pertinently. *The other cyclic plasticity models are limited to the description of metal deformation.*
3. Inelastic strain rate induced by the stress rate tangential to the yield surface, i.e. tangential-inelastic strain rate is described appropriately, fulfilling the continuity condition, which is required to describe the non-proportional loading behavior. *All the other models assume the purely-elastic domain so that they predict the tangential-inelastic strain rate induced suddenly at the moment when the stress reaches the yield surface if the tangential-inelastic strain rate is incorporated, violating the continuity condition at the moment.*
4. Viscoplastic constitutive deformation is described pertinently in a general rate from the quasi-static to the impact loading. *The other models are incapable of describing the viscoplastic deformation behavior at high rate, predicting an elastic response for an impact loading.*
5. Damage phenomenon under cyclic loading is described appropriately, which leads to a softening behavior in general. *The other cyclic plasticity models are incapable of describing the cyclic damage phenomenon with a softening pertinently.*
6. Constitutive equation of fatigue would be described pertinently, which is required to describe plastic strain rate induced in low stress level and small stress amplitude. *The other models are incapable of describing the fatigue phenomenon, predicting only an elastic strain rate in low stress level and small stress amplitude.*
7. Constitutive equation of phase-transformation of metals can be described pertinently. *The other models are incapable of describing the phase-transformation phenomenon pertinently.*

8. Friction phenomenon is described pertinently, including the transition from static to kinetic friction by the sliding, the recovery of static friction with elapse of time and the both of positive and negative rate-sensitivities. The negative and the positive rate sensitivities are relevant to the dry and the lubricated (fluid) friction, respectively. *The other model is incapable of describing these friction phenomena.*
9. Crystal plasticity analysis can be executed pertinently, in which calculation of slips in numerous number of slip systems are required. The yield judgment is unnecessary since the smooth elastic-plastic transition is described and the resolved shear stress is automatically pulled-back to the critical shear stress. *The other models are inapplicable to the crystal plasticity analysis rigorously because the yield judgment and the operation to pull back the resolved shear stresses to the critical shear stress are required in numerous number of slip systems. Then, impertinent analysis using the creep model has been performed widely after Pierce et al. (1982, 1983). It is impertinent to adopt the rate-dependent model for the rate-independent deformation phenomenon. In addition, it should be noted that the creep model is impertinent such that it predicts a creep shear strain rate even in unloading process of the resolved shear stress from the critical shear stress.*
10. The multiplicative hyperelastic-based plasticity can be formulated exactly based on the subloading surface model, which is capable of describing the cyclic loading behavior exactly for finite elastoplastic deformation/rotation. The *multiplicative hyperelastic-based plasticity cannot be formulated by the other cyclic plasticity models.*

Thus, the subloading surface model is endowed inherently with the high generality and flexibility for the description of irreversible mechanical phenomena of solids.

Eventually, it can stated that

> "*Subloading surface concept is to be the unified constitutive law which is inevitable to describe irreversible mechanical phenomena in wide classes of solids, ranging from monotonic to cyclic loadings, from quasi-static to impact loadings, from infinitesimal to finite deformations and from micro to macro phenomena*".

Then, the elastoplasticity theory will be developed steadily (breaking through the stagnation) by incorporating the exact formulation of the subloading surface model, although it has stagnated during a half century since the study on the unconventional (cyclic) elastoplasticity aiming at pertinent description of the plastic strain rate caused by the rate of stress inside the yield surface has started in the 1960s.

The subloading surface and the subloading-friction models have been implemented to the commercial FEM software Marc marketed by MSC Software Ltd. as the standard uploaded (ready-made) programs, so that Marc users can apply these models to their deformation analyses. The implementation was highly supported by Dr. Motohatu Tateishi, MSC Software Ltd., Japan. Then, the author decided to publish this book in order to provide the comprehensive explanation of the subloading surface model for readers ranging from beginners to specialists of the elastoplasticity so that they can apply it easily to their deformation analyses. In addition, the computer programs of the subloading surface and the subloading-friction models are released in Appendix J so as to use them capturing clearly the formulations and the calculation processes.

The main contents in each chapter will be delineated below in order.

As a foundation for the formulation of elastoplasticity theory, the mathematical and the physical ingredients of the continuum mechanics are provided in Chaps. 1–4. Chapter 1 addresses the vector-tensor analysis since physical quantities used in continuum mechanics are tensors; consequently, their relations are described mathematically by tensor equations. Explanations for mathematical properties and rules of tensors are presented to the extent that is sufficient to understand the subject of this book: elastoplasticity theory. Chapter 2 addresses the description of motion and strain (rate) and their related quantities. Chapter 3 presents conservation laws of mass, momentum, and angular momentum, and equilibrium equations and virtual work principles derived from them. In addition, their rate forms used for constitutive equations of inelastic deformation are explained concisely.

Chapter 4 specifically addresses the objectivity of constitutive equations, which is required for the description of deformation behavior under material rotation. The substantial physical meaning of the objective rate of tensor is explained incorporating the convected base. Then, the objectivities of various stress, strain, and their rates are described by examining their coordinate transformation rules. Then, the pullback and the push-forward operations are systematically explained, defining the Eulerian and the Lagrangian vectors and tensors. Further, all the objective and the corotational time derivatives of tensors are derived systematically from the convected (embedded) time derivative. The mathematical proof is given to the fact that the material time derivative of scalar-valued tensor function can be transformed to the corotational time derivative of that.

Chapter 5 specifically examines the description of elastic deformation. Elastic constitutive equations are classified into the hyperelasticity, the Cauchy elasticity, and the hypoelasticity depending on exactness in the description of reversibility. The mathematical and physical characteristics of these equations are explained prior to the description of elastoplastic constitutive equations in the subsequent chapters.

Elastoplastic constitutive equations are described comprehensively in Chaps. 6–9. In Chap. 6, the physical and mathematical backgrounds are first given to the additive decomposition of strain rare (symmetric part of the velocity gradient) into the elastic and the plastic parts and that of the continuum spin (antisymmetric part of the velocity gradient) into the elastic and the plastic parts from the standpoint of the multiplicative decomposition of deformation gradient which provides the exact decomposition of deformation gradient tensor into the elastic and the plastic parts by introducing the intermediate configuration as the hyper-elastically unloaded state to the stress-free state. In addition, the physical backgrounds are given to facts that the elastic spin designates the sum of the rigid-body rotational rate and the rotational rate due to the elastic distortion and that the plastic spin designates the rotational rate of the intermediate configuration. Thereafter, the basic formulations of elastoplastic constitutive equation, e.g., the elastic and the plastic strain rates, the consistency condition, the plastic flow rule, and the loading criterion, are provided based on the physical interpretations. Descriptions of anisotropy and the tangential inelastic strain rate are also incorporated. However, they fall within the framework of conventional plasticity on the premise that the interior of the yield surface is an elastic domain. Therefore, they are incapable of predicting a smooth transition from the elastic to plastic state and a cyclic loading behavior of real materials pertinently.

In Chap. 7, the continuity and the smoothness conditions are described first. They are the fundamental requirements for the constitutive equations of irreversible deformation, especially to describe cyclic loading behavior accurately. The subloading surface model is described in detail, which falls within the framework of the unconventional plasticity excluding the assumption that the interior of yield surface is an elastic domain. It satisfies both the continuity and the smoothness conditions. In Chap. 8, cyclic plasticity models are classified into the models based on the translation of the small yield surface enclosing a purely elastic domain, i.e., the cyclic kinematic hardening models, and the model based on the expansion/contraction of loading surface, i.e., the subloading surface model. Further, their mathematical structures and mechanical features are explained in detail. It is revealed that the cyclic kinematic hardening models, e.g., the multi-, the two, and the superposed nonlinear-kinematic hardening models, are the temporizing models, which do not possess a generality/pertinence and contain various serious deficiencies. First of all, the purpose for the creation of unconventional plasticity model is the description of the plastic strain rate by the rate of stress inside the yield surface which cannot be described by the conventional plasticity model assuming the yield surface enclosing a purely elastic domain. However, the defect of the conventional plasticity model cannot be solved endlessly by the cyclic kinematic hardening models incorporating the small yield surface enclosing a purely elastic domain. In addition, the mechanism for the development of plastic strain rate is substantially different from the mechanism for the development of anisotropy such as the kinematic hardening. It can be concluded that only the extended subloading surface model falling within the framework of the latter category possesses the generality and the mathematical structure capable of describing the cyclic loading behavior in wide classes of elastoplastic materials including metals and soils. In addition, the friction phenomenon of solids can be described rigorously by the friction model based on the concept of the subloading surface as will be described in Chap. 18.

In Chap. 9, the formulation of the extended subloading surface model is described in detail, in which the elastic core, i.e., the similarity center of the normal yield and the subloading surfaces, translates with a plastic deformation. Therein, the inelastic strain rate attributable to the stress rate tangential to the subloading surface is incorporated, which is indispensable for the accurate prediction of non-proportional loading behavior and the plastic instability phenomena. In addition, the cyclic stagnation of the isotropic hardening in metals is incorporated. In Chaps. 10 and 11, constitutive equations based on the extended subloading surface model are shown for metals and soils. Their validities are verified by the comparisons with various test data containing the cyclic loading.

In Chap. 12, the exact finite strain theory based on the multiplicative decomposition of deformation gradient is formulated, in which the extended Hashiguchi (subloading surface) model is incorporated, although only the initial subloading surface model was incorporated in the book of Hashiguchi and Yamakawa (2012). The author aims at formulating the constitutive equation possessing the generality and the universality to be inherited eternally, while any unconventional model, i.e., cyclic plasticity model other than the subloading surface model, has not been extended to the multiplicative finite strain theory up to the present.

In Chap. 13, the history of the development of the viscoplastic constitutive equation for describing rate-dependent deformation induced for the stress level over the yield surface is reviewed first. Then, the pertinent viscoplastic constitutive equation is described, in which the concept of the subloading surface is incorporated into the overstress model. It is applicable to the prediction of rate-dependent deformation behavior in the general rate ranging from quasi-static to impact loads, while the deformation behavior under the impact load cannot be described by the past overstress models. On the other hand, it is revealed that the creep model contains impertinence for the description of a quasi-static deformation behavior, although it has been studied widely. Further, the subloading-overstress model is extended to the multiplicative exact deformation theory.

In Chap. 14, the constitutive equation with the damage phenomenon based on the subloading surface model is described. The softening behavior is often observed, for which the description of a smooth transition from the elastic to the plastic state is required and thus, the subloading surface concept should be introduced inevitably. In Chap. 15, the plasticity for the phase-transformation phenomenon based on the Hashiguchi (subloading surface) model is described briefly.

Special issues related to elastoplastic deformation behavior are discussed in Chaps. 16 and 17. Chapter 16 specifically examines corotational rate tensors, the necessity of which is suggested in Chap. 4. Mechanical features of corotational tensors with various spins are examined comparing their simple shear deformation characteristics. The pertinence of the plastic spin is particularly explained. Chapter 17 opens with a mechanical interpretation for the localization of deformation inducing a shear band. Then, the approaches to the prediction of shear band inception condition, the inclination/thickness of shear band, and the eigenvalue analysis and the gradient theory are explained. The smeared model, i.e., the shear band-embedded model for the practical finite element analysis, is also described.

Chapter 18 addresses the prediction of friction phenomena between solid bodies. All bodies except those floating in a vacuum contact with other bodies so that the friction phenomena occur between their contact surfaces. Pertinent analyses, not only of the deformation behavior of bodies, but also of friction behavior on the contact surface, are necessary for the analyses of boundary-value problems. A constitutive equation of friction is formulated in the similar form to the elastoplastic constitutive equation by incorporating the concept of the subloading surface, which is called the subloading-friction model. It is capable of describing the transition from a static to a kinetic friction attributable to plastic softening and the recovery of the static friction attributable to creep hardening. The anisotropy based on the orthotropy and the rotation of sliding-yield surface is incorporated. The stick-slip phenomenon is analyzed by incorporating the subloading-friction model. Their validities are shown by comparison with various test data.

In Chap. 19, the crystal plasticity model based on the Hashiguchi (subloading surface) model is described, which would be physically pertinent and possess the high efficiency in numerical analyses. In contrast, note that the creep model without a yield surface is widely incorporated into the crystal plasticity analyses in order to avoid the numerical difficulty, although it is physically quite irrelevant.

The FEM analysis based on elastoplastic constitutive equations described in the former chapters requires pertinent numerical method. In Chap. 20, the return-mapping and the consistent (algorithmic) tangent modulus tensor are explained, which provides the calculation in a high accuracy and efficiency.

Finally, in Chap. 21, the formulations of elastoplastic constitutive equation from the second law of thermodynamics are commented and then it is suggested that the formulations are irrational and merely the prerequisite logic and thus, we have to formulate constitutive equation without falling into the thermodynamic formalism.

The distinguishable features and importance of this book are the comprehensive descriptions of fundamental concepts and formulations including the objectivity, the objective derivative of tensor function, the associated flow rule, the loading criterion, the continuity and smoothness conditions, and their physical interpretations in addition to the circumstantial explanations on wide classes of reversible/irreversible constitutive equations for monotonic, cyclic and non-proportional loading behavior, rate-dependent deformation behavior, and friction behavior of solids.

Most of the theories described in this book fall within the framework of the hypoelastic-based plasticity for the finite deformation under the infinitesimal elastic deformation, and the finite strain theory based on the multiplicative decomposition of deformation gradient is explained concisely. This book is the elaborated compilation of the former books: "*Elastoplasticity Theory*" by Hashiguchi (Springer, 2013) for the hypoelastic-based plasticity and "*Introduction to Finite Strain Theory for Continuum Elasto-Plasticity*" by Hashiguchi and Yamakawa (John-Wiley, 2012) for the multiplicative elastoplasticity (exact hyperelastic-based plasticity). It is recommendable for the readers to read also the latter book for the numerical calculation and the finite element method.

The author wishes to express cordial thanks to his colleagues at Kyushu University, who have discussed and collaborated with him over a long period of time: Prof. M. Ueno (currently Professor at Univ. Ryukyus) in particular, and Dr. T. Okayasu (currently Associate Prof. Kyushu Univ.), Dr. S. Tsutsumi (currently Associate Prof., Osaka Univ.), Dr. T. Ozaki of Kyushu Electric Eng. Consult. Inc., Dr. S. Ozaki (currently Associate Prof. Yokohama National Univ.) and Mr. T. Mase of Tokyo Electric Power Services Co., Ltd. In addition, Emeritus Prof. T. Tanaka of Univ. of Tokyo, Emeritus Prof. C. Yatomi of Kanazawa Univ., Emeritus Prof. F. Yoshida of Hiroshima Univ., Prof. M. Kuroda of Yamagata Univ. and Dr. I. Watanabe of Natl. Inst. Material Sci., Japan are appreciated for their valuable discussions and advices.

Further, the author would like to express his sincere gratitude to Prof. A. Asaoka and his colleagues at Nagoya University: Prof. M. Nakano and Prof. T. Noda who have appreciated and used the author's model widely in their analyses and who have offered discussion continually on deformation of geomaterials. In addition, the author thanks Prof. T. Nakai and Prof. T.F. Zhang of the Nagoya Institute of Technology, for their valuable comments.

Furthermore, the author is thankful to Dr. K. Okamura, Dr. N. Suzuki and Dr. R. Higuchi, Nippon Steel & Sumitomo Metal Corporation, Dr. M. Oka and Mr. T. Anjiki, Yanmar Co. Ltd., for the collaborations on constitutive relations of metals and the numerical calculations. He is also grateful to Mr. T. Kato (President) and Dr. M. Tateishi, MSC Software, Ltd., Japan, for the standard implementation of the Hashiguchi (subloading surface) model to the commercial FEM software Marc.

The author is deeply indebted to Prof. late Bogdan Raniecki and Prof. H. Petryk of the Inst. Fund. Tech. Research, Poland, who have visited Kyushu University several times to deliver lectures on applied mechanics. Further, the author thanks Prof. I.F. Collins of the University of Auckland, Prof. O.T. Bruhns of Ruhr Univ., Bochum, and Prof. E.C. Aifantis of Michigan Tech. Univ. and the late Prof. I. Vardoulakis of Natl. Univ. Tech. Athens, who have also stayed at Kyushu Univ., delivering lectures and engaging in valuable discussions related to continuum mechanics.

The heartfelt gratitude of the author is dedicated to Prof. Yuki Yamakawa of Tohoku University, for a lot of valuable advices and a close collaboration with detailed discussions on elastoplasticity theory, particularly on the finite strain theory based on the multiplicative decomposition of the deformation gradient.

In particular, the author expresses his sincere gratitude to Prof. Genki Yagawa, Emeritus Professor, University of Tokyo, for encouraging always the author with undeserved high appreciation of research contributions, and thus, the author was stimulated to the publication of this book.

Finally, the author would like to acknowledge the enthusiastic supports by Dr. Thomas Ditzinger and Ms. Servai Sukanya, Springer Project Coordinator, for the publication of this book.

Fukuoka-shi, Japan
June 2016

Koichi Hashiguchi
Technical Adviser, MSC Software Ltd.
(Emeritus Prof. Kyushu University)

References

Monographs on Solid Mechanics and Tensor Analysis

Hashiguchi K (2013) Elastoplasticity theory. Lecture note in applied computational mechanics, 2nd edn. Springer, Heidelberg

Hashiguchi K, Yamakawa Y (2012) Introduction to finite strain theory for continuum elasto-plasticity. Wiley Series in Computational Mechanics, Wiley, Chichester

Contents

Chapter 1
Vector and Tensor Analysis

Physical quantities appearing in continuum mechanics are mathematically expressed by tensors because they possess not only magnitudes but also directions in multi-dimensional space. Therefore, their relations are described by tensor equations. Before the explanation on the main theme of this book, i.e. elastoplasticity theory, the mathematical properties of tensors and mathematical rules on tensor operations are explained on the level necessary to understand elastoplasticity theory. The orthogonal Cartesian coordinate system is adopted in this book except for chapters 12 and some parts in chapters 4 and 21. A further advanced mathematics of tensors in the embedded curvilinear coordinate system is referred to Hashiguchi and Yamakawa (2012).

1.1 Conventions and Symbols

Some basic conventions and symbols appearing in the tensor analysis are described in this section.

1.1.1 Summation Convention

We first introduce the *Cartesian summation convention*. Repeated suffix in a term is summed over numbers that the suffix can take. For instance,

$$\left.\begin{array}{l} u_r v_r = \sum_{r=1}^{3} u_r v_r, \quad T_{ir} v_r = \sum_{r=1}^{3} T_{ir} v_r \\ T_{rr} = \sum_{r=1}^{3} T_{rr} \end{array}\right\} \tag{1.1}$$

K. Hashiguchi, *Foundations of Elastoplasticity: Subloading Surface Model*,
DOI 10.1007/978-3-319-48821-9_1

where the range of suffixes is $1, 2, 3$. Because of $u_r v_r = u_s v_s$, $T_{ir} v_r = T_{is} v_s$, $T_{rr} = T_{ss}$ a letter of the repeated suffix is arbitrary. It is therefore called as the *dummy index.*

The convention described above is also called *Einstein's summation convention.* Hereinafter, a repeated index obeys this convention unless otherwise specified by the additional remark "(no sum)".

1.1.2 Kronecker's Delta and Permutation Symbol

The symbol $\delta_{ij}(i,j = 1,2,3)$ defined in the following equation is called the *Kronecker's delta.*

$$\delta_{ij} = \begin{cases} 1: & i = j \\ 0: & i \neq j \end{cases} \tag{1.2}$$

for which one has

$$\delta_{ir}\delta_{rj} = \delta_{ij}, \quad \delta_{ii} = 3 \tag{1.3}$$

Furthermore, the symbol ε_{ijk} defined by the following equation is called the *alternating* (or *permutation*) *symbol* or *Eddington's epsilon* or *Levi-Citiva "e" tensor.*

$$\varepsilon_{ijk} = \begin{cases} 1 & \text{for} \quad \text{even permutation of } ijk \text{ from } 123 \\ -1 & \text{for} \quad \text{odd permutation of } ijk \text{ from } 123 \\ 0 & \text{for} \quad \text{others} \end{cases} \tag{1.4}$$

The number of same permutations that the suffixes i, j, k in ε_{ijk} take different values from each other is $3!$ and, needless to say, the square of $\varepsilon_{ijk}(=1 \text{ or } -1)$ is $+1$. Therefore, it holds that

$$\varepsilon_{ijk}\varepsilon_{ijk} = 3! \tag{1.5}$$

1.1.3 Matrix and Determinant

When the quantity $\mathbf{T}$ possessing 3×3 components T_{ij} is expressed in the arrangement

$$\mathbf{T} = [T_{ij}] = \begin{bmatrix} T_{11} & T_{12} & T_{13} \\ T_{21} & T_{22} & T_{23} \\ T_{31} & T_{32} & T_{33} \end{bmatrix} \tag{1.6}$$

the expression of $\mathbf{T}$ in this form is called a *matrix.* For the two matrices $\mathbf{T}$ and $\mathbf{S}$, their product $\mathbf{TS}$ is defined by the matrix having the following components.

$$(\mathbf{TS})_{ij} = T_{ir}S_{rj} \tag{1.7}$$

Further, the quantity defined by the following equation is called the *determinant* of $\mathbf{T}$ and is shown by the symbol det $\mathbf{T}$, i.e.

$$\boxed{\det \mathbf{T} = \varepsilon_{ijk}T_{1i}T_{2j}T_{3k} = \varepsilon_{ijk}T_{i1}T_{j2}T_{k3}} = \begin{vmatrix} T_{11} & T_{12} & T_{13} \\ T_{21} & T_{22} & T_{23} \\ T_{31} & T_{32} & T_{33} \end{vmatrix} \tag{1.8}$$

with

$$\det \mathbf{T}^T = \det \mathbf{T},\ \det(s\mathbf{T}) = s^3 \det(\mathbf{T}) \tag{1.9}$$

where $(\ )^T$ designates the *transpose*, i.e. the mutual replacement of columns and rows. Here, the number of permutations that the suffixes i, j, k in ε_{ijk} can take is 3!. Therefore, Eq. (1.8) can be written as

$$\boxed{\det \mathbf{T} = \frac{1}{3!}\varepsilon_{ijk}\varepsilon_{pqr}T_{ip}T_{jq}T_{kr}} \tag{1.10}$$

Equation (1.10) is rewritten as

$$\boxed{\det \mathbf{T} = \frac{1}{3}T_{rs}\Delta_{rs}, \quad \det \mathbf{T} = \frac{1}{3}\mathbf{T} : \boldsymbol{\Delta}^T = \frac{1}{3}\mathbf{T}^T : \boldsymbol{\Delta}} \tag{1.11}$$

or

$$\det \mathbf{T} = T_{1s}\Delta_{1s} = T_{2s}\Delta_{2s} = T_{3s}\Delta_{3s} = T_{r1}\Delta_{r1} = T_{r1}\Delta_{r1} = T_{r2}\Delta_{r2} = T_{r3}\Delta_{r3} \tag{1.12}$$

where

$$\Delta_{ip} \equiv \frac{1}{2!}\varepsilon_{ijk}\varepsilon_{pqr}T_{jp}T_{kr} \tag{1.13}$$

which is called the *cofactor* for the i-column and the j-row, noting

$$\frac{1}{3!}\varepsilon_{ijk}\varepsilon_{pqr}T_{ip}T_{jq}T_{kr} = \frac{1}{3}T_{ip}\left(\frac{1}{2!}\varepsilon_{ijk}\varepsilon_{pqr}T_{jq}T_{kr}\right)$$

The following lemmas hold for the properties of the determinant.

Lemma 1 *If the 1st and the 2nd rows are same, i.e.,* $T_{2j} = T_{1j}$ *for example, one has* $\varepsilon_{ijk}T_{1i}T_{1j}T_{3k} = \varepsilon_{jik}T_{1j}T_{1i}T_{3k} = -\varepsilon_{ijk}T_{1i}T_{1j}T_{3k}$. *Therefore, one has the lemma "the determinant having same lines or rows is zero".* Therefore, it follows from Eq. (1.12) that

$$T_{is}\Delta_{js} = T_{ri}\Delta_{rj} = \delta_{ij}\det\mathbf{T} \tag{1.14}$$

Lemma 2 *If the 1st and the 2nd lines are exchanged, i.e., 1↔2 for example, one has* $\varepsilon_{ijk}T_{2i}T_{1j}T_{3k} = \varepsilon_{jik}T_{1i}T_{2j}T_{3k} = -\varepsilon_{ijk}T_{1i}T_{2j}T_{3k}$. *Therefore, one has the lemma "the determinant changes only its sign by exchanging lines or rows".*

Multiplying ε_{ijk} to both sides in Eq. (1.8), one has

$$\varepsilon_{ijk}\det\mathbf{T} = \varepsilon_{ijk}\varepsilon_{pqr}T_{1p}T_{2q}T_{3r} = \varepsilon_{pqr}T_{ip}T_{jq}T_{kr} \tag{1.15}$$

The transformation from the second side to the third side in Eq. (1.14) results from the above-mentioned Lemmas 1 and 2. Here, note that the expression of the determinant in Eq. (1.10) is obtained also by multiplying ε_{ijk} to both sides in Eq. (1.14) and noting Eq. (1.5).

The additive decomposition of the components T_{2j} into $T_{2j} = A_{2j} + B_{2j}$ leads to

$$\varepsilon_{ijk}T_{1i}(A_{2j} + B_{2j})T_{2k} = \varepsilon_{ijk}T_{1i}A_{2j}T_{2k} + \varepsilon_{ijk}T_{1i}B_{2j}T_{2k} \tag{1.16}$$

Therefore, the value of determinant in which components in a line (or row) are decomposed additively is the sum of the two determinants made by exchanging the line (or row) of the original determinants into the decomposed components.

Consider the multiplicative decomposition of tensor, i.e., $\mathbf{T} = \mathbf{AB}$ ($T_{ij} = A_{ir}B_{rj}$). It follows from Eqs. (1.8) and (1.14) that

$$\begin{aligned}\varepsilon_{ijk}(A_{1p}B_{pi})(A_{2q}B_{qj})(A_{3r}B_{rk}) &= A_{1p}A_{2q}A_{3r}\varepsilon_{ijk}B_{pi}B_{qj}B_{rk}\\ &= A_{1p}A_{2q}A_{3r}\varepsilon_{pqr}\det\mathbf{B}\end{aligned}$$

noting $\sum_{p,q,r=1}^{3}\varepsilon_{ijk}B_{pi}B_{qj}B_{rk} = \sum_{p,q,r=1}^{3}\varepsilon_{pqr}\det\mathbf{B}$ due to Eq. (1.15), and thus one has the following *product law* of determinant.

$$\boxed{\det(\mathbf{AB}) = \det\mathbf{A}\det\mathbf{B}} \tag{1.17}$$

The partial derivative of determinant is given from Eq. (1.10) as

$$\begin{aligned}\frac{\partial\det\mathbf{T}}{\partial T_{ij}} &= \frac{\partial\frac{1}{3!}\varepsilon_{abc}\varepsilon_{pqr}T_{ap}T_{bq}T_{cr}}{\partial T_{ij}}\\ &= \frac{1}{3!}\varepsilon_{abc}\varepsilon_{pqr}(\delta_{ia}\delta_{jp}T_{bq}T_{cr} + T_{ap}\delta_{ib}\delta_{jq}T_{cr} + T_{ap}T_{bq}\delta_{ic}\delta_{jr})\\ &= \frac{1}{3!}(\varepsilon_{ibc}\varepsilon_{jqr}T_{bq}T_{cr} + \varepsilon_{aic}\varepsilon_{pjr}T_{ap}T_{cr} + \varepsilon_{abi}\varepsilon_{pqj}T_{ap}T_{bq})\\ &= \frac{1}{3!}(\varepsilon_{ibc}\varepsilon_{jqr}T_{bq}T_{cr} + \varepsilon_{bic}\varepsilon_{qjr}T_{bq}T_{cr} + \varepsilon_{cbi}\varepsilon_{rqj}T_{cr}T_{bq})\\ &= \frac{1}{2!}\varepsilon_{ibc}\varepsilon_{jqr}T_{bq}T_{cr} = \Delta_{ij}\end{aligned}$$

which leads to

$$\boxed{\frac{\partial \det \mathbf{T}}{\partial \mathbf{T}} = \mathbf{\Delta}}, \quad \frac{\partial \det \mathbf{T}}{\partial T_{ij}} = \Delta_{ij} \tag{1.18}$$

The permutation symbol in the third order, i.e. ε_{ijk} appears often hereinafter. It is related to Kronecker's delta by the determinants as

$$\varepsilon_{ijk} = \begin{vmatrix} \delta_{1i} & \delta_{1j} & \delta_{1k} \\ \delta_{2i} & \delta_{2j} & \delta_{2k} \\ \delta_{3i} & \delta_{3j} & \delta_{3k} \end{vmatrix} = \begin{vmatrix} \delta_{1i} & \delta_{2i} & \delta_{3i} \\ \delta_{1j} & \delta_{2j} & \delta_{3j} \\ \delta_{1k} & \delta_{2k} & \delta_{3k} \end{vmatrix} \tag{1.19}$$

which is proved as follows: Note that the second side in Eq. (1.19) is expanded as

$$\begin{aligned} \varepsilon_{ijk} = \begin{vmatrix} \delta_{1i} & \delta_{1j} & \delta_{1k} \\ \delta_{2i} & \delta_{2j} & \delta_{2k} \\ \delta_{3i} & \delta_{3j} & \delta_{3k} \end{vmatrix} &= \delta_{1i}\delta_{2j}\delta_{3k} + \delta_{1k}\delta_{2i}\delta_{3j} + \delta_{1j}\delta_{2k}\delta_{3i} \\ &\quad - \delta_{1k}\delta_{2j}\delta_{3i} - \delta_{1i}\delta_{2k}\delta_{3j} - \delta_{1j}\delta_{2i}\delta_{3k} \end{aligned}$$

which yields

$$\begin{aligned} \varepsilon_{123} &= \delta_{11}\delta_{22}\delta_{33} + \delta_{13}\delta_{21}\delta_{32} + \delta_{12}\delta_{23}\delta_{31} - \delta_{13}\delta_{22}\delta_{31} - \delta_{11}\delta_{23}\delta_{32} - \delta_{12}\delta_{21}\delta_{33} = 1 \\ \varepsilon_{213} &= \delta_{12}\delta_{21}\delta_{33} + \delta_{13}\delta_{22}\delta_{31} + \delta_{11}\delta_{23}\delta_{32} - \delta_{13}\delta_{21}\delta_{32} - \delta_{12}\delta_{23}\delta_{31} - \delta_{11}\delta_{22}\delta_{33} = -1 \end{aligned}$$

for instance. The third side in Eq. (1.19) could be confirmed as well.

The following relation holds from Eqs. (1.19) and (1.17).

$$\begin{aligned} \varepsilon_{ijk}\varepsilon_{pqr} &= \begin{vmatrix} \delta_{1i} & \delta_{2i} & \delta_{3i} \\ \delta_{1j} & \delta_{2j} & \delta_{3j} \\ \delta_{1k} & \delta_{2k} & \delta_{3k} \end{vmatrix} \begin{vmatrix} \delta_{1p} & \delta_{1q} & \delta_{1r} \\ \delta_{2p} & \delta_{2q} & \delta_{2r} \\ \delta_{3p} & \delta_{3q} & \delta_{3r} \end{vmatrix} = \left| \begin{bmatrix} \delta_{1i} & \delta_{2i} & \delta_{3i} \\ \delta_{1j} & \delta_{2j} & \delta_{3j} \\ \delta_{1k} & \delta_{2k} & \delta_{3k} \end{bmatrix} \begin{bmatrix} \delta_{1p} & \delta_{1q} & \delta_{1r} \\ \delta_{2p} & \delta_{2q} & \delta_{2r} \\ \delta_{3p} & \delta_{3q} & \delta_{3r} \end{bmatrix} \right| \\ &= \begin{vmatrix} \delta_{si}\delta_{sp} & \delta_{si}\delta_{sq} & \delta_{si}\delta_{sr} \\ \delta_{sj}\delta_{sp} & \delta_{sj}\delta_{sq} & \delta_{sj}\delta_{sr} \\ \delta_{sk}\delta_{sp} & \delta_{sk}\delta_{sq} & \delta_{sk}\delta_{sr} \end{vmatrix} = \begin{vmatrix} \delta_{ip} & \delta_{iq} & \delta_{ir} \\ \delta_{jp} & \delta_{jq} & \delta_{jr} \\ \delta_{kp} & \delta_{kq} & \delta_{kr} \end{vmatrix} \end{aligned} \tag{1.20}$$

from which further one has

$$\begin{aligned} \varepsilon_{ijk}\varepsilon_{pqk} &= \begin{vmatrix} \delta_{ip} & \delta_{iq} & \delta_{ik} \\ \delta_{jp} & \delta_{jq} & \delta_{jk} \\ \delta_{kp} & \delta_{kq} & \delta_{kk} \end{vmatrix} \\ &= \delta_{ip}\delta_{jq}\delta_{kk} + \delta_{iq}\delta_{jk}\delta_{kp} + \delta_{ik}\delta_{jp}\delta_{kq} - \delta_{ik}\delta_{jq}\delta_{kp} - \delta_{ip}\delta_{jk}\delta_{kq} - \delta_{iq}\delta_{jp}\delta_{kk} \\ &= 3\delta_{ip}\delta_{jq} + \delta_{iq}\delta_{jp} + \delta_{iq}\delta_{jp} - \delta_{ip}\delta_{jq} - \delta_{ip}\delta_{jq} - 3\delta_{iq}\delta_{jp} \\ &= \delta_{ip}\delta_{jq} - \delta_{iq}\delta_{jp} \end{aligned}$$

$$\begin{aligned}
\varepsilon_{ijp}\varepsilon_{ijq} &= \begin{vmatrix} \delta_{ii} & \delta_{ij} & \delta_{iq} \\ \delta_{ji} & \delta_{jj} & \delta_{jq} \\ \delta_{pi} & \delta_{pj} & \delta_{pq} \end{vmatrix} \\
&= \delta_{ii}\delta_{jj}\delta_{pq} + \delta_{ij}\delta_{jq}\delta_{pi} + \delta_{iq}\delta_{ji}\delta_{pj} - \delta_{ii}\delta_{jq}\delta_{pj} - \delta_{ij}\delta_{ji}\delta_{pq} - \delta_{iq}\delta_{jj}\delta_{pi} \\
&= 9\delta_{pq} + \delta_{iq}\delta_{pi} + \delta_{iq}\delta_{ip} - 3\delta_{pq} - 3\delta_{pq} - 3\delta_{pq} = 2\delta_{pq} \\
\varepsilon_{ijk}\varepsilon_{ijk} &= 2\delta_{kk} = 6
\end{aligned}$$

Consequently, the following relation holds.

$$\boxed{\begin{aligned} &\varepsilon_{ijk}\varepsilon_{pqk} = \varepsilon_{kij}\varepsilon_{kpq} = \delta_{ip}\delta_{jq} - \delta_{iq}\delta_{jp} \\ &\varepsilon_{ijp}\varepsilon_{ijq} = 2\delta_{pq}, \quad \varepsilon_{ijk}\varepsilon_{ijk} = 6 \end{aligned}} \tag{1.21}$$

The last equation can also be obtained directly from Eq. (1.5).

1.2 Vector

1.2.1 Definition of Vector

The quantity having only magnitude is defined as a *scalar*. On the other hand, a quantity having direction and sense in addition to magnitude and fulfilling the following three properties is defined as a vector. A vector is expressed using lowercase letters in boldface to distinguish it from a scalar.

Equivalence: The vectors having same magnitude, direction and sense are equivalent. Here, equivalence of two vectors $\mathbf{u}$ and $\mathbf{v}$ is expressed by $\mathbf{u} = \mathbf{v}$.

Addition: The addition of vectors is given by the *parallelogram law*.

Multiplication with scalar: The multiplication of vector and scalar induces a vector whose *magnitude* is given by the multiplication of the magnitude of the original vector by the scalar, *direction* is identical to that of the original vector, and *sense* is same and opposite to that of the original vector if the scalar is positive and negative, respectively.

By virtue of the properties presented above, the *commutative*, *distributive*, and the *associative laws* hold as follows:

$$\left.\begin{aligned} &\mathbf{u} + \mathbf{v} = \mathbf{v} + \mathbf{u}, \quad (\mathbf{u} + \mathbf{v}) + \mathbf{w} = \mathbf{u} + (\mathbf{v} + \mathbf{w}) \\ &a(b\mathbf{v}) = (ab)\mathbf{v} = b(a\mathbf{v}), (a + b)\mathbf{v} = (b + a)\mathbf{v}, \ a(\mathbf{u} + \mathbf{v}) = a\mathbf{u} + a\mathbf{v} \end{aligned}\right\} \tag{1.22}$$

where a, b are arbitrary scalars.

The magnitude of vector is denoted by $||\mathbf{v}||$. In particular, the vector whose magnitude is zero is called the *zero vector* and is shown as $\mathbf{0}$. The vector whose magnitude is unity, i.e. $||\mathbf{v}|| = 1$ is called the *unit vector*.

1.2.2 Operations of Vectors

1. Scalar product

Denoting the angle between the two vectors $\mathbf{u}, \mathbf{v}$ by θ when they are translated to the common initial point, the *scalar* (or *inner*) *product* is defined as $\|\mathbf{u}\| \|\mathbf{v}\| \cos\theta$ and it is denoted by the symbol $\mathbf{u} \cdot \mathbf{v}$, i.e.

$$\mathbf{u} \cdot \mathbf{v} \equiv \|\mathbf{u}\| \|\mathbf{v}\| \cos\theta \, (= u_i v_i) \tag{1.23}$$

The magnitude of vector is expressed by setting $\theta = 0$ in Eq. (1.23) as follows:

$$\|\mathbf{v}\| = \sqrt{\mathbf{v} \cdot \mathbf{v}} \, (= \sqrt{v_i v_i}) \tag{1.24}$$

The quantity obtained by the scalar product is a scalar and the following commutative, distributive and associative laws hold.

$$\mathbf{u} \cdot \mathbf{v} = \mathbf{v} \cdot \mathbf{u}, \quad \mathbf{u} \cdot (\mathbf{v} + \mathbf{w}) = \mathbf{u} \cdot \mathbf{v} + \mathbf{u} \cdot \mathbf{w}, \quad a(\mathbf{u} \cdot \mathbf{v}) = (a\mathbf{u}) \cdot \mathbf{v} \tag{1.25}$$

2. Vector product

The operation obtaining a vector having (1) magnitude identical to the area of the parallelogram formed by the two vectors $\mathbf{u}$ and $\mathbf{v}$, provided that they are translated to the common initial point, and (2) direction of the unit vector $\mathbf{n}$ which forms the right-hand bases $\mathbf{u}, \mathbf{v}, \mathbf{n}$ in this order is defined as the *vector* (or *cross*) *product* and is noted by the symbol $\mathbf{u} \times \mathbf{v}$. Therefore, denoting the angle between the two vectors $\mathbf{u}$ and $\mathbf{v}$ by θ when they are translated to the common initial point, it holds that

$$\mathbf{u} \times \mathbf{v} \equiv \|\mathbf{u}\| \|\mathbf{v}\| \sin\theta \, \mathbf{n} \, (= \varepsilon_{ijk} u_i v_j \mathbf{e}_k) \, (\|\mathbf{n}\| = 1) \tag{1.26}$$

The vector product is not commutative, i.e.

$$\mathbf{u} \times \mathbf{v} = -\mathbf{v} \times \mathbf{u} \tag{1.27}$$

On the other hand, the distributive and the associative laws hold as follows:

$$\mathbf{u} \times (\mathbf{v} + \mathbf{w}) = \mathbf{u} \times \mathbf{v} + \mathbf{u} \times \mathbf{w}, \quad (a\mathbf{u}) \times (b\mathbf{v}) = ab(\mathbf{u} \times \mathbf{v}) \tag{1.28}$$

3. Scalar and vector triple products

The operation defined by the following equation for the vector and the scalar products of three vectors is called *scalar triple product.*

$$[\mathbf{uvw}] \equiv (\mathbf{u} \times \mathbf{v}) \cdot \mathbf{w} \; (= \varepsilon_{ijk} u_i v_j w_k) \tag{1.29}$$

fulfilling

$$[\mathbf{uvw}] = [\mathbf{vwu}] = [\mathbf{wuv}] = -[\mathbf{vuw}] = -[\mathbf{wvu}] = -[\mathbf{uwv}] \tag{1.30}$$

The commutative law for the scalar triple product will be shown in the subsequent section.

Denoting the vector $\mathbf{u}, \mathbf{v}, \mathbf{w}$ as $\mathbf{v}_1, \mathbf{v}_2, \mathbf{v}_3$, it follows from Eq. (1.29) that

$$[\mathbf{v}_i \; \mathbf{v}_j \; \mathbf{v}_k] = \varepsilon_{ijk} [\mathbf{v}_1 \; \mathbf{v}_2 \; \mathbf{v}_3] \tag{1.31}$$

noting the fact that the term in the right hand side of this equation is $+[\mathbf{v}_1, \mathbf{v}_2, \mathbf{v}_3]$, $-[\mathbf{v}_1, \mathbf{v}_2, \mathbf{v}_3]$ and 0 when indices i,j,k are even and odd permutations and two of indices coincide with each other, respectively.

Here, the following equations hold for the scalar triple product.

$$[\mathbf{e}_i \quad \mathbf{e}_j \quad \mathbf{e}_k] = \varepsilon_{ijk} \tag{1.32}$$

$$[s\mathbf{u},\mathbf{v},\mathbf{w}] = [\mathbf{u},s\mathbf{v},\mathbf{w}] = [\mathbf{u},\mathbf{v},s\mathbf{w}] = s[\mathbf{uvw}] \tag{1.33}$$

$$[a\mathbf{u} + b\mathbf{v},\mathbf{w},\mathbf{x}] = a[\mathbf{uwx}] + b[\mathbf{vwx}] \tag{1.34}$$

We have the following relations for the *vector triple product*.

$$\left.\begin{aligned} \mathbf{u} \times (\mathbf{v} \times \mathbf{w}) &= (\mathbf{u} \cdot \mathbf{w})\mathbf{v} - (\mathbf{u} \cdot \mathbf{v})\mathbf{w} \\ (\mathbf{u} \times \mathbf{v}) \times \mathbf{w} &= (\mathbf{u} \cdot \mathbf{w})\mathbf{v} - (\mathbf{v} \cdot \mathbf{w})\mathbf{u} \end{aligned}\right\} \tag{1.35}$$

noting

$$\varepsilon_{kij}\varepsilon_{kpq} u_j v_p w_q = (\delta_{ip}\delta_{jq} - \delta_{iq}\delta_{jp}) u_j v_p w_q = v_i (u_j w_j) - w_i (u_j v_j)$$

exploiting Eqs. (1.21) and (1.56) which will be given in Sect. 1.2.3. Permuting the letters in Eq. (1.35), the following identity is derived.

$$\mathbf{u} \times (\mathbf{v} \times \mathbf{w}) + \mathbf{v} \times (\mathbf{w} \times \mathbf{u}) + \mathbf{w} \times (\mathbf{u} \times \mathbf{v}) = 0 \tag{1.36}$$

noting

$$\begin{aligned} &\mathbf{u} \times (\mathbf{v} \times \mathbf{w}) + \mathbf{v} \times (\mathbf{w} \times \mathbf{u}) + \mathbf{w} \times (\mathbf{u} \times \mathbf{v}) \\ &\quad = [(\mathbf{u} \cdot \mathbf{w})\mathbf{v} - (\mathbf{u} \cdot \mathbf{v})\mathbf{w}] + [(\mathbf{u} \cdot \mathbf{v})\mathbf{w} - (\mathbf{u} \cdot \mathbf{w})\mathbf{v}] + [(\mathbf{w} \cdot \mathbf{v})\mathbf{u} - (\mathbf{w} \cdot \mathbf{u})\mathbf{v}] = 0 \end{aligned}$$

Setting $\mathbf{t} = \mathbf{u} \times \mathbf{v}$ in $\mathbf{t} \times (\mathbf{w} \times \mathbf{x}) = (\mathbf{t} \cdot \mathbf{x})\mathbf{w} - (\mathbf{t} \cdot \mathbf{w})\mathbf{x}$ due to Eq. $(1.35)_1$ and $\mathbf{t} = \mathbf{w} \times \mathbf{x}$ in $(\mathbf{u} \times \mathbf{v}) \times \mathbf{t} = (\mathbf{u} \cdot \mathbf{t})\mathbf{w} - (\mathbf{v} \cdot \mathbf{t})\mathbf{u}$ due to $(1.35)_2$, it follows that

$$(\mathbf{u} \times \mathbf{v}) \times (\mathbf{w} \times \mathbf{x}) = [(\mathbf{u} \times \mathbf{v}) \cdot \mathbf{x}]\mathbf{w} - [(\mathbf{u} \times \mathbf{v}) \cdot \mathbf{w}]\mathbf{x}$$

$$(\mathbf{u} \times \mathbf{v}) \times (\mathbf{w} \times \mathbf{x}) = [\mathbf{u} \cdot (\mathbf{w} \times \mathbf{x})]\mathbf{v} - [\mathbf{v} \cdot (\mathbf{w} \times \mathbf{x})]\mathbf{u}$$

leading to

$$\begin{aligned}(\mathbf{u} \times \mathbf{v}) \times (\mathbf{w} \times \mathbf{x}) &= [\mathbf{uvx}]\mathbf{w} - [\mathbf{uvw}]\mathbf{x} \\ &= [\mathbf{uwx}]\mathbf{v} - [\mathbf{vwx}]\mathbf{u}\end{aligned} \tag{1.37}$$

It follows from Eq. (1.37) that

$$(\mathbf{u} \times \mathbf{v}) \times (\mathbf{v} \times \mathbf{w}) = [\mathbf{uvw}]\mathbf{v} - [\mathbf{uvv}]\mathbf{w} = [\mathbf{uvw}]\mathbf{v} \tag{1.38}$$

which leads to

$$(\mathbf{u} \times \mathbf{v}) \times (\mathbf{v} \times \mathbf{w}) \cdot (\mathbf{w} \times \mathbf{u}) = [\mathbf{uvw}]\mathbf{v} \cdot (\mathbf{w} \times \mathbf{u})$$

so that

$$[\mathbf{u} \times \mathbf{v} \quad \mathbf{v} \times \mathbf{w} \quad \mathbf{w} \times \mathbf{u}] = [\mathbf{uvw}]^2 \tag{1.39}$$

Because of

$$\begin{aligned}(\mathbf{u} \times \mathbf{v}) \cdot \{(\mathbf{w} \times \mathbf{x}) \times (\mathbf{y} \times \mathbf{z})\} &= (\mathbf{u} \times \mathbf{v}) \cdot \{[\mathbf{wxz}]\mathbf{y} - [\mathbf{wxy}]\mathbf{z}\} \\ &= [\mathbf{uvy}][\mathbf{wxz}] - [\mathbf{uvz}][\mathbf{wxy}]\end{aligned}$$

due to Eq. (1.37), it follows that

$$\begin{aligned}[\mathbf{u} \times \mathbf{v}, \mathbf{w} \times \mathbf{x}, \mathbf{y} \times \mathbf{z}] &= [\mathbf{uvy}][\mathbf{zwx}] - [\mathbf{uvz}][\mathbf{ywx}] \\ &= [\mathbf{uvx}][\mathbf{wyz}] - [\mathbf{uvw}][\mathbf{xyz}] \\ &= [\mathbf{wxu}][\mathbf{vyz}] - [\mathbf{wxv}][\mathbf{uyz}]\end{aligned} \tag{1.40}$$

In addition to the scalar and the vector products, the tensor product is defined as will be defined in Sect. 1.3.5.

4. Primary and reciprocal vectors

Arbitrary vector $\mathbf{v}$ is expressed by the linear combination of the independent vectors $\mathbf{a}, \mathbf{b}, \mathbf{c}$ as follows:

$$\mathbf{v} = v_a\mathbf{a} + v_b\mathbf{b} + v_c\mathbf{c} \tag{1.41}$$

where the coefficients v_a, v_b, v_c are given by operating the scalar products of $\mathbf{a}\times\mathbf{b}, \mathbf{b}\times\mathbf{c}, \mathbf{c}\times\mathbf{a}$ to Eq. (1.41) as follows:

$$[\mathbf{abv}] = v_c[\mathbf{abc}],\ [\mathbf{bcv}] = v_a[\mathbf{abc}],\ [\mathbf{cav}] = v_b[\mathbf{abc}]$$

i.e.

$$\boxed{v_a = \frac{[\mathbf{bcv}]}{[\mathbf{abc}]},\quad v_b = \frac{[\mathbf{cav}]}{[\mathbf{abc}]},\quad v_c = \frac{[\mathbf{abv}]}{[\mathbf{abc}]}} \tag{1.42}$$

The vector $\mathbf{v}$ is rewritten by substituting Eq. (1.42) into Eq. (1.41) as follows:

$$\boxed{\mathbf{v} = \frac{[\mathbf{bcv}]}{[\mathbf{abc}]}\mathbf{a} + \frac{[\mathbf{cav}]}{[\mathbf{abc}]}\mathbf{b} + \frac{[\mathbf{abv}]}{[\mathbf{abc}]}\mathbf{c}} \tag{1.43}$$

Then, the components v_a, v_b, v_c in Eq. (1.42) are rewritten by

$$v_a = \frac{\mathbf{b}\times\mathbf{c}}{[\mathbf{abc}]}\cdot\mathbf{v} = \mathbf{a}'\cdot\mathbf{v},\quad v_b = \frac{\mathbf{c}\times\mathbf{a}}{[\mathbf{abc}]}\cdot\mathbf{v} = \mathbf{b}'\cdot\mathbf{v},\quad v_c = \frac{\mathbf{a}\times\mathbf{b}}{[\mathbf{abc}]}\cdot\mathbf{v} = \mathbf{c}'\cdot\mathbf{v} \tag{1.44}$$

where

$$\boxed{\begin{array}{lll} \mathbf{a}' \equiv \dfrac{\mathbf{b}\times\mathbf{c}}{[\mathbf{abc}]}, & \mathbf{b}' \equiv \dfrac{\mathbf{c}\times\mathbf{a}}{[\mathbf{abc}]}, & \mathbf{c}' \equiv \dfrac{\mathbf{a}\times\mathbf{b}}{[\mathbf{abc}]} \\ \mathbf{a} \equiv \dfrac{\mathbf{b}'\times\mathbf{c}'}{[\mathbf{a}'\mathbf{b}'\mathbf{c}']}, & \mathbf{b} \equiv \dfrac{\mathbf{c}'\times\mathbf{a}'}{[\mathbf{a}'\mathbf{b}'\mathbf{c}']}, & \mathbf{c} \equiv \dfrac{\mathbf{a}'\times\mathbf{b}'}{[\mathbf{a}'\mathbf{b}'\mathbf{c}']} \end{array}} \tag{1.45}$$

$(\mathbf{a}', \mathbf{b}', \mathbf{c}')$ are called the *reciprocal vectors* of the *primary vectors* $(\mathbf{a}, \mathbf{b}, \mathbf{c})$. The lower part of Eq. (1.45) is derived analogously by considering $\mathbf{v} = v'_a\mathbf{a}' + v'_b\mathbf{b}' + v'_c\mathbf{c}'$ instead of Eq. (1.41) or noting

$$\mathbf{b}'\times\mathbf{c}' = \frac{(\mathbf{c}\times\mathbf{a})\times(\mathbf{a}\times\mathbf{b})}{[\mathbf{abc}]^2} = \frac{[\mathbf{abc}]\mathbf{a}}{[\mathbf{abc}]^2}$$

due to Eq. (1.38). Here, the following relations hold.

$$\left.\begin{array}{l} \mathbf{a}\cdot\mathbf{a}' = 1,\quad \mathbf{b}\cdot\mathbf{b}' = 1,\quad \mathbf{c}\cdot\mathbf{c}' = 1 \\ \mathbf{a}\cdot\mathbf{b}' = \mathbf{a}\cdot\mathbf{c}' = \mathbf{b}\cdot\mathbf{c}' = \mathbf{b}\cdot\mathbf{a}' = \mathbf{c}\cdot\mathbf{a}' = \mathbf{c}\cdot\mathbf{b}' = 0 \\ [\mathbf{abc}][\mathbf{a}'\mathbf{b}'\mathbf{c}'] = 1 \end{array}\right\} \tag{1.46}$$

noting

$$[\mathbf{a}'\mathbf{b}'\mathbf{c}'] = \frac{(\mathbf{b}\times\mathbf{c})\times(\mathbf{c}\times\mathbf{a})\cdot(\mathbf{a}\times\mathbf{b})}{[\mathbf{abc}]^3} = \frac{[\mathbf{bca}]^2}{[\mathbf{abc}]^3} = \frac{1}{[\mathbf{abc}]}$$

exploiting Eq. (1.39).

An arbitrary vector $\mathbf{v}$ is expressed from Eqs. (1.41) and (1.44) as follows:

$$\boxed{\mathbf{v} = (\mathbf{a}' \cdot \mathbf{v})\mathbf{a} + (\mathbf{b}' \cdot \mathbf{v})\mathbf{b} + (\mathbf{c}' \cdot \mathbf{v})\mathbf{c} = (\mathbf{a} \cdot \mathbf{v})\mathbf{a}' + (\mathbf{b} \cdot \mathbf{v})\mathbf{b}' + (\mathbf{c} \cdot \mathbf{v})\mathbf{c}'} \quad (1.47)$$

where the third side is obtained from the second side by exchanging (**a**, **b**, **c**) and ($\mathbf{a}'$, $\mathbf{b}'$, $\mathbf{c}'$).

1.2.3 Component Description of Vector

The component description of vector is explained here prior to the description of component description of general tensor.

1. Component description

Consider the *normalized orthogonal coordinate system.* Here, the "normalized" means that it has the unit base vectors, and "orthonormal" means that the coordinate axes, i.e. the base vectors are mutually orthonormal. Let it be denoted as $\{O - x_i\}$, while the unit vectors are denoted by the triad $\{\mathbf{e}_i\}$. The scalar and the vector products between the base vectors are given from Eqs. (1.2), (1.23) and (1.26) as follows:

$$\mathbf{e}_i \cdot \mathbf{e}_j = \delta_{ij} \quad (1.48)$$

$$\mathbf{e}_i \times \mathbf{e}_j = \varepsilon_{ijr}\mathbf{e}_r \quad (1.49)$$

Vector $\mathbf{v}$ is described in the linear associative form as follows:

$$\mathbf{v} = v_r\mathbf{e}_r (= v_1\mathbf{e}_1 + v_2\mathbf{e}_2 + v_3\mathbf{e}_3) \quad (1.50)$$

where v_1, v_2, v_3 are the components of $\mathbf{v}$. Denoting the angle of the direction of vector $\mathbf{v}$ from the direction of the base vector $\mathbf{e}_i$ by θ_i, $\cos\theta_i = \mathbf{n} \cdot \mathbf{e}_i$ is called the *direction cosine* by which the component of $\mathbf{v}$ is given as

$$v_i = \mathbf{v} \cdot \mathbf{e}_i = ||\mathbf{v}||\mathbf{n} \cdot \mathbf{e}_i = ||\mathbf{v}||\cos\theta_i \quad (1.51)$$

The magnitude of vector $\mathbf{v}$ and its unit direction vector $\mathbf{n}$ are given from Eqs. (1.24), (1.48) and (1.50) as follows:

$$||\mathbf{v}|| \equiv \sqrt{v_r v_r}, \quad \mathbf{n} \equiv \frac{\mathbf{v}}{||\mathbf{v}||} = \frac{v_r}{||\mathbf{v}||}\mathbf{e}_r \quad (1.52)$$

Because of $\mathbf{u}\cdot\mathbf{v} = u_r\mathbf{e}_r \cdot v_s\mathbf{e}_s = u_r v_s \delta_{rs}$ the scalar product is expressed by using the components as

$$\mathbf{u}\cdot\mathbf{v} = u_r v_r \tag{1.53}$$

The vector product is expressed from Eqs. (1.49) and (1.50) as follows:

$$\begin{aligned}\mathbf{u}\times\mathbf{v} &= u_i\mathbf{e}_i \times v_j\mathbf{e}_j = \varepsilon_{ijk}u_j v_k \mathbf{e}_i = \varepsilon_{ijk}u_i v_j \mathbf{e}_k \\ &= (u_2v_3 - u_3v_2)\mathbf{e}_1 + (u_3v_1 - u_1v_3)\mathbf{e}_2 + (u_1v_2 - u_2v_1)\mathbf{e}_3 = \begin{vmatrix} \mathbf{e}_1 & \mathbf{e}_2 & \mathbf{e}_3 \\ u_1 & u_2 & u_3 \\ v_1 & v_2 & v_3 \end{vmatrix}\end{aligned} \tag{1.54}$$

For the sake of Eq. (1.54) the scalar triple product defined in Eq. (1.29) is expressed in a component form as

$$\begin{aligned}[\mathbf{uvw}] &= (\mathbf{u}\times\mathbf{v})\cdot\mathbf{w} = \varepsilon_{ijr}u_i v_j \mathbf{e}_r \cdot w_k\mathbf{e}_k = \varepsilon_{ijr}u_i v_j w_k \delta_{rk} = \varepsilon_{ijk}u_i v_j w_k \\ &= \begin{vmatrix} u_1 & v_1 & w_1 \\ u_2 & v_2 & w_2 \\ u_3 & v_3 & w_3 \end{vmatrix}\end{aligned} \tag{1.55}$$

Further, the vector triple product in Eq. (1.35) is expressed as follows:

$$\mathbf{u}\times(\mathbf{v}\times\mathbf{w}) = \varepsilon_{ijk}\varepsilon_{kpq}u_j v_p w_q \mathbf{e}_i \tag{1.56}$$

noting

$$\mathbf{u}\times(\mathbf{v}\times\mathbf{w}) = u_j\mathbf{e}_j \times (\varepsilon_{kpq}v_p w_q \mathbf{e}_k) = u_j\varepsilon_{kpq}v_p w_q \mathbf{e}_j \times \mathbf{e}_k = \varepsilon_{kpq}u_j v_p w_q \varepsilon_{kij}\mathbf{e}_i$$

based on Eq. (1.54).

2. Coordinate transformation

Adopt the other normalized orthogonal coordinate system $\{O - x_i^*\}$ with the base $\{\mathbf{e}_i^*\}$ in addition to the normalized orthogonal coordinate system $\{O - x_i\}$ with the base $\{\mathbf{e}_i\}$ (Fig. 1.1). Noting $\mathbf{v} = v_j\mathbf{e}_j = (\mathbf{v}\cdot\mathbf{e}_j)\mathbf{e}_j$ in general, the following relations hold for the base vectors.

$$\mathbf{e}_i = (\mathbf{e}_i\cdot\mathbf{e}_j^*)\mathbf{e}_j^*, \quad \mathbf{e}_i^* = (\mathbf{e}_i^*\cdot\mathbf{e}_j)\mathbf{e}_j \tag{1.57}$$

$$\mathbf{e}_i = Q_{ri}\mathbf{e}_r^*, \quad \mathbf{e}_i^* = Q_{ir}\mathbf{e}_r \tag{1.58}$$

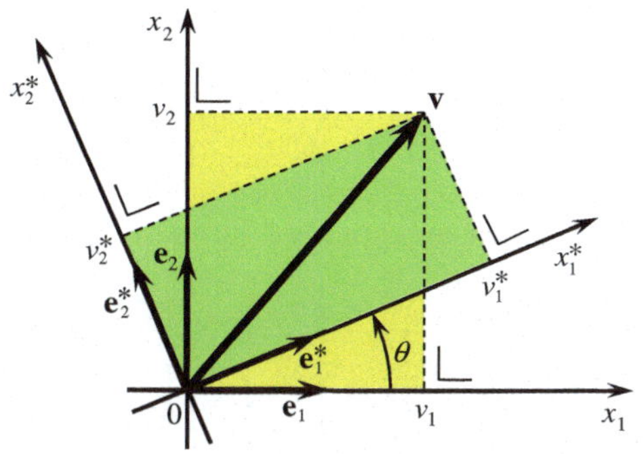

Fig. 1.1 Coordinate transformation of vector illustrated in the two-dimensional state

where the coordinate transformation operator Q_{ij} is defined by

$$\boxed{Q_{ij} \equiv \cos(\text{angle between } \mathbf{e}_i^* \text{ and } \mathbf{e}_j) = \mathbf{e}_i^* \bullet \mathbf{e}_j} \tag{1.59}$$

Moreover, because of

$$\left.\begin{aligned} Q_{ir}Q_{jr} &= (\mathbf{e}_i^* \bullet \mathbf{e}_r)(\mathbf{e}_j^* \bullet \mathbf{e}_r) = \mathbf{e}_i^* \bullet (\mathbf{e}_j^* \bullet \mathbf{e}_r)\mathbf{e}_r = \mathbf{e}_i^* \bullet \mathbf{e}_j^* \\ Q_{ri}Q_{rj} &= (\mathbf{e}_r^* \bullet \mathbf{e}_i)(\mathbf{e}_r^* \bullet \mathbf{e}_j) = (\mathbf{e}_i \bullet \mathbf{e}_r^*)\mathbf{e}_r^* \bullet \mathbf{e}_j = \mathbf{e}_i \bullet \mathbf{e}_j \end{aligned}\right\}$$

it follows that

$$\boxed{Q_{ir}Q_{jr} = Q_{ri}Q_{rj} = \delta_{ij}} \tag{1.60}$$

It is assumed for a while that the relative (parallel and rotational) motion does not exist between the above-described coordinate systems, and that their origins mutually coincide. Then, denoting the component on the base $\mathbf{e}_i^*$ by $(\,)^*$, the coordinate transformation rule, i.e. the transformation rule of the components of $\mathbf{v}$ in these coordinate systems is given by

$$v_i^* = Q_{ij}v_j \tag{1.61}$$

noting

$$v_r^*\mathbf{e}_r^* \bullet \mathbf{e}_i^* = v_j\mathbf{e}_j \bullet \mathbf{e}_i^*$$

based on

$$\mathbf{v} = v_j\mathbf{e}_j = v_r^*\mathbf{e}_r^* \tag{1.62}$$

Furthermore, noting $Q_{ri}v_r^* = Q_{ri}Q_{rs}v_s = \delta_{is}v_s$, the inverse relation of Eq. (1.61) is given as

$$v_i = Q_{ji}v_j^* \tag{1.63}$$

Equations (1.61) and (1.63) are expressed in matrix form as

$$\begin{Bmatrix} v_1^* \\ v_2^* \\ v_3^* \end{Bmatrix} = \begin{bmatrix} Q_{11} & Q_{12} & Q_{13} \\ Q_{21} & Q_{22} & Q_{23} \\ Q_{31} & Q_{32} & Q_{33} \end{bmatrix} \begin{Bmatrix} v_1 \\ v_2 \\ v_3 \end{Bmatrix}, \quad \begin{Bmatrix} v_1 \\ v_2 \\ v_3 \end{Bmatrix} = \begin{bmatrix} Q_{11} & Q_{21} & Q_{31} \\ Q_{12} & Q_{22} & Q_{32} \\ Q_{13} & Q_{23} & Q_{33} \end{bmatrix} \begin{Bmatrix} v_1^* \\ v_2^* \\ v_3^* \end{Bmatrix} \tag{1.64}$$

which are often expressed simply as

$$\{v^*\} = [Q]\{v\}, \quad \{v\} = [Q]^T\{v^*\} \tag{1.65}$$

Needless to say, an equation including ( ) and $(\ )^*$ does not describe the relation between different vectors, but describes the relations between components when a certain vector is described by two different coordinate systems.

As known from the following equation, the magnitude of vector is not influenced by the coordinate transformation, whilst it is the basic property of the scalar quantity.

$$||\mathbf{v}^*|| = \sqrt{Q_{ir}v_rQ_{is}v_s} = \sqrt{Q_{ir}Q_{is}v_rv_s} = \sqrt{\delta_{rs}v_rv_s} = \sqrt{v_rv_r} = ||\mathbf{v}||$$

The relations described above are shown below.

$$[Q] = \begin{bmatrix} \mathbf{e}_1^* \cdot \mathbf{e}_1 & \mathbf{e}_1^* \cdot \mathbf{e}_2 \\ \mathbf{e}_2^* \cdot \mathbf{e}_1 & \mathbf{e}_2^* \cdot \mathbf{e}_2 \end{bmatrix} = \begin{bmatrix} (\mathbf{e}_1^*)_1 & (\mathbf{e}_1^*)_1 \\ (\mathbf{e}_2^*)_1 & (\mathbf{e}_2^*)_2 \end{bmatrix} = \begin{bmatrix} \cos\theta & \sin\theta \\ -\sin\theta & \cos\theta \end{bmatrix}$$

$$Q_{ir}Q_{jr} = \begin{bmatrix} \cos\theta & \sin\theta \\ -\sin\theta & \cos\theta \end{bmatrix} \begin{bmatrix} \cos\theta & -\sin\theta \\ \sin\theta & \cos\theta \end{bmatrix} = \begin{bmatrix} 1 & 0 \\ 0 & 1 \end{bmatrix} = [\delta_{ij}]$$

$$\mathbf{e}_i^* = (\mathbf{e}_i^* \cdot \mathbf{e}_1)\mathbf{e}_1 + (\mathbf{e}_i^* \cdot \mathbf{e}_2)\mathbf{e}_2 = Q_{i1}\mathbf{e}_1 + Q_{i2}\mathbf{e}_2,$$

$$\left.\begin{aligned} \mathbf{e}_1^* &= \cos\theta\mathbf{e}_1 + \sin\theta\mathbf{e}_2 \\ \mathbf{e}_2^* &= -\sin\theta\mathbf{e}_1 + \cos\theta\mathbf{e}_2 \end{aligned}\right\}$$

$$\mathbf{e}_i = (\mathbf{e}_i \cdot \mathbf{e}_1^*)\mathbf{e}_1^* + (\mathbf{e}_i \cdot \mathbf{e}_2^*)\mathbf{e}_2^* = Q_{1i}\mathbf{e}_1^* + Q_{2i}\mathbf{e}_2^*$$

$$\mathbf{v} = v_1\mathbf{e}_1 + v_2\mathbf{e}_2 = v_1^*\mathbf{e}_1^* + v_2^*\mathbf{e}_2^*$$

$$\begin{Bmatrix} v_1^* \\ v_2^* \end{Bmatrix} = \begin{bmatrix} \cos\theta & \sin\theta \\ -\sin\theta & \cos\theta \end{bmatrix} \begin{Bmatrix} v_1 \\ v_2 \end{Bmatrix}, \quad \begin{Bmatrix} v_1 \\ v_2 \end{Bmatrix} = \begin{bmatrix} \cos\theta & -\sin\theta \\ \sin\theta & \cos\theta \end{bmatrix} \begin{Bmatrix} v_1^* \\ v_2^* \end{Bmatrix}$$

where θ designates the angle that the base $\{\mathbf{e}_i^*\}$ rotates in the anti-clock direction from the base $\{\mathbf{e}_i\}$.

Choosing the position vector $\mathbf{x}$ as the vector $\mathbf{v}$, it follows from Eqs. (1.61) and (1.63) that

$$\left.\begin{aligned} x_i^* &= Q_{ir}x_r \\ x_i &= Q_{ri}x_r^* \end{aligned}\right\} \tag{1.66}$$

from which one has

$$\left.\begin{aligned} \frac{\partial x_i^*}{\partial x_j} &= \frac{\partial Q_{ir}x_r}{\partial x_j} = Q_{ir}\frac{\partial x_r}{\partial x_j} = Q_{ir}\delta_{jr} = Q_{ij} \\ \frac{\partial x_j}{\partial x_i^*} &= \frac{\partial Q_{rj}x_r^*}{\partial x_i^*} = Q_{rj}\frac{\partial x_r^*}{\partial x_i^*} = Q_{rj}\delta_{ir} = Q_{ij} \end{aligned}\right\} \tag{1.67}$$

Consequently, Q_{ij} can be also described as

$$Q_{ij} = \frac{\partial x_i^*}{\partial x_j} = \frac{\partial x_j}{\partial x_i^*} \tag{1.68}$$

1.3 Tensor

The vector described in the foregoing possesses the direction in first order but there exit quantities possessing the direction in high order. They are collectively called the *tensor*. The general definition and mathematical properties of tensor are described in this section.

1.3.1 Definition of Tensor

Let the set of n^m functions be described as $T(p_1, p_2, \cdots, p_m)$ in the coordinate system $\{\mathrm{O} - x_i\}$ with the origin O and the axes $x_i (i = 1, 2, \cdots, n)$ in the n-dimensional space, where each of the indices $p_1, p_2, \cdots, p_m$ takes the number $1, 2, \cdots, n$. This set of functions is defined as the mth-order tensor in the n-dimension, if the set of functions is observed in the other coordinate system $\{\mathrm{O} - x_i^*\}$ with the origin O and the axes x_i^* as follows:

$$T^*(p_1, p_2, \cdots p_m) = Q_{p_1 q_1} Q_{p_2 q_2} \cdots Q_{p_m q_m} T(q_1, q_2, \cdots q_m) \tag{1.69}$$

or

$$T^*(p_1, p_2, \cdots p_m) = \frac{\partial x_{p_1}^*}{\partial x_{q_1}} \frac{\partial x_{p_2}^*}{\partial x_{q_2}} \cdots \frac{\partial x_{p_m}^*}{\partial x_{qm}} T(q_1, q_2, \cdots q_m) \tag{1.70}$$

provided that only the directions of axes are different but the origin is common and the relative motion does not exist. Here, Eq. (1.68) is used.

Then, designating $T(p_1, p_2, \cdots p_m)$ by the symbol $T_{p_1 p_2 \cdots p_m}$ for the simplicity of notation, Eqs. (1.69) or (1.70) is expressed as

$$\boxed{T^*_{p_1 p_2 \cdots p_m} = Q_{p_1 q_1} Q_{p_2 q_2} \cdots Q_{p_m q_m} T_{q_1 q_2 \cdots q_m}} \tag{1.71}$$

or

$$T^*_{p_1 p_2 \cdots p_m} = \frac{\partial x^*_{p_1}}{\partial x_{q_1}} \frac{\partial x^*_{p_2}}{\partial x_{q_2}} \cdots \frac{\partial x^*_{p_m}}{\partial x_{qm}} T_{q_1 q_2 \cdots q_m} \tag{1.72}$$

Noting that

$$\begin{aligned} Q_{p_1 r_1} Q_{p_2 r_2} \cdots Q_{p_m r_m} T^*_{p_1 p_2 \cdots p_m} &= Q_{p_1 r_1} Q_{p_2 r_2} \cdots Q_{p_m r_m} Q_{p_1 q_1} Q_{p_2 q_2} \cdots Q_{p_m q_m} T_{q_1 q_2 \cdots q_m} \\ &= (Q_{p_1 r_1} Q_{p_1 q_1})(Q_{p_2 r_2} Q_{p_2 q_2}) \cdots (Q_{p_m r_m} Q_{p_m q_m}) T_{q_1 q_2 \cdots q_m} \\ &= \delta_{r_1 q_1} \delta_{r_2 q_2} \cdots \delta_{r_m q_m} T_{q_1 q_2 \cdots q_m} \end{aligned}$$

the inverse relation of Eq. (1.71) is given by

$$\boxed{T_{r_1 r_2 \cdots r_m} = Q_{p_1 r_1} Q_{p_2 r_2} \cdots Q_{p_m r_m} T^*_{p_1 p_2 \cdots p_m}} \tag{1.73}$$

While the transformation rule of the first-order tensor, i.e. vector is given by Eqs. (1.61) and (1.63), the transformation rule of the second-order tensor is given by

$$T^*_{ij} = Q_{ir} Q_{js} T_{rs},\ T_{ij} = Q_{ri} Q_{js} T^*_{rs} \tag{1.74}$$

The transformation between the coordinate systems without relative motion is considered above in the definition of the tensor, whereas the transformation in the form of (1.71) or (1.73) is called as the *objective transformation.* A tensor that obeys the objective transformation even between the coordinate systems with the relative motion is called an *objective tensor.*

1.3.2 Quotient Law

One has a convenient law, called the *quotient law*, which is used to judge whether or not a quantity is a tensor as will be explained below.

Quotient law: "If a set of functions $T(p_1, p_2, \cdots, p_m)$ becomes $B_{p_{l+1} p_{l+2} \cdots p_m}$ ($m - l$-th order tensor lacking the suffices $p_1 \sim p_l$) by multiplying it by $A_{p_1 p_2 \cdots p_l}$ (l-th order tensor ($l \leq m$)), the set is an m-th order tensor".

Proof The proof can be achieved by showing that the quantity $T(p_1, p_2, \cdots, p_m)$ is the m-th order tensor when it holds that

$$T(p_1, p_2, \cdots, p_m)A_{p_1 p_2 \cdots p_l} = B_{p_{l+1} p_{l+2} \cdots p_m} \tag{1.75}$$

which is described in the coordinate system $\{O \ - x_i^*\}$ as follows:

$$T^*(p_1, p_2, \cdots, p_m)A^*_{p_1 p_2 \cdots p_m} = B^*_{p_{l+1} p_{l+2} \cdots p_m} \tag{1.76}$$

Here, the following relation holds.

$$\begin{aligned} B^*_{p_{l+1} p_{l+2} \cdots p_m} &= Q_{p_{l+1} r_{l+1}} Q_{p_{l+2} r_{l+2}} \cdots Q_{p_m r_m} B_{r_{l+1} r_{l+2} \cdots r_m} \\ &= Q_{p_{l+1} r_{l+1}} Q_{p_{l+2} r_{l+2}} \cdots Q_{p_m r_m} T(r_1, r_2, \cdots, r_m) A_{r_1 r_2 \cdots r_l} \\ &= \underbrace{Q_{p_{l+1} r_{l+1}} Q_{p_{l+2} r_{l+2}} \cdots Q_{p_m r_m}}_{l+1 \sim m} T(r_1, r_2, \cdots, r_m) \underbrace{Q_{p_1 r_1} Q_{p_2 r_2} \cdots Q_{p_l r_l}}_{1 \sim l} A^*_{p_1 p_2 \cdots p_l} \end{aligned} \tag{1.77}$$

Substituting Eq. (1.76) into Eq. (1.77) yields

$$\{T^*(p_1, p_2, \cdots, p_m) - Q_{p_1 r_1} Q_{p_2 r_2} \cdots Q_{p_m r_m} T(r_1, r_2, \cdots, r_m)\} A^*_{p_1 p_2 \cdots p_l} = 0 \tag{1.78}$$

from which it holds that

$$T^*(p_1, p_2, \cdots, p_m) = Q_{p_1 r_1} Q_{p_2 r_2} \cdots Q_{p_m r_m} T(r_1, r_2, \cdots, r_m) \tag{1.79}$$

Therefore, taking account of the definition of tensor in Eq. (1.69) into Eq. (1.79), the quantity $T(p_1, p_2, \cdots, p_m)$ is the m-th order tensor. (Q.E.D.)

According to the proof presented above, Eq. (1.75) can be written as

$$T_{p_1 p_2 \cdots p_m} A_{p_1 p_2 \cdots p_l} = B_{p_{l+1} p_{l+2} \cdots p_m} \tag{1.80}$$

For instance, if the quantity $T(i,j)$ transforms the first-order tensor, i.e. vector v_i to the vector u_i by the operation $T(i,j)v_j = u_i$, one can regard $T(i,j)$ as the second-order tensor.

Eventually, in order to prove that a certain quantity is a tensor, one needs only to show that it obeys the tensor transformation rule (1.71) or that the multiplication of a tensor to the quantity leads to a tensor by the quotient rule.

Tensors fulfill linearity as follows:

$$\left. \begin{aligned} & T_{p_1 p_2 \cdots p_m}(G_{p_1 p_2 \cdots p_l} + H_{p_1 p_2 \cdots p_l}) = T_{p_1 p_2 \cdots p_m} G_{p_1 p_2 \cdots p_l} + T_{p_1 p_2 \cdots p_m} H_{p_1 p_2 \cdots p_l} \\ & T_{p_1 p_2 \cdots p_m}(a A_{p_1 p_2 \cdots p_l}) = a T_{p_1 p_2 \cdots p_m} A_{p_1 p_2 \cdots p_l} \end{aligned} \right\}$$

where a is an arbitrary scalar. Therefore, the tensor plays the role to transform linearly a tensor to the other tensor and thus it is called the *linear transformation*. The operation that lowers the order of tensor by multiplying the other tensor is called the *contraction*.

1.3.3 Notations of Tensors

When we express the tensor **T** as

$$\boxed{\mathbf{T} = T_{p_1p_2\cdots p_m}\mathbf{e}_{p_1}\otimes\mathbf{e}_{p_2}\cdots\otimes\mathbf{e}_{p_m}} \tag{1.81}$$

in a similar form to the case of vector in Eq. (1.50), Eq. (1.81) is called the *component notation with bases*, defining $\mathbf{e}_{p_1}\otimes\mathbf{e}_{p_2}\cdots\otimes\mathbf{e}_{p_m}$ as the base of m-th order tensor. The transformation of **T** between the bases in Eq. (1.58) leads Eq. (1.81) to

$$\begin{aligned}\mathbf{T} &= T_{p_1p_2\cdots p_m}Q_{r_1p_1}\mathbf{e}^*_{r_1}\otimes Q_{r_2p_2}\mathbf{e}^*_{r_2}\otimes\cdots\otimes Q_{r_mp_m}\mathbf{e}^*_{r_m}\\ &= Q_{r_1p_1}Q_{r_2p_2}\cdots Q_{r_mp_m}T_{p_1p_2\cdots p_m}\mathbf{e}^*_{r_1}\otimes\mathbf{e}^*_{r_2}\otimes\cdots\otimes\mathbf{e}^*_{r_m}\\ &= T^*_{r_1r_2\cdots r_m}\mathbf{e}^*_{r_1}\otimes\mathbf{e}^*_{r_2}\otimes\cdots\otimes\mathbf{e}^*_{r_m}\end{aligned} \tag{1.82}$$

The following various notations are used for tensors.

Indicial (or component) notation: $T_{p_1p_2\cdots p_m}$

Component notation with base: $T_{p_1p_2\cdots p_m}\mathbf{e}_{p_1}\mathbf{e}_{p_2}\otimes\cdots\otimes\boldsymbol{e}_{p_m}$

Symbolic (or direct) notation: **T**

Matrix notation: Eq. (1.6) for second-order tensor as an example.

The matrix notation holds only for a vector or a second-order tensor or for a fourth-order tensor if it is formally expressed by two suffixes. For instance, the stress-strain relation can be expressed in matrix notation by expressing the stress and the strain of second-order tensors as a form of vector and the stiffness coefficient of fourth-order tensor as a form of second-order tensor.

Various contractions exist in the operation of higher-order tensors. and thus the symbolic notation is not useful in general. For instance, which of the following does **ST** mean: $S_{ijk}T_{jk}, S_{ijk}T_{kj}, S_{ijk}T_{ij}, S_{ijk}T_{ji}, S_{ijk}T_{jl}, S_{ijk}T_{kl}, S_{ijk}T_{il}$? In other words, the application of symbolic notation is limited to the multiplication between low order tensors. On the other hand, component notation with bases holds always without defining special rule.

Introducing the notation

$$\boxed{\begin{aligned}(\mathbf{Q}[\![\mathbf{T}]\!])_{p_1p_2\cdots p_m} &\equiv Q_{p_1q_1}Q_{p_2q_2}\cdots Q_{p_mq_m}T_{q_1q_2\cdots q_m}\\ (\mathbf{Q}^T[\![\mathbf{T}]\!])_{p_1p_2\cdots p_m} &\equiv Q_{q_1p_1}Q_{q_2p_2}\cdots Q_{q_mp_m}T_{q_1q_2\cdots q_m}\end{aligned}} \tag{1.83}$$

Equations (1.71) and (1.73) can be expressed by the symbolic notation as follows:

$$\boxed{\begin{array}{l}\mathbf{T}^* = \mathbf{Q}[\![\mathbf{T}]\!] \\ \mathbf{T} = \mathbf{Q}^T[\![\mathbf{T}^*]\!]\end{array}} \tag{1.84}$$

In particular, the transformations of the vector and the second-order tensor are expressed

$$\boxed{\mathbf{v}^* = \mathbf{Q}\mathbf{v} = \mathbf{v}\mathbf{Q}^T, \quad \mathbf{v} = \mathbf{Q}^T\mathbf{v}^* = \mathbf{v}^*\mathbf{Q}} \tag{1.85}$$

$$\boxed{\mathbf{T}^* = \mathbf{Q}\mathbf{T}\mathbf{Q}^T, \quad \mathbf{T} = \mathbf{Q}^T\mathbf{T}^*\mathbf{Q}} \tag{1.86}$$

noting $\mathbf{T}\mathbf{v} = \mathbf{v}\mathbf{T}^T(T_{ij}v_j = v_jT_{ij})$ in general for Eq. (1.85), where $\mathbf{T}^T((\mathbf{T}^T)_{ij} = T_{ji})$ is the transposed tensor defined exactly in Sect. 1.4.1.

1.3.4 Orthogonal Tensor

The coordinate transformation operator Q_{ij} which appeared in Sects. 1.2.3 and 1.3.3 plays an important role in the coordinate transformation. The component notation with bases is obtained from

$$\begin{aligned} Q_{ij}\mathbf{e}_i \otimes \mathbf{e}_j &= \mathbf{e}_i \otimes (\mathbf{e}_i^* \cdot \mathbf{e}_j)\mathbf{e}_j\,(=\mathbf{e}_i \otimes \mathbf{e}_i^*) = (\mathbf{e}_i \cdot \mathbf{e}_r^*)\mathbf{e}_r^* \otimes \mathbf{e}_i^* \\ &= Q_{ri}\mathbf{e}_r^* \otimes \mathbf{e}_i^* \end{aligned} \tag{1.87}$$

as follows:

$$\boxed{\mathbf{Q} = Q_{ij}\mathbf{e}_i \otimes \mathbf{e}_j = Q_{ij}\mathbf{e}_i^* \otimes \mathbf{e}_j^*} \tag{1.88}$$

Furthermore, substituting Eq. (1.58), the direct notation of $\mathbf{Q}$ is given by

$$\boxed{\mathbf{Q} = \mathbf{e}_i \otimes \mathbf{e}_i^*} \tag{1.89}$$

Because of

$$\left.\begin{array}{l}\mathbf{e}_i = \mathbf{e}_r\delta_{ir} = \mathbf{e}_r \otimes \mathbf{e}_r^*\mathbf{e}_i^* \\ \mathbf{e}_i^* = \mathbf{e}_r^*\delta_{ir} = \mathbf{e}_r^* \otimes \mathbf{e}_r\mathbf{e}_i\end{array}\right\}$$

it follows that

$$\boxed{\mathbf{e}_i = \mathbf{Q}\mathbf{e}_i^* = Q_{ri}\mathbf{e}_r^*, \quad \mathbf{e}_i^* = \mathbf{Q}^T\mathbf{e}_i = Q_{ir}\mathbf{e}_r} \tag{1.90}$$

Furthermore, changing Eq. (1.60) to the direct notation or noting the relation

$$\left.\begin{aligned}\mathbf{Q}\mathbf{Q}^T &= \mathbf{e}_i \otimes \mathbf{e}_i^*\mathbf{e}_j^* \otimes \mathbf{e}_j = \mathbf{e}_i\delta_{ij} \otimes \mathbf{e}_j = \mathbf{e}_i \otimes \mathbf{e}_i \\ \mathbf{Q}^T\mathbf{Q} &= \mathbf{e}_i^* \otimes \mathbf{e}_i\mathbf{e}_j \otimes \mathbf{e}_j^* = \mathbf{e}_i^*\delta_{ij} \otimes \mathbf{e}_j^* = \mathbf{e}_i^* \otimes \mathbf{e}_i^*\end{aligned}\right\}$$

obtained from Eq. (1.89), it holds that

$$\boxed{\mathbf{Q}\mathbf{Q}^T = \mathbf{Q}^T\mathbf{Q} = \mathbf{I}} \tag{1.91}$$

where **I** possesses the components of the Kronecker's delta, i.e.

$$(\mathbf{I})_{ij} = \delta_{ij} \tag{1.92}$$

which transforms vector and tensor to original vector and tensor as $\mathbf{I}\mathbf{v} = \mathbf{v}\mathbf{I} = \mathbf{v}$ and $\mathbf{I}\mathbf{T} = \mathbf{T}\mathbf{I} = \mathbf{T}$, and thus it is called the *identity tensor*.

The tensor satisfying Eq. (1.91) is defined as the *orthogonal tensor* in general, the properties of which are described below.

It follows from Eq. (1.91) that

$$\boxed{\mathbf{Q}^T = \mathbf{Q}^{-1}} \tag{1.93}$$

Moreover, from Eqs. (1.9), (1.17) and (1.91), it is obtained that

$$\boxed{\det \mathbf{Q} = \det \mathbf{Q}^T = \pm 1} \tag{1.94}$$

Further from Eq. (1.91) one obtains

$$(\mathbf{Q} - \mathbf{I})\mathbf{Q}^T = -(\mathbf{Q} - \mathbf{I})^T$$

Making the determinant of this equation and noting Eqs. (1.12), (1.17) and (1.94), it holds that

$$\det(\mathbf{Q} - \mathbf{I}) = -\det(\mathbf{Q} - \mathbf{I}) \rightarrow \det(\mathbf{Q} - \mathbf{I}) = 0 \tag{1.95}$$

Then, it is known that one of the principal values of the orthogonal tensor is unity in order to satisfy $(Q_1 - 1)(Q_2 - 1)(Q_3 - 1) = 0$ as known from the fact which will be described in Sect. 1.6.

1.3.5 Tensor Product and Component

Based on the vectors $\mathbf{v}^{(1)}, \mathbf{v}^{(2)}, \cdots, \mathbf{v}^{(m)}$, one can make the m-th order tensor as follows:

$$\mathbf{v}^{(1)} \otimes \mathbf{v}^{(2)} \cdots \otimes \mathbf{v}^{(m)} = v_{p_1}^{(1)} v_{p_2}^{(2)} \cdots v_{p_m}^{(m)} \mathbf{e}_{p_1 \otimes} \mathbf{e}_{p_2} \cdots \otimes \mathbf{e}_{p_m} \tag{1.96}$$

For two vectors, one has the second-order tensor

$$\mathbf{u} \otimes \mathbf{v} = u_i \mathbf{e}_{i \otimes} v_j \mathbf{e}_j = u_i v_j \mathbf{e}_{i \otimes} \mathbf{e}_j \tag{1.97}$$

which is expressed in the matrix form

$$[\mathbf{u} \otimes \mathbf{v}] = \begin{bmatrix} u_1 v_1 & u_1 v_2 & u_1 v_3 \\ u_2 v_1 & u_2 v_2 & u_2 v_3 \\ u_3 v_1 & u_3 v_2 & u_3 v_3 \end{bmatrix} \tag{1.98}$$

As described above, one can make a tensor from two vectors. After the scalar product $\mathbf{u} \cdot \mathbf{v}$ and the vector product $\mathbf{u} \times \mathbf{v}$ for the two vectors $\mathbf{u}$ and $\mathbf{v}$, one calls $\mathbf{u} \otimes \mathbf{v}$ as the *tensor* (*cross*) *product* or *dyad* which means "one set by two". Particularly, it holds for three arbitrary vectors $\mathbf{u}$, $\mathbf{v}$ and $\mathbf{w}$ that

$$(\mathbf{u} \otimes \mathbf{v})\mathbf{w} = (u_i \mathbf{e}_i \otimes v_j \mathbf{e}_j) w_k \mathbf{e}_k = u_i \mathbf{e}_i (v_j w_k \delta_{jk}) = u_i \mathbf{e}_i (v_j w_j) = \mathbf{u}(\mathbf{v} \cdot \mathbf{w})$$

and thus the following expression holds.

$$\boxed{\mathbf{u} \otimes \mathbf{v} = \mathbf{u}(\mathbf{v} \cdot} \tag{1.99}$$

It follows from Eq. (1.8) that

$$\det(\mathbf{u} \otimes \mathbf{v}) = \varepsilon_{ijk} u_1 v_i u_2 v_j u_3 v_k = u_1 u_2 u_3 \varepsilon_{ijk} v_i v_j v_k = u_1 u_2 u_3 [\mathbf{vvv}]$$

leading to

$$\boxed{\det(\mathbf{u} \otimes \mathbf{v}) = 0} \tag{1.100}$$

The component of vector is expressed by the direct notation in Eq. (1.51). Here, consider the component of second-order tensor in the direct notation. The second-order tensor $\mathbf{T}$ is expressed from Eq. (1.81) as

$$\mathbf{T} = T_{ij} \mathbf{e}_i \otimes \mathbf{e}_j \tag{1.101}$$

from which, noting Eq. (1.99) it follows that $\mathbf{e}_i \cdot \mathbf{T}\mathbf{e}_j = \mathbf{e}_i \cdot T_{rs}\mathbf{e}_r \otimes \mathbf{e}_s\mathbf{e}_j = T_{rs}\delta_{ir}\delta_{sj}$ and thus the component of $\mathbf{T}$ in the direct notation is given as

$$\boxed{T_{ij} = \mathbf{e}_i \cdot \mathbf{T}\mathbf{e}_j} \tag{1.102}$$

As known from Eq. (1.102), the orthogonal projection of the vector $\mathbf{T}\mathbf{e}_j$ to the base vector $\mathbf{e}_i$ is the component of the tensor $\mathbf{T}$. Especially, T_{ii}(no sum) and T_{ij} $(i \neq j)$ are called the *normal* and the *shear* components, respectively.

An arbitrary vector $\mathbf{v}$ is expressed by exploiting Eq. (1.99) in Eq. (1.47) as

$$\mathbf{v} = (\mathbf{a} \otimes \mathbf{a}' + \mathbf{b} \otimes \mathbf{b}' + \mathbf{c} \otimes \mathbf{c}')\mathbf{v} = (\mathbf{a}' \otimes \mathbf{a} + \mathbf{b}' \otimes \mathbf{b} + \mathbf{c}' \otimes \mathbf{c})\mathbf{v} \tag{1.103}$$

The identity tensor in Eq. (1.92) is expressed by virtue of (1.101) as follows:

$$\boxed{\mathbf{I} = \delta_{ij}\mathbf{e}_i \otimes \mathbf{e}_j = \mathbf{e}_i \otimes \mathbf{e}_i \, (= \mathbf{I}^T)} \tag{1.104}$$

which is described in terms of the primary vectors $(\mathbf{a}, \mathbf{b}, \mathbf{c})$ and the reciprocal vectors $(\mathbf{a}', \mathbf{b}', \mathbf{c}')$ as follows:

$$\boxed{\mathbf{I} = \mathbf{a} \otimes \mathbf{a}' + \mathbf{b} \otimes \mathbf{b}' + \mathbf{c} \otimes \mathbf{c}' = \mathbf{a}' \otimes \mathbf{a} + \mathbf{b}' \otimes \mathbf{b} + \mathbf{c}' \otimes \mathbf{c}} \tag{1.105}$$

by taking account of $\mathbf{v} = \mathbf{I}\mathbf{v}$ in general in Eq. (1.103).

Further, noting the relation $(\mathbf{T}\mathbf{u} \otimes \mathbf{T}\mathbf{v})_{ij} = T_{ir}u_r T_{js}u_s = T_{ir}u_r u_s T_{js} = (\mathbf{T}(\mathbf{u} \otimes \mathbf{v})\mathbf{T}^T)_{ij}$, one has

$$\boxed{\mathbf{T}\mathbf{u} \otimes \mathbf{T}\mathbf{v} = \mathbf{T}\mathbf{u} \otimes \mathbf{v}\mathbf{T}^T} \tag{1.106}$$

1.4 Operations of Tensors

As described in Sect. 1.3, the tensor operations must be expressed by component notation in general. However, the following direct notations of tensor contractions will be used for the simple cases throughout this book.

$$\left.\begin{array}{llll}
(\mathbf{T}\mathbf{v})_i = T_{ir}v_r, & (\mathbf{T}\mathbf{S})_{ij} = T_{ir}S_{rj}, & \mathbf{T} : \mathbf{S} = \mathrm{tr}(\mathbf{T}\mathbf{S}^T) = T_{ij}S_{ij} & \\
(\boldsymbol{\Sigma}\mathbf{v})_{ij} = \Sigma_{ijr}v_r, & (\boldsymbol{\Sigma}\mathbf{T})_{ijk} = \Sigma_{ijr}T_{rk}, & (\boldsymbol{\Sigma} : \mathbf{T})_i = \Sigma_{irs}T_{rs} & \\
(\boldsymbol{\Xi}\mathbf{v})_{ij} = \Xi_{ijr}v_r, & (\boldsymbol{\Xi} : \mathbf{T})_{ij} = \Xi_{ijrs}T_{rs}, & (\mathbf{T} : \boldsymbol{\Xi})_{ij} = T_{rs}\Xi_{rsij}, & (\boldsymbol{\Xi} : \boldsymbol{\Pi})_{ijkl} = \Xi_{ijrs}\Pi_{rskl}
\end{array}\right\} \tag{1.107}$$

where $\mathbf{v}$, $(\mathbf{T}, \mathbf{S})$, $\boldsymbol{\Xi}$ and $(\boldsymbol{\Sigma}, \boldsymbol{\Pi})$ designate the vector, the second-order, the third-order and the fourth-order tensors, respectively.

1.4.1 Trace

An operation taking the sum of the components having the same suffixes, i.e. the sum of diagonal components in the matrix notation is called the *trace* and is expressed as

$$\mathrm{tr}\mathbf{T}\ (=\mathbf{T}:\mathbf{I}) = T_{rs}\delta_{rs} = T_{rr} = T_{11} + T_{22} + T_{33} \tag{1.108}$$

$$\begin{aligned}\mathrm{tr}(\mathbf{AB})(=\mathbf{A}:\mathbf{B}^T) = A_{ir}B_{ri} &= A_{11}B_{11} + A_{12}B_{21} + A_{13}B_{31} \\ &+ A_{21}B_{12} + A_{22}B_{22} + A_{23}B_{32} \\ &+ A_{31}B_{13} + A_{32}B_{23} + A_{33}B_{33}\end{aligned} \tag{1.109}$$

The following relations hold for the trace.

$$\mathrm{tr}(\mathbf{A}+\mathbf{S}) = \mathrm{tr}\,\mathbf{A} + \mathrm{tr}\,\mathbf{B}, \quad \mathrm{tr}(a\mathbf{T}) = a\,\mathrm{tr}\,\mathbf{T}, \quad \mathrm{tr}(\mathbf{AB}) = \mathrm{tr}(\mathbf{BA}), \quad \mathrm{tr}(\mathbf{u}\otimes\mathbf{v}) = \mathbf{u}\cdot\mathbf{v} \tag{1.110}$$

1.4.2 Various Tensors

Various basic tensors used widely in tensor operations are explained in this subsection.

1. Transposed tensor

The following tensor $\mathbf{T}^T$ is called the *transposed tensor*.

$$\mathbf{T}^T = T_{ji}\mathbf{e}_i\otimes\mathbf{e}_j, \quad (\mathbf{T}^T)_{ij} = T_{ji} \tag{1.111}$$

It holds from $\{(\mathbf{AB})^T\}_{ij} = A_{jr}B_{ri}$ that

$$(\mathbf{AB})^T = \mathbf{B}^T\mathbf{A}^T \tag{1.112}$$

Further, the following relation holds for the trace.

$$\mathrm{tr}\mathbf{T}^T = \mathrm{tr}\mathbf{T}, \quad \mathrm{tr}(\mathbf{AB}) = \mathbf{A}:\mathbf{B}^T = \mathbf{A}^T:\mathbf{B} \tag{1.113}$$

The magnitude of tensor is defined as the square root of the sum of the squares of each components and thus it is expressed using Eq. (1.111) as

$$\|\mathbf{T}\| = \sqrt{T_{ij}T_{ij}} = \sqrt{\mathbf{T}:\mathbf{T}} = \sqrt{\mathrm{tr}(\mathbf{T}\mathbf{T}^T)} \tag{1.114}$$

The tensor whose magnitude is unity is called the *unit tensor*.

It follows from Eq. (1.111) that

$$\boxed{\mathbf{T}\mathbf{v} = \mathbf{v}\mathbf{T}^T}, \quad T_{ij}v_j = v_j T_{ij} \tag{1.115}$$

$$\boxed{\mathbf{T}\mathbf{u} \cdot \mathbf{v} = \mathbf{u} \cdot \mathbf{T}^T\mathbf{v}}, \quad T_{ij}u_j v_i = u_j (\mathbf{T}^T)_{ji} v_i \tag{1.116}$$

Further, it holds from Eq. (1.116) that

$$\boxed{\mathbf{A}\mathbf{u} \cdot \mathbf{B}\mathbf{v} = \mathbf{u} \cdot \mathbf{A}^T\mathbf{B}\mathbf{v}}, \ \mathbf{A} : (\mathbf{B}\mathbf{C}) = (\mathbf{B}^T\mathbf{A}) : \mathbf{C} = (\mathbf{A}\mathbf{C}^T) : \mathbf{B} \tag{1.117}$$

2. Inverse tensor

The tensor $\mathbf{T}^{-1}$ fulfilling the following relation is defined as the *inverse tensor* of the tensor $\mathbf{T}$.

$$\mathbf{T}\mathbf{T}^{-1} = \mathbf{I}, \quad T_{ir}(\mathbf{T}^{-1})_{rj} = \delta_{ij} \tag{1.118}$$

Equation (1.14) is rewritten in the direct notation as follows:

$$\mathbf{T}\Delta^T = \mathbf{I} \det \mathbf{T}$$

from which one has

$$\mathbf{T}\frac{\boldsymbol{\Delta}^T}{\det \mathbf{T}} = \mathbf{I}, \quad T_{is}\frac{\Delta_{js}}{\det \mathbf{T}} = \delta_{ij} \tag{1.119}$$

Consequently, $\mathbf{T}^{-1}$ is given by

$$\boxed{\mathbf{T}^{-1} = \frac{\boldsymbol{\Delta}^T}{\det \mathbf{T}}}, \quad (\mathbf{T}^{-1})_{ij} = \frac{\Delta_{ji}}{\det \mathbf{T}} \tag{1.120}$$

Then, $\det \mathbf{T} \neq 0$ is required in order that $\mathbf{T}^{-1}$ exists, so that the tensor fulfilling this condition is called the *non-singular* (or *invertible*) *tensor*. The partial derivative of Eq. (1.18) is rewritten by Eq. (1.120) as

$$\boxed{\frac{\partial \det \mathbf{T}}{\partial \mathbf{T}} = (\det \mathbf{T})\mathbf{T}^{-T}} \tag{1.121}$$

The derivation of Eq. (1.121) starting from the definition of the total differential equation has been often described in some literatures (cf. Leigh 1968; Hashiguchi and Yamakawa 2012) but it needs cumbersome manipulations. Compared with it, the derivation shown above would be concise and straightforward.

The following relations hold for the inverse tensor.

$$(s\mathbf{T})^{-1} = (1/s)\mathbf{T}^{-1}, (\mathbf{T}^T)^{-1} = (\mathbf{T}^{-1})^T \ (\equiv \mathbf{T}^{-T}),$$
$$(\mathbf{AB})^{-1} = \mathbf{B}^{-1}\mathbf{A}^{-1}, \ \det(\mathbf{T}^{-1}) = (\det \mathbf{T})^{-1} \tag{1.122}$$

$$(\mathbf{A}^{-1} + \mathbf{B}^{-1})^{-1} = [\mathbf{B}^{-1}(\mathbf{B} + \mathbf{A})\mathbf{A}^{-1}]^{-1} = \mathbf{A}(\mathbf{B} + \mathbf{A})^{-1}\mathbf{B} = \mathbf{B}(\mathbf{A} + \mathbf{B})^{-1}\mathbf{A} \tag{1.123}$$

because of

$$((\mathbf{T}\mathbf{T}^{-1})^T =) \ (\mathbf{T}^{-1})^T\mathbf{T}^T = \mathbf{I} = (\mathbf{T}^T)^{-1}\mathbf{T}^T$$

$$\mathbf{AB}(\mathbf{AB})^{-1} = \mathbf{I} \rightarrow \mathbf{B}(\mathbf{AB})^{-1} = \mathbf{A}^{-1}$$

Now, when we regard the transformation of the vector $\mathbf{v}$ to the vector $\mathbf{u}$ by the tensor $\mathbf{T}$, i.e.

$$\mathbf{T}\mathbf{v} = \mathbf{u}, \quad T_{ij}v_j = u_i \tag{1.124}$$

as the simultaneous equation in which the components of $\mathbf{v}$ are the unknown numbers, solution exists for $\mathbf{u} \neq \mathbf{0}$ if $\det \mathbf{T} \neq 0$ and is given by $\mathbf{v} = \mathbf{T}^{-1}\mathbf{u}$, noting Eq. (1.120), as

$$\boxed{\mathbf{v} = \frac{\mathbf{\Delta}^T}{\det \mathbf{T}}\mathbf{u}}, \quad v_i = \frac{\Delta_{ji}}{\det \mathbf{T}}u_j \tag{1.125}$$

Here, $\mathbf{T}$ must be the non-singular tensor fulfilling $\det \mathbf{T} \neq 0$ in order that the non-trivial solution $\mathbf{v} \neq \mathbf{0}$ exists for $\mathbf{u} \neq \mathbf{0}$. On the other hand, $\mathbf{T}$ must be the singular tensor fulfilling $\det \mathbf{T} = 0$ in order that the solution $\mathbf{v} \neq \mathbf{0}$ exists for $\mathbf{u} = \mathbf{0}$.

3. Symmetric and skew-symmetric tensors

Tensors $\mathbf{T}^S$ and $\mathbf{T}^A$ fulfilling the following relations are defined as the *symmetric* and the *skew*-(or *anti-*)*symmetric tensor*, respectively.

$$\mathbf{T}^{ST} = \mathbf{T}^S, \quad T^S_{ji} = T^S_{ij} \tag{1.126}$$

and

$$\mathbf{T}^{AT} = -\mathbf{T}^A, \quad T^A_{ji} = -T^A_{ij} \tag{1.127}$$

An arbitrary tensor $\mathbf{T}$ is uniquely decomposed into the symmetric and the skew (anti)-symmetric tensors.

$$\mathbf{T} = \mathbf{T}^S + \mathbf{T}^A \tag{1.128}$$

$$\mathbf{T}^S = \frac{1}{2}(\mathbf{T} + \mathbf{T}^T), \quad \mathbf{T}^A = \frac{1}{2}(\mathbf{T} - \mathbf{T}^T) \tag{1.129}$$

while the components of $\mathbf{T}^S$ and $\mathbf{T}^A$ are often denoted by $T_{(ij)}$ and $T_{[ij]}$, respectively. Equation (1.128) is called the *Cartesian decomposition*, following the decomposition of a complex number to a real and an imaginary parts. It holds that

$$\mathrm{tr}\mathbf{T}^S = \mathrm{tr}\,\mathbf{T}, \quad \mathrm{tr}\,\mathbf{T}^A = 0, \quad \mathrm{tr}(\mathbf{A}^S\mathbf{B}^A) = 0 \tag{1.130}$$

$$(\mathbf{T}^A)_{ii} = 0 \quad (\text{no sum}) \tag{1.131}$$

$$\boxed{\det \mathbf{T}^A = 0} \tag{1.132}$$

4. Mean and deviatoric parts

When the tensor $\mathbf{T}$ is decomposed as follows:

$$\boxed{\mathbf{T} = \mathbf{T}_m + \mathbf{T}'} \tag{1.133}$$

$$\left.\begin{aligned} &\mathbf{T}_m \equiv T_m\mathbf{I}, \quad T_m \equiv \frac{1}{3}(\mathrm{tr}\,\mathbf{T}) = \frac{1}{3}T_{ii} \\ &\mathbf{T}' \equiv \mathbf{T} - T_m\mathbf{I} \quad (\mathrm{tr}\,\mathbf{T}' = 0) \end{aligned}\right\} \tag{1.134}$$

while $\mathbf{T}_m$ and $\mathbf{T}'$ are called the *mean* (or *spherical*) *part* and the *deviatoric part* of the tensor $\mathbf{T}$. Noting Eq. (1.131), the skew-symmetric tensor $\mathbf{T}'^A$ of the deviatoric tensor $\mathbf{T}'$ is given by

$$\mathbf{T}'^A = \mathbf{T}^A \tag{1.135}$$

Then, the symmetric part of the deviatoric tensor is given by

$$\mathbf{T}'^S = \mathbf{T}' - \mathbf{T}^A = \mathbf{T} - T_m\mathbf{I} - \mathbf{T}^A \tag{1.136}$$

from which one has

$$\mathbf{T} = T_m\mathbf{I} + \mathbf{T}'^S + \mathbf{T}^A \tag{1.137}$$

The decomposition of $\mathbf{T}$ into the mean component $T_m\mathbf{I}$, the deviatoric symmetric component $\mathbf{T}'^S$ and the skew-symmetric component $\mathbf{T}^A$ is called *triple decomposition*.The following holds:

$$\mathrm{tr}(\mathbf{A}'\mathbf{B}) = \mathrm{tr}(\mathbf{A}\mathbf{B}') = \mathrm{tr}(\mathbf{A}'\mathbf{B}') \tag{1.138}$$

5. Axial vector

The skew-symmetric tensor $\mathbf{T}^A$ has three independent components in the three-dimensional state. Therefore, vector $\mathbf{t}^A$ having the following components is called the *axial vector*.

$$\boxed{t_i^A = -\frac{1}{2}\varepsilon_{irs}T_{rs}^A} \quad \left(\lfloor t_1^A \quad t_2^A \quad t_3^A \rfloor = \lfloor -T_{23}^A - T_{31}^A - T_{12}^A \rfloor\right) \tag{1.139}$$

Inversely from Eq. (1.139) it is obtained that

$$T_{ij}^A = -\varepsilon_{ijr}t_r^A, \quad T_{ij}^A = \begin{bmatrix} 0 & -t_3^A & t_2^A \\ & 0 & -t_1^A \\ & \text{ant.} & 0 \end{bmatrix} \tag{1.140}$$

Furthermore, noting Eq. (1.54) and the relation

$$T_{ir}^A v_r = -\varepsilon_{irs}t_s^A v_r = \varepsilon_{irs}t_r^A v_s \tag{1.141}$$

the following relation holds.

$$\boxed{\mathbf{T}^A\mathbf{v} = \mathbf{t}^A \times \mathbf{v}} \tag{1.142}$$

The relation of $\mathbf{T}^A$ and $\mathbf{t}^A$ is shown in Fig. 1.2 in the case that $\mathbf{t}^A$ is the angular velocity vector and $\mathbf{v}$ is the position vector of particle. The quantity in Eq. (1.142) designates the peripheral velocity vector, while $\mathbf{T}^A$ is called the *spin tensor* which induces the peripheral velocity by undergoing the multiplication of the position vector.

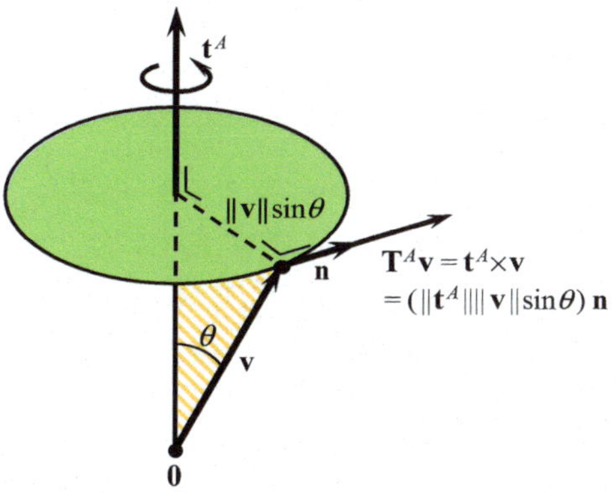

Fig. 1.2 Anti-symmetric tensor and axial vector

6. Fourth-order transformation tensors

The *fourth-order tracing identity tensor* $\mathcal{T}$ for a second-order tensor $\mathbf{I}$ is defined by

$$\boxed{\mathcal{T} \equiv \mathbf{I} \otimes \mathbf{I} = \delta_{ij}\delta_{kl}\mathbf{e}_i \otimes \mathbf{e}_j \otimes \mathbf{e}_k \otimes \mathbf{e}_l = \mathbf{e}_i \otimes \mathbf{e}_i \otimes \mathbf{e}_j \otimes \mathbf{e}_j} \tag{1.143}$$

leading to $\mathcal{T} : \mathbf{T} = \mathbf{T} : \mathcal{T} = (\mathrm{tr}\mathbf{T})\mathbf{I} = 3T_m\mathbf{I}$ for an arbitrary second-order tensor $\mathbf{T}$.

The *fourth-order identity tensor* $\mathcal{I}$ and the *fourth-order transposing tensor* $\underline{\mathcal{I}}$ for a second-order tensor are defined by

$$\boxed{\begin{aligned} \mathcal{I} &\equiv \delta_{ik}\delta_{jl}\mathbf{e}_i \otimes \mathbf{e}_j \otimes \mathbf{e}_k \otimes \mathbf{e}_l = \mathbf{e}_i \otimes \mathbf{e}_j \otimes \mathbf{e}_i \otimes \mathbf{e}_j \\ \underline{\mathcal{I}} &\equiv \delta_{il}\delta_{jk}\mathbf{e}_i \otimes \mathbf{e}_j \otimes \mathbf{e}_k \otimes \mathbf{e}_l = \mathbf{e}_i \otimes \mathbf{e}_j \otimes \mathbf{e}_j \otimes \mathbf{e}_i \end{aligned}} \tag{1.144}$$

leading to $\mathcal{I} : \mathbf{T} = \mathbf{T} : \mathcal{I} = \mathbf{T}$, $\underline{\mathcal{I}} : \mathbf{T} = \mathbf{T} : \underline{\mathcal{I}} = \mathbf{T}^T$, $\mathbf{I} : \underline{\mathcal{I}} = \underline{\mathcal{I}} : \mathbf{I} = \underline{\mathcal{I}}$ and $\partial\mathbf{T}/\partial\mathbf{T} = \mathcal{I}$, $\partial\mathbf{T}^T/\partial\mathbf{T} = \underline{\mathcal{I}}$.

The *symmetrizing tensor* $\mathcal{S}$ and the *skew-*(or *anti-*)*symmetrizing tensor* $\mathcal{A}$ are defined by

$$\begin{aligned} \mathcal{S} &\equiv \frac{1}{2}(\delta_{ik}\delta_{jl} + \delta_{il}\delta_{jk})\mathbf{e}_i \otimes \mathbf{e}_j \otimes \mathbf{e}_k \otimes \mathbf{e}_l = \frac{1}{2}(\mathcal{I} + \underline{\mathcal{I}}) \\ \mathcal{A} &\equiv \frac{1}{2}(\delta_{ik}\delta_{jl} - \delta_{il}\delta_{jk})\mathbf{e}_i \otimes \mathbf{e}_j \otimes \mathbf{e}_k \otimes \mathbf{e}_l = \frac{1}{2}(\mathcal{I} - \underline{\mathcal{I}}) \end{aligned} \tag{1.145}$$

leading to $\mathcal{S} : \mathbf{T} = \mathbf{T}^S$, $\mathcal{A} : \mathbf{T} = \mathbf{T}^A$.

The *deviatoric projection tensor* $\mathcal{I}'$ is defined by

$$\boxed{\mathcal{I}' \equiv (\delta_{ik}\delta_{jl} - \frac{1}{3}\delta_{ij}\delta_{kl})\mathbf{e}_i \otimes \mathbf{e}_j \otimes \mathbf{e}_k \otimes \mathbf{e}_l = \mathcal{I} - \frac{1}{3}\mathcal{T}} \tag{1.146}$$

leading to $\mathcal{I}' : \mathbf{T} = \mathbf{T} : \mathcal{I}' = \mathbf{T}'$.

The following relations hold from Eqs. (1.143)–(1.146).

$$\left.\begin{aligned} &\mathcal{T} : \mathcal{T} = 3\mathcal{T}, \mathcal{S} : \mathcal{S} = \mathcal{S}, \mathcal{A} : \mathcal{A} = \mathbb{O}, \mathcal{I}' : \mathcal{I}' = \mathcal{I}', \\ &\mathcal{T} : \mathcal{I} = \mathcal{I}, \mathcal{T} : \mathcal{S} = \mathcal{T}, \mathcal{T} : \mathcal{A} = \mathbb{O}, \mathcal{T} : \mathcal{I}' = \mathbb{O}, \\ &(\mathcal{T} : \mathbb{T})_{ijkl} = \delta_{ij}\,\mathbb{T}_{rrkl}, \mathcal{I} : \mathbb{T} = \mathbb{T}, \\ &(\mathcal{S} : \mathbb{T})_{ijkl} = (\mathbb{T}_{ijkl} + \mathbb{T}_{jikl})/2, (\mathcal{A} : \mathbb{T})_{ijkl} = (\mathbb{T}_{ijkl} - \mathbb{T}_{jikl})/2, \\ &(\mathcal{I}' : \mathbb{T})_{ijkl} = \mathbb{T}_{ijkl} - \frac{1}{3}\delta_{ij}\,\mathbb{T}_{rrkl}, \end{aligned}\right\} \tag{1.147}$$

where $\mathbb{T}$ is an arbitrary fourth-order tensor and $\mathbb{O}$ is the fourth-order zero tensor.

Further, the following fourth-order tensors are defined by the four types of tensor products of the second-order tensors $\mathbf{A}, \mathbf{B}, \mathbf{C}$ (cf. e.g. del Peiro 1979; Steinmann et al. 1997; Kintzel and Bazar 2006; Wang and Dui 2008).

$$\left.\begin{array}{ll}
\boxed{(\mathbf{A}\otimes\mathbf{B})_{ijkl}=A_{ij}B_{kl}} & \text{with } \mathbf{A}\otimes\mathbf{B}:\mathbf{C}=\mathbf{A}(\mathbf{B}:\mathbf{C})\ ((\mathbf{A}\otimes\mathbf{BC})_{ij}=A_{ij}(B_{kl}C_{kl})) \\
 & \mathbf{C}:\mathbf{A}\otimes\mathbf{B}=(\mathbf{C}:\mathbf{A})\mathbf{B}\ ((\mathbf{CA}\otimes\mathbf{B})_{kl}=C_{ij}A_{ij}B_{kl}) \\
\boxed{(\mathbf{A}\overline{\otimes}\mathbf{B})_{ijkl}=A_{ik}B_{jl}} & \text{with } \mathbf{A}\overline{\otimes}\mathbf{B}:\mathbf{C}=\mathbf{ACB}^T\ ((\mathbf{A}\overline{\otimes}\mathbf{B}:\mathbf{C})_{ij}=A_{ik}B_{jl}C_{kl}=A_{ik}C_{kl}B_{jl}) \\
 & \mathbf{C}:\mathbf{A}\overline{\otimes}\mathbf{B}=\mathbf{A}^T\mathbf{CB}\ ((\mathbf{C}:\mathbf{A}\overline{\otimes}\mathbf{B})_{kl}=C_{ij}A_{ik}B_{jl}=A_{ik}C_{ij}B_{jl}) \\
\boxed{(\mathbf{A}\underline{\otimes}\mathbf{B})_{ijkl}=A_{il}B_{jk}} & \text{with } \mathbf{A}\underline{\otimes}\mathbf{B}:\mathbf{C}=\mathbf{AC}^T\mathbf{B}^T\ ((\mathbf{A}\underline{\otimes}\mathbf{B}:\mathbf{C})_{ij}=A_{il}B_{jk}C_{kl}=A_{il}C_{kl}B_{jk}) \\
 & \mathbf{C}:\mathbf{A}\underline{\otimes}\mathbf{B}=\mathbf{B}^T\mathbf{C}^T\mathbf{A}\ ((\mathbf{C}:\mathbf{A}\underline{\otimes}\mathbf{B})_{kl}=C_{ij}A_{il}B_{jk}=B_{jk}C_{ij}A_{il}) \\
\boxed{(\mathbf{A}\tilde{\otimes}\mathbf{B})_{ijkl}=A_{ik}B_{lj}} & \text{with } \mathbf{A}\tilde{\otimes}\mathbf{B}:\mathbf{C}=\mathbf{ACB}\ ((\mathbf{A}\tilde{\otimes}\mathbf{B}:\mathbf{C})_{ij}=A_{ik}B_{lj}C_{kl}=A_{ik}C_{kl}B_{lj}) \\
 & \mathbf{C}:\mathbf{A}\tilde{\otimes}\mathbf{B}=\mathbf{A}^T\mathbf{CB}^T\ ((\mathbf{C}:\mathbf{A}\tilde{\otimes}\mathbf{B})_{kl}=C_{ij}A_{ik}B_{lj}=A_{ik}C_{ij}B_{lj}) \\
\boxed{(\mathbf{A}\underset{\sim}{\otimes}\mathbf{B})_{ijkl}=A_{il}B_{kj}} & \text{with } \mathbf{A}\underset{\sim}{\otimes}\mathbf{B}:\mathbf{C}=\mathbf{AC}^T\mathbf{B}\ ((\mathbf{A}\underset{\sim}{\otimes}\mathbf{B}:\mathbf{C})_{ij}=A_{il}B_{kj}C_{kl}=A_{il}C_{kl}B_{kj}) \\
 & \mathbf{C}:\mathbf{A}\underset{\sim}{\otimes}\mathbf{B}=\mathbf{BC}^T\mathbf{A}\ ((\mathbf{C}:\mathbf{A}\underset{\sim}{\otimes}\mathbf{B})_{kl}=C_{ij}A_{il}B_{kj}=B_{kj}C_{ij}A_{il})
\end{array}\right\} \tag{1.148}$$

from which it follows that

$$\left.\begin{array}{l}
\mathcal{T}=\mathbf{I}\otimes\mathbf{I},\ \mathcal{I}=\mathbf{I}\overline{\otimes}\mathbf{I},\ \underline{\mathcal{I}}=\mathbf{I}\underline{\otimes}\mathbf{I} \\
\mathbf{AB}=\mathbf{ABI}=\mathbf{A}\overline{\otimes}\mathbf{I}:\mathbf{B} \\
\mathbf{BA}^T=\mathbf{IBA}^T=\mathbf{I}\overline{\otimes}\mathbf{A}:\mathbf{B} \\
\mathbf{AB}^T=\mathbf{AB}^T\mathbf{I}=\mathbf{A}\underline{\otimes}\mathbf{I}:\mathbf{B} \\
\mathbf{A}^T\mathbf{B}^T=\mathbf{IA}^T\mathbf{B}^T=\mathbf{I}\underline{\otimes}\mathbf{B}:\mathbf{A}
\end{array}\right\} \tag{1.149}$$

1.5 Scalar Triple Products with Invariants

The following formulae of the scalar triple products related to the principal invariants hold.

$$\boxed{\begin{array}{l}
[\mathbf{Tu}\ \mathbf{v}\ \mathbf{w}]+[\mathbf{u}\ \mathbf{Tv}\ \mathbf{w}]+[\mathbf{u}\ \mathbf{v}\ \mathbf{Tw}]=\mathrm{tr}\mathbf{T}[\mathbf{uvw}]=\mathrm{I}[\mathbf{uvw}] \\
[\mathbf{u}\ \mathbf{Tv}\ \mathbf{Tw}]+[\mathbf{Tu}\ \mathbf{v}\ \mathbf{Tw}]+[\mathbf{Tu}\ \mathbf{Tv}\ \mathbf{w}]=\mathrm{II}[\mathbf{uvw}] \\
[\mathbf{Tu}\ \mathbf{Tv}\ \mathbf{Tw}]=\det\mathbf{T}[\mathbf{uvw}]=\mathrm{III}[\mathbf{uvw}]
\end{array}} \tag{1.150}$$

where $\mathrm{I},\mathrm{II},\mathrm{III}$ are *principal invariants* of $\mathbf{T}$, i.e.

$$\mathrm{I}\equiv\mathrm{tr}\mathbf{T},\quad \mathrm{II}\equiv(\mathrm{tr}^2\mathbf{T}-\mathrm{tr}\mathbf{T}^2)/2,\quad \mathrm{III}\equiv\det\mathbf{T}$$

which will be delineated in the next section.

The proof for Eq. (1.150) is given below (cf. Chadwick 1976; Kyoya 2008).

Proof of Eq. $(1.150)_1$:
We have the relation

$$
\begin{aligned}
&[\mathbf{Tu},\mathbf{v},\mathbf{w}]+[\mathbf{u},\mathbf{Tv},\mathbf{w}]+[\mathbf{u},\mathbf{v},\mathbf{Tw}]\\
&\quad= u_iv_jw_k([\mathbf{Te}_i,\mathbf{e}_j,\mathbf{e}_k]+[\mathbf{e}_i,\mathbf{Te}_j,\mathbf{e}_k]+[\mathbf{e}_i,\mathbf{e}_j,\mathbf{Te}_k])\\
&\quad= \varepsilon_{ijk}u_iv_jw_k([\mathbf{Te}_1,\mathbf{e}_2,\mathbf{e}_3]+[\mathbf{e}_1,\mathbf{Te}_2,\mathbf{e}_3]+[\mathbf{e}_1,\mathbf{e}_2,\mathbf{Te}_3])\\
&\quad= [\mathbf{uvw}]([\mathbf{Te}_1,\mathbf{e}_2,\mathbf{e}_3]+[\mathbf{e}_1,\mathbf{Te}_2,\mathbf{e}_3]+[\mathbf{e}_1,\mathbf{e}_2,\mathbf{Te}_3])
\end{aligned}
\tag{1.151}
$$

making use of Eq. (1.33) with Eq. (1.31) and noting

$$
[\mathbf{Te}_i,\mathbf{e}_j,\mathbf{e}_k]+[\mathbf{e}_i,\mathbf{Te}_j,\mathbf{e}_k]+[\mathbf{e}_i,\mathbf{e}_j,\mathbf{Te}_k]=\varepsilon_{ijk}([\mathbf{Te}_1,\mathbf{e}_2,\mathbf{e}_3]+[\mathbf{e}_1,\mathbf{Te}_2,\mathbf{e}_3]+[\mathbf{e}_1,\mathbf{e}_2,\mathbf{Te}_3])
\tag{1.152}
$$

Further, the inside of bracket in the last equation in Eq. (1.151) is described as follows:

$$
\begin{aligned}
[\mathbf{Te}_1,\mathbf{e}_2,\mathbf{e}_3]+[\mathbf{e}_1,\mathbf{Te}_2,\mathbf{e}_3]+[\mathbf{e}_1,\mathbf{e}_2,\mathbf{Te}_3] &= [T_{r1}\mathbf{e}_r,\mathbf{e}_2,\mathbf{e}_3]+[\mathbf{e}_1,T_{r2}\mathbf{e}_r,\mathbf{e}_3]+[\mathbf{e}_1,\mathbf{e}_2,T_{r3}\mathbf{e}_r]\\
&= T_{r1}[\mathbf{e}_r\mathbf{e}_2\mathbf{e}_3]+T_{r2}[\mathbf{e}_1\mathbf{e}_r\mathbf{e}_3]+T_{r3}[\mathbf{e}_1\mathbf{e}_2\mathbf{e}_r]\\
&= T_{11}[\mathbf{e}_1\mathbf{e}_2\mathbf{e}_3]+T_{22}[\mathbf{e}_1\mathbf{e}_2\mathbf{e}_3]+T_{33}[\mathbf{e}_1\mathbf{e}_2\mathbf{e}_3]\\
&= T_{11}+T_{22}+T_{33}=\mathrm{tr}\mathbf{T}
\end{aligned}
\tag{1.153}
$$

Equation $(1.150)_1$ is obtained by substituting Eq. (1.153) into Eq. (1.151).
Proof of Eq. $(1.150)_2$:
In the similar way as Eq. (1.151), we have first

$$
\begin{aligned}
&[\mathbf{u},\mathbf{Tv},\mathbf{Tw}]+[\mathbf{Tu},\mathbf{v},\mathbf{Tw}]+[\mathbf{Tu},\mathbf{Tv},\mathbf{w}]\\
&\qquad= \varepsilon_{ijk}u_iv_jw_k([\mathbf{e}_1,\mathbf{Te}_2,\mathbf{Te}_3][\mathbf{Te}_1,\mathbf{e}_2,\mathbf{Te}_3]+[\mathbf{Te}_1,\mathbf{Te}_2,\mathbf{e}_3])\\
&\qquad= [\mathbf{uvw}]([\mathbf{e}_1,\mathbf{Te}_2,\mathbf{Te}_3]+[\mathbf{Te}_1,\mathbf{e}_2,\mathbf{Te}_3]+[\mathbf{Te}_1,\mathbf{Te}_2,\mathbf{e}_3])
\end{aligned}
\tag{1.154}
$$

Here, applying Eqs. (1.31) and (1.152), the inside of bracket in this equation is described as follows:

$$
\begin{aligned}
&[\mathbf{e}_1,\mathbf{Te}_2,\mathbf{Te}_3]+[\mathbf{Te}_1,\mathbf{e}_2,\mathbf{Te}_3]+[\mathbf{Te}_1,\mathbf{Te}_2,\mathbf{e}_3]\\
&\quad= T_{r2}T_{s3}[\mathbf{e}_1\mathbf{e}_r\mathbf{e}_s]+T_{r1}T_{s3}[\mathbf{e}_r\mathbf{e}_2\mathbf{e}_s]+T_{r1}T_{s2}[\mathbf{e}_r\mathbf{e}_s\mathbf{e}_3]\\
&\quad= T_{r2}T_{s3}\varepsilon_{1rs}+T_{r1}T_{s3}\varepsilon_{r2s}+T_{r1}T_{s2}\varepsilon_{rs3}\\
&\quad= (T_{22}T_{33}-T_{23}T_{32})+(T_{11}T_{33}-T_{13}T_{31})+(T_{11}T_{22}-T_{12}T_{12})\\
&\quad= T_{11}T_{22}+T_{22}T_{33}+T_{33}T_{11}-(T_{12}T_{12}+T_{23}T_{32}+T_{31}T_{13})\\
&\quad= \{T_{11}^2+T_{22}^2+T_{33}^2+2(T_{11}T_{22}+T_{22}T_{33}+T_{33}T_{11})\}/2\\
&\qquad -\{T_{11}^2+T_{22}^2+T_{33}^2+2(T_{12}T_{12}+T_{23}T_{32}+T_{31}T_{13})\}/2\\
&\quad= (T_{rr}T_{ss}-T_{rs}T_{sr})/2
\end{aligned}
\tag{1.155}
$$

leading to

$$[\mathbf{e}_1, \mathbf{T}\mathbf{e}_2, \mathbf{T}\mathbf{e}_3] + [\mathbf{T}\mathbf{e}_1, \mathbf{e}_2, \mathbf{T}\mathbf{e}_3] + [\mathbf{T}\mathbf{e}_1, \mathbf{T}\mathbf{e}_2, \mathbf{e}_3] = [(\mathrm{tr}\,\mathbf{T})^2 - \mathrm{tr}^2\mathbf{T}]/2 \qquad (1.156)$$

Equation $(1.150)_2$ is obtained by substituting Eq. (1.156) into Eq. (1.154).

Proof of Eq. $(1.150)_3$:

Changing the representations of the tree vectors to the component-based representations and then applying Eqs. (1.31) and (1.152), we have

$$[\mathbf{T}\mathbf{u}, \mathbf{T}\mathbf{v}, \mathbf{T}\mathbf{w}] = \varepsilon_{ijk} u_i v_j w_k [\mathbf{T}\mathbf{e}_1, \mathbf{T}\mathbf{e}_2, \mathbf{T}\mathbf{e}_3] = [\mathbf{u}\mathbf{v}\mathbf{w}][\mathbf{T}\mathbf{e}_1, \mathbf{T}\mathbf{e}_2, \mathbf{T}\mathbf{e}_3] \qquad (1.157)$$

Equation $(1.150)_3$ is obtained by substituting the following equation into Eq. (1.157).

$$\mathbf{T}\mathbf{e}_p = T_{ij}\mathbf{e}_i \otimes \mathbf{e}_j \mathbf{e}_p = T_{iip}\mathbf{e}_i$$

$$\begin{aligned} |\mathbf{T}\mathbf{e}_1\ \mathbf{T}\mathbf{e}_2\ \mathbf{T}\mathbf{e}_3| &= |T_{i1}\mathbf{e}_i\ T_{j2}\mathbf{e}_j\ T_{k3}\mathbf{e}_k| = T_{i1}T_{j2}T_{k3}|\mathbf{e}_i\ \mathbf{e}_j\ \mathbf{e}_k| \\ &= T_{i1}T_{j2}T_{k3}\varepsilon_{ijk}|\mathbf{e}_1\ \mathbf{e}_2\ \mathbf{e}_3| = \det \mathbf{T} \end{aligned}$$

with the aid of Eq. (1.31). It follows from Eq. $(1.150)_3$ that

$$\varepsilon_{ijk} \det \mathbf{T} = [\mathbf{T}\mathbf{e}_i, \mathbf{T}\mathbf{e}_j, \mathbf{T}\mathbf{e}_k] \qquad (1.158)$$

1.6 Eigenvalues and Eigenvectors

The symmetric tensor **T** with real symmetric components is expressed in the component notation having only real normal (diagonal) components by choosing the orthogonal coordinate axes in particular directions. This fact will be proven below.

The unit vector **e** fulfilling

$$\boxed{\mathbf{T}\mathbf{e} = T\mathbf{e}}, \quad T_{ij}e_j = Te_i \qquad (1.159)$$

i.e.

$$(\mathbf{T} - T\mathbf{I})\mathbf{e} = \mathbf{0}, \quad (T_{ij} - T\delta_{ij})e_j = 0 \qquad (1.160)$$

for the second-order tensor is called the *eigenvector* (or *principal* or *characteristic* or *proper vector*) and the scalar T is called the *eigenvalue* (or *principal* or *characteristic* or *proper value*). Taking the determinant of Eq. (1.160), one has

$$\det(\mathbf{T} - T\mathbf{I}) \det \mathbf{e} = \det \mathbf{0}$$

noting $\det(\mathbf{T}-T\mathbf{I})\det\mathbf{e}=\det\mathbf{0}$ with $\det\mathbf{e}=1$ and $\det\mathbf{0}=0$, the necessary and sufficient condition that the simultaneous Eq. (1.160) has a non-zero solution of $\mathbf{e}$ is given by

$$\boxed{\det(\mathbf{T}-T\mathbf{I})=0},\quad \left|T_{ij}-T\delta_{ij}\right|=0 \tag{1.161}$$

Equation (1.161) is called the *characteristic equation* of the tensor, which is regarded as the cubic equation of T. Unit vectors $\mathbf{e}_1,\mathbf{e}_2,\mathbf{e}_3$ are derived for each of solutions T_1,T_2,T_3 from Eq. (1.160). Here, Eq. (1.161) is also obtained as the necessary and sufficient condition that a non-zero (non-trivial) solution of $\mathbf{e}$ exists in Eq. (1.160).

(Note) Consider the simultaneous equation for the unknown values x_1,x_2,x_3:

$$\left.\begin{aligned} a_{11}x_1+a_{12}x_2+a_{13}x_3&=c_1\\ a_{21}x_1+a_{22}x_2+a_{23}x_3&=c_2\\ a_{31}x_1+a_{32}x_2+a_{33}x_3&=c_3\end{aligned}\right\} \tag{a}$$

where $a_{ij}(i,j=1,2,3)$ are the known coefficients. Equation (a) is described in the forms:

$$\left.\begin{aligned}&\mathbf{A}\mathbf{x}=\mathbf{c}:\text{tensor form}\\ &[\mathbf{A}]\{\mathbf{x}\}=\{\mathbf{c}\}:\text{matrix form}\end{aligned}\right\} \tag{b}$$

setting

$$\mathbf{A}=\begin{bmatrix}a_{11}a_{12}a_{13}\\ a_{21}a_{22}a_{23}\\ a_{31}a_{32}a_{33}\end{bmatrix},\quad \mathbf{x}=\begin{Bmatrix}x_1\\ x_2\\ x_3\end{Bmatrix},\quad \mathbf{c}=\begin{Bmatrix}c_1\\ c_2\\ c_3\end{Bmatrix}$$

Solutions of Eq. (a) are given by

$$x_1=\frac{\begin{vmatrix}c_1a_{12}a_{13}\\ c_2a_{22}a_{23}\\ c_3a_{32}a_{33}\end{vmatrix}}{\begin{vmatrix}a_{11}a_{12}a_{13}\\ a_{21}a_{22}a_{23}\\ a_{31}a_{32}a_{33}\end{vmatrix}},\quad x_2=\frac{\begin{vmatrix}a_{11}c_1a_{13}\\ a_{21}c_2a_{23}\\ a_{31}c_3a_{33}\end{vmatrix}}{\begin{vmatrix}a_{11}a_{12}a_{13}\\ a_{21}a_{22}a_{23}\\ a_{31}a_{32}a_{33}\end{vmatrix}},\quad x_3=\frac{\begin{vmatrix}a_{11}a_{12}c_1\\ a_{21}a_{22}c_2\\ a_{31}a_{32}c_3\end{vmatrix}}{\begin{vmatrix}a_{11}a_{12}a_{13}\\ a_{21}a_{22}a_{23}\\ a_{31}a_{32}a_{33}\end{vmatrix}}$$

In the case of $c_1=c_2=c_3=0$ as seen in Eq. (1.161), only the trivial solutions $x_1=x_2=x_3=0$ can exist for $\det\mathbf{A}\neq 0$. Then, the following equation must hold in order that nontrivial solution exits.

$$\det \mathbf{A} = \begin{vmatrix} a_{11} & a_{12} & a_{13} \\ a_{21} & a_{22} & a_{23} \\ a_{31} & a_{32} & a_{33} \end{vmatrix} = 0 \rightarrow \det(\mathbf{T} - T\mathbf{I}) = \begin{vmatrix} T_{11} - T & T_{12} & T_{13} \\ T_{21} & T_{22} - T & T_{23} \\ T_{31} & T_{32} & T_{33} - T \end{vmatrix} = 0$$

In what follows, it is proven that the eigenvalues are real and the eigenvectors are mutually orthogonal in the second-order real symmetric tensor.

The complex conjugate relation for Eq. (1.159) is generally described as follows:

$$\mathbf{T}\bar{\mathbf{e}} = \bar{T}\bar{\mathbf{e}} \tag{1.162}$$

designating the conjugate quantities by the over bar ($\bar{\ }$).

The multiplication of $\bar{\mathbf{e}}$ and $\mathbf{e}$ to Eqs. (1.159) and (1.162) leads to

$$\bar{\mathbf{e}} \cdot \mathbf{T}\mathbf{e} = T\bar{\mathbf{e}} \cdot \mathbf{e}, \mathbf{e} \cdot \mathbf{T}\bar{\mathbf{e}} = \bar{T}\mathbf{e} \cdot \bar{\mathbf{e}}$$

from which we have

$$\bar{\mathbf{e}} \cdot \mathbf{T}\mathbf{e} - \mathbf{e} \cdot \mathbf{T}\bar{\mathbf{e}} = (T - \bar{T})\bar{\mathbf{e}} \cdot \mathbf{e}$$

The left-hand side of this equation is zero because of $\bar{\mathbf{e}} \cdot \mathbf{T}\mathbf{e} - \mathbf{e} \cdot T\bar{\mathbf{e}} = \bar{\mathbf{e}} \cdot (\mathbf{T} - \mathbf{T}^T)\mathbf{e} = 0$. Then, we have $T = \overline{T}$ so that T must be a real number.

Further, it follows from Eq. (1.159) that

$$\mathbf{T}\mathbf{e}_\alpha = T_\alpha \mathbf{e}_\alpha (\text{no sum}) \tag{1.163}$$

for the eigenvectors $\mathbf{e}_\alpha (\alpha = 1, 2, 3)$ of $\mathbf{T}$. By making the scalar products of Eq. (1.162) and the eigenvectors, we have

$$\left.\begin{aligned} \mathbf{e}_\beta \cdot \mathbf{T}\mathbf{e}_\alpha &= T_\alpha \mathbf{e}_\alpha \cdot \mathbf{e}_\beta \\ \mathbf{e}_\alpha \cdot \mathbf{T}\mathbf{e}_\beta &= T_\beta \mathbf{e}_\beta \cdot \mathbf{e}_\alpha \end{aligned}\right\} (\text{no sum})$$

Subtracting the lower equation from the upper equation, one has

$$\mathbf{e}_\beta \cdot \mathbf{T}\mathbf{e}_\alpha - \mathbf{e}_\alpha \cdot \mathbf{T}\mathbf{e}_\beta = (T_\alpha - T_\beta)\mathbf{e}_\alpha \cdot \mathbf{e}_\beta$$

which reads:

$$(T_\alpha - T_\beta)\mathbf{e}_\alpha \cdot \mathbf{e}_\beta = 0 \quad (\text{no sum}) \tag{1.164}$$

noting $\mathbf{e}_\beta \cdot \mathbf{T}\mathbf{e}_\alpha - \mathbf{e}_\alpha \cdot \mathbf{T}\mathbf{e}_\beta = \mathbf{e}_\beta \cdot \mathbf{T}\mathbf{e}_\alpha - \mathbf{T}^T\mathbf{e}_\alpha \cdot \mathbf{e}_\beta = \mathbf{e}_\beta \cdot (\mathbf{T} - \mathbf{T}^T)\mathbf{e}_\alpha$. The following facts can be concluded from Eq. (1.164).

1. If three principal values are all different to each other, there exist the three principal directions which are perpendicular to each other.

2. If two of three principal values are same, all directions in the plane perpendicular to the principal direction for the other principal value are the principal directions for the same principal value.
3. If all three principal values are same, all directions in the space are the principal directions.

Based on the result described above, denoting the eigenvectors by $\mathbf{e}_P$ and the corresponding eigenvalues as T_J, one can write

$$\mathbf{T}\mathbf{e}_J = T_J\mathbf{e}_J \quad (\text{no sum}) \tag{1.165}$$

In addition, noting that the shear component on the coordinate system with the base vector $\{\mathbf{e}_P\}$ is zero, i.e.

$$T_{PQ} = 0\,(P \neq Q) \tag{1.166}$$

the symmetric tensor $\mathbf{T}$ possessing orthogonal principal directions is expressed by

$$\boxed{\mathbf{T} = \sum_{P=1}^{3} T_P\mathbf{e}_P \otimes \mathbf{e}_P} \tag{1.167}$$

which is called the *spectral representation*. Then, the tensor function of tensor $\mathbf{T}$ is described in general as

$$\mathbf{f}(\mathbf{T}) \equiv \sum_{P=1}^{3} f(T_P)\mathbf{e}_P \otimes \mathbf{e}_P \tag{1.168}$$

and then one has

$$[\mathbf{f}(\mathbf{T})]^T = \mathbf{f}(\mathbf{T}^T) \tag{1.169}$$

If tensor $\widetilde{\mathbf{T}}$ having eigenvector $\widetilde{\mathbf{e}}_P$ has the same eigenvalues as tensor $\mathbf{T}$, it holds that

$$\widetilde{\mathbf{T}}\,\widetilde{\mathbf{e}}_P = T_P\widetilde{\mathbf{e}}_P\,(\text{no sum}) \tag{1.170}$$

where the orthogonal tensor $\widetilde{\mathbf{Q}}$ between the eigenvectors of these tensors is given by

$$\widetilde{Q}_{PQ} = \widetilde{\mathbf{e}}_P \cdot \mathbf{e}_Q, \quad \widetilde{\mathbf{Q}} = \mathbf{e}_P \otimes \widetilde{\mathbf{e}}_P, \qquad \mathbf{e}_J = \widetilde{\mathbf{Q}}\widetilde{\mathbf{e}}_P, \qquad \widetilde{\mathbf{e}}_P = \widetilde{\mathbf{Q}}^T\mathbf{e}_P, \tag{1.171}$$

Applying $\widetilde{\mathbf{Q}}$ to Eq. (1.165), one has

$$(\widetilde{\mathbf{Q}}^T\mathbf{T}\mathbf{e}_P =)\widetilde{\mathbf{Q}}^T\mathbf{T}\widetilde{\mathbf{Q}}\widetilde{\mathbf{Q}}^T\mathbf{e}_P = T_P\widetilde{\mathbf{Q}}^T\mathbf{e}_P \quad \text{(no sum)}$$

from which, considering Eqs. (1.170) and (1.171), it holds that

$$\widetilde{\mathbf{Q}}^T\mathbf{T}\widetilde{\mathbf{Q}}\widetilde{e}_P = T_P\widetilde{\mathbf{e}}_P = \widetilde{\mathbf{T}}\widetilde{\mathbf{e}}_P \quad \text{(no sum)} \tag{1.172}$$

Then, one obtains the relation

$$\widetilde{\mathbf{T}} = \widetilde{\mathbf{Q}}^T\mathbf{T}\widetilde{\mathbf{Q}} \tag{1.173}$$

As presented above, tensors having identical eigenvalues can be related by the orthogonal tensor; they are called the *similar tensor* mutually. The coordinate transformation rule (1.86) of a certain tensor and the relation (1.173) of two tensors having identical eigenvalues but different eigenvectors, are of mutually opposite forms.

If the function f of tensor $\mathbf{A}, \mathbf{B}, \cdots$ is observed to be identical independent of observers, i.e. if it fulfills the relation

$$f(\mathbf{A}, \mathbf{B}, \cdots) = f(\mathbf{Q}[\![\mathbf{A}]\!], \mathbf{Q}[\![\mathbf{B}]\!], \cdots) \tag{1.174}$$

using the symbol in Eq. (1.83), f is called the *isotropic scalar-valued tensor function*, which is none other than the invariant. In particular, if the isotropic scalar-valued tensor function $f(\mathbf{T})$ of single second-order tensor $\mathbf{T}$ fulfills

$$f(\mathbf{T}) = f(\mathbf{Q}\mathbf{T}\mathbf{Q}^T) \tag{1.175}$$

$f(\mathbf{T})$ can be expressed by three principal values in the three-dimensional case, including them in symmetric form so as to be identical even if they are exchanged to each other. Then, there exist three independent invariants for a single tensor. Their explicit forms are presented below. The expansion of the characteristic Eq. (1.161) of $\mathbf{T}$ leads to

$$\begin{aligned}
&\begin{vmatrix} T_{11}-T & T_{12} & T_{13} \\ T_{21} & T_{22}-T & T_{23} \\ T_{31} & T_{32} & T_{33}-T \end{vmatrix} \\
&= (T_{11}-T)(T_{22}-T)(T_{33}-T) + T_{12}T_{23}T_{31} + T_{21}T_{32}T_{13} \\
&\quad - (T_{11}-T)T_{23}T_{32} - (T_{22}-T)T_{31}T_{13} - (T_{33}-T)T_{12}T_{21} \\
&= -T^3 + (T_{11}+T_{22}+T_{33})T^2 - (T_{11}T_{22}+T_{22}T_{33}+T_{33}T_{11})T \\
&\quad + T_{11}T_{22}T_{33} + T_{12}T_{23}T_{31} + T_{21}T_{32}T_{13} \\
&\quad + (T_{12}T_{21}+T_{23}T_{32}+T_{31}T_{13})T - T_{11}T_{23}T_{32} - T_{22}T_{31}T_{13} - T_{33}T_{12}T_{21} \\
&= -T^3 + (T_{11}+T_{22}+T_{33})T^2 - (T_{11}T_{22}+T_{22}T_{33}+T_{33}T_{11} \\
&\quad - T_{12}T_{21} - T_{23}T_{32} - T_{31}T_{13})T \\
&\quad + T_{11}T_{22}T_{33} + T_{12}T_{23}T_{31} + T_{21}T_{32}T_{13} - T_{11}T_{23}T_{32} - T_{22}T_{31}T_{13} - T_{33}T_{12}T_{21}
\end{aligned} \tag{1.176}$$

from which the characteristic equation is given as

$$\boxed{T^3 - \mathrm{I}T^2 + \mathrm{II}\,T - \mathrm{III} = 0} \tag{1.177}$$

where

$$\boxed{\mathrm{I} \equiv T_{11} + T_{22} + T_{33} = T_{ii} = \mathrm{tr}\mathbf{T}} \tag{1.178}$$

$$\boxed{\mathrm{II} \equiv \begin{vmatrix} T_{11} & T_{12} \\ T_{21} & T_{22} \end{vmatrix} + \begin{vmatrix} T_{22} & T_{23} \\ T_{32} & T_{33} \end{vmatrix} + \begin{vmatrix} T_{11} & T_{13} \\ T_{31} & T_{33} \end{vmatrix} = \Delta_{ii} = \frac{1}{2}(T_{rr}T_{ss} - T_{rs}T_{sr}) = \frac{1}{2}[(\mathrm{tr}\,\mathbf{T})^2 - \mathrm{tr}\,\mathbf{T}^2]} \tag{1.179}$$

$$\boxed{\mathrm{III} \equiv \begin{vmatrix} T_{11} & T_{12} & T_{13} \\ T_{21} & T_{22} & T_{23} \\ T_{31} & T_{32} & T_{33} \end{vmatrix} = \det\mathbf{T} = \varepsilon_{rst}T_{r1}T_{s2}T_{t3} = \frac{1}{6}(\mathrm{tr}\mathbf{T})^3 - \frac{1}{2}\mathrm{tr}\mathbf{T}\mathrm{tr}\mathbf{T}^2 + \frac{1}{3}\mathrm{tr}\,\mathbf{T}^3} \tag{1.180}$$

The direct notation of III is derived by taking the trace of the expression of the Cayley-Hamilton theorem described in Sect. 1.9 and substituting Eqs. (1.178) and (1.179) into it, i.e.

$$\mathrm{tr}\mathbf{T}^3 - \mathrm{I}\,\mathrm{tr}\mathbf{T}^2 + \mathrm{II}\,\mathrm{tr}\mathbf{T} - 3\,\mathrm{III} = \mathrm{tr}\mathbf{T}^3 - (\mathrm{tr}\,\mathbf{T})\mathrm{tr}\mathbf{T}^2 + \frac{1}{2}[(\mathrm{tr}\,\mathbf{T})^2 - \mathrm{tr}\mathbf{T}^2]\mathrm{tr}\mathbf{T} - 3\,\mathrm{III} = 0$$

On the other hand, the characteristic Eq. (1.177) is expressed using the principal values as follows:

$$(T - T_1)(T - T_2)(T - T_3) = 0 \tag{1.181}$$

i.e.

$$T^3 - (T_1 + T_2 + T_3)T^2 + (T_1T_2 + T_1T_2 + \mathrm{T}_3T_1)T - T_1T_2T_3 = 0$$

Comparing Eqs. (1.177) and (1.181), coefficients I, II and III are described as

$$\boxed{\begin{aligned} &\mathrm{I} = T_1 + T_2 + T_3 \\ &\mathrm{II} = T_1T_2 + T_2T_3 + \mathrm{T}_3T_1 \\ &\mathrm{III} = T_1T_2T_3 \end{aligned}} \tag{1.182}$$

Equation (1.182) can also be derived by substituting $T_{11} = T_1, T_{22} = T_2,\ \ T_{33} = T_3, T_{12} = T_{23} = T_{31} = 0$ in Eqs. (1.178)–(1.180), while hereinafter principal value is denoted by only one suffix. Since I, II and III are the symmetric functions of principal values, they are the invariants and are called the *principal invariants*.

The following invariant are called the *moments*.

$$\bar{\mathrm{I}} \equiv \mathrm{I} = \mathrm{tr}\mathbf{T}, \quad \bar{\mathrm{II}} \equiv \mathrm{tr}\mathbf{T}^2, \quad \bar{\mathrm{III}} \equiv \mathrm{tr}\mathbf{T}^3 \tag{1.183}$$

The principal invariants are described in terms of these moments from Eqs. (1.178)–(1.180) as follows:

$$\boxed{\begin{aligned} &\mathrm{I} = \bar{\mathrm{I}} \\ &\mathrm{II} = \frac{1}{2}(\bar{\mathrm{I}}^2 - \bar{\mathrm{II}}) \\ &\mathrm{III} = \frac{1}{6}\bar{\mathrm{I}}^3 - \frac{1}{2}\bar{\mathrm{III}} + \frac{1}{3}\bar{\mathrm{III}} \end{aligned}} \tag{1.184}$$

Next, consider the deviatoric tensor $\mathbf{T}'$. The characteristic equation of $\mathbf{T}'$ is given by replacing $\mathbf{T}$ to $\mathbf{T}'$ in Eq. (1.177) as follows:

$$\boxed{T_3' - \mathrm{II}'T' - \mathrm{III}' = 0} \tag{1.185}$$

noting

$$\mathrm{I}' \equiv \mathrm{tr}\mathbf{T}' = 0 \tag{1.186}$$

where

$$\begin{aligned} &\boxed{\mathrm{II}' \equiv \Delta_{ii}' = \frac{1}{2}T_{rs}'T_{sr}' = \frac{1}{2}\mathrm{tr}\,\mathbf{T}'^2 = \frac{1}{2}\|\mathbf{T}'\|^2} \\ &= \frac{1}{2}(T_{11}'^2 + T_{22}'^2 + T_{33}'^2) + T_{12}'^2 + T_{23}'^2 + T_{31}'^2 \\ &= \frac{1}{6}\{(T_{11} - T_{22})^2 + (T_{22} - T_{33})^2 + (T_{33} - T_{11})^2\} + T_{12}'^2 + T_{23}'^2 + T_{31}' \\ &= \frac{1}{2}(T_1'^2 + T_2'^2 + T_3'^2) = \frac{1}{6}[(T_1 - T_2)^2 + (T_2 - T_3)^2 + (T_3 - T_1)^2] \end{aligned} \tag{1.187}$$

$$\begin{aligned} &\boxed{\mathrm{III}' \equiv \begin{vmatrix} T_{11}' & T_{12}' & T_{13}' \\ T_{21}' & T_{22}' & T_{23}' \\ T_{31}' & T_{32}' & T_{33}' \end{vmatrix} = \det\mathbf{T}' = \varepsilon_{ijk}T_{i1}'T_{j2}'T_{k3}'} \\ &= T_{11}'T_{22}'T_{33}' + 2T_{12}'T_{23}'T_{31}' - T_{11}'T_{23}'^2 - T_{22}'T_{31}'^2 - T_{33}'T_{12}'^2 = T_1'T_2'T_3' \end{aligned} \tag{1.188}$$

It holds by the analogous method done for Eq. (1.182) that

$$\boxed{\begin{aligned} &\mathrm{I}' = 0 \\ &\mathrm{II}' = T_1'T_2' + T_2'T_3' + \mathrm{T}_3'T_1' \\ &\mathrm{III}' = T_1'T_2'T_3' \end{aligned}} \tag{1.189}$$

1.7 Calculations of Eigenvalues and Eigenvectors

The second-order symmetric tensor can be represented by Eq. (1.167) as the spectral representation in the eigendirections. To express the tensor in the eigendirections, one must calculate the eigenvalues and the eigenvectors of the tensor. The solutions for them (cf. Hoger and Carlson 1984; Carlson and Hoger 1986) are shown in this section.

1.7.1 Eigenvalues

In order to obtain eigenvalues, one try to calculate the deviatoric components from the characteristic Eq. (1.185) which is the cubic equation having the coefficients as the functions of invariants. Now, infer the form

$$T' = \sqrt{\frac{4\mathrm{II}'}{3}}\cos\psi \tag{1.190}$$

for the eigenvalues of deviatoric part of tensor **T**. The substituting Eq. (1.190) into Eq. (1.185), we have

$$\left(\frac{4\mathrm{II}'}{3}\right)^{3/2}\cos^3\psi - \mathrm{II}'\left(\frac{4\mathrm{II}'}{3}\right)^{1/2}\cos\psi - \mathrm{III}' = 0 \tag{1.191}$$

which is reduced to

$$\frac{\sqrt{4}}{3\sqrt{3}}\mathrm{II}'^{3/2}\cos 3\psi - \mathrm{III}' = 0 \tag{1.192}$$

using the trigonometric formula

$$\cos^3\psi = \frac{1}{4}(\cos 3\psi + 3\cos\psi) \tag{1.193}$$

It is obtained from (1.192) that

$$\cos 3\psi = \frac{3\sqrt{3}\mathrm{III}'}{2\mathrm{II}'^{3/2}} \tag{1.194}$$

Noting that the cosine is the periodic function with the period 2π, the angle ψ is expressed by the following equation with a natural number J in general.

$$\psi_P = \frac{1}{3}\left[\cos^{-1}\left(\frac{3\sqrt{3}\mathrm{III}'}{2\mathrm{II}'^{3/2}}\right) - 2\pi J\right] \tag{1.195}$$

Substituting Eq. (1.195) into Eq. (1.190), one has

$$T_p' = \frac{1}{3}\sqrt{\frac{4\mathrm{II}'}{3}}\cos\left\{\frac{1}{3}\left[\cos^{-1}\left(\frac{3\sqrt{3}\mathrm{III}'}{2\mathrm{II}'^{3/2}}\right) - 2\pi J\right]\right\} \tag{1.196}$$

and adding the isotropic component I/3, the eigenvalues of $\mathbf{T}$ are given as follows:

$$\boxed{T_P = \frac{1}{3}\left(\mathrm{I} + \sqrt{\frac{4\mathrm{II}'}{3}}\cos\left\{\frac{1}{3}\left[\cos^{-1}\left(\frac{3\sqrt{3}\mathrm{III}'}{2\mathrm{II}'^{3/2}}\right) - 2\pi J\right]\right\}\right)} \tag{1.197}$$

1.7.2 Eigenvectors

Equation (1.167) can be expressed as follows:

$$\mathbf{T} = \sum_{P=1}^{3} T_P \mathbf{E}_P \tag{1.198}$$

while the tensor $\mathbf{E}_J$ is called the *eigenprojection* of $\mathbf{T}$, which is defined by

$$\boxed{\mathbf{E}_P \equiv \mathbf{e}_P \otimes \mathbf{e}_P \quad (\text{no sum})} \tag{1.199}$$

fulfilling

$$\sum_{P=1}^{3} \mathbf{E}_P (= \mathbf{e}_1 \otimes \mathbf{e}_1 + \mathbf{e}_2 \otimes \mathbf{e}_2 + \mathbf{e}_3 \otimes \mathbf{e}_3) = \mathbf{I} \tag{1.200}$$

$$\left.\begin{array}{l} \mathbf{E}_P\mathbf{E}_Q = \begin{cases} \mathbf{E}_P & \text{for } P = Q \\ \mathbf{O} & \text{for } P \neq Q \end{cases} \\ \mathbf{E}_P : \mathbf{E}_Q = \delta_{PQ} \end{array}\right\} \tag{1.201}$$

It holds that

$$\left.\begin{array}{l} \mathbf{T}\mathbf{E}_P = \left(\sum\limits_{Q=1}^{3} T_Q \mathbf{E}_Q\right)\mathbf{e}_P \otimes \mathbf{e}_P = \left(\sum\limits_{Q=1}^{3} T_Q \mathbf{e}_Q \otimes \mathbf{e}_Q\right)\mathbf{e}_P \otimes \mathbf{e}_P = T_P \mathbf{e}_P \otimes \mathbf{e}_P \\ \mathbf{E}_P\mathbf{T} = \mathbf{e}_P \otimes \mathbf{e}_P\left(\sum\limits_{Q=1}^{3} T_Q \mathbf{E}_Q\right) = \mathbf{e}_P \otimes \mathbf{e}_P\left(\sum\limits_{Q=1}^{3} T_Q \mathbf{e}_Q \otimes \mathbf{e}_Q\right) = \mathbf{e}_P \otimes \mathbf{e}_P T_P \end{array}\right\}$$

(no sum for P)

and thus one has

$$\mathbf{T}\mathbf{E}_P = \mathbf{E}_P\mathbf{T} = T_P\mathbf{E}_P \quad \text{(no sum)} \tag{1.202}$$

On the other hand, it holds from Eq. (1.200) that

$$\mathbf{T} - T_Q\mathbf{I} = \sum_{P=1}^{3} T_P\mathbf{E}_P - \mathrm{T}_Q\sum_{P=1}^{3}\mathbf{E}_P$$

and thus it is obtained that

$$\mathbf{T} - T_Q\mathbf{I} = \sum_{P=1}^{3}(T_P - \mathrm{T}_Q)\mathbf{E}_P \tag{1.203}$$

from which one has

$$\prod_{\substack{Q\neq\theta\\Q=1}}^{3}(\mathbf{T} - T_Q\mathbf{I}) = \prod_{\substack{Q\neq\theta\\Q=1}}^{3}\sum_{P=1}^{3}(T_P - \mathrm{T}_Q)\mathbf{E}_Q = \left[\prod_{\substack{Q\neq\theta\\Q=1}}^{3}(T_\theta - \mathrm{T}_Q)\right]\mathbf{E}_\theta \tag{1.204}$$

from which the following *Sylvester's formula* is obtained from Eqs. (1.204).

$$\boxed{\mathbf{E}_\theta = \prod_{\substack{Q\neq\theta\\Q=1}}^{3}\frac{\mathbf{T} - T_Q\mathbf{I}}{T_\theta - T_Q}} \tag{1.205}$$

For instance, $\mathbf{E}_2(\theta = 2)$ in the popular case of $n = 3$ is obtained from Eq. (1.205) as follows:

$$\mathbf{E}_2 = \frac{\prod_{P=1,3}(\mathbf{T} - T_P\mathbf{I})}{\prod_{P=1,3}(T_2 - T_P)} = \frac{(\mathbf{T} - T_1\mathbf{I})(\mathbf{T} - T_3\mathbf{I})}{(T_2 - \mathrm{T}_1)(T_2 - \mathrm{T}_3)}$$

1.8 Eigenvalues and Eigenvectors of Skew-Symmetric Tensor

The characteristic equation of skew-symmetric tensor is given by substituting $\mathbf{T}^A$ into Eq. (1.177) as follows:

$$T^3 - \mathrm{tr}\mathbf{T}^A T^2 + \frac{1}{2}(\mathrm{tr}^2\mathbf{T}^A - \mathrm{tr}\mathbf{T}^{A2})\mathrm{T} - \det\mathbf{T}^A = 0 \tag{1.206}$$

Noting $\mathrm{tr}\mathbf{T}^A = \det \mathbf{T}^A = 0$ in Eqs. (1.130) and (1.132), Eq. (1.206) leads to

$$(2T^2 - \mathrm{tr}\,\mathbf{T}^{A2})\frac{T}{2} = 0 \tag{1.207}$$

from which the eigenvalues are given by

$$T = \pm i\sqrt{|\mathrm{tr}\,\mathbf{T}^{A2}|/2} = \pm i\|\mathbf{t}^A\| \quad \text{and} \quad 0 \tag{1.208}$$

noting

$$\mathrm{tr}\mathbf{T}^{A2} = \mathrm{tr}\left(\begin{bmatrix} 0 & -t_3^A & t_2^A \\ & 0 & -t_1^A \\ \text{ant.} & & 0 \end{bmatrix}\begin{bmatrix} 0 & -t_3^A & t_2^A \\ & 0 & -t_1^A \\ \text{ant.} & & 0 \end{bmatrix}\right) = -2(t_1^{A2} + t_2^{A2} + t_3^{A2}) < 0$$

obtained from Eq. (1.140), where t_i^A is the axial vectors defined in Eq. (1.139). It is known from Eq. (1.208) that one principal value is zero and two principal values are pure-imaginary numbers without a real part.

Here, if one of the principal direction $\mathbf{e}_3$ is chosen to the direction of the zero principal value of $\mathbf{T}^A$ leading to $T_{13}^A = t_{23}^A = 0$, it follows by denoting $T_{12}^A = -t_{12}^A \equiv \omega$ that

$$\begin{aligned}\mathbf{T}^{A*} &= \mathbf{Q}\mathbf{T}^A\mathbf{Q}^T \\ &= \begin{bmatrix} \cos\theta & \sin\theta & 0 \\ -\sin\theta & \cos\theta & 0 \\ 0 & 0 & 1 \end{bmatrix}\begin{bmatrix} 0 & \omega & 0 \\ -\omega & 0 & 0 \\ 0 & 0 & 0 \end{bmatrix}\begin{bmatrix} \cos\theta & -\sin\theta & 0 \\ \sin\theta & \cos\theta & 0 \\ 0 & 0 & 1 \end{bmatrix} \\ &= \begin{bmatrix} -\omega\sin\theta\cos\theta + \omega\sin\theta\cos\theta & \omega\sin^2\theta + \omega\cos^2\theta & 0 \\ -\omega\cos^2\theta - \omega\sin^2\theta & \omega\sin\theta\cos\theta - \omega\sin\theta\cos\theta & 0 \\ 0 & 0 & 0 \end{bmatrix} \\ &= \begin{bmatrix} 0 & \omega & 0 \\ -\omega & 0 & 0 \\ 0 & 0 & 0 \end{bmatrix} = \mathbf{T}^A \end{aligned} \tag{1.209}$$

meaning that the components do not change in the coordinate transformation. It is caused from the fact that the independent component of skew-symmetric tensor is only one when the one of bases in the coordinate system is chosen to the principal direction of skew-symmetric tensor.

1.9 Cayley-Hamilton Theorem

Denoting the eigenvector of the tensor $\mathbf{T}$ by $\mathbf{e}$, it follows from Eq. (1.159) that

$$\mathbf{T}^r\mathbf{e} = T^r\mathbf{e} \tag{1.210}$$

by the repeated applications of Eq. (1.159), i.e. $T^r\mathbf{e} = T^{r-1}T\mathbf{e} = T^{r-1}\mathbf{T}\mathbf{e} = \mathbf{T}T^{r-2}T\mathbf{e} = \mathbf{T}T^{r-2}\mathbf{T}\mathbf{e} = \mathbf{T}^2T^{r-2}\mathbf{e} = \mathbf{T}^2T^{r-3}T\mathbf{e} = \cdots = \mathbf{T}^r\mathbf{e}$. Equation (1.210) means that the eigenvalues of the tensor $\mathbf{T}^r$ are given by T_J^r where $T_J(J = 1, 2, 3)$ are the eigenvalues of $\mathbf{T}$, and the eigenvectors of the tensor $\mathbf{T}^r$ coincide with those of $\mathbf{T}$. Tensors having an identical set of principal directions are called to be *coaxial* or said to fulfill the *coaxiality*. Then, the linear associative function $\mathbf{f}(\mathbf{T})$ of $\mathbf{T}$ is coaxial with $\mathbf{T}$ and the principal values of $\mathbf{f}(\mathbf{T})$ are given by $f(T_J)$.

The multiplication of the eigenvector $\mathbf{e}$ to the characteristic Eq. (1.177) leads to

$$(\mathbf{T}^3 - \mathrm{I}\,\mathbf{T}^2 + \mathrm{II}\,\mathbf{T} - \mathrm{III}\,\mathbf{I})\mathbf{e} = 0$$

noting Eq. (1.210). Because of $\mathbf{e} \neq \mathbf{0}$, the following *Cayley-Hamilton theorem* holds.

$$\boxed{\mathbf{T}^3 - \mathrm{I}\,\mathbf{T}^2 + \mathrm{II}\,\mathbf{T} - \mathrm{III}\,\mathbf{I} = \mathbf{O}} \tag{1.211}$$

It follows from the Cayley-Hamilton theorem that

$$\begin{aligned}\mathbf{T}^4 &= (\mathrm{I}\mathbf{T}^2 - \mathrm{II}\mathbf{T} + \mathrm{III}\mathbf{I})\mathbf{T} = \mathrm{I}\mathbf{T}^3 - \mathrm{II}\mathbf{T}^2 + \mathrm{III}\mathbf{T} = \mathrm{I}(\mathrm{I}\mathbf{T}^2 - \mathrm{II}\mathbf{T} + \mathrm{III}\mathbf{I}) - \mathrm{II}\mathbf{T}^2 + \mathrm{III}\mathbf{T}\\ &= (\mathrm{I}^2 - \mathrm{II})\mathbf{T}^2 - (\mathrm{I}\,\mathrm{II} - \mathrm{III})\mathbf{T} + \mathrm{I}\,\mathrm{III}\mathbf{I}\end{aligned} \tag{1.212}$$

$$\mathrm{III}\,\mathbf{T}^{-1} = \mathbf{T}^2 - \mathrm{I}\,\mathbf{T} + \mathrm{II}\,\mathbf{I} \tag{1.213}$$

It is concluded that the power of the tensor $\mathbf{T}$ is expressed by the linear associative of $\mathbf{T}^2, \mathbf{T}, \mathbf{I}$ with coefficients consisting of the principal invariants.

Taking the trace of Eq. (1.211) for the deviatoric tensor $\mathbf{T}'$, we have $\mathrm{tr}\,\mathbf{T}'^3 - \mathrm{III}'\mathrm{tr}\,\mathbf{I} = 0$, noting $\mathrm{I}' = \mathrm{tr}\,\mathbf{T}' = 0$, and thus the determinant of the deviatoric tensor in Eq. (1.188) is also expressed as

$$\boxed{\mathrm{III}'(= \det\mathbf{T}') = \frac{1}{3}\mathrm{tr}\,\mathbf{T}'^3} = \frac{1}{3}T'_{rs}T'_{st}T'_{tr} = \frac{1}{3}(T_1'^3 + T_2'^3 + T_3'^3) \tag{1.214}$$

The Cayley-Hamilton theorem can be also derived from the characteristic equation as follows: Eq. (1.177) holds for the three principal values T_1, T_2 and T_3, and thus the following equation must holds.

$$\begin{bmatrix} T_1^3 - \mathrm{I}\,T_1^2 + \mathrm{II}\,T_1 - \mathrm{III} & 0 & 0 \\ 0 & T_1^3 - \mathrm{I}\,T_1^2 + \mathrm{II}\,T_1 - \mathrm{III} & 0 \\ 0 & 0 & T_1^3 - \mathrm{I}\,T_1^2 + \mathrm{II}\,T_1 - \mathrm{III} \end{bmatrix}$$

$$= \begin{bmatrix} T_1^3 & 0 & 0 \\ 0 & T_2^3 & 0 \\ 0 & 0 & T_3^3 \end{bmatrix} - \mathrm{I} \begin{bmatrix} T_1^2 & 0 & 0 \\ 0 & T_2^2 & 0 \\ 0 & 0 & T_3^2 \end{bmatrix} + \mathrm{II} \begin{bmatrix} T_1 & 0 & 0 \\ 0 & T_2 & 0 \\ 0 & 0 & T_3 \end{bmatrix} - \mathrm{III} \begin{bmatrix} 1 & 0 & 0 \\ 0 & 1 & 0 \\ 0 & 0 & 1 \end{bmatrix} = \begin{bmatrix} 0 & 0 & 0 \\ 0 & 0 & 0 \\ 0 & 0 & 0 \end{bmatrix} \tag{1.215}$$

the tensor expression of which is nothing but Eq. (1.211).

1.10 Positive Definite Tensor

When the second-order tensor $\mathbf{P}$ is symmetric and fulfills

$$\boxed{\mathbf{vP} \cdot \mathbf{v} > 0} \tag{1.216}$$

for an arbitrary vector $\mathbf{v}(\neq 0)$, $\mathbf{P}$ is called the *positive-definite tensor*.

Denoting the principal value and direction of $\mathbf{P}$ as P_J and $\mathbf{e}_J$ $(J = 1, 2, 3)$, respectively, it holds that

$$\mathbf{e}_J \cdot \mathbf{P}\mathbf{e}_J = \mathbf{e}_J \cdot P_J \mathbf{e}_J = P_J \|\mathbf{e}_J\|^2 > 0 \quad (\text{no sum}) \tag{1.217}$$

noting Eq. (1.216). Then, it is known that the principal value of positive-definite tensor is positive, i.e. $P_J > 0$. Taking this fact into account for Eq. $(1.182)_3$, it follows that

$$\boxed{\det \mathbf{P} = \mathrm{III} > 0} \tag{1.218}$$

The positive definite tensor $\mathbf{U}$ having the same principal directions as those of $\mathbf{P}$ and principal values $P_J^{1/2}$ is defined as the square root of $\mathbf{P}$, i.e. $\mathbf{U}^2 = \mathbf{P}$ or $\mathbf{U} = \mathbf{P}^{1/2}$.

1.11 Polar Decomposition

Assuming that the second-order tensor $\mathbf{T}$ is not singular ($\det \mathbf{T} \neq 0$), it holds that $\mathbf{Tv} \neq \mathbf{0}$ for an arbitrary vector $\mathbf{v}(\neq \mathbf{0})$ as described in 2) of Sect. 1.4 and thus using Eq. (1.117), one obtains

$$\mathbf{v} \cdot \mathbf{T}^T\mathbf{T}\mathbf{v} = \mathbf{Tv} \cdot \mathbf{Tv} > 0 \tag{1.219}$$

where $\mathbf{T}^T\mathbf{T}$ is the symmetric tensor and thus it is the positive-definite tensor as described in Sect. 1.10. Then, $\mathbf{T}^T\mathbf{T}$ is described as $\mathbf{T}^T\mathbf{T} = \sum \Lambda_J \mathbf{u}_J \otimes \mathbf{u}_J$, where

$\mathbf{u}_J$ $(J = 1, 2, 3)$ are the unit principal vectors and $\Lambda_J (\geq 0)$ are the principal values. One can relate the tensor defined as $\mathbf{U} = \sum \lambda_J \mathbf{u}_J \otimes \mathbf{u}_J$, where $\lambda_J \equiv \sqrt{\Lambda_J}$, to $\mathbf{T}^T\mathbf{T}$ as follows:

$$\boxed{\mathbf{U} = (\mathbf{T}^T\mathbf{T})^{1/2} (= \mathbf{U}^T) \quad (\mathbf{U}^2 = \mathbf{T}^T\mathbf{T})} \tag{1.220}$$

which is also the positive definite tensor.

Further, defining the tensor $\mathbf{R}$ as

$$\mathbf{R} = \mathbf{T}\mathbf{U}^{-1}, \mathbf{U} = \mathbf{R}^T\mathbf{T} \tag{1.221}$$

and noting Eq. (1.122), one has

$$\mathbf{R}\mathbf{R}^T = (\mathbf{T}\mathbf{U}^{-1})(\mathbf{T}\mathbf{U}^{-1})^T = \mathbf{T}\mathbf{U}^{-1}\mathbf{U}^{-1}\mathbf{T}^T = \mathbf{T}(\mathbf{U}^2)^{-1}\mathbf{T}^T = \mathbf{T}\mathbf{T}^{-1}\mathbf{T}^{-T}\mathbf{T}^T = \mathbf{I} \tag{1.222}$$

Therefore, $\mathbf{R}$ is the orthogonal tensor.

Furthermore, the following tensor is proven to be the positive-definite tensor in the analogous way to the case of the tensor $\mathbf{U}$.

$$\boxed{\mathbf{V} = (\mathbf{T}\mathbf{T}^T)^{1/2} = \mathbf{T}\mathbf{R}^T (= \mathbf{V}^T)\,(\mathbf{V}^2 = \mathbf{T}\mathbf{T}^T)} \tag{1.223}$$

from which we have

$$\mathbf{V}^2 = \mathbf{T}\mathbf{T}^T = (\mathbf{R}\mathbf{U})\,(\mathbf{R}\mathbf{U})^T = \mathbf{R}\mathbf{U}\mathbf{U}\mathbf{R}^T = \mathbf{R}\mathbf{U}\mathbf{R}^T\mathbf{R}\mathbf{U}\mathbf{R}^T = (\mathbf{R}\mathbf{U}\mathbf{R}^T)^2 \tag{1.224}$$

noting Eq. (1.221), and thus it follows that

$$\boxed{\mathbf{V} = \mathbf{R}\mathbf{U}\mathbf{R}^T, \quad \mathbf{U} = \mathbf{R}^T\mathbf{V}\mathbf{R}} \tag{1.225}$$

Then, one can write

$$\boxed{\mathbf{T} = \mathbf{R}\mathbf{U} = \mathbf{V}\mathbf{R};\ \mathbf{U} = \mathbf{R}^T\mathbf{T},\ \mathbf{V} = \mathbf{T}\mathbf{R}^T} \tag{1.226}$$

Consequently, an arbitrary non-singular tensor $\mathbf{T}$ can be decomposed into the two forms in terms of the positive definite tensors $\mathbf{U}$ or $\mathbf{V}$ and the orthogonal tensor $\mathbf{R}$. Here, based on Eqs. (1.220), (1.223) and (1.226), $\mathbf{R}$ is expressed by the original tensor $\mathbf{T}$ as follows:

$$\mathbf{R} = (\mathbf{T}\mathbf{T}^T)^{-1/2}\mathbf{T} = \mathbf{T}(\mathbf{T}^T\mathbf{T})^{-1/2} \tag{1.227}$$

Based on (1.225), $\mathbf{U}$, $\mathbf{V}$ are the mutually similar tensors, as described in Sect. 1.6. For that reason, they have same positive principal values, denoted as $\lambda_J = U_J = V_J$ $(J = 1, 2, 3)$, and their unit principal vectors $\mathbf{u}_J, \mathbf{v}_J$ are mutually related by

$$\boxed{\mathbf{v}_J = \mathbf{R}\mathbf{u}_J, \quad \mathbf{R} = \sum_{J=1}^{3} \mathbf{v}_J \otimes \mathbf{u}_J} \tag{1.228}$$

and it holds that

$$\boxed{\mathbf{U} = \sum_{J=1}^{3} \lambda_J \mathbf{u}_J \otimes \mathbf{u}_J, \quad \mathbf{V} = \sum_{J=1}^{3} \lambda_J \mathbf{v}_J \otimes \mathbf{v}_J} \tag{1.229}$$

Equation (1.226) is called the *polar decomposition* in similarity to the polar form $\mathbf{Z} = |\mathbf{Z}|\mathbf{e}^{i\theta}$ (i: imaginary unit, θ: phase angle) which expresses the complex number by the decomposition into the magnitude and the direction in the polar coordinate system. Actually, $\mathbf{RU}$ and $\mathbf{VR}$ are respectively called the *right* and the *left polar decompositions*.

1.12 Isotropic Tensor-Valued Tensor Function

If the tensor-valued function $\mathbf{f}$ of tensors $\mathbf{S}, \mathbf{T}, \cdots$ fulfills the following equation, it is called the *isotropic function*.

$$\mathbf{Q}[\![\mathbf{f}(\mathbf{S}, \mathbf{T}, \cdots)]\!] = \mathbf{f}(\mathbf{Q}[\![\mathbf{S}]\!], \mathbf{Q}[\![\mathbf{T}]\!], \cdots) \tag{1.230}$$

where use is made of the symbol in Eq. (1.84). If $\mathbf{f}$ is a scalar, it is to be the invariant defined in Eq. (1.174) and if it is a tensor, it is called the *isotropic tensor-valued tensor function*.

Now, consider the isotropic second-order tensor function $\mathbf{B}$ of a single second-order tensor $\mathbf{A}$ as the simplest case of Eq. (1.230), i.e.

$$\mathbf{B} = \mathbf{f}(\mathbf{A}) \tag{1.231}$$

where $\mathbf{f}$ fulfills

$$\mathbf{f}(\mathbf{Q}\mathbf{A}\mathbf{Q}^T) = \mathbf{Q}\mathbf{f}(\mathbf{A})\mathbf{Q}^T \tag{1.232}$$

First introducing the coordinate system with the bases $\mathbf{e}_1, \mathbf{e}_2, \mathbf{e}_3$, which are the unit eigenvector of the tensor $\mathbf{A}$ and further adopting the another coordinate system rotated 180° around the base $\mathbf{e}_3$, the orthogonal tensor between the bases of these coordinate systems is given by

$$\mathbf{Q}_0 = \begin{bmatrix} -1 & 0 & 0 \\ 0 & -1 & 0 \\ 0 & 0 & 1 \end{bmatrix} \tag{1.233}$$

where $\mathbf{Q}_0$ fulfills $\mathbf{Q}_0 = \mathbf{Q}_0^T$ resulting in the symmetric tensor and it holds that

$$\begin{bmatrix} -1 & 0 & 0 \\ 0 & -1 & 0 \\ 0 & 0 & 1 \end{bmatrix} \begin{Bmatrix} 0 \\ 0 \\ 1 \end{Bmatrix} = \begin{Bmatrix} 0 \\ 0 \\ 1 \end{Bmatrix}, \text{i.e.} \quad \mathbf{Q}_0\mathbf{e}_3 = \mathbf{e}_3 \tag{1.234}$$

Then, it is known that $\mathbf{e}_3$ is one eigenvector not only of $\mathbf{A}$ but also of $\mathbf{Q}_0$. Furthermore, denoting the principal values of $\mathbf{A}$ by A_1, A_2, A_3, it holds that

$$\begin{bmatrix} -1 & 0 & 0 \\ 0 & -1 & 0 \\ 0 & 0 & 1 \end{bmatrix} \begin{bmatrix} A_1 & 0 & 0 \\ 0 & A_2 & 0 \\ 0 & 0 & A_3 \end{bmatrix} \begin{bmatrix} -1 & 0 & 0 \\ 0 & -1 & 0 \\ 0 & 0 & 1 \end{bmatrix} = \begin{bmatrix} A_1 & 0 & 0 \\ 0 & A_2 & 0 \\ 0 & 0 & A_3 \end{bmatrix},$$

i.e.

$$\mathbf{Q}_0\mathbf{A}\mathbf{Q}_0^T = \mathbf{A} \tag{1.235}$$

and thus it holds that

$$\mathbf{f}(\mathbf{Q}_0\mathbf{A}\mathbf{Q}_0^T) = \mathbf{f}(\mathbf{A}) = \mathbf{B} \tag{1.236}$$

On the other hand, from Eq. (1.232) one has

$$\mathbf{f}(\mathbf{Q}_0\mathbf{A}\mathbf{Q}_0^T) = \mathbf{Q}_0\mathbf{B}\mathbf{Q}_0^T \tag{1.237}$$

Then, one obtains $\mathbf{B} = \mathbf{Q}_0\mathbf{B}\mathbf{Q}_0^T$ leading to the commutative law

$$\mathbf{Q}_0\mathbf{B} = \mathbf{B}\mathbf{Q}_0 \tag{1.238}$$

from Eqs. (1.236) and (1.237), and further, noting Eq. (1.234), the following relation is obtained.

$$\mathbf{Q}_0\mathbf{B}\mathbf{e}_3 = \mathbf{B}\mathbf{Q}_0\mathbf{e}_3 = \mathbf{B}\mathbf{e}_3 \tag{1.239}$$

which means that $\mathbf{B}\mathbf{e}_3$ is the eigenvector of $\mathbf{Q}_0$. Then, reminding that the eigenvector of $\mathbf{Q}_0$ is $\mathbf{e}_3$, it follows that $\mathbf{B}\mathbf{e}_3$ has the same direction as $\mathbf{e}_3$. Then, it follows that

$$\mathbf{B}\mathbf{e}_3 = B_3\mathbf{e}_3 \tag{1.240}$$

where B_3 is the eigenvalue of $\mathbf{B}$ for the eigenvector $\mathbf{e}_3$. Performing the similar manipulations also for $\mathbf{e}_1$ and $\mathbf{e}_2$, it can be concluded that the tensor $\mathbf{B}$ has the same eigenvectors as the tensor $\mathbf{A}$, leading to the coaxiality. Therefore, the principal values B_1, B_2, B_3 of the tensor $\mathbf{B}$ can be represented in unique relation to the principal values A_1, A_2, A_3 of the tensor $\mathbf{A}$, i.e.

$$B_i = \hat{B}_i(A_1, A_2, A_3) \quad (i = 1,\ 2,\ 3) \tag{1.241}$$

Now, if the principal values A_i is different from each other, the most general expression of B_i is given by

$$\boxed{B_i = \hat{B}_i(A_j) = \phi_0 + \phi_1 A_i + \phi_2 A_i^2 \ (i = 1, 2, 3)} \tag{1.242}$$

i.e.

$$\left.\begin{array}{l} B_1(A_1,\ A_2,\ A_3) = \phi_0 + \phi_1 A_1 + \phi_2 A_1^2 \\ B_2(A_1,\ A_2,\ A_3) = \phi_0 + \phi_1 A_2 + \phi_2 A_2^2 \\ B_3(A_1,\ A_2,\ A_3) = \phi_0 + \phi_1 A_3 + \phi_2 A_3^2 \end{array}\right\},\quad \text{i.e.} \begin{Bmatrix} B_1 \\ B_2 \\ B_3 \end{Bmatrix} = \begin{bmatrix} 1\, A_1 A_1^2 \\ 1\, A_2 A_2^2 \\ 1\, A_3 A_3^2 \end{bmatrix} \begin{Bmatrix} \phi_0 \\ \phi_1 \\ \phi_2 \end{Bmatrix} \tag{1.243}$$

where $\phi_0,\ \phi_1,\ \phi_2$ are scalar functions of A_1, A_2, A_3. The rightness of Eq. (1.242) or (1.243) can be verified by the reason that ϕ_0, ϕ_1, ϕ_2 can be determined if A_1, A_2, A_3 and B_1, B_2, B_3 are given, because the following *Vandermonde's determinant* for the above simultaneous equation of the unknown variables ϕ_0, ϕ_1, ϕ_2 is not zero for mutually different values of A_1, A_2, A_3, i.e.

$$\begin{bmatrix} 1 & A_1 & A_1^2 \\ 1 & A_2 & A_2^2 \\ 1 & A_3 & A_2^2 \end{bmatrix} = (A_1 - A_2)(A_2 - A_3)(A_3 - A_1) \neq 0 \tag{1.244}$$

leading to the solutions

$$\phi_0 = \frac{\begin{bmatrix} B_0 & A_1 & A_1^2 \\ B_1 & A_2 & A_2^2 \\ B_2 & A_3 & A_3^2 \end{bmatrix}}{\begin{bmatrix} 1 & A_1 & A_1^2 \\ 1 & A_2 & A_2^2 \\ 1 & A_3 & A_3^2 \end{bmatrix}},\quad \phi_0 = \frac{\begin{bmatrix} 1 & B_1 & A_1^2 \\ 1 & B_2 & A_2^2 \\ 1 & B_3 & A_3^2 \end{bmatrix}}{\begin{bmatrix} 1 & A_1 & A_1^2 \\ 1 & A_2 & A_2^2 \\ 1 & A_3 & A_3^2 \end{bmatrix}},\quad \phi_2 = \frac{\begin{bmatrix} 1 & A_1 & B_1 \\ 1 & A_2 & B_2 \\ 1 & A_3 & B_3 \end{bmatrix}}{\begin{bmatrix} 1 & A_1 & A_1^2 \\ 1 & A_2 & A_2^2 \\ 1 & A_3 & A_3^2 \end{bmatrix}} \tag{1.245}$$

While Eq. (1.243) is regarded as a representation of the relation of the tensors **A** and **B** in their common principal coordinate system, it is expressed in the direct notation of tensor as

$$\boxed{\mathbf{B} = \phi_0 \mathbf{I} + \phi_1 \mathbf{A} + \phi_2 \mathbf{A}^2} \tag{1.246}$$

This fact can also be verified simply using Cayley-Hamilton theorem (Sect. 1.8) for the special case that **f** is a polynomial equation of the power of **A** in Eq. (1.231). However, for the case in which **f** is the general function of **A**, one must depend on the proof given in this section.

It follows analogously to the above-mentioned method that

$$\mathbf{B} = \phi_0 \mathbf{I} + \phi_1 \mathbf{A} \tag{1.247}$$

when the two of principal values are same, and

$$\mathbf{B} = \phi_0 \mathbf{I} \tag{1.248}$$

when all principal values are same.

In the particular case in which **f** is the linear function of the tensor **A**, Eq. (1.246) is reduced to

$$\mathbf{B} = a(\mathrm{tr}\,\mathbf{A})\mathbf{I} + b\mathbf{A} \tag{1.249}$$

where a and b are the material constants. Equation (1.249) is rewritten as

$$\mathbf{B} = \boldsymbol{\Xi} : \mathbf{A} \tag{1.250}$$

where

$$\boldsymbol{\Xi} \equiv a\boldsymbol{\mathcal{T}} + b\boldsymbol{\mathcal{S}},\ \Xi_{ijkl} \equiv a\delta_{ij}\delta_{kl} + \frac{1}{2}b(\delta_{ik}\delta_{jl} + \delta_{il}\delta_{jk}) \tag{1.251}$$

While the second-order isotropic tensor-valued tensor function of single tensor is considered above, the *representation theorem* of the second-order isotropic tensor-valued symmetric tensor function $\mathbf{f}^S$ and anti-symmetric tensor function $\mathbf{f}^A$ of the two tensors **A** and **B** are represented as follows (Wang 1970).

$$\left.\begin{aligned} \mathbf{f}^S(\mathbf{A},\mathbf{B}) &= \varphi_0\mathbf{I} + \varphi_1\mathbf{A} + \varphi_2\mathbf{B} + \varphi_3\mathbf{A}^2 + \varphi_4\mathbf{B}^2 \\ &\quad + \varphi_5(\mathbf{AB}+\mathbf{BA}) + \varphi_6(\mathbf{A}^2\mathbf{B}+\mathbf{BA}^2) + \varphi_7(\mathbf{AB}^2+\mathbf{B}^2\mathbf{A}) + \varphi_8(\mathbf{A}^2\mathbf{B}^2+\mathbf{B}^2\mathbf{A}^2) \\ \mathbf{f}^A(\mathbf{A},\mathbf{B}) &= \eta_1(\mathbf{AB}-\mathbf{BA}) + \eta_2(\mathbf{A}^2\mathbf{B}-\mathbf{BA}^2) + \eta_3(\mathbf{AB}^2-\mathbf{B}^2\mathbf{A}) \\ &\quad + \eta_4(\mathbf{ABA}^2-\mathbf{A}^2\mathbf{BA}) + \eta_5(\mathbf{BAB}^2-\mathbf{B}^2\mathbf{AB}) \end{aligned}\right\} \tag{1.252}$$

where $\varphi_0, \varphi_1, \cdots, \varphi_8$ and $\eta_1, \eta_2, \cdots, \eta_5$ are the scalar-valued isotropic functions of invariants of **A** and **B**, i.e.

$$\left.\begin{array}{l} \mathrm{tr}\,\mathbf{A}, \mathrm{tr}\mathbf{A}^2, \mathrm{tr}\,\mathbf{A}^3, \mathrm{tr}\,\mathbf{B}, \mathrm{tr}\,\mathbf{B}^2, \mathrm{tr}\,\mathbf{B}^3 \\ \mathrm{tr}(\mathbf{AB}), \mathrm{tr}(\mathbf{AB}^2), \mathrm{tr}(\mathbf{A}^2\mathbf{B}), \mathrm{tr}(\mathbf{A}^2\mathbf{B}^2) \end{array}\right\} \tag{1.253}$$

noting that all tensors can be represented by powers of tensor lower than the second power by the Cayley-Hamilton theorem in Section 1.9.

1.13 Representation of Tensor in Principal Space

Second-order tensor **T** is described by only three independent components, i.e. the principal values, in the directions of the eigenvectors $\mathbf{e}_1, \mathbf{e}_2, \mathbf{e}_3$. Designating the principal values by T_1, T_2, T_3, it can be represented as follows:

$$\mathbf{T} = T_1\mathbf{e}_1 + T_2\mathbf{e}_2 + T_3\mathbf{e}_3 \tag{1.254}$$

Equation (1.254) may be called the *representation of tensor in principal space* (Fig. 1.3) by which the second-order tensor can be visualized in the three dimensional space. Equation (1.254) is rewritten by decomposing **T** into the mean and the deviatoric components as follows:

$$\mathbf{T} = (T_m + T'_1)\mathbf{e}_1 + (T_m + T'_2)\mathbf{e}_2 + (T_m + T'_3)\mathbf{e}_3 = \mathbf{T}_m + \mathbf{T}' \tag{1.255}$$

where

$$\mathbf{T}_m \equiv T_m\mathbf{I}_m = \sqrt{3}T_m\mathbf{e}_m \;(T_m \equiv (T_1 + T_2 + T_3)/3) \tag{1.256}$$

$$\mathbf{I}_m \equiv \mathbf{e}_1 + \mathbf{e}_2 + \mathbf{e}_3, \quad \mathbf{e}_m \equiv \frac{1}{\sqrt{3}}\mathbf{I}_m \;\; (\|\mathbf{e}_m\| = 1) \tag{1.257}$$

$$\mathbf{T}' \equiv T'_1\mathbf{e}_1 + T'_2\mathbf{e}_2 + T'_3\mathbf{e}_3 = \|\mathbf{T}'\|\mathbf{t}' \tag{1.258}$$

$$\mathbf{t}' \equiv \mathbf{T}'/\|\mathbf{T}'\|(\|\mathbf{t}'\| = 1) \tag{1.259}$$

$$\left.\begin{aligned} &T'_1 \equiv T_1 - T_m, \quad T'_2 \equiv T_2 - T_m, \quad T'_3 \equiv T_3 - T_m \\ &\mathbf{T}' = \sqrt{T_1'^2 + T_2'^2 + T_3'^2} = \frac{1}{\sqrt{3}}\sqrt{(T_1 - T_2)^2 + (T_2 - T_3)^2 + (T_3 - T_1)^2} \end{aligned}\right\} \tag{1.260}$$

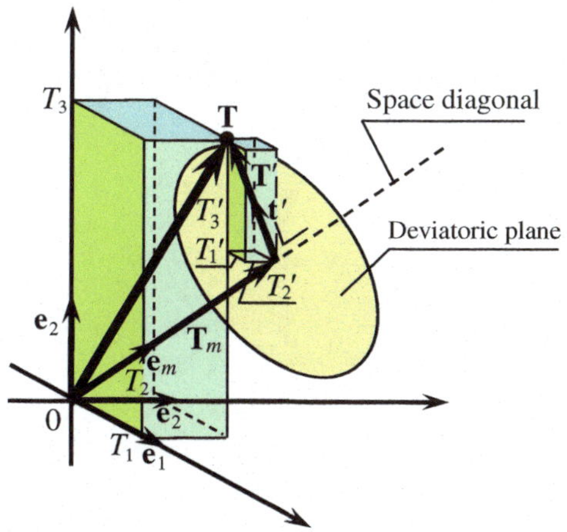

Fig. 1.3 Representation of second-order tensor in principal space

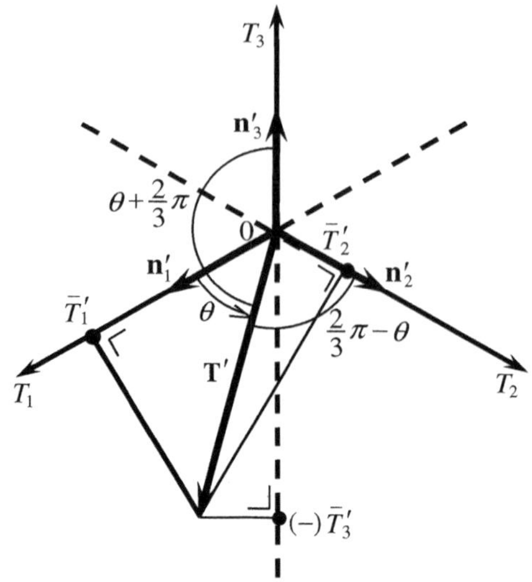

Fig. 1.4 Deviatoric part of tensor in deviatoric plane (π-plane)

Here, introduce the following unit vectors $\mathbf{n}'_1$, $\mathbf{n}'_2$ and $\mathbf{n}'_3$ directed towards the projection of the base vectors $\mathbf{e}_1$, $\mathbf{e}_2$ and $\mathbf{e}_3$, respectively, onto the deviatoric plane as shown in Fig. 1.4.

$$\left.\begin{aligned}
\mathbf{n}'_1 &\equiv \frac{1}{\sqrt{6}}(2\mathbf{e}_1 - \mathbf{e}_2 - \mathbf{e}_3) \quad (\|\mathbf{n}'_1\| = 1)\\
\mathbf{n}'_2 &\equiv \frac{1}{\sqrt{6}}(-\mathbf{e}_1 + 2\mathbf{e}_2 - \mathbf{e}_3) \quad (\|\mathbf{n}'_2\| = 1)\\
\mathbf{n}'_3 &\equiv \frac{1}{\sqrt{6}}(-\mathbf{e}_1 - \mathbf{e}_2 + 2\mathbf{e}_3) \quad (\|\mathbf{n}'_3\| = 1)
\end{aligned}\right\} \tag{1.261}$$

fulfilling

$$\mathbf{n}'_1 \cdot \mathbf{e}_1 = \mathbf{n}'_2 \cdot \mathbf{e}_2 = \mathbf{n}'_3 \cdot \mathbf{e}_3 = \sqrt{2/3} \tag{1.262}$$

It follows from Eqs. (1.258) and (1.261) for the projections of $\mathbf{T}'$ onto the directions $\mathbf{n}'_1$, $\mathbf{n}'_2$ and $\mathbf{n}'_3$ that

$$\left.\begin{aligned}
\overline{T}'_1 &\equiv \mathbf{T}' \cdot \mathbf{n}'_1 = \|\mathbf{T}'\| \mathbf{t}' \cdot \mathbf{n}'_1 = \|\mathbf{T}'\| \cos\theta\\
\overline{T}'_2 &\equiv \mathbf{T}' \cdot \mathbf{n}'_2 = \|\mathbf{T}'\| \mathbf{t}' \cdot \mathbf{n}'_2 = \|\mathbf{T}'\| \cos(\theta - \frac{2}{3}\pi)\\
\overline{T}'_3 &\equiv \mathbf{T}' \cdot \mathbf{n}'_3 = \|\mathbf{T}'\| \mathbf{t}' \cdot \mathbf{n}'_3 = \|\mathbf{T}'\| \cos(\theta + \frac{2}{3}\pi)
\end{aligned}\right\} \tag{1.263}$$

as shown in Fig. 1.4. Exploiting Eq. (1.263), the deviatoric components are expressed as

$$\left.\begin{array}{l} T_1' = (T_1'\mathbf{n}_1')\cdot\mathbf{e}_1 = \|\mathbf{T}'\|\cos\theta\mathbf{n}_1'\cdot\mathbf{e}_1 = \sqrt{\dfrac{2}{3}}\|\mathbf{T}'\|\cos\theta \\ T_2' = (T_2'\mathbf{n}_2')\cdot\mathbf{e}_2 = \|\mathbf{T}'\|\cos(\theta-\dfrac{2}{3}\pi)\mathbf{n}_2'\cdot\mathbf{e}_2 = \sqrt{\dfrac{2}{3}}\|\mathbf{T}'\|\cos(\theta-\dfrac{2}{3}\pi) \\ T_3' = (T_3'\mathbf{n}_3')\cdot\mathbf{e}_3 = \|\mathbf{T}'\|(\theta+\dfrac{2}{3}\pi)\mathbf{n}_3'\cdot\mathbf{e}_3 = \sqrt{\dfrac{2}{3}}\|\mathbf{T}'\|\cos(\theta+\dfrac{2}{3}\pi) \end{array}\right\} \tag{1.264}$$

where θ is the angle measured in the anti-clock wise direction from $\mathbf{n}_1'$ to $\mathbf{T}'$ in the deviatoric plane as shown in Fig. 1.4 and it is called the *Lode angle*.

The product of the three components is given by

$$\frac{\mathrm{III}'}{(2\mathrm{II}')^{3/2}} = \frac{T_1'}{\|\mathbf{T}'\|}\frac{T_2'}{\|\mathbf{T}'\|}\frac{T_3'}{\|\mathbf{T}'\|} = \frac{2}{3}\sqrt{\frac{2}{3}}\cos\theta\,\cos(\theta-\frac{2}{3}\pi)\cos(\theta+\frac{2}{3}\pi) = \frac{1}{3\sqrt{6}}\cos 3\theta \tag{1.265}$$

noting Eqs. (1.187) and (1.188). It follows from Eq. (1.265) that

$$\boxed{\cos 3\theta = \frac{3\sqrt{3}}{2}\frac{\mathrm{III}'}{\mathrm{II}'^{3/2}} = \sqrt{6}\mathrm{tr}\,\mathbf{t}'^3} \tag{1.266}$$

The above-mentioned method for deriving Eq. (1.266) would be most concise possessing the clear physical meaning among others (cf. e.g. Ottosen and Ristinmaa 2005; Hashiguchi 2013). Equation (1.196) for the eigenvalues of the deviatoric part in tensor can be derived also by substituting Eq. (1.266) into Eq. (1.264).

It follows from Eq. (1.264) with some mathematical operations on the trigonometric functions that

$$\mu \equiv \frac{2T_2 - T_1 - T_3}{T_1 - T_3} = \frac{2T_2' - T_1' - T_3'}{T_1' - T_3'} = \sqrt{3}\tan\left(\theta - \frac{1}{6}\pi\right) \tag{1.267}$$

which is called the *Lode variable*.

Substituting the principal value $T_1' = \sqrt{2/3}\|\mathbf{T}'\|\cos\theta$ into the characteristic equation of deviatoric tensor (1.185), one has the following equation

$$\left(\sqrt{\frac{2}{3}}\|\mathbf{T}'\|\cos\theta\right)^3 - \frac{1}{2}\|\mathbf{T}'\|^2\left(\sqrt{\frac{2}{3}}\|\mathbf{T}'\|\cos\theta\right) - \frac{1}{3}\mathrm{tr}\,\mathbf{T}'^3 = 0$$

resulting in

$$4\cos^3\theta - 3\cos\theta - \sqrt{6}\mathrm{tr}\,\mathbf{t}'^3 = 0 \tag{1.268}$$

from which one can derive Eq. (1.266), noting $4\cos^3\theta - 3\cos\theta = \cos 3\theta$.

1.14 Two-Dimensional State

Consider the two-dimensional state in which the components related to the $\mathbf{e}_3$ direction in the coordinate system (x_1, x_2, x_3) with the fixed base $(\mathbf{e}_1, \mathbf{e}_2, \mathbf{e}_3)$ are zero, i.e. $T_{33} = T_{31} = T_{23} = 0$. Furthermore, introduce the coordinate system (x_1^*, x_2^*, x_3^*) with the bases $(\mathbf{e}_1^*, \mathbf{e}_2^*, \mathbf{e}_3^*(=\mathbf{e}_3))$ which is rotated by the angle α in the counterclockwise direction around the axis x_3 as shown in Fig. 1.5, where the normal components T_{ij}^* $(i,j = 1,2;\ i = j)$ and the shear components T_{ij}^* $(i \neq j)$ in the base $(\mathbf{e}_1^*, \mathbf{e}_2^*)$ are designated by T_n and by T_t, respectively. The orthogonal tensor between these bases is given from Eq. (1.59) as follows:

$$[Q] = \begin{bmatrix} \cos\alpha & \sin\alpha & 0 \\ -\sin\alpha & \cos\alpha & 0 \\ 0 & 0 & 1 \end{bmatrix} \tag{1.269}$$

Substituting Eq. (1.269) into Eq. (1.74), one has

$$\left.\begin{aligned} T_{11}^* &= T_{11}\cos^2\alpha + T_{22}\sin^2\alpha + 2T_{12}\cos\alpha\,\sin\alpha \\ T_{22}^* &= T_{11}\sin^2\alpha + T_{22}\cos^2\alpha - 2T_{12}\sin\alpha\,\cos\alpha \\ T_{12}^* &= (T_{22} - T_{11})\cos\alpha\,\sin\alpha + T_{12}(\cos^2\alpha - \sin^2\alpha) \end{aligned}\right\} \tag{1.270}$$

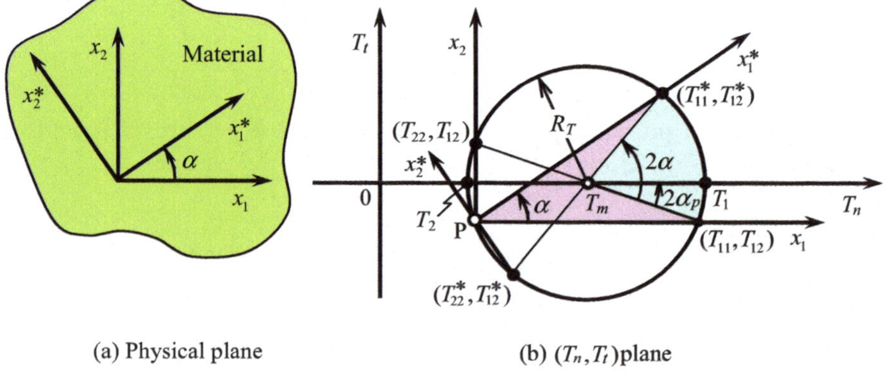

Fig. 1.5 Mohr's circle

which is rewritten as

$$\left.\begin{aligned} T_{11}^* &= T_m + \overline{T}\cos 2\alpha + T_{12}\sin 2\alpha \\ T_{22}^* &= T_m - \overline{T}\cos 2\alpha - T_{12}\sin 2\alpha \\ T_{12}^* &= -\overline{T}\sin 2\alpha + T_{12}\cos 2\alpha \end{aligned}\right\} \tag{1.271}$$

where

$$T_m \equiv \frac{T_{11} + T_{22}}{2}, \quad \overline{T} \equiv \frac{T_{11} - T_{22}}{2} \tag{1.272}$$

Furthermore, it follows from Eq. (1.271) that

$$T_{11}^* + T_{22}^* = T_{11} + T_{22} \tag{1.273}$$

$$\frac{\partial T_{11}^*}{\partial \alpha} = 2T_{12}^*, \ \frac{\partial T_{22}^*}{\partial \alpha} = -2T_{12}^* \tag{1.274}$$

While the axis $x_3(=x_3^*)$ is one of the principal directions, the other principal directions exist on the plane (x_1, x_2). Denoting the principal direction from the x_1- axis by α, it is obtained by substituting $\partial T_{11}^*/\partial\alpha = T_{12}^* = 0$ or $\partial T_{22}^*/\partial\alpha = 0$ in Eq. (1.274) into Eq. (1.271) that

$$\tan 2\alpha_p = \frac{T_{12}}{\overline{T}} \tag{1.275}$$

from which one obtains

$$\left.\begin{aligned} &\overline{T} = \pm R_T\cos 2\alpha_p, \quad T_{12} = \pm R_T \sin 2\alpha_p \\ &R_T = \sqrt{\overline{T}^2 + \mathrm{T}_{12}^2} \end{aligned}\right\} \tag{1.276}$$

Substituting Eq. (1.276) into the upper two of Eq. (1.271) with specifying α as α_p, the maximum and the minimum principal values T_1 and T_2 are described by

$$\left.\begin{aligned} T_1 \\ T_2 \end{aligned}\right\} = T_m \pm R_T \tag{1.277}$$

Equation (1.277) can also be derived directly from the quadratic equation

$$T^2 - (T_{11} + T_{22})T + T_{11}T_{22} - T_{12}^2 = 0$$

which is obtained by inserting $\mathrm{I} = T_{11} + T_{22}$, $\mathrm{II} = T_{11}T_{22} - T_{12}^2$, $\mathrm{III} = 0$ $(T_{33} = T_{31} = T_{23} = 0)$ in Eq. (1.177).

Furthermore, denoting α for the extremal value of T^*_{12} as α_s, it follows by taking $\partial T^*_{12}/\partial\alpha = 0$ in Eq. (1.271) that

$$\tan 2\alpha_s = -\frac{\overline{T}}{2T_{12}} \tag{1.278}$$

Equations (1.275) and (1.278) yield the relation $\alpha_s = \alpha_p \pm \pi/4\,(\tan 2\alpha_p \tan 2\alpha_s = -1)$ and thus there exist the two directions for the extremal values of T^*_{12} and they divide the two principal directions into two equal angles, i.e. $\pi/4$. The extremal values of T^*_{12} denoted by T_M is given from Eq. $(1.276)_2$ as follows:

$$T_M = \pm R_T \tag{1.279}$$

which is also expressed by Eq. (1.277) as

$$T_M = \pm\frac{T_1 - T_2}{2} \tag{1.280}$$

The following equation is derived from Eqs. (1.271) and $(1.276)_3$.

$$(T_n - T_m)^2 + T_t^2 = \overline{R}_T^2 \tag{1.281}$$

noting the following equations obtained from Eq. $(1.276)_{1,3}$ or $(1.276)_{2,3}$ with the replacement of $T^*_{11}\,(T^*_{22}) \to T_n$, $T^*_{12} \to T_t$.

$$\left.\begin{aligned}(T_n - T_m)^2 &= \overline{T}^2\cos^2 2\alpha + T_{12}^2\sin^2 2\alpha + 2\overline{T}T_{12}\sin 2\alpha\cos 2\alpha\\ T_t^2 &= \overline{T}^2\sin^2 2\alpha + T_{12}^2\cos^2 2\alpha - 2\overline{T}T_{12}\sin 2\alpha\cos 2\alpha\end{aligned}\right\}$$

Consequently, the components on an arbitrary plane is expressed by the point on the circle with the radius R_T centering at $(T_m, 0)$ in the two-dimensional plane (T_n, T_t) as shown in Fig. 1.6. This circle is called the *Mohr's circle*.

Substituting Eq. (1.276) into Eq. (1.271), we have the expressions

$$\left.\begin{aligned}T^*_{11} &= T_m + R_T\cos(2\alpha - 2\alpha_p),\\ T^*_{22} &= T_m - R_T\cos(2\alpha - 2\alpha_p),\\ T^*_{12} &= -R_T\cos(2\alpha - 2\alpha_p)\end{aligned}\right\} \tag{1.282}$$

Therefore, T^*_{11} and T^*_{12} are shown by the values in the ordinate and abscissa axes, respectively, of the point rotated 2α counterclockwise from point T_{11}, T_{12} on the Mohr's circle as shown in Fig. 1.5, provided that the definition for the sign of shear component is altered to be positive when it applies to the body surface in the clockwise direction, in the Mohr's circle.

As shown in Fig. 1.5, the intersecting angle of the two straight lines drawn in parallel to the physical plane x_1 and x^*_1 stemming from the points (T_{11}, T_{12}) and (T^*_{11}, T^*_{12}), respectively, on the Mohr's circle is α which is the angle of circumference of Mohr's circle and thus the intersecting point lies on the circle. This point is called the *pole*. Generally speaking, the normal component T_n and the shear component T_t applying to a certain physical plane are given by the intersecting point of Mohr's circle and the straight line drawn parallel to that physical plane from the pole P.

1.15 Tensor Functions

Some tensor functions are shown in this section. Based on Eq. (1.168) with the Taylor expansion, we have the following tensor functions

$$\begin{aligned}\sin \mathbf{T} &= \sum_{n=0}^{\infty} \frac{(-1)^n}{(2n+1)!} T_P^n \mathbf{e}_P \otimes \mathbf{e}_P = \left(1 - \frac{1}{3!} T_P + \frac{1}{5!} T_P^3 - \cdots\right) \mathbf{e}_P \otimes \mathbf{e}_P \\ &= \mathbf{T} - \frac{1}{3!}\mathbf{T}^3 + \frac{1}{5!}\mathbf{T}^5 - \cdots = \sum_{n=0}^{\infty} \frac{(-1)^n}{(2n+1)!} \mathbf{T}^{2n+1}\end{aligned} \tag{1.283}$$

$$\begin{aligned}\cos \mathbf{T} &= \sum_{n=0}^{\infty} \frac{(-1)^n}{(2n)!} T_P^{2n} \mathbf{e}_P \otimes \mathbf{e}_P = \left(1 - \frac{1}{2!} T_P^2 + \frac{1}{4!} T_P^4 - \cdots\right) \mathbf{e}_P \otimes \mathbf{e}_P \\ &= \mathbf{T} - \frac{1}{2!}\mathbf{T}^2 + \frac{1}{4!}\mathbf{T}^4 - \cdots = \sum_{n=0}^{\infty} \frac{(-1)^n}{(2n)!} \mathbf{T}^{2n}\end{aligned} \tag{1.284}$$

$$\begin{aligned}\exp \mathbf{T} &= \sum_{n=0}^{\infty} \frac{1}{n!} T_P^n \mathbf{e}_P \otimes \mathbf{e}_P = \left(1 + T_P + \frac{1}{2!} T_P^2 + \frac{1}{3!} T_P^3 + \cdots\right) \mathbf{e}_P \otimes \mathbf{e}_P \\ &= \sum_{n=0}^{\infty} \frac{1}{n!} \mathbf{T}^n = \mathbf{I} + \mathbf{T} + \frac{1}{2!}\mathbf{T}^2 + \frac{1}{3!}\mathbf{T}^3 + \cdots = \sum_{n=0}^{\infty} \frac{1}{n!} \mathbf{T}^n\end{aligned} \tag{1.285}$$

Then, it follows noting Eqs. (1.168) and (1.285) that

$$(\exp \mathbf{T}^A)^T = (\exp \mathbf{T}^{A\,T}) = \exp(-\mathbf{T}^A) \tag{1.286}$$

$$(\exp \mathbf{T})^T = \exp(\mathbf{T}^T) \tag{1.287}$$

$$(\exp \mathbf{T}^A)^T = (\exp \mathbf{T}^{AT}) = \exp(-\mathbf{T}^A) \tag{1.288}$$

$$\exp(n\mathbf{T}) = (\exp \mathbf{T})^n \tag{1.289}$$

$$\exp(\mathbf{A} + \mathbf{B}) = \exp \mathbf{A} \exp \mathbf{B} \tag{1.290}$$

$$\det[\exp(\mathbf{A} + \mathbf{B})] = \exp(\mathrm{tr}\,\mathbf{A}) \exp(\mathrm{tr}\,\mathbf{B}) \tag{1.291}$$

$$(\exp \mathbf{T}^A)(\exp \mathbf{T}^A)^T = (\exp \mathbf{T}^A)(\exp \mathbf{T}^{AT}) = \exp(\mathbf{T}^A - \mathbf{T}^A) = \mathbf{I} \tag{1.292}$$

noting

$$\sum_{n=0}^{\infty} \frac{1}{n!}(\mathbf{A}+\mathbf{B})^n = \sum_{n=0}^{\infty} \frac{1}{n!}(\mathbf{A}^n + n\mathbf{A}^{n-1}\mathbf{B} + \cdots + n\mathbf{A}\mathbf{B}^{n-1} + \mathbf{B}^n) = \left(\sum_{n=0}^{\infty} \frac{1}{n!}\mathbf{A}^n\right)\left(\sum_{n=0}^{\infty} \frac{1}{n!}\mathbf{B}^n\right)$$

$$\begin{aligned} \det[\exp \mathbf{T}] &= (\exp \mathbf{T})_1 (\exp \mathbf{T})_2 (\exp \mathbf{T})_3 = \exp T_1 \exp T_2 \exp T_3 \\ &= \exp(T_1 + T_2 + T_3) \end{aligned}$$

for Eqs. (1.290) and (1.291).

1.16 Partial Differential Calculi

Partial derivatives of symmetric tensors appearing often in elastoplasticity are shown below.

$$\begin{aligned} \frac{\partial T_{ij}}{\partial T_{kl}} &= \delta_{ik}\delta_{jl} \\ \frac{\partial (T_{ir}T_{rj})}{\partial T_{kl}} &= \delta_{ik}\delta_{rl}T_{rj} + T_{ir}\delta_{rk}\delta_{jl} = \delta_{ik}T_{lj} + T_{ik}\delta_{jl} \\ \frac{\partial (T_{ir}T_{rs}T_{sj})}{\partial T_{kl}} &= \delta_{ik}\delta_{rl}T_{rs}T_{sj} + T_{ir}\delta_{rk}\delta_{sl}T_{sj} + T_{ir}T_{rs}\delta_{sk}\delta_{jl} = \delta_{ik}T_{ls}T_{sj} + T_{ik}T_{lj} + T_{ir}T_{rk}\delta_{jl} \end{aligned}$$

$$\boxed{\frac{\partial \mathbf{T}}{\partial \mathbf{T}} = \mathcal{I} = \mathbf{I}\,\overline{\otimes}\,\mathbf{I}, \quad \frac{\partial \mathbf{T}^2}{\partial \mathbf{T}} = \mathbf{I}\,\tilde{\otimes}\,\mathbf{T} + \mathbf{T}\,\overline{\otimes}\,\mathbf{I}, \quad \frac{\partial \mathbf{T}^3}{\partial \mathbf{T}} = \mathbf{I}\,\tilde{\otimes}\,\mathbf{T}^2 + \mathbf{T}\,\tilde{\otimes}\,\mathbf{T} + \mathbf{T}^2\,\overline{\otimes}\,\mathbf{I}} \tag{1.293}$$

using made of the symbol in Eq. $(1.148)_3$.

$$\begin{aligned} \frac{\partial (T_{rs}\delta_{rs})}{\partial T_{ij}} &= \delta_{ir}\delta_{js}\delta_{rs} = \delta_{ij} \\ \frac{\partial (T_{rs}T_{sr})}{\partial T_{ij}} &= \delta_{ri}\delta_{sj}T_{sr} + T_{rs}\delta_{si}\delta_{rj} = 2T_{ji} \\ \frac{\partial (T_{rs}T_{st}T_{tr})}{\partial T_{ij}} &= \delta_{ri}\delta_{sj}T_{st}T_{tr} + T_{rs}\delta_{si}\delta_{tj}T_{tr} + T_{rs}T_{st}\delta_{ti}\delta_{rj} = T_{jt}T_{ti} + T_{ri}T_{jr} + T_{js}T_{si} = 3T_{jt}T_{ti} \\ \frac{\partial T'_{ij}}{\partial T_{kl}} &= \mathcal{I}'_{ijkl} \\ &\left(\frac{\partial T'_{ij}}{\partial T_{kl}} = \frac{\partial (T_{ij} - T_m\delta_{ij})}{\partial T_{kl}} = \delta_{ik}\delta_{jl} - \frac{1}{3}\delta_{ij}\delta_{kl}\right) \end{aligned}$$

$$\boxed{\frac{\partial \mathrm{tr}\,\mathbf{T}}{\partial \mathbf{T}} = \frac{\partial\,\mathrm{I}}{\partial \mathbf{T}} = \frac{\partial \bar{\mathrm{I}}}{\partial \mathbf{T}} = \mathbf{I}, \quad \frac{\partial\,\mathrm{tr}\,\mathbf{T}^2}{\partial \mathbf{T}} = \frac{\partial\,\overline{\mathrm{II}}}{\partial \mathbf{T}} = 2\mathbf{T}^T, \quad \frac{\partial\,\mathrm{tr}\,\mathbf{T}^3}{\partial \mathbf{T}} = \frac{\partial\,\overline{\mathrm{III}}}{\partial \mathbf{T}} = 3\mathbf{T}^{T^2}} \tag{1.294}$$

$$\boxed{\begin{aligned} \frac{\partial \mathrm{II}}{\partial \mathbf{T}} &= \frac{\partial \frac{1}{2}((\mathrm{tr}\mathbf{T})^2 - \mathrm{tr}\mathbf{T}^2)}{\partial \mathbf{T}} = \mathrm{I}\mathbf{I} - \mathbf{T}^T \\ \frac{\partial \mathrm{III}}{\partial \mathbf{T}} &= \frac{\partial\,\det \mathbf{T}}{\partial \mathbf{T}} = (\det \mathbf{T})\mathbf{T}^{-T} = \mathrm{II}\mathbf{I} - \mathrm{I}\mathbf{T}^T + \mathbf{T}^{2T} = (\mathrm{II}\mathbf{I} - \mathrm{I}\mathbf{T} + \mathbf{T}^2)^T \end{aligned}} \tag{1.295}$$

noting Eq. (1.121) and

$$\begin{aligned} \frac{\partial \mathrm{III}}{\partial \mathbf{T}} &= \frac{\partial \left[\frac{1}{6}(\mathrm{tr}\,\mathbf{T})^3 - \frac{1}{2}\mathrm{tr}\,\mathbf{T}\mathrm{tr}\,\mathbf{T}^2 + \frac{1}{3}\mathrm{tr}\,\mathbf{T}^3\right]}{\partial \mathbf{T}} = \frac{3}{6}(\mathrm{tr}\,\mathbf{T})^2\mathbf{I} - \frac{1}{2}(\mathrm{tr}\mathbf{T}^2)\mathbf{I} - \frac{1}{2}(\mathrm{tr}\,\mathbf{T})2\mathbf{T}^T + \frac{1}{3}3\mathbf{T}^{T^2} \\ &= \frac{1}{2}[(\mathrm{tr}\,\mathbf{T})^2 - \mathrm{tr}\,\mathbf{T}^2]\mathbf{I} - (\mathrm{tr}\,\mathbf{T})\mathbf{T}^T + \mathbf{T}^{T^2} \end{aligned}$$

$$\frac{\partial\sqrt{T_{rs}T_{rs}}}{\partial T_{ij}} = \frac{1}{2}(T_{rs}T_{rs})^{-1/2}(\delta_{ri}\delta_{sj}T_{sr} + T_{rs}\delta_{ri}\delta_{sj}) = \frac{T_{ij}}{\sqrt{T_{rs}T_{rs}}}, \quad \boxed{\frac{\partial\|\mathbf{T}\|}{\partial \mathbf{T}} = \frac{\mathbf{T}}{\|\mathbf{T}\|}} \tag{1.296}$$

$$\frac{\partial T'_{ij}}{\partial T_{kl}} = \frac{\partial(T_{ij} - T_m\delta_{ij})}{\partial T_{kl}} = \delta_{ik}\delta_{jl} - \frac{1}{3}\delta_{ij}\delta_{kl} = \mathcal{I}'_{ijkl}, \quad \boxed{\frac{\partial \mathbf{T}'}{\partial \mathbf{T}} = \mathcal{I}'} \tag{1.297}$$

$$\frac{\partial\sqrt{T'_{rs}T'_{rs}}}{\partial T'_{ij}} = \frac{1}{2}(T'_{rs}T'_{rs})^{-1/2}\frac{\partial(T'_{rs}T'_{rs})}{\partial T'_{ij}} = \frac{1}{2}(T'_{rs}T'_{rs})^{-1/2}2\delta_{ir}\delta_{js}T'_{rs} = t'_{ij},$$

$$\boxed{\frac{\partial\|\mathbf{T}'\|}{\partial \mathbf{T}'}\left(= \frac{\partial\|\mathbf{T}'\|}{\partial \mathbf{T}}\frac{\partial \mathbf{T}}{\partial \mathbf{T}'} = \frac{\mathbf{T}'}{\|\mathbf{T}'\|}\mathcal{I}'\right) = \frac{\mathbf{T}'}{\|\mathbf{T}'\|} = \mathbf{t}'} \tag{1.298}$$

$$\frac{\partial t'_{ij}}{\partial T'_{kl}} = \frac{\partial \dfrac{T'_{ij}}{\sqrt{T'_{pq}T'_{pq}}}}{\partial T'_{kl}} = \frac{\delta_{ik}\delta_{jl}\sqrt{T'_{pq}T'_{pq}} - T'_{ij}\dfrac{T'_{kl}}{\sqrt{T'_{pq}T'_{pq}}}}{T'_{pq}T'_{pq}} = \frac{1}{\sqrt{T'_{pq}T'_{pq}}}(\delta_{ik}\delta_{jl} - t'_{ij}t'_{kl}),$$

$$\boxed{\frac{\partial \mathbf{t}'}{\partial \mathbf{T}'} = \frac{1}{\|\mathbf{T}'\|}(\mathcal{I} - \mathbf{t}'\otimes\mathbf{t}')} \tag{1.299}$$

$$\frac{\partial t'_{ij}}{\partial T_{kl}} = \frac{\partial t'_{ij}}{\partial T'_{rs}}\frac{\partial T'_{rs}}{\partial T_{kl}} = \frac{1}{\sqrt{T'_{pq}T'_{pq}}}(\delta_{ir}\delta_{js} - t'_{ij}t'_{rs})(\delta_{rk}\delta_{sl} - \frac{1}{3}\delta_{rs}\delta_{kl})$$
$$= \frac{1}{\sqrt{T'_{pq}T'_{pq}}}(\delta_{ik}\delta_{jl} - \frac{1}{3}\delta_{ij}\delta_{kl} - t'_{ij}t'_{kl}),$$

$$\boxed{\frac{\partial \mathbf{t}'}{\partial \mathbf{T}} = \frac{1}{\|\mathbf{T}'\|}(\mathcal{I} - \frac{1}{3}\mathcal{T} - \mathbf{t}' \otimes \mathbf{t}') = \frac{1}{\|\mathbf{T}'\|}(\mathcal{I}' - \mathbf{t}' \otimes \mathbf{t}')} \tag{1.300}$$

$$\boxed{\cos 3\theta \equiv \sqrt{6}\mathrm{tr}\,\mathbf{t}'^3} \tag{1.301}$$

$$\frac{\partial \cos 3\theta}{\partial t'_{ij}} = \frac{\partial \sqrt{6} t'_{pq}t'_{qr}t'_{rp}}{\partial t'_{ij}} = 3\sqrt{6}\delta_{ip}\delta_{jq}t'_{qr}t'_{rp} = 3\sqrt{6}t'_{ir}t'_{rj},$$

$$\boxed{\frac{\partial \cos 3\theta}{\partial \mathbf{t}'} = 3\sqrt{6}\mathbf{t}'^2} \tag{1.302}$$

$$\frac{\partial \cos 3\theta}{\partial T_{ij}} = \frac{\partial \cos 3\theta}{\partial t'_{rs}}\frac{\partial t'_{rs}}{\partial T_{ij}}$$
$$= 3\sqrt{6}t'_{rt}t'_{ts}\frac{1}{\sqrt{T'_{pq}T'_{pq}}}(\delta_{ir}\delta_{js} - \frac{1}{3}\delta_{ij}\delta_{rs} - t'_{ij}t'_{rs})$$
$$= 3\sqrt{6}\frac{1}{\sqrt{T'_{pq}T'_{pq}}}(t'_{ir}t'_{rj} - \frac{1}{3}t'_{rt}t'_{tr}\delta_{ij} - t'_{rs}t'_{st}t'_{tr}t'_{ij}),$$

$$\boxed{\frac{\partial \cos 3\theta}{\partial \mathbf{T}} = -\frac{1}{\|\mathbf{T}'\|}(\sqrt{6}\|\mathbf{t}'\|^2\mathbf{I} + 3\cos 3\theta\mathbf{t}' - 3\sqrt{6}\mathbf{t}'^2)} \tag{1.303}$$

Differentiating $T^{-1}_{ir}T_{rs} = \delta_{is}$, one has

$$\frac{\partial (T^{-1}_{ir}T_{rs})}{\partial T_{kl}} = \frac{\partial T^{-1}_{ir}}{\partial T_{kl}}T_{rs} + T^{-1}_{ir}\delta_{rk}\delta_{sl} = 0 \rightarrow \frac{\partial T^{-1}_{ir}}{\partial T_{kl}}T_{rs}T^{-1}_{sj} + T^{-1}_{ir}\delta_{rk}\delta_{sl}T^{-1}_{sj} = 0$$

$$\dot{\delta}_{ij} = (T^{-1}_{ir}T_{rs})^{\bullet} = \dot{T}^{-1}_{ir}T_{rs} + T^{-1}_{ir}\dot{T}_{rs} = 0$$

which leads to

$$\frac{\partial T^{-1}_{ij}}{\partial T_{kl}} = -T^{-1}_{ik}T^{-1}_{lj} = -T^{-1}_{ir}\mathcal{I}_{rskl}T^{-1}_{sj},\ \dot{T}^{-1}_{ij} = -T^{-1}_{ir}\dot{T}_{rs}T^{-1}_{sj}$$

$$\boxed{\frac{\partial \mathbf{T}^{-1}}{\partial \mathbf{T}} = -\mathbf{T}^{-1}\tilde{\otimes}\mathbf{T}^{-1} = -\mathbf{T}^{-1}\mathcal{I}\mathbf{T}^{-1}, \quad \dot{\mathbf{T}}^{-1} = -\mathbf{T}^{-1}\dot{\mathbf{T}}\mathbf{T}^{-1}} \tag{1.304}$$

Differentiating Eq. (1.180) and noting Eq. (1.294), it holds that

$$\frac{\partial \mathrm{III}}{\partial \mathbf{T}} = \mathbf{T}^{T^2} - \mathrm{I}\mathbf{T}^T + \mathrm{II}\mathbf{I}$$

Then, noting Eq. (1.213), one has

$$\boxed{\frac{\partial \det \mathbf{T}}{\partial \mathbf{T}} = \frac{\partial \mathrm{III}}{\partial \mathbf{T}} = \mathrm{III}\,\mathbf{T}^{-T} = (\det \mathbf{T})\mathbf{T}^{-T}}$$

which was derived also in Eq. (1.121).

For symmetric tensors, the fourth-order identity tensor $\overline{\mathcal{I}}_{ijkl} = \delta_{ik}\delta_{jl}$ can be replaced to the fourth-order symmetrizing tensor $\mathcal{S}_{ijkl} = (\delta_{ik}\delta_{jl} + \delta_{il}\delta_{jk})/2$ in the above equations.

1.17 Differentiation and Integration in Tensor Field

Scalar s, vector $\mathbf{v}$, and tensor $\mathbf{T}$ are called the scalar field, the vector field, and the tensor field, respectively when they are functions of the position vector $\mathbf{x}$. Their differentiation and integration in fields are shown below, in which the following operator, called the *nabla* or *Hamilton operator*, is often used.

$$\boxed{\mathbf{\nabla} \equiv \frac{\partial}{\partial x_r}\mathbf{e}_r = \frac{\partial}{\partial \mathbf{x}}} \tag{1.305}$$

1. Gradient

Scalar field:

$$\mathrm{grad}s = \mathbf{\nabla} s = \frac{\partial s}{\partial x_r}\mathbf{e}_r \tag{1.306}$$

Vector field:

$$\mathrm{grad}\mathbf{v} = \begin{cases} \mathbf{v} \otimes \mathbf{\nabla} = v_i \mathbf{e}_i \otimes \dfrac{\partial}{\partial x_j}\mathbf{e}_j = \dfrac{\partial v_i}{\partial x_j}\mathbf{e}_i \otimes \mathbf{e}_j : \text{rear(right) form} \\ \mathbf{\nabla} \otimes \mathbf{v} = \dfrac{\partial}{\partial x_i}\mathbf{e}_i \otimes v_j \mathbf{e}_j = \dfrac{\partial v_j}{\partial x_i}\mathbf{e}_i \otimes \mathbf{e}_j : \text{front(left) form} \end{cases} \tag{1.307}$$

Second-order tensor field:

$$\mathrm{grad}\mathbf{T} = \begin{cases} \mathbf{T} \otimes \nabla = T_{ij}\mathbf{e}_i \otimes \mathbf{e}_j \otimes \dfrac{\partial}{\partial x_k}\mathbf{e}_k = \dfrac{\partial T_{ij}}{\partial x_k}\mathbf{e}_i \otimes \mathbf{e}_j \otimes \mathbf{e}_k : \text{rear(right) form} \\ \nabla \otimes \mathbf{T} = \dfrac{\partial}{\partial x_i}\mathbf{e}_i \otimes T_{jk}\mathbf{e}_j \otimes \mathbf{e}_k = \dfrac{\partial T_{jk}}{\partial x_i}\mathbf{e}_i \otimes \mathbf{e}_j \otimes \mathbf{e}_k : \text{ front(left) form} \end{cases} \tag{1.308}$$

2. Divergence

Vector field:

$$\mathrm{div}\mathbf{v} = \nabla \cdot \mathbf{v}\,(= \mathbf{v} \cdot \nabla) = v_i \mathbf{e}_i \cdot \frac{\partial}{\partial x_j}\mathbf{e}_j = \frac{\partial v_i}{\partial x_i} \tag{1.309}$$

Second-order tensor field:

$$\mathrm{div}\mathbf{T} = \begin{cases} \mathbf{T}\nabla = T_{ij}\mathbf{e}_i \otimes \mathbf{e}_j \dfrac{\partial}{\partial x_k}\mathbf{e}_k = \dfrac{\partial T_{ir}}{\partial x_r}\mathbf{e}_i : \text{ rear (right) form} \\ \nabla\mathbf{T} = \dfrac{\partial}{\partial x_i}\mathbf{e}_i T_{jk}\mathbf{e}_j \otimes \mathbf{e}_k = \dfrac{\partial T_{ri}}{\partial x_r}\mathbf{e}_i : \text{ front (left) form} \end{cases} \tag{1.310}$$

3. Rotation (or **curl**)

Vector field:

$$\mathrm{rot}\,\mathbf{v} = \begin{cases} \mathbf{v} \times \nabla = v_i\mathbf{e}_i \times \dfrac{\partial}{\partial x_j}\mathbf{e}_j = \varepsilon_{ijk}\dfrac{\partial v_i}{\partial x_j}\mathbf{e}_k : \text{rear (right) form} \\ \nabla \times \mathbf{v} = \dfrac{\partial}{\partial x_i}\mathbf{e}_i \times v_j\mathbf{e}_j = \varepsilon_{ijk}\dfrac{\partial v_j}{\partial x_i}\mathbf{e}_k : \text{front (left) form} \end{cases} \tag{1.311}$$

noting Eq. (1.49).

Second-order tensor field:

$$\mathrm{rot}\mathbf{T} = \begin{cases} \mathbf{T} \times \nabla = T_{ij}\mathbf{e}_i \otimes \mathbf{e}_j \times \dfrac{\partial}{\partial x_k}\mathbf{e}_k \\ \quad = \dfrac{\partial T_{ij}}{\partial x_k}\mathbf{e}_i \otimes (\mathbf{e}_j \times \mathbf{e}_k) = \varepsilon_{jkr}\dfrac{\partial T_{ij}}{\partial x_k}\mathbf{e}_i \otimes \mathbf{e}_r : \text{rear(right) form} \\ \nabla \times \mathbf{T} = \dfrac{\partial}{\partial x_i}\mathbf{e}_i \times T_{jk}\mathbf{e}_j \otimes \mathbf{e}_k \\ \quad = \dfrac{\partial T_{jk}}{\partial x_i}(\mathbf{e}_i \times \mathbf{e}_j) \otimes \mathbf{e}_k = \varepsilon_{ijr}\dfrac{\partial T_{jk}}{\partial x_i}\mathbf{e}_r \otimes \mathbf{e}_k : \text{front(left) form} \end{cases} \tag{1.312}$$

The symbol ∇ is the vector, and the scalar product of itself, i.e.

$$\boxed{\Delta \equiv \nabla^2 \equiv \nabla \cdot \nabla = \frac{\partial}{\partial x_r}\mathbf{e}_r \cdot \frac{\partial}{\partial x_s}\mathbf{e}_s = \frac{\partial^2}{\partial x_r \partial x_r}} \tag{1.313}$$

has the meaning of $\nabla^2(\,) \equiv \mathrm{div}(\mathrm{grad}(\,))$. The symbol Δ is called the *Laplacian* or *Laplace operator*, which is often used for scalar or vector fields as

$$\Delta s = \frac{\partial^2 s}{\partial x_r \partial x_r} \tag{1.314}$$

$$\Delta \mathbf{v} = \frac{\partial^2 v_s}{\partial x_r \partial x_r} \mathbf{e}_s \tag{1.315}$$

The following relations hold between the above-mentioned operators.

$$\left.\begin{aligned}
&\mathrm{grad}(s\mathbf{v}) = \mathbf{v} \otimes \mathrm{grad}s + s\mathrm{grad}\mathbf{v}, \\
&\mathrm{div}(s\mathbf{v}) = s\mathrm{div}\mathbf{v} + \mathbf{v} \cdot \mathrm{grad}s, \\
&\mathrm{div}(\mathbf{u} \times \mathbf{v}) = \mathbf{v} \cdot \mathrm{rot}\mathbf{u} - \mathbf{u} \cdot \mathrm{rot}\mathbf{v}, \\
&\mathrm{rot}(\mathbf{u} \times \mathbf{v}) = (\mathrm{grad}\mathbf{u})\mathbf{v} - (\mathrm{grad}\mathbf{v})\mathbf{u} + (\mathrm{div}\mathbf{v})\mathbf{u} - (\mathrm{div}\mathbf{u})\mathbf{v}, \\
&\mathrm{grad}(\mathbf{u} \cdot \mathbf{v}) = (\mathrm{grad}\mathbf{v})\mathbf{u} + (\mathrm{grad}\mathbf{u})\mathbf{v} + \mathbf{u} \times \mathrm{rot}\mathbf{v} + \mathbf{v} \times \mathrm{rot}\mathbf{u}, \\
&\mathrm{div}(s\mathbf{T}) = \mathbf{T}^T \mathrm{grad}s + s\mathrm{div}\mathbf{T}, \\
&\mathrm{div}(\mathbf{T}\mathbf{v}) = \mathbf{v} \cdot \mathrm{div}\mathbf{T} + \mathrm{tr}(\mathbf{T}\mathrm{grad}\mathbf{v}) \\
&\mathrm{div}(\mathbf{A}\mathbf{B}) = \mathbf{B} \cdot \mathrm{div}\mathbf{A} + \mathrm{div}\mathbf{A}^T \cdot \mathbf{B}
\end{aligned}\right\} \tag{1.316}$$

4. Gauss' divergence theorem

Consider the physical quantity $T(\mathbf{x})$ in the zone surrounded by a smooth surface inside a material. Then, suppose the slender prism cut by the four planes perpendicular to the x_2-axis and x_3-axis in infinitesimal intervals from a zone inside the material. The following equation holds for the prism possessing the infinitesimal volume $\partial v = dx_1 dx_2 dx_3$.

$$\int_{\partial v} \frac{\partial T}{\partial x_1} dv = \int_{\partial v} \frac{\partial T}{\partial x_1} dx_1 dx_2 dx_3 = [T]_{x_1^-}^{x_1^+} dx_2 dx_3 = (T^+ - T^-) dx_2 dx_3 \tag{1.317}$$

where $(\,)^+$ and $(\,)^-$ designate the values of physical quantity at the maximum and the minimum x_1-coordinates, respectively.

The neighborhood of the surface cut by the prism is magnified in Fig. 1.6. Consider the infinitesimal rectangular surface PQRS of the prism exposed at the surface in the maximum x_1-coordinate and the infinitesimal rectangular section $\mathrm{PQ^*R^*S^*}$ cut by the plane passing through the point P and perpendicular to the x_1-axis by the prism. Then, denoting $\overline{\mathrm{QQ^*}} = dx_Q$, $\overline{\mathrm{SS^*}} = dx_S$, the vectors $\overrightarrow{\mathrm{PQ}}$, $\overrightarrow{\mathrm{PS}}$ are given by

$$\overrightarrow{\mathrm{PQ}} = dx_2 \mathbf{e}_2 + dx_Q \mathbf{e}_1, \quad \overrightarrow{\mathrm{PS}} = dx_3 \mathbf{e}_3 + dx_S \mathbf{e}_1 \tag{1.318}$$

and thus it holds that

$$\begin{aligned}
\mathbf{n}^+ da^+ &= \overrightarrow{\mathrm{PQ}} \times \overrightarrow{\mathrm{PS}} = (dx_2 \mathbf{e}_2 + dx_Q \mathbf{e}_1) \times (dx_3 \mathbf{e}_3 + dx_S \mathbf{e}_1) \\
&= dx_2 dx_3 \mathbf{e}_2 \times \mathbf{e}_3 + dx_Q dx_3 \mathbf{e}_1 \times \mathbf{e}_3 + dx_S dx_2 \mathbf{e}_2 \times \mathbf{e}_1 \\
&= dx_2 dx_3 \mathbf{e}_1 - dx_Q dx_3 \mathbf{e}_2 - dx_S dx_2 \mathbf{e}_3
\end{aligned} \tag{1.319}$$

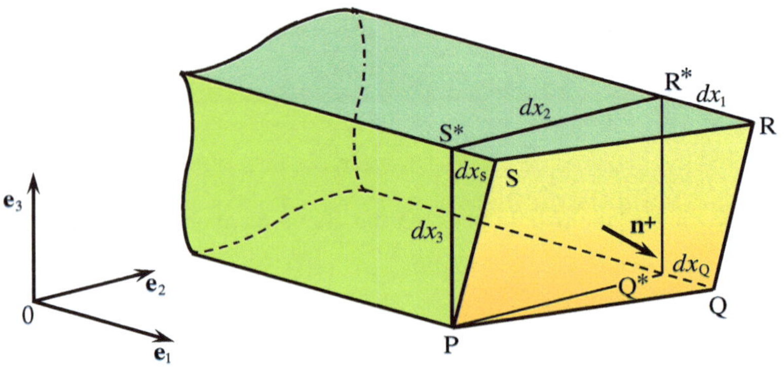

Fig. 1.6 Infinitesimal square pillar cut from a zone in material

Comparing the components in the base $\mathbf{e}_1$ on the both sides in Eq. (1.319), one has

$$n_1^+ da^+ = dx_2 dx_3 \tag{1.320}$$

In a similar manner for the surface of the prism exposed on surface in the minimum x_1-coordinate, one has

$$n_1^- da^- = -dx_2 dx_3 \tag{1.321}$$

The general expression of projected area is given in Appendix A.

Adopting Eqs. (1.320) and (1.321) in Eq. (1.317), it holds for the prism that

$$\int_{\partial v} \frac{\partial T}{\partial x_1} dv = T^+ n_1^+ da^+ - T^-(-n_1^- da^-) = T^+ n_1^+ da^+ + T^- n_1^- da^- \tag{1.322}$$

Then, the following equation holds for the whole zone.

$$\int_v \frac{\partial T}{\partial x_1} dv = \int_a T n_1 da \tag{1.323}$$

In a similar manner also for the x_2- and x_3-directions, the following Gauss' divergence theorem holds.

$$\int_v \frac{\partial T}{\partial x_i} dv = \int_a T n_i da \tag{1.324}$$

The following equations for the scalar s, the vector $\mathbf{v}$ and the tensor $\mathbf{T}$ hold from Eq. (1.324).

$$\int_v \frac{\partial s}{\partial x_i} dv = \int_a s n_i da, \quad \int_v \nabla s dv = \int_a s\mathbf{n} da \tag{1.325}$$

$$\int_v \frac{\partial v_i}{\partial x_i} dv = \int_a v_i n_i da, \quad \int_v \nabla \cdot \mathbf{v} dv = \int_a \mathbf{v} \cdot \mathbf{n} da \tag{1.326}$$

$$\int_v \frac{\partial T_{ij}}{\partial x_i} dv = \int_a T_{ij} n_i da, \quad \int_v \nabla \mathbf{T} dv = \int_a \mathbf{T}\mathbf{n} da \tag{1.327}$$

References

Carlson DE, Hoger A (1986) The derivative of a tensor-valued function of a tensor. Q Appl Math 406:409–423

Chadwick P (1976) Continuum mechanics: concise theory and problems. Dover, New York

del Peiro G (1979) Some properties of the set of fourth-order tensors, with application to elasticity. J Elast 9:245–261

Hashiguchi K (2013) Elastoplasticity theory. Lecture notes in applied and computational mechanics, 2nd edn. Springer, Heidelberg

Hashiguchi K, Yamakawa Y (2012) Introduction to finite strain theory for continuum elasto-plasticity. Wiley series in computational mechanics. Wiley, Chichester

Hoger A, Carlson DE (1984) On the derivative of the square root of a tensor and Guo's theorem. J Elast 14:329–336

Kintzel O, Bazar Y (2006) Fourth-order tensors—tensor differentiation with applications to continuum mechanics, Part I: Classical tensor analysis. ZAMM 86:291–311

Kyoya T (2008) Note on continuum mechanics. Japan Association for Nonlinear CAE, Tokyo (in Japanese)

Leigh DC (1968) Nonlinear continuum mechanics: an introduction to the continuum physics and mechanical theory of the nonlinear mechanical behavior of materials, McGraw-Hill, New York

Ottosen NS, Ristinmaa M (2005) The mechanics of constitutive modeling. Elsevier, Amsterdam

Steinmann P, Larsson R, Runesson K (1997) On the localization properties of multiplicative hyperelasto-pastic continua with strong discontinuities. Int J Solids Struct 8:969–990

Wang CCA (1970) A new representation theorem for isotropic functions: an answer to Professor G.F. Smith's criticism of my paper on representations for isotropic functions. Arch Ratl Mech Anal 36:166–223

Wang Z-Q, Dui G-S (2008) Two-point constitutive equations and integration algorithms for isotropic-hardening rate-independence elastoplastic materials in large deformation. Int J Numer Meth Eng 75:1435–1456

Chapter 2
Motion and Strain (Rate)

The tensor analysis providing the mathematical foundation for the continuum mechanics is described in Chap. 1. Basic concepts and quantities for continuum mechanics will be studied in the three chapters up to Chap. 4. The description of motion and deformation of a material body constitutes the basic introductory part of the continuum mechanics. Various expressions of motion and a variety of strain and strain rate measures are employed for the description of reversible and irreversible deformations of materials. Some selected basic expressions and measures will be explained in this chapter.

2.1 Motion of Material Point

A material body is assembly of material particles (or material elements). The map of positions of material particles in a space is referred to as the *configuration*. Here, the configurations in the initial time $t = t_0$ and the current time t are called the *initial* (or *Lagrangian*) *configuration* and the *current* (or *Eulerian*) *configuration*, respectively. Deformation is described by the change of configuration from a particular configuration which is called the *reference configuration*. Here, the reference configuration can be chosen at arbitrary intermediate time $\tau(t_0 \leq \tau \leq t)$ is called the *reference time*.

The position vectors of material particle in the initial and the current configurations are designated by $\mathbf{X}$ and $\mathbf{x}(t)$, respectively. Here, $\mathbf{X}$ is fixed and thus it can be regarded as a label of each material particle. The motion of material point during the time $t_0 \rightarrow t$ is described as

$$\mathbf{x} = \chi(\mathbf{X},t),\ \mathbf{X} = \chi^{-1}(\mathbf{x}, t) \tag{2.1}$$

Besides, the motion of material point during the time $t_0 \rightarrow \tau$ is described as

K. Hashiguchi, *Foundations of Elastoplasticity: Subloading Surface Model*,
DOI 10.1007/978-3-319-48821-9_2

$$\mathbf{x}(\tau) = \chi(\mathbf{X},\tau), \quad \mathbf{X} = \chi^{-1}(\mathbf{x}(\tau),\tau) \tag{2.2}$$

The fact that a material does not overlap or separate by the motion of material requires the existence of the one-to-one correspondence between $\mathbf{X}$ and $\mathbf{x}$ ($\mathbf{x}$ is uniquely determined by $\mathbf{x}$ and vice versa) so that $x_1(X_1,X_2,X_3)$, $x_2(X_1,X_2,X_3)$ and $x_3(X_1,X_2,X_3)$ must be mutually independent. Now, let the mathematical requirement for this fact be derived by the reductive absurdity. x_1,x_2,x_3 are not mutually independent if they satisfy the constraint

$$f(x_1(X_1,X_2,X_3),x_2(X_1,X_2,X_3),x_3(X_1,X_2,X_3)) = 0 \tag{2.3}$$

from which it follows that

$$\left.\begin{array}{l} \dfrac{\partial f}{\partial x_1}\dfrac{\partial x_1}{\partial X_1} + \dfrac{\partial f}{\partial x_2}\dfrac{\partial x_2}{\partial X_1} + \dfrac{\partial f}{\partial x_3}\dfrac{\partial x_3}{\partial X_1} = 0 \\ \dfrac{\partial f}{\partial x_1}\dfrac{\partial x_1}{\partial X_2} + \dfrac{\partial f}{\partial x_2}\dfrac{\partial x_2}{\partial X_2} + \dfrac{\partial f}{\partial x_3}\dfrac{\partial x_3}{\partial X_2} = 0 \\ \dfrac{\partial f}{\partial x_1}\dfrac{\partial x_1}{\partial X_3} + \dfrac{\partial f}{\partial x_2}\dfrac{\partial x_2}{\partial X_3} + \dfrac{\partial f}{\partial x_3}\dfrac{\partial x_3}{\partial X_3} = 0 \end{array}\right\}, \text{ i.e. } \begin{bmatrix} \dfrac{\partial x_1}{\partial X_1} & \dfrac{\partial x_2}{\partial X_1} & \dfrac{\partial x_3}{\partial X_1} \\ \dfrac{\partial x_1}{\partial X_2} & \dfrac{\partial x_2}{\partial X_2} & \dfrac{\partial x_3}{\partial X_2} \\ \dfrac{\partial x_1}{\partial X_3} & \dfrac{\partial x_2}{\partial X_3} & \dfrac{\partial x_3}{\partial X_3} \end{bmatrix} \begin{Bmatrix} \dfrac{\partial f}{\partial x_1} \\ \dfrac{\partial f}{\partial x_2} \\ \dfrac{\partial f}{\partial x_3} \end{Bmatrix} = \begin{Bmatrix} 0 \\ 0 \\ 0 \end{Bmatrix} \tag{2.4}$$

The equation

$$J = 0 \tag{2.5}$$

must hold in order that $\partial f/\partial x_1$, $\partial f/\partial x_2$, $\partial f/\partial x_3$ are determined uniquely on account of Eq. (1.125) regarding $T_{iJ} = \partial x_i/\partial X_J$ and $v_i = \partial f/\partial x_i$, where J is defined by

$$\boxed{J \equiv \det\left(\frac{\partial x_i}{\partial X_J}\right) = \varepsilon_{IJK}\frac{\partial x_1}{\partial X_I}\frac{\partial x_2}{\partial X_J}\frac{\partial x_3}{\partial X_K}} \tag{2.6}$$

and is called the *functional determinant* or *Jacobian*. In contrast, in order that they are mutually independent, it must hold that

$$J \neq 0 \tag{2.7}$$

being free from the constraint in Eq. (2.3). The transformation between $\mathbf{x}$ and $\mathbf{X}$ is called the *admissible transformation*, if f_1, f_2, f_3 in $x_1 = f_1(X_1,X_2,X_3)\, x_2 = f_2(X_1,X_2,X_3)$, $x_3 = f_3(X_1X_2X_3)$ are single-valued and continuous functions, so that the Jacobian is not zero as shown in Eq. (2.7). Further, if the Jacobian is positive, a right-hand coordinate system is transformed to other right-hand one, and it is called the *positive transformation*. Inversely, if the Jacobian is negative, a right-hand coordinate system is transformed to a left-hand one, and it is called the *negative transformation*. Admissible and positive transformation with $J > 0$ is assumed throughout this book.

Physical quantity, say $\mathbf{T}$, in the body changes generally with the position and the time. Physical quantity at current time is described $\mathbf{T}(\mathbf{X},t)$ in terms of the current configuration $\mathbf{X}$ and the current time t. This type of description of mechanical state is called the *Lagrangian* (*or material*) *description*. On the other hand, the physical quantity at current time is described $\mathbf{T}(\mathbf{x},t)$ in terms of the current configuration $\mathbf{x}$ and the current time t. This type of description of physical quantity is called the *Eulerian description* or *spatial description*. Further, the physical quantity at current time can be described in terms of the current configuration $\mathbf{x}(\tau)$ at arbitrary reference time τ and the current time t as

$$\mathbf{T}(\chi^{-1}(\mathbf{x}(\tau),\tau),t) = {}_{\tau}\mathbf{T}(\mathbf{x}(\tau),\, t) \tag{2.8}$$

${}_{\tau}(\,)$ designating to choose the reference time τ as $\tau > t_0$. Equation (2.8) is called the *relative description*. Specifically, ${}_{t}\mathbf{T}(\mathbf{x}(t),\, t)$ is called the *updated Lagrangian description*, where the reference configuration is taken as the current configuration, choosing the reference time as the current time, i.e. $\tau = t$. In contrast, the description $\mathbf{T}(\mathbf{X},t)$ will be called the *total Lagrangian description*.

2.2 Time-Derivatives

The time derivative of the tensor in the spatial description

$$\frac{\partial \mathbf{T}(\mathbf{x},t)}{\partial t} \tag{2.9}$$

describes the rate of the physical quantity at a certain spatial point and thus it is called the *spatial-time* (or *local*) derivative. In many cases of fluid mechanics, a motion and its history of individual particle from the initial state is immaterial and thus the spatial-time derivative is often adopted. In contrast, the time-derivative of the tensor in the material description

$$\frac{\partial \mathbf{T}(\mathbf{X},t)}{\partial t} \tag{2.10}$$

describes the rate of the physical quantity in a certain material particle and thus is called the *material-time derivative*. It is denoted by the symbol

$$\dot{\mathbf{T}} \equiv \frac{\partial \mathbf{T}(\mathbf{X},t)}{\partial t} \quad \text{or} \quad \frac{D\mathbf{T}}{Dt} \equiv \frac{\partial \mathbf{T}(\mathbf{X},t)}{\partial t} \tag{2.11}$$

In solid mechanics, the rate of deformation and its history of individual material particle is required and thus the material-time derivative is used usually.

The material-time derivative in Eq. (2.11) and the spatial-time derivative in Eq. (2.9) are related by

$$\boxed{\dot{\mathbf{T}} \equiv \frac{\partial \mathbf{T}(\mathbf{x},t)}{\partial t} + \frac{\partial \mathbf{T}(\mathbf{x},t)}{\partial \mathbf{x}}\mathbf{v}}, \quad \dot{T}_{ij} \equiv \frac{\partial T_{ij}(\mathbf{x},t)}{\partial t} + \frac{\partial T_{ij}(\mathbf{x},t)}{\partial x_k} v_k \tag{2.12}$$

where $\mathbf{v} \equiv \partial \mathbf{x}/\partial t$ is the velocity vector of material particle. The first term in the right-hand side signifies the *non-steady* (or *local time derivative*) *term* describing the change with time at fixed spatial point and the second term signifies the *steady* (or *convective*) *term* describing the change due to the movement of material, which results from the existence of a spatial gradient of the physical quantity $\mathbf{T}$.

Rate-type constitutive equations for the irreversible deformation of solids, e.g. the viscoelastic, the elastoplastic and the viscoplastic deformation, must be described by the material-time derivative pursuing a material particle because they must describe the relation of physical quantities in each material particle. Here, it should be noticed that the material-time derivative of physical quantity describes the rate observed by moving in parallel with material particle as known from Eq. (2.11) which concerns only with the position vector of material particle and the time. Then, the objective time-derivative based on the rate of physical quantity observed by the coordinate system deforming/rotating with a material must be used for constitutive equations of solids as will be described in Chap. 4.

2.3 Deformation Gradient and Deformation Tensors

At the initial state of deformation ($t = 0$), consider a material particle, the position vector of which is $\mathbf{X}$, and the adjacent material point, the position vector of which is $\mathbf{X} + d\mathbf{X}$. Furthermore, consider the current state ($t = t$) in which these points move to the points with position vectors $\mathbf{x}$ and $\mathbf{x} + d\mathbf{x}$, respectively. The infinitesimal line elements before and after the deformation are described as

$$d\mathbf{X} = dX_A \mathbf{e}_A, \quad d\mathbf{x}(t) = dx_i(t)\mathbf{e}_i(t) \tag{2.13}$$

where the current base $\{\mathbf{e}_i(t)\}$ rotates with the elapse of time so that it changes different from the fixed reference base $\{\mathbf{e}_A\}$, i.e. $\{\mathbf{e}_i(t)\} \neq \{\mathbf{e}_A\}$ for $t > 0$ in general. However, the same base is often used for the reference and the current bases for the sake of simplicity.

Here, based on the relation $d\mathbf{x}(t) = (\partial \mathbf{x}(t)/\partial \mathbf{X})d\mathbf{X}$, we define the deformation gradient tensor

$$\boxed{\mathbf{F}(t) \equiv \frac{\partial \mathbf{x}(t)}{\partial \mathbf{X}}} = F_{iA}(t)\mathbf{e}_i(t) \otimes \mathbf{e}_A \equiv \frac{\partial x_i(t)}{\partial X_A}\mathbf{e}_i(t) \otimes \mathbf{e}_A = x_{i,A}(t)\mathbf{e}_i(t) \otimes \mathbf{e}_A \tag{2.14}$$

$\mathbf{F}(t)$ is based in the current and the reference bases which can be chosen different to each other and thus it is called the *Eulerian-Lagrangian two-point tensor*. The infinitesimal line-element $d\mathbf{x}(t)$ is described by $d\mathbf{X}$ from Eq. (2.14) as follows:

$$d\mathbf{x}(t) = dx_i(t)\mathbf{e}_i(t) = \frac{\partial x_i(t)}{\partial X_A} dX_A \mathbf{e}_i(t) = \frac{\partial x_i(t)}{\partial X_A} \mathbf{e}_i(t) \otimes \mathbf{e}_A dX_B \mathbf{e}_B = \mathbf{F}(t) d\mathbf{X} \quad (2.15)$$

The deformation gradient tensor $\mathbf{F}(t)$ transforms the reference infinitesimal line element to the current infinitesimal line element and thus it is the most fundamental variable for the description of deformation of materials. Equation (2.6) is written in terms of the deformation gradient as

$$J = \det \mathbf{F} \quad (2.16)$$

Therefore, if $J = \det \mathbf{F} \neq 0$ holds, the inverse tensor $\mathbf{F}^{-1}$ exists by virtue of Eq. (1.120), and it is derived from $\mathbf{F}\mathbf{F}^{-1} = (\partial \mathbf{x}/\partial \mathbf{X})(\partial \mathbf{X}/\partial \mathbf{x}) = \mathbf{I}$ as

$$\mathbf{F}^{-1} = \frac{\partial \mathbf{X}}{\partial \mathbf{x}}, \quad (\mathbf{F}^{-1})_{Ai} \mathbf{e}_A \otimes \mathbf{e}_i(t) = \frac{\partial X_A}{\partial x_i} \mathbf{e}_A \otimes \mathbf{e}_i(t) = X_{A,i}(t) \mathbf{e}_A \otimes \mathbf{e}_i(t) \quad (2.17)$$

noting

$$d\mathbf{X} = dX_A \mathbf{e}_A = \frac{\partial X_A}{\partial x_i(t)} dx_i(t) \mathbf{e}_A = \frac{\partial X_A}{\partial x_i(t)} \mathbf{e}_A \otimes \mathbf{e}(t) dx_j(t) \mathbf{e}_j(t) = \mathbf{F}^{-1}(t) d\mathbf{x}(t)$$

As described above, the deformation gradient tensor $\mathbf{F}$ plays the most basic role to describe the deformation of materials. Any exact deformation (rate) measure must be represented by it. In addition, the transformation of the infinitesimal current line-element to its rate is described by the velocity gradient tensor $\boldsymbol{l}$ which is the most basic measure for deformation rate as will be described in Sect. 2.5.

Besides, consider the unit cubic cell (a parallelepiped) whose sides at the initial (reference) configuration are given by the triad $\{\mathbf{e}_I\}$ and then assume that it deforms to the cell whose sides are formed by the triad $\{\bar{\mathbf{e}}_i\}$. They are related by Eq. $(2.15)_1$ regarding $d\mathbf{x}$ and $d\mathbf{X}$ as $\bar{\mathbf{e}}_i$ and $\mathbf{e}_I$, respectively, as follows:

$$\bar{\mathbf{e}}_i = \delta_{iI} \mathbf{F} \mathbf{e}_I \quad (2.18)$$

The vectors $\bar{\mathbf{e}}_i$ are neither unit vectors nor orthonormal except for the rigid-body rotation. The curvilinear coordinate system with the base $\{\bar{\mathbf{e}}_i\}$ is referred to as the *convected coordinate system*. It is indispensable for general interpretation of deformation and rotation of materials, the detailed explanation of which can be referred to Sect. 4.4 and **Appendix B** briefly or Hashiguchi and Yamakawa (2012) in detail.

Applying the polar decomposition in Sect. 1.11 to the deformation gradient $\mathbf{F}$, we have

$$\boxed{\mathbf{F} = \mathbf{RU} = \mathbf{VR}}, \quad F_{iA} = R_{iR}U_{RA} = V_{ir}R_{rA} \tag{2.19}$$

where

$$\mathbf{U} = \mathbf{R}^T\mathbf{F} = (\mathbf{F}^T\mathbf{F})^{1/2}(= \mathbf{U}^T)(\mathbf{U}^2 = \mathbf{F}^T\mathbf{F}) \tag{2.20}$$

$$\mathbf{V} = \mathbf{F}\mathbf{R}^T = (\mathbf{F}\mathbf{F}^T)^{1/2}(= \mathbf{V}^T)(\mathbf{V}^2 = \mathbf{F}\mathbf{F}^T) \tag{2.21}$$

$$\mathbf{R} = \mathbf{F}\mathbf{U}^{-1} = \mathbf{F}(\mathbf{F}^T\mathbf{F})^{-1/2}, \quad \mathbf{R} = \mathbf{V}^{-1}\mathbf{F} = (\mathbf{F}\mathbf{F}^T)^{-1/2}\mathbf{F} \quad (\det\mathbf{R} = 1) \tag{2.22}$$

$$\mathbf{V} = \mathbf{R}\mathbf{U}\mathbf{R}^T, \quad \mathbf{U} = \mathbf{R}^T\mathbf{V}\mathbf{R} \tag{2.23}$$

U and **V** are the symmetric tensors so that there exist the two principal direction triads in which the deformation is described by the three-dimensional stretching resulting in the volume change and the shape change (shear deformation). Further, they are the similar tensors to each other, since Eq. (2.23) holds for the orthogonal tensor **R** as was described in Sect. 1.6. Therefore, they possess the same principal values, say $\lambda_\alpha(>0)(\alpha = 1, 2, 3)$. Denoting the bases for the principal directions of **U** and **V** by $\{\mathbf{N}^\alpha(t)\}$ and $\{\mathbf{n}^{(\alpha)}(t)\}$, respectively, they can be written as

$$\boxed{\mathbf{U} = \sum_{\alpha=1}^{3} \lambda_\alpha \mathbf{N}^{(\alpha)}(t) \otimes \mathbf{N}^{(\alpha)}(t), \quad \mathbf{V} = \sum_{\alpha=1}^{3} \lambda_\alpha \mathbf{n}^{(\alpha)}(t) \otimes \mathbf{n}^{(\alpha)}(t)} \tag{2.24}$$

where the relation of $\mathbf{N}^{(\alpha)}(t)$ and $\mathbf{n}^{(\alpha)}(t)$ is given from Eq. (1.228) as follows:

$$\boxed{\mathbf{n}^{(\alpha)}(t) = \mathbf{R}(t)\mathbf{N}^{(\alpha)}(t), \, \mathbf{N}^{(\alpha)}(t) = \mathbf{R}^T(t)\mathbf{n}^{(\alpha)}(t)} \tag{2.25}$$

with

$$\boxed{\mathbf{R}(t) = \sum_{\alpha=1}^{3} \mathbf{n}^{(\alpha)}(t) \otimes \mathbf{N}^{(\alpha)}(t)} \tag{2.26}$$

$\mathbf{N}^{(\alpha)}(t)$ and $\mathbf{n}^{(\alpha)}(t)$ are called the *Lagrangian triad* and the *Eulerian triad*, respectively.

Substituting Eqs. (2.24) and (2.26) into Eq. (2.19), **F** and its inverse tensor are described by

$$\boxed{\mathbf{F}(t) = \sum_{\alpha=1}^{3} \lambda_\alpha(t)\mathbf{n}^{(\alpha)}(t) \otimes \mathbf{N}^{(\alpha)}(t), \quad \mathbf{F}^{-1}(t) = \sum_{\alpha=1}^{3} \frac{1}{\lambda_\alpha}(t)\mathbf{N}^{(\alpha)}(t) \otimes \mathbf{n}^{(\alpha)}(t)} \tag{2.27}$$

Let the mechanical meanings of **U**, **V** and **R** be examined below.

The variation of infinitesimal line-element is given by the polar decomposition **F** = **RU** as follows:

$$d\mathbf{x} = \mathbf{F}d\mathbf{X} = \mathbf{RU}d\mathbf{X} = \mathbf{R}\sum_{\beta=1}^{3}\lambda_\beta \mathbf{N}^{(\beta)} \otimes \mathbf{N}^{(\beta)} \sum_{\alpha=1}^{3} dX_\alpha \mathbf{N}^{(\alpha)} = \mathbf{R}\sum_{\alpha=1}^{3}\lambda_\alpha dX_\alpha \mathbf{N}^{(\alpha)} \quad (2.28)$$

Equation (2.28) means that the infinitesimal line-elements $dX_\alpha \mathbf{N}^{(\alpha)}$ (no sum) in the principal directions $\mathbf{N}^{(\alpha)}$ are first stretched by λ_α times to $\lambda_\alpha dX_\alpha \mathbf{N}^{(\alpha)}$ (no sum) and then undergoes the rotation $\mathbf{R}$ as shown in Fig. 2.1.

On the other hand, the change of the infinitesimal line-element by the polar decomposition $\mathbf{VR}$ is described as

$$\begin{aligned} d\mathbf{x} = \mathbf{VR}d\mathbf{X} &= \sum_{\beta=1}^{3}\lambda_\beta \mathbf{n}^{(\beta)} \otimes \mathbf{n}^{(\beta)}\mathbf{R}\sum_{\alpha=1}^{3} dX_\alpha \mathbf{N}^{(\alpha)} = \sum_{\beta=1}^{3}\lambda_\beta \mathbf{n}^{(\beta)} \otimes \mathbf{n}^{(\beta)}\sum_{\alpha=1}^{3} dX_\alpha \mathbf{n}^{(\alpha)} \\ &= \sum_{\alpha=1}^{3}\lambda_\alpha dX_\alpha \mathbf{n}^{(\alpha)} = \sum_{\alpha=1}^{3}\lambda_\alpha \mathbf{R} dX_\alpha \mathbf{N}^{(\alpha)} \end{aligned} \quad (2.29)$$

Equation (2.29) means that the infinitesimal line-elements $dX_\alpha \mathbf{N}^{(\alpha)}$ (no sum) in the principal directions $\mathbf{N}^{(\alpha)}$ first becomes $dX_\alpha \mathbf{n}^{(\alpha)}$ (no sum) by rotation $\mathbf{R}$ and then are stretched by λ_α times to $\lambda_\alpha dX_\alpha \mathbf{n}^{(\alpha)}$ (no sum) (see Fig. 2.1).

As described above, $\mathbf{U}$ and $\mathbf{V}$ designates the deformation and $\mathbf{R}$ the rotation. λ_α is called the *principal stretch*, and $\mathbf{U}$ and $\mathbf{V}$ are called the *right* and *left stretch tensor*, respectively.

Letting $\mathbf{R}^L$ and $\mathbf{R}^E$ designate the rotations of the Lagrangian triad $\{\mathbf{N}^{(\alpha)}\}$ and the Eulerian triad $\{\mathbf{n}^{(\alpha)}\}$, respectively, from the fixed base $\{\mathbf{e}_\alpha\}(\alpha = 1,2,3)$, they are given by

$$\boxed{\mathbf{R}^L \equiv \sum_{\alpha=1}^{3}\mathbf{N}^{(\alpha)} \otimes \mathbf{e}_\alpha, \quad \mathbf{R}^E \equiv \sum_{\alpha=1}^{3}\mathbf{n}^{(\alpha)} \otimes \mathbf{e}_\alpha} \quad (2.30)$$

where the following relations hold.

$$\boxed{\mathbf{N}^{(\alpha)} = \mathbf{R}^L\mathbf{e}_\alpha, \quad \mathbf{n}^{(\alpha)} = \mathbf{R}^E\mathbf{e}_\alpha} \quad (2.31)$$

$$\boxed{\mathbf{R}^E = \mathbf{R}\mathbf{R}^L} \quad (2.32)$$

Considering the particle P and the adjacent particles P′ and P″, we designate their position vectors before and after the deformation by $\mathbf{X}$, $\mathbf{X}+d\mathbf{X}$, $\mathbf{X}+\delta\mathbf{X}$ and $\mathbf{x}$, $\mathbf{x}+d\mathbf{x}$, $\mathbf{x}+\delta\mathbf{x}$, respectively. Then, noting (1.116), one has

$$d\mathbf{x}\cdot\delta\mathbf{x} = \mathbf{F}d\mathbf{X}\cdot\mathbf{F}\delta\mathbf{X} = \mathbf{F}^T\mathbf{F}d\mathbf{X}\cdot\delta\mathbf{X} = \mathbf{C}d\mathbf{X}\cdot\delta\mathbf{X} \quad (2.33)$$

$$d\mathbf{X}\cdot\delta\mathbf{X} = \mathbf{F}^{-1}d\mathbf{x}\cdot\mathbf{F}^{-1}\delta\mathbf{x} = \mathbf{F}^{-T}\mathbf{F}^{-1}d\mathbf{x}\cdot\delta\mathbf{x} = (\mathbf{F}\mathbf{F}^T)^{-1}d\mathbf{x}\cdot\delta\mathbf{x} = \mathbf{b}^{-1}d\mathbf{x}\cdot\delta\mathbf{x} \quad (2.34)$$

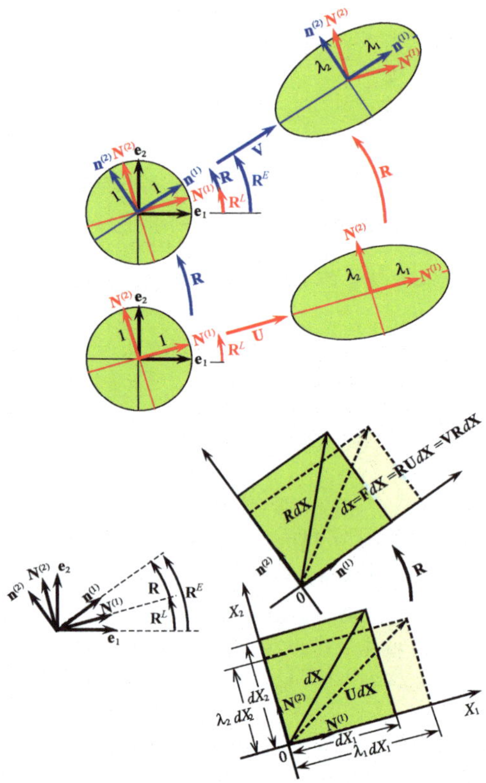

Fig. 2.1 Polar decomposition of deformation gradient

where

$$\boxed{\mathbf{C} \equiv \mathbf{F}^T\mathbf{F} = \mathbf{U}^2(= \mathbf{C}^T)}, \quad C_{AB} = F_{kA}F_{kB} \tag{2.35}$$

and

$$\boxed{\mathbf{b} \equiv \mathbf{F}\mathbf{F}^T = \mathbf{V}^2\,(= \mathbf{R}\mathbf{C}\mathbf{R}^T)(\,= \mathbf{b}^T)}, \; b_{ij} = F_{iA}F_{jA} \tag{2.36}$$

are the tensors which describe how the scalar product of two line-element vectors passing through a material point is influenced by a deformation. **C** and **b** are called the *right* and *left Cauchy-Green deformation tensor*, respectively. In accordance with Eq. (2.24) they are described by

$$\boxed{\mathbf{C} = \sum_{\alpha=1}^{3} \lambda_\alpha^2 \mathbf{N}^{(\alpha)} \otimes \mathbf{N}^{(\alpha)}}, \quad \boxed{\mathbf{b} = \sum_{\alpha=1}^{3} \lambda_\alpha^2 \mathbf{n}^{(\alpha)} \otimes \mathbf{n}^{(\alpha)}} \tag{2.37}$$

The principal values λ_α are obtained by the solutions of the characteristic equation

$$\lambda^3 - \mathrm{I}_c\lambda^2 - \mathrm{II}_c\lambda + \mathrm{III}_c = 0 \tag{2.38}$$

based on Eq. (1.177), where

$$\mathrm{I_c} \equiv \mathrm{tr}\mathbf{C},\ \mathrm{II_c} \equiv \frac{1}{2}(\mathrm{tr}\mathbf{C} - \mathrm{tr}\mathbf{C}^2),\ \mathrm{III_c} \equiv \frac{1}{6}\mathrm{tr}^3\mathbf{C} - \frac{1}{2}\mathrm{tr}\mathbf{C}\mathrm{tr}\mathbf{C}^2 + \frac{1}{3}\mathrm{tr}\mathbf{C}^3 \tag{2.39}$$

The principal values and directions are calculated by the method described in 1.6. The similar equations hold for $\mathbf{b}$ instead of $\mathbf{C}$. Using the relative description (Eq. 2.8), the *relative deformation gradient tensor* in the reference configuration $\mathbf{x}(\tau)$ is defined as

$$\boxed{{}_\tau\mathbf{F}(t) = \frac{\partial\mathbf{x}(t)}{\partial\mathbf{x}(\tau)}} \tag{2.40}$$

which is related to the deformation gradient $\mathbf{F}(t)\ (\equiv {}_0\mathbf{F}(t))$ as

$$\mathbf{F}(t) = \frac{\partial\mathbf{x}(t)}{\partial\mathbf{X}} = \frac{\partial\mathbf{x}(t)}{\partial\mathbf{x}(\tau)}\frac{\partial\mathbf{x}(\tau)}{\partial\mathbf{X}} = {}_\tau\mathbf{F}(t)\mathbf{F}(\tau) \tag{2.41}$$

and is further expressed in the polar decomposition as

$${}_\tau\mathbf{F}(t) = {}_\tau\mathbf{R}(t)_\tau\mathbf{U}(t) = {}_\tau\mathbf{V}(t)_\tau\mathbf{R}(t) \tag{2.42}$$

where ${}_\tau\mathbf{C}(t),\ {}_\tau\mathbf{b}(t)$ defined by

$$\left.\begin{aligned} {}_\tau\mathbf{C}(t) &= ({}_\tau\mathbf{F}(t))^T\,{}_\tau\mathbf{F}(t) = {}_\tau\mathbf{U}^2(t) \\ {}_\tau\mathbf{b}(t) &= {}_\tau\mathbf{F}(t)\,({}_\tau\mathbf{F}(t))^T = {}_\tau\mathbf{V}^2(t) \end{aligned}\right\} \tag{2.43}$$

which are called the *relative right* and the *left Cauchy-Green deformation tensors*.

2.4 Strain Tensors

Consider the scalar quantity which changes only by the pure deformation but is independent of the rotation. Subtracting Eq. (2.34) from Eq. (2.33), one has

$$d\mathbf{x}\cdot\delta\mathbf{x} - d\mathbf{X}\cdot\delta\mathbf{X} = \begin{cases} 2\boldsymbol{E}d\mathbf{X}\cdot\delta\mathbf{X}(= 2E_{AB}dX_A\delta X_B) \\ 2\boldsymbol{e}d\mathbf{x}\cdot\delta\mathbf{x}(= 2e_{ij}dx_i\delta x_j) \end{cases} \tag{2.44}$$

where

$$\left.\begin{array}{l} \boxed{\boldsymbol{E} \equiv \dfrac{1}{2}(\mathbf{C}-\mathbf{I}) = \dfrac{1}{2}(\mathbf{U}^2-\mathbf{I}) = \dfrac{1}{2}(\mathbf{F}^T\mathbf{F}-\mathbf{I}) = \dfrac{1}{2}\left[\left(\dfrac{\partial \mathbf{x}}{\partial \mathbf{X}}\right)^T\left(\dfrac{\partial \mathbf{x}}{\partial \mathbf{X}}\right) - \mathbf{I}\right]} \\ E_{AB} \equiv \dfrac{1}{2}(F_{kA}F_{kB} - \delta_{AB}) = \dfrac{1}{2}\left(\dfrac{\partial x_k}{\partial X_A}\dfrac{\partial x_k}{\partial X_B} - \delta_{AB}\right) \\ \boxed{\boldsymbol{e} \equiv \dfrac{1}{2}(\mathbf{I}-\mathbf{b}^{-1}) = \dfrac{1}{2}(\mathbf{I}-\mathbf{V}^{-2}) = \dfrac{1}{2}(\mathbf{I}-\mathbf{F}^{-T}\mathbf{F}^{-1}) = \dfrac{1}{2}\left[\mathbf{I}-\left(\dfrac{\partial \mathbf{X}}{\partial \mathbf{x}}\right)^T\dfrac{\partial \mathbf{X}}{\partial \mathbf{x}}\right]} \\ e_{ij} \equiv \dfrac{1}{2}[\delta_{ij}-(\mathbf{F}^{-1})_{Ki}(\mathbf{F}^{-1})_{Kj}] = \dfrac{1}{2}\left(\delta_{ij} - \dfrac{\partial X_K}{\partial x_i}\dfrac{\partial X_K}{\partial x_j}\right) \end{array}\right\} \tag{2.45}$$

which are defined by $\mathbf{C}$ and $\mathbf{b}$ describing the pure deformations. Applying the quotient law described in Sect. 1.3.2 to Eq. (2.44), it is confirmed that $\boldsymbol{E}$ and $\boldsymbol{e}$ are the second-order tensors.

If a deformation is not induced, the triangle PP′P″ keeps the same shape as in the initial state and thus the left-hand side in Eq. (2.44) is zero so that $\boldsymbol{E}$ and $\boldsymbol{e}$ are zero independent of rotation. Conversely, if $\boldsymbol{E} \neq \mathbf{O}$, $\boldsymbol{e} \neq \mathbf{O}$, the left-hand side in Eq. (2.44) is not zero so that the shape of the triangle is not same as in the initial state, resulting in a deformation. Therefore, $\boldsymbol{E}$ and $\boldsymbol{e}$ are the quantities describing the deformation independent of rigid-body rotation and called the *Green* (or *Lagrangian*) *strain tensor* and the *Almansi* (or *Eulerian*) *strain tensor*, respectively. Using the displacement vector

$$\boxed{\mathbf{u} = \mathbf{x} - \mathbf{X} = u_i\mathbf{e}_i}, \quad \frac{\partial \mathbf{u}}{\partial \mathbf{X}} = \mathbf{F} - \mathbf{I} \tag{2.46}$$

they are expressed by

$$\left.\begin{array}{ll} \boxed{\boldsymbol{E} = \dfrac{1}{2}\left[\dfrac{\partial \mathbf{u}}{\partial \mathbf{X}} + \left(\dfrac{\partial \mathbf{u}}{\partial \mathbf{X}}\right)^T + \left(\dfrac{\partial \mathbf{u}}{\partial \mathbf{X}}\right)^T\left(\dfrac{\partial \mathbf{u}}{\partial \mathbf{X}}\right)\right]}, & E_{AB} = \dfrac{1}{2}\left(\dfrac{\partial u_A}{\partial X_B} + \dfrac{\partial u_B}{\partial X_A} + \dfrac{\partial u_K}{\partial X_A}\dfrac{\partial u_K}{\partial X_B}\right) \\ \boxed{\boldsymbol{e} = \dfrac{1}{2}\left[\dfrac{\partial \mathbf{u}}{\partial \mathbf{x}} + \left(\dfrac{\partial \mathbf{u}}{\partial \mathbf{x}}\right)^T - \left(\dfrac{\partial \mathbf{u}}{\partial \mathbf{x}}\right)^T\left(\dfrac{\partial \mathbf{u}}{\partial \mathbf{x}}\right)\right]}, & e_{ij} = \dfrac{1}{2}\left(\dfrac{\partial u_i}{\partial x_j} + \dfrac{\partial u_j}{\partial x_i} - \dfrac{\partial u_k}{\partial x_i}\dfrac{\partial u_k}{\partial x_j}\right) \end{array}\right\} \tag{2.47}$$

The following relation exists between them,

$$\boxed{\boldsymbol{E} = \mathbf{F}^T\boldsymbol{e}\mathbf{F}}, \quad E_{AB} = F_{iA}F_{jB}e_{ij} \tag{2.48}$$

Now, consider the symmetric part of the displacement gradient which is the eliminations of the third terms in the brackets $\boldsymbol{E}$ and $\boldsymbol{e}$ in Eq. (2.47), i.e

$$\boxed{\boldsymbol{\varepsilon} \equiv \left(\frac{\partial \mathbf{u}}{\partial \mathbf{X}}\right)^S = \frac{1}{2}\left[\frac{\partial \mathbf{u}}{\partial \mathbf{X}} + \left(\frac{\partial \mathbf{u}}{\partial \mathbf{X}}\right)^T\right] = \frac{1}{2}(\mathbf{F}+\mathbf{F}^T) - \mathbf{I}}, \quad \varepsilon_{AB} \equiv \frac{1}{2}\left(\frac{\partial u_A}{\partial X_B} + \frac{\partial u_B}{\partial X_A}\right) \tag{2.49}$$

or

$$\boxed{\boldsymbol{\varepsilon} \equiv \left(\frac{\partial \mathbf{u}}{\partial \mathbf{x}}\right)^S = \frac{1}{2}\left[\frac{\partial \mathbf{u}}{\partial \mathbf{x}} + \left(\frac{\partial \mathbf{u}}{\partial \mathbf{x}}\right)^T\right] = \mathbf{I} - \frac{1}{2}(\mathbf{F}^{-1}+\mathbf{F}^{-T})}, \quad \varepsilon_{ij} \equiv \frac{1}{2}\left(\frac{\partial u_i}{\partial x_j} + \frac{\partial u_j}{\partial x_i}\right) \tag{2.50}$$

which describe roughly deformation, depending not only on $\mathbf{U}$ or $\mathbf{V}$ but also on the rotation tensor $\mathbf{R}$. Besides, the Lagrangian and the Eulerian tensors explained in Sect. 4.4 are mixed without the distinction of $\mathbf{X}$ and $\mathbf{x}$. Then, $\boldsymbol{\varepsilon}$ is called the *infinitesimal strain tensor*. The difference of Eqs. (2.49) and (2.50) vanishes in an infinitesimal deformation leading to $d\mathbf{x} \cong d\mathbf{X}$. It possesses various impertinence as will be described in Sect. 2.6.

In what follows, the geometrical interpretation of $\boldsymbol{E}$ and $\boldsymbol{e}$ will be given.

Considering the case that the two infinitesimal line-elements PP′ and PP″ coincide to each other, i.e. $d\mathbf{X} = \delta\mathbf{X}, d\mathbf{x} = \delta\mathbf{x}$ and denoting its direction vector in the initial configuration by $\mathbf{N}(||\mathbf{N}|| = 1)$, it follows from Eq. (2.44) that

$$||d\mathbf{x}||^2 - ||d\mathbf{X}||^2 = \begin{cases} 2\boldsymbol{E}\mathbf{N}\cdot\mathbf{N}||d\mathbf{X}||^2 \\ 2\boldsymbol{e}\mathbf{N}\cdot\mathbf{N}||d\mathbf{x}||^2 \end{cases} \tag{2.51}$$

Selecting the X_1-axis for this line-element, $(N_1, N_2, N_3) = (1, 0, 0)$ holds and thus we have

$$\left.\begin{aligned} E_{11} &= \frac{1}{2}\frac{||d\mathbf{x}||^2 - ||d\mathbf{X}||^2}{||d\mathbf{X}||^2}\left(= \frac{1}{2}\left[\left(\frac{||d\mathbf{x}||}{||d\mathbf{X}||}\right)^2 - 1\right]\right) \\ e_{11} &= \frac{1}{2}\frac{||d\mathbf{x}||^2 - ||d\mathbf{X}||^2}{||d\mathbf{x}||^2}\left(= \frac{1}{2}\left[1 - \left(\frac{||d\mathbf{X}||}{||d\mathbf{x}||}\right)^2\right]\right) \end{aligned}\right\} \tag{2.52}$$

from which the ratio of the line-elements before and after the deformation is given by

$$\frac{||d\mathbf{x}||}{||d\mathbf{X}||} = \begin{cases} \sqrt{1+2E_{11}} \\ 1/\sqrt{1-2e_{11}} \end{cases} \tag{2.53}$$

In the case that the variation of the length of the line-element is infinitesimal $(||d\mathbf{x}||/||d\mathbf{X}|| \cong 1)$, Eq. (2.52) becomes

$$\left.\begin{aligned} E_{11} &\cong \frac{||d\mathbf{x}|| - ||d\mathbf{X}||}{||d\mathbf{X}||} \\ e_{11} &\cong \frac{||d\mathbf{x}|| - ||d\mathbf{X}||}{||d\mathbf{x}||} \end{aligned}\right\} \cong \varepsilon_{11} \tag{2.54}$$

so that E_{11} and e_{11} describe the rate of elongation coinciding with the normal strain in the infinitesimal strain $\boldsymbol{\varepsilon}$.

On the other hand, denoting the direction vectors of the two distinct infinitesimal line-element PP′ and PP″ as $\mathbf{N}'$ and $\mathbf{N}''$, respectively, in the initial state and the angles contained by them as θ, it holds from Eq. (2.44) that

$$||d\mathbf{x}||||\delta\mathbf{x}||\cos\theta - ||d\mathbf{X}||||\delta\mathbf{X}||\cos\theta_0 = 2\boldsymbol{E}\mathbf{N}'\cdot\mathbf{N}''||d\mathbf{X}||||\delta\mathbf{X}|| \tag{2.55}$$

i.e.

$$\frac{||d\mathbf{x}||}{||d\mathbf{X}||}\frac{||\delta\mathbf{x}||}{||\delta\mathbf{X}||}\cos\theta - \cos\theta_0 = 2\boldsymbol{E}\mathbf{N}'\cdot\mathbf{N}'' = 2E_{ij}N'_iN''_j \tag{2.56}$$

where θ_0 is the initial value of θ. Here, assuming that the infinitesimal line-elements PP′ and PP″ were mutually perpendicular before a deformation, i.e. $\theta_0 = \pi/2$ leading to $\cos\theta_0 = 0$, and making their directions coincide to the X_1- and X_2-axes, i.e. $(N'_1, N'_2, N'_3) = (1, 0, 0)$, $(N''_1, N''_2, N''_3) = (0, 1, 0)$ leading to $E_{ij}N'_iN''_j = E_{12}$, it follows that

$$E_{12} = \frac{1}{2}\frac{||d\mathbf{x}||}{||d\mathbf{X}||}\frac{||\delta\mathbf{x}||}{||\delta\mathbf{X}||}\cos\theta = \frac{1}{2}\frac{||d\mathbf{x}||}{||d\mathbf{X}||}\frac{||\delta\mathbf{x}||}{||\delta\mathbf{X}||}\sin(\pi/2-\theta) \tag{2.57}$$

which describes the half of decrease in the sine of angle contained by the two line-elements which were mutually perpendicular before deformation when the changes in lengths of these line-elements are infinitesimal ($||d\mathbf{x}||/||d\mathbf{X}|| \cong 1$, $||\delta\mathbf{x}||/||\delta\mathbf{X}|| \cong 1$). Furthermore, when the change in the angle formed by these line-elements is infinitesimal ($\theta \cong \pi/2$), one has

$$E_{12} \cong (\pi/2 - \theta)/2 \tag{2.58}$$

Consequently, E_{12} describes half of the decrease in the angle contained by the two line-elements which were perpendicular before deformation.

In addition to the Lagrangian and Eulerian strain tensors defined above, we can define various strain tensors in terms of $\mathbf{U}$ or $\mathbf{V}$, fulfilling the condition that they are zero when $\mathbf{U} = \mathbf{V} = \mathbf{I}$ as follows (Seth 1964; Hill 1968):

$$\boxed{\boldsymbol{E}^{(m)} = \mathbf{f}(\mathbf{U}) = \begin{cases} \dfrac{1}{2m}(\mathbf{U}^{2m} - \mathbf{I}) & \text{for } m \neq 0 \\ \ln\mathbf{U} & \text{for } m = 0 \end{cases}} \tag{2.59}$$

$$\boxed{\boldsymbol{e}^{(m)} = \mathbf{f}(\mathbf{V}) = \begin{cases} \dfrac{1}{2m}(\mathbf{V}^{2m} - \mathbf{I}) & \text{for } m \neq 0 \\ \ln\mathbf{V} & \text{for } m = 0 \end{cases}} \tag{2.60}$$

where $2m$ is the integer (positive or negative). The Green strain tensor is obtained by choosing $m = 1$ in Eq. (2.59) and the Almansi strain tensor is obtained by choosing $m = -1$ in Eq. (2.60). The *Biot strain tensor* (Biot 1965) is given by choosing $m = 1/2$ in Eq. (2.59), i.e.

$$\boxed{\mathbf{B} \equiv \mathbf{U} - \mathbf{I}} \tag{2.61}$$

The generalized strain tensors in Eqs. (2.59) and (2.60) are mutually related by virtue of Eq. (2.23) as follows.

$$\boxed{\boldsymbol{E}^{(m)} = \mathbf{R}^T \boldsymbol{e}^{(m)} \mathbf{R}} \tag{2.62}$$

The strain tensors in Eqs. (2.59) and (2.60) are coaxial with $\mathbf{U}$ and $\mathbf{V}$, respectively, and their principal values are given by

$$\boxed{f(\lambda_\alpha) = \begin{cases} \dfrac{1}{2m}(\lambda_\alpha^{2m} - 1) & \text{for} \quad m \neq 0 \\ \ln \lambda_\alpha & \text{for} \quad m = 0 \end{cases}} \tag{2.63}$$

for $\alpha = 1, 2, 3$. The function $f(\lambda_\alpha)$ fulfills

$$f(1) = 0, \; f'(1) = 1 \tag{2.64}$$

and

$$f'(s) > 0 \tag{2.65}$$

where s is an arbitrary positive scalar quantity. The function $f(\lambda_\alpha)$ for several values of m is shown in Fig. 2.2.

(Note) Eq. $(2.63)_2$ for $m = 0$ is derived as follows:

$$\lim_{m \to 0} \frac{1}{m}(\lambda_\alpha^m - 1) = \lim_{m \to 0} \frac{\exp(m \ln \lambda_\alpha) - 1}{m} = \lim_{m \to 0} \frac{\exp(m \ln \lambda_\alpha) \ln \lambda_\alpha}{1} = \ln \lambda_\alpha \text{ (no sum)}$$

by the aid of l'Hôpital's.

Further, adopt the second-order tensor function $\mathbf{f}(\mathbf{U})$ which is coaxial with the right stretch tensor $\mathbf{U}$ and has the principal values $f(\lambda_\alpha)$. Therefore, we can define the general strain tensor in the spectral decomposition as follows:

$$\boxed{\mathbf{f}(\mathbf{U}) = \sum_{\alpha=1}^{3} f(\lambda_\alpha) \mathbf{N}^{(\alpha)} \otimes \mathbf{N}^{(\alpha)}} \tag{2.66}$$

In addition, for the left stretch tensor $\mathbf{V}$, we can define the following strain tensor.

$$\boxed{\mathbf{f}(\mathbf{V}) = \sum_{\alpha=1}^{3} f(\lambda_\alpha) \mathbf{n}^{(\alpha)} \otimes \mathbf{n}^{(\alpha)}} = \sum_{\alpha=1}^{3} f(\lambda_\alpha) \mathbf{R}\mathbf{N}^{(\alpha)} \otimes \mathbf{R}\mathbf{N}^{(\alpha)} = \mathbf{R}\mathbf{f}(\mathbf{U})\mathbf{R}^T \tag{2.67}$$

noting Eq. (1.106).

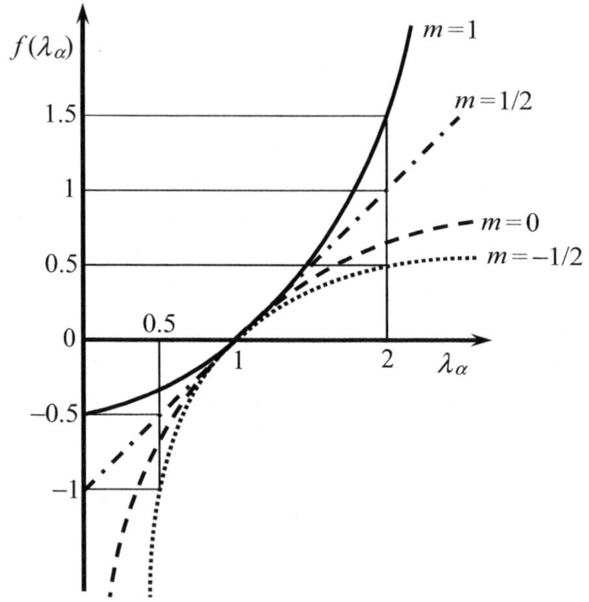

Fig. 2.2 Function of general principal strain measures

In the particular case of $m = 0$, noting $\lambda_\alpha > 0$, the strains defined by the following equation are called the *logarithmic* or *Hencky strain tensor*.

$$\text{Lagrangian-logarithmic strain tensor:}\quad \boldsymbol{E}^{(0)} = \sum_{\alpha=1}^{3} \ln \lambda_\alpha \mathbf{N}^{(\alpha)} \otimes \mathbf{N}^{(\alpha)} \equiv \ln\mathbf{U} = \frac{1}{2}\ln\mathbf{C}$$

$$\text{Eulerian-logarithmic strain tensor:}\quad \boldsymbol{e}^{(0)} = \sum_{\alpha=1}^{3} \ln \lambda_\alpha \mathbf{n}^{(\alpha)} \otimes \mathbf{n}^{(\alpha)} \equiv \ln\mathbf{V} = \frac{1}{2}\ln\mathbf{b} \tag{2.68}$$

where

$$\lambda_\alpha = U_\alpha = V_\alpha = \sqrt{C_\alpha} = \sqrt{b_\alpha} \tag{2.69}$$

$U_\alpha, C_\alpha, V_\alpha$ and b_α being the principal values of $\mathbf{U}, \mathbf{C}, \mathbf{V}$ and $\mathbf{b}$, respectively.

When the principal directions of $\mathbf{U}$ and $\mathbf{V}$ are fixed, the following equations hold in these directions.

$$\lambda_\alpha = \frac{\partial x_\alpha}{\partial X_\alpha} \text{ (no sum)} \tag{2.70}$$

$$(\dot{\boldsymbol{E}}^{(0)})_\alpha = (\dot{\boldsymbol{e}}^{(0)})_\alpha = \left(\frac{\partial x_\alpha}{\partial X_\alpha}\right)^{\bullet} \Big/ \left(\frac{\partial x_\alpha}{\partial X_\alpha}\right) = \frac{\partial \dot{x}_\alpha}{\partial x_\alpha} = (\ln\lambda_\alpha)^{\bullet} = d_{\alpha\alpha} \text{ (no sum)} \tag{2.71}$$

where $\ln \lambda_\alpha$ in Eq. (2.70) is the *logarithmic strain* and $d_{\alpha\alpha}$ (no sum) is the principal component of the strain rate tensor defined in the next section. It follows from Eq. (2.68) that

$$\left.\begin{aligned} \mathrm{tr}\boldsymbol{E}^{(0)} &= \sum_{\alpha=1}^{3} \ln \lambda_\alpha = \frac{1}{2}\sum_{\alpha=1}^{3} \ln C_\alpha = \sum_{\alpha=1}^{3} \ln U_\alpha = \ln(U_1 U_2 U_3) \\ \mathrm{tr}\boldsymbol{e}^{(0)} &= \sum_{\alpha=1}^{3} \ln \lambda_\alpha = \frac{1}{2}\sum_{\alpha=1}^{3} \ln b_\alpha = \sum_{\alpha=1}^{3} \ln V_\alpha = \ln(V_1 V_2 V_3) \end{aligned}\right\} \tag{2.72}$$

which is nothing but the *logarithmic volumetric strain*

$$\boxed{\mathrm{tr}\boldsymbol{E}^{(0)} = \mathrm{tr}\boldsymbol{e}^{(0)} = \sum_{\alpha=1}^{3} \ln \frac{\partial x_\alpha}{\partial X_\alpha} = \ln J = \ln \frac{v}{V} = \varepsilon_v} \tag{2.73}$$

The Hencky strain is relevant to the strain rate defined by the symmetric part of the velocity gradient, as the rates of principal components and volumetric part in the former coincides to the latter as will be described in Sect. 2.6.

2.5 Strain Rate and Spin Tensors

The idealized deformation process in which the deformation is uniquely determined by the state of stress independent of the loading path is called the *elastic deformation process*. To describe it, it is required to introduce the strain tensor describing the deformation from the initial state and relate it to the stress. Here, since the superposition rule does not hold in the strain tensor, the null stress state is chosen usually as the reference state of strain.

On the other hand, the deformation is not determined uniquely by the state of stress depending on the loading path and thus it cannot be related to the stress in the *irreversible deformation process*, e.g. the viscoelastic, the plastic and the viscoplastic loading processes. Therefore, it is obligatory to relate the infinitesimal changes of stress and deformation and to integrate them along the loading path in order to capture the current states of stress and deformation.

Here, introduce the *velocity gradient tensor* defined as

$$\boxed{\boldsymbol{l} \equiv \frac{\partial \mathbf{v}}{\partial \mathbf{x}}}, \quad l_{ij} \equiv \frac{\partial v_i}{\partial x_j} \equiv \partial_j v_i \tag{2.74}$$

Noting $\dot{\mathbf{F}} = \partial \dot{\mathbf{x}}/\partial \mathbf{X} = \partial \mathbf{v}/\partial \mathbf{X}$ ($d\mathbf{v} = \dot{\mathbf{F}} d\mathbf{X}$) and the chain rule of derivative, Eq. (2.74) can be rewritten as

$$\boxed{\boldsymbol{l} = \frac{\partial \mathbf{v}}{\partial \mathbf{X}}\frac{\partial \mathbf{X}}{\partial \mathbf{x}} = \dot{\mathbf{F}}\mathbf{F}^{-1}} (\dot{\mathbf{F}} = \boldsymbol{l}\mathbf{F}), \, l_{ij} = \frac{\partial v_i}{\partial X_A}\frac{\partial X_A}{\partial x_j} \tag{2.75}$$

$$\boxed{(d\mathbf{x})^{\boldsymbol{\cdot}} = \boldsymbol{l} d\mathbf{x}} \tag{2.76}$$

noting

$$(d\mathbf{x})^{\bullet} = d\mathbf{v} = \frac{\partial \mathbf{v}}{\partial \mathbf{x}} d\mathbf{x}$$

$(d\mathbf{x})^{\bullet}$, i.e. $d\mathbf{v}$ describes the rate of the infinitesimal line element, i.e. the relative velocity between the velocities of material particles in both sides of infinitesimal line element $d\mathbf{x}$. Here, we can choose the time $\tau(\leq t)$ to be arbitrary, resulting in $\boldsymbol{l} = {}_{\tau}\dot{\mathbf{F}}(t){}_{\tau}\mathbf{F}^{-1}(t)$ because the velocity gradient tensor $\boldsymbol{l}$ is substantially independent of the reference infinitesimal line element $d\mathbf{X}$ but dependent only on rates of deformation and rotation. Now, choosing the current state for the reference state leading to ${}_{t}\mathbf{F}^{-1}(t) = \mathbf{I}$, the velocity gradient tensor $\boldsymbol{l}$ can be expressed in the updated Lagrangian description as follows:

$$\boldsymbol{l} = {}_{t}\dot{\mathbf{F}}(t) \tag{2.77}$$

Further, taking the time-derivative of Eq. (2.42) and noting ${}_{t}\mathbf{R}(t) = {}_{t}\mathbf{U}(t) = {}_{t}\mathbf{V}(t) = \mathbf{I}$, it follows that

$${}_{t}\dot{\mathbf{F}}(t) = {}_{t}\dot{\mathbf{U}}(t) + {}_{t}\dot{\mathbf{R}}(t) = {}_{t}\dot{\mathbf{V}}(t) + {}_{t}\dot{\mathbf{R}}(t) \tag{2.78}$$

Decomposing $\boldsymbol{l}$ additively into the symmetric and the skew-symmetric parts and noting Eqs. (2.75)–(2.78), it is obtained that

$$\boxed{\boldsymbol{l} = \mathbf{d} + \mathbf{w}} \tag{2.79}$$

where

$$\left.\begin{array}{l} \boxed{\mathbf{d} \equiv \frac{1}{2}(\boldsymbol{l} + \boldsymbol{l}^T) = \frac{1}{2}\left[\frac{\partial \mathbf{v}}{\partial \mathbf{x}} + \left(\frac{\partial \mathbf{v}}{\partial \mathbf{x}}\right)^T\right]} = {}_{t}\dot{\mathbf{U}}(t) = {}_{t}\dot{\mathbf{V}}(t) \\ d_{ij} \equiv \frac{1}{2}\left(\frac{\partial v_i}{\partial x_j} + \frac{\partial v_j}{\partial x_i}\right) \end{array}\right\} \tag{2.80}$$

$$\left.\begin{array}{l} \boxed{\mathbf{w} \equiv \frac{1}{2}(\boldsymbol{l} - \boldsymbol{l}^T) = \frac{1}{2}\left[\frac{\partial \mathbf{v}}{\partial \mathbf{x}} - \left(\frac{\partial \mathbf{v}}{\partial \mathbf{x}}\right)^T\right]} = {}_{t}\dot{\mathbf{R}}(t) \\ w_{ij} \equiv \frac{1}{2}\left(\frac{\partial v_i}{\partial x_j} - \frac{\partial v_j}{\partial x_i}\right) \end{array}\right\} \tag{2.81}$$

where $\mathbf{d}$ is called the *strain rate tensor* or *deformation rate tensor* or *stretching* and $\mathbf{w}$ is called the (*continuum*) *rotation rate tensor* or *continuum spin tensor*. Here, note that $\mathbf{d}$ is not a time-derivative of any strain tensor but is defined independently as the rate variable although it is called the strain rate tensor. In addition, note that the time-integration of $\mathbf{d}$ cannot play any deformation measure in general, because it concerns with different material line-elements which rotate with material. Only the time-integration of axial component of **d** coincides with the axial component of the Hencky strain in Eq. (2.68) if the axial direction is fixed.

Substituting Eqs. (2.19) and (2.75) into Eqs. (2.80) and (2.81), **d** and **w** are described by $\mathbf{U}, \mathbf{R}$ as follows:

$$\left.\begin{aligned} \mathbf{d} &= \frac{1}{2}[\dot{\mathbf{F}}\mathbf{F}^{-1} + (\dot{\mathbf{F}}\mathbf{F}^{-1})^T] = \frac{1}{2}\left\{(\mathbf{RU})^{\bullet}(\mathbf{RU})^{-1} + (\mathbf{RU})^{-T}[(\mathbf{RU})^{\bullet}]^T\right\} \\ &= \frac{1}{2}\mathbf{R}(\dot{\mathbf{U}}\mathbf{U}^{-1} + \mathbf{U}^{-1}\dot{\mathbf{U}})\mathbf{R}^T \\ \mathbf{w} &= \frac{1}{2}[\dot{\mathbf{F}}\mathbf{F}^{-1} - (\dot{\mathbf{F}}\mathbf{F}^{-1})^T] = \frac{1}{2}\left\{(\mathbf{RU})^{\bullet}(\mathbf{RU})^{-1} - (\mathbf{RU})^{-T}[(\mathbf{RU})^{\bullet}]^T\right\} \\ &= \dot{\mathbf{R}}\mathbf{R}^T + \frac{1}{2}\mathbf{R}(\dot{\mathbf{U}}\mathbf{U}^{-1} - \mathbf{U}^{-1}\dot{\mathbf{U}})\mathbf{R}^T \end{aligned}\right\} \tag{2.82}$$

Consequently, we obtain

$$\left.\begin{aligned} \mathbf{d} &= \frac{1}{2}\mathbf{R}(\dot{\tilde{\mathbf{U}}} + \dot{\tilde{\mathbf{U}}}^T)\mathbf{R}^T \\ \mathbf{w} &= \boldsymbol{\Omega}^R + \frac{1}{2}\mathbf{R}(\dot{\tilde{\mathbf{U}}} - \dot{\tilde{\mathbf{U}}}^T)\mathbf{R}^T \end{aligned}\right\} \tag{2.83}$$

where

$$\dot{\tilde{\mathbf{U}}} \equiv \dot{\mathbf{U}}\mathbf{U}^{-1} \tag{2.84}$$

$$\boxed{\boldsymbol{\Omega}^R \equiv \dot{\mathbf{R}}\mathbf{R}^T} \tag{2.85}$$

$\boldsymbol{\Omega}^R$ is called the *relative* (or *polar*) *spin tensor*. Further, **d** and **w** are described by $\mathbf{V}, \mathbf{R}$ as follows:

$$\left.\begin{aligned} \mathbf{d} &= \frac{1}{2}\left\{(\mathbf{VR})^{\bullet}(\mathbf{VR})^{-1} + (\mathbf{VR})^{-T}[(\mathbf{VR})^{\bullet}]^T\right\} \\ &= \frac{1}{2}(\dot{\mathbf{V}}\mathbf{V}^{-1} + \mathbf{V}^{-T}\dot{\mathbf{V}}^T) + \frac{1}{2}(\mathbf{V}\dot{\mathbf{R}}\mathbf{R}^T\mathbf{V}^{-T} - \mathbf{V}^{-1}\dot{\mathbf{R}}\mathbf{R}^T\mathbf{V}^T) \\ \mathbf{w} &= \frac{1}{2}\left\{(\mathbf{VR})^{\bullet}(\mathbf{VR})^{-1} - (\mathbf{VR})^{-T}[(\mathbf{VR})^{\bullet}]^T\right\} \\ &= \frac{1}{2}(\mathbf{V}\dot{\mathbf{R}}\mathbf{R}^T\mathbf{V}^{-1} + \mathbf{V}^{-T}\dot{\mathbf{R}}\mathbf{R}^T\mathbf{V}^T) + \frac{1}{2}(\dot{\mathbf{V}}\mathbf{V}^{-1} - \mathbf{V}^{-T}\dot{\mathbf{V}}^T) \end{aligned}\right\} \tag{2.86}$$

and thus

$$\left.\begin{aligned} \mathbf{d} &= \frac{1}{2}(\dot{\tilde{\mathbf{V}}} + \dot{\tilde{\mathbf{V}}}^T) + \frac{1}{2}(\tilde{\boldsymbol{\Omega}}^R + \tilde{\boldsymbol{\Omega}}^{RT}) \\ \mathbf{w} &= \frac{1}{2}(\tilde{\boldsymbol{\Omega}}^R - \tilde{\boldsymbol{\Omega}}^{RT}) + \frac{1}{2}(\dot{\tilde{\mathbf{V}}} - \dot{\tilde{\mathbf{V}}}^T) \end{aligned}\right\} \tag{2.87}$$

where

$$\overset{\bullet}{\tilde{\mathbf{V}}} \equiv \dot{\mathbf{V}}\mathbf{V}^{-1} \tag{2.88}$$

$$\widetilde{\boldsymbol{\Omega}}^R \equiv \mathbf{V}\boldsymbol{\Omega}^R\mathbf{V}^{-1} \tag{2.89}$$

It follows from Eq. (2.30) that

$$\begin{aligned}\boxed{\boldsymbol{\Omega}^L \equiv \dot{\mathbf{R}}^L\mathbf{R}^{LT}} &= \sum_{\alpha,\beta=1}^{3} (\mathbf{N}^{(\alpha)} \otimes \mathbf{e}_\alpha)^{\bullet}(\mathbf{N}^{(\beta)} \otimes \mathbf{e}_\beta)^T \\ &= \sum_{\alpha,\beta=1}^{3} (\dot{\mathbf{N}}^{(\alpha)} \otimes \mathbf{e}^{(\alpha)} + \mathbf{N}^{(\alpha)} \otimes \dot{\mathbf{e}}^{(\alpha)})\mathbf{e}^{(\beta)} \otimes \mathbf{N}^{(\beta)} \\ &= \sum_{\alpha=1}^{3} \dot{\mathbf{N}}^{(\alpha)} \otimes \mathbf{N}^{(\alpha)}\end{aligned} \tag{2.90}$$

and thus one has

$$\boxed{\dot{\mathbf{N}}^{(\alpha)} = \boldsymbol{\Omega}^L\mathbf{N}^{(\alpha)}} \tag{2.91}$$

noting $\dot{\mathbf{e}}^{(\alpha)} = \mathbf{0}$ since $\{\mathbf{e}^{(\alpha)}\}$ is the fixed base. Therefore, $\boldsymbol{\Omega}^L$ describes the spin of the Lagrangian principal triad $\{\mathbf{N}^{(\alpha)}\}$ of the right stretch tensor $\mathbf{U}$ and is called the *Lagrangian spin tensor*. The components of $\boldsymbol{\Omega}^L$ in the Lagrangian principal triad $\{\mathbf{N}^{(\alpha)}\}$ are described as

$$\Omega^L_{\alpha\beta} = \mathbf{N}^{(\alpha)} \bullet \boldsymbol{\Omega}^L\mathbf{N}^{(\beta)} = \mathbf{N}^{(\alpha)} \bullet \dot{\mathbf{N}}^{(\beta)} \tag{2.92}$$

On the other hand, it follows from Eq. (2.30) that

$$\begin{aligned}\boxed{\boldsymbol{\Omega}^E \equiv \dot{\mathbf{R}}^E\mathbf{R}^{ET}} &= \sum_{\alpha,\beta=1}^{3} (\mathbf{n}^{(\alpha)} \otimes \mathbf{e}^{(\alpha)})^{\bullet}(\mathbf{n}^{(\beta)} \otimes \mathbf{e}^{(\beta)})^T \\ &= \sum_{\alpha=1}^{3} \dot{\mathbf{n}}^{(\alpha)} \otimes \mathbf{n}^{(\alpha)}\end{aligned} \tag{2.93}$$

and thus one has

$$\boxed{\dot{\mathbf{n}}^{(\alpha)} = \boldsymbol{\Omega}^E\mathbf{n}^{(\alpha)}} \tag{2.94}$$

Therefore, $\mathbf{\Omega}^E$ describes the spin of the Eulerian principal triad $\{\mathbf{n}^{(\alpha)}\}$ of the right stretch tensor $\mathbf{V}$ and is called the *Eulerian spin tensor*. The components of $\mathbf{\Omega}^E$ in the Eulerian triad $\{\mathbf{n}^{(\alpha)}\}$ are described as

$$\Omega^E_{\alpha\beta} = \mathbf{n}^{(\alpha)} \cdot \mathbf{\Omega}^E \mathbf{n}^{(\beta)} = \mathbf{n}^{(\alpha)} \cdot \dot{\mathbf{n}}^{(\beta)} \tag{2.95}$$

It follows from Eq. (2.27) that

$$\dot{\mathbf{F}} = \sum_{\alpha=1}^{3} \left[\dot{\lambda}_\alpha \mathbf{n}^\alpha \otimes \mathbf{N}^\alpha + \lambda_\alpha (\dot{\mathbf{n}}^\alpha \otimes \mathbf{N}^\alpha + \mathbf{n}^{(\alpha)} \otimes \dot{\mathbf{N}}^\alpha) \right] \tag{2.96}$$

which is rewritten by Eqs. (2.91) and (2.94) as

$$\dot{\mathbf{F}} = \sum_{\alpha=1}^{3} \dot{\lambda}_\alpha \mathbf{n}^{(\alpha)} \otimes \mathbf{N}^{(\alpha)} + \mathbf{\Omega}^E \mathbf{F} - \mathbf{F} \mathbf{\Omega}^L \tag{2.97}$$

or by Eqs. (2.92) and (2.95), Eq. (2.96) leads to

$$\dot{\mathbf{F}} = \sum_{\alpha=1}^{3} \dot{\lambda}_\alpha \mathbf{n}^{(\alpha)} \otimes \mathbf{N}^{(\alpha)} + \sum_{\alpha,\beta=1}^{3} (\lambda_\beta \Omega^E_{\alpha\beta} - \lambda_\alpha \Omega^L_{\alpha\beta}) \mathbf{n}^{(\alpha)} \otimes \mathbf{N}^{(\beta)} \tag{2.98}$$

Here, it holds that

$$\begin{aligned}
\dot{\mathbf{R}}\mathbf{R}^T &= \sum_{\alpha,\beta=1}^{3} (\mathbf{n}^{(\alpha)} \otimes \mathbf{N}^{(\alpha)})^{\bullet} (\mathbf{n}^{(\beta)} \otimes \mathbf{N}^{(\beta)})^T \\
&= \sum_{\alpha,\beta=1}^{3} (\dot{\mathbf{n}}^{(\alpha)} \otimes \mathbf{N}^{(\alpha)} + \mathbf{n}^{(\alpha)} \otimes \dot{\mathbf{N}}^{(\alpha)}) \mathbf{N}^{(\beta)} \otimes \mathbf{n}^{(\beta)} \\
&= \sum_{\alpha,\beta,\gamma=1}^{3} \left[\dot{\mathbf{n}}^{(\alpha)} \otimes \mathbf{n}^{(\alpha)} + \mathbf{n}^{(\alpha)} \otimes \mathbf{N}^{(\alpha)} (\dot{\mathbf{N}}^{(\gamma)} \otimes \dot{\mathbf{N}}^{(\gamma)}) \mathbf{N}^{(\beta)} \otimes \mathbf{n}^{(\beta)} \right] \\
&= \sum_{\alpha,\beta,\gamma=1}^{3} \left[\dot{\mathbf{n}}^{(\alpha)} \otimes \mathbf{n}^{(\alpha)} - \mathbf{n}^{(\alpha)} \otimes \mathbf{N}^{(\alpha)} (\dot{\mathbf{N}}^{(\gamma)} \otimes \mathbf{N}^{(\gamma)}) \mathbf{N}^{(\beta)} \otimes \mathbf{n}^{(\beta)} \right]
\end{aligned}$$

and thus the following relations hold.

$$\boxed{\mathbf{\Omega}^R = \mathbf{\Omega}^E - \mathbf{R}\mathbf{\Omega}^L\mathbf{R}^T,\ \mathbf{\Omega}^E = \mathbf{\Omega}^R + \mathbf{R}\mathbf{\Omega}^L\mathbf{R}^T, \mathbf{\Omega}^L = \mathbf{R}^T(\mathbf{\Omega}^E - \mathbf{\Omega}^R)\mathbf{R}} \tag{2.99}$$

The velocity gradient is described noting Eq. (2.27) as

$$
\begin{aligned}
\boldsymbol{l} &= \left(\sum_{\alpha=1}^{3} \lambda_\alpha \mathbf{n}^{(\alpha)} \otimes \mathbf{N}^{(\alpha)}\right)^{\bullet} \sum_{\beta=1}^{3} \frac{1}{\lambda_\beta} \mathbf{N}^{(\beta)} \otimes \mathbf{n}^{(\beta)} \\
&= \sum_{\alpha=1}^{3} \left(\dot{\lambda}_\alpha \mathbf{n}^{(\alpha)} \otimes \mathbf{N}^{(\alpha)} + \lambda_\alpha \dot{\mathbf{n}}^{(\alpha)} \otimes \mathbf{N}^{(\alpha)} + \lambda_\alpha \mathbf{n}^{(\alpha)} \otimes \dot{\mathbf{N}}^{(\alpha)}\right) \sum_{\beta=1}^{3} \frac{1}{\lambda_\beta} \mathbf{N}^{(\beta)} \otimes \mathbf{n}^{(\beta)} \\
&= \sum_{\alpha=1}^{3} \left(\frac{\dot{\lambda}_\alpha}{\lambda_\alpha} \mathbf{n}^{(\alpha)} \otimes \mathbf{n}^{(\alpha)} + \dot{\mathbf{n}}^{(\alpha)} \otimes \mathbf{n}^{(\alpha)}\right) + \sum_{\alpha,\beta=1}^{3} \frac{\lambda_\alpha}{\lambda_\beta} \left(\dot{\mathbf{N}}^{(\alpha)} \bullet \mathbf{N}^{(\beta)}\right) \mathbf{n}^{(\alpha)} \otimes \mathbf{n}^{(\beta)}
\end{aligned}
\tag{2.100}
$$

which is rewritten using Eqs. (2.92) and (2.95) as

$$
\boldsymbol{l} = \sum_{\alpha=1}^{3} \frac{\dot{\lambda}_\alpha}{\lambda_\alpha} \mathbf{n}^{(\alpha)} \otimes \mathbf{n}^{(\alpha)} + \sum_{\alpha,\beta=1}^{3} \left(\Omega^E_{\alpha\beta} - \frac{\lambda_\alpha}{\lambda_\beta} \Omega^L_{\alpha\beta}\right) \mathbf{n}^{(\alpha)} \otimes \mathbf{n}^{(\beta)} \tag{2.101}
$$

from which the strain rate and the continuum spin are represented as

$$
\mathbf{d} = \sum_{\alpha=1}^{3} \frac{\dot{\lambda}_\alpha}{\lambda_\alpha} \mathbf{n}^{(\alpha)} \otimes \mathbf{n}^{(\alpha)} + \sum_{\alpha,\beta=1}^{3} \frac{\lambda_\alpha^2 - \lambda_\beta^2}{2\lambda_\alpha \lambda_\beta} \Omega^L_{\alpha\beta} \mathbf{n}^{(\alpha)} \otimes \mathbf{n}^{(\beta)} \tag{2.102}
$$

$$
\mathbf{w} = \sum_{\alpha,\beta=1}^{3} \left(\Omega^E_{\alpha\beta} - \frac{\lambda_\alpha^2 + \lambda_\beta^2}{2\lambda_\alpha \lambda_\beta} \Omega^L_{\alpha\beta}\right) \mathbf{n}^{(\alpha)} \otimes \mathbf{n}^{(\beta)} \; (\alpha \neq \beta) \tag{2.103}
$$

noting

$$
\mathbf{d} = \sum_{\alpha=1}^{3} \frac{\dot{\lambda}_\alpha}{\lambda_\alpha} \mathbf{n}^{(\alpha)} \otimes \mathbf{n}^{(\alpha)} + \frac{1}{2} \sum_{\alpha,\beta=1}^{3} \left[\frac{\lambda_\alpha}{\lambda_\beta} (\dot{\mathbf{N}}^{(\alpha)} \bullet \mathbf{N}^{(\beta)}) + \frac{\lambda_\beta}{\lambda_\alpha} (\dot{\mathbf{N}}^{(\beta)} \bullet \mathbf{N}^{(\alpha)})\right] \mathbf{n}^{(\alpha)} \otimes \mathbf{n}^{(\beta)} \tag{2.104}
$$

$$
\begin{aligned}
\mathbf{w} &= \frac{1}{2} \sum_{\alpha=1}^{3} (\dot{\mathbf{n}}^{(\alpha)} \otimes \mathbf{n}^{(\alpha)} - \mathbf{n}^{(\alpha)} \otimes \dot{\mathbf{n}}^{(\alpha)}) \\
&\quad + \frac{1}{2} \sum_{\alpha,\beta=1}^{3} \left[\frac{\lambda_\alpha}{\lambda_\beta} (\dot{\mathbf{N}}^{(\alpha)} \bullet \mathbf{N}^{(\beta)}) - \frac{\lambda_\beta}{\lambda_\alpha} (\dot{\mathbf{N}}^{(\beta)} \bullet \mathbf{N}^{(\alpha)})\right] \mathbf{n}^{(\alpha)} \otimes \mathbf{n}^{(\beta)}
\end{aligned}
\tag{2.105}
$$

$$
\begin{aligned}
2\mathrm{sym}\left[\sum_{\alpha=1}^{3} \dot{\mathbf{n}}^{(\alpha)} \otimes \mathbf{n}^{(\alpha)}\right] &= \sum_{\alpha=1}^{3} (\dot{\mathbf{n}}^{(\alpha)} \otimes \mathbf{n}^{(\alpha)} + \mathbf{n}^{(\alpha)} \otimes \dot{\mathbf{n}}^{(\alpha)}) \\
&= \left(\sum_{\alpha=1}^{3} \mathbf{n}^{(\alpha)} \otimes \mathbf{n}^{(\alpha)}\right)^{\bullet} = \dot{\mathbf{I}} = \mathbf{O}
\end{aligned}
\tag{2.106}
$$

From the relation

$$d_{\alpha\beta} = \frac{\lambda_\alpha^2 - \lambda_\beta^2}{2\lambda_\alpha\lambda_\beta} \Omega_{\alpha\beta}^L \tag{2.107}$$

obtained from the anti-symmetric part in Eq. (2.102), the Lagrangian spin is written as

$$\Omega_{\alpha\beta}^L = \frac{2\lambda_\alpha\lambda_\beta}{\lambda_\beta^2 - \lambda_\alpha^2} d_{\alpha\beta}\ (\alpha \neq \beta) \tag{2.108}$$

The Eulerian spin is given from Eq. (2.103) as follows:

$$\Omega_{\alpha\beta}^E = w_{\alpha\beta} - \frac{\lambda_\alpha^2 + \lambda_\beta^2}{\lambda_\alpha^2 - \lambda_\beta^2} d_{\alpha\beta}\ (\alpha \neq \beta) \tag{2.109}$$

In the rigid-body rotation ($\dot{\mathbf{U}} = \dot{\tilde{\mathbf{U}}} = \dot{\mathbf{V}} = \dot{\tilde{\mathbf{V}}} = \mathbf{O}$, $\dot{\mathbf{N}}^{(\alpha)} = \mathbf{O}$), it follows from Eqs. (2.79), (2.83), (2.90) and (2.99) that

$$\boldsymbol{l} = \mathbf{w} = \boldsymbol{\Omega}^R = \boldsymbol{\Omega}^E,\ \boldsymbol{\Omega}^L = \mathbf{O} \tag{2.110}$$

In what follows, we consider the physical meanings of $\mathbf{d}$ and $\mathbf{w}$.

The relative velocity of the particle points P and P′, the current position vectors of which are $\mathbf{x}$ and $\mathbf{x} + d\mathbf{x}$, respectively, is given by

$$d\mathbf{v} = \boldsymbol{l} d\mathbf{x} \tag{2.111}$$

from Eq. (2.76) and it is additively decomposed as

$$d\mathbf{v} = d\mathbf{v}^d + d\mathbf{v}^w \tag{2.112}$$

where

$$d\mathbf{v}^d \equiv \mathbf{d} d\mathbf{x} \tag{2.113}$$

$$d\mathbf{v}^w \equiv \mathbf{w} d\mathbf{x} \tag{2.114}$$

The following equation is obtained for the infinitesimal line-element $d\mathbf{x} = dx_i\mathbf{e}_i$ (no sum).

$$d_{ji} = \frac{dv_j^d}{dx_i}\ \text{(no sum)} \tag{2.115}$$

noting

$$d_{ji} = \mathbf{e}_j \cdot \mathbf{d}\mathbf{e}_i = \mathbf{e}_j \cdot \mathbf{d}\frac{d\mathbf{x}}{dx_i} = \mathbf{e}_j \cdot \frac{d\mathbf{v}^d}{dx_i} = \frac{dv_j^d}{dx_i} \tag{2.116}$$

with the aid of Eq. (2.113). Therefore, d_{ji} is the $\mathbf{e}_j$−component of the relative velocity $d\mathbf{v}^d$ of the unit line element $(dx_i = 1)$ in the $\mathbf{e}_i$−direction. Consequently, the infinitesimal line-element $d\mathbf{x} = dx_i\mathbf{e}_i$ (no sum) rotates in the velocity given by the tangential component $d_{ji}(j \neq i)$ of the strain rate $\mathbf{d}$.

On the other hand, denoting the axial vector described in Eq. (1.139) for the skew-symmetric tensor $\mathbf{w}$ by $\widehat{\omega}$, it holds that

$$\widehat{\omega}_i = -\frac{1}{2}\varepsilon_{irs}w_{rs}, \quad w_{ij} = -\varepsilon_{ijr}\,\widehat{\omega}_r \tag{2.117}$$

and thus Eq. (2.114) is rewritten from Eq. (1.142) as

$$d\mathbf{v}^w = \widehat{\boldsymbol{\omega}} \times d\mathbf{x}, dv_i^w(= w_{is}dx_s = -\varepsilon_{isr}\widehat{\omega}_r dx_s) = \varepsilon_{irs}\widehat{\omega}_r dx_s \tag{2.118}$$

Therefore, the arbitrary line-element $d\mathbf{x}$ rotates in the peripheral velocity $d\mathbf{v}^w$ and angular velocity $\widehat{\boldsymbol{\omega}}$, termed often the *spin vector*, by the continuum spin $\mathbf{w}$, whereas $2\widehat{\boldsymbol{\omega}}$ is called the *vorticity*.

Noting Eqs. (2.112)–(2.118), it follows for the material line element $dx_1\mathbf{e}_1$ that

$$\begin{aligned} d\mathbf{v} &= l_{ij}\mathbf{e}_i \otimes \mathbf{e}_j dx_1\mathbf{e}_1 = l_{i1}dx_1\mathbf{e}_i = (d_{21} + w_{21})dx_1\mathbf{e}_2 + d_{11}dx_1\mathbf{e}_1 \\ &= (d_{12} - w_{12})dx_1\mathbf{e}_2 + d_{11}dx_1\mathbf{e}_1 = \varpi_3 dx_1\mathbf{e}_2 + d_{11}dx_1\mathbf{e}_1 \end{aligned} \tag{2.119}$$

as shown in Fig. 2.3, where we set

$$\varpi_3 \equiv -w_{12} + d_{12} = \widehat{\omega}_3 + d_{12} \tag{2.120}$$

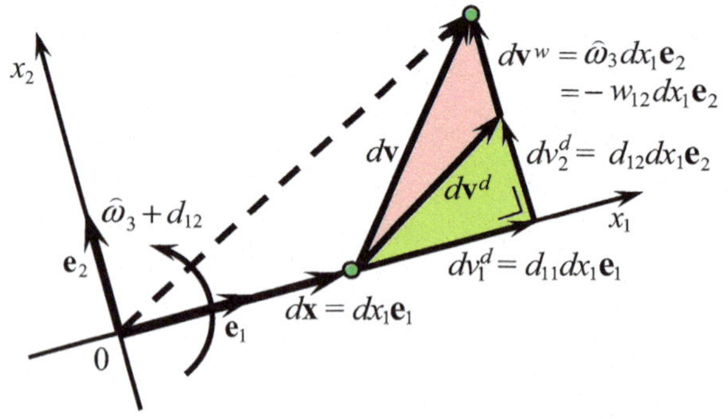

Fig. 2.3 Extension and rotation of the line-element

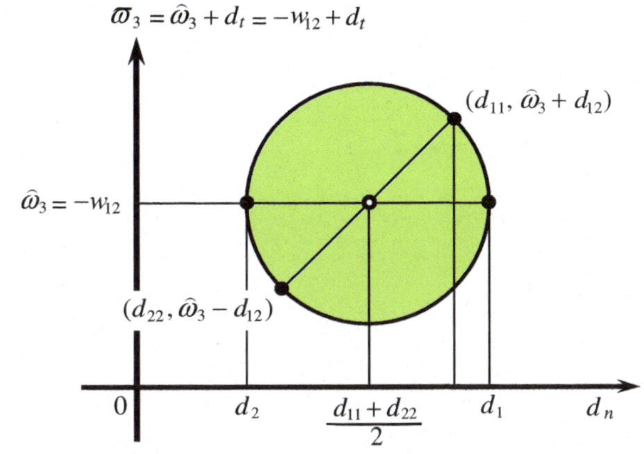

Fig. 2.4 Circle of relative velocity

ϖ_3 designates the clock-wise angular velocity of the line-element. $\widehat{\omega}_3$ denotes the average angular velocity of the line-elements in the plane, which coincides with the angular velocity of the line element in the principal directions of strain rate fulfilling $d_{12} = 0$. By choosing d_n, d_t for T_n, T_t described in Sect. 1.14, the relation of the rate of elongation and the rate of rotation is shown in Fig. 2.4. It is depicted by the *circle of relative velocity* with the radius $\sqrt{[(d_{11}+d_{22})/2]^2 + d_{12}^2}$ centering in $((d_{11}+d_{22})/2, \widehat{\omega}_3)$ in the two-dimensional plane (d_n, ϖ_3).

The parallelepiped in the principal directions of the strain rate $\mathbf{d}$ rotates by the angular velocity $\widehat{\boldsymbol{\omega}}$ as shown in Fig. 2.5.

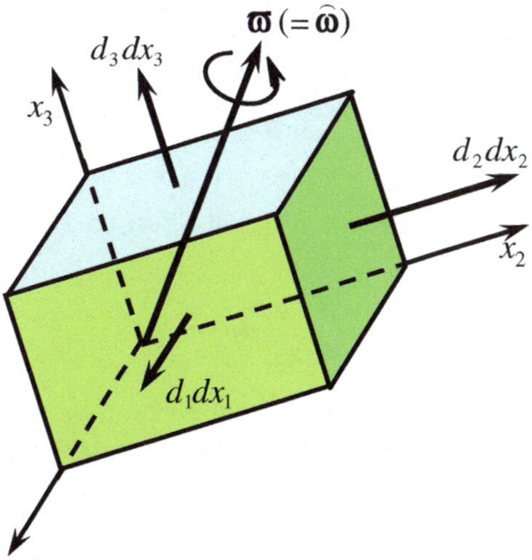

Fig. 2.5 Deformation and rotation for principal directions of strain rate

The rate of the scalar product of the vectors $d\mathbf{x}$ and $\delta\mathbf{x}$ of the infinitesimal elements connecting the three points P, P', P'' with the position vectors $\mathbf{x}, \mathbf{x}+d\mathbf{x},\ \mathbf{x}+\delta\mathbf{x}$, respectively, is given noting Eq. (1.116) as follows:

$$
\begin{aligned}
(d\mathbf{x}\cdot\delta\mathbf{x})^{\bullet} &= d\mathbf{v}\cdot\delta\mathbf{x}+d\mathbf{x}\cdot\delta\mathbf{v}\\
&= \frac{\partial\mathbf{v}}{\partial\mathbf{x}}d\mathbf{x}\cdot\delta\mathbf{x}+d\mathbf{x}\cdot\frac{\partial\mathbf{v}}{\partial\mathbf{x}}\delta\mathbf{x} = \left\{\left[\frac{\partial\mathbf{v}}{\partial\mathbf{x}}+\left(\frac{\partial\mathbf{v}}{\partial\mathbf{x}}\right)^{T}\right]d\mathbf{x}\right\}\cdot\delta\mathbf{x}
\end{aligned}
$$

leading to

$$
\boxed{(d\mathbf{x}\cdot\delta\mathbf{x})^{\bullet} = 2\mathbf{d}d\mathbf{x}\cdot\delta\mathbf{x}} \tag{2.121}
$$

If the vicinity of the particle P undergoes the rigid-body rotation, the quantity in Eq. (2.121) for an arbitrary scalar quantity $d\mathbf{x}\cdot\delta\mathbf{x}$ is zero and thus $\mathbf{d}=\mathbf{O}$ has to hold. Inversely, if $\mathbf{d}=\mathbf{O}$, the quantity in Eq. (2.121) for the scalar quantity $d\mathbf{x}\cdot\delta\mathbf{x}$ of arbitrary line-element vectors becomes zero and thus it can be stated that the vicinity of the particle P does not undergo a deformation. Then, $\mathbf{d}=\mathbf{O}$ is the necessary and the sufficient condition for the situation that a deformation is not induced, allowing only a rigid-body rotation.

Denoting the lengths of the line-elements PP′ and PP″ as ds and δs, respectively, and the angle contained by them as θ, it holds that

$$
\begin{aligned}
(d\mathbf{x}\cdot\delta\mathbf{x})^{\bullet} &= (ds\delta s\cos\theta)^{\bullet}\\
&= \left\{\left[\frac{(ds)^{\bullet}}{ds}+\frac{(\delta s)^{\bullet}}{\delta s}\right]\cos\theta-\dot{\theta}\sin\theta\right\}ds\delta s
\end{aligned} \tag{2.122}
$$

Further, denoting the unit vectors in the directions of the line-elements PP′ and PP″ as $\boldsymbol{\mu}$ and $\mathbf{v}$, respectively, and noting $d\mathbf{x}=\boldsymbol{\mu}ds$, $\delta\mathbf{x}=\mathbf{v}\delta s$, it holds from Eqs. (2.121) and (2.122) that

$$
\left[\frac{(ds)^{\bullet}}{ds}+\frac{(\delta s)^{\bullet}}{\delta s}\right]\cos\theta-\dot{\theta}\sin\theta = 2\mathbf{d}\boldsymbol{\mu}\cdot\mathbf{v}\,(=2d_{ij}\mu_i v_j) \tag{2.123}
$$

If the particles P′ and P″ chosen in same direction ($\theta=0$), it follows from Eq. (2.123) that

$$
\frac{(ds)^{\bullet}}{ds} = \mathbf{d}\boldsymbol{\mu}\cdot\boldsymbol{\mu} \tag{2.124}
$$

The left-hand side of Eq. (2.124) designates the *rate of elongation* of the line-element. Therefore, the rate of elongation is given by the normal component of $\mathbf{d}$ in the relevant direction, noting Eq. (1.102).

On the other hand, choosing the line-element PP″ to be perpendicular to the line element PP′ $(\theta = \pi/2)$, it follows from Eq. (2.123) that

$$-\dot{\theta}\,2\mathbf{d}\boldsymbol{\mu}\cdot\mathbf{v} \quad (\boldsymbol{\mu}\cdot\mathbf{v} = 0) \tag{2.125}$$

The left-hand side of Eq. (2.125) designates the decreasing rate of the angle contained by the two line-elements mutually perpendicular instantaneously and is called the *shear strain rate*.

Next, the relations of the rate $\dot{\boldsymbol{E}}$ of Green strain tensor $\boldsymbol{E}$ and the rate $\dot{\boldsymbol{e}}$ of the Almansi strain tensor $\boldsymbol{e}$ to the strain rate tensor $\mathbf{d}$ are formulated below.

The material-time derivative of Eq. (2.44) is given by

$$(d\mathbf{x}\cdot\delta\mathbf{x})^{\bullet} = 2\dot{\boldsymbol{E}}\,d\mathbf{X}\cdot\delta\mathbf{X} \tag{2.126}$$

It follows from Eqs. (2.121) and (2.126) that

$$\mathbf{d}d\mathbf{x}\cdot\delta\mathbf{x} = \mathbf{dF}d\mathbf{X}\cdot\mathbf{F}\delta\mathbf{X} = \dot{\boldsymbol{E}}\;d\mathbf{X}\cdot\delta\mathbf{X} \tag{2.127}$$

from which, noting Eq. (1.116), we have the relation of Green strain tensor to the strain rate tensor as follows:

$$\boxed{\dot{\boldsymbol{E}} = \mathbf{F}^T\mathbf{dF},\ \mathbf{d} = \mathbf{F}^{-T}\dot{\boldsymbol{E}}\mathbf{F}^{-1}} \tag{2.128}$$

which is obtained also from

$$\begin{aligned}
\dot{\boldsymbol{E}} &= \frac{1}{2}(\mathbf{F}^T\mathbf{F} - \mathbf{I})^{\bullet} = \frac{1}{2}(\dot{\mathbf{F}}^T\mathbf{F} + \mathbf{F}^T\dot{\mathbf{F}}) \\
&= \frac{1}{2}\,[\mathbf{F}^T(\mathbf{F}^{-T}\dot{\mathbf{F}}^T)\mathbf{F} + \mathbf{F}^T(\dot{\mathbf{F}}\mathbf{F}^{-1})\mathbf{F}] \\
&= \frac{1}{2}(\mathbf{F}^T\boldsymbol{l}^T\mathbf{F} + \mathbf{F}^T\boldsymbol{l}\mathbf{F}) = \frac{1}{2}\mathbf{F}^T(\boldsymbol{l}^T + \boldsymbol{l})\mathbf{F} \\
&= \mathbf{F}^T\mathbf{dF}
\end{aligned} \tag{2.129}$$

Next, the time-differentiation of Eq. $(2.45)_2$ leads to

$$\dot{\boldsymbol{e}} = -\frac{1}{2}(\dot{\mathbf{F}}^{-T}\mathbf{F}^{-1} + \mathbf{F}^{-T}\dot{\mathbf{F}}^{-1}) \tag{2.130}$$

Here, it follows from $(\mathbf{F}\mathbf{F}^{-1})^{\bullet} = \mathbf{F}\dot{\mathbf{F}}^{-1} + \dot{\mathbf{F}}\mathbf{F}^{-1} = \mathbf{O}$ with Eq. (2.75) that

$$\boxed{\dot{\mathbf{F}}^{-1} = -\mathbf{F}^{-1}\boldsymbol{l}} \tag{2.131}$$

which is derived also by

$$\begin{aligned}\dot{\mathbf{F}}^{-1} &= \frac{\partial \mathbf{F}^{-1}}{\partial t} + \frac{\partial \mathbf{F}^{-1}}{\partial \mathbf{x}}\mathbf{v} = \frac{\partial}{\partial t}\left(\frac{\partial \mathbf{X}}{\partial \mathbf{x}}\right) + \frac{\partial}{\partial \mathbf{x}}\left(\frac{\partial \mathbf{X}}{\partial \mathbf{x}}\right)\mathbf{v} \\ &= \frac{\partial}{\partial \mathbf{x}}\left(\frac{\partial \mathbf{X}}{\partial t}\right) + \frac{\partial}{\partial \mathbf{x}}\left(\frac{\partial \mathbf{X}}{\partial \mathbf{x}}\mathbf{v}\right) - \frac{\partial \mathbf{X}}{\partial \mathbf{x}}\frac{\partial \mathbf{v}}{\partial \mathbf{x}} = \frac{\partial}{\partial \mathbf{x}}\left(\frac{\partial \mathbf{X}}{\partial t} + \frac{\partial \mathbf{X}}{\partial \mathbf{x}}\mathbf{v}\right) - \frac{\partial \mathbf{X}}{\partial \mathbf{x}}\frac{\partial \mathbf{v}}{\partial \mathbf{x}} \\ &= \frac{\partial \dot{\mathbf{X}}}{\partial \mathbf{x}} - \frac{\partial \mathbf{X}}{\partial \mathbf{x}}\frac{\partial \mathbf{v}}{\partial \mathbf{x}} = -\frac{\partial \mathbf{X}}{\partial \mathbf{x}}\frac{\partial \mathbf{v}}{\partial \mathbf{x}}\end{aligned} \tag{2.132}$$

noting that the inside of the bracket ( ) in the last side of Eq. (2.132) is the material-time derivative of the initial position vector $\mathbf{X}$ and thus it is zero. Substituting Eq. (2.131) into Eq. (2.130), one has

$$\begin{aligned}\dot{\boldsymbol{e}} &= \frac{1}{2}[(\mathbf{F}^{-1}\boldsymbol{l})^T\mathbf{F}^{-1} + \mathbf{F}^{-T}\mathbf{F}^{-1}\boldsymbol{l}] \\ &= \boldsymbol{l}^T\left[\frac{1}{2}\mathbf{I} - \frac{1}{2}(\mathbf{I} - \mathbf{F}^{-T}\mathbf{F}^{-1})\right] + \frac{1}{2}\left[\mathbf{I} - \left(\frac{1}{2}\mathbf{I} - \mathbf{F}^{-T}\mathbf{F}^{-1}\right)\right]\boldsymbol{l} \\ &= \boldsymbol{l}^T\left(\frac{1}{2}\mathbf{I} - \boldsymbol{e}\right) + \left(\frac{1}{2}\mathbf{I} - \boldsymbol{e}\right)\boldsymbol{l}\end{aligned} \tag{2.133}$$

from which one has the relation of the rate of the Almansi strain tensor to the strain rate tensor:

$$\boxed{\dot{\boldsymbol{e}} = \mathbf{d} - \boldsymbol{l}^T\boldsymbol{e} - \boldsymbol{e}\boldsymbol{l}} \tag{2.134}$$

Equation (2.134) is rewritten as

$$\dot{\boldsymbol{e}} = \mathbf{d} - \frac{1}{2}[(\boldsymbol{l} + \boldsymbol{l}^T) - (\boldsymbol{l} - \boldsymbol{l}^T)]\boldsymbol{e} - \frac{1}{2}\boldsymbol{e}[(\boldsymbol{l} + \boldsymbol{l}^T) + (\boldsymbol{l} - \boldsymbol{l}^T)]$$

and thus we obtain

$$\dot{\boldsymbol{e}} - \mathbf{w}\boldsymbol{e} + \boldsymbol{e}\mathbf{w} = \mathbf{d} - \mathbf{d}\boldsymbol{e} - \boldsymbol{e}\mathbf{d} \tag{2.135}$$

where

$$\boxed{\overset{\circ}{\boldsymbol{e}}{}^{w} \equiv \dot{\boldsymbol{e}} - \mathbf{w}\boldsymbol{e} + \boldsymbol{e}\mathbf{w}} \tag{2.136}$$

is called the *Zaremba-Jaumann rate of Almansi strain tensor*, while the Zaremba-Jaumann rate will be explained in Sect. 4.4. $\dot{\boldsymbol{E}} = \dot{\boldsymbol{e}} = \overset{\circ}{\boldsymbol{e}}{}^{w} = \mathbf{d}$ holds in the initial state ($\mathbf{F} = \mathbf{I}$, $\boldsymbol{E} = \boldsymbol{e} = \mathbf{O}$) and thus all the strain rates mutually coincide by Eqs. (2.128), (2.134) and (2.136).

2.6 Logarithmic and Nominal Strains

In what follows, let **d** and **d**dt be designated as $\overset{\bullet}{\boldsymbol{\varepsilon}}$ and $d\boldsymbol{\varepsilon}$, respectively.

If the direction of the material line-element always coincides with the x_i-axis, the principal strain rate in this direction is given by

$$d_i = \overset{\bullet}{\varepsilon}_i = \frac{\partial \overset{\bullet}{u}_i}{\partial x_i} = \frac{\partial \overset{\bullet}{u}_i}{\partial (X_i + u_i)} = \frac{\partial \overset{\bullet}{u}_i}{\partial X_i} \bigg/ \left(1 + \frac{\partial u_i}{\partial X_i}\right) \ \text{(no sum)} \tag{2.137}$$

The time-integration of Eq. (2.137) leads to

$$\boxed{\varepsilon_i = \ln\left(1 + \frac{\partial u_i}{\partial X_i}\right) = \ln\frac{\partial x_i}{\partial X_i} = \ln\lambda_i = E_i^{(0)} = e_i^{(0)} = \ln(1 + \varepsilon_i) \ \text{(no sum)}} \tag{2.138}$$

Therefore, the time-integration ε_i of principal strain rate d_i does not coincide with the principal infinitesimal strain ε_i in Eq. (2.50). Setting $\partial X_i \to l^0$, $\partial x_i \to l$, $\partial u_i \to l - l^0$, where l^0 and l are the lengths of the line-element in the initial and the current states, respectively, it follows that

$$\left.\begin{aligned} \varepsilon_i &= \ln\frac{l}{l^0} = \ln(1 + \varepsilon_i) \\ \varepsilon_i &= \frac{l - l^0}{l^0} \end{aligned}\right\} \tag{2.139}$$

Consequently, the time-integration of principal component of strain rate tensor **d** and the Hencky strain tensor $\boldsymbol{E}^{(0)}$ and $\boldsymbol{e}^{(0)}$ coincide with ε_i which is called the *logarithmic* (or *natural*) *strain*, provided that their principal directions are fixed. On the other hand, the principal value of infinitesimal strain tensor $\boldsymbol{\varepsilon}$ does not coincide with them and it is called the *nominal strain*.

It follows from Eq. (2.139) that

$$\left.\begin{aligned} &\boxed{\varepsilon_i = \ln_e 2(\cong 0.693) \ \text{for} \ l = 2l^0 \ \text{and} \ \varepsilon_i = -\infty \ \text{for} \ l = 0} \\ &\varepsilon_i = +1 \ \text{for} \ l = 2l^0 \ \text{and} \ \varepsilon_i = -1 \ \text{for} \ l = 0 \end{aligned}\right\} \tag{2.140}$$

Therefore, the magnitude of nominal strain in the deformation that the material length becomes zero, i.e. the material diminishes is identical with that in the deformation that the material length becomes only twice. As a practical example, about 5 % difference is induced in the nominal strain for 10 % elongation as known from $\varepsilon_i/\varepsilon_i = 0.1/\ln(1.1) = 1.049$ for $l = 1.1 \times l_0$. This property would cause the inconvenience for the adoption in constitutive equation for the wide range of deformation.

Further, one has

$$\int_{l^0}^{l^n} \frac{dl}{l} = \int_{l^0}^{l^1} \frac{dl}{l} + \int_{l^1}^{l^2} \frac{dl}{l} + \cdots + \int_{l^{n-1}}^{l^n} \frac{dl}{l}$$

i.e.

$$1n\frac{l^n}{l^0} = 1n\frac{l^1}{l^0} + 1n\frac{l^2}{l^1} + \cdots + 1n\frac{l^n}{l^{n-1}}$$

On the other hand, one sees

$$\frac{l^n - l^0}{l^0} \neq \frac{l^1 - l^0}{l^0} + \frac{l^2 - l^1}{l^1} + \cdots + \frac{l^n - l^{n-1}}{l^{n-1}}$$

Thus, it follows for the superposition of strains that

$$\left.\begin{array}{l} \boxed{\varepsilon_i^{0\sim n} = \varepsilon_i^{0\sim 1} + \varepsilon_i^{0\sim 2} + \cdots + \varepsilon_i^{n-1\sim n}} \\ \varepsilon_i^{0\sim n} \neq \varepsilon_i^{0\sim 1} + \varepsilon_i^{0\sim 2} + \cdots + \varepsilon_i^{n-1\sim n} \end{array}\right\} \tag{2.141}$$

while $\varepsilon_i^{a\sim b}$ designates the longitudinal strain in the x_i-direction when the length of the line-element changes from l^a to l^b, provided that the principal direction of strains are fixed. Consequently, the superposition rule holds in the logarithmic strain but it does not hold in the nominal strain.

Furthermore, one has

$$1n\frac{l_1 l_2 l_3}{l_1^0 l_2^0 l_3^0} = 1n\frac{l_1}{l_1^0} + 1n\frac{l_2}{l_2^0} + 1n\frac{l_3}{l_3^0}$$

$$\frac{l_1 l_2 l_3 - l_1^0 l_2^0 l_3^0}{l_1^0 l_2^0 l_3^0} \neq \frac{l_1 - l_1^0}{l_1^0} + \frac{l_2 - l_2^0}{l_2^0} + \frac{l_3 - l_3^0}{l_3^0}$$

where l_1, l_2, l_3 are the lengths of line-elements in the directions of three fixed principal strains. Thus, it follows for the sum of the principal strains that

$$\left.\begin{array}{l} \boxed{\varepsilon_v = \ln\frac{v}{V} = \sum_{i=1}^{3} \varepsilon_i} = \ln(1 + \varepsilon_v) \\ \varepsilon_v = \frac{v - V}{V} \neq \sum_{i=1}^{3} \varepsilon_i \end{array}\right\} \tag{2.142}$$

where V and v are the initial and the current volumes, respectively, of material. Therefore, the sum of logarithmic strains in orthogonal directions coincides with the logarithmic volumetric strain but the sum of nominal strains in orthogonal directions does not coincide with the nominal volumetric strain. Further, it follows from Eqs. (2.137) and (2.142) that

$$d_v \equiv \dot{\varepsilon}_v = \mathrm{tr}\mathbf{d} = \frac{\dot{v}}{v} \neq \dot{\varepsilon}_v = \frac{\dot{v}}{V} \tag{2.143}$$

Therefore, the volumetric strain rate tr**d** coincides with the material-time derivative of the logarithmic volumetric strain ε_v but it does not coincide with that of the nominal volumetric strain ε_v.

Consequently, the nominal strain is applicable only to the description of infinitesimal deformation, so that the logarithmic strain should be adopted to constitutive equations for finite deformation.

2.7 Surface Element, Volume Element and Their Rates

Presuming that the line-elements $d\mathbf{X}^a$, $d\mathbf{X}^b$, $d\mathbf{X}^c$ change to $d\mathbf{x}^a$, $d\mathbf{x}^b$, $d\mathbf{x}^c$ by the deformation, the following relation holds for the volume element before and after the deformation by exploiting Eq. $(1.150)_3$.

$$dv = [d\mathbf{x}^a \; d\mathbf{x}^b \; d\mathbf{x}^c] = [\mathbf{F}d\mathbf{X}^a \; \mathbf{F}d\mathbf{X}^b \; \mathbf{F}d\mathbf{X}^c] = \det\mathbf{F}[d\mathbf{X}^a \; d\mathbf{X}^b \; d\mathbf{X}^c] = \det \mathbf{F}dV \tag{2.144}$$

which can be also derived from Eqs. (1.16), (1.55), (1.56), (2.6) and (2.16).

$$\begin{aligned} dv &= (d\mathbf{x}^a \times d\mathbf{x}^b)\cdot d\mathbf{x}^c = \varepsilon_{ijk} dx_i^a dx_j^b dx_k^c \\ &= \begin{vmatrix} dx_1^a & dx_1^b & dx_1^c \\ dx_2^a & dx_2^b & dx_2^c \\ dx_3^a & dx_3^b & dx_3^c \end{vmatrix} = \begin{vmatrix} F_{1R}dX_R^a & F_{1R}dX_R^b & F_{1R}dX_R^c \\ F_{1R}dX_R^a & F_{2R}dX_R^b & F_{2R}dX_R^c \\ F_{1R}dX_R^c & F_{3R}dX_R^b & F_{3R}dX_R^c \end{vmatrix} \\ &= \left| \begin{bmatrix} F_{11} & F_{12} & F_{13} \\ F_{21} & F_{22} & F_{23} \\ F_{31} & F_{32} & F_{33} \end{bmatrix} \begin{bmatrix} dX_1^a & dX_1^b & dX_1^c \\ dX_2^a & dX_2^b & dX_2^c \\ dX_3^a & dX_3^b & dX_3^c \end{bmatrix} \right| \\ &= \begin{vmatrix} F_{11} & F_{12} & F_{13} \\ F_{21} & F_{22} & F_{23} \\ F_{31} & F_{32} & F_{33} \end{vmatrix} \begin{vmatrix} dX_1^a & dX_1^b & dX_1^c \\ dX_2^a & dX_2^b & dX_2^c \\ dX_3^a & dX_3^b & dX_3^c \end{vmatrix} = JdV \end{aligned}$$

from which one has

$$\boxed{J = \det\mathbf{F} = \frac{dv}{dV} = \frac{\rho_0}{\rho}} \tag{2.145}$$

where ρ_0 and ρ are the initial and the current mass densities. On the other hand, denoting the areas and the unit normal vectors of the surface elements formed by the two line-elements $d\mathbf{X}^a, d\mathbf{X}^b$ and $d\mathbf{x}^a, d\mathbf{x}^b$ as dA , $\mathbf{N}$ and da, $\mathbf{n}$, respectively, we have

$$\left.\begin{array}{l} dV=[d\mathbf{X}^a\, d\mathbf{X}^b d\mathbf{X}^c]=(d\mathbf{X}^a\times d\mathbf{X}^b)\boldsymbol{\cdot} d\mathbf{X}^c=\mathbf{N}dA\boldsymbol{\cdot} d\mathbf{X}^c \\ dv=[d\mathbf{x}^a\, d\mathbf{x}^b\, d\mathbf{x}^c]=(d\mathbf{x}^a\times d\mathbf{x}^b)\boldsymbol{\cdot} d\mathbf{x}^c=d\mathbf{x}^c\boldsymbol{\cdot}\mathbf{n}da=\mathbf{F}d\mathbf{X}^c\boldsymbol{\cdot}\mathbf{n}da=\mathbf{F}^T\mathbf{n}da\boldsymbol{\cdot} d\mathbf{X}^c \end{array}\right\} \tag{2.146}$$

noting Eq. (1.116). The following *Nanson's formula* is derived from Eqs. (2.145) and (2.146).

$$\boxed{\mathbf{n}da=J\mathbf{F}^{-T}\mathbf{N}dA,\ \mathbf{N}dA=\frac{1}{J}\mathbf{F}^T\mathbf{n}da} \tag{2.147}$$

or

$$d\mathbf{a}=J\mathbf{F}^{-T}d\mathbf{A},\quad d\mathbf{A}=\frac{1}{J}\mathbf{F}^T d\mathbf{a} \tag{2.148}$$

where

$$d\mathbf{a}\equiv\mathbf{n}da,\quad d\mathbf{A}\equiv\mathbf{N}dA \tag{2.149}$$

It follows from Eq. (2.145) noting Eq. (1.121) that

$$\dot{J}=(\det\mathbf{F})^{\bullet}=\mathrm{tr}\left(\frac{\partial\det\mathbf{F}}{\partial\mathbf{F}}\dot{\mathbf{F}}^T\right)=\mathrm{tr}[(\det\mathbf{F})\mathbf{F}^{-T}\dot{\mathbf{F}}^T]=\mathrm{tr}[J(\dot{\mathbf{F}}\mathbf{F}^{-1})^T]=J\mathrm{tr}\boldsymbol{l}$$

Then, the following relation holds for the rate of volume element, noting $\mathrm{tr}\boldsymbol{l}=\mathrm{tr}\mathbf{d}$.

$$\boxed{\dot{\varepsilon}_v=d_v=\mathrm{tr}\mathbf{d}=\frac{\partial v_r}{\partial x_r}=\frac{(dv)^{\bullet}}{dv}=\frac{\dot{J}}{J}},\quad \text{i.e.}\quad (dv)^{\bullet}=\dot{J}dV=dv\mathrm{tr}\mathbf{d} \tag{2.150}$$

which was derived already in Eq. (2.143) exploiting the principal values under the fixed principal directions. Then, the time-integration of the volumetric strain rate leads to the logarithmic volumetric strain.

$$\varepsilon_v=\int\mathrm{tr}\mathbf{d}dt=\ln\frac{dv}{dV} \tag{2.151}$$

which was shown already in Eq. (2.142).

Moreover, it follows from Eq. (2.75) and the Nanson's formula (2.147) that

$$\begin{aligned}(\mathbf{n}da)^{\bullet}&=(J\mathbf{F}^{-T}\mathbf{N}dA)^{\bullet}=(\dot{J}\mathbf{F}^{-T}+J\dot{\mathbf{F}}^{-T})\mathbf{N}dA\\&=[(\mathrm{tr}\mathbf{d})\mathbf{I}+\dot{\mathbf{F}}^{-T}\mathbf{F}^T]\mathbf{F}^{-T}J\mathbf{N}dA\\&=[(\mathrm{tr}\mathbf{d})\mathbf{I}-\mathbf{F}^{-T}\dot{\mathbf{F}}^T]\mathbf{n}da\\&=[(\mathrm{tr}\mathbf{d})\mathbf{I}-\boldsymbol{l}^T]\mathbf{n}da\end{aligned} \tag{2.152}$$

On the other hand, noting $\dot{\mathbf{n}} \cdot \mathbf{n} = 0$ because of $\mathbf{n} \cdot \mathbf{n} = 1$ for the unit vector $\mathbf{n}$, it follows that

$$(da)^{\bullet} = \mathbf{n} \cdot \mathbf{n}(da)^{\bullet} = \mathbf{n} \cdot [(\mathbf{n}da)^{\bullet} - \dot{\mathbf{n}}\, da] = \mathbf{n} \cdot (\mathbf{n}da)^{\bullet} \tag{2.153}$$

Substituting Eq. (2.152) into Eq. (2.153), one obtains the rate of the current infinitesimal area as follows:

$$(da)^{\bullet} = \mathbf{n} \cdot [(\mathrm{tr}\mathbf{d})\mathbf{I} - \boldsymbol{l}^T]\ \mathbf{n}da \tag{2.154}$$

or

$$\boxed{(da)^{\bullet} = (\mathrm{tr}\mathbf{d} - \mathbf{n} \cdot \mathbf{dn})da} \tag{2.155}$$

Further, it holds from Eqs. (2.152) and (2.155) that

$$\begin{aligned} \dot{\mathbf{n}}\, da &= (\mathbf{n}da)^{\bullet} - \mathbf{n}(da)^{\bullet} \\ &= [(\mathrm{tr}\mathbf{d})\mathbf{I} - \boldsymbol{l}^T]\mathbf{n}da - \mathbf{n}[(\mathrm{tr}\mathbf{d}) - \mathbf{n} \cdot \mathbf{dn}]da \end{aligned} \tag{2.156}$$

Then, the rate of the unit normal vector of the current surface element is given by

$$\boxed{\dot{\mathbf{n}} = [(\mathbf{n} \cdot \mathbf{dn})\mathbf{I} - \boldsymbol{l}^T]\mathbf{n}} \tag{2.157}$$

2.8 Material-Time Derivative of Volume Integration

Supposing that the zone of material occupying the volume v at the current moment $(t = t)$ changes to occupy the volume $v + \delta v$ after the infinitesimal time $(t = t + \delta t)$, the material-time-derivative of the volume integration $\int_v T(\mathbf{x}, t)dv$ of the physical quantity $T(\mathbf{x}, t)$ involved in the volume is given by the following equation.

$$\begin{aligned} \left(\int_v T(\mathbf{x}, t)dv\right)^{\bullet} &= \lim_{\delta t \to 0} \frac{1}{\delta t}\left[\int_{v+\delta v} T(\mathbf{x}, t+\delta t)dv - \int_v T(\mathbf{x}, t)dv\right] \\ &= \lim_{\delta t \to 0} \frac{1}{\delta t}\left[\int_v \{T(\mathbf{x}, t+\delta t) - T(\mathbf{x}, t)\}dv + \int_{\delta v} T(\mathbf{x}, t+\delta t)dv\right] \end{aligned} \tag{2.158}$$

The integration of the first term in the right-hand side in Eq. (2.158) is transformed as

$$\lim_{\delta t \to 0} \frac{1}{\delta t}\int_v [T(\mathbf{x}, t+\delta t) - T(\mathbf{x}, t)]dv = \int_v \frac{\partial T(\mathbf{x}, t)}{\partial t}dv \tag{2.159}$$

On the other hand, the second term in Eq. (2.158) describes the influence caused by the change of volume during the infinitesimal time increment. Here, the volume increment δv is given by subtracting the volume flowing out from the boundary of the zone from the volume flowing into the boundary, which is the sum of $dv(= \mathbf{v} \cdot \mathbf{n} da \delta t)$ over the whole boundary surface (Fig. 2.6). Therefore, substituting the Gauss' divergence theorem in Eq. (1.324) and ignoring the second-order infinitesimal quantity, the integration of the second term in the right-hand side of Eq. (2.158) is given by

$$
\begin{aligned}
\lim_{\delta t \to 0} \frac{1}{\delta t} \int_{\delta v} T(\mathbf{x}, t + \delta t) dv &\cong \lim_{\delta t \to 0} \frac{1}{\delta t} \int_{\delta v} T(\mathbf{x}, t) dv \\
&= \lim_{\delta t \to 0} \frac{1}{\delta t} \int_a T(\mathbf{x}, t) v_r n_r da \delta t = \int_a T(\mathbf{x}, t) v_r n_r da \\
&= \int_v \frac{\partial T(\mathbf{x}, t) v_r}{\partial x_r} dv = \int_v \frac{\partial T(\mathbf{x}, t)}{\partial x_r} v_r dv + \int_v T(\mathbf{x}, t) \frac{\partial v_r}{\partial x_r} dv
\end{aligned}
$$

The sum of the first term in the right-hand side in this equation and the Eq. (2.159) is equal to the material-time derivative of $T(\mathbf{x}, t)$ by virtue of Eq. (2.12). Then, Eq. (2.158) is given by

$$
\boxed{\left(\int_v T(\mathbf{x}, t) dv \right)^{\bullet} = \int_v \left[\dot{T}(\mathbf{x}, t) + T(\mathbf{x}, t) \frac{\partial v_r}{\partial x_r} \right] dv = \int_v [\dot{T}(\mathbf{x}, t) + T(\mathbf{x}, t) \mathrm{div} \mathbf{v}] dv} \tag{2.160}
$$

which is called the *Reynolds' transportation theorem*.

Equation (2.160) can be obtained also by the following simple manner.

$$
\begin{aligned}
\left(\int_v T(\mathbf{x}, t) dv \right)^{\bullet} &= \left(\int_V T(\mathbf{X}, t) J dV \right)^{\bullet} = \int_V (\dot{T}(\mathbf{X}, t) J + T(\mathbf{X}, t) \dot{J}) dV \\
&= \int_v \left(\dot{T}(\mathbf{x}, t) + T(\mathbf{x}, t) \frac{\partial v_r}{\partial x_r} \right) dv
\end{aligned}
$$

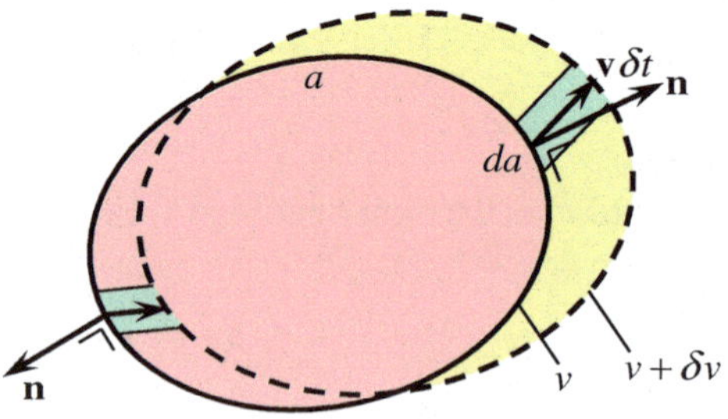

Fig. 2.6 Translation of a material in zone

where V is the initial volume. Here, Eq. (2.150), i.e. $J = dv/dV$ and $\dot{J}/J = \partial v_r/\partial x_r$ hold.

For the physical quantity T kept constant in a volume element, say a mass, Eq. (2.160) leads to

$$\int_v (\dot{T}(\mathbf{x}, t) + T(\mathbf{x}, t) div\mathbf{v}) dv = 0 \tag{2.161}$$

The local (weak) form of Eq. (2.161) is given as

$$\boxed{\dot{T}(\mathbf{x}, t) + T(\mathbf{x}, t) div\mathbf{v} = 0} \tag{2.162}$$

References

Biot MA (1965) Mechanics of incremental deformations, John-Wiley, New York

Hashiguchi K, Yamakawa Y (2012) Introduction to finite strain theory for continuum elasto-plasticity, vol Wiley Series in computational mechanics. Wiley, Chichester, UK

Hill R (1968) On the constitutive inequalities for simple materials—1. J Mech Phys Solids 16:229–242

Seth BR (1964) Generalized strain measure with applications to physical problems. In: Second-order effects inelasticity, plasticity, and fluid dynamics. Pergamon, Oxford

Chapter 3
Stress Tensors and Conservation Laws

Conservation laws of mass, momentum, angular momentum, etc. must be fulfilled during a deformation. These laws are described first in detail. Then, the Cauchy stress tensor is defined and further, based on it, various stress tensors are derived from the Cauchy stress tensor. Introducing the stress tensor, the equilibrium equations of force and moment are formulated from the conservation laws. The virtual work principle required for the analyses of boundary value problems are also described in this chapter (cf. Hashiguchi 2013).

3.1 Stress Tensor

When the infinitesimal force vector $d\mathbf{f}$ applies to the surface with infinitesimal area da and the unit normal vector $\mathbf{n}$, the *stress vector*, i.e. *traction* is given as

$$\boldsymbol{t} \equiv \frac{d\mathbf{f}}{da} \tag{3.1}$$

Now, introduce the following second-order tensor $\boldsymbol{\sigma}$ fulfilling the relation

$$\boxed{\boldsymbol{t} = \boldsymbol{\sigma}^T\mathbf{n},\ t_i = \sigma_{ji}n_j} \tag{3.2}$$

by the quotient law described in Sect. 1.3.2. The components of the tensor $\boldsymbol{\sigma}$ are given by Eq. (1.102) as

$$\sigma_{ij} = \mathbf{e}_i \cdot \boldsymbol{\sigma}\mathbf{e}_j = \boldsymbol{\sigma}\mathbf{e}_i \cdot \mathbf{e}_j = \boldsymbol{\sigma}^T\mathbf{e}_i \cdot \mathbf{e}_j, \boldsymbol{\sigma} = \sigma_{ij}\mathbf{e}_i \otimes \mathbf{e}_j \tag{3.3}$$

if $\boldsymbol{\sigma}$ is the symmetric tensor. Here, when we choose $\mathbf{e}_i$ to the unit normal vector $\mathbf{n}$ of the surface on which $\boldsymbol{t}$ applies, the following equation holds by substituting Eq. (3.2) with $\mathbf{n} = \mathbf{e}_i$ into Eq. (3.3).

$$\sigma_{ij} = \boldsymbol{t}(\mathbf{e}_i) \cdot \mathbf{e}_j \tag{3.4}$$

K. Hashiguchi, *Foundations of Elastoplasticity: Subloading Surface Model*,
DOI 10.1007/978-3-319-48821-9_3

Therefore, σ_{ij} can be interpreted as the component in the direction of $\mathbf{e}_j$ for the stress vector $\boldsymbol{t}(\mathbf{e}_i)$ applying on the surface element having the outward-normal vector $\mathbf{e}_i$. The tensor $\boldsymbol{\sigma}$ is called the *Cauchy stress tensor*. Equation (3.2) is called the *Cauchy's fundamental theorem* or *Cauchy's stress principle*. It follows from the equilibrium of angular moment described in Sect. 3.6 that

$$\boxed{\boldsymbol{\sigma} = \boldsymbol{\sigma}^T, \quad \sigma_{ij} = \sigma_{ji}} \tag{3.5}$$

which means that $\boldsymbol{\sigma}$ is the symmetric tensor.

It was shown above by the virtue of the quotient law in Sect. 1.3.2 that the Cauchy stress is the second-order tensor. In what follows, it will be verified that the Cauchy stress obeys the transformation rule of the second-order tensor.

Consider the coordinate systems $(\mathbf{e}_1, \mathbf{e}_2, \mathbf{e}_3)$ and $(\mathbf{e}_1^*, \mathbf{e}_2^*, \mathbf{e}_3^*)$ the origin of which passes through the point P in the material. The stress vectors (force vector per unit area) applying to the surface the outward-normal of which is directed to the base vectors $\mathbf{e}_1, \mathbf{e}_2, \mathbf{e}_3$ are denoted by $\mathbf{P}_1, \mathbf{P}_2, \mathbf{P}_3$, respectively. Further, the stress vector applying to the surface the outward-normal of which is directed to the base vector $\mathbf{e}_1^*$ is denoted by $\mathbf{P}_1^*$. Then, consider the infinitesimal tetrahedron possessing the surfaces perpendicular to the base vectors $\mathbf{e}_1$, $\mathbf{e}_2, \mathbf{e}_3$ and $\mathbf{e}_1^*$ as shown in Fig. 3.1, while the areas of them are designated as dS_1, dS_2, dS_3 and dS_1^*, respectively. Besides, the length of the line perpendicular to the surface perpendicular to the base $\mathbf{e}_1^*$, while the line stems from the point P and reaches to that surface, is denoted by h. The forces applying to these surfaces are given by $-\mathbf{P}_1 dS_1, -\mathbf{P}_2 dS_2, -\mathbf{P}_3 dS_3$ and $\mathbf{P}_1^* dS_1^*$. Then, the following equilibrium equation must hold.

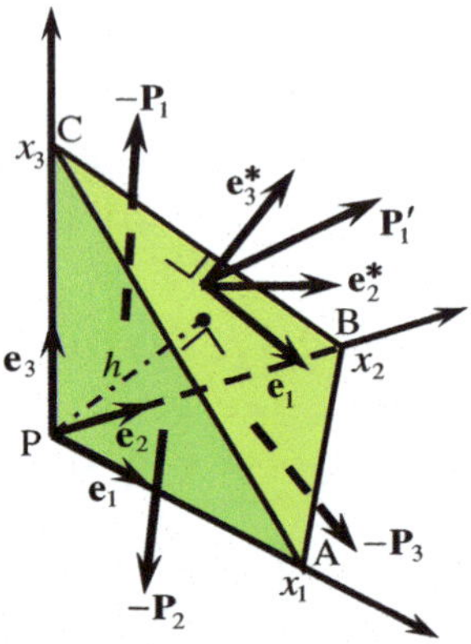

$dS_1^* = \Delta ABC$

$dS_1 = \Delta PBC,\ dS_2 = \Delta PCA,\ dS_3 = \Delta PAB$

Fig. 3.1 Stress vectors applied to tetrahedron

$$\mathbf{P}_1^* dS_1^* - \mathbf{P}_1 dS_1 - \mathbf{P}_2 dS_2 - \mathbf{P}_3 dS_3 + \rho \mathbf{b}\left(\frac{1}{3}hdS\right) = \rho \dot{\mathbf{v}}\left(\frac{1}{3}hdS\right) \tag{3.6}$$

For $h \to 0$, Eq. (3.6) is reduced to

$$\mathbf{P}_1^* dS_1^* = \mathbf{P}_1 dS_1 + \mathbf{P}_2 dS_2 + \mathbf{P}_3 dS_3 \tag{3.7}$$

Here, the area vector of the surface ΔABC perpendicular to the base vector $\mathbf{e}_1^*$ is given by

$$\begin{aligned} dS_1^* \mathbf{e}_1^* &= (\overrightarrow{\mathrm{PA}} - \overrightarrow{\mathrm{PB}}) \times (\overrightarrow{\mathrm{PB}} - \overrightarrow{\mathrm{PC}})/2 = (dx_1 \mathbf{e}_1 - dx_2 \mathbf{e}_2) \times (dx_2 \mathbf{e}_2 - dx_3 \mathbf{e}_3)/2 \\ &= (dx_2 dx_3 \mathbf{e}_1 + dx_2 dx_3 \mathbf{e}_1 + dx_2 dx_3 \mathbf{e}_1)/2. \end{aligned}$$

On the other hand, the area vector of the surface ΔPBC perpendicular to the base vector $\mathbf{e}_1$ is by

$$dS_1 \mathbf{e}_1 = \overrightarrow{\mathrm{PB}} \times \overrightarrow{\mathrm{PC}}/2 = dx_2 \mathbf{e}_2 \times dx_3 \mathbf{e}_3/2 = dx_2 dx_3 \mathbf{e}_1/2$$

Then, one has $dS_1^* \mathbf{e}_1^* \cdot \mathbf{e}_1 = dx_2 dx_3/2 = dS_1$. Further, one has $dS_1^* \mathbf{e}_1^* \cdot \mathbf{e}_2 = dS_2$ and $dS_1^* \mathbf{e}_1^* \cdot \mathbf{e}_3 = dS_3$ in the similar ways. Then, it follows that

$$dS_1^* \mathbf{e}_1^* \cdot \mathbf{e}_i = dS_i \to dS_i = dS_1^* Q_{1i} \tag{3.8}$$

The substitution of $dS_1 = dS_1^* Q_{11}, dS_2 = dS_1^* Q_{12}, dS_3 = dS_1^* Q_{13}$ obtained from Eq. (3.8) into Eq. (3.7) leads to

$$\mathbf{P}_1^* = \mathbf{P}_1 Q_{11} + \mathbf{P}_2 Q_{12} + \mathbf{P}_3 Q_{13} = \mathbf{P}_i Q_{1i} \tag{3.9}$$

The similar relations are obtained for stress vectors $\mathbf{P}_2^*$ and $\mathbf{P}_3^*$ applying to the surface perpendicular to the base vectors $\mathbf{e}_2^*$ and $\mathbf{e}_3^*$, respectively. Eventually, one has

$$\mathbf{P}_r^* = \mathbf{P}_i Q_{ri} \tag{3.10}$$

Designating the component of the stress vector $\mathbf{P}_i$ in the base vector $\mathbf{e}_j$ by σ_{ij} and considering the symmetry property of $\boldsymbol{\sigma}$ in Eq. (3.5), one has

$$\begin{aligned} \mathbf{P}_r^* &= \mathbf{P}_i Q_{ri} = \mathbf{P}_i(\mathbf{e}_r^* \cdot \mathbf{e}_i) = [(\mathbf{P}_i \cdot \mathbf{e}_j)\mathbf{e}_j](\mathbf{e}_r^* \cdot \mathbf{e}_i) = \sigma_{ij}\mathbf{e}_j(\mathbf{e}_i \cdot \mathbf{e}_r^*) \\ &= \sigma_{ij}\mathbf{e}_j \otimes \mathbf{e}_i \mathbf{e}_r^* = \sigma_{ij}\mathbf{e}_i \otimes \mathbf{e}_j \mathbf{e}_r^* \end{aligned}$$

i.e.

$$\mathbf{P}_r^* = \boldsymbol{\sigma}\mathbf{e}_r^*, \mathbf{P}_j^* \cdot \mathbf{e}_i^* = \boldsymbol{\sigma}\mathbf{e}_j^* \cdot \mathbf{e}_i^* = \sigma_{ij}^* \tag{3.11}$$

resulting in

$$\boldsymbol{\sigma} = \sigma_{ij}\mathbf{e}_i \otimes \mathbf{e}_j = \sigma^*_{ij}\mathbf{e}^*_i \otimes \mathbf{e}^*_j \tag{3.12}$$

Here, the second index of the component in the second-order tensor $\boldsymbol{\sigma}$ designates the direction of surface to which the stress vector applies and the first index does the applying direction. However, the indices can be exchanged to each other by the symmetry shown in Eq. (3.5). Eqs. (3.3) and (3.12) with the aid of Eq. $(1.90)_2$ yield the transformation rule of the components in the stress tensor $\boldsymbol{\sigma}$ itself.

$$\boldsymbol{\sigma}^* = \mathbf{Q}\boldsymbol{\sigma}\mathbf{Q}^T, \quad \sigma^*_{ij} = Q_{ir}Q_{is}\sigma_{rs}$$

Various stress tensors are defined from the Cauchy stress tensor described above. Some of them, which are often used in continuum mechanics, are presented below.

The tensor $\boldsymbol{\tau}$ defined by the following equation is called the *Kirchhoff stress tensor*.

$$\boldsymbol{\tau} = J\boldsymbol{\sigma} \tag{3.13}$$

The vector $\boldsymbol{t}_0$ defined by the following equation is called the *nominal stress vector*.

$$\boldsymbol{t}_0 \equiv \frac{d\mathbf{f}}{dA} \tag{3.14}$$

The stress tensor $\boldsymbol{\Pi}$, which is related to $\boldsymbol{t}_0$ by the following equation, is called the *first Piola-Kirchhoff stress tensor* which is the Eulerian-Lagrangian two-point tensor.

$$\boldsymbol{t}_0 \equiv \boldsymbol{\Pi}\mathbf{N} \ \ (t_{0i} \equiv \Pi_{iA}N_A) \tag{3.15}$$

Here, substituting Eqs. (2.147) and (3.15) into Eq. (3.14), we have

$$d\mathbf{f} \equiv \boldsymbol{\Pi}\mathbf{N}dA = \frac{1}{J}\boldsymbol{\Pi}\mathbf{F}^{\mathrm{T}}\mathbf{n}da, \quad df_i = \Pi_{iA}N_A dA = \frac{1}{J}\Pi_{iA}F_{rA}n_r da \tag{3.16}$$

On the other hand, the substitution of Eq. (3.2) into Eq. (3.1) yields

$$d\mathbf{f} = \boldsymbol{\sigma}^T\mathbf{n}da \tag{3.17}$$

It follows from Eqs. (3.16) and (3.17) that

$$d\mathbf{f} = \boldsymbol{\sigma}^T\mathbf{n}da = \boldsymbol{\Pi}\mathbf{N}dA \tag{3.18}$$

The relation of $\boldsymbol{\sigma}$ and $\boldsymbol{\Pi}$ is from Eqs. (2.147) and (3.18) as shown below.

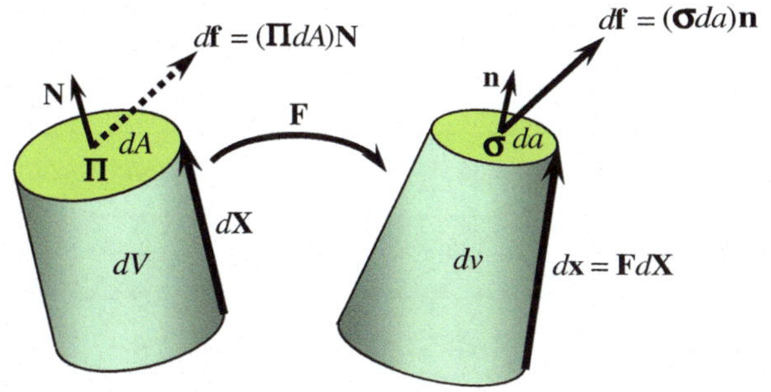

Fig. 3.2 Cauchy stress and first Piola-Kirchhoff stress

$$\left.\begin{array}{ll}\boldsymbol{\sigma}=\frac{1}{J}\mathbf{F}\boldsymbol{\Pi}^T(=\frac{1}{J}\boldsymbol{\Pi}\mathbf{F}^T), & \sigma_{ij}=\frac{1}{J}F_{iA}\Pi_{jA}=\frac{1}{J}\Pi_{iA}F_{jA}\\ \boxed{\boldsymbol{\Pi}=J\boldsymbol{\sigma}\mathbf{F}^{-T}(\neq\boldsymbol{\Pi}^T)}, & \Pi_{iA}=J\sigma_{ir}(\mathbf{F}^{-1})_{Ar}\end{array}\right\} \tag{3.19}$$

Equation (3.18) is illustrated in Fig. 3.2. The stress tensor $\mathbf{P}\equiv\boldsymbol{\Pi}^T=J\mathbf{F}^{-1}\boldsymbol{\sigma}$ is called the *nominal stress tensor*, noting $\mathbf{P}^T\mathbf{N}=d\mathbf{f}/dA$.

Further, the stress tensor $\mathbf{S}$ defined by the following equation is called the *second Piola-Kirchhoff stress tensor* which is the Lagrangian tensor.

$$\overleftarrow{\boldsymbol{t}}_0=\mathbf{S}\mathbf{N} \tag{3.20}$$

where

$$\overleftarrow{\boldsymbol{t}}_0\equiv\frac{\mathbf{F}^{-1}d\mathbf{f}}{dA}=\mathbf{F}^{-1}\boldsymbol{t}_0 \tag{3.21}$$

Using Eq. (2.147) into these equations, one has the following expression.

$$d\mathbf{f}\equiv\mathbf{F}\mathbf{S}\mathbf{N}dA=\frac{1}{J}\mathbf{F}\mathbf{S}\mathbf{F}^T\mathbf{n}da \tag{3.22}$$

Comparing Eqs. (3.17) and (3.22), it follows that

$$\left.\begin{array}{ll}\boldsymbol{\sigma}=\frac{1}{J}\mathbf{F}\mathbf{S}\mathbf{F}^T, & \sigma_{ij}=\frac{1}{J}F_{iA}S_{AB}F_{Bj}\\ \boxed{\mathbf{S}=J\mathbf{F}^{-1}\boldsymbol{\sigma}F^{-T}}(=\mathbf{S}^T), & S_{AB}=J(\mathbf{F}^{-1})_{Ai}\sigma_{ij}(\mathbf{F}^{-1})_{Bj}\end{array}\right\} \tag{3.23}$$

(Note) $\mathbf{F}^{-1}d\mathbf{f}=(\partial X_A/\partial x_i)\mathbf{e}_A\otimes\mathbf{e}_i df_j\mathbf{e}_j=(\partial X_A/\partial x_i)df_i\mathbf{e}_A$ is the pull-back, i.e. the transformation of $d\mathbf{f}$ from the current to the initial configurations. One can pull-back $d\mathbf{f}$ also by $\mathbf{F}^Td\mathbf{f}$ as will be described in Sect. 4.4 (2). However, the stress tensor $\boldsymbol{\Sigma}=\mathbf{F}^T\boldsymbol{\sigma}\mathbf{F}^{-T}/J$ obtained by setting $\mathbf{F}^Td\mathbf{f}/dA=\boldsymbol{\Sigma}\mathbf{N}$ instead of Eq. (3.20) does not satisfy the symmetry, i.e. $\boldsymbol{\Sigma}\neq\boldsymbol{\Sigma}^T$ and thus it is inconvenient for the practical use of deformation analysis.

Table 3.1 Relations of various stress tensors

Names, Notations	$\boldsymbol{\sigma}\,(=\boldsymbol{\sigma}^T)$	$\boldsymbol{\tau}\,(=\boldsymbol{\tau}^T)$	$\boldsymbol{\Pi}\,(\neq\boldsymbol{\Pi}^T)$	$\mathbf{S}\,(=\mathbf{S}^T)$	$_c\boldsymbol{\tau}\,(=\,_c\boldsymbol{\tau}^T)$
Cauchy $\boldsymbol{\sigma}$		$\frac{1}{J}\boldsymbol{\tau}$	$\frac{1}{J}\mathbf{F}\boldsymbol{\Pi}^T$	$\frac{1}{J}\mathbf{FSF}^T$	$\frac{1}{J}\mathbf{F}^{-T}\,_c\boldsymbol{\tau}\mathbf{F}^{-1}$
Kirchhoff $\boldsymbol{\tau}$	$J\boldsymbol{\sigma}$		$\mathbf{F}\boldsymbol{\Pi}^T$	$\mathbf{FSF}^T$	$\mathbf{F}^{-T}\,_c\boldsymbol{\tau}\mathbf{F}^{-1}$
1st Piola-Kirchhoff $\boldsymbol{\Pi}$ ($\mathbf{P}=\boldsymbol{\Pi}^T$: Nominal)	$J\boldsymbol{\sigma}\mathbf{F}^{-T}$	$\boldsymbol{\tau}\mathbf{F}^{-T}$		$\mathbf{FS}$	$\mathbf{F}^{-T}\,_c\boldsymbol{\tau}\mathbf{C}^{-1}$
2nd Piola-Kirchhoff $\mathbf{S}$	$J\mathbf{F}^{-1}\boldsymbol{\sigma}\mathbf{F}^{-T}$	$\mathbf{F}^{-1}\boldsymbol{\tau}\mathbf{F}^{-T}$	$\mathbf{F}^{-1}\boldsymbol{\Pi}$		$\mathbf{C}^{-1}\,_c\boldsymbol{\tau}\mathbf{C}^{-1}$
Covariant convected $_c\boldsymbol{\tau}$	$J\mathbf{F}^T\boldsymbol{\sigma}\mathbf{F}$	$\mathbf{F}^T\boldsymbol{\tau}\mathbf{F}$	$\mathbf{C}\boldsymbol{\Pi}^T\mathbf{F}$	$\mathbf{CSC}$	

Note

$$\boldsymbol{t}\equiv\frac{d\mathbf{f}}{da},\ \boldsymbol{t}_0\equiv\frac{d\mathbf{f}}{dA},\ \overleftarrow{\boldsymbol{t}}_0\equiv\frac{\mathbf{F}^{-1}d\mathbf{f}}{dA}=\mathbf{F}^{-1}\boldsymbol{t}_0$$

$$\boldsymbol{t}=\boldsymbol{\sigma}\mathbf{n},\ J\boldsymbol{t}=\boldsymbol{\tau}\mathbf{n},\ \boldsymbol{t}_0=\boldsymbol{\Pi}\mathbf{N},\ \overleftarrow{\boldsymbol{t}}_0=\mathbf{SN}$$

The relations between the above-mentioned stress tensors as the transformations between the Lagrangian and the Eulerian tensors will be described in Sect. 4.4

The following stress is called the *covariant convected stress tensor*.

$$\boxed{_c\boldsymbol{\sigma}\equiv\mathbf{F}^T\boldsymbol{\sigma}F}(=\,_c\boldsymbol{\sigma}^T),\quad _c\boldsymbol{\tau}\equiv\mathbf{F}^T\boldsymbol{\tau}F(=\,_c\boldsymbol{\tau}^T) \tag{3.24}$$

The relations of various stress tensors defined above are summarized in Table 3.1.

3.2 Conservation Law of Mass

Denoting the field of material density as $\rho(\mathbf{x},t)$, the mass in a current volume v is given as $m=\int_v\rho(\mathbf{x},\,t)dv$ which is kept constant because the mass neither flow into the volume element nor flow out from it. Therefore, the following *conservation law of mass* must hold.

$$\dot{m}=\left(\int_v\rho(\mathbf{x},t)dv\right)^{\bullet}=0 \tag{3.25}$$

from which, noting Eq. (2.162), one has the *continuity equation*.

$$\boxed{\dot{\rho}+\rho\mathrm{div}\mathbf{v}=0},\quad \dot{\rho}+\rho\frac{\partial v_r}{\partial x_r}=0 \tag{3.26}$$

Further, setting $T(\mathbf{x}, t) \equiv \rho\phi$, where ϕ is a physical quantity per unit mass, one has

$$\left(\int_v \rho\phi dv\right)^{\bullet} = \int_v \left(\rho\dot{\phi} + \dot{\rho}\phi + \rho\phi\frac{\partial v_r}{\partial x_r}\right)dv = \int_v \rho\dot{\phi}dv \tag{3.27}$$

noting the Reynolds' transportation theorem in Eqs. (2.160) and (3.26).

3.3 Conservation Law of Linear Momentum

The *linear momentum* in a current volume v is given by $\int_v \rho\mathbf{v}dv$. On the other hand, the traction applied to the surface of the region is given as $\int_a \boldsymbol{t}\,da$ and the body force applied to the region is given by $\int_v \rho\mathbf{b}dv$, denoting the *body force* per unit mass as $\mathbf{b}$. The rate of momentum has to be equivalent to the sum of the traction and the body force applied to the region. Therefore, *Euler's first law of motion* (or *conservation law of momentum*) is given as

$$\left(\int_v \rho\mathbf{v}dv\right)^{\bullet} = \int_a \boldsymbol{t}\,da + \int_v \rho\mathbf{b}dv$$

or

$$\boxed{\int_v \rho\,\dot{\mathbf{v}}\,dv = \int_a \mathbf{t}\,da + \int_v \rho\mathbf{b}dv} \tag{3.28}$$

by virtue of Eq. (3.27).

3.4 Conservation Law of Angular Momentum

The *angular momentum* in a current volume v in a current state is given as $\int_v \rho(\mathbf{x} \times \mathbf{v})dv$. On the other hand, since the angular momentum caused by the traction and the angular momentum caused by the body force are described by and $\int_v \rho(\mathbf{x} \times \mathbf{b})dv$, respectively, *Euler's second law of motion*, i.e. *conservation law of angular momentum* is described as

$$\left(\int_v \rho\mathbf{x} \times \mathbf{v}dv\right)^{\bullet} = \int_a \mathbf{x} \times \boldsymbol{t}da + \int_v \rho\mathbf{x} \times \mathbf{b}dv$$
$$\left(\int_v \rho\varepsilon_{ijk}x_j v_k dv\right)^{\bullet} = \int_a \varepsilon_{ijk}x_j t_k da + \int_v \rho\varepsilon_{ijk}x_j b_k dv$$

which is reduced to

$$\boxed{\int_v \rho\mathbf{x}\times\dot{\mathbf{v}}\,dv=\int_a \mathbf{x}\times\boldsymbol{t}\,da+\int_v \rho\mathbf{x}\times\mathbf{b}dv}$$

$$\int_v \rho\varepsilon_{ijk}x_j\dot{v}_k\,dv=\int_a \varepsilon_{ijk}x_jt_kda+\int_v \rho\varepsilon_{ijk}x_jb_kdv \tag{3.29}$$

noting $(\mathbf{x}\times\mathbf{v})^{\bullet}=\mathbf{v}\times\mathbf{v}+\mathbf{x}\times\dot{\mathbf{v}}=\mathbf{x}\times\dot{\mathbf{v}}$ and Eq. (3.27).

3.5 Equilibrium Equation

Substituting Eq. (3.2) into Eq. (3.28) for the conservation law of momentum and noting Eq. (3.5), the following equation is obtained.

$$\int_v \rho\dot{\mathbf{v}}dv=\int_a \boldsymbol{\sigma}^T\mathbf{n}da+\int_v \rho\mathbf{b}dv,\quad \int_v \rho\dot{v}_idv=\int_a \sigma_{ir}n_rda+\int_v \rho b_idv \tag{3.30}$$

The first term in the right-hand side of Eq. (3.30) is given by Eq. (1.327) of Gauss' divergence theorem as

$$\int_a \sigma_{ir}n_rda=\int_v \frac{\partial\sigma_{ir}}{\partial x_r}dv$$

By this equation the *local form* of Eq. (3.30) is given as

$$\boxed{\nabla_x\boldsymbol{\sigma}+\rho\mathbf{b}=\rho\dot{\mathbf{v}}},\quad \frac{\partial\sigma_{ij}}{\partial x_j}+\rho b_i=\rho\dot{v}_i \tag{3.31}$$

This equation is called the *Cauchy's first law of motion*, i.e. the *equilibrium equation.*

On the other hand, substituting Eqs. (2.145) and (3.18) into Eq. (3.30), one has

$$\int_V \rho_0\dot{\mathbf{v}}dV=\int_A \boldsymbol{\Pi}\mathbf{N}dA+\int_V \rho_0\mathbf{b}dV \tag{3.32}$$

which is rewritten by Gauss' divergence theorem as follows:

$$\int_V \rho_0\dot{\mathbf{v}}dV=\int_V \boldsymbol{\Pi}\nabla_XdV+\int_V \rho_0\mathbf{b}dV \tag{3.33}$$

where $\nabla_X\equiv(\partial/\partial X_A)\mathbf{e}_A=\partial/\partial\mathbf{X}$. The local form of this equation is given as

$$\boxed{\nabla_X\boldsymbol{\Pi}+\rho_0\mathbf{b}=\rho_0\dot{\mathbf{v}}},\quad \frac{\partial\Pi_{iA}}{\partial X_A}+\rho_0b_i=\rho_0\dot{v}_i \tag{3.34}$$

The equilibrium equation in a rate form is required in constitutive equations for irreversible deformation including elastoplastic deformation. The time-differentiation of Eq. (3.34) engenders the following *rate-type* (or *incremental-type*) *equilibrium equation*, provided that the acceleration does not change, i.e. $\ddot{\mathbf{v}} = \mathbf{0}$.

$$\boxed{\dot{\mathbf{\Pi}}\nabla_X + \rho_0\dot{\mathbf{b}} = \mathbf{0}}, \quad \frac{\partial\dot{\Pi}_{iA}}{\partial X_A} + \rho_0\dot{b}_i = 0 \tag{3.35}$$

In order to describe Eq. (3.35) by the Cauchy stress, specifying

$$\boxed{{}_{\Pi}\overset{\Delta}{\boldsymbol{\sigma}} \equiv \frac{1}{J}\dot{\mathbf{\Pi}}\mathbf{F}^T} \quad (\neq {}_{\Pi}\overset{\Delta}{\boldsymbol{\sigma}}^T = \frac{1}{J}\mathbf{F}\dot{\mathbf{\Pi}}^T) \tag{3.36}$$

and noting Eq. $(3.19)_2$, we have

$${}_{\Pi}\overset{\Delta}{\boldsymbol{\sigma}} = \frac{1}{J}\left(\dot{J}\boldsymbol{\sigma}F^{-T} + J\dot{\boldsymbol{\sigma}}\mathbf{F}^{-T} + J\boldsymbol{\sigma}\dot{\mathbf{F}}^{-T}\right)\mathbf{F}^T = \frac{\dot{J}}{J}\boldsymbol{\sigma} + \dot{\boldsymbol{\sigma}} + \boldsymbol{\sigma}\dot{\mathbf{F}}^{-T}\mathbf{F}^T$$

By substituting $\dot{\mathbf{I}} = (\mathbf{F}^{-T}\mathbf{F}^T)^{\bullet} = \dot{\mathbf{F}}^{-T}\mathbf{F}^T + \mathbf{F}^{-T}\dot{\mathbf{F}}^T = \dot{\mathbf{F}}^{-T}\mathbf{F}^T + \boldsymbol{l}^T = \mathbf{O}$ due to Eq. (2.75) and Eqs. (2.79) and (2.150) to this equation, we obtains

$${}_{\Pi}\overset{\Delta}{\boldsymbol{\sigma}} = \dot{\boldsymbol{\sigma}} + \boldsymbol{\sigma}\mathrm{tr}\boldsymbol{l} - \boldsymbol{\sigma}\boldsymbol{l}^T = \overset{\circ}{\boldsymbol{\sigma}}{}^{w} + \boldsymbol{\sigma}\mathrm{tr}\mathbf{d} - \boldsymbol{\sigma}\mathbf{d} + \mathbf{w}\boldsymbol{\sigma} \tag{3.37}$$

Therein, ${}_{\Pi}\overset{\Delta}{\boldsymbol{\sigma}}$ is referred to as the *nominal stress rate*, whereas $\overset{\circ}{\boldsymbol{\sigma}}{}^{w} \equiv \dot{\boldsymbol{\sigma}} - \mathbf{w}\boldsymbol{\sigma} + \boldsymbol{\sigma}\boldsymbol{w}$ is the *Zaremba-Jaumann stress rate* which will be defined in the next chapter.

It follows that

$$\frac{\partial\,{}_{\Pi}\overset{\Delta}{\boldsymbol{\sigma}}}{\partial\mathbf{x}} = \frac{1}{J}\frac{\partial\dot{\mathbf{\Pi}}}{\partial\mathbf{X}} \tag{3.38}$$

by virtue of

$$\frac{\partial\,{}_{\Pi}\overset{\Delta}{\boldsymbol{\sigma}}}{\partial\mathbf{x}} = \frac{\partial(\dot{\mathbf{\Pi}}\mathbf{F}^T/J)}{\partial\mathbf{x}} = \frac{1}{J}\frac{\partial\dot{\mathbf{\Pi}}}{\partial\mathbf{x}}\mathbf{F}^T + \dot{\mathbf{\Pi}}\frac{\partial(\mathbf{F}^T/J)}{\partial\mathbf{x}} = \frac{1}{J}\frac{\partial\dot{\mathbf{\Pi}}}{\partial\mathbf{X}} \tag{3.39}$$

$$\left(\frac{\partial\,{}_{\Pi}\overset{\Delta}{\sigma}_{ij}}{\partial x_j} = \frac{1}{J}\frac{\partial\dot{\Pi}_{iA}}{\partial x_j}F_{jA} + \dot{\Pi}_{iA}\frac{\partial(F_{jA}/J)}{\partial x_j} = \frac{1}{J}\frac{\partial\dot{\Pi}_{iA}}{\partial x_j}\frac{\partial x_j}{\partial X_A} = \frac{1}{J}\frac{\partial\dot{\Pi}_{iA}}{\partial X_A}\right)$$

with

$$\frac{\partial(\mathbf{F}^T/J)}{\partial\mathbf{x}} = \mathbf{O} \tag{3.40}$$

where $\mathbf{O}$ is the third-order zero tensor, noting

$$\frac{\partial(\mathbf{F}^T/J)}{\partial\mathbf{x}} = \frac{1}{J^2}\left(\frac{\partial\mathbf{F}^T}{\partial\mathbf{x}}J - \mathbf{F}^T\frac{\partial J}{\partial\mathbf{x}}\right) = \frac{1}{J^2}\left[\frac{\partial\mathbf{F}^T}{\partial\mathbf{x}}J - \mathbf{F}^T\frac{\partial \det\mathbf{F}}{\partial\mathbf{F}}\left(\frac{\partial\mathbf{F}}{\partial\mathbf{x}}\right)^T\right]$$
$$= \frac{1}{J^2}\left[\frac{\partial\mathbf{F}^T}{\partial\mathbf{x}}J - \mathbf{F}^T J\mathbf{F}^{-T}\left(\frac{\partial\mathbf{F}}{\partial\mathbf{x}}\right)^T\right] = \mathbf{O}$$

Substitution of Eq. (3.38) into Eq. (3.35) yields the rate-type equilibrium defined in the current configuration:

$$\boxed{{}_{\Pi}\overset{\Delta}{\boldsymbol{\sigma}}\nabla_x + \rho\dot{\mathbf{b}} = \mathbf{0}}, \quad \frac{\partial_{\Pi}\overset{\Delta}{\sigma}_{ij}}{\partial x_j} + \rho\dot{b}_i = 0 \tag{3.41}$$

This equation is derived also by the following manner. From Eq. (3.30) with $\ddot{\mathbf{v}} = \mathbf{0}$, one has

$$\left(\int_v \rho\dot{\mathbf{v}}dv\right)^{\bullet} = \left(\int_a \boldsymbol{\sigma}^T\mathbf{n}da\right)^{\bullet} + \left(\int_v \rho\mathbf{b}dv\right)^{\bullet}$$

$$\underbrace{\left(\int_V \rho_0\,\dot{\mathbf{v}}\,dV\right)^{\bullet}}_{0} = \left(\int_a \boldsymbol{\sigma}^T\mathbf{n}da\right)^{\bullet} + \int_V(\rho_0\mathbf{b}dV)^{\bullet}$$

$$\begin{aligned}\mathbf{0} &= \int_a \dot{\boldsymbol{\sigma}}^T\mathbf{n}da + \int_a \boldsymbol{\sigma}^T(\mathbf{n}da)^{\bullet} + \int_V \rho_0\,\dot{\mathbf{b}}\,dV \\ &= \int_a \dot{\boldsymbol{\sigma}}^T\mathbf{n}da + \int_a \boldsymbol{\sigma}^T[(\mathrm{tr}\mathbf{d})\mathbf{I} - \boldsymbol{l}^T]\,\mathbf{n}da + \int_v \rho\dot{\mathbf{b}}dv \\ &= \int_v (\dot{\boldsymbol{\sigma}}^T + \boldsymbol{\sigma}^T\mathrm{tr}\mathbf{d} - \boldsymbol{\sigma}^T\boldsymbol{l}^T)\nabla_x dv + \int_v \rho\dot{\mathbf{b}}dv\end{aligned}$$

which results in Eq. (3.41), noting Eqs. (1.327), (2.152) and $\boldsymbol{\sigma} = \boldsymbol{\sigma}^T$ which will be verified in the next section.

3.6 Equilibrium Equation of Angular Moment

Substituting Eq. (3.2) into Eq. (3.29) of the conservation law of angular momentum and noting Eq. (3.5), one has

$$\int_v \rho\varepsilon_{ijk}x_j\dot{v}_k dv = \int_a \varepsilon_{ijk}x_j\sigma_{kr}n_r da + \int_v \rho\varepsilon_{ijk}x_j b_k dv \tag{3.42}$$

Because the first term in the right-hand side of this equation is rewritten as

$$\int_a \varepsilon_{ijk} x_j \sigma_{kr} n_r da = \int_a \varepsilon_{ijk} \frac{\partial x_j \sigma_{kr}}{\partial x_r} dv = \int_v \left(\varepsilon_{ijk} \sigma_{kj} + \varepsilon_{ijk} x_j \frac{\partial \sigma_{kr}}{\partial x_r} \right) dv$$

Equation (3.42) leads to

$$\int_v \left\{ \varepsilon_{ijk} \sigma_{kj} + \varepsilon_{ijk} x_j \left(\frac{\partial \sigma_{kr}}{\partial x_r} + \rho b_k - \rho \dot{v}_k \right) \right\} dv = 0$$

Noting the equilibrium Eq. (3.31) to this equation, it holds that $\varepsilon_{ijk}\sigma_{kj} = 0$ (e.g. $\varepsilon_{123}\sigma_{23} + \varepsilon_{132}\sigma_{32} = \sigma_{23} - \sigma_{32} = 0$) from which we have the *symmetry of Cauchy stress tensor*, i.e. $\boldsymbol{\sigma} = \boldsymbol{\sigma}^T, \sigma_{ij} = \sigma_{ji}$.

3.7 Virtual Work Principle

The stress (rate) field fulfilling the equilibrium equation and the boundary condition of stress is called the *statically admissible filed*. On the other hand, the displacement (velocity) field fulfilling the geometrical requirement $F_{iA}(= \partial x_i/\partial X_A) = \delta_{iA} + \partial u_i/\partial X_A$ or $d_{ij} = (\partial v_i/\partial x_j + \partial v_j/\partial x_i)/2$ and the boundary condition of displacement (velocity) is called the *kinematically-admissible field*. Denoting arbitrary statically admissible stress field and kinematically-admissible velocity field by $()^{\Delta}$ and $()^{\nabla}$, one has the following equation from Eq. (3.31).

$$\int_v (\boldsymbol{\sigma}^{\Delta} \nabla_x + \rho \mathbf{b} - \rho \dot{\mathbf{v}}) \cdot \mathbf{u}^{\nabla} dv = 0, \quad \int_v \left(\frac{\partial \sigma_{ij}^{\Delta}}{\partial x_j} + \rho b_i - \rho \dot{v}_i \right) u_i^{\nabla} dv = 0 \tag{3.43}$$

Using the Eq. (1.327) of Gauss' divergence theorem, we have

$$\begin{aligned}
\int_v \sigma_{ij}^{\Delta} \frac{\partial u_i^{\nabla}}{\partial x_j} dv &= \int_v \frac{\partial (\sigma_{ij}^{\Delta} u_i^{\nabla})}{\partial x_j} dv - \int_v \frac{\partial \sigma_{ij}^{\Delta}}{\partial x_j} u_i^{\nabla} dv \\
&= \int_{a_t} \bar{\sigma}_{ij} u_i^{\nabla} n_j da + \int_{a_v} \sigma_{ij}^{\nabla} \bar{u}_i n_j da - \int_v \frac{\partial \sigma_{ij}^{\Delta}}{\partial x_i} u_i^{\nabla} dv \\
&= \int_{a_t} \bar{t}_i u_i^{\nabla} da + \int_{a_v} \sigma_{ij} n_j \bar{u}_i da - \int_v \frac{\partial \sigma_{ij}^{\Delta}}{\partial x_j} u_i^{\nabla} dv
\end{aligned}$$

where $(\bar{\ })$ designates the given boundary condition, and a_t and a_v specify the surfaces of the body on which the traction (rate) and the displacement (velocity) are given, respectively. Substituting Eq. (3.43) into this equation, the following *virtual work principle* described by the quantities in the current state is obtained.

$$\int_v \sigma_{ij}^{\triangle} \frac{\partial u_i^{\nabla}}{\partial x_j} dv = \int_{a_t} \bar{t}_i u_i^{\nabla} da + \int_{a_v} \sigma_{ij}^{\triangle} n_j \bar{u}_i da + \int_v \rho b_i u_i^{\nabla} dv - \int_v \rho \dot{v}_i u_i^{\nabla} dv \quad (3.44)$$

Similarly, the following virtual work principle can be described by the quantities in the initial state from Eq. (3.34).

$$\int_V \Pi_{iJ}^{\triangle} \frac{\partial u_i^{\nabla}}{\partial X_J} dV = \int_{A_t} \bar{t}_{0i} u_i^{\nabla} dA + \int_{A_v} \Pi_{iJ}^{\triangle} N_J \bar{u}_i dA + \int_V \rho_0 b_i u_i^{\nabla} dV - \int_V \rho_0 \dot{v}_i u_i^{\nabla} dV \quad (3.45)$$

Furthermore, one has the rate-type virtual work principle from Eqs. (3.35) and (3.41) as follows:

$$\int_v {}_{\Pi}\overset{\triangle}{\sigma}_{ij}^{\triangle} d_{ij}^{\nabla} dv = \int_{a_t} \dot{\bar{t}}_i v_i^{\nabla} da + \int_{a_v} {}_{\Pi}\overset{\circ}{\sigma}_{ij}^{\nabla} n_j \bar{v}_i da + \int_v \rho \dot{b}_i v_i^{\nabla} dv \quad (3.46)$$

$$\int_V \dot{\Pi}_{iJ}^{\triangle} \dot{F}_{iJ}^{\nabla} dV = \int_{A_t} \dot{\bar{t}}_{0i} v_i^{\nabla} dA + \int_{A_v} \dot{\Pi}_{iJ}^{\triangle} N_J \bar{v}_i dA + \int_V \rho_0 \dot{b}_i v_i^{\nabla} dV \quad (3.47)$$

3.8 Various Simple Deformations

Let various strain (rate) and stress (rate) described in the foregoing be shown explicitly and let their relation be described for various simple deformations. These deformations are often observed in experiments for measurement of material properties. Homogeneous and isotropic deformation is assumed therein.

3.8.1 Uniaxial Loading

For a cylindrical specimen with the initial length L and the initial radius R, suppose that the length and the radius changes to l and r (Khan and Huang 1995). Choosing the X_1-axis to the axial direction of cylinder, it holds that

$$x_1 = \lambda_1 X_1, \quad x_2 = \lambda_2 X_2, \quad x_3 = \lambda_3 X_3 \quad (3.48)$$

where

$$\lambda_1 = \frac{l}{L}, \quad \lambda_2 = \lambda_3 = \frac{r}{R} \quad (3.49)$$

from which one has

$$\left.\begin{array}{l}\mathbf{F}=\mathbf{U}=\mathbf{V}=\begin{bmatrix}\lambda_1 & 0 & 0\\ 0 & \lambda_2 & 0\\ 0 & 0 & \lambda_2\end{bmatrix},\quad \mathbf{F}^{-1}=\begin{bmatrix}\lambda_1^{-1} & 0 & 0\\ 0 & \lambda_2^{-1} & 0\\ 0 & 0 & \lambda_2^{-1}\end{bmatrix},\quad J=\det\mathbf{F}=\lambda_1\lambda_2^2\\ \mathbf{R}=\mathbf{I}\end{array}\right\} \tag{3.50}$$

The aforementioned measures of deformation (rate) are given as

$$\mathbf{C}=\mathbf{U}^2,\mathbf{F}^T\mathbf{F}=\begin{bmatrix}\lambda_1^2 & 0 & 0\\ 0 & \lambda_2^2 & 0\\ 0 & 0 & \lambda_2^2\end{bmatrix},\quad \mathbf{b}^{-1}=\mathbf{V}^{-2}=\mathbf{F}^{-T}\mathbf{F}^{-1}=\begin{bmatrix}\lambda_1^{-2} & 0 & 0\\ 0 & \lambda_2^{-2} & 0\\ 0 & 0 & \lambda_2^{-2}\end{bmatrix} \tag{3.51}$$

$$\boldsymbol{E}=\frac{1}{2}(\mathbf{C}-\mathbf{I})=\frac{1}{2}\begin{bmatrix}\lambda_1^2-1 & 0 & 0\\ 0 & \lambda_2^2-1 & 0\\ 0 & 0 & \lambda_2^2-1\end{bmatrix}$$

$$\boldsymbol{e}=\frac{1}{2}(\mathbf{I}-\mathbf{b}^{-1})=\frac{1}{2}\begin{bmatrix}1-\lambda_1^{-2} & 0 & 0\\ 0 & 1-\lambda_2^{-2} & 0\\ 0 & 0 & 1-\lambda_2^{-2}\end{bmatrix} \tag{3.52}$$

$$\boldsymbol{E}^{(0)}=\boldsymbol{e}^{(0)}=\begin{bmatrix}\ln\lambda_1 & 0 & 0\\ 0 & \ln\lambda_2 & 0\\ 0 & 0 & \ln\lambda_2\end{bmatrix}=\begin{bmatrix}\ln(l/L) & 0 & 0\\ 0 & \ln(r/R) & 0\\ 0 & 0 & \ln(r/R)\end{bmatrix} \tag{3.53}$$

$$\begin{aligned}\boldsymbol{l}=\dot{\mathbf{F}}\mathbf{F}^{-1}&=\begin{bmatrix}\dot\lambda_1 & 0 & 0\\ 0 & \dot\lambda_2 & 0\\ 0 & 0 & \dot\lambda_2\end{bmatrix}\begin{bmatrix}\lambda_1^{-1} & 0 & 0\\ 0 & \lambda_2^{-1} & 0\\ 0 & 0 & \lambda_2^{-1}\end{bmatrix}=\begin{bmatrix}\dot\lambda_1\lambda_1^{-1} & 0 & 0\\ 0 & \dot\lambda_2\lambda_2^{-1} & 0\\ 0 & 0 & \dot\lambda_2\lambda_2^{-1}\end{bmatrix}\\ &=\begin{bmatrix}\dot l l^{-1} & 0 & 0\\ 0 & \dot r r^{-1} & 0\\ 0 & 0 & \dot r r^{-1}\end{bmatrix}=\mathbf{d}=\dot{\boldsymbol{E}}^{(0)}=\dot{\boldsymbol{e}}^{(0)}\end{aligned} \tag{3.54}$$

$$\mathbf{w}=\mathbf{O} \tag{3.55}$$

The infinitesimal strain in Eq. (2.49) is given by

$$\boldsymbol{\varepsilon} = \begin{bmatrix} (l-L)/L & 0 & 0 \\ 0 & (r-R)/R & 0 \\ 0 & 0 & (r-R)/R \end{bmatrix} = \begin{bmatrix} \lambda_1 - 1 & 0 & 0 \\ 0 & \lambda_2 - 1 & 0 \\ 0 & 0 & \lambda_2 - 1 \end{bmatrix} = \mathbf{U} - \mathbf{I} \tag{3.56}$$

$$\dot{\boldsymbol{\varepsilon}} = \begin{bmatrix} \dot{l}/L & 0 & 0 \\ 0 & \dot{r}/R & 0 \\ 0 & 0 & \dot{r}/R \end{bmatrix} \tag{3.57}$$

Denoting the axial load as F, various stresses are shown as follows:

$$\boldsymbol{\sigma} = \begin{bmatrix} \dfrac{F}{\pi r^2} & 0 & 0 \\ 0 & 0 & 0 \\ 0 & 0 & 0 \end{bmatrix} = \begin{bmatrix} \dfrac{F}{\lambda_2^2 \pi R^2} & 0 & 0 \\ 0 & 0 & 0 \\ 0 & 0 & 0 \end{bmatrix} = \begin{bmatrix} \dfrac{F}{\lambda_2^2 A_0} & 0 & 0 \\ 0 & 0 & 0 \\ 0 & 0 & 0 \end{bmatrix} \tag{3.58}$$

$$\boldsymbol{\tau} = J\boldsymbol{\sigma} = \lambda_1 \lambda_2^2 \begin{bmatrix} \dfrac{F}{\lambda_2^2 A_0} & 0 & 0 \\ 0 & 0 & 0 \\ 0 & 0 & 0 \end{bmatrix} = \begin{bmatrix} \dfrac{F}{A_0}\lambda_1 & 0 & 0 \\ 0 & 0 & 0 \\ 0 & 0 & 0 \end{bmatrix} \tag{3.59}$$

$$\boldsymbol{\Pi} = J\mathbf{F}^{-1}\boldsymbol{\sigma} = \lambda_1 \lambda_2^2 \begin{bmatrix} \lambda_1^{-1} & 0 & 0 \\ 0 & \lambda_2^{-1} & 0 \\ 0 & 0 & \lambda_2^{-1} \end{bmatrix} \begin{bmatrix} F/(\lambda_2^2 A_0) & 0 & 0 \\ 0 & 0 & 0 \\ 0 & 0 & 0 \end{bmatrix} = \begin{bmatrix} F/A_0 & 0 & 0 \\ 0 & 0 & 0 \\ 0 & 0 & 0 \end{bmatrix} \tag{3.60}$$

$$\mathbf{S} = J\mathbf{F}^{-1}\boldsymbol{\sigma}F^{-T} = \begin{bmatrix} \dfrac{F}{\lambda_1 A_0} & 0 & 0 \\ 0 & 0 & 0 \\ 0 & 0 & 0 \end{bmatrix} \tag{3.61}$$

3.8.2 Simple Shear

Consider the simple shear in which the shear deformation is induced in parallel to the x_1-axis as shown in Fig. 3.3.

$$x_1 = X_1 + \gamma X_2, \quad x_2 = X_2, \quad x_3 = X_3 \tag{3.62}$$

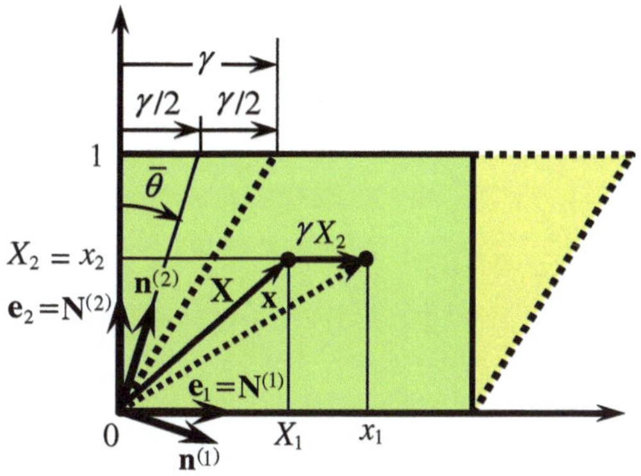

Fig. 3.3 Simple shear

where $\gamma(=2D_{12})$ is the engineering shear strain. Denoting the shear angle by $\bar{\theta}$, it holds that

$$\gamma = 2\tan\bar{\theta}, \quad \dot{\gamma} = 2\dot{\bar{\theta}}\sec^2\bar{\theta} \tag{3.63}$$

It holds in this situation that

$$\mathbf{F} = \begin{bmatrix} 1 & \gamma & 0 \\ 0 & 1 & 0 \\ 0 & 0 & 1 \end{bmatrix}, \quad \mathbf{F}^{-1} = \begin{bmatrix} 1 & -\gamma & 0 \\ 0 & 1 & 0 \\ 0 & 0 & 1 \end{bmatrix}, \quad J = \det\mathbf{F} = 1 \tag{3.64}$$

where the inverse tensor $\mathbf{F}^{-1}$ is derived using Eq. (1.120). The components in the third line and those in the third row are zero except for unity in the third line and the third row in all tensors appearing hereinafter for the simple shear deformation. Then, for simplicity, let them be expressed by the matrix with two lines and two rows.

$$\mathbf{F} = \begin{bmatrix} 1 & \gamma \\ 0 & 1 \end{bmatrix}, \quad \mathbf{F}^{-1} = \begin{bmatrix} 1 & -\gamma \\ 0 & 1 \end{bmatrix} \tag{3.65}$$

from which it is obtained that

$$\boldsymbol{l} = \dot{\mathbf{F}}\mathbf{F}^{-1} = \begin{bmatrix} 0 & \dot{\gamma} \\ 0 & 0 \end{bmatrix}\begin{bmatrix} 0 & -\gamma \\ 0 & 1 \end{bmatrix} = \begin{bmatrix} 0 & \dot{\gamma} \\ 0 & 0 \end{bmatrix} \tag{3.66}$$

$$\mathbf{d} = \begin{bmatrix} 0 & 1 \\ 1 & 0 \end{bmatrix}\frac{\dot{\gamma}}{2}, \quad \mathbf{w} = \begin{bmatrix} 0 & 1 \\ -1 & 0 \end{bmatrix}\frac{\dot{\gamma}}{2} \tag{3.67}$$

Further, it holds that

$$\mathbf{C}(=\mathbf{U}^2=\mathbf{F}^T\mathbf{F})=\begin{bmatrix}1 & \gamma\\ \gamma & 1+\gamma^2\end{bmatrix},\quad \mathbf{C}^{-1}(=\mathbf{U}^{-2}=\mathbf{F}^{-1}\mathbf{F}^T)=\begin{bmatrix}1+\gamma^2 & -\gamma\\ -\gamma & 1\end{bmatrix} \tag{3.68}$$

$$\mathbf{b}(=\mathbf{V}^2=\mathbf{F}\mathbf{F}^T)=\begin{bmatrix}1+\gamma^2 & \gamma\\ \gamma & 1\end{bmatrix},\quad \mathbf{b}^{-1}(=\mathbf{V}^{-2}=\mathbf{F}^{-T}\mathbf{F}^{-1})=\begin{bmatrix}1 & -\gamma\\ -\gamma & 1+\gamma^2\end{bmatrix} \tag{3.69}$$

$$\boldsymbol{E}=\frac{1}{2}(\mathbf{C}-\mathbf{I})=\frac{1}{2}\begin{bmatrix}1 & \gamma\\ \gamma & \gamma^2\end{bmatrix},\quad \boldsymbol{e}=\frac{1}{2}(\mathbf{I}-\mathbf{b}^{-1})=\frac{1}{2}\begin{bmatrix}0 & -\gamma\\ -\gamma & \gamma^2\end{bmatrix} \tag{3.70}$$

Next, derive the principal stretches λ_α and the eigenvectors $\mathbf{n}^{(\alpha)}$ ($\alpha = 1, 2$ in the present two dimensional state) of $\mathbf{V}$ in Eq. (2.24), i.e.,

$$\mathbf{V}\mathbf{n}^{(\alpha)}=\lambda_\alpha\mathbf{n}^{(\alpha)} \tag{3.71}$$

The principal stretches λ_α must fulfill the following characteristic equation based on Eq. (1.176).

$$\begin{aligned}\begin{vmatrix}V_{11} & -\lambda_\alpha & V_{12}\\ V_{21} & V_{22} & -\lambda_\alpha\end{vmatrix} &= (V_{11}-\lambda_\alpha)(V_{22}-\lambda_\alpha)-V_{12}V_{21}\\ &= \lambda_\alpha^2-(V_{11}+V_{22})\lambda_\alpha+V_{11}V_{22}-V_{12}V_{21}\\ &= \lambda_\alpha^2-(\mathrm{tr}\mathbf{V})\lambda_\alpha+\det\mathbf{V}=0\end{aligned} \tag{3.72}$$

where, it holds from Eq. (3.69) that

$$\det\mathbf{V}=\sqrt{\det\mathbf{V}^2}=\sqrt{1+\gamma^2-\gamma^2}=1 \tag{3.73}$$

and

$$\mathrm{tr}\mathbf{V}^2=2+\gamma^2=2+4\tan^2\bar{\theta} \tag{3.74}$$

noting Eq. (3.63). Here, denoting the principal values of the second-order tensor $\mathbf{T}$ in the two-dimensional state as T_1, T_2 with $T_3=0$, it holds that

$$\mathrm{tr}\mathbf{T}^2=T_1^2+T_2^2=(T_1+T_2)^2-2T_1T_2=(\mathrm{tr}\mathbf{T})^2-2\det\mathbf{T} \tag{3.75}$$

in general and thus we have

$$\Gamma \equiv \mathrm{tr}\mathbf{V} = \sqrt{\mathrm{tr}\mathbf{V}^2 + 2\det\mathbf{V}} = \sqrt{4+\gamma^2} = \frac{2}{\cos\bar{\theta}} \tag{3.76}$$

where it holds that

$$\dot{\Gamma} = \frac{1}{2}\Gamma^{-1}2\gamma\dot{\gamma} = \frac{\gamma}{\Gamma}\dot{\gamma} = 2\frac{\tan\bar{\theta}}{\cos\bar{\theta}}\dot{\bar{\theta}} \tag{3.77}$$

Substituting Eqs. (3.73) and (3.74) into Eq. (3.72), the principal stretches λ_+ and λ_- are given by

$$\lambda_\pm = \frac{1}{2}(\Gamma \pm \gamma) \quad (\lambda_+ = 1 \to +\infty,\ \lambda_- = 1 \to 0 \,\mathrm{for}\, \gamma = 0 \to \infty) \tag{3.78}$$

Furthermore, multiplying the identity tensor $\mathbf{I}\left(=\sum_{\alpha=1}^{3}\mathbf{n}^{(\alpha)}\otimes\mathbf{n}^{(\alpha)}\right)$ to both sides of the last equation in Eq. (3.72), it holds that

$$\mathbf{V}^2 - (\mathrm{tr}\mathbf{V})\mathbf{V} + (\det\mathbf{V})\mathbf{I} = \mathbf{O} \tag{3.79}$$

Substituting Eqs. (3.69), (3.73), (3.76) into Eq. (3.79), one has

$$\mathbf{V} = \frac{\mathbf{V}^2 + (\det\mathbf{V})\mathbf{I}}{\mathrm{tr}\mathbf{V}} = \frac{1}{\Gamma}\left(\begin{bmatrix} 1+\gamma^2 & \gamma \\ \gamma & 1 \end{bmatrix} + \begin{bmatrix} 1 & 0 \\ 0 & 1 \end{bmatrix}\right) \tag{3.80}$$

resulting in

$$\mathbf{V} = V_{ij}\mathbf{e}_i \otimes \mathbf{e}_j,\ [V_{ij}] = \frac{1}{\Gamma}\begin{bmatrix} 2+\gamma^2 & \gamma \\ \gamma & 2 \end{bmatrix} = \frac{1}{\Gamma}\begin{bmatrix} 2+\tan^2\bar{\theta} & \tan\bar{\theta} \\ \tan\bar{\theta} & 2 \end{bmatrix} \tag{3.81}$$

for which the inverse tensor of $\mathbf{V}$ is given noting Eq. (1.120) as

$$\mathbf{V}^{-1} = \frac{1}{\Gamma}\begin{bmatrix} 2 & \gamma \\ -\gamma & 2+\gamma^2 \end{bmatrix} = \frac{1}{\Gamma}\begin{bmatrix} 2 & -\tan\bar{\theta} \\ -\tan\bar{\theta} & 2+\tan^2\bar{\theta} \end{bmatrix} \tag{3.82}$$

Substituting Eqs. (3.65) and (3.82) into Eq. (2.22), $\mathbf{R}$ is described as follows:

$$\begin{aligned} \mathbf{R} = \mathbf{V}^{-1}\mathbf{F} &= \frac{1}{\Gamma}\begin{bmatrix} 2 & -\gamma \\ -\gamma & 2+\gamma^2 \end{bmatrix}\begin{bmatrix} 1 & \gamma \\ 0 & 1 \end{bmatrix} = \frac{1}{\Gamma}\begin{bmatrix} 2 & \gamma \\ -\gamma & 2 \end{bmatrix} \\ &= \begin{bmatrix} \cos\bar{\theta} & \sin\bar{\theta} \\ -\sin\bar{\theta} & \cos\bar{\theta} \end{bmatrix} (\det\mathbf{R} = 1) \end{aligned} \tag{3.83}$$

Here, setting

$$\tan\theta^R\,(=R_{12}/R_{11})=\frac{\gamma}{2}=\tan\bar{\theta}\,(\theta^R=\bar{\theta}) \tag{3.84}$$

R is also derived from Eqs. (2.26), (3.83) and (3.84) as

$$\begin{aligned}\mathbf{R}\Bigg(&=\left[\mathbf{e}_i\bullet\left(\sum_{\alpha=1}^{3}\mathbf{n}^{(\alpha)}\otimes\mathbf{N}^{(\alpha)}\right)\mathbf{e}_j\right]=\left[\sum_{\alpha=1}^{3}(\mathbf{e}_i\bullet\mathbf{n}^{(\alpha)})(\mathbf{e}_j\bullet\mathbf{N}^{(\alpha)})\right]\\&=\left[\sum_{\alpha=1}^{3}(\mathbf{e}_i\bullet\mathbf{n}^{(\alpha)})\delta_j^{(\alpha)}\right]=\left[\mathbf{e}_i\bullet\mathbf{n}^{(j)}\right]\Bigg)=\begin{bmatrix}\cos\theta^R & \sin\theta^R\\ -\sin\theta^R & \cos\theta^R\end{bmatrix}\end{aligned} \tag{3.85}$$

It holds from Eq. (3.84) that

$$\overset{\bullet}{\gamma}=\frac{2}{\cos^2\theta^R}\overset{\bullet}{\theta}{}^R=2(1+\tan^2\theta^R)\overset{\bullet}{\theta}{}^R,\quad \overset{\bullet}{\theta}{}^R=\frac{\overset{\bullet}{\gamma}}{2\{1+(\gamma/2)^2\}}=\frac{2}{\Gamma^2}\overset{\bullet}{\gamma} \tag{3.86}$$

The relative spin in Eq. (2.85) is given from Eqs. (3.85) and (3.86) as follows:

$$\begin{aligned}\mathbf{\Omega}^R&=\begin{bmatrix}-\sin\theta^R & \cos\theta^R\\ -\cos\theta^R & -\sin\theta^R\end{bmatrix}\overset{\bullet}{\theta}{}^R\begin{bmatrix}\cos\theta^R & -\sin\theta^R\\ \sin\theta^R & \cos\theta^R\end{bmatrix}\\&=\begin{bmatrix}0 & 1\\ -1 & 0\end{bmatrix}\overset{\bullet}{\theta}{}^R=\frac{2}{\Gamma^2}\begin{bmatrix}0 & 1\\ -1 & 0\end{bmatrix}\overset{\bullet}{\gamma}=\begin{bmatrix}0 & 1\\ -1 & 0\end{bmatrix}\overset{\bullet}{\bar{\theta}}\end{aligned} \tag{3.87}$$

Next, denoting the expression of **V** in the principal direction by ${}^P\mathbf{V}$, it holds from Eq. (1.59) that

$${}^P\mathbf{V}=\mathbf{Q}^E\mathbf{V}\mathbf{Q}^{ET},\ {}^PV_{ij}=(\mathbf{n}^{(i)}\bullet\mathbf{e}_r)V_{rs}(\mathbf{n}^{(s)}\bullet\mathbf{e}_j) \tag{3.88}$$

where

$${}^P\mathbf{V}=\begin{bmatrix}\lambda_+ & 0\\ 0 & \lambda_-\end{bmatrix} \tag{3.89}$$

$$\mathbf{Q}^E=\left[\mathbf{n}^{(i)}\bullet\mathbf{e}_j\right]=\begin{bmatrix}\mathbf{n}^{(1)}\bullet\mathbf{e}_1 & \mathbf{n}^{(1)}\bullet\mathbf{e}_2\\ \mathbf{n}^{(2)}\bullet\mathbf{e}_1 & \mathbf{n}^{(2)}\bullet\mathbf{e}_2\end{bmatrix}=\begin{bmatrix}\cos\theta^E & \sin\theta^E\\ -\sin\theta^E & \cos\theta^E\end{bmatrix} \tag{3.90}$$

where θ^E is the rotation angle of the eigenvector $\mathbf{n}^{(1)}$, $\mathbf{n}^{(2)}$ of **V** from the bases $\mathbf{e}_1$, $\mathbf{e}_2$ in a counterclockwise direction.

On the other hand, the components of $\mathbf{R}^E$ are given from Eq. (2.30) as follows:

$$R_{ij}^E = \mathbf{e}_i \cdot \mathbf{R}^E \mathbf{e}_j = \mathbf{e}_i \cdot \sum_{\alpha=1}^{3} \mathbf{n}^{(\alpha)} \otimes \mathbf{e}_\alpha \cdot \mathbf{e}_j = \mathbf{e}_i \cdot \mathbf{n}^{(j)} = Q_{ji}^E, \mathbf{R}^E = \mathbf{Q}^{ET} \tag{3.91}$$

The following expressions are obtained by substituting Eq. (3.91) into Eq. (3.88).

$${}^{p}\mathbf{V} = \mathbf{R}^{E_T} \mathbf{V} \mathbf{R}^E, {}^{p}V_{ij} = R_{ri}^E V_{rs} R_{sj}^E \tag{3.92}$$

$$\mathbf{V} = \mathbf{R}^{E^p} \mathbf{V} \mathbf{R}^{E^T}, V_{ij} = R_{ir}^{E^p} V_{rs} R_{js}^E \tag{3.93}$$

Since $V_1 \geq V_2$ always, choosing the maximum principal value λ_+ in the direction of $\mathbf{n}^{(1)}$, it holds from Eqs. (3.81), (3.89), (3.91) and (3.93) that

$$\begin{aligned}
\frac{1}{\Gamma}\begin{bmatrix} 2+\gamma^2 & \gamma \\ \gamma & 2 \end{bmatrix} &= \begin{bmatrix} \cos\theta^E & -\sin\theta^E \\ \sin\theta^E & \cos\theta^E \end{bmatrix} \begin{bmatrix} \lambda_+ & 0 \\ 0 & \lambda_- \end{bmatrix} \begin{bmatrix} \cos\theta^E & \sin\theta^E \\ -\sin\theta^E & \cos\theta^E \end{bmatrix} \\
&= \begin{bmatrix} \lambda_+ \cos\theta^E & -\lambda_- \sin\theta^E \\ \lambda_+ \sin\theta^E & \lambda_- \cos\theta^E \end{bmatrix} \begin{bmatrix} \cos\theta^E & \sin\theta^E \\ -\sin\theta^E & \cos\theta^E \end{bmatrix} \\
&= \begin{bmatrix} \lambda_+ \cos^2\theta^E + \lambda_- \sin^2\theta^E & (\lambda_+ - \lambda_-)\sin\theta^E \cos\theta^E \\ (\lambda_+ - \lambda_-)\sin\theta^E \cos\theta^E & \lambda_+ \sin^2\theta^E + \lambda_- \cos^2\theta^E \end{bmatrix}
\end{aligned} \tag{3.94}$$

Substituting Eq. (3.78) into Eq. (3.94), one obtains

$$\begin{aligned}
&\frac{1}{\Gamma}\begin{bmatrix} 2+\gamma^2 & \gamma \\ \gamma & 2 \end{bmatrix} \\
&= \begin{bmatrix} \frac{1}{2}(\Gamma+\gamma)\cos^2\theta^E + \frac{1}{2}(\Gamma-\gamma)\sin^2\theta^E & \gamma \sin\theta^E \cos\theta^E \\ \gamma \sin\theta^E \cos\theta^E & \frac{1}{2}(\Gamma+\gamma)\sin^2\theta^E + \frac{1}{2}(\Gamma-\gamma)\cos^2\theta^E \end{bmatrix} \\
&= \frac{1}{2}\begin{bmatrix} \Gamma+\gamma\cos 2\theta^E & \gamma \sin 2\theta^E \\ \gamma \sin 2\theta^E & \Gamma-\gamma\cos 2\theta^E \end{bmatrix}
\end{aligned} \tag{3.95}$$

from which it holds that

$$\left.\begin{array}{l}\dfrac{\gamma}{\Gamma}=\dfrac{1}{2}\gamma\ \sin 2\theta^E \rightarrow \sin 2\theta^E=\dfrac{2}{\Gamma}\\ (V_{11}-V_{22}=)\dfrac{\gamma^2}{\Gamma}=\gamma\ \cos 2\theta^E \rightarrow \cos 2\theta^E=\dfrac{\gamma}{\Gamma}\end{array}\right\}$$

and thus it can be obtained that

$$\left.\begin{array}{l}\sin\theta^E=\sqrt{\dfrac{1}{2}(1-\dfrac{\gamma}{\Gamma})}=\sqrt{\dfrac{(\Gamma-\gamma)/2}{\Gamma}}=\dfrac{\sqrt{\Gamma(\Gamma-\gamma)/2}}{\Gamma}=\dfrac{Z_-}{\Gamma}=\dfrac{1}{Z_+}\\ \cos\theta^E=\sqrt{1-(\dfrac{Z_-}{\Gamma})^2}=\dfrac{Z_+}{\Gamma}=\dfrac{1}{Z_-}\\ \tan\theta^E=\dfrac{Z_-}{Z_+}=\dfrac{\Gamma-\gamma}{2}\end{array}\right\} \tag{3.96}$$

with

$$Z_\pm \equiv \sqrt{\Gamma(\Gamma\pm\gamma)/2} \tag{3.97}$$

where the double signs $\pm$ take in the same order. The following relations hold for $Z_\pm$.

$$\left.\begin{array}{l}Z_+Z_-=\Gamma\\ \dfrac{Z_+}{Z_-}=\dfrac{\Gamma+\gamma}{2},\dfrac{Z_-}{Z_+}=\dfrac{\Gamma-\gamma}{2}\\ Z_+^2+Z_-^2=\Gamma^2,Z_+^2-Z_-^2=\gamma\Gamma\\ \dfrac{1}{Z_+^2}+\dfrac{1}{Z_-^2}=1,\dfrac{1}{Z_-^2}-\dfrac{1}{Z_+^2}=\dfrac{\gamma}{\Gamma}\end{array}\right\} \tag{3.98}$$

The substitution of Eq. (3.90), (3.91), (3.96) into Eq. (3.91) yields.

$$\mathbf{R}^E=\frac{1}{\Gamma}\begin{bmatrix}Z_+ & -Z_-\\ Z_- & Z_+\end{bmatrix}=\begin{bmatrix}\dfrac{1}{Z_-} & -\dfrac{1}{Z_+}\\ \dfrac{1}{Z_+} & \dfrac{1}{Z_-}\end{bmatrix} \tag{3.99}$$

Differentiating Eq. $(3.96)_3$ and noting Eqs. (3.77), we have

$$\frac{\dot{\theta}^E}{\cos^2\theta^E}=\frac{1}{2}\left(\frac{\gamma}{\Gamma}-1\right)\dot{\gamma},\quad \dot{\theta}^E=\frac{\Gamma(\Gamma+\gamma)/2}{\Gamma^2}\frac{1}{2}\left(\frac{\gamma}{\Gamma}-1\right)\dot{\gamma}=-\frac{1}{\Gamma^2}\dot{\gamma} \tag{3.100}$$

Substituting Eqs. (3.90), (3.91), (3.100) into Eq. (2.93), it is obtained that

$$
\begin{aligned}
\mathbf{\Omega}^E &= \begin{bmatrix} -\sin\theta^E & -\cos\theta^E \\ \cos\theta^E & -\sin\theta^E \end{bmatrix} \dot{\theta}^E \begin{bmatrix} \cos\theta^E & \sin\theta^E \\ -\sin\theta^E & \cos\theta^E \end{bmatrix} \\
&= \begin{bmatrix} 0 & -1 \\ 1 & 0 \end{bmatrix} \dot{\theta}^E = \frac{2}{\Gamma^2} \begin{bmatrix} 0 & 1 \\ -1 & 0 \end{bmatrix} \dot{\gamma} = \frac{1}{2} \mathbf{\Omega}^R
\end{aligned}
\tag{3.101}
$$

The following expression of $\mathbf{U}$ is obtained in a similar manner to that used in Eq. (3.80)

$$
\mathbf{U} = \frac{\mathbf{U}^2 + (\det \mathbf{U})\mathbf{I}}{\mathrm{tr}\mathbf{U}} = \frac{1}{\Gamma}\left(\begin{bmatrix} 1 & \gamma \\ \gamma & 1+\gamma^2 \end{bmatrix} + \begin{bmatrix} 1 & 0 \\ 0 & 1 \end{bmatrix} \right)
$$

from which it is obtained that

$$
\mathbf{U} = \frac{1}{\Gamma} \begin{bmatrix} 2 & \gamma \\ \gamma & 2+\gamma^2 \end{bmatrix} = \begin{bmatrix} \cos\bar{\theta} & \sin\bar{\theta} \\ \sin\bar{\theta} & (1+\sin^2\bar{\theta})/\cos\bar{\theta} \end{bmatrix}
\tag{3.102}
$$

In order to obtain the rotation $\mathbf{R}^L$ of the eigenvector $\mathbf{N}^{(\alpha)}$ of $\mathbf{U}$, denoting the angle measured in counterclockwise direction from $\mathbf{e}_1$, $\mathbf{e}_2$ to $\mathbf{N}^{(1)}, \mathbf{N}^{(2)}$ by θ^L, one has

$$
\mathbf{U} = \mathbf{R}^L \mathbf{V}^p \mathbf{R}^{LT}
\tag{3.103}
$$

where, setting

$$
\mathbf{R}^L = \left[\mathbf{e}_i \bullet \mathbf{N}^{(j)} \right] = \begin{bmatrix} \mathbf{e}_1 \bullet \mathbf{N}^{(1)} & \mathbf{e}_1 \bullet \mathbf{N}^{(2)} \\ \mathbf{e}_2 \bullet \mathbf{N}^{(1)} & \mathbf{e}_2 \bullet \mathbf{N}^{(2)} \end{bmatrix} = \begin{bmatrix} \cos\theta^L & -\sin\theta^L \\ \sin\theta^L & \cos\theta^L \end{bmatrix}
\tag{3.104}
$$

and substituting Eqs. (3.89), (3.102), (3.104) into Eq. (3.103), we have

$$
\frac{1}{\Gamma} \begin{bmatrix} 2 & \gamma \\ \gamma & 2+\gamma^2 \end{bmatrix} = \begin{bmatrix} \cos\theta^L & \sin\theta^L \\ -\sin\theta^L & \cos\theta^L \end{bmatrix} \begin{bmatrix} \lambda_+ & 0 \\ 0 & \lambda_- \end{bmatrix} \begin{bmatrix} \cos\theta^L & -\sin\theta^L \\ \sin\theta^L & \cos\theta^L \end{bmatrix}
\tag{3.105}
$$

The substitution of Eq. (3.78) into Eq. (3.105) leads to

$$
\frac{1}{\Gamma} \begin{bmatrix} 2 & \gamma \\ \gamma & 2+\gamma^2 \end{bmatrix} = \frac{1}{2} \begin{bmatrix} \Gamma + \gamma\cos 2\theta^L & -\gamma\sin 2\theta^L \\ -\gamma\sin 2\theta^L & \Gamma - \gamma\cos 2\theta^L \end{bmatrix}
\tag{3.106}
$$

from which one has

$$
\left.
\begin{aligned}
&\sin 2\theta^L \ (= \sin 2\theta^E) = -\frac{2}{\Gamma} \\
&(U_{11} - U_{22} =) - \frac{\gamma^2}{\Gamma} = \gamma\cos 2\theta^L \rightarrow \cos 2\theta^L = -\frac{\gamma}{\Gamma}
\end{aligned}
\right\}
\tag{3.107}
$$

and

$$\left.\begin{aligned} \sin\theta^L &= \sqrt{\frac{1}{2}\Big(1+\frac{\gamma}{\Gamma}\Big)^2} = \frac{Z_+}{\Gamma} = \frac{1}{Z_-} \\ \cos\theta^L &= \sqrt{1-\Big(\frac{Z_+}{\Gamma}\Big)^2} = \frac{Z_-}{\Gamma} = \frac{1}{Z_+} \\ \tan\theta^L &= \frac{Z_+}{Z_-} = \frac{2}{\Gamma-\gamma} = \frac{1}{\tan\theta^E} \end{aligned}\right\} \tag{3.108}$$

The substitution of Eq. (3.108) into Eq. (3.104) reads:

$$\mathbf{R}^L = \frac{1}{\Gamma}\begin{bmatrix} Z_- & -Z_+ \\ Z_+ & Z_- \end{bmatrix} = \begin{bmatrix} \dfrac{1}{Z_+} & -\dfrac{1}{Z_-} \\ \dfrac{1}{Z_-} & \dfrac{1}{Z_+} \end{bmatrix} \tag{3.109}$$

Substituting Eqs. (3.99) and (3.109) into $\mathbf{R} = \mathbf{R}^E\mathbf{R}^{LT}$ based on Eq. (2.32), it holds that

$$\begin{aligned} \mathbf{R} &= \frac{1}{\Gamma}\begin{bmatrix} Z_+ & -Z_- \\ Z_- & Z_+ \end{bmatrix}\frac{1}{\Gamma}\begin{bmatrix} Z_- & Z_+ \\ -Z_+ & Z_- \end{bmatrix} = \frac{1}{\Gamma^2}\begin{bmatrix} 2Z_+Z_- & Z_+^2 - Z_-^2 \\ -Z_+^2 + Z_-^2 & 2Z_+Z_- \end{bmatrix} \\ &= \frac{1}{\Gamma}\begin{bmatrix} 2 & \gamma \\ -\gamma & 2 \end{bmatrix} \end{aligned} \tag{3.110}$$

which coincides with Eq. (3.83) obtained by the different approach. Substituting Eqs. (3.85), (3.90), (3.91), (3.104) into $\mathbf{R} = \mathbf{R}^E\mathbf{R}^{LT}$, one obtains

$$\begin{aligned} \begin{bmatrix} \cos\theta^R & \sin\theta^R \\ -\sin\theta^R & \cos\theta^R \end{bmatrix} &= \begin{bmatrix} \cos\theta^E & -\sin\theta^E \\ \sin\theta^E & \cos\theta^E \end{bmatrix}\begin{bmatrix} \cos\theta^L & \sin\theta^L \\ -\sin\theta^L & \cos\theta^L \end{bmatrix} \\ &= \begin{bmatrix} \cos\theta^E\cos\theta^L + \sin\theta^E\sin\theta^L & \cos\theta^E\sin\theta^L - \sin\theta^E\cos\theta^L \\ \sin\theta^E\cos\theta^L - \cos\theta^E\sin\theta^L & \sin\theta^E\sin\theta^L + \cos\theta^E\cos\theta^L \end{bmatrix} \\ &= \begin{bmatrix} \cos(\theta^E-\theta^L) & -\sin(\theta^E-\theta^L) \\ \sin(\theta^E-\theta^L) & \cos(\theta^E-\theta^L) \end{bmatrix} \end{aligned}$$

from which the following relation is obtained.

$$\theta^E = \theta^L - \theta^R \tag{3.111}$$

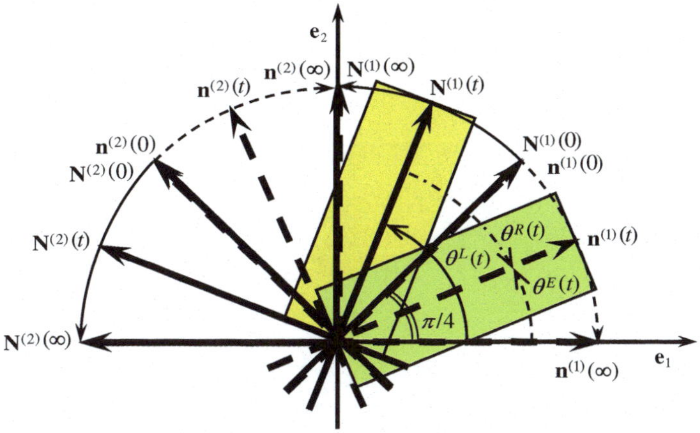

Fig. 3.4 Rotation of Lagrangian triad $\mathbf{N}^{(\alpha)}$ and Eulerian triad $\mathbf{n}^{(\alpha)}$

The rotations of $\mathbf{n}^{(\alpha)}$ and $\mathbf{N}^{(\alpha)}$ are shown in Fig. 3.4, noting

$$\left.\begin{aligned}\cos\theta^E &= \frac{1}{\sqrt{\Gamma(\Gamma-\gamma)/2}} = \frac{1}{\sqrt{\sqrt{4+\gamma^2}(\sqrt{4+\gamma^2}-\gamma)/2}}\\ \cos\theta^L &= \frac{1}{\sqrt{\Gamma(\Gamma+\gamma)/2}} = \frac{1}{\sqrt{\sqrt{4+\gamma^2}(\sqrt{4+\gamma^2}+\gamma)/2}}\end{aligned}\right\} \tag{3.112}$$

by virtue of Eqs. (3.76), (3.96), (3.97) and (3.108).

Furthermore, it is derived from Eqs. (3.77) and (3.108) that

$$\frac{\overset{\bullet}{\theta}{}^L}{\cos^2\theta^L} = \frac{1}{2}\left(\frac{\gamma}{\Gamma}+1\right)\overset{\bullet}{\gamma}, \quad \overset{\bullet}{\theta}{}^L = \frac{\Gamma(\Gamma-\gamma)/2}{\Gamma^2}\frac{1}{2}\left(\frac{\gamma}{\Gamma}+1\right)\overset{\bullet}{\gamma} = \frac{1}{\Gamma^2}\overset{\bullet}{\gamma} \tag{3.113}$$

Substituting Eqs. (3.104), (3.113) into Eq. (2.90), we obtain

$$\begin{aligned}\boldsymbol{\Omega}^L &= \begin{bmatrix} -\sin\theta^L & -\cos\theta^L \\ \cos\theta^L & -\sin\theta^L \end{bmatrix} \overset{\bullet}{\theta}{}^L \begin{bmatrix} \cos\theta^L & \sin\theta^L \\ -\sin\theta^L & \cos\theta^L \end{bmatrix} \\ &= \begin{bmatrix} 0 & -1 \\ 1 & 0 \end{bmatrix} \overset{\bullet}{\theta}{}^L = \frac{2}{\Gamma^2} \begin{bmatrix} 0 & -1 \\ 1 & 0 \end{bmatrix} \overset{\bullet}{\gamma} = -\frac{1}{2}\boldsymbol{\Omega}^R\end{aligned} \tag{3.114}$$

It can be confirmed easily that the three kinds of spins $\boldsymbol{\Omega}^R, \boldsymbol{\Omega}^E, \boldsymbol{\Omega}^L$ fulfill the relation in Eq. (2.99) by Eqs. (3.83), (3.87), (3.101) and (3.114).

Denoting $\tau = \sigma_{12}$, various stress tensors are described as follows:

$$\boldsymbol{\sigma} = \begin{bmatrix} \sigma_{11} & \tau \\ \tau & \sigma_{22} \end{bmatrix} \tag{3.115}$$

$$\boldsymbol{\Pi} = J\mathbf{F}^{-1}\boldsymbol{\sigma} = \begin{bmatrix} 1 & -\gamma \\ 0 & 1 \end{bmatrix}\begin{bmatrix} \sigma_{11} & \tau \\ \tau & \sigma_{22} \end{bmatrix} = \begin{bmatrix} \sigma_{11}-\gamma\tau & \tau-\gamma\sigma_{22} \\ \tau & \sigma_{22} \end{bmatrix} \tag{3.116}$$

$$\mathbf{S} = J\mathbf{F}^{-1}\boldsymbol{\sigma}F^{-T} = \begin{bmatrix} \sigma_{11}-\gamma\tau & \tau-\gamma\sigma_{22} \\ \tau & \sigma_{22} \end{bmatrix}\begin{bmatrix} 1 & 0 \\ -\gamma & 1 \end{bmatrix}$$

$$= \begin{bmatrix} \sigma_{11}-\gamma^2\sigma_{22}-2\gamma\tau & \tau-\gamma\sigma_{22} \\ \tau-\gamma\sigma_{22} & \sigma_{22} \end{bmatrix} \tag{3.117}$$

3.8.3 Combination of Tension and Distortion

Consider a thin cylindrical specimen subjected to the combination of tension and distortion described by the following equation in the polar coordinate system

$$r = \alpha R, \quad \theta = \Theta + \omega Z, \quad z = \lambda Z \tag{3.118}$$

where (R, Θ, Z) signifies the initial configuration, and α, ω and λ denote the proportionality factors depending on the deformation, while ω is described by the relative distortion angle ϕ between both ends as follows:

$$\omega \equiv \phi / L \tag{3.119}$$

L being the length of the specimen. The explanation in the following is referred to Khan and Huang (1995).

Variables describing a deformation are given as follows:

$$\mathbf{F} = \begin{bmatrix} F_{rR} & F_{r\Theta} & F_{rZ} \\ F_{\theta R} & F_{\theta\Theta} & F_{\theta Z} \\ F_{zR} & F_{z\Theta} & F_{zZ} \end{bmatrix} = \begin{bmatrix} \dfrac{\partial r}{\partial R} & \dfrac{\partial r}{R\partial\Theta} & \dfrac{\partial r}{\partial Z} \\ \dfrac{r\partial\theta}{\partial R} & \dfrac{r\partial\theta}{R\partial\Theta} & \dfrac{r\partial\theta}{\partial Z} \\ \dfrac{\partial z}{\partial R} & \dfrac{\partial z}{R\partial\Theta} & \dfrac{\partial z}{\partial Z} \end{bmatrix}$$
$$= \begin{bmatrix} \dfrac{\partial\alpha R}{\partial R} & \dfrac{\partial\alpha R}{R\partial\Theta} & \dfrac{\partial\alpha R}{\partial Z} \\ \dfrac{r\partial(\Theta+\omega Z)}{\partial R} & \dfrac{r\partial(\Theta+\omega Z)}{R\partial\Theta} & \dfrac{r\partial(\Theta+\omega Z)}{\partial Z} \\ \dfrac{\partial\lambda Z}{\partial R} & \dfrac{\partial\lambda Z}{R\partial\Theta} & \dfrac{\partial\lambda Z}{\partial Z} \end{bmatrix} = \begin{bmatrix} \alpha & 0 & 0 \\ 0 & \alpha & \omega\alpha\bar{R} \\ 0 & 0 & \lambda \end{bmatrix} \tag{3.120}$$

from which we have

$$\mathbf{F}^{-1} = \begin{bmatrix} \frac{1}{\alpha} & 0 & 0 \\ 0 & \frac{1}{\alpha} & -\frac{\omega \bar{R}}{\lambda} \\ 0 & 0 & \frac{1}{\lambda} \end{bmatrix} \tag{3.121}$$

and

$$\mathbf{V} = \begin{bmatrix} \alpha & 0 & 0 \\ 0 & \alpha \cos \phi + \alpha \omega \bar{R} \sin \phi & \lambda \sin s\phi \\ 0 & \lambda \sin \phi & \lambda \cos \phi \end{bmatrix} \tag{3.122}$$

$$\mathbf{U} = \begin{bmatrix} \alpha & 0 & 0 \\ 0 & \alpha \cos \phi & \alpha \sin \phi \\ 0 & \alpha \sin \phi & \alpha \omega \bar{R} \sin \phi + \lambda \cos \phi \end{bmatrix} \tag{3.123}$$

where

$$\cos \phi = \frac{\lambda + \alpha}{\sqrt{(\alpha + \lambda)^2 + (\alpha \omega \bar{R})^2}}, \quad \sin \phi = \frac{\alpha \omega \bar{R}}{\sqrt{(\alpha + \lambda)^2 + (\alpha \omega \bar{R})^2}} \tag{3.124}$$

$\bar{R}$ being the mean radius ($R \cong \bar{R}$) of the thin cylindrical specimen in the initial state.

Further, one has

$$\mathbf{C} = \mathbf{F}^T \mathbf{F} = \begin{bmatrix} \alpha^2 & 0 & 0 \\ 0 & \alpha^2 & \omega \alpha^2 \bar{R} \\ 0 & \omega \alpha^2 \bar{R} & \lambda^2 + \omega^2 \alpha^2 \bar{R}^2 \end{bmatrix} \tag{3.125}$$

$$\mathbf{b}^{-1} = \mathbf{F}^{-T} \mathbf{F}^{-1} = \begin{bmatrix} \frac{1}{\alpha^2} & 0 & 0 \\ 0 & \frac{1}{\alpha^2} & -\frac{\omega \bar{R}}{\alpha \lambda} \\ 0 & -\frac{\omega \bar{R}}{\alpha \lambda} & \frac{\omega^2 \bar{R}^2}{\lambda^2} + \frac{1}{\lambda^2} \end{bmatrix} \tag{3.126}$$

$$\boldsymbol{l} = \dot{\mathbf{F}}\mathbf{F}^{-1} = \begin{bmatrix} \frac{\dot{\alpha}}{\alpha} & -\dot{\omega}Z & 0 \\ \dot{\omega}Z & \frac{\dot{\alpha}}{\alpha} & \frac{\dot{\omega}\alpha\bar{R}}{\lambda} \\ 0 & 0 & \frac{\dot{\lambda}}{\lambda} \end{bmatrix} \tag{3.127}$$

from which we have

$$\mathbf{d} = \begin{bmatrix} \frac{\dot{\alpha}}{\alpha} & 0 & 0 \\ 0 & \frac{\dot{\alpha}}{\alpha} & \frac{\dot{\omega}\,\alpha\bar{R}}{2\lambda} \\ 0 & \frac{\dot{\omega}\,\alpha\bar{R}}{2\lambda} & \frac{\dot{\lambda}}{\lambda} \end{bmatrix}, \quad \mathbf{w} = \begin{bmatrix} 0 & -\dot{\omega}\,Z & 0 \\ \dot{\omega}\,Z & 0 & \frac{\dot{\omega}\,\alpha\bar{R}}{2\lambda} \\ 0 & -\frac{\dot{\omega}\,\alpha\bar{R}}{2\lambda} & 0 \end{bmatrix} \tag{3.128}$$

Stresses in various definitions are described by the following equations, designating the normal stress σ_{ZZ} and the shear stress $\sigma_{r\theta}$ applied to the traverse section of the cylinder by σ and τ, respectively.

$$\boldsymbol{\sigma} = \begin{bmatrix} 0 & 0 & 0 \\ 0 & 0 & \tau \\ 0 & \tau & \sigma \end{bmatrix} \tag{3.129}$$

It holds from Eqs. (3.120), (3.121) and (3.129) that

$$\begin{aligned} \boldsymbol{\Pi} = J\mathbf{F}^{-1}\boldsymbol{\sigma} &= \alpha^2\lambda \begin{bmatrix} \frac{1}{\alpha} & 0 & 0 \\ 0 & \frac{1}{\alpha} & -\frac{\omega\bar{R}}{\lambda} \\ 0 & 0 & \frac{1}{\lambda} \end{bmatrix} \begin{bmatrix} 0 & 0 & 0 \\ 0 & 0 & \tau \\ 0 & \tau & \sigma \end{bmatrix} \\ &= \begin{bmatrix} 0 & 0 & 0 \\ 0 & -\alpha^2\omega\tau\bar{R} & \alpha\lambda\tau - \alpha^2\omega\sigma\bar{R} \\ 0 & \alpha^2\tau & \alpha^2\sigma \end{bmatrix} \end{aligned} \tag{3.130}$$

$$
\mathbf{S} = \boldsymbol{\Pi} F^{-T} = \begin{bmatrix} 0 & 0 & 0 \\ 0 & -\alpha^2\omega\tau\bar{R} & \alpha\lambda\tau - \alpha^2\omega\sigma\bar{R} \\ 0 & \alpha^2\tau & \alpha^2\sigma \end{bmatrix} \begin{bmatrix} \dfrac{1}{\alpha} & 0 & 0 \\ 0 & \dfrac{1}{\alpha} & 0 \\ 0 & -\dfrac{\omega\bar{R}}{\lambda} & \dfrac{1}{\lambda} \end{bmatrix}
$$

$$
= \begin{bmatrix} 0 & 0 & 0 \\ 0 & -2\alpha\omega\tau\bar{R} + \dfrac{\alpha^2\omega^2\sigma\bar{R}^2}{\lambda} & \alpha\tau - \dfrac{\alpha^2\omega\sigma\bar{R}}{\lambda} \\ 0 & \alpha\tau - \dfrac{\alpha^2\omega\sigma\bar{R}}{\lambda} & \dfrac{\alpha^2}{\lambda} \end{bmatrix} \tag{3.131}
$$

References

Hashiguchi K (2013) Elastoplasticity theory. In: Lecture notes in applied and computational mechanics, 2nd edn. Springer, Heidelberg

Hashiguchi K, Yamakawa Y (2012) Introduction to finite strain theory for continuum elasto-plasticity. In: Wiley series in computational mechanics. Wiley, Chichester, UK

Khan AS, Huang S (1995) Continuum theory of plasticity. Wiley, New York

Chapter 4
Objectivity and Objective (Rate) Tensors

Constitutive property of material is independent of observers. Therefore, constitutive equation has to be described by variables obeying the common objective transformation rule described in Sect. 1.3.1. State variables, e.g. stress, strain and back stress tensors in the same configuration obey the common coordinate transformation rule. However, the material-time derivatives of tensors in the current configuration do not obey the objective transformation rule, since they are influenced by the rigid-body rotation. Then, instead of the material-time derivative of tensors, particular time-derivatives of tensors obeying the objective transformation rule have to be adopted in constitutive equations.

The consideration on the fulfillment of objectivity is of great importance for the hypoelastic-based constitutive equation formulated in the current configuration which is influenced directly by the rigid-body rotation, while the hypoelastic-based plasticity is comprehensively explained in this book. Then, the objectivity and the formulation of constitutive relations fulfilling the objectivity will be comprehensively described in this chapter.

4.1 Objectivity

Physical quantities except for scalar ones are observed to be different depending on the state, e.g. position, direction, velocity of observers. On the other hand, mechanical property of material is observed identically independent of the state of observers. In particular, it is observed identically independent of the rigid-body rotation of material. Therefore, a constitutive equation describing material property must be expressed in a common form independent of coordinate systems. Then, it must be described so as not to be influenced by the rigid-body rotation of material. This fact was not so obvious in the olden time and was advocated by Oldroyd (1950) in the middle of the last century. It is referred to as the *principle of material-frame indifference* (Oldroyd 1950) or *principle of objectivity* or simply *objectivity*.

K. Hashiguchi, *Foundations of Elastoplasticity: Subloading Surface Model*,
DOI 10.1007/978-3-319-48821-9_4

This would be regarded as the starting point of the modern continuum mechanics which is called sometimes as the *rational mechanics* (Truesdell and Toupin 1960; Truesdell and Noll 1965).

Here, note that components of tensor describing mechanical state of material, e.g. stress, strain and anisotropic internal variables are observed to be changed by the fixed coordinate system if the material rotates, even when the components are observed to be unchanged by the coordinate system rotating concurrently with the material itself. Therefore, the material-time derivative of tensor describing mechanical state is observed to be non-zero, by the fixed coordinate system when the material rotates even when it is observed to be zero by the observer rotating concurrently with the material itself. It is caused by the fact that the material-time derivative of tensor designates the rate of tensor observed by the coordinate system moving in parallel with material but without rotation. Then, the material-time derivative cannot be adopted for the description of constitutive equations in a current rate form.

Machine elements are often subjected not only to deformation but also to rigid-body rotation, as seen in metal forming, gears, wheels, etc. Soils near the side edges of footings, at the bottom ends of piles, etc. undergo a large rigid-body rotation. Therefore, formulations of constitutive equations which is not influenced by the rigid-body rotation are of great importance in practical engineering problems.

4.2 Influence of Rigid-Body Rotation on Various Mechanical Quantities

In order to check whether or not a constitutive equation is formulated so as to satisfy the objectivity principle, it is expedient to examine the influence of rigid-body rotation on the tensor variables used in constitutive equations. Instead, one may examine how the components of these variables are observed by the coordinate systems with the fixed base $\{\mathbf{e}_i\}$ and the rotating base $\{\mathbf{e}_i^*(t)\}$ which are related as

$$\mathbf{e}_i^*(t) = \mathbf{Q}^T(t)\mathbf{e}_i, \quad \mathbf{e}_i^*(0) = \mathbf{e}_i, \quad \dot{\mathbf{e}}_i^*(t) = \dot{\mathbf{Q}}^T(t)\mathbf{e}_i \tag{4.1}$$

provided that the rotating base $\{\mathbf{e}_i^*(t)\}$ coincides with the fixed base $\{\mathbf{e}_i\}$ at the beginning of deforming/rotation $(t = 0)$, where one has

$$\mathbf{Q}(t) = \mathbf{e}_i \otimes \mathbf{e}_i^*(t), \quad \mathbf{Q}(0) = \mathbf{I} \tag{4.2}$$

noting Eq. (1.89) with Eq. (4.1). These bases are illustrated in Fig. 4.1 for the two-dimensional state.

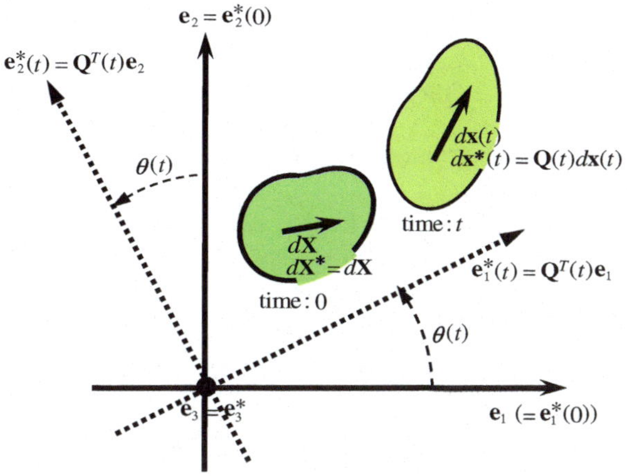

Fig. 4.1 Coordinate systems with the fixed base $\{\mathbf{e}_i\}$ and the rotating base $\{\mathbf{e}_i^*(t)\}$ which coincides with the base $\{\mathbf{e}_i\}$ at the beginning of deformation/rotation (illustrated in two-dimensional state for $\mathbf{e}_3 = \mathbf{e}_3^*$)

The components of the initial infinitesimal line element $d\mathbf{X}$ in the initial state $(t = 0)$ is observed to be identical by these bases, since $\{\mathbf{e}_i^*(t)\}$ coincides with $\{\mathbf{e}_i\}$ in the initial state. On the other hand, the components of the current infinitesimal line-element $d\mathbf{x}(t)$ is observed to be different by these bases as the rotating base $\{\mathbf{e}_i^*(t)\}$ differs from the fixed base $\{\mathbf{e}_i\}$ for $t > 0$. Here, noting $d\mathbf{x}(t) = \mathbf{F}(t)d\mathbf{X}$, we have

$$d\mathbf{X}^* = d\mathbf{X}\,(\{\mathbf{e}_i^*(0)\} = \{\mathbf{e}_i\}),\ d\mathbf{x}^*(t) = \mathbf{Q}(t)d\mathbf{x}(t) = \mathbf{Q}(t)\mathbf{F}(t)d\mathbf{X} \tag{4.3}$$

and

$$d\mathbf{x}^*(t) = \mathbf{F}^*(t)d\mathbf{X}^*(0) = \mathbf{F}^*(t)d\mathbf{X} \tag{4.4}$$

from which it follows that

$$\mathbf{F}^*(t) = \mathbf{Q}(t)\mathbf{F}(t) \tag{4.5}$$

It is known from Eq. (4.5) that the deformation gradient $\mathbf{F}(t)$ is the second-order tensor but it obeys the transformation rule of the first-order tensor. This is based on the fact that the deformation gradient is the two-point tensor as specified in Eq. (2.14).

Substituting Eq. (4.5) into Eqs. (2.20)–(2.22), (2.35), (2.36) and (2.45), the following relations are obtained for various quantities describing a deformation.

$$\mathbf{U}^* = \mathbf{U}, \quad \mathbf{C}^* = \mathbf{C}\ (\mathbf{F}^{*T}\mathbf{F}^* = (\mathbf{QF})^T\mathbf{QF} = \mathbf{F}^T\mathbf{Q}^T\mathbf{QF} = \mathbf{F}^T\mathbf{F}) \tag{4.6}$$

$$\mathbf{V}^* = \mathbf{QVQ}^T, \quad \mathbf{b}^* = \mathbf{QbQ}^T \quad (\mathbf{F}^*\mathbf{F}^{*T} = \mathbf{QF}(\mathbf{QF})^T = \mathbf{QFF}^T\mathbf{Q}^T) \tag{4.7}$$

$$\mathbf{R}^* = \mathbf{QR}\ (\mathbf{F}^* = \mathbf{QF} = \mathbf{QRU} = \mathbf{QRU}^* = \mathbf{R}^*\mathbf{U}^*) \tag{4.8}$$

$$\boldsymbol{E}^* = \boldsymbol{E}\ (= \mathbf{F}^T\boldsymbol{e}\mathbf{F}), \ \boldsymbol{e}^* = \mathbf{Q}\boldsymbol{e}\mathbf{Q}^T \tag{4.9}$$

Noting the relation

$$\dot{\mathbf{F}}^*\mathbf{F}^{*-1} = (\mathbf{QF})^{\boldsymbol{\cdot}}(\mathbf{QF})^{-1} = (\dot{\mathbf{Q}}\mathbf{F} + \mathbf{Q}\dot{\mathbf{F}})\mathbf{F}^{-1}\mathbf{Q}^{-1} = \mathbf{Q}(\dot{\mathbf{F}}\mathbf{F}^{-1} - \dot{\mathbf{Q}}^T\mathbf{Q})\mathbf{Q}^T$$

it holds for the velocity gradient in Eq. (2.75) that

$$\boldsymbol{l}^* = \mathbf{Q}(\boldsymbol{l} - \boldsymbol{\Omega})\mathbf{Q}^T = \mathbf{Q}\boldsymbol{l}\mathbf{Q}^T + \overline{\boldsymbol{\Omega}} \tag{4.10}$$

where $\boldsymbol{\Omega}$ and $\overline{\boldsymbol{\Omega}}$ are defined by

$$\left.\begin{array}{ll} \boldsymbol{\Omega} \equiv \dot{\mathbf{Q}}^T\mathbf{Q}, & \boldsymbol{\Omega} \equiv \dot{Q}_{ri}Q_{rj}\mathbf{e}_i \otimes \mathbf{e}_j \\ \overline{\boldsymbol{\Omega}} \equiv \dot{\mathbf{Q}}\mathbf{Q}^T, & \overline{\boldsymbol{\Omega}} \equiv \dot{Q}_{ir}Q_{jr}\mathbf{e}_i \otimes \mathbf{e}_j \end{array}\right\} \tag{4.11}$$

and they are related by

$$\overline{\boldsymbol{\Omega}} = -\mathbf{Q}\boldsymbol{\Omega}\mathbf{Q}^T, \ \boldsymbol{\Omega} = -\mathbf{Q}^T\overline{\boldsymbol{\Omega}}\mathbf{Q} \tag{4.12}$$

where $\dot{\mathbf{Q}}$ is given by

$$\dot{\mathbf{Q}} = \mathbf{e}_r \otimes \dot{\mathbf{e}}_r^* \tag{4.13}$$

from Eq. (4.2) because of $\dot{\mathbf{e}}_i = \mathbf{0}$. Substituting Eqs. (4.2) and (4.13) into Eq. (4.11), we have

$$\boldsymbol{\Omega} = \dot{\mathbf{e}}_r^* \otimes \mathbf{e}_r^*, \quad \overline{\boldsymbol{\Omega}} = (\dot{\mathbf{e}}_r^* \boldsymbol{\cdot} \mathbf{e}_j^*)\mathbf{e}_i \otimes \mathbf{e}_j \tag{4.14}$$

from which it follows that

$$\dot{\mathbf{e}}_i^* = \boldsymbol{\Omega}\mathbf{e}_i^* \tag{4.15}$$

It is known from Eq. (4.15) that $\boldsymbol{\Omega}$ is the spin of the base $\{\mathbf{e}_i^*\}$.

The substitution of Eq. (4.10) into Eqs. (2.80) and (2.81) yields the following transformation rules.

$$\mathbf{d}^* = \mathbf{Q}\mathbf{d}\mathbf{Q}^T \tag{4.16}$$

$$\mathbf{w}^* = \mathbf{Q}(\mathbf{w} - \mathbf{\Omega})\mathbf{Q}^T = \mathbf{Q}\mathbf{w}\mathbf{Q}^T + \overline{\mathbf{\Omega}} \tag{4.17}$$

The relative spin in Eq. (2.85), i.e.

$$\mathbf{\Omega}^R \equiv \dot{\mathbf{R}}\,\mathbf{R}^T \tag{4.18}$$

obeys the transformation identical to that of $\mathbf{w}$ as follows:

$$\mathbf{\Omega}^{R*} = \mathbf{Q}(\mathbf{\Omega}^R - \mathbf{\Omega})\mathbf{Q}^T = \mathbf{Q}\mathbf{\Omega}^R\mathbf{Q}^T + \overline{\mathbf{\Omega}} \tag{4.19}$$

noting

$$\begin{aligned}\dot{\mathbf{R}}^*\mathbf{R}^{*T} &= (\mathbf{Q}\mathbf{R})^{\bullet}(\mathbf{Q}\mathbf{R})^T = (\dot{\mathbf{Q}}\mathbf{R} + \mathbf{Q}\dot{\mathbf{R}})\mathbf{R}^T\mathbf{Q}^T \\ &= \mathbf{Q}\dot{\mathbf{R}}\mathbf{R}^T\mathbf{Q}^T + \dot{\mathbf{Q}}\mathbf{Q}^T = \mathbf{Q}\mathbf{\Omega}^R\mathbf{Q}^T + \mathbf{Q}\mathbf{Q}^T\dot{\mathbf{Q}}\mathbf{Q}^T = \mathbf{Q}\mathbf{\Omega}^R\mathbf{Q}^T - \mathbf{Q}\dot{\mathbf{Q}}^T\mathbf{Q}\mathbf{Q}^T\end{aligned}$$

We obtain the following conclusions for the influence of rigid-body rotation from Eqs. (4.6) to (4.19).

(1) The right Cauchy-Green deformation tensor $\mathbf{C}$ and the Green strain tensor $\boldsymbol{E}$ are based in the reference configuration and thus they are observed to be unchangeable, i.e. invariant, obeying the transformation rule of scalar quantities independent of the rigid-body rotation. On the other hand, the left Cauchy-Green deformation tensor $\mathbf{b}$ and the Almansi strain tensor $\boldsymbol{e}$ are based in the current configuration and thus obey the transformation rule of second-order tensor.

(2) The strain rate tensor $\mathbf{d}$ obeys the transformation rule of the second-order tensor. On the other hand, the velocity gradient tensor $\boldsymbol{l}$, the continuum spin tensor $\mathbf{w}$ and the relative spin $\mathbf{\Omega}^R$ are directly subjected to the influence of rate of rigid-body rotation, lacking the objectivity.

The following transformations hold for stress tensors described in Chap. 3.

$$\boldsymbol{\sigma}^* = \mathbf{Q}\boldsymbol{\sigma}\mathbf{Q}^T \;\; (\mathbf{t}^* = \mathbf{Q}\mathbf{t} = \mathbf{Q}\boldsymbol{\sigma}\mathbf{n} = \mathbf{Q}\boldsymbol{\sigma}\mathbf{Q}^T\mathbf{Q}\mathbf{n} = \boldsymbol{\sigma}^*\mathbf{n}^*) \tag{4.20}$$

$$\boldsymbol{\tau}^* = \mathbf{Q}\boldsymbol{\tau}\mathbf{Q}^T \tag{4.21}$$

$$\begin{gathered}\mathbf{\Pi}^* = \mathbf{Q}\mathbf{\Pi} \\ (\mathbf{\Pi}^* = J\boldsymbol{\sigma}^*\mathbf{F}^{*-T} = J\mathbf{Q}\boldsymbol{\sigma}\mathbf{Q}^T(\mathbf{Q}\mathbf{F})^{-T} = J\mathbf{Q}\boldsymbol{\sigma}\mathbf{Q}^T\mathbf{Q}^{-T}\mathbf{F}^{-T} = \mathbf{Q}J\boldsymbol{\sigma}\mathbf{F}^{-T} = \mathbf{Q}\mathbf{\Pi})\end{gathered} \tag{4.22}$$

$$\mathbf{S}^* = \mathbf{S}$$
$$(\mathbf{S}^* = \mathbf{F}^{*-1}\boldsymbol{\tau}^*\mathbf{F}^{*-T} = (\mathbf{QF})^{-1}(\mathbf{Q}\boldsymbol{\tau}\mathbf{Q}^T)(\mathbf{QF})^{-T} = \mathbf{F}^{-1}\mathbf{Q}^T\mathbf{Q}\boldsymbol{\tau}\mathbf{Q}^T\mathbf{Q}^{-T}\mathbf{F}^{-T} = \mathbf{F}^{-1}\boldsymbol{\tau}\mathbf{F}^{-T} = \mathbf{S}) \tag{4.23}$$

Then, the Cauchy stress tensor $\boldsymbol{\sigma}$ and the Kirchhoff stress tensor $\boldsymbol{\tau}$ obeys the transformation rule of the second-order tensor. On the other hand, the first Piola-Kirchhoff stress tensor $\boldsymbol{\Pi}$ obeys the transformation rule of the first-order tensor so that it is the two-point tensor. The second Piola-Kirchhoff stress tensor $\mathbf{S}$ is the invariant under the superposition of rigid-body rotation.

The consideration of objectivity is of great importance in the formulation of hypoelastic-based plastic constitutive equations since the rates of stress and anisotropic internal state variables in the current configuration are influenced directly by the rigid-body rotation as described above. Then, the time-derivatives of state variables will be further considered in the subsequent sections.

4.3 Material-Time Derivative of Tensor

The material-time derivatives of state variables is the rates of them observed by the coordinate system moving in parallel with material particle as explained in Sect. 2.2. However, it will be mathematically verified in this section that the material-time derivative of tensor does not obey the objective transformation rule and thus it cannot be used in constitutive equations.

Consider the transformation of the material-time derivative of a state variable obeying the objective transformation in Eq. (1.71) and (1.73). The material-time derivative of the tensor $\mathbf{t}$ in the current configuration reads (Hashiguchi 2007a):

$$\begin{aligned}\dot{t}^*_{p_1p_2\cdots p_m} &= \dot{Q}_{p_1q_1}Q_{p_2q_2}\cdots Q_{p_mq_m}t_{q_1q_2\cdots q_m} + Q_{p_1q_1}\dot{Q}_{p_2q_2}\cdots Q_{p_mq_m}t_{q_1q_2\cdots q_m} + \cdots \\ &\quad + Q_{p_1q_1}Q_{p_2q_2}\cdots \dot{Q}_{p_mq_m}t_{q_1q_2\cdots q_m} + Q_{p_1q_1}Q_{p_2q_2}\cdots Q_{p_mq_m}\dot{t}_{q_1q_2\cdots q_m}\end{aligned} \tag{4.24}$$

$$\begin{aligned}\dot{t}_{p_1p_2\cdots p_m} &= \dot{Q}_{q_1p_1}Q_{q_2p_2}\cdots Q_{q_mp_m}t^*_{q_1q_2\cdots q_m} + Q_{q_1p_1}\dot{Q}_{q_2p_2}\cdots Q_{q_mp_m}t^*_{q_1q_2\cdots q_m} + \cdots \\ &\quad + Q_{q_1p_1}Q_{q_2p_2}\cdots \dot{Q}_{q_mp_m}t^*_{q_1q_2\cdots q_m} + Q_{q_1p_1}Q_{q_2p_2}\cdots Q_{q_mp_m}\dot{t}^*_{q_1q_2\cdots q_m}\end{aligned} \tag{4.25}$$

Noting the relation $\dot{Q}_{p_iq_i} = \delta_{p_is}\dot{Q}_{sq_i} = Q_{p_it}Q_{st}\dot{Q}_{sq_i} = -Q_{p_it}\dot{Q}_{st}Q_{sq_i} = -Q_{p_it}\Omega_{tq_i}$ and replacing $t\rightarrow q_i$, $q_i\rightarrow r_i$, then Eqs. (4.24) and (4.25) can be rewritten as follows:

$$\begin{aligned}\dot{t}^*_{p_1p_2\cdots p_m} &= Q_{p_1q_1}Q_{p_2q_2}\cdots Q_{p_mq_m}\,(\dot{t}_{q_1q_2\cdots q_m} - \Omega_{q_1r_1}t_{r_1q_2\cdots q_m} - \Omega_{q_2r_2}t_{q_1r_2\cdots q_m}\\ &\quad - \cdots - \Omega_{q_1r_m}t_{q_1q_2\cdots r_m})\end{aligned} \tag{4.26}$$

$$\begin{aligned}\dot{t}_{p_1p_2\cdots p_m} &= Q_{q_1p_1}Q_{q_2p_2}\cdots Q_{q_mp_m}\,(\dot{t}^*_{q_1q_2\cdots q_m} - \overline{\Omega}_{q_1r_1}t^*_{r_1q_2\cdots q_m} - \overline{\Omega}_{q_2r_2}t^*_{q_1r_2\cdots q_m}\\ &\quad - \cdots - \overline{\Omega}_{q_mr_m}t^*_{q_1q_2\cdots r_m})\end{aligned} \tag{4.27}$$

It is known from Eqs. (4.26) and (4.27) that the material-time derivative does not obey the objective transformation rule, noting that the components $\dot{t}_{p_1p_2\ldots p_m}$ in the fixed coordinate system is not zero even when the components $\dot{t}^*_{p_1p_2\ldots p_m}$ in the coordinate system rotating with the material is zero. Equations (4.26) and (4.27) are expressed for the vector $\mathbf{v}$ and the second-order tensor $\mathbf{t}$ in symbolic notation as follows:

$$\dot{\mathbf{v}}^* = \mathbf{Q}(\dot{\mathbf{v}} - \boldsymbol{\Omega}\mathbf{v}),\, \dot{\mathbf{v}} = \mathbf{Q}^T(\dot{\mathbf{v}}^* - \overline{\boldsymbol{\Omega}}\mathbf{v}^*) \tag{4.28}$$

$$\dot{\mathbf{t}}^* = \mathbf{Q}(\dot{\mathbf{t}} - \boldsymbol{\Omega}\mathbf{t} + \mathbf{t}\boldsymbol{\Omega})\mathbf{Q}^T,\, \dot{\mathbf{t}} = \mathbf{Q}^T(\dot{\mathbf{t}}^* - \overline{\boldsymbol{\Omega}}\mathbf{t}^* + \mathbf{t}^*\overline{\boldsymbol{\Omega}})\mathbf{Q} \tag{4.29}$$

Consequently, the material-time derivative cannot be adopted in constitutive equations.

In order to see the irrationality for using the material-time derivative of tensor, consider the hypoelastic constitutive equation, which relates the material-time derivative of Cauchy stress tensor linearly to the strain rate tensor and to which the hypoelastic-base plastic constitutive equation described after Chap. 6 also belong, as follows:

$$\dot{\boldsymbol{\sigma}} = \mathbf{H} : \mathbf{d}$$

where the tangent modulus tensor $\mathbf{H}$ (fourth-order tensor) is the function of stress and anisotropic internal variables in general. It follows from this equation with Eq. (4.29) that

$$\begin{aligned}\mathbf{d} &= \mathbf{H}^{-1}\!:\mathbf{Q}^T(\dot{\boldsymbol{\sigma}}^* - \overline{\boldsymbol{\Omega}}\boldsymbol{\sigma}^* + \boldsymbol{\sigma}^*\overline{\boldsymbol{\Omega}})\mathbf{Q}\,(\neq \mathbf{H}^{-1}\!:\mathbf{Q}^T\dot{\boldsymbol{\sigma}}^*\mathbf{Q} = \mathbf{H}^{-1}\!:\mathbf{Q}^T(\mathbf{Q}\boldsymbol{\sigma}\mathbf{Q}^T)^{\boldsymbol{\cdot}}\mathbf{Q}\\ &= \mathbf{H}^{-1}\!:\mathring{\boldsymbol{\sigma}})\end{aligned}$$

This leads to the irrational result that the deformation is induced, i.e. $\mathbf{d} \neq \mathbf{O}$ even if the stress observed by the coordinate system rotating ($\overline{\boldsymbol{\Omega}} \neq \mathbf{O}$) with the material itself does not change, i.e. $\dot{\boldsymbol{\sigma}}^* = \mathbf{O}$. This is caused by the non-objectivity of material-time derivative of tensor, while the strain rate $\mathbf{d}$ is not a material-time derivative of tensor but is the original tensor defined so as to obey the objective transformation by excluding the continuum spin tensor $\mathbf{w}$ from the velocity gradient tensor $\boldsymbol{l}$. In the next section, the objective time-derivative of tensor will be introduced which is based on the rate of tensor observed by the coordinate system deforming/rotating with material itself, satisfying the objectivity.

4.4 Convected Time-Derivative of Tensor

The objective rate of tensor describing the physical quantity must be independent of the spin of rigid-body rotation and thus it has to be given primarily by the *convected rate*, i.e. the rate of tensors observed by the coordinate system deforming and rotating with material itself, i.e. the *convected* (or *convective* or *embedded*) *coordinate system* in which the coordinate axes are etched in material itself. The convected rate, i.e. convected time-derivative is the generalization of the *Lie derivative* (cf. e.g. Truesdell and Toupin; Marsden and Hughes 1983, 1960; Bonet and Wood 1997; Simo 1998; Belytschko et al. 2014; de Souza-Neto et al. 2008). The convected coordinate system turns to the curvilinear coordinate system in general as a deformation proceeds. Therefore, it is required first to study the mathematics on the general curvilinear coordinate system in order to capture the exact physical interpretation of the objective rate tensors. However, it is beyond the level of this book. One can refer to Hashiguchi (2012) for the comprehensive. In what follows, the explanation for the objective rate tensor will be devised so as to be understood without the detailed mathematical formulation in the curvilinear coordinate system.

(1) **Description in convected bases**

Consider the embedded primary base $\{\mathbf{G}_I\}$ in the reference configuration, which becomes $\{\mathbf{g}_i(t)\}$ in the current configuration as the deformation of material is induced. Then, let the reciprocal bases for the primary bases $\{\mathbf{G}_I\}$ and $\{\mathbf{g}_i(t)\}$ be denoted by $\{\mathbf{G}^I\}$ and $\{\mathbf{g}^i(t)\}$, respectively, noting the definition in Eq. (1.45). Here, it should be noted that the reciprocal base $\{\mathbf{g}^i(t)\}$ can be embedded under a pure rotation of material but it cannot be embedded under a deformation of material because it does not keep the reciprocal relation to the primary base $\{\mathbf{g}_i(t)\}$ if deformation is induced. They satisfy

$$\mathbf{G}_I \cdot \mathbf{G}^J = \delta_I^J, \ \mathbf{g}_i(t) \cdot \mathbf{g}^j(t) = \delta_i^j \tag{4.30}$$

by virtue of Eq. (1.46). In addition, the following tensors are the generalized expressions of the identity tensor as can be confirmed by Eq. (1.105), while the identity tensor is called *metric tensor* in the general Euclidian space described in the curvilinear coordinate system.

$$\left.\begin{aligned}
\mathbf{G} &\equiv \mathbf{G}^I \otimes \mathbf{G}_I = \mathbf{G}_I \otimes \mathbf{G}^I \\
&= G^{IJ}\mathbf{G}_I \otimes \mathbf{G}_J = G_J^I \mathbf{G}_I \otimes \mathbf{G}^J \\
&= G_I^J \mathbf{G}^I \otimes \mathbf{G}_J = G_{IJ}\mathbf{G}^I \otimes \mathbf{G}^J \\
\mathbf{g}(t) &\equiv \mathbf{g}^i(t) \otimes \mathbf{g}_i(t) = \mathbf{g}_i(t) \otimes \mathbf{g}^i(t) \\
&= g^{ij}(t)\mathbf{g}_i(t) \otimes \mathbf{g}_j(t) = g_j^i(t)\mathbf{g}_i(t) \otimes \mathbf{g}^j(t) \\
&= g_i^j(t)\mathbf{g}^i(t) \otimes \mathbf{g}_j(t) = g_{ij}(t)\mathbf{g}^i(t) \otimes \mathbf{g}^j(t)
\end{aligned}\right\} \tag{4.31}$$

setting $G^{IJ} \equiv \mathbf{G}^I \cdot \mathbf{G}^J$, $G^I_J \equiv \mathbf{G}^I \cdot \mathbf{G}_J = \delta^I_J$, $G^J_I \equiv \mathbf{G}_I \cdot \mathbf{G}^J = \delta^J_I$, $G_{IJ} \equiv \mathbf{G}_I \cdot \mathbf{G}_J$ and $g^{ij} \equiv \mathbf{g}^i \cdot \mathbf{g}^j$, $g^i_j \equiv \mathbf{g}^i \cdot \mathbf{g}_j = \delta^i_j$, $g^j_i \equiv \mathbf{g}_i \cdot \mathbf{g}^j = \delta^j_i$, $g_{ij} \equiv \mathbf{g}_i \cdot \mathbf{g}_j$. The vector and the tensor based in the reference and the current configurations are called the *Lagrangian vector* and *tensor* and the *Eulerian vector* and *tensor*, respectively. In principle, the Lagrangian and the Eulerian vectors and tensors and their indices are denoted by the uppercase and the lowercase letters, respectively. Further, the tensor based in both of the reference and the current configurations is called the *Lagrangian–Eulerian* or *Eulerian–Lagrangian two-point tensor*, and they are denoted by the uppercase letter, and their indices are denoted by using both of the uppercase and the lowercase letters so as to specify the base vectors in which they are based. The symbol (t) specifying the quantities in the current time is omitted below for the sake of simplicity. The necessity for introducing the primary and the reciprocal bases can be recognized from the typical example that the deformation gradient tensor $\mathbf{F}$, which is the most basic variable for describing the deformation, is specified by the exploiting them as will be shown below.

Regarding the infinitesimal line-element vector $d\mathbf{X}$ in the reference configuration and the infinitesimal line-element $d\mathbf{x}$ in the current configuration to be the primary base vectors $\mathbf{G}_I$ and $\mathbf{g}_i$, respectively, in Eq. (2.15), we obtain the relations between the reference and current base vectors as follows:

$$\mathbf{g}_i = \delta^I_i \mathbf{F} \mathbf{G}_I, \quad \mathbf{G}_I = \delta^i_I \mathbf{F}^{-1} \mathbf{g}_i \tag{4.32}$$

$$\mathbf{F} = \delta^i_I \mathbf{g}_i \otimes \mathbf{G}^I, \quad \mathbf{F}^T = \delta^i_I \mathbf{G}^I \otimes \mathbf{g}_i \tag{4.33}$$

$$\mathbf{F}^{-1} = \delta^I_i \mathbf{G}_I \otimes \mathbf{g}^i,\ \mathbf{F}^{-T} = \delta^I_i \mathbf{g}^i \otimes \mathbf{G}_I \tag{4.34}$$

$$\mathbf{g}^i = \delta^i_I\, \mathbf{F}^{-T} \mathbf{G}^I,\ \mathbf{G}^I = \delta^I_i \mathbf{F}^T \mathbf{g}^i \tag{4.35}$$

from which we have

$$\left.\begin{aligned} \dot{\mathbf{g}}_i &= \delta^I_i \dot{\mathbf{F}} \mathbf{G}_I = \dot{\mathbf{F}} \mathbf{F}^{-1} \mathbf{g}_i = \boldsymbol{l} \mathbf{g}_i \\ \dot{\mathbf{g}}^i &= \delta^I_i \dot{\mathbf{F}}^{-T} \mathbf{G}^I = \dot{\mathbf{F}}^{-T} \mathbf{F}^T \mathbf{g}^i = -\mathbf{F}^{-T} \dot{\mathbf{F}}^T \mathbf{g}^i = -\boldsymbol{l}^T \mathbf{g}^i \end{aligned}\right\} \tag{4.36}$$

noting Eq. (2.74) and $\dot{\mathbf{I}} = (\mathbf{F}\mathbf{F}^{-1})^{\bullet} = \dot{\mathbf{F}}\mathbf{F}^{-1} + \mathbf{F}\dot{\mathbf{F}}^{-1} = \dot{\mathbf{F}}^{-T}\mathbf{F}^T + \mathbf{F}^{-T}\dot{\mathbf{F}}^T = \mathbf{O}$. While the deformation gradient $\mathbf{F}$ is the two-point tensor based in both the current and the reference configurations, it is further regarded to be the two-point (mixed) identity tensor in the convected bases, i.e. the reference reciprocal base $\mathbf{G}^I$ and the current primary base $\mathbf{g}_i$ from Eq. (4.33).

Vector and tensor in the current base are described by Eqs. (1.44), (1.105) and (1.106) as follows:

$$\mathbf{v} = (\mathbf{v} \cdot \mathbf{g}^r)\mathbf{g}_r = (\mathbf{v} \cdot \mathbf{g}_r)\mathbf{g}^r \tag{4.37}$$

$$\mathbf{t} = (\mathbf{g}^r \cdot \mathbf{t}\mathbf{g}^s)\mathbf{g}_r \otimes \mathbf{g}_s = (\mathbf{g}^r \cdot \mathbf{t}\mathbf{g}_s)\mathbf{g}_r \otimes \mathbf{g}^s = (\mathbf{g}_r \cdot \mathbf{t}\mathbf{g}^s)\mathbf{g}^r \otimes \mathbf{g}_s = (\mathbf{g}_r \cdot \mathbf{t}\mathbf{g}_s)\mathbf{g}^r \otimes \mathbf{g}^s \tag{4.38}$$

noting

$$\mathrm{t}_{ij}\mathbf{e}_i \otimes \mathbf{e}_j = (\mathbf{e}_i \cdot \mathbf{t}\mathbf{e}_j)\mathbf{e}_i \otimes \mathbf{e}_j$$

$$= \begin{cases} \{[(\mathbf{e}_i \cdot \mathbf{g}_r)\mathbf{g}^r] \cdot \mathbf{t}[(\mathbf{e}_j \cdot \mathbf{g}_s)\mathbf{g}^s]\}(\mathbf{e}_i \cdot \mathbf{g}^p)\mathbf{g}_p \otimes (\mathbf{e}_j \cdot \mathbf{g}^q)\mathbf{g}_q \\ \quad = (\mathbf{g}^r \cdot \mathbf{t}\mathbf{g}^s)[(\mathbf{e}_i \cdot \mathbf{g}_r)\mathbf{e}_i \cdot \mathbf{g}^p]\mathbf{g}_p \otimes [(\mathbf{e}_j \cdot \mathbf{g}_s)\mathbf{e}_j \cdot \mathbf{g}^q]\mathbf{g}_q \\ \quad = (\mathbf{g}^r \cdot \mathbf{t}\mathbf{g}^s)(\mathbf{g}_r \cdot \mathbf{g}^p)\mathbf{g}_p \otimes (\mathbf{g}_s \cdot \mathbf{g}^q)\mathbf{g}_q \\ \quad = (\mathbf{g}^r \cdot \mathbf{t}\mathbf{g}^s)\mathbf{g}_r \otimes \mathbf{g}_s \\ \{[(\mathbf{e}_i \cdot \mathbf{g}_r)\mathbf{g}^r] \cdot \mathbf{t}[(\mathbf{e}_j \cdot \mathbf{g}^s)\mathbf{g}_s]\}(\mathbf{e}_i \cdot \mathbf{g}^p)\mathbf{g}_p \otimes (\mathbf{e}_j \cdot \mathbf{g}_q)\mathbf{g}^q \\ \quad = (\mathbf{g}^r \cdot \mathbf{t}\mathbf{g}_s)\mathbf{g}_r \otimes \mathbf{g}^s \\ \{[(\mathbf{e}_i \cdot \mathbf{g}^r)\mathbf{g}_r] \cdot \mathbf{t}[(\mathbf{e}_j \cdot \mathbf{g}_s)\mathbf{g}^s]\}(\mathbf{e}_i \cdot \mathbf{g}_p)\mathbf{g}^p \otimes (\mathbf{e}_j \cdot \mathbf{g}^q)\mathbf{g}_q \\ \quad = (\mathbf{g}_r \cdot \mathbf{t}\mathbf{g}^s)\mathbf{g}^r \otimes \mathbf{g}_s \\ \{[(\mathbf{e}_i \cdot \mathbf{g}^r)\mathbf{g}_r] \cdot \mathbf{t}[(\mathbf{e}_j \cdot \mathbf{g}^s)\mathbf{g}_s]\}(\mathbf{e}_i \cdot \mathbf{g}_p)\mathbf{g}^p \otimes (\mathbf{e}_j \cdot \mathbf{g}_q)\mathbf{g}^q \\ \quad = (\mathbf{g}_r \cdot \mathbf{t}\mathbf{g}_s)\mathbf{g}^r \otimes \mathbf{g}^s \end{cases}$$

where the components in the rectangular base are denoted by the roman letter t_{ij} in order to distinguish them from the components denoted by the italic letters t_{ij} in the convected bases. The following expressions in the embedded coordinate system hold from Eqs. (4.37) and (4.38).

$$\mathbf{v} = v^i\mathbf{g}_i = v_i\mathbf{g}^i \tag{4.39}$$

$$\mathbf{t} = t^{ij}\mathbf{g}_i \otimes \mathbf{g}_j = t^i_{\bullet j}\mathbf{g}_i \otimes \mathbf{g}^j = t_i^{\bullet j}\mathbf{g}^i \otimes \mathbf{g}_j = t_{ij}\mathbf{g}^i \otimes \mathbf{g}^j \tag{4.40}$$

where

$$v^i = \mathbf{v} \cdot \mathbf{g}^i, \quad v_i = \mathbf{v} \cdot \mathbf{g}_i \tag{4.41}$$

$$t^{ij} = \mathbf{g}^i \cdot \mathbf{t}\mathbf{g}^j, \quad t^i_{\bullet j} = \mathbf{g}^i \cdot \mathbf{t}\mathbf{g}_j, \quad t_i^{\bullet j} = \mathbf{g}_i \cdot \mathbf{t}\mathbf{g}^j, \quad t_{ij} = \mathbf{g}_i \cdot \mathbf{t}\mathbf{g}_j \tag{4.42}$$

Here, note that the opposite combination of the *contravariant* and the *covariant* (see **Appendix B**) holds between component and base vector of tensor in general, while this fact is obvious for vector by virtue of Eq. (1.47).

(2) **Pull-back and push-forward operations**

The objective time-derivative of tensor would be no more than the rate of variation in tensor observed from the material itself. Then, in order to derive it, we have to incorporate the tensor which changes only when the state of physical quantity is observed to change by the material itself. To this end, we incorporate the tensors defined in the following.

Eulerian tensor: The tensor based in the current configuration, i.e. standing on the current base vectors is called the *Eulerian tensor* (e.g. $\mathbf{b}$, $\mathbf{V}$, $\mathbf{e}$, $\mathbf{R}^E$, $\boldsymbol{l}$, $\boldsymbol{d}$, $\mathbf{w}$, $\boldsymbol{\omega}$, $\boldsymbol{\Omega}^E$, $\boldsymbol{\Omega}^R$, $\boldsymbol{\sigma}$, $\boldsymbol{\tau}$ obeying $\mathbf{t}^* = \mathbf{Q}\mathbf{t}\mathbf{Q}^T$).

Lagrangian tensor: The tensor based in the reference configuration, i.e. standing on the reference base vectors is called the *Lagrangian tensor* (e.g. $\mathbf{C}$, $\boldsymbol{U}$, $\boldsymbol{E}$, $\mathbf{B}$, $\mathbf{R}^L$, $\boldsymbol{\Omega}^L$, $\mathbf{S}$, $\mathfrak{B}$ obeying $\mathbf{T}^* = \mathbf{T}$).

Here, we should notice the following facts.

1) The Eulerian and the Lagrangian tensors possess the same components in the convected coordinate system. Therefore, they are derived by changing the base vectors from each other.
2) The Eulerian tensor is changeable but the Lagrangian tensor remains unaltered when the state observed from the material itself, i.e. components in the convected coordinate system does not change because the current base vectors are changeable but the reference base vectors remain unaltered in the convected coordinate system. In other words, the Eulerian tensor is influenced by the rigid-body rotation but the Lagrangian tensor is independent of that, describing the variation of state observed from the material itself. Here, on the other hand, note that one can represent tensors in any coordinate system. However, the variational rates of physical quantities observed from material itself, which can be adopted in constitutive relation, cannot be described simply in the orthogonal coordinate system.
3) Variation of physical quantity in material itself can be described by the Lagrangian tensor without the influence of superposed rigid-body rotation, while the Eulerian tensor is influenced by the superposed rigid-body rotation. This advantage of the Lagrangian tensor is utilized for the objective time-integration of tensor-valued quantities in numerical calculation as will be described in Sect. 20.10.

Two-point tensor: The tensor based in both the current and the reference configurations, i.e. standing on the current and reference base vectors is called the *two-point tensor* (e.g. $\mathbf{F}$, $\mathbf{R}$, $\boldsymbol{\Pi}(=\mathbf{P}^T)$ obeying $\mathbf{T}^* = \mathbf{QT}$ or $\mathbf{TQ}^T$).

The transformation from the Eulerian tensor to the Lagrangian tensor and its inverse are called the *pull-back* and the *push-forward* operations, respectively, and executed by multiplying the two-point tensor (Lagrangian-Eulerian tensor, e.g. $\mathbf{F}, \mathbf{F}^{-T}, \mathbf{R}$ for the pull-back and Eulerian-Lagrangian tensor, e.g. $\mathbf{F}^T, \mathbf{F}^{-1}, \mathbf{R}^T$ for the push-forward from the left and their inverse ones from the right) describing the deformation and/or rotation in general. In what follows, the concrete examples by the deformation gradient tensor are shown noting Eqs. (4.33), (4.34), (4.39) and (4.40) as follows:

$$\boxed{\begin{array}{l} \overleftarrow{\mathbf{v}}^G = \delta^I_i v^i \mathbf{G}_I = \mathbf{F}^{-1}\mathbf{v},\ \overleftarrow{\mathbf{v}}_G = \delta^i_I v_i \mathbf{G}^I = \mathbf{F}^T\mathbf{v} \\ \overrightarrow{\mathbf{V}}^g = \delta^i_I V^I \mathbf{g}_i = \mathbf{F}\mathbf{V},\ \overrightarrow{\mathbf{V}}_g = \delta^I_i V_I \mathbf{g}^i = \mathbf{F}^{-T}\mathbf{V} \end{array}} \tag{4.43}$$

$$\boxed{\begin{aligned} &\left.\begin{aligned}\overleftarrow{\mathbf{t}}^{GG} = \delta^I_i \delta^J_j t^{ij} \mathbf{G}_I \otimes \mathbf{G}_J = \mathbf{F}^{-1}\mathbf{t}\mathbf{F}^{-T},\ \overleftarrow{\mathbf{t}}^{G}_{\bullet G} = \delta^I_i \delta^j_J t^{i}_{\bullet j} \mathbf{G}_I \otimes \mathbf{G}^J = \mathbf{F}^{-1}\mathbf{t}\mathbf{F} \\ \overleftarrow{\mathbf{t}}^{\bullet G}_{G} = \delta^i_I \delta^j_J t_i^{\bullet j} \mathbf{G}_I \otimes \mathbf{G}_J = \mathbf{F}^{T}\mathbf{t}\mathbf{F}^{-T},\ \overleftarrow{\mathbf{t}}_{GG} = \delta^i_I \delta^j_J t^{ij} \mathbf{G}_I \otimes \mathbf{G}^J = \mathbf{F}^{T}\mathbf{t}\mathbf{F}\end{aligned}\right\} \\ &\left.\begin{aligned}\overrightarrow{\mathbf{T}}^{gg} = \delta^i_I \delta^j_J T^{IJ} \mathbf{g}_i \otimes \mathbf{g}_j = \mathbf{F}\mathbf{T}\mathbf{F}^{T},\ \overrightarrow{\mathbf{T}}^{g}_{\bullet g} = \delta^i_I \delta^J_j T^{I}_{\bullet J} \mathbf{g}^i \otimes \mathbf{g}^j = \mathbf{F}\mathbf{T}\mathbf{F}^{-1} \\ \overrightarrow{\mathbf{T}}^{\bullet g}_{g} = \delta^I_i \delta^j_J T_I^{\bullet J} \mathbf{g}_i \otimes \mathbf{g}_j = \mathbf{F}^{-T}\mathbf{T}\mathbf{F}^{T},\ \overrightarrow{\mathbf{T}}_{gg} = \delta^I_i \delta^J_j T^{IJ} \mathbf{g}^i \otimes \mathbf{g}^j = \mathbf{F}^{-T}\mathbf{T}\mathbf{F}^{-1}\end{aligned}\right\}\end{aligned}} \tag{4.44}$$

noting

$$\mathbf{F}^{-1}\mathbf{v} = \delta^I_i \mathbf{G}_I \otimes \mathbf{g}^i v^r \mathbf{g}_r = \delta^I_i v^i \mathbf{G}_I$$

$$\mathbf{F}^{-1}\mathbf{t}\mathbf{F}^{-T} = \delta^I_i \mathbf{G}_I \otimes \mathbf{g}^i t^{pq} \mathbf{g}_p \otimes \mathbf{g}_q \delta^J_j \mathbf{g}^j \otimes \mathbf{G}_J = \delta^I_i \mathbf{G}_I t^{ij} \delta^J_j \otimes \mathbf{G}_J$$

as the example of Eq. $(4.44)_1$. The over arrow turning left ($\leftarrow$) and right ($\rightarrow$) is added for the pull-back and the push-forward operation, respectively. Further, the uppercase letter index G is added in order to specify the replacement of the current base to the reference base in the pull-back operation and the lowercase letter index g is added in order to specify the replacement of the reference base to the current base in the push-forward operation, and they are put in the lower or upper position for the covariant or the contravariant component, respectively, while these symbols were devised by Hashiguchi (2011). Here, note that the pulled-back and push-forward operations of tensors in higher order than two cannot be expressed in the symbolic notations by the multiplications of the deformation gradient tensor but can be represented only by exchanging the current base vectors to the reference base vectors and its inverse as far as quite particular definitions of tensor operations are not adopted.

It is noteworthy that the differences between the contravariant and the covariant forms in the pull-back and the push-forward operations diminish when only rotation is taken account leading to $\mathbf{F} = \mathbf{F}^{-T} = \mathbf{R}$, $\mathbf{F}^{-1} = \mathbf{F}^{T} = \mathbf{R}^{T}$. The tensor ${}^{R}\overleftarrow{\mathbf{t}} = \mathbf{R}^{T}\mathbf{t}\mathbf{R}$ pulled back only by the rotation, regarding $\mathbf{F} = \mathbf{R}$, is called the *rotation-free tensor* or *rotation-insensitive tensor* since the rotation $\mathbf{R}$ is excluded from the Eulerian tensor.

The Lagrangian tensors $\mathbf{C}$, $\boldsymbol{E}$ and $\mathbf{S}$ described in Sect. 4.2 are derived by the pull-back from the Eulerian tensors $\mathbf{g}$, $\boldsymbol{e}$ and $\boldsymbol{\tau}$ as follows:

$$\begin{aligned} \mathbf{C} &= \overleftarrow{\mathbf{g}}_{GG} = \delta^i_I \delta^j_J g_{ij} \mathbf{G}^I \otimes \mathbf{G}^J = \mathbf{F}^{T}\mathbf{g}\mathbf{F} = \mathbf{F}^{T}\mathbf{F} \\ \boldsymbol{E} &= \overleftarrow{\boldsymbol{e}}_{GG} = \delta^i_I \delta^j_J e_{ij} \mathbf{G}^I \otimes \mathbf{G}^J = \mathbf{F}^{T}\boldsymbol{e}\mathbf{F} \\ \mathbf{S} &= \overleftarrow{\boldsymbol{\tau}}^{GG} = \delta^I_i \delta^J_j \tau^{ij} \mathbf{G}_I \otimes \mathbf{G}_J = \mathbf{F}^{-1}\boldsymbol{\tau}\mathbf{F}^{-T} \end{aligned}$$

and the two-point tensors $\mathbf{F}$ and $\boldsymbol{\Pi}$ are derived by the pull-back from the Eulerian tensors $\mathbf{g}$ and $\boldsymbol{\tau}$ as follows:

$$\mathbf{F} = \left\{ \begin{array}{l} \overleftarrow{\mathbf{g}}{}^{\hat{g}}_{\bullet G} = \delta^i_I \mathbf{g}_i \otimes \mathbf{G}^I = \mathbf{g}_i \otimes \mathbf{g}^i \delta^r_I \mathbf{g}_r \otimes \mathbf{G}^I \\ \overleftarrow{\mathbf{g}}_{\hat{g}G} = \delta^j_J g_{ij} \mathbf{g}^i \otimes \mathbf{G}^J = g_{ij} \mathbf{g}^i \otimes \mathbf{g}^j \delta^r_J \mathbf{g}_r \otimes \mathbf{G}^J \end{array} \right\} = \mathbf{gF}$$

$$\mathbf{\Pi} = \left\{ \begin{array}{l} \overleftarrow{\boldsymbol{\tau}}{}^{\hat{g}G} = \delta^J_j \tau^{ij} \mathbf{g}_i \otimes \mathbf{G}_J = \tau^{ij} \mathbf{g}_i \otimes \mathbf{g}_j \delta^J_r \mathbf{g}^r \otimes \mathbf{G}_J \\ \overleftarrow{\boldsymbol{\tau}}{}_{\hat{g}}^{\bullet G} = \delta^J_j \tau_i^{\bullet j} \mathbf{g}^i \otimes \mathbf{G}_J = \tau_i^{\bullet j} \mathbf{g}^i \otimes \mathbf{g}_j \delta^J_r \mathbf{g}^r \otimes \mathbf{G}_J \end{array} \right\} = \boldsymbol{\tau} \mathbf{F}^{-T}$$

noting Eqs. (4.32), (4.35) and (4.44). The over hat symbol $(^{\wedge})$ specifies the un-exchange of base vector. $\mathbf{\Pi}$ is the induced two-point tensor from the Kirchhoff stress $\boldsymbol{\tau}$. On the other hand, $\mathbf{F}$ is regarded as the inherent two-point tensors. Here, it can be called the identity tensor in the broad sense, since it possesses the components of the Kronecker's delta in both the current and the reference bases. $\mathbf{F}$ was called the two-point tensor in Eq. (2.14) for the orthogonal coordinate system, where it was described by the components of position vectors in the current and the reference configurations in the orthogonal coordinate systems. Then, the physical meaning of the two-point tensor would be obscure by the expression in the orthogonal coordinate system. On the other hand, the physical meaning of the two-point tensor would be captured clearly by the expression in the convected coordinate system such that it is based extending over the reference and the current bases which are not arbitrary but composed by the definite sets of the embedded base vectors.

The Eulerian tensor changes even when the state of physical quantity observed from the material itself does not change under a material rotation, since the base vectors change even by a rotation of the material. On the other hand, the Lagrangian tensor pulled-back to the ref-erence configuration with the fixed base vectors does not change in the material rotation and thus it called the *rotation-free tensor*, while the Lagrangian tensor inherits the components in the Eulerian tensor. Then, the constitutive relation described by the Lagrangian tensors is used in the deformation analysis. The Eulerian base vectors are calculated by the push-forward op-eration of the Lagrangian (reference) base vectors through the deformation gradient tensor, which is required to capture the Eulerian tensor in the current configuration from the Lagrangian tensor.

The physical and geometrical interpretations for the relations between the above-mentioned Eulerian and Lagrangian tensors can be referred to Hashiguchi and Yamakawa (2012).

(3) **Convected time-derivatives: Objective rate of tensor**

The material-time derivative of the vector $\mathbf{v}$ is described in the current primary base $\{\mathbf{g}_i\}$ and the current reciprocal base $\{\mathbf{g}^i\}$ from Eq. (4.39) by

$$\dot{\mathbf{v}} = \left\{ \begin{array}{l} (v^r \mathbf{g}_r)^{\bullet} = \dot{v}^r \mathbf{g}_r + v^r \dot{\mathbf{g}}_r \\ (v_r \mathbf{g}^r)^{\bullet} = \dot{v}_r \mathbf{g}^r + v_r \dot{\mathbf{g}}^r \end{array} \right. \tag{4.45}$$

The first terms in the right-hand sides of Eq. (4.45) represent the rates of the vector $\mathbf{v}$ observed from the embedded coordinate system and thus they are called the *convected time-derivative*. In other words, they mean the rate of physical quantity observed from the embedded coordinate system having the base vectors composed of line-elements etched in a material. Also, they are interpreted as the rates observed from material itself and thus they are independent of rigid-body rotation, possessing the *objectivity*. Here, however, note that the rotation of the embedded base is different from the rotation of the substructure of material in general as known from the fact that the movements of line-elements etched in material coincides with the deformed geometrical appearance of material but it does not necessarily coincide with the movements of material fibers representing the substructure of anisotropic material. The convected time-derivatives of vector are expressed from Eq. (4.45) as

$$\boxed{\begin{array}{l} \overset{\rightleftarrows}{\dot{\mathbf{v}}}{}^{g} \equiv \dot{v}^r\mathbf{g}_r = \dot{\mathbf{v}} - v^r\dot{\mathbf{g}}_r = \dot{\mathbf{v}} - \boldsymbol{l}\mathbf{v} = \dot{\mathbf{v}} + \mathbf{F}\dot{\mathbf{F}}^{-1}\mathbf{v} = \mathbf{F}(\mathbf{F}^{-1}\mathbf{v})^{\bullet} = (\overrightarrow{\overleftarrow{\mathbf{v}}{}^{G}})^{\bullet g} \\ \overset{\rightleftarrows}{\dot{\mathbf{v}}}_{g} \equiv \dot{v}_r\mathbf{g}^r = \dot{\mathbf{v}} - v_r\dot{\mathbf{g}}^r = \dot{\mathbf{v}} + \boldsymbol{l}^T\mathbf{v} = \dot{\mathbf{v}} + \mathbf{F}^{-T}\dot{\mathbf{F}}^{T}\mathbf{v} = \mathbf{F}^{-T}(\mathbf{F}^{T}\mathbf{v})^{\bullet} = (\overrightarrow{\overleftarrow{\mathbf{v}}_{G}})^{\bullet}_{g} \end{array}} \tag{4.46}$$

by using Eq. (4.36), noting $v^r\dot{\mathbf{g}}_r = v^r\boldsymbol{l}\mathbf{g}_r = \boldsymbol{l}\mathbf{v}$.

Analogously to the vector described above, the material-time derivative of tensor in the current base is described from Eq. (4.40) by

$$\dot{\mathbf{t}} = \begin{cases} (t^{ij}\mathbf{g}_i \otimes \mathbf{g}_j)^{\bullet} = \dot{t}^{ij}\mathbf{g}_i \otimes \mathbf{g}_j + t^{ij}\dot{\mathbf{g}}_i \otimes \mathbf{g}_j + t^{ij}\mathbf{g}_i \otimes \dot{\mathbf{g}}_j \\ (t^{i}_{\bullet j}\mathbf{g}_i \otimes \mathbf{g}^j)^{\bullet} = \dot{t}^{i}_{\bullet j}\mathbf{g}_i \otimes \mathbf{g}^j + t^{i}_{\bullet j}\dot{\mathbf{g}}_i \otimes \mathbf{g}^j + t^{i}_{\bullet j}\mathbf{g}_i \otimes \dot{\mathbf{g}}^j \\ (t_{i}^{\bullet j}\mathbf{g}^i \otimes \mathbf{g}_j)^{\bullet} = \dot{t}_{i}^{\bullet j}\mathbf{g}^i \otimes \mathbf{g}_j + t_{i}^{\bullet j}\dot{\mathbf{g}}^i \otimes \mathbf{g}_j + t_{i}^{\bullet j}\mathbf{g}^i \otimes \dot{\mathbf{g}}_j \\ (t_{ij}\mathbf{g}^i \otimes \mathbf{g}^j)^{\bullet} = \dot{t}_{ij}\mathbf{g}^i \otimes \mathbf{g}^j + t_{ij}\dot{\mathbf{g}}^i \otimes \mathbf{g}^j + t_{ij}\mathbf{g}^i \otimes \dot{\mathbf{g}}^j \end{cases} \tag{4.47}$$

Exploiting Eqs. (1.106) and (4.36) in Eq. (4.47), the following four types of convected time-derivatives are derived.

$$\boxed{\begin{array}{l} \overset{\rightleftarrows}{\dot{\mathbf{t}}}{}^{gg} \equiv \dot{t}^{ij}\mathbf{g}_i \otimes \mathbf{g}_j = \dot{\mathbf{t}} - \boldsymbol{l}\mathbf{t} - \mathbf{t}\boldsymbol{l}^T = \dot{\mathbf{t}} + \mathbf{F}\dot{\mathbf{F}}^{-1}\mathbf{t} + \mathbf{t}\dot{\mathbf{F}}^{-T}\mathbf{F}^{T} = \mathbf{F}(\mathbf{F}^{-1}\mathbf{t}\mathbf{F}^{-T})^{\bullet}\mathbf{F}^{T} = (\overrightarrow{\overleftarrow{\mathbf{t}}{}^{GG}})^{\bullet gg} \\ \overset{\rightleftarrows}{\dot{\mathbf{t}}}{}^{g}_{\bullet g} \equiv \dot{t}^{i}_{\bullet j}\mathbf{g}_i \otimes \mathbf{g}^j = \dot{\mathbf{t}} - \boldsymbol{l}\mathbf{t} + \mathbf{t}\boldsymbol{l} = \dot{\mathbf{t}} + \mathbf{F}\dot{\mathbf{F}}^{-1}\mathbf{t} + \mathbf{t}\dot{\mathbf{F}}\mathbf{F}^{-1} = \mathbf{F}(\mathbf{F}^{-1}\mathbf{t}\mathbf{F})^{\bullet}\mathbf{F}^{-1} = (\overrightarrow{\overleftarrow{\mathbf{t}}{}^{G}_{\bullet G}})^{\bullet g}_{\bullet g} \\ \overset{\rightleftarrows}{\dot{\mathbf{t}}}{}_{g}^{\bullet g} \equiv \dot{t}_{i}^{\bullet j}\mathbf{g}^i \otimes \mathbf{g}_j = \dot{\mathbf{t}} + \boldsymbol{l}^T\mathbf{t} - \mathbf{t}\boldsymbol{l}^T = \dot{\mathbf{t}} + \mathbf{F}^{-T}\dot{\mathbf{F}}^{T}\mathbf{t} + \mathbf{t}\dot{\mathbf{F}}^{-T}\mathbf{F}^{T} = \mathbf{F}^{-T}(\mathbf{F}^{T}\mathbf{t}\mathbf{F}^{-T})^{\bullet}\mathbf{F}^{T} = (\overrightarrow{\overleftarrow{\mathbf{t}}{}_{G}^{\bullet G}})^{\bullet \bullet g}_{g} \\ \overset{\rightleftarrows}{\dot{\mathbf{t}}}_{gg} \equiv \dot{t}_{ij}\mathbf{g}^i \otimes \mathbf{g}^j = \dot{\mathbf{t}} + \boldsymbol{l}^T\mathbf{t} + \mathbf{t}\boldsymbol{l} = \dot{\mathbf{t}} + \mathbf{F}^{-T}\dot{\mathbf{F}}^{T}\mathbf{t} + \mathbf{t}\dot{\mathbf{F}}\mathbf{F}^{-1} = \mathbf{F}^{-T}(\mathbf{F}^{T}\mathbf{t}\mathbf{F})^{\bullet}\mathbf{F}^{-1} = (\overrightarrow{\overleftarrow{\mathbf{t}}_{GG}})^{\bullet}_{gg} \end{array}} \tag{4.48}$$

The notations $\overset{\rightleftarrows}{\dot{\mathbf{v}}}{}^{g}$, $\overset{\rightleftarrows}{\dot{\mathbf{t}}}{}^{gg}$, $\overset{\rightleftarrows}{\dot{\mathbf{t}}}{}^{g}_{\bullet g}$, etc. were used first by Hashiguchi (2011) to specify the objective time-derivatives and their types of contravariant and covariant, where the indices "g" are added in the upper and lower positions in order to specify the contravariant and the covariant expressions (component positions), respectively, of vector and tensor (**Appendix B**). The objective time-derivative is the rate of tensor observed from material itself but it can be also interpreted from Eqs. (4.46) and (4.48) to be the *current expression of rate of Lagrangian vector* or *tensor* (transformation of vector or tensor physical quantity described in the reference base to the current base), the components of which is independent of a rotation of material. There exist the two and the four types of convected rates of vector and tensor, respectively, as shown in Eqs. (4.46) and (4.48). In particular, $\overset{\rightleftarrows}{\mathbf{t}}{}^{gg}$ and $\overset{\rightleftarrows}{\mathbf{t}}_{gg}$ are the general forms of the *Oldroyd rate* (Oldroyd, 1950)) and the *Cotter-Rivlin rate* (Cotter and Rivlin, 1955), respectively, while the former is called the *Lie derivative* based on the hyperelasticity (cf. e.g. Bonet and Wood 1997; Simo 1998; Belytschko et al. 2014; de Souza-Neto et al. 2008).

The convected derivatives in Eqs. (4.46) and (4.48) satisfy the objectivity obviously because they are based on the rates of tensor observed by a material itself. In addition, this fact can be mathematically confirmed as shown below for Eqs. $(4.46)_1$ and $(4.48)_1$ as examples, noting Eq. (4.10).

$$\overset{\rightleftarrows}{\mathbf{v}}{}^{*g} = \mathbf{F}^*(\mathbf{F}^{*-1}\mathbf{v}^*)^{\bullet} = \mathbf{QF}[(\mathbf{QF})^{-1}\mathbf{Qv}]^{\bullet} = \mathbf{Q}[\mathbf{F}(\mathbf{F}^{-1}\mathbf{v})^{\bullet}] \tag{4.49}$$

or

$$\begin{aligned}\overset{\rightleftarrows}{\mathbf{v}}{}^{*g} &= \dot{\mathbf{v}}^* - \boldsymbol{l}^*\mathbf{v}^* = (\mathbf{Qv})^{\bullet} - \mathbf{Q}(\boldsymbol{l} - \boldsymbol{\Omega})\mathbf{Q}^T\mathbf{Qv} = \dot{\mathbf{Q}}\mathbf{v} + \mathbf{Q}\dot{\mathbf{v}} - \mathbf{Q}(\boldsymbol{l} - \dot{\mathbf{Q}}^T\mathbf{Q})\mathbf{v}\\ &= \mathbf{Q}\dot{\mathbf{v}} + \dot{\mathbf{Q}}\mathbf{v} - \mathbf{Q}(\boldsymbol{l} + \mathbf{Q}^T\dot{\mathbf{Q}})\mathbf{v} = \mathbf{Q}(\dot{\mathbf{v}} - \boldsymbol{l}\mathbf{v})\end{aligned} \tag{4.50}$$

and

$$\begin{aligned}\overset{\rightleftarrows}{\mathbf{t}}{}^{*gg} &= \mathbf{F}^*(\mathbf{F}^{*-1}\mathbf{t}^*\mathbf{F}^{*-T})^{\bullet}\mathbf{F}^{*T} = \mathbf{QF}[(\mathbf{QF})^{-1}\mathbf{QtQ}^T(\mathbf{QF}^{-T})]^{\bullet}(\mathbf{QF}^T)\\ &= \mathbf{QF}(\mathbf{F}^{-1}\mathbf{Q}^T\mathbf{QtQ}^T\mathbf{QF}^{-T})^{\bullet}\mathbf{F}^T\mathbf{Q}^T = \mathbf{Q}[\mathbf{F}(\mathbf{F}^{-1}\mathbf{tF}^{-T})^{\bullet}\,\mathbf{F}^T]\mathbf{Q}^T\end{aligned} \tag{4.51}$$

or

$$\begin{aligned}\overset{\rightleftarrows}{\mathbf{t}}{}^{*gg} &= \dot{\mathbf{t}}^* - \boldsymbol{l}^*\mathbf{t}^* - \mathbf{t}^*\boldsymbol{l}^{*T} = (\mathbf{QtQ}^T)^{\bullet} - \mathbf{Q}(\boldsymbol{l}-\boldsymbol{\Omega})\mathbf{Q}^T\mathbf{QtQ}^T - \mathbf{QtQ}^T[\mathbf{Q}(\boldsymbol{l}-\boldsymbol{\Omega})\mathbf{Q}^T]^T\\ &= \dot{\mathbf{Q}}\mathbf{tQ}^T + \mathbf{Q}\dot{\mathbf{t}}\mathbf{Q}^T + \mathbf{Qt}\dot{\mathbf{Q}}^T - \mathbf{Q}(\boldsymbol{l}-\boldsymbol{\Omega})\mathbf{tQ}^T - \mathbf{QtQ}^T\mathbf{Q}(\boldsymbol{l}^T - \boldsymbol{\Omega}^T)\mathbf{Q}^T\\ &= \dot{\mathbf{Q}}\mathbf{tQ}^T + \mathbf{Q}\dot{\mathbf{t}}\mathbf{Q}^T + \mathbf{Qt}\dot{\mathbf{Q}}^T - \mathbf{Q}(\boldsymbol{l}-\dot{\mathbf{Q}}^T\mathbf{Q})\mathbf{tQ}^T - \mathbf{QtQ}^T\mathbf{Q}(\boldsymbol{l}^T - \mathbf{Q}^T\dot{\mathbf{Q}})\mathbf{Q}^T\\ &= \dot{\mathbf{Q}}\mathbf{tQ}^T + \mathbf{Q}\dot{\mathbf{t}}\mathbf{Q}^T + \mathbf{Qt}\dot{\mathbf{Q}}^T - \mathbf{Q}\boldsymbol{l}\mathbf{tQ}^T - \mathbf{QQ}^T\dot{\mathbf{Q}}^T\mathbf{tQ}^T - \mathbf{Qt}\boldsymbol{l}^T\mathbf{Q}^T - \mathbf{Qt}\dot{\mathbf{Q}}^T\mathbf{QQ}^T\\ &= \mathbf{Q}(\dot{\mathbf{t}} - \boldsymbol{l}\mathbf{t} - \mathbf{t}\boldsymbol{l}^T)\mathbf{Q}^T\end{aligned} \tag{4.52}$$

4.5 Corotational Rate Tensors

The convected time-derivatives satisfy the objectivity. However, the objectivity is satisfied even in specialized convected time-derivatives as will be shown below by the particular case in which only the rotation of material is considered.

Let the spin tensors obeying the following coordinate transformation which is seen in Eqs. (4.17), (4.19), etc. be designated by the symbol $\boldsymbol{\omega}\ (= -\boldsymbol{\omega}^T)$ collectively.

$$\boldsymbol{\omega}^* = \mathbf{Q}(\boldsymbol{\omega} - \boldsymbol{\Omega})\mathbf{Q}^T = \mathbf{Q}\boldsymbol{\omega}\mathbf{Q}^T + \overline{\boldsymbol{\Omega}} \tag{4.53}$$

It is readily known from Eqs. (4.46) and (4.48) that the rates of vector and tensor obtained by replacing the velocity gradient $\boldsymbol{l}$ to the spin tensor $\boldsymbol{\omega}$, i.e. by ignoring the rate of deformation are described as follows:

$$\boxed{\overset{\circ}{\mathbf{v}} = \dot{\mathbf{v}} - \boldsymbol{\omega}\mathbf{v}} \tag{4.54}$$

$$\boxed{\overset{\circ}{\mathbf{t}} = \dot{\mathbf{t}} - \boldsymbol{\omega}\mathbf{t} + \mathbf{t}\boldsymbol{\omega}} \tag{4.55}$$

which obey the objective transformation rules and are referred to as the *corotational rate* or *corotational time-derivative*. Their fulfillment of objectivity is obvious from the proof for the convected time-derivatives described in the foregoing, noting that $\boldsymbol{\omega}$ obeys the identical transformation rule to that of $\boldsymbol{l}$. However, Eqs. (4.54) and (4.55) are inapplicable to deformation analysis as far as $\boldsymbol{\omega}$ is not given explicitly as a physical quantity. Needless to say, they must be chosen so as to reflect the rotational rate of material appropriately. In what follows, typical explicit corotational rate vectors and tensors will be shown.

The replacement of $\mathbf{F} = \mathbf{R}$ in Eqs. (4.46) and (4.48) leads Eqs. (4.54) and (4.55) to the corotational rate with the relative spin $\boldsymbol{\omega} = \boldsymbol{\Omega}^R$ tensor in Eq. (2.85) as follows:

$$\boxed{\overset{\circ}{\mathbf{v}}{}^R \equiv \mathbf{R}(\mathbf{R}^T\mathbf{V})^{\bullet} = \dot{\mathbf{v}} - \boldsymbol{\Omega}^R\mathbf{v}} \tag{4.56}$$

$$\boxed{\overset{\circ}{\mathbf{t}}{}^R \equiv \mathbf{R}(\mathbf{R}^T\mathbf{t}\mathbf{R})^{\bullet}\mathbf{R}^T = \dot{\mathbf{t}} - \boldsymbol{\Omega}^R\mathbf{t} + \mathbf{t}\boldsymbol{\Omega}^R} \tag{4.57}$$

$\overset{\circ}{\mathbf{t}}{}^R$ in Eq. (4.57) is called the *Green-Naghdi rate* (Green-Naghdi 1965). The Green-Naghdi rate depends on the initial value of **R** describing the rotation of material. In other words, it is influenced by the estimation of initial state of rotation even in isotropic materials. Therefore, it lacks the objectivity in the broad physical sense, which requires the independence of deformation behavior on the rigid-body rotation.

Further, by choosing $\boldsymbol{\omega} = \mathbf{w}(={}_t\dot{\mathbf{R}}(t))$, i.e. the relative spin in Eq. (2.81), Eqs. (4.54) and (4.55) lead to

$$\boxed{\overset{\circ}{\mathbf{v}}{}^{w} \equiv \dot{\mathbf{v}} - \mathbf{w}\mathbf{v}} \tag{4.58}$$

$$\boxed{\overset{\circ}{\mathbf{t}}{}^{w} \equiv \dot{\mathbf{t}} - \mathbf{w}\mathbf{t} + \mathbf{t}\mathbf{w}} \tag{4.59}$$

Equation (4.59) is called the *Zaremba-Jaumann rate* (Zaremba 1903; Jaumann 1911).

The accurate numerical time-integration scheme of the corotational rate tensors will be described in Sect. 20.10.

The objectivity is the common requirement for constitutive equations. One can make various objective rates which are given by the convected rates and classified as shown in Eq. (4.46) for vector and in Eq. (4.48) for second-order tensor. However, the other consideration is required for the judgment which one of them is appropriate. In fact, the objective rates described above are determined solely by a geometrical change of outward appearance of material. On the other hand, the spin which reflects the mechanical response is the spin of substructure (microstructure) in material. However, the substructure is invisible from the outward appearance. Generally speaking, the spin of the substructure is not so large as that given by the continuum spin. An explicit form of the spin of substructure in the elastoplastic deformation will be described in Chap. 16.

4.6 Various Stress Rate Tensors

Various rates of the Cauchy stress $\boldsymbol{\sigma}$ and the Kirchhoff stress $\boldsymbol{\tau}\,(= J\boldsymbol{\sigma})$ can be obtained from the aforementioned convected and corotational time-derivatives as will be shown in this section. Corotational time-derivative with a spin tensor is designated by the symbol $(^{\circ})$ as shown in the last section. On the other and, the time-derivatives other than corotational time derivatives are designated by the symbol $(^{\Delta})$.

(a) Contravariant convected rates

Based on Eq. $(4.48)_1$, the contravariant convected rate of the Cauchy stress $\boldsymbol{\sigma}$ is given by

$$\overset{\Delta}{\boldsymbol{\sigma}}{}^{Ol} \equiv \overset{\overset{\rightarrow}{\leftarrow}\bullet}{\boldsymbol{\sigma}}{}^{gg} = \mathbf{F}(\mathbf{F}^{-1}\boldsymbol{\sigma}\mathbf{F}^{-T})^{\bullet}\mathbf{F}^{T} = \mathbf{F}\,\overset{\bullet}{\overline{\mathbf{S}/J}}\,\mathbf{F}^{T} = \dot{\boldsymbol{\sigma}} - \boldsymbol{l}\boldsymbol{\sigma} - \boldsymbol{\sigma}\boldsymbol{l}^{T}\,(= \overset{\Delta}{\boldsymbol{\sigma}}{}^{OlT}) \tag{4.60}$$

which is termed the *Oldroyd rate of Cauchy stress* (Oldroyd 1950). Likewise, it holds for the Kirchhoff stress that

$$\overset{\Delta}{\boldsymbol{\tau}}{}^{Ol} \equiv \overset{\rightleftarrows}{\boldsymbol{\tau}}{}^{gg} = \mathbf{F}(\mathbf{F}^{-1}\boldsymbol{\tau}\mathbf{F}^{-T})^{\bullet}\mathbf{F}^{T} = \mathbf{F}\dot{\mathbf{S}}\,\mathbf{F}^{T} = \dot{\boldsymbol{\tau}} - \boldsymbol{l}\boldsymbol{\tau} - \boldsymbol{\tau}\boldsymbol{l}^{T} (= \overset{\Delta}{\boldsymbol{\tau}}{}^{OlT}) \tag{4.61}$$

which is termed the Oldroyd rate of Kirchhoff stress.

Further,

$$\begin{aligned}\overset{\Delta}{\boldsymbol{\sigma}}{}^{Tr} &\equiv J^{-1}\overset{\Delta}{\boldsymbol{\tau}}{}^{Ol} = J^{-1}\overset{\rightleftarrows}{(J\boldsymbol{\sigma})}{}^{gg} = J^{-1}\mathbf{F}(\mathbf{F}^{-1}(J\boldsymbol{\sigma})\mathbf{F}^{-T})^{\bullet}\mathbf{F}^{T} = J^{-1}\mathbf{F}\,\dot{\mathbf{S}}\,\mathbf{F}^{T} = \overset{\Delta}{\boldsymbol{\sigma}}{}^{Ol} + \boldsymbol{\sigma}\,\mathrm{tr}\,\mathbf{d}\\ &= \dot{\boldsymbol{\sigma}} - \boldsymbol{l}\boldsymbol{\sigma} - \boldsymbol{\sigma}\,\boldsymbol{l}^{T} + \boldsymbol{\sigma}\,\mathrm{tr}\,\mathbf{d}\ (= \mathring{\boldsymbol{\sigma}}{}^{TrT})\end{aligned} \tag{4.62}$$

is termed the *Truesdell rate of Cauchy stress*.

(b) Covariant-contravariant convected rates

The covariant-contravariant convected rate of the Kirchhoff stress $\boldsymbol{\tau}$ is given from Eq. $(4.48)_3$ as

$$\overset{\Delta}{\boldsymbol{\tau}} \equiv \overset{\rightarrow}{\overset{\leftarrow}{\boldsymbol{\tau}}}{}_{g}^{\bullet g} = \mathbf{F}^{-T}(\mathbf{F}^{T}\boldsymbol{\tau}\mathbf{F}^{-T})^{\bullet}\,\mathbf{F}^{T} = \dot{\boldsymbol{\tau}} + \boldsymbol{l}^{T}\boldsymbol{\tau} - \boldsymbol{\tau}\,\boldsymbol{l}^{T}\ (\neq \overset{\Delta}{\boldsymbol{\tau}}{}^{T}) \tag{4.63}$$

The particular case of the rate in Eq. (4.63) is given by

$$_{\Pi}\overset{\Delta}{\boldsymbol{\tau}} \equiv \overset{\rightleftarrows}{\boldsymbol{\tau}}{}_{\hat{g}}^{\bullet g} = (\boldsymbol{\tau}\mathbf{F}^{-T})^{\bullet}\,\mathbf{F}^{T} = \dot{\boldsymbol{\Pi}}\mathbf{F}^{T} = \dot{\boldsymbol{\tau}} - \boldsymbol{\tau}\,\boldsymbol{l}^{T}\ (\neq {}_{\Pi}\overset{\Delta}{\boldsymbol{\tau}}{}^{T}) \tag{4.64}$$

which is termed the *relative 1st Piola-Kirchhoff stress rate*. The following stress rate is defined as the nominal stress rate in Eqs. (3.36) and (3.37).

$$_{\Pi}\overset{\Delta}{\boldsymbol{\sigma}} \equiv \frac{1}{J}\,{}_{\Pi}\mathring{\boldsymbol{\tau}} = \dot{\boldsymbol{\sigma}} - \boldsymbol{\sigma}\boldsymbol{l}^{T} + \boldsymbol{\sigma}\ \mathrm{tr}\,\mathbf{d}\,(\neq {}_{\Pi}\overset{\Delta}{\boldsymbol{\sigma}}{}^{T}) \tag{4.65}$$

which is the nominal stress rate and used for the equilibrium equation of rate-form in the current configuration as described in Eq. (3.41).

(c) Covariant convected rates

The *covariant convected rate* of Cauchy stress is given from Eq. $(4.48)_4$ as

$$\overset{\Delta}{\boldsymbol{\sigma}}{}^{CR} \equiv \overset{\leftleftarrows}{\boldsymbol{\sigma}}{}_{gg} = \mathbf{F}^{-T}(\mathbf{F}^{T}\boldsymbol{\sigma}\,\mathbf{F})^{\bullet}\mathbf{F}^{-1} = \dot{\boldsymbol{\sigma}} + \boldsymbol{l}^{T}\boldsymbol{\sigma} + \boldsymbol{\sigma}\boldsymbol{l}\,(= \overset{\Delta}{\boldsymbol{\sigma}}{}^{CRT}) \tag{4.66}$$

which is termed the *Cotter-Rivlin rate* of Cauchy stress (Cotter and Rivlin 1995). Likewise, the covariant convected rate of Kirchhoff stress is given by

$$\overset{\Delta}{\boldsymbol{\tau}}{}^{CR} \equiv \overset{\rightleftarrows}{\boldsymbol{\tau}}_{gg} = \mathbf{F}^{-T}(\mathbf{F}^{T}\boldsymbol{\tau}\mathbf{F})^{\bullet}\mathbf{F}^{-1} = \dot{\boldsymbol{\tau}} + \boldsymbol{l}^{T}\boldsymbol{\tau} + \boldsymbol{\tau}\boldsymbol{l}\,(= \overset{\Delta}{\boldsymbol{\tau}}{}^{CRT}) \tag{4.67}$$

(d) Corotational rates

The following stress rate based on Eq. (4.57) is termed the *Green-Naghdi rate of Cauchy stress* (Green and Naghdi 1965).

$$\overset{\circ}{\boldsymbol{\sigma}}{}^{R} \equiv \overset{\rightleftarrows}{\boldsymbol{\sigma}}{}^{R} = \mathbf{R}(\mathbf{R}^{T}\boldsymbol{\sigma}\mathbf{R})^{\bullet}\mathbf{R}^{T} = \dot{\boldsymbol{\sigma}} - \boldsymbol{\Omega}^{R}\boldsymbol{\sigma} + \boldsymbol{\sigma}\boldsymbol{\Omega}^{R}\,(= \overset{\circ}{\boldsymbol{\sigma}}{}^{RT}) \tag{4.68}$$

Similarly, the *Green-Naghdi rate of Kirchhoff stress* is given by

$$\overset{\circ}{\boldsymbol{\tau}}{}^{R} \equiv \overset{\rightleftarrows}{\boldsymbol{\tau}}{}^{R} = \mathbf{R}(\mathbf{R}^{T}\boldsymbol{\tau}R)^{\bullet}\mathbf{R}^{T} = \dot{\boldsymbol{\tau}} - \boldsymbol{\Omega}^{R}\boldsymbol{\tau} + \boldsymbol{\tau}\boldsymbol{\Omega}^{R}\,(= \overset{\circ}{\boldsymbol{\tau}}{}^{RT}) \tag{4.69}$$

The stress rate based on Eq. (4.59) is given by

$$\overset{\circ}{\boldsymbol{\sigma}}{}^{w} \equiv \dot{\boldsymbol{\sigma}} - \mathbf{w}\boldsymbol{\sigma} + \boldsymbol{\sigma}\mathbf{w}\,(= \overset{\circ}{\boldsymbol{\sigma}}{}^{wT}) \tag{4.70}$$

which is termed the *Zaremba-Jaumann rate of Cauchy stress* (Zaremba 1903; Jaumann 1911). Likewise, it follows for the Kirchhoff stress that

$$\overset{\circ}{\boldsymbol{\tau}}{}^{w} \equiv \dot{\boldsymbol{\tau}} - \mathbf{w}\boldsymbol{\tau} + \boldsymbol{\tau}\mathbf{w}\,(= \overset{\circ}{\boldsymbol{\tau}}{}^{wT}) \tag{4.71}$$

The stress rate tensors described above are listed in Table 4.1. Here, it should be noted that objective rates must be used also for all internal variables in addition to objective stress rate and strain rate.

The stress rate tensors based on the convected and the corotational time-derivatives satisfy the objectivity. Therefore, the stress rate tensors shown in this section are objective quantities.

Table 4.1 Various stress rate tensors

Oldroyd rate of Cauchy stress: $\overset{\circ}{\boldsymbol{\sigma}}{}^{Ol} \equiv \dot{\boldsymbol{\sigma}} - \boldsymbol{l}\boldsymbol{\sigma} - \boldsymbol{\sigma}\boldsymbol{l}^{T}\,(= \overset{\circ}{\boldsymbol{\sigma}}{}^{OlT})$
Truesdell rate of Cauchy stress: $\overset{\circ}{\boldsymbol{\sigma}}{}^{Tr} \equiv \overset{\circ}{\boldsymbol{\sigma}}{}^{Ol} + \boldsymbol{\sigma}\,\mathrm{tr}\,\mathbf{d} = \dot{\boldsymbol{\sigma}} - \boldsymbol{l}\boldsymbol{\sigma} - \boldsymbol{\sigma}\boldsymbol{l}^{T} + \boldsymbol{\sigma}\,\mathrm{tr}\,\mathbf{d}\,(= \overset{\circ}{\boldsymbol{\sigma}}{}^{TrT})$
Covariant-contravariant convected rate of the Kirchhoff stress: $\overset{\Delta}{\boldsymbol{\tau}} \equiv \dot{\boldsymbol{\tau}} + \boldsymbol{l}^{T}\boldsymbol{\tau} - \boldsymbol{\tau}\boldsymbol{l}^{T}\,(\neq \overset{\Delta}{\boldsymbol{\tau}}{}^{T})$
Nominal stress rate: $_{\Pi}\overset{\Delta}{\boldsymbol{\sigma}} \equiv \frac{1}{J}\,{}_{\Pi}\overset{\Delta}{\boldsymbol{\tau}} = \dot{\boldsymbol{\sigma}} - \boldsymbol{\sigma}\boldsymbol{l}^{T} + \boldsymbol{\sigma}\ \mathrm{tr}\,\mathbf{d}\,(\neq {}_{\Pi}\overset{\Delta}{\boldsymbol{\sigma}}{}^{T})$
Cotter-Rivlin rate of Cauchy stress: $\overset{\Delta}{\boldsymbol{\sigma}}{}^{CR} \equiv \dot{\boldsymbol{\sigma}} + \boldsymbol{l}^{T}\boldsymbol{\sigma} + \boldsymbol{\sigma}\boldsymbol{l}\,(= \overset{\Delta}{\boldsymbol{\sigma}}{}^{CRT})$
Green-Naghdi rate of Cauchy stress: $\overset{\circ}{\boldsymbol{\sigma}}{}^{R} \equiv \dot{\boldsymbol{\sigma}} - \boldsymbol{\Omega}^{R}\boldsymbol{\sigma} + \boldsymbol{\sigma}\boldsymbol{\Omega}^{R}\,(= \overset{\circ}{\boldsymbol{\sigma}}{}^{RT})$
Zaremba-Jaumann rate of Cauchy stress: $\overset{\circ}{\boldsymbol{\sigma}}{}^{w} \equiv \dot{\boldsymbol{\sigma}} - \mathbf{w}\boldsymbol{\sigma} + \boldsymbol{\sigma}\mathbf{w}\,(= \overset{\circ}{\boldsymbol{\sigma}}{}^{wT})$

The above-mentioned rate tensors are used for rate-type constitutive equations. In particular, the Oldroyd rate appears in the current rate form of the hyperelastic constitutive equation. The other rates are used for expressions of its variations as will be described in Chap. 5. The corotational time-derivatives are used in the derivation of the consistency condition from the yield condition.

Constitutive equation for irreversible deformation exhibiting the loading-path dependence has to be formulated in a rate-form in terms of objective stress rate and objective strain rate. On the other hand, a hyperelastic constitutive equation must be formulated in terms of objective stress and objective strain. Unfortunately, however, it is difficult to find an objective rate of strain, although various objective rates of stress have been found as explained above in detail. In this situation, the particular spin, called the *logarithmic spin*, by which the corotational rate of the Eulerian-logarithmic (Hencky) strain $\ln \mathbf{V}$ in Eq. $(2.68)_2$ coincides with the strain rate $\mathbf{d}$, was proposed and the hyperelastic constitutive equation was derived from the hypoelastic constitutive equation in terms of the logarithmic rate of Cauchy stress and the strain rate $\mathbf{d}$ by Xiao and his colleagues (Xiao 1995; Xiao et al. 1997, 1999). The formulation of the logarithmic spin is explained in **Appendix C**.

4.7 Time Derivative of Scalar-Valued Tensor Function

Scalar-valued tensor functions of stress and internal variables appear often in continuum mechanics as seen in the strain energy function and the yield function. Then, the time-derivative of scalar-valued tensor function is required in order to derive the rate-type relation of variables, e.g. the consistency condition of yield condition. The time-derivative of scalar function is independent of rigid-body rotation and thus it can be given primarily by its material-time derivative. Here, it should be noticed that the internal variables are formulated by the objective time-derivatives and thus the consistency condition must be transformed to the objective time-derivative. It can be proved that the material-time derivative of scalar-valued tensor function is transformed only to its corotational time-derivative. This fact would seem physically obvious but it must be proved mathematically. To this end, its mathematically exact proof for scalar valued function of general tensor will be given below, referring to the previous studies by Dafalias (1985, 1998; 2011) for vector and second-order tensor and Hashiguchi (2007b) for general tensor.

The corotational rate of general tensor is defined by extending Eq. (4.54) for the vector and Eq. (4.55) for the second-order tensor as follows:

$$\mathring{\mathbf{t}} = \widehat{\mathbf{R}}\left[\!\left[(\widehat{\mathbf{R}}^T[\![\mathbf{t}]\!]\ )^{\bullet}\right]\!\right] \quad \mathring{\mathbf{t}} = \widehat{\mathbf{R}}\left[\!\left[(\widehat{\mathbf{R}}^T[\![\mathbf{t}]\!]\ )^{\bullet}\right]\!\right] \tag{4.72}$$

where use is made of the symbol $[\![\;]\!]$ for general objective transformation in Eq. (1.84) with the replacement $\mathbf{Q} \to \widehat{\mathbf{R}}^T$. Here, noting $\dot{f}(\mathbf{t}) = \dot{f}(\widehat{\mathbf{R}}^T[\![\mathbf{t}]\!])$ because of the requirement $f(\mathbf{t}) = f(\widehat{\mathbf{R}}^T[\![\mathbf{t}]\!])$ for scalar variable, one has

$$\begin{aligned}\dot{f}(\mathbf{t}) &= \dot{f}(\widehat{\mathbf{R}}^T[\![\mathbf{t}]\!]) = \frac{\partial f(\widehat{\mathbf{R}}^T[\![\mathbf{t}]\!])}{\partial(\widehat{\mathbf{R}}^T[\![\mathbf{t}]\!])} * (\widehat{\mathbf{R}}^T[\![\mathbf{t}]\!])^{\cdot} = \widehat{\mathbf{R}}^T\left[\!\!\left[\frac{\partial f(\mathbf{t})}{\partial \mathbf{t}}\right]\!\!\right] * (\widehat{\mathbf{R}}^T[\![\mathbf{t}]\!])^{\cdot} \\ &= \frac{\partial f(\mathbf{t})}{\partial \mathbf{t}} * \widehat{\mathbf{R}}\left[\!\!\left[(\widehat{\mathbf{R}}^T[\![\mathbf{t}]\!])^{\cdot}\right]\!\!\right]\end{aligned} \tag{4.73}$$

$$\begin{aligned}\dot{f}(t_{w_1w_2\ldots}) &= \frac{\partial f(\widehat{R}_{u_1w_1}\widehat{R}_{u_2w_2}\cdots t_{u_1u_2\ldots})}{\partial(\widehat{R}_{s_1p_1}\widehat{R}_{s_2p_2}\cdots t_{s_1s_2\ldots})}(\widehat{R}_{v_1p_1}\widehat{R}_{v_2p_2}\cdots t_{v_1v_2\ldots})^{\cdot} \\ &= \frac{\partial f(t_{w_1w_2\ldots})}{\partial t_{s_1s_2\ldots}}\widehat{R}_{s_1p_1}\widehat{R}_{s_2p_2}\cdots(\widehat{R}_{v_1p_1}\widehat{R}_{v_2p_2}\cdots t_{v_1v_2\ldots})^{\cdot}\end{aligned} \tag{4.74}$$

where the symbol * designates the full contraction between derivative components in order between derivative components, i.e. $\mathbf{t} * \mathbf{s} = t_{p_1p_2\cdots p_m}s_{p_1p_2\cdots p_m}$. The derivation of Eq. (4.73) is shown for vector and second-order tensor as follows:

$$\dot{f}(\mathbf{v}) = \dot{f}(\widehat{\mathbf{R}}^T\mathbf{v}) = \frac{\partial f(\widehat{\mathbf{R}}^T\mathbf{v})}{\partial(\widehat{\mathbf{R}}^T\mathbf{v})}\cdot(\widehat{\mathbf{R}}^T\mathbf{v})^{\cdot} = \widehat{\mathbf{R}}^T\frac{\partial f(\mathbf{v})}{\partial \mathbf{v}}\cdot(\widehat{\mathbf{R}}^T\mathbf{v})^{\cdot} = \frac{\partial f(\mathbf{v})}{\partial \mathbf{v}}\cdot\widehat{\mathbf{R}}(\widehat{\mathbf{R}}^T\mathbf{v})^{\cdot} \tag{4.75}$$

$$\begin{aligned}\dot{f}(\mathbf{t}) &= \dot{f}(\widehat{\mathbf{R}}^T\mathbf{t}\widehat{\mathbf{R}}) = \frac{\partial f(\widehat{\mathbf{R}}^T\mathbf{t}\widehat{\mathbf{R}})}{\partial(\widehat{\mathbf{R}}^T\mathbf{t}\widehat{\mathbf{R}})}:(\widehat{\mathbf{R}}^T\mathbf{t}\widehat{\mathbf{R}})^{\cdot} = \widehat{\mathbf{R}}^T\frac{\partial f(\mathbf{t})}{\partial \mathbf{t}}\widehat{\mathbf{R}}^T:(\widehat{\mathbf{R}}^T\mathbf{t}\widehat{\mathbf{R}})^{\cdot} \\ &= \frac{\partial f(\mathbf{t})}{\partial \mathbf{t}}:\widehat{\mathbf{R}}(\widehat{\mathbf{R}}^T\mathbf{t}\widehat{\mathbf{R}})^{\cdot}\widehat{\mathbf{R}}^T\end{aligned} \tag{4.76}$$

Equations (4.73), (4.75) and (4.76) can be satisfied by the corotational rate in Eq. (4.72) amongst objective rates. Then, we have the following relation.

$$\begin{aligned}\dot{f}(\mathbf{t}) &= \frac{\partial f(\mathbf{t})}{\partial \mathbf{t}} * \dot{\mathbf{t}} = \frac{\partial f(\mathbf{t})}{\partial \mathbf{t}} * \overset{\circ}{\mathbf{t}} \\ \dot{f}(t_{q_1q_2\cdots q_m}) &= \frac{\partial f(t_{q_1q_2\cdots q_m})}{\partial t_{p_1p_2\cdots p_m}}\dot{t}_{p_1p_2\cdots p_m} = \frac{\partial f(t_{q_1q_2\cdots q_m})}{\partial t_{p_1p_2\cdots p_m}}\overset{\circ}{t}_{p_1p_2\cdots p_m}\end{aligned} \tag{4.77}$$

which is described for vector and second-order tensor as follows:

$$\overset{\bullet}{f}(\mathbf{v}) = \frac{\partial f(\mathbf{v})}{\partial \mathbf{v}} \bullet \overset{\circ}{\mathbf{v}}, \quad \overset{\bullet}{f}(v_r) = \frac{\partial f(v_r)}{\partial v_i} \overset{\circ}{v}_i \tag{4.78}$$

$$\overset{\bullet}{f}(\mathbf{t}) = \frac{\partial f(\mathbf{t})}{\partial \mathbf{t}} : \overset{\circ}{\mathbf{t}}, \quad \overset{\bullet}{f}(t_{rs}) = \frac{\partial f(t_{rs})}{\partial t_{ij}} \overset{\circ}{t}_{ij} \tag{4.79}$$

It follows from Eq. (4.74) that

$$\begin{aligned}
&\frac{\partial f(t_{w_1 w_2 w_3 \ldots})}{\partial t_{s_1 s_2 s_3 \cdots}} \widehat{R}_{s_1 p_1} \widehat{R}_{s_2 p_2} \widehat{R}_{s_3 p_3} \cdots (\overset{\bullet}{\widehat{R}}_{v_1 p_1} \widehat{R}_{v_2 p_2} \widehat{R}_{v_3 p_3} \cdots + \widehat{R}_{p_1 v_1} \overset{\bullet}{\widehat{R}}_{p_2 v_2} \widehat{R}_{p_3 v_3} \cdots + \cdots) t_{v_1 v_2 v_3 \cdots} \\
&= \frac{\partial f(t_{w_1 w_2 w_3 \cdots})}{\partial t_{s_1 s_2 s_3 \cdots}} (\widehat{R}_{s_1 p_1} \overset{\bullet}{\widehat{R}}_{v_1 p_1} \delta_{s_2 v_2} \delta_{s_3 v_3} \cdots + \delta_{s_1 v_1} \widehat{R}_{s_2 p_2} \overset{\bullet}{\widehat{R}}_{v_2 p_2} \delta_{s_3 v_3} \cdots + \cdots) t_{v_1 v_2 v_3 \cdots} \\
&= \frac{\partial f(t_{w_1 w_2 w_3 \cdots})}{\partial t_{s_1 s_2 s_3 \cdots}} (\omega_{v_1 s_1} t_{v_1 s_2 s_3 \cdots} \cdots + \omega_{s_2 v_2} t_{s_1 v_2 s_3 \cdots} \cdots + \cdots) = 0
\end{aligned} \tag{4.80}$$

which is reduced for vector and second-order tensor as follows:

$$\frac{\partial f(\mathbf{v})}{\partial \mathbf{v}} \bullet \boldsymbol{\omega} \mathbf{v} = \frac{\partial f(v_u)}{\partial v_r} \omega_{ri} v_i = v_i \frac{\partial f(v_u)}{\partial v_r} \omega_{ri} = 0, \text{ i.e. } \operatorname{tr}\left[\left(\mathbf{v} \otimes \frac{\partial f(\mathbf{v})}{\partial \mathbf{v}}\right) \boldsymbol{\omega}\right] = 0 \tag{4.81}$$

$$\operatorname{tr}\left[\left(\frac{\partial f(\mathbf{t})}{\partial \mathbf{t}} \mathbf{t}^T - \mathbf{t}^T \frac{\partial f(\mathbf{t})}{\partial \mathbf{t}}\right) \boldsymbol{\omega}\right] = 0 \tag{4.82}$$

The fulfillments of Eq. (4.81) and (4.82) require for the tensors in the brackets ( ) to be zero or symmetric tensor, while Dafalias (1998) has required for the latter to be zero tensor. The fulfillment of Eq. (4.81) is easily known for $f(\mathbf{v}) = s\mathbf{v} \bullet \mathbf{v}$ because of $\partial f(\mathbf{v}) / \partial \mathbf{v} = 2s\mathbf{v}$ with Eq. $(1.130)_3$, and that of Eq. (4.82) for the second-order symmetric tensor $\mathbf{t}$ because of $(\partial f(\mathbf{t}) / \partial \mathbf{t})\mathbf{t}^T - \mathbf{t}^T(\partial f(\mathbf{t}) / \partial \mathbf{t}) = \mathbf{O}$.

Equation (4.77) is extended for plural variables as follows:

$$\begin{aligned}
\overset{\bullet}{f}(\mathbf{t}_1, \mathbf{t}_2, \cdots) &= \frac{\partial f(\mathbf{t}_1, \mathbf{t}_2, \cdots)}{\partial \mathbf{t}_1} * \overset{\bullet}{\mathbf{t}}_1 + \frac{\partial f(\mathbf{t}_1, \mathbf{t}_2, \cdots)}{\partial \mathbf{t}_2} * \overset{\bullet}{\mathbf{t}}_2 + \cdots \\
&= \frac{\partial f(\mathbf{t}_1, \mathbf{t}_2, \cdots)}{\partial \mathbf{t}_1} * \overset{\circ}{\mathbf{t}}_1 + \frac{\partial f(\mathbf{t}_1, \mathbf{t}_2, \cdots)}{\partial \mathbf{t}_2} * \overset{\circ}{\mathbf{t}}_2 + \cdots = \overset{\circ}{f}(\mathbf{t}_1, \mathbf{t}_2, \cdots)
\end{aligned} \tag{4.83}$$

which is shown for the function of two tensors in Belytschko et al. (2014). Here, it should be noted that the mathematical property does not hold for each term, i.e.

$$\frac{\partial f(\mathbf{t}_1, \mathbf{t}_2, \cdots, \mathbf{t}_m}{\partial \mathbf{t}_i} * \overset{\bullet}{\mathbf{t}}_i \neq \frac{\partial f(\mathbf{t}_1, \mathbf{t}_2, \cdots, \mathbf{t}_m)}{\partial \mathbf{t}_i} * \overset{\circ}{\mathbf{t}}_i \text{ (no sum)} \tag{4.84}$$

Scalar-valued functions must be independent of rigid-body rotation so that their material-time derivative possess a unique value which coincides with their corotational time-derivative as can be confirmed by the above-mentioned proof. However, they do not lead to the other convected time-derivatives which depend on the rate of deformation, i.e. velocity gradient. Therefore, corotational time-derivatives can be adopted in the time-derivatives of scalar functions in constitutive relations, e.g. a strain energy function and a yield function of tensors in the current configuration but convected time-derivatives other than corotational rates cannot be adopted in them.

The most popular scalar-valued tensor functions are the principal invariants of tensor. Principal invariants of tensor are described by three independent principal invariants of tensor. Then, the material-time derivatives of the principal invariants are transformed to the corotational time-derivatives merely by replacing all the material time-derivatives of tensor to the corotational time-derivatives of tensor as will be written below.

It follows from Eqs. (4.79) and (4.83) that

$$\left.\begin{array}{l}
\overset{\bullet}{\mathrm{I}} = (\mathrm{tr}\mathbf{t})^{\bullet} = (\mathrm{tr}\mathbf{t})^{\circ} = \mathbf{I} : \overset{\circ}{\mathbf{t}} = \mathrm{tr}\,\overset{\circ}{\mathbf{t}} \\
\overset{\bullet}{\mathrm{II}} = (\mathrm{tr}\mathbf{t}^2)^{\bullet} = (\mathrm{tr}\mathbf{t}^2)^{\circ} = 2\mathbf{t}^T : \overset{\circ}{\mathbf{t}} = 2(\mathrm{tr}(\mathbf{t}\overset{\circ}{\mathbf{t}}) \\
\overset{\bullet}{\mathrm{III}} = (\mathrm{tr}\mathbf{t}^3)^{\bullet} = (\mathrm{tr}\mathbf{t}^3)^{\circ} = 3\mathbf{t}^{2T} : \overset{\circ}{\mathbf{t}} = 3(\mathrm{tr}(\mathbf{t}^2\overset{\circ}{\mathbf{t}})
\end{array}\right\} \tag{4.85}$$

$$\left.\begin{array}{l}
\dot{\mathrm{I}} = (\mathrm{tr}\,\mathbf{t})^{\bullet} = \mathrm{tr}\,\overset{\circ}{\mathbf{t}} \\
\dot{\mathrm{II}} = \dfrac{1}{2}[(\mathrm{tr}\,\mathbf{t})^2 - \mathrm{tr}\,\mathbf{t}^2]^{\bullet} = (\mathrm{tr}\,\mathbf{t})\mathrm{tr}\,\overset{\circ}{\mathbf{t}} - \mathrm{tr}\,(\mathbf{t}\overset{\circ}{\mathbf{t}}) \\
\dot{\mathrm{III}} = (\det\mathbf{t})^{\bullet} = (\det\mathbf{t})\mathbf{t}^{-T} : \overset{\circ}{\mathbf{t}} = (\det\mathbf{t})\mathrm{tr}\,(\mathbf{t}^{-1}\overset{\circ}{\mathbf{t}})
\end{array}\right\} \tag{4.86}$$

for the principal invariants in Eqs. (1.183) and (1.178)–(1.180), noting Eqs. (1.294) and (1.295). Further, it holds that

$$(\mathbf{t}_1 : \mathbf{t}_2)^{\bullet} = \overset{\circ}{\mathbf{t}}_1 : \mathbf{t}_2 + \mathbf{t}_1 : \overset{\circ}{\mathbf{t}}_2 = \mathrm{tr}(\mathbf{t}_2^T\overset{\circ}{\mathbf{t}}_1) + \mathrm{tr}(\mathbf{t}_1^T\overset{\circ}{\mathbf{t}}_2) \tag{4.87}$$

for the two tensor variables. If $\mathbf{t}_1$ and $\mathbf{t}_2$ are commutative (possessing same principal directions) leading to $\mathbf{t}_1\mathbf{t}_2 = \mathbf{t}_2\mathbf{t}_1 = \mathbf{t}_1\mathbf{t}_2^T = \mathbf{t}_2^T\mathbf{t}_1$, one has

$$\mathbf{t}_1 : \overset{\bullet}{\mathbf{t}}_2 = \mathbf{t}_1 : \overset{\circ}{\mathbf{t}}_2, \overset{\bullet}{\mathbf{t}}_1 : \mathbf{t}_2 = \overset{\circ}{\mathbf{t}}_1 : \mathbf{t}_2 \tag{4.88}$$

noting

$$\mathbf{t}_1 : (\mathbf{t}_2\omega - \omega\mathbf{t}_2) = tr\{\mathbf{t}_1(\mathbf{t}_2\boldsymbol{\omega})^T\} - tr\{\mathbf{t}_1(\omega\mathbf{t}_2)^T\} = -\mathrm{tr}(\mathbf{t}_2^T\mathbf{t}_1\boldsymbol{\omega}) + \mathrm{tr}(\mathbf{t}_1\mathbf{t}_2^T\boldsymbol{\omega}) = 0 \tag{4.89}$$

All the equations in Eqs. (4.85)–(4.87) hold for arbitrary corotational tensors as proved here, although they are written explicitly for the Zaremba-Jaumann rate in some literatures (e.g. Prager 1961; Belytschko et al. 2014).

4.8 Work Conjugacy

The work rate done for the unit volume in the current configuration is given by $\boldsymbol{\sigma}:\mathbf{d}(=\sigma_{ij}d_{ij})$. Designating the infinitesimal volumes in a specific region of material in the reference and the current configurations as dV and $dv(=JdV)$, respectively, the work rate $\overset{\bullet}{w}_0$ done per the unit reference volume, i.e. a certain volume element possessing a fixed mass is given from Eqs. (2.35), (2.45), (2.75), (2.80), (2.128), (3.13), (3.19) and (3.23) as follows:

$$
\begin{aligned}
\overset{\bullet}{w}_0 &= \boldsymbol{\sigma}:\mathbf{d}dv/dV = \mathrm{tr}(\boldsymbol{\sigma}\,\mathbf{d})J = \mathrm{tr}(\boldsymbol{\tau}\boldsymbol{l})\\
&= \begin{cases}
\mathrm{tr}[(\boldsymbol{\tau}\mathbf{F}^{-T})(\boldsymbol{l}\mathbf{F})^{T}] = \mathrm{tr}(\boldsymbol{\Pi}\overset{\bullet}{\mathbf{F}}^{T})\\
\mathrm{tr}[(\mathbf{F}^{-1}\boldsymbol{\tau}\mathbf{F}^{-T})(\mathbf{F}^{T}\mathbf{dF})] = \mathrm{tr}(\mathbf{S}\overset{\bullet}{\boldsymbol{E}}) = \mathrm{tr}(\mathbf{S}\overset{\bullet}{\mathbf{C}}/2)\\
\mathrm{tr}(\mathfrak{B}\,\overset{\bullet}{\mathbf{B}})
\end{cases}
\end{aligned}
$$

leading to

$$
\boxed{\overset{\bullet}{w}_0 = J\boldsymbol{\sigma}:\mathbf{d} = \boldsymbol{\tau}:\mathbf{d} = \boldsymbol{\Pi}:\overset{\bullet}{\mathbf{F}} = \mathbf{S}:\overset{\bullet}{\boldsymbol{E}} = \mathbf{S}:\overset{\bullet}{\mathbf{C}}/2 = \mathfrak{B}:\overset{\bullet}{\mathbf{B}}} \tag{4.90}
$$

where

$$
\boxed{\mathfrak{B} \equiv \frac{1}{2}(\mathbf{SU} + \mathbf{US})} \tag{4.91}
$$

which is called the *Biot stress tensor*. Equation $(4.90)_7$ is derived also as follows:

$$
\begin{aligned}
\mathrm{tr}(\mathbf{S}\overset{\bullet}{\boldsymbol{E}}) &= \mathrm{tr}\left[\mathbf{S}\frac{1}{2}(\mathbf{U}\overset{\bullet}{\mathbf{U}} + \overset{\bullet}{\mathbf{U}}\mathbf{U})\right] = \frac{1}{2}\mathrm{tr}(\mathbf{SU}\overset{\bullet}{\mathbf{U}} + \mathbf{US}\overset{\bullet}{\mathbf{U}})\\
&= \frac{1}{2}\mathrm{tr}(\mathbf{SU} + \mathbf{US})\overset{\bullet}{\mathbf{U}} = \mathrm{tr}(\mathfrak{B}\overset{\bullet}{\mathbf{B}})
\end{aligned} \tag{4.92}
$$

noting Eqs. (2.35), $(2.45)_1$ and (2.61).

By taking account of Eq. $(4.90)_4$ into the relation

$$
\begin{aligned}
\overset{\bullet}{w}_0 &= \sum_{i=1}^{3}\mathbf{f}_i\cdot\overset{\bullet}{\mathbf{n}}_i = \sum_{i=1}^{3}\boldsymbol{\Pi}\mathbf{N}_i\cdot(\mathbf{F}\mathbf{N}_i)^{\bullet} = \sum_{i=1}^{3}\boldsymbol{\Pi}\mathbf{N}_i\cdot\overset{\bullet}{\mathbf{F}}\mathbf{N}_i = \sum_{i=1}^{3}\mathbf{N}_i\cdot\overset{\bullet}{\mathbf{F}}^{T}\boldsymbol{\Pi}\mathbf{N}_i\\
&= (\overset{\bullet}{\mathbf{F}}^{T}\boldsymbol{\Pi})_{ii} = \mathrm{tr}\,(\boldsymbol{\Pi}\overset{\bullet}{\mathbf{F}}^{T})
\end{aligned} \tag{4.93}
$$

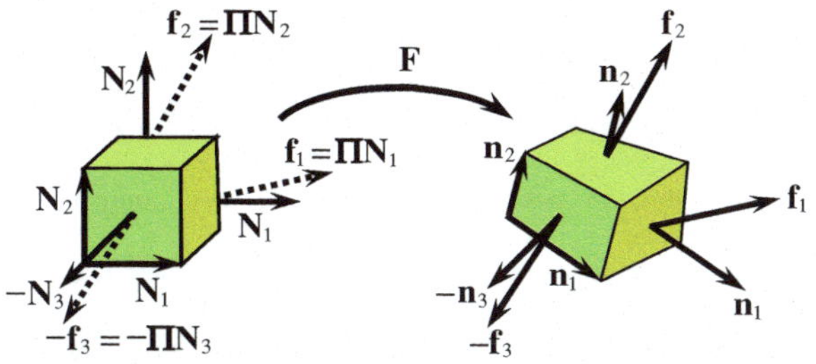

Fig. 4.2 Current cell deformed from reference orthogonal unit cell to which First Piola-Kirchhoff stress applies

we can confirm the fact that $\dot{w}_0$ designates the work rate (power) done in the current cell with the side vectors $(\mathbf{n}_1, \mathbf{n}_2, \mathbf{n}_3)$ which was the orthogonal unit cell with the side vectors $(\mathbf{N}_1, \mathbf{N}_2, \mathbf{N}_3)$ $(||\mathbf{N}_i|| = 1)$ in the reference state, noting Eqs. (1.117), (2. 15) and (3.18) with the replacements $dA \rightarrow 1$ and $d\mathbf{f} \rightarrow \mathbf{f}$, as shown in Fig. 4.2. Besides, the first Piola-Kirchhoff stress is calculated supposing that the force $\mathbf{f}_i$ on the current cell formed by the vectors $(\mathbf{n}_1, \mathbf{n}_2, \mathbf{n}_3)$ applies to the reference dell formed by the vectors $(\mathbf{N}_1, \mathbf{N}_2, \mathbf{N}_3)$.

The work rate reflecting the constitutive property is not concerned with a current unit volume but is concerned with a reference unit volume, noting that the mass in the current unit volume is variable but the mass in the reference unit volume is invariable. The pairs of stresses and strain rates (or rates of deformation gradient) shown in Eq. (4.90) are called the *work-conjugate pair*. Stress and strain rate tensors in the work-conjugacy pair have to be used for the formulation of constitutive equation.

References

Belytschko T, Liu WK, Moran B (2014) Nonlinear finite elements for continua and structures, 2nd edn. Wiley, New York

Bonet J, Wood RD (1997) Nonlinear continuum mechanics for finite element analysis. Cambridge Univ. Press, Cambridge

Cotter BA, Rivlin RS (1955) Tensors associated with time-dependent stresses. Quart Appl Math 13:177–182

Dafalias YF (1985) The plastic spin. J Appl Mech (ASME) 52:865–871

Dafalias YF (1998) Plastic spin: necessity or redundancy ? Int J Plast 14:909–931

Dafalias YF (2011) Finite elastic-plastic deformations: beyond the plastic spin. Theory Appl Mech 38:321–345

de Souza Neto EA, Perić D, Owen DJR (2008) Computational methods for plasticity. Wiley, Chichester

Green AE, Naghdi PM (1965) A general theory of an elastic-plastic continuum. Arch Ration Mech Anal 18:251–281

Hashiguchi K (2007a) General corotational rate tensor and replacement of material-time derivative to corotational derivative of yield function. Comput Model Eng Sci 17:55–62

Hashiguchi K (2007b) Anisotropic constitutive equation of friction with rotational hardening, In: Proceedings of 13th International Symposium Plasticity & its Current Applications, pp 34–36

Hashiguchi K (2011) General interpretations and tensor symbols for pull-back, push-forward and convected derivative, In: Proceedings of JSME 24th Computational Mechanics Conference, pp 669–671

Hashiguchi K (2012) Introduction to finite strain theory for continuum elasto-plasticity. Wiley Series in Computational Mechanics. Wiley, Chichester

Jaumann G (1911) Geschlossenes System physicalisher und chemischer Differentialgesetze. Sitzber Akad Wiss Wien (IIa) 120:385–530

Marsden JE, Hughes TJR (1983) Mathematical foundation of elasticity. Prentice-Hall, Englewood Cliffs

Oldroyd JG (1950) On the formulation of rheological equations of state. Proc R Soc Lond A200:523–541

Prager W (1961) Introduction to mechanics of continua. Ginn & Comp, Boston

Simo JC (1998) Numerical analysis and simulation of plasticity, In: Ciarlet PG, Lions JL (eds) Handbook of numerical analysis, vol 6, Part 3 (Numerical methods for solids). Elsevier, Amsterdam

Truesdell C, Noll W (1965) In: Flugge S (ed) The nonlinear field theories of mechanics, encyclopedia of physics, vol III/3. Springer, Berlin

Truesdell C, Toupin R (1960) In: Flugge S (ed) The classical field theories, encyclopedia of physics, vol III/1. Springer, Berlin

Xiao H (1995) Unified explicit basis-free expressions for time rate and conjugate stress of an arbitrary Hill strain. Int Solids Struct 32:3327–3340

Xiao H, Bruhns OT, Meyers A (1997) Logarithmic strain, logarithmic spin and logarithmic rate. Acta Mech 124:89–105

Xiao H, Bruhns OT, Meyers A (1999) Existence and uniqueness of the integrable-exactly hypoelastic equation $\overset{\circ}{\boldsymbol{\tau}}^{*} = \lambda(\mathrm{tr}\,\mathbf{D})\mathbf{I} + 2\mu\mathbf{D}$ and its significance to finite inelasticity. Acta Mech 138:31–50

Zaremba S (1903) Su une forme perfectionnee de la theorie de la relaxation, In: Bulletin International de l'Academie des Sciences de Cracovie, pp. 594–614 (in French)

Chapter 5
Elastic Constitutive Equations

Elastic deformation is induced by the reversible deformation of material particles themselves without a mutual slip between them. They therefore exhibit high stiffness. Elastic constitutive equations are classifiable into the three types depending on the exactness in the description of reversibility, i.e. the *hyperelasticity* (or *Green elasticity*) possessing the strain energy function, the *Cauchy elasticity* possessing the one-to-one correspondence between stress and strain and the *hypoelasticity* possessing the linear relation between stress rate and strain rate. As preparation for the study of elastoplasticity in the subsequent chapters, they are explained in this chapter.

5.1 Hyperelasticity

In the hyperelastic material, the one to one correspondence between the stress and the strain exists and further the work done during the loading process from a certain strain to another certain stain is determined uniquely independent of the loading path in that process. Then, the hyperelastic material must possess the *strain (Helmholtz) energy function* which is determined uniquely by a tensor describing deformation of material. For instance, let the deformation gradient tensor $\mathbf{F}$ be adopted for the tensor describing the deformation with the strain energy function, which is the most basic tensor describing the deformation of material. Then, letting the strain energy function per unit volume in the reference configuration be denoted by φ, the work done per the reference unit volume during the change of the deformation gradient from $\mathbf{F}_0$ to $\mathbf{F}$ must be uniquely determined by the values of deformation gradient in the reference and the current states, i.e.

K. Hashiguchi, *Foundations of Elastoplasticity: Subloading Surface Model*,
DOI 10.1007/978-3-319-48821-9_5

$$\int_{\mathbf{F}_0}^{\mathbf{F}} dw_0(\mathbf{F}) = \int_{\mathbf{F}_0}^{\mathbf{F}} \frac{\partial\varphi(\mathbf{F})}{\partial\mathbf{F}} : d\mathbf{F} = \varphi(\mathbf{F}) - \varphi(\mathbf{F}_0) \tag{5.1}$$

leading to

$$\dot{w}_0(\mathbf{F}) = \frac{\partial\varphi(\mathbf{F})}{\partial\mathbf{F}} : \dot{\mathbf{F}}$$

Then, the 1st Piola Kirchhoff stress tensor $\mathbf{\Pi}$ is given by

$$\mathbf{\Pi} = \frac{\partial\varphi(\mathbf{F})}{\partial\mathbf{F}} \tag{5.2}$$

noting Eq. $(4.90)_4$.

Substituting Eq. (5.2) into Eqs. (3.19) and (3.23), we obtain various expressions of the hyperelasticity by the deformation gradient as follows:

$$\boxed{\boldsymbol{\sigma} = \frac{1}{\det\mathbf{F}}\frac{\partial\varphi(\mathbf{F})}{\partial\mathbf{F}}\mathbf{F}^T, \quad \boldsymbol{\tau} = \frac{\partial\varphi(\mathbf{F})}{\partial\mathbf{F}}\mathbf{F}^T, \quad \mathbf{S} = \mathbf{F}^{-1}\frac{\partial\varphi(\mathbf{F})}{\partial\mathbf{F}}} \tag{5.3}$$

Furthermore, noting

$$\begin{aligned} \frac{\partial}{\partial F_{iA}} &= \frac{\partial}{\partial C_{PQ}}\frac{\partial C_{PQ}}{\partial F_{iA}} = \frac{\partial}{\partial C_{PQ}}\frac{\partial F_{rP}F_{rQ}}{\partial F_{iA}} = \frac{\partial}{\partial C_{PQ}}(\delta_{ri}\delta_{PA}F_{rQ} + F_{rP}\delta_{ri}\delta_{QA}) \\ &= \frac{\partial}{\partial C_{AQ}}F_{iQ} + \frac{\partial}{\partial C_{PA}}F_{iP} = 2F_{iP}\frac{\partial}{\partial C_{PA}} \end{aligned} \tag{5.4}$$

and denoting the strain energy function described in terms of the right Cauchy-Green tensor $\mathbf{C}$ or the Green strain $\boldsymbol{E}$ by ψ, one has

$$\left.\begin{aligned} \frac{\partial\varphi}{\partial\mathbf{F}} &= 2\mathbf{F}\frac{\partial\psi}{\partial\mathbf{C}} = \mathbf{F}\frac{\partial\psi}{\partial\boldsymbol{E}} \\ \frac{\partial\psi}{\partial\mathbf{C}} &= \frac{1}{2}\mathbf{F}^{-1}\frac{\partial\varphi}{\partial\mathbf{F}} = \frac{1}{2}\frac{\partial\psi}{\partial\boldsymbol{E}} \\ \frac{\partial\psi}{\partial\boldsymbol{E}} &= \mathbf{F}^{-1}\frac{\partial\varphi}{\partial\mathbf{F}} = 2\frac{\partial\psi}{\partial\mathbf{C}} \end{aligned}\right\} \tag{5.5}$$

Then, substituting Eq. (5.5) into Eq. (5.3), the hyperelasticity is expressed as follows:

$$\boxed{\begin{aligned} &\boldsymbol{\sigma} = 2\frac{1}{\det \mathbf{F}}\mathbf{F}\frac{\partial\psi(\mathbf{C})}{\partial \mathbf{C}}\mathbf{F}^T = \frac{1}{\det \mathbf{F}}\mathbf{F}\frac{\partial\psi(\boldsymbol{E})}{\partial \boldsymbol{E}}\mathbf{F}^T \\ &\boldsymbol{\tau} = 2\mathbf{F}\frac{\partial\psi(\mathbf{C})}{\partial \mathbf{C}}\mathbf{F}^T = \mathbf{F}\frac{\partial\psi(\boldsymbol{E})}{\partial \boldsymbol{E}}\mathbf{F}^T \\ &\boldsymbol{\Pi} = 2\mathbf{F}\frac{\partial\psi(\mathbf{C})}{\partial \mathbf{C}} = \mathbf{F}\frac{\partial\psi(\boldsymbol{E})}{\partial \boldsymbol{E}} \\ &\mathbf{S} = 2\frac{\partial\psi(\mathbf{C})}{\partial \mathbf{C}} = \frac{\partial\psi(\boldsymbol{E})}{\partial \boldsymbol{E}} \end{aligned}} \tag{5.6}$$

As known from Eq. $(5.6)_4$, the constitutive relation for isotropic elastic deformation is described through the elastic potential energy function of the right Cauchy-Green deformation tensor $\mathbf{C}$ in the initial configuration, i.e. the three-dimensional stretching resulting in the volume change and the shape change (pure shear deformation). It is based on the physical background that the elastic deformation is induced by the expansion/contraction of the intervals between material particles connected by the elastic springs. On the other hand, the plastic deformation is induced by the slips between material particles so that it cannot be formulated through the potential energy function but it must be formulated in a rate form.

It holds from Eq. (1.246) for any scalar-valued tensor function $\partial\psi(\boldsymbol{E})$ leading to the isotropic material that

$$\boxed{\frac{\partial\psi(\boldsymbol{E})}{\partial \boldsymbol{E}} = \phi_0^E\mathbf{I} + \phi_1^E\boldsymbol{E} + \phi_2^E\boldsymbol{E}^2} \tag{5.7}$$

where $\phi_0^E, \phi_1^E, \phi_2^E$ are the functions of invariants of $\boldsymbol{E}$. Equation (5.7) reduces to the following equation for the linear elastic material.

$$\frac{\partial\psi(\boldsymbol{E})}{\partial \boldsymbol{E}} = a(\mathrm{tr}\boldsymbol{E})\mathbf{I} + 2b\boldsymbol{E} \tag{5.8}$$

where a, b are the material parameters.

The function $\partial\psi(\mathbf{C})$ is described as

$$\psi(\mathbf{C}) = \psi(\mathrm{I}_C, \mathrm{II}_C, \mathrm{III}_C) \tag{5.9}$$

where

$$\left.\begin{aligned} &\mathrm{I}_C \equiv \mathrm{tr}\mathbf{C} \\ &\mathrm{II}_C \equiv \frac{1}{2}(\mathrm{tr}^2\mathbf{C} - \mathrm{tr}\mathbf{C}^2) \\ &\mathrm{III}_C \equiv \det \mathbf{C} \end{aligned}\right\} \tag{5.10}$$

with

$$\left.\begin{aligned}&\frac{\partial \mathrm{I}_C}{\partial \mathbf{C}}=\frac{\partial \mathrm{tr}\mathbf{C}}{\partial \mathbf{C}}=\mathbf{I}\\&\frac{\partial \mathrm{II}_C}{\partial \mathbf{C}}=\frac{\partial \frac{1}{2}(\mathrm{tr}^2\mathbf{C}-\mathrm{tr}\mathbf{C}^2)}{\partial \mathbf{C}}=\mathrm{I}_C\mathbf{I}-\mathbf{C}\\&\frac{\partial \mathrm{III}_C}{\partial \mathbf{C}}=\frac{\partial \det \mathbf{C}}{\partial \mathbf{C}}=\mathrm{II}_C\mathbf{I}-\mathrm{I}_C\mathbf{C}+\mathbf{C}^2=\mathrm{III}_C\mathbf{C}^{-1}\end{aligned}\right\} \tag{5.11}$$

noting Eq. (1.295). Then, substituting Eq. (5.11) into Eq. (5.6), it follows that

$$\boldsymbol{\sigma}=2\frac{1}{\sqrt{\mathrm{III}_C}}\mathbf{F}\left[\frac{\partial\psi}{\partial \mathrm{I}_C}\mathbf{I}+\frac{\partial\psi}{\partial \mathrm{II}_C}(\mathrm{I}_C\mathbf{I}-\mathbf{C})+\frac{\partial\psi}{\partial \mathrm{III}_C}(\mathrm{II}_C\mathbf{I}-\mathrm{I}_C\mathbf{C}+\mathbf{C}^2)\right]\mathbf{F}^T \tag{5.12}$$

The strain energy function of the *Mooney-Rivlin model* (Mooney 1940; Rivlin 1948) which is applicable to the elastic deformation of the incompressible rubber is given as

$$\psi=a_1(\mathrm{I}_C-3)+a_2(\mathrm{II}_C-3)\quad(\mathrm{III}_C=1) \tag{5.13}$$

where a_1 and a_2 are material parameters.

Further, the *neo-Hookean model* is given by the simplification setting $a_2=0$ as follows:

$$\psi=\frac{1}{2}v(\mathrm{I}_C-3)\quad(\mathrm{III}_C=1) \tag{5.14}$$

where v is the material parameter.

Further, the strain energy function of the *Ogden model* (Ogden 1982, 1984) which is applicable to the elastic deformation of the incompressible rubber for a large deformation is given as

$$\psi=\sum_{n=1}^{3}\frac{\beta_n}{\alpha_n}(\lambda_1^{\alpha_n}+\lambda_2^{\alpha_n}+\lambda_3^{\alpha_n}-3)\quad(\lambda_1\lambda_2\lambda_3=1) \tag{5.15}$$

where λ_i are the principal values of $\mathbf{U}=\mathbf{C}^{1/2}$, α_n and β_n are material parameters. Equation (5.15) is reduced to Eq. (5.13) for the Mooney-Rivlin model by choosing the material parameters as follows (cf. Hisada 1992):

$$\left.\begin{aligned}&\beta_1=2C_1, &&\alpha_1=2\\&\beta_2=-2C_2, &&\alpha_2=-2\\&\beta_3=0, &&\alpha_3=0\end{aligned}\right\} \tag{5.16}$$

The hyperelastic equation for soils is referred to Sect. 11.10.

The time-differentiation of Eq. $(5.6)_4$ leads to

$$\boxed{\dot{\mathbf{S}} = \frac{\partial^2 \psi(\boldsymbol{E})}{\partial \boldsymbol{E} \otimes \partial \boldsymbol{E}} : \dot{\boldsymbol{E}}} \tag{5.17}$$

Here, the symbol $\otimes$ specifies the fourth-order tensor due to the second-order partial derivative by the second-order tensor, although the expression without this symbol is widely used in a lot of literatures (e.g. Simo and Hughes 1988; Bonet and Wood 1997; Belytschko et al. 2014). Substituting Eqs. (2.128) and (5.17) into Eq. (4.61), the Truesdell rate of Kirchhoff stress $\overset{\circ}{\boldsymbol{\tau}}^{Ol}$ is rewritten as

$$\overset{\Delta}{\boldsymbol{\tau}}^{Ol} = \mathbf{F}\left[\frac{\partial^2 \psi(\boldsymbol{E})}{\partial \boldsymbol{E} \otimes \partial \boldsymbol{E}} :(\mathbf{F}^T \mathbf{dF})\right]\mathbf{F}^T \left(\overset{\Delta}{\tau}{}^{Ol}_{ij} = F_{iA} F_{jB} F_{kC} F_{lD} \frac{\partial^2 \psi(\boldsymbol{E})}{\partial E_{AB} \partial E_{CD}} d_{kl}\right) \tag{5.18}$$

which is the rate of hyperelastic equation in the current configuration. Here, $\overset{\Delta}{\boldsymbol{\tau}}^{Ol}$ is related to the Zaremba-Jaumann rate of Cauchy stress in Eq. (4.70) as

$$\boxed{\overset{\Delta}{\boldsymbol{\tau}}^{Ol} = J(\overset{\circ}{\boldsymbol{\sigma}}^{w} - \mathbf{d}\boldsymbol{\sigma} - \boldsymbol{\sigma}\mathbf{d} + \boldsymbol{\sigma}\mathrm{tr}\mathbf{d})} \tag{5.19}$$

The Zaremba-Jaumann rate of Cauchy stress is related to the strain rate from these equations as

$$\overset{\circ}{\boldsymbol{\sigma}}^{w} = \frac{1}{\det \mathbf{F}} \mathbf{F}\left[\frac{\partial^2 \psi(\boldsymbol{E})}{\partial \boldsymbol{E} \otimes \partial \boldsymbol{E}} (\mathbf{F}^T \mathbf{dF})\right]\mathbf{F}^T + \mathbf{d}\boldsymbol{\sigma} + \boldsymbol{\sigma}\mathbf{d} - \boldsymbol{\sigma}\mathrm{tr}\mathbf{d} \tag{5.20}$$

which is expressed as

$$\overset{\circ}{\boldsymbol{\sigma}}^{w} = \tilde{\mathbf{E}} : \mathbf{d} \tag{5.21}$$

where the *hyperelastic tangent modulus tensor* $\tilde{\mathbf{E}}$ in the current configuration is given by

$$\tilde{E}_{ijkl} \equiv \frac{1}{\det \mathbf{F}} F_{iA} F_{kC} F_{lD} F_{jB} \frac{\partial^2 \psi(\boldsymbol{E})}{\partial E_{AB} \partial E_{CD}} + \Sigma_{ijkl} - \sigma_{ij} \delta_{kl} \tag{5.22}$$

with

$$\Sigma_{ijkl} \equiv \frac{1}{2}(\sigma_{ik}\delta_{jl} + \sigma_{il}\delta_{jk} + \sigma_{jk}\delta_{il} + \sigma_{jl}\delta_{ik}) \quad (\Sigma_{ijkl} = \Sigma_{klij} = \Sigma_{jikl} = \Sigma_{ijlk}) \tag{5.23}$$

5.2 Infinitesimal Elastic Deformation

For the infinitesimal deformation, the hyperelastic constitutive equation can be given as

$$\boldsymbol{\sigma} = \frac{\partial \psi(\boldsymbol{\varepsilon})}{\partial \boldsymbol{\varepsilon}}, \quad \dot{\boldsymbol{\sigma}} = \frac{\partial \psi^2(\boldsymbol{\varepsilon})}{\partial \boldsymbol{\varepsilon} \otimes \partial \boldsymbol{\varepsilon}} : \dot{\boldsymbol{\varepsilon}} = \mathbf{E} : \dot{\boldsymbol{\varepsilon}}, \quad \mathbf{E} \equiv \frac{\partial \boldsymbol{\sigma}}{\partial \boldsymbol{\varepsilon}} = \frac{\partial^2 \psi(\boldsymbol{\varepsilon})}{\partial \boldsymbol{\varepsilon} \otimes \partial \boldsymbol{\varepsilon}} \tag{5.24}$$

where $\boldsymbol{\varepsilon}$ is the infinitesimal strain in Eq. (2.55).

For the particular Helmholtz free strain energy function (strain energy function)

$$\psi(\boldsymbol{\varepsilon}) = \frac{1}{2} L (\mathrm{tr}\boldsymbol{\varepsilon})^2 + G \mathrm{tr}\boldsymbol{\varepsilon}^2 \tag{5.25}$$

the stress is given by the linear relation to the elastic strain $\boldsymbol{\varepsilon}$ as

$$\boldsymbol{\sigma} = L \varepsilon_v \mathbf{I} + 2G\boldsymbol{\varepsilon} \tag{5.26}$$

i.e.

$$\boldsymbol{\sigma} = \left(L + \frac{2}{3} G \right) \varepsilon_v \mathbf{I} + 2G\boldsymbol{\varepsilon}' \tag{5.27}$$

which is referred to as the *Hooke's law*, where L and G are called the *Lamé constants*. It follows by taking the trace and the deviatoric part of Eq. (5.27) that

$$\varepsilon_v = \frac{3}{3L + 2G} \sigma_m, \quad \boldsymbol{\varepsilon}' = \frac{1}{2G} \boldsymbol{\sigma}' \tag{5.28}$$

where $\sigma_m (\equiv (\mathrm{tr}\boldsymbol{\sigma})/3)$ is the mean stress. Then, the inverse relation of Eq. (5.26) is given by

$$\boldsymbol{\varepsilon} = \frac{1}{3L + 2G} (\mathrm{tr}\boldsymbol{\sigma}) \mathbf{I} + \frac{1}{2G} \boldsymbol{\sigma}' = \frac{4G - 3L}{6G(3L + 2G)} (\mathrm{tr}\boldsymbol{\sigma}) \mathbf{I} + \frac{1}{2G} \boldsymbol{\sigma} \tag{5.29}$$

The inverse relation can be derived by first making the spherical and the deviatoric parts and then combing them as shown above.

Equations (5.26) and (5.29) are rewritten as

$$\boxed{\begin{aligned} \boldsymbol{\sigma} &= K \varepsilon_v \mathbf{I} + 2G\boldsymbol{\varepsilon}' = \left(K - \frac{2}{3} G \right) \varepsilon_v \mathbf{I} + 2G\boldsymbol{\varepsilon} \\ \boldsymbol{\varepsilon} &= \frac{1}{3K} \sigma_m \mathbf{I} + \frac{1}{2G} \boldsymbol{\sigma}' = \left(\frac{1}{3K} - \frac{1}{2G} \right) \sigma_m \mathbf{I} + \frac{1}{2G} \boldsymbol{\sigma} \end{aligned}} \tag{5.30}$$

where

$$K \equiv L + \frac{2}{3}G \tag{5.31}$$

It follows from Eq. (5.30) that

$$\sigma_m = K\varepsilon_v, \ \boldsymbol{\sigma}' = 2G\boldsymbol{\varepsilon}' \tag{5.32}$$

Then, K and G are called the *bulk elastic modulus* and the *shear elastic modulus*, respectively.

Equations (5.26), (5.29) and (5.30) are represented as

$$\boxed{\boldsymbol{\sigma} = \mathbf{E} : \boldsymbol{\varepsilon}} \tag{5.33}$$

using the elastic modulus tensor $\mathbf{E}$ given by

$$\left.\begin{aligned} \mathbf{E} &= \left(L + \frac{2}{3}G\right)\mathcal{T} + 2G\mathcal{I}', \quad E_{ijkl} \equiv \left(L + \frac{2}{3}G\right)\delta_{ij}\delta_{kl} + 2G\left(\delta_{ik}\delta_{jl} + \delta_{il}\delta_{jk} - \frac{1}{3}\delta_{ij}\delta_{kl}\right) \\ \mathbf{E}^{-1} &= \frac{1}{9L+6G}\mathcal{T} + \frac{1}{2G}\mathcal{I}', \quad (\mathbf{E}^{-1})_{ijkl} = \frac{1}{9L+6G}\delta_{ij}\delta_{kl} + \frac{1}{2G}\left(\delta_{ik}\delta_{jl} + \delta_{il}\delta_{jk} - \frac{1}{3}\delta_{ij}\delta_{kl}\right) \end{aligned}\right\} \tag{5.34}$$

and

$$\boxed{\begin{aligned} \mathbf{E} &= K\mathcal{T} + 2G\mathcal{I}', \quad E_{ijkl} \equiv K\delta_{ij}\delta_{kl} + 2G\left[\frac{1}{2}(\delta_{ik}\delta_{jl} + \delta_{il}\delta_{jk}) - \frac{1}{3}\delta_{ij}\delta_{kl}\right] \\ \mathbf{E}^{-1} &= \frac{1}{9K}\mathcal{T} + \frac{1}{2G}\mathcal{I}', \quad (\mathbf{E}^{-1})_{ijkl} = \frac{1}{9K}\delta_{ij}\delta_{kl} + \frac{1}{2G}\left[\frac{1}{2}(\delta_{ik}\delta_{jl} + \delta_{il}\delta_{jk}) - \frac{1}{3}\delta_{ij}\delta_{kl}\right] \end{aligned}} \tag{5.35}$$

where $\mathcal{T}$ is the fourth-order tracing tensor and $\mathcal{I}'$ is the fourth-order deviatoric projection tensor defined in Eq. (1.143) and (1.146), respectively. The inverse relation between the two equation in Eq. (5.35) is confirmed by $(K\mathcal{T} + 2G\mathcal{I}'):[(1/9)K\mathcal{T} + (1/2G)\mathcal{I}'] = (1/3)\mathcal{T} + \mathcal{I}' = \mathcal{I}$, noting Eq. (1.146).

It follows from Eq. $(5.35)_2$ for the uniaxial loading process ($\sigma_{ij} = 0$ for $i = j \neq 1$ and $i \neq j$), noting $\mathcal{T}_{1111} = 1$, $\mathcal{I}'_{1111} = 2/3$, $\mathcal{T}_{2211} = 0$, $\mathcal{I}'_{2211} = -1/3$ that

$$\varepsilon_{11} = \frac{1}{E}\sigma_{11}, \quad \varepsilon_{22} = -\frac{\nu}{E}\sigma_{11} \rightarrow \frac{\varepsilon_{22}}{\varepsilon_{11}} = -\nu \tag{5.36}$$

where

$$E=\frac{9KG}{3K+G},\quad \nu=\frac{3K-2G}{2(3K+G)} \tag{5.37}$$

the inverses of which are given as

$$K\equiv\frac{E}{3(1-2\nu)},\quad G\equiv\frac{E}{2(1+\nu)} \tag{5.38}$$

Here, E is the ratio of the axial stress rate to the axial strain rate and is called the *Young's modulus*, and ν is the ratio of lateral strain rate to axial strain rate and is called the *Poisson's ratio*. The strain energy function in (5.25) is expressed as

$$\psi(\boldsymbol{\varepsilon})=\frac{\nu E}{2(1+\nu)(1-2\nu)}(\mathrm{tr}\boldsymbol{\varepsilon})^2+G\mathrm{tr}\boldsymbol{\varepsilon}^2 \tag{5.39}$$

which must be positive so that the Poisson's ratio is limited in the range

$$-1<\nu<1/2 \tag{5.40}$$

The lower limit and the upper limit correspond to the similar shape and the constant volume, respectively, as known from Eq. (5.38). The former is seen in artificial structures, e.g. honeycomb.

Substituting Eq. (5.38) into Eq. (5.35), the elastic modulus tensor is also described using the Young's modulus and the Poisson's ratio as follows:

$$\boxed{\begin{aligned}&\mathbf{E}=\frac{E}{3(1-2\nu)}\mathcal{T}+\frac{E}{1+\nu}\mathcal{I}',\quad E_{ijkl}=\frac{E}{3(1-2\nu)}\delta_{ij}\delta_{kl}+\frac{E}{1+\nu}\left[\frac{1}{2}(\delta_{ik}\delta_{jl}+\delta_{il}\delta_{jk})-\frac{1}{3}\delta_{ij}\delta_{kl}\right]\\&\mathbf{E}^{-1}=\frac{1-2\nu}{3E}\mathcal{T}+\frac{1+\nu}{E}\mathcal{I}',\quad (\mathbf{E}^{-1})_{ijkl}=\frac{1-2\nu}{3E}\delta_{ij}\,\delta_{kl}+\frac{1+\nu}{E}\left[\frac{1}{2}(\delta_{ik}\delta_{jl}+\delta_{il}\delta_{jk})-\frac{1}{3}\delta_{ij}\delta_{kl}\right]\end{aligned}} \tag{5.41}$$

or

$$\boxed{\begin{aligned}&\mathbf{E}=\frac{E}{1+\nu}\left(\frac{\nu}{1-2\nu}\mathcal{T}+\mathcal{I}\right),\quad E_{ijkl}=\frac{E}{1+\nu}\left[\frac{\nu}{1-2\nu}\delta_{ij}\delta_{kl}+\frac{1}{2}(\delta_{ik}\delta_{jl}+\delta_{il}\delta_{jk})\right]\\&\mathbf{E}^{-1}=-\frac{1}{E}\left[\nu\mathcal{T}-(1+\nu)\mathcal{I}\right],\quad (\mathbf{E}^{-1})_{ijkl}=-\frac{1}{E}\left[\nu\delta_{ij}\delta_{kl}-\frac{1}{2}(1+\nu)(\delta_{ik}\delta_{jl}+\delta_{il}\delta_{jk})\right]\end{aligned}} \tag{5.42}$$

Table 5.1 Relationships between two independent elastic constants

	$E,\ \nu$	$G,\ \nu$	$E,\ G$	$E,\ K$	$G,\ K$	$L,\ G$
E	E	$2(1+\nu)G$	E	E	$\dfrac{9KG}{3K+G}$	$\dfrac{\mu(3L+2G)}{L+G}$
G	$\dfrac{E}{2(1+\nu)}$	G	G	$\dfrac{3EK}{9K-E}$	G	G
K	$\dfrac{E}{3(1-2\nu)}$	$\dfrac{2(1+\nu)G}{3(1-2\nu)}$	$\dfrac{EG}{3(3G-E)}$	K	K	$L+\dfrac{2}{3}G$
ν	ν	ν	$\dfrac{E-2G}{2G}$	$\dfrac{3K-E}{6K}$	$\dfrac{3K-2G}{2(3K+G)}$	$\dfrac{L}{2(L+G)}$
L	$\dfrac{\nu E}{(1+\nu)(1-2\nu)}$	$\dfrac{2G\nu}{1-\nu}$	$\dfrac{G(E-2G)}{3G-E}$	$\dfrac{3K(3K-E)}{9K-E}$	$K-\dfrac{2}{3}G$	L

Relationships between two independent elastic constants are listed in Table 5.1.

The Helmholtz free energy function (strain energy function) $\psi(\boldsymbol{\varepsilon})$ and the *Gibbs' free energy function* (complementary energy function) $\phi(\boldsymbol{\sigma})$ are given for the linear elasticity as

$$\psi(\boldsymbol{\varepsilon}) = \frac{1}{2}\boldsymbol{\varepsilon}:\mathbf{E}:\boldsymbol{\varepsilon} = \frac{1}{2}\frac{E}{1+\nu}\left[\varepsilon_{ij}\varepsilon_{ij} + \frac{E}{1-2\nu}(\varepsilon_{kk})^2\right] \tag{5.43}$$

$$\phi(\boldsymbol{\sigma}) = \frac{1}{2}\boldsymbol{\sigma}:\mathbf{E}^{-1}:\boldsymbol{\sigma} = \frac{1}{2E}\left[(1+\nu)\sigma_{ij}\sigma_{ij} - \nu(\sigma_{kk})^2\right] \tag{5.44}$$

from which it follows that

$$\boldsymbol{\sigma} = \frac{\partial\psi}{\partial\boldsymbol{\varepsilon}} = \mathbf{E}:\boldsymbol{\varepsilon} = \frac{E}{1+\nu}\left(\frac{\nu}{1-2\nu}\varepsilon_v\mathbf{I} + \boldsymbol{\varepsilon}\right) = E\left[\frac{1}{3(1-2\nu)}\varepsilon_v\mathbf{I} + \frac{1}{1+\nu}\boldsymbol{\varepsilon}'\right] \tag{5.45}$$

$$\boldsymbol{\varepsilon} = \frac{\partial\phi}{\partial\boldsymbol{\sigma}} = \mathbf{E}^{-1}:\boldsymbol{\sigma} = \frac{1}{E}\left[(1+\nu)\boldsymbol{\sigma} - 3\nu\sigma_m\mathbf{I}\right] = \frac{1}{E}\left[(1-2\nu)\sigma_m\mathbf{I} + (1+\nu)\boldsymbol{\sigma}'\right] \tag{5.46}$$

5.3 Cauchy Elasticity

The elastic material which does not have a strain energy function but has a one-to-one correspondence between the Cauchy stress and a strain is called the *Cauchy elastic material*. Here, the stress tensor is given by an equation of strain tensor and thus the equation includes six strain components. The equation of six strain components does not fulfill the condition of complete integration leading to the strain energy function so that it does not result in the hyperelasticity in general. Then, the work done by the stress is generally dependent on the deformation path.

For that reason, an energy dissipation/production is induced during the stress or strain cycle.

In the above-mentioned definition, the Cauchy elastic material is described as

$$\boldsymbol{\sigma} = \mathbf{f}(\boldsymbol{e}) \tag{5.47}$$

in terms of the Almansi strain tensor $\boldsymbol{e}$ in Eq. (2.45) or (2.47). Equation (5.47) reduces to the following equation by virtue of Eq. (1.246) for the isotropic material.

$$\boldsymbol{\sigma} = \phi_0^e \mathbf{I} + \phi_1^e \boldsymbol{e} + \phi_2^e \boldsymbol{e}^2 \tag{5.48}$$

where $\phi_0^e, \phi_1^e, \phi_2^e$ are functions of invariants of $\boldsymbol{e}$. Furthermore, for an isotropic linear elastic material, Eq. (5.48) reduces to

$$\boldsymbol{\sigma} = L(\mathrm{tr}\boldsymbol{e})\mathbf{I} + 2G\boldsymbol{e} \tag{5.49}$$

noting Eq. (5.26). Limiting to the infinitesimal strain leading to $\boldsymbol{e} \cong \boldsymbol{\varepsilon}$, Eq. (5.49) results in Eq. (5.26), i.e.

$$\boldsymbol{\sigma} = L(\mathrm{tr}\boldsymbol{\varepsilon})\mathbf{I} + 2G\boldsymbol{\varepsilon} \tag{5.50}$$

Here, substituting Eq. (5.50) with Eq. (2.50) into Eq. (3.31), the *Navier's equation* is obtained by replacing L and G to a and b, respectively, as follows:

$$(a+b)\frac{\partial^2 u_j}{\partial x_j \partial x_i} + b\frac{\partial^2 u_i}{\partial x_j \partial x_j} + \rho b_i = \rho \dot{v}_i$$

$$(a+b)\nabla(\nabla \cdot \mathbf{u}) + b\Delta\mathbf{u} + \rho\mathbf{b} = \rho\,\dot{\mathbf{v}} \tag{5.51}$$

noting Eqs. (1.309), (1.315) and

$$\frac{\partial\left[a\dfrac{\partial u_k}{\partial x_k}\delta_{ij} + 2b\dfrac{1}{2}\left(\dfrac{\partial u_i}{\partial x_j} + \dfrac{\partial u_j}{\partial x_i}\right)\right]}{\partial x_j} = a\frac{\partial^2 u_j}{\partial x_j \partial x_i} + b\frac{\partial^2 u_i}{\partial x_j \partial x_j} + b\frac{\partial^2 u_j}{\partial x_j \partial x_i}$$

5.4 Hypoelasticity

The following material, for which the corotational rate of stress is related linearly to the strain rate, is referred to as the *hypoelastic material* by Truesdell (1955).

$$\overset{\circ}{\boldsymbol{\sigma}} = \mathbf{H}(\boldsymbol{\sigma})[\mathbf{d}] \tag{5.52}$$

where the tensor function $\mathbf{H}(\boldsymbol{\sigma})[\mathbf{d}]$ designates the linearity in the strain rate $\mathbf{d}$ and the isotropies in $\boldsymbol{\sigma}$ and $\mathbf{d}$.

In what follows, we adopt the following elastic constitutive relation with the elastic modulus tensor $\mathbf{E}$ incorporated in the infinitesimal elastic deformation in Sect. 5.2, i.e.

$$\boxed{\overset{\circ}{\boldsymbol{\sigma}} = \mathbf{E} : \mathbf{d}} \tag{5.53}$$

which leads to the following relations for the Hooke's law, noting Eqs. (5.30), (5.45) and (5.46).

$$\left.\begin{aligned} \overset{\circ}{\boldsymbol{\sigma}} &= K d_v \mathbf{I} + 2G\mathbf{d}' = \left(K - \frac{2}{3}G\right) d_v \mathbf{I} + 2G\mathbf{d} \\ \mathbf{d} &= \frac{1}{3K}\dot{\sigma}_m \mathbf{I} + \frac{1}{2G}\overset{\circ}{\boldsymbol{\sigma}}{}' = \left(\frac{1}{3K} - \frac{1}{2G}\right)\dot{\sigma}_m \mathbf{I} + \frac{1}{2G}\overset{\circ}{\boldsymbol{\sigma}} \end{aligned}\right\} \tag{5.54}$$

$$\dot{\sigma}_m = K d_v, \quad \overset{\circ}{\boldsymbol{\sigma}}{}' = 2G\mathbf{d}' \tag{5.55}$$

$$\left.\begin{aligned} \overset{\circ}{\boldsymbol{\sigma}} &= E\left[\frac{1}{3(1-2\nu)} d_v \mathbf{I} + \frac{1}{1+\nu}\mathbf{d}'\right] = \frac{E}{1+\nu}\left(\frac{\nu}{1-2\nu} d_v \mathbf{I} + \mathbf{d}\right) \\ \mathbf{d} &= \frac{1}{E}\left[(1-2\nu)\dot{\sigma}_m \mathbf{I} + (1+\nu)\overset{\circ}{\boldsymbol{\sigma}}{}'\right] = \frac{1}{E}\left[(1+\nu)\overset{\circ}{\boldsymbol{\sigma}} - 3\nu\dot{\sigma}_m \mathbf{I}\right] \end{aligned}\right\} \tag{5.56}$$

Besides, the following equation in which the Jaumann rate of Cauchy stress is related nonlinearly to the strain rate is called the *hypoplastic material* (Kolymbas and Wu 1993).

$$\overset{\circ}{\boldsymbol{\sigma}} = \mathbf{f}(\mathbf{d}, \boldsymbol{\sigma}), \quad \overset{\circ}{\sigma}_{ij} = f_{ij}(d_{kl}, \sigma_{kl}) \tag{5.57}$$

where f_{ij} is the nonlinear function of d_{kl}, and for rate-independent deformation it is the homogeneous function of d_{kl} in degree-one fulfilling $f_{ij}(|s|d_{kl}) = |s|f_{ij}(d_{kl})$ which implies $(\partial f_{ij}/\partial d_{kl})d_{kl} = f_{ij}$ on account of Euler's theorem for homogeneous function (see **Appendix D**).

While the three popular types of elastic materials are described in this chapter, the other elastic material, called the *Cosserat elastic material*, was advocated by Cosserat and Cosserat (1909). The *couple stress* is related to the *rotational strain* in this material. It has been applied to the prediction of localized deformation (e.g. cf. Mindlin 1963; Muhlhaus and Vardoulaskis 1987).

References

Belytschko T, Liu WK, Moran B (2014) Nonlinear finite elements for continua and structures, 2nd edn. Wiley, New York

Bonet J, Wood RD (1997) Nonlinear continuum mechanics for finite element analysis. Cambridge University Press, Cambridge

Cosserat E, Cosserat F (1909) Theorie des Corps Deformation, Traite de Physique, transl. Davaux E, ed Chwolson OD, 2nd edn, 2, Paris, pp 953–1173

Hisada T (1992) Tensor analysis for nonlinear finite element method. Maruzen Publications Inc., Tokyo (in Japanese)

Kolymbas D, Wu W (1993) Introduction to plasticity. In: Modern approaches to plasticity, Elsevier, Amsterdam, pp 213–224

Mindlin RD (1963) Influence of couple-stresses on stress concentrations. Experiment Mech 3:1–7

Mooney M (1940) A theory of large elastic deformation. J Appl Phys 11(9):582–592

Muhlhaus HB, Vardoulakis I (1987) The thickness of shear bands in granular materials. Geotechnique 37:271–283

Ogden RW (1982) Elastic deformations of rubberlike solids. In: Hopkins HG, Sewell MJ (eds) Mechanics of solids: the rodney hill 60th anniversary volume. Pergamon, Oxford, pp 499–537

Ogden RW (1984) Non-linear elastic deformations. Ellis-Horwood, Chichester, UK

Rivlin RS (1948) Large elastic deformations of isotropic materials. IV. Further developments of the general theory. Philos Trans R Soc Lond Ser A 241(835):379–397

Simo JC, Hughes TJR (1998) Computational Inelasticity. Springer, New York

Truesdell C (1955) Hypo-elasticity. J Rational Mech Anal 4:83–133

Chapter 6
Basic Formulations for Elastoplastic Constitutive Equations

Elastic deformation is induced microscopically by the elastic deformations of the material particles themselves, exhibiting a one-to-one correspondence to the stress as described in Chap. 4. However, when the stress reaches an yield stress, slippages between material particles (e.g. crystal lattice in metals and soil particles in soils) are induced, which do not disappear even if the stress is removed, leading to macroscopically to the plastic deformation. Then, the one-to-one correspondence between the stress and the strain, i.e. the stress-strain relation does not hold in the elastoplastic deformation process, exhibiting the *loading-path dependence*. Therefore, one must formulate the elastoplastic constitutive equation as a relation between the stress rate and the strain rate. This chapter addresses the basic concept and formulation for elastoplastic constitutive equations in the *conventional elastoplasticity* (Drucker 1988) based on the assumption that the inside of the yield surface is a purely elastic domain as the introduction to elastoplasticity, while the formulations are given within the framework of the hypoelastic-based plasticity. The *unconventional elastoplasticity* describing the plastic strain rate induced by the rate of stress inside the yield surface will be described in the subsequent chapters, which is required to describe the cyclic loading behavior, and the exact finite strain theory based on the multiplicative decomposition will be given in Chap. 12.

6.1 Multiplicative Decomposition of Deformation Gradient Tensor

Readers who wish only to use the elastoplasticity for their analyses may omit to read Sects. 6.1 and 6.2 and start to read from Sect. 6.3.

Consider the deformation/rotation measure which is relevant to the exact description of elastoplastic deformation/rotation. The measure must satisfy the following requirements.

K. Hashiguchi, *Foundations of Elastoplasticity: Subloading Surface Model*,
DOI 10.1007/978-3-319-48821-9_6

(1) It must be based on the deformation gradient tensor $\mathbf{F}$ in Eq. (2.14), which is the basic measure to describe the deformation/rotation exactly, since it transforms the initial infinitesimal line-element $d\mathbf{X}$ to the current infinitesimal line-element $d\mathbf{x}$.
(2) It is decomposed exactly into the elastic and the plastic parts.

Responding to these requirements, the deformation gradient tensor $\mathbf{F}$ has been multiplicatively decomposed into the elastic deformation gradient tensor $\mathbf{F}^e$ and the plastic deformation gradient tensor $\mathbf{F}^p$ (Kroner 1960; Lee and Liu 1967; Lee 1969; Mandel 1971, 1973), which is referred to as the *multiplicative decomposition*, i.e.

$$\boxed{\mathbf{F} = \mathbf{F}^e\mathbf{F}^p} \tag{6.1}$$

where

$$d\mathbf{x} = \mathbf{F}^e d\overline{\mathbf{X}}, \quad d\overline{\mathbf{X}} = \mathbf{F}^p d\mathbf{X} \quad (d\mathbf{x} = \mathbf{F}^e\mathbf{F}^p d\mathbf{X} = \mathbf{F}d\mathbf{X}) \tag{6.2}$$

$\overline{\mathbf{X}}$ designates the position vector of material particle in the *intermediate configuration* obtained by the unloading to the stress-free state along the hyperelastic relation. Here, assume that the positive transformations

$$J = \det\mathbf{F} > 0, \quad J^e = \det\mathbf{F}^e > 0, \quad J^p = \det\mathbf{F}^p > 0 \tag{6.3}$$

hold, so that not only $\mathbf{F}$ but also $\mathbf{F}^e$ and $\mathbf{F}^p$ are the invertible tensors possessing the positive determinant required for the existence of inverse tensor as described in Eq. (1.120). The multiplicative decomposition is illustrated in Fig. 6.1, where the initial, the intermediate and the current configurations are specified by the symbols $\mathcal{K}_0$, $\overline{\mathcal{K}}$ and $\mathcal{K}$, respectively.

Here, it should be noticed that the intermediate configuration is not an actual configuration which is physically achievable but it is merely a virtual configuration which is conceptually introduced as will be explained in the following. Solids possess the heterogeneous substructures which are statically-indeterminate in general. Therefore, the purely-elastic deformation is induced only in the initiation of unloading (reverse loading in general) process and the plastic deformation is slightly induced (removed) in the actual unloading process to the stress-free state. In order to let all material points be released to the real stress-free state, we must give different amounts of de-stressing to individual material points by cutting a material up into pieces. Therefore, the position vector $\overline{\mathbf{X}}$ of material particle in the intermediate configuration is not actual but virtual vector and it is discontinuous function of $\mathbf{X}$ and $\mathbf{x}$. Then, it is impossible to formulate $\mathbf{F}^e = \partial\mathbf{x}/\partial\overline{\mathbf{X}}$ and $\mathbf{F}^p = \partial\overline{\mathbf{X}}/\partial\mathbf{X}$ in general. Consequently, $\mathbf{F}^e$ and $\mathbf{F}^p$ are fictitious tensor-valued functions. Therefore, the intermediate configuration $\overline{\mathcal{K}}$ is the virtual configuration calculated by the unloading to the stress-free state along the hyperelastic constitutive equation as shown in Fig. 6.2. In other words, the multiplicative decomposition in Eq. (6.1) is the fictitious concept introduced in order to formulate exactly the hyperelastic-based plastic constitutive equation.

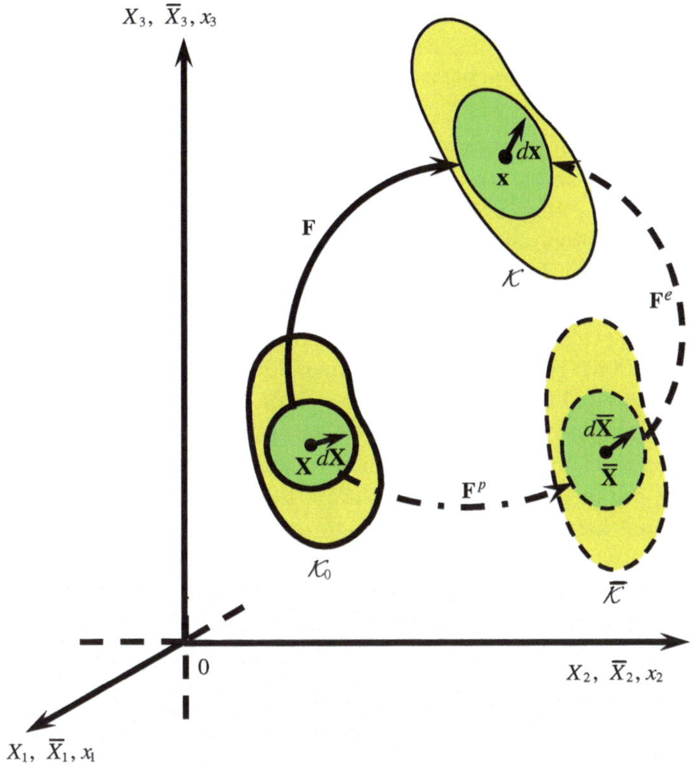

Fig. 6.1 Multiplicative decomposition of deformation gradient

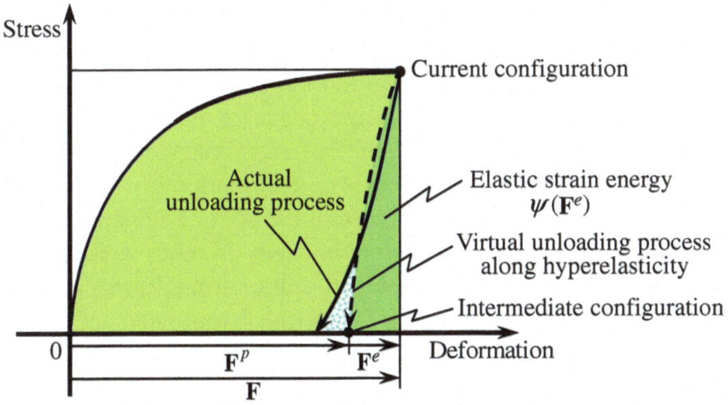

Fig. 6.2 Intermediate configuration illustrated conceptually in one-dimensional deformation

Here, note that the multiplicative decomposition in Eq. (6.1) is not attained uniquely as far as which of the elastic or the plastic deformation gradient includes the rigid-body rotation in what degree is not determined as known from $\mathbf{F} = \mathbf{F}^e\mathbf{F}^p = (\mathbf{F}^e\mathbf{R}^T)(\mathbf{R}\mathbf{F}^p) = \mathbf{F}^{e*}\mathbf{F}^{p*}$, where $\mathbf{R}$ is an arbitrary orthogonal tensor describing a rotation of the intermediate configuration and $\mathbf{F}^{e*} = \mathbf{F}^e\mathbf{R}^T$, $\mathbf{F}^{p*} = \mathbf{R}\mathbf{F}^p$. Then, the extensive debate as to which an elastic deformation gradient or a plastic deformation gradient must include the rigid-body rotation have been repeated for a long time after the proposition of the multiplicative decomposition. The inclusion of the rigid-body rotation in the plastic deformation gradient has been insisted by Lee (1969), Fardshisheh and Onat (1974), Lubarda and Lee (1981), Dafalias (1985a, b), Khan and Huang (1995), Han et al. (2003), Lubarda (2002, 2004), Wu (2004), Asaro and Lubarda (2006), Harrysson and Ristinmaa (2007), etc. On the other hand, the inclusion of the rigid-body rotation in the elastic deformation gradient has been insisted by Holsapple (1973), White (1975), Haupt (1985), Van der Giessen (1989), etc. The debates on this issue has been commented repeatedly without a definite conclusion by various authors (Dashner 1986; Lubliner 1990; Simo 1998, etc.).

The situation that the rigid-body rotation must be included in the plastic deformation gradient would be caused by worrying the fact that the elastic distortion is known from the current stress but the rigid-body rotation is unknown and thus it is possible to exclude only the elastic distortion but it is impossible to exclude both of the rigid-body rotation and the elastic distortion from the current configuration in order to get to the intermediate configuration. However, note that the deformation gradient changes from $\mathbf{F}$ to $\mathbf{F}^* = \mathbf{R}^R\mathbf{F}$ by the superposed rigid-body rotation $\mathbf{R}^R$ resulting in the change of the elastic deformation gradient from $\mathbf{F}^e$ to $\mathbf{F}^{e*} = \mathbf{R}^R\mathbf{F}^e$ because of $\mathbf{F}^* = \mathbf{F}^{e*}\mathbf{F}^{p*} = \mathbf{R}^R\mathbf{F}^e\mathbf{F}^p$, while the plastic deformation gradient does not change, i.e. $\mathbf{F}^{p*} = \mathbf{F}^p$ standing for the plastic deformation/rotation which is irrelevant to the rigid-body rotation. Therefore, the intermediate configuration is not influenced by the rigid-body rotation. The mechanically meaningful rotation is not the rotation of outside appearance of material but the rotation of substructure of material, since the anisotropic response is relevant to the anisotropic substructure of material. The plastic deformation is induced by the slips between material particles, which is induced nearly parallel to the substructure of material, causing the *plastic spin* macroscopically, and thus hardly influences on the rotation of substructure, while it causes directly the continuum spin. Therefore, the spin of substructure is given by subtracting the plastic spin from the continuum spin. In other words, the rotation of substructure is given by subtracting the plastic rotation from the macroscopic rotation observed from the outside appearance of material. Eventually, the following conclusion is obtained.

1) The intermediate configuration is the virtual configuration attained by unloading along the hyperelastic constitutive relation.
2) The substructure does not rotate during the plastic deformation gradient. It is called the *isoclinic* (constant inclination) concept (Mandel, 1973, 1974). Therefore, the rigid-body rotation is included in the elastic deformation gradient.

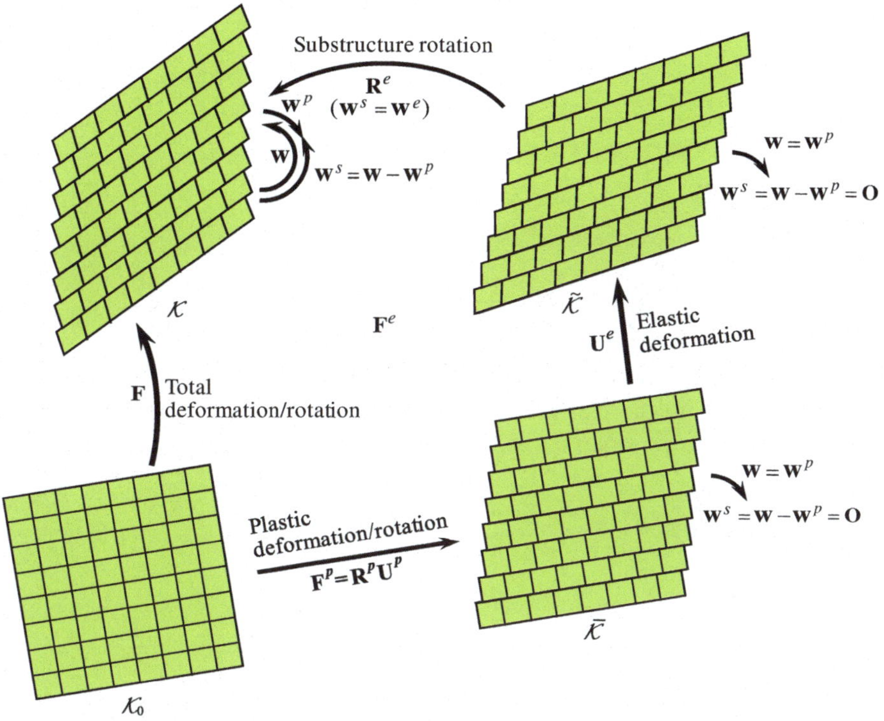

Fig. 6.3 Multiplicative decomposition, where the rigid-body rotation is included in elastic deformation gradient tensor

The multiplicative decomposition based on these notions is illustrated in Fig. 6.3. Here, the configuration excluded the substructure rotation $\mathbf{R}^e$ in $\mathbf{F}^e = \mathbf{R}^e\mathbf{U}^e$ from the current configuration is designated by the symbol $\tilde{\mathcal{K}}$.

The continuum spin is known from the outside appearance of material. In contrast, the spin of substructure is unknown only from the continuum spin as far as the plastic spin is unknown, which depends on the substructure of material. Then, it is obliged to formulate the plastic spin as the macroscopic constitutive equation.

Hereinafter, the barren discussion on the inclusion of the rigid-body rotation in the elastic or in the plastic deformation gradient should be stopped by the above-mentioned conclusion.

It holds from Eq. (6.1) with Eq. (1.17) that

$$\det \mathbf{F} = \det(\mathbf{F}^e\mathbf{F}^p) = \det\mathbf{F}^e\det\mathbf{F}^p \tag{6.4}$$

On the other hand, presuming that the line-elements $d\mathbf{X}^a, d\mathbf{X}^b, d\mathbf{X}^c$ in the initial configuration $\mathcal{K}_0$ change to $d\mathbf{x}^a, d\mathbf{x}^b, d\mathbf{x}^c$ in the current configuration $\mathcal{K}$ and to $d\overline{\mathbf{X}}^a, d\overline{\mathbf{X}}^b, d\overline{\mathbf{X}}^c$ in the intermediate configuration $\overline{\mathcal{K}}$, and denoting the initial, current and intermediate volumes as V, v and $\overline{V}$, respectively, it follows from Eq. (2.145) with Eq. (6.4) that

$$\left.\begin{array}{l} dv = JdV = J^e d\overline{V} \\ d\overline{V} = J^p dV \end{array}\right\} \tag{6.5}$$

from which one has

$$J = J^e J^p \tag{6.6}$$

where

$$J \equiv \det\mathbf{F} = \frac{dv}{dV}, \quad J^e \equiv \det\mathbf{F}^e = \frac{dv}{d\overline{V}}, \quad J^p \equiv \det\mathbf{F}^p = \frac{d\overline{V}}{dV} \tag{6.7}$$

Equation (6.6) engenders the additive decomposition of the logarithmic volumetric strain ε_v into the elastic logarithmic volumetric strain ε_v^e and the plastic logarithmic volumetric strain ε_v^p, i.e.

$$\varepsilon_v = \varepsilon_v^e + \varepsilon_v^p \tag{6.8}$$

where

$$\varepsilon_v \equiv \ln J = \ln\frac{dv}{dV}, \quad \varepsilon_v^e \equiv \ln J^e = \ln\frac{dv}{d\overline{V}}, \quad \varepsilon_v^p \equiv \ln J^p = \ln\frac{d\overline{V}}{dV} \tag{6.9}$$

It follows from Eq. (6.9) that

$$\dot{\varepsilon}_v = \mathrm{tr}\mathbf{d} = \frac{\dot{J}}{J} = \frac{(dv)^{\bullet}}{dv}, \quad \dot{\varepsilon}_v^e = \frac{\dot{J}^e}{J^e} = \frac{(dv)^{\bullet}}{dv} - \frac{(d\overline{V})^{\bullet}}{d\overline{V}}, \quad \dot{\varepsilon}_v^p = \frac{\dot{J}^p}{J^p} = \frac{(d\overline{V})^{\bullet}}{d\overline{V}} \tag{6.10}$$

Equations (6.9) and (6.10) conform to Eqs. (2.142) and (2.150).

Substitution of Eq. (6.1) into the velocity gradient $\boldsymbol{l}$ in Eq. (2.75) leads to

$$\boldsymbol{l} = \dot{\mathbf{F}}\mathbf{F}^{-1} = \dot{\mathbf{F}}^e\mathbf{F}^{e-1} + \mathbf{F}^e\dot{\mathbf{F}}^p\mathbf{F}^{p-1}\mathbf{F}^{e-1} = \boldsymbol{l}^e + \boldsymbol{l}^p \tag{6.11}$$

where

$$\boxed{\begin{array}{l} \boldsymbol{l}^e \equiv \dot{\mathbf{F}}^e\mathbf{F}^{e-1} \\ \boldsymbol{l}^p \equiv \mathbf{F}^e\dot{\mathbf{F}}^p\mathbf{F}^{p-1}\mathbf{F}^{e-1} = \mathbf{F}^e\overline{\mathbf{L}}^p\mathbf{F}^{e-1} \\ \overline{\mathbf{L}}^p \equiv \dot{\mathbf{F}}^p\mathbf{F}^{p-1} \end{array}} \tag{6.12}$$

The velocity gradients $\boldsymbol{l}, \boldsymbol{l}^e, \boldsymbol{l}^p, \overline{\mathbf{L}}^p$ can be decomposed additively into the symmetric and skew-symmetric parts as follows:

$$\left.\begin{aligned} \boldsymbol{l} &= \mathbf{d}+\mathbf{w} \\ \boldsymbol{l}^e &= \mathbf{d}^e+\mathbf{w}^e \\ \boldsymbol{l}^p &= \mathbf{d}^p+\mathbf{w}^p \\ \overline{\mathbf{L}}^p &= \overline{\mathbf{D}}^p+\overline{\mathbf{W}}^p \end{aligned}\right\} \tag{6.13}$$

and

$$\left.\begin{aligned} \mathbf{d} &= \mathbf{d}^e+\mathbf{d}^p \\ \mathbf{w} &= \mathbf{w}^e+\mathbf{w}^p \end{aligned}\right\} \tag{6.14}$$

where

$$\left.\begin{aligned} \mathbf{d} &= \mathrm{sym}[\boldsymbol{l}] = \mathrm{sym}[\dot{\mathbf{F}}\,\mathbf{F}^{-1}] \\ \mathbf{w} &= \mathrm{ant}[\boldsymbol{l}] = \mathrm{ant}[\dot{\mathbf{F}}\,\mathbf{F}^{-1}] \end{aligned}\right\} \tag{6.15}$$

$$\left.\begin{aligned} \mathbf{d}^e &= \mathrm{sym}[\boldsymbol{l}^e] = \mathrm{sym}[\dot{\mathbf{F}}^e\mathbf{F}^{e-1}] = \mathbf{R}^e\mathrm{sym}[\dot{\mathbf{U}}^e\mathbf{U}^{e-1}]\mathbf{R}^{eT} \\ \mathbf{w}^e &= \mathrm{ant}[\boldsymbol{l}^e] = \mathrm{ant}[\dot{\mathbf{F}}^e\mathbf{F}^{e-1}] = \dot{\mathbf{R}}^e\mathbf{R}^{eT}+\mathbf{R}^e\mathrm{ant}[\dot{\mathbf{U}}^e\mathbf{U}^{e-1}]\mathbf{R}^{eT} \end{aligned}\right\} \tag{6.16}$$

$$\left.\begin{aligned} \mathbf{d}^p &= \mathrm{sym}[\boldsymbol{l}^p] = \mathrm{sym}[\mathbf{F}^e\overline{\mathbf{L}}^p\mathbf{F}^{e-1}] \\ \mathbf{w}^p &= \mathrm{ant}[\boldsymbol{l}^p] = \mathrm{ant}[\mathbf{F}^e\overline{\mathbf{L}}^p\mathbf{F}^{e-1}] \end{aligned}\right\} \tag{6.17}$$

$$\left.\begin{aligned} \overline{\mathbf{D}}^p &= \mathrm{sym}[\overline{\mathbf{L}}^p] = \mathrm{sym}\dot{\mathbf{F}}^p\mathbf{F}^{p-1}] \\ \overline{\mathbf{W}}^p &= \mathrm{ant}[\overline{\mathbf{L}}^p] = \mathrm{ant}[\dot{\mathbf{F}}^p\mathbf{F}^{p-1}] \end{aligned}\right\} \tag{6.18}$$

6.2 Requirement for Additive Decomposition of Strain Rate and Spin

The additive decomposition of the strain rate $\mathbf{d}$ and the continuum spin $\mathbf{w}$ is required for the formulation of the hypoelastic-based plasticity. The requirement for the additive decomposition into the elastic and the plastic parts from the multiplicative decomposition of the deformation gradient will be considered in this section.

The elastic strain rate $\mathbf{d}^e$ based in the current configuration depends only on the elastic deformation gradient $\mathbf{F}^e$ as shown in Eq. $(6.16)_1$ and it is given by excluding the anti-symmetric part, i.e. the substructure spin $\mathbf{w}^e$ from the elastic velocity gradient. Then, let $\mathbf{d}^e$ be related to the corotational rate $\overset{\circ}{\boldsymbol{\sigma}}$ based on the substructure rotation as the hypoelastic equation (Truesdell, 1955) described in Eq. (5.53), i.e.

$$\mathbf{d}^e = \mathrm{sym}[\boldsymbol{l}^e] = \mathrm{sym}[\dot{\mathbf{F}}^e\mathbf{F}^{e-1}] = \mathbf{E}^{-1} : \overset{\circ}{\boldsymbol{\sigma}} \tag{6.19}$$

where $\mathbf{E}$ is the elastic modulus tensor. The substructure spin $\mathbf{w}^e$ will be formulated later.

On the other hand, the plastic strain rate $\mathbf{d}^p$ and the plastic spin $\mathbf{w}^p$ depend not only on the plastic deformation gradient $\mathbf{F}^p$ but also on the elastic deformation gradient $\mathbf{F}^e$ as known from Eq. (6.17). On the other hand, $\overline{\mathbf{D}}^p$ and $\overline{\mathbf{W}}^p$ depend only on the plastic deformation gradient $\mathbf{F}^p$. Then, we assume the following plastic constitutive equations with the plastic flow rules:

$$\left.\begin{aligned} \overline{\mathbf{D}}^p &= \mathrm{sym}[\dot{\mathbf{F}}^p\mathbf{F}^{p-1}] = \dot{\lambda}\,\overline{\mathbf{N}}^p(\|\overline{\mathbf{N}}^p\| = 1) \\ \overline{\mathbf{W}}^p &= \mathrm{ant}[\dot{\mathbf{F}}^p\mathbf{F}^{p-1}] = \dot{\lambda}\,\overline{\boldsymbol{\Omega}}^p(\|\overline{\boldsymbol{\Omega}}^p\| = 1) \end{aligned}\right\}(\dot{\lambda} > 0) \tag{6.20}$$

where $\overline{\mathbf{N}}^p$ is the normalized direction tensors of the plastic strain rate and $\overline{\boldsymbol{\Omega}}^p$ is the second-order anti-symmetric tensor in the intermediate configuration $\overline{\mathcal{K}}$.

Now, assume that the elastic deformation is infinitesimal and thus the elastic stretch tensors are reduced to

$$\mathbf{U}^e \cong \mathbf{I}, \quad \mathbf{V}^e \cong \mathbf{I} \tag{6.21}$$

in the polar decomposition

$$\mathbf{F}^e = \mathbf{R}^e\mathbf{U}^e = \mathbf{V}^e\mathbf{R}^e \tag{6.22}$$

where $\dot{\mathbf{U}}^e \cong \mathbf{O}$, $\dot{\mathbf{V}}^e \cong \mathbf{O}$ do not hold in general. Then, it follows that

$$\left.\begin{aligned} \mathbf{F}^e &\cong \mathbf{R}^e, \\ \mathrm{sym}[\dot{\mathbf{U}}^e\mathbf{U}^{e-1}] &\cong \mathrm{ant}[\dot{\mathbf{U}}^e] = \dot{\mathbf{U}}^e \\ \mathrm{ant}[\dot{\mathbf{U}}^e\mathbf{U}^{e-1}] &\cong \mathrm{ant}[\dot{\mathbf{U}}^e] = \mathbf{O} \end{aligned}\right\} \tag{6.23}$$

and from Eqs. (6.17) and (6.18) with Eq. (6.23) that

$$\left.\begin{aligned} \mathbf{d}^e &\cong \mathbf{R}^e\mathrm{sym}[\dot{\mathbf{U}}^e]\mathbf{R}^{eT} = \mathbf{R}^e\,\dot{\mathbf{U}}^e\mathbf{R}^{eT} \\ \mathbf{w}^e &\cong \dot{\mathbf{R}}^e\mathbf{R}^{eT} + \mathbf{R}^e\mathrm{ant}[\dot{\mathbf{U}}^e]\mathbf{R}^{eT} = \dot{\mathbf{R}}^e\mathbf{R}^{eT} \\ \mathbf{d}^p &\cong \mathbf{R}^e\mathrm{sym}[\overline{\mathbf{L}}^p]\mathbf{R}^{eT} = \mathbf{R}^e\,\overline{\mathbf{D}}^p\mathbf{R}^{eT} \\ \mathbf{w}^p &\cong \mathbf{R}^e\mathrm{ant}[\overline{\mathbf{L}}^p]\mathbf{R}^{eT} = \mathbf{R}^e\,\overline{\mathbf{W}}^p\mathbf{R}^{eT} \end{aligned}\right\} \tag{6.24}$$

where the substructure spin $\mathbf{w}^e$ is not infinitesimal in general.

It follows from Eq. (6.23) with Eq. (6.24) that

$$\left.\begin{aligned} \mathbf{d}^p &= \dot{\lambda}\,\mathbf{n}^p(\|\mathbf{n}^p\| = 1) \\ \mathbf{w}^p &= \dot{\lambda}\,\boldsymbol{\omega}^p \end{aligned}\right\} \tag{6.25}$$

where

$$\left.\begin{aligned}\mathbf{n}^p &= \mathbf{R}^e\overline{\mathbf{N}}^p\mathbf{R}^{eT}\\ \boldsymbol{\omega}^p &= \mathbf{R}^e\overline{\boldsymbol{\Omega}}^p\mathbf{R}^{eT}\end{aligned}\right\} \tag{6.26}$$

The substitution of Eqs. (6.19) and (6.25) into Eq. (6.14) leads to the following additive decompositions of the strain rate **d** and the continuum spin **w** into the purely-elastic and the purely-plastic parts:

$$\left.\begin{aligned}\boxed{\mathbf{d} = \mathbf{d}^e + \mathbf{d}^p} &= \mathbf{E}^{-1} : \overset{\circ}{\boldsymbol{\sigma}} + \dot{\lambda}\,\mathbf{n}^p\\ \boxed{\mathbf{w} = \mathbf{w}^e + \mathbf{w}^p} &= \dot{\mathbf{R}}^e\mathbf{R}^{eT} + \dot{\lambda}\,\boldsymbol{\omega}^p\end{aligned}\right\} \tag{6.27}$$

which is called the *hypoelasto-plasticity* or the *hypoelastic-based plasticity*. The physical background for the additive decomposition of the strain rate and the spin into the elastic and the plastic parts is given above, which has been adopted a priori (cf. e.g. Lubliner 1990; Simo and Hughes 1998; Belytschko et al. 2014; Lubarda 2002; Asaro and Lubarda 2006) or without an exact proof (cf. e.g. Dafalias 1984, 1985b).

The rotation of substructure (material fibers) , $\boldsymbol{\omega}_s$, is different from the rotation of the outside appearance of material, i.e. the rotation of line elements depicted on the material surface but it is suppressed by the plastic deformation in plastically anisotropic material. It is given by subtracting the plastic spin $\mathbf{w}^p$ from the continuum spin $\mathbf{w}$, resulting in $\boldsymbol{\omega}_s = \mathbf{w} - \mathbf{w}^p$. Here, the plastic spin $\mathbf{w}^p$ is determined depending on the constitutive property of plastic anisotropy, while the continuum spin $\mathbf{w}$ is determined only by the geometrical variation of outside appearance. Consequently, $\boldsymbol{\omega}_s = \mathbf{w}^e$ holds, i.e.

$$\boldsymbol{\omega}_s = \mathbf{w}^e = \dot{\mathbf{R}}^e\mathbf{R}^{eT} = \mathbf{w} - \mathbf{w}^p \tag{6.28}$$

from Eq. $(6.27)_2$ (Mandel 1971; Kratochvil 1971; Dafalias 1983, 1985a; Loret 1983). The physical interpretation of this fact will be given in Chaps. 16 and 19. The explicit formulation of the plastic spin $\mathbf{w}^p$ in Eqs. $(6.25)_2$ or $(6.27)_2$ will be given in Sect. 9.9 and Chap. 16.

Based on Eq. (6.27), the hypoelastic-based plastic constitutive equation will be formulated throughout this book, which holds for finite deformation and rotation under the limitation of the infinitesimal elastic deformation.

6.3 Conventional Hypoelastic-Based Plastic Constitutive Equations

The elastic strain rate is given by Eqs. (5.53), $(6.14)_1$ and (6.19), i.e.

$$\boxed{\overset{\circ}{\boldsymbol{\sigma}} = \mathbf{E} : \mathbf{d}^e = \mathbf{E} : (\mathbf{d} - \mathbf{d}^p), \quad \mathbf{d}^e = \mathbf{E}^{-1} : \overset{\circ}{\boldsymbol{\sigma}}} \tag{6.29}$$

In what follows, the plastic strain rate will be formulated limiting to the conventional plasticity (Drucker 1988) as was described at the beginning of this chapter.

Now, consider first the following isotropic *yield condition* exhibiting the isotropic hardening/softening.

$$\boxed{f(\boldsymbol{\sigma}) = F(H)} \tag{6.30}$$

where $F(\geq 0)$ is the function of the *isotropic hardening variable* $H(\geq 0)$ and is called the *hardening function* which describes the isotropic hardening or softening, i.e. the expansion or contraction of yield surface. Here, we choose the yield stress function $f(\boldsymbol{\sigma})(\geq 0)$ in Eq. (6.30) to be the homogeneous function of $\boldsymbol{\sigma}$ in degree-one. Therefore, it follows that

$$f(|s|\boldsymbol{\sigma}) = |s|f(\boldsymbol{\sigma}) \tag{6.31}$$

for an arbitrary scalar s and

$$\frac{\partial f(\boldsymbol{\sigma})}{\partial \boldsymbol{\sigma}} : \boldsymbol{\sigma} = f(\boldsymbol{\sigma}) = F(H) \tag{6.32}$$

for the sake of *Euler's theorem for homogeneous function* in degree-one (see **Appendix D**). Then, it follows from Eq. (6.32) that

$$1/\left\|\frac{\partial f(\boldsymbol{\sigma})}{\partial \boldsymbol{\sigma}}\right\| = \frac{\dfrac{\partial f(\boldsymbol{\sigma})}{\partial \boldsymbol{\sigma}} : \boldsymbol{\sigma}}{f(\boldsymbol{\sigma})} / \left\|\frac{\partial f(\boldsymbol{\sigma})}{\partial \boldsymbol{\sigma}}\right\| = \frac{\mathbf{n} : \boldsymbol{\sigma}}{F} \tag{6.33}$$

where $\mathbf{n}$ is the normalized outward-normal of the yield surface (see **Appendix E**).

$$\mathbf{n} \equiv \frac{\partial f(\boldsymbol{\sigma})}{\partial \boldsymbol{\sigma}} \bigg/ \left\|\frac{\partial f(\boldsymbol{\sigma})}{\partial \boldsymbol{\sigma}}\right\| \quad (\|\mathbf{n}\| = 1) \tag{6.34}$$

Taking account of the corotational time-derivative in Eq. (4.79), the material-time derivative of Eq. (6.30) leads to the *consistency condition* in the corotational time-derivative:

$$\frac{\partial f(\boldsymbol{\sigma})}{\partial \boldsymbol{\sigma}} : \overset{\circ}{\boldsymbol{\sigma}} - F' \dot{H} = 0 \tag{6.35}$$

$$F' \equiv dF/dH \tag{6.36}$$

$$\overset{\bullet}{H}(\boldsymbol{\sigma}, H; \mathbf{d}^p) = f_{Hd}(\boldsymbol{\sigma}, H; \mathbf{d}^p/||\mathbf{d}^p||)||\mathbf{d}^p|| \tag{6.37}$$

$\overset{\bullet}{H}$ has to be the homogeneous function of $\mathbf{d}^p$ in degree-one for the rate-independent deformation behavior because it evolves only in the plastic loading process ($\mathbf{d}^p \neq \mathbf{O}$) and possesses the dimension of time in minus one, while, needless to say, it is a nonlinear equation of the components d_{ij}^p in general as seen in metals, i.e. $\overset{\bullet}{H} = \sqrt{2/3}||\mathbf{d}^p|| = \sqrt{2/3}\sqrt{d_{rs}^p d_{rs}^p}$ described in Sect. 6.4.

Now, assume the *associated flow rule* in which the *plastic potential* function is given by the yield function:

$$\boxed{\mathbf{d}^p = \overset{\bullet}{\lambda}\mathbf{n}} \quad (\overset{\bullet}{\lambda} = ||\mathbf{d}^p|| > 0, ||\mathbf{n}|| = 1) \tag{6.38}$$

adopting $\mathbf{n}$ for $\mathbf{n}^p$ in Eq. (6.25), where $\overset{\bullet}{\lambda}$ is the magnitude of plastic strain rate, called often *plastic positive proportionality factor* or *plastic multiplier*. The expression of the flow rule in Eq. (6.38) possesses the clear physical meaning as definitely divided into the magnitude $\overset{\bullet}{\lambda}$ and the pure direction tensor $\mathbf{n}$. Then, the magnitude of plastic strain rate, i.e. $||\mathbf{d}^p||$ appearing often in the hardening variable can be replaced to $\overset{\bullet}{\lambda}$ so that the physical interpretation of plastic constitutive relation can be provided definitely. On the other hand, the magnitude cannot be represented only by $\overset{\bullet}{\lambda}$ in the flow rule with the expression $\mathbf{d}^p = \overset{\bullet}{\lambda}\,\partial f(\boldsymbol{\sigma})/\partial\boldsymbol{\sigma}$ except for the particular yield function given by the magnitude of stress or deviatoric stress (Mises yield function) leading to $||\partial f(\boldsymbol{\sigma})/\partial\boldsymbol{\sigma}|| = ||\partial||\boldsymbol{\sigma}||/\partial\boldsymbol{\sigma}|| = ||\partial||\boldsymbol{\sigma}'||/\partial\boldsymbol{\sigma}|| = 1$.

Substituting the plastic flow rule in Eq. (6.38) into the consistency condition (6.35), one has

$$\mathbf{n} : \overset{\circ}{\boldsymbol{\sigma}} - F'\overset{\bullet}{H}\frac{\mathbf{n}:\boldsymbol{\sigma}}{F} = \mathbf{n} : \overset{\circ}{\boldsymbol{\sigma}} - \frac{F'}{F}\overset{\bullet}{\lambda} f_{Hn}(\boldsymbol{\sigma}, H;\ \mathbf{n})\mathbf{n}:\boldsymbol{\sigma} = 0 \tag{6.39}$$

by virtue of Eq. (6.33), which leads to

$$\boxed{\mathbf{n} : \overset{\circ}{\boldsymbol{\sigma}} - \overset{\bullet}{\lambda} M^p = 0} \tag{6.40}$$

where M^p is called the *plastic modulus* and is given by

$$\boxed{M^p \equiv \frac{F'}{F} f_{Hn}(\boldsymbol{\sigma}, H;\ \mathbf{n})\mathbf{n}:\boldsymbol{\sigma}} \tag{6.41}$$

$$f_{Hn}(\boldsymbol{\sigma}, H;\ \mathbf{n}) = \overset{\bullet}{H} \,/\, \overset{\bullet}{\lambda} \tag{6.42}$$

Here, the yield surface in Eq. (6.30) retains the similar shape and orientation with respect to the origin of stress space by virtue of homogeneity of function $f(\boldsymbol{\sigma})$.

It follows from Eq. (6.39) that

$$\boxed{\dot{\lambda} = \frac{\mathbf{n} : \overset{\circ}{\boldsymbol{\sigma}}}{M^p}}, \quad \boxed{\mathbf{d}^p = \frac{\mathbf{n} : \overset{\circ}{\boldsymbol{\sigma}}}{M^p}\mathbf{n}} \tag{6.43}$$

Substituting Eqs. (6.29) and (6.43) into Eq. $(6.14)_1$, the strain rate is given by

$$\boxed{\mathbf{d} = \mathbf{E}^{-1} : \overset{\circ}{\boldsymbol{\sigma}} + \frac{\mathbf{n} : \overset{\circ}{\boldsymbol{\sigma}}}{M^p}\mathbf{n} = \left(\mathbf{E}^{-1} + \frac{\mathbf{n} \otimes \mathbf{n}}{M^p}\right) : \overset{\circ}{\boldsymbol{\sigma}}} \tag{6.44}$$

The scalar product of $\mathbf{n}:\mathbf{E}$ to Eq. (6.44) leads to

$$\mathbf{n}:\mathbf{E}:\mathbf{d}\left(= \mathbf{n} : \overset{\circ}{\boldsymbol{\sigma}} + \frac{\mathbf{n} : \overset{\circ}{\boldsymbol{\sigma}}}{M^p}\mathbf{n}:\mathbf{E}:\mathbf{n} = (M^p + \mathbf{n}:\mathbf{E}:\mathbf{n})\frac{\mathbf{n} : \overset{\circ}{\boldsymbol{\sigma}}}{M^p} = (M^p + \mathbf{n}:\mathbf{E}:\mathbf{n})\frac{\mathbf{n} : \overset{\circ}{\boldsymbol{\sigma}}}{M^p}\right)$$
$$= (M^p + \mathbf{n}:\mathbf{E}:\mathbf{n})\,\dot{\lambda}$$

from which the magnitude of plastic strain rate described in terms of strain rate, denoted by $\dot{\Lambda}$ instead of $\dot{\lambda}$, in the flow rule (6.38) is described as follows:

$$\boxed{\dot{\Lambda} = \frac{\mathbf{n}:\mathbf{E}:\mathbf{d}}{M^p + \mathbf{n}:\mathbf{E}:\mathbf{n}}}, \quad \boxed{\mathbf{d}^p = \frac{\mathbf{n}:\mathbf{E}:\mathbf{d}}{M^p + \mathbf{n}:\mathbf{E}:\mathbf{n}}\mathbf{n}} \tag{6.45}$$

Incidentally, Eq. $(6.45)_1$ can be also derived directly from the following relation obtained by substituting Eqs. $(6.14)_1$, (6.29) and (6.38) into the consistency condition (6.40).

$$\mathbf{n} : \overset{\circ}{\boldsymbol{\sigma}} - \dot{\Lambda}M^p = \mathbf{n}:\mathbf{E}:\mathbf{d}^e - \dot{\Lambda}M^p = \mathbf{n}:\mathbf{E}:(\mathbf{d} - \mathbf{d}^p) - \dot{\Lambda}M^p$$
$$= \mathbf{n}:\mathbf{E}:(\mathbf{d} - \dot{\Lambda}\,\mathbf{n}) - \dot{\Lambda}M^p = \mathbf{n}:\mathbf{E}:\mathbf{d} - (\mathbf{n}:\mathbf{E}:\mathbf{n} + M^p)\,\dot{\Lambda} = 0 \tag{6.46}$$

The stress rate is given from Eq. (6.29) with Eq. (6.45) as follows:

$$\boxed{\overset{\circ}{\boldsymbol{\sigma}} = \overset{\circ}{\boldsymbol{\sigma}}{}^e + \overset{\circ}{\boldsymbol{\sigma}}{}^p = \mathbf{E}:\mathbf{d} - \frac{\mathbf{n}:\mathbf{E}:\mathbf{d}}{M^p + \mathbf{n}:\mathbf{E}:\mathbf{n}}\mathbf{E}:\mathbf{n} = \left(\mathbf{E} - \frac{\mathbf{E}:\mathbf{n} \otimes \mathbf{n}:\mathbf{E}}{M^p + \mathbf{n}:\mathbf{E}:\mathbf{n}}\right):\mathbf{d}}$$

$$\overset{\circ}{\sigma}_{ij} = \overset{\circ}{\sigma}{}^e_{ij} + \overset{\circ}{\sigma}{}^p_{ij} = E_{ijkl}d_{kl} - \frac{n_{pq}E_{pqkl}d_{kl}}{M^p + n_{ab}E_{abcd}n_{cd}}E_{ijrs}n_{rs} \tag{6.47}$$

where

$$\overset{\circ}{\boldsymbol{\sigma}}{}^e \equiv \mathbf{E}:\mathbf{d},\ \overset{\circ}{\boldsymbol{\sigma}}{}^p \equiv -\mathbf{E}:\mathbf{d}^p = -\mathbf{K}^{pr}:\mathbf{d} \tag{6.48}$$

$$\mathbf{K}^{pr} \equiv \frac{\mathbf{E}:\mathbf{n} \otimes \mathbf{n}:\mathbf{E}}{M^p + \mathbf{n}:\mathbf{E}:\mathbf{n}}, \quad K^{pr}_{ijkl} \equiv \frac{E_{ijrs}n_{rs}n_{pq}E_{pqkl}}{M^p + n_{ab}E_{abcd}n_{cd}} \tag{6.49}$$

$\overset{\circ}{\boldsymbol{\sigma}}{}^e$ is called the *elastic stress rate* since it is calculated supposing that a purely elastic deformation is induced by the strain rate $\mathbf{d}$, and $\overset{\circ}{\boldsymbol{\sigma}}{}^p$ is called the *plastic relaxation stress rate*. The fourth-order tensor $\mathbf{K}^{pr}$ is called the *plastic relaxation modulus*.

Furthermore, using the *elastoplastic stiffness modulus tensor*

$$\boxed{\mathbf{K}^{ep} \equiv \mathbf{E} - \mathbf{K}^{pr} = \mathbf{E} - \frac{\mathbf{E}:\mathbf{n} \otimes \mathbf{n}:\mathbf{E}}{M^p + \mathbf{n}:\mathbf{E}:\mathbf{n}}}, \quad K^{ep}_{ijkl} \equiv E_{ijkl} - K^{pr}_{ijkl} = E_{ijkl} - \frac{E_{ijrs} n_{rs} n_{pq} E_{pqkl}}{M^p + n_{ab} E_{abcd} n_{cd}} \tag{6.50}$$

the stress rate can be described as

$$\overset{\circ}{\boldsymbol{\sigma}} = \mathbf{K}^{ep} : \mathbf{d} \tag{6.51}$$

Here, note that $\mathbf{E}$ possesses not only the minor symmetry but also the major symmetry $E_{ijkl} = E_{klij}$ providing $(\mathbf{E}:\mathbf{n})_{ij} = (\mathbf{n}:\mathbf{E})_{ij} = E_{ijkl}n_{kl}$, so that the symmetries of $\mathbf{E}:\mathbf{n} \otimes \mathbf{n}:\mathbf{E} = [(\mathbf{E}:\mathbf{n}) \otimes (\mathbf{n}:\mathbf{E})]^T$ and $\mathbf{K}^{ep} = \mathbf{K}^{ep\,T}$ hold.

For the non-associated flow rule

$$\mathbf{d}^p = \dot{\lambda}\, \mathbf{m} \quad (\|\mathbf{m}\| = 1,\ \mathbf{m} \neq \mathbf{n}) \tag{6.52}$$

where $\mathbf{m}$ is the normalized second-order function of stress and internal variables. The magnitude of plastic strain rate $\dot{\Lambda}$ is given instead of Eq. (6.45) as follows:

$$\dot{\Lambda} = \frac{\mathbf{n}:\mathbf{E}:\mathbf{d}}{M^p + \mathbf{n}:\mathbf{E}:\mathbf{m}} \tag{6.53}$$

Then, the elastoplastic stiffness modulus tensor is given by

$$\mathbf{K}^{ep} = \mathbf{E} - \mathbf{K}^{pr} = \mathbf{E} - \frac{\mathbf{E}:\mathbf{m} \otimes \mathbf{n}:\mathbf{E}}{M^p + \mathbf{n}:\mathbf{E}:\mathbf{m}} \tag{6.54}$$

Therefore, the plastic relaxation modulus tensor and the elastoplastic stiffness modulus tensor are not the symmetric tensors, i.e.

$$\mathbf{K}^p \neq \mathbf{K}^{pT}, \quad \mathbf{K}^{ep} \neq \mathbf{K}^{epT}$$

in the non-associated flow rule.

6.4 Constitutive Equation of Metals

The elastoplastic constitutive equations for isotropic materials are described above. The constitutive equation of metals is shown below, which has made an important contribution to the development of elastoplasticity. The following *von Mises yield condition* with the associated flow rule is assumed for metals.

$$f(\boldsymbol{\sigma}) = \sigma^{eq}, \quad \sigma^{eq} \equiv \sqrt{\frac{3}{2}}||\boldsymbol{\sigma}'|| \tag{6.55}$$

$$\left.\begin{array}{l} H = \varepsilon^{eqp} \equiv \int \sqrt{\frac{2}{3}}||\mathbf{d}^p||dt, \quad \dot{H} = \dot{\varepsilon}^{eqp} = \sqrt{\frac{2}{3}}||\mathbf{d}^p||, \quad h = \sqrt{\frac{2}{3}}||\mathbf{n}|| = \sqrt{\frac{2}{3}} \\ F(H) = F_0\{1 + h_1[1 - \exp(-h_2 H)]\}, \quad F' = F_0\, h_1\, h_2 \exp(-h_2 H) \end{array}\right\} \tag{6.56}$$

In the monotonic uniaxial loading with the assumption of the plastically-pressure independence [$\sigma_{ij} = 0$ except for $i = j = 1$; $d_2^p = d_3^p = -d_1^p/2$ ($\mathrm{tr}\mathbf{d}^p = 0$), $d_{ij}^p = 0 (i \neq j)$], it holds that

$$\left.\begin{array}{l} \sigma^{eq} \equiv \sqrt{3/2}\sqrt{\sigma_1'^2 + 2\sigma_2'^2} = \sqrt{3/2}\sqrt{(\sigma_1 - \sigma_1/3)^2 + 2(0 - \sigma_1/3)^2} = \sigma_1 \\ \varepsilon^{eqp} \equiv \int \sqrt{2/3}\sqrt{d_1^{p\prime 2} + 2d_2^{p\prime 2}}dt = \int \sqrt{2/3}\sqrt{d_1^{p\,2} + 2(-d_1^p/2)^2}dt = \int d_1^p dt = \varepsilon_1^p \end{array}\right\} \tag{6.57}$$

Then, σ^{eq} and ε^{eqp} coincide with the axial stress and the axial plastic strain in that loading and thus are called the *equivalent stress* and the *equivalent* (or *accumulated*) *plastic strain*, respectively. Substituting Eq. (6.56) into Eq. (6.44) and using the relations

$$\left.\begin{array}{l} \dfrac{\partial f(\boldsymbol{\sigma})}{\partial \boldsymbol{\sigma}} = \sqrt{\dfrac{3}{2}}\dfrac{\boldsymbol{\sigma}'}{||\boldsymbol{\sigma}'||}, \quad \mathbf{n} = \dfrac{\boldsymbol{\sigma}'}{||\boldsymbol{\sigma}'||}, \quad \mathbf{n}:\boldsymbol{\sigma} = ||\boldsymbol{\sigma}'|| \\ \mathring{\sigma}^{eq} = \sqrt{\dfrac{3}{2}}\dfrac{\boldsymbol{\sigma}'}{||\boldsymbol{\sigma}'||}:\mathring{\boldsymbol{\sigma}}' = \dfrac{3}{2}\dfrac{\boldsymbol{\sigma}':\mathring{\boldsymbol{\sigma}}}{\sigma^{eq}} = \sqrt{\dfrac{3}{2}}\mathbf{n}:\mathring{\boldsymbol{\sigma}} \\ \mathbf{d}^p = \mathbf{d}^{p\prime} \\ M^p = \dfrac{F'}{F}\sqrt{\dfrac{2}{3}}||\boldsymbol{\sigma}'|| = \dfrac{2}{3}F' \end{array}\right\} \tag{6.58}$$

the constitutive equation of the isotropic Mises material is given as follows:

$$\mathbf{d} = \mathbf{E}^{-1}:\mathring{\boldsymbol{\sigma}} + \frac{\mathbf{n}:\mathring{\boldsymbol{\sigma}}}{\frac{2}{3}F'}\mathbf{n} = \mathbf{E}^{-1}:\mathring{\boldsymbol{\sigma}} + \frac{\sqrt{\frac{2}{3}}\mathring{\sigma}^{eq}}{\frac{2}{3}F'}\frac{\boldsymbol{\sigma}'}{||\boldsymbol{\sigma}'||} = \mathbf{E}^{-1}:\mathring{\boldsymbol{\sigma}} + \frac{3}{2}\frac{1}{F'}\frac{\mathring{\sigma}^{eq}}{\sigma^{eq}}\boldsymbol{\sigma}' \tag{6.59}$$

which is called the *Prandtl-Reuss equation*. The plastic work rate of this material is described as

$$\boldsymbol{\sigma}:\mathbf{d}^p = \boldsymbol{\sigma}:\dot{\lambda}\frac{\boldsymbol{\sigma}'}{||\boldsymbol{\sigma}'||} = \dot{\lambda}||\boldsymbol{\sigma}'|| = ||\boldsymbol{\sigma}'||||\mathbf{d}^p|| = \sigma^{eq}\dot{\varepsilon}^{eqp} = F(\varepsilon^{eqp})\dot{\varepsilon}^{eqp} \tag{6.60}$$

which is the product of the hardening function and the equivalent plastic strain and thus the hardening attributable to the equivalent plastic strain is called the *work hardening*, too.

The traction $\boldsymbol{t}$ acting on the octahedral plane is expressed by

$$\begin{aligned}\boldsymbol{t} = \boldsymbol{\sigma}\,\mathbf{e}_m &= (\sigma_1\mathbf{e}_1\otimes\mathbf{e}_1+\sigma_2\mathbf{e}_2\otimes\mathbf{e}_2+\sigma_3\mathbf{e}_3\otimes\mathbf{e}_3)\left(\frac{1}{\sqrt{3}}\mathbf{e}_1+\frac{1}{\sqrt{3}}\mathbf{e}_2+\frac{1}{\sqrt{3}}\mathbf{e}_3\right)\\ &=\frac{1}{\sqrt{3}}(\sigma_1\mathbf{e}_1+\sigma_2\mathbf{e}_2+\sigma_3\mathbf{e}_3)\end{aligned}\tag{6.61}$$

on the principal base, noting Eq. (3.2) with $\mathbf{n} = \mathbf{e}_m$. Then, the normal component σ_{oct} and the tangential component τ_{oct} of the traction $\boldsymbol{t}$ are given as

$$\begin{aligned}\sigma_{\text{oct}} \equiv \boldsymbol{t}\cdot\mathbf{e}_m &= \frac{1}{\sqrt{3}}(\sigma_1\mathbf{e}_1+\sigma_2\mathbf{e}_2+\sigma_3\mathbf{e}_3)\cdot\left(\frac{1}{\sqrt{3}}\mathbf{e}_1+\frac{1}{\sqrt{3}}\mathbf{e}_2+\frac{1}{\sqrt{3}}\mathbf{e}_3\right)\\ &=\frac{1}{3}(\sigma_1+\sigma_2+\sigma_3)=\sigma_m\end{aligned}\tag{6.62}$$

$$\begin{aligned}\tau_{\text{oct}} &\equiv \sqrt{\|\boldsymbol{t}\|^2-\sigma_{\text{oct}}^2}\\ &=\sqrt{\frac{1}{3}(\sigma_1^2+\sigma_2^2+\sigma_3^2)-\frac{1}{9}(\sigma_1+\sigma_2+\sigma_3)^2}\\ &=\sqrt{\frac{2}{9}\{\sigma_1^2+\sigma_2^2+\sigma_3^2-(\sigma_1\sigma_2+\sigma_2\sigma_3+\sigma_3\sigma_1)\}}\\ &=\frac{1}{3}\sqrt{(\sigma_1-\sigma_2)^2+(\sigma_2-\sigma_3)^2+(\sigma_3-\sigma_1)^2}=\sqrt{\frac{2}{3}}\sqrt{J_2}\end{aligned}\tag{6.63}$$

where τ_{oct} is called the *octahedral shear stress* and J_2 is given by choosing the deviatoric Cauchy tensor $\boldsymbol{\sigma}'$ as the second invariant of the deviatoric tensor $\mathbf{T}'$ in Eq. (1.187) as follows:

$$\begin{aligned}J_2 &\equiv \frac{1}{2}\|\boldsymbol{\sigma}'\|^2=\frac{1}{2}\mathrm{tr}\boldsymbol{\sigma}'^2=\frac{1}{2}\sigma'_{rs}\sigma'_{sr}\\ &=\frac{1}{2}(\sigma'^2_{11}+\sigma'^2_{22}+\sigma'^2_{33})+\sigma'^2_{12}+\sigma'^2_{23}+\sigma'^2_{31}\\ &=\frac{1}{2}(\sigma'^2_1+\sigma'^2_2+\sigma'^2_3)=\frac{1}{6}\{(\sigma_1-\sigma_2)^2+(\sigma_2-\sigma_3)^2+(\sigma_3-\sigma_1)^2\}\end{aligned}\tag{6.64}$$

It is interpreted from Eqs. (6.56), (6.63) and (6.64) for the Mises yield condition that the yielding is induced when the octahedral shear stress reaches a certain value. Equations (6.62) and (6.63) are also derived as the components in the directions $\mathbf{I}_m(=\sqrt{3}\mathbf{e}_m)$ and $\mathbf{t}'(=\mathbf{T}'/\|\mathbf{T}'\|)$, i.e. T_m and $\|\mathbf{T}'\|$ of $\mathbf{T}_m$ and $\mathbf{T}'$, respectively, regarding $\mathbf{T}$ as $\boldsymbol{\sigma}$ in Eq. (1.255).

6.5 Formulation of Loading Criterion

The judgment of whether or not the plastic strain rate is induced for a given incremental loading is required for the elastoplastic deformation analysis. The criterion for this judgment is called the *loading criterion*. In what follows, this criterion is formulated (Hashiguchi 2000).

1. It is required that

$$\left.\begin{aligned} \dot{\lambda} &= \frac{\mathbf{n}:\overset{\circ}{\boldsymbol{\sigma}}}{M^p} > 0 \\ \dot{\Lambda} &= \frac{\mathbf{n}:\mathbf{E}:\mathbf{d}}{M^p + \mathbf{n}:\mathbf{E}:\mathbf{m}} > 0 \end{aligned}\right\} \tag{6.65}$$

in the loading (plastic deformation) process $\mathbf{d}^p \neq \mathbf{O}$.

2. It holds that

$$\mathbf{n}:\overset{\circ}{\boldsymbol{\sigma}} \leq 0 \tag{6.66}$$

in the unloading (elastic deformation) process $\mathbf{d}^p = \mathbf{O}$. Further, substituting $\mathbf{d} = \mathbf{d}^e$ leading to $\mathbf{n}:\mathbf{E}:\mathbf{d} = \mathbf{n}:\mathbf{E}:\mathbf{d}^e = \mathbf{n}:\overset{\circ}{\boldsymbol{\sigma}}$ into Eq. (6.45), $\dot{\Lambda}$ is described as

$$\dot{\Lambda} = \frac{\mathbf{n}:\overset{\circ}{\boldsymbol{\sigma}}}{M^p + \mathbf{n}:\mathbf{E}:\mathbf{m}}. \tag{6.67}$$

in this process.

3. The plastic modulus M^p takes both positive (hardening) and negative (softening) signs in general. On the other hand, the elastic modulus $\mathbf{E}$ is the positive definite tensor fulfilling $\mathbf{n}:\mathbf{E}:\mathbf{n} > 0$ for an arbitrary $\mathbf{n}$ as defined in Eq. (1.216). Here, we may assume $\mathbf{n}:\mathbf{E}:\mathbf{m} \gg |M^p|$, provided that $\mathbf{m}$ is not far different from $\mathbf{n}$. Eventually, it follows that

$$M^p + \mathbf{n}:\mathbf{E}:\mathbf{m} > 0. \tag{6.68}$$

Also, note that the infinite rate of plastic relaxation is induced so that the stress decreases in the infinite rate if the denominator becomes zero, i.e. $M^p + \mathbf{n}:\mathbf{E}:\mathbf{m} \to 0$ as known from Eq. (6.47) and illustrated in Fig. 6.4 for the uniaxial loading process.

Then, in the unloading process ($\mathbf{d}^p = \mathbf{O}, \mathbf{d}^e \neq \mathbf{O}$), the following inequalities hold from Eqs. (6.43) and (6.65)–(6.68), depending on the sign of the plastic modulus M^p leading to the hardening, perfectly-plastic and softening state.

$$\left.\begin{array}{l}\dot{\lambda} < 0 \text{ and } \dot{\Lambda} < 0 \text{ when } M^p > 0 \\ \dot{\lambda} \to -\infty \text{ or indeterminate and } \dot{\Lambda} < 0 \text{ when } M^p = 0 \\ \dot{\lambda} > 0 \text{ and } \dot{\Lambda} < 0 \text{ when } M^p < 0\end{array}\right\} \quad (6.69)$$

Consequently, the sign of $\dot{\lambda}$ at the moment of unloading from the state $M^p \leq 0$ can be positive or indeterminate in the unloading process. On the other hand, $\dot{\Lambda}$ is definitely negative in the unloading process. Thus, the distinction between a loading and an unloading process cannot be judged by the sign of $\dot{\lambda}$ but it can be done by that of $\dot{\Lambda}$. Therefore, the loading criterion is given as either

$$\left.\begin{array}{l}\mathbf{D}^p \neq \mathbf{O}: f(\boldsymbol{\sigma}) = F(H) \text{ and } \dfrac{\mathbf{n}:\mathbf{E}:\mathbf{d}}{M^p + \mathbf{n}:\mathbf{E}:\mathbf{m}} > 0 \\ \mathbf{D}^p = \mathbf{O}: \text{otherwise}\end{array}\right\} \quad (6.70)$$

or

$$\boxed{\begin{array}{l}\mathbf{d}^p \neq \mathbf{O}: f(\boldsymbol{\sigma}) = F(H) \text{ and } \mathbf{n}:\mathbf{E}:\mathbf{d} > 0 \\ \mathbf{d}^p = \mathbf{O}: \text{otherwise}\end{array}} \quad (6.71)$$

in lieu of Eq. (6.68). Equation (6.71) has been shown a priori by Hill (1967, 1983). Limiting to the hardening process with $M^p > 0$, Eq. (6.71) leads to

$$\left.\begin{array}{l}\mathbf{d}^p \neq \mathbf{O}: f(\boldsymbol{\sigma}) = F(H) \text{ and } \mathbf{n}:\mathring{\boldsymbol{\sigma}} > 0 \\ \mathbf{d}^p = \mathbf{O}: \text{otherwise}\end{array}\right\} \quad (6.72)$$

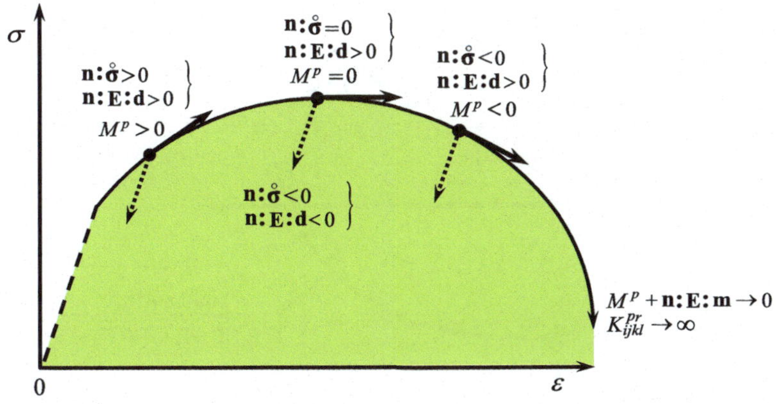

Fig. 6.4 Signs of $\mathbf{n}:\mathring{\boldsymbol{\sigma}}$ and $\mathbf{n}:\mathbf{E}:\mathbf{d}$ in uniaxial loading process

The plastic loading is interpreted as the process that the relaxation stress rate $\overset{\circ}{\boldsymbol{\sigma}}{}^p$ is induced, noting Eq. (6.47) with Eq. (6.70) (Fig. 6.4).

The loading criterion in Eq. (6.71) is rewritten as

$$\left.\begin{array}{l}\mathbf{d}^p \neq \mathbf{O}: f(\boldsymbol{\sigma}) = F(H) \;\text{ and }\; \mathbf{n}:\overset{\circ}{\boldsymbol{\sigma}}{}^e > 0\\ \mathbf{d}^p = \mathbf{O}: \text{otherwise}\end{array}\right\} \tag{6.73}$$

where

$$\overset{\circ}{\boldsymbol{\sigma}}{}^e \equiv \mathbf{E}:\mathbf{d} \tag{6.74}$$

which is called the *elastic stress rate*. Equation (6.73) is interpreted as follows: The loading, the neutral loading and the unloading are defined as processes that the elastic stress rate is directed outward, tangential and inward direction, respectively, of the yield surface as shown in Fig. 6.5, where $d\boldsymbol{\varepsilon}$ stands for $\mathbf{d}dt$.

In elastoplastic deformation analysis, suppose to calculate first $\overset{\circ}{\boldsymbol{\sigma}}$ and $\mathbf{d}$ by either of the elastic or the elastoplastic constitutive equation. Then, check the sign of $\mathbf{n}:\mathbf{E}:\mathbf{d}$. If the sign conflicts with the loading criterion, it is required to recalculate them using another constitutive equation. Here, it would be efficient to calculate first by the elastoplastic constitutive equation since the monotonic loading process in which the elastoplastic deformation process continues is seen often in practical engineering problems.

The loading judgment by the direction of strain rate, i.e. the sign of $\mathbf{n}:\mathbf{E}:\mathbf{d}$ (or $\mathbf{n}:\overset{\circ}{\boldsymbol{\sigma}}$ for the hardening state) is not required for the numerical calculation exploiting the return-mapping projection for hyperelastic-based constitutive equations as will be described in Chap. 20.

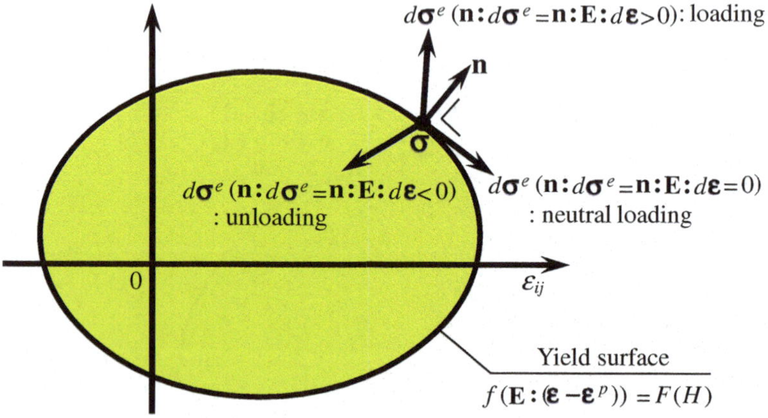

Fig. 6.5 Loading criterion by the direction of the elastic stress increment $d\boldsymbol{\sigma}^e \equiv \mathbf{E}:d\boldsymbol{\varepsilon}$ in strain space

6.6 Physical Backgrounds of Associated Flow Rule

The *associated flow rule* holds under some assumptions for a wide range of materials. Some mechanical interpretations for the associated flow rule are described in this section.

6.6.1 Positiveness of Second-Order Plastic Work Rate: Prager's Interpretation

Prager (1949) reported that the associated flow rule must hold to fulfill the positivity of the *second-order plastic work rate*, i.e.

$$\overset{\circ}{\boldsymbol{\sigma}} : \mathbf{d}^p \geq 0 \tag{6.75}$$

The direction of the plastic strain rate must be outward-normal to the yield surface for any stress increments directing outwards the yield surface, assuming that the direction of plastic strain rate depends on the state of stress but independent of the rate of stress as shown in Fig. 6.6. However, Eq. (6.75) holds only for hardening materials in which the stress rate is directed outwards the yield surface but does not hold for softening materials.

6.6.2 Principle of Maximum Plastic Work

The postulate, called the *principle of maximum plastic work*, was proposed by Mises for rigid-plastic materials and Hill (1948b, 1950) and Mandel (1964) for

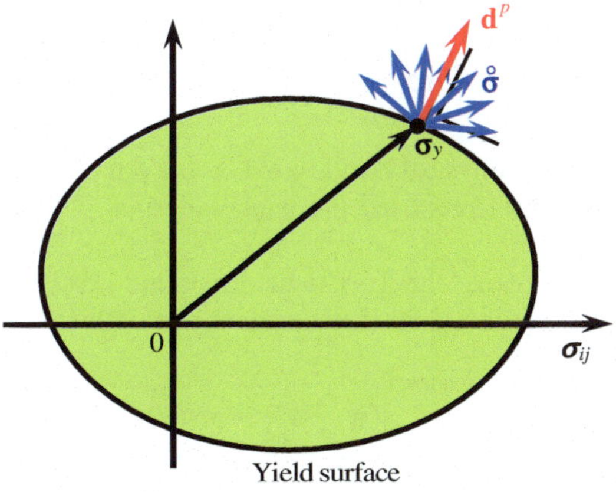

Fig. 6.6 Prager's positive plastic work rate for hardening materials

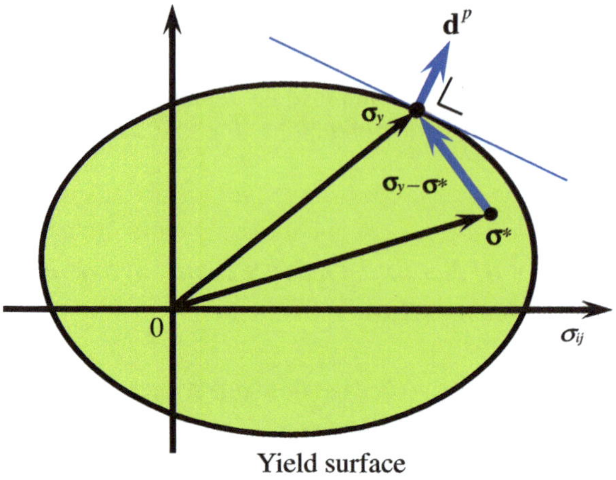

Fig. 6.7 Principle of maximum plastic work

elastoplastic materials It insists that the plastic work rate done by the actual stress $\boldsymbol{\sigma}_y$ on the yield surface is greater than a plastic work done by any statically-admissible stress $\boldsymbol{\sigma}^*$ inside the convex yield surface, leading to

$$\boldsymbol{\sigma}_y : \mathbf{d}^p > \boldsymbol{\sigma}^* : \mathbf{d}^p, \quad \text{i.e.}(\boldsymbol{\sigma}_y - \boldsymbol{\sigma}^*) : \mathbf{d}^p > 0 \tag{6.76}$$

as depicted in Fig. 6.7. Then, the plastic strain rate must be directed to the outward-normal of convex yield surface, so that the associated flow rule must hold in order to satisfy the principle of maximum plastic work under the assumption that the direction of plastic strain rate $\mathbf{d}^p$ depends only on the state of stress but it is independent of the rate of stress. It was discussed for the finite strain theory by Lubliner (1984).

6.6.3 Positiveness of Work Done During Stress Cycle: Drucker's Postulate

Hill did not comment any physical background on the principle of maximum plastic work rate. Drucker (1951) proposed the background as will be described in the following.

Drucker (1951) postulated "the work done during the stress cycle by the external agency is positive". It is described mathematically as follows:

$$\int_{t_0(\boldsymbol{\sigma}_0)}^{t(\boldsymbol{\sigma}_0)} (\boldsymbol{\sigma} - \boldsymbol{\sigma}_0) : \mathbf{d}dt \geq 0 \tag{6.77}$$

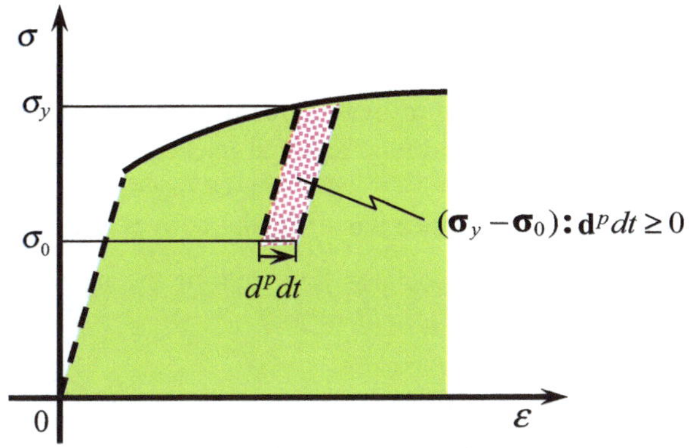

Fig. 6.8 Positive work done by external agency in Drucker's (1951) postulate (illustration in uniaxial loading process)

where $\boldsymbol{\sigma}_0$ stands for the initial stress at the initial time $t_0(\boldsymbol{\sigma}_0)$, and $t(\boldsymbol{\sigma}_0)$ designates the time that the stress returns to the initial stress. The following inequality is obtained from Eq. (6.77) under the assumption that the inside of the yield surface is elastic domain (see Fig. 6.8).

$$\boxed{(\boldsymbol{\sigma}_y - \boldsymbol{\sigma}_0) : \mathbf{d}^p \geq 0} \tag{6.78}$$

where $\boldsymbol{\sigma}_y$ designates the stress on the yield surface in which the plastic strain rate $\mathbf{d}^p$ is induced.

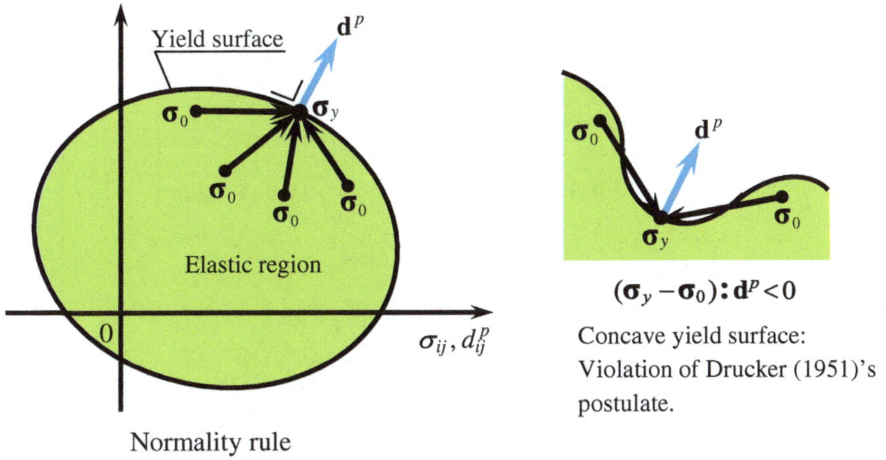

Fig. 6.9 Associated flow (normality) rule and convexity of yield surface based on the Drucker's (1951) postulate

The followings should hold in order to fulfill Eq. (6.78).

(1) The plastic strain rate is directed outward-normal of the yield surface. Then, the associated flow rule must hold, provided that the direction of plastic strain rate is determined solely by the current stress and internal variable but independent of stress rate.
(2) In this occasion the yield surface has to be the convex surface (see Fig. 6.9).

The result (1) is called the associated flow rule or the normality rule and the result (2) is called the *convexity of yield surface*.

6.6.4 Positiveness of Second-Order Plastic Relaxation Work Rate

Ilyushin (1961) postulated that "*the work done during the strain cycle is positive*". Limiting to the infinitesimal deformation process, it leads to the postulate "*the second-order work increment d^2w is not larger than the second-order elastic stress work increment d^2w^{es} calculated by presuming that the strain increment is induced elastically*" or "*the second-order plastic relaxation work increment d^2w^{pr}, i.e. the work increment done during the infinitesimal strain cycle is not negative*" (Hill 1968; Petryk 1991, 1997; Hashiguchi 1993; see Fig. 6.10). It is described mathematically as follows:

$$d^2w = d^2w^{es} - d^2w^{pr}; \quad d^2w \le d^2w^{es},\ d^2w^{pr} \ge 0 \tag{6.79}$$

where

$$\left.\begin{aligned} d^2w &\equiv \frac{1}{2}d\boldsymbol{\sigma} : \mathbf{d}dt = \frac{1}{2}\mathbf{d}^e dt : \mathbf{E} : \mathbf{d}dt \\ d^2w^{es} &\equiv \frac{1}{2}d\boldsymbol{\sigma}^e : \mathbf{d}dt = \frac{1}{2}\mathbf{d}dt : \mathbf{E} : \mathbf{d}dt \\ d^2w^{pr} &\equiv -\frac{1}{2}d\boldsymbol{\sigma}^p : \mathbf{d}dt = \frac{1}{2}\mathbf{d}^p dt : \mathbf{E} : \mathbf{d}dt = \frac{1}{2}\dot{\lambda}\, dt\mathbf{m} : \mathbf{E} : \mathbf{d}dt \end{aligned}\right\} \tag{6.80}$$

with

$$d\boldsymbol{\sigma}^e \equiv \mathbf{E} : \mathbf{d}dt, \quad d\boldsymbol{\sigma}^p \equiv -\mathbf{E} : \mathbf{d}^p dt \tag{6.81}$$

$d\boldsymbol{\sigma}^e$ and $d\boldsymbol{\sigma}^p$ are called the *elastic stress increment* and the *plastic relaxation stress increment*, respectively, following $\mathring{\boldsymbol{\sigma}}^e$ and $\mathring{\boldsymbol{\sigma}}^p$ in Eq. (6.48).

It should be noted that the associated flow rule with $\mathbf{m} = \mathbf{n}$ in Eq. (6.52) must hold in order that Eq. $(6.79)_3$ conforms to the loading condition in Eq. (6.71) which is not restricted to the associated flow rule.

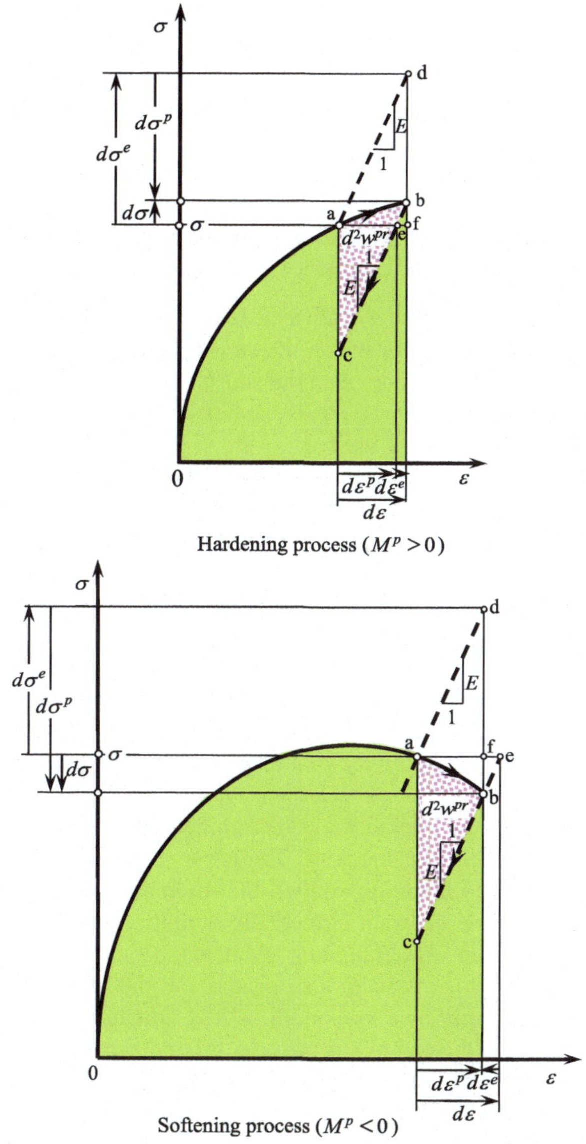

Fig. 6.10 Positiveness of second-order work rate (illustration in uniaxial loading)

6.6.5 *Comparison of Interpretations for Associated Flow Rule*

Prager's (1949) interpretation of the associated flow rule is concerned only with hardening materials as described previously. Hill's (1948b, 1950) principle of maximum plastic work is not limited to the hardening behavior but does not provide a clear physical background. On the other hand, the interpretations of the positivity of work done by the external agency, i.e. the additional stress during a stress cycle by Drucker (1951) and of the positivity of the second-order plastic relaxation work rate during a strain cycle are based on the conceivable physical postulates of the dissipation energy of materials unlimited to the hardening behavior.

Here, Drucker's (1951) postulate is related to the stress cycle but the postulate of the second-order plastic relaxation work rate is related with the infinitesimal strain cycle. Now, compare below the pertinence of these postulates.

(1) The strain cycle can be realized always. However, the stress cycle cannot be made in the softening state in which the stress cannot be returned to the initial state if the plastic strain rate is induced. It is based on the fact that any deformation can be given but a stress cannot be given arbitrarily to materials since strength of materials is limited.
(2) Limiting to the infinitesimal cycles, consider the stress and strain cycles. The second-order work increment done during the infinitesimal stress cycle is given by $(1/2)d\boldsymbol{\sigma}:\mathbf{d}^p dt$ (Δabe in Fig. 6.10a). On the other hand, the additional work increment $(1/2)\mathbf{d}^p dt:\mathbf{E}:\mathbf{d}^p dt$ (Δaec in Fig. 6.10a) must be done to close the strain cycle, whilst $(1/2)\mathbf{d}^p dt:\mathbf{E}:\mathbf{d}^p dt \geq 0$ holds because of the positive-definiteness of the elastic modulus tensor $\mathbf{E}$. Therefore, the work done during the infinitesimal stress cycle is far smaller than the work during the infinitesimal strain cycle. In other words, Drucker's (1951) postulate holds for the materials fulfilling a more restricted condition, i.e., more particular materials than the materials fulfilling the positivity of the second-order plastic relaxation work rate.
(3) Strain (increment) in any definition is determined uniquely by the displacement (increment) induced in the material. Therefore, the configuration of material returns to the initial configuration only if the strain returns to the initial value. In other words, if a cycle of strain in a certain definition closes, cycle of strain in any other definition (Lagrangian, Almansi, logarithmic, nominal and infinitesimal strains for instance) also closes. On the other hand, the stress is defined by the force per unit area and thus it is influenced by the deformation. The configuration in the end of stress cycle differs from the initial configuration depending on the loading path chosen during the cycle and on the definition of stress (Cauchy, Kirchhoff, nominal and second Piola-Kirchhoff stresses for instance). Then, even if a cycle of stress in a certain definition closes, the cycle in the stress in the other definition does not close. Eventually, the strain cycle possesses the objectivity, but the stress cycle does not possess it.
(4) The assumption that the interior of the yield surface is the purely elastic domain is adopted in Drucker's postulate. On the other hand, it is not required by the postulate of the positivity of second-order plastic relaxation work rate, which holds on the quite natural premise that the purely elastic deformation is induced at the moment of reverse loading.

Eventually, it can be stated that postulate of the positivity of second-order plastic relaxation work rate is more general than Drucker's postulate. However, even the former is based on the premise that the direction of the plastic strain rate is dependent on the normal component but independent of the tangential component of stress rate to the yield surface. It is observed in the test result that the inelastic deformation is induced even by the tangential component. The inelastic strain rate induced by the component of stress rate tangential to yield surface will be described in Sect. 7.7

6.7 Anisotropy

The plastic strain rate described in Sect. 6.3 concerns the yield condition with the function of stress invariants and scalar-valued internal variables. Therefore, it is limited to the materials exhibiting the isotropy in the plastic deformation behavior. In what follows, first the isotropy in constitutive equation is defined. Then, the plastic strain rate extended to the anisotropy will be explained in this section.

6.7.1 Definition of Isotropy

An isotropic material is defined as one exhibiting identical mechanical response that is independent of the chosen direction of material element or of the coordinate system by which the response is observed. Here, the input/output variables are the stress rate and the strain rate in the irreversible deformation.

The rate-type constitutive equation is described in general as follows:

$$\mathbf{f}(\mathring{\boldsymbol{\sigma}},\ \boldsymbol{\sigma}, H_i,\ \mathbf{d}) = \mathbf{O} \tag{6.82}$$

where H_i ($i = 1, 2, 3, \cdots$) denotes collectively scalar-valued or tensor-valued internal state variables. When the following equation holds by giving coordinate transformations only for stress (rate) and strain rate tensors in the function $\mathbf{f}$, it can be stated that Eq. (6.82) describes the constitutive equation of isotropic material.

$$\mathbf{f}(\mathbf{Q}\mathring{\boldsymbol{\sigma}}\mathbf{Q}^T,\ \mathbf{Q}\boldsymbol{\sigma}\mathbf{Q}^T, H_i,\ \mathbf{Q}\mathbf{d}\mathbf{Q}^T) = \mathbf{Q}\mathbf{f}(\mathring{\boldsymbol{\sigma}},\ \boldsymbol{\sigma}, H_i,\ \mathbf{d})\mathbf{Q}^T \tag{6.83}$$

In the plastic constitutive equation formulated incorporating the yield and/or plastic potential function, the isotropy holds if the yield and/or plastic potential function is given by the function of stress invariants and scalar internal variables. Then, designating these functions by f, it must fulfill the equation.

$$f(\mathbf{Q}\boldsymbol{\sigma}\mathbf{Q}^T, H_i) = f(\boldsymbol{\sigma}, H_i) \tag{6.84}$$

In contrast, the anisotropic plastic constitutive equation is described by incorporating the yield and/or plastic potential function including tensor-valued internal variable in addition to the stress invariants and scalar internal variables. Then, Eqs. (6.83) and (6.84) do not hold in anisotropic constitutive equations.

6.7.2 Elastoplastic Constitutive Equation with Kinematic Hardening

If the monotonic loading proceeds towards a certain direction in the stress space, the hardening develops in that direction but the yield stress lowers in the opposite direction. This phenomenon is induced by the statically-indeterminable deformation of internal structure and is called the *Bauschinger effect.* To reflect this effect in the elastoplastic constitutive equation, the translation or the rotation of the yield surface is adopted widely. The translation of the yield surface, called the *kinematic hardening*, is realized by introducing the back stress translating towards the loading direction and replacing the stress tensor with the tensor given by subtracting the back stress tensor from the stress tensor. On the other hand, soils, which is the assembly of particles with weak adhesion among them, can bear a far larger compression stress than the tensile stress. Therefore, they exhibit a strong frictional property that the deviatoric yield stress increases with the pressure, while the yield surface only slightly includes the origin of the stress space. Therefore, once the yield surface translates leaving the origin, it can never come back to include the origin again because the yield surface contracts with the plastic volume expansion leading to the softening. Therefore, the kinematic hardening cannot be applied but the rotation of the yield surface around the origin of stress space, i.e. the *rotational hardening*, is pertinent to soils as will be described in Chap. 11.

Now, let the yield condition (6.30) be extended to describe the anisotropy by introducing the internal variables of second-order tensors as follows:

$$\boxed{f(\hat{\boldsymbol{\sigma}}) = F(H)} \tag{6.85}$$

where

$$\hat{\boldsymbol{\sigma}} \equiv \boldsymbol{\sigma} - \boldsymbol{\alpha} \tag{6.86}$$

$\boldsymbol{\alpha}(= \boldsymbol{\alpha}')$ being the *back stress* (*kinematic hardening variable*) proposed by Prager (1956) in order to describe the induced anisotropy, the evolution rule of which depends on stress and internal variables as will be described in the next section. Note that an anisotropic hardening variable is deviatoric in general. Here, it is assumed that f in Eq. (6.85) is the function of $\hat{\boldsymbol{\sigma}}$ in the homogeneous degree-one fulfilling $f(|s|\hat{\boldsymbol{\sigma}}) = |s|f(\hat{\boldsymbol{\sigma}})$. Then, the yield surface (6.85) maintains the similar shape and orientation with respect to $\boldsymbol{\sigma} = \boldsymbol{\alpha}$.

The material time-derivative of Eq. (6.85) leads to the consistency condition in the corotational time-derivative.

$$\frac{\partial f(\hat{\boldsymbol{\sigma}})}{\partial \hat{\boldsymbol{\sigma}}} : \overset{\circ}{\boldsymbol{\sigma}} - \frac{\partial f(\hat{\boldsymbol{\sigma}})}{\partial \hat{\boldsymbol{\sigma}}} : \overset{\circ}{\boldsymbol{\alpha}} - F' \dot{H} = 0 \tag{6.87}$$

where

$$\left.\begin{array}{l} \dot{H} = f_{Hd}(\boldsymbol{\sigma}, H;\, \mathbf{d}^p) = f_{Hd}(\boldsymbol{\sigma}, H;\, \mathbf{d}^p/||\mathbf{d}^p||)||\mathbf{d}^p|| \\ \overset{\circ}{\boldsymbol{\alpha}} = \mathbf{f}_{kd}(\boldsymbol{\sigma}, F, \boldsymbol{\alpha};\, \mathbf{d}^{p'}) = \mathbf{f}_{kd}(\boldsymbol{\sigma}, F, \boldsymbol{\alpha};\, \mathbf{d}^{p'}/||\mathbf{d}^p||)||\mathbf{d}^p|| \end{array}\right\} \tag{6.88}$$

noting that they are homogeneous functions of $\mathbf{d}^p$ in degree-one since they are induced only in the plastic loading process $\mathbf{d}^p \neq \mathbf{O}$ and the first-order time-differential quantities and that anisotropic hardening variables evolve only with the deviatoric strain rate.

Substituting the associated flow rule

$$\mathbf{d}^p = \dot{\lambda}\, \hat{\mathbf{n}} \quad (\dot{\lambda} > 0, ||\hat{\mathbf{n}}|| = 1, ||\mathbf{d}^p|| = \dot{\lambda}) \tag{6.89}$$

with

$$\hat{\mathbf{n}} \equiv \frac{\partial f(\hat{\boldsymbol{\sigma}})}{\partial \boldsymbol{\sigma}} / || \frac{\partial f(\hat{\boldsymbol{\sigma}})}{\partial \boldsymbol{\sigma}} || \tag{6.90}$$

into Eq. (6.87) and noting $\partial f(\hat{\boldsymbol{\sigma}})/\partial \boldsymbol{\sigma} = \partial f(\hat{\boldsymbol{\sigma}})/\partial \hat{\boldsymbol{\sigma}}$, it follows that

$$\hat{\mathbf{n}} : \overset{\circ}{\boldsymbol{\sigma}} - \hat{\mathbf{n}} : \overset{\circ}{\boldsymbol{\alpha}} - \frac{\hat{\mathbf{n}} : \hat{\boldsymbol{\sigma}}}{F} F' \dot{H} = 0$$

i.e.

$$\hat{\mathbf{n}} : \overset{\circ}{\boldsymbol{\sigma}} - \dot{\lambda} M^p = 0 \tag{6.91}$$

where

$$M^p \equiv \hat{\mathbf{n}} : \left(\frac{F'}{F} f_{Hn}(\boldsymbol{\sigma}, H; \hat{\mathbf{n}}) \hat{\boldsymbol{\sigma}} + \mathbf{f}_{kn}(\boldsymbol{\sigma}, \boldsymbol{\alpha},\, F;\, \hat{\mathbf{n}}'/||\hat{\mathbf{n}}||) \right) \tag{6.92}$$

$$\begin{array}{c} f_{Hn}(\boldsymbol{\sigma}, H;\, \hat{\mathbf{n}}) = \dot{H} / \dot{\lambda}, \quad \mathbf{f}_{kn}(\boldsymbol{\sigma}, \boldsymbol{\alpha},\, F;\, \hat{\mathbf{n}}'/||\hat{\mathbf{n}}||) = \overset{\circ}{\boldsymbol{\alpha}} / \dot{\lambda} \\ (\mathrm{tr}\mathbf{f}_{kn}(\boldsymbol{\sigma}, \boldsymbol{\alpha},\, F;\, \hat{\mathbf{n}}'/||\hat{\mathbf{n}}||) = 0) \end{array} \tag{6.93}$$

taking account of the Euler's theorem for homogeneous function $f(\hat{\boldsymbol{\sigma}})$ in order-one leading to

$$\left.\begin{array}{l} \dfrac{\partial f(\hat{\boldsymbol{\sigma}})}{\partial \hat{\boldsymbol{\sigma}}} : \hat{\boldsymbol{\sigma}} = f(\hat{\boldsymbol{\sigma}}) = F \\[3mm] 1/\left\| \dfrac{\partial f(\hat{\boldsymbol{\sigma}})}{\partial \boldsymbol{\sigma}} \right\| \hat{\mathbf{n}} = \dfrac{\dfrac{\partial f(\hat{\boldsymbol{\sigma}})}{\partial \boldsymbol{\sigma}} : \hat{\boldsymbol{\sigma}}}{f(\hat{\boldsymbol{\sigma}})} / \left\| \dfrac{\partial f(\hat{\boldsymbol{\sigma}})}{\partial \boldsymbol{\sigma}} \right\| = \dfrac{\hat{\mathbf{n}} : \hat{\boldsymbol{\sigma}}}{F} \end{array}\right\} \qquad (6.94)$$

It follows from Eq. (6.91) that

$$\dot{\lambda} = \dfrac{\hat{\mathbf{n}} : \overset{\circ}{\boldsymbol{\sigma}}}{M^p}, \quad \mathbf{d}^p = \dfrac{\hat{\mathbf{n}} : \overset{\circ}{\boldsymbol{\sigma}}}{M^p} \hat{\mathbf{n}} \qquad (6.95)$$

and

$$\mathbf{d} = \mathbf{E}^{-1} : \overset{\circ}{\boldsymbol{\sigma}} + \dfrac{\hat{\mathbf{n}} : \overset{\circ}{\boldsymbol{\sigma}}}{M^p} \hat{\mathbf{n}} = \left(\mathbf{E}^{-1} + \dfrac{\hat{\mathbf{n}} \otimes \hat{\mathbf{n}}}{M^p} \right) : \overset{\circ}{\boldsymbol{\sigma}} \qquad (6.96)$$

from which it follows that

$$\dot{\Lambda} = \dfrac{\hat{\mathbf{n}} : \mathbf{E} : \mathbf{d}}{M^p + \hat{\mathbf{n}} : \mathbf{E} : \hat{\mathbf{n}}} \qquad (6.97)$$

$$\overset{\circ}{\boldsymbol{\sigma}} = \mathbf{E} : \mathbf{d} - \dfrac{\hat{\mathbf{n}} : \mathbf{E} : \mathbf{d}}{M^p + \hat{\mathbf{n}} : \mathbf{E} : \hat{\mathbf{n}}} \mathbf{E} : \hat{\mathbf{n}} = \left(\mathbf{E} - \dfrac{\mathbf{E} : \hat{\mathbf{n}} \otimes \mathbf{E} : \hat{\mathbf{n}}}{M^p + \hat{\mathbf{n}} : \mathbf{E} : \hat{\mathbf{n}}} \right) : \mathbf{d} \qquad (6.98)$$

The loading criterion is given by

$$\left.\begin{array}{l} \mathbf{d}^p \neq \mathbf{O} : f(\hat{\boldsymbol{\sigma}}) = F(H) \ \text{ and } \ \hat{\mathbf{n}} : \mathbf{E} : \mathbf{d} > 0 \\[1mm] \mathbf{d}^p = \mathbf{O} : \text{otherwise} \end{array}\right\} \qquad (6.99)$$

6.7.3 Kinematic Hardening Rules

The evolution rule of the back stress for the plastically-incompressible metals was given by Prager (1956) as follows:

$$\overset{\circ}{\boldsymbol{\alpha}} = c_k \mathbf{d}^{p\prime} \qquad (6.100)$$

where c_k is the material constant with the dimension of stress. Equation (6.100) is described in the uniaxial loading process by

$$\dot{\alpha}_a = c_k d_a^{p\prime} = c_k d_a^p \qquad (6.101)$$

where $(\ )_a$ designates the axial component. If c_k is extended to be a monotonic decreasing function of the equivalent plastic strain, the nonlinear behavior is

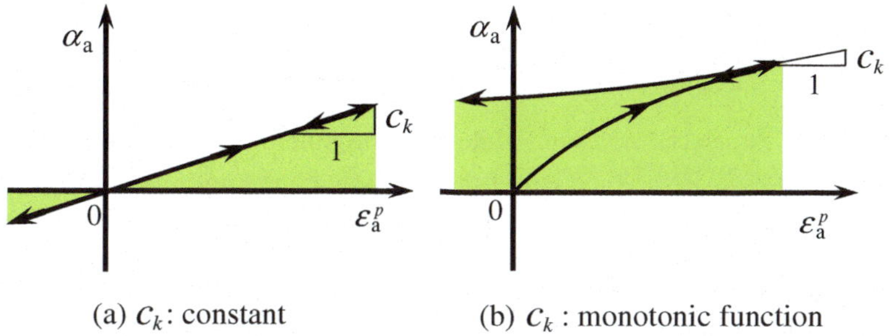

Fig. 6.11 Linear kinematic hardening rules illustrated in one-dimensional state

described in the initial monotonic loading process but the peculiar behavior is described in the inverse loading process as shown in Fig. 6.11b. It should be noted that this extension is not accepted, resulting in the worsening, although it has been recommended in some literatures (e.g. de Souza Neto et al. 2008)

Shield and Ziegler (1958) and Ziegler (1959) proposed the following modification of Prager's (1956) linear kinematic hardening rule in Eq. (6.100), insisting that Prager's rule possesses the mathematical inconvenience such that the components of back stress tensor are induced also in the directions in which the components of stress tensor are zero.

$$\overset{\circ}{\boldsymbol{\alpha}} = c_z \hat{\boldsymbol{\sigma}}' ||\mathbf{d}^{p\prime}|| \tag{6.102}$$

where c_z is the material constant, noting that $\overset{\circ}{\alpha}_i = 0$ for $\hat{\sigma}'_i = 0$ which is seen in a plane stress condition. The directions of the translations of the back stresses in Eqs. (6.100) and (6.102) are identical to each other for the Mises yield surface whose section cut by the deviatoric plane is circle for which $\hat{\mathbf{n}} = \hat{\boldsymbol{\sigma}}'/||\hat{\boldsymbol{\sigma}}'||$ holds, while they are obviously different from each other for the Tresca yield surface for instance. However, Eq. (6.102) is physically unacceptable in general since it is irrelevant to the hyperelasticity as will be explained in Sect. 6.9, so that it is impossible to be extended to the multiplicative finite elastoplasticity which will be described in Chap. 12. Eventually, it is merely the disturbance in the history of plasticity, although it has been introduced in a lot of literatures (e.g. Chakrabarty 1987; Duszek and Perzyna 1991; Khan and Huang 1995; Lubarda 2002; Asaro and Lubarda 2006).

The nonlinear-kinematic hardening rule of metals was proposed by Armstrong and Frederick (1966). It can be extended to describe the general kinematic hardening behavior including also the rotational hardening, i.e. the rotation of yield surface in soils (Sekiguchi and Ohta 1977; Hashiguchi and Chen 1998; Hashiguchi 1977, 2001, 2013) of plastically-compressible soils by the following equation, noting that the kinematic hardening is induced only by the deviatoric deformation.

$$\boxed{\overset{\circ}{\boldsymbol{\alpha}} = c_k\left(\mathbf{d}^{p\prime} - \frac{1}{b_k}||\mathbf{d}^{p\prime}||\boldsymbol{\alpha}\right)} = \dot{\lambda}\mathbf{f}_{kn}, \quad \mathbf{f}_{kn} = c_k\left(\hat{\mathbf{n}}' - \frac{1}{b_k}||\hat{\mathbf{n}}'||\boldsymbol{\alpha}\right) \tag{6.103}$$

where c_k is the material constant. Note that the kinematic hardening saturates as $\boldsymbol{\alpha} \to b_k\hat{\mathbf{n}}'/||\hat{\mathbf{n}}'||$. b_k is the function of the isotropic hardening function F in metals. On the other hand, note that the kinematic hardening variable $\boldsymbol{\alpha}$ does not designate the stress-valued variable but it designates the rotation of the central axis of yield surface in the rotational hardening in soils, which possesses the non-dimension of the stress ratio $||\boldsymbol{\sigma}'||/p$ ($p \equiv -\mathrm{tr}\boldsymbol{\sigma}/3$: pressure). Besides, b_k is the function of the Lode angle θ_σ (Eq. 1.266) for the stress in the rotational hardening, i.e. $b_k = M_r(\theta_\sigma)$ (Hashiguchi 2013).

Now, Eq. (6.103) reduces to the following equation for the uniaxial loading process of the Mises metals.

$$\overset{\circ}{\alpha}_{\mathrm{a}} = c_k\left(d_{\mathrm{a}}^{p\prime} - \frac{1}{b_k}\sqrt{d_{\mathrm{a}}^{p\prime 2} + 2d_l^{p\prime 2}}\,\alpha_{\mathrm{a}}\right) = c_k\left(1 \mp \sqrt{\frac{3}{2}}\frac{1}{b_k}\alpha_{\mathrm{a}}\right)d_{\mathrm{a}}^p \tag{6.104}$$

leading to

$$\overset{\circ}{\alpha}_{\mathrm{a}} = \begin{cases} \left.\begin{array}{l} 0 \text{ for } d_{\mathrm{a}}^p > 0 \\ 2c_k \text{ for } d_{\mathrm{a}}^p < 0 \end{array}\right\} \text{ in } \alpha_{\mathrm{a}} = \sqrt{2/3}b_k \\ c_k \text{ in } \alpha_{\mathrm{a}} = 0 \\ \left.\begin{array}{l} 2c_k \text{ for } d_{\mathrm{a}}^p > 0 \\ 0 \text{ for } d_{\mathrm{a}}^p < 0 \end{array}\right\} \text{in } \alpha_{\mathrm{a}} = -\sqrt{2/3}b_k \end{cases} \tag{6.105}$$

noting $d_l^p = d_{\mathrm{a}}^p/2$, where $(\ )_l$ designates the lateral component. The relation of the axial components of the back stress α_{a} versus the plastic strain $\varepsilon_{\mathrm{a}}^p$ in the uniaxial

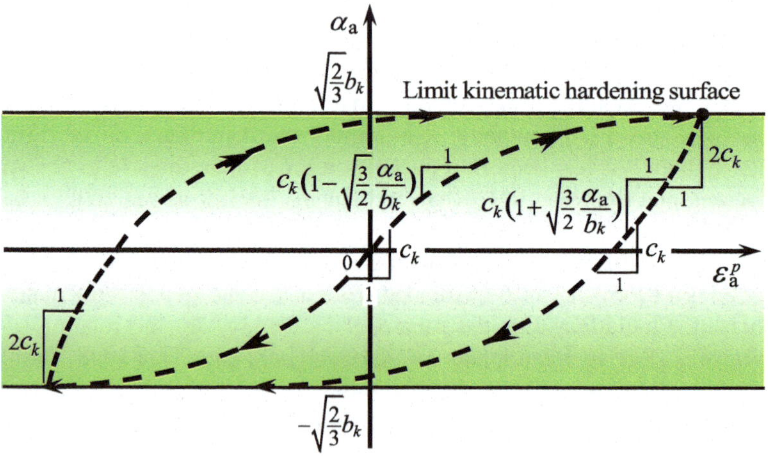

Fig. 6.12 Nonlinear kinematic hardening rule illustrated for uniaxial loading process

loading for the Mises material is illustrated in Fig. 6.12 in which the kinematic hardening gradually saturates in the monotonic loading process but the abrupt increase of the kinematic hardening rate occurs at the moment of reverse loading. The kinematic hardening saturates at $\sqrt{2/3}b_k$, where the material parameter b_k would be the function of the isotropic hardening variable F, i.e. $b_k = \sqrt{3/2}\zeta F$ (ζ: material constant) as will be described in Sect. 10.1.

6.8 Plastic Spin

The plastic spin in Eq. (6.25) is given specifically following Zbib and Aifantis (1988) as follows:

$$\mathbf{w}^p = \eta^p(\boldsymbol{\sigma}\mathbf{d}^p - \mathbf{d}^p\boldsymbol{\sigma}) = \eta^p\,\dot{\lambda}(\boldsymbol{\sigma}\hat{\mathbf{n}} - \hat{\mathbf{n}}\boldsymbol{\sigma}), \quad \boldsymbol{\omega}^p = \eta^p(\boldsymbol{\sigma}\hat{\mathbf{n}} - \hat{\mathbf{n}}\boldsymbol{\sigma}) \tag{6.106}$$

where η^p is the material parameter. Equation (6.106) is based on the non-coaxiality between the stress and the plastic strain rate, which is generally accompanied with the anisotropy. Obviously, the plastic spin is not induced in isotropic materials in which $\boldsymbol{\alpha} = \mathbf{O}$ holds.

6.9 Infinitesimal Hyperelastic-Based Plasticity and Physical Interpretation of Nonlinear Kinematic Hardening Rule

The subloading surface model has been formulated in the hypoelastic-based plasticity in the preceding sections. The formulation of the subloading surface model is presented in the framework of the infinitesimal hyperelastic-based plasticity for the infinitesimal deformation/rotation in this section. Besides, the formulation of the infinitesimal hypoelastic-based plasticity is obtained simply by replacing the strain rate $\mathbf{d}$ to the infinitesimal strain rate $\dot{\boldsymbol{\varepsilon}}$ and the corotational rate to the material-time derivative in the previous sections.

Let the infinitesimal strain $\boldsymbol{\varepsilon}$ in Eqs. (2.49) or (2.50) be assumed to be additively decomposed into the elastic strain $\boldsymbol{\varepsilon}^e$ and the plastic strain $\boldsymbol{\varepsilon}^p$, i.e.

$$\boldsymbol{\varepsilon} = \boldsymbol{\varepsilon}^e + \boldsymbol{\varepsilon}^p \tag{6.107}$$

Further, let the plastic strain $\boldsymbol{\varepsilon}^p$ be additively decomposed into the storage part $\boldsymbol{\varepsilon}^p_{ks}$ which induces the kinematic hardening and the dissipative part $\boldsymbol{\varepsilon}^p_{kd}$, i.e.

$$\boldsymbol{\varepsilon}^p = \boldsymbol{\varepsilon}^p_{ks} + \boldsymbol{\varepsilon}^p_{kd} \tag{6.108}$$

The stress $\boldsymbol{\sigma}$ and the kinematic hardening variable $\boldsymbol{\alpha}$ which describe the variation of substructures in material are given by the partial derivatives of the potential functions of the elastic strain and the storage parts of the plastic strain so that the following relations hold.

$$\boldsymbol{\sigma} = \frac{\partial \psi^e(\boldsymbol{\varepsilon}^e)}{\partial \boldsymbol{\varepsilon}^e}, \quad \boldsymbol{\alpha} = \frac{\partial \psi^k(\boldsymbol{\varepsilon}^p_{ks})}{\partial \boldsymbol{\varepsilon}^p_{ks}} \tag{6.109}$$

Now, assuming

$$\psi^e(\boldsymbol{\varepsilon}^e) = \frac{1}{2}\boldsymbol{\varepsilon}^e : \mathbf{E} : \boldsymbol{\varepsilon}^e, \quad \psi^k(\boldsymbol{\varepsilon}^p_{ks}) = \frac{1}{2}c_k \boldsymbol{\varepsilon}^{p\prime}_{ks} : \boldsymbol{\varepsilon}^{p\prime}_{ks} \tag{6.110}$$

it follows from Eq. (6.109) that

$$\boldsymbol{\sigma} = \mathrm{E} : \varepsilon^e = \mathrm{E} :(\boldsymbol{\varepsilon} - \boldsymbol{\varepsilon}^p), \quad \boldsymbol{\alpha} = c_k \boldsymbol{\varepsilon}^{p\prime}_{ks} = c_k(\boldsymbol{\varepsilon}^{p\prime} - \boldsymbol{\varepsilon}^{p\prime}_{kd}) \tag{6.111}$$

The time-differentiation of Eq. (6.111) reads:

$$\dot{\boldsymbol{\sigma}} = \mathrm{E} :(\dot{\boldsymbol{\varepsilon}} - \dot{\boldsymbol{\varepsilon}}^p), \quad \dot{\boldsymbol{\alpha}} = c_k(\dot{\boldsymbol{\varepsilon}}^{p\prime} - \dot{\boldsymbol{\varepsilon}}^{p\prime}_{kd}) \tag{6.112}$$

while the elastic modulus tensor E, and the elastic-like scalar c_k is assumed to be constant.

Further, assume the rates of the plastic strain and of the dissipative parts in the kinematic hardening as

$$\dot{\boldsymbol{\varepsilon}}^p = \dot{\lambda}\,\hat{\mathbf{n}}, \quad \dot{\boldsymbol{\varepsilon}}^p_{kd} = \frac{1}{b_k}||\dot{\boldsymbol{\varepsilon}}^{p\prime}||\boldsymbol{\alpha} = \frac{1}{b_k}\dot{\lambda}\,||\hat{\mathbf{n}}'||\boldsymbol{\alpha} \tag{6.113}$$

The dissipative part of the kinematic hardening in Eq. $(6.113)_2$ satisfies the positivity of the dissipation energy, i.e. $\boldsymbol{\alpha} : \dot{\boldsymbol{\varepsilon}}^p_{kd} = (\boldsymbol{\alpha} : \boldsymbol{\alpha})||\dot{\boldsymbol{\varepsilon}}^{p\prime}||/b_k \geq 0$. Substituting Eq. (6.113) into Eq. (6.112), one obtains the Armstrong-Frederick type nonlinear kinematic-hardening rule as follows:

$$\dot{\boldsymbol{\alpha}} = c_k\left(\dot{\boldsymbol{\varepsilon}}^{p\prime} - \frac{1}{b_k}||\dot{\boldsymbol{\varepsilon}}^{p\prime}||\boldsymbol{\alpha}\right) \tag{6.114}$$

The elastic relation and the kinematic hardening rule in Eqs. (6.29) and (6.103) are obtained by replacing the strain rates $\dot{\boldsymbol{\varepsilon}}$, $\dot{\boldsymbol{\varepsilon}}^e, \dot{\boldsymbol{\varepsilon}}^p$ to $\mathbf{d}$, $\mathbf{d}^e, \mathbf{d}^p$ and the material-time derivative $(\dot{\ })$ to the corotational rate $(^\circ)$.

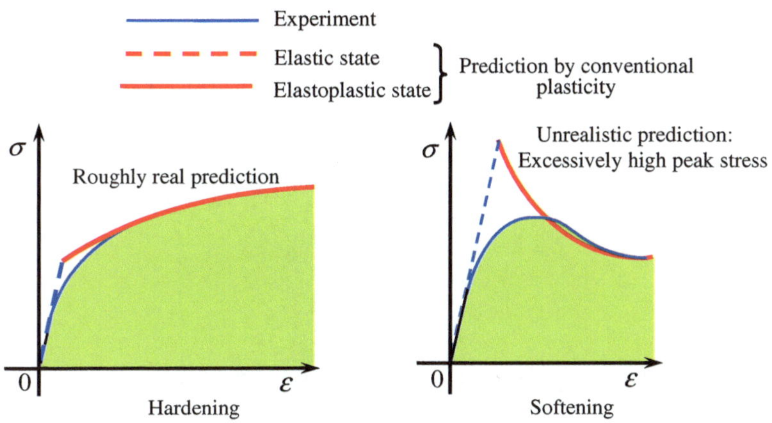

Fig. 6.13 Prediction of monotonic loading behavior by conventional plasticity

The stress $\boldsymbol{\sigma}$ and the kinematic hardening variable $\boldsymbol{\alpha}$ can be calculated directly from Eq. (6.111) as the hyperelasticity by substituting $\boldsymbol{\varepsilon}^p$, $\boldsymbol{\varepsilon}^p_{kd}$ obtained by the time-integrations of Eq. (6.113) without the time-integrations of Eq. (6.114). However, in the forward-Euler calculation method within the framework of the infinitesimal theory for which the spin of material is ignored, the accuracies in calculations of stress and internal variables using the hyperelastic and hyperelastic-like relations are identical to those by the direct time-integrations of the rates themselves.

6.10 Limitations of Conventional Elastoplasticity

The conventional elastoplasticity described in this chapter is premised on the assumption that the interior of yield surface is a purely elastic domain. Therefore, the relation of stress rate versus strain rate is predicted to change abruptly at the moment when the stress reaches the yield surface. Therefore, the smooth stress-strain curve observed in real materials is not predicted as shown in Fig. 6.13. This results in the defects: a determination of *offset value*, i.e. plastic strain at yield point is required, which is influenced by an arbitrariness since a plastic deformation develops gradually and thus a stress-strain curve is smooth usually in real materials, and a peak stress value is predicted to be excessively high in a softening behavior. Further, only an elastic deformation is repeated for the cyclic loading of stress below the yield stress. In real materials, however, plastic deformation is accumulated for stress cycles less than the yield stress and the strain is amplified leading to the failure as depicted in Fig. 6.14. Therefore, the conventional plasticity possesses various limitations in the application to the mechanical design of machines and structures in engineering practice.

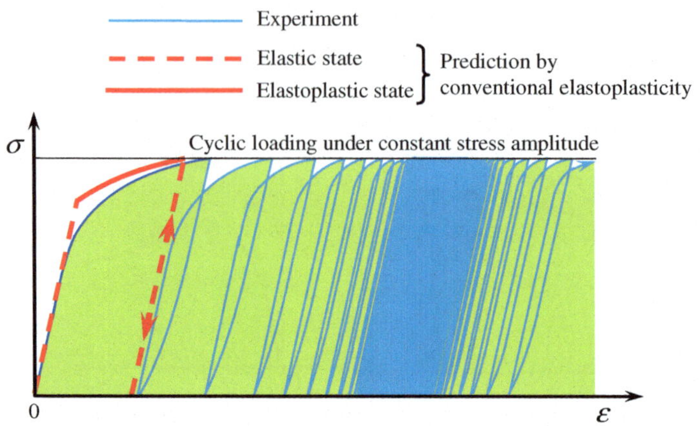

Fig. 6.14 Cyclic loading behavior: inability of conventional plasticity

References

Armstrong PJ, Frederick CO (1966) A mathematical representation of the multiaxial Bauschinger effect. CEGB Report RD/B/N 731 (or in Mater High Temp 24:1–26 (2007))

Asaro RJ, Lubarda V (2006) Mechanics of solids and materials. Cambridge University Press, Cambridge

Belytschko T, Liu WK, and Moran B (2014) Nonlinear Finite Elements for Continua and Structures. 2nd edn, Wiley, New York

Chakrabarty J (1987) Theory of plasticity. McGraw-Hill, New York

Dafalias YF (1983) Corotational rates for kinematic hardening at large plastic deformations. J Appl Mech (ASME) 50:561–565

Dafalias YF (1984) The plastic spin concept and a simple illustration of its role in finite plastic transformation. Mech Mater 3:223–233

Dafalias YF (1985a) The plastic spin. J Appl Mech (ASME) 52:865–871

Dafalias YF (1985b) A missing link in the macroscopic constitutive formulation of large plastic deformations. In: Sawczuk A, Bianchi G (eds) Plasticity today, international symposium on recent trends and results in plasticity. Elsevier Publ., pp 135–151

Dashner PA (1986) Invariance considerations in large strain elasto-plasticity. J Appl Mech (ASME) 53:55–60

de Souza Neto EA, Perić D, Owen DJR (2008) Computational methods for plasticity. Wiley, Chichester

Drucker DC (1951) A more fundamental approach to plastic stress-strain relations. In: Proceedings of 1st US national congress applied mechanics (ASME), vol 1, pp 487–491

Drucker DC (1988) Conventional and unconventional plastic response and representation. Appl Meek Rev (ASME) 41:151–167

Duszek MK, Perzyna P (1991) On combined isotropic and kinematic hardening effects in plastic flow process. Int J Plast 9:351–363

Fardshisheh F, Onat ET (1974) Representations of elastoplastic behavior by means of state variables. In: Sawczuk A (ed) Problems of plasticity, pp 89–115

Han C-S, Chung K, Wagoner RH, Oh S-I (2003) A multiplicative finite elasto-plastic formulation with anisotropic yield functions. Int J Plast 19:197–211

Harrysson M, Ristinmaa M (2007) Description of evolving anisotropy at large strains. Mech Mater 39:267–282

Hashiguchi K (1977) An expression of anisotropy in a plastic constitutive equation of soils. In: Murayama S, Schofield AN (eds) Constitutive equations of soils (Proc 9th Int Conf Soil Mech Found Eng Spec Session 9), Tokyo, JSSMFE, pp 302–305

Hashiguchi K (1993) Fundamental requirements and formulation of elastoplastic constitutive equations with tangential plasticity. Int J Plast 9:525–549

Hashiguchi K (2000) Fundamentals in constitutive equation: continuity and smoothness conditions and loading criterion. Soils Found 40(3):155–161

Hashiguchi K (2001) Description of inherent/induced anisotropy of soils: rotational hardening rule with objectivity. Soils Found 41(6):139–145

Hashiguchi K (2013) Elastoplasticity theory. Lecture note in Appl Comput Mech, 2nd edn. Springer, Heidelberg

Hashiguchi K, Chen Z-P (1998) Elastoplastic constitutive equations of soils with the subloading surface and the rotational hardening. Int J Numer Anal Meth Geomech 22:197–227

Haupt P (1985) On the concept of an intermediate configuration and its application to a representation of viscoelastic-plastic material behavior. Int J Plast 1:303–316

Hill R (1948) A variational principle of maximum plastic work in classical plasticity. Q J Mech Appl Math 1:18–28

Hill R (1950) The mathematical theory of plasticity. Oxford University Press, London

Hill R (1967) On the classical constitutive relations for elastic/plastic solids. In: Recent progress in applied mechanics, pp 241–249

Hill R (1968) On the constitutive inequalities for simple materials—I. J Mech Phys Solids 16:229–242

Hill R (1983) On the intrinsic eigenstates in plasticity with generalized variables. Math Proc Camb Phil Soc 93:177–189

Holsapple KA (1973) A finite elastic-plastic theory and invariance requirements. Acta Mech 17:277–290

Ilyushin AA (1961) On the postulate of plasticity. Appl Math Meek 25:746–752 (translation of O postulate plastichnosti, Prikladnaya Mathematika i Mekkanika 25:503–507)

Khan AS, Huang S (1995) Continuum theory of plasticity. Wiley, New York

Kratochvil J (1971) Finite-strain theory of crystalline elastic-inelastic materials. J Appl Phys 42:1104–1108

Kroner E (1960) Allgemeine Kontinuumstheoreie der Versetzungen und Eigenspannnungen. Arch Ration Mech Anal 4:273–334

Lee EH (1969) Elastic-plastic deformation at finite strain. J Appl Mech (ASME) 36:1–6

Lee EH, Liu DT (1967) Finite-strain elastic-plastic theory with application to plane-wave analysis. J Appl Phys 38:19–27

Loret B (1983) On the effects of plastic rotation in the finite deformation of anisotropic elastoplastic materials. Mech Mater 2:287–304

Lubarda VA (2002) Elastoplasticity theory. CRC Press, Boca Ranton

Lubarda VA (2004) Constitutive theories based on the multiplicative decomposition of deformation gradient: thermoplasticity, elastoplasticity, and biomechanics. Appl Mech Rev 57:95–108

Lubarda VA, Lee EH (1981) A correct definition of elastic and plastic deformation and its computational significance. J Appl Mech (ASME) 48:35–40

Lubliner J (1984) A maximum-dissipation principle in generalized plasticity. Acta Mech 52:225–237

Lubliner J (1990) Plasticity theory. Macmillan, New York

Mandel J (1964) Contribution theorique a l'eude de l'ecrouissage et des lois de l'ecoulement plastique. In: Proceedings of 11th international congress on applied mechanics, pp 502–509

Mandel J (1971) Plastidite classique et viscoplasticite. Course and lectures. No. 97, International Center for Mechanical Sciences, Udine, Springer, Heidelberg

Mandel J (1973) Equations constitutives directeurs dans les milieux plastiques at viscoplastiques. Int J Solids Struct 9:725–740

Mandel J (1974) Director vectors and constitutive equations for plastic and viscoplastic media. Problems of Plasticity (Proc. Int. Symp Foundation of Plasticity). (Ed. A. Sawczuk). Noordhoff Int. Publ., Leyden, Netherland. pp.135–141

Petryk H (1991) On the second-order work in plasticity. Arch Mech 43:377–397

Petryk H (1997) Plastic instability: criteria and computational approaches. Arch Comput Approach Meth Eng 4:111–151

Prager W (1949) Recent development in the mathematical theory of plasticity. J Appl Mech (ASME) 20:235–241

Prager W (1956) A new methods of analyzing stresses and strains in work hardening plastic solids. J Appl Mech (ASME) 23:493–496

Sekiguchi H, Ohta H (1977) Induced anisotropy and its time dependence in clays. In: Constitutive equations of soils (Proc Spec Session 9, 9th ICSFME), Tokyo, pp 229–238

Shield RT, Ziegler H (1958) On Prager's hardening rule. Z ang Math Mech 9:260–276

Simo JC (1998) Numerical analysis and simulation of plasticity. In: Ciarlet PG, Lions JL (eds) Handbook of numerical analysis, vol 6, part 3 (numerical methods for solids). Elsevier, Amsterdam

Simo JC, Hughes TJR (1998) Computational inelasticity. Springer, New York

Truesdell C (1955) Hypo-elasticity. J Ration Mech Anal 4:83–133

Van der Giessen E (1989) Micromechanical and thermodynamic aspects of the plastic spin. Int J Plast 7:365–386

White PS (1975) Elastic-plastic solids as simple materials. Q J Mech Appl Math 28:483–496

Wu H-C (2004) Continuum mechanics and plasticity. Chapman & Hall/CRC, New York

Zbib, HM, Aifantis EC (1988) On the concept of relative and plastic spins and its implications to large deformation theories. Part I: Hypoelasticity and vertex-type plasticity, Acta Mech., 75:15–33

Ziegler H (1959) A modification of Prager's hardening rule. Q Appl Phys 17:55–60

Chapter 7
Unconventional Elastoplasticity Model: Subloading Surface Model

Elastoplastic constitutive equations with the yield surface enclosing the elastic domain possess many limitations in the description of elastoplastic deformation, as explained in the last chapter. They are called the *conventional model* in the Drucker's (1988) classification of plasticity models. Various *unconventional elastoplasticity* models have been proposed, which are intended to describe the plastic strain rate induced by the rate of stress inside the yield surface. Among them, the subloading surface model is the only pertinent model fulfilling the mechanical requirements for unconventional models. These mechanical requirements are first described and then the subloading surface model is explained in detail.

7.1 Mechanical Requirements

There exist various mechanical requirements, e.g., the thermodynamic restriction and the objectivity for constitutive equations. Among them, the continuity and the smoothness conditions are violated in many elastoplasticity models, while their importance for formulation of constitutive equations has not been sufficiently recognized to date. Before formulation of the plastic strain rate, these conditions will be explained below (Hashiguchi 1993a, b, 1997, 2000).

7.1.1 Continuity Condition

It is observed in experiments that "*stress rate changes continuously for a continuous change of strain rate*". This fact is called the *continuity condition* and is expressed mathematically as follows (Hashiguchi 1993a, b, 1997, 2000).

K. Hashiguchi, *Foundations of Elastoplasticity: Subloading Surface Model*,
DOI 10.1007/978-3-319-48821-9_7

$$\boxed{\lim_{\delta \mathbf{d} \to \mathbf{O}} [\overset{\circ}{\boldsymbol{\sigma}}(\boldsymbol{\sigma}, H_i;\ \mathbf{d} + \delta\mathbf{d}) - \overset{\circ}{\boldsymbol{\sigma}}(\boldsymbol{\sigma}, H_i;\ \mathbf{d})] \to \mathbf{O}} \tag{7.1}$$

where H_i ($i = 1, 2, 3, \cdots$) denotes collectively scalar-valued or tensor-valued internal state variables. In addition, $\delta(\ )$ stands for an infinitesimal variation. The response of the stress rate to the input of strain rate in the current state of stress and internal variables is designated by $\overset{\circ}{\boldsymbol{\sigma}}(\boldsymbol{\sigma}, H_i;\ \mathbf{d})$. *Uniqueness of solution* is not guaranteed in constitutive equations violating the continuity condition, predicting different stresses or deformations for identical input loading. The violation of this condition is schematically shown in Fig. 7.1. Ordinary elastoplastic constitutive equations, in which the plastic strain rate is derived obeying the consistency condition, fulfill the continuity condition. As described later, however, no elastoplastic constitutive equation fulfills it except for the subloading surface model when they are extended to describe the tangential inelastic strain rate.

The concept of the continuity condition was first advocated by Prager (1949). However, a mathematical expression of this condition was not given. The condition was defined as the continuity of strain rate to the input of stress rate by Prager (1949) inversely to the definition given above. However, identical stress rate directing inwards the yield surface can induce different strain rates in loading and unloading states for softening materials. Also, identical stress rate along the yield surface can induce different strain rates in a perfectly-plastic material as illustrated in Fig. 7.2. Besides, it is noteworthy that a stress rate cannot be given arbitrarily since there exists a limitation in strength of materials although a strain rate can be given arbitrarily. For that reason, the Prager's (1949) notion does not hold in the general loading state including softening and the perfectly plastic states.

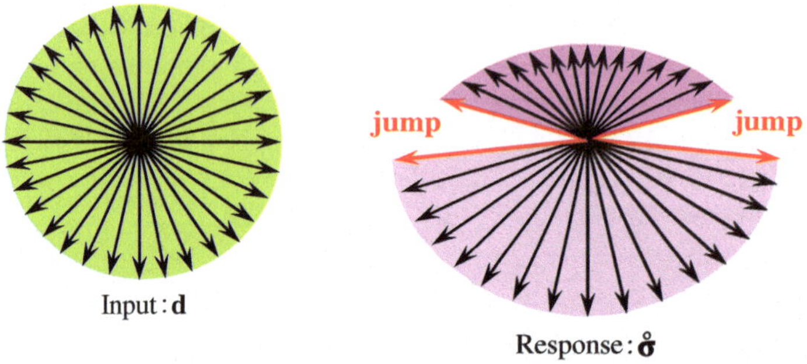

Fig. 7.1 Violation of continuity condition

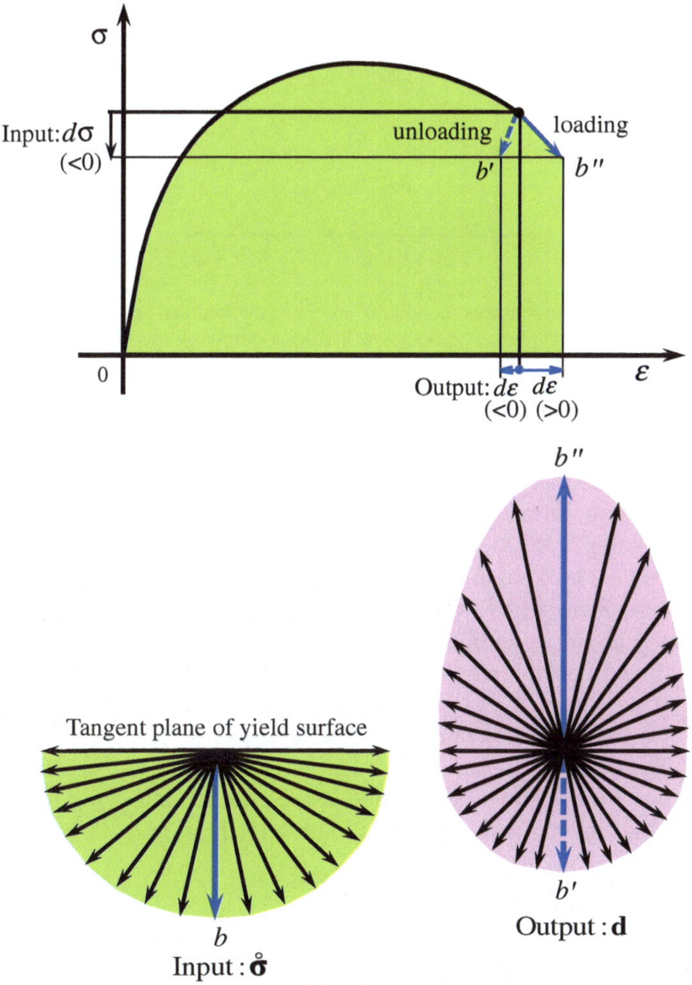

Fig. 7.2 Impertinence of Prager's (1949) continuity condition in softening state

7.1.2 *Smoothness Condition*

It is observed in experiments that "*the stress rate induced by the identical strain rate changes continuously for a continuous change of stress state*". This fact is called the *smoothness condition* and is expressed mathematically as follows:

$$\boxed{\lim_{\delta\boldsymbol{\sigma}\to\mathbf{O}} \overset{\circ}{\boldsymbol{\sigma}}(\boldsymbol{\sigma} + \delta\boldsymbol{\sigma}, H_i; \mathbf{d}) - \overset{\circ}{\boldsymbol{\sigma}}(\boldsymbol{\sigma}, H_i\ ; \mathbf{d}) \to \mathbf{O}} \tag{7.2}$$

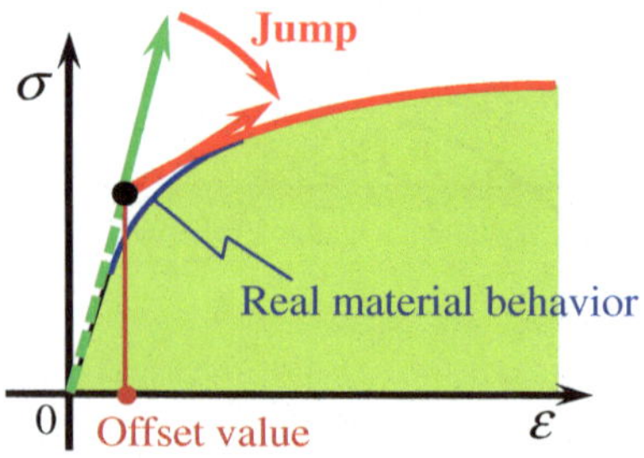

Fig. 7.3 Violation of smoothness condition in the conventional plasticity model and the kinematic hardening cyclic plasticity models assuming a purely elastic domain

A smooth response of stress–strain relation is not described by constitutive equations violating the smoothness condition, causing discontinuous change of tangent modulus, as illustrated in Fig. 7.3 for the conventional elastoplastic constitutive equations assuming the yield surface enclosing a purely-elastic domain. Then, constitutive equations violating the smoothness condition exhibit abrupt change of tangent modulus from the elastic to the elastoplastic state, so that it is required to determine the offset (permanent strain) value, i.e. the plastic stain at yield point, which is accompanied with an arbitrariness (although 0.2 % plastic strain is often used).

The rate-linear constitutive equation is described as

$$\overset{\circ}{\boldsymbol{\sigma}} = \mathbf{K}^{ep}(\boldsymbol{\sigma}, H_i) : \mathbf{d} \tag{7.3}$$

where the fourth-order tensor $\mathbf{K}^{ep}$ is the elastoplastic modulus, which is a function of the stress and internal variables, can be described generally as

$$\mathbf{K}^{ep} = \frac{\partial \overset{\circ}{\boldsymbol{\sigma}}}{\partial \mathbf{d}} \tag{7.4}$$

Consequently, Eq. (7.2) can be rewritten as

$$\boxed{\lim_{\delta\boldsymbol{\sigma}\to\mathbf{O}} [\mathbf{K}^{ep}(\boldsymbol{\sigma} + \delta\boldsymbol{\sigma}, H_i) - \mathbf{K}^{ep}(\boldsymbol{\sigma}, H_i)] \to \mathbf{O}} \tag{7.5}$$

where $\mathbf{O}$ designates the second-order and fourth-order zero tensors.

Constitutive equations violating the smoothness condition cannot predict a smooth stress-strain curve. Therefore, they cannot describe softening behavior pertinently, as depicted in Fig. 6.13. Further, they cannot predict the strain accumulation for cyclic loading of stress amplitude less than the yield stress, as depicted in Fig. 6.14. The smoothness condition is of great importance in the description of cyclic loading behavior for which an accurate description of plastic strain rate induced by the rate of stress inside the yield surface is required. Among the existing constitutive models only the subloading surface model always fulfills the smoothness condition.

7.2 Subloading Surface (Hashiguchi) Model

The basic concept and equations for the subloading surface model (Hashiguchi and Ueno 1977; Hashiguchi 1978, 1980, 1989) are described below. This is the only model fulfilling the mathematical requirements described in 7.1.

In order to describe the plastic strain rate induced by the rate of stress inside the yield surface, let the following postulate be incorporated based on the concept of the subloading surface (Hashiguchi 1980, 1989, 2013b).

Fundamental postulate of unconventional elastoplasticity (Subloading surface concept): *The stress approaches the yield surface when the plastic strain rate is induced, exhibiting a continuous variation of tangent modulus, but it recedes from the yield surface when only the elastic strain rate is induced.*

In this context, it is first required to incorporate the general measure which describes the approaching degree of the stress to the yield surface, renamed the *normal-yield surface*, in order to formulate the plastic strain rate. Then, let the following ***subloading surface*** which always passes through the current stress and maintains a similar shape and an orientation to the normal-yield surface be introduced (see Fig. 7.4).

$$\boxed{f(\boldsymbol{\sigma}) = RF(H)} \tag{7.6}$$

where $R(0 \le R \le 1)$ is the ratio of the size of the subloading surface to that of the normal-yield surface and called the *normal-yield ratio*, playing the role for the measure of approaching degree to the normal-yield surface.

Based on the above-mentioned fundamental postulate of elastoplasticity, the rate of the normal-yield ratio must satisfy the following conditions (see Fig. 7.5).

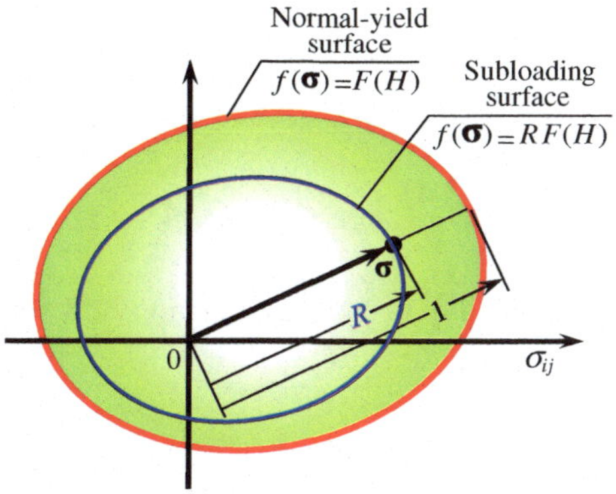

Fig. 7.4 Normal-yield and subloading surfaces

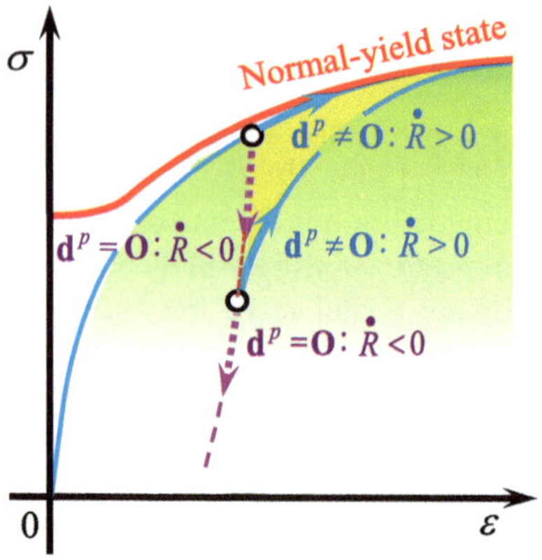

Fig. 7.5 Plastic strain rate based on the subloading surface concept

$$\dot{R}\begin{cases} \to \infty \text{ for } R = 0 \\ > 0 \text{ for } R < 1 \\ = 0 \text{ for } R = 1 \\ (< 0 \text{ for } R > 1) \end{cases} \quad \text{for } \mathbf{d}^p \neq \mathbf{O} \tag{7.7}$$

$$\dot{R}\begin{cases} = 0 \text{ for } \mathbf{d}^e = \mathbf{O} \\ < 0 \text{ for } \mathbf{d}^e \neq \mathbf{O} \end{cases} \quad \text{for } \mathbf{d}^p = \mathbf{O} \tag{7.8}$$

Here, the rate of the normal-yield ratio evolves with the plastic strain rate, obeying Eq. (7.7) but it is calculated from the equation of the subloading surface in Eq. (7.6), substituting a stress changing by the elastic constitutive relation under fixed internal variables when only the elastic strain rate is induced. Then, it follows that

$$\boxed{\dot{R} = U(R)\|\,\mathbf{d}^p\| = U(R)\,\dot{\bar{\lambda}} \quad \text{for } \mathbf{d}^p \neq \mathbf{O}} \tag{7.9}$$

$$R = \frac{f(\boldsymbol{\sigma})}{F} \quad \text{for } \mathbf{d}^e \neq \mathbf{O},\, \mathbf{d}^p = \mathbf{O} \tag{7.10}$$

where $U(R)$ is the monotonically-decreasing function of R fulfilling the conditions (see Fig. 7.6).

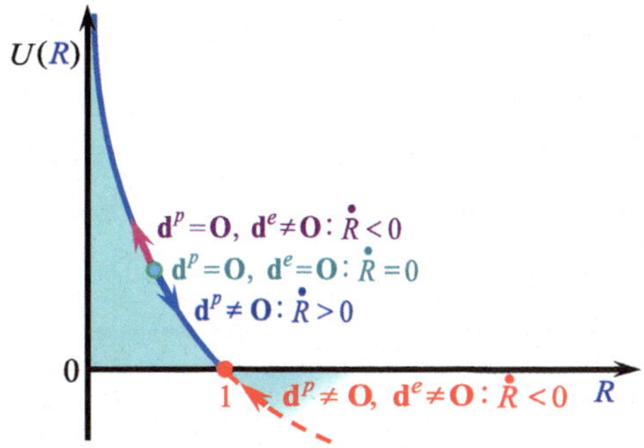

Fig. 7.6 Function $U(R)$ in rate of normal-yield ratio R

$$U(R)\begin{cases} \to +\infty \ \text{for}\ R = 0\ \text{(elastic state)} \\ > 0\ \text{for}\ R < 1\ \text{(subyield state)} \\ = 0\ \text{for}\ R = 1\ \text{(normal-yield state)} \\ < 0\ \text{for}\ R > 1\ \text{(over normal-yield state)} \end{cases} \tag{7.11}$$

The function $U(R)$ in Eq. (7.9) with Eq. (7.11) is schematically shown in Fig. 7.6.

The explicit form of the function $U(R)$ is given by

$$U(R) = u\ \cot[(\pi/2)R] \tag{7.12}$$

and the other examples of the function $U(R)$ are shown as

$$U(R) = -u\ \ln R,\ U(R) = u\left(\frac{1}{R} - 1\right) \tag{7.13}$$

where u is the material parameters. Here, note that the normal-yield ratio R increases obeying the evolution rule in Eq. (7.9) formulated by the plastic strain rate in the plastic-loading process. On the other hand, it decreases obeying Eq. (7.10) where R is calculated using the stress calculated by the elastic constitutive relation in Eq. (6.29) in the elastic-unloading process, where the internal variable F is fixed.

Equation (7.9) with Eqs. (7.12) and (7.13) is analytically integrated as follows:

$$\left.\begin{array}{ll} R = \dfrac{2}{\pi}\cos^{-1}\left(\cos\left\{\dfrac{2}{\pi}R_0\exp\left[-\dfrac{2}{\pi}u(\varepsilon^p - \varepsilon_0^p)\right]\right\}\right) & \text{for}\ U = u\ \cot\left(\dfrac{2}{\pi}R\right) \\ R(\ln R - 1) - R_0(\ln R_0 - 1) = u(\varepsilon^p - \varepsilon_0^p) & \text{for}\ U = -u\ln R \\ -R(\ln R + 1) + R_0(\ln R_0 + 1) = u(\varepsilon^p - \varepsilon_0^p) & \text{for}\ U = u\left(\dfrac{1}{R} - 1\right) \end{array}\right\} \tag{7.14}$$

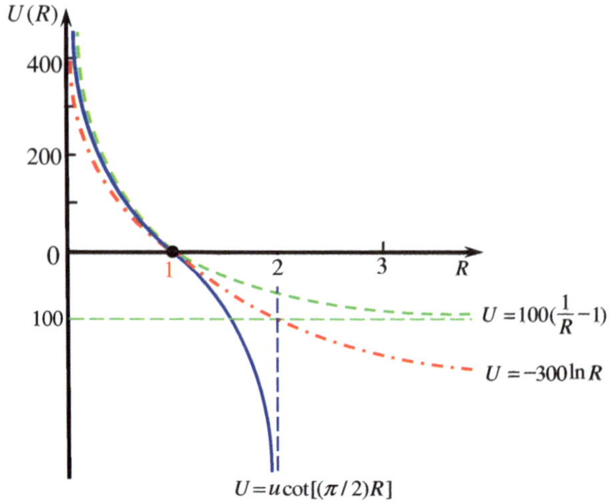

Fig. 7.7 Three types of function $U(R)$ in the evolution rule of the normal friction-yield ratio R

under the initial condition $\varepsilon^p = \varepsilon_0^p : R = R_0$, where $\varepsilon^p \equiv \int ||\mathbf{d}^p||dt(t:$ time). Here, note that analytical expression of R holds only for the cotangent function in Eq. $(7.14)_1$, although inversely ε^p can be expressed analytically by R for all of these functions.

Further, note that the following inequality holds as depicted in Fig. 7.7 in which $u = 30$ is used for $u\cot[(\pi/2)R]$ and $u[(1/R) - 1]$ and $u = 100$ for $-u\ln R$.

$$u \cot[(\pi/2)R] < -u \ln R < u(\frac{1}{R} - 1) < 0 \quad \text{For } R > 1 \tag{7.15}$$

when the material parameter u is chosen such that the values of $U(R)$ in these functions are almost identical in the range $R < 1$, noting

$$\left.\begin{array}{ll} R \to 2: & \dot{R} \to -\infty \text{ for } U = u \cot[(\pi/2)R] \\ R \to \infty: & \dot{R} \to -\infty \text{ for } U = -\mathrm{u} \ln R \\ R \to \infty: & \dot{R} \to -u \quad \text{for } U = u\left(\dfrac{1}{R} - 1\right) \end{array}\right\} \tag{7.16}$$

The cotangent function in Eq. (7.12) would be most beneficial among the three functions shown above by the following reasons.

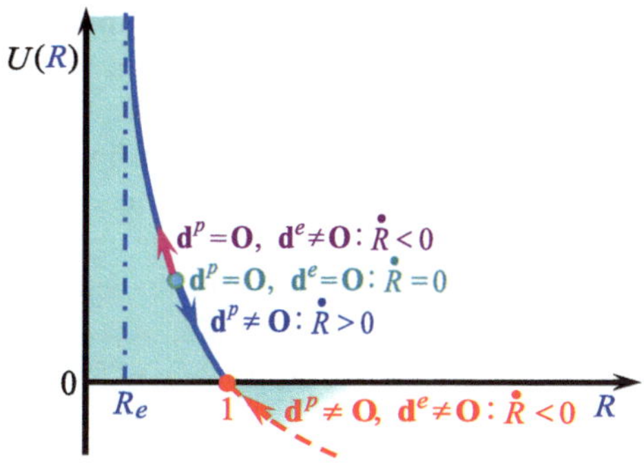

Fig. 7.8 Function $U(R)$ with purely-elastic domain for evolution rule of normal-yield ratio

(1) The analytical expression of R can be obtained only for the cotangent function in Eq. (7.12) as shown in Eq. $(7.14)_1$.
(2) It possesses the largest negative value in the range $R > 1$ among these three kinds of function $U(R)$. Then, it provides the most intense controlling function to pull back the stress to the normal-yield surface when the stress jumps out from that surface in numerical calculations as will be explained in Sect. 7.3 with Fig. 7.13.

Here, note that there exist a lot of materials containing usual metals in which the plastic strain rate is hardly induced in a wide range of the normal-yield ratio. Then, let the following relation be assumed instead of Eq. (7.11), in which the plastic strain rate is not induced until the normal-yield ratio R reaches a certain value of the material parameter $R_e(<1)$ (see Fig. 7.8).

$$U(R)\begin{cases} \to +\infty \text{ for } 0 \leq R \leq R_e \text{ (elastic state)} \\ > 0 \text{ for } R_e < R < 1 \text{ (subyield state)} \\ = 0 \text{ for } R = 1 \text{ (normal-yield state)} \\ < 0 \text{ for } R > 1 \text{ (over normal-yield state)} \end{cases} \tag{7.17}$$

The material parameter R_e is interpreted to be the ratio of the (half) stress amplitude σ_{fl} at the fatigue (or endurance) limit, i.e. the fatigue limit stress to the yield stress σ_y under the zero value of average stress $\bar{\sigma}$, i.e. $R_e = \sigma_{fl}/\sigma_y|_{\bar{\sigma}=0}$. Fatigue limit is observed in steels, titanium, etc. but it is not observed in other materials involving non-ferrous metals.

Note here that the incorporation of the material parameter R_e does not mean the incorporation of the yield surface enclosing a purely-elastic domain as known from the fact: The plastic strain rate is predicted for the cyclic loading with a small stress amplitude under a high average stress by the subloading surface model with the incorporation of R_e but it cannot be predicted if the yield surface enclosing a purely-elastic domain is incorporated as seen in the cyclic kinematic hardening model, i.e. the multi surface, the two surface and the superposed kinematic hardening models which will be described in Sect. 8.2.

Equation (7.12) is modified to satisfy Eq. (7.17) as follows:

$$U(R) = u \cot\left(\frac{\pi}{2}\frac{\langle R - R_e\rangle}{1 - R_e}\right) \tag{7.18}$$

where $\langle\,\rangle$ is the Macaulay's bracket defined by $\langle s\rangle = (s + |s|)/2$, i.e. $s < 0 : \langle s\rangle = 0$ and $s \geq 0 : \langle s\rangle = s$ (s: arbitrary scalar variable). Equation (7.18) conforms to the fulfillment of the smoothness condition since it decreases continuously from infinite value in $R = R_e$. If u is fixed to be constant, Eq. (7.9) with Eq. (7.18) can be integrated analytically as

$$\left.\begin{aligned} R &= \frac{2}{\pi}(1 - R_e)\cos^{-1}\left[\cos\left(\frac{\pi}{2}\frac{R_0 - R_e}{1 - R_e}\right)\exp\left(-u\frac{\pi}{2}\frac{\varepsilon^p - \varepsilon^p_0}{1 - R_e}\right)\right] + R_e \\ \varepsilon^p - \varepsilon^p_0 &= \frac{2}{\pi}\frac{1 - R_e}{u}\ln\frac{\cos\left(\dfrac{\pi}{2}\dfrac{R_0 - R_e}{1 - R_e}\right)}{\cos\left(\dfrac{\pi}{2}\dfrac{R - R_e}{1 - R_e}\right)} \end{aligned}\right\} \text{for } R_0 \geq R_e \tag{7.19}$$

whilst one must set $R_0 = R_e$ for $R_0 < R_e$. However, the judgment whether of $R < R_e$ or $R \geq R_e$ is required in Eq. (7.18). The judgment whether R reaches R_e is required in Eq. (7.18), although the yield judgment is not required.

Equation (7.18) with the purely-elastic limit R_e is given above. However, a monotonic loading behaviour and a low cycle loading behavior with a large stress amplitude cannot be realistically described resulting in an excessively large plastic strain accumulation if we choose $R_e \cong 0$ and a high cycle loading behaviour with a small stress amplitude cannot be described resulting in no accumulation of plastic strain if we choose $R_e \gg 0$. Therefore, Eq. (7.18) is not applicable to a general cyclic loading behaviour with variable (fluctuating) stress amplitudes.

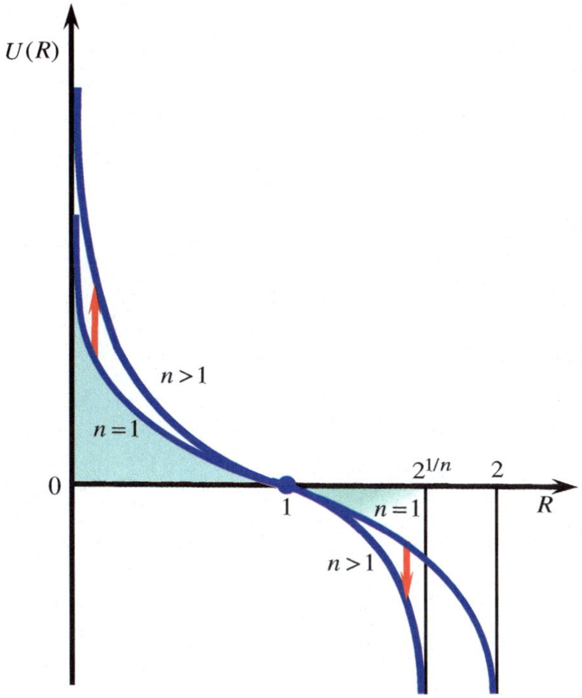

Fig. 7.9 Extended function $U(R) = u\ \cot[(\pi/2)R^n]$

Eventually, any of Eqs. (7.12), (7.13) and (7.18) is inapplicable to the prediction of cyclic loading behavior under variable stress amplitudes in materials without a fatigue limit. The following functions would be applicable to general cyclic loading behavior in wide classes of materials.

$$U(R) = u\cot\left(\frac{\pi}{2}\left\langle\frac{R-R_e}{1-R_e}\right\rangle^n\right) \tag{7.20}$$

$$U(R) = u\left(\left\langle\frac{1-R_e}{R-R_e}\right\rangle^n - 1\right) \tag{7.21}$$

where $n(\geq 1)$ is the material parameter. The value of the function $U(R)$ is larger throughout a whole region of R for a lager value of u, and it is larger in a region of small value of R for a larger value of n. The function in Eq. (7.12) is schematically illustrated in Fig. 7.9 for $R_e = 0$.

The time-differentiation of Eq. (7.6) of the subloading surface leads to

$$\frac{\partial f(\boldsymbol{\sigma})}{\partial \boldsymbol{\sigma}} : \overset{\circ}{\boldsymbol{\sigma}} - R\dot{F} - \dot{R}F = 0 \tag{7.22}$$

Substituting Eq. (7.6) into Eq. (6.32), one has

$$\frac{\partial f(\boldsymbol{\sigma})}{\partial \boldsymbol{\sigma}} : \boldsymbol{\sigma} = RF \tag{7.23}$$

which yields

$$\mathbf{n} : \boldsymbol{\sigma} = \frac{\dfrac{\partial f(\boldsymbol{\sigma})}{\partial \boldsymbol{\sigma}} : \boldsymbol{\sigma}}{\left\| \dfrac{\partial f(\boldsymbol{\sigma})}{\partial \boldsymbol{\sigma}} \right\|} = \frac{RF}{\left\| \dfrac{\partial f(\boldsymbol{\sigma})}{\partial \boldsymbol{\sigma}} \right\|}, \quad \frac{1}{\left\| \dfrac{\partial f(\boldsymbol{\sigma})}{\partial \boldsymbol{\sigma}} \right\|} = \frac{\mathbf{n} : \boldsymbol{\sigma}}{RF} \tag{7.24}$$

where

$$\mathbf{n} \equiv \frac{\partial f(\boldsymbol{\sigma})}{\partial \boldsymbol{\sigma}} / \left\| \frac{\partial f(\boldsymbol{\sigma})}{\partial \boldsymbol{\sigma}} \right\| \ (\|\mathbf{n}\| = 1) \tag{7.25}$$

Equation (7.22) with Eq. (7.24) results in

$$\mathbf{n} : \left[\overset{\circ}{\boldsymbol{\sigma}} - \left(\frac{\dot{F}}{F} + \frac{\dot{R}}{R} \right) \boldsymbol{\sigma} \right] = 0 \tag{7.26}$$

Now, adopt the associated flow rule

$$\mathbf{d}^p = \dot{\bar{\lambda}} \mathbf{n} \ (\dot{\bar{\lambda}} = \|\mathbf{d}^p\| > 0) \tag{7.27}$$

The evolution rule of the normal-yield ratio is described as the equation $\dot{R} = U(R)\dot{\bar{\lambda}}$ in addition to the Eq. (7.9) for the expression of the flow rule in Eq. (7.27). Note, however, that the equation $\dot{R} = U(R)\dot{\bar{\lambda}}$ does not hold for the expression of the flow rule $\mathbf{d}^p = \dot{\bar{\lambda}}\, \partial f(\boldsymbol{\sigma})/\partial \boldsymbol{\sigma}$ because of $\|\partial f(\boldsymbol{\sigma})/\partial \boldsymbol{\sigma}\| \neq 1$ in general.

Substituting Eqs. (6.42) and $(7.9)_1$ with Eq. (7.27), one has

$$\mathbf{n}:\left[\overset{\circ}{\boldsymbol{\sigma}}-\left(\frac{F'}{F}f_{Hn}(\boldsymbol{\sigma},\,H;\mathbf{n})\dot{\bar{\lambda}}\;+\;\frac{U(R)}{R}\dot{\bar{\lambda}}\right)\boldsymbol{\sigma}\right]=0 \tag{7.28}$$

It follows from Eqs. (7.27) and (7.28) that

$$\dot{\bar{\lambda}}=\frac{\mathbf{n}:\overset{\circ}{\boldsymbol{\sigma}}}{\bar{M}^p},\ \mathbf{d}^p=\frac{\mathbf{n}:\overset{\circ}{\boldsymbol{\sigma}}}{\bar{M}^p}\mathbf{n} \tag{7.29}$$

where

$$\boxed{\bar{M}^p\equiv\left(\frac{F'}{F}f_{Hn}(\boldsymbol{\sigma},\,H;\mathbf{n})+\frac{U(R)}{R}\right)\mathbf{n}:\boldsymbol{\sigma}} \tag{7.30}$$

which is reduced to the plastic modulus of the conventional elastoplasticity, i.e.

$$\bar{M}^p=\frac{F'}{F}f_{Hn}(\boldsymbol{\sigma},\,H;\mathbf{n})\mathbf{n}:\boldsymbol{\sigma}=M^p \tag{7.31}$$

in the normal-yield state ($R=1\rightarrow U(R)=0$).

The strain rate is described from Eqs. $(6.27)_1$, (6.29) and (7.29) as

$$\boxed{\mathbf{d}=\mathbf{E}^{-1}:\overset{\circ}{\boldsymbol{\sigma}}+\frac{\mathbf{n}:\overset{\circ}{\boldsymbol{\sigma}}}{\bar{M}^p}\mathbf{n}=\left(\mathbf{E}^{-1}\;+\;\frac{\mathbf{n}\otimes\mathbf{n}}{\bar{M}^p}\right):\overset{\circ}{\boldsymbol{\sigma}}} \tag{7.32}$$

from which the magnitude of plastic strain $\dot{\bar{\lambda}}$ in terms of strain rate, denoted by the symbol $\dot{\bar{\Lambda}}$, is derived as follows:

$$\dot{\bar{\Lambda}}\;=\;\frac{\mathbf{n}:\mathbf{E}:\mathbf{d}}{\bar{M}^p+\mathbf{n}:\mathbf{E}:\mathbf{n}},\ \mathbf{d}^p=\frac{\mathbf{n}:\mathbf{E}:\mathbf{d}}{\bar{M}^p+\mathbf{n}:\mathbf{E}:\mathbf{n}}\mathbf{n} \tag{7.33}$$

The stress rate is described from Eq. (6.29) with Eq. (7.33) as

$$\boxed{\overset{\circ}{\boldsymbol{\sigma}}\;=\mathbf{E}:\mathbf{d}-\frac{\mathbf{n}:\mathbf{E}:\mathbf{d}}{\bar{M}^p+\mathbf{n}:\mathbf{E}:\mathbf{n}}\mathbf{E}:\mathbf{n}=\left(\mathbf{E}-\frac{\mathbf{E}:\mathbf{n}\otimes\mathbf{n}:\mathbf{E}}{\bar{M}^p+\mathbf{n}:\mathbf{E}:\mathbf{n}}\right):\mathbf{d}} \tag{7.34}$$

The loading criterion is given by

$$\left.\begin{array}{l}\mathbf{d}^p\neq\mathbf{O}\ \text{for}\ \dot{\bar{\Lambda}}\;>0\\ \mathbf{d}^p=\mathbf{O}\ \text{for}\ \dot{\bar{\Lambda}}\;\leq 0\end{array}\right\} \tag{7.35}$$

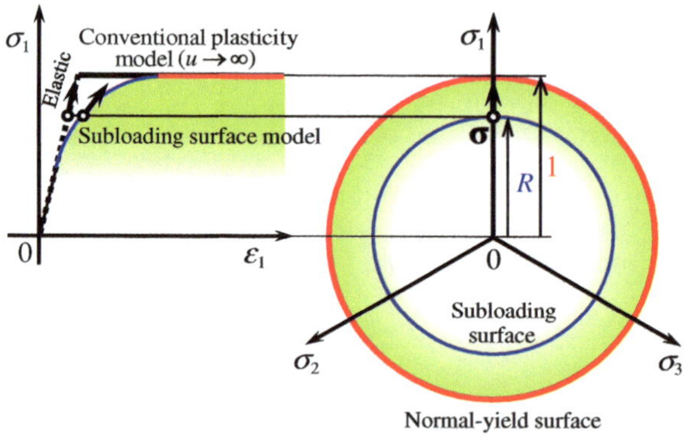

Fig. 7.10 Smooth stress–strain curve predicted by the subloading surface

or

$$\boxed{\begin{array}{l}\mathbf{d}^p \neq \mathbf{O} : \mathbf{n} : \mathbf{E} : \mathbf{d} > 0 \\ \mathbf{d}^p = \mathbf{O} : \text{otherwise}\end{array}} \tag{7.36}$$

where the judgment whether or not the stress reaches the yield surface is not required since the plastic strain rate is induced continuously as the stress approaches the normal-yield surface.

There exists the risk that the subloading surface once contracts and then expands, so that a plastic strain rate is induced at the moment of stress reversal event in the neighborhood of the similarity-center of the normal-yield and the subloading surfaces even when $\mathbf{n} : \mathbf{E} : \mathbf{d} < 0$ holds. Therefore, the magnitude of input loading increment must be so small as to avoid the risk in the numerical calculation method.

The stress vs. strain curve by the subloading surface model is illustrated for the simplest case of the perfectly-plastic material in Fig. 7.10.

7.3 Salient Features of Subloading Surface Model

This model possesses the following distinguished abilities.

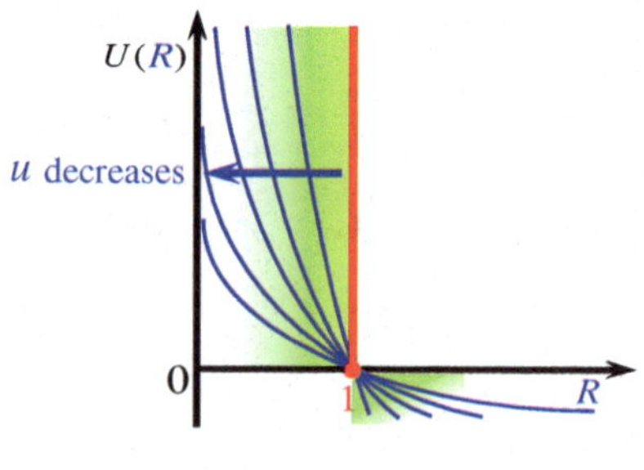

Fig. 7.11 Influence of u on function $U(R)$

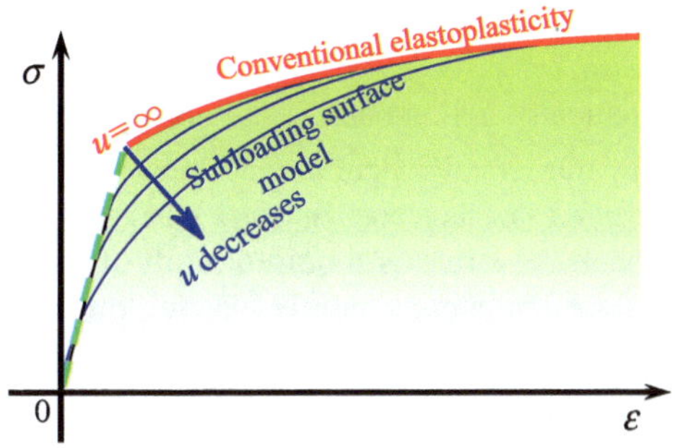

Fig. 7.12 Influence of u on stress–strain curve

(1) Smooth transition from elastic to plastic state is described, which is observed in real material behavior. Then, we don't need suffer from the determination of an offset value (plastic strain value in yield point). In contrast, the determination is required in all of the other elastoplastic models since they assume a surface enclosing a purely-elastic domain leading to the abrupt elastic-plastic transition, while the determination is accompanied with an arbitrariness. The influences of the material parameter u on the function $U(R)$ and the stress-strain curve are depicted in Figs. 7.11 and 7.12, respectively. The larger the material parameter u, the more rapidly the normal-yield ratio R increases causing the more rapid increase of stress, i.e. approaching the behavior of the conventional plasticity.
(2) Plastic strain rate can be described even for low stress level and for cyclic loading process under small stress amplitudes since a purely-elastic domain is not assumed.
(3) The yield-judgment whether or not the stress reaches the yield surface is unnecessary since the plastic strain rate develops continuously as the stress approaches the normal-yield surface. In contrast, the yield judgment is required

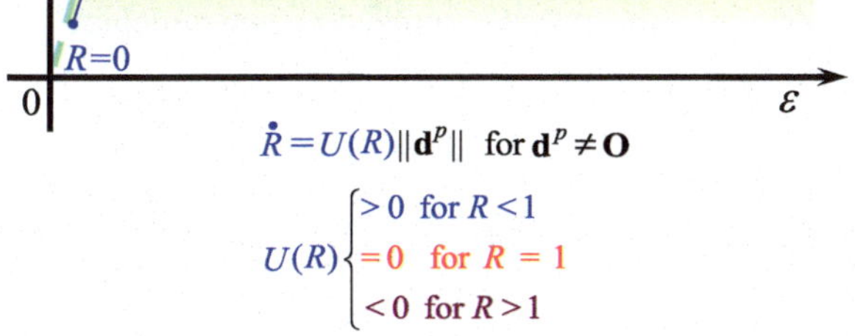

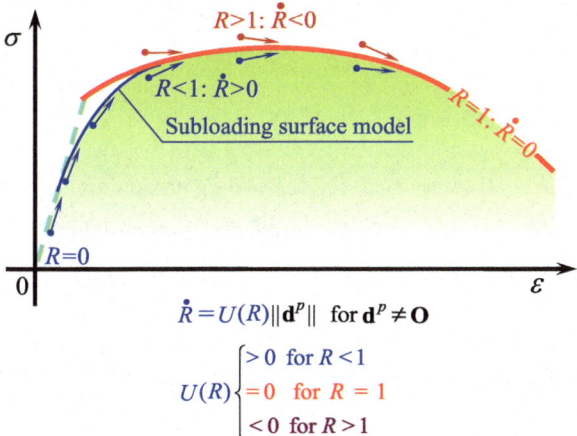

Fig. 7.13 Stress is automatically controlled to be attracted to the normal-yield surface in the subloading surface model

in all of the other elastoplastic models since they assume a surface enclosing a purely-elastic domain.

(4) The stress is automatically pulled back to the normal-yield surface when it goes over the surface in numerical calculation because of $\dot{R} < 0$ for $R > 1$ from Eq. (7.7) with Eq. $(7.11)_4$ as seen in Fig. 7.13. In contrast, the particular operation to pull back the stress is required in all of the other models because they assume a surface enclosing a purely-elastic domain.

For the normal-yield state $R = 1$ ($U = 0$), the plastic strain rate in Eq. (7.29) with Eq. (7.30) is reduced to Eq. (6.43) with Eq. (6.41) for the conventional plasticity, i.e.

$$\mathbf{d}^p = \frac{\mathbf{n} : \overset{\circ}{\boldsymbol{\sigma}}}{\frac{F'}{F} f_{Hn}(\boldsymbol{\sigma}, H; \mathbf{n})\mathbf{n} : \boldsymbol{\sigma}} \mathbf{n} \left(= \frac{\dfrac{\partial f(\boldsymbol{\sigma})}{\partial \boldsymbol{\sigma}} : \overset{\circ}{\boldsymbol{\sigma}}}{F' f_{Hn}\left(\boldsymbol{\sigma}, H; \dfrac{\partial f(\boldsymbol{\sigma})}{\partial \boldsymbol{\sigma}}\right)} \frac{\partial f(\boldsymbol{\sigma})}{\partial \boldsymbol{\sigma}} \right)$$

For $u \to \infty$ leading to the sudden decrease of the function U from $U \to \infty$ for $R < 1$ to $U = 0$ for $R = 1$ in Eq. (7.11), the plastic modulus $\bar{M}^p$ in Eq. (7.30) drops suddenly from the infinite value to the value M^p in Eq. (6.41) so that the present model behavior is reduced to the conventional elastoplasticity model behavior by choosing a large value of the material parameter u, thereby exhibiting an sudden transition from the elastic to plastic state. On the other hand, as u becomes smaller, a gentler transition from the elastic to plastic state is described. Therefore, u plays the role to alleviate the sudden transition from the elastic to plastic state.

It follows from Eq. (6.43) and Eq. (7.29) in the plastic loading process, fulfilling $\dot{\lambda} \geq 0$ or $\dot{\bar{\lambda}} \geq 0$, that

$$\left.\begin{array}{l} M^p > 0 \to \mathbf{n} : \overset{\circ}{\boldsymbol{\sigma}} > 0,\ \dot{F} > 0 : \text{normal hardening} \\ M^p = 0 \to \mathbf{n} : \overset{\circ}{\boldsymbol{\sigma}} = 0,\ \dot{F} = 0 : \text{normal nonhardening} \\ M^p < 0 \to \mathbf{n} : \overset{\circ}{\boldsymbol{\sigma}} < 0,\ \dot{F} < 0 : \text{normal softening} \end{array}\right\} \tag{7.37}$$

for the conventional model and

$$\left.\begin{array}{l} \bar{M}^p > 0 \to \mathbf{n} : \overset{\circ}{\boldsymbol{\sigma}} > 0 : \text{subloading hardening} \\ \bar{M}^p = 0 \to \mathbf{n} : \overset{\circ}{\boldsymbol{\sigma}} = 0 : \text{subloading nonhardening} \\ \bar{M}^p < 0 \to \mathbf{n} : \overset{\circ}{\boldsymbol{\sigma}} < 0 : \text{subloading softening} \end{array}\right\} \tag{7.38}$$

for the subloading surface model. Here, it should be noted that the signs of M^p or $\bar{M}^p$ and $\mathbf{n} : \overset{\circ}{\boldsymbol{\sigma}}$ coincide with each other in both models but they do not necessarily coincide with the sign of $\dot{F}$ in the subloading surface model.

The distinguished advantages of the subloading surface model in the descriptions of irreversible mechanical phenomena can be obtained by the simple modification of existing computer program for the conventional elastoplasticity model to add only one material parameter u for the evolution rule of the normal-yield ratio without any expense.

7.4 Numerical Performance of Subloading Surface Model

The stress controlling function of the subloading surface model is described in Sect. 7.3. This fact will be shown below by the numerical calculation for the response of the uniaxial loading behavior, adopting the simplest subloading surface model for the isotropic Mises material with the evolution rule of the normal-yield ratio in Eq. (7.9) with Eq. (7.12). The response of the conventional elastoplastic constitutive model is also shown for the comparison.

The relations of the axial stress σ_a and the normal-yield ratio R versus the axial strain ε_a are depicted in Fig. 7.14. The responses adopting the linear isotropic hardening $F = F_0 + h_c \varepsilon^{ep}$ (h_c: material constant) are depicted in Fig. 7.14a and those for the nonlinear isotropic hardening in Eq. (6.56) are shown in Fig. 7.14b. The two levels of axial strain increment $d\varepsilon_a = 0.0006$ and 0.0055 are input in the numerical calculations. Here, any special stress controlling algorithm to pull it back to the yield surface is not introduced. The material parameters are chosen as follows:

Material constants:

$$
\begin{aligned}
&\text{Youg's modulus: } E = 100000\text{MPa},\\
&\text{Hardening}\begin{cases} \text{Linear isotropic: } h_c = 7000\text{MPa} \\ \text{Nonlinar isotropic: } h_1 = 0.8,\ h_2 = 50, \end{cases}\\
&\text{Evolution of normal - yield ratio : } u = 200.
\end{aligned}
$$

Initial values:

$$
\begin{aligned}
&\text{Hardening function: } F_0 = 500\text{MPa},\\
&\text{Stress : } \boldsymbol{\sigma}_0 = \mathbf{O}\text{MPa}
\end{aligned}
$$

The nonsmooth curves bent at the yield stress are expressed by the conventional model. Moreover, the stress deviates from the exact curve of conventional elasto-plasticity. The deviation becomes large with the increases in the nonlinearity of hardening and in the increase of input strain increment. On the other hand, the stress is automatically attracted to the normal-yield surface in the subloading surface model even for the quite large strain increment $d\varepsilon_a = 0.0055\,(0.55\%)$. The zigzag lines tracing the exact curve are calculated such that the stress rises up when the it lies below the normal-yield surface but it drops down immediately if it goes

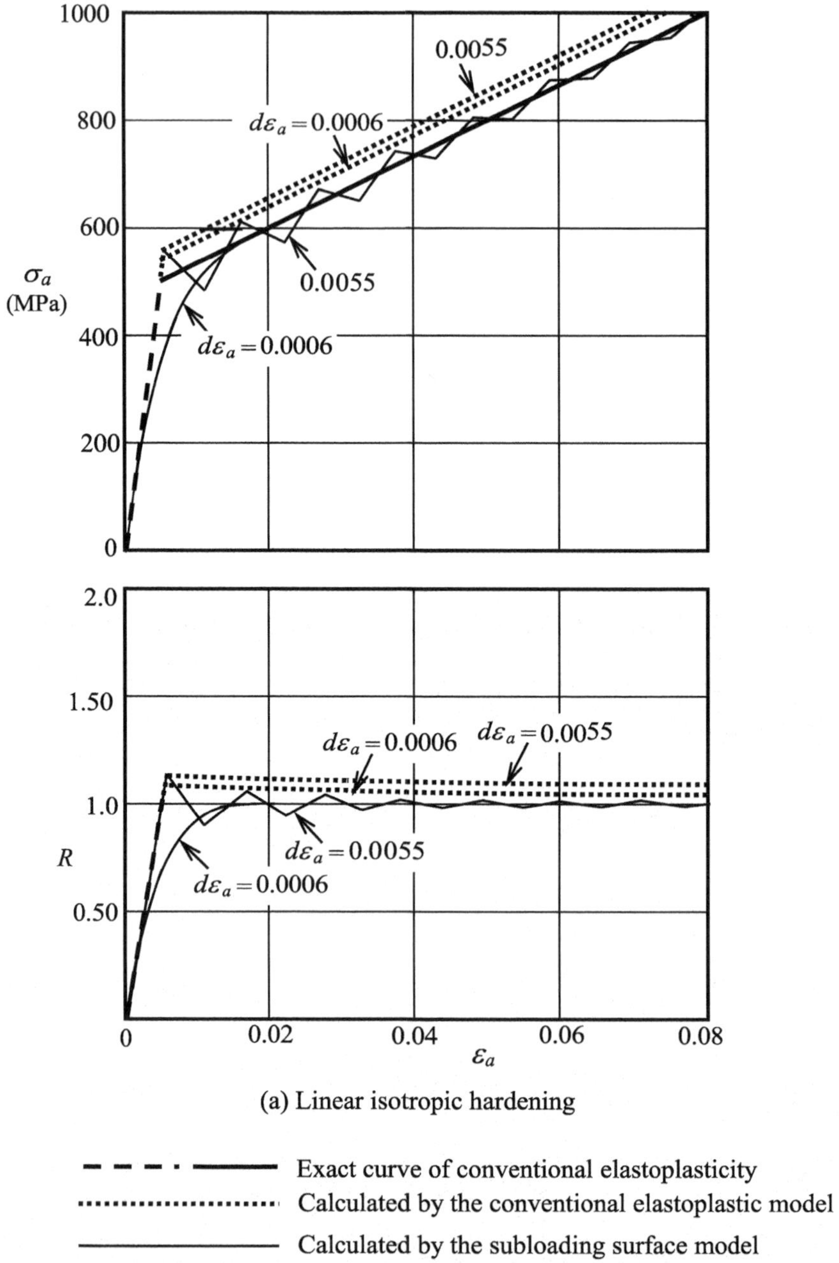

Fig. 7.14 Numerical accuracies of the conventional elastoplastic and the subloading surface model: Uniaxial loading behavior of Mises material with isotropic hardening. **a** Linear isotropic hardening **b** nonlinear isotropic hardening (continued)

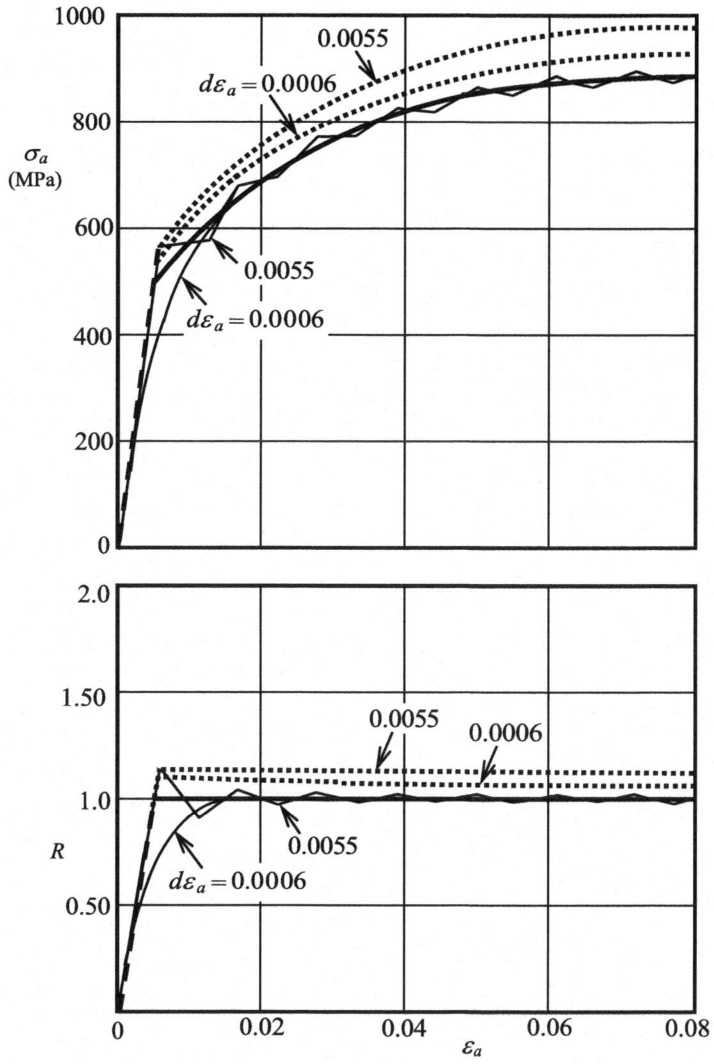

(b) Nonlinear isotropic hardening

Fig. 7.14 (continued)

over the normal-yield surface, obeying the evolution rule of normal-yield ratio in Eq. (7.9) with Eq. (7.12), i.e. $\dot{R} > 0$ for $R < 1$ and $\dot{R} < 0$ for $R > 1$. The plastic modulus $\bar{M}^p$ lowers than that in the conventional one and further it can be negative at the over normal-yield state $R > 1$ leading to $U < 0$, while $\mathbf{n} : \overset{\circ}{\boldsymbol{\sigma}} < 0$ (subloading softening defined in Eq. (7.38)) holds for $\bar{M}^p < 0$ because of $\dot{\bar{\lambda}} > 0$ as known from Eqs. (7.29–7.31). The amplitude of zigzag decreases gradually in the monotonic loading process, while, needless to say, the amplitude is smaller for a smaller input increment of strain. Eventually, the subloading surface model posseses the distinguished high ability for numerical calculation as verified also quantitatively in these concrete examples, which has not been attained in any other elastoplastic constitutive equations including the multi, the two, the infinite, superposed kinematic hardening and the bounding surface models assuming a purely-elastic domain as will be described in Chap. 8.

7.5 On Bounding Surface and Bounding Surface Model

The terms *bounding surface* and *bounding surface model* are widely used for models falling within the framework of unconventional plasticity describing the plastic strain rate induced by the rate of stress inside the yield surface. They were named by Y. F. Dafalias, who also coined the terms *plastic spin* (Dafalias 1985a) and *hypoplasticity* (Dafalias 1986). The only concrete model proposed by Dafalias as the bounding surface model is the two-surface model (Dafalias and Popov 1975), in which a small subyield surface is assumed inside the yield surface. The small subyield surface encloses the purely elastic domain and translates maintaining the size which maintains constant ratio to the size of the bounding surface (Dafalias and Popov 1977). On the other hand, the basic structure of the *bounding surface model with a radial mapping* used later by Dafalias and Herrmann (1980) falls within the framework of the subloading surface model, as has been recognized by Dafalias himself in his statement "*It appears that the first time a radial mapping formulation was proposed, it was in reference to granular materials by Hashiguchi and Ueno (1977)*" which is the original sentence in Dafalias (1986, p. 980).

However, note the following facts.

(1) The bounding surface is no more than the yield surface that has been assumed historically in the field of plasticity. The term yield surface has a clear physical meaning that the plastic deformation begins when stress reaches it; it also has the geometrical meaning that the stress cannot go out from it in the quasi-static deformation process. In contrast, the phrase bounding surface has only a geometrical meaning but has no physical meaning.

(2) The yield surface always exists. However, the stress goes over the yield surface in the deformation process at a high rate as represented by the *overstress model* describing a viscoplastic deformation. Therefore, no surface exists which bounds the stress except for the quasi-static deformation process. Consequently, the phrase "bounding surface" has no generality.

(3) The term bounding surface model induces the confusion as if all unconventional plasticity models inheriting the yield surface belong to the bounding surface model. Krieg (1975) uses the term *limit surface* in his two surface

model, Mroz (1967) uses *outmost surface* in his *multi surface model*, and Hashiguchi (1989) uses *normal-yield surface* in his *subloading surface model* instead of yield surface. However, they use these words only in a limited sense for naming elements in their models: they never use these words as names of their proposed models such as the *limit surface model*, the *outmost surface model*, or the *normal-yield surface model*. The term bounding surface should be used only for the two surface model of Dafalias and Popov (1975) in order to avoid the confusion.

Furthermore, Dafalias uses the phrase *bounding surface model with a radial mapping* (Dafalias and Herrmann 1980). Nevertheless, it possesses physical and mathematical structure which differs from the two-surface model but it is based on the identical basic structure to the subloading surface model proposed in 1977 three years earlier than 1980 when Dafalias began to write the articles on the bounding surface model with a radial mapping. Furthermore, it includes various the immaturity and the impertinence in the explicit formulations as described below.

In the bounding surface model with a radial mapping, the following ratio is adopted as the measure to describe the degree of approaching the yield (bounding) surface.

$$b \equiv ||\boldsymbol{\sigma}_y||/||\boldsymbol{\sigma}||\ (1 \leq b \leq \infty) \tag{7.39}$$

which is the ratio of the magnitude of conjugate stress $\boldsymbol{\sigma}_y$ on the yield surface to the magnitude of current stress $\boldsymbol{\sigma}$. Then, the plastic modulus M^p in the plastic strain rate of the conventional plasticity, i.e. Eq. (6.41) is modified merely by the interpolation method without incorporation of the consistency condition as

$$M^p \rightarrow M^p + \hat{H}\langle\frac{b-1}{b_e - b}\rangle \left(= \begin{cases} \rightarrow \infty & \text{for } b \geq b_e \\ M^p & \text{for } b = 1 \end{cases} \right) \tag{7.40}$$

where $\hat{H}$ is the function of stress and internal variables and b_e is the value of the variable b at the elastic limit.

Here, note the following facts.

(i) The variable $b(\infty \geq b \geq 1)$ is merely the inverse of normal-yield ratio $R(0 \leq R \leq 1)$ in the subloading surface model, whilst $b \rightarrow \infty$ and $b = 1$ correspond to $R = 0$ and $R = 1$, respectively.
(ii) The plastic modulus M^p in the bounding surface model with a radial mapping is given by the interpolation method between the stress in an elastic limit and the stress on the yield (bounding) surface, where no consistency condition is introduced as can be confirmed from the statement "*No consistency condition* $\dot{f} = 0$ *is required for stress points inside* $F = 0$, *since now* $f = 0$ *is always defined at any* σ_{ij}." (Dafalias 1986, p. 978), whereas the consistency condition for the subloading surface is introduced into the subloading surface model. Various equations other than Eq. (7.40) can be

assumed for the plastic modulus if an easy-going interpolation method is adopted. In fact, Eq. (7.40) differs substantially from the plastic modulus of Eq. (7.30) in the subloading surface model which is derived rigorously from the consistency condition formulated based on the assumption that the normal-yield ratio approaches unity in the plastic loading process.

(iii) Therefore, it is not guaranteed that the stress approaches the yield surface in the plastic loading process. On the other hand, the subloading surface model possesses a stress controlling function to attract the stress to the yield surface in the plastic loading process even if the stress goes out from the yield surface in the numerical calculation by the finite strain or stress increments.

(iv) A formulation for describing cyclic loading behavior has not been given for the bounding surface model with a radial mapping. On the other hand, it was attained in the subloading surface model as the *extended subloading surface model* (Hashiguchi 1989) by making the similarity-center of the normal-yield and the subloading surfaces move with the plastic strain rate as will be described in detail in the next chapter.

Eventually, it can be concluded for the bounding surface model with radial mapping as follows:

(I) The bounding surface is substantially the synonym of the yield surface although it does not express any physical meaning. Therefore, the term: bounding surface model would cause confusion as if all models adopting the yield surface belong to this model, while in fact it is insisted by Dafalias that *the model with "stress reversal surfaces" (infinite surface model of* Mroz et al. 1981) *proposed for soils can be classified as a radial mapping model* (Dafalias 1986, p. 981) in addition to the impertinent assessment on the subloading surface model. It is desirable to make effort for concrete formulation of pertinent model rather than only coining new terms. Eventually, the term bounding surface should be used only for the two surface model formulated by Dafalias himself.

(II) The bounding surface model with radial mapping falls within the framework of the subloading surface model but it is not formulated rationally, whereas the rigorous formulations including the description of cyclic loading behavior have been given by the subloading surface model.

Eventually, the ones using the bounding surface with radial mapping should abandon its use and instead they should notice the subloading surface model for rigorous deformation analyses and sound development of plasticity.

7.6 Incorporation of Kinematic Hardening

The subloading surface based on the yield surface in Eq. (6.85) with the kinematic hardening is described as

$$f(\hat{\boldsymbol{\sigma}}) = RF(H) \tag{7.41}$$

The material-time derivative of Eq. (7.41) leads to

$$\frac{\partial f(\hat{\boldsymbol{\sigma}})}{\partial \hat{\boldsymbol{\sigma}}} : \overset{\circ}{\boldsymbol{\sigma}} - \frac{\partial f(\hat{\boldsymbol{\sigma}})}{\partial \hat{\boldsymbol{\sigma}}} : \overset{\circ}{\boldsymbol{\alpha}} - RF' \dot{H} - \dot{R} F = 0 \tag{7.42}$$

Substituting the associated flow rule

$$\mathbf{d}^p = \dot{\bar{\lambda}} \hat{\mathbf{n}} \ (\dot{\bar{\lambda}} = ||\mathbf{d}^p|| \geq 0) \tag{7.43}$$

Eq. (7.42) is rewritten substituting Eq. (6.88) as

$$\frac{\partial f(\hat{\boldsymbol{\sigma}})}{\partial \hat{\boldsymbol{\sigma}}} : \overset{\circ}{\boldsymbol{\sigma}} - \frac{\partial f(\hat{\boldsymbol{\sigma}})}{\partial \hat{\boldsymbol{\sigma}}} : \dot{\bar{\lambda}} \mathbf{f}_{kn}(\boldsymbol{\sigma}, \boldsymbol{\alpha}, F; \hat{\mathbf{n}}') - RF' \dot{\bar{\lambda}} f_{Hn}(\boldsymbol{\sigma}, H; \hat{\mathbf{n}}) - U \dot{\bar{\lambda}} F = 0 \tag{7.44}$$

Noting

$$\left.\begin{aligned} &\frac{\partial f(\hat{\boldsymbol{\sigma}})}{\partial \hat{\boldsymbol{\sigma}}} : \hat{\boldsymbol{\sigma}} = f(\hat{\boldsymbol{\sigma}}) = RF \\ &\frac{\partial f(\hat{\boldsymbol{\sigma}})}{\partial \hat{\boldsymbol{\sigma}}} = ||\frac{\partial f(\hat{\boldsymbol{\sigma}})}{\partial \hat{\boldsymbol{\sigma}}}|| \hat{\mathbf{n}} \\ &1/||\frac{\partial f(\hat{\boldsymbol{\sigma}})}{\partial \hat{\boldsymbol{\sigma}}}|| = \frac{\dfrac{\partial f(\hat{\boldsymbol{\sigma}})}{\partial \hat{\boldsymbol{\sigma}}} : \hat{\boldsymbol{\sigma}}}{f(\hat{\boldsymbol{\sigma}})} / ||\frac{\partial f(\hat{\boldsymbol{\sigma}})}{\partial \hat{\boldsymbol{\sigma}}}|| = \frac{\hat{\mathbf{n}} : \hat{\boldsymbol{\sigma}}}{RF} \end{aligned}\right\} \tag{7.45}$$

Equation (7.44) is rewritten as

$$\hat{\mathbf{n}} : \overset{\circ}{\boldsymbol{\sigma}} - \hat{\mathbf{n}} : \dot{\bar{\lambda}} \mathbf{f}_{kn}(\boldsymbol{\sigma}, \boldsymbol{\alpha}, F; \hat{\mathbf{n}}') - \hat{\mathbf{n}} : \hat{\boldsymbol{\sigma}} \left(\frac{F'}{F} \dot{\bar{\lambda}} f_{Hn}(\boldsymbol{\sigma}, H; \hat{\mathbf{n}}) + \frac{U}{R} \dot{\bar{\lambda}} \right) = 0 \tag{7.46}$$

from which one has

$$\dot{\bar{\lambda}} = \frac{\hat{\mathbf{n}} : \overset{\circ}{\boldsymbol{\sigma}}}{\bar{M}^p}, \ \mathbf{d}^p = \frac{\hat{\mathbf{n}} : \overset{\circ}{\boldsymbol{\sigma}}}{\bar{M}^p} \hat{\mathbf{n}} \tag{7.47}$$

where

$$\boxed{\bar{M}^p \equiv \hat{\mathbf{n}} : \left[\left(\frac{F'}{F} f_{Hn}(\boldsymbol{\sigma}, H; \hat{\mathbf{n}}) + \frac{U}{R} \right) \hat{\boldsymbol{\sigma}} + \mathbf{f}_{kn}(\boldsymbol{\sigma}, \boldsymbol{\alpha}, F; \hat{\mathbf{n}}') \right]} \tag{7.48}$$

The loading criterion is given by

$$\left. \begin{array}{l} \mathbf{d}^p \neq \mathbf{O} : \hat{\mathbf{n}} : \mathbf{E} : \mathbf{d} > 0 \\ \mathbf{d}^p = \mathbf{O} : \text{ otherwise} \end{array} \right\} \tag{7.49}$$

7.7 Incorporation of Tangential-Inelastic Strain Rate

As presented in Eqs. (6.43) and (6.95), the inelastic strain rate in the traditional constitutive equation has the following limitations.

(i) The inelastic strain rate depends solely on the stress rate component normal to the yield surface, called the *normal stress rate*, but is independent of the component tangential to the yield surface, called the *tangential stress rate*, since it is derived merely based on the consistency condition.
(ii) The direction of inelastic strain rate is determined solely by the current state of stress and internal variables but it is independent of the stress rate.
(iii) The principal directions of inelastic strain rate tensor coincide with those of stress tensor, exhibiting the so-called *coaxiality*, in the case of isotropy in which the direction of plastic strain rate depends only on the direction of stress by the fact described in Sect. 1.12.

On the other hand, it has been verified by experiments that an inelastic strain rate induced by the deviatoric part of the tangential stress rate, called the *deviatoric tangential stress rate*, influences considerably on a deformation in the non-proportional loading process deviating from the proportional loading path normal to the yield surface, which is called the *tangential inelastic strain rate*. Here, the spherical part of the tangential stress rate does not induce an inelastic strain rate, as Rudnicki and Rice (1975) verified based on the fissure model. In addition, the tangential inelastic strain rate is induced considerably in the plastic instability phenomena with the strain localization induced by the generation of the shear band and it influences on the macroscopic deformation and strength characteristics. To remedy these insufficiencies of the traditional plastic constitutive equation, various models have been proposed to date as follows:

(1) *Intersection of plural yield surfaces*: Various models assuming the intersection of plural yield surfaces have been proposed (Batdorf and Budiansky 1949; Koiter 1953; Bland 1957; Mandel 1965; Hill 1966; Sewell 1973, 1974). The Koiter's (1953) model was adopted by Sewell (1973, 1974), but it is indicated that the applicability of the model is limited to the inception of uniaxial loading. Models in this category cannot describe the latent hardening pertinently and are

not readily applicable to general loading processes (cf. Christoffersen and Hutchinson (1979)).

(2) *Corner theory*: The singularity of outward-normal of the yield surface is introduced by assuming the conical corner or vertex at the stress point on the yield surface. Therefore, the direction of plastic strain rate can take a wide range surrounded by the outward-normal of the yield surface (Christoffersen and Hutchinson 1979; Ito 1979; Gotoh 1985; Goya and Ito 1991; Petryk and Thermann 1997). There exist the two kinds of models: One kind is based on the assumption of an imaginary infinitesimal vertex and the other subsumes a finite projecting cone. The evolution rule of the cone cannot be formulated and the reloading from the cone surface after partial unloading cannot be described pertinently in the latter models. It was described by Hecker (1972) and Ikegami (1979) that the yield surface projects towards the loading direction generally but the formation of the so-called vertex is doubtful.

(3) *Hypoplasticity*: This term was first used by Dafalias (1986) in the analogy to the term hypoelasticity introduced by Truesdell (1955) described in Sect. 5.4. Models in this category are classified into the two kind of models in which the direction of plastic strain rate depends on the direction of the stress rate $\overset{\circ}{\boldsymbol{\sigma}}/\|\overset{\circ}{\boldsymbol{\sigma}}\|$ (Mroz 1966; Dafalias and Popov 1977; Hughes and Shakib 1986; Wang et al. 1990; Hashiguchi 1993a) and the models in which the direction of the plastic strain rate depends on the direction of strain rate $\mathbf{d}/\|\mathbf{d}\|$ (Hill 1959; Simo 1987; Hashiguchi 1997). The singularity in the field of direction of plastic strain rate is introduced in the algebraic ways into these models, although it is done geometrically in the models described in (1) and (2). However, the magnitude of the plastic strain rate is derived from the consistency condition. Therefore, the plastic strain rate diminishes when the stress rate is directed tangentially to the yield surface, as in the traditional constitutive equations without the vertex.

The constitutive equations described in (1)–(3) possess the following problems.

(i) A formulation of pertinent model which fulfills the consistency condition and is applicable to the general loading process is difficult.

(ii) The stress rate vs. strain rate relation becomes nonlinear. Therefore, the inverse expression cannot be derived, which renders deformation analysis as difficult.

Differently from the above-mentioned models, the following linear relation between the stress rate vs. strain rate with the tangential-inelastic strain rate, called the J_2- *deformation theory*, has been formulated by Budiansky (1959) and later Rudnicki and Rice (1975) by extending Eq. (6.59) with the isotropic Mises yield condition as follows:

$$\mathbf{d} = \mathbf{E}^{-1} : \overset{\circ}{\boldsymbol{\sigma}} + \frac{3}{2}\frac{1}{F'}\frac{\dot{\sigma}^{eq}}{\sigma^{eq}}\boldsymbol{\sigma}' + \phi(\sigma^{eq})\left[\overset{\circ}{\boldsymbol{\sigma}}{}' - \left(\frac{\boldsymbol{\sigma}'}{\|\boldsymbol{\sigma}'\|} : \overset{\circ}{\boldsymbol{\sigma}}{}'\right)\frac{\boldsymbol{\sigma}'}{\|\boldsymbol{\sigma}'\|}\right] \tag{7.50}$$

which can be rewritten as

$$
\begin{aligned}
\mathbf{d} &= \mathbf{E}^{-1} : \mathring{\boldsymbol{\sigma}} + \frac{3}{2}\frac{1}{F'}\frac{\overset{\bullet}{\sigma}^{eq}}{\sigma^{eq}}\boldsymbol{\sigma}' + \phi(\sigma^{eq})\left(\mathring{\boldsymbol{\sigma}}' - \sqrt{2/3}\overset{\bullet}{\sigma}^{eq}\frac{\boldsymbol{\sigma}'}{\sqrt{2/3}\sigma^{eq}}\right) \\
&= \mathbf{E}^{-1} : \mathring{\boldsymbol{\sigma}} + \left(\frac{3}{2}\frac{1}{F'} - \phi(\sigma^{eq})\right)\frac{\overset{\bullet}{\sigma}^{eq}}{\sigma^{eq}}\boldsymbol{\sigma}' + \phi(\sigma^{eq})\mathring{\boldsymbol{\sigma}}'
\end{aligned}
\tag{7.51}
$$

where the rate-linearity is retained.

On the other hand, *Hencky's deformation theory* (Hencky 1924) is described as

$$
\boldsymbol{\varepsilon} = \mathbf{E}^{-1} : \sigma + \phi(\sigma^{eq})\boldsymbol{\sigma}' \tag{7.52}
$$

The corotational time-derivative of Eq. (7.52) leads to

$$
\mathring{\boldsymbol{\varepsilon}} = \mathbf{E}^{-1} : \mathring{\boldsymbol{\sigma}} + \phi'(\sigma^{eq})\overset{\bullet}{\sigma}^{eq}\boldsymbol{\sigma}' + \phi(\sigma^{eq})\mathring{\boldsymbol{\sigma}}' \tag{7.53}
$$

Comparing Eq. (7.51) with Eq. (7.53), choosing $F(\sigma^{eq})$ so as to fulfill

$$
F'(\varepsilon^{eqp}(\sigma^{eq})) = \frac{3}{2}\frac{1}{\phi(\sigma^{eq}) + \phi'(\sigma^{eq})\sigma^{eq}} \tag{7.54}
$$

and regarding $\mathbf{d}$ as $\mathring{\boldsymbol{\varepsilon}}$, it is known that the J_2- deformation theory coincides with Hencky's deformation theory (7.52). However, it possesses crucial limitations as described in below.

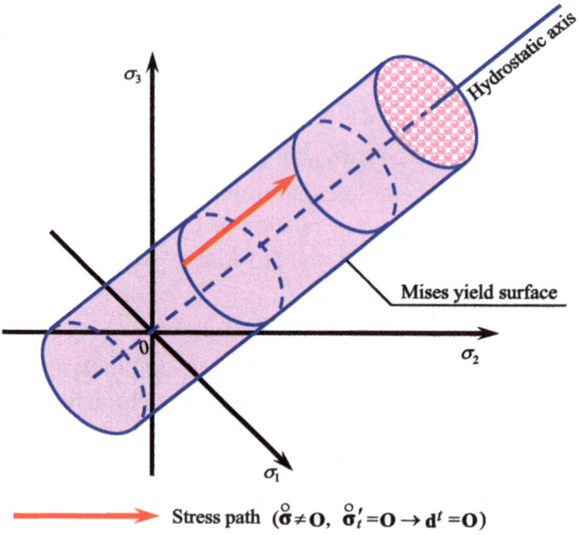

Fig. 7.15 Example showing the fact that tangential-inelastic strain rate is not induced by spherical stress rate since inelastic volumetric change is not induced in metals

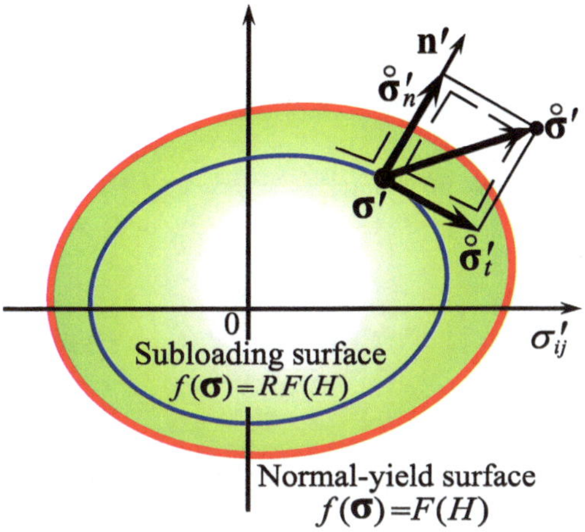

Fig. 7.16 Normal and tangential stress rates for subloading surface model in deviatoric stress plane

In what follows, let the tangential inelastic strain rate be incorporated into the above-mentioned subloading surface model in the following (Hashiguchi 1998, 2005; Hashiguchi and Tsutsumi 2003; Hashiguchi and Protasov 2004; Khojastepour and Hashiguchi 2004a, b; Khojastehpour et al. 2006).

Inelastic strain rate is induced even by the deviatoric stress rate component tangential to the loading surface in real material behavior, while it would not be induced by spherical stress rate as would be inferred from the example for the fact that inelastic volumetric change would not be induced in metals (see Fig. 7.15). However, this fact has been ignored in the traditional plasticity described in the preceding sections in which the inelastic strain rate is given only by the plastic strain rate derived from the consistency condition of the subloading surface so that it is depends only on the stress rate component normal to the subloading surface.

In order to describe the inelastic strain rate induced by the deviatoric stress rate component tangential to the subloading surface, assume that the strain rate is additively composed of the elastic strain rate, the plastic strain rate and the *tangential-inelastic strain rate* $\mathbf{d}^t$ as follows (Hashiguchi, 1998, 2013b, 2016):

$$\mathbf{d} = \mathbf{d}^e + \mathbf{d}^p + \mathbf{d}^t \tag{7.55}$$

Further, assume that the tangential-inelastic strain rate $\mathbf{d}^t$ is induced by the tangential component of the stress rate $\overset{\circ}{\boldsymbol{\sigma}}$ to the subloading surface in the deviatoric stress space, which is denoted by $\overset{\circ}{\boldsymbol{\sigma}}{}'_t$ (Fig. 7.16) defined by

$$\left.\begin{aligned} \overset{\circ}{\boldsymbol{\sigma}}{}'_t &\equiv \boldsymbol{\mathfrak{T}}'_t : \overset{\circ}{\boldsymbol{\sigma}} = \overset{\circ}{\boldsymbol{\sigma}}{}' - \overset{\circ}{\boldsymbol{\sigma}}{}'_n (\mathbf{n} : \overset{\circ}{\boldsymbol{\sigma}}{}'_t = \mathbf{n}' : \overset{\circ}{\boldsymbol{\sigma}}{}'_t = 0) \\ \overset{\circ}{\boldsymbol{\sigma}}{}'_n &\equiv (\mathbf{n}' : \overset{\circ}{\boldsymbol{\sigma}})\mathbf{n}' = (\mathbf{n}' \otimes \mathbf{n}') : \overset{\circ}{\boldsymbol{\sigma}} \end{aligned}\right\} \tag{7.56}$$

$$\mathbf{n}' \equiv \frac{\mathbf{n}'}{\|\mathbf{n}'\|} = \left(\frac{\partial f(\boldsymbol{\sigma})}{\partial \boldsymbol{\sigma}}\right)' \Big/ \left\|\left(\frac{\partial f(\boldsymbol{\sigma})}{\partial \boldsymbol{\sigma}}\right)'\right\| \left(\neq \mathbf{n}' = \left(\frac{\partial f(\boldsymbol{\sigma})}{\partial \boldsymbol{\sigma}}\right)' \Big/ \left\|\frac{\partial f(\boldsymbol{\sigma})}{\partial \boldsymbol{\sigma}}\right\|\right) (\|\mathbf{n}'\| = 1) \tag{7.57}$$

$$\boldsymbol{\mathfrak{T}}'_t \equiv \mathcal{I}' - \mathbf{n}' \otimes \mathbf{n}', \ \mathfrak{T}'_{ijkl} \equiv \mathcal{I}'_{ijkl} - \mathrm{n}'_{ij}\mathrm{n}'_{kl} \tag{7.58}$$

fulfilling

$$\mathbf{n}'_t \equiv \boldsymbol{\mathfrak{T}}'_t : \mathbf{n} = \mathbf{n}' - (\mathbf{n}' : \mathbf{n})\mathbf{n}' = \mathbf{O} \tag{7.59}$$

The fourth-order tensor $\mathcal{I}'$ is the *deviatoric projection tensor* in Eq. (1.146). The fourth-order tensor $\boldsymbol{\mathfrak{T}}'_t$ is the *deviatoric-tangential projection tensor* which transforms an arbitrary second-order tensor to the tangential part to the subloading surface in the deviatoric stress space and the second-order tensor subjected to this projection is designated by $(\ )'_t$, i.e. $\mathbf{t}'_t \equiv \boldsymbol{\mathfrak{T}}'_t : \mathbf{t}$ leading further to $\boldsymbol{\mathfrak{T}}'_t : \mathbf{t}'_t = \mathbf{t}'_t$. Then, $\overset{\circ}{\boldsymbol{\sigma}}{}'_t$ is the deviatoric-tangential projection tensor of the stress rate $\overset{\circ}{\boldsymbol{\sigma}}$, which is called the *deviatoric-tangential stress rate*.

Now, assume that the tangential-inelastic strain rate $\mathbf{d}^t$ is related linearly to the tangential-deviatoric stress rate $\overset{\circ}{\boldsymbol{\sigma}}{}'_t$ by the extended hypoelastic relation (Truesdell 1955) in the normal-yield state ($R = 1$) as follows:

$$\mathbf{d}^t = \mathbf{E}^{-1} : \overset{\circ}{\boldsymbol{\sigma}}{}'_t \tag{7.60}$$

where $\mathbf{E}$ is the fourth-order tensor which is a function of stress and internal variables in general. Let Eq. (7.60) be extended for the sub-yield state as follows:

$$\mathbf{d}^t = T(R)\mathbf{E}^{-1} : \overset{\circ}{\boldsymbol{\sigma}}{}'_t \tag{7.61}$$

In this equation $T(R)$ is the monotonically-increasing function of R given by

$$\boxed{T(R) = \tilde{c}R^{\tilde{n}}} \tag{7.62}$$

or

$$T(R) = \tilde{c}\left\langle \frac{R - \tilde{R}_e}{1 - \tilde{R}_e} \right\rangle^{\tilde{n}} \left(\begin{cases} = 0 \ \text{for} \ R \leq \tilde{R}_e \\ > 0 \ \text{for} \ \tilde{R}_e < R < 1 \\ = \tilde{c} \ \text{for} \ R = 1 \end{cases} \right) \tag{7.63}$$

where $\tilde{c}$, $\tilde{n}(\geq 1)$ and $\tilde{R}_e$ (<1) are the material constants. The tangential-inelastic strain rate is induced increasingly as the stress approaches the normal-yield surface, always fulfilling the continuity and the smoothness conditions (Hashiguchi 1993a,b, 1997, 2000). On the other hand, if the tangential-inelastic strain rate is incorporated into the plasticity model assuming the yield surface enclosing a purely-elastic domain, both of the continuity and the smoothness conditions are violated, since the tangential-inelastic strain rate is induced suddenly at the moment when the stress reaches the yield surface.

Adding the tangential-inelastic strain rate in Eq. (7.62) to Eq. (7.32), the strain rate is given by

$$\begin{aligned}\mathbf{d} &= \mathbf{E}^{-1}:\overset{\circ}{\boldsymbol{\sigma}} + \frac{\mathbf{n}:\overset{\circ}{\boldsymbol{\sigma}}}{\bar{M}^p}\mathbf{n} + T(R)\mathbf{E}^{-1}:\overset{\circ}{\boldsymbol{\sigma}}{}'_t \\ &= \left(\mathbf{E}^{-1} + \frac{\mathbf{n}\otimes\mathbf{n}}{\bar{M}^p} + T(R)\mathbf{E}^{-1}:\mathfrak{T}'_t\right):\overset{\circ}{\boldsymbol{\sigma}}\end{aligned} \tag{7.64}$$

In what follows, we assume the elastic modulus tensor $\mathbf{E}$ fulfilling

$$\mathfrak{T}'_t:\mathbf{E}:\mathbf{n} = \mathbf{O} \tag{7.65}$$

which holds in the Hooke's law in (Eq. 5.35) for example. Taking account of Eq. (7.65), it follows from Eq. (7.64) that

$$\mathfrak{T}'_t:\mathbf{E}:\mathbf{d} = (1+T(R))\overset{\circ}{\boldsymbol{\sigma}}{}'_t \tag{7.66}$$

leading to

$$\overset{\circ}{\boldsymbol{\sigma}}{}'_t = \frac{1}{1+T(R)}\mathfrak{T}'_t:\mathbf{E}:\mathbf{d} \tag{7.67}$$

Substituting Eq. (7.67) into Eq. (7.61), one obtains

$$\mathbf{d}^t - \frac{T(R)}{1+T(R)}\mathbf{E}^{-1}\mathfrak{T}'_t:\mathbf{E}:\mathbf{d} \tag{7.68}$$

The expression in Eq. (7.33) itself for the plastic multiplier in terms of strain rate is obtained from Eq. (7.64), noting Eq. (7.59). Then, the stress rate is given by substituting Eq. (7.55) with Eq. (7.34) and Eq. (7.68) into the relation $\overset{\circ}{\boldsymbol{\sigma}} = \mathbf{E}:\mathbf{d}^e$ in Eq. (6.29) as follows:

$$\begin{aligned}\overset{\circ}{\boldsymbol{\sigma}} &= \mathbf{E}:(\mathbf{d}-\mathbf{d}^p-\mathbf{d}^t) = \mathbf{E}:\mathbf{d} - \frac{\mathbf{n}:\mathbf{E}:\mathbf{d}}{\bar{M}^p+\mathbf{n}:\mathbf{E}:\mathbf{n}}\mathbf{E}:\mathbf{n} - \frac{T(R)}{1+T(R)}\mathfrak{T}'_t:\mathbf{E}:\mathbf{d} \\ &= \left(\mathbf{E} - \frac{\mathbf{E}:\mathbf{n}\otimes\mathbf{n}:\mathbf{E}}{\bar{M}^p+\mathbf{n}:\mathbf{E}:\mathbf{n}} - \frac{T(R)}{1+T(R)}\mathfrak{T}'_t:\mathbf{E}\right):\mathbf{d}\end{aligned} \tag{7.69}$$

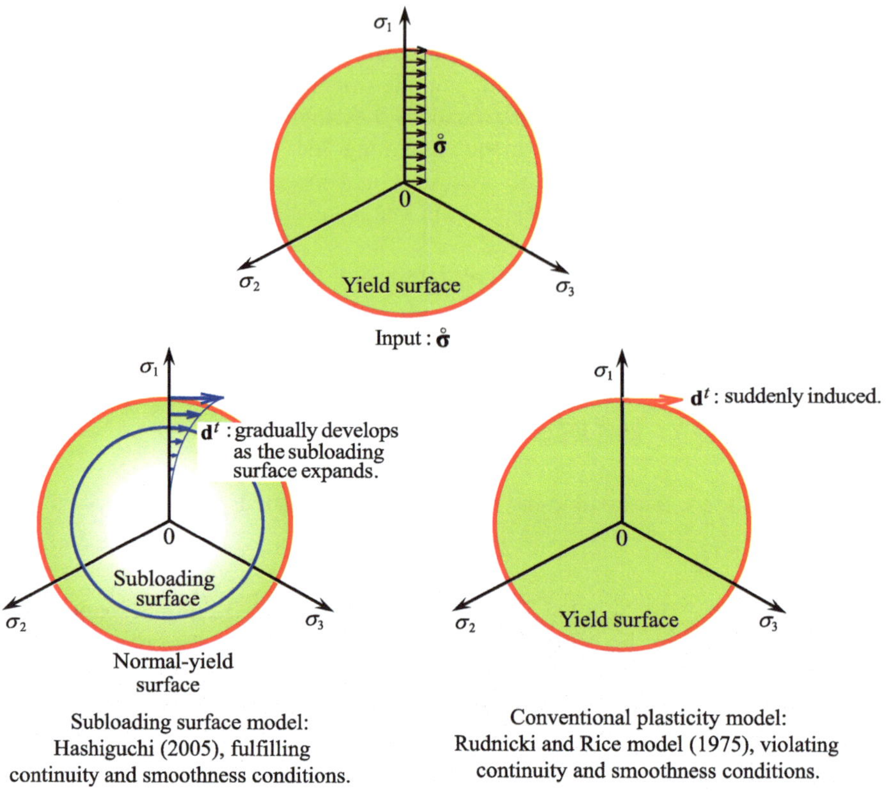

Fig. 7.17 Incorporation of tangential inelastic strain rate illustrated for von Mises yield surface

Equations (7.64) and (7.69) are given for the Hooke's law in Eq. (5.35) as follows:

$$\boxed{\mathbf{d} = \left(\mathbf{E}^{-1} + \frac{\mathbf{n} \otimes \mathbf{n}}{\bar{M}^p} + \frac{T(R)}{2G}\mathfrak{T}'_t\right) : \overset{\circ}{\boldsymbol{\sigma}}} \tag{7.70}$$

$$\boxed{\overset{\circ}{\boldsymbol{\sigma}} = \left(\mathbf{E} - \frac{\mathbf{E} : \mathbf{n} \otimes \mathbf{n} : \mathbf{E}}{\bar{M}^p + \mathbf{n} : \mathbf{E} : \mathbf{n}} - 2G\frac{T(R)}{1 + T(R)}\mathfrak{T}'_t\right) : \mathbf{d}} \tag{7.71}$$

Here, it is known that the bulk modulus K is irrelevant to the tangential inelastic strain rate which is induced only by the deviatoric part of stress rate.

The loading criterion is given by the equation identical to that without the tangential-inelastic strain rate, since the tangential-inelastic strain rate is always induced by the tangential stress rate.

The tangential-inelastic strain rate $\mathbf{d}^t$ develops gradually as the current stress approaches the normal-yield surface, i.e. the subloading surface expands fulfilling the continuity and the smoothness condition in the subloading surface model as shown in Fig. 7.17 for the isotropic Mises material. The validity of Eq. (7.70) or (7.71) has been verified by Hashiguchi and Protasov (2004) for metals and Hashiguchi and Tsutsumi (2001, 2003, 2007) and Tsutsumi and Hashiguchi (2005) and for geomaterials. On the other hand, all models other than the subloading surface model violate the smoothness condition. Therefore, they violate also the continuity condition in Eq. (7.1) as illustrated for the J_2-deformation model of Rudnicki and Rice (1975) in Fig. 10. Note that the tangential-inelastic strain rate does not affect the yield surface and thus the consistency condition.

The subloading surface model has been applied to metals (Hashiguchi 1980, 1989; Hashiguchi and Yoshimaru 1995; Hashiguchi and Tsutsumi 2001; Hashiguchi and Protasov 2004; Khojastehpor et al. 2006; Tsutsumi et al. 2006; Hashiguchi et al. 2012) and soils (Hashiguchi and Ueno 1977; Hashiguchi 1978; Topolnicki 1990; Kohgo et al. 1993; Asaoka et al. 1997; Hashiguchi and Chen 1998; Chowdhury et al. 1999; Hashiguchi et al. 2002; Khojastehpor and Hashiguchi 2004a, b; Khojastehpor et al. 2006; Nakai and Hinokio 2004; Hashiguchi and Tsutsumi 2006; Hashiguchi and Mase 2007, 2011; Wongsaroj et al. 2007). Consequently, its capability has been verified widely.

References

Asaoka A, Nakano M, Noda T (1997) Soil-water coupled behaviour of heavily overconsolidated clay near/at critical state. Soils Found 37(1):13–28

Batdorf SB, Budiansky B (1949) A mathematical theory of plasticity based on the concept of slip. NACA TC1871, pp 1–31

Bland DR (1957) The associated flow rule of plasticity. J Mech Phys Solids 6:71–78

Budiansky B (1959) A reassessment of deformation theories of plasticity. J Appl Mech (ASME) 20:259–264

Chowdhury EQ, Nakai T, Tawada M, Yamada S (1999) A model for clay using modified stress under various loading conditions with the application of subloading concept. Soils Found 39 (6):103–116

Christoffersen J, Hutchinson JW (1979) A class of phenomenological corner theories of plasticity. J Mech Phys Solids 27:465–487

Dafalias YF (1985) The plastic spin. J Appl Mech (ASME) 52:865–871

Dafalias YF (1986) Bounding surface plasticity. I: mathematical foundation and hypoplasticity. J Eng Mech (ASCE) 112:966–987

Dafalias YF, Herrmann LR (1980) A bounding surface soil plasticity model. Proc. Int. Symp. Soils Cyclic Trans. Load, Swansea, pp 335–345

Dafalias YF, Popov EP (1975) A model of nonlinearly hardening materials for complex loading. Acta Mech 23:173–192

Dafalias YF, Popov EP (1977) Cyclic loading for materials with a vanishing elastic domain. Nucl Eng Des 41:293–302

Drucker DC (1988) Conventional and unconventional plastic response and representation. Appl Meek Rev (ASME) 41:151–167

Gotoh M (1985) A class of plastic constitutive equations with vertex effect. Int J Solids Struct 21:1101–1163

Goya M, Ito K (1991) An expression of elastic-plastic constitutive laws incorporating vertex formulation and kinematic hardening. J Appl Mech (ASME) 58:617–622

Hashiguchi K (1978) Plastic constitutive equations of granular materials. In: Cowin SC, Satake M (eds) Proceedings of US-Japan seminar on continuum mech stastical approach mechanics granular materials, Sendai, pp 321–329
Hashiguchi K (1980) Constitutive equations of elastoplastic materials with elastic-plastic transition. J Appl Mech (ASME) 47:266–272
Hashiguchi K (1989) Subloading surface model in unconventional plasticity. Int J Solids Struct 25:917–945
Hashiguchi K (1993a) Fundamental requirements and formulation of elastoplastic constitutive equations with tangential plasticity. Int J Plasticity 9:525–549
Hashiguchi K (1993b) Mechanical requirements and structures of cyclic plasticity models. Int J Plasticity 9:721–748
Hashiguchi K (1997) The extended flow rule in plasticity. Int J Plasticity 13:37–58
Hashiguchi K (1998) The tangential plasticity. Met Mater 4:652–656
Hashiguchi K (2000) Fundamentals in constitutive equation: continuity and smoothness conditions and loading criterion. Soils Found 40(3):155–161
Hashiguchi K (2005) Subloading surface model with tangential relaxation. Proc Int Symp Plasticity '05:259–261
Hashiguchi K (2013b) Elastoplasticity theory, lecture note in applied comptational mechanics, 2nd edn. Springer-Verlag, Heidelberg
Hashiguchi K (2016) Exact formulation of subloading surface model: unified constitutive law for irreversible mechanical phenomena in solids. Arch Compt Meth Eng 23:417–447
Hashiguchi K, Chen Z-P (1998) Elastoplastic constitutive equations of soils with the subloading surface and the rotational hardening. Int J Numer Anal Meth Geomech 22:197–227
Hashiguchi K, Mase T (2007) Extended yield condition of soils with tensile strength and rotational hardening. Int J Plasticity 23:1939–1956
Hashiguchi K, Mase T (2011) Physical interpretation and quantitative prediction of cyclic mobility by the subloading surface model. Jpn Geotech J 6:225–241 (in Japanese)
Hashiguchi K, Protasov A (2004) Localized necking analysis by the subloading surface model with tangential-strain rate and anisotropy. Int J Plasticity 20:1909–1930
Hashiguchi K, Tsutsumi S (2001) Elastoplastic constitutive equation with tangential stress rate effect. Int J Plasticity 17:117–145
Hashiguchi K, Tsutsumi S (2003) Shear band formation analysis in soils by the subloading surface model with tangential stress rate effect. Int J Plasticity 19:1651–1677
Hashiguchi K, Tsutsumi S (2006) Gradient plasticity with the tangential subloading surface model and the prediction of shear band thickness of granular materials. Int J Plasticity 22:767–797
Hashiguchi K, Ueno M (1977) Elastoplastic constitutive laws of granular materials, Constitutive equations of soils. In: Murayama S, Schofield AN (eds) Proceedings of 9th international conference soil mechanics and foundation engineering Spec. Ses. vol 9, Tokyo, JSSMFE, pp 73–82
Hashiguchi K, Yoshimaru T (1995) A generalized formulation of the concept of nonhardening region. Int J Plasticity 11:347–365
Hashiguchi K, Saitoh K, Okayasu T, Tsutsumi S (2002) Evaluation of typical conventional and unconventional plasticity models for prediction of softening behavior of soils. Geotechnique 52:561–573
Hashiguchi K, Ueno M, Ozaki T (2012) Elastoplastic model of metals with smooth elastic-plastic transition. Acta Mech 223:985–1013
Hecker SS (1972) Experimental investigation of corners in yield surface. Acta Mech 13:69–86
Hencky H (1924) Zur Theorie plastischer Deformationen und der hierdurch im Material herforgerufenen Nachspannungen. Z A M M 4:323–334
Hill R (1959) Some basic principles in the mechanics of solids without a natural time. J Mech Phys Solids 7:225–229
Hill R (1966) Generalized constitutive relations for incremental deformation of metal crystals. J Mech Phys Solids 14:95–102
Hughes TJR, Shakib F (1986) Pseudo-corner theory: A simple enhancement of J_2-flow theory for applications involving non-proportional loading. Eng Comput 3:116–120

Ikegami K (1979) Experimental plasticity on the anisotropy of metals. Proc Euromech Colloquium 115:201–242

Ito K (1979) New flow rule for elastic-plastic solids based on KBW model with a view to lowering the buckling stress of plates and shells. Tech Report Tohoku Univ 44:199–232

Khojastehpour M, Hashiguchi K (2004a) The plane strain bifurcation analysis of soils by the tangential-subloading surface model. Int J Solids Struct 41:5541–5563

Khojastehpour M, Hashiguchi K (2004b) Axisymmetric bifurcation analysis in soils by the tangential-subloading surface model. J Mech Phys Solids 52:2235–2262

Khojastehpour M, Murakami Y, Hashiguchi K (2006) Antisymmetric bifurcation in a circular cylinder with tangential plasticity. Mech Mater 38:1061–1071

Kohgo Y, Nakano M, Miyazaki T (1993) Verification of the generalized elastoplastic model for unsaturated soils. Soil Found 33(4):64–73

Koiter WT (1953) Stress-strain relations, uniqueness and variational theories for elastic-plastic materials with a singular yield surface. Quart Appl Math 11:350–354

Krieg RD (1975) A practical two surface plasticity theory. J Appl Mech (ASME) 42:641–646

Mandel J (1965) Generalisation de la theorie de plasticite de W.T. Koiter. Int J Solids Struct 1:273–295

Mroz Z (1966) On forms of constitutive laws for elastic-plastic solids. Arch Mech Stos 18:3–35

Mroz Z (1967) On the description of anisotropic workhardening. J Mech Phys Solids 15:163–175

Mroz Z, Norris VA, Zienkiewicz OC (1981) An anisotropic, critical state model for soils subject to cyclic loading. Geotechnique 31:451–469

Nakai T, Hinokio M (2004) A simple elastoplastic model for normally and over consolidated soils with unified material parameters. Soils Found 44(2):53–70

Petryk H (1991) On the second-order work in plasticity. Arch Mech 43:377–397

Petryk H (1997) Plastic instability: criteria and computational approaches. Arch Comput Approach Meth Eng 4:111–151

Petryk H, Thermann K (1997) A yield-vertex modification of two-surface models of metal plasticity Arch Mech 49:847–863

Prager W (1949) Recent development in the mathematical theory of plasticity. J Appl Mech (ASME) 20:235–241

Rudnicki JW, Rice JR (1975) Conditions for the localization of deformation in pressure-sensitive dilatant materials. J Mech Phys Solids 23:371–394

Sewell MJ (1973) A yield-surface comer lowers the buckling stress of an elastic-plastic plate under compression. J Mech Phys Solids 21:19–45

Sewell MJ (1974) A plastic flow at a yield vertex. J Mech Phys Solids 22:469–490

Simo JC (1987) A J_2-flow theory exhibiting a corner-like effect and suitable for large-scale computation. Comput Meth Appl Mech Eng 62:169–194

Topolnicki M (1990) An elasto-plastic suboading surface model for clay with isotropic and kinematic mixed hardening parameters. Soils Found 30(2):103–113

Truesdell C (1955) Hypo-elasticity. J Rational Mech Anal 4:83–133

Tsutsumi S, Hashiguchi K (2005) General non-proportional loading behavior of soils. Int J Plasticity 21:1941–1969

Tsutsumi S, Toyosada M, Hashiguchi K (2006) Extended subloading surface model incorporating elastic limit concept. In: Proceedings of plasticity *'06*, Halifax, pp 217–219

Wang Z-L, Dafalias YF, Shen C-K (1990) Bounding surface hypoplasticity model for sand. J Eng Mech (ASCE) 116:983–1001

Wongsaroj J, Soga K, Mair RJ (2007) Modeling of long-term ground response to tunneling under St James' Park, London. Geotechnique 57:75–90

Chapter 8
Cyclic Plasticity Models: Critical Reviews and Assessments

Accurate description of plastic deformation induced during a cyclic loading process is required for the mechanical design of machinery subjected to vibration and buildings and soil structures subjected to earthquakes since the middle of the last century. Elastoplastic constitutive model formulated for this aim is called the *cyclic plasticity model*. Substantially, the key of the pertinence in cyclic plasticity model is how to describe appropriately a small plastic strain rate induced by the rate of stress inside the yield surface. Therefore, a quite delicate formulation of plastic strain rate developing gradually as the stress approaches the yield surface is required to this end. Here, needless to say, the continuity and the smoothness conditions described in Sect. 7.1 would have to be fulfilled in a cyclic plasticity model.

Various cyclic plasticity models have been proposed to date, while most of them violate the continuity and the smoothness conditions unfortunately. Then, the beginners for the cyclic plasticity model would be perplexed as to which model is most pertinent and should be chosen for their study and analyses. In order to avoid their perplexity and missed selections, the cyclic plasticity models will be classified from the mathematical structures and their distinctive physical features will be examined in this chapter. Then, their pertinences/impertinences will be critically assessed in detail.

8.1 Classification of Cyclic Plasticity Models

Cyclic plasticity models proposed to date are classifiable into the two types described in the following.

K. Hashiguchi, *Foundations of Elastoplasticity: Subloading Surface Model*,
DOI 10.1007/978-3-319-48821-9_8

The one type is based on the concept of the kinematic hardening, i.e., the translation of subyield surface(s) assumed inside the conventional yield surface or the small single yield surface translating rapidly with a plastic deformation, while the innermost surface encloses a purely elastic domain. Several cyclic plasticity models in this type have been proposed to date. The other type is based on the natural concept that the plastic strain rate develops as the stress approaches the yield surface, i.e. the extension of the subloading surface model described in Chap. 7. The cyclic loading behavior is described rigorously by extending the subloading surface model such that the similarity-center of the normal-yield and the subloading surfaces translates with the plastic deformation. It is called the *extended subloading surface model*, while the subloading surface model described in Chap. 7 for which the similarity-center is fixed is renamed the *initial subloading surface model*.

The classification of the cyclic plasticity models is shown schematically in Fig. 8.1. Their mathematical and mechanical features and pertinences/impertinences to the description of cyclic loading behavior will be revealed in detail in the subsequent sections.

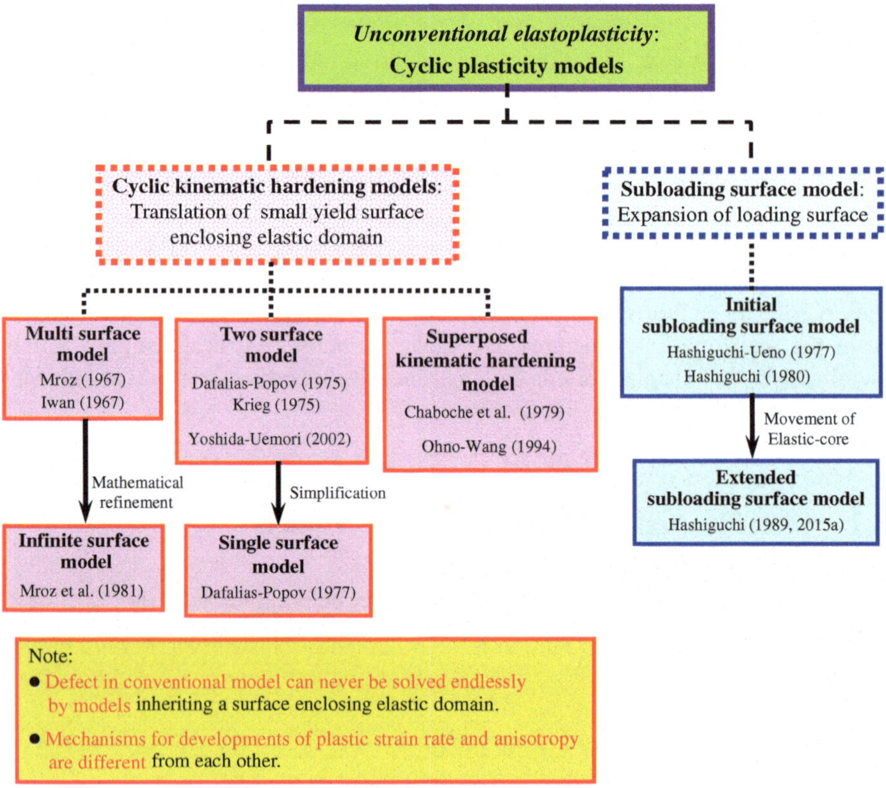

Fig. 8.1 Classification of cyclic plasticity models

8.2 Cyclic Kinematic Hardening Models: Improper Use of Kinematic Hardening

The conventional plasticity postulates that a plastic strain rate is induced by the rate of stress on the yield surface but it is required for the cyclic plasticity model to describe the plastic strain rate induced by the rate of stress inside the yield surface. The description of plastic deformation induced in the cyclic loading process was initiated by exploiting the kinematic hardening concept with the assumption of existence of a purely elastic domain. It is called the *cyclic kinematic hardening model*. However, we should notice the following crucial defects.

1. The stress region, in which a plastic strain rate is not induced but only an elastic strain rate is induced by cyclic change of stress, does not exist in general as known from the fact that the plastic strain rate is induced during a cyclic loading with any small stress amplitude. Therefore, the yield surface enclosing a purely-elastic domain must not be incorporated in the modelling of cyclic elastoplastic constitutive equation.
2. The purpose for the creation of unconventional plasticity model is the description of the plastic strain rate by the rate of stress inside the yield surface which cannot be described by the conventional plasticity model assuming the yield surface enclosing a purely-elastic domain. However, the cyclic kinematic hardening models are incapable of describing the plastic strain rate in the cyclic loading by the stress amplitude inside the small yield surface enclosing a purely-elastic domain. Then, the models inheriting the yield surface enclosing a purely-elastic domain are required to incorporate smaller and smaller yield surfaces one after another endlessly depending on the stress level and the stress amplitude, like an infinity mirror, a nest of boxes, a matryoshka doll in Russia, etc. so that they fall into the endless pit without the fundamental resolution of the problem. In addition, note that it cannot be physically accepted to incorporate an infinitely small yield surface, resulting in the indetermination of the direction of plastic strain rate, since the normal direction to the infinitesimal surface is indeterminate. Eventually, ***the defect of the conventional plasticity model cannot be solved by the cyclic kinematic hardening models*** inheriting a (small) yield surface enclosing a purely-elastic domain.
3. The ***mechanism for the development of plastic strain rate is substantially different from the mechanism for the development of anisotropy*** such as the kinematic hardening,
4. A purely elastic state would not move up to a high stress in a fully-yield state.
5. Plastic deformation behavior of material with a fatigue limit smaller than a size of yield surface enclosing a purely-elastic domain cannot be described for the stress amplitude smaller than the size of yield surface.

Besides, remind that the kinematic hardening (Prager 1956; Armstrong and Frederick 1966) is merely the simple method proposed primarily to describe the induced anisotropy of non-frictional, i.e. plastically-pressure independent metals.

The kinematic hardening is inapplicable but the rotational hardening rule (cf. Chap. 11) has to be adopted to describe the anisotropy of the plastically-pressure dependent, i.e. frictional materials (e.g. soils, rocks, concretes and friction behavior). The cyclic kinematic hardening models would not possess the rationality and the generality as will be explained in detail for each model.

8.2.1 Multi-surface Model

Mroz (1966, 1967, 1976) and Iwan (1967) proposed the *multi surface model* based on the following basic assumptions.

(a) Plural encircled subyield surfaces are incorporated, while the ratios of the sizes of these surfaces to the outer-most surface (conventional yield surface) are kept constant throughout a deformation.
(b) Interior of the innermost subyield surface is a purely-elastic domain.
(c) Subyield surfaces are pushed out by the current stress point. Then, plural surfaces contact at a point.
(d) Plastic modulus is prescribed by the size of the subyield surface on which the current stress lies, while it is smaller for a more outer subyield surface.

The uniaxial deformation behavior of the multi-surface model is illustrated in Fig. 8.2 for the simple material without a variation of the outmost surface.

However, this model possesses the following serious defects.

1. Plastic modulus decreases suddenly at the moment when the stress reaches a larger subyield surface, so that the smoothness condition in Eq. (7.2) is violated at that moment. Smooth stress-strain curve cannot be described but piece-wise linear curve is predicted. Therefore, it is required to determine the offset value, i.e. the plastic stain at yield point, which is accompanied with an arbitrariness.
2. Plural subyield surfaces contact at the current stress point and there exist plural plastic moduli at that contact point so that the singular point in the field of plastic modulus is induced there. Numerical calculation becomes unstable for the cyclic loading behavior in the vicinity of contact point.
3. Plastic deformation cannot be predicted at all for the cyclic loading inside the innermost surface even in a high stress level.
4. It is physically impertinent that the innermost subyield surface contacts with the outermost surface, so that the purely-elastic domain reaches the fully-plastic state.
5. Stress transfers to a larger subyield surface by moving half of the difference of the sizes of subyield surfaces in the initial loading process as shown in Fig. 8.2b. On the other hand, it transfers to a larger subyield surface by moving just the difference of the sizes of subyield surfaces in the unloading-reverse loading process as shown in Fig. 8.2d. Therefore, the *Masing rule* (Masing 1926) meaning that the curvature of stress-strain curve in the unloading-reverse loading decreases to a half of the curvature of initial loading curve is described

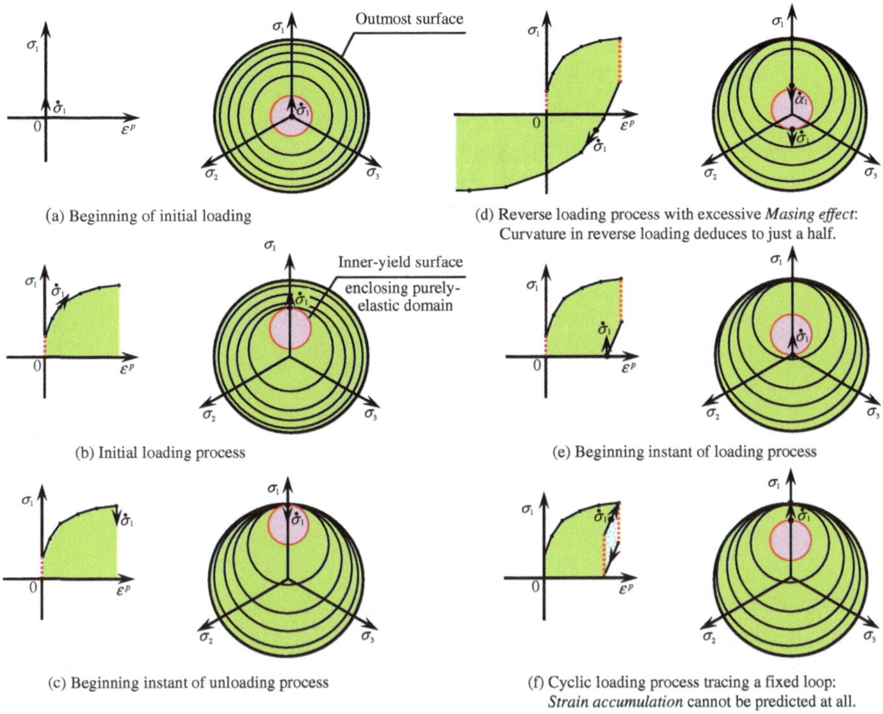
Outmost surface
Inner-yield surface enclosing purely-elastic domain
(a) Beginning of initial loading
(b) Initial loading process
(c) Beginning instant of unloading process
(d) Reverse loading process with excessive *Masing effect*: Curvature in reverse loading deduces to just a half.
(e) Beginning instant of loading process
(f) Cyclic loading process tracing a fixed loop: *Strain accumulation* cannot be predicted at all.

Fig. 8.2 One-dimensional loading behavior predicted by multi-surface model

exactly and simply as shown in Fig. 8.2d. By virtue of this mechanical feature, this model has been used widely. However, the variation of curvature observed in real material behavior is not so large as described by the Masing rule.

6. In the cyclic loading process under a constant amplitude, the *plastic shakedown* is induced immediately for non-hardening (sub)yield surfaces and after several cycles for hardening (sub)yield surfaces, tracing the fixed loop cyclically (Figs. 8.2f and 8.3). In other words, the accumulation of plastic strain during a pulsating stress loading, called the *mechanical ratcheting effect*, cannot be described at all by this model. In fact, however, the remarkable mechanical ratcheting is observed in real metal behavior which can be simulated accurately by the extended subloading surface model as will be described in Sect. 10.4 for metals. Therefore, the deformations of machinery and structures subjected to cyclic loading are predicted to be unrealistically small by the multi surface model, resulting in a risky mechanical design of machinery.
7. The continuity condition in Eq. (7.1) is also violated at the moment when the stress transfers to a larger subyield surface if the tangential inelastic strain rate described in Sect. 7.7 is incorporated, which is induced discontinuously.

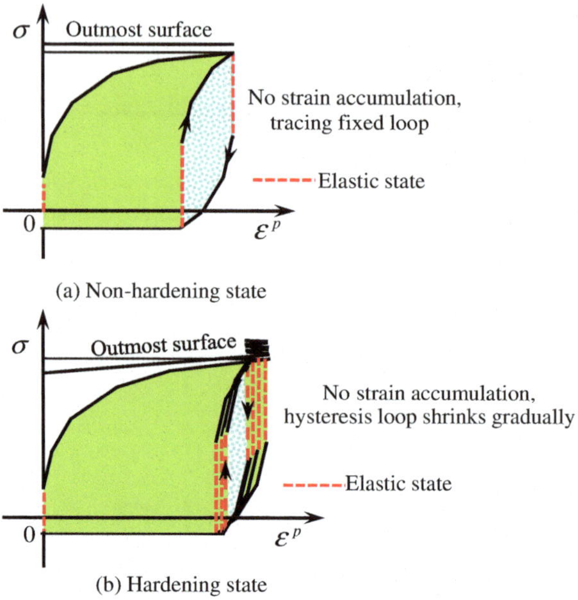

Fig. 8.3 Prediction of cyclic loading behavior by the multi surface model under a constant stress amplitude

8. Judgment on which subyield surface among multiple subyield surfaces the current stress lies is required in the loading criterion. In addition, deformation analysis by this model is complicated because it is necessary to calculate all movements of multi-subyield surfaces. It results in the increases of memory usage and calculations necessary for numerical analysis. It becomes more serious in the analysis of cyclic loading behavior in fluctuating unsteady loading process.
9. Numerical calculation of cyclic loading behavior in the vicinity of innermost yield surface is unstable, the tangent modulus jumping from the elastic to the elastoplastic ones and vice versa.

8.2.2 *Infinite Surface Model*

Modification of the multi surface model was proposed by Mroz et al. (1981), in which infinite number of subyield surfaces are incorporated inside the yield surface in contrast to the two surface model described in the next subsection. It is called the *infinite surface model*. The smoothness condition in Eq. (7.2) is fulfilled so that the smooth stress-strain curve is described in the initial monotonic loading process but it is violated at the moment when the stress passes through the starting point of reverse loading, called the *stress reversal point*, in the reloading process after the partial unloading, whereas the singularity of the plastic modulus is induced at that point, since subyield surfaces with different sizes contact mutually. It is caused by the inherent characteristic of the multi surface model that the plural yield surfaces contact at one point, resulting in the singularity of the plastic modulus. All the defects (1)–(9) described in the multi surface model except for (3) and the fulfillment of smoothness condition in the initial monotonic loading process are retained in the infinite surface model. Further, note that the incorporation of an infinitely small yield surface cannot be physically accepted since it causes the singularity in the direction of plastic strain rate as was described in the item 2 in Section 8.2.

8.2.3 Two-Surface Model

Dafalias and Popov (1975, 1976) and Krieg (1975) proposed the *two-surface model* based on the following assumptions:

(a) Only one subyield surface enclosing a purely-elastic domain is incorporated inside the conventional-yield surface which is renamed as the "*bounding surface*" by Dafalias and Popov (1975) and "limit surface" by Krieg (1975).
(b) The ratio of the size of the subyield surface to that of the bounding surface is kept constant throughout the deformation.
(c) The subyield surface is pushed out by the current stress point and translates toward the conjugate point on the bounding surface, while the outward-normal at conjugate point on the bounding surface is identical with that at the current stress point on the subyield surface.
(d) The plastic modulus is determined by the distance from the current stress point to the conjugate stress point on the bounding surface.

Here, it is required that the subyield surface must translate so as not to intersect with the bounding surface because the direction of plastic strain rate becomes indeterminate at the intersecting point of these surfaces. The rigorous translation rule was derived by Hashiguchi (1981, 1988).

This model has been adopted widely for the prediction of deformation behavior of metals (cf. e.g. Dafalias and Popov 1976; McDowell 1985, 1989; Ohno and Kachi 1986; Ellyin 1989; Hassan and Kyriakides 1992; Yoshida and Uemori 2002a, b, 2003).

The uniaxial loading behavior of the two-surface model is illustrated in Fig. 8.4 for the simple material without a variation of the bounding surface. The mechanical response of this model is opposite to that of the multi-surface model, although they would seem similar since only the numbers of subyield surfaces are different, as follows:

1. The tangent modulus changes suddenly from the elastic to the elastoplastic modulus at the moment when the stress reaches the subyield surface. Therefore, the smoothness condition in Eq. (7.2) is violated at that moment. Needless to say, the smooth stress-strain curve is not described but the suddenly-bent stress-strain curve is predicted. Therefore, it is required to determine the offset value, which is accompanied with an arbitrariness.

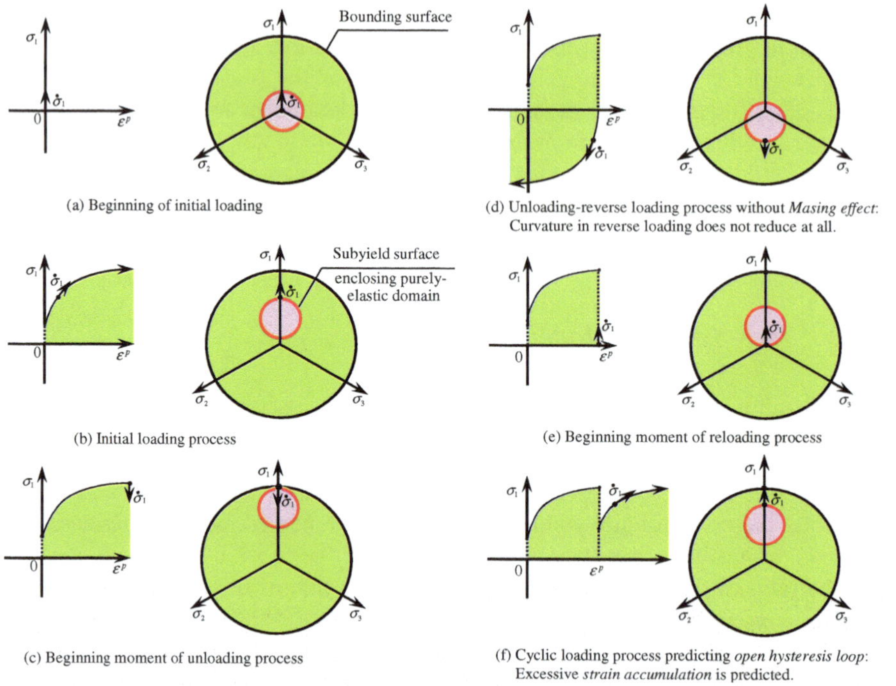
Bounding surface
(a) Beginning of initial loading
(d) Unloading-reverse loading process without *Masing effect*: Curvature in reverse loading does not reduce at all.
Subyield surface enclosing purely-elastic domain
(b) Initial loading process
(e) Beginning moment of reloading process
(c) Beginning moment of unloading process
(f) Cyclic loading process predicting *open hysteresis loop*: Excessive *strain accumulation* is predicted.

Fig. 8.4 Uniaxial loading behavior predicted by two-surface model

2. The *singular point* of plastic modulus is induced at the contact point of the bounding and the subyield surfaces, while the elastic and the elastoplastic tangent moduli are induced at that one point. Numerical calculation of cyclic loading behavior in the vicinity of contact point becomes unstable.
3. Plastic deformation cannot be predicted at all for the cyclic loading inside the cyclic stress inside the subyield surface even in a high stress level.
4. It is physically impertinent that the subyield surface enclosing the purely-elastic domain contacts directly with the bounding surface describing the fully-plastic state.
5. The plastic modulus depends on the distance from the current stress to the conjugate stress irrespective of the loading process, i.e. the initial, the reverse and the reloading processes and thus the curvature of stress-strain curves are identical irrespective of these processes. Then, the Masing effect cannot be described at all contrary to the multi-surface model. The ratio of the size of the inner-yield surface to that of the bounding surface is chosen to be 1/3–2/5

(Yoshida and Uemori 2002a, b, 2003) so that the inner yield surface contains the zero stress state as far as a kinematic hardening does not highly develop. Based on this fact too, it is obvious that plastic strain rate cannot be predicted in the unloading (stress reducing) process by this model and the size ratio must be chosen less than $1/4$ in order that the inner yield surface does not contain the zero stress state in general. Eventually, the unloading behavior becomes unrealistically elastic in order that the initial loading curve matches to real behavior. Consequently, the closed hysteresis loop cannot be depicted in the pulsating loading process (positive or negative one side cyclic loading process) so that the unrealistically large mechanical ratchetting is predicted. Stress versus strain curves are depicted in reloading process but unloading process have never been depicted in the pulsating loading in literatures (Yoshida and Uemori 2002a, b, 2003). Nevertheless, Yoshida and Uemori (2002a, b, 2003) insist that their model can predict the spring-back phenomenon. In their formulation, the Young's modulus is formulated to decrease but saturate to the limited value with the plastic equivalent strain in order to improve the simulation of the test data of *spring-back* behavior of metals. In fact, however, if once the Young's modulus decreases in the tension loading, it decreases acceleratingly to zero, since the decrease of the Young's modulus is caused by the growth of cracks which develop increasingly in the continuing tension loading process as has been revealed in the damage mechanics described in Chap. 14. Therefore, the method for prediction of springback due to the decrease of Young's modulus (Yoshida and Uemori 2003) is physically unacceptable, which also leads to the unrealistic prediction of deformation behavior after the springback. In other words, it would be quite irrational idea invented deceiving the inherent defect of the two surface model that only the elastic deformation is induced in the unloading (stress reducing) process as will be examined in detail in Sect. 10.5.

6. In the cyclic loading process with the constant amplitude of the positive or negative one side stress, the open hysteretic loop is described and thus the excessive strain accumulation in the cyclic loading, i.e. the excessive mechanical ratcheting is predicted contrary to the multi-surface model as shown in Fig. 8.5.
7. The continuity condition in Eq. (7.1) is also violated at the moment when the stress reaches the bounding surface if the tangential inelastic strain rate is incorporated.
8. Judgment whether or not the current stress reaches the subyield and/or the bounding surface is required in the loading criterion.
9. Numerical calculation of cyclic loading behavior in the vicinity of subyield surface is unstable, the tangent modulus jumping from the elastic to the elastoplastic ones and vice versa.

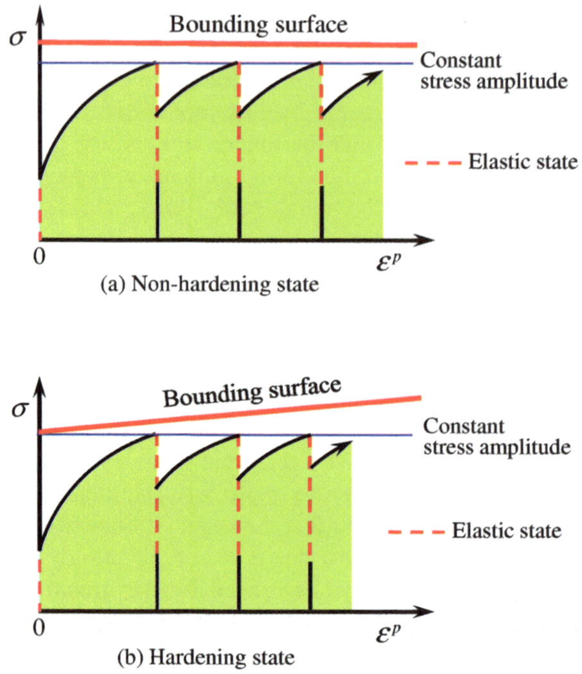

Fig. 8.5 Prediction of cyclic loading behavior by the two surface model under a constant stress amplitude

8.2.4 *Single Surface Model*

The *single surface model* is proposed by Dafalias and Popov (1977), in which the subyield surface shrinks to a point in the two surface model. It would describe a smooth response unless the shrinking surface does not lie on the bounding surface. Besides, the intense dependence of the direction of plastic strain rate on the direction of stress rate leading to the rate-nonlinearity which would be impertinent physically and mathematically. Further, this model cannot be free from the basic defects contained in the two-surface model.

8.2.5 *Superposed Kinematic Hardening Model*

The cyclic plasticity model assuming the small single yield surface which translates by the superposition of plural non-linear kinematic rules, excluding the conventional yield surface, was proposed by Chaboche et al. (1979) and Chaboche and Rousselier (1983). It may be called the *superposed kinematic hardening model.* Its alteration was proposed by Ohno and Wang (1993), which is called the *combined nonlinear kinematic-isotropic hardening model* by Ohno et al. (2013). It is not an physical model but merely an empirical model in the polynomial approximation of test data by using a lot of material constants lacking physical meanings.

(a) Chaboche model

The following small Mises type yield surface with the isotropic and the Armstrong-Frederick (1966) kinematic hardenings is introduced by Chaboche et al. (1979) and improved by Chaboche and Rousselier (1983).

$$\sqrt{\frac{3}{2}}||\hat{\boldsymbol{\sigma}}'|| = F_0 + \bar{F} \tag{8.1}$$

where the increase of the isotropic hardening function, $\bar{F}$, evolves by the following equation.

$$\dot{\bar{F}} = c(\bar{F}_s - \bar{F})\dot{\varepsilon}^{eqp} \tag{8.2}$$

c is the material constant and $\bar{F}_s$ is the material constant describing the saturation value of $\bar{F}$.

The kinematic hardening is given by the superposition of the several non-linear kinematic hardening rules of Armstrong and Frederick (1966) as follows:

$$\dot{\boldsymbol{\alpha}} = \sum_{i=1}^{n} \dot{\boldsymbol{\alpha}}_i \tag{8.3}$$

where

$$\dot{\boldsymbol{\alpha}}_i = \left(A_i\hat{\mathbf{n}} - \sqrt{\frac{2}{3}}b_i\boldsymbol{\alpha}_i\right)||\dot{\boldsymbol{\varepsilon}}^p||\ \text{(no sum)} \tag{8.4}$$

which is based on Eq. (6.114). A_i and b_i $(i = 1, 2, \ldots, n)$ are the material constants, while n is chosen usually $4 \sim 8$. Equation (8.3) is integrated for the uniaxial loading process as follows:

$$\alpha_a = \sum_{i=1}^{n} \frac{A_i}{b_i}[1 - \exp(-b_i\varepsilon_a^p)] \tag{8.5}$$

for $\varepsilon_a^p > 0$ under the initial condition $\alpha_a = 0$ for $\varepsilon_a^p = 0$.

The uniaxial loading behavior of Chaboche model is illustrated in Fig. 8.6, where the isotropic hardening is not incorporated by setting $c = 0$.

(b) Ohno-Wang model

Ohno and Wang (1993) introduced the small Mises type yield surface which translates by the superposition of the several bilinear kinematic hardening rules composed of the linear kinematic hardening and the isotropic non-hardening as follows:

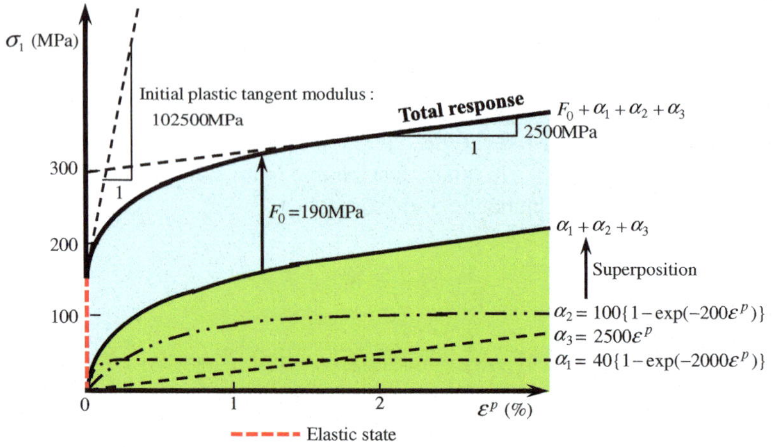

Fig. 8.6 Chaboche model (after Lamaitre and Chaboche 1990)

$$\sqrt{\frac{3}{2}}||\hat{\boldsymbol{\sigma}}'|| = F_c \tag{8.6}$$

where F_c is the material constant and thus the isotropic hardening is not induced. The kinematic hardening rule is given by

$$\dot{\boldsymbol{\alpha}} = \sum_{i=1}^{n} \dot{\boldsymbol{\alpha}}_i \tag{8.7}$$

$$\dot{\boldsymbol{\alpha}}_i = h_i\left(\frac{2}{3}\dot{\boldsymbol{\varepsilon}}^p - H[f_i]\left\langle \dot{\boldsymbol{\varepsilon}}^p : \frac{\boldsymbol{\alpha}_i}{\bar{a}_i}\right\rangle \frac{\boldsymbol{\alpha}_i}{r_i}\right) \quad \text{(no sum)} \tag{8.8}$$

where h_i and r_i $(i = 1, 2, \ldots, n)$ are the material constants, while n is chosen $4 \sim 8$ usually. The linear kinematic hardenings proceed but they stop suddenly when the following condition is satisfied in each of them.

$$f_i \equiv \bar{a}_i^2 - r_i^2 \tag{8.9}$$

where

$$\bar{a}_i \equiv \sqrt{\frac{3}{2}}||\boldsymbol{\alpha}_i|| \tag{8.10}$$

$H[\ ]$ is the Heaviside step function, i.e., $H[s] = 1$ for $s \geq 0$, $H[s] = 0$ for $s < 0$. Then, the kinematic hardening proceeds when the plastic strain rate is induced directing outwards the surface described by Eq. (8.9) but it ceases when $\boldsymbol{\alpha}_i$ reaches the surface $f_i \equiv \bar{a}_i^2 - r_i^2 = 0$ $(H[f_i] = 1)$ as ascertained by

$$\begin{aligned}
\dot{f}_i &= 2\bar{a}_i\dot{\bar{a}}_i = 2\bar{a}_i\sqrt{\frac{3}{2}}(||\boldsymbol{\alpha}_i||)^{\bullet} = 2\bar{a}_i\sqrt{\frac{3}{2}}\frac{\boldsymbol{\alpha}_i}{||\boldsymbol{\alpha}_i||} : \overset{\circ}{\boldsymbol{\alpha}}_i \\
&= 2\bar{a}_i\sqrt{\frac{3}{2}}\frac{\boldsymbol{\alpha}_i}{||\boldsymbol{\alpha}_i||} : h_i\left(\frac{2}{3}\overset{\bullet}{\boldsymbol{\varepsilon}}{}^p - H[f_i]\left\langle \overset{\bullet}{\boldsymbol{\varepsilon}}{}^p : \frac{\boldsymbol{\alpha}_i}{\bar{a}_i}\right\rangle \frac{\boldsymbol{\alpha}_i}{r_i}\right) \\
&= 2\bar{a}_i\sqrt{\frac{3}{2}}\frac{\boldsymbol{\alpha}_i}{||\boldsymbol{\alpha}_i||} : h_i\left(\frac{2}{3}\dot{\lambda}\frac{\boldsymbol{\alpha}_i}{||\boldsymbol{\alpha}_i||} - \left\langle \dot{\lambda}\frac{\boldsymbol{\alpha}_i}{||\boldsymbol{\alpha}_i||} : \frac{\boldsymbol{\alpha}_i}{\bar{a}_i}\right\rangle \frac{\boldsymbol{\alpha}_i}{\bar{a}_i}\right) \\
&= 2\bar{a}_i\sqrt{\frac{3}{2}}\frac{\boldsymbol{\alpha}_i}{||\boldsymbol{\alpha}_i||} : h_i\left(\frac{2}{3}\dot{\lambda}\frac{\boldsymbol{\alpha}_i}{||\boldsymbol{\alpha}_i||} - \left\langle \dot{\lambda}\frac{\boldsymbol{\alpha}_i}{||\boldsymbol{\alpha}_i||} : \frac{\boldsymbol{\alpha}_i}{\sqrt{\frac{3}{2}}||\boldsymbol{\alpha}_i||}\right\rangle \frac{\boldsymbol{\alpha}_i}{\sqrt{\frac{3}{2}}||\boldsymbol{\alpha}_i||}\right) = 0
\end{aligned}$$

The uniaxial loading behavior of this model is illustrated in Fig. 8.7, exhibiting the piecewise linear relation requiring the cumbersome judgments for the Heaviside step function in numerical calculation.

Ohno and Wang (1993) showed that the model with Eq. (8.8) exhibits similar behavior to the multi surface model. They extended Eq. (8.8) by replacing the Heaviside step function to the continuous function as follows:

$$\overset{\circ}{\underset{i}{\boldsymbol{\alpha}}} = h_i\left[\frac{2}{3}\mathbf{d}^p - \left(\frac{\bar{a}_i}{r_i}\right)^{m_i}\left\langle \mathbf{d}^p : \frac{\boldsymbol{\alpha}_i}{\bar{a}_i}\right\rangle \frac{\boldsymbol{\alpha}_i}{r_i}\right] \text{ (no sum)} \tag{8.11}$$

where m_i $(i = 1, 2, \cdots, n)$ are material constants. On account of this modification, the stress vs. strain curve in the monotonic loading process becomes smooth after the stress reached the yield surface. Here, note that Eq. (8.11) for $m_i \to \infty$ is reduced to Eq. (8.8) exhibiting the bilinear curve so that the piecewise linear stress versus strain curve with the completely closed hysteresis loop resulting in a

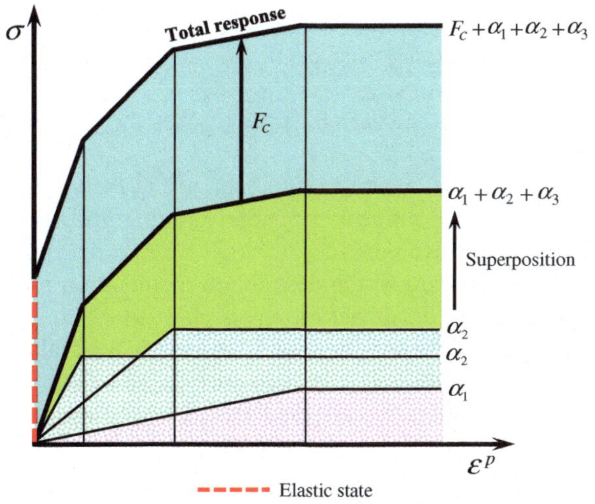

Fig. 8.7 Ohno-Wang model (Ohno and Wang 1993)

no-ratcheting is predicted. On the other hand, Eq. (8.11) for $m_i = 0$ is reduced to the Armstrong-Frederick nonlinear kinematic hardening rule so that the excessively large ratcheting is predicted. Further, the model was extended by Ohno and Abdel-Karim (2000) as follows:

$$\dot{\boldsymbol{\alpha}}_i = h_i\left(\frac{2}{3}\mathbf{d}^p - \mu_i\boldsymbol{\alpha}_i||\dot{\boldsymbol{\varepsilon}}^p|| - H[f_i]\left\langle \mathbf{d}^p:\frac{\boldsymbol{\alpha}_i}{\bar{a}_i}\right\rangle\frac{\boldsymbol{\alpha}_i}{r_i}\right) \text{ (no sum)} \tag{8.12}$$

where μ_i are the material constants. Equation (8.12) is the combination of the Armstrong-Frederick kinematic hardening and the piecewise kinematic hardening in Eq. (8.8) so that it approaches the behavior of the aforementioned Chaboche model. The Ohno model would be regarded as an impertinent modification of the Chaboche model.

The superposed kinematic hardening model possesses the following defects.

1. It is not a physical model but the empirical model due to the polynomial approximation of test data.
2. A lot of material constants without clear physical meaning are incorporated.
3. The yield surface is limited to the Mises yield surface or cylindrical yield surface in the principal stress space.
4. The tangent modulus changes suddenly from the elastic to the elastoplastic modulus at the moment when the stress reaches the yield surface, violating the smoothness condition in Eq. (7.2). Needless to say, the smooth stress-strain curve is not described but the suddenly-bent stress-strain curve is predicted. Therefore, it is required to determine the offset value, i.e. the plastic stain at yield point, which is accompanied with an arbitrariness.
5. Plastic deformation cannot be predicted for the cyclic loading inside the yield surface even in a high stress level.
6. It is physically impertinent that the small yield surface enclosing a purely-elastic domain reaches a high stress, i.e. full yield state.
7. The continuity condition in Eq. (7.1) is also violated at the moment when the stress reaches the bounding surface if the tangential inelastic strain rate is incorporated.
8. Judgment whether or not the current stress reaches the yield surface is required in the loading criterion.
9. Numerical calculation of cyclic loading behavior in the vicinity of small yield surface is unstable, the tangent modulus jumping from the elastic to the elastoplastic ones and vice versa.
10. The applicability is limited to the description of the deformation behavior for the variation of stress in the deviatoric stress plane so that it is limited to metals only with the Mises yield condition and the plastic equivalent hardening. On the other hand, the multi- and the two-surface and the subloading surface models have been applied to soils, and the multi-surface and the subloading

surface model has been further applied to friction phenomena (e.g. Mroz and Stupkiewicz 1994; Hashiguchi et al. 2005b, 2016; Hashiguchi, 2013a; Hashiguchi and Ozaki, 2008). Consequently, the superposed kinematic hardening model lacks the generality markedly.

11. The explicit formulations of this model are concerned with infinitesimal deformation up to several percent strain without a rotation, which are based on the infinitesimal strain in Eq. (2.49) possessing the deficiencies described in Sect. 2.6 and the material-time derivatives of the stress and the kinematic hardening violating the objectivity as described in Sect. 4.4. Therefore, it ignores even the fundamentals of modern continuum mechanics started by Oldroyd (1950) at the middle of the last century. Needless to say, it is inapplicable to the deformation analyses under a material rotation as seen in a metal forming process for instance. On the other hand, the other models, e.g. the subloading loading and the two surface models are formulated in the hypoelastic-based plasticity which holds for the finite deformation up to 100 % strain under a finite rotation even if the Jaumann rate is adopted.

Nevertheless, the ad hoc Chaboche model was officially implemented in the commercial software Marc and Abaqus so that it is used widely by metal engineers because it can be understood even by the beginners of elastoplasticity theory possessing the elementary knowledge only of the Mises yield condition, the kinematic hardening, the infinitesimal strain and the material-time derivative lacking the objectivity.

8.2.6 Common Drawbacks in Cyclic Kinematic Hardening Models

The cyclic plasticity models based on the kinematic hardening concept possess the common drawbacks as follows:

1. The original purpose for the creation of unconventional plasticity model is the description of the plastic strain rate by the rate of stress inside the yield surface which cannot be described by the conventional model assuming the yield surface enclosing a purely-elastic domain. The purpose cannot be attained endlessly by the methods which incorporate the small yield surface enclosing a purely-elastic domain.
2. It is premised that the development of plastic strain rate (decrease of plastic tangent modulus) proceeds by the development of the kinematic hardening. In fact, however, the main source for the development of plastic strain rate would be different from the kinematic hardening, i.e. anisotropy because the plastic

strain rate develops as the stress increases to overcome the friction resistance between material particles, i.e. as the stress approaches the yield surface. Therefore, these models would lack the physical rationality.

3. The tangent modulus lowers suddenly from the elastic one to the elastoplastic one at the yield point so that the smoothness condition in Eq. (7.2) is violated and thus a smooth stress-strain curve cannot be described. Therefore, it is required to determine the offset value, i.e. the plastic stain at yield point, which is accompanied with an arbitrariness.
4. The strain accumulation in a small cyclic loading inside a small yield surface cannot be described since a small yield surface enclosing an elastic domain is assumed. Note that even the cyclic loading in a high stress near full yield state does not cause the strain accumulation as shown in Fig. 8.8. This kind of loading situation is often observed in practical engineering, e.g. the phenomenon that the cantilever supporting a weight near yielding load is subjected to a cyclic loading with a small stress amplitude by a wind or a sea waves for instance. This defect leads to the risky mechanical design for the cyclic loading by which a large strain accumulation is induced in real materials. In addition, the spring-back phenomenon cannot be described as described for the two-surface model.
5. The judgment whether or not the stress reaches conventional-yield (outmost, bounding, limit, small yield) and/or subyield surface is fulfilled is required in the loading criterion.
6. The continuity condition in Eq. (7.1) is also violated if the tangential-inelastic strain rate is incorporated since it is induced suddenly when the stress reaches the surface(s). Therefore, the non-proportional loading behavior and the plastic instability phenomena cannot be described pertinently.

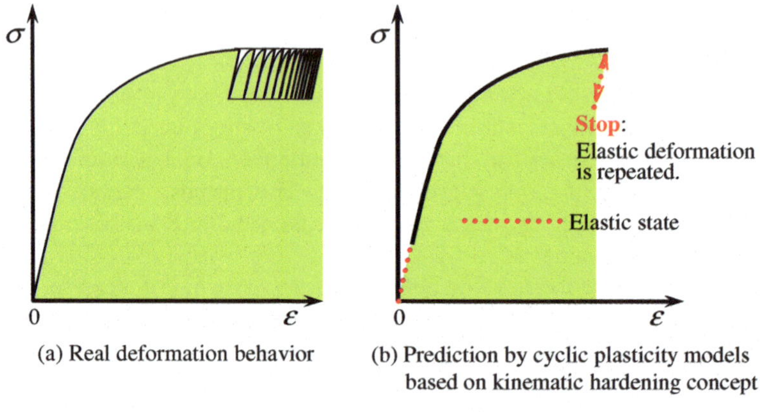

Fig. 8.8 Unrealistic prediction of cyclic loading behavior after partial unloading-reloading by the multi, the two and the superposed kinematic hardening models postulating purely elastic domain

7. The mathematical structures of these models differ basically from the conventional plasticity, assuming the plural surfaces (the multi and the two surface models) or the small yield surface (the superposed kinematic hardening model). On the other hand, the mathematical structure in the subloading surface model is the natural extension of that in the conventional one, assuming only the conventional yield surface as an independent surface as was shown in Chap. 7 and will be shown elaborately in Chap. 9.
8. Quite small loading increments must be input resulting in inefficient numerical calculation as far as any particular computer subroutine to pulled-back the stress to the yield surface (Kobayashi and Ohno, 2002; Ghaei and Green, 2010) is not incorporated, because the automatic controlling function to pull-back the stress to the yield surface contained in the subloading surface model is not furnished. Besides, numerical calculation of cyclic loading behavior in the vicinity of small yield surface is unstable, the tangent modulus jumping from the elastic to the elastoplastic ones and vice versa.
9. Phenomena which can be implemented to the return-mapping projection are limited. In fact, it would be difficult to incorporate the cyclic isotropic hardening-stagnation which will be described in Sect. 10.2 since the pull-back of the plastic strain or the back stress to the isotropic hardening surface cannot be executed readily in the return-mapping projection. It is quite peculiar that the cyclic isotropic hardening-stagnation is abandoned by the proposers themselves (Chaboche et al. 1979; Chaboche 2008; Ohno 1982; Kobayashi and Ohno 2002) in the FEM analyses.
10. The exact finite strain theory based on the multiplicative decomposition of deformation gradient, called the *multiplicative finite strain theory*, cannot be formulated for the kinematic hardening models. On the other hand, the exact formulation was attained for the subloading surface model as will be described in detail in Chap. 12.
11. They have been formulated for the description of elastoplastic deformation of pressure-independent Mises metals and thus it is difficult or impossible for these models to be applied to materials other than pressure-independent metals, e.g. pressure-dependent metals, soils, rocks and concretes. Therefore, the cyclic kinematic hardening models would be the ad hoc models which are applicable only to quite limited material and phenomena.

As known from the facts revealed in this section, the cyclic kinematic hardening models possess various serious defects. Nevertheless, they are installed widely in the commercial FEM software (Chaboche model: Marc, Abaqus; Yoshida-Uemori model: PAM-STAMP, LS-DYNA (Japan)). The proposers of the cyclic kinematic-hardening models, i.e. the multi surface, the two surface and the superposed-kinematic hardening models should accept sincerely the intrinsic limits and defects of these models and also the users of these models should recognize these serious limits and defects for the sound development of elastoplastic theory and deformation analysis.

On the other hand, the subloading surface model described in the last chapter, the next section and the subsequent chapters would be the universal model which will become widespread with the passage of time. The overall assessment of cyclic plasticity models is summarized in Table 8.1.

8.3 Expansion of Loading Surface: Extended Subloading Surface Model

The subloading surface model formulated in Chap. 7, called the *initial subloading surface model* hereinafter, is incapable of describing cyclic loading behavior appropriately, predicting an open hysteresis loop in an unloading-reloading process and thus overestimating a mechanical ratcheting phenomenon. The insufficiency is caused by the fact that the similarity-center of the normal-yield and the subloading surfaces is fixed at the origin of stress space and thus a purely-elastic deformation is described in the unloading process, resulting in the open hysteresis loop. Here, it should be noted that purely-elastic response is induced only in an initiation of reverse loading process in general. Then, the insufficiency was remedied by making the similarity-center of the normal-yield and the subloading surfaces translate with the plastic deformation (Hashiguchi 1985b, 1986, 1989).

The uniaxial loading behavior is depicted in Fig. 8.9 for the simple material behavior without a variation of the normal-yield surface. The similarity-center goes up following the stress by the plastic strain rate in the initial loading process as seen in Fig. 8.9a, b. The subloading surface shrinks and thus only elastic strain rate is induced until the stress goes down to the similarity-center in the unloading process as seen in Fig. 8.9c. After that the subloading surface begins to expand and thus the plastic strain rate in the compression is induced in the unloading-inverse loading process whilst the similarity-center goes down following the stress by the plastic strain rate as seen in Fig. 8.9d. Again only the elastic strain rate is induced until the stress goes up to the similarity-center in the reloading process from the complete unloading as seen in Fig. 8.9e. After that the subloading surface begins to expand and thus the plastic strain rate is induced whilst the similarity-center goes up following the stress by the plastic strain rate as seen in Fig. 8.9f. The expanded figure of Fig. 8.9f is also shown at the lowest part in Fig. 8.9. Consequently, the closed hysteresis loop is depicted realistically as shown in this figure.

Table 8.1 Assessment of cyclic plasticity models

Cyclic plasticity models		Smoothness condition	Judgment of yielding in loading criterion	Automatic control to pull-back stress to yield surface	Description of plastic strain accumulation under cyclic loading for small strain amplitude	Continuity condition in incorporation of tangential inelastic strain rate	Applicability to materials other than Mises metals	Formulation of multiplicative finite strain theory
Cyclic kinematic hardening model (Translation of small yield surface)	Multi surface model (Mroz)	Violate	Necessary	Impossible	Impossible	Violate	Difficult	Impossible
	Two surface model (Dafalias)							
	Superposed Kinematic hardening model (Chaboche)						Impossible	
Expansion of loading surface	Subloading surface model (Hashiguchi)	Fulfills	Un-necessary	Possible	Possible	Fulfills	Applicable generally	Already formulated

The extended subloading surface model would describe the cyclic loading behavior realistically as illustratively shown in Fig. 8.10. It does not contain any drawbacks in the cyclic plasticity models based on the kinematic hardening concept, while the continuity and the smoothness conditions in Eqs. (7.1) and (7.2) are satisfied only in this model. Then, it has been applied to the descriptions of rate-independent and rate-dependent elastoplastic deformation behavior and plastic-instability phenomena of not only metals but also geomaterials and further the friction phenomena between solids as will be described in detail in the subsequent chapters.

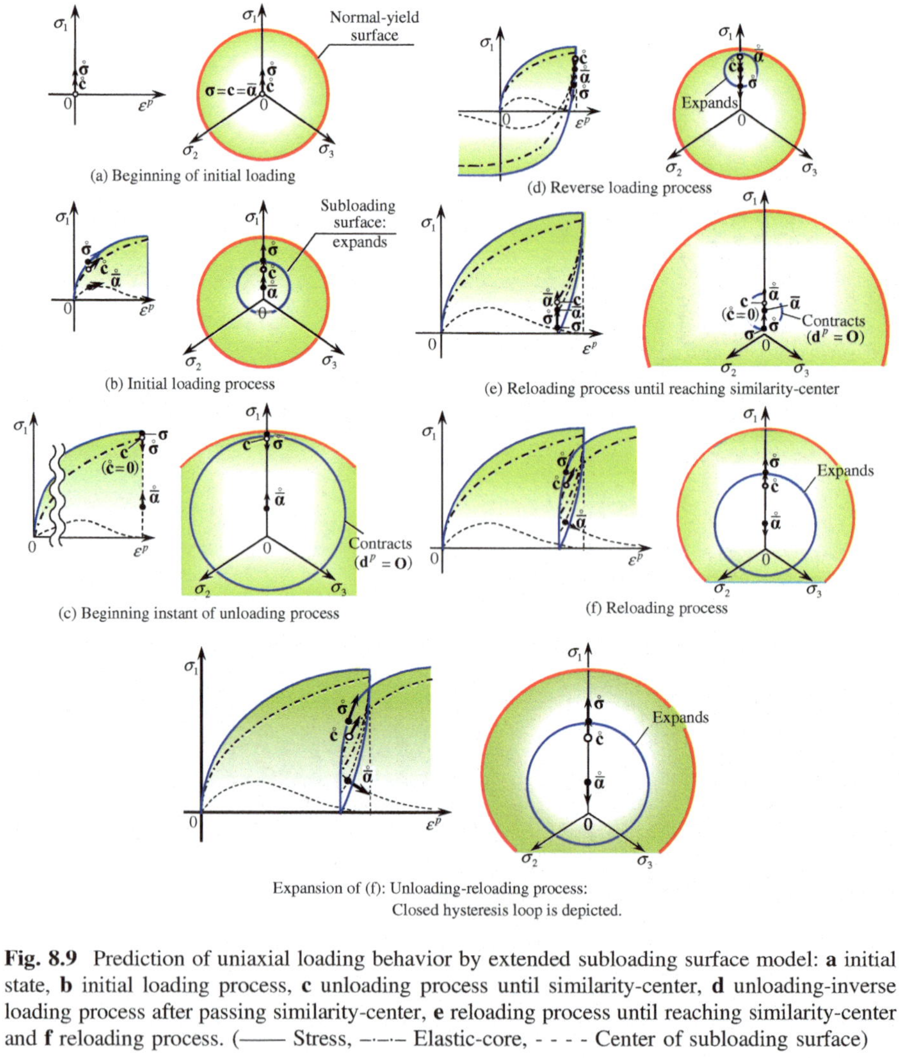

Fig. 8.9 Prediction of uniaxial loading behavior by extended subloading surface model: **a** initial state, **b** initial loading process, **c** unloading process until similarity-center, **d** unloading-inverse loading process after passing similarity-center, **e** reloading process until reaching similarity-center and **f** reloading process. (—— Stress, –·–·– Elastic-core, - - - - Center of subloading surface)

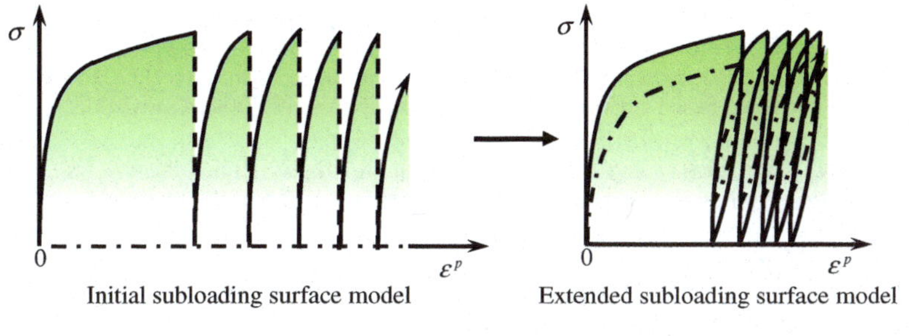

Fig. 8.10 Modification of subloading surface model to describe cyclic loading behavior

References

Armstrong PJ, Frederick CO (1966) A mathematical representation of the multiaxial Bauschinger effect. CEGB Report RD/B/N 731 (or in Materials at High Temperature 24:1–26 (2007))

Chaboche JL (2008) A review of some plasticity and viscoplasticity constitutive theories. Int J Plast 24:1642–1693

Chaboche JL, Rousselier G (1983) On the plastic and viscoplastic constitutive equations, Parts I and II. J Pressure Vessel Tech (ASME) 165:153–164

Dafalias YF, Popov EP (1975) A model of nonlinearly hardening materials for complex loading. Acta Mech 23:173–192

Dafalias YF, Popov EP (1976) Plastic internal variables formalism of cyclic plasticity. J Appl Mech (ASME) 43:645–651

Dafalias YF, Popov EP (1977) Cyclic loading for materials with a vanishing elastic domain. Nucl Eng Design 41:293–302

Chaboche JL, Dang-Van, K, Cordier G (1979) Modelization of the strain memory effect on the cyclic hardening of 316 stainless steel, Transactions on 5th international conference of SMiRT, Berlin, Division L., Paper No. L. 11/3

Ellyin F (1989) An anisotropic hardening rule for elastoplastic solids based on experimental observations. J Appl Mech (ASME) 56:499–507

Ghaei A, Green DE (2010) Numerical implementation of Yoshida-Uemori two-surface plasticity model using a fully implicit integration scheme. Compt Mater Sci 48:195–205

Hashiguchi K (1981) Constitutive equations of elastoplastic materials with anisotropic hardening and elastic-plastic transition. J Appl Mech (ASME) 48:297–301

Hashiguchi K (1985b) Subloading surface model of plasticity. In: Constitutive laws of soils (Proc. Discuss. Ses. 1A, 11th Int. Conf. Soil Mech. Found. Eng.), San Francisco, pp 127–130

Hashiguchi K (1986) Elastoplastic constitutive model with a subloading surface. In: Proceedings of international conference on computer mechanics, pp IV65–70

Hashiguchi K (1988) A mathematical modification of two surface model formulation in plasticity. Int J Solids Struct 24:987–1001

Hashiguchi K (1989) Subloading surface model in unconventional plasticity. Int J Solids Struct 25:917–945

Hashiguchi K (2013) General description of elastoplastic deformation/sliding phenomena of solids in high accuracy and numerical efficiency: subloading surface concept. Arch Compt Meth Eng 20:361–417

Hashiguchi K, Ozaki S (2008) Constitutive equation for friction with transition from static to kinetic friction and recovery of static friction. Int. J. Plasticity 24:2102–2124

Hashiguchi K, Ozaki S, Okayasu T (2005) Unconventional friction theory based on the subloading surface concept. Int J Solids Struct 42:1705–1727

Hashiguchi K, Ueno M, Kuwayama T, Suzuki N, Yonemura S, Yoshikawa N (2016) Constitutive equation of friction based on the subloading surface concept. Proc. Royal Soc., London A: 472:1–24 http://dx.doi.org/10.1098/rspa.2016.0212

Hashiguchi K, Yamakawa Y (2012) Introduction to finite strain theory for continuum elasto-plasticity, wiley series in computational mechanics. Wiley, Chichester

Hassan S, Kyriakides S (1992) Ratcheting in cyclic plasticity. Part I: uniaxial behavior. J Appl Mech (ASME) 8:91–116

Iwan WD (1967) On a class of models for the yielding behavior of continuous and composite systems. J Appl Mech (ASME) 34:612–617

Kobayashi M, Ohno N (2002) Implementation of cyclic plasticity models based on a general form of kinematic hardening. Int J Numer Meth Eng 53:2217–2238

Krieg RD (1975) A practical two surface plasticity theory. J Appl Mech (ASME) 42:641–646

Lemaitre JA, Chaboche J-L (1990) Mechanics of solid materials. Cambridge University Press, Cambridge

Masing G (1926) Eigenspannungen und Verfestigung beim Messing. In: Proceedings of the 2nd international congress of applied mechanics, Zurich, pp 332–335

McDowell DL (1985) An experimental study of the structure of constitutive equations for nonproportional cyclic plasticity. J Eng Mater Tech (ASME) 107:307–315

McDowell DL (1989) Evaluation of intersection conditions for two-surface plasticity theory. Int J Plast 5:29–50

Mroz Z (1966) On forms of constitutive laws for elastic-plastic solids. Arch Mech Stos 18:3–35

Mroz Z (1967) On the description of anisotropic workhardening. J Mech Phys Solids 15:163–175

Mroz Z (1976) A non-linear hardening model and its application to cyclic plasticity. Acta Mech 25:51–61

Mroz Z, Stupkiewicz S (1994) An anisotropic friction and wear model. Int J Solids Struct 31:1113–1131

Mroz Z, Norris VA, Zienkiewicz OC (1981) An anisotropic, critical state model for soils subject to cyclic loading. Geotechnique 31:451–469

Ohno N (1982) A constitutive model of cyclic plasticity with a non-hardening strain region. J Appl Mech (ASME) 49:721–727

Ohno N, Abdel-Karim M (2000) Uniaxial ratchetting of 316FR steel at room temperature—part II: constitutive modeling and simulation. J Eng Mater Tech (ASME) 122:35–41

Ohno N, Kachi Y (1986) A constitutive model of cyclic plasticity for nonlinearly hardening materials. J Appl Mech (ASME) 53:395–403

Ohno N, Wang JD (1993) Kinematic hardening rules with critical state of dynamic recovery, Part I: formulation and basic features for ratcheting behavior. Part II: application to experiments of ratcheting behavior. Int J Plast 9:375–403

Ohno N, Tsuda M, Kamei T (2013) Elastoplastic implicit integration algorithm applicable to both plane stress and three-dimensional stress states. Finite Elements Anal Design 66:1–11

Oldroyd JG (1950) On the formulation of rheological equations of state. Proc Roy Soc Lond A200:523–541

Prager W (1956) A new methods of analyzing stresses and strains in work hardening plastic solids. J Appl Mech (ASME) 23:493–496

Yoshida F, Uemori T (2002a) Elastic-plastic behavior of steel sheets under in-plane cyclic tension-compression at large strain. Int J Plast 18:633–659

Yoshida F, Uemori T (2002b) A model of large-strain cyclic plasticity describing the Bauschinger effect and workhardening stagnation. Int J Plast 18:661–686

Yoshida F, Uemori T (2003) A model of large-strain cyclic plasticity and its application to springback simulation. Int J Mech Sci 45:1687–1702

Chapter 9
Extended Subloading Surface Model

As was deliberated in Chap. 8, it can be presumed that only the extended subloading surface model is capable of describing the general loading behavior of materials appropriately. The explicit formulation of the extended model is shown in this chapter. Then, this model will be applied to metals and soils and their validities will be verified by comparisons with test data of metals in Chap. 10 and soils in Chap. 11.

9.1 Normal-Yield and Subloading Surfaces

The normal-yield surface with the isotropic and the kinematic hardening is described as

$$f(\hat{\boldsymbol{\sigma}})=F(H) \tag{9.1}$$

as shown in Eq. (6.85) already. The extended subloading surface for the normal-yield surface in Eq. (9.1) is given as follows (see Fig. 9.1).

$$\boxed{f(\bar{\boldsymbol{\sigma}})=RF(H)} \tag{9.2}$$

where $R(0 \leq R \leq 1)$ is the normal-yield ratio and

$$\bar{\boldsymbol{\sigma}} \equiv \boldsymbol{\sigma} - \bar{\boldsymbol{\alpha}} \tag{9.3}$$

$\bar{\boldsymbol{\alpha}}$ stands for the conjugate (similar) point in the subloading surface to the point $\boldsymbol{\alpha}$ in the normal-yield surface. By letting $\mathbf{c}$ denote the center of similarity of the normal-yield and the subloading surfaces, i.e. the similarity-center, which is called *elastic-core* since the most elastic deformation behavior is induced when the stress

K. Hashiguchi, *Foundations of Elastoplasticity: Subloading Surface Model*,
DOI 10.1007/978-3-319-48821-9_9

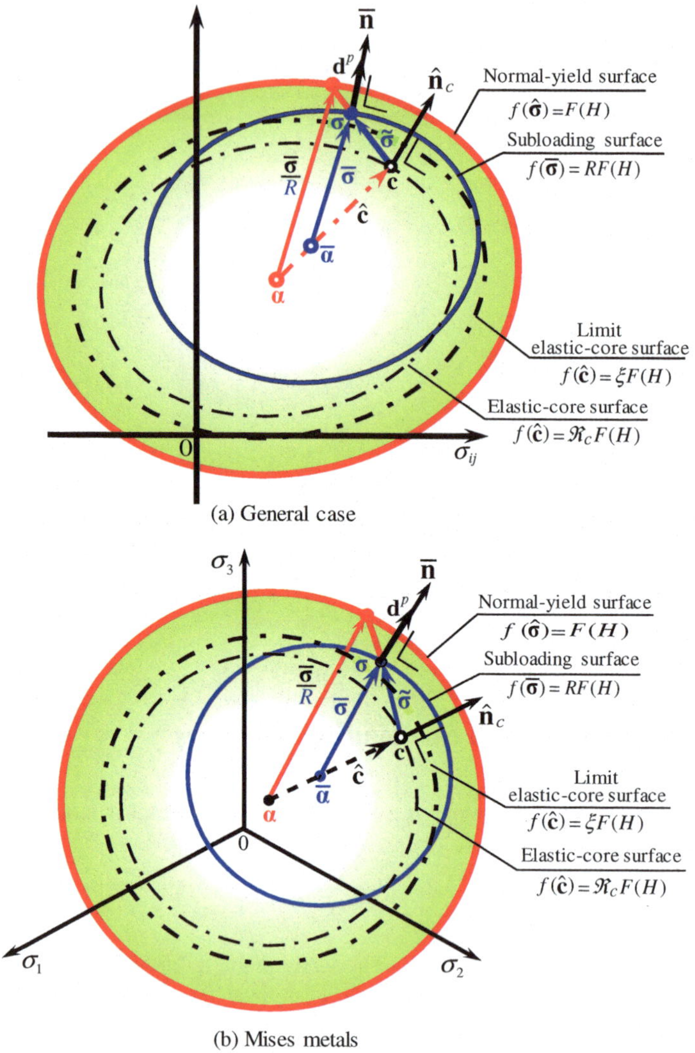

Fig. 9.1 Normal-yield, subloading and elastic-core surfaces

lies on it fulfilling $R = 0$ as will be explained later, the following relation holds (see Fig. 9.1).

$$\mathbf{c} - \bar{\boldsymbol{\alpha}} = R(\mathbf{c} - \boldsymbol{\alpha}) \tag{9.4}$$

which yields

$$\bar{\boldsymbol{\alpha}} = \mathbf{c} - R\hat{\mathbf{c}} \tag{9.5}$$

$$\bar{\boldsymbol{\sigma}} = \tilde{\boldsymbol{\sigma}} + R\hat{\mathbf{c}} \tag{9.6}$$

where

$$\left.\begin{array}{l} \hat{\mathbf{c}} \equiv \mathbf{c} - \boldsymbol{\alpha} \\ \tilde{\boldsymbol{\sigma}} \equiv \boldsymbol{\sigma} - \mathbf{c} \end{array}\right\} \tag{9.7}$$

All the relations in Eqs. (9.4–9.6) hold by virtue of the similarity of the subloading surface to the normal-yield surface.

Adopt the associated flow rule for the subloading surface:

$$\mathbf{d}^p = \dot{\bar{\lambda}}\,\bar{\mathbf{n}} \ (\dot{\bar{\lambda}} = ||\mathbf{d}^p|| > 0) \tag{9.8}$$

where

$$\bar{\mathbf{n}} \equiv \frac{\partial f(\bar{\boldsymbol{\sigma}})}{\partial \bar{\boldsymbol{\sigma}}} \bigg/ \left\|\frac{\partial f(\bar{\boldsymbol{\sigma}})}{\partial \bar{\boldsymbol{\sigma}}}\right\| \ (||\bar{\mathbf{n}}|| = 1) \tag{9.9}$$

The rate of the isotropic hardening variable is described by using the function f_{Hn} given from Eqs. (6.37) and (6.42) with Eq. (9.8) and the replacement of $\hat{\mathbf{n}}$ to $\bar{\mathbf{n}}$ as follows:

$$\dot{H}(\boldsymbol{\sigma}, H; \mathbf{d}^p) = f_{Hn}(\boldsymbol{\sigma}, H; \bar{\mathbf{n}})\,\dot{\bar{\lambda}} \tag{9.10}$$

The rate of the kinematic hardening variable is described by using the function $\mathbf{f}_{kn}$ given from (6.103) with Eq. (9.8) and the replacement of $\hat{\mathbf{n}}$ to $\bar{\mathbf{n}}$ as follows:

$$\mathring{\boldsymbol{\alpha}} = \dot{\bar{\lambda}}\,\bar{\mathbf{f}}_{kn}, \quad \bar{\mathbf{f}}_{kn} = c_k\left(\bar{\mathbf{n}}' - \frac{1}{b_k}||\mathbf{n}'||\boldsymbol{\alpha}\right) \tag{9.11}$$

9.2   Evolution Rule of Elastic-Core

The most elastic deformation behavior is induced in the state that the stress lies on the similarity-center, i.e. $\boldsymbol{\sigma} = \mathbf{c}$ leading to $R = 0$. Then, the similarity-center $\mathbf{c}$ is interpreted as the most elastic stress state so that let it be called the *elastic-core* or *elastic-center*. Here, note that the elastic-core $\mathbf{c}$ moves following the stress $\boldsymbol{\sigma}$ in the plastic loading process. However, from the physical point of view, the elastic-core should not approach the normal-yield surface without limitation in order that the smooth transition from the elastic to the plastic state is described always, although the small yield surface enclosing a purely-elastic region goes up unlimitedly contacting to the yield surface describing the fully-plastic state in the cyclic kinematic-hardening models, e.g. the multi surface, the two surface and the superposed kinematic hardening models. Further, from the mathematical point of view, the subloading surface is not determined uniquely if the stress coincides with the similarity-center lying just on the normal-yield surface.

Now, let the following *elastic-core surface* be introduced, which always passes through the elastic-core $\mathbf{c}$ and maintains a similarity to the normal-yield surface with respect to the kinematic-hardening variable $\boldsymbol{\alpha}$.

$$\boxed{f(\hat{\mathbf{c}}) = \Re_c F(H)}, \quad \text{i.e. } \Re_c = f(\hat{\mathbf{c}})/F(H) \tag{9.12}$$

where $\Re_c$ designates the ratio of the size of the elastic-core surface to the normal-yield surface (see Fig. 9.1) so that let it be called the *elastic-core yield ratio*. Then, let it be postulated that the elastic-core can never reach the normal-yield surface designating the fully-plastic stress state so that the elastic-core does not go over the following *limit elastic-core surface*.

$$f(\hat{\mathbf{c}}) = \xi F(H) \tag{9.13}$$

where $\xi(<1)$ is material constant or function in general and the following inequality must be satisfied.

$$f(\hat{\mathbf{c}}) \leq \xi F(H), \quad \text{i.e. } \Re_c \leq \xi \tag{9.14}$$

Let the translation rule of elastic-core be formulated as

$$\boxed{\overset{\circ}{\mathbf{c}} = c\left(\mathbf{d}^p - \frac{\Re_c}{\xi}\|\mathbf{d}^p\|\hat{\mathbf{n}}_c\right)} = \dot{\bar{\lambda}}\,\bar{\mathbf{f}}_{cn} \tag{9.15}$$

$$\bar{\mathbf{f}}_{cn} \equiv c\left(\bar{\mathbf{n}} - \frac{\Re_c}{\xi}\hat{\mathbf{n}}_c\right) \tag{9.16}$$

where c is a material parameter and

$$\hat{\mathbf{n}}_c \equiv \frac{\partial f(\hat{\mathbf{c}})}{\partial \mathbf{c}} \Bigg/ \left\|\frac{\partial f(\hat{\mathbf{c}})}{\partial \mathbf{c}}\right\| (\|\hat{\mathbf{n}}_c\| = 1) \tag{9.17}$$

Equation (9.15) is nonlinear in $\mathbf{d}^p$ as well as the nonlinear kinematic hardening rule in Eq. (9.11). The translation of the elastic-core from various point for input of various plastic strain rates is illustrated in Fig. 9.2.

The following facts for the translation of the elastic-core are recognized from Eq. (9.15).

(1) The elastic-core translates more inward direction of the elastic-core surface than the direction of the plastic strain rate, except for the state that the elastic-core lies just on the back-stress ($\Re_c = 0$).
(2) The following relations hold for Eq. (9.15).

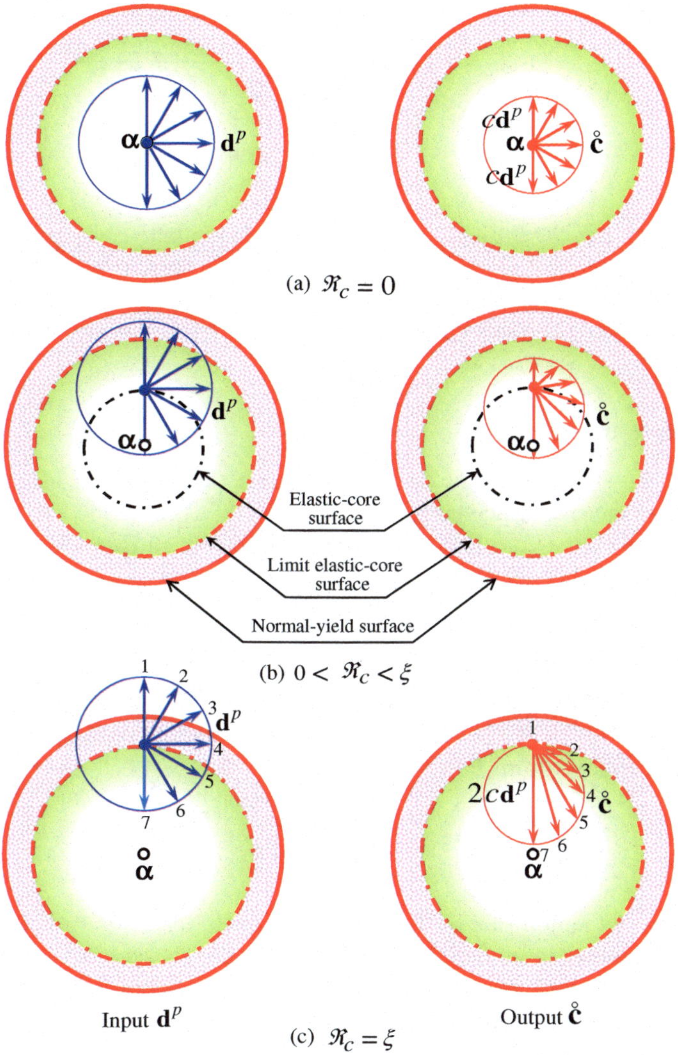

Fig. 9.2 Translation of elastic-core

$$\mathring{\mathbf{c}} = \begin{cases} c\mathbf{d}^p \text{ for } \mathscr{R}_c = 0 \\ \left.\begin{array}{l} \mathbf{O} \text{ for } \bar{\mathbf{n}} = \hat{\mathbf{n}}_c \\ c(\mathbf{d}^p - \|\mathbf{d}^p\|\hat{\mathbf{n}}_c) = c\|\mathbf{d}^p\|(\bar{\mathbf{n}} - \hat{\mathbf{n}}_c) \\ 2c\mathbf{d}^p \text{ for } \bar{\mathbf{n}} = -\hat{\mathbf{n}}_c \end{array}\right\} \text{for } \mathscr{R}_c = \xi \end{cases} \tag{9.18}$$

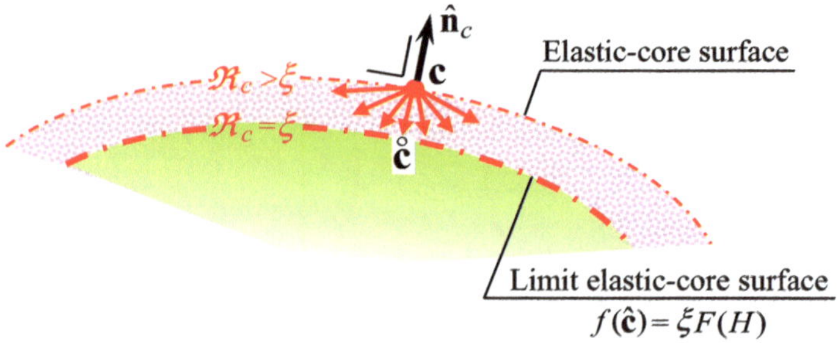

Fig. 9.3 Elastic-core is automatically pulled-back to the limit elastic-core surface when it goes out from the surface

which leads to the facts:

(a) When the elastic-core lies on the back-stress ($\mathcal{R}_c = 0$), it translates towards the direction of plastic strain rate.
(b) When the elastic-core lies on the limit elastic-core surface ($\mathcal{R}_c = \xi$),
  (i) it does not translate when the plastic strain rate is induced in the outward-normal of the limit elastic-core surface ($\bar{\mathbf{n}} = \hat{\mathbf{n}}_c$).
  (ii) it translates inward-normal of the limit elastic-core surface in the highest speed $2c\mathbf{d}^p$, when the plastic strain rate is induced in the inward-normal of the limit elastic-core surface ($\bar{\mathbf{n}} = -\hat{\mathbf{n}}_c$).

Further, it follows from Eq. (9.15) that

$$\hat{\mathbf{n}}_c : \overset{\circ}{\mathbf{c}} = c\|\mathbf{d}^p\|\left(\hat{\mathbf{n}}_c : \bar{\mathbf{n}} - \frac{\mathcal{R}_c}{\xi}\right) < 0 \quad \text{for } \mathcal{R}_c > \xi \tag{9.19}$$

Therefore, the evolution rule of the elastic-core in Eq. (9.15) is furnished with the distinguished ability in numerical calculation that the elastic-core is automatically pulled-back to the limit elastic-core surface when it goes out from that surface by the input of finite numerical increment as shown in Fig. 9.3.

The following relation holds in the uniaxial loading process for the Mises material ($f(\hat{\boldsymbol{\sigma}}) = \sqrt{3/2}\|\hat{\boldsymbol{\sigma}}'\|$) from Eq. (9.15), denoting the axial components of $\mathbf{c}, \boldsymbol{\alpha}$ and $\mathbf{d}^p$ as c_a, α_a and $\dot{\varepsilon}^p_a$.

$$\begin{aligned}\dot{c}_a &= c\left[\dot{\varepsilon}^p_a - \frac{\dfrac{|c_a-\alpha_a|}{F}}{\xi}\sqrt{\frac{3}{2}}|\dot{\varepsilon}^p_a|\frac{\dfrac{2}{3}(c_a-\alpha_a)}{\sqrt{\dfrac{2}{3}}|c_a-\alpha_a|}\right]\\ &= c\left[\dot{\varepsilon}^p_a - \frac{\pm(c_a-\alpha_a)}{\xi F}(\pm\dot{\varepsilon}^p_a)\frac{c_a-\alpha_a}{\pm(c_a-\alpha_a)}\right] \quad (\text{upper: } \dot{\varepsilon}^p_a > 0,\ \text{lower: } \dot{\varepsilon}^p_a < 0)\end{aligned}$$

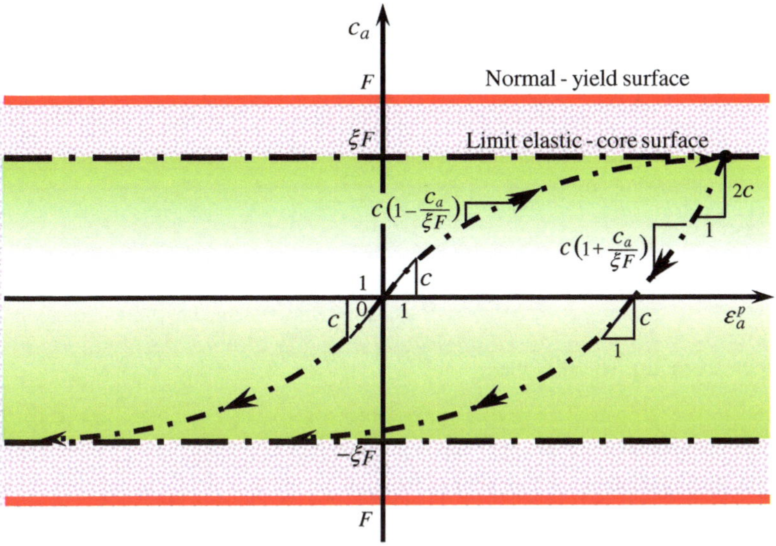

Fig. 9.4 Translation of elastic-core in uniaxial loading for Mises material without hardening

leading to

$$\dot{c}_a = c\left[1 \mp \frac{c_a - \alpha_a}{\xi F}\right]\dot{\varepsilon}_a^p \quad \left(\text{upper: } \dot{\varepsilon}_a^p > 0, \ \ \text{lower: } \dot{\varepsilon}_a^p < 0\right) \tag{9.20}$$

which is time-integrated for $F = \text{const.}$ and $\alpha_a = 0$ as follows:

$$\frac{\xi F - c_a}{\xi F - c_{a0}} = \exp\left[\mp \frac{c}{\xi F}\left(\varepsilon_a^p - \varepsilon_{a0}^p\right)\right] \tag{9.21}$$

where c_{a0} and ε_{a0}^p are the initial values of c_a and ε_a^p. The relation of c_a and ε_a^p is shown in Fig. 9.4.

The physical interpretations for the hyperelasticity and the kinematic hardening rule were given in Sect. 6.9. The analogous interpretation will be given for the translation rule of the elastic-core in this section.

Let the plastic strain $\boldsymbol{\varepsilon}^p$ be additively decomposed into the storage part $\boldsymbol{\varepsilon}_{cs}^p$ which induces the translation of the elastic-core and the dissipative part $\boldsymbol{\varepsilon}_{cd}^p$, i.e.

$$\boldsymbol{\varepsilon}^p = \boldsymbol{\varepsilon}_{cs}^p + \boldsymbol{\varepsilon}_{cd}^p \tag{9.22}$$

The elastic-core $\mathbf{c}$ which describes the variation of substructures in material is given by the partial derivatives of the potential function of the storage parts of the plastic strain so that the following relation holds.

$$\mathbf{c} = \frac{\partial \psi^c(\boldsymbol{\varepsilon}^p_{cs})}{\partial \boldsymbol{\varepsilon}^p_{cs}} \tag{9.23}$$

Now, assuming

$$\psi^c(\boldsymbol{\varepsilon}^p_{cs}) = \frac{1}{2} c \boldsymbol{\varepsilon}^p_{cs} : \boldsymbol{\varepsilon}^p_{cs} \tag{9.24}$$

it follows from Eq. (6.109) that

$$\mathbf{c} = c\boldsymbol{\varepsilon}^p_{cs} = c(\boldsymbol{\varepsilon}^p - \boldsymbol{\varepsilon}^p_{cd}) \tag{9.25}$$

The time-differentiation of Eq. (6.111) reads:

$$\dot{\mathbf{c}} = c(\dot{\boldsymbol{\varepsilon}}^p - \dot{\boldsymbol{\varepsilon}}^p_{cd}) \tag{9.26}$$

Further, assume the rates of the dissipative parts in the elastic-core as follows:

$$\dot{\boldsymbol{\varepsilon}}^p_{cd} = \frac{\Re_c}{\xi} ||\dot{\boldsymbol{\varepsilon}}^p|| \hat{\mathbf{n}}_c = \frac{\Re_c}{\xi} \dot{\bar{\lambda}} \hat{\mathbf{n}}_c \tag{9.27}$$

The dissipative part in the rate of elastic-core in Eq. (9.27) satisfies the positivity of the dissipation energy, i.e. $\hat{\mathbf{c}} : \dot{\boldsymbol{\varepsilon}}^p_{cd} = (\hat{\mathbf{c}} : \hat{\mathbf{n}}_c)(\Re_c/\xi)||\dot{\boldsymbol{\varepsilon}}^p|| \geq 0$. Substituting Eq. (9.27) into Eq. (9.26), one obtains the translation rule of the elastic-core which is similar to the kinematic hardening rule in Eq. (6.114) as follows:

$$\dot{\mathbf{c}} = c\left(\dot{\boldsymbol{\varepsilon}}^p - \frac{\Re_c}{\xi} ||\dot{\boldsymbol{\varepsilon}}^p|| \hat{\mathbf{n}}_c\right) \tag{9.28}$$

The translation rule of the elastic-core in Eq. (9.15) is obtained from Eq. (9.28) by replacing the plastic strain rate $\dot{\boldsymbol{\varepsilon}}^p$ to $\mathbf{d}^p$ and the material-time derivative $(^\bullet)$ to the corotational rate $(^\circ)$.

The elastic-core $\mathbf{c}$ can be calculated directly from Eq. (9.25) in the hyperelastic-like relation by substituting $\boldsymbol{\varepsilon}^p$ and $\boldsymbol{\varepsilon}^p_{cd}$ obtained by the time-integration of Eq. (9.27) without the time-integration of Eq. (9.28). However, in the forward-Euler calculation method within the framework of the infinitesimal theory for which the spin of material is ignored, the accuracy in calculation of elastic-core using the hyperelastic-like relation is identical to that by the direct time-integration of the rate itself.

9.3 Plastic Strain Rate

The material-time derivative of Eq. (9.2) leads to the consistency condition of the subloading surface in the corotational time-derivative:

$$\frac{\partial f(\bar{\boldsymbol{\sigma}})}{\partial \bar{\boldsymbol{\sigma}}} : \overset{\circ}{\boldsymbol{\sigma}} - \frac{\partial f(\bar{\boldsymbol{\sigma}})}{\partial \bar{\boldsymbol{\sigma}}} : \overset{\circ}{\bar{\boldsymbol{\alpha}}} - R\dot{F} - \dot{R}F = 0 \tag{9.29}$$

Here, one has

$$\frac{\partial f(\bar{\boldsymbol{\sigma}})}{\partial \bar{\boldsymbol{\sigma}}} : \bar{\boldsymbol{\sigma}} = f(\bar{\boldsymbol{\sigma}}) = RF \tag{9.30}$$

based on the homogeneous function $f(\bar{\boldsymbol{\sigma}})$ of $\bar{\boldsymbol{\sigma}}$ in degree-one by the Euler's theorem. Then, it follows that

$$\bar{\mathbf{n}} : \bar{\boldsymbol{\sigma}} = \frac{\dfrac{\partial f(\bar{\boldsymbol{\sigma}})}{\partial \bar{\boldsymbol{\sigma}}} : \bar{\boldsymbol{\sigma}}}{\left\| \dfrac{\partial f(\bar{\boldsymbol{\sigma}})}{\partial \bar{\boldsymbol{\sigma}}} \right\|} = \frac{RF}{\left\| \dfrac{\partial f(\bar{\boldsymbol{\sigma}})}{\partial \bar{\boldsymbol{\sigma}}} \right\|}, \quad \frac{1}{\left\| \dfrac{\partial f(\bar{\boldsymbol{\sigma}})}{\partial \bar{\boldsymbol{\sigma}}} \right\|} = \frac{\bar{\mathbf{n}} : \bar{\boldsymbol{\sigma}}}{RF} \tag{9.31}$$

The substitution of Eq. (9.31) into Eq. (9.29) leads to

$$\bar{\mathbf{n}} : \overset{\circ}{\boldsymbol{\sigma}} - \bar{\mathbf{n}} : \left[\left(\frac{\dot{F}}{F} + \frac{\dot{R}}{R} \right) \bar{\boldsymbol{\sigma}} + \overset{\circ}{\bar{\boldsymbol{\alpha}}} \right] = 0 \tag{9.32}$$

Here, the corotational rate variable $\overset{\circ}{\bar{\boldsymbol{\alpha}}}$ is described from Eq. (9.5) as

$$\overset{\circ}{\bar{\boldsymbol{\alpha}}} = R\overset{\circ}{\hat{\boldsymbol{\alpha}}} + (1 - R)\,\overset{\circ}{\mathbf{c}} - \dot{R}\,\hat{\mathbf{c}} \tag{9.33}$$

The substitution of Eq. (9.33) into Eq. (9.32) leads to

$$\bar{\mathbf{n}} : \overset{\circ}{\boldsymbol{\sigma}} - \bar{\mathbf{n}} : \left[\frac{\dot{F}}{F} \bar{\boldsymbol{\sigma}} + R\overset{\circ}{\hat{\boldsymbol{\alpha}}} + (1 - R)\overset{\circ}{\mathbf{c}} + \frac{\dot{R}}{R}(\bar{\boldsymbol{\sigma}} - R\hat{\mathbf{c}}) \right] = 0 \tag{9.34}$$

Noting the relation

$$\bar{\boldsymbol{\sigma}} - R\hat{\mathbf{c}} = \boldsymbol{\sigma} - \bar{\boldsymbol{\alpha}} - (\mathbf{c} - \bar{\boldsymbol{\alpha}}) = \tilde{\boldsymbol{\sigma}} \tag{9.35}$$

it follows from Eq. (9.34) that

$$\bar{\mathbf{n}} : \overset{\circ}{\boldsymbol{\sigma}} - \bar{\mathbf{n}} : \left[\frac{\dot{F}}{F} \bar{\boldsymbol{\sigma}} + R\,\overset{\circ}{\hat{\boldsymbol{\alpha}}} + (1 - R)\,\overset{\circ}{\mathbf{c}} + \frac{\dot{R}}{R} \tilde{\boldsymbol{\sigma}} \right] = 0 \tag{9.36}$$

The substitutions of Eqs. (7.9), (9.10), (9.11) and (9.15) into Eq. (9.36) leads to

$$\bar{\mathbf{n}}:\overset{\circ}{\boldsymbol{\sigma}}-\bar{\mathbf{n}}:\left[\frac{F'}{F}\dot{\bar{\lambda}}f_{Hn}\bar{\boldsymbol{\sigma}}+R\dot{\bar{\lambda}}\bar{\mathbf{f}}_{kn}+(1-R)\dot{\bar{\lambda}}\bar{\mathbf{f}}_{cn}+\frac{U}{R}\dot{\bar{\lambda}}\tilde{\boldsymbol{\sigma}}\right]=0 \tag{9.37}$$

from which the plastic multiplier $\dot{\bar{\lambda}}$ and the plastic strain rate $\mathbf{d}^p$ are given as follows:

$$\dot{\bar{\lambda}}=\frac{\bar{\mathbf{n}}:\overset{\circ}{\boldsymbol{\sigma}}}{\bar{M}^p},\quad \mathbf{d}^p=\frac{\bar{\mathbf{n}}:\overset{\circ}{\boldsymbol{\sigma}}}{\bar{M}^p}\bar{\mathbf{n}} \tag{9.38}$$

where

$$\boxed{\bar{M}^p=\bar{\mathbf{n}}:\left[\frac{F'}{F}f_{Hn}\bar{\boldsymbol{\sigma}}+R\bar{\mathbf{f}}_{kn}+(1-R)\bar{\mathbf{f}}_{cn}+\frac{U}{R}\tilde{\boldsymbol{\sigma}}\right]} \tag{9.39}$$

The plastic modulus in Eq. (9.39) reduces to Eq. (6.92) for the conventional model in the normal-yield state ($R = 1$). In addition, it should be noted that the exactly same calculation result to that due to the conventional model can be obtained by merely setting the material parameter $u \to \infty$, i.e. using a quite large number for u in the evolution rule of the normal-yield ratio in Eq. (7.20). It is quite convenient in the comparison of mechanical response with the conventional model.

9.4 Stain Rate Versus Stress Rate Relations

The strain rate is given by substituting Eqs. (6.29) and (9.38) into Eq. (6.27) as follows:

$$\mathbf{d}=\mathbf{E}^{-1}:\overset{\circ}{\boldsymbol{\sigma}}+\frac{\bar{\mathbf{n}}:\overset{\circ}{\boldsymbol{\sigma}}}{\bar{M}^p}\bar{\mathbf{n}}=\left(\mathbf{E}^{-1}+\frac{\bar{\mathbf{n}}\otimes:\bar{\mathbf{n}}}{\bar{M}^p}\right):\overset{\circ}{\boldsymbol{\sigma}} \tag{9.40}$$

from which the magnitude of plastic strain rate described in terms of the strain rate, denoted by $\dot{\bar{\Lambda}}$ instead of $\dot{\bar{\lambda}}$, in the flow rule of Eq. (9.8) is given as follows:

$$\dot{\bar{\Lambda}}=\frac{\bar{\mathbf{n}}:\mathbf{E}:\mathbf{d}}{\bar{M}^p+\bar{\mathbf{n}}:\mathbf{E}:\bar{\mathbf{n}}},\quad \mathbf{d}^p=\frac{\bar{\mathbf{n}}:\mathbf{E}:\mathbf{d}}{\bar{M}^p+\bar{\mathbf{n}}:\mathbf{E}:\bar{\mathbf{n}}}\bar{\mathbf{n}} \tag{9.41}$$

The stress rate is given from Eq. (6.29) with Eq. (9.41) as follows:

$$\overset{\circ}{\boldsymbol{\sigma}}=\mathbf{E}:\mathbf{d}-\frac{\bar{\mathbf{n}}:\mathbf{E}:\mathbf{d}}{\bar{M}^p+\bar{\mathbf{n}}:\mathbf{E}:\bar{\mathbf{n}}}\mathbf{E}:\bar{\mathbf{n}}=\left(\mathbf{E}-\frac{\mathbf{E}:\bar{\mathbf{n}}\otimes\bar{\mathbf{n}}:\mathbf{E}}{\bar{M}^p+\bar{\mathbf{n}}:\mathbf{E}:\bar{\mathbf{n}}}\right):\mathbf{d} \tag{9.42}$$

9.5 Loading Criterion

The loading criterion is given as follows (Hashiguchi 2000, 2013):

$$\left.\begin{array}{ll}\mathbf{d}^p \neq \mathbf{O} & \text{for } \overset{\bullet}{\Lambda} > 0 \\ \mathbf{d}^p = \mathbf{O} & \text{for } \overset{\bullet}{\Lambda} \leq 0\end{array}\right\} \tag{9.43}$$

or

$$\left\{\begin{array}{l}\mathbf{d}^p \neq \mathbf{O} \ \text{ for } \bar{\mathbf{n}}:\mathbf{E}:\mathbf{d} > 0 \\ \mathbf{d}^p = \mathbf{O} \ \text{ for } \bar{\mathbf{n}}:\mathbf{E}:\mathbf{d} \leq 0\end{array}\right. \tag{9.44}$$

premising $\bar{M}^p + \bar{\mathbf{n}}:\mathbf{E}:\bar{\mathbf{n}} > 0$ based on the same physical background to that for Eq. (6.68) described in Sect. 6.5, where the judgment whether or not the stress reaches the yield surface is not required since the plastic strain rate is induced continuously as the stress approaches the normal-yield surface. Equations (9.43) and (9.44) are applicable not only to the hardening state but also to the perfectly-plastic and softening state (Hashiguchi 2000, 2013). A small loading increment must be input in order to avoid the risk that the subloading surface once contracts and then expands, so that a plastic strain rate is induced at the moment of stress reversal event even when $\bar{\mathbf{n}}:\mathbf{E}:\mathbf{d} < 0$ holds as was described in Sect. 7.2.

9.6 Calculation of Normal-Yield Ratio

Substituting Eq. (9.6) into Eq. (9.2), the subloading surface is described as follows:

$$f(\tilde{\boldsymbol{\sigma}} + R\hat{\mathbf{c}}) = RF(H) \tag{9.45}$$

from which the normal-yield ratio R is calculated by substituting the updated values of $\boldsymbol{\sigma}, \boldsymbol{\alpha}, \mathbf{c}, H$.

The normal-yield ratio can be calculated by the following two methods:

(1) Calculating it by Eq. (9.45) in both of the plastic (loading) and the elastic (unloading) processes after all the other variables $\boldsymbol{\sigma}, \boldsymbol{\alpha}, \mathbf{c}, H$ were updated.
(2) Calculating it by the time-integration of Eq. (7.9) in the plastic (loading) process. Here, the analytical time-integration in Eq. (7.19) is beneficial to the enhancement of numerical analysis in the return-mapping projection (cf. e.g. Simo and Hughes 1998; Hashiguchi 2013b). However, its use spoils the

controlling function to pull-back the stress to the normal-yield surface in numerical analysis based on the forward-Euler method.

The second method (2) would be superior to the first method (1), since the normal-yield ratio is calculated directly from the plastic strain rate.

9.7 Improvement of Inverse and Reloading Responses

The unique relation $\varepsilon^p - \varepsilon^p_0 = f(R - R_0)$ holds under the initial condition $\varepsilon^p = \varepsilon^p_0$: $R = R_0$ in the monotonic loading process if U in Eq. (7.9) is the function of only the normal-yield ratio R. Therefore, ε^p induced during a certain change of R in the monotonic loading process is identical irrespective of the difference of loading processes, e.g. the initial loading, the reloading and the inverse loading and of the proportional and non-proportional loadings. This property causes the description that the returning of the reloading stress–strain curve to the previous loading curve is unrealistically gentle as shown in Fig. 9.5a. Therefore, it engenders the impertinent prediction of cyclic loading behavior, i.e. the prediction of the unrealistically large plastic strain accumulation during the cyclic loading process as shown in the upper part of Fig. 9.5b. This insufficiency in the past formulation of the subloading surface model was criticized intensely by Dafalias, stating "*the predictions reported in* Hashiguchi (1980) *for the uniaxial loading of metals were quite unrealistic, basically due to the strong undershooting phenomenon*" (Dafalias 1986, p. 980). However, the insufficiency in the description of deformation behavior by the past formulation of the subloading surface model would not originate from the intrinsic nature of this model contrary to the criticism by Dafalias (1986). In what follows, the past formulation of subloading surface model will be modified so as to remedy the insufficiency in the description of reloading behavior.
First, note the following facts:

(1) The difference between the curvatures in the reloading and the inverse loading curves becomes larger as the plastic deformation proceeds continuously. This fact is known as the Masing rule (Masing 1926).
(2) The similarity-center corresponding to the most elastic stress state approaches the normal-yield surface, following the current stress, as the plastic deformation proceeds continuously, and the approaching degree of the similarity-center to the normal-yield surface is expressed by the elastic-core yield ratio $\mathfrak{R}_c$ in Eq. (9.12) as described in Sect. 9.1
(3) The transition from the elastic to plastic state is more abrupt, i.e. the curvature of stress–strain curve is greater for a larger value of the material parameter u in the function $U(R)$ in Eqs. (7.12), (7.13) and (7.20), as described in Sect. 7.2. Therefore, the increase in the curvature of stress-strain curve can be described by giving a larger value to the material parameter u.
(4) By the facts (1–3), the difference between the values of u for the reloading and the inverse loading states should be greater for the larger value of $\mathfrak{R}_c$.

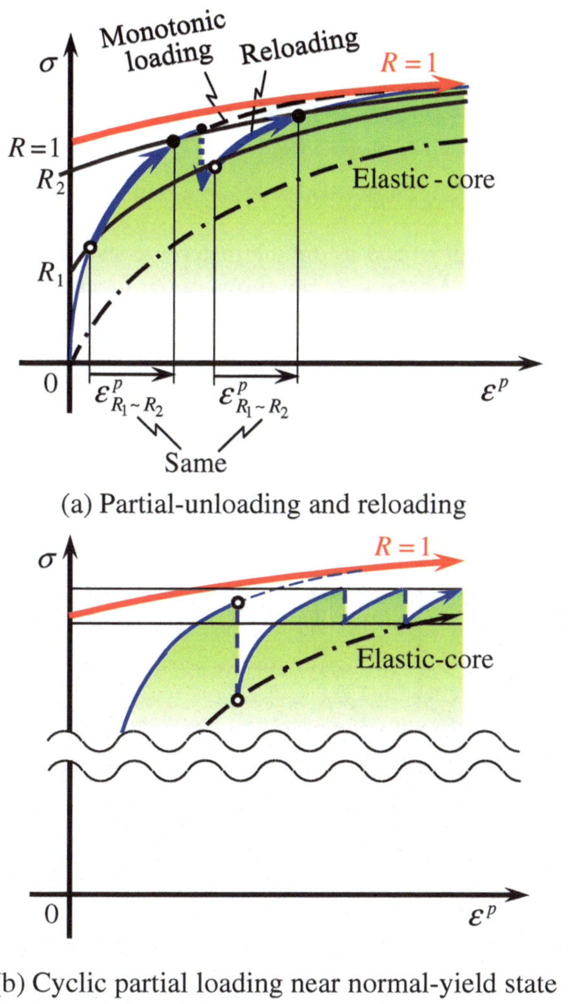

Fig. 9.5 The defect of past subloading surface model: Unrealistically gentle returning to preceding loading curve

(5) The direction $\bar{\mathbf{n}}$ of plastic strain rate is near the outward-normal $\hat{\mathbf{n}}_c$ (Eq. 9.17) of the elastic-core surface in the reloading process but it is far from $\hat{\mathbf{n}}_c$ in the unloading process. Then, the degree to which the process is near the reloading process can be expressed by the following scalar product of these unit tensors:

$$\boxed{C_\sigma \equiv \hat{\mathbf{n}}_c : \bar{\mathbf{n}} \quad (-1 \leq C_\sigma \leq 1)} \tag{9.46}$$

Eventually, introducing the variables $\mathfrak{R}_c$ and C_σ, let the material parameter u in Eq. (7.20) be extended as follows:

$$\boxed{u = \bar{u}\exp(u_c \mathfrak{R}_c C_\sigma)} \quad (\bar{u}\exp(-u_c\xi) \leq u \leq \bar{u}\exp(u_c\xi))$$

$$\left(= \begin{cases} \bar{u}\exp(u_c\xi) \text{ (largest) for } \mathfrak{R}_c = \xi \text{ and } C_\sigma = 1 \\ \bar{u} \quad \text{(average) for } \mathfrak{R}_c = 0 \text{ or } C_\sigma = 0 \\ \bar{u}\exp(-u_c\xi) \text{ (smallest) for } \mathfrak{R}_c = \xi \text{ and } C_\sigma = -1 \end{cases} \right) \tag{9.47}$$

where $\bar{u}$ (average value of u) and u_c are the material constant. u is the continuous function of the variables $\mathfrak{R}_c$ and C_σ. The forms of the function u for the particular states are shown in the bracket. $C_\sigma = 1$, 0 and -1 designate the states that the

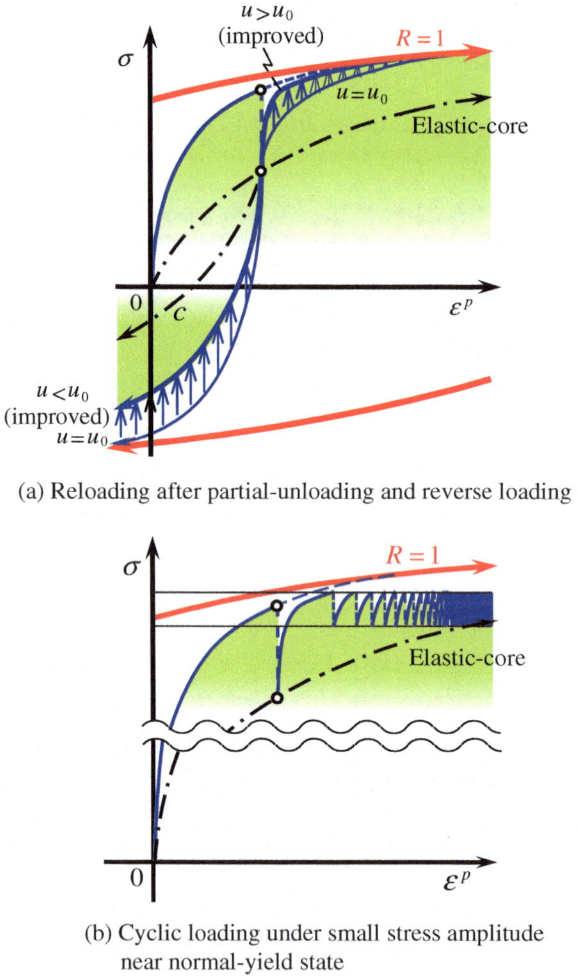

Fig. 9.6 Stress-plastic strain curve predicted by the modified evolution rule of normal-yield ratio: rapid recovery to preceding monotonic loading curve

plastic strain rate is directed outward-normal, tangential and inward-normal, respectively, to the elastic-core surface. $\mathfrak{R}_c = 0$ and $\mathfrak{R}_c = \xi$ designate the states that the similarity-center lies on the center of the normal-yield surface and on the limit elastic-core surface, respectively. By this modification, the phenomenon that the reloading curve after a partial unloading returns rapidly to the preceding loading curve and the curvature of inverse loading curve decreases can be described realistically as shown in Fig. 9.6a. Besides, the plastic strain accumulation for the cyclic loading process in the neighborhood of yield surface is suppressed as shown in the lower part of Fig. 9.6b.

9.8 Plastic Spin

The plastic spin in Eq. (6.25) is given analogously to Eq. (6.106) as follows:

$$\boxed{\mathbf{w}^p = \eta^p(\boldsymbol{\sigma}\mathbf{d}^p - \mathbf{d}^p\boldsymbol{\sigma})} = \eta^p\,\dot{\lambda}(\boldsymbol{\sigma}\bar{\mathbf{n}} - \bar{\mathbf{n}}\boldsymbol{\sigma}), \quad \boldsymbol{\omega}^p = \eta^p(\boldsymbol{\sigma}\bar{\mathbf{n}} - \bar{\mathbf{n}}\boldsymbol{\sigma}) \tag{9.48}$$

9.9 Incorporation of Tangential-Inelastic Strain Rate

The tangential-inelastic strain rate described in Sect.7.7 is extended for the extended subloading surface model as shown in this section.

The tangential-deviatoric stress rate $\bar{\overset{\circ}{\boldsymbol{\sigma}}}{}'_t$ (Fig. 9.7) is defined by

$$\left.\begin{array}{l} \bar{\overset{\circ}{\boldsymbol{\sigma}}}{}'_t \equiv \bar{\mathfrak{T}}'_t : \overset{\circ}{\boldsymbol{\sigma}} = \overset{\circ}{\boldsymbol{\sigma}}{}' - \bar{\overset{\circ}{\boldsymbol{\sigma}}}{}'_n \quad \left(\bar{\mathbf{n}} : \bar{\overset{\circ}{\boldsymbol{\sigma}}}{}'_t = \bar{\mathfrak{n}}' : \bar{\overset{\circ}{\boldsymbol{\sigma}}}{}'_t = 0\right) \\ \bar{\overset{\circ}{\boldsymbol{\sigma}}}{}'_n \equiv (\bar{\mathfrak{n}}' : \overset{\circ}{\boldsymbol{\sigma}})\bar{\mathfrak{n}}' = (\bar{\mathfrak{n}}' \otimes \bar{\mathfrak{n}}') : \overset{\circ}{\boldsymbol{\sigma}} \end{array}\right\} \tag{9.49}$$

$$\bar{\mathfrak{n}}' \equiv \frac{\bar{\mathbf{n}}'}{\|\bar{\mathbf{n}}'\|} = \left(\frac{\partial f(\bar{\boldsymbol{\sigma}})}{\partial \bar{\boldsymbol{\sigma}}}\right)' \Bigg/ \left\| \left(\frac{\partial f(\bar{\boldsymbol{\sigma}})}{\partial \bar{\boldsymbol{\sigma}}}\right)' \right\| \left(\neq \bar{\mathbf{n}}' = \left(\frac{\partial f(\bar{\boldsymbol{\sigma}})}{\partial \bar{\boldsymbol{\sigma}}}\right)' \Bigg/ \left\| \frac{\partial f(\bar{\boldsymbol{\sigma}})}{\partial \bar{\boldsymbol{\sigma}}} \right\| \right) (\|\bar{\mathfrak{n}}'\| = 1) \tag{9.50}$$

$$\bar{\mathfrak{T}}'_t \equiv \mathcal{I}' - \bar{\mathfrak{n}}' \otimes \bar{\mathfrak{n}}', \quad \bar{\mathcal{T}}'_{t\,ijkl} \equiv \mathcal{I}'_{ijkl} - \bar{n}'_{ij}\bar{n}'_{kl} \tag{9.51}$$

fulfilling

$$\bar{\mathbf{n}}'_t \equiv \bar{\mathfrak{T}}'_t : \bar{\mathbf{n}} = \bar{\mathbf{n}}' - (\bar{\mathfrak{n}}' : \bar{\mathbf{n}})\bar{\mathfrak{n}}' = \mathbf{O} \tag{9.52}$$

The fourth-order tensor $\bar{\mathfrak{T}}'_t$ is the *tangential-deviatoric projection tensor* which transforms an arbitrary second-order tensor to the tangential part to the subloading

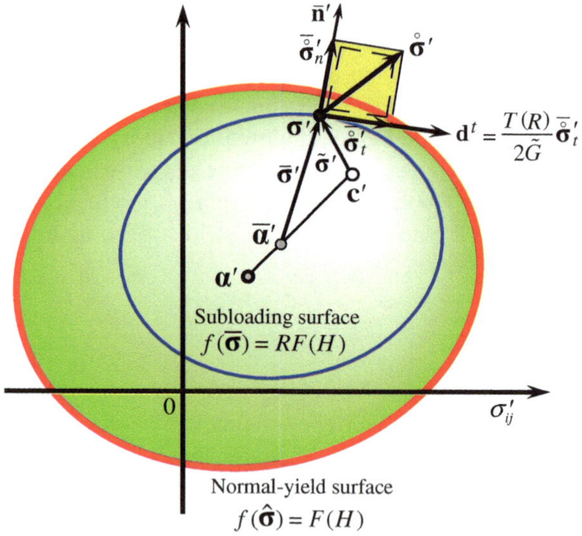

Fig. 9.7 Tangential-deviatoric stress rate and tangential-inelastic strain rate in the deviatoric stress plane

surface in the deviatoric stress space and the second-order tensor subjected to this projection is designated as $(\bar{\ \ })'_t$, i.e. $\bar{\mathbf{t}}'_t \equiv \bar{\mathfrak{T}}'_t \mathbf{:}\, \mathbf{t}$ leading further to $\bar{\mathfrak{T}}'_t \mathbf{:} \bar{\mathbf{t}}'_t = \bar{\mathbf{t}}'_t$.

All the relations in Eqs. (7.61–7.71) hold under the replacements of $\mathbf{n} \to \bar{\mathbf{n}}, \mathfrak{n} \to \bar{\mathfrak{n}}, (\ \ )'_t \to (\bar{\ \ })'_t, M^p \to \bar{M}^p$. Consequently, it follows that

$$\mathbf{d}^t = T(R)\mathbf{E}^{-1} \mathbf{:} \overset{\circ}{\bar{\boldsymbol{\sigma}}}{}'_t \tag{9.53}$$

$$\boxed{\mathbf{d} = \left(\mathbf{E}^{-1} + \frac{\bar{\mathbf{n}} \otimes \bar{\mathbf{n}}}{\bar{M}^p} + \frac{T(R)}{2G}\bar{\mathfrak{T}}'_t\right) \mathbf{:} \overset{\circ}{\boldsymbol{\sigma}}} \tag{9.54}$$

$$\boxed{\overset{\circ}{\boldsymbol{\sigma}} = \left(\mathbf{E} - \frac{\mathbf{E} \mathbf{:} \bar{\mathbf{n}} \otimes \bar{\mathbf{n}} \mathbf{:} \mathbf{E}}{\bar{M}^p + \bar{\mathbf{n}} \mathbf{:} \mathbf{E} \mathbf{:} \bar{\mathbf{n}}} - 2G\frac{T(R)}{1+T(R)}\bar{\mathfrak{T}}'_t\right) \mathbf{:} \mathbf{d}} \tag{9.55}$$

The loading criterion in Eq. (9.43) or (9.44) which was given for the equation without the tangential-inelastic strain rate holds as it is, as was described in Sect. 7.7.

The importance for the introduction of the tangential-inelastic strain rate will be verified by comparison with test data on a metal in Sect. 10.4.

Further, limiting to the plastically-incompressible materials satisfying $\bar{\mathfrak{n}}' = \bar{\mathbf{n}}' = \bar{\mathbf{n}}$, the tangential-inelastic strain rate can be given instead of Eq. (9.53) as follows:

$$\mathbf{d}^t = T(R)\mathbf{E}^{-1} : \overset{\circ}{\bar{\boldsymbol{\sigma}}}'_{tn}, \quad \overset{\circ}{\bar{\boldsymbol{\sigma}}}'_{tn} \equiv (\mathcal{I}' - c_t \bar{\mathbf{n}} \otimes \bar{\mathbf{n}}) : \overset{\circ}{\boldsymbol{\sigma}} \tag{9.56}$$

where c_t $(0 \le c_t \le 1)$ is the material parameter. The stress rate $\overset{\circ}{\bar{\boldsymbol{\sigma}}}'_{tn}$ is normal and tangential to the subloading surface for $c_t = 0$ and $c_t = 1$, respectively as illustrated in Fig. 9.8. The strain rate is given for Eq. (9.56) by

$$\mathbf{d} = \mathbf{E}^{-1} : \overset{\circ}{\boldsymbol{\sigma}} + \frac{\bar{\mathbf{n}} : \overset{\circ}{\boldsymbol{\sigma}}}{\bar{M}^p} \bar{\mathbf{n}} + T(R)\mathbf{E}^{-1} : (\mathcal{I}' - c_t \bar{\mathbf{n}} \otimes \bar{\mathbf{n}}) : \overset{\circ}{\boldsymbol{\sigma}} \tag{9.57}$$

from which it follows that

$$\begin{aligned} \bar{\mathbf{n}} : \mathbf{E} : \mathbf{d} &= \bar{\mathbf{n}} : \overset{\circ}{\boldsymbol{\sigma}} + \frac{\bar{\mathbf{n}} : \mathbf{E} : \bar{\mathbf{n}}}{\bar{M}^p} \bar{\mathbf{n}} : \overset{\circ}{\boldsymbol{\sigma}} + T(R)\bar{\mathbf{n}} : \overset{\circ}{\boldsymbol{\sigma}} - c_t T(R)\bar{\mathbf{n}} : \overset{\circ}{\boldsymbol{\sigma}} \\ &= [\bar{M}^p + \bar{\mathbf{n}} : \mathbf{E} : \bar{\mathbf{n}} + (1 - c_t)T(R)\bar{M}^p] \frac{\bar{\mathbf{n}} : \overset{\circ}{\boldsymbol{\sigma}}}{\bar{M}^p} \end{aligned}$$

leading to

$$\dot{\Lambda}\left(= \frac{\bar{\mathbf{n}} : \overset{\circ}{\boldsymbol{\sigma}}}{\bar{M}^p} \right) = \frac{\bar{\mathbf{n}} : \mathbf{E} : \mathbf{d}}{[1 + (1 - c_t)T(R)]\bar{M}^p + \bar{\mathbf{n}} : \mathbf{E} : \bar{\mathbf{n}}} \tag{9.58}$$

It follows taking account of the plastic multiplier in Eq. (9.58) that

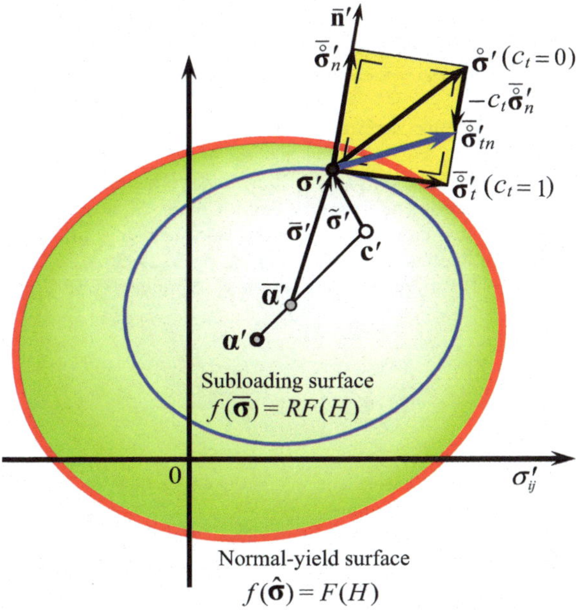

Fig. 9.8 Normal and tangential stress rates for subloading surface model in deviatoric stress plane for plastically-compressible materials.

$$
\begin{aligned}
\overset{\circ}{\boldsymbol{\sigma}} &= \mathbf{E} : (\mathbf{d} - \mathbf{d}^p - \mathbf{d}^t) \\
&= \mathbf{E} : \left\{ \mathbf{d} - \frac{\bar{\mathbf{n}} : \mathbf{E} : \mathbf{d}}{[1 + (1 - c_t)T(R)]\bar{M}^p + \bar{\mathbf{n}} : \mathbf{E} : \bar{\mathbf{n}}} \bar{\mathbf{n}} - T(R)\mathbf{E}^{-1} : (\mathcal{I}' - c_t \bar{\mathbf{n}} \otimes \bar{\mathbf{n}}) : \overset{\circ}{\boldsymbol{\sigma}} \right\}
\end{aligned}
$$

from which one has

$$
[\mathcal{I} + T(R)(\mathcal{I}' - c_t \bar{\mathbf{n}} \otimes \bar{\mathbf{n}})] : \overset{\circ}{\boldsymbol{\sigma}} = \mathbf{E} : \left\{ \mathbf{d} - \frac{\bar{\mathbf{n}} : \mathbf{E} : \mathbf{d}}{[1 + (1 - c_t)T(R)]\bar{M}^p + \bar{\mathbf{n}} : \mathbf{E} : \bar{\mathbf{n}}} \bar{\mathbf{n}} \right\}
$$

Consequently, the stress rate is given as follows:

$$
\overset{\circ}{\boldsymbol{\sigma}} = [\mathcal{I} + T(R)(\mathcal{I}' - c_t \bar{\mathbf{n}} \otimes \bar{\mathbf{n}})]^{-1} : \left\{ \mathbf{E} - \frac{\mathbf{E} : \bar{\mathbf{n}} \otimes \bar{\mathbf{n}} : \mathbf{E}}{[1 + (1 - c_t)T(R)]\bar{M}^p + \bar{\mathbf{n}} : \mathbf{E} : \bar{\mathbf{n}}} \right\} : \mathbf{d} \tag{9.59}
$$

References

Dafalias YF (1986) Bounding surface plasticity. I: Mathematical foundation and hypoplasticity. J Eng Mech (ASCE) 112:966–987

Hashiguchi K (1980) Constitutive equations of elastoplastic materials with elastic-plastic transition. J Appl Mech (ASME) 47:266–272

Hashiguchi K (2000) Fundamentals in constitutive equation: continuity and smoothness conditions and loading criterion. Soils Found 40(3):155–161

Hashiguchi K (2013) Elastoplasticity theory. Lecture note in applied computational mechanics, 2 edn. Springer, Heidelberg

Hashiguchi K (1989) Subloading surface model in unconventional plasticity. Int. J. Solids Structure 25:917–945

Hashiguchi K (2016) Exact formulation of subloading surface model: unified constitutive law for irreversible mechanical phenomena in solids, Arch. Compt. Meth Eng 23:86–112

Masing G (1926) Eigenspannungen und Verfestigung beim Messing. In: Proceedings of the 2nd international congress of applied mechanics, Zurich, pp 332–335

Simo JC, Hughes TJR (1998) Computational inelasticity. Springer, New York

Chapter 10
Constitutive Equations of Metals

The plasticity theory has highly developed through the prediction of deformation of metals up to date. The reason would be caused by the fact that, among various materials exhibiting plastic deformation, metals are used most widely as engineering materials and exhibit the simplest plastic deformation behavior without a pressure dependence, a plastic compressibility, a dependence on the third invariant of deviatoric stress and a softening. Nevertheless, metals exhibit various particular aspects, e.g., the kinematic hardening and the stagnation of isotropic hardening in a cyclic loading. Explicit constitutive equations of metals will be delineated in this chapter, which are based on the extended subloading surface model described in the preceding chapters.

10.1 Isotropic and Kinematic Hardening

The yield function for the Mises yield condition is extended to incorporate kinematic hardening by replacing $\boldsymbol{\sigma}'$ to $\hat{\boldsymbol{\sigma}}'$ in Eqs. (6.55), (6.58) and (9.50) as follows:

$$f(\hat{\boldsymbol{\sigma}}) = \sqrt{\frac{3}{2}}\|\hat{\boldsymbol{\sigma}}'\|,\ \hat{\mathbf{n}} = \hat{\mathbf{n}}' = \bar{\mathbf{n}}' = \frac{\hat{\boldsymbol{\sigma}}'}{\|\hat{\boldsymbol{\sigma}}'\|} \tag{10.1}$$

while the subloading function $f(\overline{\boldsymbol{\sigma}})$ for Eq. (10.1) is given by

$$\boxed{f(\overline{\boldsymbol{\sigma}})=\sqrt{\frac{3}{2}}\|\overline{\boldsymbol{\sigma}}'\|},\ \bar{\mathbf{n}} = \bar{\mathbf{n}}' = \bar{\mathbf{n}}' = \frac{\overline{\boldsymbol{\sigma}}'}{\|\overline{\boldsymbol{\sigma}}'\|} \tag{10.2}$$

Further, the elastic-core function in Eq. (9.12) for Eq. (10.1) is given by

K. Hashiguchi, *Foundations of Elastoplasticity: Subloading Surface Model*,
DOI 10.1007/978-3-319-48821-9_10

$$\boxed{f(\hat{\mathbf{c}}) = \sqrt{\frac{3}{2}}\|\hat{\mathbf{c}}'\|}, \quad \frac{\partial f(\hat{\mathbf{c}})}{\partial \hat{\mathbf{c}}} = \sqrt{\frac{3}{2}}\frac{\mathbf{c}'}{\|\hat{\mathbf{c}}'\|} \tag{10.3}$$

It follows from Eq. (9.12) with Eq. (10.3) that

$$\boxed{\Re_c = \sqrt{\frac{3}{2}}\frac{\|\hat{\mathbf{c}}'\|}{F}}, \quad \hat{\mathbf{n}}_c = \frac{\hat{\mathbf{c}}'}{\|\hat{\mathbf{c}}'\|} \tag{10.4}$$

The isotropic hardening function is given by Eq. (6.56), i.e.

$$\boxed{F(H) = F_0\{1+h_1[1-\exp(-h_2H)]\}}, \quad F' = F_0\, h_1\, h_2 \exp(-h_2H) \tag{10.5}$$

$$\overset{\bullet}{H} = \sqrt{\frac{2}{3}}\|\mathbf{d}^p\| = \overset{\bullet}{\bar{\lambda}} f_{Hn}(\boldsymbol{\sigma}, H;\, \bar{\mathbf{n}}) \tag{10.6}$$

$$f_{Hn}(\boldsymbol{\sigma}, H;\, \bar{\mathbf{n}}) = \sqrt{\frac{2}{3}} \tag{10.7}$$

where F_0, h_1, h_2 are the material constants. The hardening function F in Eq. (10.7) increases from the initial value F_0 by the equivalent plastic strain ε^{eqp} and saturates when it reaches the maximum value $(1+h_1)F_0$ as shown in Fig. 10.1.

Let the following evolution rule of the nonlinear kinematic hardening based on Eq. (6.103) be given for metals as follows:

$$\boxed{\overset{\circ}{\boldsymbol{\alpha}} = c_k\left(\mathbf{d}^p - \frac{1}{\sqrt{3/2}\zeta F}\|\mathbf{d}^p\|\boldsymbol{\alpha}\right) = \overset{\bullet}{\bar{\lambda}}\, \bar{\mathbf{f}}_{kn}(\boldsymbol{\sigma}, \boldsymbol{\alpha};\, \bar{\mathbf{n}})} \tag{10.8}$$

$$\bar{\mathbf{f}}_{kn}(\boldsymbol{\sigma}, \boldsymbol{\alpha};\, \bar{\mathbf{n}}) = c_k\left(\bar{\mathbf{n}} - \frac{1}{\sqrt{3/2}\zeta F}\boldsymbol{\alpha}\right) \tag{10.9}$$

where ζ is the material constant. As shown in Fig. 10.2, $\boldsymbol{\alpha}$ translates toward the conjugate point, i.e. $\boldsymbol{\alpha} \to \sqrt{3/2}\zeta F\overline{\boldsymbol{\sigma}}'/\|\overline{\boldsymbol{\sigma}}'\|$ on the limit surface $\|\boldsymbol{\alpha}\| = \sqrt{3/2}\zeta F$ of kinematic hardening.

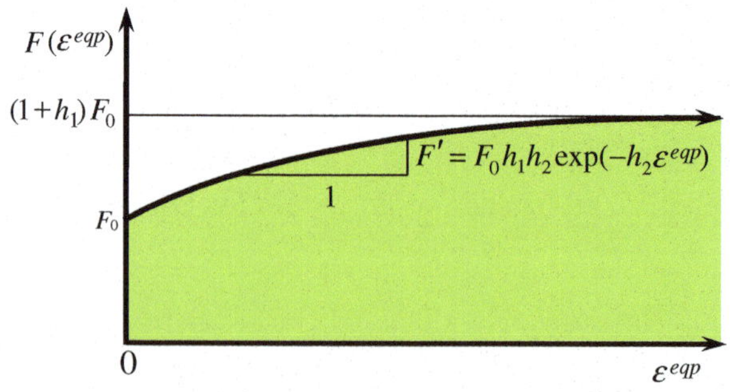

Fig. 10.1 Isotropic hardening function in the uniaxial loading process

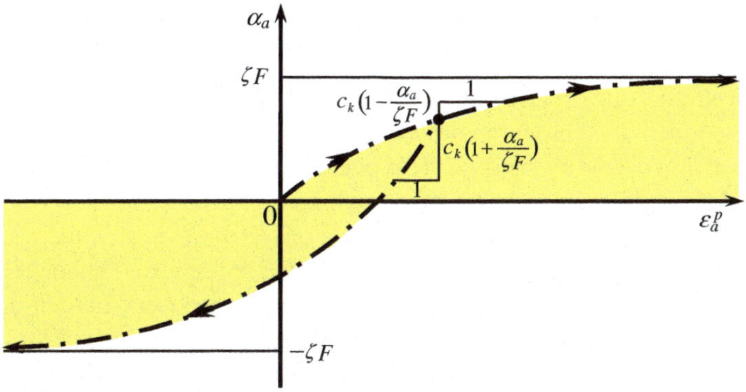

Fig. 10.2 Kinematic hardening rule illustrated for uniaxial loading process

The following relation holds in the uniaxial loading process ($\alpha_a = \sqrt{2/3}\,\|\boldsymbol{\alpha}\|, \varepsilon_a^p = \sqrt{2/3}\,\|\boldsymbol{\varepsilon}^p\|$) from Eq. (10.8).

$$\dot{\alpha}_a = c_k\left(\dot{\varepsilon}_a^p - \frac{\alpha_a}{\zeta F}|\dot{\varepsilon}_a^p|\right) = c_k\left(1 \mp \frac{\alpha_a}{\zeta F}\right)\dot{\varepsilon}_a^p \;(\text{upper} : \dot{\varepsilon}_a^p > 0, \text{lower} : \dot{\varepsilon}_a^p < 0) \tag{10.10}$$

which is time-integrated for $F=$ const. as follows:

$$\frac{\zeta F - \alpha_a}{\zeta F - \alpha_{a0}} = \exp\left[\mp \frac{c_k}{\zeta F}\left(\varepsilon_a^p - \varepsilon_{a0}^p\right)\right] \tag{10.11}$$

where α_{a0} is the initial value of α_a. The relation of α_a versus ε_a^p is shown in Fig. 10.2. The axial back stress saturates at $\alpha_a = \alpha'_a = \zeta F$. Here, it should be noted that the saturation value of the axial stress is not $\sigma_a = (1+\zeta)F$ but it is given by $\sigma_a = [1+(3/2)\zeta]F$, substituting $\hat{\sigma}'_a = \alpha'_a - \alpha'_a = (2/3)\sigma_a - \zeta F$ into $\hat{\sigma}^{eq} = \sqrt{3/2}\|\hat{\boldsymbol{\sigma}}'\| = (3/2)\hat{\sigma}'_a = F$.

The plastic modulus is given by substituting Eqs. (9.16), (10.2), (10.3), (10.7) and (10.9) into Eq. (9.39) as follows:

$$\boxed{\bar{M}^p \equiv \frac{\overline{\boldsymbol{\sigma}}'}{\|\overline{\boldsymbol{\sigma}}'\|}:\left[\sqrt{\frac{2}{3}}\frac{F'}{F}\overline{\boldsymbol{\sigma}} + c_k R\left(\frac{\overline{\boldsymbol{\sigma}}'}{\|\overline{\boldsymbol{\sigma}}'\|} - \frac{1}{\sqrt{3/2}\zeta F}\boldsymbol{\alpha}\right) + c(1-R)\left(\frac{\overline{\boldsymbol{\sigma}}'}{\|\overline{\boldsymbol{\sigma}}'\|} - \frac{\Re_c}{\xi}\frac{\hat{\mathbf{c}}'}{\|\hat{\mathbf{c}}'\|}\right) + \frac{U}{R}\widetilde{\boldsymbol{\sigma}}\right]}$$
$$= \frac{2}{3}RF' + c_k R\left(1 - \frac{1}{\sqrt{3/2}3\zeta F}\frac{\overline{\boldsymbol{\sigma}}'}{\|\overline{\boldsymbol{\sigma}}'\|}:\boldsymbol{\alpha}\right) + c(1-R)\left(1 - \frac{\Re_c}{\xi}\frac{\overline{\boldsymbol{\sigma}}'}{\|\overline{\boldsymbol{\sigma}}'\|}:\frac{\hat{\mathbf{c}}'}{\|\hat{\mathbf{c}}'\|}\right) + \frac{U}{R}\frac{\overline{\boldsymbol{\sigma}}'}{\|\overline{\boldsymbol{\sigma}}'\|}:\widetilde{\boldsymbol{\sigma}} \tag{10.12}$$

10.2 Cyclic Stagnation of Isotropic Hardening

It is observed through experiments for metals that the isotropic hardening stagnates and only the kinematic hardening proceeds in a certain period of reverse deformation starting from the reverse re-yielding. This phenomenon considerably affects the cyclic loading behavior in which the reverse loading is repeated. To describe this phenomenon, the concept of the *cyclic stagnation of isotropic hardening*, i.e. *nonhardening region* was proposed by Chaboche et al. (1979; see also Chaboche 1989) and studied also by Ohno (1982). The concept insists that isotropic hardening does not proceed when the plastic strain given by the time-integration of plastic strain rate lies inside a certain region, called the *nonhardening region*, in the plastic strain space. The non-hardening region expands and translates when the plastic strain lies on the boundary of the region and the plastic strain rate is induced directing outwards the region. It is similar to the notion of the yield surface based on the assumption that the plastic strain rate is induced only when the stress lies on that surface, while the plastic strain and the rate of isotropic hardening variable for the nonhardening region correspond to the stress and the plastic strain rate, respectively, for the yield surface. Thereafter, the other formulation that the isotropic hardening stagnates when the back stress lies inside the certain region of stress space was proposed by Yoshida and Uemori (2002, 2003), where the nonlinear kinematic hardening rule is adopted. Here, it should be noted that the time-integration of the plastic strain rate has no physical meaning in a general state under a material rotation, so that it is not a state variable, and the formulation of the isotropic hardening stagnation in terms of the time-integration of the plastic strain rate cannot be unified as a stress space formulation (cannot be depicted in the stress space). In contrast, the back stress is obviously the state variable and the isotropic hardening stagnation in terms of the back stress can be unified as a stress space formulation. Then, the rigorous formulation for the isotropic hardening stagnation was provided based on the notion of the subloading surface model by Hashiguchi (2015c), adopting the back stress instead of the plastic strain following Yoshida and Uemori (2002, 2003), as will be described in this section.

Assuming that the isotropic hardening stagnates when the back stress $\boldsymbol{\alpha}$ lies inside a certain region, let the following surface, called the *normal-isotropic hardening surface*, be introduced.

$$\mathrm{g}(\widetilde{\boldsymbol{\alpha}}) = \widetilde{K} \tag{10.13}$$

where

$$\widetilde{\boldsymbol{\alpha}} \equiv \boldsymbol{\alpha} - \boldsymbol{\rho} \tag{10.14}$$

$\widetilde{K}$ and $\boldsymbol{\rho}\,(=\boldsymbol{\rho}')$ designate the size and the center, respectively, of the normal-isotropic hardening surface, the evolution rules of which will be formulated later. Furthermore, we introduce the surface, called the *subloading-isotropic hardening surface*, which always passes through the current back stress $\boldsymbol{\alpha}$ and which has a similar shape and an orientation to the normal-isotropic hardening surface (see Fig. 10.3). It is expressed by the following equation.

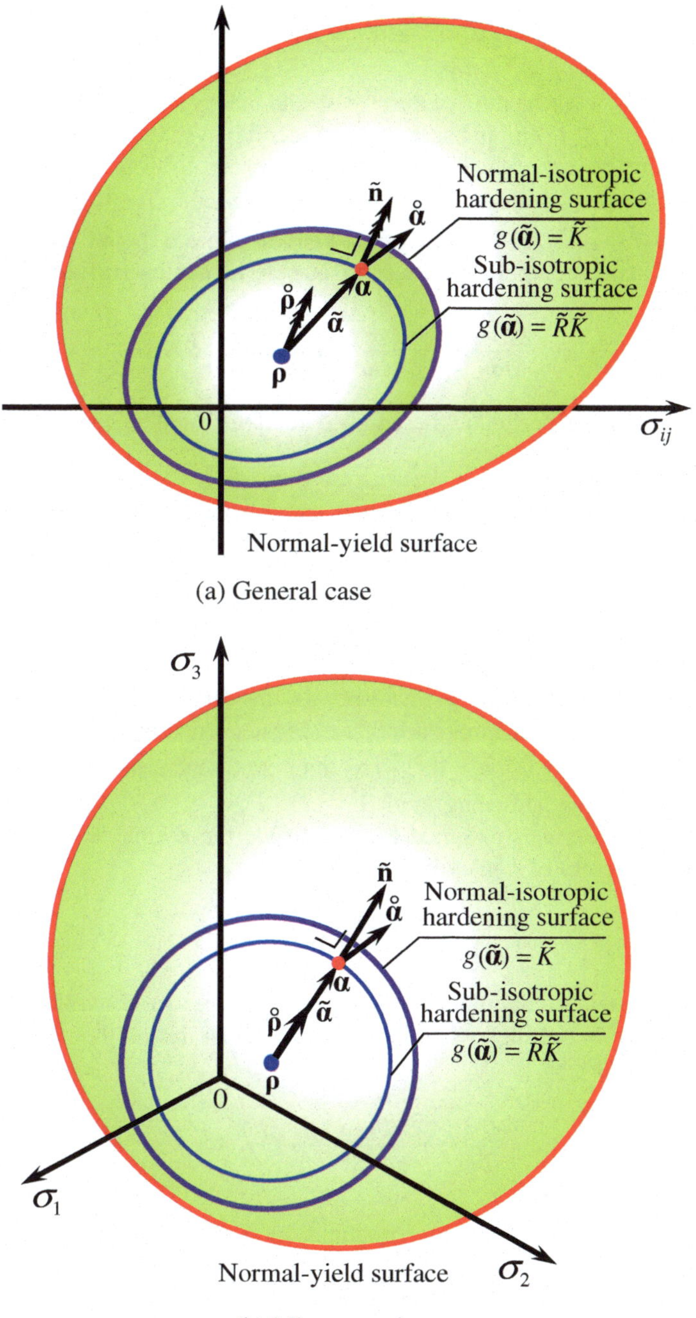

Fig. 10.3 Normal- and sub-isotropic hardening surfaces

$$\boxed{g(\widetilde{\boldsymbol{\alpha}}) = \widetilde{R}\widetilde{K}} \tag{10.15}$$

where $\widetilde{R}(0 \leq \widetilde{R} \leq 1)$ is the ratio of the size of subloading-isotropic hardening surface to that of the normal-isotropic hardening surface. It plays the role as the measure for the approaching degree of the back stress to the normal-isotropic hardening surface. Then, $\widetilde{R}$ is referred to as the *normal-isotropic hardening ratio*. It is calculable from the equation $\widetilde{R} = g(\widetilde{\boldsymbol{\alpha}})/\widetilde{K}$ in terms of the known values $\boldsymbol{\alpha}$, $\boldsymbol{\rho}$ and $\widetilde{K}$.

The consistency condition of the sub-isotropic hardening surface is given by

$$\frac{\partial g(\widetilde{\boldsymbol{\alpha}})}{\partial \widetilde{\boldsymbol{\alpha}}} : \overset{\circ}{\boldsymbol{\alpha}} - \frac{\partial g(\widetilde{\boldsymbol{\alpha}})}{\partial \widetilde{\boldsymbol{\alpha}}} : \overset{\circ}{\boldsymbol{\rho}} = \widetilde{R}\dot{\widetilde{K}} + \dot{\widetilde{R}}\widetilde{K}. \tag{10.16}$$

Let the following postulates be adopted for the formulations of the evolution rules of $\widetilde{K}$ and $\boldsymbol{\rho}$.

(1) $\widetilde{K}$ and $\boldsymbol{\rho}$ evolve when the back stress rate $\overset{\circ}{\boldsymbol{\alpha}}$ is induced directing outwards the subloading-isotropic hardening surface, fulfilling

$$\frac{\partial g(\widetilde{\boldsymbol{\alpha}})}{\partial \widetilde{\boldsymbol{\alpha}}} : \overset{\circ}{\boldsymbol{\alpha}} > 0 \tag{10.17}$$

(2) The rates of $\widetilde{K}$ and $\boldsymbol{\rho}$ increase as the back stress approaches the normal-isotropic hardening surface, i.e. as the normal-isotropic hardening ratio $\widetilde{R}$ increases. Therefore, they are monotonic-increasing function of the normal-isotropic hardening ratio $\widetilde{R}$.

(3) The back stress $\boldsymbol{\alpha}$ is assumed to exist inside the normal-isotropic hardening surface. Therefore, it must hold that

$$\dot{\widetilde{R}} = 0 \text{ for } \widetilde{R} = 1 \tag{10.18}$$

(4) The consistency condition in Eq. (10.16) reduces to the following relation which must be fulfilled when the back stress just lies on the normal-isotropic hardening surface.

$$\frac{\partial g(\widetilde{\boldsymbol{\alpha}})}{\partial \widetilde{\boldsymbol{\alpha}}} : \overset{\circ}{\boldsymbol{\alpha}} - \frac{\partial g(\widetilde{\boldsymbol{\alpha}})}{\partial \widetilde{\boldsymbol{\alpha}}} : \overset{\circ}{\boldsymbol{\rho}} = \dot{\widetilde{K}} \text{ for } \widetilde{R} = 1. \tag{10.19}$$

Then, we assume the following equations so as to fulfill all these postulates.

$$\boxed{\dot{\widetilde{K}} = C\widetilde{R}^{\varsigma}\langle \widetilde{\mathbf{n}} : \overset{\circ}{\boldsymbol{\alpha}} \rangle \left\| \frac{\partial g(\widetilde{\boldsymbol{\alpha}})}{\partial \widetilde{\boldsymbol{\alpha}}} \right\|}, \tag{10.20}$$

$$\boxed{\mathring{\boldsymbol{\rho}} = (1-C)\widetilde{R}^{\varsigma}\langle \widetilde{\mathbf{n}} : \mathring{\boldsymbol{\alpha}} \rangle \widetilde{\mathbf{n}}}, \tag{10.21}$$

where $0 \leq C \leq 1$ and $\varsigma(\geq 1)$ are the material constants and

$$\widetilde{\mathbf{n}} \equiv \frac{\partial g(\widetilde{\boldsymbol{\alpha}})}{\partial \widetilde{\boldsymbol{\alpha}}} \bigg/ \left\| \frac{\partial g(\widetilde{\boldsymbol{\alpha}})}{\partial \widetilde{\boldsymbol{\alpha}}} \right\|. \tag{10.22}$$

Substituting Eqs. (10.21) and (10.22) for the evolution rules of $\widetilde{K}$ and $\boldsymbol{\rho}$ into Eq. (10.17), the rate of the normal-isotropic hardening ratio is given by

$$\begin{aligned} \dot{\widetilde{R}} &= \frac{1}{\widetilde{\widetilde{K}}} \left[\left\langle \frac{\partial g(\widetilde{\boldsymbol{\alpha}})}{\partial \widetilde{\boldsymbol{\alpha}}} : \mathring{\boldsymbol{\alpha}} \right\rangle - (1-C)\widetilde{R}^{\varsigma} \frac{\partial g(\widetilde{\boldsymbol{\alpha}})}{\partial \widetilde{\boldsymbol{\alpha}}} : \langle \widetilde{\mathbf{n}} : \mathring{\boldsymbol{\alpha}} \rangle \widetilde{\mathbf{n}} - \widetilde{R}C\widetilde{R}^{\varsigma} \langle \widetilde{\mathbf{n}} : \mathring{\boldsymbol{\alpha}} \rangle \left\| \frac{\partial g(\widetilde{\boldsymbol{\alpha}})}{\partial \widetilde{\boldsymbol{\alpha}}} \right\| \right] \\ &= \frac{1}{\widetilde{\widetilde{K}}} \left\langle \frac{\partial g(\widetilde{\boldsymbol{\alpha}})}{\partial \widetilde{\boldsymbol{\alpha}}} : \mathring{\boldsymbol{\alpha}} \right\rangle \{1 - [1 - C(1 - \widetilde{R})]\widetilde{R}^{\varsigma}\} \end{aligned} \tag{10.23}$$

which is the monotonically-decreasing function of $\widetilde{R}$ fulfilling

$$\dot{\widetilde{R}} \begin{cases} = \dfrac{1}{\widetilde{\widetilde{K}}} \left\langle \dfrac{\partial g(\widetilde{\boldsymbol{\alpha}})}{\partial \widetilde{\boldsymbol{\alpha}}} : \mathring{\boldsymbol{\alpha}} \right\rangle (>0) \quad \text{for} \quad \widetilde{R} = 0 \\ < \dfrac{1}{\widetilde{\widetilde{K}}} \left\langle \dfrac{\partial g(\widetilde{\boldsymbol{\alpha}})}{\partial \widetilde{\boldsymbol{\alpha}}} : \mathring{\boldsymbol{\alpha}} \right\rangle (>0) \quad \text{for} \quad \widetilde{R} < 1 \\ = 0 \text{ for } \widetilde{R} = 1 \\ < 0 \text{ for } \widetilde{R} > 1 \end{cases} \tag{10.24}$$

as shown in Fig. 10.4. Therefore, the normal-isotropic hardening ratio increases when the back stress moves to the outward of the sub-isotropic hardening surface

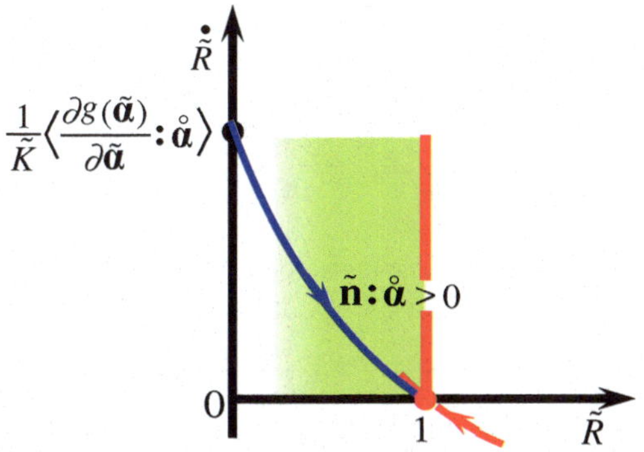

Fig. 10.4 Evolution of normal-isotropic hardening ratio: back stress is attracted to the normal-isotropic hardening surface

but it decreases such that the normal-isotropic hardening surface involves the back stress when the back stress goes out from the normal-isotropic hardening surface by virtue of the inequality $\dot{\widetilde{R}} < 0$ for $\widetilde{R} > 1$ as shown in Eq. (10.24). Furthermore, needless to say, the judgment of whether the back stress reaches the normal-isotropic hardening surface is not necessary in the present formulation.

It is assumed that the isotropic hardening variable H evolves under the following conditions.

(1) The isotropic hardening is induced when the back stress rate $\overset{\circ}{\boldsymbol{\alpha}}$ is induced directing outwards the sub-isotropic hardening surface, i.e.

$$\dot{H}\begin{cases} > 0 & \text{for } \widetilde{\mathbf{n}} : \overset{\circ}{\boldsymbol{\alpha}} > 0 \\ = 0 & \text{for } \widetilde{\mathbf{n}} : \overset{\circ}{\boldsymbol{\alpha}} \leq 0 \end{cases} \tag{10.25}$$

(2) The isotropic hardening rate increases as the back stress approaches the normal-isotropic hardening surface, i.e. as the normal-isotropic hardening ratio $\widetilde{R}$ increases. Then, $\dot{H}$ is the monotonically-increasing function of $\widetilde{R}$. Here, in order that the isotropic hardening develops continuously, its rate must be zero, i.e. $\dot{H} = 0$ for $\widetilde{R} = 0$, i.e. when the back stress lies just on the center of the normal-isotropic hardening surface because the rate is zero during the process in which the back stress moves towards the inside of the sub-isotropic stagnation surface.

(3) The isotropic hardening rule of Eq. (10.6) in the monotonic loading process holds when the back stress lies on the normal-isotropic hardening surface ($\widetilde{R} = 1$) and the plastic strain rate is induced in the outward-direction of that surface.

Eventually, let the following evolution rule of isotropic hardening be assumed by extending Eq. (10.6).

$$\boxed{\dot{H} = \sqrt{\frac{2}{3}}\widetilde{R}^{\upsilon}\langle \widetilde{\mathbf{n}} : \overline{\mathbf{n}}_{kn} \rangle \|\mathbf{d}^{p}\|} \tag{10.26}$$

where υ is the material constant and $\overline{\mathbf{n}}_{kn}$ is the normalized direction of increment of the back stress, noting Eq. (10.9).

$$\overline{\mathbf{n}}_{kn} \equiv \frac{\overline{\mathbf{f}}_{kn}}{\|\overline{\mathbf{f}}_{kn}\|} \tag{10.27}$$

Employing the extended isotropic hardening rule in Eq. (10.27) instead of Eq. (10.7) into Eq. (10.12), the plastic modulus is modified as follows:

$$\boxed{\overline{M}^{p} \equiv \overline{\mathbf{n}} : \left[\sqrt{\frac{2}{3}}\frac{F'}{F}\widetilde{R}\langle \widetilde{\mathbf{n}} : \overline{\mathbf{n}}_{kn} \rangle \overline{\boldsymbol{\sigma}} + R\overline{\mathbf{f}}_{kn} + \frac{U}{R}\widetilde{\boldsymbol{\sigma}} + (1 - R)\overline{\mathbf{f}}_{cn} \right]} \tag{10.28}$$

The normal-isotropic hardening surface evolves such that the boundary of the surface always approaches the back stress and moves so as to involve it even if the back stress goes out from the boundary by the inequality $\dot{\widetilde{R}} < 0$ for $\widetilde{R} > 1$ as shown in Eq. (10.24). Furthermore, the judgment of whether the back stress reaches the normal-isotropic hardening surface is not necessary in the present formulation. In contrast, the judgment whether the plastic strain or the back stress reaches the isotropic hardening (stagnation) surface is required in the other models (Chaboche et al. 1979; Chaboche 1991; Ohno 1982; Yoshida and Uemori 2002). In addition, the boundary of the isotropic (stagnation) surface does not approach the plastic strain or the back stress and does not move so as to involve it even if they go out from the surface. Therefore, these models would be obliged to abandon the incorporation of isotropic stagnation (Chaboche 2008; Kobayashi and Ohno 2002) except for the calculation by the forward-Euler method with infinitesimal loading increments, although the isotropic stagnation formulations have been proposed by the proposers of these cyclic kinematic models themselves.

The function $g(\tilde{\boldsymbol{\alpha}})$ is given in the simplest form as follows:

$$g(\tilde{\boldsymbol{\alpha}}) = ||\tilde{\boldsymbol{\alpha}}||, \quad \tilde{\mathbf{n}} = \frac{\tilde{\boldsymbol{\alpha}}}{||\tilde{\boldsymbol{\alpha}}||}, ||\frac{\partial g(\tilde{\boldsymbol{\alpha}})}{\partial \tilde{\boldsymbol{\alpha}}}|| = ||\tilde{\mathbf{n}}|| = 1 \tag{10.29}$$

which will be used in the subsequent sections for the comparisons with test data. It follows from Eq. (10.28) that

10.3 Calculation of Normal-Yield Ratio

The normal-yield ratio R must be calculated from the equation of the subloading surface in the unloading process ($\mathbf{d}^p = \mathbf{O}$). It can be calculated directly by $R = f(\hat{\boldsymbol{\sigma}})/F$ in the initial subloading surface model. However, it has to be calculated by solving the equation of the subloading surface in the extended subloading surface model as described below.

Substituting Eq. (9.6) into Eq. (10.2), the extended subloading surface is described as follows:

$$\sqrt{\frac{3}{2}}||\widetilde{\boldsymbol{\sigma}}' + R\hat{\mathbf{c}}'|| = RF(H) \tag{10.30}$$

i.e.

$$\text{tr}\,(\widetilde{\boldsymbol{\sigma}}' + R\hat{\mathbf{c}}')^2 = \frac{2}{3}R^2F^2 \tag{10.31}$$

The normal-yield ratio R is derived from the quadratic Eq. (10.31) as follows:

$$\boxed{R = \frac{\tilde{\boldsymbol{\sigma}}' : \hat{\mathbf{c}}' + \sqrt{(\tilde{\boldsymbol{\sigma}}' : \hat{\mathbf{c}}')^2 + \left(\frac{2}{3}F^2 - \|\hat{\mathbf{c}}'\|^2\right)\|\tilde{\boldsymbol{\sigma}}'\|^2}}{\frac{2}{3}F^2 - \|\hat{\mathbf{c}}'\|^2}} \tag{10.32}$$

10.4 Material Parameters and Comparisons with Test Data

Material parameters will be collectively shown and the capability of the subloading surface model to describe various loading behavior will be verified by the comparisons with test data in this section.

10.4.1 Material Parameters

Material parameters are shown collectively below for three versions of the subloading surface model. Eqs. (7.20) and (7.62) will be used as the functions in the evolution equation of the normal-yield ratio and the tangential-inelastic strain rate in the following.

i) The simplest subloading surface model, which is the improvement of the conventional elastoplasticity model only with the isotropic and the kinematic hardenings, contains the following 7 material constants and 2 initial values.

Material constants:

Elastic moduli: E, ν

$$\text{Hardening}\begin{cases}\text{isotropic: } h_1\ h_2\\ \text{kinematic: } c_k, \zeta\end{cases}$$

Evolution of normal-yield ratio : u

Initial values:

$$\text{Normal-yield surface}\begin{cases}\text{size: } F_0\\ \text{center: } \boldsymbol{\alpha}_0\end{cases}$$

The computer program is shown in **Appendix J** (a) i).

ii) The simplified subloading surface model, in which the tangential-inelastic strain rate and the isotropic hardening stagnation are ignored, contains the following 12 material constants and 3 initial values.

Material constants:

Elastic moduli: E, v

$$\text{Hardening}\begin{cases}\text{isotropic: } h_1 \ h_2\\ \text{kinematic: } c_k, \zeta\end{cases}$$

Evolution of normal-yield ratio : $\bar{u}, u_c, R_e(\geq 1), n(\geq 1)$

Translation of elastic-core: $c,\ \xi(<1)$

Initial values:

$$\text{Normal-yield surface}\begin{cases}\text{size: } F_0\\ \text{center: } \boldsymbol{\alpha}_0\end{cases}$$

Elastic-core : $\mathbf{c}_0$

The computer program is shown in **Appendix J** (a) ii).

iii) The most general subloading surface model, which possesses all the behavior involving the tangential-inelastic strain rate and the isotropic stagnation, contains the following 17 material constants and 5 initial values at most, while the full version of computer program is shown in **Appendix J**(a) iii).

Material constants:

Elastic moduli: E, v

$$\text{Hardening}\begin{cases}\text{isotropic: } h_1 \ h_2\\ \text{kinematic: } c_k, \zeta\end{cases}$$

Evolution of normal-yield ratio : $\bar{u}, u_c, R_e(<1), n(\geq 1)$

Translation of elastic-core: $c,\ \xi(<1)$

Tangential inelasticity: $\widetilde{c}(\leq 1), \widetilde{n} = (\geq 1)$

Stagnation of isotropic hardening: $C(0 \leq C \leq 1), \varsigma(>1),\ \upsilon$

Initial values:

$$\text{Normal-yield surface}\begin{cases}\text{size: } F_0\\ \text{center: } \boldsymbol{\alpha}_0\end{cases}$$

Elastic-core : $\mathbf{c}_0$

$$\text{Normal-isotropic hardening surface}\begin{cases}\text{size: } \widetilde{K}_0\\ \text{center: } \boldsymbol{\rho}_0\end{cases}$$

The computer program is shown in **Appendix J** (a) iii).

The determination of these material parameters is explained below in brief.

(1) Young's modulus E and Poisson's ratio v are determined from the slope and the ratio of lateral to axial strains in the initial part of stress–strain curve.
(2) h_1, h_2 and F_0 for the isotropic hardening and c_k, ζ and $\boldsymbol{\alpha}_0$ for the kinematic hardening are determined from stress–strain curves in the initial and the inverse loadings.
(3) $\bar{u}$, u_c, $R_e(<1)$ and n for the evolution of the normal-yield ratio are determined from the stress–strain curve in the subyield state, i.e. the elastic-plastic transitional state.
(4) c, ξ and $\mathbf{c}_0$ for the elastic-core are determined from the stress–strain curves in cyclic loading.
(5) $\widetilde{c}$ and $\widetilde{n}$ for the tangential-inelastic strain rate are determined by the difference of the strain in the non-proportional loading from that in the proportional loading.
(6) C, ς, υ, $\widetilde{K}_0$ and $\boldsymbol{\rho}_0$ for the elastic-core are determined from the stress–strain curves in cyclic loading under a constant strain amplitude.

All of these material parameters except for $\widetilde{c}$ and $\widetilde{n}$ for the tangential-inelastic strain rate can be determined only by the stress–strain curves in the uniaxial loading for initial isotropic materials. One can put $\boldsymbol{\alpha}_0 = \mathbf{c}_0 = \boldsymbol{\rho}_0 = \mathbf{O}$ for the initial isotropy, which is assumed in all the subsequent simulations. We may calculate $\widetilde{K}_0$ by $\widetilde{K}_0 = \|\widetilde{\boldsymbol{\alpha}}_0\| (\cong 0)$ leading to $\widetilde{R}_0 = 1$, by input of a small value of $\boldsymbol{\alpha}_0$ and $\boldsymbol{\rho}_0 = \mathbf{O}$ in order that the isotropic hardening rule in Eq. (10.5) with Eq. (10.6) holds in an initial loading process. Tangential inelasticity is irrelevant to the proportional loading.

10.4.2 Comparison with Test Data

The capability of the present model for describing the deformation behavior of metals is verified through comparisons with several basic test data in this section, referring to Hashiguchi et al. (2012) and Hashiguchi and Ueno (2017). Capability of unconventional plasticity model aimed at describing plastic strain rate induced by a rate of stress inside yield surface must be evaluated by a degree in which cyclic loading behavior can be described appropriately. Then, various cyclic loading test data in uniaxial loading are first simulated and thereafter a circular strain path test datum is simulated to verify capability for describing non-proportional loading behavior.

The cyclic loading behavior under the stress amplitude to both positive and negative sides can be predicted to some extent by any models, including even the conventional plasticity model. On the other hand, the prediction of the cyclic loading behavior under the stress amplitude in positive or negative one side, i.e. the pulsating loading inducing the so-called *mechanical ratcheting effect* requires a high ability for the description of plastic strain rate induced by the rate of stress inside the yield surface. Furthermore, it is noteworthy that we often encounter the

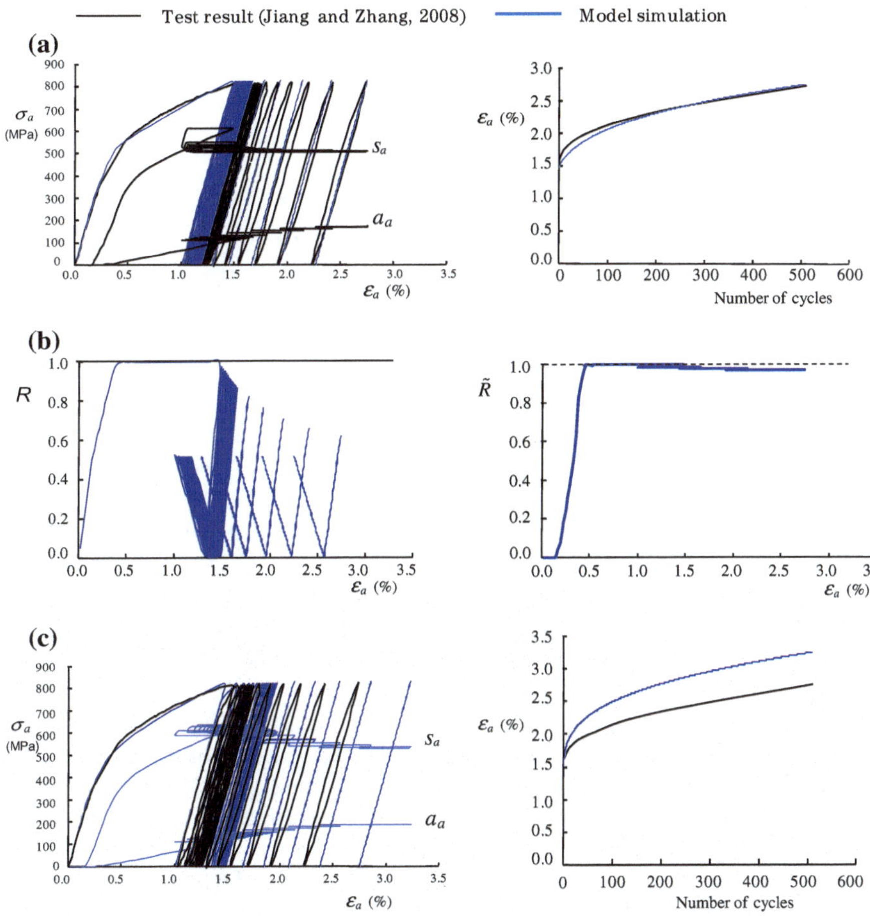

Fig. 10.5 Uniaxial cyclic loading behavior under the pulsating loading between 0 and 830 MPa of 1070 steel (Test data after Jiang and Zhang 2008): **a** Test result and simulation result and simulation without stagnation of isotropic hardening, **b** Variations of normal-yield ratio and normal-isotropic hardening ratio and **c** Test result and simulation without improvement of reloading behavior.

pulsating loading phenomena in the boundary-value problems in engineering practice, e.g. railways and gears. The comparison with the test data for the 1070 steel under the cyclic loading of axial stress between 0 and +830 MPa after Jiang and Zhang (2008) is depicted in Fig. 10.5, where the material parameters are selected as follows:

Material constants:

$$\text{Elastic moduli: } E = 160,000\,\text{MPa},\ \nu = 0.3,$$
$$\text{Hardening}\begin{cases}\text{isotropic: } h_1 = 0.58,\ h_2 = 170,\\ \text{kinematic: } c_k = 5000\ \text{MPa},\ \zeta = 0.5,\end{cases}$$
$$\text{Evolution of normal-yield ratio: } \bar{u} = 200, u_c = 6, R_e = 0.5, n = 1,$$
$$\text{Translation of elastic-core: } c = 7000\ \text{MPa},\ \xi = 0.7,$$
$$\text{Stagnation of isotropic hardening: } C = 0.5, \varsigma = 5, \upsilon = 0.1,$$

Initial values:

$$\text{Isotropic hardening function: } F_0 = 507\,\text{MPa}.$$

The relation of the axial stress and the axial components of back stress and similarity-center versus the axial strain and the relation of the axial strain versus the number of cycles are depicted in Fig. 10.5a, where the axial components are designated by $(\ )_a$. The accumulation of axial strain is simulated closely by the present model. The calculation is controlled automatically such that the stress and the back stress are attracted to the normal-yield and the normal-isotropic hardening surfaces, respectively, as known from the variations of the normal-yield ratio R and the normal-isotropic hardening ratio $\widetilde{R}$ depicted in Fig. 10.5b. Accumulation of axial strain is overestimated as depicted in Fig. 10.5c if the reloading behavior is not improved by setting $u_c = 0$ ignoring the Masing effect. Despite of the improvement for reloading behavior, however, hysteresis loops are simulated as narrower than those in the test result in order to fit the strain accumulation in the test result. A further improvement is desirable for this insufficiency.

Next, examine the uniaxial cyclic loading behavior under the constant stress amplitude to both positive and negative sides with different magnitudes. Comparison with the test data for the 304L steel under the cyclic loading of axial stress between $+250$ and -150 MPa after Hassan et al. (2008) is depicted in Fig. 10.6 where the material parameters are selected as shown below.

Material constants:

$$\text{Elastic moduli: } E = 200,000\text{MPa},\ \nu = 0.3,$$
$$\text{Hardening}\begin{cases}\text{isotropic: } h_1 = 0.3,\ h_2 = 30,\\ \text{kinematic: } c_k = 130\ \text{MPa},\ \zeta = 0.9,\end{cases}$$
$$\text{Evolution of normal-yield ratio: } \bar{u} = 2, u_c = 10, R_e = 0.5, n = 1,$$
$$\text{Translation of elastic-core: } c = 10{,}000\ \text{MPa},\ \xi = 0.7,$$
$$\text{Stagnation of isotropic hardening: } C = 0.5, \varsigma = 15, \upsilon = 1,$$

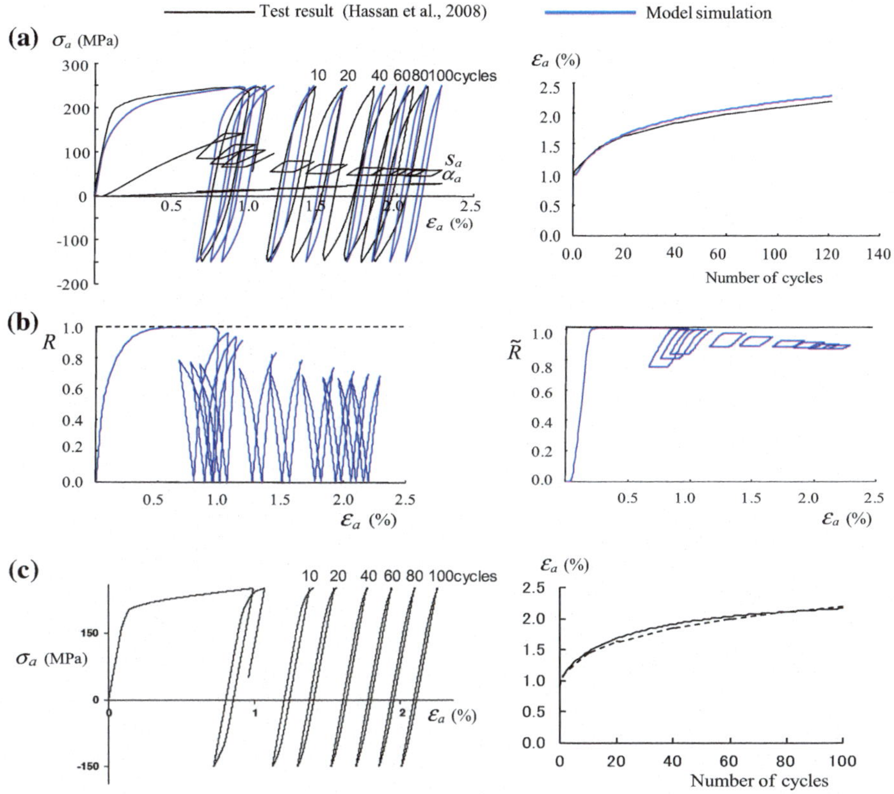

Fig. 10.6 Uniaxial cyclic loading behavior under the constant stress amplitude between −150 and 250 MPa of 304L steel (Test data after Hassan et al. 2008): **a** Test result and simulation, **b** Variations of normal-yield ratio and normal-isotropic hardening ratio and **c** Simulation by modified Chaboche model (cf. Hassan et al. 2008)

Initial values:

Isotropic hardening function: $F_0 = 232$ MPa.

The relation of the axial stress and the axial components of back stress and similarity-center versus the axial strain and the relation of the axial strain versus the number of cycles are depicted in Fig. 10.6a. Both the accumulation of strain and the hysteresis loops are simulated closely by the present model. The calculation is controlled automatically such that the stress and the back stress are attracted to the normal-yield and the normal-isotropic hardening surfaces, respectively, as known from the variations of the normal-yield ratio R and the normal-isotropic hardening ratio $\widetilde{R}$ depicted in Fig. 10.6b. The relations of the axial stress versus the axial strain

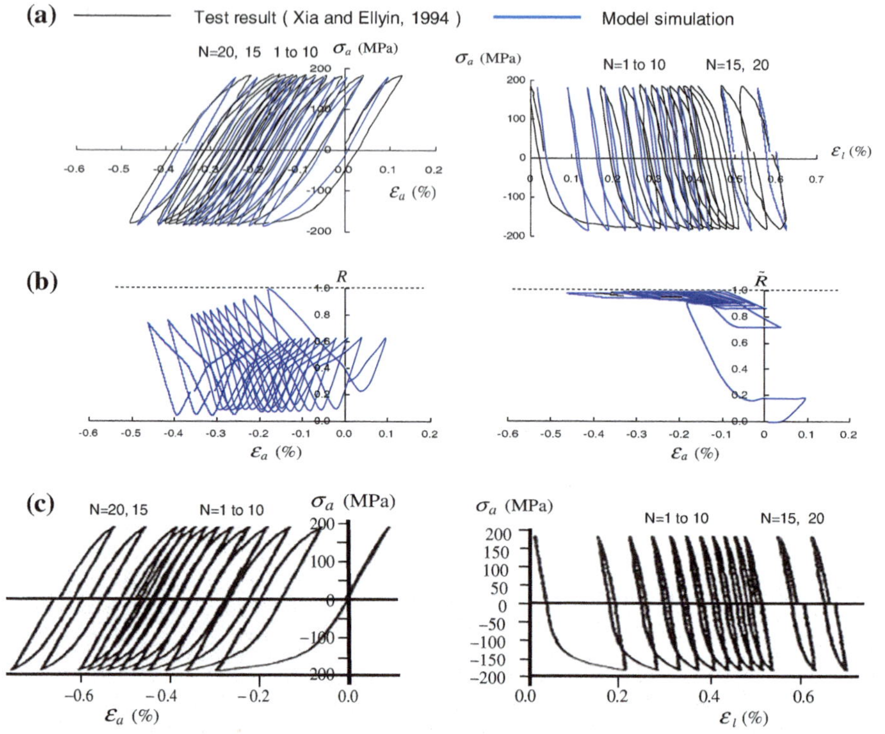

Fig. 10.7 Uniaxial cyclic loading behavior under the constant stress amplitude between −182 and +182 MPa of 304L steel (Test data after Xia and Ellyin 1994): **a** Test result and simulation, **b** Variations of normal-yield ratio and normal-isotropic hardening ratio and **c** Simulation by Xia and Ellyin (1994)

and the relation of the axial strain versus the number of cycles simulated using the modified Chaboche model (Chaboche 1991) are also depicted in Fig. 10.6c in which the strain is simulated as larger than the test result and the hysteresis loops are simulated as narrower than the test data. The prediction of this steel deformation behavior will be improved by incorporating the rate-dependence.

Further, we examine the uniaxial cyclic loading behavior for constant symmetric stress amplitude to both positive and negative sides. Comparison with test data of the 304 steel under the cyclic loading of axial stress between $+182$ and -182 MPa under the constant hoop stress 80 MPa after Xia and Ellyin (1994) is depicted in Fig. 10.7 where the material parameters are selected as follows:

Material constants:

Elastic moduli: $E = 190,000\,\text{MPa}$, $\nu = 0.3$,

$$\text{Hardening}\begin{cases}\text{isotropic: } h_1 = 1.3, h_2 = 100,\\ \text{kinematic: } c_k = 25\ \text{MPa},\ \zeta = 0.3,\end{cases}$$

Evolution of normal-yield ratio: $\bar{u} = 200, u_c = 6, R_e = 0.5, n = 1$,

Translation of elastic-core: $c = 5000$ MPa, $\xi = 0.7$,

Stagnation of isotropic hardening: $C = 0.5, \varsigma = 8, \upsilon = 5$,

Initial values:

Isotropic hardening function: $F_0 = 212\,\text{MPa}$.

The relation of the axial stress and the axial components of back stress and similarity-center versus the axial strain and the circumferential strain ε_l with the number of cycles are shown in Fig. 10.7a, while the back stress is induced quite slightly so that it is invisible in this figure. The simulations for the accumulation of axial strain and the hysteresis loops agree well with the test result, except for the prediction of hysteresis loops as narrower than the test result in the initial stage. Here, the axial strain and the lateral strain are accumulated to the compression side and the extension side, respectively, by the application of the hoop stress 80 MPa. The calculation is automatically controlled such that the stress and the back stress are attracted to the normal-yield and the normal-isotropic hardening surfaces, respectively, as known from the variations of the normal-yield ratio R and the normal-isotropic hardening ratio $\widetilde{R}$ depicted in Fig. 10.7b. The relations of the axial stress versus the axial and lateral strains and the relation with the number of cycles simulated by Xia and Ellyin (1994; cf. also Ellyin 1997) are also depicted in Fig. 10.7c where both the axial and the circumferential strains are overestimated.

Furthermore, examine the uniaxial cyclic loading behavior under the constant symmetric strain amplitudes to both positive and negative sides. Comparison with the test data of the 316 steel under the cyclic loading with the increasing axial strain amplitudes $\pm 1.0, \pm 1.5, \pm 2.0, \pm 2.5, \pm 3.0\,\%$ after Chaboche et al. (1979) is depicted in Fig. 10.8 where the material parameters are selected as follows:

Material constants:

Elastic moduli: $E = 170,000\,\text{MPa}, \nu = 0.3$,

$$\text{Hardening}\begin{cases}\text{isotropic: } h_1 = 0.85, h_2 = 5,\\ \text{kinematic: } c_k = 2000\ \text{MPa},\ \zeta = 0.5,\end{cases}$$

Evolution of normal-yield ratio: $\bar{u} = 100, u_c = 3, R_e = 0.5, n = 1$,

Translation of elastic-core: $c = 2000$ MPa, $\xi = 0.7$,

Stagnation of isotropic hardening: $C = 0.5, \varsigma = 5, \upsilon = 1$,

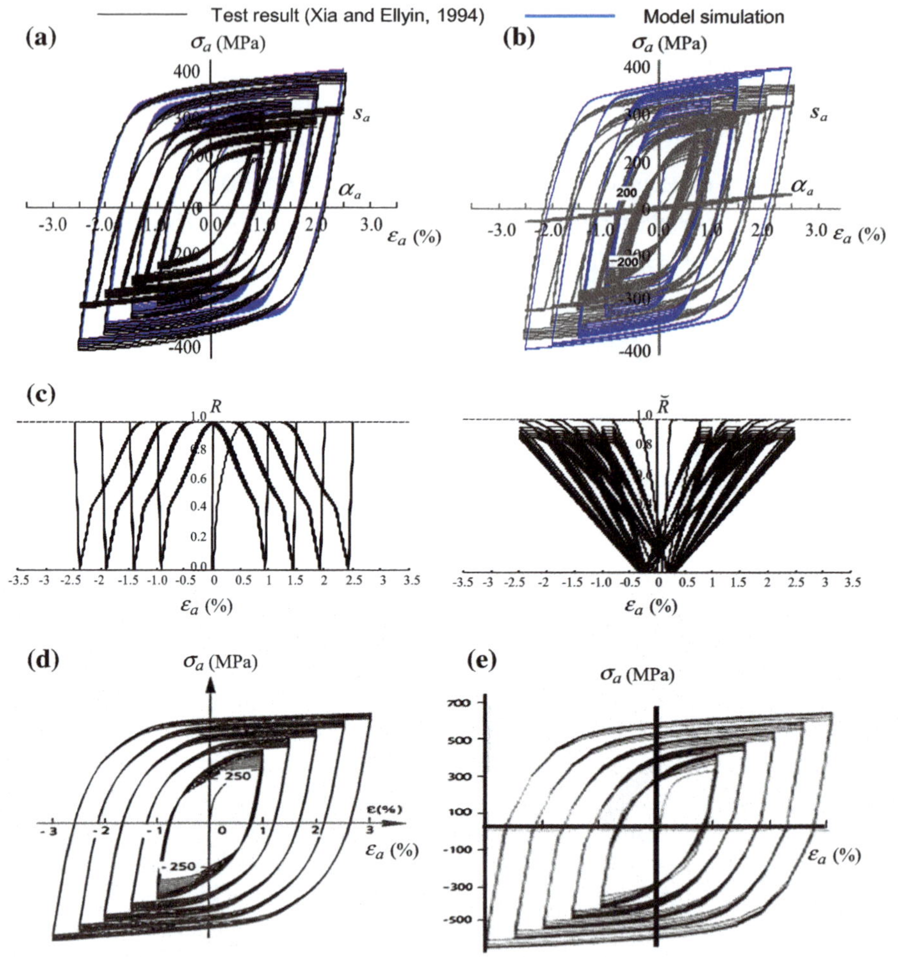

Fig. 10.8 Uniaxial cyclic loading behavior under the constant strain amplitude the 5 levels increasing strain amplitudes of 316 steel (Test data after Chaboche et al. 1979): **a** Test result and simulation by present model, **b** Test result and simulation without stagnation of isotropic hardening, **c** Variations of normal-yield ratio and normal-isotropic hardening, **d** Simulation by Chaboche (2008) and **e** Simulation by Ellyin and Xia (1989)

Initial values:

Isotropic hardening function: $F_0 = 320\,\text{MPa}$.

The relation of the axial stress and the axial components of back stress and similarity-center versus the axial strain are shown in Fig. 10.8a. The hysteresis loops and the stagnation of isotropic hardening are simulated closely by the present model. On the other hand, the calculated result without the cyclic stagnation of isotropic hardening overestimates the hardening behavior as shown in Fig. 10.8b. The calculation is controlled automatically such that the stress and the back stress are attracted to the normal-yield and the normal-isotropic hardening surfaces, respectively, as known from the variations of the normal-yield ratio R and the normal-isotropic hardening ratio $\widetilde{R}$ depicted in Fig. 10.8c. The relations of the axial stress and the axial strain simulated by Chaboche (1991) and Ellyin and Xia (1989) are depicted in Fig. 10.8d, e, respectively. The strain in the initial stage is simulated as larger than the test result by the former and the curves predicted by the latter is not smooth but piece-wise linear.

Finally, we examine the non-proportional loading behavior. Comparison with the test data of the austenitic 17–12 Mo SPH carbon stainless steel subjected to the approximately circular strain path in the strain plane $(\varepsilon_a,\ \varepsilon_{a\theta})$ by the inputs of the axial strain $\varepsilon_a = 0.004\sin(\alpha - \pi/2)$ and the axial-circumferential shear strain $\varepsilon_{a\theta} = 0.0036\sin\alpha$ in the sinusoidal waves under the constant circumferential normal stress $\sigma_\theta = 50$ MPa during 40 cycles after the uniaxial loading to $\varepsilon_a = 0.004$ after Delobelle et al. (1995) is depicted in Fig. 10.9. Here, α is the angle measured from the axis of ε_a in the strain plane $(\varepsilon_a, \varepsilon_{a\theta})$. Consequently, the cyclic loadings of the axial strain ε_a under the constant amplitude $\varepsilon_a = \pm 0.004$ and of the axial-circumferential shear strain $\varepsilon_{a\theta}$ under the constant amplitude $\varepsilon_{a\theta} = \pm 0.0036$ are executed simultaneously, while the phase of $\varepsilon_{a\theta}$ is later than that of ε_a by $\pi/2$. The material parameters are selected as follows:

Material constants:

$$
\begin{aligned}
&\text{Elastic moduli: } E = 170{,}000\,\text{MPa},\ \nu = 0.3 (G = 65{,}385\ \text{MPa}),\\
&\text{Hardening}\begin{cases}\text{isotropic: } h_1 = 1.7, h_2 = 40,\\ \text{kinematic: } c_k = 200\ \text{MPa},\ \zeta = 0.9,\end{cases}\\
&\text{Evolution of normal-yield ratio: } \bar{u} = 800, u_c = 3, R_e = 0.5, n = 1,\\
&\text{Translation of elastic - core: } c = 7000\ \text{MPa}, \xi = 0.7,\\
&\text{Tangential inelasticity: } \widetilde{c} = 0.6,\ \widetilde{n} = 3,\\
&\text{Stagnation of isotropic hardening: } C = 0.5, \varsigma = 5, \upsilon = 1,
\end{aligned}
$$

Initial values:

$$\text{Isotropic hardening function: } F_0 = 240\,\text{MPa}.$$

The strain path $(\varepsilon_a, \varepsilon_\theta)$ (ε_θ: circumferential normal strain) and the stress path $(\sigma_a, \sqrt{3}\sigma_{a\theta})$ ($\sigma_{a\theta}$: axial-circumferential shear stress) are shown for the test result and the model simulation in Fig. 10.9a, b, respectively. The simulation of the stress path and the accumulation of lateral strain are in good agreement with the test result. The stress and the back stress are attracted to the normal-yield and the normal-isotropic hardening surfaces, respectively, as known from the variations of the normal-yield ratio R and the normal-isotropic hardening ratio $\widetilde{R}$ depicted in Fig. 10.9b.

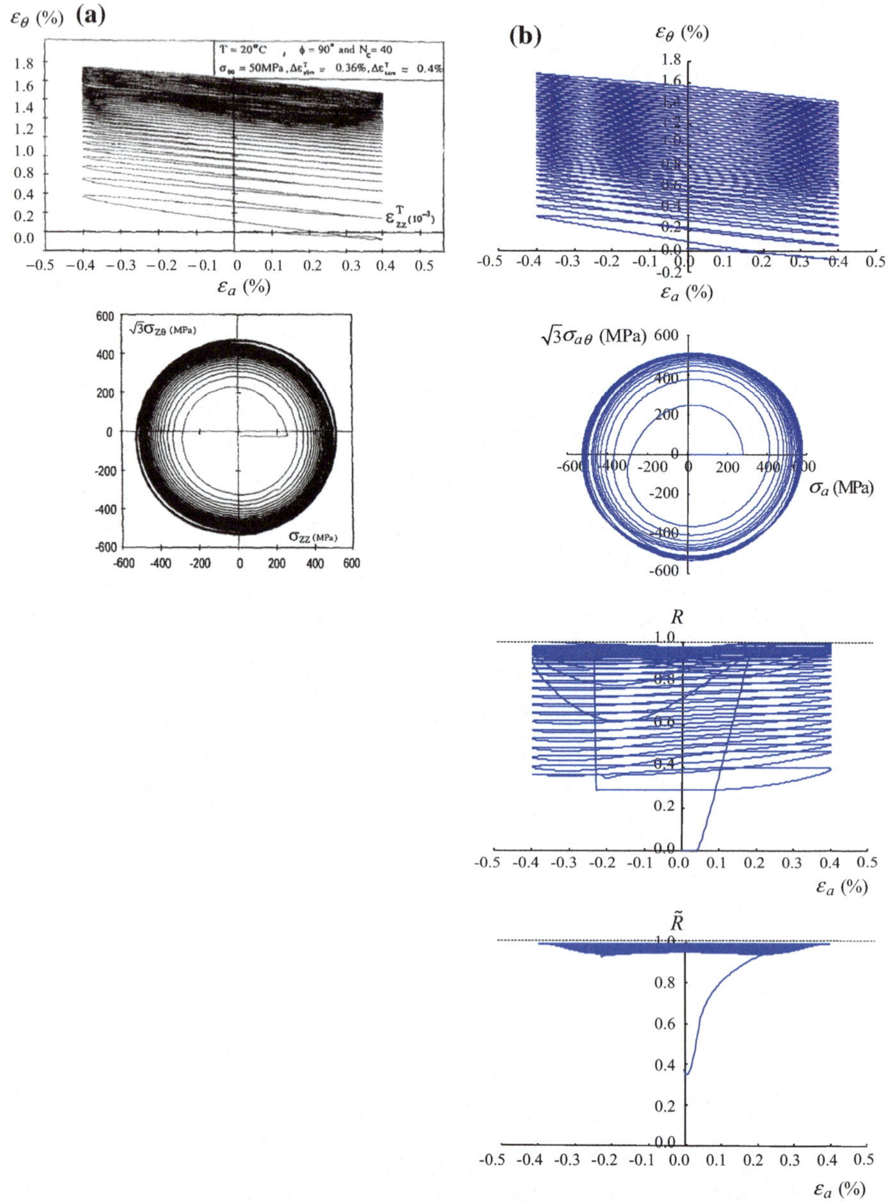

Fig. 10.9 Circular strain path loading given by the axial strain and the axial-circumferential engineering shear strain during 40 cycles after the uniaxial loading of austenitic 17–12 Mo SPH carbon stainless steel (Test data after Delobelle et al. 1995): **a** Test result, **b** Model simulation

The circular strain path in this test produces the spiral stress path approaching the circular stress path along the Mises yield surface which is expressed by the circle $\sqrt{\sigma_a^2+(\sqrt{3}\sigma_{a\theta})^2}=F$ in the two-dimensional stress plane $(\sigma_a, \sqrt{3}\sigma_{a\theta})$. Here, note the following facts.

(1) The tangential-deviatoric stress rate is induced almost only in the component $((\boldsymbol{\sigma}'_t)_a, (\boldsymbol{\sigma}'_t)_{a\theta})$, and thus the tangential-inelastic strain rate is induced almost only in the component $((\mathbf{d}^t)_a, (\mathbf{d}^t)_{a\theta})$ in this test, if the isotropy is roughly assumed.
(2) The axial strain rate $(\mathbf{d})_a$ is composed of the axial elastoplastic strain rate $(\mathbf{d}^{ep})_a(=(\mathbf{d}^e)_a+(\mathbf{d}^p)_a)$ and the axial tangential-inelastic strain rate $(\mathbf{d}^t)_a$.
(3) The ratio of the axial component of tangential-inelastic strain rate, $(\mathbf{d}^t)_a$, to that of the elastoplastic strain rate, $(\mathbf{d}^{ep})_a$, is larger when the ratio $(\mathbf{d})_a/(\mathbf{d})_{a\theta}$ is larger in the strain space $(\varepsilon_a, \varepsilon_{a\theta})$. In other words, it is largest when the strain path passes through the top or the bottom of the circle in the strain space.
(4) The circumferential strain component ε_θ is almost independent of the tangential-inelastic strain rate.
(5) Then, the inclination of the $\varepsilon_\theta-\varepsilon_a$ curve, i.e. $\varepsilon_\theta/\varepsilon_a$ is smaller for the larger value of $\widetilde{c}$ in the strain plane $(\varepsilon_\theta, \varepsilon_a)$. Then, the strain path rises up higher compared with the test data in the strain plane $(\varepsilon_\theta, \varepsilon_a)$ if the tangential-strain rate is ignored by setting $\widetilde{c}=0$ as shown in Fig. 10.10.

The importance of the introduction of the tangential-inelastic strain rate would have been definitely recognized by the verification shown above.

10.5 Springback and Residual Stress Analyses

The metal forming analyses are of importance in the industrial production. The analyses of the two typical phenomena, i.e. the springback analysis and the residual stress analysis by use of the subloading surface model will be described in this section.

(1) **Springback analysis**

The high tensile (strength) steel sheets and aluminum sheets exhibiting far larger springback than ordinary mild steel sheets are widely used in automobile industries. The springback cannot be described by the constitutive models which use the yield surface enclosing a purely-elastic domain, i.e. the conventional model and the cyclic kinematic hardening models (multi-surface, two-surface and superposed kinematic hardening models), since a plastic strain rate in the unloading process is not described appropriately by these models. Based on the two surface model, however, the method for the springback analysis was proposed by Yoshida and Uemori (2003) in which the Young's modulus is formulated to decrease but approach the saturated value with the equivalent plastic strain as shown in

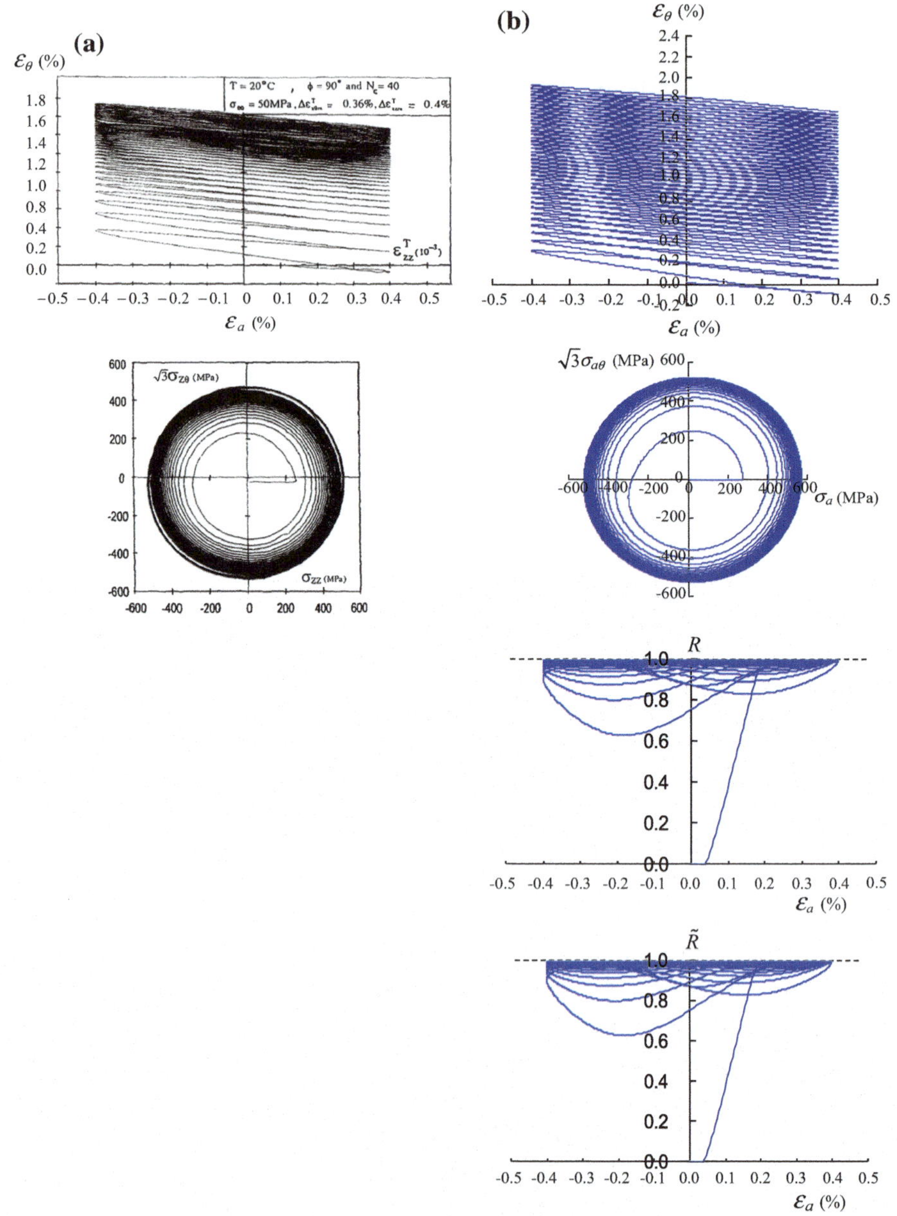

Fig. 10.10 Circular strain path loading given by the axial strain and the axial-circumferential engineering shear strain during 40 cycles after the uniaxial loading of austenitic 17–12 Mo SPH carbon stainless steel (Test data after Delobelle et al. 1995): **a** Test result, **b** Model simulation without tangential-inelasticity

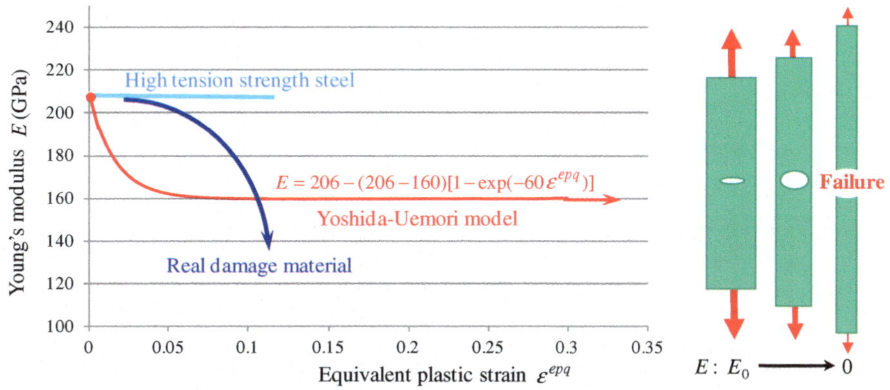

Fig. 10.11 Variation of Young's modulus with equivalent plastic strain: Physically-unacceptable prediction by Yoshida and Uemori (2003)

Fig. 10.11, which is calculated by the following equation for the *high tensile strength steel* (Yoshida and Uemori 2003).

$$E = E_0 - (E_0 - E_a)[1 - \exp(-\xi\varepsilon^{epq})] \text{ with } E_0 = 206,\ E_a = 160,\ \xi = 60$$

In fact, however, the purely elastic deformation is induced only at the initiation of stress reversal event and the plastic deformation induced in the unloading process to the stress free state would increase with the preceding plastic strain history. Nevertheless, the unloading process is regarded to be the purely elastic deformation process and then the Young's modulus is calculated from the inclination of straight line connecting the initial and the final points of the uniaxial unloading curve in the Yoshida and Uemori's model. Besides, the Young's moduli of real materials decreases acceleratingly to zero with the equivalent plastic strain in the continuing tension loading process if once it decreases as shown in Fig. 10.11. It is caused by the fact that the cracks grow increasingly so that not only the Young's modulus but also the hardening function F decreases in that loading process as has been revealed in the damage mechanics which will be described in Chapt. 14. Nevertheless, this fact based on the damage mechanics is also ignored and the so-called *prohibited technique* is used in the Yoshida and Uemori's model. Further, the incorporation of the *quasi-plastic-elastic strain* in the unloading process by Wagoner's group (cf. Sun and Wagoner 2011; Wagoner et al., 2013) is unacceptable physically in addition to causing the unnecessary complexity in formulation of constitutive relation. Besides, Barlat's group (cf. Lee et al., 2015) proposed the irrational friction model in which the friction coefficient is given by the multiplicative form consisting of the functions of sliding velocity and contact pressure, although the friction phenomenon is described rationally by the subloading-friction model which will be described in detail in Chapter 18. It is impertinent that the physically unacceptable model is adopted in the commercial softwares, i.e. the PAM-STAMP

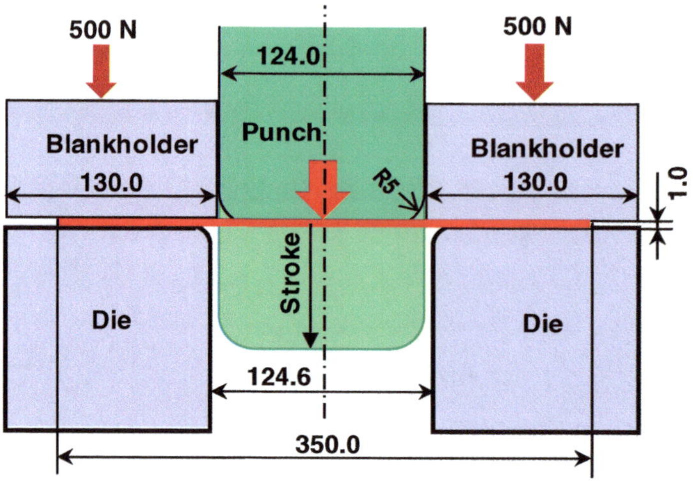

Fig. 10.12 Schematic illustration of the set-up of hat-bending after Yoshida and Uemori (2003)

and the LS-DYNA (Japan) and used widely in automobile industries. Constitutive equation capable of describing the plastic deformation in the unloading process should be incorporated for the springback analysis.

The pertinent calculation result of the springback is shown below, which was analyzed by Dr. Motoharu Tateishi (MSC Software, Ltd., Japan) by implementing the subloading surface model to the commercial software Marc (MSC Software, Ltd.).

The schematic illustration of the draw-bending (so-called *hat-bending*) is shown in Fig. 10.12, which was adopted by Yoshida and Uemori (2003).

The calculation results of the shapes of the sheet after the springback are shown in Fig. 10.13, choosing the die diameter 5 mm and using the following values of material parameters.

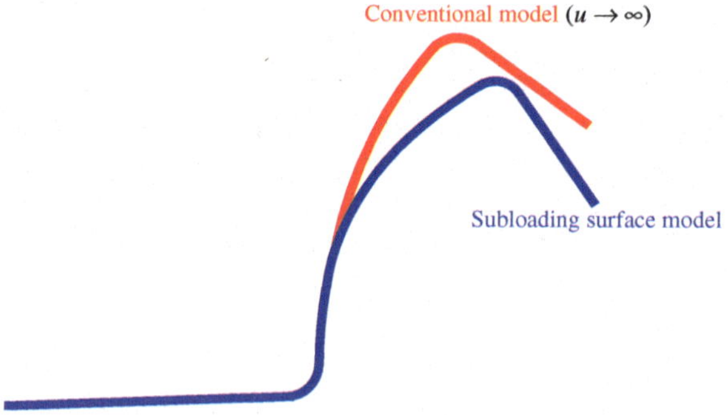

Fig. 10.13 Springback analysis

Material constants:

$$\text{Elastic moduli: } E = 205{,}000\,\text{MPa},\ \nu = 0.3,$$
$$\text{Hardening}\begin{cases}\text{isotropic: } h_1 = 0.5,\ h_2 = 15,\\ \text{kinematic: } c_k = 3{,}000\ \text{MPa},\ \zeta = 0.5,\end{cases}$$
$$\text{Evolution of normal-yield ratio: } \bar{u} = 200,\ u_c = 3.5,\ R_e = 0.5,\ n = 1,$$
$$\text{Translation of elastic-core: } c = 3{,}000\ \text{MPa},\ \xi = 0.7,$$
$$\text{Stagnation of isotropic hardening: } C = 0.5,\ \varsigma = 20,\ \upsilon = 1,$$

Initial values:

$$\text{Isotropic hardening function: } F_0 = 400\,\text{MPa}.$$

The enough springback is predicted, which is caused by the plastic deformation in the stress-releasing process by virtue of the advantage of the subloading surface model describing the plastic strain rate due to the rate of stress inside the yield surface. In contrast, the spingback is predicted just slightly by the conventional elastoplastic model which is realized by using the large value for the material constant in the evolution of the normal-yield ratio $\bar{u} = 100{,}000$ only in the springback process. Then, the importance is recognized for the introduction of the rigorous elastoplastic model, i.e. the subloading surface model capable of describing the plastic strain rate in the stress-reducing process appropriately. Hereinafter, it is desirable that the prediction of springback behavior will be executed by the pertinent analysis exploiting the subloading surface model, aiming at the epochal improvement of the prediction of the springback behavior in industries.

(1) **Residual stress analysis** (Higuchi and Okamura 2016)

The prediction of the residual stress is of importance in the metal forming process. The estimations of residual stress change due to cyclic loading by the conventional elastoplasticity model and the subloading surface model are shown below, which was examined by Higuchi and Okamura (2016).

A four-point cyclic bending test was conducted to examine the change of the residual stress in the cyclic loading process. A specimen with a width of 13 mm, a thickness of 13 mm and a length of 100 mm was cut from a seamless steel pipe P110. The specimen was loaded under a four-point bend configuration with the intervals of 20 mm between the inner rollers and 80 mm between the outer rollers as shown in Fig. 10.14. At first, the specimen was plastically deformed by static bending load corresponding to the maximum bending stress of 900 MPa. After the unloading of the static bending load, the compressive residual stress was generated in the side of outer rollers and the tensile residual stress in the side of inner rollers. The distribution of the residual stress was measured by the X-ray stress measurement method. Then, the specimen was turned upside down, so that the side of outer rollers was in the tensile residual stress state. Sinusoidal waveform load between 5 and 100 % of 900 MPa was applied 20 times to the specimen. The maximum

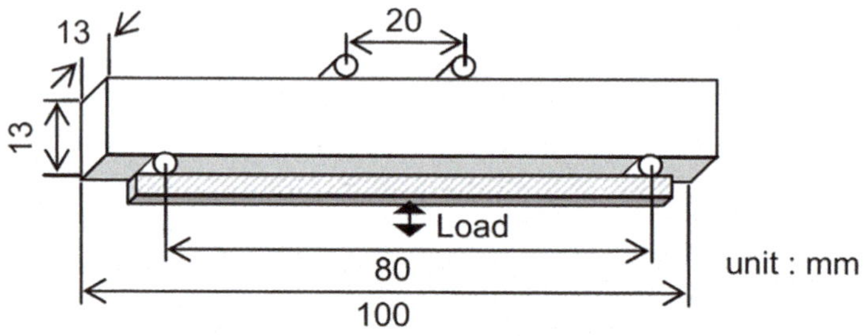

Fig. 10.14 Configuration of the four-point bending test

bending stress was 500 MPa. The distributions of the residual stress after cyclic loading were measured by the X-ray stress measurement method.

The simulations by the Chaboche model described in Sect. 8.2.5 and the subloading surface model described in this chapter are executed, while the commercial software Abaqus is used in the simulation by the Chaboche model. First, material constants of these two models were determined so as to fit to the test data of uniaxial cyclic loading of a round bar specimen of P110 as shown in Fig. 10.15. The test data can be simulated accurately by the subloading surface model. On the other hand, the simulation by the Chaboche model is not in agreement with the test data. Especially, it was difficult to simulate appropriately the smooth elastic-plastic transition in the reverse loading processes by the Chaboche model.

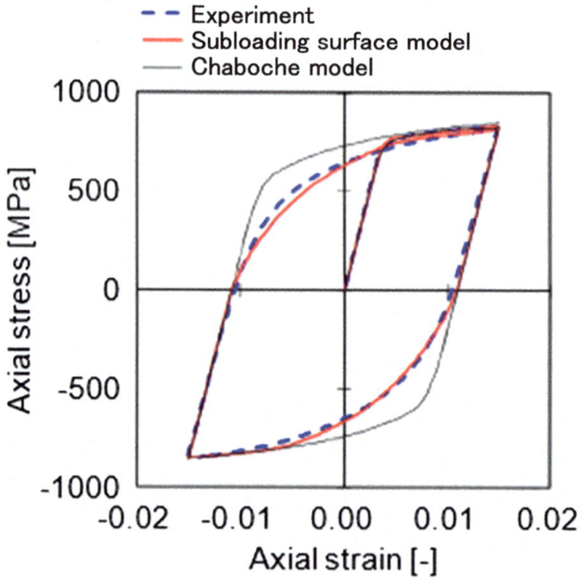

Fig. 10.15 Comparison of stress-strain curves between experiment and simulations in uniaxial loading

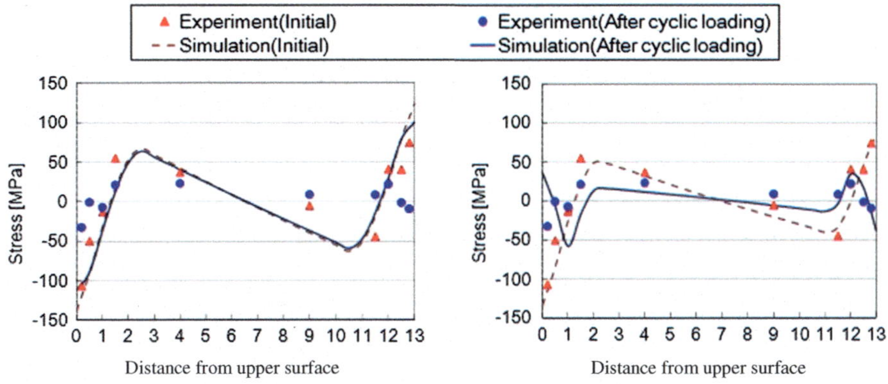

(a) Chaboche model (b) Subloading surface model

Fig. 10.16 Comparison of distributions of residual stresses before and after cyclic loading between experiment and simulations

Measured distributions of the residual stresses after the initial loading and the 20 times cyclic loading are simulated by the two models as shown in Fig. 10.16. The horizontal axis denotes the distance from upper surface to lower surface at center of the specimen. The initial residual stress distributions are simulated well by both of these models. As for the residual stress distribution after cyclic loading, however, the decreases of the residual stresses near the upper and the lower surfaces of the specimen is accurately simulated by the subloading surface model, whereas they are not simulated by the Chaboche model. This would be caused by the fact that the subloading surface model is capable of describing the cyclic loading behavior more accurately than the other constitutive models.

10.6 Orthotropic Anisotropy

The kinematic hardening incorporated in the foregoing is regarded to be the induced anisotropy. On the other hand, various inherent anisotropies are induced in the manufacturing process of metals. The typical inherent anisotropy is the *orthotropic anisotropy* formulated by Hill (1948).

Now, consider the general yield function in the quadratic form shown as follows:

$$\boxed{f(\sigma_{ij}) = \sqrt{\frac{1}{2} C_{ijkl}\sigma_{ij}\sigma_{kl}}} \tag{10.33}$$

where C_{ijkl} is the fourth-order anisotropic tensor having eighty-one components fulfilling the symmetry

$$C_{ijkl} = C_{ijlk} = C_{jikl} = C_{jilk} = C_{klij} = C_{klji} = C_{lkij} = C_{lkji} \tag{10.34}$$

by the minor symmetries $C_{ijkl} = C_{ijlk} = C_{jikl}$ based on the symmetry of the stress tensor $\sigma_{ij} = \sigma_{ji}$ and the major symmetries $C_{ijkl} = C_{klij}$ based on $C_{ijkl}\sigma_{ij}\sigma_{kl}= C_{klij}\sigma_{kl}\sigma_{ij} = C_{klij}\sigma_{ij}\sigma_{kl}$. Then, the independent components is reduced to twenty-one leading to

$$\begin{aligned}
&C_{ijkl}\sigma_{ij}\sigma_{kl}\\
&= C_{1111}\sigma_{11}^2 + 2C_{1122}\sigma_{11}\sigma_{22} + 2C_{1133}\sigma_{11}\sigma_{33} + 2C_{1112}\sigma_{11}\sigma_{12} + 2C_{1123}\sigma_{11}\sigma_{23} + 2C_{1131}\sigma_{11}\sigma_{33}\\
&\quad + C_{2222}\sigma_{22}^2 + 2C_{2233}\sigma_{22}\sigma_{33} + 2C_{2212}\sigma_{22}\sigma_{12} + 2C_{2223}\sigma_{22}\sigma_{23} + 2C_{2231}\sigma_{22}\sigma_{31}\\
&\quad + C_{3333}\sigma_{33}^2 + 2C_{3312}\sigma_{33}\sigma_{12} + 2C_{3323}\sigma_{33}\sigma_{23} + 2C_{3331}\sigma_{33}\sigma_{31}\\
&\quad + C_{1212}\sigma_{12}^2 + 2C_{1223}\sigma_{12}\sigma_{23} + 2C_{1231}\sigma_{12}\sigma_{31}\\
&\quad + C_{2323}\sigma_{23}^2 + 2C_{2331}\sigma_{23}\sigma_{31}\\
&\quad + C_{3131}\sigma_{31}^2
\end{aligned} \tag{10.35}$$

which is the general form of yield function in the quadratic form.

Here, assuming the plastic incompressibility, it holds that

$$\begin{aligned}
(\partial(2f^2)/\partial\sigma_{pq})\delta_{pq} &= (\partial C_{ijkl}\sigma_{ij}\sigma_{kl}/\partial\sigma_{pq})\delta_{pq}\\
&= C_{ijkl}\delta_{pi}\delta_{qj}\sigma_{kl}\delta_{pq} + C_{ijkl}\sigma_{ij}\delta_{pk}\delta_{ql}\delta_{pq}\\
&= C_{ppkl}\sigma_{kl} + C_{ijpp}\sigma_{ij}=C_{ppkl}\sigma_{kl} + C_{ppij}\sigma_{ij}\\
&= 2C_{ppkl}\sigma_{kl}= 0
\end{aligned}$$

This relation must hold for any σ_{ij} and thus one obtains

$$C_{ppkl} = C_{ijqq} = 0 \tag{10.36}$$

which leads to

$$\left.\begin{array}{l}
C_{1111} + C_{1122} + C_{1133} = 0\\
C_{2211} + C_{2222} + C_{2233} = 0\\
C_{3311} + C_{3322} + C_{3333} = 0
\end{array}\right\} \tag{10.37}$$

$$\left.\begin{array}{l}
C_{1112} + C_{2212} + C_{3312} = 0\\
C_{1123} + C_{2223} + C_{3323} = 0\\
C_{1131} + C_{2231} + C_{3331} = 0
\end{array}\right\} \to \left.\begin{array}{l}
C_{3312} = -(C_{1112} + C_{2212})\\
C_{1123} = -(C_{2223} + C_{3323})\\
C_{2231} = -(C_{1131} + C_{3331})
\end{array}\right\} \tag{10.38}$$

The substitution of Eq. (10.38) into Eq. (10.35) gives the expression

$$
\begin{aligned}
& C_{ijkl}\sigma_{ij}\sigma_{kl} \\
& = C_{1111}\sigma_{11}^2 + 2C_{1122}\sigma_{11}\sigma_{22} + 2C_{1133}\sigma_{11}\sigma_{33} + 2C_{1112}\sigma_{11}\sigma_{12} - 2(C_{2223} \\
& \quad + C_{3323})\sigma_{11}\sigma_{23} + 2C_{1131}\sigma_{11}\sigma_{31} \\
& \quad + C_{2222}\sigma_{22}^2 + 2C_{2233}\sigma_{22}\sigma_{33} + 2C_{2212}\sigma_{22}\sigma_{12} + 2C_{2223}\sigma_{22}\sigma_{23} - 2(C_{1131} + C_{3331})\sigma_{22}\sigma_{33} \\
& \quad + C_{3333}\sigma_{33}^2 - 2(C_{1112} + C_{2212})\sigma_{33}\sigma_{12} + 2C_{3323}\sigma_{33}\sigma_{23} + 2C_{3331}\sigma_{33}\sigma_{31} \\
& \quad + C_{1212}\sigma_{12}^2 + 2C_{1223}\sigma_{123}\sigma_{23} + 2C_{1231}\sigma_{12}\sigma_{31} \\
& \quad + C_{2323}\sigma_{23}^2 + 2C_{2331}\sigma_{23}\sigma_{31} \\
& \quad + C_{3131}\sigma_{31}^2
\end{aligned}
\tag{10.39}
$$

Further, noting Eq. (10.37), the terms in the form $C_{iijj}\sigma_{ii}\sigma_{jj}$ (no sum) are written as

$$
\begin{aligned}
& C_{1111}\sigma_{11}^2 + C_{2222}\sigma_{22}^2 + C_{3333}\sigma_{33}^2 + 2C_{1122}\sigma_{11}\sigma_{22} + 2C_{2233}\sigma_{22}\sigma_{33} + 2C_{1133}\sigma_{11}\sigma_{33} \\
& \quad = C_{1111}\sigma_{11}^2 + C_{2222}\sigma_{22}^2 + C_{3333}\sigma_{33}^2 \\
& \qquad - C_{1122}(\sigma_{11} - \sigma_{22})^2 + C_{1122}\sigma_{11}^2 + C_{1122}\sigma_{22}^2 \\
& \qquad - C_{2233}(\sigma_{22} - \sigma_{33})^2 + C_{2233}\sigma_{22}^2 + C_{2233}\sigma_{33}^2 \\
& \qquad - C_{1133}(\sigma_{33} - \sigma_{11})^2 + C_{1133}\sigma_{33}^2 + C_{2233}\sigma_{11}^2 \\
& \quad = (C_{1111} + C_{1122} + C_{2233})\sigma_{11}^2 + (C_{1122} + C_{2222} + C_{2233})\sigma_{22}^2 \\
& \qquad + (C_{1133} + C_{2233} + C_{3333})\sigma_{33}^2 \\
& \qquad - C_{1122}(\sigma_{11} - \sigma_{22})^2 - C_{2233}(\sigma_{22} - \sigma_{33})^2 - C_{1133}(\sigma_{33} - \sigma_{11})^2 \\
& \quad = -C_{1122}(\sigma_{11} - \sigma_{22})^2 - C_{2233}(\sigma_{22} - \sigma_{33})^2 - C_{1133}(\sigma_{33} - \sigma_{11})^2
\end{aligned}
\tag{10.40}
$$

Then, by setting

$$
\left.
\begin{array}{l}
a_1 \equiv -C_{1122}, a_2 \equiv -C_{2233}\ a_3 \equiv -C_{1133} \\
a_4 \equiv -2C_{1112}, a_5 \equiv -2C_{2212}, a_6 \equiv -C_{2223} \\
a_7 \equiv -2C_{3323}\ a_8 \equiv -2C_{3331}, a_9 \equiv -2C_{1131} \\
a_{10} \equiv 2C_{1223}, a_{11} \equiv 2C_{2331}\ a_{12} \equiv 2C_{1231} \\
a_{13} \equiv C_{1212}, a_{14} \equiv C_{2323}\ a_{15} \equiv C_{3131}
\end{array}
\right\}
\tag{10.41}
$$

and substituting Eqs. (10.39) and (10.40) with Eq. (10.41) into Eq. (10.39) reads:

$$
\begin{aligned}
C_{ijkl}\sigma_{ij}\sigma_{kl} = & \ a_1(\sigma_{11} - \sigma_{22})^2 + a_2(\sigma_{22} - \sigma_{33})^2 + a_3(\sigma_{33} - \sigma_{11})^2 \\
& + \{a_4(\sigma_{33} - \sigma_{11}) + a_5(\sigma_{33} - \sigma_{22})\}\sigma_{12} \\
& + \{a_6(\sigma_{11} - \sigma_{22}) + a_7(\sigma_{22} - \sigma_{33})\}\sigma_{23} \\
& + \{a_8(\sigma_{22} - \sigma_{33}) + a_9(\sigma_{22} - \sigma_{11})\}\sigma_{31} \\
& + a_{10}\sigma_{12}\sigma_{23} + a_{11}\sigma_{23}\sigma_{31} + a_{12}\sigma_{31}\sigma_{12} \\
& + a_{13}\sigma_{12}^2 + a_{14}\sigma_{23}^2 + a_{15}\sigma_{31}^2
\end{aligned}
\tag{10.42}
$$

Equation (10.42) is the general yield function for the plastically-incompressible materials in the quadratic form.

Now, assume orthotropic anisotropy. Then, if we describe the yield surface by the coordinate axes selected to the principal axes $\{\widehat{\mathbf{e}}_i\}$ of orthotropic anisotropy, the yield function is independent of the sign of shear stress components in this coordinate system. Therefore, it must hold that

$$a_4 = a_5 = a_6 = a_7 = a_8 = a_9 = a_{10} = a_{11} = a_{12} = 0$$

Here, replacing the symbols a_i as

$$F = a_1, G = a_2, H = a_3, L = a_{13}/2, M = a_{14}/2, H = a_{15}/2$$

used by Hill (1948), Eq. (10.42) leads to the Hill's yield condition with orthotropic anisotropy:

$$\boxed{\frac{1}{2}\sqrt{F(\sigma_{11}-\sigma_{22})^2+G(\sigma_{22}-\sigma_{33})^2+H(\sigma_{33}-\sigma_{11})^2+6(L\sigma_{12}^2+M\sigma_{23}^2+N\sigma_{31}^2)} = F(H)} \tag{10.43}$$

Here, note that for the isotropic material all the material parameters are unity, i.e. $F = G = H = L = M = N = 1$ holds and thus Eq. (10.43) is reduced to $\sqrt{3/2}\|\hat{\boldsymbol{\sigma}}'\| = F(H)$ which is the equivalent stress, noting Eq. (1.187). While Eq. (10.43) is the expression on the principal axis $\{\widehat{\mathbf{e}}_i\}$ of orthotropic anisotropy, it is rewritten by the following equation stipulating this fact.

$$\frac{1}{2}\sqrt{F(\widehat{\sigma}_{11}-\widehat{\sigma}_{22})^2+G(\widehat{\sigma}_{22}-\widehat{\sigma}_{33})^2+H(\widehat{\sigma}_{33}-\widehat{\sigma}_{11})^2+6(L\widehat{\sigma}_{12}^2+M\widehat{\sigma}_{23}^2+N\widehat{\sigma}_{31}^2)}$$
$$= F(H) \tag{10.44}$$

or

$$\bar{F}(\widehat{\sigma}_{11}-\widehat{\sigma}_{22})^2+\bar{G}(\widehat{\sigma}_{22}-\widehat{\sigma}_{33})^2+\bar{H}(\widehat{\sigma}_{33}-\widehat{\sigma}_{11})^2+6(\bar{L}\widehat{\sigma}_{12}^2+\bar{M}\widehat{\sigma}_{23}^2+\bar{N}\widehat{\sigma}_{31}^2) = 1 \tag{10.45}$$

where

$$\left.\begin{array}{l} \bar{H} \equiv \dfrac{H}{[2F(H)]^2}, \ \bar{G} \equiv \dfrac{G}{[2F(H)]^2}, \ \bar{F} \equiv \dfrac{F}{[2F(H)]^2} \\ \bar{L} \equiv \dfrac{L}{[2F(H)]^2}, \ \bar{M} \equiv \dfrac{L}{[2F(H)]^2}, \ \bar{N} \equiv \dfrac{L}{[2F(H)]^2} \end{array}\right\} \tag{10.46}$$

Denoting $\widehat{\sigma}_{11}, \widehat{\sigma}_{22}, \widehat{\sigma}_{33}, \widehat{\sigma}_{12}, \widehat{\sigma}_{23}$ and $\widehat{\sigma}_{31}$ by the yield stress by $\widehat{\sigma}^y_1, \widehat{\sigma}^y_2, \widehat{\sigma}^y_3$, $\widehat{\sigma}^y_{12}, \widehat{\sigma}^y_{23}$ and $\widehat{\sigma}^y_{31}$, respectively, when only each stress applies, one has

$$\left.\begin{array}{l}(\bar{H}+\bar{F})\widehat{\sigma}_1^{y2} = 1, (\bar{G}+\bar{F})\widehat{\sigma}_2^{y2} = 1, (\bar{G}+\bar{H})\widehat{\sigma}_3^{y2} = 1 \\ 6\bar{L}\widehat{\tau}_{12}^{y2} = 1, 6\bar{M}\widehat{\tau}_{23}^{y2} = 1, 6\bar{N}\widehat{\tau}_{31}^{y2} = 1\end{array}\right\} \tag{10.47}$$

from which the material parameters are given by

$$\left.\begin{array}{l}\bar{F} = \dfrac{1}{2}\left(\dfrac{1}{\widehat{\sigma}_1^{y2}} + \dfrac{1}{\widehat{\sigma}_2^{y2}} + \dfrac{1}{\widehat{\sigma}_3^{y2}}\right) \\ \bar{G} = \dfrac{1}{2}\left(\dfrac{1}{\widehat{\sigma}_2^{y2}} + \dfrac{1}{\widehat{\sigma}_3^{y2}} + \dfrac{1}{\widehat{\sigma}_1^{y2}}\right) \\ \bar{H} = \dfrac{1}{2}\left(\dfrac{1}{\widehat{\sigma}_3^{y2}} + \dfrac{1}{\widehat{\sigma}_1^{y2}} + \dfrac{1}{\widehat{\sigma}_2^{y2}}\right) \\ \bar{L} = \dfrac{1}{6\widehat{\tau}_{12}^{y2}}, \bar{M} = \dfrac{1}{6\widehat{\tau}_{23}^{y2}}, \bar{N} = \dfrac{1}{6\widehat{\tau}_{23}^{y2}}\end{array}\right\} \tag{10.48}$$

Equation (10.44) is reduced under the uniaxial loading in the sheet metal forming as follows:

$$\frac{1}{2}\sqrt{(F+H)}\widehat{\sigma}_{11} = F(H) \quad \text{or} \quad (\bar{F}+\bar{H})\widehat{\sigma}_{11}^2 = F(H) \tag{10.49}$$

Further, under the plane stress condition observed in the sheet metal forming it holds that $\widehat{\sigma}_{23} = \widehat{\sigma}_{31} = \widehat{\sigma}_{33} = 0$ and thus Eqs. (10.44) and (10.45) are reduced to

$$\frac{1}{2}\sqrt{(F+H)\widehat{\sigma}_{11}^2 - 2F\widehat{\sigma}_{11}\widehat{\sigma}_{22} + (F+G)\widehat{\sigma}_{22}^2 + 6L\widehat{\sigma}_{12}^2} = F(H) \tag{10.50}$$

i.e.

$$(\bar{F}+\bar{H})\widehat{\sigma}_{11}^2 - 2\bar{F}\widehat{\sigma}_{11}\widehat{\sigma}_{22} + (\bar{F}+\bar{G})\widehat{\sigma}_{22}^2 + 6\bar{L}\widehat{\sigma}_{12}^2 = 1 \tag{10.51}$$

Equation (10.51) is described in the state that the principal stress directions coincide with the orthogonal anisotropy axes as follows:

$$(\bar{F}+\bar{H})\widehat{\sigma}_1^2 - 2\bar{F}\widehat{\sigma}_1\widehat{\sigma}_2 + (\bar{F}+\bar{G})\widehat{\sigma}_2^2 = 1 \tag{10.52}$$

The plastic strain rate is given for as follows:

$$\left.\begin{array}{l}d_1^p/\dot{\lambda} = 2(\bar{F}+\bar{H})\widehat{\sigma}_1 - 2\bar{F}\widehat{\sigma}_2 \\ d_2^p/\dot{\lambda} = -2\bar{F}\widehat{\sigma}_1 + 2(\bar{F}+\bar{G})\widehat{\sigma}_2 \\ d_3^p/\dot{\lambda} = -2\bar{H}\widehat{\sigma}_1 - 2\bar{G}\widehat{\sigma}_2\end{array}\right\} \tag{10.53}$$

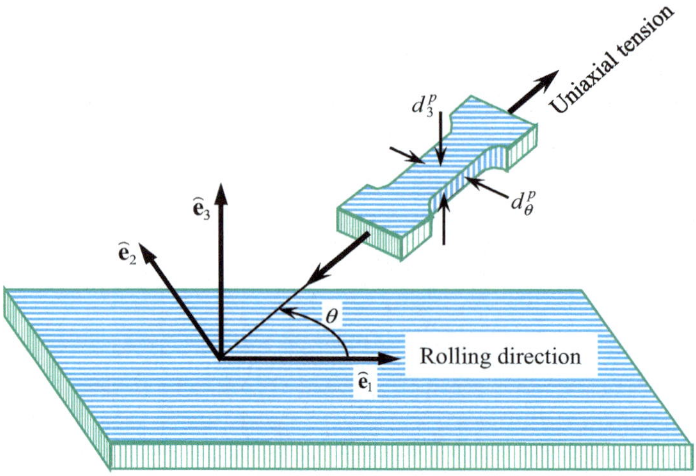

Fig. 10.17 Uniaxial tension test for R-value for metal formed by rolling process

The orthogonal anisotropy is induced seriously in the rolling process for the sheet metal forming. Choosing the axes $\widehat{\mathbf{e}}_1, \widehat{\mathbf{e}}_2, \widehat{\mathbf{e}}_3$ to the rolling, the traverse, and the thickness directions, respectively, the following R-value is adopted widely in order to evaluate the intensity of the orthotropy.

$$R_\theta = \frac{d_\theta^p}{d_3^p} \tag{10.54}$$

where $d_\theta^p(<0)$ is the lateral strain rate measured from the uniaxial tension test of the test specimen cut out at the angle θ measured counterclockwise from the rolling direction. The plastic strain rate $d_3^p(<0)$ in the thickness direction is calculated from the axial and the lateral strain rates by the assumption of plastic incompressibility (see Fig. 10.17). $R = 1$ means the isotropy. A small R-value means that the produced sheet metal is easily thinned resulting in an easy failure. It follows from Eq. (10.53) that

$$\left.\begin{aligned} R_0 &= \frac{d_2^p}{d_3^p} = \frac{\bar{F}}{\bar{H}} \quad \text{for } \widehat{\sigma}_2 = \widehat{\sigma}_3 = 0 \\ R_{90} &= \frac{d_1^p}{d_3^p} = \frac{\bar{F}}{\bar{G}} \quad \text{for } \widehat{\sigma}_1 = \widehat{\sigma}_3 = 0 \end{aligned}\right\} \tag{10.55}$$

Substituting $\bar{F} = R_0\bar{H}$ due to Eq. (10.55) into Eq. (10.52), one has

$$(\bar{H} + R_0\bar{H})\widehat{\sigma}_1^2 + (\bar{G} + R_0\bar{H})\widehat{\sigma}_2^2 - 2R_0\bar{H}\widehat{\sigma}_1\widehat{\sigma}_2 = 1$$

leading to

$$\widehat{\sigma}_1^2 + \frac{\bar{G} + R_0\bar{H}}{(1+R_0)\bar{H}}\widehat{\sigma}_2^2 - \frac{2R_0}{1+R_0}\widehat{\sigma}_1\widehat{\sigma}_2 = \frac{1}{(1+R_0)\bar{H}} \tag{10.56}$$

Substituting Eq. (10.48) into Eq. (10.55), it follows that

$$\left.\begin{aligned} (1+R_0)\frac{1}{\widehat{\sigma}_3^{y2}} - (1+R_0)\frac{1}{\widehat{\sigma}_2^{y2}} = (1-R_0)\frac{1}{\widehat{\sigma}_1^{y2}} \\ (1+R_{90})\frac{1}{\widehat{\sigma}_3^{y2}} - (1-R_{90})\frac{1}{\widehat{\sigma}_2^{y2}} = (1+R_{90})\frac{1}{\widehat{\sigma}_1^{y2}} \end{aligned}\right\}$$

By solving this equation, one has

$$\left.\begin{aligned} \frac{1}{\widehat{\sigma}_2^{y2}} = \frac{R_0(1+R_{90})}{(1+R_0)R_{90}}\frac{1}{\widehat{\sigma}_1^{y2}} \\ \frac{1}{\widehat{\sigma}_3^{y2}} = \frac{R_0+R_{90}}{(1+R_0)R_{90}}\frac{1}{\widehat{\sigma}_1^{y2}} \end{aligned}\right\} \tag{10.57}$$

Substituting Eq. (10.57) into Eq. (10.48), it follows that

$$\left.\begin{aligned} \bar{G} = \frac{R_0}{(1+R_0)R_{90}}\frac{1}{\widehat{\sigma}_1^{y2}} \\ \bar{H} = \frac{1}{1+R_0}\frac{1}{\widehat{\sigma}_1^{y2}} \end{aligned}\right\} \tag{10.58}$$

from which we have

$$\left.\begin{aligned} \frac{\bar{G}+R_0\bar{H}}{(1+R_0)\bar{H}} = \frac{R_0(1+R_{90})}{(1+R_0)R_{90}} \\ \frac{1}{(1+R_0)\bar{H}} = \widehat{\sigma}_1^{y2} \end{aligned}\right\} \tag{10.59}$$

Substituting Eq. (10.59) into Eq. (10.56), it follows that

$$\boxed{\widehat{\sigma}_1^2 + \frac{R_0(1+R_{90})}{(1+R_0)R_{90}}\widehat{\sigma}_2^2 - \frac{2R_0}{1+R_0}\widehat{\sigma}_1\widehat{\sigma}_2 = \widehat{\sigma}_1^{y2}} \tag{10.60}$$

Equation (10.51) is rewritten as

$$\left[\frac{1}{4}(\bar{G}+\bar{H})+\frac{1}{4}(\bar{G}+\bar{H}+4\bar{F})-\frac{1}{2}(\bar{G}-\bar{H})\right]\widehat{\sigma}_{11}^2+\left[\frac{2}{4}(\bar{G}+\bar{H})-\frac{2}{4}(\bar{G}+\bar{H}+4\bar{F})\right]\widehat{\sigma}_{11}\widehat{\sigma}_{22}$$
$$+\left[\frac{1}{4}(\bar{G}+\bar{H})+\frac{1}{4}(\bar{G}+\bar{H}+4\bar{F})+\frac{1}{2}(\bar{G}-\bar{H})\right]\widehat{\sigma}_{22}^2+6\bar{L}\widehat{\sigma}_{12}^2=1$$

which is arranged as follows:

$$\frac{1}{4}(\bar{G}+\bar{H})(\widehat{\sigma}_{11}+\widehat{\sigma}_{22})^2+\frac{1}{4}(\bar{G}+\bar{H}+4\bar{F})(\widehat{\sigma}_{11}-\widehat{\sigma}_{22})^2$$
$$-\frac{1}{2}(\bar{G}-\bar{H})(\widehat{\sigma}_{11}^2-\widehat{\sigma}_{22}^2)+6\bar{L}\widehat{\sigma}_{12}^2=1 \tag{10.61}$$

Denoting the angle measured in the counterclockwise direction from the principal axes of anisotropy to the principal stress as α and substituting the relations

$$\widehat{\sigma}_{11}+\widehat{\sigma}_{22}=\sigma_1+\sigma_2,\ \widehat{\sigma}_{11}-\widehat{\sigma}_{22}=(\sigma_1-\sigma_2)\cos 2\alpha,\ 2\widehat{\sigma}_{12}=(\sigma_1-\sigma_2)\sin 2\alpha \tag{10.62}$$

into Eq. (10.61), one has

$$\frac{1}{4}(\bar{G}+\bar{H})(\sigma_1+\sigma_2)^2+\frac{1}{4}(\bar{G}+\bar{H}+4\bar{F})(\sigma_1-\sigma_2)^2\cos^2 2\alpha$$
$$-\frac{1}{2}(\bar{G}-\bar{H})(\sigma_1^2-\sigma_2^2)\cos 2\alpha+\frac{3}{2}\bar{L}(\sigma_1-\sigma_2)^2\sin^2 2\alpha=1$$

which is rewritten as

$$(\sigma_1+\sigma_2)^2-2a(\sigma_1^2-\sigma_2^2)\cos 2\alpha+b(\sigma_1-\sigma_2)^2\cos^2 2\alpha$$
$$+6\frac{\bar{L}}{\bar{G}+\bar{H}}(\sigma_1-\sigma_2)^2=\frac{4}{\bar{G}+\bar{H}} \tag{10.63}$$

where

$$a\equiv\frac{\bar{G}-\bar{H}}{\bar{G}+\bar{H}},\ b\equiv\frac{\bar{G}+\bar{H}+4\bar{F}-6\bar{L}}{\bar{G}+\bar{H}} \tag{10.64}$$

Here, denoting the yielding strength in the equi-two axis tension as σ and that of the pure shear as τ, it follows from Eq. (10.51) that

$$\sigma\equiv(\bar{H}+\bar{G})^{-1/2},\ \tau\equiv(6\bar{L})^{-1/2} \tag{10.65}$$

The substitution of Eq. (10.65) into Eq. (10.63) leads to

$$(\sigma_1+\sigma_2)^2+\left(\frac{\sigma}{\tau}\right)^2(\sigma_1-\sigma_2)^2-2a(\sigma_1^2-\sigma_2^2)\cos 2\alpha+b(\sigma_1$$
$$-\sigma_2)^2\cos^2 2\alpha=(2\sigma)^2 \tag{10.66}$$

Equation (10.66) is extended to the following equation for the in-plane isotropy with the material constant $m(\geq 1)$.

$$|\sigma_1+\sigma_2|^m+\left(\frac{\sigma}{\tau}\right)^m|\sigma_1-\sigma_2|^m=(2\sigma)^m \tag{10.67}$$

Hill (1990) proposed the following extended orthotropic yield condition from Eqs. (10.66) and (10.67).

$$\begin{aligned}&|\sigma_1+\sigma_2|^m+\left(\frac{\sigma}{\tau}\right)^m|\sigma_1-\sigma_2|^m\\&\quad+|\sigma_1^2+\sigma_2^2|^{(m/2)-1}[-2a(\sigma_1^2-\sigma_2^2)+b(\sigma_1-\sigma_2)^2\cos 2\alpha]\cos 2\alpha=(2\sigma)^m\end{aligned} \tag{10.68}$$

Equation (10.68) includes the five material constants, i.e. the yield stress σ,τ and the dimensionless number a,b,m. It is reduced to Eq. (10.66) for $m=2$ and to Eq. (10.67) for $a=b=0$ (or $\alpha=\pi/4$). By use of Eq. (10.62), Eq. (10.68) is rewritten in the anisotropic axes as follows:

$$\begin{aligned}&|\widehat{\sigma}_{11}+\widehat{\sigma}_{22}|^m+\left(\frac{\sigma}{\tau}\right)^m\left|(\widehat{\sigma}_{11}-\widehat{\sigma}_{22})^2+4\widehat{\sigma}_{12}^2\right|^{m/2}\\&\quad+\left|(\widehat{\sigma}_{11}+\widehat{\sigma}_{22})^2+4\widehat{\sigma}_{12}^2\right|^{(m/2)-1}[-2a(\widehat{\sigma}_{11}^2-\widehat{\sigma}_{22}^2)+b(\widehat{\sigma}_{11}-\widehat{\sigma}_{22})^2]=(2\sigma)^m\end{aligned} \tag{10.69}$$

Generally, the yield surface is described in the principal axes of anisotropy as follows:

$$f(\widehat{\sigma}_{ij})=F(H) \tag{10.70}$$

where

$$\widehat{\boldsymbol{\sigma}}=\widehat{\mathbf{R}}^T\boldsymbol{\sigma}\widehat{\mathbf{R}},\widehat{\sigma}_{ij}=\widehat{R}_{ri}\widehat{R}_{sj}\sigma_{rs} \tag{10.71}$$

$$\widehat{R}_{ij}(t)\equiv\mathbf{e}_i\cdot\widehat{\mathbf{e}}_j(t)\ (=\cos(\mathbf{e}_i,\widehat{\mathbf{e}}_j(t))) \tag{10.72}$$

$\widehat{\mathbf{e}}_i(t)$ are the base vectors taken to the directions of the principal axes of the orthotropic anisotropy. Needless to say, Eq. (10.70) is not a general tensor expression but is merely the expression by the components. The variation of $\widehat{\mathbf{e}}_i$ is calculated using the following equation with the initial value of $\widehat{\mathbf{e}}_{i0}$.

$$\widehat{\mathbf{e}}_i=\widehat{\mathbf{e}}_{i0}+\int\dot{\widehat{\mathbf{e}}}_i dt \tag{10.73}$$

where $\dot{\widehat{\mathbf{e}}}_i$ is given by

$$\overset{\bullet}{\widehat{\mathbf{e}}}_i = \boldsymbol{\omega}_a \widehat{\mathbf{e}}_i \; (\boldsymbol{\omega}_a = \overset{\bullet}{\widehat{\mathbf{e}}}_i \otimes \widehat{\mathbf{e}}_i) \tag{10.74}$$

Here, the stress rate $\overset{\bullet}{\widehat{\sigma}}_{ij}$ is given by

$$\overset{\bullet}{\widehat{\sigma}}_{ij} = \overset{\circ}{\widehat{\sigma}}_{ij} = \widehat{R}_{ri}\widehat{R}_{sj}\overset{\circ}{\sigma}_{rs} = \widehat{R}_{ri}\widehat{R}_{sj}(\overset{\bullet}{\sigma}_{rs} - \omega_{arp}\sigma_{ps} + \sigma_{rp}\omega_{aps}) \tag{10.75}$$

noting Eqs. (1.86) and (4.55) with $\mathbf{Q} = \widehat{\mathbf{R}}^T$.

Various anisotropic yield surfaces in the plane stress state are proposed by the Barlat's group (e.g. Barlat et al. 2007), Yoshida (2015), etc.

10.7 Representation of Isotropic Mises Yield Condition

The isotropic yield function described by Eq. (6.57) can be expressed in the following various forms.

$$\begin{aligned}
f(\boldsymbol{\sigma}) = \sigma^{eq} &= \sqrt{\frac{3}{2}}\|\boldsymbol{\sigma}'\| = \sqrt{\frac{3}{2}}\sqrt{\sigma'_{rs}\sigma'_{rs}} \\
&= \sqrt{\frac{3}{2}}\sqrt{\sigma'^2_{11} + \sigma'^2_{22} + \sigma'^2_{33} + 2(\sigma'^2_{12} + \sigma'^2_{23} + \sigma'^2_{31})} \\
&= \sqrt{\frac{1}{2}}\sqrt{(\sigma_{11} - \sigma_{22})^2 + (\sigma_{22} - \sigma_{33})^2 + (\sigma_{33} - \sigma_{11})^2 + 6(\sigma^2_{12} + \sigma^2_{23} + \sigma^2_{31})} \\
&= \sqrt{\frac{3}{2}}\sqrt{\sigma'^2_1 + \sigma'^2_2 + \sigma'^2_3} \\
&= \sqrt{\frac{1}{2}}\sqrt{(\sigma_1 - \sigma_2)^2 + (\sigma_2 - \sigma_3)^2 + (\sigma_3 - \sigma_1)^2} = F
\end{aligned} \tag{10.76}$$

The combined test of the tensile stress $\sigma(=\sigma_{11})$ and the distortional stress $\tau(=\sigma_{12})$ for a thin wall cylinder specimen is widely adopted for metal. In this case Eq. (10.76) is rewritten as

$$\sigma^2 + (\sqrt{3}\tau)^2 = F^2 \tag{10.77}$$

Then, the Mises yield condition is shown by a circle of radius F in the $(\sigma, \sqrt{3}\tau)$ plane.

The visualization of the stress state can be realized in the space of three and less dimension. The stress state can be represented completely in the principal stress space when principal stress directions are fixed to materials and only the principal stress values change. In general, however, one must use the six-dimensional space or memorize the variation of the principal stress direction if the directions change. However, in the cases for which the number of independent variable components is less than three, such as the tension-distortion test described above and the plane stress and strain tests, the state of stress can be represented in the three and less dimensional stress space. The *Ilyushin's isotropic stress space* (Ilyushin 1963) is convenient to depict the Mises yield surface, which depends only on the deviatoric stress, as explained below.

The deviatoric stress tensor includes the five independent variables and thus the Mises yield surface in Eq. (10.76) is described by the independent components as follows:

$$f(\boldsymbol{\sigma}) = \sqrt{3\sigma'^2_{11} + 3\sigma'^2_{22} + 3\sigma'_{11}\sigma'_{22} + 3(\sigma'^2_{12} + \sigma'^2_{23} + \sigma'^2_{31})}$$
$$= \sqrt{\left(\frac{3}{2}\sigma'_{11}\right)^2 + 3\left(\frac{1}{2}\sigma'_{11} + \sigma'_{22}\right)^2 + 3(\sigma'^2_{12} + \sigma'^2_{23} + \sigma'^2_{31})} = F$$

and thus it can be rewritten as

$$S_1^2 + S_2^2 + S_3^2 + S_4^2 + S_5^2 = F^2 \qquad (10.78)$$

in the five-dimensional space with the axes

$$S_1 = \frac{3}{2}\sigma'_{11},\ S_2 = \sqrt{3}\left(\frac{1}{2}\sigma'_{11} + \sigma'_{22}\right),\ S_3 = \sqrt{3}\sigma'_{12},\ S_4 = \sqrt{3}\sigma'_{23},\ S_5 = \sqrt{3}\sigma'_{31} \qquad (10.79)$$

Equation (10.78) exhibits the *five-dimensional spherical super surface*. Further, consider the expression of the Mises yield surface for the plane stress and strain conditions in the following.

10.7.1 Plane Stress State

The plane stress state fulfilling $\sigma_{3j} = 0$ can be described in the three-dimensional space $(\sigma_{11}, \sigma_{22}, \sigma_{12})$ and thus the Mises yield condition (10.76) is described by the following equation.

$$\sqrt{\sigma_{11}^2 - \sigma_{11}\sigma_{22} + \sigma_{22}^2 + 3\sigma_{12}^2} = F \qquad (10.80)$$

On the other hand, Eq. (10.80) can be described in the two-dimensional principal stress plane as follows:

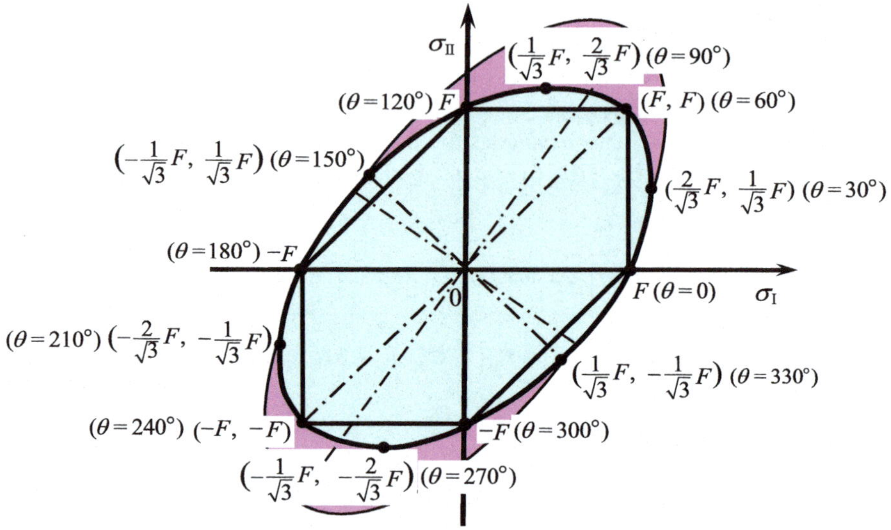

Fig. 10.18 Mises yield surface in the plane stress condition (Thin curve describes Hill's orthotropic Mises yield condition)

$$\sqrt{\sigma_1^2 - \sigma_1\sigma_2 + \sigma_2^2} = F \tag{10.81}$$

which is the section of the Mises yield condition cut by the plane $\sigma_3 = 0$ and exhibits *Mises's ellipse* in the principal stress plane (σ_1, σ_2) as shown in Fig. 10.18. It follows from the third equation of Eq. (1.264) that

$$\sigma_m = -\sqrt{\frac{2}{3}}\|\boldsymbol{\sigma}'\| F\cos\left(\theta + \frac{2}{3}\pi\right) = -\frac{2}{3}F\cos\left(\theta + \frac{2}{3}\pi\right) \tag{10.82}$$

because of $\sigma_m + \sigma'_3 = 0$ leading to $\sigma_m = -\sigma'_3$. Substituting Eq. (1.264) and Eq. (10.82) with $\|\boldsymbol{\sigma}'\| = \sqrt{2/3}F$ into Eq. (1.260), one obtains

$$\left.\begin{aligned}\sigma_1 &= -\frac{2}{3}F\cos\left(\theta + \frac{2}{3}\pi\right) + \frac{2}{3}F\cos\theta = \frac{2}{\sqrt{3}}F\sin\left(\theta + \frac{\pi}{3}\right)\\ \sigma_2 &= -\frac{2}{3}F\cos\left(\theta + \frac{2}{3}\pi\right) + \frac{2}{3}F\cos\left(\theta - \frac{2}{3}\pi\right) = \frac{2}{\sqrt{3}}F\sin\theta\end{aligned}\right\} \tag{10.83}$$

from which the coordinates of main points on the Mises's ellipse are calculated as shown in Fig. 10.18. The thin curve shows the Hill's orthotropic Mises yield surface in Eq. (10.50), which is rotated the principal axes of ellipse with the changes of its long and short radii from the isotropic Mises yield surface.

Next, consider the *Ilyushin's isotropic stress space* in which the variables in Eq. (10.79) are used. Here, in the present case fulfilling $\sigma_{3j} = 0$ leading to $S_4 =$

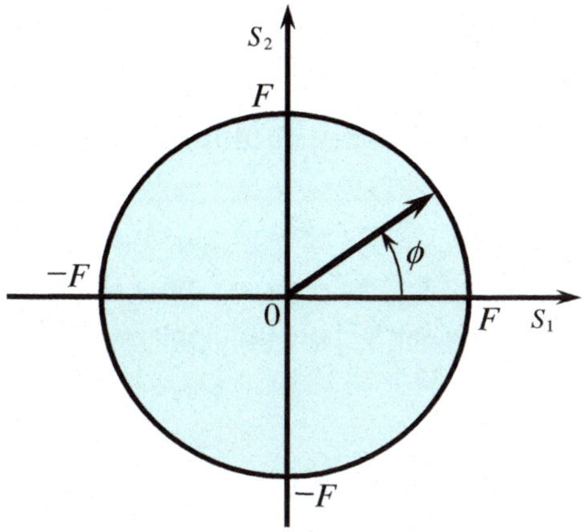

Fig. 10.19 Mises yield surface in plane stress state without shear stress

$S_5 = 0$ the Mises yield surface is represented by the sphere in the (S_1, S_2, S_3) space, while it holds that

$$\left.\begin{aligned} S_1 &= (3/2)\sigma'_{11} = (3/2)[\sigma_{11} - (\sigma_{11} + \sigma_{22})/3] = \sigma_{11} - \sigma_{22}/2 \\ S_2 &= \sqrt{3}[(1/2)\sigma'_{11} + \sigma'_{22}] \\ &= \sqrt{3}\{(1/2)[\sigma_{11} - (\sigma_{11} + \sigma_{22})/3] + [\sigma_{22} - (\sigma_{11} + \sigma_{22})/3]\} = (\sqrt{3}/2)\sigma_{22} \end{aligned}\right\} \tag{10.84}$$

Furthermore, in the case fulfilling $\sigma_{12} = 0$, the Mises yield surface is represented by the circle in the (S_1, S_2) plane (Fig. 10.19). Here, setting

$$S_1 = F\cos\phi, \quad S_2 = F\sin\phi \tag{10.85}$$

and substituting them into Eq. (10.84), it holds that

$$\sigma_{11} = \frac{2}{\sqrt{3}}F\sin\left(\phi + \frac{\pi}{3}\right), \quad \sigma_{22} = \frac{2}{\sqrt{3}}F\sin\phi \tag{10.86}$$

10.7.2 Plane Strain State

If the elastic strain rate can be ignored compared with the plastic strain rate in the plane strain state, the following relation holds by substituting $D^p_{33} = \dot{\lambda}\,\sigma'_{33} = 0$ into $\sigma'_{rr} = 0$.

$$\sigma_{33} = \frac{1}{2}(\sigma_{11} + \sigma_{22}) \tag{10.87}$$

Then, the Mises yield surface is described from Eq. $(10.76)_3$ by

$$\sqrt{3}\sqrt{\left(\frac{\sigma_{11} - \sigma_{22}}{2}\right)^2 + \sigma_{12}^2} = F \tag{10.88}$$

which is represented by the Mohr's circle in the plane of the normal and the shear stresses.

References

Barlat F, Yoon JW, Cazacu O (2007) On linear transformations of stress tensors for the description of plastic anisotropy. Int J Plast 23:876–896

Chaboche JL (1989) Constitutive equations for cyclic plasticity and cyclic viscoplasticity. Int J Plast 5:247–302

Chaboche JL (1991) On some modifications of kinematic hardening to improve the description of ratcheting effects. Int J Plast 7:661–678

Chaboche JL (2008) A review of some plasticity and viscoplasticity constitutive theories. Int J Plast 24:1642–1693

Chaboche JL, Dang-Van K, Cordier G (1979) Modelization of the strain memory effect on the cyclic hardening of 316 stainless steel, Transaction on 5th International Conference SMiRT, Berlin, Division L., Paper No. L. 11/3

Delobelle P, Robinet P, Bocher L (1995) Experimental study and phenomenological modelization of ratchet under uniaxial and biaxial loading on austenitic stainless steel. Int J Plast 11:295–330

Ellyin F (1997) Fracture damage, crack growth and life prediction. Chapman & Hall, London

Ellyin F, Xia Z (1989) A rate-independent constitutive model for transient non-proportional loading. J Mech Phys Solids 37:71–91

Ghaei A, Green DE (2010) Numerical implementation of Yoshida-Uemori two-surface plasticity model using a fully implicit integration scheme. Compt Mater Sci 48:195–205

Hashiguchi K (2015c) Cyclic stagnation of isotropic hardening in metals, In: Proceedings of 2nd Science Meeting of Kyushu Branch of Society Material Science, Japan, B18

Hashiguchi K, Ueno M (2017) Elastoplastic constitutive equation of metals under cyclic loading. Int J Eng Sci 111:86–112

Hashiguchi K, Ueno M, Ozaki T (2012) Elastoplastic model of metals with smooth elastic-plastic transition. Acta Mech 223:985–1013

Hashiguchi K, Yamakawa Y (2012) Introduction to finite strain theory for continuum elasto-plasticity, Wiley series in computational mechanics. Wiley, Chichester

Hassan T, Taleb T, Krishna S (2008) Influence of non-proportional loading on ratcheting responses and simulations by two recent cyclic plasticity models. Int J Plast 24:1863–1889

Higuchi R, Okamura K (2016) Prediction of residual stress change due to cyclic loading ~ validation of advantage of sub-loading surface model ~

Hill R (1948) Theory of yielding and plastic flow of anisotropic metals. Proc Royal Soc Lond A193:281–297

Hill R (1990) Constitutive modeling of orthotropic plasticity in sheet metals. J Mech Phys Solids 38:241–249

Ilyushin AA (1963) Plasticity—foundation of the general mathematical theory, Izdatielistbo Akademii Nauk CCCR (Publisher of the Russian Academy of Sciences), Moscow

Jiang Y, Zhang J (2008) Benchmark experiments and characteristic cyclic plasticity deformation. Int J Plast 24:1481–1515

Kobayashi M, Ohno N (2002) Implementation of cyclic plasticity models based on a general form of kinematic hardening. Int J Numer Meth Eng 53:2217–2238

Lee JY, Barlat F, Lee MG (2015) Constitutive and friction modeling for accurate springback analysis of advanced high strength steel sheets. Int J Pasticity 71:113–135

Lemaitre JA (1992) A course on damage mechanics. Springer, Heidelberg

Murakami S (2012) Continuum damage mechanics: a continuum mechanics approach to the analysis of damage and fracture. Springer, Dordrecht

Ohno N (1982) A constitutive model of cyclic plasticity with a non-hardening strain region. J Appl Mech (ASME) 49:721–727

Sun L, Wagoner RH (2011) Complex unloading behavior: nature of the deformation and its consistent constitutive representation. Int J Pasticity 27:1126–1144

Wagoner RH, Lim H, Lee MG (2013) Advanced issues in springback. Int J Pasticity 45:3–20

Xia Z, Ellyin F (1994) Biaxial ratcheting under strain or stress-controlled axial cycling with constant hoop stress. J Appl Mech (ASME) 61:422–428

Yoshida F, Uemori T (2002) A model of large-strain cyclic plasticity describing the Bauschinger effect and workhardening stagnation. Int J Plast 18:661–686

Yoshida F, Uemori T (2003) A model of large-strain cyclic plasticity and its application to springback simulation. Int J Mech Sci 45:1687–1702

Yoshida F, Hamasaki H, Uemori T (2015) Modeling of anisotropic hardening of sheet metals including description of the Bauschinger effect. Int J Plast 75:170–188

Chapter 11
Constitutive Equations of Soils

The history of plasticity started in the study of deformation behavior of soils by Coulomb (1773) when he proposed the yield condition of soils by applying the friction law proposed by himself. Thereafter, the soil plasticity was superseded the leadership by the metal plasticity. One of the reasons would be caused by the fact that soils exhibit various complex plastic deformation behaviors, e.g., the pressure-dependence, the plastic compressibility, the dependence on the third invariant of deviatoric stress, the softening and the rotational hardening. Explicit constitutive equations of soils will be described in this chapter, based on the elastoplastic constitutive equations in Chaps. 6–9. All the stresses in this chapter mean the so-called *effective stress* excluded the pore pressure.

11.1 Isotropic Consolidation Characteristics

The isotropic consolidation characteristics is the one of the fundamentals in the constitutive equations of soils, which provides the base of isotropic hardening of soils.

(a) Linear relation between double-logarithm of volume and pressure

The $\ln v - \ln p$ *linear relation* ($p \equiv -(\mathrm{tr}\boldsymbol{\sigma})/3$: pressure) for the *isotropic consolidation* characteristics of soils was proposed by Hashiguchi (1974, 1977, 1985, 1995, 2008; Hashiguchi and Ueno 1977) and later its conformity to test data was shown by Butterfield (1979). It is schematically shown in Fig. 11.1.

$$\left.\begin{array}{ll}\text{Normal-consolidation:} & \\ \text{(Elastoplastic state)} & \ln\dfrac{v_y}{V_y} = -\tilde{\lambda}\ln\dfrac{p_y+p_e}{p_{y0}+p_e} \\ \text{Swelling line:} & \\ \text{(Elastic state)} & \ln\dfrac{v}{\overline{\overline{V}}} = -\tilde{\kappa}\ln\dfrac{p+p_e}{p_0+p_e}\end{array}\right\} \tag{11.1}$$

K. Hashiguchi, *Foundations of Elastoplasticity: Subloading Surface Model*,
DOI 10.1007/978-3-319-48821-9_11

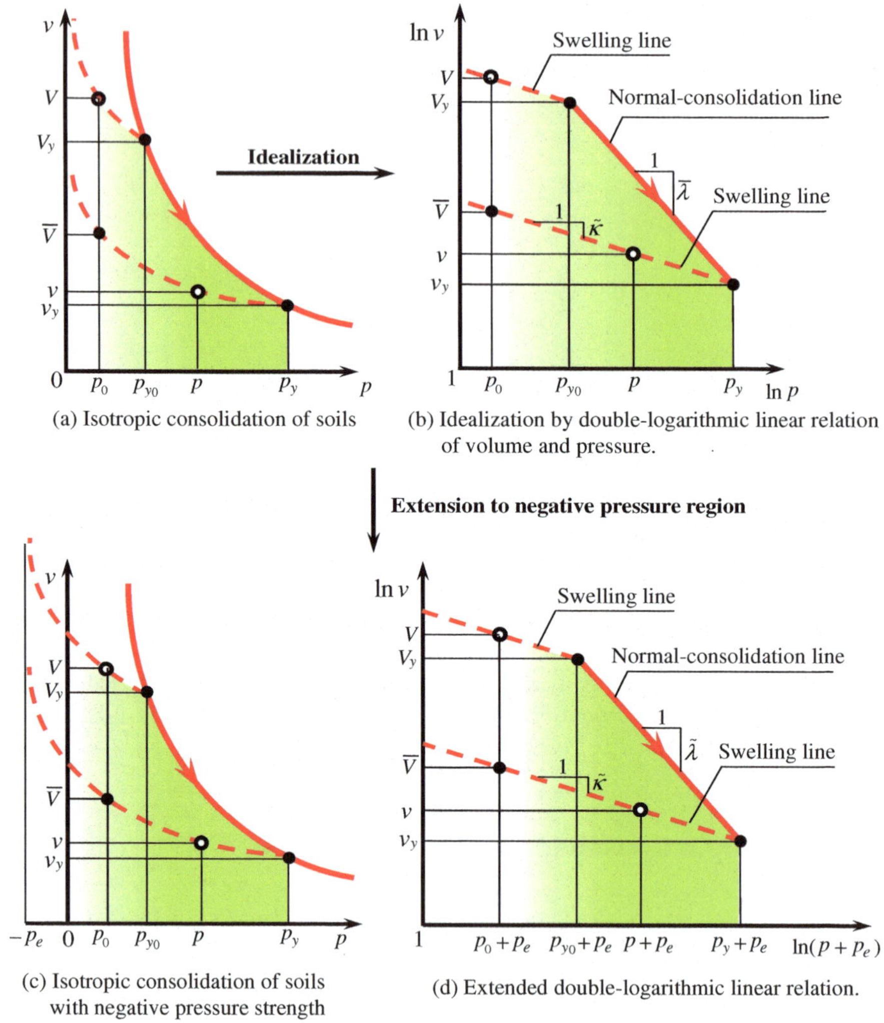

(a) Isotropic consolidation of soils

(b) Idealization by double-logarithmic linear relation of volume and pressure.

(c) Isotropic consolidation of soils with negative pressure strength

(d) Extended double-logarithmic linear relation.

Fig. 11.1 Linear relation between double-logarithms of volume and pressure for isotropic consolidation of soils

where (V, p_0), (V_y, p_{y0}) and (v_y, p_y) are the volumes and the pressures in the initial state, the initial yield state and the current yield state, respectively. p_y is the so-called *preconsolidated pressure*. In addition, $\overline{V}$ is the volume in the unloaded state to the initial pressure p_0, while the unloaded state corresponds to the intermediate configuration described in Sect. 6.1. Further, $\tilde{\lambda}$ and $\tilde{\kappa}$ are the material constants prescribing the slopes of *normal-consolidation line* and *swelling line*, respectively, in the $(\ln v, \ln p)$ plane, while in the range of infinitesimal deformation they approximately coincide to the values of λ and κ in the $e - \ln p$ *linear relation* (e: void ratio) used in the Cam-clay model as will be described subsequently in this section. The extension so as to be applicable to the negative pressure range is shown in the lower part in Fig. 11.1, where $p_e\ (\geq 0)$ is the material constant prescribing the negative pressure for which the volume becomes infinite, i.e. $v \to \infty$ for $p \to -p_e$.

The logarithmic volumetric strain and its elastic and plastic parts are given from Eqs. (6.8), (6.9) and (11.1) as follows:

$$\begin{aligned}
\varepsilon_v &= \ln\frac{v}{V} = \varepsilon_v^e + \varepsilon_v^p = \ln\frac{v}{\overline{V}} + \ln\frac{\overline{V}}{V} = \ln\frac{v}{\overline{V}} + \left(\ln\frac{V_y}{V} + \ln\frac{v_y}{V_y} + \ln\frac{\overline{V}}{v_y}\right) \\
&= -\tilde{\kappa}\ln\frac{p+p_e}{p_0+p_e} + \left(-\tilde{\kappa}\ln\frac{p_{y0}+p_e}{p_0+p_e} - \tilde{\lambda}\ln\frac{p_y+p_e}{p_{y0}+p_e} - \tilde{\kappa}\ln\frac{p_0+p_e}{p_y+p_e}\right) \\
&= -\tilde{\kappa}\ln\frac{p+p_e}{p_0+p_e} - (\tilde{\lambda}-\tilde{\kappa})\ln\frac{p_y+p_e}{p_{y0}+p_e}
\end{aligned} \tag{11.2}$$

leading to

$$\left.\begin{aligned}
\varepsilon_v^e &= \ln\frac{v}{\overline{V}} = -\tilde{\kappa}\ln\frac{p+p_e}{p_0+p_e} \\
\varepsilon_v^p &= \ln\frac{\overline{V}}{V} = -(\tilde{\lambda}-\tilde{\kappa})\ln\frac{p_y+p_e}{p_{y0}+p_e}
\end{aligned}\right\} \tag{11.3}$$

Here, let it be assumed that the isotropic hardening/softening is induced by the plastic volume contraction/expansion and further let the value of the isotropic hardening function F in Eq. (6.30) be chosen to coincide with the pre-consolidation pressure p_y, i.e.

$$F(H) = p_y \tag{11.4}$$

$$H = -\varepsilon_v^p \tag{11.5}$$

where the minus sign is added so that the plastic volume contraction/expansion, i.e. the minus/plus plastic volumetric strain causes the hardening/softening.

Strictly speaking from the physical point of view, the negative pressure p_e, up to which a soil can bear, would increase in proportion to the preconsolidated pressure p_y. Then, let the elastic and the plastic volumetric strains be extended by substituting $p_e = \vartheta p_y = \vartheta F$ in the general state and $p_e = \vartheta p_{y0} = \vartheta F_0$ in the initial state into Eq. (11.3) as follows:

$$\left.\begin{aligned}
\varepsilon_v^e &= -\tilde{\kappa}\ln\frac{p+\vartheta F}{p_0+\vartheta F_0} \\
\varepsilon_v^p &= -(\tilde{\lambda}-\tilde{\kappa})\ln\frac{F+\vartheta F}{F_0+\vartheta F_0} = -(\tilde{\lambda}-\tilde{\kappa})\ln\frac{F}{F_0}
\end{aligned}\right\} \tag{11.6}$$

where ϑ is the material constant leading to $\varepsilon^e \to \infty$ for $p \to -\vartheta F$. Here, the pressure will be formulated to keep $p + \vartheta F > 0$ inside the yield surface in Subsect. 11.4.1. It follows from Eq. (11.6) that

$$\left.\begin{aligned} \overset{\bullet}{\varepsilon}{}^e_v &= d^e_v = \mathrm{tr}\mathbf{d}^e = -\tilde{\kappa}\,\frac{p_0 + \vartheta F_0}{p + \vartheta F}(\overset{\bullet}{p} + \vartheta \overset{\bullet}{F}). \\ \overset{\bullet}{\varepsilon}{}^p_v &= d^p_v = \mathrm{tr}\mathbf{d}^p = -(\tilde{\lambda} - \tilde{\kappa})\frac{\overset{\bullet}{F}}{F} \end{aligned}\right\} \tag{11.7}$$

Adopting Eq. $(11.7)_1$ in the explicit Eq. (5.35) for the elastic modulus tensor $\mathbf{E}$ in Eq. (6.29), let the elastic bulk modulus K and the elastic shear modulus G in the hypoelastic relation be given, noting that the elastic strain rate is far smaller than the plastic strain rate and further the elastic strain rate induced through the rate of hardening variable $\vartheta \overset{\bullet}{F}$ is far more smaller in Eq. (11.7) and taking account of the dependence of G on the pressure and the pre-consolidation pressure, as follows:

$$\boxed{K = \frac{\overset{\bullet}{\sigma}_m}{\overset{\bullet}{\varepsilon}{}^e_v} = \frac{-\overset{\bullet}{p}}{d^e_v} = \frac{1}{\tilde{\kappa}}\frac{p + \vartheta F}{p_0 + \vartheta F_0}, \quad G = \frac{\|\overset{\circ}{\boldsymbol{\sigma}}{}'\|}{2\|\mathbf{d}^{e'}\|} = G_0\left(\frac{p + \vartheta F}{p_0 + \vartheta F_0}\right).} \tag{11.8}$$

noting Eq. (5.55), where G_0 is the initial value of G and $n\,(\leq |)$ is the material constant. Therefore, the elastic tangent moduli depend on the pressure p and the isotropic hardening function F. The hyperelastic relation based on Eq. (11.8) will be described in Subsect. 20.9.2.

The hardening function is given from Eqs. (11.5) and $(11.6)_2$ as follows:

$$\boxed{F(H) = F_0 \exp\left(\frac{H}{\tilde{\lambda} - \tilde{\kappa}}\right) = F_0 \exp\left(\frac{-\varepsilon^p_v}{\tilde{\lambda} - \tilde{\kappa}}\right)} \tag{11.9}$$

(b) **Linear relation between void ratio and logarithm of pressure**

The following $e - \ln p$ *linear relation* (e: void ratio) for the isotropic consolidation has been widely adopted for constitutive equation of soils after the Cam-clay models (Roscoe and Burland 1968; Schofield and Wroth 1968), although it possesses a lot of serious deficiencies described later in detail but unfortunately these deficiencies have not been recognized definitely.

$$\left.\begin{aligned} &\text{Normal-consolidation line: } e_y - e_{y0} = -\lambda \ln\frac{p_y}{p_{y0}} \\ &\text{Swelling line: } e - \bar{e} = -\kappa \ln\frac{p}{p_0} \end{aligned}\right\} \tag{11.10}$$

where the material constants λ and κ are the slopes of normal-consolidation and swelling lines, respectively, in the $(e, \ln p)$ plane as shown in Fig. 11.2, where (e_0, p_0), (e_{y0}, p_{y0}) and (e_y, p_y) are the void ratios and the pressures in the initial, the initial yield and the current yield states, respectively. In addition, $\bar{e}$ is the void ratio in the unloaded state to the initial pressure p_0.

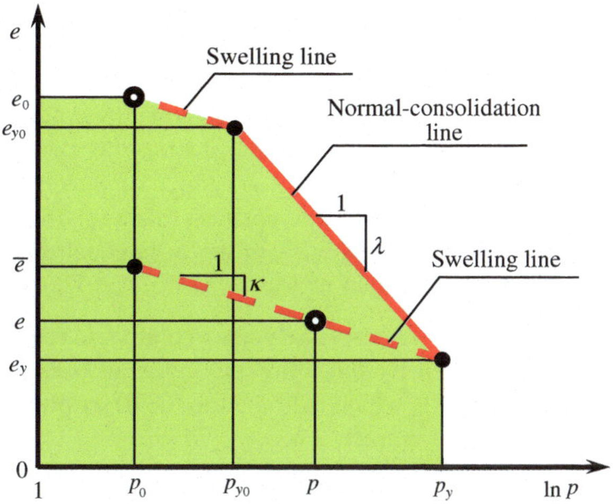

Fig. 11.2 Linear relation between void ratio and logarithm of pressure

However, the $e - \ln p$ linear relation has the following physical impertinence.

1. The change of void ratio induced during a change of pressure from a certain pressure to other certain pressure (e.g. p_0 to p_{y0} in Fig. 11.2) along the swelling line is identical in spite of the plastic decrease of void ratio by the increase of a pre-consolidation pressure p_y. This defect is caused from the fact that the void ratio itself is adopted in vertical axis in the $e - \ln p$ linear relation, although the logarithm of volume is adopted the $\ln v - \ln p$ linear relation.
2. The void ratio becomes negative if the pressure becomes large as $p_y > p_{y0} \exp(e_{y0}/\lambda)$ or $p > p_0 \exp(\bar{e}/\kappa)$, noting Eq. (11.10).
3. The void ratio becomes infinitely large when the pressure approaches zero because of $e \to \infty$ due to $-\ln p \to +\infty$ for $p \to 0$ in Eq. $(11.10)_2$. This defect causes the serious problem in the deformation analysis in which pressure decreases as seen in the footing-settlement analysis: Pressure in soils in the periphery of footing decreases to zero in non-cohesive soils and even to negative in cohesive soils and the cyclic mobility analysis for liquefaction in which an accurate prediction of deformation under a quite low pressure is required.
4. The pressure is not related to the volume but to the void ratio or specific volume (void ratio plus unity), although the volumetric strain is not defined by the change of void ratio but by the change of volume. Note here that the ratio of the current volume to the initial volume required for the derivation of the logarithmic volumetric strain can be transformed to the ratio of the current specific volume to the initial specific volume as $v/V = (1+e)v_s(p)/(1+e_0)v_s(p_0) = (1+e)/(1+e_0)$ under the assumption of the incompressibility of soil particles themselves, i.e. $v_s = \text{const.}$ where v_s designates the volume occupied by the soil particles.
5. The $e - \ln p$ linear relation is not formulated by the logarithm of ratio of volumes but it is given by the difference of void ratios. Therefore, it does not fit to the logarithmic strain but it fits to the nominal strain, so that the nominal strain is

obliged to be adopted in the derivation of volumetric strain from the $e - \ln p$ linear relation. However, the nominal volumetric strain does not coincide with the time-integration of trace of the strain rate **d** used in the elastoplastic constitutive equations in the preceding chapters as known from Eq. (2.143) and it is impertinent to the description of large deformation, because (a) the strain is merely minus one even when the material vanishes completely as shown in Eq. $(2.140)_2$, (b) it does not satisfy the superposition rule as shown in Eq. $(2.141)_2$, (c) the sum of three longitudinal strains in the orthogonal directions does not coincide with the volumetric strain as shown in Eq. $(2.142)_2$.

Because of the deficiency in 5, a pressure versus volumetric relation based on the $e - \ln p$ linear relation is given by the following equation based on the nominal volumetric strain in Eq. $(2.142)_2$ which is limited to the description of infinitesimal deformation as was delineated in Sect. 2.6.

$$\mathrm{\varepsilon}_v = \mathrm{\varepsilon}_v^e + \mathrm{\varepsilon}_v^p \tag{11.11}$$

where

$$\mathrm{\varepsilon}_v = \frac{v - V}{V},\ \mathrm{\varepsilon}_v^e = \frac{v - \overline{V}}{V},\ \mathrm{\varepsilon}_v^p = \frac{\overline{V} - V}{V} \tag{11.12}$$

Here, the italic letter ε is used for the logarithmic strain but the roman letter $\mathrm{\varepsilon}$ is used for the nominal strain in order to distinguish them as described in Sect. 2.6. It should be noted that the nominal strain cannot be related to the strain rate **d** in the exact sense as known by $\dot{\varepsilon}_v = d_v = \mathrm{tr}\mathbf{d} = \dot{v}/v \neq \dot{v}/V = \dot{\mathrm{\varepsilon}}_v$. The following relations are used by substituting the specific volume instead of the volume into Eq. (11.12) on the approximation at the sacrifice of the above-mentioned deficiency 4.

$$\left.\begin{aligned}
\mathrm{\varepsilon}_v &\cong \frac{e - e_0}{1 + e_0} \\
\mathrm{\varepsilon}_v^e &\cong \frac{e - \bar{e}}{1 + e_0} \\
\mathrm{\varepsilon}_v^p &\cong \frac{\bar{e} - e_0}{1 + e_0} = \frac{(e_{y0} - e_0) + (\bar{e}_y - e_{y0}) + (\bar{e} - \bar{e}_y)}{1 + e_0}
\end{aligned}\right\} \tag{11.13}$$

where the nearly equal symbol $\cong$ is used to specify the approximation by the incompressibility of soil particles, noting that the symbols v and V in Eq. (11.12) are not the specific volumes but the volumes. Substituting Eq. (11.10) into Eq. (11.13), one has

$$\left.\begin{aligned}
\mathrm{\varepsilon}_v &\cong -\frac{\kappa}{1+e_0}\ln\frac{p}{p_0} - \frac{\lambda - \kappa}{1+e_0}\ln\frac{p_y}{p_{y0}} \\
\mathrm{\varepsilon}_v^e &\cong -\frac{\kappa}{1+e_0}\ln\frac{p}{p_0} \\
\mathrm{\varepsilon}_v^p &\cong -\frac{\kappa}{1+e_0}\ln\frac{p_{y0}}{p_0} - \frac{\lambda}{1+e_0}\ln\frac{p_y}{p_{y0}} - \frac{\kappa}{1+e_0}\ln\frac{p_0}{p_y} = -\frac{\lambda - \kappa}{1+e_0}\ln\frac{p_y}{p_{y0}}
\end{aligned}\right\} \tag{11.14}$$

from which the nominal volumetric strain rate is given by

$$\left.\begin{aligned}
\overset{\bullet}{\varepsilon}_v &= \frac{\overset{\bullet}{v}}{V} \cong \frac{\overset{\bullet}{e}}{1+e_0} = -\frac{\kappa}{1+e_0}\frac{\overset{\bullet}{p}}{p} - \frac{\lambda-\kappa}{1+e_0}\frac{\overset{\bullet}{p}_y}{p_y} \\
\overset{\bullet}{\varepsilon}{}^e_v &= \frac{\overset{\bullet}{v}}{V} - \frac{\overset{\bullet}{\bar{V}}}{V} \cong \frac{\overset{\bullet}{e}-\overset{\bullet}{\bar{e}}}{1+e_0} = -\frac{\kappa}{1+e_0}\frac{\overset{\bullet}{p}}{p} \\
\overset{\bullet}{\varepsilon}{}^p_v &= \frac{\overset{\bullet}{\bar{V}}}{V} \cong \frac{\overset{\bullet}{\bar{e}}}{1+e_0} = -\frac{\lambda-\kappa}{1+e_0}\frac{\overset{\bullet}{p}_y}{p_y}
\end{aligned}\right\} \tag{11.15}$$

Equation (11.15) for the rates of volumetric strain rate and pressure is adopted widely for elastoplastic constitutive equations of soils after the Cam-clay model which is quite primitive model, while its basic concept was already clarified by Drucker and Prager (1952) and Drucker et al. (1957). Nevertheless, it causes the further deficiencies as follows:

6. It is not derived exactly but approximately from the $e - \ln p$ linear relation.
7. It cannot be adopted to describe finite deformation since it is derived based on the definition of nominal strain.
8. The tangent elastic bulk modulus is given by

$$K \equiv \frac{-\overset{\bullet}{p}}{\overset{\bullet}{\varepsilon}{}^e_v} = \frac{1+e_0}{\kappa}p \tag{11.16}$$

from Eq. (11.15). Here, the tangent elastic bulk modulus K depends on the initial void ratio e_0. It has a crucial physical impertinence: "*The larger the initial void ratio, the larger is the elastic bulk modulus. In other words, the looser the soil, the more difficult to be compressed*".

Eventually, it is concluded that the $e - \ln p$ linear relation is inadequate physically and mathematically for formulation of constitutive equations for finite deformation of soils. On the other hand, all the deficiencies in the $e - \ln p$ linear relation can be excluded in the $\ln v - \ln p$ linear relation.

The nominal volumetric strain and its elastic and plastic parts are related to the strain rate and its elastic and plastic parts used in the elastoplastic constitutive equation as follows:

$$\left.\begin{aligned}
\overset{\bullet}{\varepsilon}_v &= \left(\frac{v-V}{V}\right)^{\bullet} = \frac{\overset{\bullet}{v}}{V} = \frac{v}{V}\frac{\overset{\bullet}{v}}{v} = J\frac{\overset{\bullet}{v}}{v} = (\det\mathbf{F})\mathrm{tr}\mathbf{d} \\
\overset{\bullet}{\varepsilon}{}^e_v &= \frac{\overset{\bullet}{v}}{V} - \frac{\overset{\bullet}{\bar{V}}}{V} = \frac{v}{V}\frac{\overset{\bullet}{v}}{v} - \frac{\bar{V}}{V}\frac{\overset{\bullet}{\bar{V}}}{\bar{V}} = J\frac{\overset{\bullet}{v}}{v} - J^p\mathrm{tr}\mathbf{d}^p = (\det\mathbf{F})\mathrm{tr}\mathbf{d} - (\det\mathbf{F}^p)\mathrm{tr}\mathbf{d}^p \\
\overset{\bullet}{\varepsilon}{}^p_v &= \frac{\overset{\bullet}{\bar{V}}}{V} = \frac{\bar{V}}{V}\frac{\overset{\bullet}{\bar{V}}}{\bar{V}} = J^p\mathrm{tr}\mathbf{d}^p = (\det\mathbf{F}^p)\mathrm{tr}\mathbf{d}^p
\end{aligned}\right\} \tag{11.17}$$

which are of complicated forms containing the Jacobian and its elastic and plastic parts. Equation (11.17) have been used for constitutive equations under the approximation of $J^e \cong 1, J \cong J^p$ by some workers (cf. Asaoka et al. 1997; Zhang et al. 2007). However, it is yet merely the approximation and the physical property cannot be remedied, so that it is limited to the description of infinitesimal deformation in spite of the complexity. Then, this modification causes no good and much harmful.

A lot of constitutive equations of soils adopt the $e - \ln p$ linear relation, so that unfortunately they are applicable to the description of infinitesimal deformation. On the other hand, the $\ln v - \ln p$ linear relation is applicable to the description of finite deformation. It has been adopted not only in hypoelastic-based plastic constitutive equations of soils (cf. Hashiguchi 1974, 1978; Hashiguchi and Ueno 1977; Hashiguchi and Chen 1998) but also in hyperelastic-based plastic constitutive equations of soils based on the additive decomposition of infinitesimal strain into elastic and plastic parts (e.g. Houlsby 1985; Collins and Hilder 2002; Coombs and Crouch 2011; Coombs et al. 2013). Further, it has been adopted in hyperelastic-based plastic constitutive equations of soils under the multiplicative decomposition of deformation gradient (e.g. Borja and Tamagnini 1998; Yamakawa et al. 2010) as will be described in Sect. 20.9.2. The material constants $\tilde{\lambda}$ and $\tilde{\kappa}$ in the $\ln v - \ln p$ linear relation are approximately related to λ and κ in the $e - \ln p$ linear relation by $\tilde{\lambda} = \lambda/(1+e_0)$, $\tilde{\kappa} = \kappa/(1+e_0)$ as described in Appendix F, while numerous test data have been accumulated for the latter.

11.2 Yield Conditions

Various yield conditions of soils have been proposed to date. The functions $f(\boldsymbol{\sigma})$ in Eq. (6.30) can be reduced to the following common form for soils.

$$\boxed{f(\boldsymbol{\sigma}) = pg(\eta/M)} \tag{11.18}$$

or

$$\boxed{f(\boldsymbol{\sigma}) = pg(\eta_m)} \tag{11.19}$$

where

$$\boldsymbol{\eta} \equiv \frac{\boldsymbol{\sigma}'}{p}, \quad \eta \equiv \|\boldsymbol{\eta}\|, \quad \eta_m \equiv \frac{\eta}{M} \tag{11.20}$$

M is the stress ratio $\|\boldsymbol{\eta}\|$ in the maximum state of $\|\boldsymbol{\sigma}'\|$ in the fixed yield surface, i.e. the *critical state* and is called the *critical state stress ratio*.

It is premised in Eq. (11.19) that the yield surface passes through the isotropic compression state and the null stress point. Then, the function $g(\eta_m)$ fulfill the following conditions

$$\left.\begin{array}{ll} g = 1 & \text{for } \eta = 0 \\ g \to \infty & \text{for } \eta \to \infty \end{array}\right\} \tag{11.21}$$

Furthermore, note that the following equality hold in the critical state (cf. **Appendix G**).

$$g'(1) = g(1) \tag{11.22}$$

which is illustrated in Fig. 11.3.

Denoting the p-value in the critical state as p_{cr}, the following equation is obtained by substituting Eq. (11.19) into Eq. (6.30) at $\eta_m = 1$.

$$p_{cr} = F/g(1) \tag{11.23}$$

The function $g(\eta_m)$ for the yield surfaces proposed in the past are given as follows:

(1) *Original Cam-clay model* (Schofield and Wroth 1968)

$$g = \exp(\eta_m), \quad \text{i.e. } p\exp\left(\frac{\|\boldsymbol{\sigma}'\|/p}{M}\right) = F \tag{11.24}$$

$$p_{cr} = F/\mathrm{e} \tag{11.25}$$

where e is the exponent, i.e. $\mathrm{e} = 2.71828\cdots$.

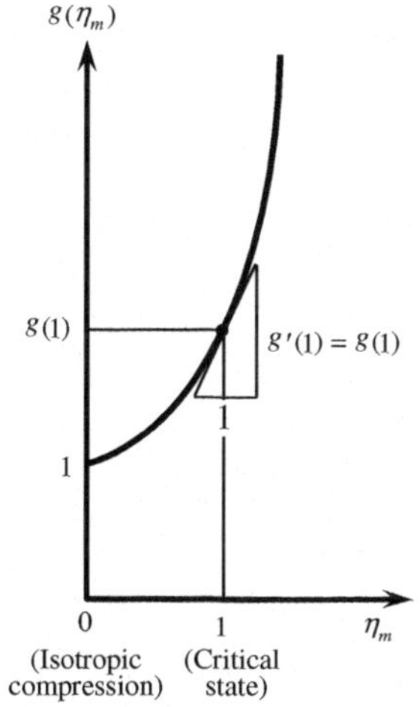

Fig. 11.3 Function $g(\eta_m)$ in yield surface of soils

(2) *Modified Cam-clay model* (Burland 1965; Roscoe and Burland 1968)

$$\boxed{g = 1 + \eta_m^2, \quad \text{i.e. } p\left[1 + \left(\frac{\|\boldsymbol{\sigma}'\|/p}{M}\right)^2\right] = F} \tag{11.26}$$

$$P_{cr} = F/2 \tag{11.27}$$

(3) *Hashiguchi model* (Hashiguchi 1972, 1985)

$$g = \exp\left(\frac{\eta_m^2}{2}\right), \quad \text{i.e. } p\exp\left[\frac{1}{2}\left(\frac{\|\boldsymbol{\sigma}'\|/p}{M}\right)^2\right] = F \tag{11.28}$$

$$p_{cr} = F/\sqrt{\mathrm{e}} \tag{11.29}$$

The functions in (1) and (2) are unified as follows:

$$g = \exp\left(\frac{\eta_m^n}{n}\right), \quad \text{i.e. } p\exp\left[\frac{1}{n}\left(\frac{\|\boldsymbol{\sigma}'\|/p}{M}\right)^n\right] = F \tag{11.30}$$

$$p_{cr} = F/\exp(1/n) \tag{11.31}$$

where Eqs. (11.24) and (11.28) are given for $n = 1$ and 2, respectively.

The above-mentioned three yield surfaces are shown in Fig. 11.4, where

$$\left.\begin{array}{l} p \equiv -\dfrac{1}{3}\mathrm{tr}\boldsymbol{\sigma} = -\dfrac{1}{3}(\sigma_a + 2\sigma_l) \\ q \equiv \sigma_l - \sigma_a = \begin{cases} -\dfrac{3}{2}\left[\sigma_a - \dfrac{1}{3}(\sigma_a + 2\sigma_l)\right] = -\dfrac{3}{2}\sigma'_a \\ 3\left[\sigma_l - \dfrac{1}{3}(\sigma_a + 2\sigma_l)\right] = 3\sigma'_l \end{cases} \\ \|\boldsymbol{\sigma}'\| = \sqrt{\dfrac{2}{3}}|\sigma_l - \sigma_a| = \sqrt{\dfrac{2}{3}}|q| \end{array}\right\} \tag{11.32}$$

where σ_a and σ_l are the axial and the lateral stress, respectively, in the triaxial compression/extension state.

Among the above-mentioned yield surfaces, only the yield surface of the modified Cam-clay model in Eq. (11.26) does not include any corner so that the singularity of the normal direction of the surface is not induced. Equation (6.30) with Eqs. (11.19) and (11.26) is rewritten as

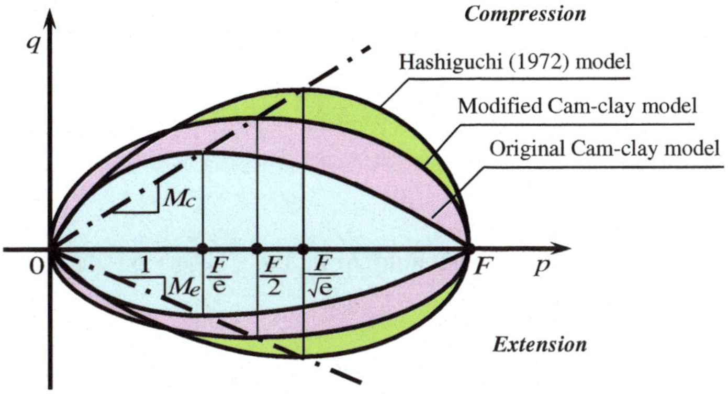

Fig. 11.4 Various yield surfaces of soils

$$\left[\frac{p-(F/2)}{F/2}\right]^2+\left[\frac{||\boldsymbol{\sigma}'||}{MF/2}\right]^2=1 \tag{11.33}$$

In what follows, we will find the pertinent function M for the stress ratio in the critical state. Then, assume the following equation.

$$M=\frac{2\sqrt{6}\sin\phi_c}{3-(1-c)\sin\phi_c+c\sin\phi_c\cos 3\theta_\sigma} \tag{11.34}$$

where c $(0\leq c\leq 1)$ is the material parameter, ϕ_c is the angle of internal friction in the critical state for the axisymmetric compression stress state, i.e. the so-called triaxial compression state and

$$\cos 3\theta_\sigma\equiv\sqrt{6}\mathrm{tr}\mathbf{t}'^3_\sigma,\quad \mathbf{t}'_\sigma\equiv\frac{\boldsymbol{\sigma}'}{||\boldsymbol{\sigma}'||} \tag{11.35}$$

The convexity condition (cf. Eq. (A41) in **Appendix H**) for Eq. (11.34) is given by

$$\begin{aligned}\Xi&\equiv\frac{1}{M}+\frac{d}{d\theta_\sigma^2}\left(\frac{1}{M}\right)^2\\&=\frac{1}{2\sqrt{6}\sin\phi_c}[3-(1-c)\sin\phi_c+c\sin\phi_c\cos 3\theta_\sigma-9c\sin\phi_c\cos 3\theta_\sigma]\\&=\frac{1}{2\sqrt{6}\sin\phi_c}[3-(1-c)\sin\phi_c-8\mathrm{c}\sin\phi_c\cos 3\theta_\sigma]\geq 0\end{aligned} \tag{11.36}$$

noting

$$\left.\begin{aligned} \frac{d}{d\theta_\sigma}\left(\frac{1}{M}\right) &= \frac{d}{d\theta_\sigma}\left(\frac{3-(1-c)\sin\phi_c + c\sin\phi_c\cos 3\theta_\sigma}{2\sqrt{6}\sin\phi_c}\right) = -\frac{3\,c\sin 3\theta_\sigma}{2\sqrt{6}} \\ \frac{d}{d\theta_\sigma^2}\left(\frac{1}{M}\right)^2 &= -\frac{9\,c\cos 3\theta_\sigma}{2\sqrt{6}} \end{aligned}\right\}$$

The following inequality must hold from Eq. (11.36) on the convexity for Eq. (11.34).

$$c \leq \frac{3-\sin\phi_c}{7\sin\phi_c} \tag{11.37}$$

The function M in Eq. (11.34) possesses the following properties.

1) It coincides always to the value of the Coulomb-Moher criterion in the triaxial compression state ($\theta_\sigma = (1 \pm 2n)(\pi/3)$) but it is different from the value of the Coulomb-Mohr criterion in the triaxial extension state ($\theta_\sigma = \pm 2n(\pi/3)$), i.e.

$$\left.\begin{aligned} M &= M_c \text{ for } \theta_\sigma = (1 \pm 2n(\pi/3) \\ M &= \frac{2\sqrt{6}\sin\phi_c}{3+(2c-1)\sin\phi_c} \neq M_e \text{ for } \theta_\sigma = \pm 2n(\pi/3) \end{aligned}\right\}$$

where M_c and M_e are the values of M in the triaxial compression and extension states, respectively, in the Coulomb-Moher criterion, i.e.

$$\left.\begin{aligned} M_c &\equiv \frac{2\sqrt{6}\sin\phi_c}{3-\sin\phi_c} \text{ for } \theta_\sigma = (1 \pm 2n)(\pi/3) \\ M_e &\equiv \frac{2\sqrt{6}\sin\phi_c}{3+\sin\phi_c} \text{ for } \theta_\sigma = \pm 2n(\pi/3) \end{aligned}\right\} \tag{11.38}$$

and n is a natural number.

2) It follows for $c = 0$ that M is constant coinciding with the value of M in the triaxial compression state, i.e. $M = M_c$.
3) It coincides also to the value of M for the Coulomb-Moher criterion in the triaxial extension state for $c = 1$ which Eq. (11.34) is reduced to

$$M = \frac{2\sqrt{6}\sin\phi_c}{3+\sin\phi_c\cos 3\theta_\sigma} \tag{11.39}$$

(see Fig. 11.5). Eq. (11.39) was proposed by Satake (1972) and later by Gudehus (1973) and Argyris et al. (1973) as the failure criterion of soils. The convexity in Eq. (11.37) leads to the following inequality for $c = 1$.

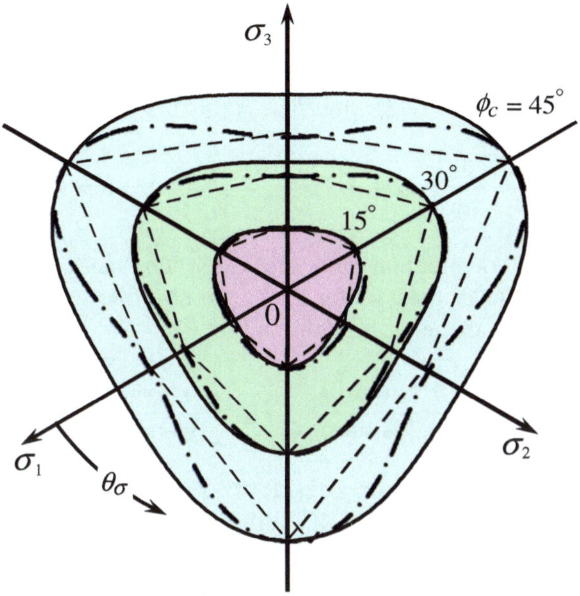

Fig. 11.5 Sections of the critical state surfaces of soils in p-constant plane (Hashiguchi 2002)

$$\sin\phi_c \leq \frac{3}{8}, \text{ i.e. } \phi_c \leq \phi_{ch} \equiv 22.02^\circ \tag{11.40}$$

4) The value of M in the triaxial extension state is lower for a larger value of the material parameter c because of $\partial M/\partial c = -4\sqrt{6}\,\sin^2\phi_c/[3 + (2c-1)\,\sin\,\phi_c]^2 < 0 \; for\, \theta_\sigma \; = \; \pm 2n\,(\pi/3)$.
Then, choosing c to be the largest value satisfying the equality in Eq. (11.37) for the convexity condition, M is reduced to

$$M = \frac{2\sqrt{6}\sin\phi_c}{3 - (1 - \frac{3-\sin\phi_c}{7\sin\phi_c})\sin\phi_c + \frac{3-\sin\phi_c}{7\sin\phi_c}\sin\phi_c\cos 3\theta_\sigma} = \frac{7}{8+\cos 3\theta_\sigma}\frac{2\sqrt{6}\sin\phi_c}{3-\sin\phi_c}$$

i.e.

$$\boxed{M = \frac{7}{8+\cos 3\theta_\sigma}\frac{2\sqrt{6}\sin\phi_c}{3-\sin\phi_c} = \frac{7}{8+\cos 3\theta_\sigma}M_c} \tag{11.41}$$

which was Proposed by Hashiguchi (2002) (see Fig. 11.5), being reduced to

$$M = \begin{cases} \dfrac{2\sqrt{6}\sin\phi_c}{3-\sin\phi_c} = M_c \ \text{for} \ \theta_\sigma = (1\pm 2n)(\pi/3) \\ \dfrac{7}{9}\dfrac{2\sqrt{6}\sin\phi_c}{3-\sin\phi_c} = \dfrac{7}{9}M_c = \dfrac{7}{9}\dfrac{3+\sin\phi_c}{3-\sin\phi_c}M_e \ \text{for} \ \theta_\sigma = \pm 2n(\pi/3) \end{cases}$$

in the triaxial compression and the extension states. Therefore, M for Eq. (11.41) is smaller than M_e for the Coulomb-Mohr criterion in the triaxial extension state for the angle ϕ_c fulfilling

$$\frac{7}{9}\frac{2\sqrt{6}\sin\phi_c}{3-\sin\phi_c} \leq M_e = \frac{2\sqrt{6}\sin\phi_c}{3+\sin\phi_c}$$

which leads to the inequality in Eq. (11.40). Therefore, the upper-limit angle of ϕ_c, i.e. ϕ_{ch} up to which Eq. (11.40) fulfills the convexity condition coincides with that up to which the value of M given by Eq. (11.41) in the triaxial extension state is smaller than M_e.

11.3 Subloading Surface Model

The initial subloading surface model with the isotropic hardening is described in Chap. 7. In what follows, the simple initial subloading surface model with the modified Cam-clay yield surface will be formulated and compared with the other soil models.

The following functions are adopted based on Eqs. (11.5), (11.9), (11.19), (11.26) and (11.34).

$$\left.\begin{aligned} & f(\boldsymbol{\sigma}) = p\left[1+\left(\frac{\|\boldsymbol{\sigma}'\|/p}{M}\right)^2\right] = p\left[1+\left(\frac{\eta}{M}\right)^2\right] \\ & F(H) = F_0\exp\frac{H}{\tilde{\lambda}-\tilde{\kappa}}, \quad F' \equiv \frac{dF}{dH} = \frac{F}{\tilde{\lambda}-\tilde{\kappa}} \\ & H = -\varepsilon_v^p, \quad \dot{H} = -\mathrm{tr}\mathbf{d}^p, \quad h(=\dot{H}\,/\,\dot{\lambda}) = -\mathrm{tr}\mathbf{n} \\ & M(\cos 3\theta_\sigma) = \frac{7}{8+\cos 3\theta_\sigma}M_c \end{aligned}\right\} \tag{11.42}$$

The plastic modulus $\overline{M}^p$ in Eq. (7.30) is given for Eq. (11.42) as

$$\overline{M}^p \equiv \left(\frac{-\mathrm{tr}\mathbf{n}}{\tilde{\lambda}-\tilde{\kappa}} + \frac{U(R)}{R}\right)\mathbf{n}:\boldsymbol{\sigma} \tag{11.43}$$

and thus the plastic strain rate is described by

$$\mathbf{d}^p = \frac{\mathbf{n}:\overset{\circ}{\boldsymbol{\sigma}}}{\overline{M}^p}\mathbf{n} = \frac{\mathbf{n}:\overset{\circ}{\boldsymbol{\sigma}}}{\left(\dfrac{-\mathrm{tr}\mathbf{n}}{\tilde{\lambda}-\tilde{\kappa}} + \dfrac{U(R)}{R}\right)\mathbf{n}:\boldsymbol{\sigma}}\mathbf{n} \tag{11.44}$$

and thus

$$\mathbf{d} = \mathbf{E}^{-1}:\overset{\circ}{\boldsymbol{\sigma}} + \frac{\mathbf{n}:\overset{\circ}{\boldsymbol{\sigma}}}{\left(\dfrac{-\mathrm{tr}\mathbf{n}}{\tilde{\lambda}-\tilde{\kappa}} + \dfrac{U(R)}{R}\right)\mathbf{n}:\boldsymbol{\sigma}}\mathbf{n} \tag{11.45}$$

Here, note that the subloading hardening, i.e. $\mathbf{n}:\overset{\circ}{\boldsymbol{\sigma}} > 0$ in Eq. (7.38) can be induced over the critical state line, fulfilling $\mathrm{tr}\mathbf{n} \geq 0$ since the positive quantity $U(R)/R$ is contained in the plastic modulus $\overline{M}^p$.

The partial derivatives of the yield function in Eq. (11.42) is given as follows:

$$\begin{aligned} &\frac{\partial f(p,\eta,M)}{\partial p} = 1 + \left(\frac{\eta}{M}\right)^2, \quad \frac{\partial f(p,\eta,M)}{\partial \eta} = 2\frac{p\eta}{M^2}, \\ &\frac{\partial f(p,\eta,M)}{\partial M} = -2\frac{p}{M}\left(\frac{\eta}{M}\right)^2 \end{aligned} \tag{11.46}$$

$$\frac{\partial \eta}{\partial \boldsymbol{\sigma}} = \frac{\partial \|\boldsymbol{\sigma}'\|/p}{\partial \boldsymbol{\sigma}} = \frac{1}{p^2}\left(p\mathbf{t}'_\sigma + \frac{1}{3}\|\boldsymbol{\sigma}'\|\mathbf{I}\right) = \frac{1}{p}\left(\mathbf{t}'_\sigma + \frac{1}{3}\eta\mathbf{I}\right) \tag{11.47}$$

$$\begin{aligned} \frac{\partial M}{\partial \cos 3\theta_\sigma} &= \frac{14\sqrt{6}\sin\phi_c}{(3-\sin\phi_c)(8+\cos 3\theta_\sigma)^2} = -\frac{M}{8+\cos 3\theta_\sigma} \\ &= -\frac{3-\sin\phi_c}{14\sqrt{6}\sin\phi_c}M^2 \end{aligned} \tag{11.48}$$

$$\frac{\partial \cos 3\theta_\sigma}{\partial \boldsymbol{\sigma}'} = \frac{3}{\|\boldsymbol{\sigma}'\|}\left(\sqrt{6}\mathbf{t}'^2_\sigma - \cos 3\theta_\sigma \mathbf{t}'_\sigma\right) \tag{11.49}$$

(cf. Eq. (1.302))

$$\begin{aligned} \frac{\partial f(\boldsymbol{\sigma})}{\partial \boldsymbol{\sigma}} &= \frac{\partial f(p,\eta,M)}{\partial \boldsymbol{\sigma}} = \frac{\partial f(p,\eta,M)}{\partial p}\frac{\partial p}{\partial \boldsymbol{\sigma}} + \frac{\partial f(p,\eta,M)}{\partial \eta}\frac{\partial \eta}{\partial \boldsymbol{\sigma}} + \frac{\partial f(p,\eta,M)}{\partial M}\frac{\partial M}{\partial \boldsymbol{\sigma}} \\ &= -\frac{1}{3}\left[1+\left(\frac{\eta}{M}\right)^2\right]\mathbf{I} + 2\frac{p}{M}\frac{\eta}{M}\frac{1}{p}\left(\mathbf{t}'_\sigma + \frac{1}{3}\eta\mathbf{I}\right) \\ &\quad -2\frac{p}{M}\left(\frac{\eta}{M}\right)^2\left(-\frac{3-\sin\phi_c}{14\sqrt{6}\sin\phi_c}M^2\right)\frac{3}{\|\boldsymbol{\sigma}'\|}\left(\sqrt{6}\mathbf{t}'^2_\sigma - \cos 3\theta_\sigma\mathbf{t}'_\sigma\right):\mathcal{I}' \\ &= -\frac{1}{3}\left[1-\left(\frac{\eta}{M}\right)^2\right]\mathbf{I} + \frac{2\eta}{M^2}\mathbf{t}'_\sigma + 3\frac{\eta}{M}\frac{3-\sin\phi_c}{7\sqrt{6}\sin\phi_c}\left(\sqrt{6}\mathbf{t}'^2_\sigma:\mathcal{I}' - \cos 3\theta_\sigma\mathbf{t}'_\sigma\right) \end{aligned} \tag{11.50}$$

noting the partial derivative formulae in Sect. 1.16.

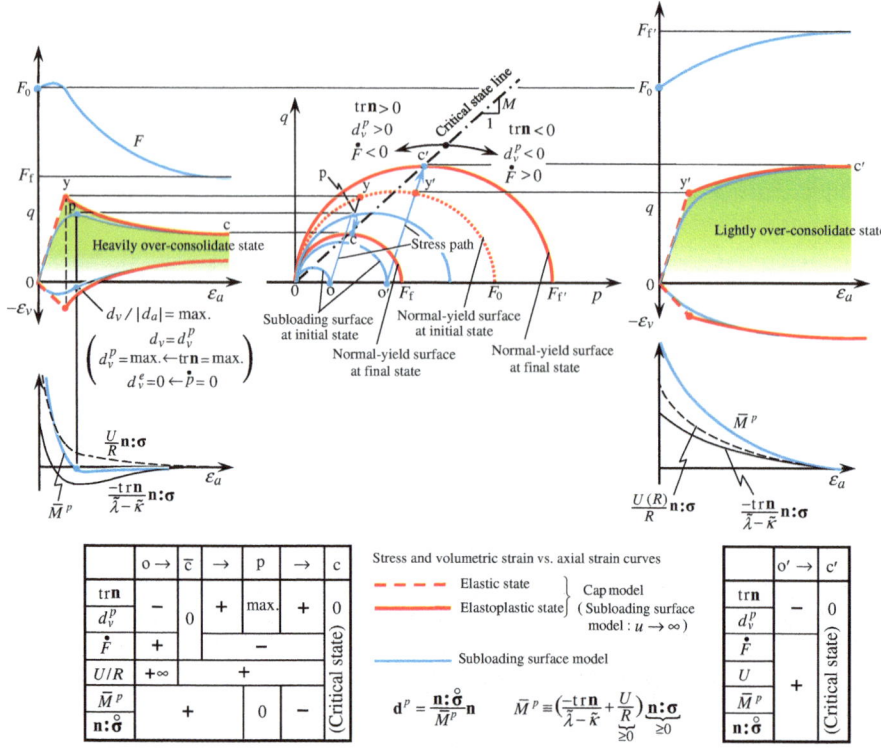

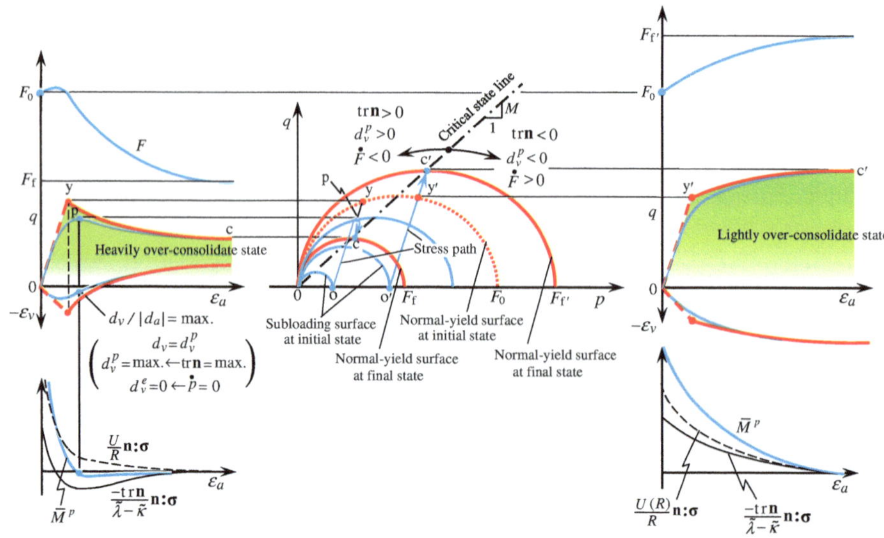

	o →	c̄	→	p	→	c
$\mathrm{tr}\mathbf{n}$	−	0	+	max.	+	0
d_v^p						
$\overset{\bullet}{F}$	+		−			(Critical state)
U/R	+∞	+				
$\bar{M}^p$	+			0	−	
$\mathbf{n}:\overset{\circ}{\boldsymbol{\sigma}}$						

Stress and volumetric strain vs. axial strain curves

- - - - Elastic state } Cap model
——— Elastoplastic state } (Subloading surface model : $u \to \infty$)

——— Subloading surface model

$$\mathbf{d}^p = \frac{\mathbf{n}:\overset{\circ}{\boldsymbol{\sigma}}}{\bar{M}^p}\mathbf{n} \qquad \bar{M}^p \equiv \Big(\underbrace{\frac{-\mathrm{tr}\mathbf{n}}{\tilde{\lambda}-\tilde{\kappa}} + \frac{U}{R}}_{\geq 0}\Big)\underbrace{\mathbf{n}:\boldsymbol{\sigma}}_{\geq 0}$$

	o′ →	c′
$\mathrm{tr}\mathbf{n}$	−	0
d_v^p		
$\overset{\bullet}{F}$	+	(Critical state)
U		
$\bar{M}^p$		
$\mathbf{n}:\overset{\circ}{\boldsymbol{\sigma}}$		

Fig. 11.6 Comparison of predictions of triaxial compression behavior under constant lateral stress by the conventional Cam-clay model and the subloading surface model

The predictions of the drained triaxial compression behavior of soils under the constant lateral stress by the subloading surface model in Eq. (11.45) and the conventional *Cap model* (Roscoe and Burland 1968; Schofield and Wroth 1968) are depicted in Fig. 11.6, where ε_a is the axial strain. Here, the curves of axial stress and volumetric strain versus the axial strain in the loading from the heavily and the lightly over-consolidated states, i.e. points o and o′, respectively are shown in this figure.

In the loading from the lightly over-consolidated state o′, the volume contraction proceeds and the axial stress increases monotonically up to the critical state. The abrupt transition from the elastic to the plastic state is predicted by the conventional Cam-clay model. On the other hand, the smooth behavior is predicted always by the subloading surface model as observed in experiments.

Next, consider the loading from the heavily over-consolidated state o. The elastic behavior is predicted until the stress reaches the yield surface and then the stress is predicted to decrease abruptly toward the critical state, exhibiting the intense softening by the Cam-clay model. On the other hand, the following realistic deformation behavior is predicted by the subloading surface model.

(1) The first term in the plastic modulus in Eq. (11.42) decreases from the positive value to zero because of $\mathrm{tr}\mathbf{n} < 0$ in the process from the initial state up to the critical state line. On the other hand, the second term is always positive.

Therefore, the plastic modulus is kept to be positive in this process, i.e. $-\mathrm{tr}\mathbf{n}/(\tilde{\lambda}-\tilde{\kappa}) > 0 \to 0$, $U/R > 0$, $\overline{M}^p > 0$ for o $\to$ c̄. Both of the normal hardening $\dot{F} > 0$ in Eq. (7.37) and the subloading hardening $\mathbf{n}:\overset{\circ}{\boldsymbol{\sigma}} > 0$ in Eq. (7.38) proceed in this process.

(2) The first term becomes zero but the second term is positive at the point on the critical state line, i.e. $-\mathrm{tr}\mathbf{n}/(\tilde{\lambda}-\tilde{\kappa}) = 0$, $U/R > 0$, $\overline{M}^p > 0$ at the point c̄. The normal and the subloading hardenings proceed continuously rising up along this line.

(3) The first term tends to be negative but the second term is positive, so that the plastic modulus is kept to be positive until the stress reaches the peak, i.e. $-\mathrm{tr}\mathbf{n}/(\tilde{\lambda}-\tilde{\kappa}) < 0$, $U/R > 0$, $\overline{M}^p > 0$ for c̄ $\to$ p. The normal softening $\dot{F} < 0$ but the subloading hardening $\mathbf{n}:\overset{\circ}{\boldsymbol{\sigma}} > 0$ proceed in this process.

(4) The first term reaches the minimum value while the second term is always positive so that the sum of them becomes zero, canceling each other at the peak stress state, i.e. $-\mathrm{tr}\mathbf{n}/(\tilde{\lambda}-\tilde{\kappa}) = \mathrm{minimum}(<0)$, $U/R > 0$, $\overline{M}^p = 0$ at the peak point p. The normal softening proceeds continuously but the subloading hardening ceases at this point.

(5) The first term is negative but tends to increase toward zero while the second term is always positive but decreases continuously so that the sum of them becomes negative, exhibiting the subloading softening, i.e. $-\mathrm{tr}\mathbf{n}/(\tilde{\lambda}-\tilde{\kappa}) < 0$, $U/R > 0$, $\overline{M}^p < 0$ and then reducing to the critical (residual) state ($\overline{M}^p = 0$) as p $\to$ c. Both of the normal softening $\dot{F} < 0$ and the subloading softening $\mathbf{n}:\overset{\circ}{\boldsymbol{\sigma}} < 0$ proceed in this process.

(6) The first and the second terms reach zero at the residual (critical) state, i.e. $-\mathrm{tr}\mathbf{n}/(\tilde{\lambda}-\tilde{\kappa}) = 0$, $U/R = 0$, $\overline{M}^p = 0$ at the final point c. Both of the normal and the subloading softening cease in this state.

(7) The sign of the plastic volumetric strain rate is identical to that of tr**n**, whilst the sign of the elastic volumetric strain rate is opposite to that of the rate of pressure. The volume contraction is induced by the elastic volume contraction due to the increase of pressure in the initial stage of loading. Thereafter, the rate of plastic volume expansion tends to be larger than the rate of elastic volume contraction, so that the rate of volume expansion proceeds. However, reaching the peak stress ($\mathbf{n}:\overset{\circ}{\boldsymbol{\sigma}} = \overline{M}^p = d_v^e = 0, \dot{p} = 0$) at which the subloading surface expands at most and thus tr**n** becomes maximum leading to $d_v = d_v^p = \max$. by virtue of the "associated flow rule", the maximum ratio of volume expansion strain rate to axial strain rate, i.e. $d_v/|d_a| = \max$. are induced, where d_a is the rate of axial strain, i.e. $d_a = \dot{\varepsilon}_a$. This fact was indicated by Taylor (1948) based on the experimental evidence.

Eventually, these typical deformation behavior in normally-consolidated and over-consolidated states can be described pertinently by the initial subloading surface model. Here, it should be emphasized that these realistic predictions are attained by the subloading surface model with the adoption of the associated flow rule.

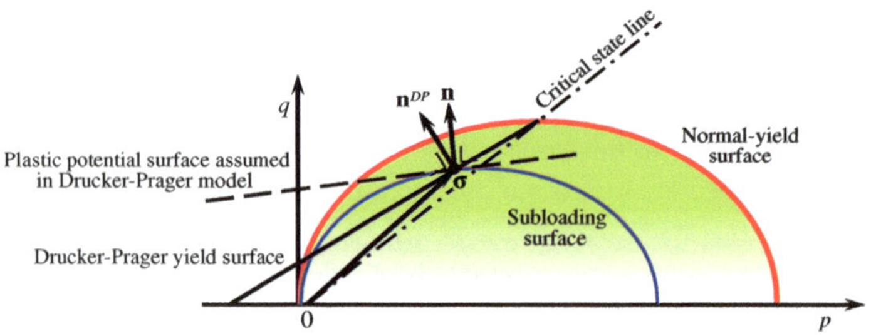

Fig. 11.7 Outward-normal of subloading surface coinciding approximately with the plastic potential surface assumed in the Drucker-Prager model

As described above, the conventional Cam-clay model predicts unrealistically high yield stress in the over-consolidated state. Then, the *Cap model* in which the over-consolidated side of Cam-clay yield surface is replaced by the Drucker-Prager yield surface (Drucker and Prager 1952) in Fig. 11.7 is widely used. However, the Cap model possesses various drawbacks described in the following.

1. The Cap model falls within the framework of conventional plasticity with the yield surface enclosing the purely elastic domain. Therefore, it predicts the stress-strain curve which rises up steeply (elastically) to the peak stress, and subsequently the stress decreases suddenly exhibiting a strong softening. In contrast, the subloading surface model can describe the realistic stress-strain curve with the smooth elastic-plastic transition since the plastic strain rate develops gradually as the stress approaches the yield surface.
2. The Cap model necessitates the yielding judgment, i.e. the judgment whether or not the stress lies on the yield surface in addition to the judgment on the direction of strain rate in the loading criterion as shown in Eq. (6.71). In contrast, the yielding judgment is not required in the subloading surface model since the stress lies always on the subloading surface playing the role of the loading surface as shown in Eq. (7.36).
3. The Cap model requires the particular computer algorithm for pulling-back the stress to the yield surface when the increments of stress or strain with finite magnitudes are input in numerical calculations. Otherwise, it is obliged to perform numerical calculations with quite infinitesimal loading increments. In contrast, the subloading surface model enables numerical calculations with finite loading increments without incorporation of particular computer algorithm since it possesses an automatic controlling function to attract the stress to the yield surface in the loading process as was described in Sect. 7.3.
4. The cap model additionally adopts the plastic potential surface of the conical shape to predict a dilatancy angle lower than that predicted by applying the associated flow rule to the Drucker-Prager yield surface, avoiding an unrealistically large plastic volume expansion. Then, the constitutive equation becomes complicated, including additional material parameters. In contrast, the subloading surface model can use the associated flow rule, whereas the outward-normal $\mathbf{n}$ of the subloading surface in the current stress is

approximately identical to the outward-normal $\mathbf{n}^{DP}$ of the plastic potential surface adopted in the Drucker-Prager model as shown in Fig. 11.7.

5. The cap model is obliged to adopt the non-associativity for the Drucker-Prager yield surface as described in 4. Therefore, it is accompanied with the asymmetry of the elastoplastic stiffness modulus tensor $\mathbf{K}^{ep}$ as shown in Eq. (6.54). This fact engenders the complexity in the formulation of variational principle and thus the difficulty in the analysis of boundary value problems. In contrast, the subloading surface model adopts the associativity leading to the symmetry of the elastoplastic stiffness modulus tensor.
6. The cap model predicts the *failure surface* which is determined uniquely by the Drucker-Prager yield surface itself, independent of the loading paths, because the interior of the yield surface is assumed to be a purely elastic domain and only the softening is induced when the stress reaches the Drucker-Prager yield surface. However, the surface depicted by connecting the peak stresses depends on the loading paths and its meridian section for the constant Lode angle is not straight but curved in real soils. In contrast, these facts can be described pertinently by the subloading surface model (cf. Hashiguchi et al. 2002).
7. The cap model is required the *tension cut* for the Drucker-Prager yield surface, which runs out sharply into the negative pressure range, when it is applied to the description of deformation in vicinity of zero pressure. In contrast, the subloading surface model is not required the tension cut because it adopts the normal-yield surface passing through the vicinity of the null stress state.
8. The cap model is accompanied with the singularity in the direction of plastic strain rate on the intersecting lines of the Drucker-Prager yield surface with the Cam-clay and the tension-cut yield surfaces. It results in unrealistic description of deformation behavior and would induce the difficulty in deformation analysis. In contrast, the subloading surface model adopts a single smooth normal-yield surface so that it is not accompanied with the singularity of the plastic modulus.
9. The cap model predicts the simultaneous occurrence of the peak stress and the maximum volume compression in over-consolidated clays and dense sands, in contradiction to experimental facts. In contrast, the subloading surface model provides the realistic prediction that the peak stress and the maximum ratio of volume expansion strain rate versus axial strain rate occur simultaneously as observed in real soils and was described in the benefit (7) for the subloading surface model.
10. The cap model is required to incorporate at least two more material constants describing the inclinations of yield and plastic potential surfaces in addition to the material constants in the Cam-clay model. In contrast, the subloading surface model is required to incorporate only one more material constant u in the evolution rule of the normal-yield ratio despite the distinctively accurate description.

In what follows, some comparisons of the simulations of typical triaxial test data by the Cap model and the subloading surface model are shown (Hashiguchi et al. 2002).

The simulations of the test data measured by Skempton and Brown (1961) for Weald clay subjected to the drained triaxial compression with a constant lateral stress are shown in Fig. 11.8 where the material constants and the initial value are selected as follows:

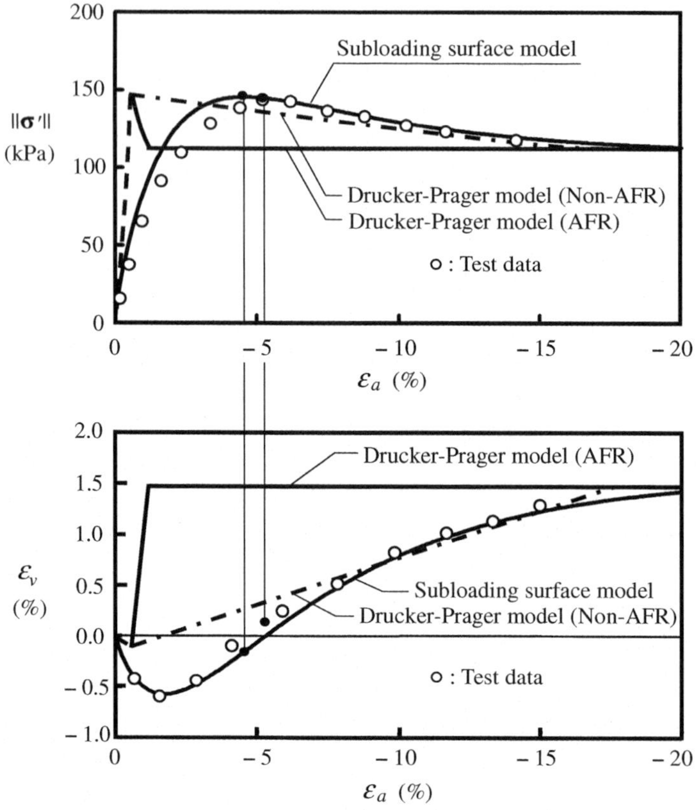

Fig. 11.8 Comparison of the calculated results by the Drucker-Prager and the subloading surface models with the test data (after Skempton and Brown 1961) of Weald clay for the drained triaxial compression with the constant lateral pressure

$$
\begin{aligned}
&\tilde{\lambda} = 0.045, \quad \tilde{\kappa} = 0.002, \quad \nu = 0.37, \quad M = 1.2,\\
&M_y = 0.574, \quad M_p = 0.071 \quad \text{for the Drucker-Prager yield surface},\\
&u = 33.0 \quad \text{for the subloading surface model},\\
&F_0 = 330.0\,\text{kPa},
\end{aligned}
$$

whilst the initial stress state is $\boldsymbol{\sigma}_0 = -67.0\mathbf{I}\,\text{kPa}$. Here, the function M_y and M_p are the inclinations of yield and the plastic potential surface, respectively, of the Drucker-Prager model in the $(p, ||\boldsymbol{\sigma}'||)$ plane. The associated flow rule and the nonassociated flow rule are abbreviated as AFR and Non-AFR, respectively, in this figure. On the other hand, the logarithmic function in Eq. $(7.13)_1$ is used for the evolution rule of the normal-yield ratio in the subloading surface model. The similar simulations for the test data of kaolinite-silt mixtures measured by Stark et al. (1994) are shown in Fig. 11.9 where the material constants and the initial value are selected as follows:

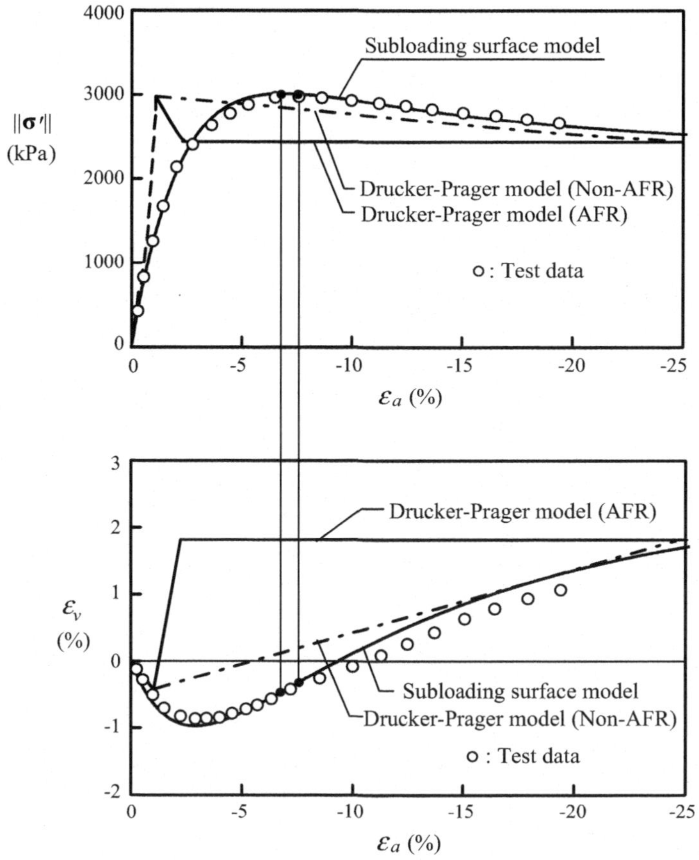

Fig. 11.9 Comparison of the calculated results by the Drucker-Prager and the subloading surface models with the test data (after Stark et al. 1994) of kaolinite-silt mixtures for the drained triaxial compression with the constant lateral pressure

$$\tilde{\lambda} = 0.1, \quad \tilde{\kappa} = 0.006, \quad \nu = 0.3, \quad M = 1.051,$$
$$M_y = 0.528, \quad M_p = 0.093 \quad \text{for the Drucker-Prager model},$$
$$u = 35.0 \quad \text{for the subloading surface model},$$
$$F_0 = 6000.0\,\text{kPa},$$

whilst the initial stress state is $\boldsymbol{\sigma}_0 = -1275.0\mathbf{I}$ kPa. The subloading surface model gives rise to the clearly better prediction than the Drucker-Prager model for both the axial stress-axial strain and the volumetric strain-axial strain curves. The curves predicted by the Drucker-Prager model are not smooth, which are formed by the three segments, i.e. the elastic, the elastoplastic and the critical state segments, whilst the former two form the concave curves of the 'Eiffel-tower' shape. Intense softening is induced rapidly lowering to the critical state immediately after the stress reaches the Drucker-Prager yield surface. However, note that the adoption of the non-associated flow rule in the Drucker-Prager model does not lead to the substantial improvement in simulation, whilst the subloading surface model adopting the associated flow rule gives the realistic prediction even for the volumetric strain. The parameter u is determined such that the stress-strain curve fit to the gentleness in the elastic-plastic transition.

The simulations of the stress paths and the stress-strain curves to the test data measured by Bishop et al. (1965) for London clay subjected to the undrained triaxial compression are shown in Fig. 11.10 where the material constants and the initial value are selected as follows:

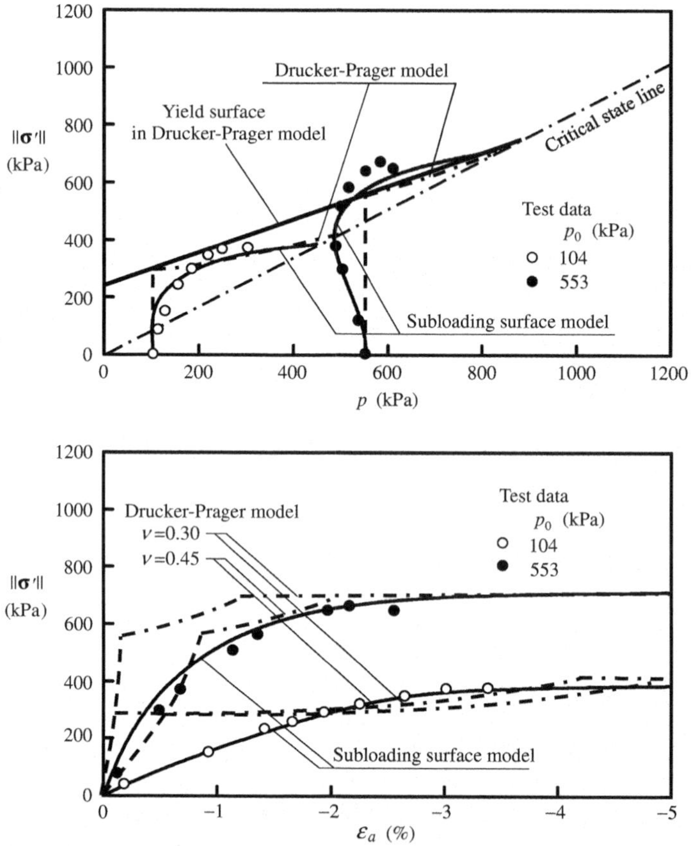

Fig. 11.10 Comparison of the calculated results by the Drucker-Prager and the subloading surface models with the test data (after Bishop et al. 1965) for the undrained triaxial compression with the constant lateral pressure

$\tilde{\lambda} = 0.022, \quad \tilde{\kappa} = 0.0063, \quad M = 0.82,$

$\nu = 0.3$ and $\nu = 0.45$, $M_y = 0.62$, $M_p = 0.21$ for the Drucker-Prager model,

$\nu = 0.3$, $u = 70.0$ for the subloading surface model,

$F_0 = 1700.0\,\mathrm{kPa}$.

The similar simulations for the test data of red clay measured by Wesley (1990) are shown in Fig. 11.11 where the material constants and the initial value are selected as follows:

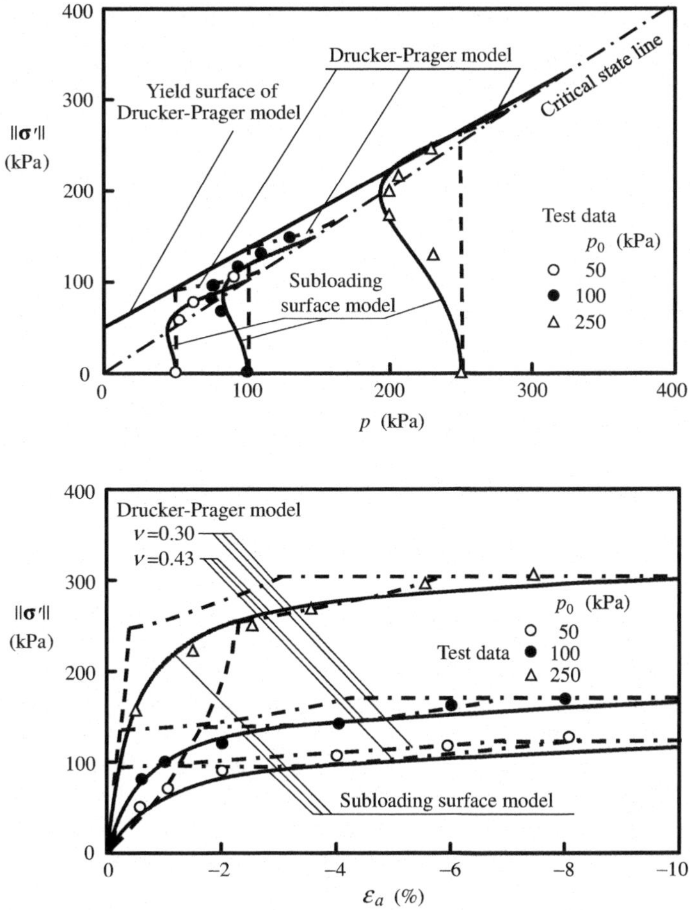

Fig. 11.11 Comparison of the calculated results by the Drucker-Prager and the subloading surface models with the test data (after Wesley 1990) for the undrained triaxial compression with the constant lateral pressure

$\tilde{\lambda} = 0.035, \quad \tilde{\kappa} = 0.012, \quad M = 1.015,$
$\nu = 0.3$ and $\nu = 0.43, \quad M_y = 0.767, \quad M_p = 0.24$ for the Drucker-Prager model,
$\nu = 0.3, \quad u = 20.0$ for the subloading surface model,
$F_0 = 300.0\,\mathrm{kPa}.$

The test data are predicted fairly well by the subloading surface model. On the other hand, both the stress paths and the stress-strain curves predicted by the Drucker-Prager model are quite different from the test data, which are not smooth being formed by the three segments, where the Poisson's ratio is selected two levels of $\nu = 0.30$ and 0.45 in Fig. 11.10 and $\nu = 0.30$ and 0.43 in Fig. 11.11.

The simple (initial) subloading surface model has been widely applied to the analyses of soil deformation behavior (e.g. Hashiguchi and Ueno 1977; Hashiguchi 1978; Topolnicki 1990; Kohgo et al. 1993; Asaoka et al. 1997; Hashiguchi and Chen 1998; Chowdhury et al. 1999; Hashiguchi et al. 2002; Khojastehpour and Hashiguchi 2004a, b; Khojastehpour et al. 2006; Nakai and Hinokio 2004; Hashiguchi and Tsutsumi 2006; Hashiguchi and Mase 2007; Wongsaroj et al. 2007).

11.4 Extension of Material Functions

Material functions contained in constitutive equation of soils formulated in the last section will be extended in order to describe the deformation behavior more realistically for the negative pressure range and the isotropic and anisotropic hardening behavior.

11.4.1 Yield Surface with Tensile Strength

Consider the extended yield surface fulfilling the following conditions.

(1) It includes not only positive but also negative pressure ranges. Here, note that the subloading surface is indeterminate at the null stress point when the stress reaches the null stress and thus the singular point of plastic modulus is induced since the normal-yield and the subloading surfaces pass through the null stress point which is thus to be the similarity-center of these surfaces in the initial subloading surface model described in Sect. 11.3. On the other hand, this problem is not induced in the extended subloading surface model because the similarity-center of the normal-yield and the subloading surfaces, i.e. elastic-core is not fixed at the null stress point and thus the subloading surface does not pass through the null stress point in general. The exclusion of the singularity of plastic modulus at the null stress point is of importance for the engineering design of soil structures because soils near the side edges of footings, soils at the pointed ends of piles, etc. are exposed to the null or further

negative stress state. In addition, the incorporation of tensile yield strength is of importance for the engineering design of structures of natural soils such as soft rocks and cement-treated soils widely used recently, which have the tensile yield strength.

(2) In the case that the anisotropy does not change, the yield surface expands/contracts maintaining a similarity with respect to the origin of stress space so that the yield stress increases/decreases in all directions in the space.

(3) For the sake of mathematical simplicity, the yield condition is described by a separate form consisting of the function of the stress and the internal variable, i.e. $f(\boldsymbol{\sigma}, \boldsymbol{\beta})$ and the function including the isotropic hardening variable, i.e. the isotropic hardening function $F(H)$ which describes the size of the yield surface.

Here, the function $f(\boldsymbol{\sigma}, \boldsymbol{\beta})$ must be a homogeneous function of the stress tensor $\boldsymbol{\sigma}$ in order to fulfill the above-mentioned conditions (2) and (3).

Equation (11.33) becomes the following equation through the translation of the yield surface to the negative pressure range by $\xi_h F(p \to p + \xi_h F)$ (Hashiguchi 2007; Hashiguchi and Mase 2007).

$$\left[\frac{p - ((1/2) - \xi_h)F}{F/2}\right]^2 + \left[\frac{\|\boldsymbol{\sigma}'\|}{MF/2}\right]^2 = 1 \tag{11.51}$$

leading to

$$(1 - \xi_h)\xi_h F^2 + (1 - 2\xi_h)pF - (p^2 + \chi^2) = 0 \tag{11.52}$$

where

$$\chi \equiv \frac{\|\boldsymbol{\sigma}'\|}{M} \tag{11.53}$$

ξ_h is the material constant, while it must fulfill $\xi_h \leq 1/2$ since the tensile yield stress is smaller than the compression yield stress and further the inequality $\xi_h < \vartheta$ is required, since the volume does not become infinite by the elastic deformation inside the yield surface, i.e. for $p > -\xi_h F$. The yield surface in Eq. (11.53), i.e. (11.54) is depicted in Fig. 11.12 for the axisymmetric stress state. $\widetilde{M}$ is given from the relation $\|\boldsymbol{\sigma}'\| = MF/2 = \widetilde{M}(1 - 2\xi_h)F/2$ as follows:

$$\widetilde{M} = \frac{1}{1 - 2\xi_h} M \tag{11.54}$$

ϕ_c is described from Eqs. (11.38) and (11.54) as follows:

$$\phi_c = \sin^{-1}\left(\frac{3M_c}{2\sqrt{6} + M_c}\right) = \sin^{-1}\left(\frac{3(1 - 2\xi_h)\widetilde{M}_c}{2\sqrt{6} + (1 - 2\xi_h)\widetilde{M}_c}\right) \tag{11.55}$$

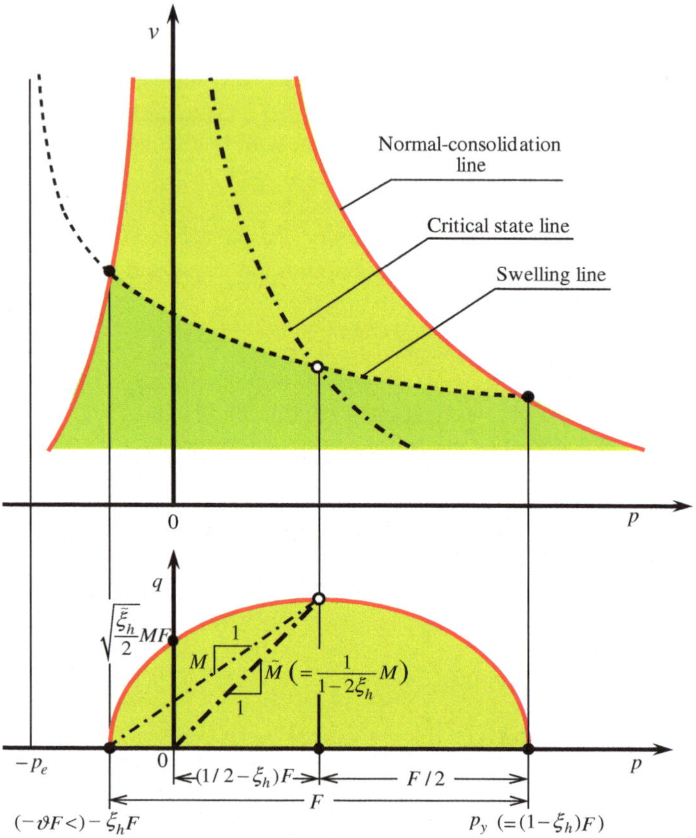

Fig. 11.12 Yield surface of soils with tensile strength

where M_c and $\widetilde{M}_c$ are the values of M and $\widetilde{M}$ respectively in the axisymmetric compression stress state.

Equation (11.51) can be expressed in the separated form of the function $f(p, \chi)$ of the stress and internal variable and the hardening function F, i.e.

$$
\boxed{f(p,\chi) = F, \quad f(p,\chi) = \begin{cases} p[1+(\chi/p)^2] & \text{for } \xi_h = 0 \\ \dfrac{1}{\bar{\xi}_h}(p_\chi - \bar{\xi}_h p) & \text{for } \xi_h \neq 0 \end{cases}} \tag{11.56}
$$

where

$$
\tilde{\xi}_h \equiv 2(1-\xi_h)\xi_h, \quad \bar{\xi}_h \equiv 1 - 2\xi_h, \quad p_\chi \equiv \sqrt{p^2 + 2\tilde{\xi}_h\chi^2} \tag{11.57}
$$

In the above, the yield surface of soils is formulated so as to fulfill the conditions (1)–(3) based on the modified Cam-clay model. It is difficult to derive the other yield surface fulfilling the conditions (1)–(3). For instance, consider the translation of the original Cam clay model to the negative pressure range by $p \to p + \xi_h F$.

$$(p + \xi_h F)\exp\left(\frac{\|\boldsymbol{\sigma}'\|}{p + \xi_h F}/M\right) = F \tag{11.58}$$

However, a separated form into the function of stress and internal variables and the hardening function cannot be derived from this equation. On the other hand, the translation of the yield surface to the negative pressure range by the constant value C_y $(p \to p + C_y)$ is adopted for constitutive equations for unsaturated soils (e.g. Alonso et al. 1990; Simo and Meschke 1993; Borja 2004). The modified Cam-clay model, for instance, is described by this translation as follows:

$$\left[\frac{p - (1/2)F + C_y}{F/2}\right]^2 + \left[\frac{\|\boldsymbol{\sigma}'\|}{MF/2}\right]^2 = 1 \tag{11.59}$$

In this equation, the yield surface expands/contracts from/to the fixed point $\boldsymbol{\sigma} = C_y\mathbf{I}$ $(p = C_y)$ on the hydrostatic axis and thus it does not fulfill the condition (2). The incorporation of this yield condition into the subloading surface model leads to the physical impertinence that the unloading is induced against the fact that a large plastic deformation would be induced when the stress changes towards the negative pressure direction.

11.4.2 Extended Isotropic Hardening Function with Deviatoric Hardening

Clays possess high void ratio and thus the hardening/softening is induced by the plastic volumetric change. On the other hand, metals are highly condensed solids without void and thus the plastic volumetric change is hardly induced so that the hardening/softening is induced by the plastic deviatoric strain. Sands possess void ratio far lower than clays and thus exhibit properties of not only clays but also metals so that the hardening/softening is induced not only by the plastic volumetric change but also the plastic deviatoric strain. It was assumed that the plastic deviatoric strain causes the isotropic hardening and softening when the stress ratio is higher and lower, respectively, than a certain value M_d (Nova 1977; Wilde 1977) in addition to the plastic volumetric hardening/softening, leading to

$$\dot{H} = -d_v^p + \mu_d\|\mathbf{d}^{p\prime}\|\frac{\|\boldsymbol{\sigma}'\| - M_d p}{F} \tag{11.60}$$

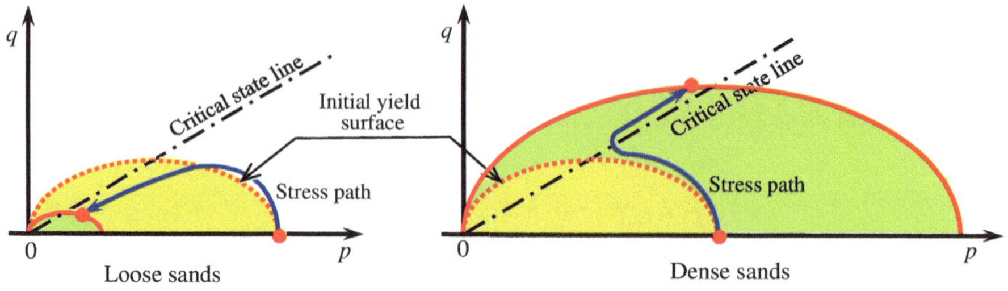

Fig. 11.13 Stress paths under constant volume or undrained condition

where μ_d is the material constant. Here, let the function M_d be extended as follows (Hashiguchi and Chen 1998):

$$M_d(\cos 3\theta_\sigma) \equiv \frac{14\sqrt{6}\sin\phi_d}{(3-\sin\phi_d)(8+\cos 3\theta_\sigma)} \tag{11.61}$$

where ϕ_d is the material constant, following Eqs. (11.38) and (11.41). The conical surface $\|\boldsymbol{\sigma}'\| = M_d p$ is called the *deviatoric hardening(/softening) surface* describing the boundary of the deviatoric hardening and softening region in the stress space. The stress paths for sands under the constant volume or undrained condition can be predicted realistically by the isotropic hardening rate in Eq. (11.59) as follows:

(1) The stress path goes down toward the origin of stress space by the deviatoric softening below the critical state line in loose sands,
(2) The stress path goes up over the critical state line and along it by the deviatoric hardening in dense sands

as illustrated in Fig. 11.13. Here, it can be stated that the denser the sand, the smaller is the angle ϕ_d and that ϕ_d is larger than ϕ_c, i.e. $\phi_d > \phi_c (M_d > M)$ in loose sands by the fact (1) but ϕ_d is smaller than ϕ_c, i.e. $\phi_d < \phi_c (M_d < M)$ in dense sands by the fact (2).

11.4.3 Rotational Hardening

The inherent anisotropy represented in the orthotropic anisotropy described in Sect. 10.6 cannot be ignored in metals and woods. On the other hand, the induced anisotropy is more dominant in soils since soils are assemblies of particles with weak cohesions between them and thus the rearrangement of soil particles is induced easily. Here, the yield surface of soils must always include the origin of stress space but does very slightly because of the weak cohesion. Besides, the remarkable softening (contraction of the yield surface) by the plastic volume expansion is induced as the stress approaches the null stress state. Then, the stress can never return to the origin of stress space, once the yield surface translates so as not to include the origin, as illustrated on the (p, q) plane for the axisymmetric stress state in Fig. 11.14. Therefore, the kinematic hardening described in Sect. 6.7.3 is not applicable to soils.

General speaking, the *stress* in pressure-independent materials would correspond to the *stress ratio*, i.e. the ratio of the deviatoric stress versus pressure in pressure-dependent, i.e. frictional materials such as soils and further the translation of yield surface, i.e. the kinematic hardening in the former would correspond to the *rotation of yield surface* in the latter. The description of anisotropy of soils by the rotation of the Cam-clay yield surface was proposed by Sekiguchi and Ohta (1977), replacing the deviatoric stress $\boldsymbol{\sigma}'$ to the novel variable $\boldsymbol{\sigma}' - p\boldsymbol{\beta}\,(\mathrm{tr}\boldsymbol{\beta} = 0)$ in the yield condition. This concept was called the *rotational hardening* in contrast to the kinematic hardening for pressure-independent materials and the second-order deviatoric dimensionless tensor $\boldsymbol{\beta}$ was called the *rotational hardening variable* by Hashiguchi (1977). Then, the yield condition in Eqs. (11.51) or (11.56) is extended as follows (Hashiguchi and Mase 2007):

$$\left[\frac{p - ((1/2) - \xi_h)F}{F/2}\right]^2 + \left[\frac{\|\widehat{\boldsymbol{\sigma}}'\|}{\widehat{M}F/2}\right]^2 = 1 \tag{11.62}$$

i.e.

$$f(p, \widehat{\chi}) = F, \quad f(p, \widehat{\chi}) = \begin{cases} p\left[1 + \left(\widehat{\chi}/p\right)^2\right] & \text{for } \xi_h = 0 \\ \dfrac{1}{\bar{\xi}_h}\left(\widehat{p}_\chi - \bar{\xi}_h p\right) & \text{for } \xi_h \neq 0 \end{cases} \tag{11.63}$$

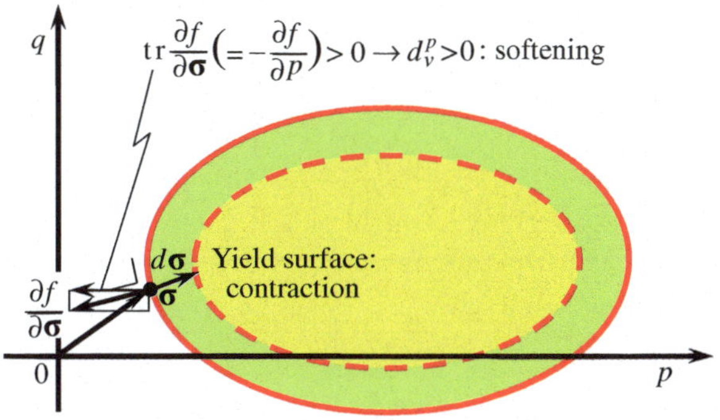

Fig. 11.14 Inadequacy of kinematic hardening for description of anisotropy of soils: stress can never return to null state

where

$$\boldsymbol{\sigma}' \equiv \boldsymbol{\sigma}' - p\boldsymbol{\beta} \tag{11.64}$$

$$\widehat{\chi} \equiv \frac{\|\widehat{\boldsymbol{\sigma}}'\|}{\widehat{M}} \tag{11.65}$$

$$\widehat{p}_{\chi} \equiv \sqrt{\widehat{p}^{2} + 2\tilde{\xi}_h \widehat{\chi}^{2}} \tag{11.66}$$

$$\widehat{M}(\cos 3\theta_{\widehat{\sigma}}) = \frac{14\sqrt{6}\sin\phi_c}{(3-\sin\phi_c)(8+\cos 3\theta_{\widehat{\sigma}})} \tag{11.67}$$

$$\cos 3\theta_{\widehat{\sigma}} \equiv \sqrt{6}\mathrm{tr}\mathbf{t}'^{3}_{\widehat{\sigma}}, \quad \mathbf{t}'_{\widehat{\sigma}} \equiv \frac{\widehat{\boldsymbol{\sigma}}'}{\|\widehat{\boldsymbol{\sigma}}'\|} \tag{11.68}$$

The yield surface in Eq. (11.62), i.e. (11.63) is depicted in Fig. 11.15 for the axisymmetric stress state.

The evolution rule of rotational hardening tensor $\boldsymbol{\beta}$ is given below (Hashiguchi and Chen 1998; Hashiguchi 2001). The following assumptions are adopted for the formulation of the evolution rule.

(1) Rotation of the yield surface is induced only by the deviatoric component of the plastic strain rate independent of the mean component.
(2) The rotation ceases when the central axis of yield surface reaches the surface, called the *rotational limit surface*, which exhibits the conical surface having the summit at the origin of stress space. Let the rotational hardening limit surface be given by

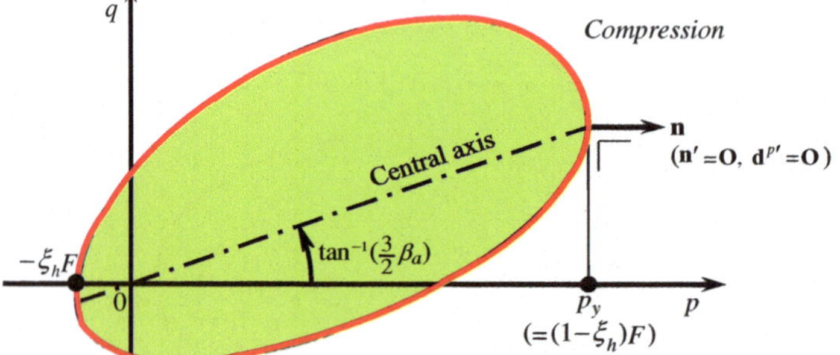

Fig. 11.15 Rotated yield surface in the $(p,\, q)$ plane

$$\|\boldsymbol{\eta}\| = M_r \tag{11.69}$$

where M_r is the stress ratio in the rotational hardening limit surface, called the *rotational limit stress ratio*, and let it be given following Eq. (11.41) by

$$M_r(\cos 3\theta_{\widehat{\sigma}}) = \frac{14\sqrt{6}\sin\phi_r}{(3-\sin\phi_r)(8+\cos 3\theta_\sigma)} \tag{11.70}$$

ϕ_r being the material constant, called the *rotational limit angle*.

(3) The central axis of yield surface $\boldsymbol{\eta} = \boldsymbol{\beta}$ rotates towards the conjugate line $\boldsymbol{\eta} = M_r \mathbf{t}_{\widehat{\sigma}}$ on the rotational limit surface, where the conjugate line is the generating line of the rotational limit surface which is observed from the hydrostatic axis in the same direction observed from the central axis $\boldsymbol{\eta} = \boldsymbol{\beta}$ of the yield surface to the current stress (see Fig. 11.16).

Based on the above-mentioned assumptions, let the following evolution rule of rotational hardening be postulated in the form in Eq. (6.103) with the replacements of $\boldsymbol{\alpha} \to \boldsymbol{\beta}$, $c_k \to c_r$, $b_k \to M_r$ as follows:

$$\begin{aligned} \overset{\circ}{\boldsymbol{\beta}} &= c_r\left(\mathbf{d}^{p\prime} - \frac{1}{M_r}\|\mathbf{d}^{p\prime}\|\boldsymbol{\beta}\right) = \dot{\lambda}\,\mathbf{f}_{\beta n}(M_r, \boldsymbol{\beta}; \mathbf{n}') \\ \mathbf{f}_{\beta n}(M_r, \boldsymbol{\beta}; \mathbf{n}') &\equiv c_r\left(\mathbf{n}' - \frac{1}{M_r}\|\mathbf{n}'\|\boldsymbol{\beta}\right)\ (\mathrm{tr}\mathbf{f}_{\beta n} = 0) \end{aligned} \tag{11.71}$$

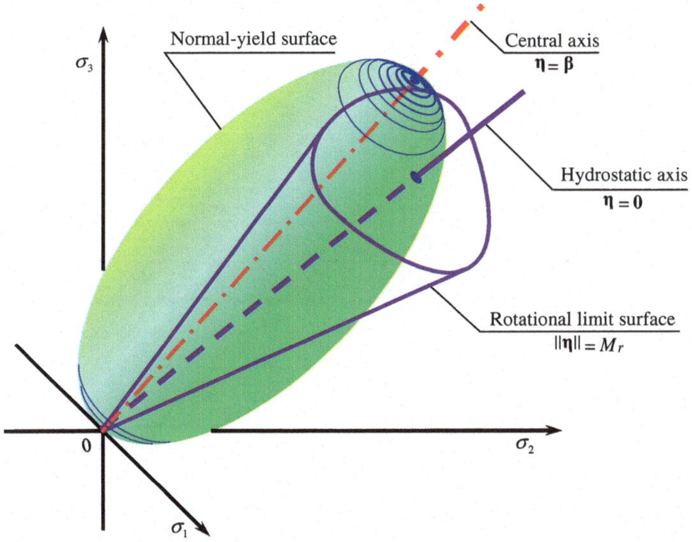

Fig. 11.16 Yield surface and rotational limit surfaces (illustrated in the principal stress space)

where c_r is the material constant. Here, it is noteworthy that the rotational hardening is not induced since the deviatoric strain rate is not induced when the stress lies on the central axis of the subloading surface in the modified Cam-clay model, i.e. $\mathbf{d}^{p\prime} = \mathbf{O} \rightarrow \overset{\circ}{\boldsymbol{\beta}} = \mathbf{O}$ for $\boldsymbol{\eta} = \boldsymbol{\beta}$ as illustrated in Fig. 11.15. Here, needless to say, the deviatoric plastic strain rate is adopted in Eq. (11.71) since the anisotropic hardening is independent of plastic volumetric strain rate. Equation (11.71) is described in one-dimensional state as follows:

$$\dot{\beta} = c_r\left(\pm 1 - \frac{\beta}{M_r}\right)(\pm\dot{\varepsilon}^p) \tag{11.72}$$

which is shown in Fig. 11.17.

11.5 Extended Subloading Surface Model

Elastoplastic constitutive equation for describing cyclic loading behavior of soils is formulated below by incorporating the rotational hardening instead of the kinematic hardening into the extended subloading surface model shown in Chap. 9.

11.5.1 Normal-Yield and Subloading Surfaces

Setting $\boldsymbol{\alpha} = \mathbf{O}$ in Eqs. (9.1) and (9.2) and incorporating the rotational hardening $\boldsymbol{\beta}$, the normal-yield and the subloading surfaces for soils are given as follows (see Fig. 11.18):

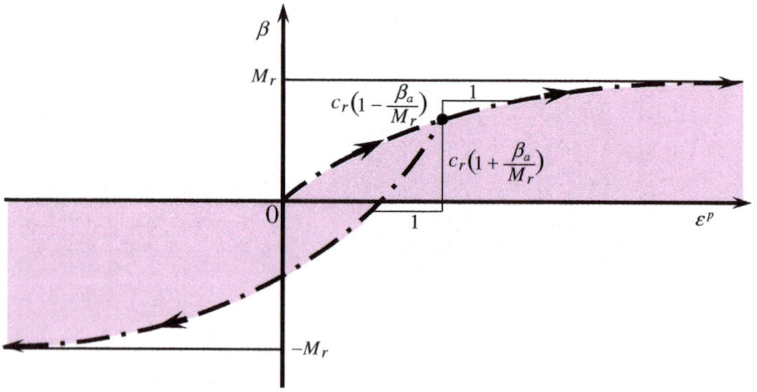

Fig. 11.17 Relation of axial component of rotational hardening variable versus axial plastic strain in the axisymmetric stress state

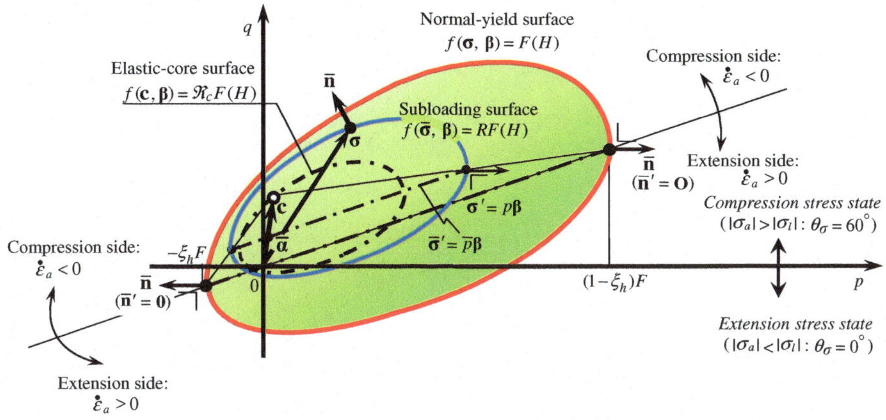

Fig. 11.18 Rotated normal-yield, subloading and similarity-center surfaces in the (p, q) plane

$$f(\boldsymbol{\sigma}, \boldsymbol{\beta}) = F(H) \tag{11.73}$$

$$\boxed{f(\bar{\boldsymbol{\sigma}}, \boldsymbol{\beta}) = RF(H)} \tag{11.74}$$

with

$$\left.\begin{array}{c} \hat{\mathbf{c}} = \mathbf{c}, \quad \bar{\mathbf{c}} = R\mathbf{c} \\ \bar{\boldsymbol{\alpha}} = (1-R)\mathbf{c} \end{array}\right\} \tag{11.75}$$

$$\bar{\boldsymbol{\sigma}} = \tilde{\boldsymbol{\sigma}} + R\mathbf{c} \tag{11.76}$$

The subloading stress function $f(\bar{\boldsymbol{\sigma}}, \boldsymbol{\beta})$ in Eq. (11.74) with a tensile strength and the rotational hardening is given from Eq. (11.63) as follows:

$$\boxed{f(\bar{\boldsymbol{\sigma}}, \boldsymbol{\beta}) = f(\bar{p}, \widehat{\bar{\chi}}) = \begin{cases} \bar{p}\left[1 + \left(\dfrac{\widehat{\bar{\chi}}}{\bar{p}}\right)^2\right] & \text{for} \quad \xi_h = 0 \\ \dfrac{1}{\bar{\xi}_h}(\widehat{\bar{p}}_\chi - \bar{\xi}_h \bar{p}) & \text{for} \quad \xi_h \neq 0 \end{cases}} \tag{11.77}$$

where

$$\bar{p} \equiv -(\mathrm{tr}\bar{\boldsymbol{\sigma}})/3 \tag{11.78}$$

$$\widehat{\bar{p}}_{\chi} \equiv \sqrt{\bar{p}^2 + 2\tilde{\xi}_h \widehat{\bar{\chi}}^2} \tag{11.79}$$

$$\widehat{\bar{\chi}} \equiv \frac{\|\widehat{\bar{\boldsymbol{\sigma}}}'\|}{\widehat{\bar{M}}}, \quad \widehat{\bar{\boldsymbol{\sigma}}}' \equiv \bar{\boldsymbol{\sigma}}' - \bar{p}\boldsymbol{\beta} \tag{11.80}$$

$$\widehat{\bar{M}}(\cos 3\theta_{\widehat{\bar{\sigma}}}) = \frac{14\sqrt{6}\sin\phi_c}{(3 - \sin\phi_c)(8 + \cos 3\theta_{\widehat{\bar{\sigma}}})} \tag{11.81}$$

$$\cos 3\theta_{\widehat{\bar{\sigma}}} \equiv \sqrt{6}\mathrm{tr}\mathbf{t}'^3_{\widehat{\bar{\sigma}}}, \quad \mathbf{t}'_{\widehat{\bar{\sigma}}} \equiv \frac{\widehat{\bar{\boldsymbol{\sigma}}}'}{\|\widehat{\bar{\boldsymbol{\sigma}}}'\|} \tag{11.82}$$

The rate of isotropic hardening is given from Eq. (11.60) as follows:

$$\dot{H} = f_{Hn}(\boldsymbol{\sigma}, H; \bar{\mathbf{n}})\,\dot{\bar{\lambda}} \tag{11.83}$$

where

$$f_{Hn}(\boldsymbol{\sigma}, H; \bar{\mathbf{n}}) = -\mathrm{tr}\bar{\mathbf{n}} + \mu_d\|\bar{\mathbf{n}}'\|\frac{\|\boldsymbol{\sigma}'\| - M_d p}{F} \tag{11.84}$$

The rate of rotational hardening is given from Eq. (11.71) as follows:

$$\mathring{\boldsymbol{\beta}} = \dot{\bar{\lambda}}\,\bar{\mathbf{f}}_{\beta n}(\widehat{\bar{M}}_r, \boldsymbol{\beta}; \bar{\mathbf{n}}') \tag{11.85}$$

where

$$\bar{\mathbf{f}}_{\beta n}(\widehat{\bar{M}}_r, \boldsymbol{\beta};\, \bar{\mathbf{n}}') = c_r\left(\bar{\mathbf{n}}' - \frac{1}{\widehat{\bar{M}}_r}\|\bar{\mathbf{n}}'\|\boldsymbol{\beta}\right) \tag{11.86}$$

$$\widehat{\bar{M}}_r(\cos 3\theta_{\widehat{\bar{\sigma}}}) = \frac{14\sqrt{6}\sin\phi_r}{(3 - \sin\phi_r)(8 + \cos 3\theta_{\widehat{\bar{\sigma}}})} \tag{11.87}$$

where ϕ_r is the material constant.

The translation rule of the elastic-core is given from Eq. (9.28) as follows:

$$\overset{\circ}{\mathbf{c}} = \dot{\lambda} c\left(\bar{\mathbf{n}} - \frac{\Re_c}{\xi}\mathbf{n}_c\right) = \dot{\lambda}\bar{\mathbf{f}}_{cn} \tag{11.88}$$

where

$$\bar{\mathbf{f}}_{cn} = c\left(\bar{\mathbf{n}} - \frac{\Re_c}{\xi}\mathbf{n}_c\right) \tag{11.89}$$

$$\mathbf{n}_c \equiv \frac{\partial f(\mathbf{c},\, \boldsymbol{\beta})}{\partial \mathbf{c}} \Big/ \left\| \frac{f(\mathbf{c},\, \boldsymbol{\beta})}{\partial \mathbf{c}} \right\| (\|\mathbf{n}_c\| = 1) \tag{11.90}$$

$$\Re_c = \frac{f(\mathbf{c},\, \boldsymbol{\beta})}{F(H)} \tag{11.91}$$

The scalar variable C_σ in Eq. (9.47) for the modification of reloading-unloading behavior is given by

$$C_\sigma \equiv \mathbf{n}_c\mathbf{:}\bar{\mathbf{n}}\ (-1 \le C_\sigma \le 1) \tag{11.92}$$

The rate of $\bar{\boldsymbol{\alpha}}$ is given from Eqs. (9.5) and (11.88) with $\boldsymbol{\alpha} = \mathbf{O}$ as follows:

$$\begin{aligned} \overset{\circ}{\bar{\boldsymbol{\alpha}}} &= (1 - R)\, \overset{\circ}{\mathbf{c}} - \dot{R}\,\mathbf{c} \\ &= (1 - R)\, \bar{\mathbf{f}}_{cn}\, \dot{\lambda} - U\, \dot{\lambda}\, \mathbf{c} \end{aligned} \tag{11.93}$$

The plastic modulus is given from Eq. (11.73), noting Eq. (9.39), as follows:

$$\bar{M}^p \equiv \bar{\mathbf{n}}\mathbf{:}\left[\frac{F'}{F} f_{Hn}\bar{\boldsymbol{\sigma}} + (1 - R)\bar{\mathbf{f}}_{cn} - \frac{1}{RF}\left(\frac{\partial f(\bar{\boldsymbol{\sigma}},\, \boldsymbol{\beta})}{\partial \boldsymbol{\beta}}\mathbf{:}\bar{\mathbf{f}}_{\beta n}\right)\bar{\boldsymbol{\sigma}} + \frac{U}{R}\tilde{\boldsymbol{\sigma}}\right] \tag{11.94}$$

noting

$$\frac{\partial f(\bar{\boldsymbol{\sigma}},\, \boldsymbol{\beta})}{\partial \bar{\boldsymbol{\sigma}}}\mathbf{:}\overset{\circ}{\boldsymbol{\sigma}} - \frac{\partial f(\bar{\boldsymbol{\sigma}},\, \boldsymbol{\beta})}{\partial \bar{\boldsymbol{\alpha}}}\mathbf{:}\overset{\circ}{\bar{\boldsymbol{\alpha}}} + \frac{\partial f(\bar{\boldsymbol{\sigma}},\, \boldsymbol{\beta})}{\partial \bar{\boldsymbol{\sigma}}}\mathbf{:}\overset{\circ}{\boldsymbol{\beta}} - R\dot{F} - \dot{R}F = 0 \tag{11.95}$$

11.5.2 Partial Derivatives of Subloading Surface Function

The partial derivatives of the function in Eq. (11.77) are shown below.

$$\frac{\partial f(\bar{p},\hat{\bar{\chi}})}{\partial \bar{p}}=\begin{cases}1+(\hat{\bar{\chi}}/\bar{p})^2+\bar{p}(-2\bar{p}^{-3})\hat{\bar{\chi}}^2 & \text{for } \xi_h=0\\ \dfrac{1}{\tilde{\bar{\xi}}_h}\left(\dfrac{1}{2}\dfrac{2\bar{p}}{\hat{\bar{p}}_\chi}-\bar{\xi}_h\right) & \text{for } \xi_h\neq 0\end{cases}$$

$$=\begin{cases}1-\left(\dfrac{\hat{\bar{\chi}}}{\bar{p}}\right)^2 & \text{for } \xi_h=0\\ \dfrac{1}{\tilde{\bar{\xi}}_h}\left(\dfrac{\bar{p}}{\hat{\bar{p}}_\chi}-\bar{\xi}_h\right) & \text{for } \xi_h\neq 0\end{cases} \tag{11.96}$$

$$\frac{\partial f(\bar{p},\hat{\bar{\chi}})}{\partial \hat{\bar{\chi}}}=\begin{cases}2\dfrac{\hat{\bar{\chi}}}{\bar{p}} & \text{for } \xi_h=0\\ \dfrac{1}{\tilde{\bar{\xi}}}\dfrac{1}{2}\dfrac{4\tilde{\xi}_h\hat{\bar{\chi}}}{\hat{\bar{p}}_\chi} & \text{for } \xi_h\neq 0\end{cases}=\begin{cases}2\dfrac{\hat{\bar{\chi}}}{\bar{p}} & \text{for } \xi_h=0\\ 2\dfrac{\hat{\bar{\chi}}}{\hat{\bar{p}}_\chi} & \text{for } \xi_h\neq 0\end{cases} \tag{11.97}$$

$$\frac{\partial \bar{p}}{\partial \boldsymbol{\sigma}}=-\frac{1}{3}\mathbf{I} \tag{11.98}$$

$$\frac{\partial \bar{\boldsymbol{\sigma}}'}{\partial \boldsymbol{\sigma}}=\mathcal{I}' \tag{11.99}$$

$$\left(\frac{\partial \bar{\sigma}'_{ij}}{\partial \sigma_{kl}}=\frac{\partial(\bar{\sigma}_{ij}+\bar{p}\delta_{ij})}{\partial \sigma_{kl}}=\frac{1}{2}(\delta_{ik}\delta_{jl}+\delta_{il}\delta_{jk})-\frac{1}{3}\delta_{ij}\delta_{kl}\right)$$

$$\frac{\partial \hat{\bar{\boldsymbol{\sigma}}}'}{\partial \boldsymbol{\sigma}}=\mathcal{I}'+\frac{1}{3}\boldsymbol{\beta}\otimes\mathbf{I} \tag{11.100}$$

$$\left(\frac{\partial \hat{\bar{\sigma}}'_{ij}}{\partial \sigma_{kl}}=\frac{\partial(\bar{\sigma}'_{ij}-\bar{p}\beta_{ij})}{\partial \sigma_{kl}}=\frac{1}{2}(\delta_{ik}\delta_{jl}+\delta_{il}\delta_{jk})-\frac{1}{3}\delta_{ij}\delta_{kl}+\frac{1}{3}\beta_{ij}\delta_{kl}\right)$$

$$\frac{\partial \hat{\bar{M}}}{\partial \cos 3\theta_{\hat{\bar{\sigma}}}}=-\frac{14\sqrt{6}\sin\phi_c}{(3-\sin\phi_c)(\cos 3\theta_{\hat{\bar{\sigma}}})^2}$$

$$=-\frac{\hat{\bar{M}}}{8+\cos 3\theta_{\hat{\bar{\sigma}}}}\left(=-\frac{3-\sin\phi_c}{14\sqrt{6}\sin\phi_c}\hat{\bar{M}}^2\right) \tag{11.101}$$

$$\frac{\partial \mathbf{t}'_{\widehat{\bar{\sigma}}}}{\partial \widehat{\bar{\boldsymbol{\sigma}}}'} = \frac{1}{\|\partial \widehat{\bar{\boldsymbol{\sigma}}}'\|}\left(\mathcal{I}' - \mathbf{t}'_{\widehat{\bar{\sigma}}} \otimes \mathbf{t}'_{\widehat{\bar{\sigma}}}\right) \tag{11.102}$$

$$\left(\frac{\partial \widehat{\bar{t}}'_{ij}}{\partial \widehat{\bar{\sigma}}'_{kl}} = \frac{\partial \dfrac{\widehat{\bar{\sigma}}'_{ij}}{\sqrt{\widehat{\bar{\sigma}}'_{rs}\widehat{\bar{\sigma}}'_{sr}}}}{\partial \widehat{\bar{\sigma}}'_{kl}} = \frac{\dfrac{\partial \widehat{\bar{\sigma}}'_{ij}}{\partial \widehat{\bar{\sigma}}'_{kl}}\sqrt{\widehat{\bar{\sigma}}'_{rs}\widehat{\bar{\sigma}}'_{sr}} - \widehat{\bar{\sigma}}'_{ij}\dfrac{\partial\sqrt{\widehat{\bar{\sigma}}'_{rs}\widehat{\bar{\sigma}}'_{sr}}}{\partial \widehat{\bar{\sigma}}'_{kl}}}{\widehat{\bar{\sigma}}'_{rs}\widehat{\bar{\sigma}}'_{sr}}\right.$$

$$\left. = \frac{\frac{1}{2}(\delta_{ik}\delta_{jl} + \delta_{il}\delta_{jk})\sqrt{\widehat{\bar{\sigma}}'_{rs}\widehat{\bar{\sigma}}'_{sr}} - \hat{\bar{\sigma}}'_{ij} t'_{\widehat{\bar{\sigma}}kl}}{\widehat{\bar{\sigma}}'_{rs}\widehat{\bar{\sigma}}'_{sr}} = \frac{1}{\sqrt{\widehat{\bar{\sigma}}'_{rs}\widehat{\bar{\sigma}}'_{sr}}}\left\{\frac{1}{2}(\delta_{ik}\delta_{jl} + \delta_{il}\delta_{jk}) - t'_{\widehat{\bar{\sigma}}ij} t'_{\widehat{\bar{\sigma}}kl}\right\}\right)$$

$$\frac{\partial(\mathrm{tr}\mathbf{t}'^3_{\widehat{\bar{\sigma}}})}{\partial \mathbf{t}'_{\widehat{\bar{\sigma}}}} = 3\mathbf{t}'^2_{\widehat{\bar{\sigma}}} \tag{11.103}$$

$$\left(\frac{\partial t'_{\widehat{\bar{\sigma}}rs} t'_{\widehat{\bar{\sigma}}st} t'_{\widehat{\bar{\sigma}}tr}}{\partial \widehat{\bar{t}}'_{ij}} = \delta_{ir}\delta_{js} t'_{\widehat{\bar{\sigma}}st} t'_{\widehat{\bar{\sigma}}tr} + t'_{\widehat{\bar{\sigma}}rs}\delta_{is}\delta_{jt} t'_{\widehat{\bar{\sigma}}tr} + t'_{\widehat{\bar{\sigma}}rs} t'_{\widehat{\bar{\sigma}}st}\delta_{it}\delta_{jr} = 3t'_{\widehat{\bar{\sigma}}ir} t'_{\widehat{\bar{\sigma}}rj}\right)$$

$$\frac{\partial \cos 3\theta_{\widehat{\bar{\sigma}}}}{\partial \widehat{\bar{\boldsymbol{\sigma}}}'} = \frac{3}{\|\widehat{\bar{\boldsymbol{\sigma}}}'\|}\left(\sqrt{6}\mathbf{t}'^2_{\widehat{\bar{\sigma}}} - \cos 3\theta_{\widehat{\bar{\sigma}}}\mathbf{t}'_{\widehat{\bar{\sigma}}}\right) \tag{11.104}$$

$$\left(\frac{\partial\sqrt{6}t'_{\widehat{\bar{\sigma}}lm} t'_{\widehat{\bar{\sigma}}mn} t'_{\widehat{\bar{\sigma}}nl}}{\partial \widehat{\bar{\sigma}}'_{ij}} = \sqrt{6}\frac{\partial t'_{\widehat{\bar{\sigma}}lm} t'_{\widehat{\bar{\sigma}}mn} t'_{\widehat{\bar{\sigma}}nl}}{\partial t'_{\widehat{\bar{\sigma}}rs}}\frac{\partial t'_{\widehat{\bar{\sigma}}rs}}{\partial \widehat{\bar{\sigma}}'_{ij}} = 3\sqrt{6}t'_{\widehat{\bar{\sigma}}sn} t'_{\widehat{\bar{\sigma}}nr}\frac{\partial t'_{\widehat{\bar{\sigma}}rs}}{\partial \widehat{\bar{\sigma}}'_{ij}}\right.$$

$$= 3\sqrt{6}t'_{\widehat{\bar{\sigma}}sn} t'_{\widehat{\bar{\sigma}}nr}\frac{1}{\sqrt{\widehat{\bar{\sigma}}'_{pq}\widehat{\bar{\sigma}}'_{qp}}}\left\{\frac{1}{2}(\delta_{ri}\delta_{sj} + \delta_{rj}\delta_{si}) - t'_{\widehat{\bar{\sigma}}rs} t'_{\widehat{\bar{\sigma}}ij}\right\}$$

$$= 3\sqrt{6}\frac{1}{\sqrt{\widehat{\bar{\sigma}}'_{pq}\widehat{\bar{\sigma}}'_{qp}}}\left(t'_{\widehat{\bar{\sigma}}in} t'_{\widehat{\bar{\sigma}}nj} - t'_{\widehat{\bar{\sigma}}sn} t'_{\widehat{\bar{\sigma}}nr} t'_{\widehat{\bar{\sigma}}rs} t'_{\widehat{\bar{\sigma}}ij}\right)$$

$$\left. = \frac{3}{\sqrt{\widehat{\bar{\sigma}}'_{pq}\widehat{\bar{\sigma}}'_{qp}}}\left(\sqrt{6}t'_{\widehat{\bar{\sigma}}in} t'_{\widehat{\bar{\sigma}}nj} - \cos 3\theta_{\widehat{\bar{\sigma}}} t'_{\widehat{\bar{\sigma}}ij}\right)\right)$$

$$\boxed{\frac{\partial \widehat{\bar{\chi}}}{\partial \widehat{\bar{\boldsymbol{\sigma}}}'} = \frac{1}{\widehat{\overline{M}}}\left[\mathbf{t}'_{\widehat{\bar{\sigma}}} + \frac{3}{8 + \cos 3\theta_{\widehat{\bar{\sigma}}}}(\sqrt{6}\mathbf{t}'^2_{\widehat{\bar{\sigma}}} - \cos 3\theta_{\widehat{\bar{\sigma}}}\mathbf{t}'_{\widehat{\bar{\sigma}}})\right]} \tag{11.105}$$

$$
\left(\frac{\partial\widehat{\bar{\chi}}}{\partial\widehat{\bar{\sigma}}'_{ij}}=\frac{\partial\widehat{\bar{\chi}}}{\partial\sqrt{\widehat{\bar{\sigma}}'_{pq}\widehat{\bar{\sigma}}'_{qp}}}\frac{\partial\sqrt{\widehat{\bar{\sigma}}'_{pq}\widehat{\bar{\sigma}}'_{qp}}}{\partial\widehat{\bar{\sigma}}'_{ij}}+\frac{\partial\widehat{\bar{\chi}}}{\partial\widehat{\bar{M}}}\frac{\partial\widehat{\bar{M}}}{\partial\cos 3\theta_{\widehat{\bar{\sigma}}}}\frac{\partial\cos 3\theta_{\widehat{\bar{\sigma}}}}{\partial\widehat{\bar{\sigma}}'_{ij}}\right.
$$

$$
=\frac{1}{\widehat{\bar{M}}}\frac{\widehat{\bar{\sigma}}'_{ij}}{\sqrt{\widehat{\bar{\sigma}}'_{pq}\widehat{\bar{\sigma}}'_{qp}}}-\frac{\sqrt{\widehat{\bar{\sigma}}'_{pq}\widehat{\bar{\sigma}}'_{qp}}}{\widehat{\bar{M}}^2}\frac{\partial\widehat{\bar{M}}}{\partial\cos 3\theta_{\widehat{\bar{\sigma}}}}\frac{\partial\cos 3\theta_{\widehat{\bar{\sigma}}}}{\partial\widehat{\bar{\sigma}}'_{ij}}
$$

$$
\left.=\frac{1}{\widehat{\bar{M}}}\left[t'_{\widehat{\bar{\sigma}}ij}+\frac{3}{8+\cos 3\theta_{\widehat{\bar{\sigma}}}}\left(\sqrt{6}t'_{\widehat{\bar{\sigma}}ir}t'_{\widehat{\bar{\sigma}}rj}-\cos 3\theta_{\widehat{\bar{\sigma}}}t'_{\widehat{\bar{\sigma}}ij}\right)\right]\right)
$$

$$
\begin{aligned}
\frac{\partial f(\bar{p},\widehat{\bar{\chi}})}{\partial\boldsymbol{\sigma}}&=\frac{\partial f(\bar{p},\widehat{\bar{\chi}})}{\partial\bar{p}}\frac{\partial\bar{p}}{\partial\boldsymbol{\sigma}}+\frac{\partial f(\bar{p},\widehat{\bar{\chi}})}{\partial\widehat{\bar{\chi}}}\frac{\partial\widehat{\bar{\chi}}}{\partial\widehat{\bar{\boldsymbol{\sigma}}}'}\frac{\partial\widehat{\bar{\boldsymbol{\sigma}}}'}{\partial\boldsymbol{\sigma}}\\
&=-\frac{1}{3}\frac{\partial f(\bar{p},\widehat{\bar{\chi}})}{\partial\bar{p}}\mathbf{I}+\frac{\partial f(\bar{p},\widehat{\bar{\chi}})}{\partial\widehat{\bar{\chi}}}\frac{\partial\widehat{\bar{\chi}}}{\partial\widehat{\bar{\boldsymbol{\sigma}}}'}\left(\mathcal{I}'+\frac{1}{3}\boldsymbol{\beta}\otimes\mathbf{I}\right)\\
&=-\frac{1}{3}\frac{\partial f(\bar{p},\widehat{\bar{\chi}})}{\partial\bar{p}}\mathbf{I}+\frac{\partial f(\bar{p},\widehat{\bar{\chi}})}{\partial\widehat{\bar{\chi}}}\left\{\frac{\partial\widehat{\bar{\chi}}}{\partial\widehat{\bar{\boldsymbol{\sigma}}}'}-\frac{1}{3}\mathrm{tr}\left[\frac{\partial\widehat{\bar{\chi}}}{\partial\widehat{\bar{\boldsymbol{\sigma}}}'}(\mathbf{I}-\boldsymbol{\beta})\right]\mathbf{I}\right\}
\end{aligned}
\tag{11.106}
$$

$$
\left(\frac{\partial f(\bar{p},\widehat{\bar{\chi}})}{\partial\sigma_{ij}}=-\frac{1}{3}\frac{\partial f(\bar{p},\widehat{\bar{\chi}})}{\partial\bar{p}}\frac{\partial\bar{p}}{\partial\sigma_{ij}}+\frac{\partial f(\bar{p},\widehat{\bar{\chi}})}{\partial\widehat{\bar{\chi}}}\frac{\partial\widehat{\bar{\chi}}}{\partial\widehat{\bar{\sigma}}'_{rs}}\frac{\partial\widehat{\bar{\sigma}}'_{rs}}{\partial\sigma_{ij}}\right.
$$

$$
=-\frac{1}{3}\frac{\partial f(\bar{p},\widehat{\bar{\chi}})}{\partial\bar{p}}\frac{\partial\bar{p}}{\partial\sigma_{ij}}+\frac{\partial f(\bar{p},\widehat{\bar{\chi}})}{\partial\widehat{\bar{\chi}}}\frac{\partial\widehat{\bar{\chi}}}{\partial\widehat{\bar{\sigma}}'_{rs}}\left[\frac{1}{2}(\delta_{ri}\delta_{sj}+\delta_{rj}\delta_{si})-\frac{1}{3}\delta_{rs}\delta_{ij}+\frac{1}{3}\beta_{rs}\delta_{ij}\right]
$$

$$
\left.=-\frac{1}{3}\frac{\partial f(\bar{p},\widehat{\bar{\chi}})}{\partial\bar{p}}\delta_{ij}+\frac{\partial f(\bar{p},\widehat{\bar{\chi}})}{\partial\widehat{\bar{\chi}}}\frac{\partial\widehat{\bar{\chi}}}{\partial\widehat{\bar{\sigma}}'_{rs}}\left[\frac{1}{2}(\delta_{ri}\delta_{sj}+\delta_{rj}\delta_{si})-\frac{1}{3}\delta_{rs}\delta_{ij}+\frac{1}{3}\beta_{rs}\delta_{ij}\right]\right)
$$

$$
\boxed{\frac{\partial f(\bar{p},\widehat{\bar{\chi}})}{\partial\boldsymbol{\sigma}}=\begin{cases}-\dfrac{1}{3}\left[1-\left(\dfrac{\widehat{\bar{\chi}}}{\bar{p}}\right)^2\right]\mathbf{I}+2\dfrac{\widehat{\bar{\chi}}}{\bar{p}}\left\{\dfrac{\partial\widehat{\bar{\chi}}}{\partial\widehat{\bar{\boldsymbol{\sigma}}}'}-\dfrac{1}{3}\left[\dfrac{\partial\widehat{\bar{\chi}}}{\partial\widehat{\bar{\boldsymbol{\sigma}}}'}:(\mathbf{I}-\boldsymbol{\beta})\right]\mathbf{I}\right\} & \text{for}\quad \xi_h=0\\[2ex] -\dfrac{1}{3}\dfrac{1}{\widetilde{\xi}_h}\left(\dfrac{\bar{p}}{\widehat{\bar{p}}_\chi}-\widetilde{\xi}_h\right)\mathbf{I}+2\dfrac{\widehat{\bar{\chi}}}{\widehat{\bar{p}}_\chi}\left\{\dfrac{\partial\widehat{\bar{\chi}}}{\partial\widehat{\bar{\boldsymbol{\sigma}}}'}-\dfrac{1}{3}\left[\dfrac{\partial\widehat{\bar{\chi}}}{\partial\widehat{\bar{\boldsymbol{\sigma}}}'}:(\mathbf{I}-\boldsymbol{\beta})\right]\mathbf{I}\right\} & \text{for}\quad \xi_h\neq 0\end{cases}}
\tag{11.107}
$$

$$
\frac{\partial f(\bar{p},\widehat{\bar{\chi}})}{\partial\boldsymbol{\beta}}=-\bar{p}\frac{\partial f(\bar{p},\widehat{\bar{\chi}})}{\partial\widehat{\bar{\chi}}}\frac{\partial\widehat{\bar{\chi}}}{\partial\widehat{\bar{\boldsymbol{\sigma}}}'}
\tag{11.108}
$$

$$\left(\frac{\partial f(\bar{p}, \widehat{\bar{\chi}})}{\partial \beta_{ij}} = \frac{\partial f(\bar{p}, \widehat{\bar{\chi}})}{\partial \widehat{\bar{\chi}}} \frac{\partial \widehat{\bar{\chi}}}{\partial \widehat{\bar{\sigma}}'_{rs}} \frac{\partial \widehat{\bar{\sigma}}'_{rs}}{\partial \beta_{ij}} = \frac{\partial f(\bar{p}, \widehat{\bar{\chi}})}{\partial \widehat{\bar{\chi}}} \frac{\partial \widehat{\bar{\chi}}}{\partial \widehat{\bar{\sigma}}'_{rs}} (-\bar{p} \bar{I}_{rsij}) \right.$$

$$\left. = -\bar{p} \frac{\partial f(\bar{p}, \widehat{\bar{\chi}})}{\partial \widehat{\bar{\chi}}} \frac{\partial \widehat{\bar{\chi}}}{\partial \widehat{\bar{\sigma}}'_{rs}} \frac{1}{2} (\delta_{rj}\delta_{sj} + \delta_{rj}\delta_{si}) = -\bar{p} \frac{\partial f(\bar{p}, \widehat{\bar{\chi}})}{\partial \widehat{\bar{\chi}}} \frac{\partial \widehat{\bar{\chi}}}{\partial \widehat{\bar{\sigma}}'_{rs}} \right)$$

$$\boxed{\frac{\partial f(\bar{p}, \widehat{\bar{\chi}})}{\partial \boldsymbol{\beta}} = \begin{cases} -2\widehat{\bar{\chi}} \dfrac{\partial \widehat{\bar{\chi}}}{\partial \widehat{\bar{\boldsymbol{\sigma}}}'} & \text{for} \quad \xi_h = 0 \\ -2\dfrac{\bar{p}}{\widehat{\bar{p}}_\chi} \widehat{\bar{\chi}} \dfrac{\partial \widehat{\bar{\chi}}}{\partial \widehat{\bar{\boldsymbol{\sigma}}}'} & \text{for} \quad \xi_h \neq 0 \end{cases}} \tag{11.109}$$

$$f(\mathbf{c},\, \boldsymbol{\beta}) = f(p_c, \widehat{\chi}_c) = \begin{cases} p_c[1 + (\widehat{\chi}_c/p_c)^2] & \text{for} \quad \xi_h = 0 \\ \dfrac{1}{\bar{\xi}_h} (\widehat{p}_{\chi_c} - \bar{\xi}_h p_c) & \text{for} \quad \xi_h \neq 0 \end{cases} \tag{11.110}$$

where

$$p_c \equiv -\frac{1}{3} \mathrm{tr}\mathbf{c}, \quad \mathbf{c}' \equiv \mathbf{c} + p_c \mathbf{I} \tag{11.111}$$

$$\widehat{\mathbf{c}}' \equiv \mathbf{c}' - p_c \boldsymbol{\beta} \tag{11.112}$$

$$\widehat{\chi}_c \equiv \frac{\|\widehat{\mathbf{c}}'\|}{\widehat{M}_c} \tag{11.113}$$

$$\widehat{M}_c(\cos 3\theta_{\widehat{c}}) = \frac{14\sqrt{6} \sin \phi_c}{(3 - \sin \phi_c)(8 + \cos 3\theta_{\widehat{c}})} \tag{11.114}$$

$$\cos 3\theta_{\widehat{c}} \equiv \sqrt{6} \mathrm{tr} \mathbf{t}'^3_{\widehat{c}}, \quad \mathbf{t}'_{\widehat{c}} \equiv \frac{\widehat{\mathbf{c}}'}{\|\widehat{\mathbf{c}}'\|} \tag{11.115}$$

$$\widehat{p}_{\chi_c} \equiv \sqrt{p_c^2 + 2\bar{\xi}_h \widehat{\chi}_c^2} \tag{11.116}$$

$$\boxed{\frac{\partial \widehat{\chi}_c}{\partial \widehat{\mathbf{c}}'} = \frac{1}{\widehat{M}_c} \left[\mathbf{t}'_{\widehat{c}} + \frac{3}{8 + \cos 3\theta_{\widehat{c}}} \left(\sqrt{6} \mathbf{t}'^2_{\widehat{c}} - \cos 3\theta_{\widehat{c}} \mathbf{t}'_{\widehat{c}} \right) \right]} \tag{11.117}$$

$$\boxed{\frac{\partial f(\mathbf{c},\, \boldsymbol{\beta})}{\partial \boldsymbol{\beta}} = \begin{cases} -2\widehat{\chi}_c \dfrac{\partial \widehat{\chi}_c}{\partial \widehat{\mathbf{c}}'} & \text{for } \xi_h = 0 \\ -2\dfrac{p_c}{\widehat{p}_{\chi_c}} \widehat{\chi}_c \dfrac{\partial \widehat{\chi}_c}{\partial \mathbf{c}'} & \text{for } \xi_h \neq 0 \end{cases}} \tag{11.118}$$

$$\boxed{\frac{\partial f(\mathbf{c},\, \boldsymbol{\beta})}{\partial \mathbf{c}} = -\frac{1}{3}\frac{1}{\tilde{\xi}_h}\left(\frac{p}{\widehat{p}_{\chi_c}} - \bar{\xi}_h\right)\mathbf{I} + 2\frac{\widehat{\chi}_c}{\widehat{p}_{\chi_c}}\left\{\frac{\partial \widehat{\chi}_c}{\partial \widehat{\mathbf{c}}'} - \frac{1}{3}\left[\frac{\partial \widehat{\chi}_c}{\partial \widehat{\mathbf{c}}'} : (\mathbf{I} - \boldsymbol{\beta})\right]\mathbf{I}\right\}} \tag{11.119}$$

11.5.3 Calculation of Normal-Yield Ratio

The normal-yield ratio R must be calculated from the equation of the subloading surface in the unloading process. It can be calculated directly from $R = f(\boldsymbol{\sigma},\, \boldsymbol{\beta})/F$ in the initial subloading surface model but it has to be calculated by the numerical method for the extended subloading surface model.

(a) **Semi-analytical method**

In general, R is calculated numerically by solving the nonlinear equation obtained by substituting the current known values of $\boldsymbol{\sigma}$, $\boldsymbol{\beta}$ and $\mathbf{c}$ into Eq. (11.74) with Eq. (11.77), noting $\bar{\boldsymbol{\sigma}} = \tilde{\boldsymbol{\sigma}} + R\mathbf{c}$ in Eq. (9.6). The Newton-Raphson method would be useful for the calculation.

The other numerical method is shown here. First, one has

$$\bar{p} = -\frac{1}{3}\mathrm{tr}(\tilde{\boldsymbol{\sigma}} + R\mathbf{c}) = -(\tilde{\sigma}_m + Rc_m), \quad \bar{\boldsymbol{\sigma}}' = \tilde{\boldsymbol{\sigma}}' + R\mathbf{c}' \tag{11.120}$$

from Eq. (11.76). Substituting Eq. (11.120) into Eq. (11.80), one has

$$\widehat{\bar{\boldsymbol{\sigma}}}'(= \bar{\boldsymbol{\sigma}}' - \bar{p}\boldsymbol{\beta}) = \tilde{\boldsymbol{\sigma}}' + R\mathbf{c}' + \frac{1}{3}[\mathrm{tr}(\tilde{\boldsymbol{\sigma}} + R\mathbf{c})]\boldsymbol{\beta} = \tilde{\boldsymbol{\sigma}}' + R\mathbf{c}' + (\tilde{\sigma}_m + Rc_m)\boldsymbol{\beta} \tag{11.121}$$

$$\widehat{\bar{\chi}} \equiv \frac{\|\widehat{\bar{\boldsymbol{\sigma}}}'\|}{\widehat{M}} = \frac{\|\tilde{\boldsymbol{\sigma}}' + R\mathbf{c}' + (\tilde{\sigma}_m + Rc_m)\boldsymbol{\beta}\|}{\widehat{M}} \tag{11.122}$$

Further, substituting Eqs. (11.120)–(11.122) into Eq. (11.74) with Eq. (11.77) of the extended subloading surface, one has the following equation and can transform it in turn.

$$
\left.\begin{aligned}
&-(\tilde{\sigma}_m+Rc_m)\left[1+\left(\frac{\dfrac{\|\tilde{\boldsymbol{\sigma}}'+R\mathbf{c}'+(\tilde{\sigma}_m+Rc_m)\boldsymbol{\beta}\|}{\widehat{\overline{M}}}}{-(\tilde{\sigma}_m+Rc_m)}\right)^2\right]=RF \quad \text{for } \xi_h=0\\
&\frac{1}{\xi_h}\left\{\sqrt{(\tilde{\sigma}_m+Rc_m)^2+2\xi_h\left[\frac{\|\tilde{\boldsymbol{\sigma}}'+R\mathbf{c}'+(\tilde{\sigma}_m+Rc_m)\boldsymbol{\beta}\|}{\widehat{\overline{M}}}\right]^2}+\xi_h(\tilde{\sigma}_m+Rc_m)\right\}\\
&=RF \quad \text{for } \xi_h\neq 0
\end{aligned}\right\}
\tag{11.123}
$$

$$
\left.\begin{aligned}
&-(\tilde{\sigma}_m+Rc_m)^2-\left[\frac{\|(\tilde{\boldsymbol{\sigma}}'+R\mathbf{c}')+(\tilde{\sigma}_m+Rc_m)\boldsymbol{\beta}\|}{\widehat{\overline{M}}}\right]^2=(\tilde{\sigma}_m+Rc_m)RF\\
&\text{for } \xi_h=0\\
&\sqrt{[-(\tilde{\sigma}_m+Rc_m)]^2+2\tilde{\xi}_h\left[\frac{\|\tilde{\boldsymbol{\sigma}}'+R\mathbf{c}'+(\tilde{\sigma}_m+Rc_m)\boldsymbol{\beta}\|}{\widehat{\overline{M}}}\right]^2}\\
&=\tilde{\xi}_hRF-\bar{\xi}_h(\tilde{\sigma}_m+Rc_m) \quad \text{for } \xi_h\neq 0
\end{aligned}\right\}
$$

$$
\left.\begin{aligned}
&\widehat{\overline{M}}^2(\tilde{\sigma}_m+Rc_m)^2+\|\tilde{\boldsymbol{\sigma}}'+R\mathbf{c}'\|^2+2(\tilde{\sigma}_m+Rc_m)[(\tilde{\boldsymbol{\sigma}}'+R\mathbf{c}'):\boldsymbol{\beta}]\\
&\quad+(\tilde{\sigma}_m+Rc_m)^2\|\boldsymbol{\beta}\|^2+\widehat{\overline{M}}^2(\tilde{\sigma}_m+Rc_m)RF=0 \quad \text{for } \xi_h=0\\
&\widehat{\overline{M}}^2(\tilde{\sigma}_m+Rc_m)^2+2\tilde{\xi}_h\|\tilde{\boldsymbol{\sigma}}'+R\mathbf{c}'\|^2+4\tilde{\xi}_h(\tilde{\sigma}_m+Re_m)[(\tilde{\boldsymbol{\sigma}}'+R\mathbf{c}'):\boldsymbol{\beta}]\\
&\quad+2\tilde{\xi}(\tilde{\sigma}_m+Rc_m)^2\|\boldsymbol{\beta}\|^2-\{\widehat{\overline{M}}[\tilde{\xi}_hRF-\bar{\xi}_h(\tilde{\sigma}_m+Rc_m)]\}^2=0 \quad \text{for } \xi_h\neq 0
\end{aligned}\right\}
$$

$$
\left.\begin{aligned}
&\widehat{\overline{M}}^2\tilde{\sigma}_m^2+2\widehat{\overline{M}}^2\tilde{\sigma}_mc_mR+\widehat{\overline{M}}^2c_m^2R^2+\|\tilde{\boldsymbol{\sigma}}'\|^2+2(\tilde{\boldsymbol{\sigma}}':\mathbf{c}')R+\|\mathbf{c}'\|^2R^2\\
&\quad+2(\tilde{\sigma}_m+Rc_m)(\tilde{\boldsymbol{\sigma}}':\boldsymbol{\beta})+2(\tilde{\sigma}_mR+c_mR^2)(\mathbf{c}':\boldsymbol{\beta})\\
&\quad+(\tilde{\sigma}_m^2+2R\tilde{\sigma}_mc_m+c_m^2R^2)\|\boldsymbol{\beta}\|^2+\widehat{\overline{M}}^2(\tilde{\sigma}_mFR+c_mFR^2)=0\\
&\quad\text{for } \xi_h=0\\
&\widehat{\overline{M}}^2\tilde{\sigma}_m^2+2\widehat{\overline{M}}^2\tilde{\sigma}_mc_mR+\widehat{\overline{M}}^2c_m^2R^2+2\tilde{\xi}_h\|\tilde{\boldsymbol{\sigma}}'\|^2\\
&\quad+4\tilde{\xi}(\tilde{\boldsymbol{\sigma}}':\mathbf{c}')R+2\tilde{\xi}_h\|\mathbf{c}'\|^2R^2+4\tilde{\xi}_h(\tilde{\sigma}_m+c_mR)(\tilde{\boldsymbol{\sigma}}':\boldsymbol{\beta})\\
&\quad+4\tilde{\xi}(\tilde{\sigma}_mR+c_mR^2)(\mathbf{c}':\boldsymbol{\beta})+2\tilde{\xi}_h(\tilde{\sigma}_m^2+2\tilde{\sigma}_mc_mR+c_m^2R^2)\|\boldsymbol{\beta}\|^2\\
&\quad-\widehat{\overline{M}}^2[(\tilde{\xi}_hF-\bar{\xi}_hc_m)^2R^2-2\bar{\xi}_h\tilde{\sigma}_m(\tilde{\xi}_hF-\bar{\xi}_hc_m)R+\bar{\xi}_h^2\tilde{\sigma}_m^2]=0\\
&\quad\text{for } \xi_h\neq 0
\end{aligned}\right\}
$$

$$\left(\begin{array}{l}[\tilde{\xi}_h RF - \bar{\xi}_h(\tilde{\sigma}_m + Rc_m)]^2 = [(\tilde{\xi}_h F - \bar{\xi} c_m)R - \bar{\xi}_h\tilde{\sigma}_m]^2 \\ = (\tilde{\xi}_h F - \bar{\xi}_h c_m)^2 R^2 - 2\bar{\xi}_h\tilde{\sigma}_m(\tilde{\xi}_h F - \bar{\xi}_h c_m)R + \bar{\xi}_h^2\tilde{\sigma}_m^2\end{array}\right)$$

$$\left.\begin{array}{l}\overline{\widehat{M}}^2 c_m^2 R^2 + ||\mathbf{c}'||^2 R^2 + 2c_m(\mathbf{c}' : \boldsymbol{\beta})R^2 + c_m^2||\boldsymbol{\beta}||^2 R^2 + \overline{\widehat{M}}^2 c_m F R^2 \\ \quad + 2\overline{\widehat{M}}^2 \tilde{\sigma}_m c_m R + 2(\tilde{\boldsymbol{\sigma}}' : \mathbf{c}')R + 2e_m(\tilde{\boldsymbol{\sigma}}' : \boldsymbol{\beta})R \\ \quad + 2\tilde{\sigma}_m(\mathbf{c}' : \boldsymbol{\beta})R + \overline{\widehat{M}}^2 \tilde{\sigma}_m F R + 2\tilde{\sigma}_m c_m ||\boldsymbol{\beta}||^2 R \\ \quad + \overline{\widehat{M}}^2 \tilde{\sigma}_m^2 + ||\tilde{\boldsymbol{\sigma}}'||^2 + 2\tilde{\sigma}_m(\tilde{\boldsymbol{\sigma}}' : \boldsymbol{\beta}) + \tilde{\sigma}_m^2||\boldsymbol{\beta}||^2 = 0 \quad \text{for } \xi_h = 0 \\ \overline{\widehat{M}}^2 c_m^2 R^2 + 2\tilde{\xi}_h||\mathbf{c}'|^2 R^2 + 4\tilde{\xi}_h c_m(\mathbf{e}' : \boldsymbol{\beta})R^2 + 2\tilde{\xi}_h c_m^2||\boldsymbol{\beta}||^2 R^2 \\ \quad - \overline{\widehat{M}}^2 R^2(\tilde{\xi}_h F - \bar{\xi}_h c_m)^2 + 2\overline{\widehat{M}}^2 \tilde{\sigma}_m c_m R + 4\tilde{\xi}_h(\tilde{\boldsymbol{\sigma}}' : \mathbf{c}')R \\ \quad + 4\tilde{\xi}_h c_m(\tilde{\boldsymbol{\sigma}}' : \boldsymbol{\beta})R + 4\tilde{\xi}_h\tilde{\sigma}_m(\mathbf{c}' : \boldsymbol{\beta})R + 4\tilde{\xi}_h\tilde{\sigma}_m c_m||\boldsymbol{\beta}||^2 R \\ \quad + 2\overline{\widehat{M}}^2 \bar{\xi}_h\tilde{\sigma}_m(\tilde{\xi}_h F - \bar{\xi}_h e_m)R + \overline{\widehat{M}}^2 \tilde{\sigma}_m^2 + 2\tilde{\xi}_h||\tilde{\boldsymbol{\sigma}}'||^2 + 4\tilde{\xi}_h\tilde{\sigma}_m(\tilde{\boldsymbol{\sigma}}' : \boldsymbol{\beta}) \\ \quad + 2\tilde{\xi}_h\tilde{\sigma}_m^2||\boldsymbol{\beta}||^2 - \overline{\widehat{M}}^2 \bar{\xi}_h^2\tilde{\sigma}_m^2 = 0 \quad \text{for } \xi_h \neq 0\end{array}\right\}$$

$$\left.\begin{array}{l}[\overline{\widehat{M}}^2 c_m^2 + ||\mathbf{c}'||^2 + 2c_m(\mathbf{c}' : \boldsymbol{\beta}) + c_m^2||\boldsymbol{\beta}||^2 + \overline{\widehat{M}}^2 c_m F]R^2 \\ \quad + [2\overline{\widehat{M}}^2 \tilde{\sigma}_m c_m + 2(\tilde{\boldsymbol{\sigma}}' : \mathbf{c}') + 2c_m(\tilde{\boldsymbol{\sigma}}' : \boldsymbol{\beta}) + 2\tilde{\sigma}_m(\mathbf{c}' : \boldsymbol{\beta}) \\ \quad\quad + 2\tilde{\sigma}_m c_m||\boldsymbol{\beta}||^2 + \overline{\widehat{M}}^2 \tilde{\sigma}_m F]R + \overline{\widehat{M}}^2 \tilde{\sigma}_m^2 + ||\tilde{\boldsymbol{\sigma}}'||^2 \\ \quad\quad + 2\tilde{\sigma}_m(\tilde{\boldsymbol{\sigma}}' : \boldsymbol{\beta}) + \tilde{\sigma}_m^2||\boldsymbol{\beta}||^2 = 0 \quad \text{for } \xi_h = 0 \\ [\overline{\widehat{M}}^2 c_m^2 + 2\tilde{\xi}_h||\mathbf{c}'||^2 + 4\tilde{\xi}_h c_m(\mathbf{c}' : \boldsymbol{\beta}) + 2\tilde{\xi}_h c_m^2||\boldsymbol{\beta}||^2 + \overline{\widehat{M}}^2(\tilde{\xi}_h F - \tilde{\xi}_h c_m)^2]R^2 \\ \quad + 2[\overline{\widehat{M}}^2 \tilde{\sigma}_m c_m + 2\tilde{\xi}_h(\tilde{\boldsymbol{\sigma}}' : \mathbf{c}') + 2\tilde{\xi}_h e_m(\tilde{\boldsymbol{\sigma}}' : \boldsymbol{\beta}) + 2\tilde{\xi}_h\tilde{\sigma}_m(\mathbf{c}' : \boldsymbol{\beta}) \\ \quad + 2\tilde{\xi}_h\tilde{\sigma}_m c_m||\boldsymbol{\beta}||^2 + \overline{\widehat{M}}^2 \bar{\xi}_h\tilde{\sigma}_m(\tilde{\xi}_h F - \bar{\xi}_h c_m)]R + \overline{\widehat{M}}^2(1 - \bar{\xi}_h^2)\tilde{\sigma}_m^2 \\ \quad + 2\tilde{\xi}_h||\tilde{\boldsymbol{\sigma}}'||^2 + 4\tilde{\xi}_h\tilde{\sigma}_m(\tilde{\boldsymbol{\sigma}}' : \boldsymbol{\beta}) + 2\tilde{\xi}_h\tilde{\sigma}_m^2||\boldsymbol{\beta}||^2 = 0 \quad \text{for } \xi_h \neq 0\end{array}\right\}$$

Solving this quadratic equation, the normal-yield ratio R is expressed as follows:

$$\boxed{R = \begin{cases}\dfrac{\sqrt{B^2 - AC} - B}{A} & \text{for } \xi_h = 0 \\[2ex] \dfrac{\sqrt{\widetilde{B}^2 - \widetilde{A}\widetilde{C}} - \widetilde{B}}{\widetilde{A}} & \text{for } \xi_h \neq 0\end{cases}} \tag{11.124}$$

where

$$
\left.
\begin{aligned}
&A \equiv \widehat{\overline{M}}^2 c_m^2 + \|\mathbf{c}'\|^2 + 2c_m(\mathbf{c}':\boldsymbol{\beta}) + c_m^2\|\boldsymbol{\beta}\|^2 + \widehat{\overline{M}}^2 c_m F \\
&B \equiv 2\widehat{\overline{M}}^2 \tilde{\sigma}_m c_m + 2\mathrm{tr}(\tilde{\boldsymbol{\sigma}}'\mathbf{c}') + 2c_m(\tilde{\boldsymbol{\sigma}}':\boldsymbol{\beta}) + 2\tilde{\sigma}_m(\mathbf{c}':\boldsymbol{\beta}) \\
&\quad + 2\tilde{\sigma}_m c_m\|\boldsymbol{\beta}\|^2 + \widehat{\overline{M}}^2 \tilde{\sigma}_m F \\
&C \equiv \widehat{\overline{M}}^2 \tilde{\sigma}_m^2 + \|\tilde{\boldsymbol{\sigma}}'\|^2 + 2\tilde{\sigma}_m(\tilde{\boldsymbol{\sigma}}':\boldsymbol{\beta}) + \tilde{\boldsymbol{\sigma}}_m^2\|\boldsymbol{\beta}\|^2 \\
&\widetilde{A} \equiv \widehat{\overline{M}}^2 c_m^2 + 2\bar{\xi}_h\|\mathbf{c}'\|^2 + 4\bar{\xi}_h c_m(\mathbf{c}':\boldsymbol{\beta}) + 2\bar{\xi} c_m^2\|\boldsymbol{\beta}\|^2 \\
&\quad - \widehat{\overline{M}}(\bar{\xi}_h F - \bar{\xi}_h e_m)^2 \\
&\widetilde{B} \equiv \widehat{\overline{M}}^2 \tilde{\sigma}_m c_m + 2\bar{\xi}_h(\tilde{\boldsymbol{\sigma}}':\mathbf{c}') + 2\bar{\xi}_h c_m(\tilde{\boldsymbol{\sigma}}':\boldsymbol{\beta}) + 2\bar{\xi}_h\tilde{\sigma}_m(\mathbf{c}':\boldsymbol{\beta}) \\
&\quad + 2\bar{\xi}_h\tilde{\sigma}_m e_m\|\boldsymbol{\beta}\|^2 + \widehat{\overline{M}}^2 \bar{\xi}_h\tilde{\sigma}_m(\bar{\xi}_h F - \bar{\xi}_h e_m) \\
&\widetilde{C} \equiv \widehat{\overline{M}}^2 (1 - \bar{\xi}_h^2)\tilde{\sigma}_m^2 + 2\bar{\xi}_h\|\tilde{\boldsymbol{\sigma}}'\|^2 + 4\bar{\xi}_h\tilde{\sigma}_m(\tilde{\boldsymbol{\sigma}}':\boldsymbol{\beta}) + 2\bar{\xi}_h\tilde{\sigma}_m^2\|\beta\|^2
\end{aligned}
\right\} \tag{11.125}
$$

Explicit numerical calculation processes:

(1) First step (beginning of calculation):

Calculate the normal-yield ratio R by Eq. (11.124), substituting the trial value $\widehat{\overline{M}} = 2\sqrt{6}\sin\phi_c/3$ which is the average of $\widehat{\overline{M}} = 2\sqrt{6}\sin\phi_c/(3+\sin\phi_c)$ and $2\sqrt{6}\sin\phi_c/(3-\sin\phi_c)$.

(2) Second step:

Recalculate R by substituting the value

$$
\widehat{\overline{M}}(\cos 3\theta_{\widehat{\bar{\sigma}}})\left(=\frac{14\sqrt{6}\sin\phi_c}{(3-\sin\phi_c)(8+\cos 3\theta_{\widehat{\bar{\sigma}}})}\right)=\frac{14\sqrt{6}\sin\phi_c}{(3-\sin\phi_c)\left(8+\sqrt{6}\mathrm{tr}\mathbf{t}_{\widehat{\bar{\sigma}}}^3\right)} \tag{11.126}
$$

into Eq. (11.124), while the value of R obtained in the former step is used in Eq. (11.126).

(3) Repeat the process (2) until R will reaches the convergence within a prescribed tolerance.

(b) **Numerical method**

In what follows, the numerical calculation of R by the Newton-Raphson method is described below.

Consider the following equation of R given from Eq. $(11.123)_2$ for the subloading surface.

$$\mathrm{g}(R) = \frac{1}{\bar{\xi}_h}\left[\sqrt{(\tilde{\sigma}_m + Rc_m)^2 + 2\tilde{\xi}_h\left(\frac{\chi_c}{\widehat{\overline{M}}}\right)^2} + \bar{\xi}_h(\tilde{\sigma}_m + Rc_m)\right] - RF = 0 \quad (11.127)$$

where

$$\chi_c \equiv \left\|\tilde{\boldsymbol{\sigma}}' + R\mathbf{c}' + (\tilde{\sigma}_m + Rc_m)\boldsymbol{\beta}\right\| \quad (11.128)$$

Differentiation of Eq. (11.127) leads to

$$\frac{\partial g(R)}{\partial R} = \frac{1}{2\bar{\xi}_h}\left[(\tilde{\sigma}_m + Rc_m)^2 + 2\tilde{\xi}_h\left(\frac{\chi_c}{\widehat{\overline{M}}}\right)^2\right]^{-1/2}\left[2(\tilde{\sigma}_m + Rc_m)c_m + 4\tilde{\xi}_h\frac{\chi_c}{\widehat{\overline{M}}}\frac{\frac{\partial\chi_c}{\partial R}\widehat{\overline{M}} - \chi_c\frac{\partial\widehat{\overline{M}}}{\partial R}}{\widehat{\overline{M}}^2}\right] + \frac{\bar{\xi}_h}{\bar{\xi}_h}c_m - F \quad (11.129)$$

$$g'(R) = \frac{1}{\bar{\xi}_h}\left[(\tilde{\sigma}_m + Rc_m)^2 + 2\tilde{\xi}_h\left(\frac{\chi_c}{\widehat{\overline{M}}}\right)^2\right]^{-1/2}\left[(\tilde{\sigma}_m + Rc_m)c_m + 2\tilde{\xi}_h\chi_c - \frac{\frac{\partial\chi_c}{\partial R} - \frac{\chi_c}{\widehat{\overline{M}}}\frac{\partial\widehat{\overline{M}}}{\partial R}}{\widehat{\overline{M}}^2}\right] + \frac{\bar{\xi}_h}{\bar{\xi}_h}c_m - F \quad (11.130)$$

where

$$\frac{\partial\chi_c}{\partial R} = \chi_c^{-1}[\tilde{\boldsymbol{\sigma}}' + R\mathbf{c}' + (\tilde{\sigma}_m + Rc_m)\boldsymbol{\beta}] : (\mathbf{c}' + c_m\boldsymbol{\beta}) \quad (11.131)$$

$$\left(\frac{\partial\|\mathbf{T}\|}{\partial R} = \frac{\mathbf{T}}{\|\mathbf{T}\|} : \frac{\partial\mathbf{T}}{\partial R}\right)$$

$$\frac{\partial\widehat{\overline{M}}}{\partial R} = -\frac{3 - \sin\phi_c}{14\sqrt{6}\sin\phi_c}\widehat{\overline{M}}^2\left[\frac{3}{\|\widehat{\overline{\boldsymbol{\sigma}}}'\|}\left(\sqrt{6}\mathbf{t}^2_{\widehat{\overline{\sigma}}} - \cos 3\theta_{\widehat{\overline{\sigma}}}\mathbf{t}_{\widehat{\overline{\sigma}}}\right)\right] : (\mathbf{c}' - \boldsymbol{\beta}\mathrm{tr}\mathbf{c}) \quad (11.132)$$

noting

$$\begin{aligned}\frac{\partial\widehat{\overline{M}}}{\partial R} &= \frac{\partial\widehat{\overline{M}}}{\partial\cos 3\theta_{\widehat{\overline{\sigma}}}}\frac{\partial\cos 3\theta_{\widehat{\overline{\sigma}}}}{\partial\widehat{\overline{\boldsymbol{\sigma}}}'} : \frac{\partial\widehat{\overline{\boldsymbol{\sigma}}}'}{\partial R}\\ \frac{\partial\widehat{\overline{\boldsymbol{\sigma}}}'}{\partial R} &= \frac{\partial[(\tilde{\boldsymbol{\sigma}} + R\mathbf{c})' - \mathrm{tr}(\tilde{\boldsymbol{\sigma}} + R\mathbf{c})\boldsymbol{\beta}]}{\partial R} = \mathbf{c}' - \boldsymbol{\beta}\mathrm{tr}\mathbf{c}\end{aligned} \quad (11.133)$$

Here, noting

$$g'(R^n) = \frac{g(R^{n+1}) - g(R^n)}{R^{n+1} - R^n} = \frac{0 - g(R^n)}{R^{n+1} - R^n}$$

we have

$$\boxed{R^{n+1} = R^n - \frac{f(R^n)}{f'(R^n)}} \tag{11.134}$$

from which R is obtained by repeating calculation until Eq. (11.134) converges within a given tolerance. In addition, adopt the following equation instead of Eq. (11.134) if the convergence is not obtained.

$$R^{n+1} = R^n - \mu^n \frac{g(R^n)}{g'(R^n)} \tag{11.135}$$

Calculate first by setting $\mu^n = 1$. If the inequality

$$|g(R^{n+1})| < \left(1 - \frac{\mu^n}{4}\right)|g'(R^n)| \tag{11.136}$$

is fulfilled, continue the calculation. However, if Eq. (11.136) is not fulfilled, calculate by reducing μ^n to the half and repeat again the calculation by reducing μ^n to the further half until Eq. (11.136) is fulfilled.

The constitutive equation of soils based on the extended subloading surface model is described above in detail. The tangential-inelastic strain rate has to be incorporated for the analyses of non-proportional loading behavior as shown in the analyses by Hashiguchi and Tsutsumi (2001, 2003, 2006) and Khojastehpour and Hashiguchi (2004a, b).

11.6 Simulations of Test Results

Some simulations of test data are given below in order to show the capability of the subloading surface model to reproduce the real deformation behavior of soils (Hashiguchi and Chen 1998).

The simulation of the test data (after Saada and Bianchini 1989) for Hostun sand subjected to the drained triaxial compression with a constant lateral stress, which includes the unloading-reloading process, is shown in Fig. 11.19 where the material constants and the initial values are selected as follows:

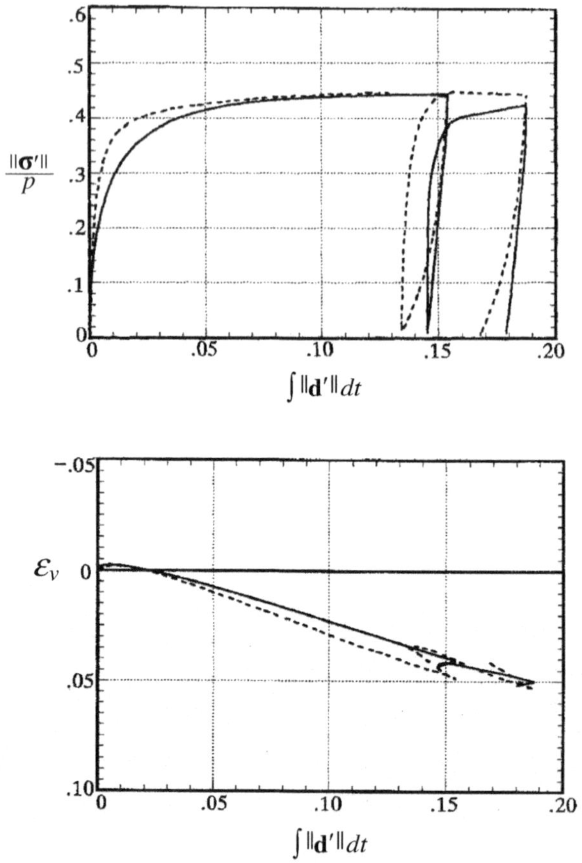

Fig. 11.19 Drained behavior of Hostun sand (data from Saada and Bianchini 1989). Measured and calculated results are shown by the *dashed* and *solid line*, respectively

Material constants:

Yield surface (ellipsoid): $\phi_c = 27^\circ$,

$$\text{Hardening/softening}\begin{cases}\text{isotropic}\begin{cases}\text{volumetric: } \tilde{\lambda} = 0.008,\ \tilde{\kappa} = 0.003,\ \vartheta = 0.025,\\ \text{deviatoric: } \mu_d = 0.6,\ \phi_d = 25^\circ,\end{cases}\\ \text{rotational: } b_r = 10,\ \phi_r = 20^\circ,\end{cases}$$

Evolution of normal-yield ratio: $u_1 = 1.5, m_1 = 3.8$,

Translationon of elastic-core: $c = 20$,

Elastic shear modulus: $G = 200{,}000$ kPa,

Initial values:

$$\text{Hardening function: } F_0 = 400\,\text{kPa},$$
$$\text{Rotational hardening variable: } \boldsymbol{\beta}_0 = \mathbf{O},$$
$$\text{Elastic-core: } \mathbf{c}_0 = -50\mathbf{I}\,\text{kPa},$$
$$\text{Stress: } \boldsymbol{\sigma}_0 = -100\mathbf{I}\ \text{kPa}$$

where the hyperbolic equation $U(R) = u_1(1/R^{m_1} - 1)$ (u_1, m_1 : material constants) is used for the evolution rule of the normal-yield ratio.

The simulation of the test data (after Saada and Bianchini 1989) for Hostun sand subjected to the drained proportional loading with $b(= (\sigma_2 - \sigma_3)/(\sigma_1 - \sigma_3)) = 0.666$ ($\theta_{\boldsymbol{\sigma}} = 19°09'$) from $\boldsymbol{\sigma}_0 = -500\mathbf{I}\,\text{kPa}$ by the true triaxial test apparatus is shown in Fig. 11.20. The material parameters are the same as those for the above-mentioned drained triaxial compression, while the sample was preliminarily loaded the isotropic compression from $\boldsymbol{\sigma} = -100\mathbf{I}\,\text{kPa}$ to $-500\mathbf{I}\,\text{kPa}$ before the test.

The simulations of the test data (after Castro 1969) for Banding sand subjected to the undrained triaxial compression with a constant lateral stress are shown in Fig. 11.21 where the material constants and the initial values are selected as follows:

Material constants:

$$\text{Yield surface (ellipsoid): } \phi_c = 26°,\ 30°,\ 31°,\ 32°,$$

$$\text{Hardening/softening}\begin{cases}\text{isotropic}\begin{cases}\text{volumetric}\begin{cases}\tilde{\lambda} = 0.025,\ 0.018,\ 0.014,\ 0.010,\\ \tilde{\kappa} = 0.0067,\ 0.0065,\ 0.0060,\ 0.0058,\\ \vartheta = 0, 0.021,\ 0.058, 0.138,\end{cases}\\ \text{deviatoric}\begin{cases}\mu_d = 1.00, 0.65, 0.30, 0.10,\\ \phi_d = 40°, 33°, 30°, 20°,\end{cases}\end{cases}\\ \text{rotational: } b_r = 10,\ \phi_r = 20°,\end{cases}$$

$$\text{Evolution of normal-yield ratio}\begin{cases}u_1 = 0.1, 0.3, 0.5, 1.0, 33.0,\\ m_1 = 0.1, 0.4, 0.5, 0, 7,\end{cases}$$

$$\text{Translationon of elastic-core: } c = 20,\ 18,\ 14,\ 8,$$
$$\text{Elastic shear modulus: } G = 18{,}000,\ 23{,}000,\ 25{,}000,\ 35{,}000\,\text{kPa},$$

Initial values:

$$\text{Hardening function: } F_0 = 410,\ 480,\ 520,\ 580\,\text{kPa},$$
$$\text{Rotational hardening variable: } \boldsymbol{\beta}_0 = \mathbf{O},$$
$$\text{Elastic-core: } \mathbf{c}_0 = -200,\ -110,\ -100,\ -80\mathbf{I}\,\text{kPa},$$
$$\text{Stress: } \boldsymbol{\sigma}_0 = -67.0\mathbf{I}\,\text{kPa}$$

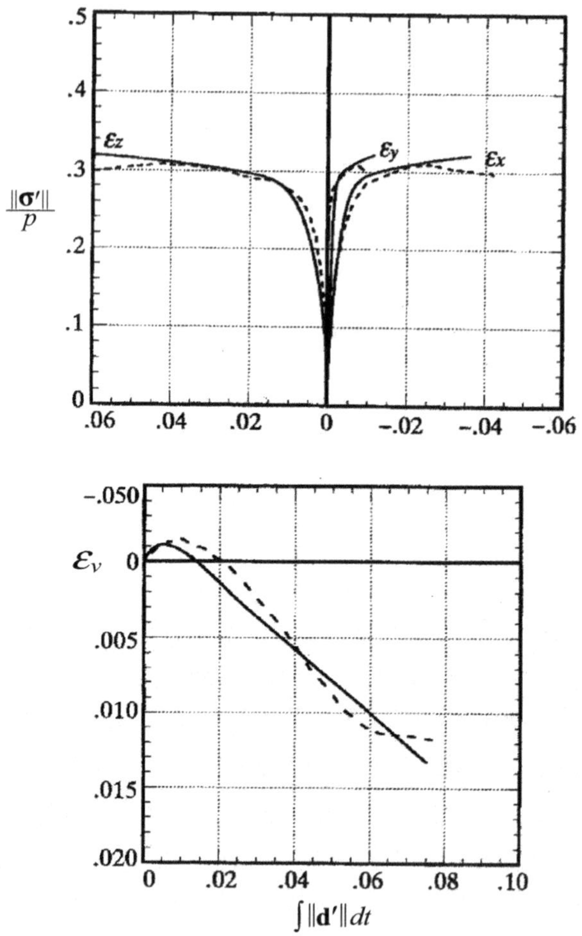

Fig. 11.20 Drained proportional loading behavior of Hostun sand (data from Saada and Bianchini 1989). Measured and calculated results are shown by the *dashed* and *solid line*, respectively

where the four values correspond to the initial relative densities $D_r = 0.29, 0.44, 0.47, 0.64$, respectively, in this order.

11.7 Description of Cyclic Mobility

Cyclic mobility occurring prior to the *liquefaction* in sands is a peculiar phenomenon exhibiting a butterfly-shaped stress loops and a *S*-shaped stress-strain loops under undrained cyclic loading. Accurate prediction of this phenomenon is of great importance for the earthquake-resistant design of soil structures. The description of the cyclic mobility has been studied after the Chile earthquake in

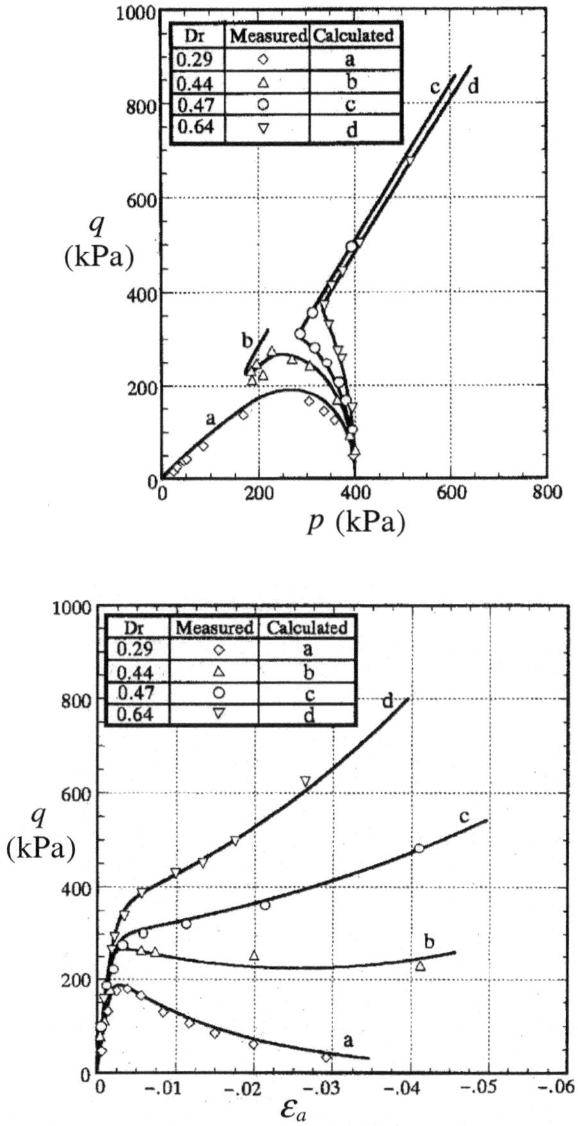

Fig. 11.21 Undrained behavior of banding sand (data from Castro 1969). Calculated results are shown by the *solid lines*

1960 and the Alaska and the Niigata earthquakes in 1964. However, it has not been attained up to recent years since the cyclic mobility is quite peculiar mechanical behavior the description of which requires the formulation of elastoplastic constitutive equation in a high level.

Various empirical nonlinear stress-strain relations (Fukutake et al. 1990; Commercial FEM code *FLIP* by Iai and Ohtsuki 2005) have been applied to the earthquake-resistant design of soil structures. However, they are inapplicable to the general undrained deformation behavior and the drained deformation behavior and possess various fundamental impertinences, e.g. the ignorance of loading path dependence, the difficulty of loading-unloading judgment with the generality and the description of irreversible deformation history because yield surface is not incorporated and the deformation is not decomposed to the elastic and the plastic parts. Needless to say, they cannot be accepted from the scientific viewpoint of mechanics.

The elastoplastic constitutive equation proposed by Oka et al. (1999) is installed to the computer program LIQCA for the liquefaction analysis. It incorporates the Drucker-Prager yield surface rotating with the current stress around the origin of stress apace into the original Cam-clay model and it ignores the influence of the third invariant of deviatoric stress. It possesses the following fundamental defects.

(1) The description of the monotonic loading behavior would be spoiled by the addition of the Drucker-Prager model, disturbing the Cam-clay model behavior.
(2) The deformation behavior in the triaxial compression and extension states cannot be described pertinently by a same set of material parameters because the influence of the third principal deviatoric stress invariant is ignored.
(3) The cyclic loading behavior cannot be described for the proportional cyclic loading in which the stress oscillates in the radial direction from the origin of stress space.

The constitutive equation based on the initial subloading surface model is applied to the liquefaction analysis by Zhang et al. (2007). The normal-yield and the subloading surfaces are assumed to become flat with their rotation and the influence of the third principal deviatoric stress invariant is ignored. Further, the rotational hardening rule was changed for the worse as was described in Sect. 11.4.3. It would spoil the description of the monotonic loading behavior and it cannot describe uniformly the deformation behavior in the triaxial compression and extension states. The cyclic mobility cannot be described rigorously by the initial subloading surface model which is limited to the description of monotonic loading behavior since the elastic-core is fixed as was described in the beginning of Sect. 8.3. Further, it should be noted that in the cyclic mobility the stress decreases to zero stress which is the elastic-core, i.e. similarity-center and thus the subloading surface is indeterminable in the initial subloading surface model as described in (1) in Sect. 11.4.1.

The extended subloading surface model has been applied to the description of the cyclic mobility by elaborating the material functions by Hashiguchi and Mase (2011) as will be deliberated in this section.

11.7.1 Physical Interpretation for Mechanism of Cyclic Mobility and Its Description by Subloading Surface Model

Physical background for the mechanism of cyclic mobility induced for the cyclic loading with the constant amplitude of deviatoric stress under the undrained condition in the compression/extension triaxial test is examined below together with how the cyclic mobility is predicted by the extended subloading surface model of soils formulated in the preceding sections. Deformation behavior in compression and extension sides are qualitatively identical so that only the behavior in the compression side will be described. Here, note that the axial compression strain rate $\dot{\varepsilon}_\alpha < 0$ and extension strain rate $\dot{\varepsilon}_\alpha > 0$ are induced in the compression side and extension side, respectively, divided by the central axis of the subloading surface as shown in Fig. 11.18. In addition, note that the decrease of effective pressure $dp < 0$ and the increase $dp > 0$ are induced to cause the elastic volume expansion $\mathrm{tr}\mathbf{d}^e > 0$ and the contraction $\mathrm{tr}\mathbf{d}^e < 0$ in order to maintain constant volume responding to the plastic volume contraction $\mathrm{tr}\mathbf{d}^p < 0$ induced inside the critical state line of the subloading surface, i.e. *looser than critical state* $(\mathrm{tr}\bar{\mathbf{n}} < 0)$ and the plastic volume expansion $\mathrm{tr}\mathbf{d}^p > 0$ induced outside, i.e. *denser than critical state* $(\mathrm{tr}\bar{\mathbf{n}} > 0)$, respectively. The looser and denser than critical states are abbreviated as LC and DC, respectively, in the following.

Process 1: *Cyclic loading of deviatoric stress under high effective confining pressure* (Fig. 11.22)

Deviatoric stress varies under a high effective pressure in the initial stage of cyclic loading so that the plastic volume contraction is induced leading to a denser arrangement of sand particles. To maintain the volume constant, elastic volume expansion is induced by the decrease of effective confining pressure. Then, the effective stress path moves gradually to the lower effective pressure, depicting the zigzag curves, as cyclic loading proceeds.

This process can be simulated by the subloading surface model as follows: The effective stress varies in the LC state $(\mathrm{tr}\bar{\mathbf{n}} < 0)$ causing the plastic volume contraction so that the effective pressure decreases responding to the elastic volume expansion in order to maintain the volume constant as represented at the point $\bar{\mathbf{a}}$ in Fig. 11.22. In this process, the volumetric hardening because of $\mathrm{tr}\bar{\mathbf{n}} < 0$ and the deviatoric softening because of $\|\boldsymbol{\sigma}'\|/p < M_d$ are induced. However, the deviatoric softening develops exceeding the volumetric hardening because of the increase of deviatoric strain rate as the stress approaches the critical state, and thus the normal-yield surface contracts.

Process 2: *Increase of deviatoric stress from state of low effective confining pressure* (Fig. 11.23)

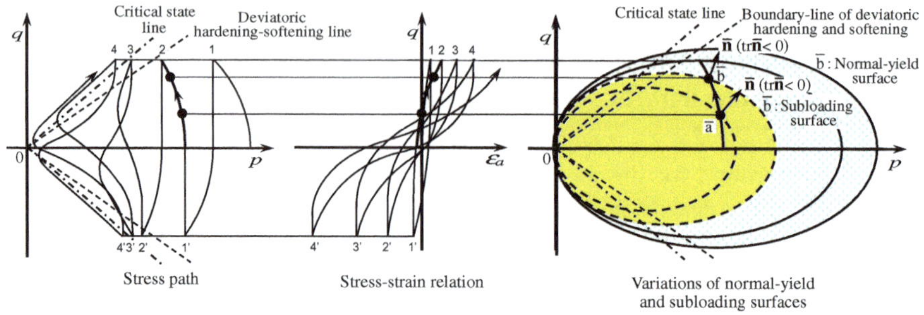

Fig. 11.22 Process 1: physical interpretation of cyclic loading behavior under high effective pressure

Arrangement of soil particles tends to looser so that the plastic volume expansion is induced by the application of deviatoric stress after the effective pressure decreased to a low level. Then, to maintain the volume constant, the elastic volume contraction must be induced by an increase of effective pressure. Eventually, the effective stress path goes up almost straight from the origin in the (p, q) plane (Castro 1969; Ishihara et al. 1975).

This process can be simulated by the subloading surface model as follows: After the effective pressure decreases as represented at the point $\bar{\mathbf{d}}$ in Fig. 11.23, the deviatoric stress increases over the critical sate line by the deviatoric hardening for $\|\boldsymbol{\sigma}'\|/p > \tan\phi_d(\phi_d \leq \phi_c)$ and reaches the DC state ($\mathrm{tr}\bar{\mathbf{n}} > 0$) causing the plastic volume expansion so that the effective pressure increases responding to the elastic volume contraction in order to maintain the volume constant.

The term *phase-transformation line* has been often used for the state that the effective pressure changes from decrease to increase after Ishihara et al. (1975). On the other hand, it can be interpreted by the elastoplasticity theory that the effective pressure changes from decrease to increase in the transition from the LC state to the DC state which is separated by the critical state line fulfilling $\mathrm{tr}\bar{\mathbf{n}} = 0\,(\mathrm{tr}\mathbf{d}^p = 0)$ on which plastic volume changes from the contraction to the expansion. Therefore, the phase-transformation line is nothing else the critical state line in the interpretation by the elastoplastic constitutive theory. Consequently, the term phase-transformation line would have to be unified to the critical state line hereinafter. In addition, the *inflection point* appears in the $q - \varepsilon_a$ curve on the critical state line as illustrated in Fig. 11.23 since the maximum deviatoric strain rate is induced on that line where $\mathrm{tr}\bar{\mathbf{n}} = 0$ and $\mathring{p} = 0$ hold leading to $d_v = d_v^e = d_v^p = 0, \mathbf{d} = \mathbf{d}'$, while q increases by the deviatoric hardening.

As the DC state with the deviatoric hardening because of $\|\boldsymbol{\sigma}'\|/p > M_c > M_d$ proceeds, the effective stress rises up at almost constant effective stress ratio as represented at the point $\bar{\mathbf{f}}$ in Fig. 11.23. Consequently, the effective stress path goes up straightly from the origin in the (p, q) plane. The normal-yield surface expands markedly so that the strain rate decreases gradually in this process. Then, the $q - \varepsilon_a$ curve gets warped to the upper as shown in this figure.

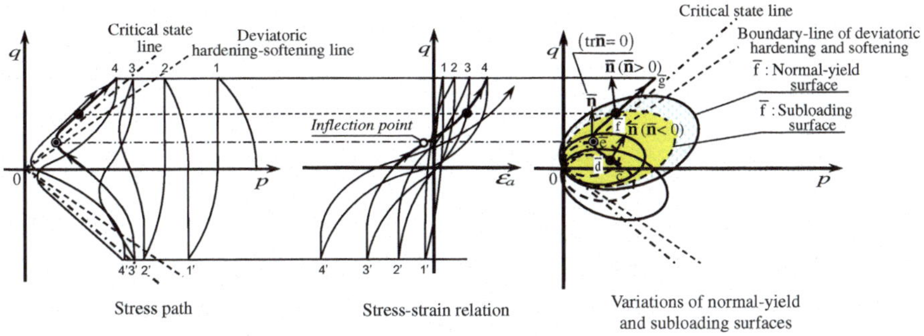

Fig. 11.23 Processes 2: increase of deviatoric stress from state of low effective confining pressure: physical interpretation of cyclic mobility, i.e. stress path of *butterfly-shape* and stress-strain relation of *S-shape*

Process 3: *Decrease of deviatoric stress from state of high effective stress ratio* (Fig. 11.24)

After the effective pressure increased to a high level and the plastic volume expansion proceeded much as the deviatoric stress increased in the process 2, a plastic volume contraction is induced as the deviatoric stress decreases. Here, to maintain the volume constant, the elastic volume expansion must be induced by the decrease of effective pressure. Then, the effective stress path goes down to the origin in the (p, q) plane (Castro 1969; Ishihara et al. 1975).

This process can be simulated by the subloading surface model as follows: The subloading surface contracts as the deviatoric stress decreases so that the unloading $(\mathbf{d}^p = \mathbf{O})$ proceeds without the elastic strain rate also because of the constant volume condition leading to the constant effective pressure in the initial stage of the decrease of deviatoric stress as represented at the point $\bar{\mathbf{h}}$ in Fig. 11.24. However, as the deviatoric stress decreases further, the subloading surface begins to expand, whilst the deviatoric softening because of $||\boldsymbol{\sigma}'||/p < M_d$ proceeds in the LC state resulting in the decrease of the effective pressure as represented at the point $\bar{\mathbf{i}}$ in Fig. 11.24. Here, the effective stress path goes down to the origin in the (p, q) plane. In this process, the deviatoric softening causes the contraction of normal-yield surface and thus the increase of strain rate so that the $q - \varepsilon_a$ curve gets warped to the upper as shown in Fig. 11.24. In this process, the effective stress passes through the critical state and reaches the DC state in the extension side. Thereafter, the identical phenomenon described in the process 2 is repeated.

Eventually, the butterfly-shaped effective stress path and the S-shaped stress-strain curve are described through the processes 2 and 3. Strain amplitude increases gradually since the effective pressure decreases with the cyclic loading.

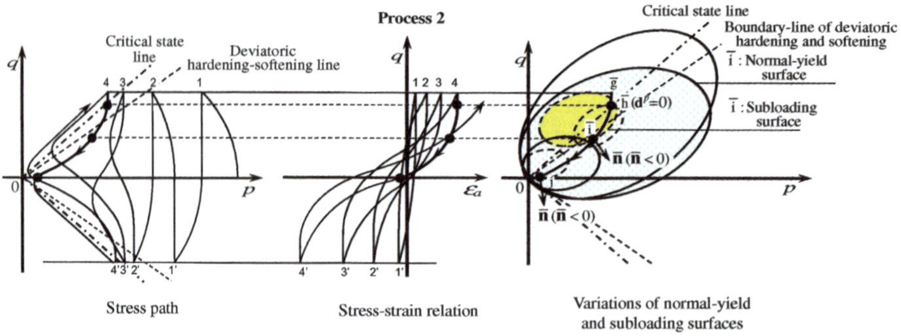

Fig. 11.24 Processes 3: decrease of deviatoric stress from state of high effective stress ratio: physical interpretation of cyclic mobility, i.e. stress path of *butterfly-shape* and stress-strain relation of *S-shape*

11.7.2 Material Functions

The material functions included in the subloading surface model described above will be formulated for the wide range of soils including sands in this section, modifying the past formulations (Hashiguchi 1995, 2002; Hashiguchi and Chen 1998; Hashiguchi and Mase 2007) in order to achieve a quantitative description of cyclic mobility.

(a) **Isotropic hardening**

The rate of isotropic hardening/softening variable H was extended to incorporate the influence of the deviatoric plastic strain rate in Sect. 11.4.2. Here, aiming at the quantitative description of the cyclic mobility, let it be further extended as follows:

$$\dot{H} = -d_v^p + d_s^p \tag{11.137}$$

$$d_v^p \equiv \mathrm{tr}\mathbf{d}^p,\ \mathbf{d}^{p\prime} \equiv \mathbf{d}^p - (d_v^p/3)\mathbf{I} \tag{11.138}$$

$$d_s^p = \mu_d \|\mathbf{d}^{p\prime}\| \frac{[\chi_d/(p+\zeta F)]^a - 1}{[\chi_d/(p+\zeta F)]^a - 1 + b} \tag{11.139}$$

$$\left(= \mu_d \|\mathbf{d}_{p'}\| \begin{cases} 1 & \text{for } \chi_d/(p+\zeta F) \to \infty \\ 0 & \text{for } \chi_d/(p+\zeta F) = 1 \\ -1(b-1)(<0) & \text{for } \chi_d/(p+\zeta F) = 0 \end{cases} \right) \tag{11.140}$$

$$\chi_d \equiv \frac{\|\boldsymbol{\sigma}'\|}{M_d} \tag{11.141}$$

$$M_d(\cos 3\theta_\sigma) \equiv \frac{14\sqrt{6}\sin\phi_d}{(3-\sin\phi_d)(8+\cos 3\theta_\sigma)} \tag{11.142}$$

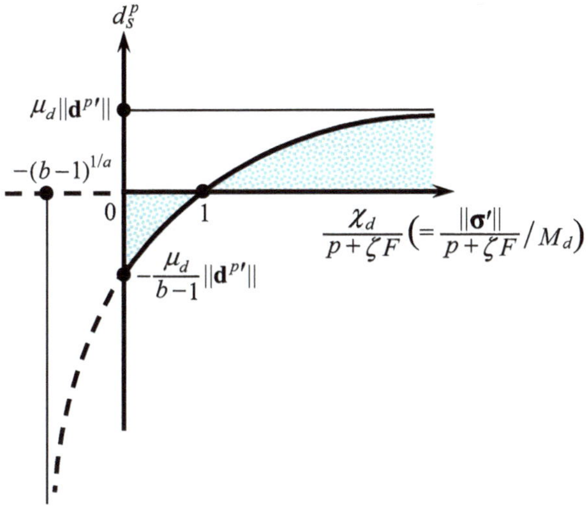

Fig. 11.25 Relation of deviatoric hardening strain rate versus modified stress ratio

$$\cos 3\theta_\sigma \equiv \sqrt{6}\mathrm{tr}\boldsymbol{\tau}^3, \quad \boldsymbol{\tau} \equiv \frac{\boldsymbol{\sigma}'}{\|\boldsymbol{\sigma}'\|} \tag{11.143}$$

$$h = -\mathrm{tr}\bar{\mathbf{n}} + \mu_d\|\bar{\mathbf{n}}'\| \frac{[\chi_d/(p+\zeta F)]^a - 1}{[\chi_d/(p+\zeta F)]^a - 1 + b} \tag{11.144}$$

where μ_d, ϕ_d, $a(\geq 1)$, $b(\geq 1)$ and $\zeta(\xi<\zeta<0.5)$ are material constants. The hardening and the softening are induced outside and inside, respectively, the conical surface $\|\boldsymbol{\sigma}'\| = M_d(p+\zeta F)$, i.e. the deviatoric hardening surface. The deviatoric hardening rate depends nonlinearly on the modified stress ratio $\chi_d/(p+\zeta F)$ (see Fig. 11.25).

The equation of stress ratio $\|\boldsymbol{\sigma}'\|/p$ in the past formulation (Hashiguchi and Chen 1998) engenders a singularity at $p=0$. Therefore, it cannot be applicable to the state of stress in non-positive pressure. Introducing a material constant $\zeta(>\xi)$ leading to $p+\zeta F>0$, $\|\boldsymbol{\sigma}'\|/p$ is replaced to the modified stress ratio $\|\boldsymbol{\sigma}'\|/(p+\zeta F)(>0)$ for which the singularity is not induced in the whole range of normal-yield surface including the zero/negative pressure reached in the cyclic mobility.

(b) **Evolution of normal-yield ratio**

The material parameter $\bar{u}$ in Eq. (9.48) for the normal-yield ratio is extended as follows:

$$\bar{u} = \frac{u_0}{\overset{\frown}{\overline{M}}{}^{\upsilon} \exp(\varsigma \varepsilon^{p\prime})} \tag{11.145}$$

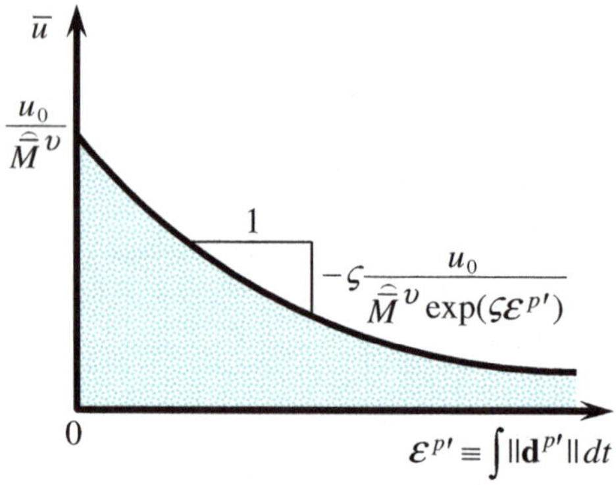

Fig. 11.26 Relation of material parameter versus accumulation of deviatoric plastic strain rate

where

$$\varepsilon^{p'} \equiv \int \|\mathbf{d}^{p'}\| dt \tag{11.146}$$

u_0, υ and ς are material constants (see Fig. 11.26). The material function $\bar{u}$ becomes smaller inducing a larger plastic strain rate as the accumulation of deviatoric plastic strain rate proceeds, and this trend is more remarkable for a larger value of ς. In addition, $\bar{u}$ is inversely proportional to the value of $\widehat{\overline{M}}$ so that the plastic strain rate is induced larger in the compression side than the tension side, and this trend is more remarkable for a larger value of υ.

11.7.3 Simulation of Cyclic Mobility

The constitutive equation of soils formulated in the previous sections is applied to the simulation of various test data of cyclic mobility in this section (Hashiguchi and Mase 2011).

(a) **Material parameters**

The material parameters included in the present model are shown below. Nineteen material constants and three initial values are included for the accurate prediction of cyclic mobility.

Material constants:

Yield surface (ellipsoid): ϕ_c, $\xi_h(<\vartheta)$

Isotropic hardening/softening: $\begin{cases} \text{volumetric: } \tilde{\lambda}, \tilde{\kappa}, \vartheta \\ \text{deviatoric: } \mu_d, \phi_d(<\phi_c), a(\geq 1), b(\geq 1), \zeta(\xi<\zeta<0.5) \end{cases}$

Anisotropic (rotational) hardening: b_r, ϕ_r

Normal-yield ratio: $u_0, u_s, \upsilon, \varsigma, n(\geq 1)$

Elastic-core: $c, \chi(<1)$

Poisson's ratio: v

Elastic shear modulus: G_0, η

Initial values:

Isotropic hardening function: F_0
Rotational hardening: $\boldsymbol{\beta}_0$
Elastic-core: $\mathbf{c}_0$

F_0, ξ, ϕ_c, u_0, c and G_0 are larger but ϕ_d, $\tilde{\lambda}$ and $\tilde{\kappa}$ are smaller in sands with higher strength which is usually observed for denser sands if arrangement of particles (same preparation of specimen in tests) is same.

Main influences of material constants on the deformation behavior are described below.

1. The transition from the elastic to plastic state is gentler for smaller value of u_0.
2. Strain is induced more intensely in the compression side for larger value of υ.
3. Strain rate increases more rapidly with the accumulated deviatoric plastic strain for larger value of ς.
4. The difference between the stiffness moduli in the reloading and the reverse loading is larger and thus the difference between the curvatures of the stress-strain curves in them is larger for larger value of u_s.
5. Plastic deformation begins sooner after unloading for larger value of c for which the closed hysteresis loop is depicted so that the strain accumulation is suppressed. On the other hand, the open hysteresis loop is depicted for $c = 0$.
6. The opening angle of the conical surface $\|\boldsymbol{\sigma}'\| = M_d(p + \zeta F)$, which regulates deviatoric hardening and softening induced its outside and inside, respectively, is larger for a larger value of ϕ_d, leading to the wider range of deviatoric softening.
7. The normal-yield and subloading surfaces rotate in a wider range for a larger value of ϕ_r.

(b) **Comparisons with test data**

All test data adopted for the simulations were obtained by the triaxial compression/extension tests with the symmetric constant deviatoric stress amplitudes from the isotropic stress state under constant total confining pressures denoted

Table 11.1 Physical properties of sands in test data used for simulations of liquefaction

Tested sands	Maximum void ratio e_{max}	Minimum void ratio e_{min}	Initial void ratio (relative density Dr (%))	Figure numbers
Edo river sand (Kiyota et al. 2009b)	1.132	0.714	0.762 (88.4)	Figure 11.27
Tone river sand (Kiyota et al. 2009a)	1.066	0.675	0.739 (83.7)	Figure 11.28
Toyoura sand (A) (Yamada et al. 2010)	0.985	0.639	0.718 (77.1)	Figure 11.29
			0.686 (86.3)	Figure 11.30
Toyoura sand (B) (Yamamoto 1998)	0.973	0.635	0.727 (72.7)	Figure 11.31
			0.672 (89.1)	Figure 11.32

by p_c. The initial isotropy, i.e. $\boldsymbol{\beta}_0 = \mathbf{O}$ is assumed and the common values $\nu = 0.3$ (Poisson ratio) and $\chi = 0.7$ (limit elastic-core yield ratio) are used for all test data in the present calculations. Physical properties of tested sands are shown in Table 11.1.

Stress paths in the (p, q) plane and stress-strain relations in the (q, ε_a) plane are compared between test results and calculated results. Test results and calculated results are depicted dotted by gray and black curve, respectively.

Results for Edo river sand (initial void ratio: 0.762, confining pressure: 160 kPa, stress ratio $q/p_c = 0.600$, cycle number: 20) after Kiyota et al. (2009b) are shown in Fig. 11.27. The material parameters were selected as follows:

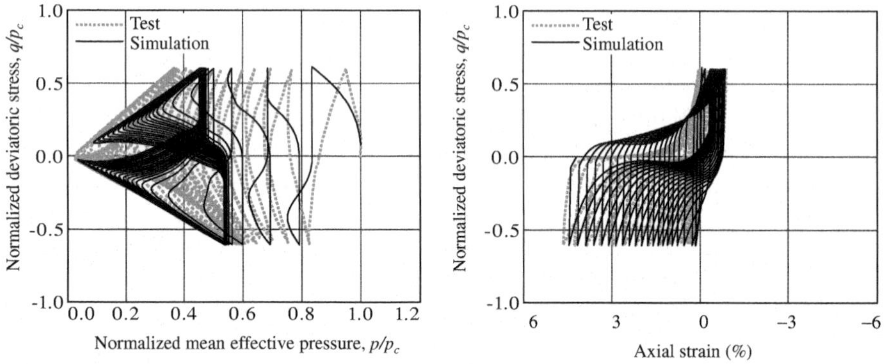

Fig. 11.27 Simulation of test data for the medium Edo river sand (initial void ratio: 0.762, confining pressure: 160 kPa, stress ratio $q/p_c = 0.600$, cycle number: 20) after Kiyota et al. (2009b)

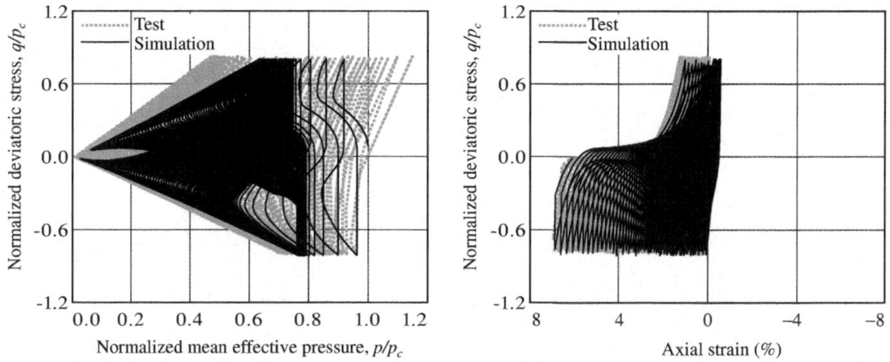

Fig. 11.28 Simulation of test data for the medium Tone river sand (initial void ratio: 0.739, confining pressure: 100 kPa, stress ratio $q/p_c = 0.800$, cycle number: 87) after Kiyota et al. (2009a)

$$
\begin{aligned}
&\phi_c = 32^\circ,\ \xi_h = 0.01;\ \tilde{\lambda} = 0.002,\ \tilde{\kappa} = 0.001,\ \vartheta = 0.043,\\
&\mu_d = 3,\ \phi_d = 22^\circ,\ a = 2.8,\ b = 22,\ \zeta = 0.08,\\
&b_r = 5\ \phi_b = 20^\circ,\\
&u_0 = 30,\ u_c = 2,\ \upsilon = 3.7,\ \varsigma = 1.17,\ n = 1,\\
&c = 20, \xi = 0.7,\\
&F_0 = 420\,\text{kPa},\ \mathbf{c}_0 = -60\mathbf{I}\ \text{kPa}
\end{aligned}
$$

Results for Tone river sand (initial void ratio: 0.739, confining pressure: 100 kPa, stress ratio $q/p_c = 0.800$, cycle number: 87) after Kiyota et al. (2009a) are shown in Fig. 11.28. The material parameters were selected as follows:

$$
\begin{aligned}
&\phi_c = 32^\circ,\ \xi_h = 0.05;\ \tilde{\lambda} = 0.002,\ \tilde{\kappa} = 0.001,\ \vartheta = 0.05,\\
&\mu_d = 3,\ \phi_d = 20^\circ,\ a = 3,\ b = 24,\ \zeta = 0.06,\\
&b_r = 1, \phi_b = 2^\circ,\\
&u_0 = 47,\ u_c = 1,\ \upsilon = 2.75,\ \varsigma = 0.95,\ n = 1,\\
&c = 20,\ \xi = 0.7,\\
&F_0 = 400\ \text{kPa},\ \mathbf{c}_0 = -60\mathbf{I}\,\text{kPa}
\end{aligned}
$$

Results for the medium Toyoura sand (A) (initial void ratio: 0.718, confining pressure: 98.1 kPa, stress ratio $q/p_c = 0.400$, cycle number: 7) after Yamada et al. (2010) are shown in Fig. 11.29. The material parameters were selected as follows:

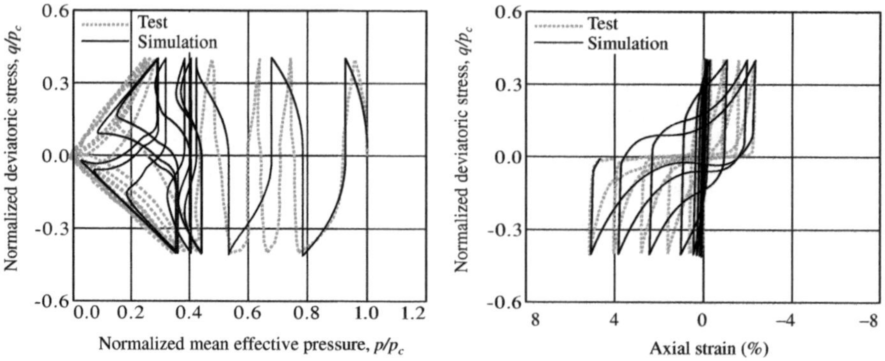

Fig. 11.29 Simulation of test data for the medium Toyoura sand (A) (initial void ratio: 0.718, confining pressure: 98.1 kPa, stress ratio $q/p_c = 0.400$, cycle number: 7) after Yamada et al. (2010)

$$
\begin{aligned}
&\phi_c = 30^\circ,\ \xi_h = 0.05;\ \tilde{\lambda} = 0.0065,\ \tilde{\kappa} = 0.0022,\ \vartheta = 0.102,\\
&\mu_d = 3,\ \phi_d = 25^\circ,\ a = 3,\ b = 14,\ \zeta = 0.06,\\
&b_r = 1,\ \phi_b = 20^\circ,\\
&u_0 = 16,\ u_c = 3,\ \xi = 0.7,\ \upsilon = 2.4,\ \varsigma = 2,\ n = 1,\\
&c = 13,\ \xi = 0.7,\\
&F_0 = 197\ \text{kPa},\ \mathbf{c}_0 = -10\mathbf{I}\ \text{kPa}
\end{aligned}
$$

Results for the dense Toyoura sand (A) (initial void ratio: 0.686, confining pressure: 300 kPa, stress ratio $q/p_c = 0.980$, cycle number: 2) after Yamada et al. (2010) are shown in Fig. 11.30. The material parameters were selected as follows:

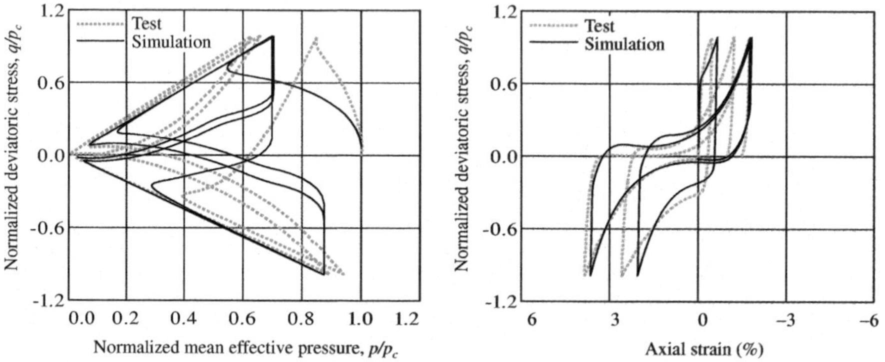

Fig. 11.30 Simulation of test data for the dense Toyoura sand (A) (initial void ratio: 0.686, confining pressure: 300 kPa, stress ratio $q/p_c = 0.980$, cycle number: 2) after Yamada et al. (2010)

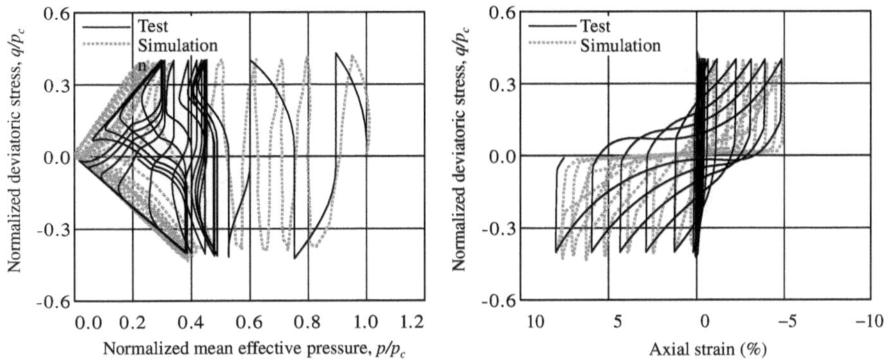

Fig. 11.31 Simulation of test data for the medium Toyoura sand (B) (initial void ratio: 0.727, confining pressure: 100 kPa, stress ratio $q/p_c = 0.400$, cycle number: 12) after Yamamoto (1998)

$$\begin{aligned}
&\phi_c = 30^\circ,\ \xi_h = 0.05;\ \tilde{\lambda} = 0.0065,\ \tilde{\kappa} = 0.001,\ \vartheta = 0.059,\\
&\mu_d = 3,\ \phi_d = 24^\circ,\ a = 3,\ b = 5,\ \zeta = 0.1,\\
&b_r = 5, \phi_b = 30^\circ,\\
&u_0 = 20,\ u_c = 3,\ \upsilon = 1,\ \varsigma = 7,\ n = 1,\\
&c = 40,\ \xi \equiv 0.7,\\
&F_0 = 340\ \text{kPa},\ \mathbf{c}_0 = -40\mathbf{I}\ \text{kPa}
\end{aligned}$$

Results for the medium Toyoura sand (B) (initial void ratio: 0.727, confining pressure: 100 kPa, stress ratio $q/p_c = 0.400$, cycle number: 12) after Yamamoto (1998) are shown in Fig. 11.31. The material parameters were selected as follows:

$$\begin{aligned}
&\phi_c = 30^\circ,\ \xi_h = 0.05;\ \tilde{\lambda} = 0.007,\ \tilde{\kappa} = 0.0013,\ \vartheta = 0.054,\\
&\mu_d = 3,\ \phi_d = 25^\circ,\ a = 3,\ b = 15,\ \zeta = 0.08,\\
&b_r = 2, \phi_b = 30^\circ,\\
&u_0 = 10,\ u_c = 3,\ \upsilon = 5,\ \varsigma = 3,\ n = 1,\\
&c = 20,\ \xi \equiv 0.7,\\
&F_0 = 370\ \text{kPa},\ \mathbf{c}_0 = -20\mathbf{I}\ \text{kPa}
\end{aligned}$$

Results for the dense Toyoura sand (B) (initial void ratio: 0.672, confining pressure: 100 kPa, stress ratio $q/p_c = 0.586$, cycle number: 29) after Yamamoto (1998) are shown in Fig. 11.32. The material parameters were selected as follows:

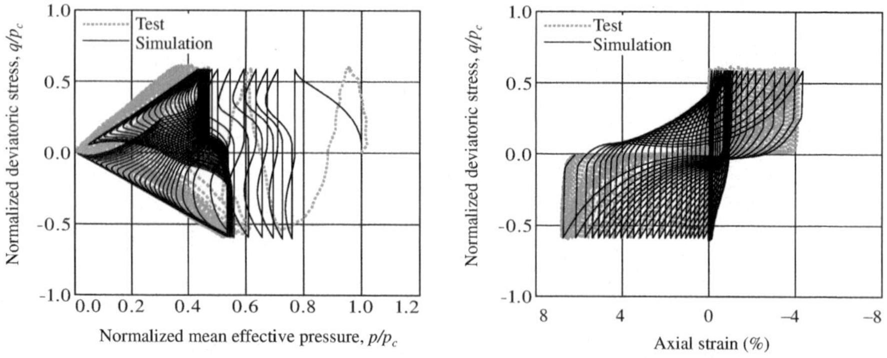

Fig. 11.32 Simulation of test data for the dense Toyoura sand (B) (initial void ratio: 0.672, confining pressure: 100 kPa, stress ratio $q/p_c = 0.586$, cycle number: 29) after Yamamoto (1998)

$$
\begin{aligned}
&\phi_c = 37^\circ,\ \xi_h = 0.05;\ \tilde{\lambda} = 0.007,\ \tilde{\kappa} = 0.0013,\ \vartheta = 0.038,\\
&\mu_d = 7,\ \phi_d = 30^\circ,\ a = 1,\ b = 15,\ \zeta = 0.07,\\
&b_r = 1,\ \phi_b = 33^\circ,\\
&u_0 = 70,\ u_c = 3,\ \upsilon = 6.85,\ \varsigma = 0.58,\ n = 1,\\
&c = 20,\ \xi \equiv 0.7,\\
&F_0 = 650\ \text{kPa},\ \mathbf{c}_0 = -50\mathbf{I}\ \text{kPa}
\end{aligned}
$$

The simulations of the drained and undrained monotonic loading behavior under the confining pressure 100 kPa are shown in Fig. 11.33, where the same sand in the same void ratio and thus same values of material parameters are used as in Fig. 11.32.

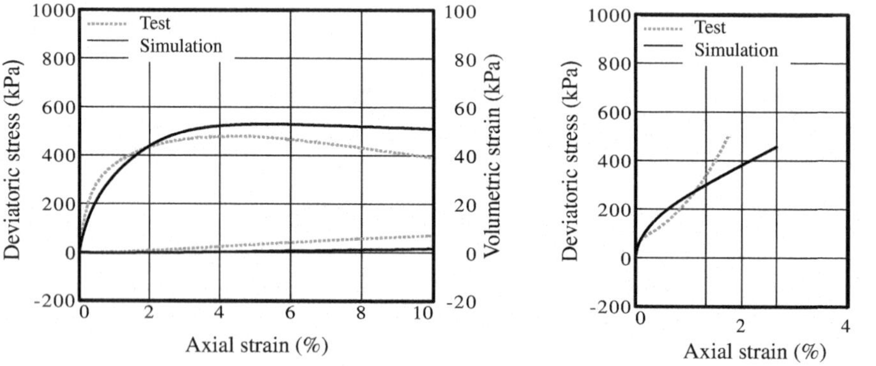

(a) Drained triaxial compression

(a) Undrained triaxial compression

Fig. 11.33 Simulation of test data for the monotonic loading behavior of same material shown in Fig. 11.32

11.8 Numerical Analysis of Footing Settlement Problem

Numerical analysis of footing settlement problem will be shown in this section (Mase and Hashiguchi 2009). The prediction of peak load and post-peak behavior for the footing-settlement problem on sands having the high friction and dilatancy cannot be attained in fact by the usual implicit finite element method requiring the repeated calculations of total stiffness equation which needs quite large calculation time. On the other hand, it can be attained by the explicit dynamic relaxation method in which the dynamic equilibrium equation is solved directly without solving the total stiffness equation so that the calculation time is drastically reduced. The *FLAC3D* (Fast Lagrangian Analysis of Continua in 3 Dimensions; Cundall and Board 1988; Itasca Consulting Group 2006) based on the explicit dynamic relaxation method is adopted in the present analysis, in which the initial subloading surface model of soils with the automatic controlling function to attract the stress to the normal-yield surface is implemented as the constitutive equation. The calculation is executed by the forward Euler method without iteration calculation for convergence in this program by adopting small incremental steps so as not to influence on the calculation, while this fact is examined prior to the calculation. The finite elements are composed of eight-noded cuboidal elements. Each cuboidal element is divided into the two kinds of overlays, i.e. assembly of five tetrahedral sub-elements having different directions. Then, the deviatoric variables are analyzed using individual values in each tetrahedral sub-element. On the other hand, isotropic variables are analyzed using averaging values in ten tetrahedral sub-elements in order to avoid the over-constraint problems common in finite element calculations for dilatant materials, i.e. the *dilatancy locking*.

Test data used for numerical simulation

The test data of footing settlement phenomenon on sand layers under the plane strain condition are used for the present analysis. The sizes of the test apparatus of type A (Tatsuoka et al. 1984) and type B (Tani 1986) have the same height 49 cm and depth 40 cm and the different widths 122 and 183 cm, respectively. The size of type C (Okahara et al. 1989) has the height 400 cm and depth 350 cm and the widths 7002 cm. The footings width, denoted as B, is taken 10 cm for the types A and B and 50 cm for the type C. The sand layers has been prepared carefully by the air-pluviation method for the dried Toyoura sand in order to obtain the same homogeneous layers but the test data exhibit dispersion more or less test by test despite of the laborious preparation work.

Numerical analysis and comparison with test data

The finite element meshes in the present analyses for the simulations of the test data are shown in Fig. 11.34. The nodal points of soil layer contacting with the footing and the bottom of soil bin are fixed to them, respectively. On the other hand, the nodal points at the side walls can move freely in the vertical direction. The right half of soil layer is analyzed in order to reduce the calculation time as has been done

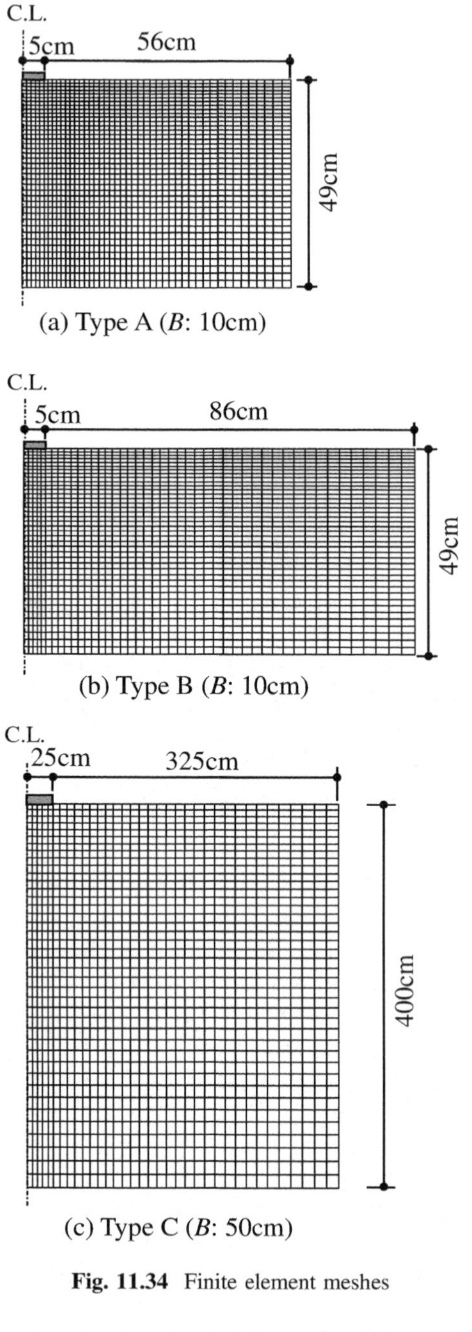

Fig. 11.34 Finite element meshes

widely even for searching the localized deformation (cf. e.g. Sloan and Randolph 1982; Pietruszczak and Niu 1993; Stallebrass and Taylor 1997; Borja and Tamagnini 1998; Siddiquee et al. 1999; de Borst and Groen 1999; Sheng et al.

2000; Borja et al. 2003). First, the analysis of deformation caused by the gravity force was performed. Then, the vertical displacement of footing is given by incremental steps of $10^{-5} \sim 5 \times 10^{-4}$ cm.

The material parameters in the subloading surface model are selected as

$$
\begin{aligned}
&F_0 = 350\ \text{kPa},\ \phi_c = 30^\circ,\ \xi_h = 0.001,\ \vartheta = 0.00003,\\
&\tilde{\lambda} = 0.0015,\ \tilde{\kappa} = 0.00015,\\
&\nu = 0.3,\\
&\phi_d = 29^\circ,\ \mu_d = 0.2\\
&u = 15.0,\ n = 1,
\end{aligned}
$$

where the logarithmic function in Eq. $(7.13)_1$ is used for the evolution rule of the normal-yield ratio in the subloading surface model. The values of material parameters listed above are used for all the following numerical calculations because Toyoura sands having the same initial void ratio 0.66 are used in these tests.

The comparisons of test and calculated results are shown in Fig. 11.35, where the prediction by Siddiquee et al. (1999) is also depicted in (c). In this figure q_m is the average footing pressure, γ_d is the unit dry weight, N_γ is the normalized footing pressure and S is the settlement. The qualitative trends of test results and the quantitative simulation to some extent are captured and the ultimate loads, i.e. bearing capacities are predicted well by the present analyses, although the analyses are performed for the sand with the high friction and dilatancy. Here, the post-peak behavior, i.e. the increase of load after exhibiting once the minimal value is also predicted well qualitatively. It would be provided by the adoption of the up-dated Lagrangian calculation realizing the accumulation of displacements by updating the positions of nodal points, which results in the upsurge of soils around the footing and thus the increase of footing load. However, the quantitative prediction of post-peak behavior would require the further study taking account of the tangential inelastic strain rate due to the stress rate tangential to the loading surface (Hashiguchi and Tsutsumi 2001) and the gradient effect (cf. Hashiguchi and Tsutsumi 2006) by introducing the shear-embedded model (cf. Pietruszczak and Mroz 1983; Tanaka and Kawamoto 1988; Tanaka and Sakai 1993) for example, which will be described in Sect. 14.3.

The displacements of nodal points from the initiation of settlement are shown in Fig. 11.36 at the settlement 11, 15, 80 mm for Type A, B and C, respectively, which are the final stage of calculation. The Prantdl's slip line solution with the triangle wedge, the logarithmic spiral zone and the passive Rankine zone is observed clearly in this figure.

On the other hand, the soils in the periphery of footing inevitably experience the null or further negative pressure since they are pulled into the vertical direction as the footing settlement proceeds (see Fig. 11.37). It causes the singularity of plastic modulus for the normal-yield surface passing through the origin of stress space at which the normal-yield and the subloading surfaces contact with each other. This

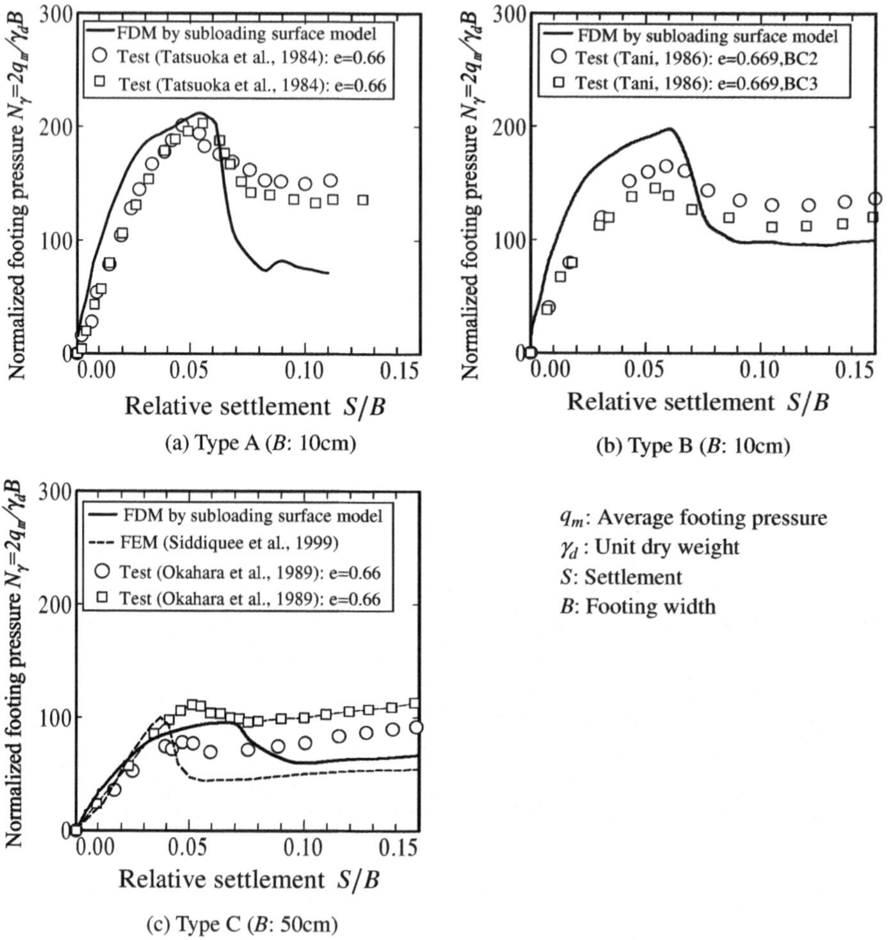

Fig. 11.35 Comparisons of test and calculated results for footing settlement phenomenon

defect is improved in the present model by making the normal-yield surface translate to the region of negative pressure as shown in Fig. 11.12, whilst the numerical difficulty can be avoided although the translation was taken quite small as 1/1000 in size of the normal-yield surface. In addition, the impertinence that the volume becomes infinite elastically is avoided by shifting the isotropic consolidation characteristic into the negative range of pressure as shown in Fig. 11.1. It should be emphasized that the stable analysis cannot be executed without these improvements.

The pertinent result for the footing-settlement problem on the sand with a high friction, one of the difficult problems in soil mechanics, is obtained in the present study as described above. Here, the peak, the subsequent reduction and the final increase of footing load are predicted well qualitatively and quantitatively to some extent. The reasons for succession are summarized as follows:

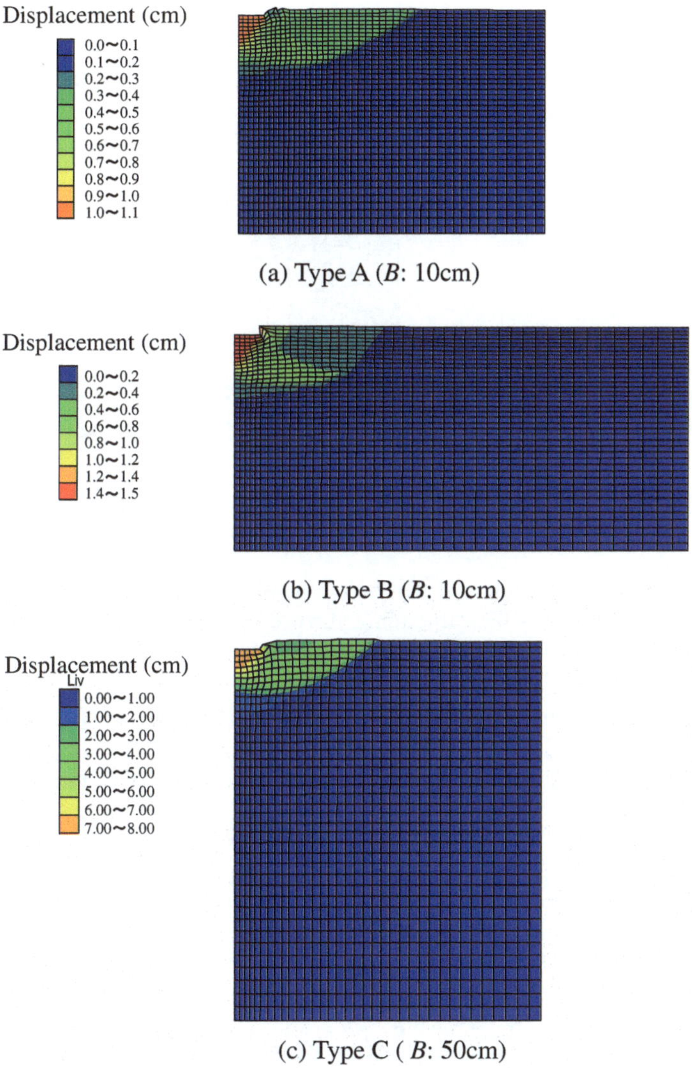

Fig. 11.36 Deformed finite element meshes at final step

(1) The subloading surface model applied in the present analysis has the advantages: (i) It is furnished with the automatic controlling function to attract the stress to the yield surface, whilst all other elastoplastic constitutive models are required to incorporate a return-mapping algorithm to pull back the stress to the yield surface in the plastic deformation process: (ii) It is capable of describing the softening behavior and dilatancy characteristics quite realistically, predicting the simultaneous occurrence of the peak load and the highest dilatancy rate as was found experimentally by Taylor (1948). (iii) It has the

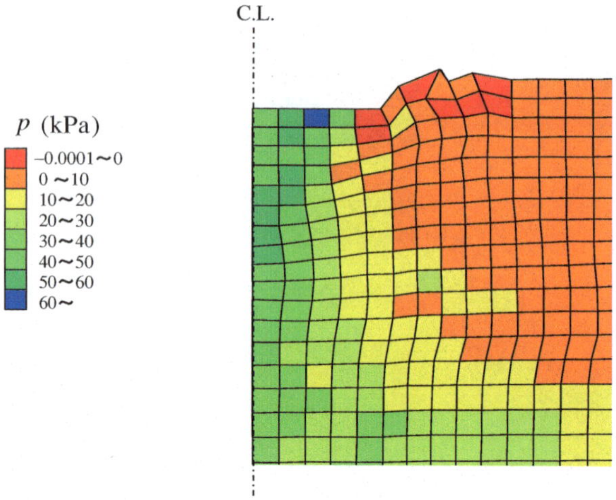

Fig. 11.37 Distribution of mean pressure for type A at final step

full regularity since the normal-yield surface does not pass through the zero stress point and thus the subloading surface is always determined uniquely. In addition, the elastic property is improved such that the elastic bulk modulus does not become zero for the stress inside the normal-yield surface.

(2) The finite difference program FLAC3D adopted in the present study is based on the explicit-relaxation method which enables us to shorten the calculation time drastically since it is not required to solve the total stiffness matrix.

11.9 On Isotropic Hardening Stagnation

The isotropic hardening stagnation in the cyclic loading process is not introduced for soils in the preceding sections. It will be formulated noting the equations for metals in Sect. 10.2 in the following.

The normal-isotropic hardening surface which is the conical surface stemming from the origin of stress space is given by

$$g(\tilde{\boldsymbol{\beta}}) = \tilde{K}_\beta \tag{11.147}$$

where

$$\tilde{\boldsymbol{\beta}} \equiv \boldsymbol{\beta} - \boldsymbol{\gamma} \tag{11.148}$$

$\tilde{K}_\beta$ and $\boldsymbol{\gamma}(= \boldsymbol{\gamma}\prime)$ designate the size and the center, respectively, of the normal isotropic hardening surface. The subloading-isotropic hardening surface is given as follows:

$$\boxed{g(\tilde{\boldsymbol{\beta}}) = \tilde{R}\tilde{K}_\beta} \tag{11.149}$$

where $\tilde{R}\,(0 \le \tilde{R} \le 1)$ is the normal-isotropic hardening ratio. It is calculated from the equation $\tilde{R} = g(\tilde{\boldsymbol{\beta}})/\tilde{K}_\beta$ in terms of the known values $\boldsymbol{\beta}, \boldsymbol{\gamma}$ and $\tilde{K}_\beta$.

The evolution rules of $\tilde{K}_\beta$ and $\boldsymbol{\gamma}$ are given anologously to Eqs. (10.21) and (10.22), as follow:

$$\boxed{\dot{\tilde{K}}_\beta = C\tilde{R}^{\varsigma}\,\langle\, \tilde{\mathbf{n}}_\beta : \overset{\circ}{\boldsymbol{\beta}}\,\rangle \left\| \frac{\partial g(\tilde{\boldsymbol{\beta}})}{\partial \tilde{\boldsymbol{\beta}}} \right\|} \tag{11.150}$$

$$\boxed{\overset{\circ}{\boldsymbol{\gamma}} = (1 - C)\tilde{R}^{\varsigma}\,\langle \tilde{\mathbf{n}}_\beta : \overset{\circ}{\boldsymbol{\beta}}\,\rangle\, \tilde{\mathbf{n}}_\beta} \tag{11.151}$$

where

$$\tilde{\mathbf{n}}_\beta \equiv \frac{\partial g(\tilde{\boldsymbol{\beta}})}{\partial \tilde{\boldsymbol{\beta}}} / \left|\left| \frac{\partial g(\tilde{\boldsymbol{\beta}})}{\partial \tilde{\boldsymbol{\beta}}} \right|\right| \tag{11.152}$$

The isotropic hardening is given analogously to Eq. (10.28), noting Eqs. (11.83) and (11.85), as follows:

$$\boxed{\dot{H} = \tilde{R}^{\upsilon} \left\langle \tilde{\mathbf{n}}_\beta \mathbf{:} \frac{\bar{\mathbf{f}}_{\beta n}}{||\bar{\mathbf{f}}_{\beta n}||} \right\rangle f_{Hn}\, \dot{\bar{\lambda}}} \tag{11.153}$$

where υ is the material constant.

The plastic modulus is given by taking account of Eq. (11.153) into Eq. (11.94) as follows:

$$\bar{M}^p \equiv \bar{\mathbf{n}} : \left[\frac{F'}{F} \tilde{R}^{\upsilon} \langle \tilde{\mathbf{n}}_\beta \mathbf{:} \frac{\bar{\mathbf{f}}_{\beta n}}{||\bar{\mathbf{f}}_{\beta n}||} \rangle f_{Hn} \bar{\boldsymbol{\sigma}} + (1 - R)\bar{\mathbf{f}}_{cn} - \frac{1}{RF} \left(\frac{\partial f(\bar{\boldsymbol{\sigma}},\ \boldsymbol{\beta})}{\partial \boldsymbol{\beta}} \mathbf{:} \bar{\mathbf{f}}_{\beta n} \right) \bar{\boldsymbol{\sigma}} + \frac{U}{R} \tilde{\boldsymbol{\sigma}} \right] \tag{11.154}$$

The function $g(\tilde{\boldsymbol{\beta}})$ depending on the third deviatoric invariant is given as follows:

$$g(\tilde{\boldsymbol{\beta}}) = \widetilde{G}_\beta(\cos 3\theta_{\tilde{\beta}})||\tilde{\boldsymbol{\beta}}|| \tag{11.155}$$

leading to the following equation of the normal-isotropic hardening surface.

$$||\tilde{\boldsymbol{\beta}}|| = \frac{\tilde{K}_\beta}{\widetilde{G}_\beta(\cos 3\theta_{\tilde{\beta}})} \tag{11.156}$$

where

$$\cos 3\theta_{\tilde{\beta}} \equiv \sqrt{6}\mathrm{tr}\mathbf{t}_{\tilde{\beta}}^{\prime 3},\ \mathbf{t}_{\tilde{\beta}}^{\prime} \equiv \frac{\tilde{\boldsymbol{\beta}}}{||\tilde{\boldsymbol{\beta}}||} \tag{11.157}$$

An example of $\widetilde{G}_{\beta}(\boldsymbol{cos}\, 3\theta_{\tilde{\beta}})$ is given noting Eq. (11.40) as

$$\widetilde{G}_{\beta}(\cos 3\theta_{\tilde{\beta}}) = 1 + (1/8)\cos 3\theta_{\tilde{\beta}} \tag{11.158}$$

The partial differential $\partial g(\tilde{\boldsymbol{\beta}})/\partial\tilde{\boldsymbol{\beta}}$ is given exploiting Eq. (1.302) as

$$\frac{\partial g(\tilde{\boldsymbol{\beta}})}{\partial\tilde{\boldsymbol{\beta}}} = \frac{\partial \widetilde{G}_{\beta}(\cos 3\theta_{\tilde{\beta}})}{\partial\tilde{\boldsymbol{\beta}}}||\tilde{\boldsymbol{\beta}}|| + \widetilde{G}_{\beta}(\cos 3\theta_{\tilde{\beta}})\frac{\tilde{\boldsymbol{\beta}}}{||\tilde{\boldsymbol{\beta}}||} \tag{11.159}$$

for Eq. (11.155), which is reduced for Eq. (11.158) to

$$\frac{\partial g(\tilde{\boldsymbol{\beta}})}{\partial\tilde{\boldsymbol{\beta}}} = (3\sqrt{6}/8)||\tilde{\boldsymbol{\beta}}||\mathbf{t}_{\tilde{\beta}}^{\prime 2} + [1 + (1/8)\cos 3\theta_{\tilde{\beta}}]||\tilde{\boldsymbol{\beta}}|| \tag{11.160}$$

It is expected that the prediction of cyclic loading behavior including the liquefaction will be improved by incorporating the isotropic hardening stagnation.

11.10 Hyperelastic Constitutive Equation of Soils

The hyperelastic equation of soils will be provided below within the frameworks of the infinitesimal and multiplicative finite strain theories.

(a) **Infinitesimal strain theory**

The isotropic hyperelastic constitutive equation for soils independent of the third invariant of stress and strain is given by

$$\boldsymbol{\sigma} = \frac{\partial\psi(\varepsilon_v^e, \varepsilon_d^e)}{\partial\boldsymbol{\varepsilon}^e} = \frac{\partial\psi(\varepsilon_v^e, \varepsilon_d^e)}{\partial\varepsilon_v^e}\frac{\partial\varepsilon_v^e}{\partial\boldsymbol{\varepsilon}^e} + \frac{\partial\psi(\varepsilon_v^e, \varepsilon_d^e)}{\partial\varepsilon_d^e}\frac{\partial\varepsilon_d^e}{\partial\boldsymbol{\varepsilon}^e} \tag{11.161}$$

with

$$\frac{\partial\varepsilon_v^e}{\partial\boldsymbol{\varepsilon}^e} = \frac{\partial\mathrm{tr}\boldsymbol{\varepsilon}^e}{\partial\boldsymbol{\varepsilon}^e} = \mathbf{I},\ \frac{\partial\varepsilon_d^e}{\partial\boldsymbol{\varepsilon}^e} = \frac{\boldsymbol{\varepsilon}^{e\prime}}{||\boldsymbol{\varepsilon}^{e\prime}||} = \frac{\boldsymbol{\varepsilon}^{e\prime}}{\varepsilon_d^e} \tag{11.162}$$

where

$$\varepsilon_v^e \equiv \mathrm{tr}\boldsymbol{\varepsilon}^e, \quad \varepsilon_d^e \equiv ||\boldsymbol{\varepsilon}^{e'}|| \tag{11.163}$$

Equation (11.161) is rewritten by Eq. (11.162) as

$$\boldsymbol{\sigma} = \frac{\partial\psi(\varepsilon_v^e, \varepsilon_d^e)}{\partial\varepsilon_v^e}\mathbf{I} + \frac{\partial\psi(\varepsilon_v^e, \varepsilon_d^e)}{\partial\varepsilon_d^e}\frac{\boldsymbol{\varepsilon}^{e'}}{\varepsilon_d^e} \tag{11.164}$$

The strain energy function of soils was first proposed by Houlsby (1985) and subsequently modified by Borja and Tamagnini (1998), Tamagnini et al. (2002), in which the $\ln v - \ln p$ linear relation in Eq. (11.1) under $p_e = 0$ in Sect. 11.1 is incorporated and the dependence of the shear modulus on the pressure is taken account. Mechanical and mathematical properties of this model have been studied by Callari et al. (1998), Niemunis and Cudny (1998), Houlsby et al. (2005), Amorosi et al. (2007), etc., while they have been formulated in terms of the Hencky (logarithmic) strain in Eq. (2.68) in Sect. 2.4 for the multiplicative elastoplasticity, which is described in the principal direction and thus it is unfortunately premised on the co-axiality limited to the isotropic constitutive equation.

The pertinent hyperlatic relation of soils will be formulated in the following. Firstly, we incorporate the following strain energy function based of Eqs. $(11.6)_1$ and $(11.8)_2$.

$$\boxed{\begin{aligned}\psi(\varepsilon_v^e, \varepsilon_d^e) = \vartheta F\varepsilon_v^e + \left[\tilde{\kappa}(p_0 + \vartheta F_0) + \exp - \left(\frac{\varepsilon_c^e}{\tilde{\kappa}}\right)\right] \\ + G_0\left[n\left(-\frac{\varepsilon_v^e}{\tilde{\kappa}}\right)\right]\varepsilon_d^{e2}\end{aligned}} \tag{11.165}$$

G_0 and η are material constants defined in Eq. (11.8), while the latter designates the pressure-dependence of shear modulus. It follows for Eq. (11.165) that

$$\left.\begin{aligned}\frac{\partial\psi(\varepsilon_v^e, \varepsilon_d^e)}{\partial\varepsilon_v^e} = \vartheta F - (p_0 + \vartheta F_0)\exp\left(-\frac{\varepsilon_v^e}{\tilde{\kappa}}\right)\left(\frac{n}{\tilde{\kappa}}G_0\exp\left[\mathrm{n}\left(-\frac{\varepsilon_\mathrm{v}^\mathrm{e}}{\kappa}\right)\right]\varepsilon_\mathrm{d}^\mathrm{e2}\right. \\ \frac{\partial\psi(\varepsilon_v^e, \varepsilon_d^e)}{\partial\varepsilon_d^e} = 2G_0\exp\left[n\left(-\frac{\varepsilon_v^e}{\tilde{\kappa}}\right)\right]\varepsilon_d^{e2}\end{aligned}\right\} \tag{11.166}$$

Substituting Eq. (11.166) into Eq. (11.164) leads to

$$\boxed{\boldsymbol{\sigma} = \left[\vartheta F - (p_0 + \vartheta F_0)\exp -\frac{\varepsilon_v^e}{\tilde{\kappa}} - \frac{n}{\tilde{\kappa}}G_0\exp\left[n\left(-\frac{\varepsilon_v^e}{\tilde{\kappa}}\right)\right]\varepsilon_d^{e2} + 2G_0\exp\left(-\frac{\varepsilon_v^e}{\tilde{\kappa}}\right)\right]\boldsymbol{\varepsilon}^{e'}} \tag{11.167}$$

The hydrostatic and deviatoric parts in Eq. (11.167) are given by

$$\frac{p+\vartheta F}{p_0+\vartheta F_0}=\exp\left(-\frac{\varepsilon_v^e}{\tilde{\kappa}}\right) \text{ for } \varepsilon^{e'}=\mathbf{O} \tag{11.168}$$

i.e.

$$\varepsilon_v^e=-\tilde{\kappa}\ln\left(\frac{p+\vartheta F}{p_0+\vartheta F_0}\right) \text{ for } \varepsilon^{e'}=\mathbf{O} \tag{11.169}$$

$$\boldsymbol{\sigma}'=2G_0\exp\left[n\left(-\frac{\varepsilon_v^e}{\tilde{\kappa}}\right)\right]\boldsymbol{\varepsilon}^{e'}=2G_0\left(\frac{p+\vartheta F}{p_0+\vartheta F_0}\right)^n\boldsymbol{\varepsilon}^{e'} \tag{11.170}$$

which coincide with Eq. $(11.6)_1$ and Eq. $(11.8)_2$ in the rate form, respectively, while Eq. (11.169) is used for deriving Eq. (11.170) from the second term in the right-hand side of Eq. (11.167).

The above-mentioned hyperelastic constitutive equation based on Eq. (11.6) is the improvement and the generalization of the afore-mentioned past formulations and their modification by Yamakawa et al. (2010a) incorporating p_e in Eq. (11.3).

(b) **Multiplicative finite strain theory**

Various immature multiplicative hyperelastic equations for soils have been proposed by Borja and Tamagnini (1998), Callari et al. (1998), etc. using the Hencky strain in the current configuration and by Yamakawa et al. (2010) using the second Piola-Kirchhoff stress in the intermediate configuration. The exact hyperelastic equation within the framework of the multiplicative finite strain theory for soils will be formulated below.

The isotropic hyperelastic equation in the intermediate configuration is given noting Eq. (5.6) and replacing ε_v^e and ε_d^e to $\ln J^e$ and $\mathrm{tr}\underline{\bar{\mathbf{C}}}^e$ ($\mathrm{tr}\underline{\bar{\mathbf{C}}}^e-3$ in detailed expression), respectively, in Eq. (11.161) as follows:

$$\bar{\mathbf{S}}=2\frac{\partial\psi(\ln J^e,\ \mathrm{tr}\underline{\bar{\mathbf{C}}}^e)}{\partial\bar{\mathbf{C}}^e}=2\frac{\partial\psi(\ln J^e,\ \mathrm{tr}\underline{\bar{\mathbf{C}}}^e)}{\partial\ln J^e}\frac{\partial\ln J^e}{\partial\bar{\mathbf{C}}^e}+2\frac{\partial\psi(\ln J^e,\ \mathrm{tr}\underline{\bar{\mathbf{C}}}^e)}{\partial\mathrm{tr}\underline{\bar{\mathbf{C}}}^e}\frac{\partial\mathrm{tr}\underline{\bar{\mathbf{C}}}^e}{\partial\bar{\mathbf{C}}^e} \tag{11.171}$$

where

$$\left.\begin{array}{l}
J^e=\det\mathbf{F}^e,\ \bar{\mathbf{C}}^e=\mathbf{F}^{eT}\mathbf{F}^e,\det\overline{\mathbf{C}}^e=J^{e2}\\
\mathbf{F}^e=\mathbf{F}^e_{vol}\underline{\mathbf{F}}^e,\mathbf{F}^e_{vol}\equiv J^{e1/3}\mathbf{I},\ \underline{\mathbf{F}}^e\equiv J^{e-1/3}\mathbf{F}^e\\
\underline{\bar{\mathbf{C}}}^e\equiv\underline{\mathbf{F}}^{eT}\underline{\mathbf{F}}^e=J^{e-2/3}\bar{\mathbf{C}}^e,\ \mathrm{tr}\underline{\bar{\mathbf{C}}}^e=J^{e-2/3}\mathrm{tr}\bar{\mathbf{C}}^e\\
\left(\begin{array}{l}\det\underline{\mathbf{F}}^e=\det\underline{\bar{\mathbf{C}}}^e=1,\underline{\bar{\mathbf{C}}}^{e\prime}=0\\ \underline{\bar{\mathbf{C}}}^e=\mathbf{I},\ \mathrm{tr}\underline{\bar{\mathbf{C}}}^e=3\quad\text{for }\mathbf{F}^e=\mathbf{F}^e_{vol}\end{array}\right)
\end{array}\right\} \tag{11.172}$$

$\mathbf{F}^e_{vol}$ is the volumetric part and $\mathbf{F}^e$ is the so-called *unimodular tensor* designating the isochoric (distortional under constant volume) part of $\mathbf{F}^e$. In addition, $\mathrm{tr}\underline{\bar{\mathbf{C}}}^e = 3\,(\mathbf{F}^e = \mathbf{F}^e_{vol}\ \ln J^e = 0\,(\mathbf{F}^e = \underline{\mathbf{F}}^e)$ is required in and the pure volumetric deformation and the pure deviatoric deformation, respectively.

The following partial derivatives hold.

$$\left.\begin{aligned}\frac{\partial \ln J^e}{\partial \bar{\mathbf{C}}^e} &= \frac{1}{2J^e\sqrt{\det \bar{\mathbf{C}}^e}}(\det \bar{\mathbf{C}}^e)\bar{\mathbf{C}}^{e-1} = \frac{1}{2}\bar{\mathbf{C}}^{e-1}\\ \frac{\partial \mathrm{tr}\underline{\bar{\mathbf{C}}}^e}{\partial \bar{\mathbf{C}}^e} &= J^{e-2/3}\left[\mathbf{I} - \frac{1}{3}(\mathrm{tr}\bar{\mathbf{C}}^e)\bar{\mathbf{C}}^{e-1}\right]\end{aligned}\right\} \tag{11.173}$$

noting

$$\left.\begin{aligned}\frac{\partial \ln J^e}{\partial \bar{\mathbf{C}}^e} &= \frac{\partial \ln J^e}{\partial J^e}\frac{\partial J^e}{\partial \bar{\mathbf{C}}^e} = \frac{1}{J^e}\frac{\partial\sqrt{\det \bar{\mathbf{C}}^e}}{\partial \bar{\mathbf{C}}^e} = \frac{1}{J^e}\frac{1}{2\sqrt{\det \bar{\mathbf{C}}^e}}\frac{\partial \det \bar{\mathbf{C}}^e}{\partial \bar{\mathbf{C}}^e}\\ &= \frac{1}{2J^e\sqrt{\det \bar{\mathbf{C}}^e}}(\det \bar{\mathbf{C}}^e)\bar{\mathbf{C}}^{e-1}\\ \frac{\partial \mathrm{tr}\underline{\bar{\mathbf{C}}}^e}{\partial \bar{\mathbf{C}}^e} &= \frac{\partial \mathrm{tr}[(\det \bar{\mathbf{C}}^e)^{-1/3}\bar{\mathbf{C}}^e]}{\partial \bar{\mathbf{C}}^e} = \frac{\partial[(\det \bar{\mathbf{C}}^e)^{-1/3}\mathrm{tr}\bar{\mathbf{C}}^e]}{\partial \bar{\mathbf{C}}^e}\\ &= (\det \bar{\mathbf{C}}^e)^{-1/3}[\mathbf{I} - \frac{1}{3}(\mathrm{tr}\bar{\mathbf{C}}^e)\bar{\mathbf{C}}^{e-1}]\end{aligned}\right\}$$

The substitution of Eq. (11.173) into Eq. (11.171) reads:

$$\bar{\mathbf{S}} = \frac{\partial\psi(\ln J^e,\ \mathrm{tr}\underline{\bar{\mathbf{C}}}^e)}{\partial \ln J^e}\bar{\mathbf{C}}^{e-1} + 2\frac{\partial\psi(\ln J^e,\ \mathrm{tr}\underline{\bar{\mathbf{C}}}^e)}{\partial \mathrm{tr}\underline{\bar{\mathbf{C}}}^e}(J^{e-2/3}\left[\mathbf{I} - \frac{1}{3}(\mathrm{tr}\bar{\mathbf{C}}^e)\bar{\mathbf{C}}^{e-1}\right] \tag{11.174}$$

Let the following strain energy function be assumed.

$$\boxed{\begin{aligned}\psi(\mathrm{In}\,J^e, \mathrm{tr}\underline{\bar{\mathbf{C}}}^e) = \vartheta F \mathrm{In} J^e + \tilde{\kappa}(\overline{P}M_0 + \vartheta F_0)J^{e-1/\tilde{\kappa}}\\ +\ G_0 J^{e-n/k}\,(\mathrm{tr}\underline{\bar{\mathbf{C}}}^e - 3)\end{aligned}} \tag{11.175}$$

noting

$$J^{e-n/\tilde{\kappa}} = \exp\left(-\frac{n}{\tilde{\kappa}}\ln \mathrm{J^e}\right)$$

where $\overline{P}_{M0}$ is the initial value of the pressure defined in terms of the Mandel stress $\overline{\mathbf{M}}$ by $\overline{P}_M \equiv -(1/3)\mathrm{tr}\overline{\mathbf{M}}$.

The following partial derivatives hold for Eq. (11.175).

$$\left.\begin{aligned} \frac{\partial\psi(\ln J^e,\ \mathrm{tr}\underline{\bar{\mathbf{C}}}^e)}{\partial \ln J^e} &= \vartheta F - (\bar{P}_{M0} + \vartheta F_0) + J^{e-1/\tilde{\kappa}} - \frac{\eta}{\tilde{\kappa}} G_0 J^{e-n/\tilde{\kappa}} (\mathrm{tr}\underline{\bar{\mathbf{C}}}^e - 3) \\ \frac{\partial\psi(\ln J^e,\ \mathrm{tr}\underline{\bar{\mathbf{C}}}^e)}{\partial \mathrm{tr}\underline{\bar{\mathbf{C}}}^e} &= G_0 + \eta\, J^{e-1/\tilde{\kappa}} \end{aligned}\right\}$$

Equation (11.171) with Eqs. (11.173) and (11.176) reads:

$$\begin{aligned} \bar{\mathbf{S}} = &\left[\vartheta F - (\bar{P}_{M0} + \vartheta F_0) J^{e-1/\tilde{\kappa}} - \frac{\eta}{\tilde{\kappa}} G_0 J^{e-n/\tilde{\kappa}} (\mathrm{tr}\underline{\bar{\mathbf{C}}}^e - 3)\right] \bar{\mathbf{C}}^{e-1} \\ &+ 2G_0\, J^{e-n/\tilde{\kappa}} J^{e-2/3} \left[\mathbf{I} - \frac{1}{3}(\mathrm{tr}\bar{\mathbf{C}}^e)\bar{\mathbf{C}}^{e-1}\right] \end{aligned} \tag{11.177}$$

from which the Mandel stress is given as follows:

$$\boxed{\begin{aligned} \overline{\mathbf{M}} = \overline{\mathbf{C}}^e \overline{\mathbf{S}} = &\left[\vartheta F - (\bar{P}_{M0} + \vartheta F_0) + J^{e-1/\tilde{\kappa}} - \frac{\eta}{\tilde{\kappa}} G_0 J^{e-n/\tilde{\kappa}} (\mathrm{tr}\underline{\bar{\mathbf{C}}}^e - 3)\right] \mathbf{I} \\ &+ 2G_0\, J^{e-n/\tilde{\kappa}} J^{e-2/3} \overline{\mathbf{C}}^{e\prime} \end{aligned}} \tag{11.178}$$

It is followed from Eq. (11.178) that

$$\frac{\bar{P}_M + \vartheta F}{\bar{P}_{M0} + \vartheta F_0} = J^{e-1/\tilde{\kappa}} = \exp\left(-\frac{1}{\tilde{\kappa}} \ln J^e\right) \quad \text{for } \mathbf{F}^e = \mathbf{F}^e_{vol} \tag{11.179}$$

i.e.

$$\ln J^e = -\tilde{\kappa} \ln \frac{\bar{P}_M + \vartheta F}{\bar{P}_{M0} + \vartheta F_0} \quad \text{for } \mathbf{F}^e = \mathbf{F}^e_{vol} \tag{11.180}$$

and

$$\bar{\mathbf{M}}' = 2G_0\, J^{e-1/\tilde{\kappa}} J^{e-2/3} \bar{\mathbf{C}}^{e\prime} = 2G_0 \left(\frac{\bar{P}_M + \vartheta F}{\bar{P}_{M0} + \vartheta F_0}\right)^n J^{e-2/3} \bar{\mathbf{C}}^{e\prime} \tag{11.181}$$

References

Alonso EE, Gens A, Josa A (1990) A constitutive model for partially saturated soils. Geotechnique 40:405–430

Amorosi A, Boldini D, Germano V (2007) Implicit integration of a mixed isotropic–kinematic hardening plasticity model for structured clays. Int J Numer Anal Methods Geomech 32:1173–1203

Argyris JH, Faust G, Szimma J, Warnke EP, William KJ (1973) Recent developments in the finite element analysis of PCRV. In: Proceedings of 2nd international conference SMIRT, Berlin

Asaoka A, Nakano M, Noda T (1997) Soil-water coupled behaviour of heavily over consolidated clay near/at critical state. Soils Found 37(1):13–28

Bishop AW, Webb DL, Lewin PI (1965) Undisturbed samples of London clay from the Ashford common shaft: strength-effective stress relationships. Geotechnique 15:1–31
Borja RI (2004) Cam-clay plasticity. Part V: a mathematical framework for three-phase deformation and strain localization analyses of partially saturated porous media. Comp Meth Appl Mech Eng 193:5301–5338
Borja RI, Tamagnini C (1998) Cam-clay plasticity, part III: extension of the infinitesimal model to include finite strains. Comp Meth Appl Mech Eng 155:73–95
Borja RI, Sama KM, Sanz PF (2003) On the numerical integration of three-invariant elastoplastic constitutive models. Comp Meth Appl Mech Eng 192:1227–1258
Burland JB (1965) The yielding and dilatation of clay, correspondence. Geotechnique 15:211–214
Butterfield R (1979) A natural compression law for soils (an advance on e-log p'). Geotechnique 29:469–480
Callari C, Auricchio F, Sacco E (1998) A finite-strain Cam-clay model in the framework of multiplicative elasto-plasticity. Int J Plast 14:1155–1187
Castro G (1969) Liquefaction of sands. Ph.D. thesis, Harvard Soil Mech. Series 81
Chowdhury EQ, Nakai T, Tawada M, Yamada S (1999) A model for clay using modified stress under various loading conditions with the application of subloading concept. Soils Found 39 (6):103–116
Collins IF, Hilder T (2002) A theoretical framework for constructing elastic/plastic constitutive models of triaxial tests. Int J Nemer Anal Meth Geomech 26:1313–1347
Coombs WM, Crouch RS (2011) Algorithmic issues for three-invariant hyperplastic critical state models. Comp Meth Appl Mech Eng 200:2297–2318
Coombs WM, Crouch RS, Augarde CE (2013) A unique critical state two-surface hyperplasticity model for fine-grained particulate media. J Mech Phys Solis 61:175–189
Cundall P, Board M (1988) A microcomputer program for modeling large-strain plasticity problems. In: Prepare for the 6th international conference numerical methods geomechanics, Innsbruck, Austria, pp 2101–2108
De Borst R, Groen AE (1999) Towards Efficient and robust elements for 3-D plasticity. Comput Struct Dynamics 19:977–992
Drucker DC, Prager W (1952) Soil mechanics and plastic analysis or limit design. Quart Appl Math 10:157–165
Drucker DC, Gibson Re and Henkel DJ (1957) Soil mechanics and workhardening theories of plasticity. Trans Amer Soc Civil Eng 122:338–346
Fukutake K, Ohtsuki M and Sato M (1990) Analysis of saturated dense sandstructure system and comparison with results from shaking table test. Earthquake eng Struct Dynamics. 19:977–992
Gudehus G (1973) Elastoplastische Stoffgleichungen fur trockenen Sand. Ing Arch 42:151–169 (in German)
Hashiguchi K (1972) On a yielding of frictional materials—a hardening law. In: Proceedings of 27th annual meeting, JSCE, pp 105–108 (in Japanese)
Hashiguchi K (1974) Isotropic hardening theory of granular media. Proc Japan Soc Civil Eng (227):45–60 (in Japanese)
Hashiguchi K (1977) An expression of anisotropy in a plastic constitutive equation of soils. In: Murayama S, Schofield AN (eds) Constitutive equations of soils (Proceedings of 9th international conference on soil mechanical and foundation engineering, Spec Ses 9). JSSMFE, Tokyo, pp 302–305
Hashiguchi K (1978) Plastic constitutive equations of granular materials. In: Cowin SC, Satake M (eds) Proceedings US-Japan seminar on continuum mechanical statistical approaches in the mechanics of granular materials. Sendai, pp 321–329
Hashiguchi K (1985) Macrometric approaches—static—intrinsically time-independent. In: Constitutive laws of soils (Proceedings discussions Ses 1A, 11th international conference soil mechanical foundation engineering). San Francisco, pp 25–65
Hashiguchi K (1995) On the linear relations of $V - \ln p$ and $\ln v - \ln p$ for isotropic consolidation of soils. Int J Numer Anal Meth Geomech 19:367–376

Hashiguchi K (2001) Description of inherent/induced anisotropy of soils: rotational hardening rule with objectivity. Soils Found 41(6):139–145

Hashiguchi K (2002) A proposal of the simplest convex-conical surface for soils. Soils Found 42 (3):107–113

Hashiguchi K (2007) Yield condition of soils with tensile yield strength and rotational hardening. Proc Int Conf Compt Exp Eng Sci 07:1441–1446

Hashiguchi K (2008) Verification of compatibility of isotropic consolidation characteristics of soils to multiplicative decomposition of deformation gradient. Soils Found 48:597–602

Hashiguchi K, Chen Z-P (1998) Elastoplastic constitutive equations of soils with the subloading surface and the rotational hardening. Int J Numer Anal Meth Geomech 22:197–227

Hashiguchi K, Mase T (2007) Extended yield condition of soils with tensile strength and rotational hardening. Int J Plast 23:1939–1956

Hashiguchi K, Mase T (2011) Physical interpretation and quantitative prediction of cyclic mobility by the subloading surface model. Japanese Geotech J 6:225–241 (in Japanese)

Hashiguchi K, Tsutsumi S (2001) Elastoplastic constitutive equation with tangential stress rate effect. Int J Plast 17:117–145

Hashiguchi K, Tsutsumi S (2003) Shear band formation analysis in soils by the subloading surface model with tangential stress rate effect. Int J Plast 19:1651–1677

Hashiguchi K, Tsutsumi S (2006) Gradient plasticity with the tangential subloading surface model and the prediction of shear band thickness of granular materials. Int J Plast 22:767–797

Hashiguchi K, Ueno M (1977) Elastoplastic constitutive laws of granular materials. In: Murayama S, Schofield AN (eds) Constitutive equations of soils (Proceedings 9th international conference soil mechanical foundation engineering, Spec Ses 9). JSSMFE, Tokyo, pp 73–82

Hashiguchi K, Saitoh K, Okayasu T, Tsutsumi S (2002) Evaluation of typical conventional and unconventional plasticity models for prediction of softening behavior of soils. Geotechnique 52:561–573

Houlsby GT (1985) The use of a variable shear modulus in elastic-plastic models for clays. Comput Geotech 1:3–13

Houlsby GT, Amorosi A, Rojas E (2005) Elastic moduli of soils dependent on pressure: a hyperelastic formulation. Geotechnique 55:383–392

Iai S, Ohtsuki O (2005) Yield and cyclic behaviour of a strain space multiple mechanism model for granular materials. Int J Numer Anal Meth Goemech 29:417–442

Ishihara K, Tatsuoka F, Yasuda S (1975) Undrained deformation and liquefaction of sand under cyclic stresses. Soils Found 15:29–44

Itasca Consulting Group (2006) FLAC3D, "Fast Lagrangian Analysis of Continua in 3 Dimensions". Minneapolis, Minnesota, USA

Khojastehpour M, Hashiguchi K (2004a) The plane strain bifurcation analysis of soils by the tangential-subloading surface model. Int J Solids Struct 41:5541–5563

Khojastehpour M, Hashiguchi K (2004b) Axisymmetric bifurcation analysis in soils by the tangential-subloading surface model. J Mech Phys Solids 52:2235–2262

Khojastehpour M, Murakami Y, Hashiguchi K (2006) Antisymmetric bifurcation in a circular cylinder with tangential plasticity. Mech Mater 38:1061–1071

Kiyota T, Kozeki J, Sato T, Tsutsumi Y (2009a) Effects of sample disturbance on small strain characteristics and liquefaction properties of Holocene and pleistocene sandy soils. Soils Found 49:509–523

Kiyota T, Kozeki J, Sato T, Kuwano S (2009b) Aging effects on small strain shear moduli and liquefaction properties of in-situ frozen and reconstituted sandy soils. Soils Found 49:259–274

Kohgo Y, Nakano M, Miyazaki T (1993) Verification of the generalized elastoplastic model for unsaturated soils. Soil Found 33(4):64–73

Mase T, Hashiguchi K (2009) Numerical analysis of footing settlement problem by subloading surface model. Soils Found 49:207–220

Matsuoka H, Nakai T (1974) Stress-deformation and strength characteristics of soil under three different principal stress. Proc Japan Soc Civil Eng (232):59–70

Matsuoka H, Yao YP, Sun DA (1999) The Cam-clay model revised by SMP criterion. Soils Found 39(1):81–95
Nakai T, Hinokio M (2004) A simple elastoplastic model for normally and over consolidated soils with unified material parameters. Soils Found 44(2):53–70
Nakai T, Mihara Y (1984) A new mechanical quantity for soils and its application to elastoplastic constitutive models. Soils Found 24(2):82–941
Niemunis A, Cudny M (1998) On hyperelasticity for clays. Comput Geotech 23:221–236
Nova R (1977) On the hardening of soils. Arch Mech Stos 29:445–458
Oka F, Yashima A, Taguchi A, Yamashita S (1999) A cyclic elasto-plastic constitutive model for sand considering a plastic-strain dependence of the shear modulus. Geotechnique 49:661–680
Okahara M, Takagi S, Mori H, Koike S, Tatsuda M, Tatsuoka F, Morimoto H (1989) Largescale plane strain bearing capacity tests of shallow foundation on sand (part 1). In: Proceedings of 24th annual meeting japanese society geotechnical engineering, pp 1239–1242 (in Japanese)
Pietruszczak ST, Mroz Z (1983) On hardening anisotropy of Ko-consolidated clays Int J Numer Anal Meth Geomech 7:19–38
Pietruszczak ST, Niu X (1993) On the description of localized deformation. Int J Numer Anal Meth Geomech 17:791–805
Roscoe KH, Burland JB (1968) On the generalized stress-strain behaviour of 'wet' clay. In: Engineering plasticity. Cambridge University Press, Cambridge, pp 535–608
Saada AS, Bianchini G (1989) Proceedings of the international workshop on constitutive equations for granular non-cohesive soils, Balkema, Amsterdam
Satake M (1972) A proposal of new yield criterion for soils. Proc Japan Soc Civil Eng 189:79–88 (in Japanese)
Schofield AN, Wroth CP (1968) Critical state soil mechanics. McGraw-Hill, London
Sekiguchi H, Ohta H (1977) Induced anisotropy and its time dependence in clays. In: Constitutive equations of soils (Proceedings of Spec Session 9, 9th ICSFME). Tokyo, pp 229–238
Sheng D, Sloan SW, Yu HS (2000) Aspects of finite element implementation of critical state models. Comput Mech 26:185–196
Siddiquee MSA, Tanaka T, Tatsuoka F, Tani K, Morimoto T (1999) Numerical simulation of bearing capacity characteristics of strip footing on sand. Soils Found 39(4):93–109
Simo JC, Meschke G (1993) A new class of algorithms for classical plasticity extended to finite strains. Application to geomaterials. Comput Mech 11:253–278
Skempton AW, Brown JD (1961) A landslide in boulder clay at Selset, Yorkshire. Geotechnique 11:280–293
Sloan SW, Randolph MF (1982) Numerical prediction of collapse loads using finite element methods. Int J Numer Anal Meth Geomech 6:47–76
Stallebrass SE, Taylor RN (1997) The development and evaluation of a constitutive model for the prediction of ground movements in over consolidated clay. Geotechnique 47:235–253
Stark TD, Ebeling RM, Vettel JJ (1994) Hyperbolic stress-strain parameters for silts. J Geotech Eng (ASCE) 120:420–441
Tanaka T, Kawamoto O (1988) Three dimensional finite element collapse analysis for foundations and slopes using dynamic relaxation. Proc Numer Meth Geomech Innsbruck pp 1213–1218
Tanaka T, Sakai T (1993) Progressive failure effect of trap-door problems with granular materials. Soils Found 33(1):11–22
Tani K (1986) Mechanism of bearing capacity of shallow foundation. Master thesis, University of Tokyo (in Japanese)
Tatsuoka F, Ikuhara O, Fukushima S, Kawamura T (1984) On the relation of bearing capacity of shallow footing on model sand ground and element test strength. In: Proceedings of Symp. Asses. Deform. Fail. Strength of Sandy Soils and Sand Grounds, Japan. Soc. Geotech. Eng., pp. 141–148 (in Japanese)
Taylor DW (1948) Fundamentals of soil mechanics. Wiley, Chichester, UK
Topolnicki M (1990) An elasto-plastic subloading surface model for clay with isotropic and kinematic mixed hardening parameters. Soils Found 30(2):103–113

Wesley LD (1990) Influence of structure and composition on residual soils. J Geotech Eng (ASCE) 116:589–603
Wilde P (1977) Two invariants depending models of granular media. Arch Mech Stos 29:799–809
Wongsaroj J, Soga K, Mair RJ (2007) Modeling of long-term ground response to tunneling under St James' Park, London. Geotechnique 57:75–90
Yamada S, Takamori T, Sato K (2010) Effects on reliquefaction resistance produced by changes in anisotropy during liquefaction. Soils Found 50:9–25
Yamakawa Y, Hashiguchi K, Ikeda K (2010) Implicit stress-update algorithm for isotropic Cam-clay model based on the subloading surface concept at finite strains. Int J Plast 26:634–658
Yamamoto Y (1998) Evaluation of seismic behavior of clay and sand grounds. Ph.D. thesis, Yamaguchi University
Zhang F, Ye B, Noda T, Nakano M, Nakai K (2007) Explanation of cyclic mobility of soils: approach by stress-induced anisotropy. Soils Found 47:635–648

Chapter 12
Multiplicative Elastoplasticity: Subloading Finite Strain Theory

The subloading surface model was formulated in the Chaps. 6–11 within the frameworks of the finite hypoelastic-based plasticity in detail and of the infinitesimal hyperelastic-based plasticity (Sect. 6.9) in brief. Finite deformation and rotation cannot be described in the exact sense by these formulations. The multiplicative elastoplastic constitutive equation will be formulated for the subloading surface model with the translation of the elastic-core, although the multiplicative constitutive equation for the initial subloading surface model, in which the elastic-core is fixed in the back stress point, was formulated in an immature form by Hashiguchi and Yamakawa (2012). One must formulate the constitutive equation possessing the generality and the universality to be inherited eternally, while any unconventional model, i.e. cyclic plasticity model other than the subloading surface model has not been extended to the multiplicative finite strain theory. The exact formulation of the multiplicative finite strain theory based on the extended subloading surface model has been attained by Hashiguchi (2016a, b, c, d), which will be explained in detail in this chapter.

12.1 Classification of Elastoplastic Constitutive Equation

The basic frameworks of elastoplasticity are classified as follows:

Infinitesimal elastoplasticity

(1) The infinitesimal strain and its material-time derivative are additively decomposed into the elastic and the plastic parts,
(2) The Cauchy stress is used as the stress measure,
(3) The elastic deformation is formulated in the hyperelastic relation,
(4) The initial and the current configurations are not distinguished,
(5) The infinitesimal elastic and plastic deformation is described, ignoring a rotation,

K. Hashiguchi, *Foundations of Elastoplasticity: Subloading Surface Model*,
DOI 10.1007/978-3-319-48821-9_12

Finite elastoplasticity

There are the following two frameworks for the description of finite deformation/rotation.

Hypoelastic-based plasticity

(1) The symmetric and anti-symmetric parts of the velocity gradient are defined to be the strain rate and the spin, respectively and further they are is additively decomposed into the elastic and the plastic parts,
(2) The Cauchy stress and its corotational rate (also for internal variables) are used,
(3) The elastic strain rate is formulated in the hypoelastic relation,
(4) The pertinent time-integration of stress rate is required,
(5) The formulation is executed in the current configuration which is influenced by the material rotation,
(6) The finite plastic deformation and the finite rotation are described under the restriction of the infinitesimal elastic deformation.

Multiplicative hyperelastic-based plasticity

(1) The multiplicative decomposition of the deformation gradient tensor is used consistently,
(2) The additive decomposition of the strain rate and the spin tensors in the intermediate configuration into the elastic and the plastic parts are used, which are decomposed definitely into these parts,
(3) The Mandel stress in the intermediate configuration is used as the stress measure,
(4) The elastic deformation is formulated in the hyperelastic relation,
(5) The formulation is executed in the intermediate configuration which is not influenced by the material rotation,
(6) The finite elastic and the plastic deformation and rotation are described exactly.

Then, it realizes the exact description of the finite deformation/rotation.

The formulation of the subloading surface model in the multiplicative hyperelastic-based plasticity was given by Hashiguchi and Yamakawa (2012) in the immature form for the initial subloading surface model in which the elastic-core is fixed so that it is limited to the description of the monotonic loading behavior. The subloading surface model with the translation of the elastic-core will be formulated in this section within the framework of the multiplicative hyperelastic-based plasticity. It is to be the first cyclic (unconventional) elastoplasticity model in the multiplicative hyperelastic-based plasticity for the exact description of finite elastoplastic deformation/rotation.

12.2 Further Multiplicative Decomposition of Plastic Deformation Gradient

The deformation gradient $\mathbf{F}$ is multiplicatively decomposed into the elastic deformation gradient $\mathbf{F}^e$ and the plastic deformation gradient $\mathbf{F}^e$ as described in Sect. 6.1. Further, decompose $\mathbf{F}^p$ into the plastic storage part $\mathbf{F}^p_{ks}$ causing the kinematic hardening and its plastic dissipative part $\mathbf{F}^p_{kd}$ multiplicatively (Lion 2000). Analogously, decompose $\mathbf{F}^p$ into the plastic storage part $\mathbf{F}^p_{cs}$ causing the translation of elastic-core and its plastic dissipative part $\mathbf{F}^p_{cd}$ multiplicatively as follows:

$$\mathbf{F} = \mathbf{F}^e\mathbf{F}^p, \quad \mathbf{F}^p = \mathbf{F}^p_{ks}\mathbf{F}^p_{kd}, \quad \mathbf{F}^p = \mathbf{F}^p_{cs}\mathbf{F}^p_{cd} \tag{12.1}$$

The configurations based on these decompositions are illustrated in Fig. 12.1.

Based on the right Cauchy-Green deformation tensor

$$\mathbf{C} \equiv \mathbf{F}^T\mathbf{F} \tag{12.2}$$

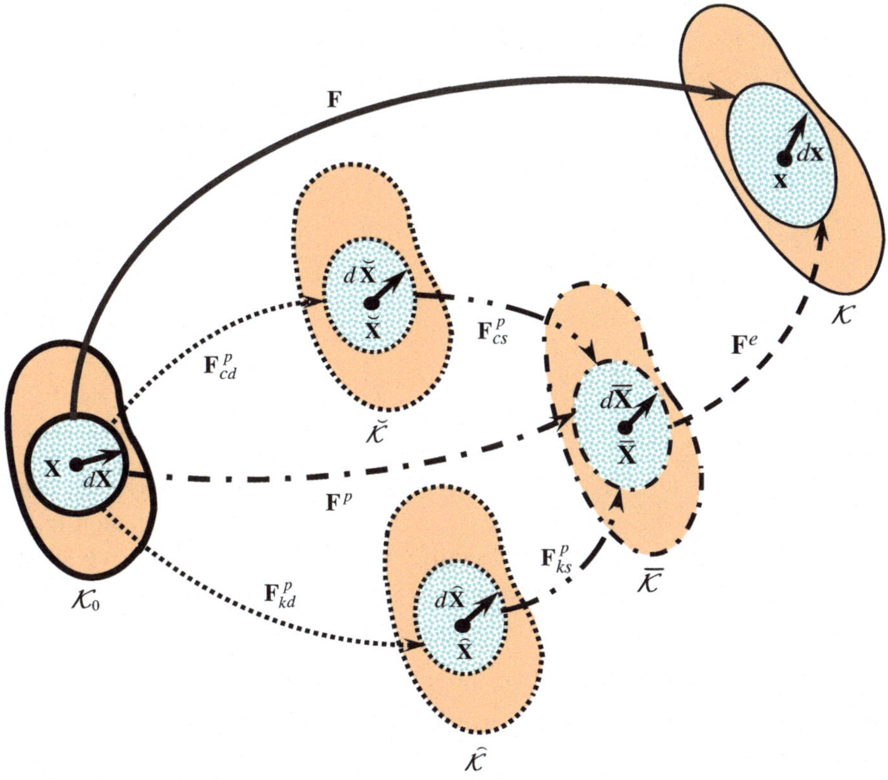

Fig. 12.1 Multiplicative decompositions of deformation gradient for elastoplastic material with translations of kinematic hardening and elastic-core

the following tensors of the storage parts $\overline{\mathbf{C}}^{e}, \widehat{\mathbf{C}}^{p}_{ks}, \breve{\mathbf{C}}^{p}_{cs}$ and the dissipative parts $\mathbf{C}^{p}, \widehat{\mathbf{C}}^{p}_{kd}, \breve{\mathbf{C}}^{p}_{cd}$ are defined.

$$\left.\begin{array}{l}\overline{\mathbf{C}}^{e} \equiv \mathbf{F}^{eT}\mathbf{F}^{e} = (\mathbf{R}^{e}\overline{\mathbf{U}}^{e})^{T}\mathbf{R}^{e}\overline{\mathbf{U}}^{e} = \overline{\mathbf{U}}^{e2}, \ \mathbf{C}^{p} \equiv \mathbf{F}^{pT}\mathbf{F}^{p}, \\ \widehat{\mathbf{C}}^{p}_{ks} \equiv \mathbf{F}^{pT}_{ks}\mathbf{F}^{p}_{ks} = \widehat{\mathbf{U}}^{p\,2}_{ks}, \mathbf{C}^{p}_{kd} \equiv \mathbf{F}^{pT}_{kd}\mathbf{F}^{p}_{kd}, \\ \breve{\mathbf{C}}^{p}_{cs} \equiv \mathbf{F}^{pT}_{cs}\mathbf{F}^{p}_{cs} = \widehat{\mathbf{U}}^{p\,2}_{cs}, \mathbf{C}^{p}_{cd} \equiv \mathbf{F}^{pT}_{cd}\mathbf{F}^{p}_{cd}\end{array}\right\} \tag{12.3}$$

where one has

$$\left.\begin{array}{l}\overline{\mathbf{C}}^{p}_{ks} \equiv {}^{p}_{ks}\widehat{\mathbf{C}}^{p}_{ks\bar{G}\bar{G}} = \mathbf{F}^{p-T}_{ks}\widehat{\mathbf{C}}^{p}_{ks}\mathbf{F}^{p-1}_{ks} \\ \overline{\mathbf{C}}^{p}_{cs} \equiv {}^{p}_{cs}\breve{\mathbf{C}}^{p}_{cs\bar{G}\bar{G}} = \mathbf{F}^{p-T}_{cs}\breve{\mathbf{C}}^{p}_{cs}\mathbf{F}^{p-1}_{cs}\end{array}\right\} = \overline{\mathbf{G}} \tag{12.4}$$

$\overline{\mathbf{G}}$ is the metric tensors defined in Eq. (4.31) in the intermediate configuration. The hat symbols $(\bar{\ })$, $(\frown)$ and $(\smile)$ are added to the variables based in the intermediate configuration $\overline{\mathcal{K}}$, the kinematic hardening intermediate configuration $\widehat{\mathcal{K}}$ and the elastic-core intermediate configuration $\breve{\mathcal{K}}$, respectively. The superscript and/or subscript is (are) added in the left side in order to specify the pull-back or push-forward due to the elastic (or plastic) deformation gradient.

In order to explain clearly, the equations described already in Eqs. (6.11) to (6.18) in Sect. 6.1 are again written in the following Eqs. (12.5)–(12.11).

The velocity gradient $\boldsymbol{l}$ in the current configuration $\mathcal{K}$ is additively decomposed into the elastic and the plastic parts:

$$\boldsymbol{l} = \boldsymbol{l}^{e} + \boldsymbol{l}^{p} \tag{12.5}$$

where

$$\left.\begin{array}{l}\boldsymbol{l} \equiv \dot{\mathbf{F}}\mathbf{F}^{-1}, \\ \boldsymbol{l}^{e} \equiv \dot{\mathbf{F}}^{e}\mathbf{F}^{e-1}, \quad \boldsymbol{l}^{p} \equiv \mathbf{F}^{e}\dot{\mathbf{F}}^{p}\mathbf{F}^{p-1}\mathbf{F}^{e-1} = {}^{e}\overrightarrow{\overline{\mathbf{L}}}{}^{p\,g}_{\bullet g} = \mathbf{F}^{e}\overline{\mathbf{L}}^{p}\mathbf{F}^{e-1} \\ \overline{\mathbf{L}}^{p} \equiv \dot{\mathbf{F}}^{p}\mathbf{F}^{p-1}\end{array}\right\} \tag{12.6}$$

Further, the velocity gradient $\overline{\mathbf{L}}$ defined as the *contravariant-covariant* pull-back (Eq. (4.44)) of the velocity gradient tensor $\boldsymbol{l}$ in the current configuration to the intermediate configuration $\overline{\mathcal{K}}$ can be additively decomposed into the purely elastic and the purely plastic parts as follows:

$$\overline{\mathbf{L}} = \overline{\mathbf{L}}^{e} + \overline{\mathbf{L}}^{p} \tag{12.7}$$

where

$$\left.\begin{array}{l}\overline{\mathbf{L}} \equiv {}^{e}\overleftarrow{\boldsymbol{l}}{}^{\overline{G}}_{\bullet\overline{G}} = \mathbf{F}^{e-1}\boldsymbol{l}\mathbf{F}^{e} \\ \overline{\mathbf{L}}^{e} \equiv {}^{e}\overleftarrow{\boldsymbol{l}}{}^{\overline{G}}_{\bullet\overline{G}} = \mathbf{F}^{e-1}\boldsymbol{l}^{e}\mathbf{F}^{e} = \mathbf{F}^{e-1}\dot{\mathbf{F}}^{e}, \quad \overline{\mathbf{L}}^{p} \equiv {}^{e}\overleftarrow{\boldsymbol{l}^{p}}{}^{\overline{G}}_{\bullet\overline{G}} = \mathbf{F}^{e-1}\boldsymbol{l}^{p}\mathbf{F}^{e} = \dot{\mathbf{F}}^{p}\mathbf{F}^{p-1}\end{array}\right\} \tag{12.8}$$

Therefore, $\overline{\mathbf{L}}$ and $\overline{\mathbf{L}}^e$, $\overline{\mathbf{L}}^p$ can be pertinently adopted in the formulation of elasto-plastic constitutive equation. It follows from Eq. (12.8) that

$$\left.\begin{array}{l}\overline{\mathbf{L}}=\overline{\mathbf{D}}+\overline{\mathbf{W}}\\ \overline{\mathbf{L}}^e=\overline{\mathbf{D}}^e+\overline{\mathbf{W}}^e,\quad \overline{\mathbf{L}}^p=\overline{\mathbf{D}}^p+\overline{\mathbf{W}}^p\end{array}\right\}\tag{12.9}$$

$$\overline{\mathbf{D}}=\overline{\mathbf{D}}^e+\overline{\mathbf{D}}^p,\quad \overline{\mathbf{W}}=\overline{\mathbf{W}}^e+\overline{\mathbf{W}}^p\tag{12.10}$$

where

$$\left.\begin{array}{ll}\overline{\mathbf{D}}=\mathrm{sym}\left[\overline{\mathbf{L}}\right], & \overline{\mathbf{W}}=\mathrm{ant}\left[\overline{\mathbf{L}}\right]\\ \overline{\mathbf{D}}^e=\mathrm{sym}\left[\overline{\mathbf{L}}^e\right], & \overline{\mathbf{W}}^e=\mathrm{ant}\left[\overline{\mathbf{L}}^e\right]\\ \overline{\mathbf{D}}^p=\mathrm{sym}\left[\overline{\mathbf{L}}^p\right], & \overline{\mathbf{W}}^p=\mathrm{ant}\left[\overline{\mathbf{L}}^p\right]\end{array}\right\}\tag{12.11}$$

The rate of $\overline{\mathbf{C}}^e$ is given from Eqs. $(12.3)_1$ and (12.7) as

$$\dot{\overline{\mathbf{C}}}{}^e=2\mathrm{sym}\left[\overline{\mathbf{C}}^e\overline{\mathbf{L}}^e\right]=2\mathrm{sym}\left[\overline{\mathbf{C}}^e\left(\overline{\mathbf{L}}-\overline{\mathbf{L}}^p\right)\right]\tag{12.12}$$

noting

$$\begin{aligned}\dot{\overline{\mathbf{C}}}{}^e&=(\mathbf{F}^{eT}\mathbf{F}^e)^{\bullet}=\mathbf{F}^{eT}\dot{\mathbf{F}}^e+\dot{\mathbf{F}}^{eT}\mathbf{F}^e=\mathbf{F}^{eT}\mathbf{F}^e(\mathbf{F}^{e-1}\dot{\mathbf{F}}^e)+(\dot{\mathbf{F}}^{eT}\mathbf{F}^{e-T})\mathbf{F}^{eT}\mathbf{F}^e\\&=\overline{\mathbf{C}}^e\overline{\mathbf{L}}^e+\overline{\mathbf{L}}^{eT}\overline{\mathbf{C}}^e\end{aligned}$$

Further, the plastic velocity gradient $\overline{\mathbf{L}}^p$ is additively decomposed for the kinematic hardening as follows:

$$\overline{\mathbf{L}}^p=\overline{\mathbf{L}}^p_{ks}+\overline{\mathbf{L}}^p_{kd}\tag{12.13}$$

where

$$\overline{\mathbf{L}}^p_{ks}\equiv\dot{\mathbf{F}}^p_{ks}\mathbf{F}^{p-1}_{ks},\ \overline{\mathbf{L}}^p_{kd}\equiv{}^{\ p}_{ks}\overrightarrow{\widehat{\mathbf{L}}}{}^{p\ \bar{G}}_{kd\bullet\bar{G}}=\mathbf{F}^p_{ks}\widehat{\mathbf{L}}^p_{kd}\mathbf{F}^{p-1}_{ks}\tag{12.14}$$

$$\left.\begin{array}{l}\widehat{\mathbf{L}}^p_{kd}=\dot{\mathbf{F}}^p_{kd}\mathbf{F}^{p-1}_{kd}\equiv{}^{\ p}_{ks}\overleftarrow{\overline{\mathbf{L}}}{}^{p\ \widehat{G}}_{kd\bullet\widehat{G}}=\mathbf{F}^{p-1}_{ks}\overline{\mathbf{L}}^p_{kd}\mathbf{F}^p_{ks}=\widehat{\mathbf{D}}^p_{kd}+\widehat{\mathbf{W}}^p_{kd}\\ \widehat{\mathbf{D}}^p_{kd}=\mathrm{sym}\left[\widehat{\mathbf{L}}^p_{kd}\right],\quad \widehat{\mathbf{W}}^p_{kd}=\mathrm{ant}\left[\widehat{\mathbf{L}}^p_{kd}\right]\end{array}\right\}\tag{12.15}$$

noting

$$\begin{aligned}\overline{\mathbf{L}}^p&=\left(\mathbf{F}^p_{ks}\mathbf{F}^p_{kd}\right)^{\bullet}\left(\mathbf{F}^p_{ks}\mathbf{F}^p_{kd}\right)^{-1}=(\dot{\mathbf{F}}^p_{ks}\mathbf{F}^p_{kd}+\mathbf{F}^p_{ks}\dot{\mathbf{F}}^p_{kd})\mathbf{F}^{p-1}_{kd}\mathbf{F}^{p-1}_{ks}\\&=\dot{\mathbf{F}}^p_{ks}\mathbf{F}^{p-1}_{ks}+\mathbf{F}^p_{ks}\dot{\mathbf{F}}^p_{kd}\mathbf{F}^{p-1}_{kd}\mathbf{F}^{p-1}_{ks}\end{aligned}$$

Analogously, the following additive decomposition of the velocity gradient holds for the elastic-core.

$$\overline{\mathbf{L}}^p = \overline{\mathbf{L}}^p_{cs} + \overline{\mathbf{L}}^p_{cd} \tag{12.16}$$

where

$$\overline{\mathbf{L}}^p_{cs} \equiv \dot{\mathbf{F}}^p_{cs}\mathbf{F}^{p-1}_{cs}, \quad \overline{\mathbf{L}}^p_{cd} \equiv {}^p_{cs}\overleftarrow{\breve{\mathbf{L}}}{}^{p\,\bar{G}}_{cd\bullet\bar{G}} = \mathbf{F}^p_{cs}\breve{\mathbf{L}}^p_{cd}\mathbf{F}^{p-1}_{cs} \tag{12.17}$$

$$\left.\begin{aligned} &\breve{\mathbf{L}}^p_{cd} = \dot{\mathbf{F}}^p_{cd}\mathbf{F}^{p-1}_{cd} = {}^p_{cs}\overleftarrow{\overline{\mathbf{L}}}{}^{p\,\bar{G}}_{cd\bullet\bar{G}} = \mathbf{F}^{p-1}_{cs}\overline{\mathbf{L}}^p_{cd}\mathbf{F}^p_{cs} = \breve{\mathbf{D}}^p_{cd} + \breve{\mathbf{W}}^p_{cd} \\ &\breve{\mathbf{D}}^p_{cd} = \mathrm{sym}\left[\breve{\mathbf{L}}^p_{cd}\right], \quad \breve{\mathbf{W}}^p_{cd} = \mathrm{ant}\left[\breve{\mathbf{L}}^p_{cd}\right] \end{aligned}\right\} \tag{12.18}$$

The time-derivative of $\widehat{\mathbf{C}}^p_{ks}$ in Eq. (12.4) is given by

$$\dot{\widehat{\mathbf{C}}}{}^p_{ks} = 2{}^p_{ks}\overleftarrow{\overline{\mathbf{D}}}{}^p_{ks\widehat{G}\widehat{G}} = 2\mathbf{F}^{pT}_{ks}\overline{\mathbf{D}}^p_{ks}\mathbf{F}^p_{ks} = 2\mathbf{F}^{pT}_{ks}(\overline{\mathbf{D}}^p - \overline{\mathbf{D}}^p_{kd})\mathbf{F}^p_{ks} \tag{12.19}$$

noting

$$\begin{aligned} \dot{\widehat{\mathbf{C}}}{}^p_{ks} &= (\mathbf{F}^{pT}_{ks}\mathbf{F}^p_{ks})^{\bullet} = \mathbf{F}^{pT}_{ks}\dot{\mathbf{F}}^p_{ks} + \dot{\mathbf{F}}^{pT}_{ks}\mathbf{F}^p_{ks} = \mathbf{F}^{pT}_{ks}\dot{\mathbf{F}}^p_{ks}\mathbf{F}^{p-1}_{ks}\mathbf{F}^p_{ks} + \mathbf{F}^{pT}_{ks}\mathbf{F}^{p-T}_{ks}\dot{\mathbf{F}}^{pT}_{ks}\mathbf{F}^p_{ks} \\ &= \mathbf{F}^{pT}_{ks}\dot{\mathbf{F}}^p_{ks}\mathbf{F}^{p-1}_{ks}\mathbf{F}^p_{ks} + \mathbf{F}^{pT}_{ks}(\dot{\mathbf{F}}^p_{ks}\mathbf{F}^{p-1}_{ks})^T\mathbf{F}^p_{ks} = \mathbf{F}^{pT}_{ks}\overline{\mathbf{L}}^p_{ks}\mathbf{F}^p_{ks} + \mathbf{F}^{pT}_{ks}\overline{\mathbf{L}}^{pT}_{ks}\mathbf{F}^p_{ks} = 2\mathbf{F}^{pT}_{ks}\overline{\mathbf{D}}^p_{ks}\mathbf{F}^p_{ks} \end{aligned}$$

Analogously, one has

$$\dot{\widehat{\mathbf{C}}}{}^p_{cs} = 2{}^p_{cs}\overleftarrow{\overline{\mathbf{D}}}{}^p_{cs\widehat{G}\widehat{G}} = 2\mathbf{F}^{pT}_{cs}\overline{\mathbf{D}}^p_{cs}\mathbf{F}^p_{cs} = 2\mathbf{F}^{pT}_{cs}(\overline{\mathbf{D}}^p - \overline{\mathbf{D}}^p_{cd})\mathbf{F}^p_{cs} \tag{12.20}$$

Further, it follows for the dissipative parts that

$$\left.\begin{aligned} &\dot{\mathbf{C}}^p = 2{}^p\overleftarrow{\overline{\mathbf{D}}}{}^p{}_{GG} = 2\mathbf{F}^{pT}\overline{\mathbf{D}}^p\mathbf{F}^p, \\ &\dot{\mathbf{C}}^p_{kd} = 2{}^p_{kd}\overleftarrow{\overline{\mathbf{D}}}{}^p_{kd\,GG} = 2\mathbf{F}^{pT}_{kd}\overline{\mathbf{D}}^p_{kd}\mathbf{F}^p_{kd}, \dot{\mathbf{C}}^p_{cd} = 2{}^p_{cd}\overleftarrow{\overline{\mathbf{D}}}{}^p_{cd\,GG} = 2\mathbf{F}^{pT}_{cd}\overline{\mathbf{D}}^p_{cd}\mathbf{F}^p_{cd} \end{aligned}\right\} \tag{12.21}$$

noting

$$\begin{aligned} \dot{\mathbf{C}}^p &= (\mathbf{F}^{pT}\mathbf{F}^p)^{\bullet} = \mathbf{F}^{pT}\dot{\mathbf{F}}^p + \dot{\mathbf{F}}^{pT}\mathbf{F}^p = \mathbf{F}^{pT}(\dot{\mathbf{F}}^p\mathbf{F}^{p-1} + \mathbf{F}^{p-T}\dot{\mathbf{F}}^{pT})\mathbf{F}^p \\ &= \mathbf{F}^{pT}[\dot{\mathbf{F}}^p\mathbf{F}^{p-1} + (\dot{\mathbf{F}}^p\mathbf{F}^{p-1})^T]\mathbf{F}^p \end{aligned}$$

12.3 Stress Measures

Introduce the second Piola-Kirchhoff stress tensor in the intermediate configuration, which is the contravariant pulled-back (Eq. (4.44)) of the Kirchhoff stress tensor, i.e.

$$\overline{\mathbf{S}}\left(=\overline{\mathbf{S}}^T\right)\equiv {}^p\overset{\rightarrow\bar{G}\bar{G}}{\mathbf{S}}=\mathbf{F}^p\mathbf{S}\mathbf{F}^{pT}=\mathbf{F}^{e-1}\left(\mathbf{F}\mathbf{S}\mathbf{F}^T\right)\mathbf{F}^{e-T}\equiv {}^e\overset{\leftarrow\bar{G}\bar{G}}{\boldsymbol{\tau}}=\mathbf{F}^{e-1}\boldsymbol{\tau}\mathbf{F}^{e-T} \tag{12.22}$$

and the *Mandel stress*

$$\overline{\mathbf{M}}\equiv\overline{\mathbf{C}}^e\overline{\mathbf{S}}=\mathbf{F}^{eT}\boldsymbol{\tau}\mathbf{F}^{e-T}(\neq\overline{\mathbf{M}}^T) \tag{12.23}$$

noting

$$\overline{\mathbf{C}}^e\overline{\mathbf{S}}=\left(\mathbf{F}^{eT}\mathbf{F}^e\right)\left(\mathbf{F}^{e-1}\boldsymbol{\tau}\mathbf{F}^{e-T}\right)=\mathbf{F}^{eT}\boldsymbol{\tau}\mathbf{F}^{e-T} \tag{12.24}$$

Here, note that the work-conjugate stress measure with the strain rare $\overline{\mathbf{L}}$ in the intermediate configuration is the Mandel stress $\overline{\mathbf{M}}$ as known from

$$\begin{aligned}\boldsymbol{\tau}:\boldsymbol{l}&=\mathrm{tr}\left[\left(\mathbf{F}^e\overline{\mathbf{S}}\mathbf{F}^{eT}\right)\left(\mathbf{F}^e\overline{\mathbf{L}}\mathbf{F}^{e-1}\right)^T\right]=\mathrm{tr}(\mathbf{F}^e\overline{\mathbf{S}}\mathbf{F}^{eT}\mathbf{F}^{e-T}\overline{\mathbf{L}}^T\mathbf{F}^{eT})\\&=\mathrm{tr}(\mathbf{F}^{eT}\mathbf{F}^e\overline{\mathbf{S}}\overline{\mathbf{L}}^T)=\mathrm{tr}(\overline{\mathbf{C}}^e\overline{\mathbf{S}}\overline{\mathbf{L}}^T)=\overline{\mathbf{C}}^e\overline{\mathbf{S}}:\overline{\mathbf{L}}=\overline{\mathbf{M}}:\overline{\mathbf{L}}\end{aligned}$$

Further, the *contravariant* push-forward (Eq. (4.44)) of the kinematic hardening variable $\hat{\mathbf{S}}_k$ and the elastic-core $\hat{\mathbf{S}}_c$ to the intermediate configuration $\overline{\mathcal{K}}$ is given by

$$\left.\begin{aligned}\overline{\mathbf{S}}_k&\equiv {}^p_{ks}\overset{\rightarrow\bar{G}\bar{G}}{\hat{\mathbf{S}}_k}=\mathbf{F}^p_{ks}\hat{\mathbf{S}}_k\mathbf{F}^{pT}_{ks}(=\overline{\mathbf{S}}_k^T),\quad \hat{\mathbf{S}}_k\equiv {}^p_{cs}\overset{\leftarrow\hat{G}\hat{G}}{\overline{\mathbf{S}}_k}=\mathbf{F}^{p-1}_{ks}\overline{\mathbf{S}}_k\mathbf{F}^{p-T}_{ks}(=\hat{\mathbf{S}}_k^T)\\ \overline{\mathbf{S}}_c&\equiv {}^p_{cs}\overset{\rightarrow\bar{G}\bar{G}}{\breve{\mathbf{S}}_c}=\mathbf{F}^p_{cs}\hat{\mathbf{S}}_c\mathbf{F}^{pT}_{cs}(=\overline{\mathbf{S}}_c^T),\quad \hat{\mathbf{S}}_c\equiv {}^p_{cs}\overset{\leftarrow\breve{G}\breve{G}}{\overline{\mathbf{S}}_c}=\mathbf{F}^{p-1}_{cs}\overline{\mathbf{S}}_c\mathbf{F}^{p-T}_{cs}(=\breve{\mathbf{S}}_c^T)\end{aligned}\right\} \tag{12.25}$$

Further, the Mandel-like variables $\overline{\mathbf{M}}_k$ and $\overline{\mathbf{M}}_c$ for the kinematic hardening variable and the elastic-core, respectively, are defined as

$$\left.\begin{aligned}\overline{\mathbf{M}}_k=\overline{\mathbf{C}}^p_{ks}\overline{\mathbf{S}}_k=\overline{\mathbf{G}}\overline{\mathbf{S}}_k=\overline{\mathbf{S}}_k=\mathbf{F}^p_{ks}\hat{\mathbf{S}}_k\mathbf{F}^{pT}_{ks}={}^p_{ks}\overset{\rightarrow}{\hat{\mathbf{M}}}{}_k{}^{\bullet\bar{G}}_{\hat{G}}=\mathbf{F}^{p-T}_{ks}\hat{\mathbf{M}}_k\mathbf{F}^{pT}_{ks}(\neq\overline{\mathbf{M}}_k^T)\\ \hat{\mathbf{M}}_k=\hat{\mathbf{C}}^p_{ks}\hat{\mathbf{S}}_k={}^p_{ks}\overset{\rightarrow}{\overline{\mathbf{M}}}{}_k{}^{\bullet\hat{G}}_{\bar{G}}=\mathbf{F}^{pT}_{ks}\overline{\mathbf{M}}_k\mathbf{F}^{p-T}_{ks}(\neq\hat{\mathbf{M}}_k^T)\end{aligned}\right\} \tag{12.26}$$

$$\left.\begin{aligned}\overline{\mathbf{M}}_c&=\overline{\mathbf{C}}^p_{cs}\overline{\mathbf{S}}_c=\bar{\mathbf{G}}\overline{\mathbf{S}}_c=\overline{\mathbf{S}}_c={}^p_{cs}\overset{\rightarrow}{\breve{\mathbf{M}}}{}_c{}^{\bullet\bar{G}}_{\breve{G}}=\mathbf{F}^{p-T}_{cs}\breve{\mathbf{M}}_c\mathbf{F}^{pT}_{cs}\left(\neq\overline{\mathbf{M}}_c^T\right)\\ \breve{\mathbf{M}}_c&=\breve{\mathbf{C}}^p_{cs}\breve{\mathbf{S}}_c={}^p_{cs}\overset{\leftarrow}{\overline{\mathbf{M}}}{}_c{}^{\bullet\breve{G}}_{\bar{G}}=\mathbf{F}^{pT}_{cs}\overline{\mathbf{M}}_c\mathbf{F}^{p-T}_{cs}\left(\neq\breve{\mathbf{M}}_c^T\right)\end{aligned}\right\} \tag{12.27}$$

noting Eq. (12.4), (12.25) and $\overline{\mathbf{S}}_k=\mathbf{F}^p_{ks}\hat{\mathbf{S}}_k\mathbf{F}^{pT}_{ks}=\mathbf{F}^p_{ks}\mathbf{C}^{p-1}_{ks}\hat{\mathbf{M}}_k\mathbf{F}^{pT}_{ks}=\mathbf{F}^{p-T}_{ks}\hat{\mathbf{M}}_k\mathbf{F}^{pT}_{ks}$.

The material-time derivative of the kinematic hardening variable $\overline{\mathbf{S}}_k$ in the intermediate configuration is given by

$$\dot{\overline{\mathbf{S}}}_k = \mathbf{F}^p_{ks}\dot{\hat{\mathbf{S}}}_k\mathbf{F}^{pT}_{ks} + 2\mathrm{sym}\left[\overline{\mathbf{L}}^p_{ks}\overline{\mathbf{S}}_k\right] \tag{12.28}$$

from Eqs. (12.14) and (12.25), noting

$$\begin{aligned}
\dot{\overline{\mathbf{S}}}_k &= \mathbf{F}^p_{ks}\dot{\hat{\mathbf{S}}}_k\mathbf{F}^{pT}_{ks} + \dot{\mathbf{F}}^p_{ks}\hat{\mathbf{S}}_k\mathbf{F}^{pT}_{ks} + \mathbf{F}^p_{ks}\hat{\mathbf{S}}_k\dot{\mathbf{F}}^{pT}_{ks} \\
&= \mathbf{F}^p_{ks}\dot{\hat{\mathbf{S}}}_k\mathbf{F}^{pT}_{ks} + \dot{\mathbf{F}}^p_{ks}\mathbf{F}^{p-1}_{ks}\overline{\mathbf{S}}_k\mathbf{F}^{p-T}_{ks}\mathbf{F}^{pT}_{ks} + \mathbf{F}^p_{ks}\mathbf{F}^{p-1}_{ks}\overline{\mathbf{S}}_k\mathbf{F}^{p-T}_{ks}\dot{\mathbf{F}}^{pT}_{ks} \\
&= \mathbf{F}^p_{ks}\dot{\hat{\mathbf{S}}}_k\mathbf{F}^{pT}_{ks} + \dot{\mathbf{F}}^p_{ks}\mathbf{F}^{p-1}_{ks}\overline{\mathbf{S}}_k + \overline{\mathbf{S}}_k\mathbf{F}^{p-T}_{ks}\dot{\mathbf{F}}^{pT}_{ks} \\
&= \mathbf{F}^p_{ks}\dot{\hat{\mathbf{S}}}_k\mathbf{F}^{pT}_{ks} + \overline{\mathbf{L}}^p_{ks}\overline{\mathbf{S}}_k + (\overline{\mathbf{L}}^p_{ks}\overline{\mathbf{S}}_k)^T
\end{aligned}$$

Further, the material-time derivative of $\overline{\mathbf{M}}_k$ is given from Eq. (12.28) with Eqs. (12.13) and (12.26) by

$$\dot{\overline{\mathbf{M}}}_k = \dot{\overline{\mathbf{S}}}_k = \mathbf{F}^p_{ks}\dot{\hat{\mathbf{S}}}_k\mathbf{F}^{pT}_{ks} + 2\mathrm{sym}\left[\overline{\mathbf{L}}^p_{ks}\overline{\mathbf{M}}_k\right] = \mathbf{F}^p_{ks}\dot{\hat{\mathbf{S}}}_k\mathbf{F}^{pT}_{ks} + 2\mathrm{sym}\left[(\overline{\mathbf{L}} - \overline{\mathbf{L}}^p_{kd})\overline{\mathbf{M}}_k\right] \tag{12.29}$$

Analogously, the following relation holds for Mandel-like elastic-core stress.

$$\dot{\overline{\mathbf{M}}}_c = \dot{\overline{\mathbf{S}}}_c = \mathbf{F}^p_{cs}\dot{\breve{\mathbf{S}}}_c\mathbf{F}^{pT}_{cs} + 2\mathrm{sym}\left[\overline{\mathbf{L}}^p_{cs}\overline{\mathbf{M}}_c\right] = \mathbf{F}^p_{cs}\dot{\breve{\mathbf{S}}}_c\mathbf{F}^{pT}_{cs} + 2\mathrm{sym}\left[(\overline{\mathbf{L}} - \overline{\mathbf{L}}^p_{cd})\overline{\mathbf{M}}_c\right] \tag{12.30}$$

12.4 Hyperelastic Constitutive Equations

Now, the 2nd Piola-Kirchhoff stress push-forwarded to the intermediate configuration, $\overline{\mathbf{S}}$, is given by the following equation with the strain energy function $\psi(\overline{\mathbf{C}}^e)$, noting Eq. $(5.6)_4$, where $\overline{\mathbf{C}}^e$ stands for the purely elastic deformation because of $\overline{\mathbf{C}}^e = \overline{\mathbf{U}}^{e2}$ as shown in Eq. (12.3).

$$\overline{\mathbf{S}} = 2\frac{\partial\psi^e(\overline{\mathbf{C}}^e)}{\partial\overline{\mathbf{C}}^e} \tag{12.31}$$

and the Mandel stress is given by

$$\overline{\mathbf{M}} \equiv \overline{\mathbf{C}}^e\overline{\mathbf{S}} = 2\overline{\mathbf{C}}^e\frac{\partial\psi^e(\overline{\mathbf{C}}^e)}{\partial\overline{\mathbf{C}}^e}(\neq \overline{\mathbf{M}}^T) \tag{12.32}$$

The rate of the Mandel stress is given noting Eq. (12.12) as

$$\overset{\bullet}{\overline{\mathbf{M}}} = (\overline{\mathbf{C}}^e\overline{\mathbf{S}})^\bullet = \overline{\mathbb{L}}^e : \overset{\bullet}{\overline{\mathbf{C}}}^e = \overline{\mathbb{L}}^e : \mathrm{sym}[\overline{\mathbf{C}}^e(\overline{\mathbf{L}} - \overline{\mathbf{L}}^p)] \tag{12.33}$$

where $\overline{\mathbb{L}}^e$ is the fourth-order hyperelastic tangent modulus tensor given by

$$\overline{\mathbb{L}}^e \equiv \frac{\partial \overline{\mathbf{M}}}{\partial \overline{\mathbf{C}}^e} = \overline{\mathbf{S}} + \frac{1}{2}\overline{\mathbf{C}}^e : \overline{\mathbb{C}}^e \tag{12.34}$$

with

$$\overline{\mathbb{C}}^e \equiv 2\frac{\partial \overline{\mathbf{S}}}{\partial \overline{\mathbf{C}}^e} = 4\frac{\partial^2 \psi^e(\overline{\mathbf{C}}^e)}{\partial \overline{\mathbf{C}}^e \otimes \partial \overline{\mathbf{C}}^e} \tag{12.35}$$

Further, let the 2nd Piola-Kirchhoff stress-like variables for the kinematic hardening variable $\widehat{\mathbf{L}}_k$ based in $\widehat{\mathcal{K}}$ and for the elastic-core $\breve{\mathbf{S}}_c$ based in $\breve{\mathcal{K}}$ be formulated by the potential energy functions $\psi^k(\widehat{\mathbf{C}}^p_{ks})$ and $\psi^c(\breve{\mathbf{C}}^p_{cs})$, noting Eq. (12.25) with Eq. (12.4), as follows:

$$\widehat{\mathbf{S}}_k = 2\frac{\partial \psi^k(\widehat{\mathbf{C}}^p_{ks})}{\partial \widehat{\mathbf{C}}^p_{ks}}, \quad \breve{\mathbf{S}}_c = 2\frac{\partial \psi^c(\breve{\mathbf{C}}^p_{cs})}{\partial \breve{\mathbf{C}}^p_{cs}} \tag{12.36}$$

$$\overline{\mathbf{S}}_k = {}^p_{ks}\overset{\rightarrow}{\widehat{\mathbf{S}}}_k{}^{\overline{G}\overline{G}} = 2\mathbf{F}^p_{ks}\frac{\partial \psi^k(\widehat{\mathbf{C}}^p_{ks})}{\partial \widehat{\mathbf{S}}^p_{ks}}\mathbf{F}^{pT}_{ks}, \quad \overline{\mathbf{S}}_c = {}^p_{cs}\overset{\rightarrow}{\breve{\mathbf{S}}}_c{}^{\overline{G}\overline{G}} = 2\mathbf{F}^p_{cs}\frac{\partial \psi^c(\breve{\mathbf{C}}^p_{cs})}{\partial \breve{\mathbf{C}}^p_{cs}}\mathbf{F}^{pT}_{cs} \tag{12.37}$$

$$\widehat{\mathbf{M}}_k = \widehat{\mathbf{C}}^p_{ks}\widehat{\mathbf{S}}_k = 2\widehat{\mathbf{C}}^p_{ks}\frac{\partial \psi^k(\widehat{\mathbf{C}}^p_{ks})}{\partial \widehat{\mathbf{C}}^p_{ks}}, \quad \breve{\mathbf{M}}_c = \breve{\mathbf{C}}^p_{cs}\breve{\mathbf{S}}_c = 2\breve{\mathbf{C}}^p_{cs}\frac{\partial \psi^c(\breve{\mathbf{C}}^p_{cs})}{\partial \breve{\mathbf{C}}^p_{cs}} \tag{12.38}$$

$$\overline{\mathbf{M}}_k = \overline{\mathbf{C}}^p_{ks}\overline{\mathbf{S}}_k = \overline{\mathbf{S}}_k = 2\mathbf{F}^p_{ks}\frac{\partial \psi^k(\widehat{\mathbf{C}}^p_{ks})}{\partial \widehat{\mathbf{C}}^p_{ks}}\mathbf{F}^{pT}_{ks}, \quad \overline{\mathbf{M}}_c = \overline{\mathbf{C}}^p_{cs}\overline{\mathbf{S}}_c = \overline{\mathbf{S}}_c = 2\mathbf{F}^p_{cs}\frac{\partial \psi^c(\breve{\mathbf{C}}^p_{cs})}{\partial \breve{\mathbf{C}}^p_{cs}}\mathbf{F}^{pT}_{cs} \tag{12.39}$$

The tensors $\overline{\mathbf{M}}, \overline{\mathbf{M}}_k, \overline{\mathbf{M}}_c$ satisfy the symmetries, i.e. $\overline{\mathbf{M}} = \overline{\mathbf{M}}^T$, $\overline{\mathbf{M}}_k = \overline{\mathbf{M}}_k^T$, $\overline{\mathbf{M}}_c = \overline{\mathbf{M}}_c^T$ for particular cases of the strain energy functions ψ^e, ψ^k, ψ^c only of the tensors $\overline{\mathbf{C}}^e$, $\widehat{\mathbf{C}}^p_{ks}$, $\breve{\mathbf{C}}^p_{cs}$, respectively, while the elastic-isotropy is caused only from the first one.

The rates of $\widehat{\mathbf{S}}_k$ and $\breve{\mathbf{S}}_c$ are given from Eq. (12.36) as

$$\overset{\bullet}{\widehat{\mathbf{S}}}_k = \widehat{\mathbb{C}}^k : \frac{1}{2}\overset{\bullet}{\widehat{\mathbf{C}}}{}^p_{ks},\ \overset{\bullet}{\breve{\mathbf{S}}}_c = \breve{\mathbb{C}}^c : \frac{1}{2}\overset{\bullet}{\breve{\mathbf{C}}}{}^p_{cs} \tag{12.40}$$

where

$$\widehat{\mathbb{C}}^k \equiv 2\frac{\partial \widehat{\mathbf{S}}^p_{ks}}{\partial \widehat{\mathbf{C}}^p_{ks}} = 4\frac{\partial^2 \psi^k(\widehat{\mathbf{C}}^p_{ks})}{\partial \widehat{\mathbf{C}}^p_{ks} \otimes \partial \widehat{\mathbf{S}}^p_{ks}},\quad \breve{\mathbb{C}}^c \equiv 2\frac{\partial \breve{\mathbf{S}}^p_{ks}}{\partial \breve{\mathbf{C}}^p_{cs}} = 4\frac{\partial \psi^c(\breve{\mathbf{C}}^p_{cs})}{\partial \breve{\mathbf{C}}^p_{cs} \otimes \partial \breve{\mathbf{C}}^p_{cs}} \tag{12.41}$$

Substituting Eq. (12.40) with Eq. (12.19) into Eq. (12.29), $\overset{\bullet}{\overline{\mathbf{M}}}_k$ is given as follows:

$$\overset{\bullet}{\overline{\mathbf{M}}}_k = \mathbf{F}^p_{ks}\widehat{\mathbb{C}}^k : \mathbf{F}^{pT}_{ks}(\overline{\mathbf{D}}^p - \overline{\mathbf{D}}^p_{kd})\mathbf{F}^p_{ks}\mathbf{F}^{pT}_{ks} + 2\mathrm{sym}[(\overline{\mathbf{L}}^p - \overline{\mathbf{L}}^p_{kd})\overline{\mathbf{M}}_k] \tag{12.42}$$

Analogously, it follows for the elastic-core that

$$\overset{\bullet}{\overline{\mathbf{M}}}_c = \mathbf{F}^p_{cs}\breve{\mathbb{C}}^c : \mathbf{F}^{pT}_{cs}(\overline{\mathbf{D}}^p - \overline{\mathbf{D}}^p_{cd})\mathbf{F}^p_{cs}\mathbf{F}^{pT}_{cs} + 2\mathrm{sym}[(\overline{\mathbf{L}}^p - \overline{\mathbf{L}}^p_{cd})\overline{\mathbf{M}}_c] \tag{12.43}$$

12.5 Normal-Yield and Subloading Surfaces

The *normal-yield surface* with the isotropic and the kinematic-hardenings is described following Eq. (9.1) in the intermediate configuration by

$$f(\hat{\overline{\mathbf{M}}}) = F(H) \tag{12.44}$$

and the *subloading surface* following Eq. (9.2) by

$$f(\underline{\overline{\mathbf{M}}}) = RF(H) \tag{12.45}$$

in the intermediate configuration, which are depicted in Fig. 12.2, where

$$\underline{\overline{\mathbf{M}}} \equiv \overline{\mathbf{M}} - \underline{\overline{\mathbf{M}}}_k (\neq \underline{\overline{\mathbf{M}}}^T) \tag{12.46}$$

$$\left.\begin{array}{c} \hat{\overline{\mathbf{M}}} \equiv \overline{\mathbf{M}} - \overline{\mathbf{M}}_k(\neq \hat{\overline{\mathbf{M}}}^T),\quad \tilde{\overline{\mathbf{M}}} \equiv \overline{\mathbf{M}} - \overline{\mathbf{M}}_c(\neq \tilde{\overline{\mathbf{M}}}^T), \\ \hat{\overline{\mathbf{M}}}_c \equiv \overline{\mathbf{M}}_c - \overline{\mathbf{M}}_k(\neq \overline{\mathbf{M}}_c^T) \end{array}\right\} \tag{12.47}$$

$\underline{\overline{\mathbf{M}}}_k$ is the conjugate point in the subloading surface to $\overline{\mathbf{M}}_k$ in the normal-yield surface. The following relations hold.

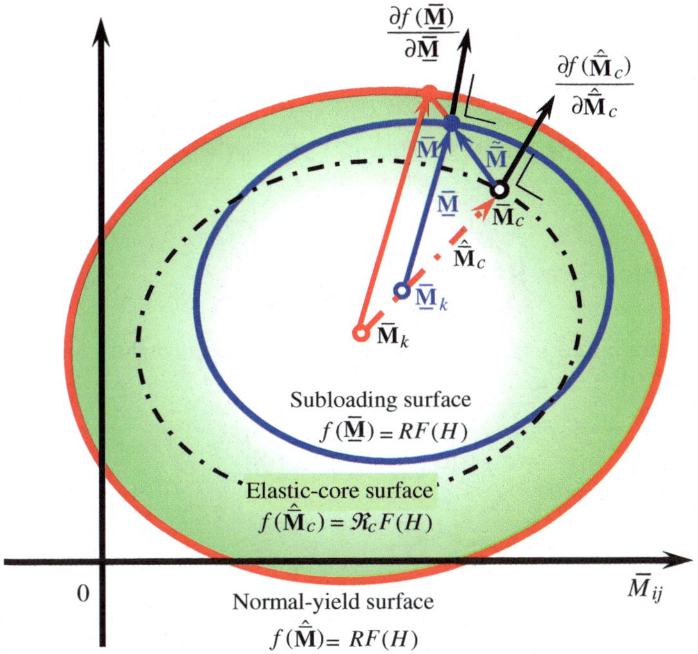

Fig. 12.2 Normal-yield, subloading and elastic-core surfaces in the intermediate configuration in multiplicative elastoplasticity theory

$$\underline{\overline{\mathbf{M}}}_k = \overline{\mathbf{M}}_c - R\hat{\overline{\mathbf{M}}}_c \;\; (\underline{\overline{\mathbf{M}}}_k - \overline{\mathbf{M}}_c = R(\overline{\mathbf{M}}_c - \overline{\mathbf{M}}_k)) \tag{12.48}$$

the rate of which is given by

$$\dot{\underline{\overline{\mathbf{M}}}}_k = R\dot{\overline{\mathbf{M}}}_k + (1-R)\dot{\overline{\mathbf{M}}}_c - \dot{R}\hat{\overline{\mathbf{M}}}_c \tag{12.49}$$

leading to

$$\dot{\underline{\overline{\mathbf{M}}}} \equiv \dot{\overline{\mathbf{M}}} - R\dot{\overline{\mathbf{M}}}_k - (1-R)\dot{\overline{\mathbf{M}}}_c + \dot{R}\hat{\overline{\mathbf{M}}}_c \tag{12.50}$$

The variables in the hypoelastic-based plasticity described in the previous sections correspond to the following variables in the intermediate configuration for the multiplicative hyperelastic-based plasticity.

$$\left.\begin{array}{l} \boldsymbol{\sigma} \rightarrow \overline{\mathbf{M}}, \; \overline{\boldsymbol{\sigma}} \rightarrow \underline{\overline{\mathbf{M}}} = \overline{\mathbf{M}} - \underline{\overline{\mathbf{M}}}_k \\ \boldsymbol{\alpha} \rightarrow \overline{\mathbf{M}}_k, \; \overline{\boldsymbol{\alpha}} \rightarrow \underline{\overline{\mathbf{M}}}_k \\ \mathbf{c} \rightarrow \overline{\mathbf{M}}_c, \; \hat{\mathbf{c}} \rightarrow \hat{\overline{\mathbf{M}}}_c = \overline{\mathbf{M}}_c - \overline{\mathbf{M}}_k \\ \tilde{\boldsymbol{\sigma}} \rightarrow \tilde{\overline{\mathbf{M}}} = \overline{\mathbf{M}} - \overline{\mathbf{M}}_c \end{array}\right\} \tag{12.51}$$

The *elastic-core surface* which passes through the elastic-core $\overline{\mathbf{M}}_c$ and is similar to the normal-yield surface with respect to the back-stress $\overline{\mathbf{M}}_k$ in the hyperelastic-based-plasticity is given following Eq. (9.17) as follows:

$$f(\hat{\overline{\mathbf{M}}}_c) = \mathfrak{R}_c F(H), \quad \text{i.e. } \mathfrak{R}_c = f(\hat{\overline{\mathbf{M}}}_c)/F(H) \tag{12.52}$$

12.6 Plastic Flow Rules

The plastic strain rate is given in the following associated flow rule proposed by Hashiguchi (2016a, b, c, d).

$$\boxed{\overline{\mathbf{D}}^p = \dot{\bar{\lambda}}\underline{\overline{\mathbf{N}}}\;(\dot{\bar{\lambda}} \geq 0)} \tag{12.53}$$

where $\dot{\bar{\lambda}}$ is the plastic multiplier and

$$\underline{\overline{\mathbf{N}}} \equiv \text{sym}\left[\frac{\partial f(\underline{\overline{\mathbf{M}}})}{\partial \overline{\mathbf{M}}}\right] \Big/ \left\| \text{sym}\left[\frac{\partial f(\underline{\overline{\mathbf{M}}})}{\partial \overline{\mathbf{M}}}\right] \right\| \quad (\|\underline{\overline{\mathbf{N}}}\| = 1) \tag{12.54}$$

which is the normalized and symmetrized tensor. If the symmetries of the Mandel stress and Mandel-like kinematic hardening variable, i.e. $\overline{M} = \overline{M}^T$ and $\overline{M}_k = \overline{M}_k^T$ hold, which are provided by the strain energy functions ψ^e and ψ^k of only $\overline{\mathbf{C}}^e$ and $\widehat{\mathbf{C}}^p_{ks}$, respectively, one obtains the symmetry $\partial f(\underline{\overline{\mathbf{M}}})/\partial \overline{\mathbf{M}} = (\partial f(\underline{\overline{\mathbf{M}}})/\partial \overline{\mathbf{M}})^T$.

The symmetric plastic dissipative parts of the plastic velocity gradients for the kinematic hardening variable and the elastic-core in Eqs. (12.15) and (12.18) are assumed following Eqs. (9.11) and (9.15) as follows:

$$\boxed{\overline{\mathbf{D}}^p_{kd} = \frac{1}{b_k}\|\overline{\mathbf{D}}^{p\prime}\|\text{sym}[\overline{\mathbf{M}}_k]} = \frac{1}{b_k}\dot{\bar{\lambda}}\,\|\underline{\overline{\mathbf{N}}}'\|\text{sym}[\overline{\mathbf{M}}_k] \tag{12.55}$$

$$\boxed{\overline{\mathbf{D}}^p_{cd} = \frac{\mathfrak{R}_c}{\xi}\|\overline{\mathbf{D}}^p\|\hat{\overline{\mathbf{N}}}_c} = \frac{\mathfrak{R}_c}{\xi}\dot{\bar{\lambda}}\,\hat{\overline{\mathbf{N}}}_c \tag{12.56}$$

where

$$\hat{\overline{\mathbf{N}}}_c \equiv \text{sym}\left[\frac{\partial f(\hat{\overline{\mathbf{M}}}_c)}{\partial \hat{\overline{\mathbf{M}}}_c}\right] \Big/ \left\| \text{sym}\left[\frac{\partial f(\hat{\overline{\mathbf{M}}}_c)}{\partial \hat{\overline{\mathbf{M}}}_c}\right] \right\| \quad (\|\hat{\overline{\mathbf{N}}}_c\| = 1) \tag{12.57}$$

$$\hat{\overline{\mathbf{M}}}_c \equiv \overline{\mathbf{M}}_c - \overline{\mathbf{M}}_k \tag{12.58}$$

In the material parameter $u = \bar{u}\exp(u_c \mathfrak{R}_c C_\sigma)$ in Eq. (9.47) for the Masing effect, $\mathfrak{R}_c$ is given by Eq. (12.51) and C_σ is given by

$$C_\sigma \equiv \hat{\underline{\mathrm{N}}}_c : \overline{\underline{\mathrm{N}}} (-1 \leq C_\sigma \leq 1) \tag{12.59}$$

Let the spins $\overline{\mathbf{W}}^p$, $\widehat{\mathbf{W}}^p_{kd}$ and $\breve{\mathbf{W}}^p_{cd}$ in Eqs. (12.11), (12.15) and (12.18), which are induced by the plastic and the dissipative parts, be given by extending Eq. (9.48) as follows:.

$$\boxed{\begin{aligned}
\overline{\mathbf{W}}^p &= \eta^p(\overline{\mathbf{M}}\,\overline{\mathbf{D}}^p - \overline{\mathbf{D}}^p\,\overline{\mathbf{M}}) = \eta^p\,\dot{\bar{\lambda}}(\overline{\mathbf{M}}\,\overline{\underline{\mathrm{N}}} - \overline{\underline{\mathrm{N}}}\,\overline{\mathbf{M}})\\
\overline{\mathbf{W}}^p_{kd} &= \eta^p_k(\overline{\mathbf{M}}\,\overline{\mathbf{D}}^p_{kd} - \overline{\mathbf{D}}^p_{kd}\overline{\mathbf{M}}) = (\eta^p_k/b_k)\,\dot{\bar{\lambda}}\,||\overline{\underline{\mathrm{N}}}'||(\overline{\mathbf{M}}\,\mathrm{sym}[\overline{\mathbf{M}}_k] - \mathrm{sym}[\overline{\mathbf{M}}_k]\overline{\mathbf{M}})\\
\overline{\mathbf{W}}^p_{cd} &= \eta^p_c(\overline{\mathbf{M}}\,\overline{\mathbf{D}}^p_{cd} - \overline{\mathbf{D}}^p_{cd}\overline{\mathbf{M}}) = \eta^p_c(\mathfrak{R}_c/\xi)\,\dot{\bar{\lambda}}(\overline{\mathbf{M}}\,\hat{\underline{\mathrm{N}}}_c - \hat{\underline{\mathrm{N}}}_c\overline{\mathbf{M}})
\end{aligned}} \tag{12.60}$$

where η^p_k and η^p_c are the material parameters, while the flow rules in Eqs. (12.53), (12.55) and (12.56) are exploited. The plastic spin tensor $\overline{\mathbf{W}}^p$ diminishes if the symmetry of the Mandel stress, i.e. $\overline{\mathbf{M}} = \overline{\mathbf{M}}^T$ due to the elastic isotropy and the plastic isotropy due to $\overline{\mathbf{M}}_k = \overline{\mathbf{M}}_c = \mathbf{O}$ hold. Further, the spin tensors $\overline{\mathbf{W}}^p_{kd}$ and $\overline{\mathbf{W}}^p_{cd}$ diminish for the plastic-isotropy due to $\overline{\mathbf{M}}_k = \overline{\mathbf{M}}_c = \mathbf{O}$.

The velocity gradients are given by substituting Eqs. (12.53), (12.55), (12.56) and (12.60) into Eqs. (12.9), (12.15) and (12.18) as follows:

$$\left.\begin{aligned}
\overline{\mathbf{L}}^p &= \dot{\bar{\lambda}}[\overline{\underline{\mathrm{N}}} + \eta^p(\overline{\mathbf{M}}\,\overline{\mathrm{N}} - \overline{\underline{\mathrm{N}}}\,\overline{\mathbf{M}})]\\
\overline{\mathbf{L}}^p_{kd} &= (1/b_k)\dot{\bar{\lambda}}||\overline{\underline{\mathrm{N}}}'||\{\mathrm{sym}[\overline{\mathbf{M}}_k] + \eta^p_k(\overline{\mathbf{M}}\,\mathrm{sym}[\overline{\mathbf{M}}_k] - \mathrm{sym}[\overline{\mathbf{M}}_k]\overline{\mathbf{M}})\}\\
\overline{\mathbf{L}}^p_{cd} &= (\mathfrak{R}_c/\xi)\,\dot{\bar{\lambda}}[\hat{\underline{\mathrm{N}}}_c + \eta^p_c(\overline{\mathbf{M}}\,\hat{\underline{\mathrm{N}}}_c - \hat{\underline{\mathrm{N}}}_c\overline{\mathbf{M}})]
\end{aligned}\right\} \tag{12.61}$$

The substitutions of Eq. (12.61) into Eqs. (12.33), (12.42) and (12.43) yield:

$$\dot{\overline{\mathbf{M}}} = \overline{\mathbb{L}}^e : \mathrm{sym}[\overline{\mathbf{C}}^e\{\overline{\mathbf{L}} - \dot{\bar{\lambda}}\,[\overline{\underline{\mathrm{N}}} + \eta^p(\overline{\mathbf{M}}\,\overline{\underline{\mathrm{N}}} - \overline{\underline{\mathrm{N}}}\,\overline{\mathbf{M}})]\}] \tag{12.62}$$

$$\begin{aligned}
\dot{\overline{\mathbf{M}}}_k = \dot{\bar{\lambda}}\{&\mathbf{F}^p_{ks}\widehat{\mathbb{C}}^k : \mathbf{F}^{pT}_{ks}(\overline{\underline{\mathrm{N}}} - (1/b_k)||\overline{\underline{\mathrm{N}}}'||\mathrm{sym}[\overline{\mathbf{M}}_k])\mathbf{F}^p_{ks}\mathbf{F}^{pT}_{ks} + 2\mathrm{sym}[(\overline{\underline{\mathrm{N}}} + \eta^p(\overline{\mathbf{M}}\,\overline{\underline{\mathrm{N}}} - \overline{\underline{\mathrm{N}}}\,\overline{\mathbf{M}})\\
&- (1/b_k)||\overline{\underline{\mathrm{N}}}'||\{\mathrm{sym}[\overline{\mathbf{M}}_k] + \eta^p_k(\overline{\mathbf{M}}\,\mathrm{sym}[\overline{\mathbf{M}}_k] - \mathrm{sym}[\overline{\mathbf{M}}_k]\overline{\mathbf{M}})\})\overline{\mathbf{M}}_k]\}
\end{aligned} \tag{12.63}$$

$$\begin{aligned}
\dot{\overline{\mathbf{M}}}_c = \dot{\bar{\lambda}}\{&\mathbf{F}^p_{cs}\widehat{\mathbb{C}}^c : \mathbf{F}^{pT}_{cs}(\overline{\underline{\mathrm{N}}} - (\mathfrak{R}_c/\xi)\hat{\underline{\mathrm{N}}}_c)\mathbf{F}^p_{cs}\mathbf{F}^{pT}_{cs}\\
&+ 2\mathrm{sym}[(\overline{\underline{\mathrm{N}}} + \eta^p(\overline{\mathbf{M}}\,\overline{\underline{\mathrm{N}}} - \overline{\underline{\mathrm{N}}}\,\overline{\mathbf{M}}) - (\mathfrak{R}_c/\xi)[\hat{\underline{\mathrm{N}}}_c + \eta^p_c(\overline{\mathbf{M}}\,\hat{\underline{\mathrm{N}}}_c - \hat{\underline{\mathrm{N}}}_c\overline{\mathbf{M}})])\overline{\mathbf{M}}_c]\}
\end{aligned} \tag{12.64}$$

12.7 Plastic Strain Rate

The elastic constitutive equation is given by Eqs. (12.31)–(12.33). The plastic strain rate formulated by Hashiguchi (2016a, b, c, d) will be shown in this section. The formulations given in this section is not necessary in the numerical calculation by the return-mapping based on the closet-point projection described in Chap. 20.

The time-differentiation of Eq. (12.45) leads to the consistency condition of the subloading surface as follows:

$$\frac{\partial f(\underline{\overline{\mathbf{M}}})}{\partial \underline{\overline{\mathbf{M}}}} : \underline{\overline{\dot{\mathbf{M}}}} - \dot{R}F - R\dot{F} = 0 \tag{12.65}$$

It holds from Eq. (12.45) that

$$\frac{\partial f(\underline{\overline{\mathbf{M}}})}{\partial \underline{\overline{\mathbf{M}}}} : \underline{\overline{\mathbf{M}}}(=f(\underline{\overline{\mathbf{M}}})) = RF \tag{12.66}$$

by the Euler's theorem for the homogenous function $f(\underline{\overline{\mathbf{M}}})$ of $\underline{\overline{\mathbf{M}}}$ in degree-one, and then it follows that

$$\underline{\overline{\mathbf{N}}} : \underline{\overline{\mathbf{M}}} = \frac{\partial f(\underline{\overline{\mathbf{M}}})}{\partial \underline{\overline{\mathbf{M}}}} : \underline{\overline{\mathbf{M}}} / \left|\left| \frac{\partial f(\underline{\overline{\mathbf{M}}})}{\partial \underline{\overline{\mathbf{M}}}} \right|\right| = f(\underline{\overline{\mathbf{M}}}) / \left|\left| \frac{\partial f(\underline{\overline{\mathbf{M}}})}{\partial \underline{\overline{\mathbf{M}}}} \right|\right| = RF / \left|\left| \frac{\partial f(\underline{\overline{\mathbf{M}}})}{\partial \underline{\overline{\mathbf{M}}}} \right|\right|$$

which leads to

$$1 / \left|\left| \frac{\partial f(\underline{\overline{\mathbf{M}}})}{\partial \underline{\overline{\mathbf{M}}}} \right|\right| = \frac{\underline{\overline{\mathbf{N}}} : \underline{\overline{\mathbf{M}}}}{RF} \tag{12.67}$$

where

$$\underline{\overline{\mathbf{N}}} \equiv \frac{\partial f(\underline{\overline{\mathbf{M}}})}{\partial \underline{\overline{\mathbf{M}}}} / \left|\left| \frac{\partial f(\underline{\overline{\mathbf{M}}})}{\partial \underline{\overline{\mathbf{M}}}} \right|\right| (\neq \underline{\overline{\mathbf{N}}}^T, ||\underline{\overline{\mathbf{N}}}|| = 1) \tag{12.68}$$

The substitution of Eq. (12.67) into Eq. (12.65) leads to

$$\underline{\overline{\mathbf{N}}} : \underline{\overline{\dot{\mathbf{M}}}} - \left(\frac{\dot{F}}{F} + \frac{\dot{R}}{R}\right) \underline{\overline{\mathbf{N}}} : \underline{\overline{\mathbf{M}}} = 0 \tag{12.69}$$

The substitution of Eq. (12.50) into Eq. (12.69) leads to

$$\underline{\overline{\mathbf{N}}} : \underline{\overline{\dot{\mathbf{M}}}} - \underline{\overline{\mathbf{N}}} : \left[\frac{\dot{F}}{F} \underline{\overline{\mathbf{M}}} + \frac{\dot{R}}{R} (\underline{\overline{\mathbf{M}}} - R\hat{\underline{\overline{\mathbf{M}}}}_c) + R \underline{\overline{\dot{\mathbf{M}}}}_k + (1 - R) \underline{\overline{\dot{\mathbf{M}}}}_c\right] = 0 \tag{12.70}$$

Further, substituting the relation

$$\underline{\overline{\mathbf{M}}} - R\hat{\overline{\mathbf{M}}}_c = \overline{\mathbf{M}} - \underline{\overline{\mathbf{M}}}_k - (\overline{\mathbf{M}}_c - \underline{\overline{\mathbf{M}}}_k) = \tilde{\overline{\mathbf{M}}} \tag{12.71}$$

Eq. (12.70) is rewritten as

$$\underline{\overline{\mathbf{N}}}:\dot{\overline{\mathbf{M}}} - \underline{\overline{\mathbf{N}}}:\left[\frac{F'\dot{H}}{F}\underline{\overline{\mathbf{M}}} + \frac{\dot{R}}{R}\tilde{\overline{\mathbf{M}}} + R\dot{\overline{\mathbf{M}}}_k + (1-R)\dot{\overline{\mathbf{M}}}_c\right] = 0 \tag{12.72}$$

where

$$\dot{H} = f_{Hd}(\overline{\mathbf{M}}, H, \overline{\mathbf{D}}^p/||\overline{\mathbf{D}}^p||)||\overline{\mathbf{D}}^p|| = f_{Hn}(\overline{\mathbf{M}}, H, \underline{\overline{\mathbf{N}}})\dot{\bar{\lambda}} \tag{12.73}$$

$$\dot{R} = U(R)||\overline{\mathbf{D}}^p|| = U(R)\dot{\bar{\lambda}} \quad \text{for} \quad \overline{\mathbf{D}}^p \neq \mathbf{O} \tag{12.74}$$

based on Eqs. (6.37) and (7.9) with Eq. (12.53). The normal-yield ratio is calculated in general from Eq. (9.45) as follows:

$$f(\tilde{\overline{\mathbf{M}}} + R\hat{\overline{\mathbf{M}}}_c) = RF(H) \tag{12.75}$$

which is explicitly described for the Mises metals from Eq. (10.32) as

$$R = \frac{\tilde{\overline{\mathbf{M}}}' : \hat{\overline{\mathbf{M}}}'_c + \sqrt{(\tilde{\overline{\mathbf{M}}}' : \hat{\overline{\mathbf{M}}}'_c)^2 + \left(\frac{2}{3}F^2 - ||\hat{\overline{\mathbf{M}}}'_c||^2\right)||\tilde{\overline{\mathbf{M}}}'||^2}}{\frac{2}{3}F^2 - ||\hat{\overline{\mathbf{M}}}'_c||^2} \tag{12.76}$$

The substitutions of Eqs. (12.63), (12.64), (12.73) and (12.74) into Eq. (12.72) lead to the consistency condition:

$$\underline{\overline{\mathbf{N}}} : \dot{\overline{\mathbf{M}}} - \overline{M}^p \dot{\bar{\lambda}} = 0 \tag{12.77}$$

from which it follows that

$$\dot{\bar{\lambda}} = \frac{\underline{\overline{\mathbf{N}}} : \dot{\overline{\mathbf{M}}}}{\overline{M}^p}, \quad \overline{\mathbf{D}}^p = \frac{\underline{\overline{\mathbf{N}}} : \dot{\overline{\mathbf{M}}}}{\overline{M}^p}\underline{\overline{\mathbf{N}}} \tag{12.78}$$

where

$$
\begin{aligned}
\overline{M}^p \equiv \underline{\overline{\mathrm{N}}}: \Bigg[& \frac{F' f_{Hn}(\overline{\mathbf{M}}, F, \underline{\overline{\mathrm{N}}})}{F} \underline{\overline{\mathbf{M}}} + \frac{U(R)}{R} \tilde{\overline{\mathbf{M}}} \\
& + R\{\mathbf{F}^p_{ks} \widehat{\mathbb{C}}^k : \mathbf{F}^{pT}_{ks} (\underline{\overline{\mathrm{N}}} - (1/b_k)\|\underline{\overline{\mathrm{N}}}'\| \mathrm{sym}[\overline{\mathbf{M}}_k]) \mathbf{F}^p_{ks} \mathbf{F}^{pT}_{ks} + 2\mathrm{sym}[(\underline{\overline{\mathrm{N}}} + \eta^p(\overline{\mathbf{M}}\underline{\overline{\mathrm{N}}} - \underline{\overline{\mathrm{N}}}\overline{\mathbf{M}}) \\
& - (1/b_k)\|\underline{\overline{\mathrm{N}}}'\| \{\mathrm{sym}[\overline{\mathbf{M}}_k] + \eta^p_k(\overline{\mathbf{M}}\mathrm{sym}[\overline{\mathbf{M}}_k] - \mathrm{sym}[\overline{\mathbf{M}}_k]\overline{\mathbf{M}})\}) \overline{\mathbf{M}}_k]\} \\
& + (1 - R)\mathbf{F}^p_{cs} \breve{\mathbb{C}}^c : \mathbf{F}^{pT}_{cs} (\underline{\overline{\mathrm{N}}} - (\Re_c/\xi)\hat{\overline{\mathrm{N}}}_c) \mathbf{F}^p_{cs} \mathbf{F}^{pT}_{cs} \\
& + 2\mathrm{sym}[(\underline{\overline{\mathrm{N}}} + \eta^p(\overline{\mathbf{M}}\underline{\overline{\mathrm{N}}} - \underline{\overline{\mathrm{N}}}\overline{\mathbf{M}}) - (\Re_c/\xi)[\hat{\overline{\mathrm{N}}}_c + \eta^p_c(\overline{\mathbf{M}}\hat{\overline{\mathrm{N}}}_c - \hat{\overline{\mathrm{N}}}_c\overline{\mathbf{M}})])\overline{\mathbf{M}}_c]\} \Bigg]
\end{aligned}
\tag{12.79}
$$

The substitutions of Eq. (12.62) into Eq. (12.77) lead to the consistency condition:

$$
\underline{\overline{\mathrm{N}}}: 2\overline{\mathbb{L}}^e : \mathrm{sym}[\overline{\mathbf{C}}^e \overline{\mathbf{L}}] - \{\underline{\overline{\mathrm{N}}}: \overline{\mathbb{L}}^e : \mathrm{sym}[\overline{\mathbf{C}}^e \{\underline{\overline{\mathrm{N}}} + \eta^p(\overline{\mathbf{M}}\underline{\overline{\mathrm{N}}} - \underline{\overline{\mathrm{N}}}\overline{\mathbf{M}})\}] + \overline{M}^p\} \dot{\overline{\Lambda}} = 0
\tag{12.80}
$$

using the symbol $\dot{\overline{\Lambda}}$ for the plastic multiplier in terms of the strain rate instead of $\dot{\overline{\lambda}}$ in terms of the stress rate. The plastic multiplier is given from Eq. (12.80) as follows:

$$
\dot{\overline{\Lambda}} = \frac{2\underline{\overline{\mathrm{N}}} : \overline{\mathbb{L}}^e : \mathrm{sym}[\overline{\mathbf{C}}^e \overline{\mathbf{L}}]}{\overline{M}^p + \underline{\overline{\mathrm{N}}} : \overline{\mathbb{L}}^e : \mathrm{sym}[\overline{\mathbf{C}}^e \{\underline{\overline{\mathrm{N}}} + \eta^p(\overline{\mathbf{M}}\underline{\overline{\mathrm{N}}} - \underline{\overline{\mathrm{N}}}\overline{\mathbf{M}})\}]}
\tag{12.81}
$$

The loading criterion is given by

$$
\left.
\begin{array}{ll}
\overline{\mathbf{D}}^p \neq \mathbf{O} & \text{for } \dot{\overline{\Lambda}} > 0 \\
\overline{\mathbf{D}}^p = \mathbf{O} & \text{for others}
\end{array}
\right\}
\tag{12.82}
$$

which can be given actually as

$$
\left.
\begin{array}{ll}
\overline{\mathbf{D}}^p \neq \mathbf{O} & \text{for } \underline{\overline{\mathrm{N}}} : \overline{\mathbb{L}}^e : \mathrm{sym}[\overline{\mathbf{C}}^e \overline{\mathbf{L}}] > 0 \\
\overline{\mathbf{D}}^p = \mathbf{O} & \text{for others}
\end{array}
\right\}
\tag{12.83}
$$

12.8 Calculation Procedures

The calculation procedure by the above-mentioned formulations is described in this section.

First, the plastic multiplier $\dot{\overline{\Lambda}}$ is calculated by the input of the velocity gradient $\overline{\mathbf{L}}$ into Eq. (12.81). The forward-Euler method or the return-mapping projection can be adopted to this calculation. Then, substituting it into Eq. (12.61), the plastic and the dissipative parts $\overline{\mathbf{L}}^p$, $\overline{\mathbf{L}}^p_{kd}$ and $\overline{\mathbf{L}}^p_{cd}$ are calculated. Thereafter, the stress and the tensor-valued internal variables are calculated by the method described below.

The rates of the plastic gradient and its dissipative parts are given from Eqs. $(12.6)_3$, $(12.15)_1$ and $(12.18)_1$ as follows:

$$\left.\begin{aligned}\dot{\mathbf{F}}^p &= \overline{\mathbf{L}}^p\mathbf{F}^p\\ \dot{\mathbf{F}}^p_{kd} &= \widehat{\mathbf{L}}^p_{kd}\mathbf{F}^p_{kd} = \mathbf{F}^{p-1}_{ks}\overline{\mathbf{L}}^p_{kd}\mathbf{F}^p_{ks}\mathbf{F}^p_{kd}\\ \dot{\mathbf{F}}^p_{cd} &= \breve{\mathbf{L}}^p_{cd}\mathbf{F}^p_{cd} = \mathbf{F}^{p-1}_{cs}\overline{\mathbf{L}}^p_{cd}\mathbf{F}^p_{cs}\mathbf{F}^p_{cd}\end{aligned}\right\} \tag{12.84}$$

where $\overline{\mathbf{L}}^p$, $\overline{\mathbf{L}}^p_{kd}$ and $\overline{\mathbf{L}}^p_{cd}$ are given by Eq. (12.61). The storage parts $\mathbf{F}^e$, $\mathbf{F}^p_{ks}$ and $\mathbf{F}^p_{cs}$ of the deformation gradient are given by substituting the time-integrations of Eq. (12.84) into

$$\mathbf{F}^e = \mathbf{F}\mathbf{F}^{p-1},\quad \mathbf{F}^p_{ks} = \mathbf{F}^p\mathbf{F}^{p-1}_{kd},\quad \mathbf{F}^p_{cs} = \mathbf{F}^p\mathbf{F}^{p-1}_{cd} \tag{12.85}$$

Further, $\overline{\mathbf{C}}^e$, $\widehat{\mathbf{C}}^p_{ks}$ and $\breve{\mathbf{C}}^p_{cs}$ are calculated by substituting Eq. (12.85) into Eq. (12.3). Further, the stress $\overline{\mathbf{S}}$, the kinematic hardening variable $\widehat{\mathbf{S}}_k$ and the elastic-core $\breve{\mathbf{S}}_c$ are calculated by substituting $\overline{\mathbf{C}}^e$, $\widehat{\mathbf{C}}^p_{ks}$ and $\breve{\mathbf{C}}^p_{cs}$ into Eqs. (12.31) and (12.36). The isotropic hardening variable and the normal-yield ratio are calculated by the time-integration of Eqs. (12.73), (12.74) and (12.75).

The plastic constitutive equation with the plastic modulus in Eq. (12.78) is not necessary to be used in the numerical calculation by the return-mapping in which the plastic strain rate is calculated by use of only the plastic flow rule in Eq. (12.53) and then the stress and internal variables are calculated.

The time-integrations of Eq. (12.84) for the deformation gradient tensors $\mathbf{F}^p$, $\mathbf{F}^p_{kd}$ and $\mathbf{F}^p_{cd}$ can be executed in high efficiency by the tensor exponential method (Miehe 1996; Weber and Anand 1990; Hashiguchi and Yamakawa 2012).

The numerical calculation scheme for the multiplicative elastoplasticity adopting the initial subloading surface model ($\overline{\mathbf{M}}_c = \mathbf{O}$) without the plastic spins ($\overline{\mathbf{W}}^p = \overline{\mathbf{W}}^p_{kd} = \overline{\mathbf{W}}^p_{cd} = \mathbf{O}$) can be referred to Hashiguchi and Yamakawa (2012).

12.9 Cyclic Stagnation of Isotropic Hardening of Metals

The stagnation of the isotropic hardening for a while after the reverse of loading was described in Sect. 10.2. It will be extended to the framework of the multiplicative finite strain theory in this section.

The normal-isotropic hardening surface in the intermediate configuration is given by

$$g(\tilde{\overline{\mathbf{M}}}_k) = \tilde{K} \tag{12.86}$$

where

$$\tilde{\overline{\mathbf{M}}}_k \equiv \overline{\mathbf{M}}_k - \overline{\mathbf{\Theta}}(\neq \tilde{\overline{\mathbf{M}}}_k^T) \tag{12.87}$$

The scalar variable K and the second-order tensor variable $\overline{\mathbf{\Theta}}(\neq \overline{\mathbf{\Theta}}^T, \mathrm{tr}\overline{\mathbf{\Theta}} = 0)$ designate the size and the center, respectively, of the normal-isotropic hardening surface, the evolution rules of which will be formulated later. Further, the sub-isotropic hardening surface, which always passes through the back stress $\overline{\mathbf{M}}_k$ in the intermediate configuration and has a similar shape and a same orientation to the normal-isotropic hardening surface is expressed by the following equation (see Fig. 12.3).

$$g(\tilde{\overline{\mathbf{M}}}_k) = \tilde{R}\tilde{K} \tag{12.88}$$

The normal-isotropic hardening ratio $\tilde{R}$ is calculable from the equation $\tilde{R} = g(\tilde{\overline{\mathbf{M}}}_k)/\tilde{K}$ in terms of the known values $\overline{\mathbf{M}}_k$, $\overline{\mathbf{\Theta}}$ and $\tilde{K}$.

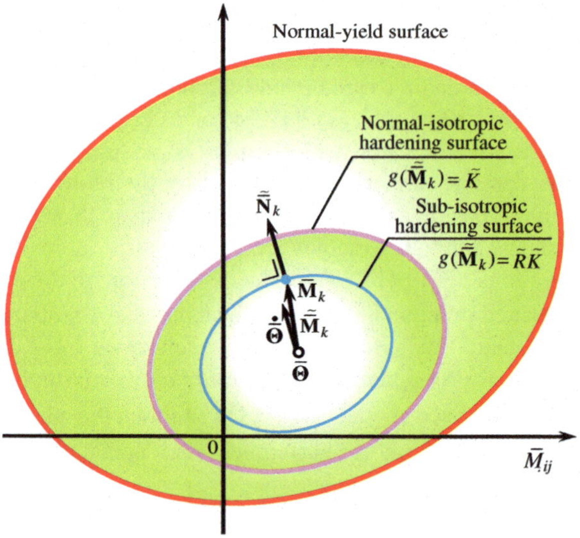

Fig. 12.3 Normal- and sub-isotropic hardening surfaces in multiplicative elastoplasticity

The consistency condition of the sub-isotropic hardening surface is given by

$$\frac{\partial g(\tilde{\overline{\mathbf{M}}}_k)}{\partial \tilde{\overline{\mathbf{M}}}_k} : \dot{\overline{\mathbf{M}}}_k - \frac{\partial g(\tilde{\overline{\mathbf{M}}}_k)}{\partial \tilde{\overline{\mathbf{M}}}_k} : \dot{\overline{\boldsymbol{\Theta}}} = \tilde{R}\dot{\tilde{K}} + \dot{\tilde{R}}\tilde{K} \tag{12.89}$$

The rates of $\tilde{K}$ and $\overline{\boldsymbol{\Theta}}$ are given by the following equations based on Eqs. (10.21) and (10.22).

$$\dot{\tilde{K}} = C\tilde{R}^{\varsigma}\langle \partial f \tilde{\overline{\mathbf{N}}}_k : \dot{\overline{\mathbf{M}}}_k \rangle \left\| \frac{\partial g(\tilde{\overline{\mathbf{M}}}_k)}{\partial \tilde{\overline{\mathbf{M}}}_k} \right\| \tag{12.90}$$

$$\dot{\overline{\boldsymbol{\Theta}}} = (1-C)\tilde{R}^{\varsigma}\langle \tilde{\overline{\mathbf{N}}}_k : \dot{\overline{\mathbf{M}}}_k \rangle \tilde{\overline{\mathbf{N}}}_k \tag{12.91}$$

where are the material constants and

$$\tilde{\overline{\mathbf{N}}}_k \equiv \frac{\partial g(\tilde{\overline{\mathbf{M}}}_k)}{\partial \tilde{\overline{\mathbf{M}}}_k} / || \frac{\partial g(\tilde{\overline{\mathbf{M}}}_k)}{\partial \tilde{\overline{\mathbf{M}}}_k} || \quad (\neq \tilde{\overline{\mathbf{N}}}_k^T) \tag{12.92}$$

Substituting Eqs. (12.91) and (12.92) for the evolution rules of K and $\boldsymbol{\Theta}$ into Eq. (12.90), the rate of the normal-isotropic hardening ratio is given by

$$\begin{aligned} \dot{\tilde{R}} &= \frac{1}{\tilde{K}}[\langle \frac{\partial g(\tilde{\overline{\mathbf{M}}}_k)}{\partial \tilde{\overline{\mathbf{M}}}_k} : \dot{\overline{\mathbf{M}}}_k \rangle - \frac{\partial g(\tilde{\overline{\mathbf{M}}}_k)}{\partial \tilde{\overline{\mathbf{M}}}_k} : (1-C)\tilde{R}^{\varsigma}\langle \overline{\mathbf{N}}_k : \dot{\overline{\mathbf{M}}}_k \rangle \overline{\mathbf{N}}_k - \tilde{R}C\tilde{R}^{\varsigma}\langle \frac{\partial g(\tilde{\overline{\mathbf{M}}}_k)}{\partial \tilde{\overline{\mathbf{M}}}_k} : \dot{\overline{\mathbf{M}}}_k \rangle] \\ &= \frac{1}{\tilde{K}}\langle \frac{\partial g(\tilde{\overline{\mathbf{M}}}_k)}{\partial \tilde{\overline{\mathbf{M}}}_k} : \dot{\overline{\mathbf{M}}}_k \rangle \{1 - [1 - C(1-\tilde{R})]\tilde{R}^{\zeta}\} \end{aligned} \tag{12.93}$$

which is the monotonically-decreasing function of $\tilde{R}$ fulfilling

$$\dot{\tilde{R}} \begin{cases} = \dfrac{1}{\tilde{K}}\langle \dfrac{\partial f(\tilde{\overline{\mathbf{M}}}_k)}{\partial \tilde{\overline{\mathbf{M}}}_k} : \dot{\overline{\mathbf{M}}}_k \rangle \ (>0) \ \text{ for } \tilde{R} = 0 \\ < \dfrac{1}{\tilde{K}}\langle \dfrac{\partial f(\tilde{\overline{\mathbf{M}}}_k)}{\partial \tilde{\overline{\mathbf{M}}}_k} : \dot{\overline{\mathbf{M}}}_k \rangle \ (>0) \ \text{ for } \tilde{R} < 1 \\ = 0 \ \text{ for } \tilde{R} = 1 \\ < 0 \ \text{ for } \tilde{R} > 1 \end{cases} \tag{12.94}$$

Therefore, $\overline{\mathbf{M}}_k$ is attracted automatically to the normal-isotropic hardening surface even if it goes out from that surface by virtue of the inequality $\dot{\tilde{R}} < 0$ for $\tilde{R} > 1$ as shown in Eq. (12.96). Furthermore, the judgment of whether $\overline{\mathbf{M}}_k$ lies on the normal-isotropic hardening surface is not required in the present formulation.

The evolution rule of isotropic hardening is given analogously to Eqs. (10.27) and (10.28), noting Eq. (12.63) as follows:

$$\boxed{\dot{H} = \tilde{R}^{\upsilon}\langle \tilde{\overline{\mathbf{N}}}_k : \overline{\mathbf{A}}/\|\overline{\mathbf{A}}\|\rangle f_{Hn}\,\dot{\bar{\lambda}}} = f_{Hsn}\,\dot{\bar{\lambda}} \tag{12.95}$$

where υ is the material constant and

$$f_{Hsn} \equiv \tilde{R}^{\upsilon}\langle \tilde{\overline{\mathbf{N}}}_k : \overline{\mathbf{A}}/\|\overline{\mathbf{A}}\|\rangle f_{Hn} \tag{12.96}$$

$$\begin{aligned}\overline{\mathbf{A}} \equiv\; & \mathbf{F}^{p}_{ks}\widehat{\mathcal{C}}^{k} : \mathbf{F}^{pT}_{ks}(\underline{\overline{\mathbf{N}}} - (1/b_k)\|\underline{\overline{\mathbf{N}}}\|\mathrm{sym}[\overline{\mathbf{M}}_k])\mathbf{F}^{p}_{ks}\,\mathbf{F}^{pT}_{ks} + 2\mathrm{sym}[(\underline{\overline{\mathbf{N}}} + \eta^{p}(\overline{\mathbf{M}}\,\underline{\overline{\mathbf{N}}} - \underline{\overline{\mathbf{N}}}\,\overline{\mathbf{M}})\\ & - (1/b_k)\|\underline{\overline{\mathbf{N}}}'\|\{\mathrm{sym}[\overline{\mathbf{M}}_k] + \eta^{p}_{k}(\overline{\mathbf{M}}_k\,\mathrm{sym}[\overline{\mathbf{M}}_k] - \mathrm{sym}[\overline{\mathbf{M}}_k]\overline{\mathbf{M}}_k)\})\overline{\mathbf{M}}_k]\end{aligned} \tag{12.97}$$

The plastic modulus is given by replacing f_{Hn} to f_{Hsn} in Eq. (12.79).

The function $g(\tilde{\overline{\mathbf{M}}}_k)$ is given in the simplest form as follows:

$$g(\tilde{\overline{\mathbf{M}}}_k) = \|\tilde{\overline{\mathbf{M}}}_k\| \tag{12.98}$$

which will be used in the subsequent sections for the comparisons with test data. It follows from Eqs. (12.92) and (12.98) that

$$\tilde{\overline{\mathbf{N}}}_k \equiv \frac{\partial g(\tilde{\overline{\mathbf{M}}}_k)}{\partial \tilde{\overline{\mathbf{M}}}_k} = \frac{\tilde{\overline{\mathbf{M}}}_k}{\|\tilde{\overline{\mathbf{M}}}_k\|} \tag{12.99}$$

The incorporation of the tangential-inelastic strain rate described in Sect. 9.10 into the multiplicative elastoplasticity requires a further study.

References

Hashiguchi K (2016a) Exact formulation of subloading surface model: unified constitutive law for irreversible mechanical phenomena in solids. Arch Compt Meth Eng 23:417–447

Hashiguchi K (2016b) Multiplicative finite strain theory based on subloading surface model. Proc Comput Eng Conf JSCE:B-8-3

Hashiguchi K (2016c) Loading criterion in return-mapping for subloading surface model. Proc Comput Mech Div JSME:03-6

Hashiguchi K (2016d) Exact multiplicative finite strain theory based on subloading surface model. Proc Mater Mech Div JSME:GS-26

Hashiguchi K, Yamakawa Y (2012) Introduction to finite strain theory for continuum elasto-plasticity. Wiley series in computational mechanics. Wiley, Chichester

Lion A (2000) Constitutive modeling in finite thermoviscoplasticity: a physical approach based on nonlinear rheological models. Int J Plast 16:469–494

Miehe C (1996) Numerical computation of algorithmic (consistent) tangent moduli in large-strain computational inelasticity. Comput Methods Appl Mech Eng 134:223–240

Weber G, Anand L (1990) Finite deformation constitutive equations and a integration procedure for isotropic, hyperelastic-viscoplastic solids. Comput Mech Appl Mech Eng 79:173–202

Chapter 13
Viscoplastic Constitutive Equations

Plastic deformations of solids depend on the rate of loading or deformation, exhibiting the time-dependence or rate-dependence in general, which is called the viscoplastic deformation. Constitutive equations for the viscoplastic deformation is described in this chapter. The physical background of rate-dependent plastic deformation, i.e. viscoplastic deformation and the history of the viscoplastic constitutive equation are reviewed first. Then, the pertinent formulation of the viscoplastic constitutive equation is formulated based on the subloading surface model, which is capable of describing the smooth elastic-viscoplastic transition and the rate-dependence for the general rate ranging from the quasi-static to the impact loadings.

13.1 Rate-Dependent Deformation of Solids

The elastic and the plastic deformations of solids are induced by the deformation of solid particles themselves (crystals in metals, soil particles in soils, etc.) and the mutual slips between them, respectively. Therefore,

(1) Elastic tangent modulus is high.
(2) High stress has to apply to solids in order that plastic deformation is induced, which overcomes the friction between solid particles. The stress inducing the plastic deformation is macroscopically the so-called yield stress.
(3) Tangent modulus in the elastoplastic deformation process lowers from that in the elastic deformation process.

The rate-dependent elastic deformation induced in the stress lower than the yield stress is called the *viscoelastic deformation*. On the other hand, the rate-dependent plastic deformation induced over the yield stress is called the *viscoplastic deformation*. Then, they are classified as

K. Hashiguchi, *Foundations of Elastoplasticity: Subloading Surface Model*,
DOI 10.1007/978-3-319-48821-9_13

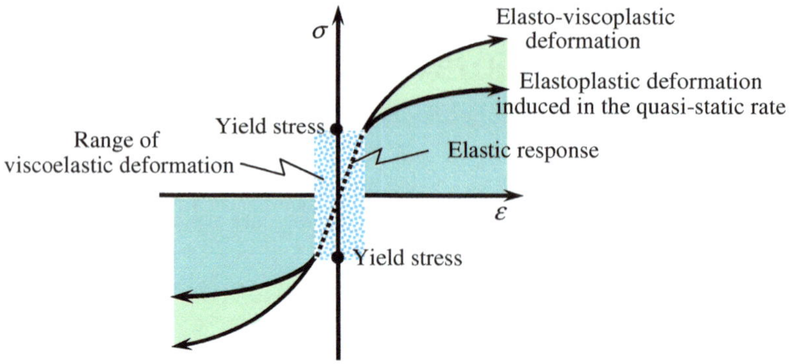

Fig. 13.1 Rate-dependent deformation of solids

Deformation of solids	Deformation of solid particles	Rate-independence: Elastic constitutive equation
		Rate-dependence: Viscoelastic constitutive equation
	Mutual slips of solid particles	Rate-independence: Elastoplastic constitutive equation
		Rate-dependence: Viscoplastic constitutive equation

which is illustrated in Fig. 13.1.

13.2 History of Viscoplastic Constitutive Equations

The most pertinent viscoplastic model would be the *overstress model*. The development of this model is reviewed in this section, while the overview of the history is portrayed in Fig. 13.2.

The elastic constitutive equation extended so as to describe the rate-dependence is called the viscoelastic constitutive equation and one of the typical models is the *Maxwell model*, in which the spring and the dashpot are connected in series. Therefore, the strain rate $\overset{\bullet}{\varepsilon}$ is additively decomposed into the elastic strain rate $\overset{\bullet}{\varepsilon}{}^{e} = E^{-1}\overset{\bullet}{\sigma}$ and the viscous strain rate $\overset{\bullet}{\varepsilon}{}^{v} = \mu^{-1}\sigma$, where σ designates the stress, E is the elastic modulus and μ is the viscous coefficient, leading to

$$\overset{\bullet}{\varepsilon} = \overset{\bullet}{\varepsilon}{}^{e} + \overset{\bullet}{\varepsilon}{}^{v} = E^{-1}\overset{\bullet}{\sigma} + \mu^{-1}\sigma \tag{13.1}$$

This model is concerned with the rate-dependent deformation at the low stress level below the yield stress.

On the other hand, the elastoplastic constitutive equation can be schematically expressed by the Prandtl model in which the dashpot is replaced with the slider in the Maxwell model, whereas the slider begins to move in the state that the stress σ

Maxwell's viscoelastic model

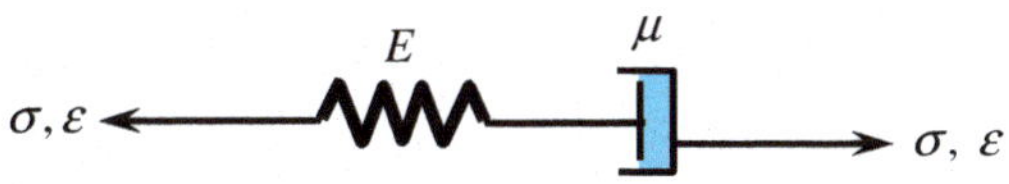

$$\dot{\varepsilon} = \dot{\varepsilon}^e + \dot{\varepsilon}^v = E^{-1}\dot{\sigma} + \mu^{-1}\sigma$$

Viscous fluid:

$$\tau = \mu\dot{\gamma}$$

$$\downarrow \tau \rightarrow \sigma, \quad \dot{\gamma} \rightarrow \dot{\varepsilon}^v$$

$$\dot{\varepsilon}^v = \mu^{-1}\sigma$$

+

Prandtl's elastoplastic model

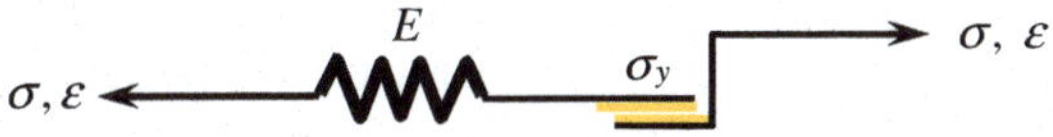

$$\dot{\varepsilon} = \begin{cases} \dot{\varepsilon}^e = E^{-1}\dot{\sigma} & \text{for } \sigma < \sigma_y \\ \dot{\varepsilon}^e + \dot{\varepsilon}^p = E^{-1}\dot{\sigma} + M^{p\,-1}\dot{\sigma} & \text{for } \sigma = \sigma_y \end{cases}$$

↓

Bingham's viscoplastic model (Bingham, 1922)**: Original concept of overstress model**

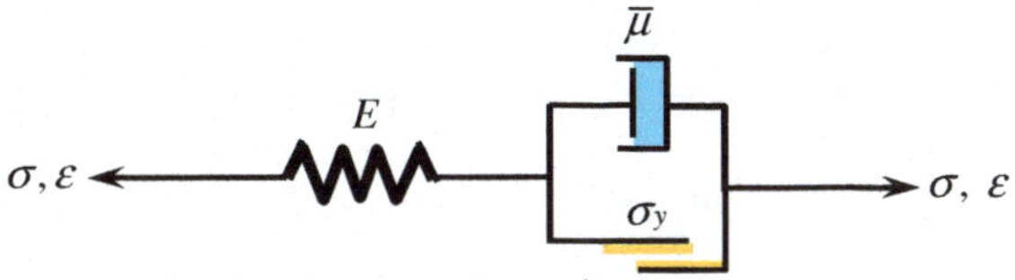

$$\dot{\varepsilon} = \dot{\varepsilon}^e + \dot{\varepsilon}^{vp} = E^{-1}\dot{\sigma} + \bar{\mu}^{-1} < \sigma - \sigma_y >^n$$

↓

Prager's overstress model (Prager, 1961)**: Overstress model for von Mises metals**

$$\mathbf{d} = \mathbf{d}^e + \mathbf{d}^{vp} = \mathbf{E}^{-1} : \mathring{\boldsymbol{\sigma}} + \bar{\mu}^{-1}\left\langle \frac{\sigma^{eq}}{F(\varepsilon^{eqvp})} - 1 \right\rangle^n \frac{\boldsymbol{\sigma}'}{\|\boldsymbol{\sigma}'\|}$$

↓

Generalized overstress model (Perzyna, 1963)**: Overstress model for generic materials**

$$\mathbf{d} = \mathbf{d}^e + \mathbf{d}^{vp} = \mathbf{E}^{-1} : \mathring{\boldsymbol{\sigma}} + \bar{\mu}^{-1}\left\langle \frac{f(\boldsymbol{\sigma})}{F(H)} - 1 \right\rangle^n \mathbf{n}$$

Fig. 13.2 History of viscoplastic model

reaches the yield stress σ_y, by which the plastic strain rate is induced (see Fig. 13.2). Then, the strain rate is additively composed of the elastic and the plastic strain rates, i.e.

$$\overset{\bullet}{\varepsilon} = \begin{cases} \overset{\bullet}{\varepsilon}{}^e = E^{-1}\overset{\bullet}{\sigma} \quad \text{for} \quad \sigma < \sigma_y \\ \overset{\bullet}{\varepsilon}{}^e + \overset{\bullet}{\varepsilon}{}^p = E^{-1}\overset{\bullet}{\sigma} + M^{p-1}\overset{\bullet}{\sigma} \quad \text{for} \quad \sigma = \sigma_y \end{cases} \tag{13.2}$$

where M^p is the plastic modulus.

Furthermore, the model which describes the rate-dependent plastic strain rate $\overset{\bullet}{\varepsilon}{}^{vp}$ induced for the state of stress over the yield stress, called the *viscoplastic strain rate*, was introduced by Bingham (1922), combining the above-mentioned Maxwell model and Prandtl model so as to connect the dashpot and the slider in parallel as shown in Fig. 13.2, where $\bar{\mu}$ is the *viscoplastic coefficient* and n is the material constant, while n is chosen to be 4–8 in practice. Then, the strain rate is given by

$$\overset{\bullet}{\varepsilon} = \overset{\bullet}{\varepsilon}{}^e + \overset{\bullet}{\varepsilon}{}^{vp} = E^{-1}\overset{\bullet}{\sigma} + \bar{\mu}^{-1}\langle\sigma - \sigma_y\rangle^n \tag{13.3}$$

The Bingham model for the elasto-viscoplastic deformation is the origin of the *overstress model* based on the concept that the viscoplastic strain rate is induced by the *overstress*, i.e. stress over the yield stress.

The above-mentioned Bingham model for the one-dimensional deformation was extended by Hohenemser and Prager (1932) and Prager (1961) to describe the three-dimensional deformation of metals, adopting the Mises yield condition for the slider as shown in Fig. 13.2. In this model, the strain rate $\mathbf{d}$ is additively decomposed into the elastic strain rate $\mathbf{d}^e$ and the viscoplastic strain rate $\mathbf{d}^{vp}$, i.e.

$$\mathbf{d} = \mathbf{d}^e + \mathbf{d}^{vp} \tag{13.4}$$

with

$$\mathbf{d}^{vp} = \frac{1}{\bar{\mu}}\left\langle\frac{\sigma^{eq}}{F(\varepsilon^{eqvp})} - 1\right\rangle^n \frac{\boldsymbol{\sigma}'}{||\boldsymbol{\sigma}'||} \tag{13.5}$$

where $\varepsilon^{eqvp} \equiv \sqrt{2/3}\int||\mathbf{d}^{vp\prime}||dt$ is the *equivalent viscoplastic strain* given by replacing the plastic strain rate $\mathbf{d}^p$ to the viscoplastic strain rate $\mathbf{d}^{vp}$ in the plastic equivalent strain $\varepsilon^{eqp} \equiv \sqrt{2/3}\int||\mathbf{d}^{p\prime}||dt$ in Eq. (6.56). The viscoplastic coefficient $\bar{\mu}$ depends on stress, internal variables and temperature in general.

Furthermore, the viscoplastic strain rate in the Prager's overstress model was extended by Perzyna (1963, 1966) for materials having the general yield condition unlimited to the Mises yield condition as

$$\mathbf{d}^{vp} = \frac{1}{\bar{\mu}}\left\langle\frac{f(\boldsymbol{\sigma})}{F(H)} - 1\right\rangle^n \mathbf{n}, \quad \mathbf{n} \equiv \frac{\partial f(\boldsymbol{\sigma})}{\partial\boldsymbol{\sigma}} \bigg/ \left\|\frac{\partial f(\boldsymbol{\sigma})}{\partial\boldsymbol{\sigma}}\right\| \tag{13.6}$$

Then, substituting Eqs. (6.29) and (13.6) into Eq. (13.4), we have

$$\mathbf{d} = \mathbf{E}^{-1}:\overset{\circ}{\boldsymbol{\sigma}} + \frac{1}{\bar{\mu}}\left\langle \frac{f(\boldsymbol{\sigma})}{F(H)} - 1 \right\rangle^{n} \mathbf{n} \tag{13.7}$$

and thus

$$\overset{\circ}{\boldsymbol{\sigma}} = \mathbf{E}:\mathbf{d} - \frac{1}{\bar{\mu}}\left\langle \frac{f(\boldsymbol{\sigma})}{F(H)} - 1 \right\rangle^{n} \mathbf{E}:\mathbf{n} \tag{13.8}$$

where the hardening variable H evolves by

$$\dot{H} = \dot{H}(\boldsymbol{\sigma}, H;\, \mathbf{d}^{vp}) \tag{13.9}$$

by replacing the plastic strain rate $\mathbf{d}^{p}$ to the viscoplastic strain rate $\mathbf{d}^{vp}$ in the evolution rule of the isotropic hardening variable in Eq. (6.37) for the plastic constitutive equation. Therefore, $\dot{H}$ is the homogeneous function of $\mathbf{d}^{vp}$ in degree-one. In what follows, the isotropic yield condition $f(\boldsymbol{\sigma}) = F(H)$ in Eq. (6. 30) is used below for the sake of simplicity in explanation up to Sect. 13.4.

13.3 On the Creep Model

Based on a concept different from the overstress model, the *creep model*, which also aims at describing the viscoplastic deformation, has been studied widely. The typical one is the *Norton law* (Norton 1929) in which the *creep strain rate* is given as follows:

$$\mathbf{d}^{c} = d_0^c \|\boldsymbol{\sigma}\|^{m} \mathbf{n} \tag{13.10}$$

where d_0^c and m ($\gg 1$) are the material constants. Further, it is modified for the crystal plasticity by Nakada and Keh (1966), Hutchinson (1976), Pan and Rice (1983), Peirce et al. (1983), etc. as follows:

$$\mathbf{d}^{c} = \|\mathbf{d}^{c}\|\mathbf{n} = d_0^c \left(\frac{f(\boldsymbol{\sigma})}{F(H)} \right)^{m} \mathbf{n} \tag{13.11}$$

where the rate of the isotropic hardening variable is given by

$$\dot{H} = \dot{H}(\boldsymbol{\sigma}, H; \mathbf{d}^{c}) \tag{13.12}$$

instead of Eq. (13.9).

The strain rate is given by

$$\mathbf{d} = \mathbf{d}^e + \mathbf{d}^c = \mathbf{E}^{-1} : \overset{\circ}{\boldsymbol{\sigma}} + d_0^c \|\boldsymbol{\sigma}\|^m \mathbf{n} \tag{13.13}$$

for Eq. (13.10) and

$$\mathbf{d} = \mathbf{d}^e + \mathbf{d}^c = \mathbf{E}^{-1} : \overset{\circ}{\boldsymbol{\sigma}} + d_0^c \left(\frac{f(\boldsymbol{\sigma})}{F(H)} \right)^m \mathbf{n} \tag{13.14}$$

for Eq. (13.11).

The creep model described above has different structures from the overstress model because Eqs. (13.10) and (13.11) possesses no threshold value for the generation of the creep strain rate. Especially, its deformation behavior is not reduced to that of the elastoplastic constitutive equation at quasi-static deformation. Therefore, this model cannot describe appropriately the deformation behavior at a low rate. In fact, the creep strain rate does not diminish even if the stress decreases into the inside of yield surface, exhibiting the overrunning stress-strain curve and the creep deformation proceeds unlimitedly under a constant stress state in the creep

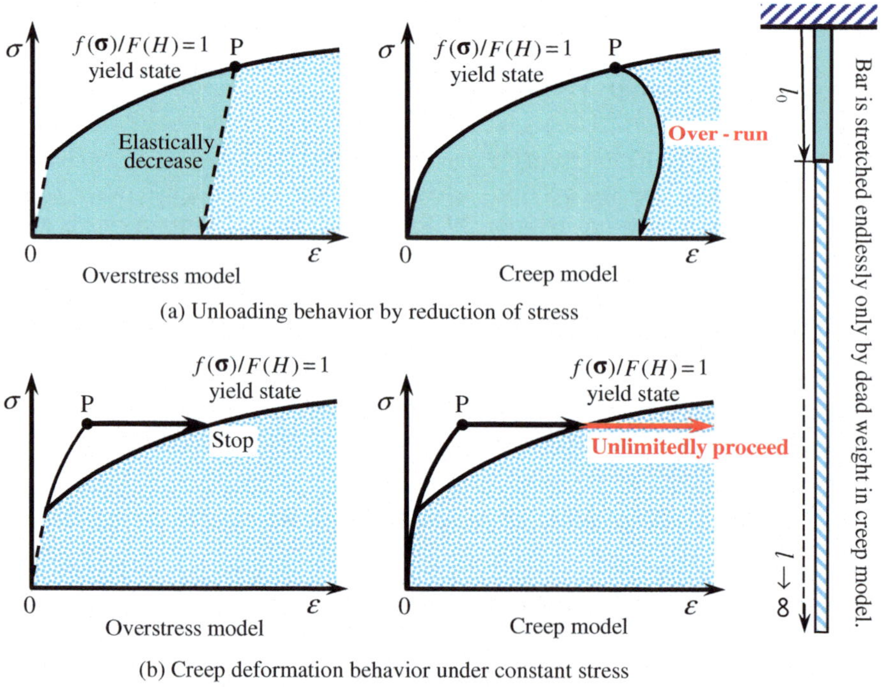

Fig. 13.3 Comparison of overstress model and creep model: the latter exhibits unrealistic behavior predicting always creep strain rate

model as shown in Fig. 13.3. As a concrete example, it is quite unnatural that a bar subjected to any low tension continues to be elongated endlessly as a time elapse. In addition, it is incapable of describing appropriately the impact loading behavior as it describes an elastic deformation behavior with an infinite strength. On the other hand, the viscoplastic strain rate diminishes immediately after the stress decreases into the inside of yield surface in the overstress model as is observed in real materials. Consequently, the creep model is not physically accepted but the overstress model is appropriate for the description of the rate-dependent plastic behavior.

Furthermore, various constitutive models including the time itself elapsed after a loading mode (loading/unloading) changed have been proposed to date. Here, note that the judgment whether or not the loading mode changes contains an ambiguity depending on the subjectivity of observers, especially in a fluctuating state of deformation rate. Therefore, these models are impertinent, lacking an objectivity.

13.4 Mechanical Response of Past Overstress Model

The development of rate-dependent elastoplastic constitutive equation is reviewed above and it is described that the overstress model would have a pertinent basic structure. Here, let the mechanical responses at the infinitesimal and the infinite rates of deformation be examined in order to clarify the basic property of this model.

The past overstress model advocated by Bingham (1922) and extended by Prager (1961) and Perzyna (1963, 1966) describes the elastoplastic deformation in the quasi-static loading since the viscous resistance of the dashpot diminishes but the elastic deformation in the impact loading since the viscous resistance of the dashpot becomes infinite. Therefore, it cannot be applied to the deformation behavior in a high rate as an impact loading. In what follows, we will show this fact on the past overstress model.

Equations (13.7) and (13.8) are expressed in the following equations for the incremental forms.

$$d\boldsymbol{\varepsilon} = \mathbf{E}^{-1} : d\boldsymbol{\sigma} + \frac{1}{\bar{\mu}}\left\langle \frac{f(\boldsymbol{\sigma})}{F(H)} - 1 \right\rangle^{n} \mathbf{n}\, dt \tag{13.15}$$

and thus

$$d\boldsymbol{\sigma} = \mathbf{E} : d\boldsymbol{\varepsilon} - \frac{1}{\bar{\mu}}\left\langle \frac{f(\boldsymbol{\sigma})}{F(H)} - 1 \right\rangle^{n} \mathbf{E} : \mathbf{n}\, dt \tag{13.16}$$

designating $d\boldsymbol{\varepsilon} \equiv \mathbf{d}dt$.

Equation (13.16) reduces to the following relation for the infinitesimal rate of deformation (quasi-static deformation) fulfilling $dt \to \infty$: $d\boldsymbol{\sigma}/dt \to \mathbf{O}$ and $\mathbf{d} = d\boldsymbol{\varepsilon}/dt \to \mathbf{O}$.

$$\mathbf{O} \cong \mathbf{O} - \frac{1}{\bar{\mu}}\left\langle \frac{f(\boldsymbol{\sigma})}{F(H)} - 1 \right\rangle^{n} \mathbf{E} : \mathbf{n} \tag{13.17}$$

leading to

$$\frac{f(\boldsymbol{\sigma})}{F(H)} - 1 \to 0 \tag{13.18}$$

Then, the stress changes fulfilling the yield condition $f(\boldsymbol{\sigma}) = F(H)$, i.e. obeying the plastic constitutive relation, while the elastic deformation is given by the change of stress, so that Eqs. (13.7) and (13.8) exhibit the response of the elastoplastic constitutive relation in the quasi-static deformation as shown in Fig. 13.4. Then, the overstress model is the extension of the elastoplastic constitutive equation for the non-zero rate of deformation. In fact, the elastoplastic constitutive relation is reproduced for the quasi-static deformation. Here, it is reproduced easily for the non-viscous material, i.e. the material with a small viscoplastic coefficient $\bar{\mu} \cong 0$. It should be noted that the viscoplastic strain rate depends on the rate of stress but it depends always on the state of stress in the creep model as described in Sect. 13.3.

On the other hand, in the infinite rate of deformation fulfilling $dt \cong 0$: $d\boldsymbol{\sigma}/dt \to \infty$ and $\mathbf{d} = d\boldsymbol{\varepsilon}/dt \to \infty$, Eq. (13.8) is reduced to

$$d\boldsymbol{\sigma} = \mathbf{E} : d\boldsymbol{\varepsilon} - \mathbf{O}, \quad \text{i.e. } \overset{\circ}{\boldsymbol{\sigma}} = \mathbf{E} : \mathbf{d} - \mathbf{O} \tag{13.19}$$

approaching the elastic response as shown in Fig. 13.4. Therefore, it predicts the unrealistic response that the material can bear an infinite load. Equation (13.19) is

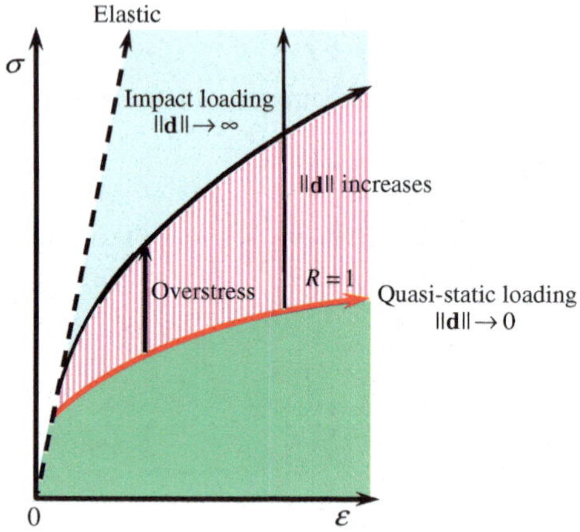

Fig. 13.4 Past overstress model which is inapplicable to prediction of deformation behavior at high rate

also known directly from Eq. (13.8) by setting $\bar{\mu} \to \infty$ resulting in the material with an infinite viscoplastic coefficient. The past overstress model is inapplicable to the prediction of deformation at high rate in general. In order to modify this defect, Lemaitre and Chaoboche (1990) and Chaboche (2008) proposed to add the creep strain rate in addition to the elastic and the viscoplastic strain rate irrationally, although the viscoplastic strain rate describes the time-dependent plastic strain rate. The rigorous modification of the overstress model so as to describe the elasto-viscoplastic behavior in the general rate ranging from the quasi-static to the impact loading will be given in the next section.

Eventually, the existing formulation of overstress model in Eq. (13.7), i.e. (13.8) is inapplicable to the prediction of deformation at a high rate. The material constant n included as the power form in Eq. (13.7), i.e. (13.8) is usually selected to be larger than five, but the fitting to the test data for impact load is impossible even if n is selected as one hundred which, needless to say, results in the inappropriate prediction of deformation in a slow loading process. In addition, the inclusion of a high power in the equation induces difficulty in numerical calculations.

The so-called *flow stress model* was proposed by Johnson and Cook (1983) in which the yield stress, i.e. flow stress depends not only on the viscoplastic strain rate but also on the viscoplastic strain rate. It is the empirical model without a generality, which is concerned with the fast loading behavior but inapplicable to the slow loading behavior. However, unfortunately it is used widely for the deformation analyses in the fast loading behavior by adopting the commercial FEM software, e.g. Abaqus and LD-DYNA. Hereinafter, it is desirable to adopt the subloading-overstress model for the analyses of the general loading behavior ranging from the quasi-static to the impact loadings, which will be described in the next section.

13.5 Extension to General Rate of Deformation: Subloading Overstress Model

Equation (13.7) can be rewritten concisely in terms of the normal-yield ratio R incorporated in the subloading surface model as follows:

$$\mathbf{d} = \mathbf{d}^e + \mathbf{d}^{vp} = \mathbf{E}^{-1} : \overset{\circ}{\boldsymbol{\sigma}} + \frac{1}{\bar{\mu}} \langle R - 1 \rangle^n \mathbf{n} \tag{13.20}$$

where $R = f(\boldsymbol{\sigma})F(H)$ takes value larger than unity when the stress goes over the yield surface so that $R \geq 1$ in general, and thus let it be renamed as the *dynamic-loading ratio*. The surface which passes through the current stress and is similar to the yield surface, called the *dynamic-loading surface*, is described by

$$f(\boldsymbol{\sigma}) = RF(H) \tag{13.21}$$

Equation (13.21) is formally identical to Eq. (7.6) for the subloading surface. Further, Eq. (13.21) is extended by incorporating the kinematic hardening as follows:

$$f(\hat{\boldsymbol{\sigma}}) = RF(H) \tag{13.22}$$

where

$$\dot{H} = \dot{H}(\boldsymbol{\sigma}, H; \mathbf{d}^{vp}), \quad \overset{\circ}{\boldsymbol{\alpha}} = \overset{\circ}{\boldsymbol{\alpha}}(\boldsymbol{\sigma}, \boldsymbol{\alpha}, F; \mathbf{d}^{vp}) \tag{13.23}$$

by replacing the plastic strain rate $\mathbf{d}^p$ to the viscoplastic strain rate $\mathbf{d}^{vp}$ in the plastic evolution equations.

Equation (13.20) can be extended so as to be applicable to the description of deformation in the general rate ranging from the quasi-static to the impact loading by incorporating the limit for the dynamic-loading ratio as follows (Hashiguchi 2007c):

$$\mathbf{d} = \mathbf{E}^{-1} \colon \overset{\circ}{\boldsymbol{\sigma}} + \frac{1}{\bar{\mu}} \frac{\langle R - 1 \rangle^n}{R_m - R} \mathbf{n} \tag{13.24}$$

where R_m ($\gg 1$) is the material constant, called the *limit dynamic-loading ratio*. By virtue of this modification, the stress cannot increase over the *limit dynamic-loading surface* described by $f(\boldsymbol{\sigma}) = R_m F(H)$ to which the stress reaches at an infinite rate of deformation, i.e. impact loading. The response of Eq. (13.24) is illustrated in Fig. 13.5.

The power equation has been used for the viscoplastic strain rate after Prager (1961) and Perzyna (1963, 1966). However, the numerical problem is caused by the

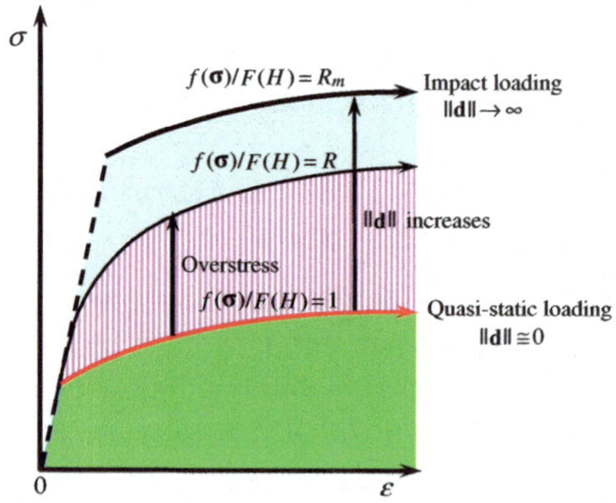

Fig. 13.5 Stress–strain curve predicted by the modified overstress model

calculation of high power of infinitesimal number. It can be remedied by using the exponential function as follows:

$$\mathbf{d} = \mathbf{E}^{-1} : \overset{\circ}{\boldsymbol{\sigma}} + \frac{1}{\bar{\mu}} \frac{\langle \exp[n(R-1)] - 1 \rangle}{R_m - R} \mathbf{n} \tag{13.25}$$

The necessity of incorporation of the plastic strain rate in addition to the elastic and the viscoplastic strain rates has been insisted in order to relax the elastic response in Eq. (13.19) such that the inelastic deformation is induced even in the infinite strain rate (cf. Lemaitre and Chaboche 1990; Hashiguchi et al. 2005a; Chaboche 2008). However, it would lead to the physical contradiction that the inelastic strain rate is described redundantly by both terms of the viscoplastic and the plastic strain rates.

Further, let the above-mentioned constitutive equation be extended such that the deformation behavior in the subloading surface model is induced in the quasi-static deformation, exhibiting the smooth elastic-viscoplastic transition. Then, assume that the subloading surface develops by the evolution rule of the normal-yield ratio in Eq. (7.9) in the viscoplastic state, provided that the plastic strain rate is replaced to the viscoplastic strain rate, i.e.,

$$\boxed{\dot{R}_s = \begin{cases} U(R_s)\|\mathbf{d}^{vp}\| & \text{for } \mathbf{d}^{vp} \neq \mathbf{O} \\ \dot{R}\,(R_s = R < 1) & \text{for } \mathbf{d}^{vp} = \mathbf{O} \end{cases}} \tag{13.26}$$

where the normal-yield ratio R in Eq. (7.6) is renamed as the *subloading-yield ratio* and denoted by the symbol R_s $(0 \leq R_s \leq 1)$. Let the function $U(R_s)$ be given by Eq. (7.18) with the replacement of R to R_s, i.e.

$$\boxed{U(R_s) = u \cot\left(\frac{\pi}{2}\frac{\langle R_s - R_e \rangle}{1 - R_e}\right)} \tag{13.27}$$

R_s can be calculated analytically through the integration of Eq. $(13.26)_1$ with Eq. (13.27) in the viscoplastic deformation process $\mathbf{d}^{vp} \neq \mathbf{O}$ similarly to Eq. (7.19) as

$$R_s = \frac{2}{\pi}(1 - R_e)\cos^{-1}\left[\cos\left(\frac{\pi}{2}\frac{R_{s0} - R_e}{1 - R_e}\right)\exp\left(-\frac{\pi}{2}u\frac{\varepsilon^{vp} - \varepsilon_0^{vp}}{1 - R_e}\right)\right] + R_e \tag{13.28}$$

under the initial condition $\varepsilon^{vp} = \varepsilon_0^{vp}$: $R_s = R_{s0}$, defining $\varepsilon^{vp} \equiv \int \|\mathbf{d}^{vp}\| dt$. Then, introducing the subloading-yield ratio, Eqs. (13.24) and (13.25) can be extended to describe the smooth elastic-plastic transition as follows:

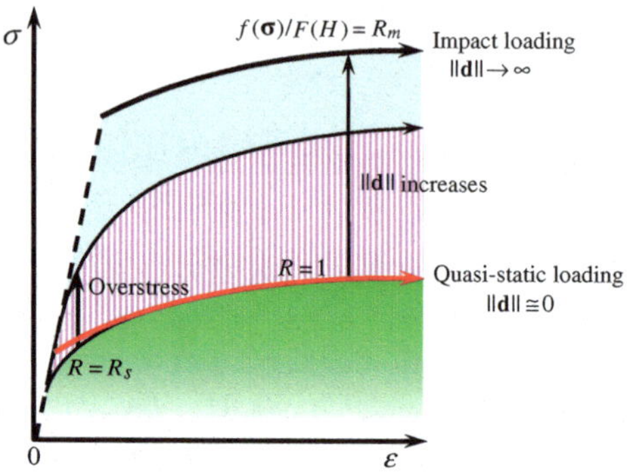

Fig. 13.6 Stress–strain curve predicted by the subloading-overstress model

$$\mathbf{d} = \mathbf{E}^{-1}\mathbf{:}\overset{\circ}{\boldsymbol{\sigma}} + \frac{1}{\bar{\mu}}\frac{\langle R-R_s\rangle^n}{R_m - R}\mathbf{n},\ \overset{\circ}{\boldsymbol{\sigma}} = \mathbf{E} : \mathbf{d} - \frac{1}{\bar{\mu}}\frac{\langle R-R_s\rangle^n}{R_m - R}\mathbf{E} : \mathbf{n} \tag{13.29}$$

$$\boxed{\mathbf{d} = \mathbf{E}^{-1}\mathbf{:}\overset{\circ}{\boldsymbol{\sigma}} + \frac{1}{\bar{\mu}}\frac{\langle \exp[n(R-R_s)] - 1\rangle}{R_m - R}\mathbf{n},\quad \overset{\circ}{\boldsymbol{\sigma}} = \mathbf{E} : \mathbf{d} - \frac{1}{\bar{\mu}}\frac{\langle \exp[n(R-R_s)] - 1\rangle}{R_m - R}\mathbf{E} : \mathbf{n}} \tag{13.30}$$

The response of the subloading-overstress model is illustrated in Fig. 13.6.

Incorporating the tangential inelastic strain rate formulated in Sect. 9.9 into Eqs. (13.29) and (13.30), the strain rate and the stress rate for the extended subloading surface model are given as follows:

$$\mathbf{d} = \mathbf{E}^{-1}\mathbf{:}\overset{\circ}{\boldsymbol{\sigma}} + \frac{1}{\bar{\mu}}\frac{\langle R-R_s\rangle^n}{R_m - R}\bar{\mathbf{n}} + \frac{T(R)}{2G}\overset{\circ}{\bar{\boldsymbol{\sigma}}}_t{}' \tag{13.31}$$

$$\overset{\circ}{\boldsymbol{\sigma}} = \mathbf{E} : \mathbf{d} - \frac{1}{\bar{\mu}}\frac{\langle R-R_s\rangle^n}{R_m - R}\mathbf{E} : \bar{\mathbf{n}} - 2G\frac{T(R)}{1+T(R)}\bar{\mathbf{d}}_t' \tag{13.32}$$

or

$$\mathbf{d} = \mathbf{E}^{-1}\mathbf{:}\overset{\circ}{\boldsymbol{\sigma}} + \frac{1}{\bar{\mu}}\frac{\langle \exp[n(R-R_s)] - 1\rangle}{R_m - R}\bar{\mathbf{n}} + \frac{T(R)}{2G}\overset{\circ}{\bar{\boldsymbol{\sigma}}}_t{}' \tag{13.33}$$

$$\overset{\circ}{\boldsymbol{\sigma}} = \mathbf{E} : \mathbf{d} - \frac{1}{\bar{\mu}}\frac{\langle \exp[n(R-R_s)] - 1\rangle}{R_m - R}\mathbf{E} : \bar{\mathbf{n}} - 2G\frac{T(R)}{1+T(R)}\bar{\mathbf{d}}_t' \tag{13.34}$$

where all the evolution rules of internal variables H, $\boldsymbol{\alpha}$, $\mathbf{c}$ (in $\bar{\mathbf{n}}$) and R_s are given by those in the elastoplastic constitutive equation (Chapter 9) with the replacements of the plastic strain rate $\mathbf{d}^p$ to the viscoplastic strain rate $\mathbf{d}^{vp}$. The subloading ratio R is calculated in general from Eq. (9.45) which is explicitly described by Eq. (10.32) for the Mises metals.

There is the other overstress model (Duvaut and Lions 1972; Simo et al. 1988; Simo and Hughes 1998) in which the tensor connecting the current stress point and its closest point on the yield surface is used instead of the ordinary overstress which is the scalar variable described in this section. It would be beneficial for the viscoplastic constitutive models assuming the intersecting yield surfaces but the calculation becomes complicated. However, the existence of the intersecting yield surfaces leads the constitutive model to be quite complicated, and its pertinence would be doubtful.

13.6 Subloading-Viscoplastic Model Based on Multiplicative Decomposition

The deformation gradient $\mathbf{F}$ is multiplicatively decomposed into the elastic deformation gradient $\mathbf{F}^e$ and the plastic deformation gradient $\mathbf{F}^{vp}$ instead of the plastic deformation gradient $\mathbf{F}^p$ in the multiplicative elastoplasticity described in Chap. 12. Then, we first adopt the following equation instead of Eq. (12.1).

$$\mathbf{F} = \mathbf{F}^e\mathbf{F}^{vp} \tag{13.35}$$

The velocity gradient $\boldsymbol{l}$ in the current configuration is additively decomposed into the elastic and the viscoplastic parts:

$$\boldsymbol{l} = \boldsymbol{l}^e + \boldsymbol{l}^{vp} \tag{13.36}$$

where

$$\left.\begin{array}{l} \boldsymbol{l} \equiv \dot{\mathbf{F}}\mathbf{F}^{-1}, \\ \boldsymbol{l}^e \equiv \dot{\mathbf{F}}^e\mathbf{F}^{e-1}, \boldsymbol{l}^p \equiv \mathbf{F}^e\dot{\mathbf{F}}^{vp}\mathbf{F}^{vp-1}\mathbf{F}^{e-1} = \mathbf{F}^e\overline{\mathbf{L}}^{vp}\mathbf{F}^{e-1} \\ \overline{\mathbf{L}}^{vp} \equiv \dot{\mathbf{F}}^{vp}\mathbf{F}^{vp-1} \end{array}\right\} \tag{13.37}$$

Further, the velocity gradient $\overline{\mathbf{L}}$ in the intermediate configuration is additively decomposed into the elastic and the plastic parts as follows:

$$\overline{\mathbf{L}} = \overline{\mathbf{L}}^e + \overline{\mathbf{L}}^{vp} \tag{13.38}$$

where

$$\left.\begin{array}{l} \overline{\mathbf{L}} \equiv \mathbf{F}^{e-1}\boldsymbol{l}\mathbf{F}^e \\ \overline{\mathbf{L}}^e \equiv \mathbf{F}^{e-1}\boldsymbol{l}^e\mathbf{F}^e = \mathbf{F}^{e-1}\dot{\mathbf{F}}^e, \overline{\mathbf{L}}^{vp} \equiv \mathbf{F}^{e-1}\boldsymbol{l}^{vp}\mathbf{F}^e = \dot{\mathbf{F}}^{vp}\mathbf{F}^{vp-1} \end{array}\right\} \tag{13.39}$$

from which it follows that

$$\left.\begin{array}{l}\overline{\mathbf{L}}=\overline{\mathbf{D}}+\overline{\mathbf{W}}\\ \overline{\mathbf{L}}^{e}=\overline{\mathbf{D}}^{e}+\overline{\mathbf{W}}^{e},\overline{\mathbf{L}}^{vp}=\overline{\mathbf{D}}^{vp}+\overline{\mathbf{W}}^{vp}\end{array}\right\} \tag{13.40}$$

$$\overline{\mathbf{D}}=\overline{\mathbf{D}}^{e}+\overline{\mathbf{D}}^{vp}, \quad \overline{\mathbf{W}}=\overline{\mathbf{W}}^{e}+\overline{\mathbf{W}}^{vp} \tag{13.41}$$

where

$$\left.\begin{array}{l}\overline{\mathbf{D}}=\operatorname{sym}[\overline{\mathbf{L}}],\ \overline{\mathbf{W}}=\operatorname{ant}[\overline{\mathbf{L}}]\\ \overline{\mathbf{D}}^{e}=\operatorname{sym}[\overline{\mathbf{L}}^{e}],\ \overline{\mathbf{W}}^{e}=\operatorname{ant}[\overline{\mathbf{L}}^{e}]\\ \overline{\mathbf{D}}^{vp}=\operatorname{sym}[\overline{\mathbf{L}}^{vp}],\ \overline{\mathbf{W}}^{vp}=\operatorname{ant}[\overline{\mathbf{L}}^{vp}]\end{array}\right\} \tag{13.42}$$

The viscoplastic strain rate is given noting Eqs. (13.31) or (13.33) by

$$\overline{\mathbf{D}}^{vp}=\frac{1}{\bar{\mu}}\frac{\langle R-R_s\rangle^n}{R_m-R}\underline{\overline{\mathbf{N}}} \tag{13.43}$$

or

$$\overline{\mathbf{D}}^{vp}=\frac{1}{\bar{\mu}}\frac{\langle\exp[n(R-R_s)]-1\rangle}{R_m-R}\underline{\overline{\mathbf{N}}} \tag{13.44}$$

where $\underline{\overline{\mathbf{N}}}$ is given by Eq. (12.54).

The viscoplastic spin $\overline{\mathbf{W}}^{vp}$ is given analogously to Eq. (12.60) as follows:

$$\overline{\mathbf{W}}^{vp}=\eta^{vp}(\overline{\mathbf{M}}\,\overline{\mathbf{D}}^{vp}-\overline{\mathbf{D}}^{vp}\overline{\mathbf{M}})=\eta^{vp}\frac{1}{\bar{\mu}}\frac{\langle R-R_s\rangle^n}{R_m-R}(\overline{\mathbf{M}}\,\underline{\overline{\mathbf{N}}}-\underline{\overline{\mathbf{N}}}\,\overline{\mathbf{M}}) \tag{13.45}$$

where η^{vp} is the material parameter.

The viscoplastic velocity gradient is given by substituting Eqs. (13.43) and (13.45) into Eqs. (13.40) as follows:

$$\overline{\mathbf{L}}^{vp}=\frac{1}{\bar{\mu}}\frac{\langle R-R_s\rangle^n}{R_m-R}[\underline{\overline{\mathbf{N}}}+\eta^{vp}(\overline{\mathbf{M}}\,\underline{\overline{\mathbf{N}}}-\underline{\overline{\mathbf{N}}}\,\overline{\mathbf{M}})] \tag{13.46}$$

The rate of the viscoplastic gradient is given from Eq. $(13.37)_3$ as follows:

$$\dot{\mathbf{F}}^{vp}=\overline{\mathbf{L}}^{vp}\mathbf{F}^{vp} \tag{13.47}$$

The time-integration for $\mathbf{F}^{vp}$ can be performed effectively by the tensor exponential method as described for the plastic deformation gradient $\mathbf{F}^{p}$ in the end of Sect. 12.8. The elastic deformation gradient $\mathbf{F}^{e}$ is given by substituting the time-integration $\mathbf{F}^{vp}$ of Eq. (13.47) into

$$\mathbf{F}^e = \mathbf{F}\mathbf{F}^{vp-1} \tag{13.48}$$

Then, $\overline{\mathbf{C}}^e$ is calculated by Eq. (12.3) and further the stresses $\overline{\mathbf{S}}$ and $\overline{\mathbf{M}}$ are calculated by Eqs. (12.31) and (12.32) as the hyperelastic relation as described in Sect. 12.8.

Needless to say, the internal variables H $\overline{M}_k$, $\overline{M}_c$ (in $\underline{\overline{\mathbf{N}}}$) and R_s evolve by the viscoplastic strain rate $\overline{\mathbf{D}}^{vp}$ as described in Sect. 13.5.

References

Bingham EC (1922) Fluidity and plasticity. McGraw-Hill, New York
Chaboche JL (2008) A review of some plasticity and viscoplasticity constitutive theories. Int J Plast 24:1642–1693
Duvaut G, Lions JL (1972) Les Inequations en Mechanique et en Physique. Dunod, Paris (Duvaut G, Lions JL (1976) Inequalities in Mechanics and Physics, Springer, New York)
Hashiguchi K (2007c) Extended overstress model for general rate of deformation including impact load. In: Proceedings of 13th international symposium on plasticity and its current application, pp 37–39
Hashiguchi K, Okayasu T, Saitoh K (2005) Rate-dependent inelastic constitutive equation: the extension of elastoplasticity. Int J Plast 21:463–491
Hohenemser K, Prager W (1932) Uber die Ansatze der Mechanik isotroper Kontinua. Z. A. M. M. 12:216–226
Hutchinson JW (1976) Bounds and self-consistent estimates for creep of polycrystalline materials. Proc Roy Soc Lon A 348:101–127
Johnson GR, Cook WH (1983) A constitutive model and data for metals subjected to large strain, high strain rates and high temperatures. Proc 7th Int. Symp. Ballistics, The Hague, pp 541–547
Lemaitre JA, Chaboche J-L (1990) Mechanics of solid materials. Cambridge University Press, Cambridge
Nakada Y, Keh AS (1966) Latent hardening in iron single crystals. Acta Metall 14:961–973
Norton FH (1929) Creep of steel at high temperature. McGraw-Hill, New York
Pan J, Rice JR, Rice (1983) Rate sensitivity of plastic flow and implications for yield surface vertices. J Mech Phys Solids 19:973–987
Peirce D, Asaro JR, Needleman A (1983) Overview 32: material rate dependence and localized deformation in crystal solids. Act Metall 31:1951–1976
Perzyna P (1963) The constitutive equations for rate sensitive plastic materials. Quart Appl Math 20:321–332
Perzyna P (1966) Fundamental problems in viscoplasticity. Adv Appl Mech 9:243–377
Prager W (1961) Linearization in visco-plasticity. Ing Archiv 15:152–157
Simo JC, Hughes TJR (1998) Computational inelasticity. Springer-Verlag, New York
Simo JC, Kennedy JG, Govindjee S (1988) Non-smooth multisurface plasticity and viscoplasticity —loading unloading conditions and numerical algorithms. Int J Numer Meth Eng 26:2161–2185

Chapter 14
Damage Model

The elastic deformation due to the deformation of material particles themselves is induced even when the stress is low, the elastoplastic deformation due to the slips between material particles (dislocations of crystal lattices in case of metals and slips between soils particles in soils) is induced when the stress increases up to a certain stress (yield stress) and the damage due to the separations of material particles is induced when the stress further increases. The phenomenological formulation of the deformation up to the failure induced in the damage process within the framework of the continuum mechanics is called the *continuum damage mechanics*.

This chapter addresses the elastoplastic constitutive equation with damage which is extended by incorporating the subloading surface model within the continuum damage mechanics. It is extended further to the description of the elasto-viscoplastic deformation with the damage.

The second law of thermodynamics, i.e. the Clausius-Duhem inequality has been used for the formulation of constitutive relation (e.g. Lemaitre and Chaboche 1990; Lemaitre 1992; Lemaitre and Desmoral 2005; Murakami 2012) but it would be helpless or harmful for the formulation of plastic constitutive relation as will be explained in Chap. 21 and Sect. 21.8 in detail. In addition, the plastic constitutive relation has been formulated originally in terms of the current stress in the actual damaged configuration (e.g. Lemaitre and Chaboche 1990; Lemaitre 1992; Lemaitre and Desmoral 2005; Murakami 2012), although it should be formulated rigorously in terms of the effective stress in the fictitious undamaged configuration. Constitutive equations will be first formulated consistently in terms of the effective stress and thereafter it is transformed to the damaged actual configuration, irrelevantly to the second law of thermodynamics in this chapter.

K. Hashiguchi, *Foundations of Elastoplasticity: Subloading Surface Model*,
DOI 10.1007/978-3-319-48821-9_14

14.1 Damage Phenomenon

The conventional elastoplastic model premises on the postulate that the yield surface encloses a purely-elastic domain. Therefore, it describes the abrupt transition from the elastic to the plastic state so that it cannot describe realistically the softening behavior which is observed in the damage phenomenon. Further, it requires the yield-judgment whether or not the stress reaches the yield surface and the operation to pull-back the stress to the yield surface when it goes out from the yield surface in numerical calculation with finite loading increments. The existing damage models (cf. e.g. Kachanov 1958; Rabotnov 1969; Lemaitre and Chaboche 1990; Lemaitre 1992; Lemaitre and Desmoral 2005; de Sauza Neto et al. 2008; Murakami 2012) are based on the conventional elastoplastic model. On the other hand, the subloading surface model (1980, 1989) describes the smooth elastic-plastic transition fulfilling always the smoothness condition in Eq. (7.2). Then, it possesses the distinguished ability as it does not require the yield-judgment and possesses the automatic controlling function to attract the stress to the yield surface in the plastic deformation process so that the stress is automatically pulled-back to the yield surface when it goes out from the yield surface by finite loading increments in numerical calculation.

The extended elastoplastic constitutive equation for the damage phenomenon is formulated by incorporating the concept of subloading surface in this chapter. The formulation is given originally in terms of the effective stress applied in the fictitious undamaged configuration, which is of the identical form to the original subloading surface model formulation without the damage. However, the hypoelasticity cannot be adopted but the hyperelasticity should be adopted for the damage model since we cannot assume that the elastic deformation is far small compared with the plastic deformation in the damage phenomena. Therefore, the formulation is modified to the hyperelastic-based plasticity in terms of the so-called infinitesimal strain. Further, it will be extended to be taken account of the rate-dependent plastic deformation by incorporating the subloading-overstress model described in Chap. 13.

14.2 Damage Variable

Let the relation of the current area da to the area $dA(\leq da)$ in the undamaged initial state be described through the damage variable D in the one-dimensional deformation process as follows:

$$da = \frac{dA}{1-D} \tag{14.1}$$

leading to

$$\sigma = \frac{df}{da} = (1 - D)\underset{\sim}{\sigma}, \underset{\sim}{\sigma} = \frac{df}{dA} = \frac{\sigma}{1 - D} \tag{14.2}$$

where $1/(1 - D)$ (≥ 1; $0 \leq D < 1$) designates the increase of area caused by the damage and df is the traction. Mechanical quantities in the fictitious undamaged configuration are specified by adding the wave under them, i.e. ($\underset{\sim}{\,}$). Let Eq. (14.1) be extended to the three-dimensional deformation process through the second-order positive-definite tensor $\mathcal{D}(=\mathcal{D}^T)$ as follows:

$$\mathbf{n}da = (\mathbf{I} - \mathcal{D})^{-1}\mathbf{N}dA, \quad \mathbf{N}dA = (\mathbf{I} - \mathcal{D})\mathbf{n}da \tag{14.3}$$

leading to

$$\boldsymbol{\sigma} = \frac{d\mathbf{f}}{da} = \underset{\sim}{\boldsymbol{\sigma}}\mathbf{N}\frac{dA}{da}, \quad \underset{\sim}{\boldsymbol{\sigma}} = \frac{d\mathbf{f}}{dA} = \boldsymbol{\sigma}\mathbf{n}\frac{da}{dA} \tag{14.4}$$

noting Eq. (3.1), where $\mathbf{N}$ and $\mathbf{n}$ are the unit outward-normal vectors of the initial and the current surface, respectively. $\mathcal{D}$ is referred to as the *damage tensor* and plays the most basic role for the description of constitutive equation of damage.

14.3 Hyperelastic Relation

In what follows, we adopt the *hypothesis of strain equivalence* (Lemaitre 1971) insisting that the strain and its elastic and plastic parts in the fictitious undamaged configuration are equivalent to those in the actual damaged configuration. As described in Sect. 6.9, the infinitesimal strain is additively decomposed as follows:

$$\boldsymbol{\varepsilon} = \boldsymbol{\varepsilon}^e + \boldsymbol{\varepsilon}^p \tag{14.5}$$

Now, adopting the linear elasticity in the Hooke's law in Eq. (5.42), i.e.

$$\left.\begin{aligned} E_{0\,ijkl} &= \frac{E_0}{1+\nu}\left[\frac{1}{2}(\delta_{ik}\delta_{jl} + \delta_{il}\delta_{jk}) + \frac{\nu}{1-2\nu}\delta_{ij}\delta_{kl}\right] \\ E^{-1}_{0\,ijkl} &= \frac{1}{E_0}\left[\frac{1}{2}(1+\nu)(\delta_{ik}\delta_{jl} + \delta_{il}\delta_{jk}) - \nu\delta_{ij}\delta_{kl}\right] \end{aligned}\right\} \tag{14.6}$$

for which the Helmholtz free energy function $\underset{\sim}{\psi}(\boldsymbol{\varepsilon}^e)$ and the Gibbs' free energy $\underset{\sim}{\phi}(\underset{\sim}{\boldsymbol{\sigma}})$ in the undamaged configuration are given noting Eqs. (5.43) and (5.44) as

$$\underset{\sim}{\psi}(\boldsymbol{\varepsilon}^e) = \frac{1}{2}\boldsymbol{\varepsilon}^e : \mathbf{E}_0 : \boldsymbol{\varepsilon}^e\left(= \frac{1}{2}\underset{\sim}{\boldsymbol{\sigma}} : \boldsymbol{\varepsilon}^e\right) = \frac{1}{2}\frac{E_0}{1+\nu}\left[\varepsilon^e_{ij}\varepsilon^e_{ij} + \frac{\nu}{1-2\nu}(\varepsilon^e_{kk})^2\right] \tag{14.7}$$

$$\underset{\sim}{\phi}(\underset{\sim}{\boldsymbol{\sigma}}) = \frac{1}{2}\underset{\sim}{\boldsymbol{\sigma}} : \mathbf{E}_0^{-1} : \underset{\sim}{\boldsymbol{\sigma}}\left(= \frac{1}{2}\underset{\sim}{\boldsymbol{\sigma}} : \boldsymbol{\varepsilon}^e \right) = \frac{1}{2E_0}\left[(1+\nu)\underset{\sim ij}{\sigma}\underset{\sim ij}{\sigma} - \nu(\underset{\sim kk}{\sigma})^2 \right] \tag{14.8}$$

where the fourth-order tensor $\mathbf{E}_0$ is assumed to be the constant tensor in the fictitious undamaged configuration, using the Young's modulus E_0 and the Poisson ratio ν in the fictitious undamaged configuration.

The effective stress $\underset{\sim}{\boldsymbol{\sigma}}$ and the elastic strain $\boldsymbol{\varepsilon}^e$ are derived from Eqs. (14.7) and (14.8), noting Eqs. (5.43) and (5.44) as follows:

$$\underset{\sim}{\boldsymbol{\sigma}} = \frac{\partial \underset{\sim}{\psi}(\boldsymbol{\varepsilon}^e)}{\partial \boldsymbol{\varepsilon}^e} = \mathbf{E}_0 : \boldsymbol{\varepsilon}^e, \quad \underset{\sim}{\sigma}_{ij} = \frac{\partial \underset{\sim}{\psi}(\boldsymbol{\varepsilon}^e)}{\partial \varepsilon_{ij}^e} = E_{0\,ijkl}\varepsilon_{kl}^e = \frac{E_0}{1+\nu}\left(\varepsilon_{ij} + \frac{\nu}{1-2\nu}\varepsilon_{kk}\delta_{ij} \right) \tag{14.9}$$

$$\boldsymbol{\varepsilon}^e = \frac{\partial \underset{\sim}{\phi}(\underset{\sim}{\boldsymbol{\sigma}})}{\partial \underset{\sim}{\boldsymbol{\sigma}}} = \mathbf{E}_0^{-1} : \underset{\sim}{\boldsymbol{\sigma}}, \quad \varepsilon_{ij}^e = \frac{\partial \underset{\sim}{\phi}(\underset{\sim}{\boldsymbol{\sigma}})}{\partial \underset{\sim ij}{\sigma}} = \frac{1}{E_0}\left[(1+\nu)\underset{\sim ij}{\sigma} - \nu \underset{\sim kk}{\sigma}\delta_{ij} \right] \tag{14.10}$$

The elastic strain rate is given from Eq. (14.10) as follows:

$$\dot{\boldsymbol{\varepsilon}}^e = \mathbf{E}_0^{-1} : \underset{\sim}{\dot{\boldsymbol{\sigma}}}, \quad \underset{\sim}{\dot{\boldsymbol{\sigma}}} = \mathbf{E}_0 : \dot{\boldsymbol{\varepsilon}}^e \tag{14.11}$$

14.4 Subloading-Damage Model

The plastic strain rate will be formulated based on the concept of subloading surface in this section (cf. Hashiguchi 2015a).

14.4.1 *Normal-Yield and Subloading Surfaces*

The yield condition is given by

$$f(\underset{\sim}{\hat{\boldsymbol{\sigma}}}) = \underset{\sim}{F}(\underset{\sim}{H}) \tag{14.12}$$

where $\underset{\sim}{H}$ is the isotropic hardening variable in the fictitious undamaged configuration and

$$\underset{\sim}{\hat{\boldsymbol{\sigma}}} \equiv \underset{\sim}{\boldsymbol{\sigma}} - \underset{\sim}{\boldsymbol{\alpha}} \tag{14.13}$$

$\underset{\sim}{\boldsymbol{\alpha}}$ is the kinematic hardening variable in the fictitious undamaged configuration. The rates of these variables are be described as follows:

$$\left.\begin{array}{l}\overset{\bullet}{\underset{\sim}{H}} = \underset{\sim}{f}_{H\varepsilon}(\underset{\sim}{\boldsymbol{\sigma}}, \underset{\sim}{H}; \dot{\boldsymbol{\varepsilon}}^p) = \underset{\sim}{f}_{H\varepsilon}(\underset{\sim}{\boldsymbol{\sigma}}, \underset{\sim}{H}; \dot{\boldsymbol{\varepsilon}}^p/\|\dot{\boldsymbol{\varepsilon}}^p\|)\|\dot{\boldsymbol{\varepsilon}}^p\| \\ \overset{\bullet}{\underset{\sim}{\boldsymbol{\alpha}}} = \underset{\sim}{\mathbf{f}}_{k\varepsilon}(\underset{\sim}{\boldsymbol{\sigma}}, \underset{\sim}{\boldsymbol{\alpha}}, \underset{\sim}{F}; \dot{\boldsymbol{\varepsilon}}^{p\prime}) = \underset{\sim}{\mathbf{f}}_{k\varepsilon}(\underset{\sim}{\boldsymbol{\sigma}}, \underset{\sim}{\boldsymbol{\alpha}}, \underset{\sim}{F}; \dot{\boldsymbol{\varepsilon}}^{p\prime}/\|\dot{\boldsymbol{\varepsilon}}^p\|)\|\dot{\boldsymbol{\varepsilon}}^p\|\end{array}\right\} \quad (14.14)$$

which are homogeneous functions of $\dot{\boldsymbol{\varepsilon}}^p$ in degree-one since they are induced only in the plastic loading process $\dot{\boldsymbol{\varepsilon}}^p \neq \mathbf{O}$ and the first-order time-differential quantities. Here, let it be assumed that the function $f(\hat{\underset{\sim}{\boldsymbol{\sigma}}})$ is the homogeneous function of $\hat{\underset{\sim}{\boldsymbol{\sigma}}}$ in degree-one so that the following relation holds.

$$\frac{\partial f(\hat{\underset{\sim}{\boldsymbol{\sigma}}})}{\partial \hat{\underset{\sim}{\boldsymbol{\sigma}}}} : \hat{\underset{\sim}{\boldsymbol{\sigma}}} = f(\hat{\underset{\sim}{\boldsymbol{\sigma}}}) \quad (14.15)$$

Therefore, the yield surface retains the similar shape.

Hereinafter, the yield surface in Eq. (14.12) is called the normal-yield surface in the fictitious undamaged configuration. Further, incorporate the following subloading surface in the fictitious undamaged configuration (see Fig. 14.1).

$$f(\hat{\underset{\sim}{\boldsymbol{\sigma}}}) = R\underset{\sim}{F}(\underset{\sim}{H}) \quad (14.16)$$

where R is the normal-yield ratio.

The rate of the normal-yield ratio is given by

$$\dot{R} = U(R)\|\dot{\boldsymbol{\varepsilon}}^p\| \quad \text{for} \quad \dot{\boldsymbol{\varepsilon}}^p \neq \mathbf{O} \quad (14.17)$$

based on Eq. (7.9).

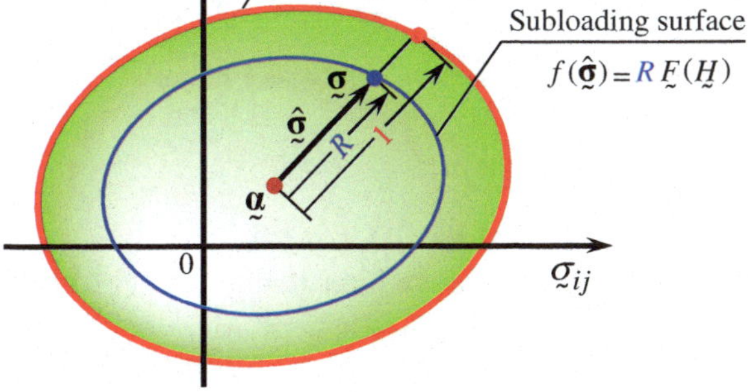

Fig. 14.1 Normal-yield and subloading surfaces in the virtual undamaged configuration

The time-differentiation of Eq. (14.16) leads to

$$\frac{\partial f(\hat{\boldsymbol{\sigma}})}{\partial \hat{\boldsymbol{\sigma}}}:\overset{\circ}{\boldsymbol{\sigma}} - \frac{\partial f(\hat{\boldsymbol{\sigma}})}{\partial \hat{\boldsymbol{\sigma}}}:\overset{\circ}{\boldsymbol{\alpha}} - R\dot{F} - \dot{R}F = 0 \tag{14.18}$$

It follows from Eqs. (14.15) and (14.16) that

$$\frac{\partial f(\hat{\boldsymbol{\sigma}})}{\partial \hat{\boldsymbol{\sigma}}}:\hat{\boldsymbol{\sigma}} = RF \tag{14.19}$$

which yields

$$\hat{\mathbf{n}}:\hat{\boldsymbol{\sigma}} = \frac{\partial f(\hat{\boldsymbol{\sigma}})}{\partial \hat{\boldsymbol{\sigma}}}:\hat{\boldsymbol{\sigma}} \Big/ \left\| \frac{\partial f(\hat{\boldsymbol{\sigma}})}{\partial \hat{\boldsymbol{\sigma}}} \right\| = RF \Big/ \left\| \frac{\partial f(\hat{\boldsymbol{\sigma}})}{\partial \hat{\boldsymbol{\sigma}}} \right\| \tag{14.20}$$

leading to

$$1 \Big/ \left\| \frac{\partial f(\hat{\boldsymbol{\sigma}})}{\partial \hat{\boldsymbol{\sigma}}} \right\| = \frac{\hat{\mathbf{n}}:\hat{\boldsymbol{\sigma}}}{RF} \tag{14.21}$$

where

$$\hat{\mathbf{n}} \equiv \frac{\partial f(\hat{\boldsymbol{\sigma}})}{\partial \hat{\boldsymbol{\sigma}}} \Big/ \left\| \frac{\partial f(\hat{\boldsymbol{\sigma}})}{\partial \hat{\boldsymbol{\sigma}}} \right\| \quad (\|\hat{\mathbf{n}}\| = 1) \tag{14.22}$$

The substitution of Eq. (14.21) into Eq. (14.18) leads to

$$\hat{\mathbf{n}}: \left[\overset{\circ}{\boldsymbol{\sigma}} - \overset{\circ}{\boldsymbol{\alpha}} - \left(\frac{\dot{F}}{F} + \frac{\dot{R}}{R} \right) \hat{\boldsymbol{\sigma}} \right] = 0 \tag{14.23}$$

Now, adopt the associated flow rule

$$\mathbf{D}^p = \dot{\bar{\lambda}}\, \hat{\mathbf{n}} \; (\dot{\bar{\lambda}} \geq 0) \tag{14.24}$$

where $\dot{\bar{\lambda}}$ is the positive plastic multiplier. Substituting Eqs. (14.14) and (14.17) with Eq. (14.24) into Eq. (14.23), one has

$$\hat{\underset{\sim}{\mathbf{n}}}\mathbf{:}\left[\overset{\circ}{\underset{\sim}{\boldsymbol{\sigma}}} - \underset{\sim}{\mathbf{f}}_{kn}(\underset{\sim}{\boldsymbol{\sigma}}, \underset{\sim}{\boldsymbol{\alpha}}, \underset{\sim}{F}; \hat{\underset{\sim}{\mathbf{n}}}')\,\dot{\bar{\lambda}} - \left(\frac{\underset{\sim}{F}'}{\underset{\sim}{F}}\underset{\sim}{f}_{Hn}(\underset{\sim}{\boldsymbol{\sigma}}, \underset{\sim}{H}; \hat{\underset{\sim}{\mathbf{n}}})\,\dot{\bar{\lambda}} + \frac{U(R)}{R}\dot{\bar{\lambda}}\right)\hat{\underset{\sim}{\boldsymbol{\sigma}}}\right] \tag{14.25}$$

where

$$\underset{\sim}{f}_{Hn}(\underset{\sim}{\boldsymbol{\sigma}}, \underset{\sim}{H}; \hat{\underset{\sim}{\mathbf{n}}}) = \dot{\underset{\sim}{H}}/\dot{\bar{\lambda}}, \quad \underset{\sim}{\mathbf{f}}_{kn}(\underset{\sim}{\boldsymbol{\sigma}}, \underset{\sim}{\boldsymbol{\alpha}}, \underset{\sim}{F}; \hat{\underset{\sim}{\mathbf{n}}}') = \dot{\underset{\sim}{\boldsymbol{\alpha}}}/\dot{\bar{\lambda}} \tag{14.26}$$

noting the homogeneities of $\dot{\underset{\sim}{H}}$ and $\dot{\underset{\sim}{\boldsymbol{\alpha}}}$ in degree-one of $\dot{\boldsymbol{\varepsilon}}^p$. It follows from Eqs. (14.24) and (14.25) that

$$\dot{\bar{\lambda}}\frac{\hat{\underset{\sim}{\mathbf{n}}} : \overset{\circ}{\underset{\sim}{\boldsymbol{\sigma}}}}{\underset{\sim}{\overline{M}^p}}, \dot{\varepsilon}^p = \frac{\hat{\underset{\sim}{\mathbf{n}}} : \overset{\circ}{\underset{\sim}{\boldsymbol{\sigma}}}}{\underset{\sim}{\overline{M}^p}}\hat{\underset{\sim}{\mathbf{n}}} \tag{14.27}$$

where

$$\underset{\sim}{\overline{M}^p} \equiv \hat{\underset{\sim}{\mathbf{n}}}\mathbf{:}\left[\underset{\sim}{\mathbf{f}}_{kn}(\underset{\sim}{\boldsymbol{\sigma}}, \underset{\sim}{\boldsymbol{\alpha}}, \underset{\sim}{F}; \hat{\underset{\sim}{\mathbf{n}}}') + \left(\frac{\underset{\sim}{F}'}{\underset{\sim}{F}}\underset{\sim}{f}_{Hn}(\underset{\sim}{\boldsymbol{\sigma}}, \underset{\sim}{H}; \hat{\underset{\sim}{\mathbf{n}}}) + \frac{U(R)}{R}\right)\hat{\underset{\sim}{\boldsymbol{\sigma}}}\right] \tag{14.28}$$

14.4.2 Stress Rate Versus Strain Rate Relations

The strain rate is described from Eqs. (14.5), (14.11) and (14.27) as

$$\dot{\boldsymbol{\varepsilon}} = \mathbf{E}_0^{-1} : \overset{\circ}{\underset{\sim}{\boldsymbol{\sigma}}} + \frac{\hat{\underset{\sim}{\mathbf{n}}}:\overset{\circ}{\underset{\sim}{\boldsymbol{\sigma}}}}{\underset{\sim}{\overline{M}^p}}\hat{\underset{\sim}{\mathbf{n}}} = (\mathbf{E}_0^{-1} + \frac{\hat{\underset{\sim}{\mathbf{n}}} \otimes \hat{\underset{\sim}{\mathbf{n}}}}{\underset{\sim}{\overline{M}^p}}):\overset{\circ}{\underset{\sim}{\boldsymbol{\sigma}}} \tag{14.29}$$

from which the plastic multiplier in terms of strain rate is derived as follows:

$$\dot{\bar{\Lambda}} = \frac{\hat{\underset{\sim}{\mathbf{n}}} : \mathbf{E}_0 : \dot{\boldsymbol{\varepsilon}}}{\underset{\sim}{\overline{M}^p} + \hat{\underset{\sim}{\mathbf{n}}} : \mathbf{E}_0 : \hat{\underset{\sim}{\mathbf{n}}}} \tag{14.30}$$

The stress rate is described from Eqs. (14.11), (14.24) and (14.30) as

$$\overset{\circ}{\underset{\sim}{\boldsymbol{\sigma}}} = \mathbf{E}_0 : \dot{\boldsymbol{\varepsilon}} - \frac{\hat{\underset{\sim}{\mathbf{n}}} : \mathbf{E}_0 : \dot{\boldsymbol{\varepsilon}}}{\underset{\sim}{\overline{M}^p} + \hat{\underset{\sim}{\mathbf{n}}} : \mathbf{E}_0 : \hat{\underset{\sim}{\mathbf{n}}}}\mathbf{E}_0 : \hat{\underset{\sim}{\mathbf{n}}} \tag{14.31}$$

The loading criterion is given by

$$\left.\begin{array}{ll}\dot{\boldsymbol{\varepsilon}}^p \neq \mathbf{O} & \text{for } \overset{\bullet}{\bar{\Lambda}} > 0 \\ \dot{\boldsymbol{\varepsilon}}^p \neq \mathbf{O} & \text{for } \overset{\bullet}{\bar{\Lambda}} \leq 0\end{array}\right\} \tag{14.32}$$

14.5 Hardening Rules

The evolution rules of the isotropic and the kinematic hardening variables are given in this section.

14.5.1 Isotropic Hardening Rule

The isotropic hardening is described by

$$\underset{\sim}{F}(\underset{\sim}{H}) = \underset{\sim}{F}_0[1 + h_1\{1 - \exp(-h_2\underset{\sim}{H})\}], \quad \underset{\sim}{F}' = h_1 h_2 \underset{\sim}{F}_0 \exp(-h_2\underset{\sim}{H}) \tag{14.33}$$

$$\dot{\underset{\sim}{H}} = \sqrt{\frac{2}{3}}\|\dot{\boldsymbol{\varepsilon}}^p\| = \dot{\bar{\lambda}} \underset{\sim}{f}_{Hn}(\underset{\sim}{\boldsymbol{\sigma}}, \underset{\sim}{H}; \underset{\sim}{\hat{\mathbf{n}}}) \tag{14.34}$$

$$\underset{\sim}{f}_{Hn}(\underset{\sim}{\boldsymbol{\sigma}}, \underset{\sim}{H}; \underset{\sim}{\hat{\mathbf{n}}}) = \sqrt{\frac{2}{3}} \tag{14.35}$$

14.5.2 Nonlinear Kinematic Hardening Rule

Internal variables describe state of substructure of material so that tensor-valued internal variables must be given by the hyperelastic-like relation, i.e. the partial differential of the energy storage function of conjugate strain measure.

As described in Sect. 6.9, the plastic strain rate $\boldsymbol{\varepsilon}^p$ be decomposed into the storage part $\boldsymbol{\varepsilon}^p_{ks}$ and the dissipation part $\boldsymbol{\varepsilon}^p_{kd}$, i.e.

$$\boldsymbol{\varepsilon}^p = \boldsymbol{\varepsilon}^p_{ks} + \boldsymbol{\varepsilon}^p_{kd} \tag{14.36}$$

for the kinematic hardening. In addition, incorporate the energy storage function $\psi^k(\boldsymbol{\varepsilon}^p_{ks})$ of the storage part of the plastic strain for the kinematic hardening variable. Then, the kinematic hardening variable is given as follows:

$$\underset{\sim}{\boldsymbol{\alpha}} = \frac{\partial \psi^k(\underset{\sim}{\boldsymbol{\varepsilon}}^{p\prime}_{ks})}{\partial \boldsymbol{\varepsilon}^p_{ks}} \tag{14.37}$$

Now, adopt the explicit function

$$\underset{\sim}{\psi}^k(\boldsymbol{\varepsilon}^p_{ks}) = \frac{1}{2} c_k \boldsymbol{\varepsilon}^{p\prime}_{ks} : \boldsymbol{\varepsilon}^{p\prime}_{ks} \tag{14.38}$$

where c_k is the material constant. The kinematic hardening variable is given from Eq. (14.37) with Eq. (14.38) by

$$\underset{\sim}{\boldsymbol{\alpha}} = c_k \boldsymbol{\varepsilon}^{p\prime}_{ks} \tag{14.39}$$

Further, let the dissipative part of plastic strain rate be given, noting Eq. (6.103), as

$$\overset{\bullet}{\boldsymbol{\varepsilon}}{}^{p\prime}_{kd} = \frac{1}{b_k(\underset{\sim}{\boldsymbol{\sigma}}, \underset{\sim}{F})} \underset{\sim}{\boldsymbol{\alpha}} \| \overset{\bullet}{\boldsymbol{\varepsilon}}{}^{p\prime} \| = \frac{1}{b_k(\underset{\sim}{\boldsymbol{\alpha}}, \underset{\sim}{F})} \underset{\sim}{\boldsymbol{\alpha}} \| \underset{\sim}{\hat{\mathbf{n}}}' \| \overset{\bullet}{\lambda} \tag{14.40}$$

noting Eq. (14.24), where $b_k(>0)$ is the material function of $\underset{\sim}{\boldsymbol{\sigma}}$ and $\underset{\sim}{F}$ in general. Equation (14.40) satisfies the positivity of the dissipation energy, i.e. $\underset{\sim}{\boldsymbol{\alpha}} : \overset{\bullet}{\boldsymbol{\varepsilon}}{}^p_{kd} = (\underset{\sim}{\boldsymbol{\alpha}} : \underset{\sim}{\boldsymbol{\alpha}}) \| \overset{\bullet}{\boldsymbol{\varepsilon}}{}^{p\prime} \| / b_k \geq 0$. It follows from Eqs. (14.24), (14.36) and (14.40) that

$$\overset{\bullet}{\boldsymbol{\varepsilon}}{}^{p\prime}_{ks} = \overset{\bullet}{\boldsymbol{\varepsilon}}{}^{p\prime} - \overset{\bullet}{\boldsymbol{\varepsilon}}{}^{p\prime}_{kd} = \overset{\bullet}{\boldsymbol{\varepsilon}}{}^{p\prime} - \frac{1}{b_k(\underset{\sim}{\boldsymbol{\sigma}}, \underset{\sim}{F})} \| \overset{\bullet}{\boldsymbol{\varepsilon}}{}^{p\prime} \| \underset{\sim}{\boldsymbol{\alpha}} = \overset{\bullet}{\lambda} \left(\underset{\sim}{\hat{\mathbf{n}}}' - \frac{1}{b_k(\underset{\sim}{\boldsymbol{\sigma}}, \underset{\sim}{F})} \underset{\sim}{\boldsymbol{\alpha}} \| \underset{\sim}{\hat{\mathbf{n}}}' \| \right) \tag{14.41}$$

The following equation is derived from Eqs. (14.39) and (14.41).

$$\left.\begin{aligned} \overset{\bullet}{\underset{\sim}{\boldsymbol{\alpha}}} &= c_k \overset{\bullet}{\boldsymbol{\varepsilon}}{}^{p\prime}_{ks} = c_k \left(\overset{\bullet}{\boldsymbol{\varepsilon}}{}^{p\prime} - \frac{1}{b_k(\underset{\sim}{\boldsymbol{\sigma}}, \underset{\sim}{F})} \| \overset{\bullet}{\boldsymbol{\varepsilon}}{}^{p\prime} \| \underset{\sim}{\boldsymbol{\alpha}} \right) = \overset{\bullet}{\lambda} \, \underset{\sim}{\mathbf{f}}_{kn} \\ \underset{\sim}{\mathbf{f}}_{kn} &= c_k \overset{\bullet}{\lambda} \left(\underset{\sim}{\hat{\mathbf{n}}}' - \frac{1}{b_k(\underset{\sim}{\boldsymbol{\sigma}}, \underset{\sim}{F})} \underset{\sim}{\boldsymbol{\alpha}} \| \underset{\sim}{\hat{\mathbf{n}}}' \| \right) \end{aligned}\right\} \tag{14.42}$$

which is the modification of the nonlinear kinematic hardening rule by Armstrong and Frederick (1966).

The constitutive relation for the effective stress was described in the previous and this sections. Further, we have to formulate the calculation method of the current stress in the actual configuration from the effective stress, which will be attained through the damage tensor as described in the subsequent sections.

14.6 Damage Tensor

The damage variable is the tensor which transforms the elastic response in the fictitious undamaged configuration to the one in the actual damaged configuration in general. However, assume that the elastic response in the actual damaged configuration is also given by the linear relation between the Cauchy stress and the elastic strain as well as the one in the fictitious undamaged configuration shown in Eq. (14.10), i.e.

$$\boldsymbol{\sigma} = \mathbf{E} : \boldsymbol{\varepsilon}^e, \quad \boldsymbol{\varepsilon}^e = \mathbf{E}^{-1} : \boldsymbol{\sigma} \tag{14.43}$$

based on the hypothesis of strain equivalence (Lemaitre 1971), where $\mathbf{E}$ is elastic modulus tensor in the actual damaged configuration. It follows from Eqs. (14.10) and (14.43) that

$$\boldsymbol{\sigma} = \mathbf{E} : \mathbf{E}_0^{-1} : \underset{\sim}{\boldsymbol{\sigma}}, \quad \underset{\sim}{\boldsymbol{\sigma}} = \mathbf{E}_0 : \mathbf{E}^{-1} : \boldsymbol{\sigma} \tag{14.44}$$

The relation of $\mathbf{E}$ to $\mathbf{E}_0$ is described using the damage tensor $\mathfrak{D}$ in general as follows:

$$\mathbf{E} = \mathbf{E}(\mathbf{E}_0, \mathfrak{D}) \tag{14.45}$$

where the evolution rule of the damage tensor $\mathfrak{D}$ can be generally given as follows:

$$\dot{\mathfrak{D}} = \mathbf{f}_D(\underset{\sim}{\boldsymbol{\sigma}}, \varepsilon^{eqp}; \dot{\boldsymbol{\varepsilon}}^p) \tag{14.46}$$

where $\mathbf{f}_D$ is the homogeneous function of $\dot{\boldsymbol{\varepsilon}}^p$ in degree-one.

Now, if we assume the linear relation between the fourth-order tensors $\mathbf{E}_0$ and $\mathbf{E}$, i.e.

$$E_{ijkl} = \mathfrak{D}_{ijklpqrs} E_{0pqrs} \tag{14.47}$$

the damage variable $\mathfrak{D}$ is the eight-order tensor in general.

The following simple transformation rule in terms of the fourth-order tensor $\mathbb{D}$ is adopted by Chaboche (1982).

$$\mathbf{E} = (\mathcal{I} - \mathbb{D}) : \mathbf{E}_0 \tag{14.48}$$

which yields

$$\boldsymbol{\sigma} = (\mathcal{I} - \mathbb{D}) : \underset{\sim}{\boldsymbol{\sigma}}, \quad \underset{\sim}{\boldsymbol{\sigma}} = (\mathcal{I} - \mathbb{D})^{-1} : \boldsymbol{\sigma} \tag{14.49}$$

However, unfortunately the current stress $\boldsymbol{\sigma}$ is generally no longer symmetric even for the effective stress tensor $\underset{\sim}{\boldsymbol{\sigma}}$ which is the symmetric tensor. Besides, it cannot be used as far as the tensor $\mathbb{D}$ is described explicitly by the known physical variable.

Various explicit damage tensors have been proposed as will be described in the following.

14.6.1 Isotropic Damage Tensor

Assume the following isotropic damage tensor with the scalar variable D in Eq. (14.1):

$$\mathbb{D} = D\mathcal{I} \tag{14.50}$$

Substituting Eq. (14.50) into Eq. (14.48), it follows that

$$\mathbf{E} = (\mathcal{I} - \mathbb{D}):\mathbf{E}_0 = (1 - D)\mathcal{I}:\mathbf{E}_0 = (1 - D)\mathbf{E}_0, \quad \mathbf{E}_0 = \mathbf{E}/(1 - D) \tag{14.51}$$

$$\mathbf{E}^{-1} = \mathbf{E}_0^{-1}/(1 - D), \quad \mathbf{E}_0^{-1} = (1 - D)\mathbf{E}^{-1} \tag{14.52}$$

which is described for the Hooke's law as follows:

$$\left.\begin{aligned} E_{ijkl} &= (1 - D)E_{0ijkl} = (1 - D)\frac{E_0}{1 + \nu}\left[\frac{\nu}{1 - 2\nu}\delta_{ij}\delta_{kl} + \frac{1}{2}(\delta_{ik}\delta_{jl} + \delta_{il}\delta_{jk})\right] \\ E_{ijkl}^{-1} &= \frac{1}{(1 - D)}E_{0ijkl}^{-1} = \frac{1}{(1 - D)E_0}\left[\frac{1}{2}(1 + \nu)(\delta_{ik}\delta_{jl} + \delta_{il}\delta_{jk}) - \nu\delta_{ij}\delta_{kl}\right] \end{aligned}\right\} \tag{14.53}$$

resulting in

$$E = (1 - D)E_0, \quad D = 1 - E/E_0 \tag{14.54}$$

Then, substituting Eq. (14.52) into Eq. (14.44), one has

$$\boldsymbol{\sigma} = (1 - D)\underset{\sim}{\boldsymbol{\sigma}}, \quad \underset{\sim}{\boldsymbol{\sigma}} = \boldsymbol{\sigma}/(1 - D) \tag{14.55}$$

from which it follows that

$$\dot{\boldsymbol{\sigma}} = (1-D)\dot{\underset{\sim}{\boldsymbol{\sigma}}} - \dot{D}\underset{\sim}{\boldsymbol{\sigma}}, \quad \dot{\underset{\sim}{\boldsymbol{\sigma}}} = \frac{\dot{\boldsymbol{\sigma}} + \dot{D}\underset{\sim}{\boldsymbol{\sigma}}}{1 - D} = \frac{\dot{\boldsymbol{\sigma}} + \dot{D}\dfrac{\boldsymbol{\sigma}}{1-D}}{1 - D} \tag{14.56}$$

The following relations hold from Eqs. (14.51) and (14.52).

$$\boldsymbol{\sigma} = \mathbf{E} : \boldsymbol{\varepsilon}^e = (1-D)\mathbf{E}_0 : \boldsymbol{\varepsilon}^e, \quad \underset{\sim}{\boldsymbol{\sigma}} = \mathbf{E}_0 : \boldsymbol{\varepsilon}^e = \frac{1}{1-D}\mathbf{E} : \boldsymbol{\varepsilon}^e \tag{14.57}$$

$$\boldsymbol{\varepsilon}^e = \mathbf{E}^{-1} : \boldsymbol{\sigma} = \frac{1}{1-D}\mathbf{E}_0^{-1} : \boldsymbol{\sigma} = \mathbf{E}_0^{-1} : \underset{\sim}{\boldsymbol{\sigma}} = (1-D)\mathbf{E}^{-1} : \boldsymbol{\sigma} \tag{14.58}$$

The Helmholtz free energy function $\psi(\boldsymbol{\varepsilon}^e)$ and the Gibbs' free energy $\phi(\boldsymbol{\sigma})$ in the damaged state are given for Eq. (14.53) noting Eqs. (5.40) as

$$\begin{aligned}\psi(\boldsymbol{\varepsilon}^e, D)\left(=\frac{1}{2}\boldsymbol{\sigma} : \boldsymbol{\varepsilon}^e\right) &= \frac{1}{2}\boldsymbol{\varepsilon}^e : \mathbf{E} : \boldsymbol{\varepsilon}^e = \frac{1}{2}(1-D)\boldsymbol{\varepsilon}^e : \mathbf{E}_0 : \boldsymbol{\varepsilon}^e \\ &= \frac{1}{2}(1-D)\frac{E_0}{1+\nu}\left[\varepsilon^e_{ij}\varepsilon^e_{ij} + \frac{\nu}{1-2\nu}(\varepsilon^e_{kk})^2\right]\end{aligned} \tag{14.59}$$

$$\begin{aligned}\phi(\boldsymbol{\sigma}, D)\left(=\frac{1}{2}\boldsymbol{\sigma} : \boldsymbol{\varepsilon}^e\right) &= \frac{1}{2}\boldsymbol{\sigma} : \mathbf{E}^{-1} : \boldsymbol{\sigma} = \frac{1}{2(1-D)}\boldsymbol{\sigma} : \mathbf{E}_0^{-1} : \boldsymbol{\sigma} \\ &= \frac{1}{2(1-D)E_0}[(1+\nu)\sigma_{ij}\sigma_{ij} - \nu(\sigma_{kk})^2]\end{aligned} \tag{14.60}$$

from which it follows that

$$\sigma_{ij} = \frac{\partial\psi(\boldsymbol{\varepsilon}^e, D)}{\partial\varepsilon^e_{ij}} = (1-D)E_{0ijkl}\varepsilon^e_{kl} = (1-D)\frac{E_0}{1+\nu}\left(\varepsilon_{ij} + \frac{\nu}{1-2\nu}\varepsilon_{kk}\delta_{ij}\right) \tag{14.61}$$

$$\begin{aligned}\varepsilon^e_{ij} &= \frac{\partial\phi(\boldsymbol{\sigma}, D)}{\partial\sigma_{ij}} = \frac{1}{1-D}E^{-1}_{0ijkl}\sigma_{kl} = \frac{1}{(1-D)E_0}[(1+\nu)\sigma_{ij} - \nu\sigma_{kk}\delta_{ij}] \\ &= \frac{1}{E_0}[(1+\nu)\underset{\sim}{\sigma}_{ij} - \nu\underset{\sim}{\sigma}_{kk}\delta_{ij}]\end{aligned} \tag{14.62}$$

The example of the evolution rule of the isotropic damage variable D is given by Lemaitre and Chaboche (1990) as follows:

$$\dot{D} = \left(\frac{Y}{\zeta}\right)^a \frac{H(\varepsilon^{eqp} - \varepsilon^{eqp}_d)}{1-D}\dot{\varepsilon}^{eqp} \tag{14.63}$$

where ζ and a are the material constants, and ε^{eqp}_d is the threshold value of ε^{eqp}, and $H(\ )$ is the Heaviside step function, i.e. $H(s) = 0$ for $s \le 0$ and $H(s) = 1$ for $s > 0$ for a scalar variable s. Y is defined by

$$Y = -\frac{\partial\psi(\boldsymbol{\varepsilon}^e, D)}{\partial D} = \frac{\psi(\boldsymbol{\varepsilon}^e, D)}{1-D} = \frac{1}{2}E_{0\,ijkl}\varepsilon^e_{ij}\varepsilon^e_{kl} = \frac{1}{2}\boldsymbol{\varepsilon}^e : \mathbf{E}_0 : \boldsymbol{\varepsilon}^e = \frac{1}{2}\underset{\sim}{\boldsymbol{\sigma}} : \boldsymbol{\varepsilon}^e \tag{14.64}$$

noting Eq. (14.59), which designates the releasing rate of the strain energy due to the damage under the constant strain ($\partial\psi(\boldsymbol{\varepsilon}^e, D)/\partial D < 0$), i.e. the rate of energy dissipated in the crack extension and is called the *strain energy density release rate* in the failure mechanics or the *damage associated variable* (Chaboche 1988). Equation (14.64) is rewritten as follows:

$$Y = \frac{1}{2E_0}[(1+\nu)\underset{\sim}{\sigma}_{ij}\underset{\sim}{\sigma}_{ij} - \nu(\underset{\sim}{\sigma}_{rr})^2] = \frac{1+\nu}{2(1-D)^2E_0}[(1+\nu)\sigma_{ij}\sigma_{ij} - \nu(\sigma_{rr})^2] \tag{14.65}$$

or

$$Y = \frac{2}{3}\frac{1+\nu}{2E_0}\underset{\sim}{\sigma}^{eq\,2} + 3\frac{1-2\nu}{2E_0}\underset{\sim}{\sigma}_m^2 \tag{14.66}$$

noting

$$\begin{aligned}
\frac{1}{2}\underset{\sim}{\sigma}_{ij}\varepsilon^e_{ij} &= \frac{1}{2}\underset{\sim}{\sigma}_{ij}\frac{1}{E_0}[(1+\nu)\underset{\sim}{\sigma}_{ij} - \nu\underset{\sim}{\sigma}_{kk}\delta_{ij})] = \frac{1}{2E_0}\left[(1+\nu)\underset{\sim}{\sigma}_{ij}\underset{\sim}{\sigma}_{ij} - \nu(\underset{\sim}{\sigma}_{kk})^2\right] \\
&= \frac{1}{2E_0}\left[(1+\nu)\left(\underset{\sim}{\sigma}'_{ij} + \frac{1}{3}\underset{\sim}{\sigma}_{kk}\delta_{ij}\right)\left(\underset{\sim}{\sigma}'_{ij} + \frac{1}{3}\underset{\sim}{\sigma}_{kk}\delta_{ij}\right) - \nu(\underset{\sim}{\sigma}_{kk})^2\right] \\
&= \frac{1}{2E_0}\left[(1+\nu)\underset{\sim}{\sigma}'_{ij}\underset{\sim}{\sigma}'_{ij} + \frac{1}{3}(1-2\nu)(\underset{\sim}{\sigma}_{kk})^2\right] = \frac{1}{2E_0}\left[(1+\nu)\frac{2}{3}\underset{\sim}{\sigma}^{eq\,2} + 3(1-2\nu)(\underset{\sim}{\sigma}_m)^2\right]
\end{aligned}$$

Equation (14.64) is further rewritten as

$$Y = \frac{\underset{\sim}{\sigma}^{eq\,2}}{2E_0}R_\nu \tag{14.67}$$

where R_ν is defined by the following equation and called the *stress triaxiality function*

$$R_\nu \equiv \frac{2}{3}(1+\nu) + 3(1-2\nu)\left(\frac{\underset{\sim}{\sigma}_m}{\underset{\sim}{\sigma}^{eq}}\right)^2 \tag{14.68}$$

$\underset{\sim}{\sigma}_m/\underset{\sim}{\sigma}^{eq}$ is referred to as the *stress triaxiality*.

Analogously to Eqs. (14.55) and (14.56), the relation of the isotropic hardening, the kinematic hardening variables and their rates in the actual damaged and the fictitious undamaged configurations are given by

$$F = (1-D)\underset{\sim}{F}, \quad \underset{\sim}{F} = F/(1-D) \tag{14.69}$$

$$\dot{F} = (1-D)\dot{\underset{\sim}{F}} - \dot{D}\underset{\sim}{F} \tag{14.70}$$

$$\boldsymbol{\alpha} = (1 - D)\underset{\sim}{\boldsymbol{\alpha}}, \quad \underset{\sim}{\boldsymbol{\alpha}} = \boldsymbol{\alpha}/(1 - D) \tag{14.71}$$

$$\dot{\boldsymbol{\alpha}} = (1 - D)\dot{\underset{\sim}{\boldsymbol{\alpha}}} - \dot{D}\underset{\sim}{\boldsymbol{\alpha}} \tag{14.72}$$

Equation (14.50) is widely employed in deformation analyses. However, it would be inapplicable to damage behavior with a strong anisotropy.

14.6.2 On Strain Energy Density Release Rate

Physical interpretation of the strain energy density release rate is given concisely in this section referring to Chaboche (1988).

Consider the elastic deformation with crack extension under the uniaxial loading in Fig. 14.3. It follows in the deformation process from the point a to the point b that

Energy release increment dE_r (from elastic body with crack extension)
= Input energy increment – Strain energy increment
$= \square a'abb' - (\Delta 0bb' - \Delta 0aa')$
$= \square a'abb' - [(\Delta 0aa' + \square a'abb' - \Delta 0ab) - \Delta 0aa']$
$= \Delta 0ab$

Now, suppose the state that the stress increment is small to be negligible, i.e. $d\sigma \cong 0$ and thus one has

$$\begin{aligned} d\psi &= \Delta 0bb' - \Delta 0aa' = (\Delta 0aa' + \square a'abb' - \Delta 0ab) - \Delta 0aa' \\ &= \square a'abb' - \Delta 0ab \cong 2\Delta 0ab - \Delta 0ab = \Delta 0ab \\ &= dE_r \end{aligned} \tag{14.73}$$

Therefore, the half of the input energy increment transforms to the energy release increment dE_r and the other half transfers to the strain energy increment $d\psi$.

Now, one has

$$d\varepsilon^e \cong \frac{\varepsilon^e}{1 - D} dD \tag{14.74}$$

noting

$$d\sigma = d(E\varepsilon^e) = d[(1 - D)E_0\varepsilon^e] = -dDE_0\varepsilon^e + (1 - D)E_0 d\varepsilon^e \cong 0$$

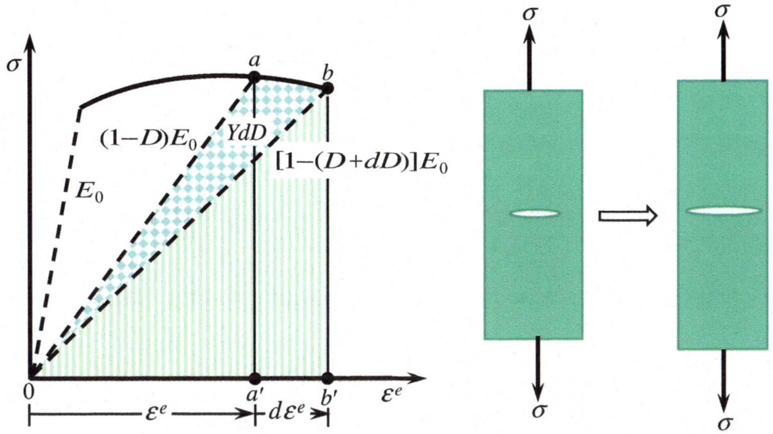

Fig. 14.2 Strain energy release due to crack extension

Then, it follows from Eqs. (14.59) and (14.64) with Eq. (14.74) that

$$d\psi = d\left(\frac{1}{2}\sigma\varepsilon^e\right) \cong \frac{1}{2}\sigma d\varepsilon^e \simeq \frac{1}{2}\frac{\sigma}{1-D}\varepsilon^e dD = \frac{1}{2}\underset{\sim}{\sigma}\varepsilon^e dD = YdD \qquad (14.75)$$

Therefore, YdD is shown by Δ0ab under the constant stress state in Fig. 14.2.

14.6.3 Unilateral Damage: Microcrack Closure Effect

The degrees of damage in directions subjected to the tension and the compression stresses are different in some materials, e.g. cast iron, rocks and concretes. It is called the *unilateral damage*, while the identical damage generation in directions subjected to the tension and the compression stresses described in the last section is called the *bilateral damage*. The unilateral damage is formulated by Ladeveze and Lemaitre (1984) as will be described below.

The tensor is described in the spectral representation in Eq. (1.170), i.e.

$$\mathbf{T} = \sum_{P=1}^{3} T_P \mathbf{e}_P \otimes \mathbf{e}_P$$

where T_P are the principal values and $\mathbf{e}_P$ are the principal vectors. The components are described as follows:

$$T_{ij} = \sum_{P=1}^{3} T_P e_{Pi} e_{Pj}, \; e_{Pi} = e_P \cdot e_i \tag{14.76}$$

noting

$$T_{ij} = \mathbf{e}_i \cdot \mathbf{T}\mathbf{e}_j = \mathbf{e}_i \cdot \sum_{P=1}^{3} T_P \mathbf{e}_P \otimes \mathbf{e}_P \mathbf{e}_j \tag{14.77}$$

The components are decomposed into the positive and the negative parts:

$$T_{ij} = \langle \mathbf{T} \rangle_{ij}^{+} + \langle \mathbf{T} \rangle_{ij}^{-} \tag{14.78}$$

where

$$\left. \begin{aligned} \langle \mathbf{T} \rangle_{ij}^{+} &\equiv \sum_{P=1}^{3} \langle T_P \rangle e_{Pi} e_{Pj} \qquad \text{for } T_P \geq 0 \\ \langle \mathbf{T} \rangle_{ij}^{-} &\equiv \sum_{P=1}^{3} -\langle -T_P \rangle e_{Pi} e_{Pj} \quad \text{for } T_P < 0 \end{aligned} \right\} \tag{14.79}$$

with

$$\langle \mathbf{T} \rangle_{ij}^{+} \langle \mathbf{T} \rangle_{ij}^{-} = 0 \tag{14.80}$$

noting

$$\langle T_P \rangle \langle -T_P \rangle = 0 \tag{14.81}$$

It follows noting Eq. (14.80) from Eq. (14.78) that

$$T_{ij} T_{ij} = \langle \mathbf{T} \rangle_{ij}^{+} \langle \mathbf{T} \rangle_{ij}^{+} + \langle \mathbf{T} \rangle_{ij}^{-} \langle \mathbf{T} \rangle_{ij}^{-} \tag{14.82}$$

$$\langle T_{kk} \rangle^2 = \langle T_{kk} \rangle^2 + \langle -T_{kk} \rangle^2 \tag{14.83}$$

$$\langle T_{kk} \rangle = \langle T_{kk} \rangle - \langle -T_{kk} \rangle \tag{14.84}$$

with

$$\langle \mathbf{T} \rangle_{kl}^{+} \delta_{kl} \neq \langle T_{kk} \rangle, \quad \langle \mathbf{T} \rangle_{kl}^{-} \delta_{kl} \neq -\langle -T_{kk} \rangle \tag{14.85}$$

and the following derivatives hold.

$$\frac{\partial}{\partial T_{ij}}\left(\frac{1}{2}\langle T_{kk}\rangle^2\right) = \langle T_{kk}\rangle\delta_{ij} \tag{14.86}$$

$$\frac{\partial}{\partial T_{ij}}\left(\frac{1}{2}\langle \mathbf{T}\rangle_{rs}^{+}\langle \mathbf{T}\rangle_{rs}^{+}\right) = \langle \mathbf{T}\rangle_{ij}^{+}, \quad \frac{\partial}{\partial T_{ij}}\left(\frac{1}{2}\langle \mathbf{T}\rangle_{rs}^{-}\langle \mathbf{T}\rangle_{rs}^{-}\right) = \langle \mathbf{T}\rangle_{ij}^{-} \tag{14.87}$$

Equations (14.60), (14.62) and (14.65) are described by Eqs. (14.79), (14.82), (14.83) and (14.84) as follows:

$$\phi(\boldsymbol{\sigma}, D) = \frac{1 + v}{2E_0}\frac{\langle\boldsymbol{\sigma}\rangle_{ij}^{+}\langle\boldsymbol{\sigma}\rangle_{ij}^{+} + \langle\boldsymbol{\sigma}\rangle_{ij}^{-}\langle\boldsymbol{\sigma}\rangle_{ij}^{-}}{1 - D} - \frac{v}{2E_0}\frac{\langle\sigma_{kk}\rangle^2 - \langle-\sigma_{kk}\rangle^2}{1 - D} \tag{14.88}$$

$$Y = \frac{1 + v}{2E_0}\frac{\langle\boldsymbol{\sigma}\rangle_{ij}^{+}\langle\boldsymbol{\sigma}\rangle_{ij}^{+} + \langle\boldsymbol{\sigma}\rangle_{ij}^{-}\langle\boldsymbol{\sigma}\rangle_{ij}^{-}}{(1 - D)^2} - \frac{v}{2E_0}\frac{\langle\sigma_{kk}\rangle^2 + \langle-\sigma_{kk}\rangle^2}{(1 - D)^2} \tag{14.89}$$

$$\varepsilon_{ij}^{e} = \frac{\partial\phi(\boldsymbol{\sigma}, D)}{\partial\boldsymbol{\sigma}} = \frac{1 + v}{E_0}\frac{\langle\boldsymbol{\sigma}\rangle_{ij}^{+} + \langle\boldsymbol{\sigma}\rangle_{ij}^{-}}{1 - D} - \frac{v}{E_0}\frac{\langle\sigma_{kk}\rangle - \langle-\sigma_{kk}\rangle}{1 - D}\delta_{ij} \tag{14.90}$$

Here, noting

$$\begin{aligned}
&(1+v)\langle\boldsymbol{\sigma}\rangle_{ij}^{+} - v\langle\sigma_{rr}\rangle\delta_{ij} \\
&\quad = (1+v)\langle\boldsymbol{\sigma}\rangle_{ij}^{+} + \frac{1}{1-2v}[(v+v^2)\langle\boldsymbol{\sigma}\rangle_{kl}^{+}\delta_{kl} - v(1-2v)\langle\boldsymbol{\sigma}\rangle_{kl}^{+}\delta_{kl} \\
&\quad\quad - 3v^2\langle\boldsymbol{\sigma}\rangle_{rs}^{+}\delta_{rs} - (v+v^2)\langle\sigma_{rr}\rangle + 3v^2\langle\sigma_{rr}\rangle]\delta_{ij} \\
&\quad = (1+v)\langle\boldsymbol{\sigma}\rangle_{ij}^{+} + \frac{1}{1-2v}[(v+v^2)\langle\boldsymbol{\sigma}\rangle_{kl}^{+}\delta_{kl} - (v+v^2)\langle\sigma_{rr}\rangle - v(1-2v)\langle\boldsymbol{\sigma}\rangle_{kl}^{+}\delta_{kl} \\
&\quad\quad - 3v^2\langle\boldsymbol{\sigma}\rangle_{rs}^{+}\delta_{rs} + 3v^2\langle\sigma_{rr}\rangle]\delta_{ij} \\
&\quad = (1+v)\left[\langle\boldsymbol{\sigma}\rangle_{ij}^{+} + \frac{v}{1-2v}(\langle\boldsymbol{\sigma}\rangle_{kl}^{+}\delta_{kl} - \langle\sigma_{rr}\rangle)\delta_{ij}\right] \\
&\quad\quad - v\left[\langle\boldsymbol{\sigma}\rangle_{kl}^{+} + \frac{v}{1-2v}(\langle\boldsymbol{\sigma}\rangle_{rs}^{+}\delta_{rs} - \langle\sigma_{rr}\rangle)\delta_{kl}\right]\delta_{kl}\delta_{ij}
\end{aligned}$$

Eq. (14.90) is rewritten as

$$\begin{aligned}
\varepsilon_{ij}^{e} &= \frac{1+v}{E_0}\left[\frac{\langle\boldsymbol{\sigma}\rangle_{ij}^{+} + \frac{v}{1-2v}(\langle\boldsymbol{\sigma}\rangle_{kl}^{+}\delta_{kl} - \langle\sigma_{rr}\rangle)\delta_{ij}}{1-D} - \frac{\langle\boldsymbol{\sigma}\rangle_{ij}^{-} + \frac{v}{1-2v}(\langle\boldsymbol{\sigma}\rangle_{kl}^{-}\delta_{kl} - \langle-\sigma_{rr}\rangle)\delta_{ij}}{1-D}\right] \\
&\quad - \frac{v}{E_0}\left[\frac{\langle\boldsymbol{\sigma}\rangle_{kl}^{+} + \frac{v}{1-2v}(\langle\boldsymbol{\sigma}\rangle_{rs}^{+}\delta_{rs} - \langle\sigma_{rr}\rangle)\delta_{kl}}{1-D} - \frac{\langle\boldsymbol{\sigma}\rangle_{kl}^{-} + \frac{v}{1-2v}(\langle\boldsymbol{\sigma}\rangle_{rs}^{+}\delta_{rs} - \langle-\sigma_{rr}\rangle)\delta_{kl}}{1-D}\right]\delta_{kl}\delta_{ij}
\end{aligned} \tag{14.91}$$

On the other hand, the elastic strain is described in terms of the effective stress by Eq. (5.44) as follows:

$$\varepsilon^e_{ij} = \frac{1+\nu}{E_0}\underset{\sim}{\sigma}_{ij} - \frac{\nu}{E_0}\underset{\sim}{\sigma}_{kk}\delta_{ij} \tag{14.92}$$

The effective stress is described in terms of the current stress by comparing Eqs. (14.91) and (14.92) as follows:

$$\underset{\sim}{\sigma}_{ij} = \frac{\langle\boldsymbol{\sigma}\rangle^+_{ij} + \langle\boldsymbol{\sigma}\rangle^-_{ij}}{1-D} + \frac{\nu}{1-2\nu}\frac{(\langle\boldsymbol{\sigma}\rangle^+_{ij} + \langle\boldsymbol{\sigma}\rangle^-_{ij})\delta_{kl} + \langle\sigma_{rr}\rangle - \langle-\sigma_{rr}\rangle}{1-D}\delta_{ij} \tag{14.93}$$

Equations (14.88), (14.89), (14.90) and (14.93) are extended to the unilateral equations as follows:

$$\phi(\boldsymbol{\sigma}, D) = \frac{1+\nu}{2E_0}\left(\frac{\langle\boldsymbol{\sigma}\rangle^+_{ij}\langle\boldsymbol{\sigma}\rangle^+_{ij}}{1-D} + \frac{\langle\boldsymbol{\sigma}\rangle^-_{ij}\langle\boldsymbol{\sigma}\rangle^-_{ij}}{1-hD}\right) - \frac{\nu}{2E_0}\left(\frac{\langle\sigma_{kk}\rangle^2}{1-D} - \frac{\langle-\sigma_{kk}\rangle^2}{1-hD}\right) \tag{14.94}$$

$$\begin{aligned} Y &= -\frac{\partial\psi(\boldsymbol{\varepsilon}^e, D)}{\partial D} \\ &= \frac{1+\nu}{2E_0}\left(\frac{\langle\boldsymbol{\sigma}\rangle^+_{ij}\langle\boldsymbol{\sigma}\rangle^+_{ij}}{(1-D)^2} + h\frac{\langle\boldsymbol{\sigma}\rangle^-_{ij}\langle\boldsymbol{\sigma}\rangle^-_{ij}}{(1-hD)^2}\right) - \frac{\nu}{2E_0}\left(\frac{\langle\sigma_{kk}\rangle^2}{(1-D)^2} + h\frac{\langle-\sigma_{kk}\rangle^2}{(1-hD)^2}\right) \end{aligned} \tag{14.95}$$

$$\varepsilon^e_{ij} = \frac{\partial\phi(\boldsymbol{\sigma}, D)}{\partial\boldsymbol{\sigma}} = \frac{1+\nu}{E_0}\left(\frac{\langle\boldsymbol{\sigma}\rangle^+_{ij}}{1-D} + \frac{\langle\boldsymbol{\sigma}\rangle^-_{ij}}{1-hD}\right) - \frac{\nu}{E_0}\left(\frac{\langle\sigma_{kk}\rangle}{1-D} - \frac{\langle-\sigma_{kk}\rangle}{1-hD}\right)\delta_{ij} \tag{14.96}$$

$$\underset{\sim}{\sigma}_{ij} = \frac{\langle\boldsymbol{\sigma}\rangle^+_{ij}}{1-D} + \frac{\langle\boldsymbol{\sigma}\rangle^-_{ij}}{1-hD} + \frac{\nu}{1-2\nu}\left(\frac{\langle\boldsymbol{\sigma}\rangle^+_{kl}\delta_{kl} - \langle-\sigma_{rr}\rangle}{1-D} + \frac{\langle\boldsymbol{\sigma}\rangle^-_{kl}\delta_{kl} - \langle-\sigma_{rr}\rangle}{1-hD}\right)\delta_{ij} \tag{14.97}$$

where $h(0 \leq h \leq 1)$ is the material parameter corresponding to the *bilateral* and the *unilateral crack effect* for $h = 1$ and $h = 0$, respectively.

Let the following variables be introduced for more concise description.

$$\sigma_m \equiv \sigma_{kk}/3 \tag{14.98}$$

$$\sigma^+_m \equiv \langle\boldsymbol{\sigma}\rangle^+_{kl}\delta_{kl}/3, \quad \sigma^-_m \equiv \langle\boldsymbol{\sigma}\rangle^-_{ij}\delta_{kl}/3 \tag{14.99}$$

$$\sigma^+_{ij} \equiv \langle\boldsymbol{\sigma}\rangle^+_{ij} + \frac{3\nu}{1-2\nu}(\sigma^+_m - \langle\sigma_m\rangle)\delta_{ij}, \quad \sigma^-_{ij} \equiv \langle\boldsymbol{\sigma}\rangle^-_{ij} + \frac{3\nu}{1-2\nu}(\sigma^-_m - \langle-\sigma_m\rangle)\delta_{ij} \tag{14.100}$$

by which the effective stress in Eq. (14.97) is described as

$$\underset{\sim}{\sigma}_{ij} = \frac{\sigma_{ij}^{+}}{1-D} + \frac{\sigma_{ij}^{-}}{1-hD} \tag{14.101}$$

where the factor $3v/(1-2v)$ is coupling term which accounts for shear effect.

Equations (14.98) to (14.101) are reduced in the one-dimensional state which can be interpreted concisely as follows (Lemaitre 1992):

$$\boldsymbol{\sigma} = \begin{bmatrix} \sigma & 0 & 0 \\ 0 & 0 & 0 \\ 0 & 0 & 0 \end{bmatrix}$$

$$\begin{array}{ll} \text{Tension } \sigma > 0 & \text{Compression } \sigma < 0 \\ \sigma_m = \sigma/3 & \sigma_m = \sigma/3(<0) \\ \left.\begin{array}{l} \sigma_m^{+} = \sigma/3 \\ \sigma_m^{-} = 0 \end{array}\right\} & \left.\begin{array}{l} \sigma_m^{+} = 0 \\ \sigma_m^{-} = -\sigma/3(>0) \end{array}\right\} \\ \left.\begin{array}{l} \sigma_m^{+} - \langle \sigma_m \rangle \\ \sigma_m^{-} - \langle -\sigma_m \rangle \end{array}\right\} = 0 & \left.\begin{array}{l} \sigma_m^{+} - \langle \sigma_m \rangle \\ \sigma_m^{-} - \langle -\sigma_m \rangle \end{array}\right\} = 0 \\ \left.\begin{array}{l} \sigma_{11}^{+} = \sigma \\ \sigma_{11}^{-} = 0 \end{array}\right\} & \left.\begin{array}{l} \sigma_{11}^{+} = 0 \\ \sigma_{11}^{-} = -\sigma \end{array}\right\} \\ \underset{\sim}{\sigma}_{11} = \dfrac{\sigma}{1-D} & \underset{\sim}{\sigma}_{11} = \dfrac{\sigma}{1-hD}(<0) \end{array} \tag{14.102}$$

Therefore, the difference between the actual stress and the effective stress in the compression state is smaller than that in the tension state if $h < 1$.

14.6.4 Anisotropic (Orthotropic) Damage Tensor

The following asymmetric effective stress was proposed by Murakami and Ohno (1981) and Murakami (1988).

$$\underset{\sim}{\boldsymbol{\sigma}} = (\mathbf{I} - \boldsymbol{\mathcal{D}})^{-1}\boldsymbol{\sigma} \tag{14.103}$$

However, an asymmetric stress tensor makes the mathematical formulation and mechanical analysis very complicated. Then, various symmetrized effective stress tensors have been proposed. For instance,

$$\underset{\sim}{\boldsymbol{\sigma}} = [\boldsymbol{\sigma}(\mathbf{I} - \boldsymbol{\mathcal{D}})^{-1} + (\mathbf{I} - \boldsymbol{\mathcal{D}})^{-1}\boldsymbol{\sigma}]/2 \tag{14.104}$$

and

$$\underset{\sim}{\boldsymbol{\sigma}} = (\mathbf{I} - \mathcal{D})^{-1}\boldsymbol{\sigma}(\mathbf{I} - \mathcal{D})^{-1} \tag{14.105}$$

have been proposed by Murakami and Ohno (1981) and Betton (1986), respectively. However, there would not exist potential function leading to them.

Cordebois and Sidoroff (1982a, b) proposed the effective stress tensor

$$\underset{\sim}{\boldsymbol{\sigma}} = \mathbf{H}\boldsymbol{\sigma}\mathbf{H} \tag{14.106}$$

where

$$\mathbf{H} \equiv (\mathbf{I} - \mathcal{D})^{-1/2} (= \mathbf{H}^T) \tag{14.107}$$

Equation (14.106) is derived from the potential function as follows:

$$\psi = C\mathrm{tr}(\mathbf{H}\boldsymbol{\sigma}\mathbf{H}\boldsymbol{\sigma}) \tag{14.108}$$

$$\boldsymbol{\varepsilon}^e = \frac{\partial \psi}{\partial \boldsymbol{\sigma}} = 2C\mathbf{H}\boldsymbol{\sigma}\mathbf{H} = 2C\underset{\sim}{\boldsymbol{\sigma}} \tag{14.109}$$

However, Eq. (14.109) is physically impertinent in the present form.

Extending Eq. (14.59) to the anisotropic damage, Lemaitre et al. (2000) assumed the following Gibbs energy.

$$\psi = \frac{1+v}{2E}\mathrm{tr}(\mathbf{H}\boldsymbol{\sigma}'\mathbf{H}\boldsymbol{\sigma}') + \frac{3(1-2v)}{2E}\frac{\sigma_m^2}{1-\eta\mathcal{D}_m} \tag{14.110}$$

where

$$\mathcal{D}_m \equiv \frac{1}{3}\mathrm{tr}\mathcal{D} \tag{14.111}$$

η is an hydrostatic sensitivity parameter concerning the Poisson's ratio with damage, while $\eta \cong 3$ is used most often. The particular case chosen as $\mathcal{D} = D\mathbf{I}$ and $\eta = 1$ corresponds to the isotropic damage.

The elastic strain is derived from Eq. (14.110) as follows:

$$\boldsymbol{\varepsilon}^e = \frac{\partial \psi}{\partial \boldsymbol{\sigma}} = \frac{1+v}{E}\underset{\sim}{\boldsymbol{\sigma}} - \frac{3v}{E}\underset{\sim}{\sigma}_m\mathbf{I} \tag{14.112}$$

which is of identical form to the Hooke's law but the effective stress is related to the actual stress as follows (Lemaitre et al., 2000):

$$\underset{\sim}{\boldsymbol{\sigma}} \equiv (\mathbf{H}\boldsymbol{\sigma}'\mathbf{H})' + \frac{\sigma_m}{1-\eta\mathcal{D}_m}\mathbf{I} \tag{14.113}$$

noting

$$\begin{aligned}
\frac{\partial\psi}{\partial\sigma_{ij}} &= \frac{\partial}{\partial\sigma_{ij}}\left[\frac{1+v}{2E}H_{pq}\sigma'_{qr}H_{rs}\sigma'_{sp} + \frac{3(1-2v)}{2E}\frac{\sigma_m^2}{1-\eta\mathcal{D}_m}\right] \\
&= \frac{1+v}{2E}\left[H_{pq}\frac{\partial\sigma'_{qr}}{\partial\sigma_{ij}}H_{rs}\sigma'_{sp} + H_{pq}\sigma'_{qr}H_{rs}\frac{\partial\sigma'_{sp}}{\partial\sigma_{ij}}\right] + \frac{3(1-2v)}{2E}\frac{2\sigma_m}{1-\eta\mathcal{D}_m}\frac{1}{3}\delta_{ij} \\
&= \frac{1+v}{2E}\left[H_{pq}(\delta_{iq}\delta_{jr} - \frac{1}{3}\delta_{qr}\delta_{ij})H_{rs}\sigma'_{sp} + H_{pq}\sigma'_{qr}H_{rs}(\delta_{is}\delta_{jp} - \frac{1}{3}\delta_{sp}\delta_{ij})\right] \\
&\quad + \frac{1-2v}{E}\frac{\sigma_m}{1-\eta\mathcal{D}_m}\delta_{ij} \\
&= \frac{1+v}{2E}\left[(H_{pi}\sigma'_{sp}H_{js} - \frac{1}{3}H_{pq}\sigma'_{sp}H_{qs}\delta_{ij}) + (H_{jq}\sigma'_{qr}H_{ri} - \frac{1}{3}H_{pq}\sigma'_{qr}H_{rp}\delta_{ij})\right] \\
&\quad + \frac{1-2v}{E}\frac{\sigma_m}{1-\eta\mathcal{D}_m}\delta_{ij} \\
&= \frac{1+v}{E}(H_{pi}\sigma'_{sp}H_{js})' + \frac{1-2v}{E}\frac{\sigma_m}{1-\eta\mathcal{D}_m}\delta_{ij} \\
&= \frac{1+v}{E}\left[(H_{pi}\sigma'_{sp}H_{js})' + \frac{\sigma_m}{1-\eta\mathcal{D}_m}\delta_{ij}\right] - \frac{v}{E}\frac{\sigma_m}{1-\eta\mathcal{D}_m}3\delta_{ij} \\
&= \frac{1+v}{E}\left[(H_{pi}\sigma'_{sp}H_{js})' + \frac{\sigma_m}{1-\eta\mathcal{D}_m}\delta_{ij}\right] - \frac{v}{E}[\underbrace{(H_{pi}\sigma'_{sp}H_{js})'_{aa}}_{0} + \frac{\sigma_m}{1-\eta\mathcal{D}_m}\delta_{aa}]\delta_{ij}
\end{aligned}$$

Equation (14.113) is regarded to be the modification of Eq. (14.106) proposed by Cordebois and Sidoroff (1982a, b) so as to conform to the Hooke's elastic behavior.

Equation (14.113) is rewritten as follows:

$$\underset{\sim}{\boldsymbol{\sigma}} = \mathbf{M}(\mathcal{D}) : \boldsymbol{\sigma} \tag{14.114}$$

i.e.

$$\underset{\sim}{\boldsymbol{\sigma}} = \mathbf{M}(\mathcal{D}) : \boldsymbol{\sigma}' + \sigma_m\mathbf{M}(\mathcal{D}) : \mathbf{I} \tag{14.115}$$

where

$$\mathbf{M}(\mathcal{D}) = \mathbf{H}\tilde{\otimes}\mathbf{H} - \frac{1}{3}(\mathbf{H}^2 \otimes \mathbf{I} + \mathbf{I} \otimes \mathbf{H}^2) + \frac{1}{9}(\mathrm{tr}\mathbf{H}^2)\mathbf{I} \otimes \mathbf{I} + \frac{\mathbf{I} \otimes \mathbf{I}}{3(1-\eta\mathcal{D}_m)} \tag{14.116}$$

with the notation $(\mathbf{A}\tilde{\otimes}\mathbf{B})_{ij} \equiv T_{ik}T_{lj}$ in Eq. (1.151$_4$), noting

$$\begin{aligned}
(\mathbf{H}\boldsymbol{\sigma}'\mathbf{H})'_{ij} &= (H_{ik}\sigma'_{kl}H_{lj})' = H_{ik}\sigma'_{kl}H_{lj} - H_{rk}\sigma'_{kl}H_{lr}\delta_{ij}/3 \\
&= H_{ik}(\sigma_{kl} - \sigma_m\delta_{kl})H_{lj} - H_{rk}(\sigma_{kl} - \sigma_m\delta_{kl})H_{lr}\delta_{ij}/3 \\
&= H_{ik}H_{lj}\sigma_{kl} - H_{ik}H_{kj}\sigma_m - H_{lr}H_{rk}\sigma_{kl}\delta_{ij}/3 + H_{rk}H_{kr}\delta_{ij}\sigma_m/3 \\
&= H_{ik}H_{lj}\sigma_{kl} - H_{ir}H_{rj}\sigma_m - H_{lr}H_{rk}\sigma_{kl}\delta_{ij}/3 + H_{rk}H_{kr}\delta_{ij}\sigma_m/3 \\
&= [H_{ik}H_{lj} - (H_{ir}H_{rj}\delta_{kl} + \delta_{ij}H_{lr}H_{rk})/3 + H_{rk}H_{kr}\delta_{ij}\delta_{kl}/9]\sigma_{kl} \\
&= \{[\mathbf{H}\tilde{\otimes}\mathbf{H} - (\mathbf{H}^2\otimes\mathbf{I} + \mathbf{I}\otimes\mathbf{H}^2)/3 + (\mathrm{tr}\mathbf{H}^2)\mathbf{I}\otimes\mathbf{I}/9]\boldsymbol{\sigma}\}_{ij}
\end{aligned}$$

$$\mathbf{M}(\mathcal{D}):\mathbf{I} = \left[\mathbf{H}\tilde{\otimes}\mathbf{H} - \frac{1}{3}(\mathbf{H}^2\otimes\mathbf{I} + \mathbf{I}\otimes\mathbf{H}^2) + \frac{1}{9}(\mathrm{tr}\mathbf{H}^2)\mathbf{I}\otimes\mathbf{I} + \frac{\mathbf{I}\otimes\mathbf{I}}{3(1-\eta\mathcal{D}_m)}\right]:\mathbf{I} = \frac{\mathbf{I}}{1-\eta\mathcal{D}_m}$$

$$\mathbf{H}\tilde{\otimes}\mathbf{H}:\mathbf{I} = \mathbf{H}^2,\ \mathbf{H}^2\otimes\mathbf{I}:\mathbf{I} = 3\mathbf{H}^2,\ \mathbf{I}\otimes\mathbf{H}^2:\mathbf{I} = (\mathrm{tr}\mathbf{H}^2)\mathbf{I}$$

Equations (14.114) are inverted as follows (Mengoni and Ponthot 2015):

$$\boldsymbol{\sigma} = \mathbf{H}^{-1}\underset{\sim}{\boldsymbol{\sigma}}'\mathbf{H}^{-1} - \frac{\underset{\sim}{\boldsymbol{\sigma}}':\mathbf{H}^2}{\mathrm{tr}\mathbf{H}^{-2}}\mathbf{H}^{-2} + (1-\eta\mathcal{D}_m)\underset{\sim}{\sigma}_m\mathbf{I} \tag{14.117}$$

$$\boldsymbol{\sigma} = \mathbf{M}^{-1}(\mathcal{D}):\underset{\sim}{\boldsymbol{\sigma}} \tag{14.118}$$

where

$$\mathbf{M}^{-1}(\boldsymbol{D}) = \mathbf{H}^{-1}\underline{\otimes}\mathbf{H}^{-1} - \frac{\mathbf{H}^{-2}\otimes\mathbf{H}^{-2}}{\mathrm{tr}\mathbf{H}^{-2}} + \frac{1}{3}(1-\eta\mathcal{D}_m)\mathbf{I}\otimes\mathbf{I} \tag{14.119}$$

The rate of the damage tensor is given by

$$\dot{\mathcal{D}} = \left(\frac{Y}{S}\right)^m |\mathbf{d}^p| \tag{14.120}$$

where S and m are material parameters and $|\mathbf{d}^p|$ is defined as

$$|\mathbf{d}^P| \equiv \sum_{P=1}^{3} |d_P^p|\mathbf{n}^P\otimes\mathbf{n}^P \tag{14.121}$$

d_P^p and $\mathbf{n}^P$ are the principal values and the normalized principal direction vectors of plastic strain rate. Y is already shown in Eq. (14.67). Then, the principal directions of the damage rate coincide with those of the plastic strain rate.

The plastic strain rate and the fictitious effective stress rate are calculated by the fictitious elastoplastic constitutive relations by inputs of strain rate. Then, the internal variables in the fictitious undamaged configuration and the damage tensor are calculated by the plastic strain rate, and then the current stress is calculated though the damage tensor from the fictitious undamaged stress.

In the uniaxial loading, the damage tensors $\mathcal{D}$ and $\mathbf{H}$ are described in the orthotropic coordinate system the axes of which coincide with the principal directions of the damage tensor as

$$\boldsymbol{\mathcal{D}} = \begin{bmatrix} \mathcal{D}_1 & 0 & 0 \\ 0 & \mathcal{D}_2 & 0 \\ 0 & 0 & \mathcal{D}_2 \end{bmatrix} \tag{14.122}$$

$$\mathbf{H} = \begin{bmatrix} \dfrac{1}{\sqrt{1-\mathcal{D}_1}} & 0 & 0 \\ 0 & \dfrac{1}{\sqrt{1-\mathcal{D}_2}} & 0 \\ 0 & 0 & \dfrac{1}{\sqrt{1-\mathcal{D}_3}} \end{bmatrix} \tag{14.123}$$

The effective equivalent stress $\underset{\sim}{\sigma}^{eq}$ is different from the effective stress $\underset{\sim}{\sigma}_1$ as follows:

$$\left.\begin{aligned} \underset{\sim}{\boldsymbol{\sigma}}^{eq} &= \frac{2}{3}\frac{\sigma_1}{1-\mathcal{D}_1} + \frac{1}{3}\frac{\sigma_1}{1-\mathcal{D}_2} \\ \underset{\sim}{\boldsymbol{\sigma}}_1 &= \frac{4}{9}\frac{\sigma_1}{1-\mathcal{D}_1} + \frac{2}{9}\frac{\sigma_1}{1-\mathcal{D}_2} + \frac{1}{3}\frac{\sigma_1}{1-\eta\mathcal{D}_m} \end{aligned}\right\} \tag{14.124}$$

The elastic strain is described by Eq. (14.112) with Eq. (14.113), noting Eqs. (14.122) and (14.123) as follows (Lemaitre et al. 2000):

$$\begin{aligned}
&\begin{bmatrix} \varepsilon_1^e & 0 & 0 \\ 0 & \varepsilon_2^e & 0 \\ 0 & 0 & \varepsilon_3^e \end{bmatrix} \\
&= \frac{1+v}{E}\left(\begin{bmatrix} \dfrac{1}{\sqrt{1-\mathcal{D}_1}} & 0 & 0 \\ 0 & \dfrac{1}{\sqrt{1-\mathcal{D}_2}} & 0 \\ 0 & 0 & \dfrac{1}{\sqrt{1-\mathcal{D}_3}} \end{bmatrix} \begin{bmatrix} \frac{2}{3}\sigma_1 & 0 & 0 \\ 0 & -\frac{1}{3}\sigma_1 & 0 \\ 0 & 0 & -\frac{1}{3}\sigma_1 \end{bmatrix} \begin{bmatrix} \dfrac{1}{\sqrt{1-\mathcal{D}_1}} & 0 & 0 \\ 0 & \dfrac{1}{\sqrt{1-\mathcal{D}_2}} & 0 \\ 0 & 0 & \dfrac{1}{\sqrt{1-\mathcal{D}_3}} \end{bmatrix} \right)' \\
&\quad + \frac{1+v}{E}\frac{\sigma_m}{1-\eta\mathcal{D}_m}\begin{bmatrix} 1 & 0 & 0 \\ 0 & 1 & 0 \\ 0 & 0 & 1 \end{bmatrix} - \frac{v}{E}\frac{3\sigma_m}{1-\eta\mathcal{D}_m}\begin{bmatrix} 1 & 0 & 0 \\ 0 & 1 & 0 \\ 0 & 0 & 1 \end{bmatrix} \\
&= \frac{1+v}{E}\begin{bmatrix} \dfrac{2}{3}\dfrac{1}{1-\mathcal{D}_1} & 0 & 0 \\ 0 & -\dfrac{1}{3}\dfrac{1}{1-\mathcal{D}_2} & 0 \\ 0 & 0 & -\dfrac{1}{3}\dfrac{1}{1-\mathcal{D}_3} \end{bmatrix}' \sigma_1 + \frac{1+v}{E}\frac{\sigma_1/3}{1-\eta\mathcal{D}_m}\begin{bmatrix} 1 & 0 & 0 \\ 0 & 1 & 0 \\ 0 & 0 & 1 \end{bmatrix} - \frac{v}{E}\frac{3\sigma_1/3}{1-\eta\mathcal{D}_m}\begin{bmatrix} 1 & 0 & 0 \\ 0 & 1 & 0 \\ 0 & 0 & 1 \end{bmatrix} \\
&\qquad \left(\mathcal{D}_m = \frac{1}{3}\left(\frac{2}{3}\frac{1}{1-\mathcal{D}_1} - \frac{1}{3}\frac{1}{1-\mathcal{D}_2} - \frac{1}{3}\frac{1}{1-\mathcal{D}_3} \right) = \frac{1}{9}\left(\frac{2}{1-\mathcal{D}_1} - \frac{1}{1-\mathcal{D}_2} - \frac{1}{1-\mathcal{D}_3} \right) \right) \\
&= \frac{1+v}{9E}\begin{bmatrix} \dfrac{4}{1-\mathcal{D}_1} + \dfrac{1}{1-\mathcal{D}_2} + \dfrac{1}{1-\mathcal{D}_3} & 0 & 0 \\ 0 & -\dfrac{2}{1-\mathcal{D}_1} - \dfrac{2}{1-\mathcal{D}_2} + \dfrac{1}{1-\mathcal{D}_3} & 0 \\ 0 & 0 & -\dfrac{2}{1-\mathcal{D}_1} + \dfrac{1}{1-\mathcal{D}_2} - \dfrac{2}{1-\mathcal{D}_3} \end{bmatrix}\sigma_1 \\
&\quad + \frac{1-2v}{E}\frac{\sigma_1}{3(1-\eta\mathcal{D}_m)}\begin{bmatrix} 1 & 0 & 0 \\ 0 & 1 & 0 \\ 0 & 0 & 1 \end{bmatrix}
\end{aligned} \tag{14.125}$$

Then, setting

$$\underset{\sim}{E}_1 \equiv \frac{\sigma_1}{\varepsilon_1^e}, \quad \underset{\sim}{\nu}_{12} \equiv -\frac{\varepsilon_2^e}{\varepsilon_1^e}, \quad \underset{\sim}{\nu}_{13} \equiv -\frac{\varepsilon_3^e}{\varepsilon_1^e} \tag{14.126}$$

one has

$$\left.\begin{aligned}
\frac{E}{\underset{\sim}{E}_1} &= \frac{1+\nu}{9}\left(\frac{4}{1-\mathcal{D}_1}+\frac{1}{1-\mathcal{D}_2}+\frac{1}{1-\mathcal{D}_3}\right)+\frac{1-2\nu}{3(1-\eta\mathcal{D}_m)} \\
\underset{\sim}{\nu}_{12}\frac{E}{\underset{\sim}{E}_1} &= \frac{1+\nu}{9}\left(\frac{2}{1-\mathcal{D}_1}+\frac{2}{1-\mathcal{D}_2}-\frac{1}{1-\mathcal{D}_3}\right)-\frac{1-2\nu}{3(1-\eta\mathcal{D}_m)} \\
\underset{\sim}{\nu}_{13}\frac{E}{\underset{\sim}{E}_1} &= \frac{1+\nu}{9}\left(\frac{2}{1-\mathcal{D}_1}-\frac{1}{1-\mathcal{D}_2}+\frac{2}{1-\mathcal{D}_3}\right)-\frac{1-2\nu}{3(1-\eta\mathcal{D}_m)}
\end{aligned}\right\} \tag{14.127}$$

The unilateral formulation for the anisotropic damage can be referred to Ladeveze and Lemaitre (1984) Lemaitre and Desmora (2005).

14.7 Subloading-Overstress Damage Model

The subloading-damage model formulated in the preceding sections will be extended to be taken account of the rate-dependent plastic deformation by incorporating the subloading-overstress model (Hashiguchi 2013a) in the following.

Equations (14.29) and (14.31) in the subloading-damage model is extended by incorporating the subloading-overstress model in Eq. (13.29) as follows:

$$\dot{\boldsymbol{\varepsilon}} = \dot{\boldsymbol{\varepsilon}}^e + \dot{\boldsymbol{\varepsilon}}^{vp} = \mathbf{E}_0^{-1} : \overset{\circ}{\underset{\sim}{\boldsymbol{\sigma}}} + \frac{1}{\bar{\mu}}\frac{\langle R - R_s\rangle^n}{R_m - R}\underset{\sim}{\hat{\mathbf{n}}} \tag{14.128}$$

$$\overset{\circ}{\underset{\sim}{\boldsymbol{\sigma}}} = \mathbf{E}_0 : \dot{\boldsymbol{\varepsilon}} - \frac{1}{\bar{\mu}}\frac{\langle R - R_s\rangle^n}{R_m - R}\mathbf{E}_0 : \underset{\sim}{\hat{\mathbf{n}}} \tag{14.129}$$

where $\dot{\boldsymbol{\varepsilon}}^{vp}$ is the viscoplastic strain rate for which the loading criterion is imposed by incorporating the Macaulay's bracket. $\bar{\mu}$ and n are the material parameters, while $\bar{\mu}$ stands for the viscoplastic coefficient. The surface which passes through the current stress and is similar to the normal-yield surface is called the *dynamic-loading surface* and the ratio of the size of the dynamic-loading surface to the normal-yield surface is described by $R(=f(\underset{\sim}{\hat{\boldsymbol{\sigma}}})/\tilde{F})$ which can be larger than unity and is called the *dynamic-loading ratio*. $R_m(\gg 1)$ is the material constant designating the maximum value of the dynamic-loading ratio, called the *limit dynamic-loading ratio*. The rates of the internal state variables $\underset{\sim}{H}, \underset{\sim}{\boldsymbol{\alpha}}, D, \mathcal{D}$ are given by replacing the plastic strain rate $\dot{\boldsymbol{\varepsilon}}^p$ to the viscoplastic strain rate $\dot{\boldsymbol{\varepsilon}}^{vp}$ in Eqs. (14.14), (14.42), (14.63) and (14.120) as follows:

$$\dot{\underset{\sim}{H}} = f_{\underset{\sim}{H\varepsilon}}(\underset{\sim}{\boldsymbol{\sigma}}, \underset{\sim}{H}; \dot{\boldsymbol{\varepsilon}}^{vp}) \tag{14.130}$$

$$\dot{\underset{\sim}{\boldsymbol{\alpha}}} = \mathbf{f}_{\underset{\sim}{k\varepsilon}}(\underset{\sim}{\boldsymbol{\sigma}}, \underset{\sim}{\boldsymbol{\alpha}}, \underset{\sim}{F}; \dot{\boldsymbol{\varepsilon}}^{vp\prime}) = c_k\left(\dot{\boldsymbol{\varepsilon}}^{vp\prime} - \frac{1}{b_k(\underset{\sim}{\boldsymbol{\sigma}}, \underset{\sim}{F})} \underset{\sim}{\boldsymbol{\alpha}} \left\| \dot{\boldsymbol{\varepsilon}}^{vp\prime} \right\| \right) \tag{14.131}$$

$$\dot{D} = \left(\frac{Y}{\zeta} \right)^a \frac{\hat{H}[\varepsilon^{eqvp} - \varepsilon_d^{eqvp}]}{1 - D} \dot{\varepsilon}^{eqvp} \tag{14.132}$$

$$\dot{\boldsymbol{D}} = \left(\frac{\overline{Y}}{S} \right)^m |\mathbf{d}^{vp}| = \left(\frac{\overline{Y}}{S} \right)^m \sum_{P=1}^{3} |d_P^{vp}| \mathbf{n}^P \otimes \mathbf{n}^P \tag{14.133}$$

$$\varepsilon^{eqvp} \equiv \sqrt{2/3} \int \left\| \dot{\boldsymbol{\varepsilon}}^{vp} \right\| dt \tag{14.134}$$

The rate of the subloading ratio $R_s (0 \le R_s \le 1)$ is given by replacing the normal-yield ratio R to R_s and the plastic strain increment $\dot{\boldsymbol{\varepsilon}}^p$ to the viscoplastic strain rate $\dot{\boldsymbol{\varepsilon}}^{vp}$ in Eq. (14.17) for the plastic sliding process and the subloading ratio R_s is identical to the normal sliding-yield ratio R for the elastic sliding process as follows:

$$\dot{R}_s = U(R_s) \left\| \dot{\boldsymbol{\varepsilon}}^{vp} \right\| \quad \text{for } \dot{\boldsymbol{\varepsilon}}^{vp} \ne \mathbf{O} \tag{14.135}$$

$$R_s = R \quad \text{for } \dot{\boldsymbol{\varepsilon}}^{vp} = \mathbf{O} \tag{14.136}$$

The smooth transition from the elastic to the viscoplastic state is described by incorporating R_s instead of unity. The response of the subloading-overstress damage model is shown in Fig. 14.3.

14.8 Subloading-Gruson Model

Plastic deformation is induced under hydrostatic stress in porous media even if the base material is the Mises material the plastic deformation behavior of which is independent of hydrostatic stress. The elastoplastic constitutive model taken account of the nucleation and the growth of round voids was proposed first by Gurson (1977) and further studied by Needleman and Rice (1978), Tvergaard and Needleman (1984), Needleman and Tvergaard (1985), etc. The yield surface is

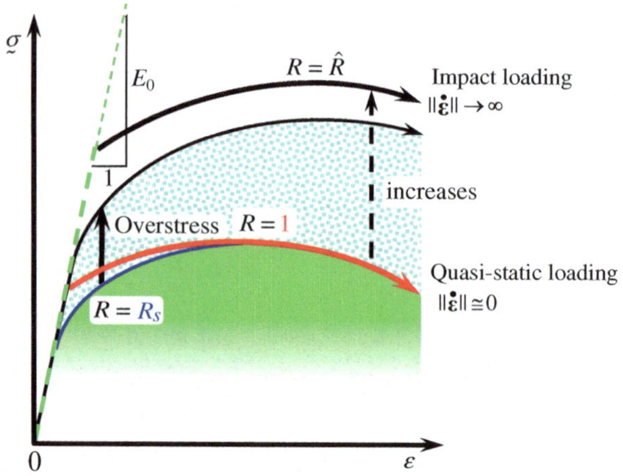

Fig. 14.3 Stress-strain curve predicted by the subloading-overstress damage model

introduced, which is taken account of the void volume fraction and the mean stress with the evolution rule of the void volume fraction. It is often called the *Gurson model* and its elaboration taken account of the void coalescence is called the *GTN (Gurson-Tvergaard-Needleman) model*. The subloading-void(Gurson) model will be described in this section.

The following yield condition is derived by Gurson (1977) by the symmetric deformation analysis of the rigid-plastic Mises material containing a spherical cavity.

$$\psi(\boldsymbol{\sigma}, F, \xi) = \left(\frac{\sigma^{eq}}{F}\right)^2 + 2\xi \cosh\left(\frac{3}{2}\frac{\sigma_m}{F}\right) - \xi^2 - 1 = 0 \tag{14.137}$$

where ξ is the void volume fraction. Equation (14.137) is reduced to the von Mises yield condition, i.e. $\sigma^{eq} = F$ for $\xi = 0$. The dependence of the yield function in Eq. (14.137) is shown in Fig. 14.4.

The rate of the void volume fraction $\dot{\xi}$ is given by sum of the growth rate $\dot{\xi}_{grow}$ and the nucleation rate of new void $\dot{\xi}_{nucl}$ as follows (Needleman and Rice 1978):

$$\dot{\xi} = \dot{\xi}_{grow} + \dot{\xi}_{nucl} \tag{14.138}$$

where

$$\left.\begin{aligned} \dot{\xi}_{grow} &= (1-\xi)\operatorname{tr}\dot{\boldsymbol{\varepsilon}}_v^p \\ \dot{\xi}_{nucl} &= a_1(\dot{F} + \dot{\sigma}_m) + a_2\,\dot{\varepsilon}^{eqp} \end{aligned}\right\} \tag{14.139}$$

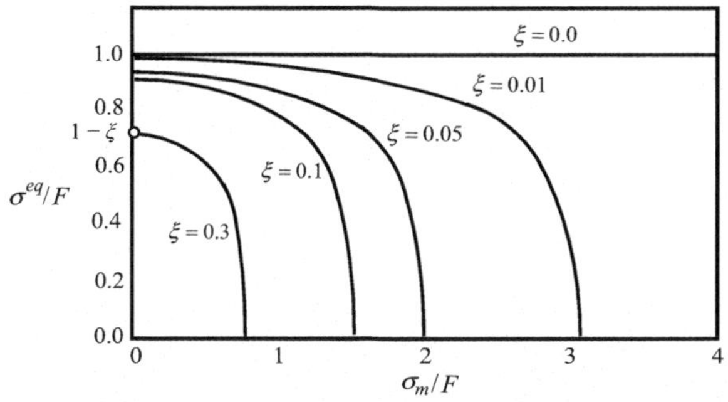

Fig. 14.4 Effect of void volume fraction in Gurson yield surface

The coefficient a_1 and a_2 are given by Chu and Needleman (1980) as follows:

$$\left.\begin{aligned} a_1 &= \frac{f_n}{\sqrt{2\pi} s_n}\exp\left[-\frac{1}{2}\left(\frac{F+\sigma_m-\sigma_n}{s_n}\right)^2\right] \\ a_2 &= \frac{f_n}{\sqrt{2\pi} s_n}\exp\left[-\frac{1}{2}\left(\frac{\varepsilon^{eqp}-\varepsilon_n}{s_n}\right)^2\right] \end{aligned}\right\} \tag{14.140}$$

which is derived postulating that the voids nucleates according to the probability distribution with the stress σ_n and the strain ε_n as their mean values together with s_n as their standard deviation, and f_n is the volume fraction of void nucleating particles.

The subloading surface for the normal-yield surface in Eq. (14.137) is given by replacing F to RF in Eq. (14.137) as follows:

$$\psi(\boldsymbol{\sigma}, F, \xi) = \left(\frac{\sigma^{eq}}{RF}\right)^2 + 2\xi\cosh\left(\frac{3}{2}\frac{\sigma_m}{RF}\right) - \xi^2 - 1 = 0 \tag{14.141}$$

The time-differentiation of Eq. (14.141) is given by

$$\begin{aligned} \dot{\psi}(\boldsymbol{\sigma}, F, \xi) = {} & 2\left(\frac{\sigma^{eq}}{RF}\right)\frac{\dot{\sigma}^{eq}RF - \sigma^{eq}(R\dot{F} + \dot{R}F)}{R^2F^2} \\ & + 2\xi\sinh\left(\frac{3}{2}\frac{\sigma_m}{RF}\right)\frac{3}{2}\frac{\dot{\sigma}_m RF - \sigma_m(R\dot{F} + \dot{R}F)}{R^2F^2} - 2\dot{\xi} = 0 \end{aligned} \tag{14.142}$$

from which one has

$$2\left(\frac{\sigma^{eq}}{RF}\right)\left[\dot{\sigma}^{eq} - \sigma^{eq}\left(\frac{\dot{F}}{F} + \frac{\dot{R}}{R}\right)\right] + 3\xi\sinh\left(\frac{3}{2}\frac{\sigma_m}{RF}\right)\left[\dot{\sigma}_m - \sigma_m\left(\frac{\dot{F}}{F} + \frac{\dot{R}}{R}\right)\right] - 2RF\dot{\xi} = 0 \tag{14.143}$$

Assume the associated flow rule

$$\overset{\bullet}{\boldsymbol{\varepsilon}}^p = \overset{\bullet}{\bar{\lambda}}\,\mathbf{n}^\psi \quad (\overset{\bullet}{\bar{\lambda}} > 0) \tag{14.144}$$

where

$$\mathbf{n}^\psi \equiv \frac{\partial \psi}{\partial \boldsymbol{\sigma}} / \left\| \frac{\partial \psi}{\partial \boldsymbol{\sigma}} \right\| \quad (\|\mathbf{n}^\psi\| = 1) \tag{14.145}$$

It follows adopting the associated flow rule in Eq. (14.144) with Eqs. (6.42) and (7.9) that

$$2\left(\frac{\sigma^{eq}}{RF}\right)\left[\overset{\bullet}{\sigma}^{eq} - \sigma^{eq}\left(\frac{F'\overset{\bullet}{\bar{\lambda}}h^\psi}{F} + \frac{U\overset{\bullet}{\bar{\lambda}}}{R}\right)\right] + 3\xi \sinh\left(\frac{3}{2}\frac{\sigma_m}{RF}\right)\left[\overset{\bullet}{\sigma}_m - \sigma_m\left(\frac{F'\overset{\bullet}{\bar{\lambda}}h^\psi}{F} + \frac{U\overset{\bullet}{\bar{\lambda}}}{R}\right)\right]$$
$$- 2RF\left[(1-\xi)\overset{\bullet}{\bar{\lambda}}\,\mathrm{tr}\mathbf{n}^\psi + a_1(F'\overset{\bullet}{\bar{\lambda}}h^\psi + \overset{\bullet}{\sigma}_m) + a_2\sqrt{\frac{2}{3}}\overset{\bullet}{\bar{\lambda}}\right] = 0$$

resulting in

$$\begin{aligned} &2\left(\frac{\sigma^{eq}}{RF}\right)\overset{\bullet}{\sigma}^{eq} + \left[3\xi \sinh\left(\frac{3}{2}\frac{\sigma_m}{RF}\right) - 2a_1RF\right]\overset{\bullet}{\sigma}_m \\ &- \left\{\left[2\sigma^{eq}\left(\frac{\sigma^{eq}}{RF}\right) + 3\xi\sigma_m \sinh\left(\frac{3}{2}\frac{\sigma_m}{RF}\right)\right]\left(\frac{F'}{F}h^\psi + \frac{U}{R}\right)\right] \\ &+ 2RF\left[(1-\xi)\mathrm{tr}\mathbf{n}^\psi + a_1F'h^\psi + \sqrt{\frac{2}{3}}a_2\right]\right\}\overset{\bullet}{\bar{\lambda}} = 0 \end{aligned} \tag{14.146}$$

where

$$h^\psi \equiv \dot{H} \,/\, \overset{\bullet}{\bar{\lambda}}\,(=\sqrt{2/3}) \tag{14.147}$$

Noting

$$\overset{\bullet}{\sigma}_m = \frac{1}{3}\mathbf{I} : \overset{\bullet}{\boldsymbol{\sigma}},\ \overset{\bullet}{\sigma}^{eq} = \left(\sqrt{\frac{3}{2}}\|\boldsymbol{\sigma}'\|\right)^{\bullet} = \sqrt{\frac{3}{2}}\frac{\boldsymbol{\sigma}' : \overset{\bullet}{\boldsymbol{\sigma}}}{\|\boldsymbol{\sigma}'\|} = \frac{3}{2}\frac{\boldsymbol{\sigma}' : \overset{\bullet}{\boldsymbol{\sigma}}}{\sigma^{eq}} \tag{14.148}$$

one has

$$
\begin{aligned}
&2\left(\frac{\sigma^{eq}}{RF}\right)\dot{\sigma}^{eq}+\left[3\xi\sinh\left(\frac{3}{2}\frac{\sigma_m}{RF}\right)-2a_1RF\right]\dot{\sigma}_m\\
&=\left\{3\left(\frac{\boldsymbol{\sigma}'}{RF}\right)+\left[\xi\sinh\left(\frac{3}{2}\frac{\sigma_m}{RF}\right)-\frac{2}{3}a_1RF\right]\mathbf{I}\right\}:\dot{\boldsymbol{\sigma}}
\end{aligned}
\tag{14.149}
$$

Substituting Eq. (14.149) into Eq. (14.146), the plastic multiplier is derived as follows:

$$
\dot{\bar{\lambda}}=\frac{\mathbf{t}^{SG}:\dot{\boldsymbol{\sigma}}}{\bar{M}^{SG}} \tag{14.150}
$$

where

$$
\mathbf{t}^{SG}\equiv 3\frac{\boldsymbol{\sigma}'}{RF}+\left[\xi\sinh\left(\frac{3}{2}\frac{\sigma_m}{RF}\right)-\frac{2}{3}a_1RF\right]\mathbf{I} \tag{14.151}
$$

$$
\bar{M}^{SG}\equiv\left[2\sigma^{eq}\left(\frac{\sigma^{eq}}{RF}\right)+3\xi\sigma_m\sinh\left(\frac{3}{2}\frac{\sigma_m}{RF}\right)\right]\left(\frac{F'}{F}h+\frac{U}{R}\right)+2RF\left[(1-\xi)\mathrm{n}^{\psi}+a_1F'h^{\psi}+\sqrt{\frac{2}{3}}a_2\right] \tag{14.152}
$$

The strain rate is given from Eqs. (14.5), (14.144) and (14.150) as

$$
\dot{\boldsymbol{\varepsilon}}=\mathbf{E}^{-1}:\dot{\boldsymbol{\sigma}}+\frac{\mathbf{t}^{SG}:\dot{\boldsymbol{\sigma}}}{\bar{M}^{SG}}\mathbf{n}^{\psi}=\left(\mathbf{E}^{-1}+\frac{\mathbf{n}^{\psi}\otimes\mathbf{t}^{SG}}{\bar{M}^{SG}}\right):\dot{\boldsymbol{\sigma}} \tag{14.153}
$$

from which the plastic multiplier in terms of strain rate is derived as follows:

$$
\dot{\bar{\Lambda}}=\frac{\mathbf{t}^{SG}:\mathbf{E}:\dot{\boldsymbol{\varepsilon}}}{\bar{M}^{SG}+\mathbf{t}^{SG}:\mathbf{E}:\mathbf{n}^{\psi}} \tag{14.154}
$$

The stress rate is described from Eqs. (14.5), (14.144) and (14.154) as

$$
\dot{\boldsymbol{\sigma}}=\mathbf{E}:\dot{\boldsymbol{\varepsilon}}-\frac{\mathbf{t}^{SG}:\mathbf{E}:\dot{\boldsymbol{\varepsilon}}}{\bar{M}^{SG}+\mathbf{t}^{SG}:\mathbf{E}:\mathbf{n}^{\psi}}\mathbf{E}:\mathbf{n}^{\psi}=\left[\mathbf{E}-\frac{(\mathbf{E}:\mathbf{n}^{\psi})\otimes(\mathbf{t}^{SG}:\mathbf{E})}{\bar{M}^{SG}+\mathbf{t}^{SG}:\mathbf{E}:\mathbf{n}^{\psi}}\right]:\dot{\boldsymbol{\varepsilon}} \tag{14.155}
$$

The loading criterion is given by the equation same as Eq. (14.32).

The elaboration of the Gurson model was proposed by Tvergaard (1982) (see also Tvergaard and Needleman 1984) by introducing the void coalescence into the yield condition in Eq. (14.137). It is further extended by the concept of the subloading surface as follows:

$$\psi(\boldsymbol{\sigma}, F, \xi) = \left(\frac{\sigma^{eq}}{RF}\right) + 2\xi^{*} q_1 \cosh\left(\frac{3}{2}\frac{q_2 \sigma_m}{RF}\right) - q_3 \xi^{*2} - 1 = 0 \tag{14.156}$$

where $\xi^{*}(\xi)$ is the extension of the void volume fraction ξ introduced so as to represent the loss of the load-carrying capacity due to the void coalescence, i.e.

$$\xi^{*} = \begin{cases} \xi & \text{for } \xi \leq \xi_c \\ \xi_c + \left(\dfrac{1}{q_1} - \xi_c\right) \dfrac{\xi - \xi_c}{\xi_f - \xi_c} & \text{for } \xi > \xi_c \end{cases} \tag{14.157}$$

ξ_c and ξ_f are the critical void volume fraction at the initiation of void coalescence and the void volume fraction at failure (complete loss of load-carrying capacity), respectively. q_1, q_2 and q_3 are the material parameters for the enforcement of the accuracy which are usually chosen as $q_1 = 1.5$, $q_2 = 1.0$ and $q_3 = q_1^2$. The constitutive model with the yield condition in Eq. (14.153) is called the *GTN (Gurson-Tvergaard-Needleman) model.*

The plastic volumetric strain rate is considered in the Gurson model, while it is not considered in the damage model explained in the preceding sections. On the other hand, the decrease of the elastic modulus is not considered in the Gurson model, while it is considered in the damage model. The extended model taken account of both of them would have to be formulated in feature.

References

Armstrong PJ, Frederick CO (1966) A mathematical representation of the multiaxial Bauschinger effect. CEGB report RD/B/N 731 [or in Materials at High Temperature 24:1–26 (2007)]

Betton J (1986) Application of tensor functions to the formulation of constitutive equations involving damage and initial anisotropy. Eng Fract Mech 25:573–584

Chaboche JL (1982) The concept of effective stress applied to elasticity and viscoplasticity in the presence of anisotropic damage. In: Boehler JP (ed) Mechanical behavior of anisotropic solids. Matrinus Nijhoff Publ., Hague, Netherland

Chaboche JL (1988) Continuum damage mechanics, Pert I General concept; Part II Damage growth, crack initiation, and crack growth. J Appl Mech (ASME) 55:59–72

Chu CC, Needleman A (1980) Void nucleation effects in biaxially stretched sheets. J Eng Technol ASME 102:249–256

Cordebois JP, Sidoroff F (1982a) Damage induced elastic anisotropy. In: Boehler JP (ed) Mechanical behavior of anisotropic solids. Martinuus Nijhoff Publ., pp 761–774

Cordebois JP, Sidoroff F (1982b) Endommagement anisotrope en elasticite et plasticite. J de Mech Theor et Appl Numero Spec 45–60

de Souza Neto EA, Perić D, Owen DJR (2008) Computational methods for plasticity. Wiley, Chichester, UK

Gurson AL (1977) Continuum theory of ductile rupture by void nucleation and growth: Part I—Yield criteria and flow rules for porous media. J Eng Mater Technol (ASME) 99:2–15

Hashiguchi K (2015a) Subloading-damage constitutive equation. Proc Compt Eng Conf Japan 20: D-2-4

Kachanov LM (1958) On rupture time under condition of creep. Izvestia Akademi Nauk SSSR, Otd Tekhn Nauk 8:26–31 (in Russian)

Ladevéze P, Lemaitre JA (1984) Damage effective stress in quasi unilateral conditions. In: 16th international congress of theoretical and applied mechanics, Lyngby, Denmark

Lemaitre JA (1971) Evaluation of dissipation and damage in metals subjected to dynamic loading. In: Proceedings of the International Congress on Mechanical Behavior of Materials 1 (ICM 1), Kyoto

Lemaitre JA (1992) A course on damage mechanics. Springer, Heidelberg

Lemaitre JA, Chaboche J-L (1990) Mechanics of solid materials. Cambridge University Press, Cambridge

Lemaitre JA, Desmoral R (2005) Engineering damage mechanics. Springer, Heidelberg

Lemaitre JA, Dosmorat R, Sauzay M (2000) Anisotropic damage law of evolution. Eur J Mech A/Solids 19:182–208

Mengoni M, Ponthot JP (2015) A generic anisotropic continuum damage model integration scheme adaptable to both ductile damage and biological damage-like situations. Int J Plast 66:46–70

Murakami S (1988) Mechanical modelling of material damage. J Appl Mech (ASME) 55:280–286

Murakami S (2012) Continuum damage mechanics: a continuum mechanics approach to the analysis of damage and fracture. Springer, Dordrecht, Netherland

Murakami S, Ohno N (1981) A continuum theory of creep and creep damage. In: Proceedings of the 3rd IUTAM symposium on creep in structures, pp 422–444

Needleman A, Rice JR (1978) Limits to ductility set by plastic flow localization. In: Koistinen DP, Wang N-M (eds) Mechanics of sheet metal forming. Plenum press, New York, pp 237–265

Needleman A, Tvergaard V (1985) Material strain-rate sensitivity in round tensile bar. In: Proceedings of the international symposium on plastic instability. Pressure de l'cole nationale des Ponts et Shausseses, Paris, pp 251–262

Rabotnov YN (1969) Creep problems in structural members. North-Holland, Amsterdam

Tvergaard V (1982) On localization in ductile materials containing spherical voids. Int J Fract 18:237–252

Tvergaard V, Needleman A (1984) Analysis of the cup-cone fracture in a round tensile bar. Acta Metall 32:157–169

Chapter 15
Plasticity for Phase Transformation

The phase transformation analysis is of importance for the thermo-mechanical treatment of metals. The elastoplastic constitutive equation with the phase transformation has been developed by Inoue and Raniecki (1978), Inoue et al. (2007), Rohde and Jeppsson (2000), Okamura and Kawashima (1988a, b), Okamura et al. (2005), Okamura (2006a, b), etc. However, the existing formulation falls within the framework of the conventional plasticity assuming the yield surface enclosing a purely-elastic domain. Therefore, it requires the yield judgment whether or not the stress reaches the yield surface and the operation to pull-back the stress to the yield surface in the plastic loading process. The modified constitutive equation of phase transformation will be formulated by incorporating the subloading surface model into the constitutive equations of phase transformation formulated by Okamura and his colleagues within the framework of the conventional elasoplasticity (Okamura and Kawashima 1988a, b; Okamura et al. 2005; Okamura 2006a, b). The subloading phase-transformation model described in this chapter is based on the article by Hashiguchi and Okamura (2014), and the description in this chapter depends highly on the valuable advices by Dr. Kazuo Okamura, Nippon Steel & Sumitomo Metal Corporation, and his articles on this subject.

15.1 Constitutive Equation

The strain increment $d\boldsymbol{\varepsilon}$ splits additively into the elastic strain increment $d\boldsymbol{\varepsilon}^e$, the plastic strain increment $d\boldsymbol{\varepsilon}^p$, the thermal and transformation strain increment $d\boldsymbol{\varepsilon}^\theta$ and the transformation-plastic strain increment $d\boldsymbol{\varepsilon}^{Tp}$ as follows (Okamura and Kawashima 1988a, b; Okamura et al. 2005; Okamura 2006a, b):

$$d\boldsymbol{\varepsilon} = d\boldsymbol{\varepsilon}^e + d\boldsymbol{\varepsilon}^p + d\boldsymbol{\varepsilon}^\theta + d\boldsymbol{\varepsilon}^{Tp} \tag{15.1}$$

These strain increments are formulated in the following.

K. Hashiguchi, *Foundations of Elastoplasticity: Subloading Surface Model*,
DOI 10.1007/978-3-319-48821-9_15

15.1.1 Elastic Strain Increment

Taking account of the dependence of the elastic modulus $\mathbf{E}$ on the temperature, the elastic strain increment $d\boldsymbol{\varepsilon}^e$ is given by

$$d\boldsymbol{\varepsilon}^e = \mathbf{E}^{-1} : d\boldsymbol{\sigma} + \frac{\partial \mathbf{E}^{-1}}{\partial \theta} d\theta : \boldsymbol{\sigma} \tag{15.2}$$

where θ is the temperature. Here, it is assumed that the elastic modulus tensor is independent of stress.

15.1.2 Plastic Strain Increment Based on Subloading Surface Model

Taking account of the fact that the hardening is given by the assemble of the hardenings of each phase so that the yield surface with the isotropic hardening is described by

$$f(\boldsymbol{\sigma}) = F \tag{15.3}$$

where the hardening function F is expressed by

$$F = \sum_{I=1}^{\mathrm{N}} F_{\mathrm{I}}(H, \theta)\xi_{\mathrm{I}} \tag{15.4}$$

The increment of the isotropic hardening variable is given by

$$dH = f_{H\varepsilon}(\boldsymbol{\sigma}, H; d\boldsymbol{\varepsilon}^p) \tag{15.5}$$

which is the homogeneous function of $d\boldsymbol{\varepsilon}^p$ in degree-one since dH is induced only for $d\boldsymbol{\varepsilon}^p \neq \mathbf{O}$ and the first-order time-differential quantity. ξ_{I} ($\mathrm{I} = 1, \cdots, \mathrm{N}$) designates the ratio of the volume V_{I} of each phase to the volume V of whole body, i.e.

$$\xi_{\mathrm{I}} = \frac{V_{\mathrm{I}}}{V} \left(\sum_{\mathrm{I}=1}^{\mathrm{N}} \xi_{\mathrm{I}} = 1 \right) \tag{15.6}$$

The subloading surface is described as

$$f(\boldsymbol{\sigma}) = RF \tag{15.7}$$

where the evolution rule of the normal-yield ratio is given by

$$dR = U(R)||d\boldsymbol{\varepsilon}^p|| \quad \text{for } d\boldsymbol{\varepsilon}^p \neq \mathbf{O} \tag{15.8}$$

$$R = f(\boldsymbol{\sigma})/F \quad \text{for } d\boldsymbol{\varepsilon}^p = \mathbf{O} \tag{15.9}$$

The time-differentiation of Eq. (15.7), noting Eqs. (15.4), (15.5) and (15.8), is given by

$$\frac{\partial f(\boldsymbol{\sigma})}{\partial \boldsymbol{\sigma}} : d\boldsymbol{\sigma} - R\sum_{\mathrm{I}=1}^{\mathrm{N}}\left[\left(\frac{\partial F_{\mathrm{I}}(H,\theta)}{\partial H} f_{H\varepsilon}(\boldsymbol{\sigma}, H; d\boldsymbol{\varepsilon}^p) + \frac{\partial F_{\mathrm{I}}(H,\theta)}{\partial \theta} d\theta\right)\xi_{\mathrm{I}} + F_{\mathrm{I}}(H,\theta) d\xi_{\mathrm{I}}\right] + U(R)||d\boldsymbol{\varepsilon}^p||F = 0 \tag{15.10}$$

Equation (15.10) is rewritten exploiting Eq. (6.33) based on the homogeneous degree-one of the function $f(\boldsymbol{\sigma})$ for $\boldsymbol{\sigma}$ as follows:

$$\mathbf{n} : d\boldsymbol{\sigma} - \left\{\frac{1}{F}\sum_{\mathrm{I}=1}^{\mathrm{N}}\left[\left(\frac{\partial F_{\mathrm{I}}(H,\theta)}{\partial H} f_{H\varepsilon}(\boldsymbol{\sigma}, H; d\boldsymbol{\varepsilon}^p) + \frac{\partial F_{\mathrm{I}}(H,\theta)}{\partial \theta} d\theta\right)\xi_{\mathrm{I}} + F_{\mathrm{I}}(H,\theta) d\xi_{\mathrm{I}}\right] + \frac{U(R)||d\boldsymbol{\varepsilon}^p||}{R}\right\}\mathbf{n} : \boldsymbol{\sigma} = 0 \tag{15.11}$$

Now, adopt the associated flow rule

$$d\boldsymbol{\varepsilon}^p = d\bar{\lambda}\mathbf{n} \ (d\bar{\lambda} \geq 0, ||\mathbf{n}|| = 1) \tag{15.12}$$

where $d\bar{\lambda}$ and $\mathbf{n}$ are the magnitude of the plastic strain increment and the normalized outward-normal of the subloading surface.

Substituting Eq. (15.12) into Eq. (15.11), one has

$$\mathbf{n} : d\boldsymbol{\sigma} - d\bar{\lambda}\overline{M}^p - \frac{1}{F}\sum_{\mathrm{I}=1}^{\mathrm{N}}\left[\left(\frac{\partial F_{\mathrm{I}}(H,\theta)}{\partial \theta} d\theta\right)\xi_{\mathrm{I}} + F_{\mathrm{I}}(H,\theta) d\xi_{\mathrm{I}}\right]\mathbf{n} : \boldsymbol{\sigma} = 0 \tag{15.13}$$

where

$$\overline{M}^p \equiv \left(\frac{1}{F}\sum_{\mathrm{I}=1}^{\mathrm{N}}\frac{\partial F_{\mathrm{I}}(H,\theta)}{\partial H}\xi_{\mathrm{I}} f_{Hn}(\boldsymbol{\sigma}, H; \mathbf{n}) + \frac{U(R)}{R}\right)\mathbf{n} : \boldsymbol{\sigma} \tag{15.14}$$

with

$$f_{Hn}(\boldsymbol{\sigma}, H; \mathbf{n}) = dH/d\bar{\lambda} \tag{15.15}$$

It follows from Eqs. (15.12) and (15.13) that

$$d\bar{\lambda} = \frac{\mathbf{n} : d\boldsymbol{\sigma} - \sum_{\mathrm{I}=1}^{\mathrm{N}} \left[\left(\dfrac{\partial F_{\mathrm{I}}(H,\theta)}{\partial \theta} d\theta \right) \xi_{\mathrm{I}} + F_{\mathrm{I}}(H,\theta) d\xi_{\mathrm{I}} \right] \dfrac{\mathbf{n} : \boldsymbol{\sigma}}{F}}{\overline{M}^{p}} \tag{15.16}$$

$$d\boldsymbol{\varepsilon}^{p} = \frac{\mathbf{n} : d\boldsymbol{\sigma} - \sum_{\mathrm{I}=1}^{\mathrm{N}} \left[\left(\dfrac{\partial F_{\mathrm{I}}(H,\theta)}{\partial \theta} d\theta \right) \xi_{\mathrm{I}} + F_{\mathrm{I}}(H,\theta) d\xi_{\mathrm{I}} \right] \dfrac{\mathbf{n} : \boldsymbol{\sigma}}{F}}{\overline{M}^{p}} \mathbf{n} \tag{15.17}$$

15.2 Thermal and Transformation Strain Increments

The inelastic strain increments in the phase-transformation process, i.e. heat-transformation strain and transformation-plastic strain increments is described in this section.

15.2.1 Heat-Transformation Strain Increment

The heat-transformation strain increment is given as follows (Okamura 2006a, b):

$$d\boldsymbol{\varepsilon}^{\theta} = -\frac{1}{3\rho}\left(\frac{{}^{0}\rho}{\rho}\right)^{1/3} d\rho \mathbf{I}$$

i.e.

$$d\boldsymbol{\varepsilon}^{\theta} = -\frac{1}{3\rho}\left(\frac{{}^{0}\rho}{\rho}\right)^{1/3} \left(\frac{\partial \rho}{\partial \theta} d\theta + \sum_{\mathrm{I}=1}^{\mathrm{N}} \frac{\partial \rho}{\partial \xi_{\mathrm{I}}} d\xi_{\mathrm{I}} \right) \mathbf{I} \tag{15.18}$$

where ρ is the mass density and ${}^{0}\rho$ is its initial value.

15.2.2 Transformation-Plastic Strain Increment

Transformation-plastic strain increment is given as follows (Denis et al. 1985):

$$d\boldsymbol{\varepsilon}^{Tp} = \sum_{\mathrm{I}=2}^{\mathrm{N}} 3K_{\mathrm{I}}(1-X)d\xi_{\mathrm{I}}\boldsymbol{\sigma}' = d\mathfrak{h}\boldsymbol{\sigma}' \tag{15.19}$$

where K_{I} is the transformation-plastic coefficient in each phase, and X and $d\mathfrak{h}$ are given by

$$X = \sum_{\mathrm{I}=2}^{\mathrm{N}} \xi_{\mathrm{I}}, \quad d\mathfrak{h} \equiv \sum_{\mathrm{I}=2}^{\mathrm{N}} 3K_{\mathrm{I}}(1-X)d\xi_{\mathrm{I}} \tag{15.20}$$

15.3 Stress Rate Versus Strain Rate Relation

Substituting Eqs. (15.2), (15.17), (15.18) and (15.19) into Eq. (15.1), we have the constitutive equation as follows:

$$\begin{aligned} d\boldsymbol{\varepsilon} = \mathbf{E}^{-1} : d\boldsymbol{\sigma} + \frac{\partial \mathbf{E}^{-1}}{\partial \theta} d\theta : \boldsymbol{\sigma} + \frac{\mathbf{n} : d\boldsymbol{\sigma} - \sum_{\mathrm{I}=1}^{\mathrm{N}} \left[\left(\frac{\partial F_{\mathrm{I}}}{\partial \theta} d\theta \right) \xi_{\mathrm{I}} + F_{\mathrm{I}} \xi_{\mathrm{I}} \right] \frac{\mathbf{n} : \boldsymbol{\sigma}}{F}}{\overline{M}^{p}} \mathbf{n} \\ - \frac{1}{3\rho} \left(\frac{{}^{0}\rho}{\rho} \right)^{1/3} \left(\frac{\partial \rho}{\partial \theta} d\theta + \sum_{I=1}^{N} \frac{\partial \rho}{\partial \xi_I} d\xi_I \right) \mathbf{I} + \sum_{I=2}^{N} 3K_I (1-X) d\xi_I \, \boldsymbol{\sigma}' \end{aligned} \tag{15.21}$$

It follows from Eq. (15.21) that

$$\begin{aligned} \mathbf{n} : \mathbf{E} : d\boldsymbol{\varepsilon} = \mathbf{n} : d\boldsymbol{\sigma} + \mathbf{n} : \mathbf{E} : \frac{\partial \mathbf{E}^{-1}}{\partial \theta} d\theta : \boldsymbol{\sigma} \\ + \frac{\mathbf{n} : d\boldsymbol{\sigma} - \sum_{\mathrm{I}=1}^{\mathrm{N}} \left[\left(\frac{\partial F_{\mathrm{I}}}{\partial \theta} d\theta \right) \xi_{\mathrm{I}} + F_{\mathrm{I}} \xi_{\mathrm{I}} \right] \frac{\mathbf{n} : \boldsymbol{\sigma}}{F}}{\bar{M}^{p}} \mathbf{n} : \mathbf{E} : \mathbf{n} \\ - \frac{1}{3\rho} \left(\frac{{}^{0}\rho}{\rho} \right)^{1/3} \left(\frac{\partial \rho}{\partial \theta} d\theta + \sum_{\mathrm{I}=1}^{\mathrm{N}} \frac{\partial \rho}{\partial \xi_{\mathrm{I}}} d\xi_{\mathrm{I}} \right) \mathbf{n} : \mathbf{E} : \mathbf{I} \\ + \sum_{\mathrm{I}=2}^{\mathrm{N}} 3K_{\mathrm{I}} (1-X) d\xi_{\mathrm{I}} \mathbf{n} : \mathbf{E} : \boldsymbol{\sigma}' \end{aligned}$$

from which the magnitude of plastic strain increment, denoted by $d\Lambda$ instead of $d\lambda$, is derived as follows:

$$d\bar{\Lambda} = \frac{1}{\bar{M}^p + \mathbf{n}:\mathbf{E}:\mathbf{n}}\left[\mathbf{n}:\mathbf{E}:d\boldsymbol{\varepsilon} - \sum_{I=1}^{N}\left[\left(\frac{\partial F_I}{\partial \theta}d\theta\right)\xi_I + F_I\xi_I\right]\frac{\mathbf{n}:\boldsymbol{\sigma}}{F}\frac{\mathbf{n}:\mathbf{E}:\mathbf{n}}{\bar{M}^p}\right.$$
$$-\mathbf{n}:\mathbf{E}:\frac{\partial \mathbf{E}^{-1}}{\partial \theta}d\theta:\boldsymbol{\sigma} + \frac{1}{3\rho}\left(\frac{{}^0\rho}{\rho}\right)^{1/3}\left(\frac{\partial \rho}{\partial \theta}d\theta + \sum_{I=1}^{N}\frac{\partial \rho}{\partial \xi_I}d\xi_I\right)\mathbf{n}:\mathbf{E}:\mathbf{I}$$
$$\left. - \sum_{I=2}^{N} 3K_I(1-X)d\xi_I\mathbf{n}:\mathbf{E}:\boldsymbol{\sigma}'\right] \tag{15.22}$$

It follows from Eqs. (15.1) and (15.2) that

$$d\boldsymbol{\sigma} = \mathbf{E}:\left(d\boldsymbol{\varepsilon} - \frac{\partial \mathbf{E}^{-1}}{\partial \theta}d\theta:\boldsymbol{\sigma} - d\boldsymbol{\varepsilon}^p - d\boldsymbol{\varepsilon}^\theta - d\boldsymbol{\varepsilon}^{Tp}\right) \tag{15.23}$$

which leads to the following equation by substituting Eqs. (15.12), (15.18), (15.20) and (15.22) into Eq. (15.23).

$$d\boldsymbol{\sigma} = \mathbf{E}:\left\{d\boldsymbol{\varepsilon} - \frac{\partial \mathbf{E}^{-1}}{\partial \theta}d\theta:\boldsymbol{\sigma} - \frac{1}{\bar{M}^p + \mathbf{n}:\mathbf{E}:\mathbf{n}}\left[\mathbf{n}:\mathbf{E}:d\boldsymbol{\varepsilon} - \sum_{\mathrm{I}=1}^{\mathrm{N}}\left[\left(\frac{\partial F_{\mathrm{I}}}{\partial \theta}d\theta\right)\xi_{\mathrm{I}} + F_{\mathrm{I}}\xi_{\mathrm{I}}\right]\frac{\mathbf{n}:\boldsymbol{\sigma}}{F}\frac{\mathbf{n}:\mathbf{E}:\mathbf{n}}{\bar{M}^p}\right.\right.$$
$$-\mathbf{n}:\mathbf{E}:\frac{\partial \mathbf{E}^{-1}}{\partial \theta}d\theta:\boldsymbol{\sigma} + \frac{1}{3\rho}\left(\frac{{}^0\rho}{\rho}\right)^{1/3}\left(\frac{\partial \rho}{\partial \theta}d\theta + \sum_{\mathrm{I}=1}^{\mathrm{N}}\frac{\partial \rho}{\partial \xi_{\mathrm{I}}}d\xi_{\mathrm{I}}\right)\mathbf{n}:\mathbf{E}:\mathbf{I}$$
$$\left. - \sum_{\mathrm{I}=2}^{\mathrm{N}} 3K_{\mathrm{I}}(1-X)d\xi_{\mathrm{I}}\mathbf{n}:\mathbf{E}:\boldsymbol{\sigma}'\right]\mathbf{n}$$
$$\left. + \frac{1}{3\rho}\left(\frac{{}^0\rho}{\rho}\right)^{1/3}\left(\frac{\partial \rho}{\partial \theta}d\theta + \sum_{\mathrm{I}=1}^{\mathrm{N}}\frac{\partial \rho}{\partial \xi_{\mathrm{I}}}d\xi_{\mathrm{I}}\right)\mathbf{I} - \sum_{\mathrm{I}=2}^{\mathrm{N}} 3K_{\mathrm{I}}(1-X)d\xi_{\mathrm{I}}\boldsymbol{\sigma}'\right\} \tag{15.24}$$

The loading criterion for the plastic strain increment is given by

$$\left.\begin{array}{ll} d\boldsymbol{\varepsilon}^p \neq \mathbf{O} & \text{for } d\bar{\Lambda} > 0 \\ d\boldsymbol{\varepsilon}^p = \mathbf{O} & \text{for } d\bar{\Lambda} \leq 0 \end{array}\right\} \tag{15.25}$$

The subloading phase-transformation model was formulated within the framework of the initial subloading surface model in this section. It can be generalized to the extended subloading surface model (Hashiguchi 1989, 2013) which incorporates the evolution rule of the elastic-core, i.e. the similarity-center of the subloading surface and the normal-yield surface so that the cyclic loading behavior can be also described appropriately. Further, it can be extended to the finite strain theory based on the multiplicative decomposition (Hashiguchi and Yamakawa 2012).

References

Denis S, Gautier E, Simon A, Beck G (1985) Stress-phase-transformation basic principles, modelling, and calculation of internal stresses. Mater Sci Tech 1:805–814

Hashiguchi K (1989) Subloading surface model in unconventional plasticity. Int J Solids Structures 25:917–945

Hashiguchi K (2013) Elastoplasticity theory. Lecture note in applied and computational mechanics, 2nd edn. Springer, Heidelberg

Hashiguchi K, Yamakawa Y (2012) Introduction to finite strain theory for continuum elasto-plasticity. Wiley series in computational mechanics. Wiley Chichester

Hashiguchi K, Okamura K (2014) Subloading-phase transformation model. In: Proceedings of 27th JSME on computational mechanical division conference, pp OS17–1707

Inoue T, Raniecki B (1978) Determination of thermal-hardening stress in steels by use of thermoplasticity theory. J Mech Phys Solids 26:187–212

Inoue T, Watanabe Y, Okamura K, Narazaki M, Shichino H, Ju D-Y, Kanamori H, Ichitani K (2007) Metallo-thermo-mechanical simulation of carburized quenching process by several codes—a benchmark project. Key Eng Mater 340–341:1061–1066

Okamura K (2006a) Reviews and perspective on hardening simulation. In: Proceedings of Japan Institute on Metals Materials and Iron Steel Institute, Japan, Kyushu-branch, pp 1–12 (in Japanese)

Okamura K (2006b) Actuarity and scope on simulation of heat treatment: I material properties and database. J Soc Mater Sci Jpn 55:529–535 (in Japanese)

Okamura K, Kawashima H (1988a) Finite element analysis of thermal stress in heat treatment. Netu-shori (Heat Treatment) 28:141–148 (in Japanese)

Okamura K, Kawashima H (1988b) Analysis of residual deformation of a gear during quenching. In: Proceedings of 32nd Japan congress materials research, pp 323–329 (in Japanese)

Okamura K, Yamamoto K, Fukumoto M (2005) Material properties for quenching simulation and assessment on computational results. In: Proceedings of 3rd Asian conference on heat treatment material, pp 353–355

Rohde J, Jeppsson A (2000) Literature review of heat treatment simulations with respect to phase transformation, residual stresses and distortion. Scand J Metall 29:47–62

Chapter 16
Corotational Rate Tensor

It was studied in Chap. 4 that the material-time derivatives of state variables, e.g. stress and internal variables in elastoplasticity do not possess the objectivity and thus, instead of them, we must adopt their objective time-derivatives. The responses of simple constitutive equations introducing corotational rates with various spins including the plastic spin will be examined in this chapter.

16.1 Hypoelasticity

Consider the hypoelastic constitutive equation in Eq. (5.54), i.e.

$$\overset{\circ}{\boldsymbol{\sigma}} = K(\mathrm{tr}\mathbf{d})\mathbf{I} + 2G\mathbf{d}' \tag{16.1}$$

Equation (16.1) is described by the following equation for Eq. (4.55), noting $\sigma_{12} = \sigma_{21}$, $\omega_{12} = -\omega_{21}$ and using Eqs. (3.63) and $(3.67)_1$ for the simple shear deformation described in Subsection 3.8.2.

$$\begin{bmatrix} \dot{\sigma}_{11} - 2\sigma_{12}\omega_{12} & \dot{\sigma}_{12} + (\sigma_{11} - \sigma_{22})\omega_{12} \\ \text{Sym.} & \dot{\sigma}_{12} + 2\sigma_{12}\omega_{12} \end{bmatrix} = G\begin{bmatrix} 0 & 1 \\ 1 & 0 \end{bmatrix}\dot{\gamma} = G\frac{2}{\cos^2\bar{\theta}}\begin{bmatrix} 0 & 1 \\ 1 & 0 \end{bmatrix}\dot{\bar{\theta}} \tag{16.2}$$

where the spin of material is represented by the symbol $\boldsymbol{\omega}$ as shown in Eq. (4.55).

16.1.1 Zaremba-Jaumann Rate

When the Zaremba-Jaumann rate in Eq. (4.59) is adopted for the corotational rate, Eq. (16.2) leads to the following equation by setting $\boldsymbol{\omega} = \mathbf{w}$ with Eq. $(3.67)_2$.

K. Hashiguchi, *Foundations of Elastoplasticity: Subloading Surface Model*,
DOI 10.1007/978-3-319-48821-9_16

$$\begin{bmatrix} \dot{\sigma}_{11} - \dot{\gamma}\sigma_{12} & \dot{\sigma}_{12} + \dfrac{\dot{\gamma}}{2}(\sigma_{11} - \sigma_{22}) \\ \text{sym.} & \dot{\sigma}_{22} + \dot{\gamma}\sigma_{12} \end{bmatrix} = G\begin{bmatrix} 0 & 1 \\ 1 & 0 \end{bmatrix}\dot{\gamma} \tag{16.3}$$

from which we have

$$\left.\begin{aligned} &\dot{\sigma}_{11} - \dot{\gamma}\sigma_{12} = 0 \\ &\dot{\sigma}_{12} + \frac{\dot{\gamma}}{2}(\sigma_{11} - \sigma_{22}) = G\dot{\gamma} \\ &\dot{\sigma}_{22} + \dot{\gamma}\sigma_{12} = 0 \end{aligned}\right\} \tag{16.4}$$

Substituting

$$\sigma_{22} = -\sigma_{11}, \quad \dot{\gamma} = \frac{\dot{\sigma}_{11}}{\sigma_{12}} \tag{16.5}$$

obtained from the first and the third equations into the second equation in Eq. (16.4), yields

$$\dot{\sigma}_{12} + \frac{\dot{\sigma}_{11}}{\sigma_{12}}\sigma_{11} = G\frac{\dot{\sigma}_{11}}{\sigma_{12}} \tag{16.6}$$

the time-integration of which is given as

$$\sigma_{12} = \sqrt{2G\sigma_{11} - \sigma_{11}^2} \tag{16.7}$$

Substituting this equation into the second equation of Eq. (16.4), we have

$$\frac{\dfrac{\dot{\sigma}_{11}}{G}}{\sqrt{1 - \left(1 - \dfrac{\sigma_{11}}{G}\right)^2}} = \dot{\gamma}$$

the integration of which is given by

$$\cos^{-1}\left(1 - \frac{\sigma_{11}}{G}\right) = \gamma$$

i.e.

$$\sigma_{11} = -\sigma_{22} = G(1 - \cos\gamma) \tag{16.8}$$

The substitution of Eq. (16.8) into Eq. (16.7) leads to

$$\sigma_{12} = G \sin \gamma \tag{16.9}$$

The continuum spin $\mathbf{w}$ designates the instantaneous rate of rotation of the principal directions of strain rate, i.e. the instantaneous rate of rotation of the cross depicted momentarily on the material surface. Therefore, if it is used in the simple shear deformation with the constant shear strain rate, i.e. $\overset{\bullet}{\gamma} = \text{const.}$ leading to $\mathbf{w} = \text{const.}$, the material is regarded to rotate in a constant angular velocity, while the strain rate $\mathbf{d}$ is also kept constant. Then, the oscillatory shear stress is predicted by the hypoelastic constitutive equation using the Zaremba-Jaumann rate with the continuum spin as shown in Eq. (16.9) and depicted in Fig. 16.1 (cf. e.g. Dienes 1979).

16.1.2 Green-Naghdi Rate

Consider the Green-Naghdi rate for the corotational rate with the relative spin $\boldsymbol{\omega} = \boldsymbol{\Omega}^R$, i.e. Eq. (4.57). It follows from Eq. (3.87) that

$$\boldsymbol{\omega} = \boldsymbol{\Omega}^R = \overset{\bullet}{\mathbf{R}}\mathbf{R}^T = \frac{2}{4+\gamma^2}\begin{bmatrix} 0 & 1 \\ -1 & 0 \end{bmatrix}\overset{\bullet}{\gamma} = \begin{bmatrix} 0 & 1 \\ 1 & 0 \end{bmatrix}\overset{\bullet}{\bar{\theta}} \tag{16.10}$$

The substitution of Eq. (16.10) into Eq. (16.2) reads:

$$\begin{bmatrix} \overset{\bullet}{\sigma}_{11} - 2\sigma_{12}\overset{\bullet}{\bar{\theta}} & \overset{\bullet}{\sigma}_{12} + (\sigma_{11} - \sigma_{22})\overset{\bullet}{\bar{\theta}} \\ \text{Sym.} & \overset{\bullet}{\sigma}_{22} + 2\sigma_{12}\overset{\bullet}{\bar{\theta}} \end{bmatrix} = G\frac{2}{\cos^2\bar{\theta}}\begin{bmatrix} 0 & 1 \\ 1 & 0 \end{bmatrix}\overset{\bullet}{\bar{\theta}} \tag{16.11}$$

from which we have

$$\left.\begin{aligned} &\overset{\bullet}{\sigma}_{11} - 2\sigma_{12}\overset{\bullet}{\bar{\theta}} = 0 \\ &\overset{\bullet}{\sigma}_{12} + (\sigma_{11} - \sigma_{22})\overset{\bullet}{\bar{\theta}} = G\frac{2}{\cos^2\bar{\theta}}\overset{\bullet}{\bar{\theta}} \\ &\overset{\bullet}{\sigma}_{22} + 2\sigma_{12}\overset{\bullet}{\bar{\theta}} = 0 \end{aligned}\right\} \tag{16.12}$$

It is obtained that

$$\sigma_{22} = -\sigma_{11} \quad \left(\overset{\bullet}{\bar{\theta}} = \frac{1}{2}\frac{\overset{\bullet}{\sigma}_{11}}{\sigma_{12}}\right) \tag{16.13}$$

from the first and the third equations, and

$$\frac{d\sigma_{11}}{d\bar{\theta}} = 2\sigma_{12} \rightarrow \frac{d\sigma_{12}}{d\bar{\theta}} = \frac{1}{2}\frac{d^2\sigma_{11}}{d\bar{\theta}^2} \tag{16.14}$$

from the first equation in Eq. (16.12). Substituting Eqs. (16.13) and (16.14) into the second equation in Eq. (16.12), i.e. $d\sigma_{12}/d\bar{\theta} + 2\sigma_{11} = 2G/\cos^2\bar{\theta}$, we have the ordinary differential equation

$$\frac{d^2\sigma_{11}}{d\bar{\theta}^2} + 4\sigma_{11} = \frac{4G}{\cos^2\bar{\theta}} \tag{16.15}$$

The roots of the characteristic equation of the second-order homogeneous linear differential equation for Eq. (16.15) are given by

$$m^2 + 4m = 0 \rightarrow m = \frac{\pm\sqrt{-16}}{2} = \pm 2i$$

Thus, the complementary function of Eq. (16.15) is given by the following equation.

$$\sigma_{11} = A\cos 2\bar{\theta} + B\sin 2\bar{\theta} = \sqrt{A^2 + B^2}\sin[2\bar{\theta} + \arctan(A/B)] \tag{16.16}$$

where A, B are the integral constants. Further, adding the particular solution for Eq. (16.15) itself, the general solution of Eq. (16.15) is obtained as follows:

$$\sigma_{11} = \sqrt{A^2 + B^2}\sin(2\bar{\theta} + \arctan(A/B)) + 4G(\cos 2\bar{\theta}\ln\cos\bar{\theta} + \bar{\theta}\sin 2\bar{\theta} - \sin^2\bar{\theta}) \tag{16.17}$$

Assuming that the initial stress is zero, Eq. (16.17) becomes

$$\sigma_{11} = 4G(\cos 2\bar{\theta}\ln\cos\bar{\theta} + \bar{\theta}\sin 2\bar{\theta} - \sin^2\bar{\theta}) \tag{16.18}$$

Furthermore, substituting Eq. (16.18) into Eq. (16.14), we have

$$\sigma_{12} = \frac{1}{2}\frac{d\sigma_{11}}{d\bar{\theta}} = 2G\cos 2\bar{\theta}(2\bar{\theta} - 2\tan^2\bar{\theta}\ln\cos\bar{\theta} - \tan\bar{\theta}) \tag{16.19}$$

These equations have been derived by Dienes (1979).

The relative spin $\mathbf{\Omega}^R$ designates the mean rate of rotation of the cross depicted on the material surface at the beginning of deformation. Therefore, it coincides with the continuum spin $\mathbf{w}$ at the initial state but it decreases gradually with the shear deformation. Then, the oscillation of shear stress observed in Jaumann rate is not predicted if the Green-Naghdi rate is adopted as the corotational rate as seen in Fig. 16.1 calculated by Eq. (16.19).

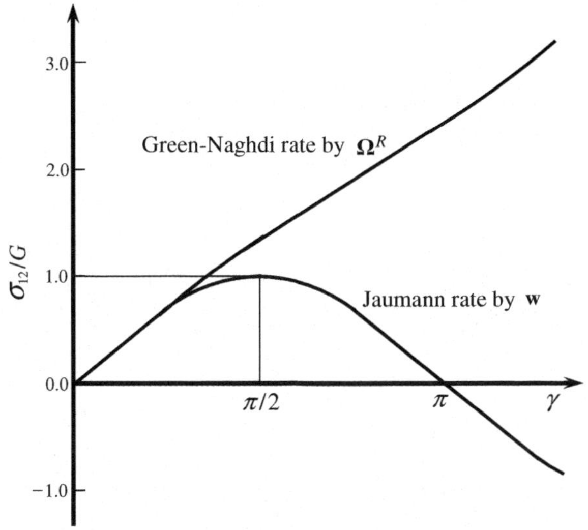

Fig. 16.1 Description of simple shear deformation of hypoelastic material by Jaumann rate and Green-Naghdi rate (Dienes 1979)

16.2 Kinematic Hardening Material

For the sake of simplicity, consider the response of a rigid plastic material fulfilling $\mathbf{d} = \mathbf{d}^p$ and assume the Mises material with linear kinematic hardening in Eq. (6.100), i.e.

$$\overset{\circ}{\boldsymbol{\alpha}} = \frac{2}{3} h_a \hat{\mathbf{n}} ||\mathbf{d}^{p\prime}|| = \frac{2}{3} h_a \mathbf{d}^{p\prime} \tag{16.20}$$

setting $c_k = (2/3)h_a$, which was analyzed by Dafalias (1983).

Then, it holds for the simple shear that

$$\begin{bmatrix} \overset{\circ}{\alpha}_{11} & \overset{\circ}{\alpha}_{12} \\ \overset{\circ}{\alpha}_{21} & \overset{\circ}{\alpha}_{22} \end{bmatrix} = \frac{1}{3} h_a \begin{bmatrix} 0 & 1 \\ 1 & 0 \end{bmatrix} \dot{\gamma} \tag{16.21}$$

by substituting Eq. $(3.67)_1$ into Eq. (16.20). Equation (16.21) is described for Eq. (4.55) as follows:

$$\begin{bmatrix} \dot{\alpha}_{11} - 2\alpha_{12}\omega_{12} & \dot{\alpha}_{12} + (\alpha_{11} - \alpha_{22})\omega_{12} \\ \text{Sym.} & \dot{\alpha}_{22} + 2\alpha_{12}\omega_{12} \end{bmatrix} = \frac{1}{3} h_a \begin{bmatrix} 0 & 1 \\ 1 & 0 \end{bmatrix} \dot{\gamma} \tag{16.22}$$

where

$$\boldsymbol{\omega} = \begin{bmatrix} 0 & 1 \\ -1 & 0 \end{bmatrix} z(\gamma)\,\dot{\gamma} \tag{16.23}$$

$$z(\gamma) = \begin{cases} 1/2 & \text{for } \boldsymbol{\omega} = \mathbf{w} \\ 2/(4+\gamma^2) & \text{for } \boldsymbol{\omega} = \boldsymbol{\Omega}^p \end{cases} \tag{16.24}$$

noting Eqs. $(3.67)_2$ and (16.10). The substitution of Eq. (16.23) into Eq. (16.22) leads to

$$\begin{bmatrix} \overset{\circ}{\alpha}_{11} - 2\alpha_{12} z(\gamma)\,\dot{\gamma} & \overset{\circ}{\alpha}_{12} + (\alpha_{11} - \alpha_{22}) z(\gamma)\,\dot{\gamma} \\ \text{Sym.} & \overset{\circ}{\alpha}_{22} + 2\alpha_{12} z(\gamma)\,\dot{\gamma} \end{bmatrix} = \frac{1}{3} h_a \begin{bmatrix} 0 & 1 \\ -1 & 0 \end{bmatrix} \dot{\gamma} \tag{16.25}$$

from which we have

$$\left.\begin{aligned} &\overset{\circ}{\alpha}_{11} - 2\alpha_{12} z(\gamma)\,\dot{\gamma} = 0 \\ &\overset{\circ}{\alpha}_{12} + (\alpha_{11} - \alpha_{22}) z(\gamma)\,\dot{\gamma} = \frac{1}{3} h_a \dot{\gamma} \\ &\overset{\circ}{\alpha}_{22} + 2\alpha_{12} z(\gamma)\,\dot{\gamma} = 0 \end{aligned}\right\} \tag{16.26}$$

In addition, noting $\alpha_{11} = -\alpha_{22}$, we have

$$\left.\begin{aligned} &\alpha'_{11} = -\alpha'_{22} = 2z(\gamma)\alpha_{12} \\ &\alpha'_{12} + 2z(\gamma)\alpha_{11} = \frac{1}{3} h_a \end{aligned}\right\} \tag{16.27}$$

where $(\,)' = d(\,)/d\gamma$. Differentiating Eq. (16.27), we have

$$\left.\begin{aligned} &\alpha''_{11} - 2z\alpha'_{12} - 2\alpha_{12} z' = 0 \\ &\alpha''_{12} + 2\alpha'_{11} z + 2\alpha_{11} z' = 0 \end{aligned}\right\} \tag{16.28}$$

which, noting Eq. (16.27), becomes

$$\left.\begin{aligned} &\alpha''_{11} - 2z\left(\frac{1}{3} h_a - 2z\alpha_{11}\right) - 2\frac{\alpha'_{11}}{2z} z' = 0 \\ &\alpha''_{12} + 4z\alpha_{12} z + 2\frac{1}{2z}\left(\frac{1}{3} h_a - \alpha'_{12}\right) z' = 0 \end{aligned}\right\}$$

Then, it is obtained that

$$\left.\begin{aligned} &\alpha''_{11} - \frac{z'}{z}\alpha'_{11} + 4z^2\alpha_{11} - \frac{2}{3} h_a z = 0 \\ &\alpha''_{12} - \frac{z'}{z}\alpha'_{12} + 4z^2\alpha_{12} + \frac{1}{3} h_a \frac{z'}{z} = 0 \end{aligned}\right\} \tag{16.29}$$

16.2.1 Zaremba-Jaumann Rate

The substitution of Eq. (16.24)$_1$ into Eq. (16.29) leads to

$$\left.\begin{aligned} &\alpha''_{11}+\alpha_{11}-\frac{1}{3}h_a=0\\ &\alpha''_{12}+\alpha_{12}=0 \end{aligned}\right\} \tag{16.30}$$

from which, noting the initial condition $\alpha_{11}=0$, $\alpha_{12}=0$ for $\gamma=0$, we have

$$\left.\begin{aligned} &\alpha_{11}=-\alpha_{22}=\frac{1}{3}h_a(1-\cos\gamma)\\ &\alpha_{12}=\frac{1}{3}h_a\sin\gamma \end{aligned}\right\} \tag{16.31}$$

It is obtained from Eq. (16.31) that

$$\left.\begin{aligned} &\sigma_{11}=\alpha_{11}=-\sigma_{22}=-\alpha_{22}=\frac{1}{3}h_a(1-\cos\gamma)\\ &\sigma_{12}=\frac{1}{\sqrt{3}}F+\alpha_{12}=\frac{1}{\sqrt{3}}F+\frac{1}{3}h_a\sin\gamma \end{aligned}\right\} \tag{16.32}$$

noting $\sqrt{3/2}\sqrt{(\sigma_{11}-\alpha_{11})^2+(\sigma_{22}-\alpha_{22})^2+2(\sigma_{12}-\alpha_{12})^2}=\sqrt{3}(\sigma_{12}-\alpha_{12})=F$ with $d_{ii}=d^p_{ii}=\dot{\lambda}\,\hat{n}_{ii}=\dot{\lambda}\,(\sigma_{ii}-\alpha_{ii})/||\hat{\boldsymbol{\sigma}}||=0$ (no sum; $i=1,2$). Both σ_{11} and σ_{12} oscillates in sine curves as shown in Fig. 16.2.

The above-mentioned fact that the kinematic hardening model with the Zaremba-Jaumann rate exhibits the oscillation was indicated first by Nagtegaal and De Jong (1982).

16.2.2 Green-Naghdi Rate

Substituting Eq. (16.24)$_2$ and

$$z'=\frac{-4\gamma}{(4+\gamma^2)^2} \tag{16.33}$$

into Eq. (16.29), we have

$$\left.\begin{aligned} &\alpha''_{11}-\frac{\dfrac{-4\gamma}{(4+\gamma^2)^2}}{\dfrac{2}{4+\gamma^2}}\alpha'_{11}+4\frac{4}{(4+\gamma^2)^2}\alpha_{11}-\frac{2}{3}h_a\frac{2}{4+\gamma^2}=0\\ &\alpha''_{12}-\frac{\dfrac{-4\gamma}{(4+\gamma^2)^2}}{\dfrac{2}{4+\gamma^2}}\alpha'_{12}+4\frac{4}{(4+\gamma^2)^2}\alpha_{12}+\frac{1}{3}h_a\frac{\dfrac{-4\gamma}{(4+\gamma^2)^2}}{\dfrac{2}{4+\gamma^2}}=0 \end{aligned}\right\}$$

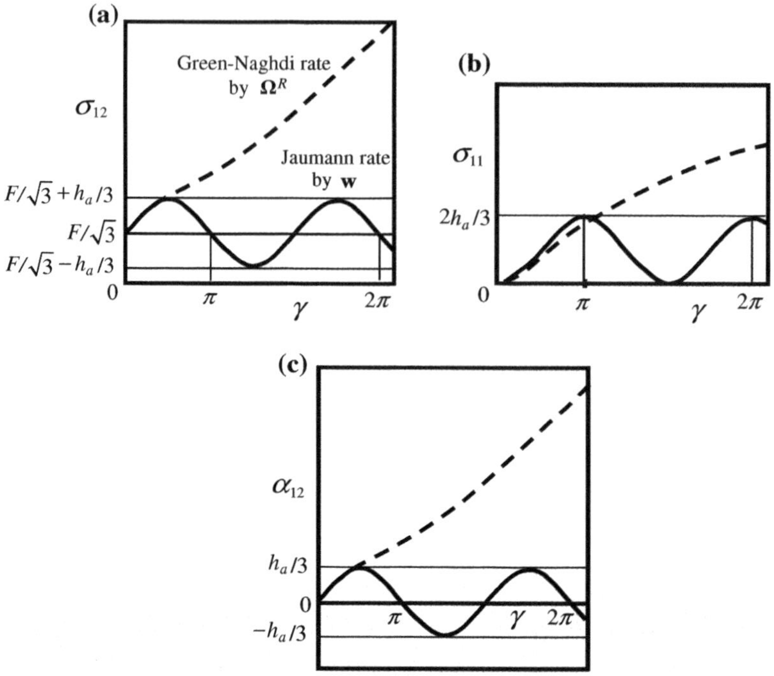

Fig. 16.2 Description of simple shear deformation of kinematic hardening material by Zaremba-Jaumann and Green-Naghdi rates (Dafalias 1983)

i.e.

$$\left.\begin{aligned}\alpha_{11}'' + \frac{2\gamma}{4+\gamma^2}\alpha_{11}' + \frac{16}{(4+\gamma^2)^2}\alpha_{11} - h_a\frac{4}{3(4+\gamma^2)} = 0\\ \alpha_{12}'' + \frac{2\gamma}{4+\gamma^2}\alpha_{12}' + \frac{16}{(4+\gamma^2)^2}\alpha_{12} - h_a\frac{2\gamma}{3(4+\gamma^2)}h_a = 0\end{aligned}\right\} \tag{16.34}$$

the general solution of which is derived as the following equation by the method of variable coefficients (Dafalias 1983).

$$\left.\begin{aligned}\alpha_{11} = \frac{1}{3}h_a\frac{1}{4+\gamma^2}\left[4\gamma\left\{4\tan^{-1}\left(\frac{\gamma}{2}\right) - \gamma\right\} - 4(\gamma^2-4)\ln\frac{2}{\sqrt{4+\gamma^2}}\right]\\ \alpha_{12} = \frac{1}{3}h_a\frac{1}{4+\gamma^2}\left[\gamma^3 - 4(\gamma^2-4)4\tan^{-1}\left(\frac{\gamma}{2}\right) - 4\gamma\left(1+4\ln\frac{2}{\sqrt{4+\gamma^2}}\right)\right]\end{aligned}\right\} \tag{16.35}$$

The relation of σ_{11}, σ_{12} to α_{11}, α_{12} is given by Eq. (16.32) also in this case. An oscillation is not predicted in the simple shear deformation as shown in Fig. 16.2.

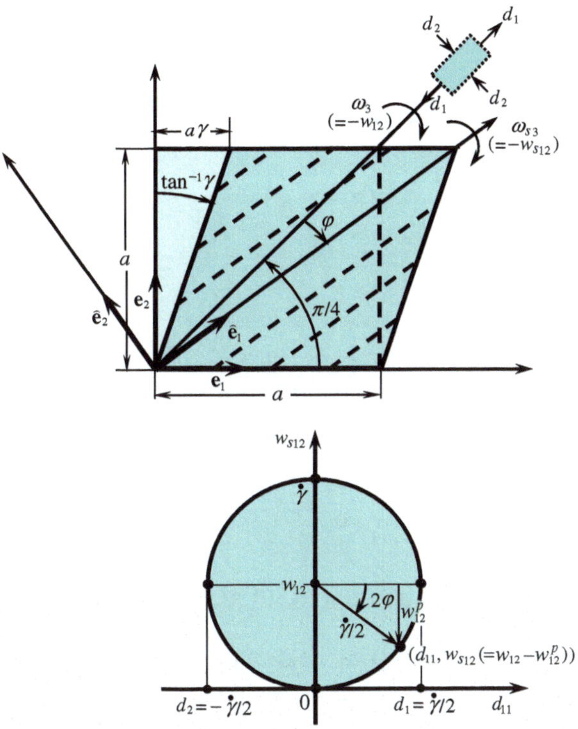

Fig. 16.3 Substructure spin in traverse isotropic material

16.3 Plastic Spin

The above-mentioned Zaremba-Jaumann rate and the Green-Naghdi rate do not reflect the substructure of material but they are uniquely determined only by the change of external appearance of the material. However, the mechanically meaningful rotation would be the spin of substructure, as known presuming the crystals of metals or the annual ring of woods, which would be the rotation of the principal direction of anisotropy (Kratochvil 1971; Mandel 1971). The concept of the plastic spin is proposed in order to incorporate such rotation into elastoplastic constitutive equations (Dafalias 1983, 1985a, b; Loret 1983).

In what follows, in order to interpret the mechanical meaning of the plastic spin, assume the rigid plasticity and the simplest anisotropy, i.e. the *traverse anisotropic material* (Fig. 16.3) with the parallel line-elements of substructure having the direction $\hat{\mathbf{e}}_1$ inclined $\pi/4$ from the fixed base $\mathbf{e}_1$ in the initial state of deformation and rotates by the angle φ in the clockwise direction with the increase of shear strain (Dafalias 1984). Then, it holds that

$$\widehat{\mathbf{e}}_1 = \begin{Bmatrix} \cos(\pi/4 - \varphi) \\ \sin(\pi/4 - \varphi) \end{Bmatrix}, \quad \dot{\widehat{\mathbf{e}}}_1 = \begin{Bmatrix} \sin(\pi/4 - \varphi) \\ -\cos(\pi/4 - \varphi) \end{Bmatrix} \dot{\varphi} \tag{16.36}$$

Here, referring to Fig. 16.3, one has

$$\tan(\pi/4 - \varphi) = \frac{1}{1+\gamma} \tag{16.37}$$

from which one has

$$\frac{-\dot{\varphi}}{\cos^2(\pi/4 - \varphi)} = \frac{-\dot{\gamma}}{(1+\gamma)^2} = -\dot{\gamma}\tan^2(\pi/4 - \varphi) \tag{16.38}$$

Then it holds that

$$\dot{\varphi} = \dot{\gamma}\sin^2(\pi/4 - \varphi)\left(= \frac{\dot{\gamma}}{2}[1 - \cos\{2(\pi/4 - \varphi)\}]\right) = (1 - \sin 2\varphi)\frac{\dot{\gamma}}{2} \tag{16.39}$$

Using Eq. (16.39) along with Eq. (16.36), it is obtained that

$$\dot{\widehat{\mathbf{e}}}_1 = \begin{Bmatrix} \sin(\pi/4 - \varphi) \\ -\cos(\pi/4 - \varphi) \end{Bmatrix}(1 - \sin 2\varphi)\frac{\dot{\gamma}}{2} \tag{16.40}$$

which is rewritten as

$$\begin{Bmatrix} \sin(\pi/4 - \varphi) \\ -\cos(\pi/4 - \varphi) \end{Bmatrix}(1 - \sin 2\varphi)\frac{\dot{\gamma}}{2} = \left(\begin{bmatrix} 0 & \dot{\gamma}/2 \\ -\dot{\gamma}/2 & 0 \end{bmatrix} - \begin{bmatrix} 0 & (\dot{\gamma}/2)\sin 2\varphi \\ -(\dot{\gamma}/2)\sin 2\varphi & 0 \end{bmatrix}\right) \begin{Bmatrix} \cos(\pi/4 - \varphi) \\ \sin(\pi/4 - \varphi) \end{Bmatrix}$$

i.e

$$\dot{\widehat{\mathbf{e}}}_1 = \boldsymbol{\omega}_s \widehat{\mathbf{e}}_1$$

Then, it can be confirmed from Eqs. $(3.67)_2$ that Eq. (6.28) for the spin of substructure holds as follows:

$$\left.\begin{aligned} \boldsymbol{\omega}_s &= \mathbf{w}^e = \mathbf{w} - \mathbf{w}^p = \begin{bmatrix} 0 & \dot{\gamma}/2 \\ -\dot{\gamma}/2 & 0 \end{bmatrix} - \begin{bmatrix} 0 & (\dot{\gamma}/2)\sin 2\varphi \\ -(\dot{\gamma}/2)\sin 2\varphi & 0 \end{bmatrix} \\ \mathbf{w}^p &= \frac{1}{2}\begin{bmatrix} 0 & \sin 2\varphi \\ -\sin 2\varphi & 0 \end{bmatrix}\dot{\gamma} \end{aligned}\right\} \tag{16.41}$$

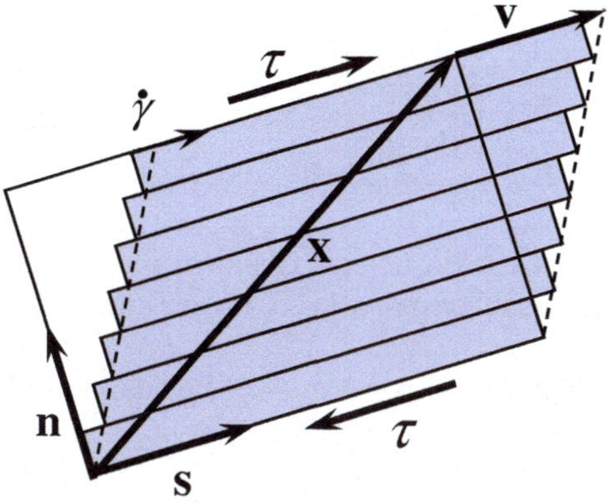

Fig. 16.4 Slip system and slip deformation

as illustrated in Fig. 16.3. Kuroda (1997) applied the above-mentioned formulation for the traverse isotropic material to the orthotropic material described in Eq. (10.44) in Sect. 10.5 and showed the numerical calculation results for the rotation of the principal axes of orthotropic yield surface.

Next, consider the same problem by the deformation of metal crystals. If the substructure does not rotate, it holds for the slip system in Fig. 16.4 that

$$\mathbf{v} = \dot{\gamma}(\mathbf{x}\cdot\mathbf{n})\mathbf{s} = \dot{\gamma}(\mathbf{s}\otimes\mathbf{x})\mathbf{n},\, v_i = \dot{\gamma}(x_r n_r)s_i \tag{16.42}$$

$$\boldsymbol{l}^p = \frac{\partial \mathbf{v}}{\partial \mathbf{x}} = \dot{\gamma}\,\mathbf{s}\otimes\mathbf{n},\, l_{ij}^p = \frac{\partial v_i}{\partial x_j} = \dot{\gamma}\,\delta_{jr}n_r s_i = \dot{\gamma}\,s_i n_j \tag{16.43}$$

$$\mathbf{d}^p = \frac{1}{2}\dot{\gamma}\,(\mathbf{s}\otimes\mathbf{n}+\mathbf{n}\otimes\mathbf{s}),\, d_{ij}^p = \frac{1}{2}\left(\frac{\partial v_i}{\partial x_j}+\frac{\partial v_j}{\partial x_i}\right) = \frac{1}{2}\dot{\gamma}(s_i n_j + n_i s_j) \tag{16.44}$$

$$\mathbf{w}^p = \frac{1}{2}\dot{\gamma}\,(\mathbf{s}\otimes\mathbf{n}-\mathbf{n}\otimes\mathbf{s}),\, w_{ij}^p = \frac{1}{2}\left(\frac{\partial v_i}{\partial x_j}-\frac{\partial v_j}{\partial x_i}\right) = \frac{1}{2}\dot{\gamma}(s_i n_j - n_i s_j) \tag{16.45}$$

Equations (16.42)–(16.45) are extended for multi slip systems of number n as follows:

$$\boldsymbol{l}^p = \sum_{\alpha=1}^{n}\dot{\gamma}^{(\alpha)}\mathbf{s}^{(\alpha)}\otimes\mathbf{n}^{(\alpha)} \tag{16.46}$$

$$\mathbf{d}^p = \sum_{\alpha=1}^{n}\dot{\gamma}^{(\alpha)}\left\{\frac{1}{2}\left(\mathbf{s}^{(\alpha)}\otimes\mathbf{n}^{(\alpha)}+\mathbf{n}^{(\alpha)}\otimes\mathbf{s}^{(\alpha)}\right)\right\} = \sum_{\alpha=1}^{n}\dot{\gamma}^{(\alpha)}\mathbf{p}^{(\alpha)} \tag{16.47}$$

$$\mathbf{p}^{(\alpha)} \equiv \frac{1}{2}\left(\mathbf{s}^{(\alpha)}\otimes\mathbf{n}^{(\alpha)}+\mathbf{n}^{(\alpha)}\otimes\mathbf{s}^{(\alpha)}\right) \tag{16.48}$$

$$\mathbf{w}^p = \sum_{\alpha=1}^{n} \frac{1}{2}\dot{\gamma}^{(\alpha)}\left(\mathbf{s}^{(\alpha)} \otimes \mathbf{n}^{(\alpha)} - \mathbf{n}^{(\alpha)} \otimes \mathbf{s}^{(\alpha)}\right) = \sum_{\alpha=1}^{n} \dot{\gamma}^{(\alpha)}\mathbf{q}^{(\alpha)} \tag{16.49}$$

$$\mathbf{q}^{(\alpha)} \equiv \frac{1}{2}\left(\mathbf{s}^{(\alpha)} \otimes \mathbf{n}^{(\alpha)} - \mathbf{n}^{(\alpha)} \otimes \mathbf{s}^{(\alpha)}\right) \tag{16.50}$$

The simple example of the plastic spin is shown above. Dafalias (1985a, b) provided the general mechanical interpretation of the plastic spin based on Eq. (6.28) as follows.

The substitution of Eq. (6.28) into Eq. (4.55) reads:

$$\overset{\circ}{\mathbf{t}} = \dot{\mathbf{t}} - (\mathbf{w} - \mathbf{w}^p)\mathbf{t} + \mathbf{t}(\mathbf{w} - \mathbf{w}^p) = \overset{\circ}{\mathbf{t}}{}^{w} + \mathbf{w}^p\mathbf{t} - \mathbf{t}\mathbf{w}^p \tag{16.51}$$

The relation of the corotational rate and the Zaremba-Jaumann rate of Cauchy stress is given from Eqs. (4.59), (6.106) and (16.51) as

$$\begin{aligned}\overset{\circ}{\boldsymbol{\sigma}} &= \overset{\circ}{\boldsymbol{\sigma}}{}^{w} + \mathbf{w}^p\boldsymbol{\sigma} - \boldsymbol{\sigma}\mathbf{w}^p \\ &= \overset{\circ}{\boldsymbol{\sigma}}{}^{w} + \eta^p\,\dot{\lambda}\{(\boldsymbol{\sigma}\hat{\mathbf{n}} - \hat{\mathbf{n}}\boldsymbol{\sigma})\boldsymbol{\sigma} - \boldsymbol{\sigma}(\boldsymbol{\sigma}\hat{\mathbf{n}} - \hat{\mathbf{n}}\boldsymbol{\sigma})\}\end{aligned}$$

i.e.

$$\overset{\circ}{\boldsymbol{\sigma}} = \overset{\circ}{\boldsymbol{\sigma}}{}^{w} + \dot{\lambda}\,\boldsymbol{\sigma}_n \tag{16.52}$$

where

$$\boldsymbol{\sigma}_n \equiv \eta^p\left(2\boldsymbol{\sigma}\hat{\mathbf{n}}\boldsymbol{\sigma} - \hat{\mathbf{n}}\boldsymbol{\sigma}^2 - \boldsymbol{\sigma}^2\hat{\mathbf{n}}\right) \tag{16.53}$$

Now, we derive the elastoplastic constitutive equation. Substituting Eq. (16.52) into Eq. (6.95), it follows that

$$\dot{\lambda} = \frac{\hat{\mathbf{n}} : \overset{\circ}{\boldsymbol{\sigma}}}{M^p} = \frac{\hat{\mathbf{n}} : (\overset{\circ}{\boldsymbol{\sigma}}{}^{w} + \dot{\lambda}\,\boldsymbol{\sigma}_n)}{M^p} \tag{16.54}$$

from which it is obtained that

$$\dot{\lambda} = \frac{\hat{\mathbf{n}} : \overset{\circ}{\boldsymbol{\sigma}}{}^{w}}{\widetilde{M}^p},\ \mathbf{d}^p = \frac{\hat{\mathbf{n}} : \overset{\circ}{\boldsymbol{\sigma}}{}^{w}}{\widetilde{M}^p}\hat{\mathbf{n}} \tag{16.55}$$

where

$$\widetilde{M}^p = M^p - \hat{\mathbf{n}} : \boldsymbol{\sigma}_n \tag{16.56}$$

The substitution of Eq. (16.55) into Eq. (16.52), one has

$$\overset{\circ}{\boldsymbol{\sigma}} = \overset{\circ}{\boldsymbol{\sigma}}{}^{w} + \frac{\hat{\mathbf{n}} : \overset{\circ}{\boldsymbol{\sigma}}{}^{w}}{\widetilde{M}^p}\boldsymbol{\sigma}_n \tag{16.57}$$

Then, the strain rate is given by

$$\mathbf{d} = \mathbf{E}^{-1} : \overset{\circ}{\boldsymbol{\sigma}} + \mathbf{d}^p = \mathbf{E}^{-1} : \left(\overset{\circ}{\boldsymbol{\sigma}}^w + \frac{\hat{\mathbf{n}} : \overset{\circ}{\boldsymbol{\sigma}}^w}{\widetilde{M}^p} \boldsymbol{\sigma}_n \right) + \frac{\hat{\mathbf{n}} : \overset{\circ}{\boldsymbol{\sigma}}^w}{\widetilde{M}^p} \hat{\mathbf{n}}$$

leading to

$$\mathbf{d} = \mathbf{E}^{-1} : \overset{\circ}{\boldsymbol{\sigma}}^w + \frac{\hat{\mathbf{n}} : \overset{\circ}{\boldsymbol{\sigma}}^w}{\widetilde{M}^p} (\hat{\mathbf{n}} + \mathbf{E}^{-1} : \boldsymbol{\sigma}_n) \tag{16.58}$$

The plastic multiplier $\dot{\Lambda}$ in terms of strain rate is given by Eq. (6.97) as it is since Eq. (6.95), i.e. (16.54) holds even in the present formulation. Then, the Zaremba-Jaumann rate of Cauchy stress is given from Eq. (16.52) with $\overset{\circ}{\boldsymbol{\sigma}} = \mathbf{E} : \mathbf{d}^e$ by

$$\begin{aligned}\overset{\circ}{\boldsymbol{\sigma}}^w &= \mathbf{E} : (\mathbf{d} - \mathbf{d}^p) - \dot{\Lambda}\, \boldsymbol{\sigma}_n \\ &= \mathbf{E} : \mathbf{d} - \mathbf{E} : \hat{\mathbf{n}} \frac{\hat{\mathbf{n}} : \mathbf{E} : \mathbf{d}}{M^p + \hat{\mathbf{n}} : \mathbf{E} : \hat{\mathbf{n}}} - \frac{\hat{\mathbf{n}} : \mathbf{E} : \mathbf{d}}{M^p + \hat{\mathbf{n}} : \mathbf{E} : \hat{\mathbf{n}}} \boldsymbol{\sigma}_n\end{aligned}$$

i.e.

$$\overset{\circ}{\boldsymbol{\sigma}}^w = \left[\mathbf{E} - \frac{(\mathbf{E} : \hat{\mathbf{n}} + \boldsymbol{\sigma}_n) \otimes (\hat{\mathbf{n}} : \mathbf{E})}{M^p + \hat{\mathbf{n}} : \mathbf{E} : \hat{\mathbf{n}}} \right] : \mathbf{d} \tag{16.59}$$

which is related by the non-symmetric tangent modulus tensor. The stress and the kinematic hardening variable are updated by the time-integrations of

$$\dot{\boldsymbol{\sigma}} = \overset{\circ}{\boldsymbol{\sigma}}^w + \mathbf{w}\boldsymbol{\sigma} - \boldsymbol{\sigma}\mathbf{w} \tag{16.60}$$

$$\dot{\boldsymbol{\alpha}} = \|\mathbf{d}^p\| \mathbf{f}_{kn} + (\mathbf{w} - \mathbf{w}^p)\, \boldsymbol{\alpha} - \boldsymbol{\alpha}(\mathbf{w} - \mathbf{w}^p) \tag{16.61}$$

noting Eq. (16.51).

Hereinafter, limit to the Mises yield condition with the kinematic hardening. Then, substituting Eq. (6.89) with Eq. (10.1) into Eq. (6.106), the plastic spin is reduced to the following equation.

$$\begin{aligned}\mathbf{w}^p &= \eta^p\, \dot{\lambda}(\boldsymbol{\sigma}\hat{\mathbf{n}} - \hat{\mathbf{n}\boldsymbol{\sigma}}) \\ &= \eta^p\, \dot{\lambda}\left(\boldsymbol{\sigma} \frac{\hat{\boldsymbol{\sigma}}'}{\|\hat{\boldsymbol{\sigma}}'\|} - \frac{\hat{\boldsymbol{\sigma}}'}{\|\hat{\boldsymbol{\sigma}}'\|} \boldsymbol{\sigma} \right) = \eta^p\, \dot{\lambda} \left\{ (\hat{\boldsymbol{\sigma}} + \boldsymbol{\alpha}) \frac{\hat{\boldsymbol{\sigma}}'}{\|\hat{\boldsymbol{\sigma}}'\|} - \frac{\hat{\boldsymbol{\sigma}}'}{\|\hat{\boldsymbol{\sigma}}'\|} (\hat{\boldsymbol{\sigma}} + \boldsymbol{\alpha}) \right\} \\ &= \eta^p\, \dot{\lambda}\, (\boldsymbol{\alpha}\boldsymbol{\sigma}' - \boldsymbol{\sigma}'\boldsymbol{\alpha}) / \|\hat{\boldsymbol{\sigma}}'\|\end{aligned} \tag{16.62}$$

which was first proposed by Dafalias (1985a, b), where Eq. (6.106) is regarded as the generalized form of Eq. (16.62) which is limited to the Mises material with the kinematic hardening. Then, considering the simple shear deformation and assuming the rigid plasticity, the initial isotropy and the linear kinematic hardening as in Sect. 16.2, Eq. (16.62) leads to

$$\begin{aligned}\mathbf{w}^p &= \eta^p\left(\begin{bmatrix}\alpha_{11} & \alpha_{12}\\ \alpha_{12} & -\alpha_{11}\end{bmatrix}\begin{bmatrix}0 & 1\\ 1 & 0\end{bmatrix}\frac{\dot{\gamma}}{2}-\begin{bmatrix}0 & 1\\ 1 & 0\end{bmatrix}\frac{\dot{\gamma}}{2}\begin{bmatrix}\alpha_{11} & \alpha_{12}\\ \alpha_{12} & -\alpha_{11}\end{bmatrix}\right)\\ &= \eta^p\left(\begin{bmatrix}\alpha_{12} & \alpha_{11}\\ -\alpha_{11} & \alpha_{12}\end{bmatrix}-\begin{bmatrix}\alpha_{12} & -\alpha_{11}\\ \alpha_{11} & \alpha_{12}\end{bmatrix}\right)\frac{\dot{\gamma}}{2}=\eta^p\begin{bmatrix}0 & \alpha_{11}\\ -\alpha_{11} & 0\end{bmatrix}\dot{\gamma}\end{aligned} \tag{16.63}$$

Substituting Eqs. (3.67), (16.21) and (16.63) into Eq. (16.61), we have

$$\begin{aligned}\begin{bmatrix}\dot{\alpha}_{11} & \dot{\alpha}_{12}\\ \dot{\alpha}_{12} & -\dot{\alpha}_{11}\end{bmatrix} &= \frac{2}{3}h_a\begin{bmatrix}0 & 1\\ 1 & 0\end{bmatrix}\frac{\dot{\gamma}}{2}+\left(\begin{bmatrix}0 & 1\\ -1 & 0\end{bmatrix}\frac{\dot{\gamma}}{2}-\eta^p\alpha_{11}\begin{bmatrix}0 & 1\\ -1 & 0\end{bmatrix}\dot{\gamma}\right)\begin{bmatrix}\alpha_{11} & \alpha_{12}\\ \alpha_{12} & -\alpha_{11}\end{bmatrix}\\ &\quad-\begin{bmatrix}\alpha_{11} & \alpha_{12}\\ \alpha_{12} & -\alpha_{11}\end{bmatrix}\left(\begin{bmatrix}0 & 1\\ -1 & 0\end{bmatrix}\frac{\dot{\gamma}}{2}-\eta^p\alpha_{11}\begin{bmatrix}0 & 1\\ -1 & 0\end{bmatrix}\dot{\gamma}\right)\\ &= \frac{1}{3}h_a\begin{bmatrix}0 & 1\\ 1 & 0\end{bmatrix}\dot{\gamma}+\begin{bmatrix}\alpha_{12} & \alpha_{11}\\ -\alpha_{11} & -\alpha_{12}\end{bmatrix}\frac{\dot{\gamma}}{2}-\eta^p\alpha_{11}\begin{bmatrix}\alpha_{12} & -\alpha_{11}\\ -\alpha_{11} & -\alpha_{12}\end{bmatrix}\dot{\gamma}\\ &\quad-\begin{bmatrix}-\alpha_{12} & \alpha_{11}\\ \alpha_{11} & \alpha_{12}\end{bmatrix}\frac{\dot{\gamma}}{2}+\eta^p\alpha_{11}\begin{bmatrix}-\alpha_{12} & \alpha_{11}\\ \alpha_{11} & \alpha_{12}\end{bmatrix}\dot{\gamma}\\ &= \frac{1}{3}h_a\begin{bmatrix}0 & 1\\ 1 & 0\end{bmatrix}\dot{\gamma}+\begin{bmatrix}\alpha_{12} & -\alpha_{11}\\ -\alpha_{11} & -\alpha_{12}\end{bmatrix}\dot{\gamma}-2\eta^p\alpha_{11}\begin{bmatrix}\alpha_{12} & -\alpha_{11}\\ -\alpha_{11} & -\alpha_{12}\end{bmatrix}\dot{\gamma}\\ &= \begin{bmatrix}(1-2\eta^p\alpha_{11})\alpha_{12} & \frac{1}{3}h_a-(1-2\eta^p\alpha_{11})\alpha_{11}\\ \frac{1}{3}h_a-(1-2\eta^p\alpha_{11})\alpha_{11} & -(1-2\eta^p\alpha_{11})\alpha_{12}\end{bmatrix}\dot{\gamma}\end{aligned}$$

from which we obtain

$$\left.\begin{aligned}\frac{d\alpha_{11}}{d\gamma} &= (1-2\eta^p\alpha_{11})\alpha_{12}\\ \frac{d\alpha_{12}}{d\gamma} &= \frac{1}{3}h_a-(1-2\eta^p\alpha_{11})\alpha_{11}\end{aligned}\right\} \tag{16.64}$$

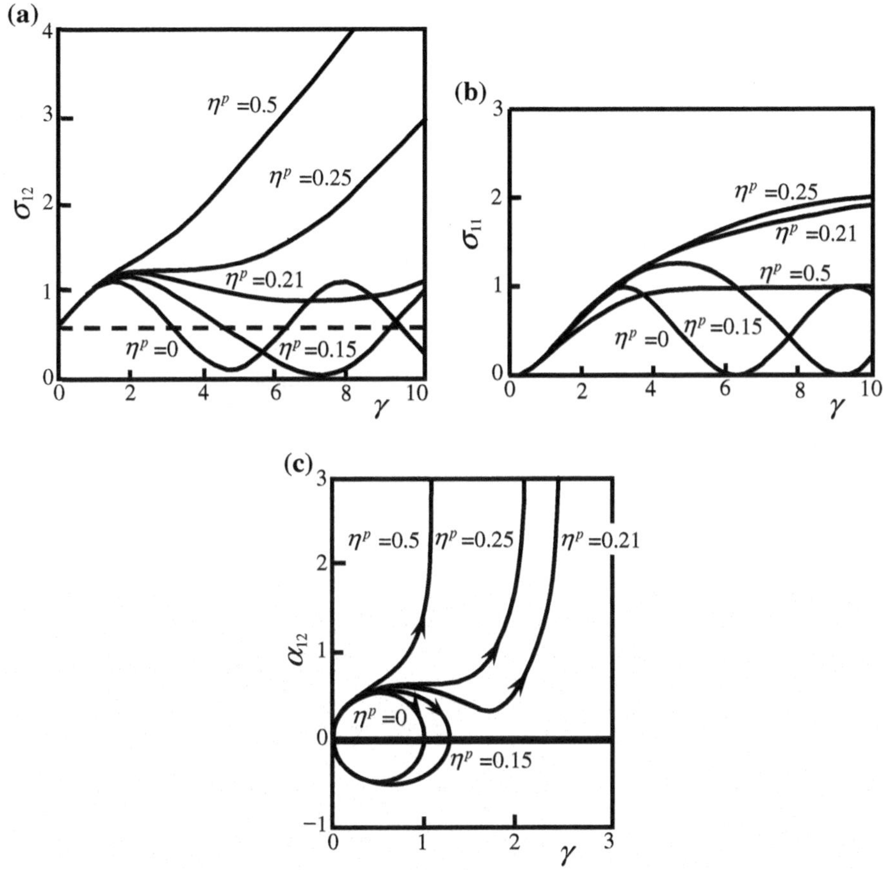

Fig. 16.5 Description of simple shear deformation of kinematic hardening material by the corotational rate with a plastic spin (Dafalias 1985b)

The nonlinear differential Eq. (16.64) is solved numerically by Dafalias (1985a). The calculation result is shown in Fig. 16.5. As seen in this figure, the non-oscillation curve is obtained by choosing the material parameter η^p appropriately. When choosing $\eta^p > 0.25$, $\sigma_{11} = -\sigma_{22}$ and σ_{12} increase monotonically with the increase of shear strain γ. On the other hand, the Zaremba-Jaumann and the Green-Naghdi rates are independent of material property and thus they would lack the physical exactness.

Oscillatory shear stress is predicted in the simple shear by the hypoelasto-plastic material with the kinematic hardening if the Jaumann rates are adopted for the stress and the back stress as described above. In the plastically-isotropic material, however, it should be noted that the plastic strain rate is independent of the types of stress rate, e.g. material-time derivative, Green-Naghdi rate, Zaremba-Jaumann rate, etc. as known from Eq. (6.43) because of

$$\mathbf{n}(\boldsymbol{\sigma}):\overset{\circ}{\boldsymbol{\sigma}} = \mathbf{n}(\boldsymbol{\sigma}):\dot{\boldsymbol{\sigma}} \tag{16.65}$$

by Eq. (4.83) since only one variable $\boldsymbol{\sigma}$ is involved in Eq. (16.65) exhibiting the isotropy, while the elastic response is influenced by the types of stress rate. In the plastically-anisotropic material, on the other hand, the response depends on the difference of the time-derivative as follows:

(1) The plastic strain rate is influenced by the difference of stress rate because of $\hat{\mathbf{n}}(\boldsymbol{\sigma},\boldsymbol{\alpha}):\overset{\circ}{\boldsymbol{\sigma}} \neq \hat{\mathbf{n}}(\boldsymbol{\sigma},\boldsymbol{\alpha}):\dot{\boldsymbol{\sigma}}$ by Eq. (4.84), which is included in the constitutive relation of plastic strain rate in Eq. (6.93).
(2) The development of the anisotropic hardening variable, i.e. the back stress is influenced by the difference of the time-derivative, so that the yield surface on which the stress lies is also influenced by it. Therefore, the stress vs. strain relation is influenced by the difference of the time-derivative.

It should be recognized that the accurate description of elastoplastic deformation can be attained up to around one hundred percent shear strain by the Jaumann rate without the plastic spin even for plastically-anisotropic materials, since identical deformation behavior is described by using any corotational rate up to that percent as known from Figs. 16.2 and 16.5.

References

Dafalias YF (1983) Corotational rates for kinematic hardening at large plastic deformations. J Appl Mech (ASME) 50:561–565

Dafalias YF (1984) The plastic spin concept and a simple illustration of its role in finite plastic transformation. Mech Mater 3:223–233

Dafalias YF (1985a) The plastic spin. J Appl Mech (ASME) 52:865–871

Dafalias YF (1985b) A missing link in the macroscopic constitutive formulation of large plastic deformations. Plasticity today. In: Sawczuk A, Bianchi G (eds) International symposium on recent trends and results in plasticity. Elsevier Publications, pp 135–151

Dienes JK (1979) On the analysis of rotation and stress rate in deforming bodies. Acta Mech 32:217–232

Kratochvil J (1971) Finite-strain theory of crystalline elastic-inelastic materials. J Appl Phys 42:1104–1108

Kuroda M (1997) Interpretation of the behavior of metals under large plastic shear deformations: a macroscopic approach. Int J Plasticity 13:359–383

Loret B (1983) On the effects of plastic rotation in the finite deformation of anisotropic elastoplastic materials. Mech Mater 2:287–304

Mandel J (1971) Plastidite classique et viscoplasticite. Course & Lectures, no 97, International Center for Mechanical Sciences, Udine. Springer, Heidelberg

Nagtegaal JC, De Jong JE (1982) Some aspects of non-isotropic work hardening in finite strain plasticity. In: Lee EH, Mallett RL (eds) Plasticity of metals and finite strain: theory, experiment and computation. Division of Applied Mechanics, Stanford University and Department of Mechanical Engineering, Rensseler Polytechnic Institute, pp 65–102

Chapter 17
Localization of Deformation

Even if material is subjected to a homogeneous stress, the deformation concentrates in a quite narrow strip zone as the deformation becomes large and finally the material results in failure. Such a concentration of deformation is called the *localization* of deformation and the strip zone is called the *shear band*. The shear band thickness is the order of several microns in metals and ten and several times of particle diameter in soils. Therefore, a large shear deformation inside the shear band is hardly reflected in the change of external appearance of the whole body, although the stress is estimated by the external load and the outer appearance of material. Therefore, a special care is required for the interpretation of element test data and the analysis taken account of the inception of shear band is indispensable when a large deformation is induced. The localization phenomenon of deformation and its pertinent analysis are described in this chapter.

17.1 Element Test

The purpose of the element test of material is to find the constitutive property of material and thus the element test is premised that a homogeneous deformation proceeds reflecting the constitutive property. However, the deformation becomes heterogeneous when a large deformation accompanying with the shear band is induced. Then, the strain (rate) inside the shear band is far larger than the one based on the variation in the outer appearance of the whole material element. Here, the stress (rate) applied to the material element is relevant to the strain (rate) inside the shear band. In other words, the strain (rate) relevant to the constitutive property is not a strain (rate) based on the variation of the outer appearance of material. This fact is illustratively depicted in Fig. 17.1 for the softening material. The strain rate observed in the element test is designated by $d\bar{\varepsilon}$ which is the average strain rate in the whole of the element, while the strain rate inside the shear band is designated by

K. Hashiguchi, *Foundations of Elastoplasticity: Subloading Surface Model*,
DOI 10.1007/978-3-319-48821-9_17

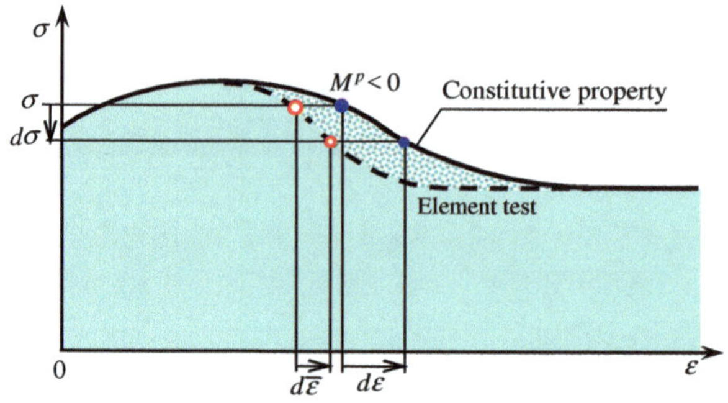

Fig. 17.1 Stress-strain curve in the constitutive property and the element test with a localization in a softening state

$d\bar{\varepsilon}$. Here, note that the actual stress increment $d\sigma$ is not relevant to $d\bar{\varepsilon}$ but relevant to $d\varepsilon$, where $d\sigma$ and $d\varepsilon$ are related by constitutive property. It must be considered when we determine the material parameters from element test data.

17.2 Gradient Theory

The stress is determined locally by the deformation at each material point when the deformation is small moderately. However, it is influenced also by the deformations in surrounding material particles, i.e. by the gradient of deformation when the deformation develops to be inhomogeneous as observed inside the shear band, in the necking part, etc. The former can be described by the *local theory* taking account of only the deformation in each material point but the latter can be described by incorporating the *nonlocal theory* taking account of the deformations not only in each material point but also in the surrounding material particles. Here, there exists the limitation in the gradient of deformation so that the shear band thickness does not decrease less than a certain limitation. The limitation is regulated by the material constant, called the *characteristic length.* In the finite element method ignoring this fact, the deformation concentrates in the narrow band zone corresponding to the size of one element. Then, if the finite elements are downsized aiming at obtaining an accurate solution, the shear band thickness is reduced infinitely, resulting in the *mesh-size dependence* losing the reliability of solution, i.e. the *ill-posedness.* The non-local theory, called the gradient theory, for the elasto-plastic deformation, in which the gradients of strain rate, stress rate and internal variables are incorporated, was proposed by Aifantis (1984). It will be described briefly in this section.

Adopting the yield condition with the isotropic hardening for the sake of simplicity and introducing the spatial gradient of the isotropic hardening variable, let the yield condition (6.30) be extended as follows:

$$f(\boldsymbol{\sigma}) = \langle\langle F(H) \rangle\rangle \tag{17.1}$$

where $\langle\langle\rangle\rangle$ designates the second-order gradient, i.e.

$$\langle\langle\rangle\rangle \equiv 1 + c_2^2\nabla^2, \quad \nabla^2 \equiv \frac{\partial^2}{\partial x_1^2} + \frac{\partial^2}{\partial x_2^2} + \frac{\partial^2}{\partial x_3^2} \tag{17.2}$$

c_2 is the material constant reflecting the effect of the gradient of the mechanical state. Here, for the sake of simplicity, the higher order gradient is not incorporated. The first-order gradient is not incorporated because the odd-order gradients in opposite directions are canceled each other.

The material-time derivative of Eq. (17.1) leads to

$$\frac{\partial f(\boldsymbol{\sigma})}{\partial \boldsymbol{\sigma}} : \overset{\circ}{\boldsymbol{\sigma}} = F'(1 + c_2^2\nabla^2)[\dot{H}] \tag{17.3}$$

Substituting Eq. (6.37) into Eq. (17.3), one has

$$\frac{\partial f(\boldsymbol{\sigma})}{\partial \boldsymbol{\sigma}} : \overset{\circ}{\boldsymbol{\sigma}} = F' f_{Hn}(\boldsymbol{\sigma}, H; \mathbf{n})(1 + c_2^2\nabla^2)[\dot{\lambda}] \tag{17.4}$$

It follows from Eq. (17.4), noting $(1 + c_2^2\nabla^2)(1 - c_2^2\nabla^2) = 1 - c_2^4\nabla^4 \cong 1$, that

$$\dot{\lambda} = \frac{\mathbf{n} : (1 - c_2^2\nabla^2)[\overset{\circ}{\boldsymbol{\sigma}}]}{M^p}, \quad \mathbf{d}^p = \frac{\mathbf{n} : (1 - c_2^2\nabla^2)[\overset{\circ}{\boldsymbol{\sigma}}]}{M^p}\mathbf{n} \tag{17.5}$$

$$\mathbf{d} = \mathbf{E}^{-1} : \overset{\circ}{\boldsymbol{\sigma}} + \frac{\mathbf{n} : (1 - c_2^2\nabla^2)[\overset{\circ}{\boldsymbol{\sigma}}]}{M^p}\mathbf{n} \tag{17.6}$$

$$\begin{aligned}
\mathbf{n} : \mathbf{E} : \mathbf{d} &= \mathbf{n} : \overset{\circ}{\boldsymbol{\sigma}} + \mathbf{n} : \mathbf{E} : \mathbf{n}\frac{\mathbf{n} : (1 - c_2^2\nabla^2)[\overset{\circ}{\boldsymbol{\sigma}}]}{M^p} \\
&\cong [M^p(1 + c_2^2\nabla^2) + \mathbf{n} : \mathbf{E} : \mathbf{n}]\left[\frac{\mathbf{n} : (1 - c_2^2\nabla^2)[\overset{\circ}{\boldsymbol{\sigma}}]}{M^p}\right] \\
&= (M^p + \mathbf{n} : \mathbf{E} : \mathbf{n})\left[1 + c_2^2\frac{M^p}{M_p + \mathbf{n} : \mathbf{E} : \mathbf{n}}\nabla^2\right]\left[\frac{\mathbf{n} : (1 - c_2^2\nabla^2)[\overset{\circ}{\boldsymbol{\sigma}}]}{M^p}\right] \\
&= (M^p + \mathbf{n} : \mathbf{E} : \mathbf{n})(1 + \widetilde{\nabla})\left[\frac{\mathbf{n} : (1 - c_2^2\nabla)[\overset{\circ}{\boldsymbol{\sigma}}]}{M^p}\right]
\end{aligned} \tag{17.7}$$

where

$$\widetilde{\nabla} \equiv c_2^2 \vartheta \nabla^2, \quad \vartheta \equiv \frac{M^p}{M^p + \mathbf{n}:\mathbf{E}:\mathbf{n}} \tag{17.8}$$

The plastic modulus M^p is given by Eq. (6.41) as it is.

On the other hand, it is obtained from Eq. (17.7) that

$$\dot{\Lambda} = (1 - \widetilde{\nabla}) \left[\frac{\mathbf{n}:\mathbf{E}:\mathbf{d}}{M^p + \mathbf{n}:\mathbf{E}:\mathbf{n}} \right] \cong \frac{\mathbf{n}:\mathbf{E}:(1 - \widetilde{\nabla})\mathbf{d}}{M^p + \mathbf{n}:\mathbf{E}:\mathbf{n}} \tag{17.9}$$

$$\overset{\circ}{\boldsymbol{\sigma}} = \mathbf{E}:\mathbf{d} - \mathbf{E}:\mathbf{n} \frac{\mathbf{n}:\mathbf{E}:(1 - \widetilde{\nabla})\mathbf{d}}{M^p + \mathbf{n}:\mathbf{E}:\mathbf{n}} \tag{17.10}$$

In what follows, the above-mentioned equations are extended for the subloading surface model with the isotropic and the anisotropic hardening. Incorporating the gradient into the internal variables, the subloading surface with the kinematic and the rotational hardening in addition to the isotropic hardening is given as (Hashiguchi and Tsutsumi 2006):

$$f(\boldsymbol{\sigma} - \langle\langle \boldsymbol{\alpha} \rangle\rangle, \langle\langle \boldsymbol{\beta} \rangle\rangle) = \langle\langle RF(H) \rangle\rangle \tag{17.11}$$

Considering Eqs. (17.2) and (17.11) leads to

$$f(\boldsymbol{\sigma} - (1 + c_2^2 \nabla^2)[\boldsymbol{\alpha}],\ (1 + c_2^2 \nabla^2)[\boldsymbol{\beta}]) = (1 + c_2^2 \nabla^2)[RF(H)] \tag{17.12}$$

The material-time derivative of Eq. (17.12) leads to

$$\begin{aligned} &\frac{\partial f(\boldsymbol{\sigma} - (1 + c_2^2 \nabla^2)[\boldsymbol{\alpha}],\ (1 + c_2^2 \nabla^2)[\boldsymbol{\beta}])}{\partial \boldsymbol{\sigma}} : \overset{\circ}{\boldsymbol{\sigma}} \\ &\quad - \frac{\partial f(\boldsymbol{\sigma} - (1 + c_2^2 \nabla^2)[\boldsymbol{\alpha}],\ (1 + c_2^2 \nabla^2)[\boldsymbol{\beta}])}{\partial \boldsymbol{\sigma}} : (1 + c_2^2 \nabla^2)[\overset{\circ}{\boldsymbol{\alpha}}] \\ &\quad + \frac{\partial f(\boldsymbol{\sigma} - (1 + c_2^2 \nabla^2)[\boldsymbol{\alpha}], (1 + c_2^2 \nabla^2)[\boldsymbol{\beta}])}{\partial (1 + c_2^2 \nabla^2)[\boldsymbol{\beta}]} : (1 + c_2^2 \nabla^2)[\overset{\circ}{\boldsymbol{\beta}}] \\ &= (1 + c_2^2 \nabla^2)[\dot{R} F + R \dot{F}] \end{aligned} \tag{17.13}$$

The gradients of internal variables can be ignored since they are small compared to the gradient of their rates and thus Eq. (17.13) is reduced approximately to

$$\begin{aligned} &\frac{\partial f(\hat{\boldsymbol{\sigma}}, \boldsymbol{\beta})}{\partial \boldsymbol{\sigma}} : \overset{\circ}{\boldsymbol{\sigma}} - \frac{\partial f(\hat{\boldsymbol{\sigma}}, \boldsymbol{\beta})}{\partial \boldsymbol{\sigma}} : (1 + c_2^2 \nabla^2)[\overset{\circ}{\boldsymbol{\alpha}}] + \frac{\partial f(\hat{\boldsymbol{\sigma}}, \boldsymbol{\beta})}{\partial \boldsymbol{\beta}} : (1 + c_2^2 \nabla^2)\,[\overset{\circ}{\boldsymbol{\beta}}] \\ &= F(1 + c_2^2 \nabla^2)[\dot{R}] + R F'(1 + c_2^2 \nabla^2)[\dot{H}] \end{aligned} \tag{17.14}$$

Substituting Eq. (7.9) for the evolution rule of normal-yield ratio, the consistency condition is derived from Eq. (17.14) as follows:

$$\frac{\partial f(\hat{\boldsymbol{\sigma}},\boldsymbol{\beta})}{\partial \hat{\boldsymbol{\sigma}}}:\overset{\circ}{\boldsymbol{\sigma}}-\frac{\partial f(\hat{\boldsymbol{\sigma}},\boldsymbol{\beta})}{\partial \hat{\boldsymbol{\sigma}}}:(1+c_2^2\nabla^2)[\overset{\circ}{\boldsymbol{\alpha}}]+\frac{\partial f(\hat{\boldsymbol{\sigma}},\boldsymbol{\beta})}{\partial \boldsymbol{\beta}}:(1+c_2^2\nabla^2)\,[\overset{\circ}{\boldsymbol{\beta}}]$$
$$=UF(1+c_2^2\nabla^2)[||\mathbf{d}^p||]+RF'(1+c_2^2\nabla^2)[\dot{H}] \tag{17.15}$$

Further, substituting the associated flow rule (7.25) into Eq. (17.15), it is obtained that

$$\begin{aligned}&\frac{\partial f(\hat{\boldsymbol{\sigma}},\boldsymbol{\beta})}{\partial \boldsymbol{\sigma}}:\overset{\circ}{\boldsymbol{\sigma}}-\frac{\partial f(\hat{\boldsymbol{\sigma}},\boldsymbol{\beta})}{\partial \boldsymbol{\sigma}}:(1+c_2^2\nabla^2)[\dot{\bar{\lambda}}\,\mathbf{f}_{kn}(\boldsymbol{\sigma},\boldsymbol{\alpha},F;\hat{\mathbf{n}})]\\&\quad+\frac{\partial f(\hat{\boldsymbol{\sigma}},\boldsymbol{\beta})}{\partial \boldsymbol{\beta}}:(1+c_2^2\nabla^2)[\dot{\bar{\lambda}}\,\mathbf{f}_{\beta n}(M_r,\boldsymbol{\beta},F;\mathbf{n}')]\\&=UF(1+c_2^2\nabla^2)[\dot{\bar{\lambda}}]+RF'(1+c_2^2\nabla^2)[\dot{\bar{\lambda}}f_{Hn}(\boldsymbol{\sigma},H;\mathbf{n})]\end{aligned} \tag{17.16}$$

which can be approximately given by

$$\begin{aligned}&\frac{\partial f(\hat{\boldsymbol{\sigma}},\boldsymbol{\beta})}{\partial \boldsymbol{\sigma}}:\overset{\circ}{\boldsymbol{\sigma}}-\frac{\partial f(\hat{\boldsymbol{\sigma}},\boldsymbol{\beta})}{\partial \boldsymbol{\sigma}}:\mathbf{f}_{kn}(\boldsymbol{\sigma},\boldsymbol{\alpha},F;\hat{\mathbf{n}})(1+c_2^2\nabla^2)[\dot{\bar{\lambda}}]\\&\quad+\frac{\partial f(\hat{\boldsymbol{\sigma}},\boldsymbol{\beta})}{\partial \boldsymbol{\beta}}:\mathbf{f}_{\beta n}(M_r,\boldsymbol{\beta},F;\mathbf{n}')(1+c_2^2\nabla^2)[\dot{\bar{\lambda}}]\\&=(UF+RF'f_{Hn}(\boldsymbol{\sigma},H;\mathbf{n}))(1+c_2^2\nabla^2)[\dot{\bar{\lambda}}]\end{aligned} \tag{17.17}$$

The plastic multiplier is derived from Eq. (17.17) as

$$\dot{\bar{\lambda}}=(1-c_2^2\nabla^2)\left[\frac{\mathbf{n}:\overset{\circ}{\boldsymbol{\sigma}}}{\bar{M}^p}\right]\cong\frac{\mathbf{n}:(1-c_2^2\nabla^2)[\overset{\circ}{\boldsymbol{\sigma}}]}{\bar{M}^p} \tag{17.18}$$

where

$$\bar{M}^p\equiv\mathbf{n}:\left[\left(\frac{F'}{F}f_{Hn}(\boldsymbol{\sigma},H;\mathbf{n})+\frac{U}{R}-\frac{1}{RF}\frac{\partial f(\hat{\boldsymbol{\sigma}},\boldsymbol{\beta})}{\partial \boldsymbol{\beta}}:\mathbf{f}_{\beta n}(M_r,\boldsymbol{\beta},F;\mathbf{n}')\right)\boldsymbol{\sigma}+\mathbf{f}_{kn}(\boldsymbol{\sigma},\boldsymbol{\alpha};\hat{\mathbf{n}})\right] \tag{17.19}$$

Consequently, the plastic strain rate is given as

$$\mathbf{d}^p=\frac{\mathbf{n}:(1-c_2^2\nabla^2)[\overset{\circ}{\boldsymbol{\sigma}}]}{\bar{M}_p}\mathbf{n} \tag{17.20}$$

The shear band thickness of softening soil was predicted adopting Eq. (17.20) by Hashiguchi and Tsutsumi (2006).

Here, it is noteworthy that we must use quite small elements with the size of several tens of shear band thickness to take the effect of the gradient into account correctly in the finite element analysis. Therefore, it is nearly impossible to apply the gradient theory to the finite element analysis of boundary value problems in engineering practice at least at present. The gradient theory is used widely for prediction of shear band thickness, size effects, etc. using fine meshes for very small specimens.

17.3 Shear-Band Embedded Model: Smeared Crack Model

Although the gradient theory is not applicable to the analysis of practical engineering problems at present, the practical model for the finite element analysis for softening materials has been proposed as described below.

As the deformation becomes large and the shear band is formed, the plastic deformation concentrates inside the shear band and thus the softening is accelerated leading to the rapid reduction of stress. As the result, inversely, the unloading leading to the elastic state occurs outside the shear band. Consequently, the elastoplastic constitutive equation holds only inside the shear band. Then, denoting the strain rate and the plastic strain rate calculated from the external appearance by $\bar{\mathbf{d}}$ and $\bar{\mathbf{d}}^p$, respectively, called the *apparent strain rate* and the *apparent plastic strain rate*, respectively, the elastoplastic constitutive equation in terms of the apparent strain rate is proposed. It is called the *shear-band embedded model* or *smeared crack model* (Pietrueszczak and Mroz 1983; Bazant and Cedolin 1991).

Denoting the ratio of the area of a shear band to the area of a two-dimensional finite element by $S(\ll 1)$, the following relations are postulated.

$$\mathbf{d}^p = S\bar{\mathbf{d}}^p \tag{17.21}$$

$$\bar{\mathbf{d}} = \mathbf{d}^e + \bar{\mathbf{d}}^p = \mathbf{d}^e + S\mathbf{d}^p = \mathbf{E}^{-1}:\overset{\circ}{\boldsymbol{\sigma}} + S\frac{\mathbf{n}:\overset{\circ}{\boldsymbol{\sigma}}}{M^p}\mathbf{n} \tag{17.22}$$

Tanaka and Kawamoto (1988) proposed the simple equation of S for the plane strain condition as follows:

$$S = (w \times l)/(l \times l) = w/\sqrt{F_e} \tag{17.23}$$

supposing simply the square finite element with the side-length l and the shear band having the thickness w, where $F_e(= l \times l)$ is the area of the finite element.

The plastic multiplier is expressed in terms of the apparent strain rate from Eq. (17.22) as follows:

$$\dot{\Lambda}\left(= S\frac{\mathbf{n}:\overset{\circ}{\boldsymbol{\sigma}}}{M^p}\right) = \frac{\mathbf{n}:\mathbf{E}:\bar{\mathbf{d}}}{\dfrac{M^p}{S} + \mathbf{n}:\mathbf{E}:\mathbf{n}} \tag{17.24}$$

Then, the stress rate is given by

$$\overset{\circ}{\boldsymbol{\sigma}} = \mathbf{E}:\bar{\mathbf{d}} - \frac{\mathbf{n}:\mathbf{E}:\bar{\mathbf{d}}}{\dfrac{M^p}{S} + \mathbf{n}:\mathbf{E}:\mathbf{n}}\mathbf{E}:\mathbf{n} \tag{17.25}$$

Then, we have

$$\mathbf{n}:\overset{\circ}{\boldsymbol{\sigma}} = \frac{M^p}{M^p + S\mathbf{n}:\mathbf{E}:\mathbf{n}}\mathbf{n}:\mathbf{E}:\bar{\mathbf{d}} \tag{17.26}$$

from which it is known that the stress reduction is larger for smaller value of S, i.e. smaller thickness of shear band, noting $M^p < 0$, $\mathbf{n}:\mathbf{E}:\bar{\mathbf{d}} > 0$ and $S < 1$. It is desirable to choose material parameters such that Eqs. (17.22) or (17.25) fits to a measured stress-strain curve, using the value of S predicted by a pertinent method, if we determine them from element test data.

17.4 Necessary Condition for Shear Band Inception

Discontinuity of the velocity gradient is induced at the shear band boundary. Here, incorporate the coordinate system in which the coordinate axes x_1^* and x_2^* are taken to be normal and parallel, respectively, to the shear band as shown in Fig. 17.2. The discontinuity of velocity gradient can be induced only in the x_1^*-direction. Therefore, only the following quantities are not zero, designating the discontinuous quantity by $\Delta(\,)$.

$$g_{1*}^1 \equiv \Delta\left(\frac{\partial v_1}{\partial x_1^*}\right), \quad g_{1*}^2 \equiv \Delta\left(\frac{\partial v_2}{\partial x_1^*}\right) \tag{17.27}$$

Then, the discontinuity of strain rate is given by

$$\begin{aligned}
\Delta d_{ij} &= \frac{1}{2}\left\{\Delta\left(\frac{\partial v_i}{\partial x_j}\right) + \Delta\left(\frac{\partial v_j}{\partial x_i}\right)\right\} = \frac{1}{2}\left\{\Delta\left(\frac{\partial v_i}{\partial x_r^*}\right)\frac{\partial x_r^*}{\partial x_j} + \Delta\left(\frac{\partial v_j}{\partial x_r^*}\right)\frac{\partial x_r^*}{\partial x_i}\right\} \\
&= \frac{1}{2}\left\{\Delta\left(\frac{\partial v_i}{\partial x_1^*}\right)\frac{\partial x_1^*}{\partial x_j} + \Delta\left(\frac{\partial v_j}{\partial x_1^*}\right)\frac{\partial x_1^*}{\partial x_i}\right\} = \frac{1}{2}\{g_{1*}^i(\mathbf{n}\bullet\mathbf{e}_j) + g_{1*}^j(\mathbf{n}\bullet\mathbf{e}_i)\} \\
&= \frac{1}{2}(g_{1*}^i n_j + g_{1*}^j n_i)
\end{aligned} \tag{17.28}$$

where $\mathbf{n}$ is the unit vector in the direction normal to the shear band, i.e. the x_1^*-direction.

On the other hand, the discontinuity in the rate of traction vector $\mathbf{t}_n$ applying to the discontinuity surface of velocity gradient, i.e. shear band having the direction vector $\mathbf{n}$ is described by

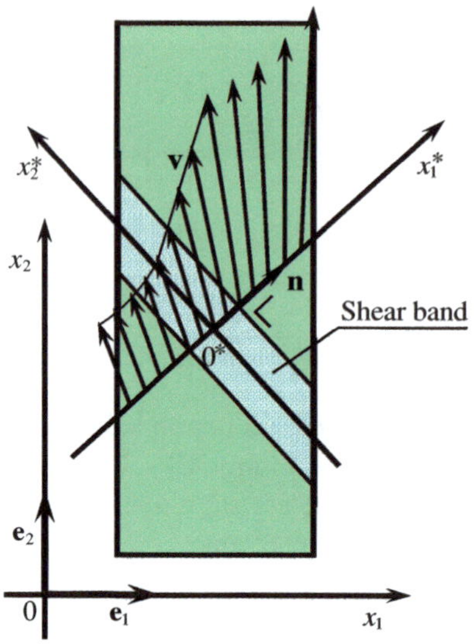

Fig. 17.2 Discontinuity of velocity gradient induced in the direction normal to the shear band

$$\left.\begin{aligned}\Delta\dot{t}_{n1} &= \Delta\sigma_{j1}n_j = C^{ep}_{j1kl}\Delta d_{kl}n_j = C^{ep}_{j1kl}\frac{1}{2}(g^k_{1*}n_l + g^l_{1*}n_k)n_j \\ \Delta\dot{t}_{n2} &= \Delta\sigma_{j2}n_j = C^{ep}_{j2kl}\Delta d_{kl}n_j = C^{ep}_{j2kl}\frac{1}{2}(g^k_{1*}n_l + g^l_{1*}n_k)n_j\end{aligned}\right\} \tag{17.29}$$

Noting

$$\left.\begin{aligned}C^{ep}_{j1kl}\frac{1}{2}(g^k_{1*}n_l + g^l_{1*}n_k)n_j &= C^{ep}_{i1kj}g^k_{1*}n_in_j \\ C^{ep}_{j2kl}\frac{1}{2}(g^k_{1*}n_l + g^l_{1*}n_k)n_j &= C^{ep}_{i2kj}g^k_{1*}n_in_j\end{aligned}\right\}$$

Equation (17.29) is expressed as

$$\begin{Bmatrix}\Delta\dot{t}_{n1} \\ \Delta\dot{t}_{n2}\end{Bmatrix} = \begin{bmatrix}C^{ep}_{i11j}n_in_j & C^{ep}_{i12j}n_in_j \\ C^{ep}_{i21j}n_in_j & C^{ep}_{i22j}n_in_j\end{bmatrix}\begin{Bmatrix}g^1_{1*} \\ g^2_{1*}\end{Bmatrix}$$

That is to say,

$$\begin{Bmatrix}\Delta\dot{t}_{n1} \\ \Delta\dot{t}_{n2}\end{Bmatrix} = \begin{bmatrix}A_{11} & A_{12} \\ A_{21} & A_{22}\end{bmatrix}\begin{Bmatrix}g^1_{1*} \\ g^2_{1*}\end{Bmatrix} \quad (\Delta\dot{t}_{ni} = A_{ij}g^j_{1*}, \quad \Delta\dot{\mathbf{t}}_n = \mathbf{A}\mathbf{g}_{1*}) \tag{17.30}$$

where A_{ij} is given by the following equation and called the *acoustic tensor*.

$$\mathbf{A} \equiv \mathbf{n}\mathbf{C}^{ep}\mathbf{n}, \quad A_{ij} \equiv C^{ep}_{rijs} n_r n_s \tag{17.31}$$

Here, noting that the traction rate vector must be continuous, i.e. $\Delta\mathbf{t}_n = \mathbf{0}$ by the equilibrium and thus it must hold from Eq. (17.30) that

$$\begin{bmatrix} A_{11} & A_{12} \\ A_{21} & A_{22} \end{bmatrix} \begin{Bmatrix} g^1_{1*} \\ g^2_{1*} \end{Bmatrix} = \begin{Bmatrix} 0 \\ 0 \end{Bmatrix} \quad (A_{ij} g^j_{1*} = 0, \quad \mathbf{A}\mathbf{g}_{1*} = \mathbf{0}) \tag{17.32}$$

In order that Eq. (17.32) has a solution other than the non-trivial solution $\mathbf{g}_{1*} = \mathbf{0}$, i.e. that the discontinuity of velocity gradient is induced, the following equation must hold, noting $\mathbf{g}_{1*} = \mathbf{A}^{-1}\mathbf{0}$ with Eq. (1.125).

$$\det \mathbf{A} = \begin{vmatrix} A_{11} & A_{12} \\ A_{21} & A_{22} \end{vmatrix} = 0 \tag{17.33}$$

At least one of eigenvalue of the acoustic tensor $\mathbf{A}$ is zero when (17.33) holds. The search for the occurrence of $\mathbf{n}$ fulfilling Eq. (17.33), i.e. the inception of the shear band and its direction, is called the *eigenvalue analysis*. Equation (17.33) is given explicitly as (Hashiguchi and Protasov 2004)

$$\begin{aligned}
\det(\mathbf{n}\mathbf{C}^{ep}\mathbf{n}) &= \det\left(C^{ep}_{1ij1} n_1 n_1 + C^{ep}_{1ij2} n_1 n_2 + C^{ep}_{2ij1} n_2 n_1 + C^{ep}_{2ij2} n_2 n_2\right) \\
&= \begin{vmatrix} C^{ep}_{1111} n_1 n_1 + C^{ep}_{1112} n_1 n_2 + C^{ep}_{2111} n_2 n_1 + C^{ep}_{2112} n_2 n_2 & C^{ep}_{1121} n_1 n_1 + C^{ep}_{1122} n_1 n_2 + C^{ep}_{2121} n_2 n_1 + C^{ep}_{2122} n_2 n_2 \\ C^{ep}_{1211} n_1 n_1 + C^{ep}_{1212} n_1 n_2 + C^{ep}_{2211} n_2 n_1 + C^{ep}_{2212} n_2 n_2 & C^{ep}_{1221} n_1 n_1 + C^{ep}_{1222} n_1 n_2 + C^{ep}_{2221} n_2 n_1 + C^{ep}_{2222} n_2 n_2 \end{vmatrix} \\
&= \begin{vmatrix} C^{ep}_{1111} n_1^2 + (C^{ep}_{1112} + C^{ep}_{2111}) n_1 n_2 + C^{ep}_{2112} n_2^2 & C^{ep}_{1121} n_1^2 + (C^{ep}_{1122} + C^{ep}_{2121}) n_1 n_2 + C^{ep}_{2122} n_2^2 \\ C^{ep}_{1211} n_1^2 + (C^{ep}_{1212} + C^{ep}_{2211}) n_1 n_2 + C^{ep}_{2212} n_2^2 & C^{ep}_{1221} n_1^2 + (C^{ep}_{1222} + C^{ep}_{2221}) n_1 n_2 + C^{ep}_{2222} n_2^2 \end{vmatrix} \\
&= \{C^{ep}_{1111} n_1^2 + (C^{ep}_{1112} + C^{ep}_{2111}) n_1 n_2 + C^{ep}_{2112} n_2^2\}\{C^{ep}_{1221} n_1^2 + (C^{ep}_{1222} + C^{ep}_{2221}) n_1 n_2 + C^{ep}_{2222} n_2^2\} \\
&\quad - \{C^{ep}_{1121} n_1^2 + (C^{ep}_{1122} + C^{ep}_{2121}) n_1 n_2 + C^{ep}_{2122} n_2^2\}\{C^{ep}_{1211} n_1^2 + (C^{ep}_{1212} + C^{ep}_{2211}) n_1 n_2 + C^{ep}_{2212} n_2^2\} \\
&= (C^{ep}_{1111} C^{ep}_{1221} - C^{ep}_{1121} C^{ep}_{1211}) n_1^4 \\
&\quad + (C^{ep}_{1111} C^{ep}_{1222} + C^{ep}_{1111} C^{ep}_{2221} + C^{ep}_{1112} C^{ep}_{1221} + C^{ep}_{2111} C^{ep}_{1221} \\
&\quad - C^{ep}_{1121} C^{ep}_{1212} - C^{ep}_{1121} C^{ep}_{2211} - C^{ep}_{1122} C^{ep}_{1211} - C^{ep}_{2121} C^{ep}_{1211}) n_1^3 n_2 \\
&\quad + (C^{ep}_{1111} C^{ep}_{2222} + C^{ep}_{1221} C^{ep}_{2112} - C^{ep}_{1121} C^{ep}_{1211} - C^{ep}_{1211} C^{ep}_{2122}) n_1^2 n_2^2 \\
&\quad + (C^{ep}_{1122} C^{ep}_{2222} + C^{ep}_{2111} C^{ep}_{2222} + C^{ep}_{1222} C^{ep}_{2112} + C^{ep}_{2221} C^{ep}_{2112} \\
&\quad - C^{ep}_{1122} C^{ep}_{2212} - C^{ep}_{2121} C^{ep}_{2212} - C^{ep}_{2122} C^{ep}_{1212} - C^{ep}_{2122} C^{ep}_{2211}) n_1 n_2^3 \\
&\quad + (C^{ep}_{2112} C^{ep}_{2222} - C^{ep}_{2122} C^{ep}_{2212}) n_2^4 \\
&= \left(C^{ep}_{1111} C^{ep}_{1212} - C^{ep2}_{1112}\right) n_1^4 \\
&\quad + (C^{ep}_{1111} C^{ep}_{1222} + C^{ep}_{1111} C^{ep}_{2212} - C^{ep}_{1112} C^{ep}_{1122} - C^{ep}_{1122} C^{ep}_{1211}) n_1^3 n_2 \\
&\quad + (C^{ep}_{1111} C^{ep}_{2222} + C^{ep}_{1212} C^{ep}_{1212} - C^{ep}_{1112} C^{ep}_{1211} - C^{ep}_{1211} C^{ep}_{1222}) n_1^2 n_2^2 \\
&\quad + (C^{ep}_{1122} C^{ep}_{2222} + C^{ep}_{12111} C^{ep}_{2222} - C^{ep}_{1122} C^{ep}_{1222} - C^{ep}_{1222} C^{ep}_{1122}) n_1 n_2^3 \\
&\quad + \left(C^{ep}_{1212} C^{ep}_{2222} - C^{ep2}_{1222}\right) n_2^4
\end{aligned}$$

which is reduced to

$$\det \mathbf{A} = a_1 n_1^4 + a_2 n_1^3 n_2 + a_3 n_1^2 n_2^2 + a_4 n_1 n_2^3 + a_5 n_2^4 = 0 \tag{17.34}$$

where

$$\left.\begin{aligned} a_1 &\equiv C^{ep}_{1111} C^{ep}_{1221} - C^{ep\,2}_{1121}, \\ a_2 &\equiv C^{ep}_{1111} C^{ep}_{1222} + C^{ep}_{1111} C^{ep}_{2221} - C^{ep}_{1121} C^{ep}_{2211} - C^{ep}_{1122} C^{ep}_{1211}, \\ a_3 &\equiv C^{ep}_{1111} C^{ep}_{2222} + C^{ep}_{1221} C^{ep}_{2112} - C^{ep}_{1121} C^{ep}_{1211} - C^{ep}_{1211} C^{ep}_{2122}, \\ a_4 &\equiv C^{ep}_{1122} C^{ep}_{2222} + C^{ep}_{2111} C^{ep}_{2222} - C^{ep}_{1122} C^{ep}_{2212} - C^{ep}_{2122} C^{ep}_{2211}, \\ a_5 &\equiv C^{ep}_{2112} C^{ep}_{2222} - C^{ep\,2}_{2122} \end{aligned}\right\} \tag{17.35}$$

Setting

$$n_1 = \cos\theta, \quad n_2 = \sin\theta \tag{17.36}$$

Equation (17.34) is rewritten as

$$g(\theta) = a_5 \tan^4\theta + a_4 \tan^3\theta + a_3 \tan^2\theta + a_2 \tan\theta + a_0 = 0 \tag{17.37}$$

which, noting the symmetry $g(\theta) = g(-\theta)$, leads to

$$g(\theta) = a_5 \tan^4\theta + a_3 \tan^2\theta + a_0 = 0 \tag{17.38}$$

There exists the possibility that a shear band occurs in the direction θ fulfilling Eq. (17.38). Here, note that Eq. (17.33) is the only necessary condition for the inception of the shear band.

We searched above the discontinuity of the velocity gradient in the direction normal to the shear band, while the traction rate vector must be continuous in that direction. Inversely, on the other hand, the search for the discontinuity in the normal stress rate component applied to the surface normal to the shear band, i.e. $\Delta\overset{\bullet}{\sigma}{}^*_{22} \neq 0$, while the discontinuity in the normal strain rate component in the direction parallel to the shear band must be zero, i.e. $\Delta\overset{\bullet}{\varepsilon}{}^*_{22} = 0$, is called the *compliance method* (cf. Mandel 1964). Here, note that the normal strain rate component in the direction normal to the shear band can be discontinuous, i.e. $\Delta\overset{\bullet}{\varepsilon}{}^*_{11} \neq 0$ in dilative materials. It was applied to the prediction of the direction of shear band formation in soils (Vermeer 1982) (Fig. 17.2).

References

Aifantis EC (1984) On the microstructural origin of certain inelastic models. J Eng Mater Tech (ASME) 106:326–330

Bazant ZP, Cedolin L (1991) Stability of structures: elastic, inelastic, fracture and damage theories. Oxford University Press, New York

Hashiguchi K, Protasov A (2004) Localized necking analysis by the subloading surface model with tangential-strain rate and anisotropy. Int J Plast 20:1909–1930

Hashiguchi K, Tsutsumi S (2006) Gradient plasticity with the tangential subloading surface model and the prediction of shear band thickness of granular materials. Int J Plast 22:767–797

Mandel J (1964) Contribution theorique a l'eude de l'ecrouissage et des lois de l'ecoulement plastique. In: Proceedings of the 11th international congress of applied mechanics, pp 502–509

Pietruszczak St, Mroz Z (1983) On hardening anisotropy of Ko-consolidated clays. Int J Numer Anal Meth Geomech 7:19–38

Tanaka T, Kawamoto O (1988) Three dimensional finite element collapse analysis for foundations and slopes using dynamic relaxation. In: Proceedings of the numerical methods in geomechanics, Innsbruck, pp 1213–1218

Vermeer PA (1982) A simple shear band analysis using compliances, In: Proceedings of the IUTAM symposium deformation and failure of granular materials, Balkema, Amsterdam, pp 493–499

Chapter 18
Constitutive Equation for Friction: Subloading-Friction Model

All bodies in the natural world are exposed to friction phenomena, contacting with other bodies, except for bodies floating in a vacuum. Therefore, it is indispensable to analyze friction phenomena rigorously in addition to the deformation behavior of bodies themselves in analyses of boundary value problems. The friction phenomenon can be formulated as a constitutive relation in a similar form to the elastoplastic constitutive equation of materials. A constitutive equation for friction with the transition from the static to the kinetic friction and vice versa and the orthotropic and rotational anisotropy is described in this chapter. The stick-slip phenomenon is also delineated, which is an unstable and intermittent motion caused by the friction and thus important for the prediction of earthquake, vibration of machinery, etc.

18.1 History of Constitutive Equation for Friction

Constitutive equations of friction within the framework of elastoplasticity were formulated as rigid-plasticity (Seguchi et al. 1974; Fredriksson 1976). Subsequently, they were extended to elasto-perfect-plasticity (Michalowski and Mroz 1978; Oden and Pires 1983a, b; Curnier 1984; Cheng and Kikuchi 1985; Kikuchi and Oden 1988; Wriggers et al. 1990; Peric and Owen 1992; Anand 1993; Zhong 1993; Mroz and Stupkiewicz 1994; Wriggers 2003). Further, isotropic hardening was introduced (Oden and Martines 1986; Gearing et al. 2001) to describe the test results (cf. e.g. Courtney-Pratt and Eisner 1957). However, the interior of the sliding-yield surface was assumed as an elastic domain. Therefore, the plastic sliding velocity induced by the rate of traction inside the sliding-yield surface cannot be described. Needless to say, the accumulation of plastic sliding displacement induced by the cyclic loading of contact traction within the sliding-yield surface cannot be described by these models. They might be called the *conventional friction model* in accordance with the classification of plastic

K. Hashiguchi, *Foundations of Elastoplasticity: Subloading Surface Model*,
DOI 10.1007/978-3-319-48821-9_18

constitutive models by Drucker (1988). Therefore, they are incapable of describing an accumulation of sliding displacement during cyclic loading of contact stress in addition to the incapability of describing the transition from the static to the kinetic friction and the recovery of the static friction.

It is widely known that when bodies at rest begin to slide, a high friction coefficient appears first, which is called the *static friction*. Subsequently, a friction coefficient decreases approaching a stationary value, called the *kinetic friction*. Furthermore, if the sliding ceases for a while and then starts again, the friction coefficient recovers and similar behavior to that of the initial sliding is reproduced (Dokos 1946; Rabinowicz 1951, 1958; Howe et al. 1955; Derjaguin et al. 1957; Brockley and Davis 1968; Kato et al. 1972; Horowitz and Ruina 1989; Ferrero and Barrau 1997; Bureau et al. 2001). The recovery of friction coefficient has been formulated using equations including a time elapsed after the stop of sliding (cf. Rabinowicz 1951; Howe et al. 1955; Brockley and Davis 1968; Kato et al. 1972; Horowitz and Ruina 1989; Bureau et al. 2001). However, the inclusion of time itself leads to the loss of objectivity in constitutive equations, as is evident from the fact that the evaluation of elapsed time is accompanied with the ambiguity in the judgment about when the sliding commences and ceases, especially in the state that the sliding velocity varies in a low-velocity regime. On the other hand, generally speaking, the variation of material property has to be described in terms of the sliding velocity, the contact traction and internal variables without the inclusion of time itself.

The reduction of the friction coefficient from the static to kinetic friction and the recovery of the friction coefficient as described above are the fundamental characteristics in friction phenomena, which have been widely recognized for a long time. Difference of the static and kinetic frictions often reaches up to several tens of percent, and thus the formulation taken account of these characteristics is of importance for analyses of practical problems in engineering. The constitutive equation describing these fundamental friction behavior has been formulated based on the subloading surface model and thus it is called the *subloading-friction model* (Hashiguchi et al. 2005b; Hashiguchi and Ozaki 2008a).

In addition, the difference of friction coefficients is observed in opposite sliding directions. It can be described by the rotation of a sliding-yield surface in the traction vector space, noting that the anisotropy of soils has been described by the rotation of a yield surface, as described in Sect. 11.4. Further, the range of friction coefficient depends on the sliding direction. It would be describable by the concept of orthotropy of the sliding-yield surface (Mroz and Stupkiewicz 1994). The subloading-friction model has been extended so as to describe these anisotropy (Hashiguchi 2006; Hashiguchi and Ozaki 2008b; Ozaki et al. 2012).

The *rate-and-state friction model* (cf. e.g. Dieterich 1978, 1979; Ruina 1980, 1983; Rice and Ruina 1983; Scholz 1998; Rice et al. 2001; Kame 2013) the basic idea of which is the dependence of the contact shear stress or friction coefficient on the rate of sliding and some state variables based on experimental data (cf. Dieterich 1979; Ruina 1980) has been widely used for the prediction of earthquake phenomena. An earthquake is a typical irreversible phenomenon which can be described appropriately by elasoplasticity but the rate-and-state friction model is not based on elastoplasticity which is premised on (1) decomposition of the rate of deformation or sliding into the reversible, i.e. elastic part and the irreversible, i.e. plastic part, (2) incorporation of the yield condition and (3) the potential flow rule of plastic strain rate or plastic sliding rate. Therefore, the rate-and-state friction model possesses various difficulties as follows:

1. It is limited to the description of one-dimensional sliding behavior and inapplicable to two-dimensional sliding behavior in which the sliding direction changes, because whether the rate of plastic sliding is induced cannot be judged rigorously as far as the yield surface is not incorporated.
2. The hardening/softening of friction coefficient cannot be described appropriately so that the transition from the static to the kinetic friction and the recovery of the static friction cannot be described appropriately.

The rate-and-state model would be easily understood and felt familiar by the workers who are unfamiliar to the elastoplasticity. It has been limited almost to the prediction of earthquakes but hardly applied to the description of the other general friction behavior of solids. It is just an empirical law or an *ad hoc* model. The earthquake disaster prevention cannot be attained forever as far as the rate-and-state model is used for the prediction of earthquakes.

It should be emphasized that the irreversible deformation phenomena can be formulated rigorously within the framework of the elastoplasticity which has been developed to describe the irreversible deformation of solids in the history of the applied mechanics. Consequently, the subloading-friction model is formulated within the framework of the elastoplasticty in this article aiming at establishing the rate-independent and -dependent constitutive equation of friction with the generality and the universality in the three-dimensional sliding under monotonic and cyclic frictional loadings.

18.2 Decomposition of Sliding Velocity and Contract Traction

The sliding velocity vector $\bar{\mathbf{v}}$, which is defined as the relative velocity of the counter (slave) body to the main (master) body, is orthogonally decomposed into the normal sliding velocity vector $\bar{\mathbf{v}}_n$ and the tangential sliding velocity vector $\bar{\mathbf{v}}_t$ relative to the contact surface as follows (Fig. 18.1):

$$\bar{\mathbf{v}} = \bar{\mathbf{v}}_n + \bar{\mathbf{v}}_t = -\bar{v}_n\mathbf{n} + \bar{v}_t\mathbf{t}_v \tag{18.1}$$

where

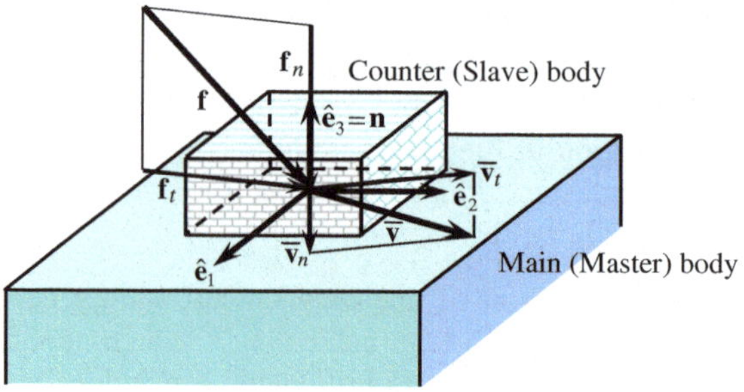

Fig. 18.1 Contact traction and sliding velocity

$$\left.\begin{array}{l}\bar{\mathbf{v}}_n = (\bar{\mathbf{v}}\cdot\mathbf{n})\mathbf{n} = (\mathbf{n}\otimes\mathbf{n})\bar{\mathbf{v}} = -\bar{v}_n\mathbf{n}\\ \bar{\mathbf{v}}_t = \bar{\mathbf{v}} - \bar{\mathbf{v}}_n = (\mathbf{I}-\mathbf{n}\otimes\mathbf{n})\bar{\mathbf{v}} = \bar{v}_t\mathbf{t}_v\end{array}\right\} \tag{18.2}$$

whilst $\mathbf{n}$ is the unit outward-normal vector of the main body and

$$\left.\begin{array}{l}\bar{v}_n \equiv -\mathbf{n}\cdot\bar{\mathbf{v}}_n = -\mathbf{n}\cdot\bar{\mathbf{v}}\\ \bar{v}_t = \|\bar{\mathbf{v}}_t\|,\ \mathbf{t}_v \equiv \dfrac{\bar{\mathbf{v}}_t}{\|\bar{\mathbf{v}}_t\|}(\mathbf{n}\cdot\mathbf{t}_v = 0, \|\mathbf{t}_v\| = 1)\end{array}\right\} \tag{18.3}$$

A minus sign is added to the definition of $\bar{v}_n$ so that friction increases as the counter (slave) body approaches the main (master) body. The velocity of a material particle depends on the observer's reference frame and it is thus not an objective quantity. Nevertheless, the observed velocities of different material particles at the contact point are identically affected by the relative velocity of the observer because the velocities coincide instantaneously at the contact point. Consequently, the sliding velocity is an objective quantity; see Fig. 18.2.

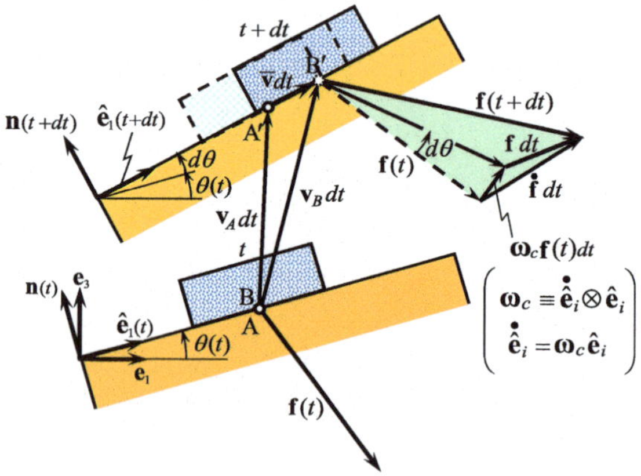

Fig. 18.2 Objectivities of the sliding velocity vector $\bar{\mathbf{v}}$ and the corotational traction rate vector $\overset{\circ}{\mathbf{f}}$ illustrated in two-dimensional space under the constant normal traction, i.e. $\overset{\circ}{\mathbf{f}}_n = \mathbf{0}$ leading to $\overset{\circ}{\mathbf{f}} = \overset{\circ}{\mathbf{f}}_t$ ($\mathbf{e}_1$, $\mathbf{e}_3$: fixed base vectors)

Here, it is further assumed that $\bar{\mathbf{v}}$ is additively decomposed into elastic sliding velocity $\bar{\mathbf{v}}^e$ and plastic sliding velocity $\bar{\mathbf{v}}^p$, i.e.

$$\bar{\mathbf{v}} = \bar{\mathbf{v}}^e + \bar{\mathbf{v}}^p \tag{18.4}$$

The elastic sliding velocity is infinitesimal because it is induced by the elastic deformations of asperities on the contact surface; it is uniquely related to the rate of contact stress. On the other hand, the plastic sliding velocity is induced by the mutual slips between surfaces of the main and the counter bodies and thus it is finite usually. Then, the direction of plastic sliding velocity is given by the plastic potential function of contact stress and thus it depends on the contact stress but is independent of the rate of contact stress as it will be formulated in Sect. 18.4.4. Hence, the rate of contact stress is not determined uniquely by the sliding velocity, making it difficult to carry out the analysis, if the elastic sliding velocity is not given. Therefore, we incorporate the elastic sliding velocity analogously to the elastic strain rate in ordinary elastoplastic constitutive equations.

Furthermore, $\bar{\mathbf{v}}_n$ and $\bar{\mathbf{v}}_t$ are expressed by the elastic and the plastic parts as follows:

$$\left.\begin{aligned} \bar{\mathbf{v}}_n &= \bar{\mathbf{v}}_n^e + \bar{\mathbf{v}}_n^p \\ \bar{\mathbf{v}}_t &= \bar{\mathbf{v}}_t^e + \bar{\mathbf{v}}_t^p \end{aligned}\right\} \tag{18.5}$$

and thus

$$\left.\begin{aligned} \bar{\mathbf{v}}^e &= \bar{\mathbf{v}}_n^e + \bar{\mathbf{v}}_t^e = -\bar{v}_n^e \mathbf{n} + \bar{v}_t^e \mathbf{t}_v^e \\ \bar{\mathbf{v}}^p &= \bar{\mathbf{v}}_n^p + \bar{\mathbf{v}}_t^p = -\bar{v}_n^p \mathbf{n} + \bar{v}_t^p \mathbf{t}_v^p \end{aligned}\right\} \tag{18.6}$$

where

$$\left.\begin{aligned} \bar{\mathbf{v}}_n^e &= (\bar{\mathbf{v}}^e \cdot \mathbf{n})\mathbf{n} = (\mathbf{n} \otimes \mathbf{n})\bar{\mathbf{v}}^e = -\bar{v}_n^e \mathbf{n} \\ \bar{\mathbf{v}}_t^e &= \bar{\mathbf{v}}^e - \bar{\mathbf{v}}_n^e = (\mathbf{I} - \mathbf{n} \otimes \mathbf{n})\bar{\mathbf{v}}^e = \bar{v}_t^e \mathbf{t}_v^e \end{aligned}\right\} \tag{18.7}$$

$$\left.\begin{aligned} \bar{\mathbf{v}}_n^p &= (\bar{\mathbf{v}}^p \cdot \mathbf{n})\mathbf{n} = (\mathbf{n} \otimes \mathbf{n})\bar{\mathbf{v}}^p = -\bar{v}_n^p \mathbf{n} \\ \bar{\mathbf{v}}_t^p &= \bar{\mathbf{v}}^p - \bar{\mathbf{v}}_n^p = (\mathbf{I} - \mathbf{n} \otimes \mathbf{n})\bar{\mathbf{v}}^p = \bar{v}_t^p \mathbf{t}_v^p \end{aligned}\right\} \tag{18.8}$$

setting

$$\left.\begin{aligned} &\bar{v}_n^e \equiv -\mathbf{n} \cdot \bar{\mathbf{v}}_n^e = -\mathbf{n} \cdot \bar{\mathbf{v}}^e \\ &\bar{v}_t^e = \|\bar{\mathbf{v}}_t^e\|,\ \mathbf{t}_v^e \equiv \frac{\bar{\mathbf{v}}_t^e}{\|\bar{\mathbf{v}}_t^e\|} (\mathbf{n} \cdot \mathbf{t}_v^e = 0,\ \|\mathbf{t}_v^e\| = 1) \end{aligned}\right\} \tag{18.9}$$

$$\left.\begin{array}{l} \bar{v}_n^p \equiv -\mathbf{n}\cdot\bar{\mathbf{v}}_n^p = -\mathbf{n}\cdot\bar{\mathbf{v}}^p \\ \bar{v}_t^p = \|\bar{\mathbf{v}}_t^p\|,\ \mathbf{t}_v^p \equiv \dfrac{\bar{\mathbf{v}}_t^p}{\|\bar{\mathbf{v}}_t^p\|}\ (\mathbf{n}\cdot\mathbf{t}_v^p = 0,\ \|\mathbf{t}_v^p\| = 1) \end{array}\right\} \quad (18.10)$$

The *contact traction vector* **f** acting on the body is expressed by the *normal traction vector* $\mathbf{f}_n$ and the *tangential traction vector* $\mathbf{f}_t$ as follows:

$$\mathbf{f} = \mathbf{f}_n + \mathbf{f}_t = -f_n\mathbf{n} + f_t\mathbf{t}_f \quad (18.11)$$

where

$$\left.\begin{array}{l} \mathbf{f}_n \equiv (\mathbf{n}\cdot\mathbf{f})\mathbf{n} = (\mathbf{n}\otimes\mathbf{n})\mathbf{f} = -f_n\mathbf{n} \\ \mathbf{f}_t \equiv \mathbf{f} - \mathbf{f}_n = (\mathbf{I} - \mathbf{n}\otimes\mathbf{n})\mathbf{f} = f_t\mathbf{t}_f \end{array}\right\} \quad (18.12)$$

$$\left.\begin{array}{l} f_n \equiv -\mathbf{n}\cdot\mathbf{f} \\ f_t = \mathbf{t}_f\cdot\mathbf{f} = \|\mathbf{f}_t\|,\ \mathbf{t}_f \equiv \dfrac{\mathbf{f}_t}{\|\mathbf{f}_t\|}(\mathbf{n}\cdot\mathbf{t}_f = 0,\ \|\mathbf{t}_f\| = 1) \end{array}\right\} \quad (18.13)$$

The minus sign is added for f_n to be positive when the compression is applied to the main body by the counter body. The contact traction vector **f**, the normal traction vector $\mathbf{f}_n$ and the tangential traction vector $\mathbf{f}_t$ are calculated from the Cauchy stress applied to the contact surface by the Cauchy's fundamental theorem in Eq. (3.2) as follows:

$$\left.\begin{array}{l} \mathbf{f} = \boldsymbol{\sigma}\mathbf{n} \\ \mathbf{f}_n = (\mathbf{n}\cdot\boldsymbol{\sigma}\mathbf{n})\mathbf{n} = (\mathbf{n}\otimes\mathbf{n})\boldsymbol{\sigma}\mathbf{n} \\ \mathbf{f}_t = (\mathbf{I} - \mathbf{n}\otimes\mathbf{n})\boldsymbol{\sigma}\mathbf{n} \end{array}\right\} \quad (18.14)$$

where $\boldsymbol{\sigma}$ is the stress tensor applied to the main body.

Here, note that the directions of the tangential contact traction and the tangential sliding velocity are different, i.e. $\mathbf{t}_v \neq \mathbf{t}_f$ in general, except for the isotropic sliding behavior in a sliding yield surface having a circular section.

The following equalities hold for the directions of the normal parts from the first equations in Eqs. (18.2), (18.7), (18.8) and (18.12).

$$\mathbf{n} = \frac{\bar{\mathbf{v}}_n}{\|\bar{\mathbf{v}}_n\|} = \frac{\bar{\mathbf{v}}_n^e}{\|\bar{\mathbf{v}}_n^e\|} = \frac{\bar{\mathbf{v}}_n^p}{\|\bar{\mathbf{v}}_n^p\|} = \frac{\mathbf{f}_n}{\|\mathbf{f}_n\|} \quad (18.15)$$

The corotational rate $\mathring{\mathbf{f}}$ based on Eq. (4.54) is decomposed to the normal and the tangential parts, i.e.

$$\mathring{\mathbf{f}} = \mathring{\mathbf{f}}_n + \mathring{\mathbf{f}}_t \quad (18.16)$$

where

$$\overset{\circ}{\mathbf{f}} = \overset{\bullet}{\mathbf{f}} - \boldsymbol{\omega}_c\mathbf{f}, \quad \overset{\circ}{\mathbf{f}}_n = \overset{\bullet}{\mathbf{f}}_n - \boldsymbol{\omega}_c\mathbf{f}_n, \quad \overset{\circ}{\mathbf{f}}_t = \overset{\bullet}{\mathbf{f}}_t - \boldsymbol{\omega}_c\mathbf{f}_t \tag{18.17}$$

noting

$$\overset{\circ}{\mathbf{f}} = (\mathbf{f}_n + \mathbf{f}_t)^{\bullet} - \boldsymbol{\omega}_c(\mathbf{f}_n + \mathbf{f}_t) = \overset{\bullet}{\mathbf{f}}_n - \boldsymbol{\omega}_c\mathbf{f}_n + \overset{\bullet}{\mathbf{f}}_t - \boldsymbol{\omega}_c\mathbf{f}_t \tag{18.18}$$

The second-order tensor $\boldsymbol{\omega}_c$ is the spin of the contact surface, which is given by

$$\boldsymbol{\omega}_c \equiv \overset{\bullet}{\hat{\mathbf{e}}}_i \otimes \hat{\mathbf{e}}_i \; (\overset{\bullet}{\hat{\mathbf{e}}}_i = \boldsymbol{\omega}_c \hat{\mathbf{e}}_i) \tag{18.19}$$

where $(\hat{\mathbf{e}}_1, \hat{\mathbf{e}}_2, \hat{\mathbf{e}}_3) = (\hat{\mathbf{e}}_1, \hat{\mathbf{e}}_2, \mathbf{n})$ is the base vectors fixed to the contact surface, while $\hat{\mathbf{e}}_3$ is taken to be $\mathbf{n}$ (see Fig. 18.2). $\boldsymbol{\omega}_c$ obeys the following coordinate transformation between the observation bases $\{\mathbf{e}_i\}$ and $\{\mathbf{e}_i^*\}$ with the relation $\mathbf{e}_i^* = \mathbf{Q}^T\mathbf{e}_i$ as shown already in Eq. (4.17).

$$\boldsymbol{\omega}_c^* = \mathbf{Q}(\boldsymbol{\omega}_c - \boldsymbol{\Omega})\mathbf{Q}^T \tag{18.20}$$

noting

$$\begin{aligned}
\overset{\bullet}{\hat{\mathbf{e}}}{}_i^* \otimes \hat{\mathbf{e}}_i^* &= (\mathbf{Q}\hat{\mathbf{e}}_i)^{\bullet} \otimes \mathbf{Q}\hat{\mathbf{e}}_i = (\dot{\mathbf{Q}}\,\hat{\mathbf{e}}_i + \mathbf{Q}\overset{\bullet}{\hat{\mathbf{e}}}_i) \otimes \mathbf{Q}\hat{\mathbf{e}}_i = \mathbf{Q}(\overset{\bullet}{\hat{\mathbf{e}}}_i \otimes \hat{\mathbf{e}}_i + \mathbf{Q}^T\dot{\mathbf{Q}}\,\hat{\mathbf{e}}_i \otimes \hat{\mathbf{e}}_i)\mathbf{Q}^T \\
&= \mathbf{Q}(\overset{\bullet}{\hat{\mathbf{e}}}_i \otimes \hat{\mathbf{e}}_i - \dot{\mathbf{Q}}^T\mathbf{Q}\hat{\mathbf{e}}_i \otimes \hat{\mathbf{e}}_i)\mathbf{Q}^T = \mathbf{Q}(\overset{\bullet}{\hat{\mathbf{e}}}_i \otimes \hat{\mathbf{e}}_i - \boldsymbol{\Omega})\mathbf{Q}^T = \mathbf{Q}(\boldsymbol{\omega}_c - \boldsymbol{\Omega})\mathbf{Q}^T
\end{aligned}$$

where $\boldsymbol{\Omega}$ is defined in Eq. (4.11). It follows exploiting Eq. (18.20) that

$$\overset{\circ}{\mathbf{f}}{}^* = \mathbf{Q}\overset{\circ}{\mathbf{f}}\mathbf{Q}^T \tag{18.21}$$

noting

$$\begin{aligned}
\overset{\bullet}{\mathbf{f}}{}^* - \boldsymbol{\omega}_c^*\mathbf{f}^* &= (\mathbf{Q}\mathbf{f})^{\bullet} - \mathbf{Q}(\boldsymbol{\omega}_c - \boldsymbol{\Omega})\mathbf{Q}^T\mathbf{Q}\mathbf{f} = \dot{\mathbf{Q}}\,\mathbf{f} + \mathbf{Q}\,\overset{\bullet}{\mathbf{f}} - \mathbf{Q}(\boldsymbol{\omega}_c - \boldsymbol{\Omega})\mathbf{Q}^T\mathbf{Q}\mathbf{f} \\
&= \mathbf{Q}(\overset{\bullet}{\mathbf{f}} - \boldsymbol{\omega}_c\mathbf{f})\mathbf{Q}^T
\end{aligned}$$

Therefore, the corotational contact stress rate $\overset{\circ}{\mathbf{f}}$ obeys the objective transformation.

18.3 Elastic Sliding Velocity

Assuming that solids is not separated under moderate deformations, a hyperelastic constitutive relation is adopted for the elastic part of deformation. In contrast, asperities of solid surfaces adhere to different counterparts one after another in the friction phenomenon. Recalling the hypoelastic relation in terms of the Cauchy stress and the elastic strain rate (the symmetric part of the elastic velocity gradient) for the elastic deformation of a solid (Truesdell 1955; Truesdell and Noll 1955), let the elastic sliding velocity be formulated as

$$\left.\begin{aligned} \bar{\mathbf{v}}^e &= \mathbf{C}^{e-1}\mathring{\mathbf{f}} \\ \mathring{\mathbf{f}} &= \mathbf{C}^e\bar{\mathbf{v}}^e \end{aligned}\right\} \tag{18.22}$$

where the second-order tensor $\mathbf{C}^e$ is the contact elastic modulus tensor given by

$$\left.\begin{aligned} \mathbf{C}^e &= \alpha_t(\mathbf{I}-\mathbf{n}\otimes\mathbf{n})+\alpha_n\mathbf{n}\otimes\mathbf{n} \\ \mathbf{C}^{e-1} &= \frac{1}{\alpha_t}(\mathbf{I}-\mathbf{n}\otimes\mathbf{n})+\frac{1}{\alpha_n}\mathbf{n}\otimes\mathbf{n} \end{aligned}\right\} \tag{18.23}$$

where α_n and α_t are the normal and tangential *contact elastic moduli*, respectively, of the contact surface. Their values are set large because the elastic sliding is caused by elastic deformations of the surface asperities. Actually, we may choose α_n, $\alpha_t = 10^2 - 10^5\,\text{GPa/mm}^3$ currently as the penalty parameters. Equation (18.22) with Eq. (18.23) leads to

$$\left.\begin{aligned} \bar{\mathbf{v}}^e &= \frac{1}{\alpha_n}\mathring{\mathbf{f}}_n+\frac{1}{\alpha_t}\mathring{\mathbf{f}}_t \\ \mathring{\mathbf{f}} &= \alpha_n\bar{\mathbf{v}}^e_n+\alpha_t\bar{\mathbf{v}}^e_t \end{aligned}\right\} \tag{18.24}$$

which are described by the following matrix form, choosing the rectangular coordinate system $(\hat{\mathbf{e}}_1, \hat{\mathbf{e}}_2, \hat{\mathbf{e}}_3) = (\hat{\mathbf{e}}_1, \hat{\mathbf{e}}_2, \mathbf{n})$ fixed to the contact surface as shown in Figs. 18.1 and 18.3.

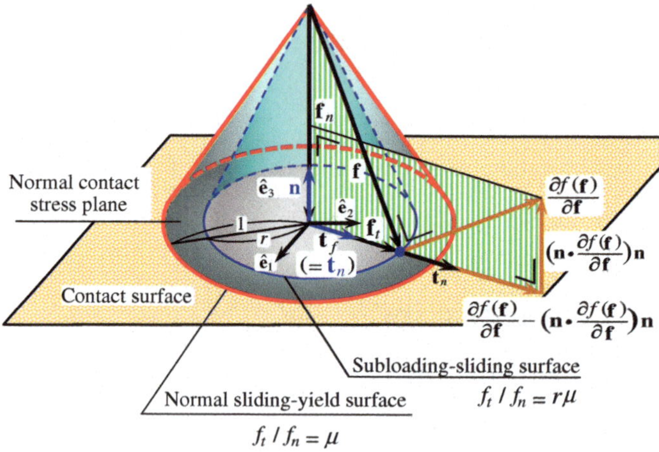

Fig. 18.3 Coulomb-type normal- and subloading-sliding surfaces

$$\begin{Bmatrix} \bar{v}_1^e \\ \bar{v}_2^e \\ \bar{v}_n^e \end{Bmatrix} = \begin{bmatrix} 1/\alpha_t & 0 & 0 \\ 0 & 1/\alpha_t & 0 \\ 0 & 0 & 1/\alpha_n \end{bmatrix} \begin{Bmatrix} \overset{\circ}{f}_1 \\ \overset{\circ}{f}_2 \\ \overset{\circ}{f}_n \end{Bmatrix}, \quad \begin{Bmatrix} \overset{\circ}{f}_1 \\ \overset{\circ}{f}_2 \\ \overset{\circ}{f}_n \end{Bmatrix} = \begin{bmatrix} \alpha_t & 0 & 0 \\ 0 & \alpha_t & 0 \\ 0 & 0 & \alpha_n \end{bmatrix} \begin{Bmatrix} \bar{v}_1^e \\ \bar{v}_2^e \\ \bar{v}_n^e \end{Bmatrix}. \tag{18.25}$$

noting

$$\begin{aligned} C_{11}^e &= \hat{\mathbf{e}}_1 \cdot (\mathbf{C}^e \hat{\mathbf{e}}_1) = \hat{\mathbf{e}}_1 \cdot \{[\alpha_n \mathbf{n} \otimes \mathbf{n} + \alpha_t (\mathbf{I} - \mathbf{n} \otimes \mathbf{n})] \hat{\mathbf{e}}_1\} = \alpha_t \\ C_{33}^e &= \mathbf{n} \cdot (\mathbf{C}^e \mathbf{n}) = \mathbf{n} \cdot \{[\alpha_n \mathbf{n} \otimes \mathbf{n} + \alpha_t (\mathbf{I} - \mathbf{n} \otimes \mathbf{n})] \mathbf{n}\} = \alpha_n \\ C_{12}^e &= \hat{\mathbf{e}}_1 \cdot (\mathbf{C}^e \hat{\mathbf{e}}_2) = \hat{\mathbf{e}}_1 \cdot \{[\alpha_n \mathbf{n} \otimes \mathbf{n} + \alpha_t (\mathbf{I} - \mathbf{n} \otimes \mathbf{n})] \hat{\mathbf{e}}_2\} = 0 \end{aligned}$$

The elastic constitutive relation is described by the simple form in Eq. (18.25) by using the coordinate system $\{\hat{\mathbf{e}}_i\}$ fixed to the contact surface.

Practically, we may assume that $\alpha \equiv \alpha_n = \alpha_t$ for the sake of simplicity, leading to

$$\mathbf{C}^e = \alpha \mathbf{I}, \quad \mathbf{C}^{e-1} = \frac{1}{\alpha} \mathbf{I} \tag{18.26}$$

since the elastic sliding displacement is quite small in general.

In the linear sliding on the fixed contact plane fulfilling $\boldsymbol{\omega} = \mathbf{O}$, the time-integration of Eq. (18.22) leads to

$$\left. \begin{aligned} \bar{\mathbf{u}}^e &= \mathbf{C}^{e-1} \mathbf{f} \\ \mathbf{f} &= \mathbf{C}^e \bar{\mathbf{u}}^e \end{aligned} \right\} \tag{18.27}$$

where $\bar{\mathbf{u}}^e = \int \bar{\mathbf{v}}^e dt$. Equation (18.27) will be used for the return-mapping projection in Chap. 20.

18.4 Elastoplastic Sliding Velocity

The plastic sliding velocity is formulated based on the subloading concept and the plastic potential theory in this section.

18.4.1 Normal Sliding-Yield and Sliding-Subloading Surfaces

Assume the following sliding-yield surface with the isotropic hardening/softening, which describes the sliding-yield condition.

$$f(\mathbf{f}) = \mu \tag{18.28}$$

μ is the isotropic hardening/softening function denoting the variation of the size of the sliding-yield surface. The friction-yield stress function $f(\mathbf{f})$ for the Coulomb friction law is given by

$$f(\mathbf{f}) = \frac{f_t}{f_n} \tag{18.29}$$

for which μ specifies the coefficient of friction. Equation (18.29) is adopted for actual calculations described in later sections.

The abrupt transition from the elastic to the plastic sliding state is described if the interior of the friction-yield surface is assumed to be a purely elastic domain. However, the plastic sliding velocity is induced even by the rate of contact stress inside the friction-yield surface and it develops gradually as contact stress approaches the friction-yield surface, thereby exhibiting the smooth elastic–plastic transition. In order to describe the plastic sliding velocity induced by the rate of contact stress inside the sliding-yield surface, let the following postulate be incorporated based on the concept of the subloading surface described for the elastoplastic deformation in Chap. 7.

Fundamental postulate of elastoplastic sliding: *The plastic sliding-velocity is induced when the contact stress approaches the sliding-yield surface but only the elastic sliding-velocity is induced when the contact stress parts from the sliding-yield surface*. In other words, the contact stress approaches the sliding-yield surface when a plastic sliding-velocity is induced but it parts from the sliding-yield surface when only an elastic sliding-velocity is induced.

Then, in order to introduce the measure of approaching degree to the sliding-yield surface, renamed the *normal sliding-yield surface*, let the following *subloading-sliding surface* passing through the current contact stress and maintaining a similarity to the normal sliding-yield surface be introduced, which plays the general measure of approaching degree of the contact stress to the normal sliding-yield surface.

$$\boxed{f(\mathbf{f}) = r\mu} \tag{18.30}$$

where $r(0 \leq r \leq 1)$ is the ratio of the size of the subloading surface to that of the normal-yield surface and called the *normal sliding-yield ratio*, playing the role of the measure of the approaching degree to the normal-yield surface. The normal friction-yield ratio r for the Coulomb friction law in Eq. (18.29) is given by the ratio of the size of the cross section of the subloading friction surface to that of the normal friction-yield surface on the constant normal contact stress plane as shown in Fig. 18.3.

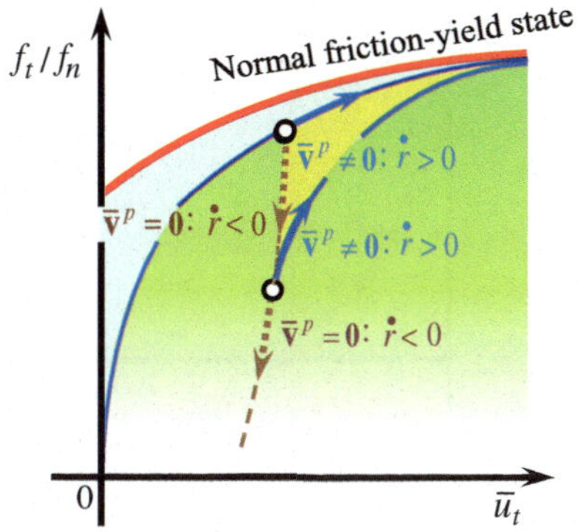

Fig. 18.4 Plastic sliding velocity based on the subloading surface concept

18.4.2 *Evolution Rule of Normal Sliding-Yield Ratio*

On the basis of the above-mentioned fundamental postulate of elastoplastic sliding, the rate of the normal friction-yield ratio r must satisfy the following conditions (see Fig. 18.4):

$$\dot{r}\begin{cases} \rightarrow +\infty & \text{for } r=0\text{: quasi-elastic sliding state} \\ >0 & \text{for } 0<r<1\text{: sub-sliding yield state} \\ =0 & \text{for } r=1\text{: normal-sliding yield state} \\ (<0 & \text{for } r>1\text{: over normal-sliding yield state)} \end{cases} \quad \text{for } \bar{\mathbf{v}}^p \neq \mathbf{0}, \tag{18.31}$$

$$\dot{r}\begin{cases} =0 & \text{for } \bar{\mathbf{v}}^e = \mathbf{0} \\ <0 & \text{for } \bar{\mathbf{v}}^e \neq \mathbf{0} \end{cases} \quad \text{for } \bar{\mathbf{v}}^p = \mathbf{0}. \tag{18.32}$$

Here, the rate of the normal sliding-yield ratio evolves with the plastic sliding-velocity, obeying Eq. (18.31) but it is calculated by substituting a contact stress changing obeying the elastic sliding-constitutive relation with fixed internal hardening variable μ into Eq. (18.30) when only the elastic sliding-velocity is induced. Then, it follows that

$$\boxed{\dot{r} = \overline{U}(r)\|\bar{\mathbf{v}}^p\| \text{for } \bar{\mathbf{v}}^p \neq \mathbf{0}} \tag{18.33}$$

$$r = \frac{f(\mathbf{f})}{\mu} \text{ for } \bar{\mathbf{v}}^p = \mathbf{0} \tag{18.34}$$

where $U(r)$ is the monotonically-increasing function of r fulfilling the conditions (see Fig. 18.5).

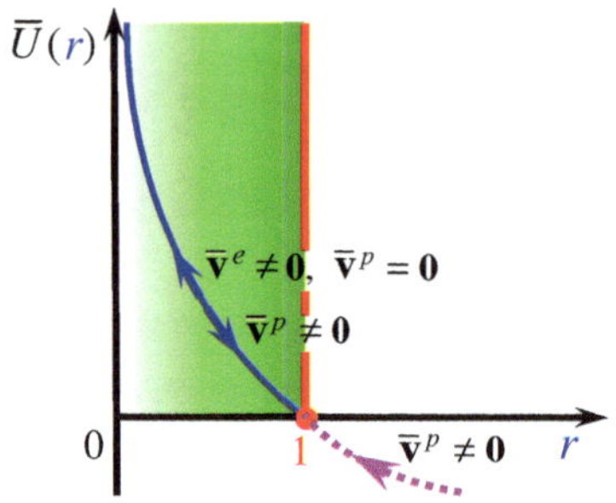

Fig. 18.5 Function $\bar{U}(r)$ for the evolution rule

$$\overline{U}(r)\begin{cases} \to +\infty \ \text{for } r = 0 (\text{quasi-elastic sliding state}) \\ > 0 \ \text{for } 0 < r < 1 (\text{sub-sliding yield state}) \\ = 0 \ \text{for } r = 1 (\text{normal-sliding yield state}) \\ (< 0 \ \text{for } r > 1 : \text{over normal-sliding yield state}) \end{cases} \tag{18.35}$$

An explicit example for $\bar{U}(r)$ is

$$\boxed{\overline{U}(r) = \tilde{u}\cot\left(\frac{\pi}{2}r\right)}, \tag{18.36}$$

where $\tilde{u}$ is a material constant. Equation (18.33) with Eq. (18.36) can be analytically integrated in the case of a monotonic sliding process as

$$\left.\begin{aligned} r &= \frac{2}{\pi}\cos^{-1}\{\cos\left(\frac{\pi}{2}r_0\right)\exp[-\frac{\pi}{2}\tilde{u}(\bar{u}^p - \bar{u}_0^p)]\} \\ \bar{u}^p - \bar{u}_0^p &= \frac{2}{\pi}\frac{1}{\tilde{u}}\ln\frac{\cos\left(\frac{\pi}{2}r_0\right)}{\cos\left(\frac{\pi}{2}r\right)} \end{aligned}\right\}, \tag{18.37}$$

where $\bar{u}_t(= \int \bar{v}_t dt)$ is the sliding displacement, and r_0 and $\bar{u}_0^p$ are the initial values of r and $\bar{u}^p$, respectively. The analytical integration would be beneficial in numerical calculations (Hashiguchi 2013b). The general trend of the effect of $\tilde{u}$ on $\overline{U}(r)$ is shown schematically in Figs. 18.6 and 18.7.

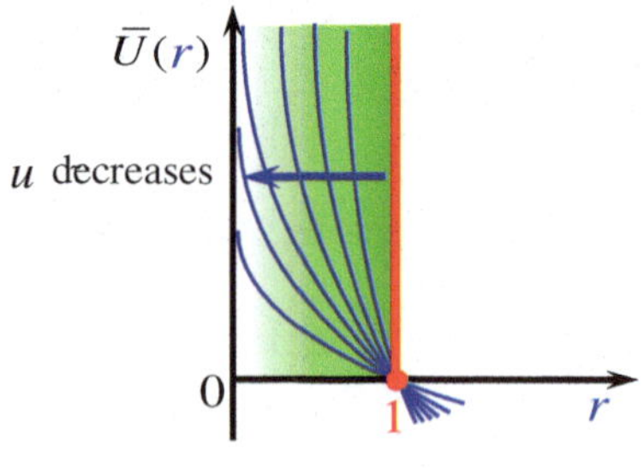

Fig. 18.6 Influence of material parameter $\tilde{u}$ on the function $\overline{U}(r)$

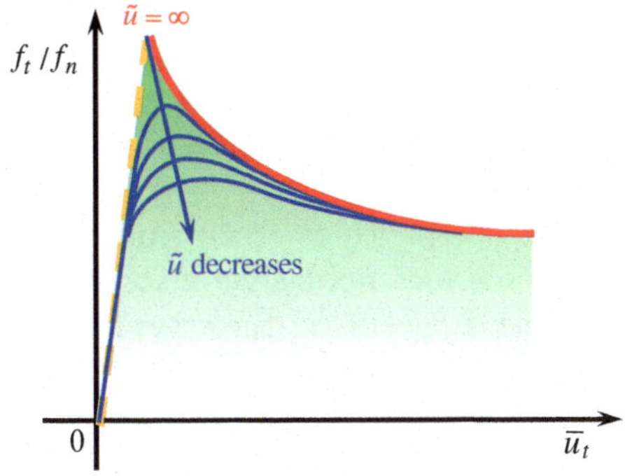

Fig. 18.7 Influence of the material parameter $\tilde{u}$ on contact stress versus sliding distance curve

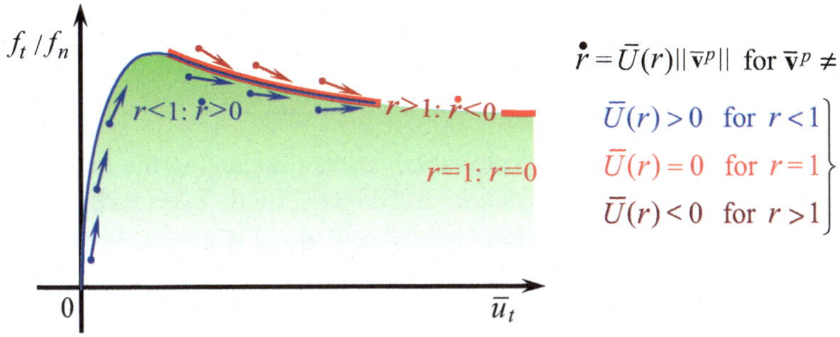

Fig. 18.8 Contact stress controlling function in subloading-friction model: constant stress is automatically attracted to yield surface in plastic-sliding process

The contact stress is automatically attracted to the normal sliding-yield surface in the plastic sliding process and it is pulled back to that surface even when it goes over the surface in numerical calculation because of $\dot{r} < 0$ for $r > 1$ from Eq. (18.33) with Eq. $(18.35)_4$ as seen in Fig. 18.8.

18.4.3 Evolution Rule of Sliding-Hardening Function

The followings might be stated from the results of experiments.

(i) The friction coefficient first reaches the maximal value of static-friction and then decreases to the minimum stationary value of kinetic-friction. Physically, this phenomenon might be interpreted to result from separations of the adhesions of surface asperities between contact bodies because of the sliding (cf. Bowden and Tabor 1958). Note here that a real contact area between tips of asperities is far smaller than apparent contact area between bodies (cf. Bay and Wanheim 1976). Then, let it be assumed that the plastic sliding leading to the separations of the adhesions of surface asperities causes the contraction of the normal sliding-yield surface, i.e., the plastic softening.

(ii) The friction coefficient recovers gradually with the elapse of time and the identical behavior as the initial sliding behavior exhibiting the static friction is reproduced if sufficient time elapses after the sliding ceases. Physically, this phenomenon might be interpreted to result from the reconstructions of the adhesions of surface asperities during the elapsed time under a quite high contact pressure between edges of surface asperities. Then, let it be assumed that the recovery results from the viscoplastic hardening.

Taking account of these facts, let the evolution rule of the isotropic hardening/softening function μ be postulated as follows (Hashiguchi 2006; Hashiguchi and Ozaki 2008a):

$$\boxed{\mathring{\mu} = \underbrace{-\kappa\left(\frac{\mu}{\mu_k} - 1\right)||\bar{\mathbf{v}}^p||}_{\text{Negative}} + \underbrace{\xi\left(1 - \frac{\mu}{\mu_s}\right)}_{\text{Positive}}} \tag{18.38}$$

where μ_s and $\mu_k (\mu_s \geq \mu \geq \mu_k)$ are material constants designating the maximum and minimum values of μ for static friction and kinetic friction, respectively. κ is a material constant specifying the rate of decrease of μ in the plastic sliding process, and ξ is a material constant specifying the rate of recovery of μ as time elapses. The first and second terms in Eq. (18.38) are relevant to the destruction and reconstruction, respectively, of the adhesion between surface asperities.

Equation (18.38) is rewritten in the incremental form as follows:

$$d\mu = -\kappa(\frac{\mu}{\mu_k} - 1)d\bar{u}^p + \xi(1 - \frac{\mu}{\mu_s})dt. \tag{18.39}$$

from which the following characteristics for the evolution of the friction coefficient are deduced.

(a) In the quasi-static sliding process ($d\mu/dt \cong 0, d\bar{u}^p/dt \cong 0$) for which the terms other than the creep part of the second term in the right-hand side are negligible in Eq. (18.39), the friction coefficient μ increases obeying the following exponential equation.

$$\mu = \mu_s - (\mu_s - \mu_0)\exp\left[-\frac{\xi}{\mu_s}(t - t_0)\right] = \begin{cases} \mu_0 & \text{for } t = t_0 \\ \mu_s & \text{for } t \to \infty \end{cases}, \tag{18.40}$$

$$\frac{d\mu}{dt} = \xi\left(1 - \frac{\mu_0}{\mu_s}\right)\exp\left[-\frac{\xi}{\mu_s}(t - t_0)\right] = \begin{cases} \xi\left(1 - \frac{\mu_0}{\mu_s}\right) & \text{for } t = t_0 \\ 0 & \text{for } t \to \infty \end{cases}, \tag{18.41}$$

denoting the initial values of t and μ as t_0 ad μ_0, respectively.

(b) In the fast sliding process $(d\mu/dt \rightarrow \infty, d\bar{u}^p/dt \rightarrow \infty)$ for which the creep part of the second term in the right-hand side is negligible in Eq. (18.39), the friction coefficient μ decreases obeying the following exponential equation.

$$\mu = (\mu_0 - \mu_k)\exp\left[-\frac{\kappa}{\mu_k}(\bar{u}^p - \bar{u}_0^p)\right] + \mu_k = \begin{cases} \mu_0 & \text{for } \bar{u}^p = \bar{u}_0^p \\ \mu_k & \text{for } \bar{u}^p \rightarrow \infty \end{cases}, \tag{18.42}$$

$$\frac{d\mu}{d\bar{u}^p} = -\kappa\left(\frac{\mu_s}{\mu_k} - 1\right)\exp\left(-\frac{\kappa}{\mu_k}\bar{u}^p\right) = \begin{cases} -\kappa\left(\dfrac{\mu_s}{\mu_k} - 1\right) & \text{for } \bar{u}^p = 0 \\ 0 & \text{for } \bar{u}^p \rightarrow \infty \end{cases}. \tag{18.43}$$

The analytical integration is beneficial for the formulation of the return-mapping method in numerical calculations as will be described in Chap. 20.

18.4.4 Plastic Sliding Velocity

The material-time derivative of Eq. (18.30) leads to the consistency condition

$$\frac{\partial f(\mathbf{f})}{\partial \mathbf{f}} \bullet \overset{\circ}{\mathbf{f}} = r\,\dot{\mu} + \dot{r}\,\mu \tag{18.44}$$

noting that the material-time derivative of scalar-valued vector function can be transformed to the corotational time-derivative as explained in Sect. 4.7.

The partial derivative of the sliding-yield function is given by

$$\frac{\partial f(\mathbf{f})}{\partial \mathbf{f}} = \frac{\partial f(\mathbf{f})}{\partial \mathbf{f}_t}\frac{\partial \mathbf{f}_t}{\partial \mathbf{f}} + \frac{\partial f(\mathbf{f})}{\partial \mathbf{f}_n}\frac{\partial \mathbf{f}_n}{\partial \mathbf{f}} = \frac{\partial f(\mathbf{f})}{\partial \mathbf{f}_t}(\mathbf{I} - \mathbf{n}\otimes\mathbf{n}) + \frac{\partial f(\mathbf{f})}{\partial \mathbf{f}_n}\mathbf{n}\otimes\mathbf{n} \tag{18.45}$$

noting

$$\left.\begin{aligned} \frac{\partial \mathbf{f}_n}{\partial \mathbf{f}} &= \frac{\partial[(\mathbf{n}\otimes\mathbf{n})\mathbf{f}]}{\partial \mathbf{f}} = \mathbf{n}\otimes\mathbf{n} \\ \frac{\partial \mathbf{f}_t}{\partial \mathbf{f}} &= \frac{\partial[(\mathbf{I} - \mathbf{n}\otimes\mathbf{n})\mathbf{f}]}{\partial \mathbf{f}} = (\mathbf{I} - \mathbf{n}\otimes\mathbf{n})\mathbf{I} = \mathbf{I} - \mathbf{n}\otimes\mathbf{n} \end{aligned}\right\} \tag{18.46}$$

and it follows from Eq. (18.13) that

$$\left.\begin{aligned} \frac{\partial f_n}{\partial \mathbf{f}} &= \frac{\partial(-\mathbf{f}\bullet\mathbf{n})}{\partial \mathbf{f}} = -\mathbf{n}\mathbf{I} = -\mathbf{n} \\ \frac{\partial f_t}{\partial \mathbf{f}} &= \frac{\partial||\mathbf{f}_t||}{\partial \mathbf{f}} = \frac{\partial||\mathbf{f}_t||}{\partial \mathbf{f}_t}\frac{\partial \mathbf{f}_t}{\partial \mathbf{f}} = \frac{\mathbf{f}_t}{||\mathbf{f}_t||}\frac{\partial[(\mathbf{I} - \mathbf{n}\otimes\mathbf{n})\mathbf{f}]}{\partial \mathbf{f}} = \frac{\mathbf{f}_t}{||\mathbf{f}_t||}(\mathbf{I} - \mathbf{n}\otimes\mathbf{n}) \equiv \mathbf{t}_f \end{aligned}\right\} \tag{18.47}$$

The substitution of Eqs. (18.33) and (18.38) into Eq. (18.44) leads to

$$\frac{\partial f(\mathbf{f})}{\partial \mathbf{f}} \cdot \overset{\circ}{\mathbf{f}} = r\left[-\kappa\left(\frac{\mu}{\mu_k} - 1\right)||\bar{\mathbf{v}}^p|| + \xi\left(1 - \frac{\mu}{\mu_s}\right)\right] + \bar{U}(r)||\bar{\mathbf{v}}^p||\mu \tag{18.48}$$

Assume that the direction of plastic sliding velocity is tangential to the contact plane and outward-normal to the curve generated by the intersection of the sliding-yield surface and the constant normal traction plane $\mathbf{f}_n =$ const., leading to the *tangent associated flow rule*

$$\boxed{\bar{\mathbf{v}}^p = \dot{\lambda}\mathbf{t}_n} \; (\dot{\lambda} \geq 0) \; (||\bar{\mathbf{v}}^p|| = \dot{\lambda}, \; \mathbf{n} \cdot \bar{\mathbf{v}}^p = 0) \tag{18.49}$$

by specializing $\mathbf{t}_v^p$ in Eq. (18.10) as

$$\mathbf{t}_n \equiv \frac{\left(\frac{\partial f(\mathbf{f})}{\partial \mathbf{f}}\right)_t}{||\left(\frac{\partial f(\mathbf{f})}{\partial \mathbf{f}}\right)_t||} \; (||\mathbf{t}_n|| = 1, \mathbf{n} \cdot \mathbf{t}_n = 0) \tag{18.50}$$

with

$$\left(\frac{\partial f(\mathbf{f})}{\partial \mathbf{f}}\right)_t \equiv \frac{\partial f(\mathbf{f})}{\partial \mathbf{f}} - \left(\mathbf{n} \cdot \frac{\partial f(\mathbf{f})}{\partial \mathbf{f}}\right)\mathbf{n} = (\mathbf{I} - \mathbf{n} \otimes \mathbf{n})\frac{\partial f(\mathbf{f})}{\partial \mathbf{f}} \tag{18.51}$$

where $\dot{\lambda}$ and $\mathbf{t}_n$ are the magnitude and direction, respectively, of the plastic sliding velocity. The vector $(\partial f(\mathbf{f})/\partial \mathbf{f})_t$ in Eq. (18.50) is the tangential projection (part) of the outward-normal vector $\partial f(\mathbf{f})/\partial \mathbf{f}$ of the subloading friction surface to the contact surface (see Fig. 18.3), and the vector $\mathbf{t}_n$ is its normalisation. It follows from Eqs. (18.8) and (18.49) with Eq. (18.50) that

$$\bar{\mathbf{v}}_n^p = \mathbf{0}, \quad \bar{\mathbf{v}}^p = \bar{\mathbf{v}}_t^p \tag{18.52}$$

The substitution of Eq. (18.49) into Eq. (18.48) reads:

$$\frac{\partial f(\mathbf{f})}{\partial \mathbf{f}} \cdot \overset{\circ}{\mathbf{f}} = r\left[-\kappa\left(\frac{\mu}{\mu_k} - 1\right)\dot{\lambda} + \xi\left(1 - \frac{\mu}{\mu_s}\right)\right] + \bar{U}(r)\,\dot{\lambda}\,\mu \tag{18.53}$$

i.e.

$$\frac{\partial f(\mathbf{f})}{\partial \mathbf{f}} \cdot \overset{\circ}{\mathbf{f}} = \dot{\lambda}\, m^p + m^c \tag{18.54}$$

where

$$\boxed{m^p \equiv -\kappa\left(\frac{\mu}{\mu_k} - 1\right)r + \bar{U}(r)\mu} \tag{18.55}$$

$$\boxed{m^c \equiv \xi\left(1-\frac{\mu}{\mu_s}\right)r(\geq 0)} \tag{18.56}$$

are relevant to the plastic and the creep sliding velocity, respectively.

It is obtained from Eqs. (18.49) and (18.54) that

$$\boxed{\overset{\bullet}{\lambda} = \frac{\dfrac{\partial f(\mathbf{f})}{\partial \mathbf{f}} \bullet \overset{\circ}{\mathbf{f}} - m^c}{m^p}, \quad \bar{\mathbf{v}}^p = \frac{\dfrac{\partial f(\mathbf{f})}{\partial \mathbf{f}} \bullet \overset{\circ}{\mathbf{f}} - m^c}{m^p}\mathbf{t}_n} \tag{18.57}$$

18.4.5 Relations Between Contact Traction Rate and Sliding Velocity

Substituting Eqs. $(18.22)_1$ and (18.57) into Eq. (18.4), the sliding velocity is given by

$$\boxed{\bar{\mathbf{v}} = \mathbf{C}^{e-1}\overset{\circ}{\mathbf{f}} + \frac{\dfrac{\partial f(\mathbf{f})}{\partial \mathbf{f}} \bullet \overset{\circ}{\mathbf{f}} - m^c}{m^p}\mathbf{t}_n} \tag{18.58}$$

The plastic multiplier in terms of the sliding velocity, denoted by the symbol $\overset{\bullet}{\Lambda}$, is given from Eqs. (18.58) as

$$\boxed{\overset{\bullet}{\Lambda} = \frac{\dfrac{\partial f(\mathbf{f})}{\partial \mathbf{f}} \bullet \mathbf{C}^e\bar{\mathbf{v}} - m^c}{m^p + \dfrac{\partial f(\mathbf{f})}{\partial \mathbf{f}} \bullet \mathbf{C}^e\mathbf{t}_n}} \tag{18.59}$$

noting

$$\begin{aligned}
\frac{\partial f(\mathbf{f})}{\partial \mathbf{f}} \bullet \mathbf{C}^e\bar{\mathbf{v}} &= \frac{\partial f(\mathbf{f})}{\partial \mathbf{f}} \bullet \overset{\circ}{\mathbf{f}} + \frac{\dfrac{\partial f(\mathbf{f})}{\partial \mathbf{f}} \bullet \overset{\circ}{\mathbf{f}} - m^c}{m^p}\frac{\partial f(\mathbf{f})}{\partial \mathbf{f}} \bullet \mathbf{C}^e\mathbf{t}_n \\
&= \left(m^p + \frac{\partial f(\mathbf{f})}{\partial \mathbf{f}} \bullet \mathbf{C}^e\mathbf{t}_n\right)\frac{\dfrac{\partial f(\mathbf{f})}{\partial \mathbf{f}} \bullet \overset{\circ}{\mathbf{f}} - m^c}{m^p} + m^c \\
&= \left(m^p + \frac{\partial f(\mathbf{f})}{\partial \mathbf{f}} \bullet \mathbf{C}^e\mathbf{t}_n\right)\overset{\bullet}{\lambda} + m^c
\end{aligned}$$

which is obtained by multiplying Eq. (18.58) by $(\partial f(\mathbf{f})/\partial \mathbf{f})\mathbf{C}^e$.

The rate of contact stress vector is described by Eqs. (18.4), $(18.22)_2$ and (18.59) as follows:

$$\boxed{\overset{\circ}{\mathbf{f}} = \mathbf{C}^e(\bar{\mathbf{v}} - \langle \frac{\frac{\partial f(\mathbf{f})}{\partial \mathbf{f}} \cdot \mathbf{C}^e\bar{\mathbf{v}} - m^c}{m^p + \frac{\partial f(\mathbf{f})}{\partial \mathbf{f}} \cdot \mathbf{C}^e\mathbf{t}_n} \rangle \mathbf{t}_n) = (\mathbf{C}^e - \frac{\mathbf{C}^e\mathbf{t}_n \otimes \frac{\partial f(\mathbf{f})}{\partial \mathbf{f}}\mathbf{C}^e}{m^p + \frac{\partial f(\mathbf{f})}{\partial \mathbf{f}} \cdot \mathbf{C}^e\mathbf{t}_n})\bar{\mathbf{v}} + \frac{m^c}{m^p + \frac{\partial f(\mathbf{f})}{\partial \mathbf{f}} \cdot \mathbf{C}^e\mathbf{t}_n}\mathbf{C}^e\mathbf{t}_n} \tag{18.60}$$

where

$$\mathbf{C}^e\mathbf{t}_n = [\alpha_n\mathbf{n} \otimes \mathbf{n} + \alpha_t(\mathbf{I} - \mathbf{n} \otimes \mathbf{n})]\mathbf{t}_n = \alpha_t\mathbf{t}_n \tag{18.61}$$

$$\frac{\partial f(\mathbf{f})}{\partial \mathbf{f}} \cdot \mathbf{C}^e\mathbf{t}_n = \alpha_t \frac{\partial f(\mathbf{f})}{\partial \mathbf{f}} \cdot \mathbf{t}_n \tag{18.62}$$

18.4.6 Isotropic Sliding-Yield Surface

The traction function for the isotropic sliding-yield surface is described as

$$\boxed{f(\mathbf{f}) = f(f_t, f_n)} \tag{18.63}$$

for which the following partial derivatives hold, noting Eq. (18.47).

$$\frac{\partial f(\mathbf{f})}{\partial \mathbf{f}} = \frac{\partial f(f_t, f_n)}{\partial \mathbf{f}} = \frac{\partial f(f_t, f_n)}{\partial f_n}\frac{\partial f_n}{\partial \mathbf{f}} + \frac{\partial f(f_t, f_n)}{\partial f_t}\frac{\partial f_t}{\partial \mathbf{f}} = -\frac{\partial f(f_t, f_n)}{\partial f_n}\mathbf{n} + \frac{\partial f(f_t, f_n)}{\partial f_t}\mathbf{t}_f \tag{18.64}$$

It follows by substituting Eq. (18.64) into Eq. (18.50) that

$$\mathbf{t}_n = \mathbf{t}_f \tag{18.65}$$

Therefore, the direction of the plastic sliding velocity coincides with the direction of the tangential traction.

As a particular form of $f(\mathbf{f})$ in Eq. (18.63), let the following Coulomb sliding-yield function be adopted.

$$\boxed{f(\mathbf{f}) = \frac{f_t}{f_n}} \tag{18.66}$$

It follows for Eqs. (18.64), (18.65) and (18.66) that

$$\frac{\partial f(\mathbf{f})}{\partial f_n} = -\frac{f_t}{f_n^2}, \quad \frac{\partial f(\mathbf{f})}{\partial f_t} = \frac{1}{f_n} \tag{18.67}$$

$$\frac{\partial f(\mathbf{f})}{\partial \mathbf{f}} = \frac{1}{f_n}\left(\mathbf{t}_f + \frac{f_t}{f_n}\mathbf{n}\right) \tag{18.68}$$

Then, we have

$$\mathbf{C}^e\mathbf{t}_n = \mathbf{C}^e\mathbf{t}_f = \alpha_t\mathbf{t}_f \tag{18.69}$$

$$\mathbf{C}^e\frac{\partial f(\mathbf{f})}{\partial \mathbf{f}} = [\alpha_t(\mathbf{I} - \mathbf{n}\otimes\mathbf{n}) + \alpha_n\mathbf{n}\otimes\mathbf{n}]\frac{1}{f_n}\left(\mathbf{t}_f + \frac{f_t}{f_n}\mathbf{n}\right) = \frac{1}{f_n}\left(\alpha_t\mathbf{t}_f + \alpha_n\frac{f_t}{f_n}\mathbf{n}\right) \tag{18.70}$$

$$\frac{\partial f(\mathbf{f})}{\partial \mathbf{f}}\cdot\mathbf{C}^e\mathbf{t}_f = \frac{1}{f_n}\left(\mathbf{t}_f + \frac{f_t}{f_n}\mathbf{n}\right)\cdot\alpha_t\mathbf{t}_f = \frac{1}{f_n}\left(\mathbf{t}_f + \frac{f_t}{f_n}\mathbf{n}\right)\cdot\alpha_t\mathbf{t}_f = \frac{\alpha_t}{f_n} \tag{18.71}$$

$$\frac{\partial f(\mathbf{f})}{\partial \mathbf{f}}\mathbf{C}^e\cdot\bar{\mathbf{v}} = \frac{1}{f_n}\left(\mathbf{t}_f + \frac{f_t}{f_n}\mathbf{n}\right)\cdot[\alpha_t(\mathbf{I} - \mathbf{n}\otimes\mathbf{n}) + \alpha_n\mathbf{n}\otimes\mathbf{n}]\bar{\mathbf{v}} = \frac{1}{f_n}\left(\alpha_t\mathbf{t}_f + \alpha_n\frac{f_t}{f_n}\mathbf{n}\right)\cdot\bar{\mathbf{v}} \tag{18.72}$$

Substituting Eqs. (18.69)–(18.72), Eqs. (18.58) and (18.60) reduces to

$$\bar{\mathbf{v}} = \left[\frac{1}{\alpha_t}(\mathbf{I} - \mathbf{n}\otimes\mathbf{n}) + \frac{1}{\alpha_n}\mathbf{n}\otimes\mathbf{n}\right]\mathring{\mathbf{f}} + \frac{\frac{1}{f_n}\left(\mathbf{t}_f + \frac{f_t}{f_n}\mathbf{n}\right)\cdot\mathring{\mathbf{f}} - m^c}{m^p}\mathbf{t}_n \tag{18.73}$$

$$\mathring{\mathbf{f}} = [\alpha_t(\mathbf{I} - \mathbf{n}\otimes\mathbf{n}) + \alpha_n\mathbf{n}\otimes\mathbf{n}]\left[\bar{\mathbf{v}} - \left\langle\frac{\frac{1}{f_n}(\alpha_t\mathbf{t}_f + \alpha_n\frac{f_t}{f_n}\mathbf{n})\cdot\bar{\mathbf{v}} - m^c}{m^p + \frac{\alpha_t}{f_n}}\right\rangle\mathbf{t}_f\right] \tag{18.74}$$

or

$$\mathring{\mathbf{f}} = \mathbf{K}^{ep}\bar{\mathbf{v}} + \frac{m^c}{m^p + \frac{\alpha_t}{f_n}}\alpha_t\mathbf{t}_f \tag{18.75}$$

$$\mathbf{K}^{ep} \equiv \alpha_t(\mathbf{I} - \mathbf{n}\otimes\mathbf{n}) + \alpha_n\mathbf{n}\otimes\mathbf{n} - \frac{\alpha_t\mathbf{t}_f\otimes\frac{1}{f_n}\left(\alpha_t\mathbf{t}_f + \alpha_n\frac{f_t}{f_n}\mathbf{n}\right)}{m^p + \frac{\alpha_t}{f_n}} \tag{18.76}$$

In the coordinate system with the base $(\hat{\mathbf{e}}_1, \hat{\mathbf{e}}_2, \mathbf{n})$, noting

$$\mathbf{I} - \mathbf{n} \otimes \mathbf{n} = \hat{\mathbf{e}}_1 \otimes \hat{\mathbf{e}}_1 + \hat{\mathbf{e}}_2 \otimes \hat{\mathbf{e}}_2 + \mathbf{n} \otimes \mathbf{n} - \mathbf{n} \otimes \mathbf{n} = \hat{\mathbf{e}}_1 \otimes \hat{\mathbf{e}}_1 + \hat{\mathbf{e}}_2 \otimes \hat{\mathbf{e}}_2 \tag{18.77}$$

$$\mathbf{t}_f = t_{f1}\hat{\mathbf{e}}_1 + t_{f2}\hat{\mathbf{e}}_2 \left(t_{f1} = \frac{f_1}{||\mathbf{f}_t||} = \frac{f_1}{\sqrt{f_1^2 + f_2^2}}, t_{f2} = \frac{f_2}{||\mathbf{f}_t||} = \frac{f_2}{\sqrt{f_1^2 + f_2^2}} \right) \tag{18.78}$$

$$\mathbf{C}^e = \alpha_t(\hat{\mathbf{e}}_1 \otimes \hat{\mathbf{e}}_1 + \hat{\mathbf{e}}_2 \otimes \hat{\mathbf{e}}_2) + \alpha_n \mathbf{n} \otimes \mathbf{n} \tag{18.79}$$

Equations (18.68)–(18.72) are expressed as

$$\frac{\partial f(\mathbf{f})}{\partial \mathbf{f}} = \frac{1}{f_n}\left(t_{f1}\hat{\mathbf{e}}_1 + t_{f2}\hat{\mathbf{e}}_2 + \frac{f_t}{f_n}\mathbf{n} \right) \tag{18.80}$$

$$\mathbf{C}^e \mathbf{t}_f = \alpha_t(t_{f1}\hat{\mathbf{e}}_1 + t_{f2}\hat{\mathbf{e}}_2) \tag{18.81}$$

$$\begin{aligned} \mathbf{C}^e \frac{\partial f(\mathbf{f})}{\partial \mathbf{f}} &= [\alpha_t(\hat{\mathbf{e}}_1 \otimes \hat{\mathbf{e}}_1 + \hat{\mathbf{e}}_2 \otimes \hat{\mathbf{e}}_2) + \alpha_n \mathbf{n} \otimes \mathbf{n}] \frac{1}{f_n}\left(t_{f1}\hat{\mathbf{e}}_1 + t_{f2}\hat{\mathbf{e}}_2 + \frac{f_t}{f_n}\mathbf{n} \right) \\ &= \frac{1}{f_n}\left(\alpha_t(t_{f1}\hat{\mathbf{e}}_1 + t_{f2}\hat{\mathbf{e}}_2) + \alpha_n \frac{f_t}{f_n}\mathbf{n} \right) \end{aligned} \tag{18.82}$$

$$\mathbf{C}^e \bar{\mathbf{v}} = [\alpha_t(\hat{\mathbf{e}}_1 \otimes \hat{\mathbf{e}}_1 + \hat{\mathbf{e}}_2 \otimes \hat{\mathbf{e}}_2) + \alpha_n \mathbf{n} \otimes \mathbf{n}]\bar{\mathbf{v}} \tag{18.83}$$

$$\frac{\partial f(\mathbf{f})}{\partial \mathbf{f}} \mathbf{C}^e \bullet \bar{\mathbf{v}} = \frac{1}{f_n}\left[\alpha_t(t_{f1}\hat{\mathbf{e}}_1 + t_{f2}\hat{\mathbf{e}}_2) + \alpha_n \frac{f_t}{f_n}\mathbf{n} \right] \bullet \bar{\mathbf{v}} \tag{18.84}$$

Substituting Eqs. (18.77) and (18.78), Eq. (18.74) reduces to

$$\begin{aligned} \overset{\circ}{\mathbf{f}} = {} & [\alpha_t(\hat{\mathbf{e}}_1 \otimes \hat{\mathbf{e}}_1 + \hat{\mathbf{e}}_2 \otimes \hat{\mathbf{e}}_2) + \alpha_n \mathbf{n} \otimes \mathbf{n}] \\ & \left[\bar{\mathbf{v}} - \left\langle \frac{\frac{1}{f_n}\left[\alpha_t(t_{f1}\hat{\mathbf{e}}_1 + t_{f2}\hat{\mathbf{e}}_2) + \alpha_n \frac{f_t}{f_n}\mathbf{n} \right] \bullet \bar{\mathbf{v}} - m^c}{m^p + \frac{\alpha_t}{f_n}} \right\rangle (t_{f1}\hat{\mathbf{e}}_1 + t_{f2}\hat{\mathbf{e}}_2) \right] \end{aligned}$$

$$
\begin{aligned}
&= [\alpha_t(\hat{\mathbf{e}}_1 \otimes \hat{\mathbf{e}}_1 + \hat{\mathbf{e}}_2 \otimes \hat{\mathbf{e}}_2) + \alpha_n \mathbf{n} \otimes \mathbf{n}]\bar{\mathbf{v}} \\
&\quad -\alpha_t(t_{f1}\hat{\mathbf{e}}_1 + t_{f2}\hat{\mathbf{e}}_2)\left\langle \frac{\dfrac{1}{f_n}\left[\alpha_t(t_{f1}\hat{\mathbf{e}}_1 + t_{f2}\hat{\mathbf{e}}_2) + \alpha_n \dfrac{f_t}{f_n}\mathbf{n}\right] \cdot \bar{\mathbf{v}} - m^c}{m^p + \dfrac{\alpha_t}{f_n}} \right\rangle \\
&= [\alpha_t(\hat{\mathbf{e}}_1 \otimes \hat{\mathbf{e}}_1 + \hat{\mathbf{e}}_2 \otimes \hat{\mathbf{e}}_2) + \alpha_n \mathbf{n} \otimes \mathbf{n}]\bar{\mathbf{v}} \\
&\quad - S\alpha_t(t_{f1}\hat{\mathbf{e}}_1 + t_{f2}\hat{\mathbf{e}}_2)\frac{\dfrac{1}{f_n}\left[\alpha_t(t_{f1}\hat{\mathbf{e}}_1 + t_{f2}\hat{\mathbf{e}}_2) + \alpha_n \dfrac{f_t}{f_n}\mathbf{n}\right] \cdot \bar{\mathbf{v}}}{m^p + \dfrac{\alpha_t}{f_n}} \\
&\quad + S\alpha_t(t_{f1}\hat{\mathbf{e}}_1 + t_{f2}\hat{\mathbf{e}}_2)\frac{m^c}{m^p + \dfrac{\alpha_t}{f_n}}
\end{aligned}
\tag{18.85}
$$

where

$$
S = \begin{cases} 1 & \text{for } \dot{\Lambda} > 0 \\ 0 & \text{for } \dot{\Lambda} \leq 0 \end{cases}
\tag{18.86}
$$

18.5 Loading Criterion

While the loading criterion for the plastic sliding velocity is similar to that for the plastic strain rate described in Sect. 6.5, it will be described below.

First, note the following facts:

1. It is necessary that

$$
\dot{\bar{\lambda}} = \dot{\Lambda} > 0
\tag{18.87}
$$

in the loading (plastic sliding) process $\bar{\mathbf{v}}^p \neq \mathbf{0}$.

2. The following inequality must hold in the unloading (elastic sliding) process $\bar{\mathbf{v}}^p = \mathbf{0}$.

$$
\frac{\partial f(\mathbf{f})}{\partial \mathbf{f}} \cdot \mathring{\mathbf{f}} \leq 0
\tag{18.88}
$$

Further, because of $\bar{\mathbf{v}} = \bar{\mathbf{v}}^e$ leading to $[\partial f(\mathbf{f})/\partial \mathbf{f}] \cdot \mathbf{C}^e \bar{\mathbf{v}} = [\partial f(\mathbf{f})/\partial \mathbf{f}] \cdot \mathbf{C}^e \bar{\mathbf{v}}^e = [\partial f(\mathbf{f})/\partial \mathbf{f}] \cdot \mathring{\mathbf{f}}$ in this process it follows from Eq. (18.59) that

$$\overset{\bullet}{\Lambda} = \frac{\dfrac{\partial f(\mathbf{f})}{\partial \mathbf{f}} \bullet \overset{\circ}{\mathbf{f}} - m^c}{m^p + \dfrac{\partial f(\mathbf{f})}{\partial \mathbf{f}} \bullet \mathbf{C}^e \mathbf{t}_v^p} \tag{18.89}$$

while it should be noted that $m^c \geq 0$ (Eq. (18.56)).

3. The plastic modulus m^p takes both positive and negative signs. However, the elastic modulus $\mathbf{C}^e$ is the positive definite tensor and thus $[\partial f(\mathbf{f})/\partial \mathbf{f}] \bullet \mathbf{C}^e \mathbf{t}_v^p \gg m^p$ holds provided that $\mathbf{t}_v^p$ is not far different from $\partial f(\mathbf{f})/\partial \mathbf{f}$, leading to

$$m^p + \frac{\partial f(\mathbf{f})}{\partial \mathbf{f}} \bullet \mathbf{C}^e \mathbf{t}_v^p > 0 \tag{18.90}$$

Therefore, the infinite plastic relaxation, i.e. infinite softening, is not induced.

Then, in the unloading process $\bar{\mathbf{v}}^p = \mathbf{0}$, the following inequalities hold depending on the sign of the plastic modulus m^p, i.e. the hardening, perfectly-plastic and softening states from Eqs. (18.57) and (18.88)–(18.90).

$$\left.\begin{array}{l} \overset{\bullet}{\lambda} \leq 0 \text{ and } \overset{\bullet}{\Lambda} \leq 0 \text{ when } m^p > 0 \\ \overset{\bullet}{\lambda} \to -\infty \text{ or indeterminate and } \overset{\bullet}{\Lambda} \leq 0 \text{ when } m^p = 0 \\ \overset{\bullet}{\lambda} \geq 0 \text{ and } \overset{\bullet}{\Lambda} \leq 0 \text{ when } m^p < 0 \end{array}\right\} \tag{18.91}$$

Therefore, the sign of $\overset{\bullet}{\lambda}$ at the moment of unloading from the state $m^p \leq 0$ is not necessarily negative. On the other hand, $\overset{\bullet}{\Lambda}$ is negative in the unloading process. Consequently, the distinction between a loading and an unloading processes cannot be judged by the sign of $\overset{\bullet}{\lambda}$ but can be done by the sign of $\overset{\bullet}{\Lambda}$. Therefore, the loading criterion is given as follows:

$$\left.\begin{array}{l} \bar{\mathbf{v}}^p \neq \mathbf{0} : \overset{\bullet}{\Lambda} > 0, \\ \bar{\mathbf{v}}^p = \mathbf{0} : \text{ otherwise.} \end{array}\right\} \tag{18.92}$$

or

$$\boxed{\begin{array}{l} \bar{\mathbf{v}}^p \neq \mathbf{0} : \dfrac{\partial f(\mathbf{f})}{\partial \mathbf{f}} \bullet \mathbf{C}^e \bar{\mathbf{v}} - m^c > 0 \\ \bar{\mathbf{v}}^p = 0 : \text{otherwise} \end{array}} \tag{18.93}$$

on account of Eq. (18.90). Here, note that the infinite plastic relaxation is generated so that the contact stress infinitely decreases if the denominator becomes zero as known from Eq. (18.60).

18.6 Calculation of Normal Friction-Yield Ratio

The normal friction-yield ratio r can be calculated using one of the following two methods.

(1) We calculate r using the following equation derived from the subloading friction surface in Eq. (18.30) for both of the elastoplastic sliding and elastic sliding processes after the contact stress vector $\mathbf{f}$ and the hardening variable μ are updated.

$$r = \frac{f(\mathbf{f})}{\mu}. \tag{18.94}$$

Needless to say, the hardening variable μ is calculated by the plastic sliding velocity in Eq. (18.57) including the plastic modulus m^p which depends on the evolution rule of the normal friction-yield ratio r in Eq. (18.33).

(2) We calculate r using method (1) for the elastic sliding process but calculate r by the time integration of Eq. (18.33) for the plastic sliding process. Actually, we adopt the function $\overline{U}(r)$ in Eq. (18.36) of the cotangent form. Here, it is beneficial to employ the analytical time integration in Eq. (18.37) for the plastic loading process in implicit numerical calculations. However, we must not use the analytical integration because the controlling function to attract the contact stress to the normal friction-yield surface is lost in numerical calculation by the forward-Euler method if we use it.

The method (2) would be superior to the method (1) since the normal friction-yield ratio is calculated directly from the plastic sliding velocity.

18.7 Fundamental Mechanical Behavior of Subloading-Friction Model

We examine below the fundamental response of the isotropic subloading-friction model by numerical experiments and comparison with test data for the linear sliding phenomenon without a rigid-body rotation under a constant normal traction and with the fixed direction of tangential contact traction on the assumption of isotropy for the sake of simplicity. The followings hold in this situation.

$$\left.\begin{array}{l} \mathbf{f}_n = \text{const.},\ \mathbf{t}_f = \text{const.},\ \boldsymbol{\omega}_c = \mathbf{O} \\ \mathring{\mathbf{f}}_n = \mathbf{0},\ \mathring{\mathbf{f}} = \mathring{\mathbf{f}}_t = \dot{f}_t \mathbf{t}_f \\ \bar{\mathbf{v}}_n^e = \mathbf{0},\ \bar{\mathbf{v}}^e = \bar{v}_t^e \mathbf{t}_f,\ \bar{\mathbf{v}} = \bar{v}_t \mathbf{t}_f \end{array}\right\} \tag{18.95}$$

18.7.1 Relation of Tangential Contact Traction Rate and Sliding Velocity

Equations (18.73) and (18.74) reduce to the following relations in the one-dimensional sliding under the conditions in Eq. (18.95), denoting $\overset{\circ}{f}_{t1} \to \overset{\circ}{f}_t, \bar{v}_{t1} \to \bar{v}_t \, (\overset{\circ}{f}_{t2} = \overset{\circ}{f}_n = 0, \bar{v}_{t2} = \bar{v}_n = 0)$ and noting Eqs. (18.55) and (18.56).

$$\bar{v}_t = \frac{1}{\alpha_t}\overset{\bullet}{f}_t + \frac{\dfrac{\overset{\bullet}{f}_t}{f_n} - m^c}{m^p} = \frac{1}{\alpha_t}\overset{\bullet}{f}_t + \frac{\dfrac{\overset{\bullet}{f}_t}{f_n} - \xi\left(1 - \dfrac{\mu}{\mu_s}\right)r}{-\kappa\left(\dfrac{\mu}{\mu_k} - 1\right)r + \overline{U}(r)\mu} \tag{18.96}$$

$$\overset{\bullet}{f}_t = \alpha_t\bar{v}_t - \alpha_t\left\langle \frac{\dfrac{\alpha_t}{f_n}\bar{v}_t - m^c}{m^p + \dfrac{\alpha_t}{f_n}} \right\rangle = \alpha_t\bar{v}_t - \alpha_t\left\langle \frac{\dfrac{\alpha_t}{f_n}\bar{v}_t - \xi\left(1 - \dfrac{\mu}{\mu_s}\right)r}{-\kappa\left(\dfrac{\mu}{\mu_k} - 1\right)r + \overline{U}(r)\mu + \dfrac{\alpha_t}{f_n}} \right\rangle \tag{18.97}$$

For the particular case that the sliding velocity is high, so that the creep term can be ignored, i.e. $m^c = 0$, Eqs. (18.96) and (18.97) are reduced to

$$\bar{v}_t = \left(\frac{1}{\alpha_t} + \frac{1}{f_n m^p}\right)\overset{\bullet}{f}_t = \left(\frac{1}{\alpha_t} + \frac{1}{f_n\left[-\kappa\left(\dfrac{\mu}{\mu_k} - 1\right)r + \overline{U}(r)\right]}\right)\overset{\bullet}{f}_t \tag{18.98}$$

$$\begin{aligned}\overset{\bullet}{f}_t &= \alpha_t\bar{v}_t - \alpha_t^2\left\langle \frac{\bar{v}_t}{m^p f_n + \alpha_t} \right\rangle = \alpha_t\left(1 - \frac{1}{1 + (f_n/\alpha_t)m^p}\right)\bar{v}_t \\ &= \alpha_t\left(1 - \frac{1}{1 + \dfrac{f_n}{\alpha_t}\left[-\kappa\left(\dfrac{\mu}{\mu_k} - 1\right)r + \overline{U}(r)\right]}\right)\bar{v}_t\end{aligned} \tag{18.99}$$

The relation between the tangential components of the contact stress vector and the displacement vector is schematically shown in Fig. 18.9 for Eq. (18.99) concerning with a high sliding velocity process in which the creep hardening of the second term in Eq. (18.97) is negligible. The relation by the conventional friction model with the sliding-yield surface enclosing an elastic domain is also shown as bold curves $0 - \boldsymbol{y} - \mathrm{k}$. In the subloading-friction model, the softening term $-\kappa(\mu/\mu_k - 1)r(\le 0)$ increases monotonically from the negative value to zero and

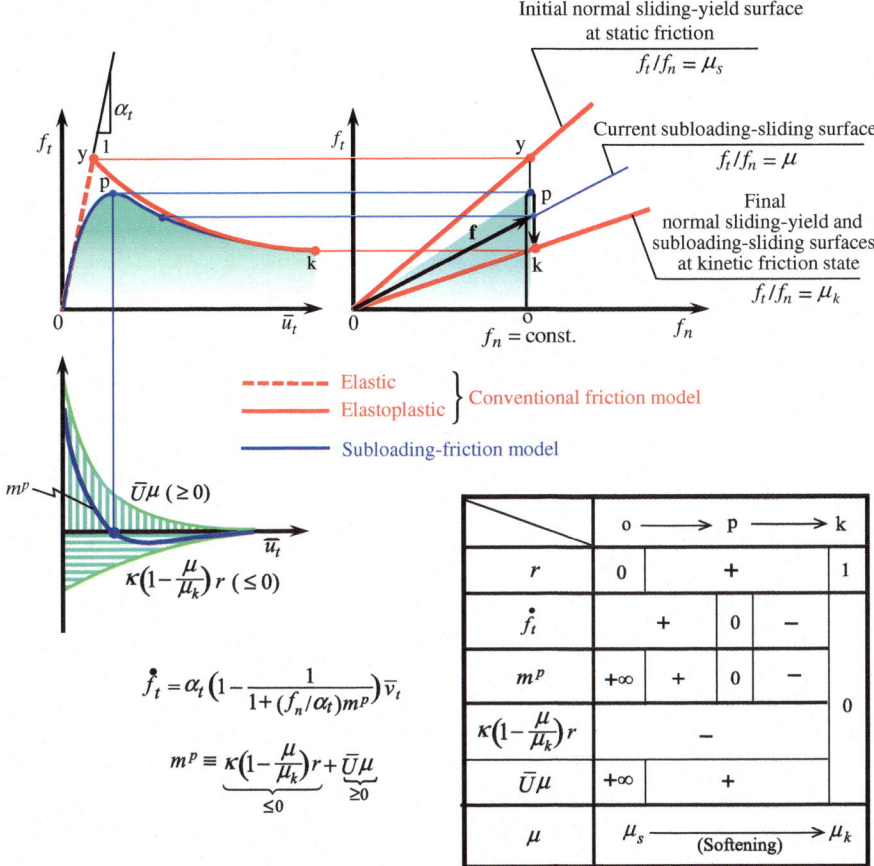

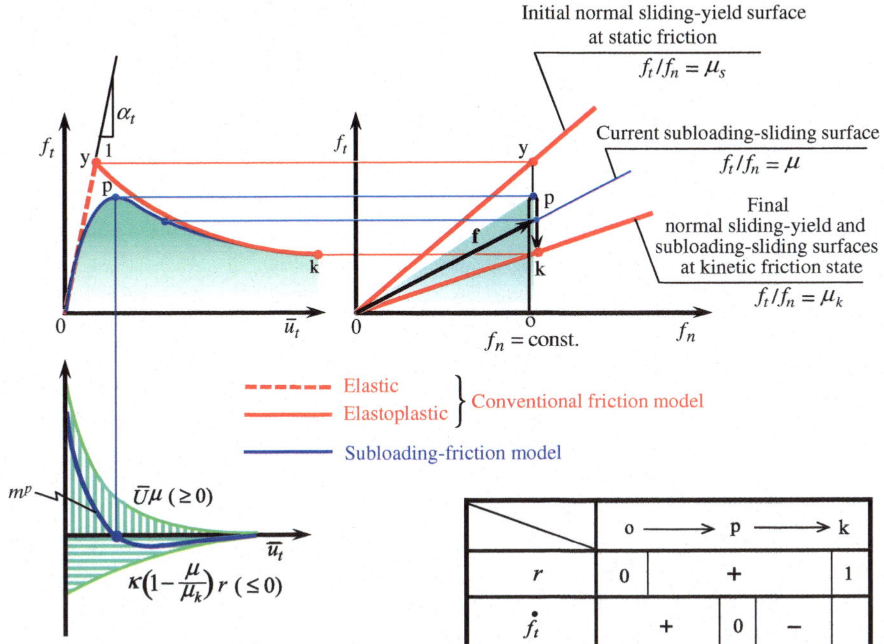

$$\mathring{f}_t = \alpha_t\left(1 - \frac{1}{1+(f_n/\alpha_t)m^p}\right)\bar{v}_t$$

$$m^p \equiv \underbrace{\kappa\left(1-\frac{\mu}{\mu_k}\right)r}_{\leq 0} + \underbrace{\bar{U}\mu}_{\geq 0}$$

	o →		p	→	k
r	0	+			1
$\mathring{f}_t$	+		0	−	0
m^p	$+\infty$	+	0	−	
$\kappa(1-\frac{\mu}{\mu_k})r$	−				
$\bar{U}\mu$	$+\infty$	+			
μ	μ_s → (Softening) → μ_k				

Fig. 18.9 Prediction of linear sliding behavior from the static to the kinetic friction by the conventional friction and the subloading friction models at a high sliding rate without the creep hardening

inversely the normal sliding-yield term $\overline{U}\mu(\geq 0)$ decreases monotonically from the infinite value to zero in the denominator of the plastic sliding velocity in second term in the bracket in Eq. (18.99). In the initial stage of sliding, the plastic modulus is positive, i.e. $m^p > 0$ so that the tangential contact traction increases but thereafter these terms cancel mutually leading to $m^p = 0$ at which the tangential contact traction reaches the peak, i.e. the static friction point p. Thereafter, the softening term increases gradually from negative to zero but the normal sliding-yield term decreases rapidly resulting in $m^p < 0$ so that the tangential contact traction decreases to the kinetic friction point k.

18.7.2 Numerical Experiments and Comparisons with Test Data

Numerical experiments and comparisons with test data for the subloading-friction model are shown below for Eqs. (18.98) and (18.99) with Eq. (18.36).

The seven material constants of μ_s, μ_k, κ, ξ, $\tilde{u}$, α_n and α_t and the initial value μ_0 of the friction coefficient are included in the present model. Material parameters are selected as follows:

$$\begin{aligned}\mu_0 &= \mu_s = 0.4,\ \mu_k = 0.2\\ \kappa &= 10\,\text{mm}^{-1},\ \xi = 0.01/\text{s}\\ \tilde{u} &= 1000\,\text{mm}^{-1}\\ \alpha_n &= \alpha_t = 1000\,\text{kN/mm}^3\end{aligned}$$

under the condition.

The influence of the sliding velocity on the relation of the traction ratio f_t/f_n versus the tangential sliding displacement $\bar{u}_t$ are shown in Fig. 18.10. Smooth transitions from the static friction to the kinetic friction and the decreases of the friction coefficient are shown. Faster decrease of friction coefficient is shown for higher sliding velocity.

The recovery of the static friction coefficient from the kinetic friction with the elapsed time t after the stop of sliding is shown in Fig. 18.11. In the calculation, the constant sliding velocity $\bar{v}_t = 0.1$ mm/s is given in the first stage reaching the kinetic friction and then the tangential contact traction is unloaded to zero. After the cessation of sliding for several elapsed times, the same sliding velocity in the first stage is given again. The recovery is larger for a longer stationary time.

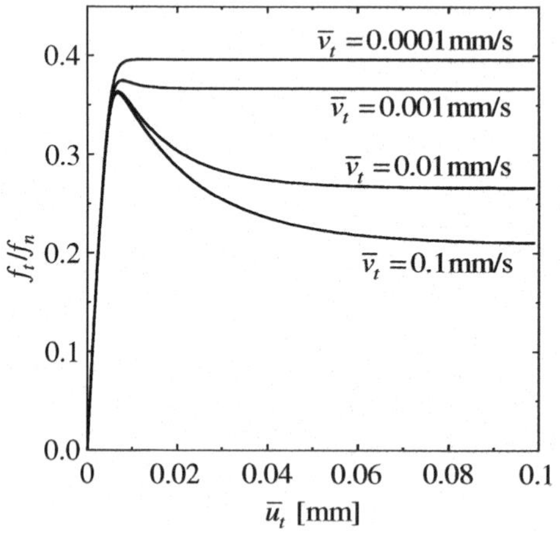

Fig. 18.10 Influence of sliding velocity

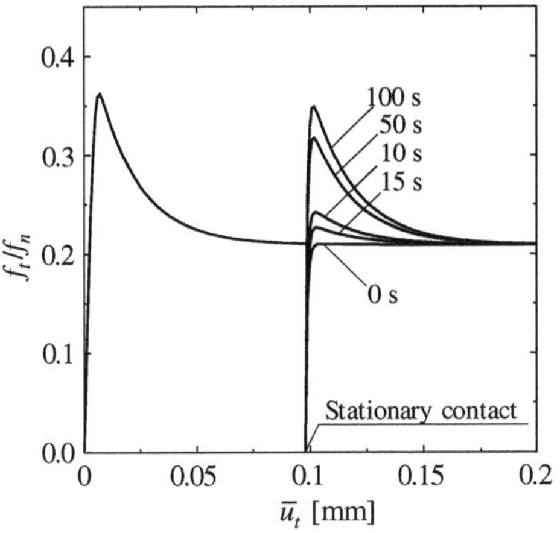

Fig. 18.11 Influence of stationary time

The influence of the material constant $\tilde{u}$ in Eq. (18.36) for the evolution rule of the normal-sliding ratio r in Eq. (18.33) on the variation of the traction ratio f_t/f_n is shown in Fig. 18.12, where other material constants and the initial value are the same as those in Fig. 18.10. The calculated results for low and high sliding velocities are shown in Fig. 18.12a, b, respectively. As shown in this figure, smoother transition from the elastic to the plastic state is shown for smaller values of $\tilde{u}$.

The accumulations of sliding displacement under the cyclic loading of tangential contact traction of 80 %, i.e., $f_t = 0{-}0.8\mu_s f_n$ for the two levels of sliding velocities are shown in Fig. 18.13, where the material parameters are chosen same as in Figs. 18.10 and 18.11. The sliding displacement increases as the velocity is larger since the recovery of friction requires time. In particular, the generation of infinite sliding under a high rate of sliding is shown in Fig. 18.13(b). In contrast, the accumulation cannot be predicted at all by the conventional friction model capable of predicting only elastic sliding for the contact stress lower than the friction-yield condition.

The serious accidents of automobiles, rail ways, airplanes, etc. and also the defects in atomic power plants occur frequently by the loosening of fastening elements, e.g. bolts and nuts. It should be emphasized that the incorporation of the subloading-friction model is inevitable in the mechanical designs of machine elements for prevention of these accidents.

The comparison with test data for the reduction process of friction coefficient from the static- to kinetic-friction is shown in Fig. 18.14. The test curve for sliding between roughly polished steel surfaces (Ferrero and Barrau 1997) under the quite low sliding velocity $\bar{v}_t \leq 0.0002$ mm/s is simulated well enough by the present model, where the material parameters are selected as follows:

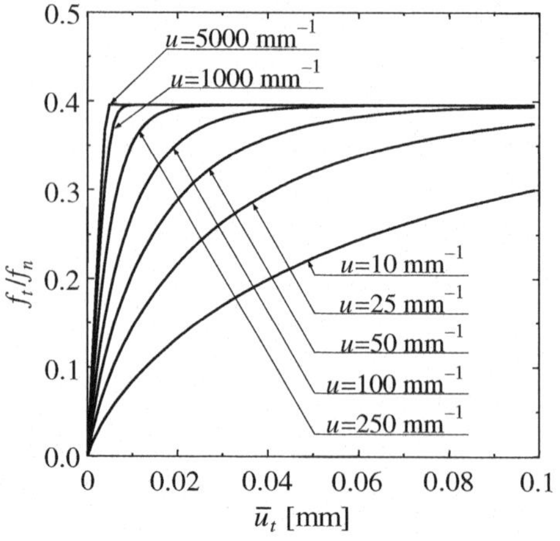

(*a*) $\bar{v}_t = 0.0001$ mm/s

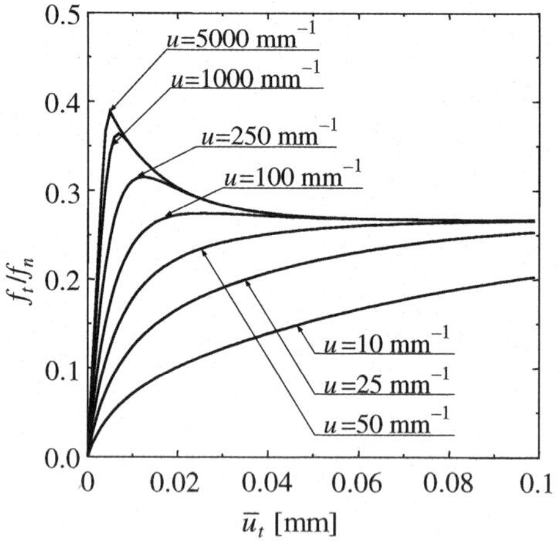

(*b*) $\bar{v} = 0.01$ mm/s

Fig. 18.12 Influences of the material constant in the evolution rule of normal sliding-yield surface on the relation of friction coefficient versus tangential sliding displacement for three levels of tangential sliding velocity

$$\mu_0 = \mu_s = 0.58,\ \mu_k = 0.38$$
$$\kappa = 35\,\text{mm}^{-1},\ \xi = 0.0005/\text{s}$$
$$\tilde{u} = 2000\,\text{mm}^{-1}$$
$$\alpha_n = \alpha_t = 1000\,\text{kN/mm}^3$$

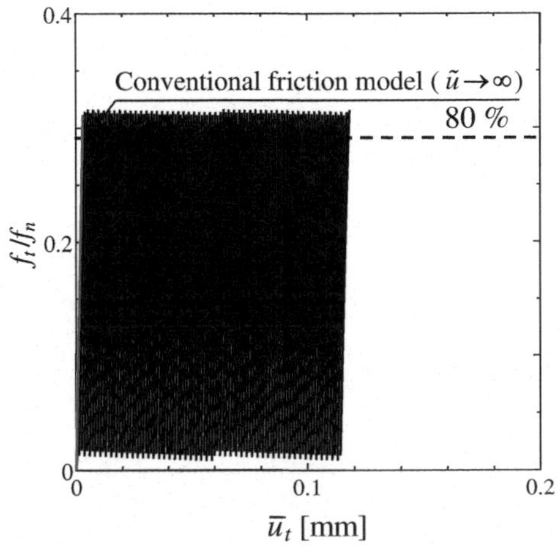

(a) $\bar{v}_t$ = 0.0001 mm/s

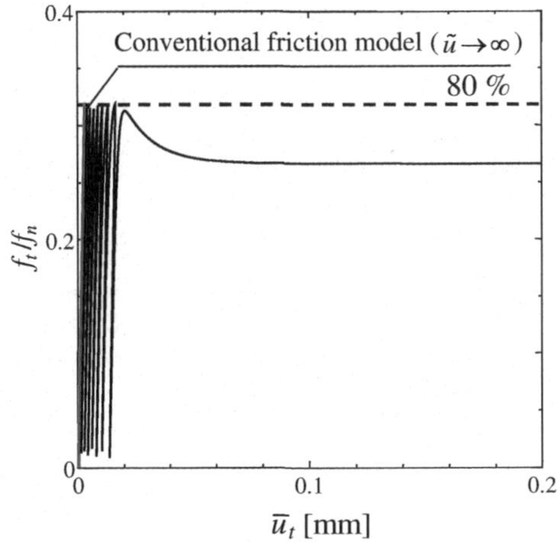

(b) $\bar{v}_t$ = 0.01 mm/s

Fig. 18.13 Influence of sliding velocity on accumulation of sliding displacement under cyclic loading

under the condition

$$f_n = 10\,\text{MPa}, \quad \bar{v}_t = 0.0002\,\text{mm/s}$$

The comparison with test data for the recovery of friction coefficient by the stop of sliding on the way of the reduction process from the static- to kinetic-friction is depicted in Fig. 18.15. The test curves for sliding between roughly polished steel surfaces (Ferrero and Barrau 1997) under the infinitesimal sliding velocity

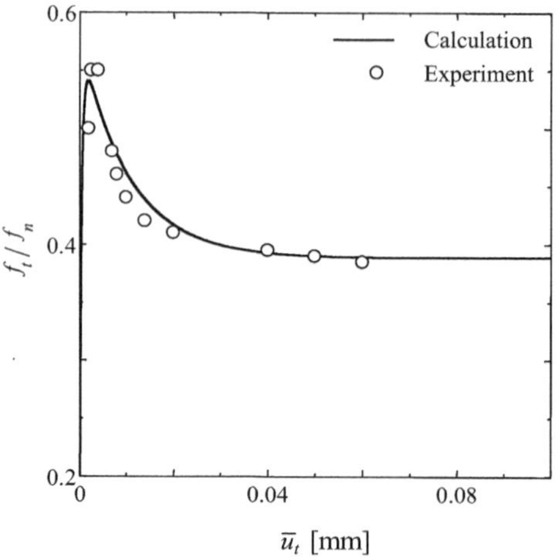

Fig. 18.14 Comparison with test data (Ferrero and Barrau 1997) for reduction of friction coefficient under linear sliding

$\bar{v}_t \leq 0.0002$ mm/s and the stationary time 20 and 400 s are simulated sufficiently well by the present model, where the material parameters are selected same as for the calculation in Fig. 18.14.

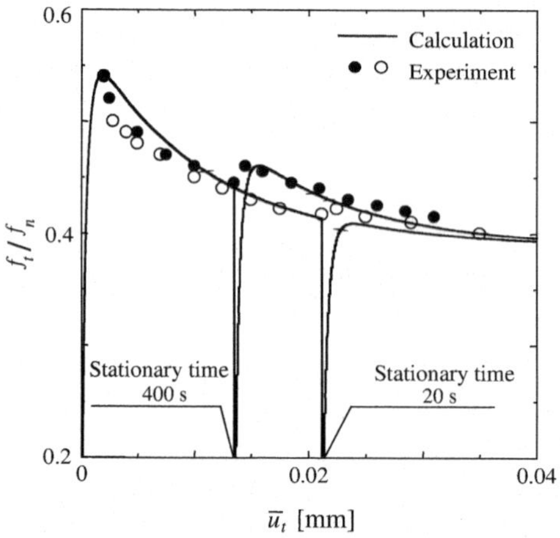

Fig. 18.15 Comparison with test data (Ferrero and Barrau 1997) for influence of stationary time on recovery of friction coefficient

18.8 Extension to Rotational and Orthotropic Anisotropy

The constitutive equation of friction explained in the preceding sections has been extended to describe the anisotropy by the rotation and orthotropy of sliding-yield surface (Hashiguchi 2007; Hashiguchi and Ozaki 2007) and its validity was verified by comparisons with experiments (Ozaki et al. 2012). The variation of friction behavior responding to the sliding direction is predicted pertinently by the extended constitutive equation of friction as will be described in this section.

The simple surface asperity model is illustrated in order to obtain the insight for the anisotropy in Fig. 18.16. Here, the inclination of surface asperities to a particular direction would lead to rotational anisotropy. In addition, the anisotropic shapes and intervals of surface asperities would lead to the orthotropic anisotropy. Now, choosing the bases $\mathbf{e}_1^*$ and $\mathbf{e}_2^*$ in the directions of the maximum and the minimum principal directions of anisotropy, respectively, and letting $\mathbf{e}_3^*$ coincide with $\mathbf{n}$ to make the right-hand coordinate system $\{\mathbf{e}_i^*\}$, the contact stress vector $\mathbf{f}$ and the rotational anisotropy vector $\bar{\boldsymbol{\beta}}$ can be written as

$$\left.\begin{aligned}\mathbf{f} &= f_1^*\mathbf{e}_1^* + f_2^*\mathbf{e}_2^* + f_3^*\mathbf{e}_3^*\\ \bar{\boldsymbol{\beta}} &= \bar{\beta}_1{}^*\mathbf{e}_1^* + \bar{\beta}_2{}^*\mathbf{e}_2^* + \bar{\beta}_3{}^*\mathbf{e}_3^*\end{aligned}\right\} \tag{18.100}$$

In what follows, it is assumed that the rotational anisotropy vector $\bar{\boldsymbol{\beta}}$ is fixed on the contact stress surface, i.e. $\dot{\bar{\beta}}_1^* = \dot{\bar{\beta}}_2^* = \dot{\bar{\beta}}_3^* = 0$, leading to $\dot{\mathbf{e}}_i^* = \boldsymbol{\omega}_c\,\mathbf{e}_i^*$.

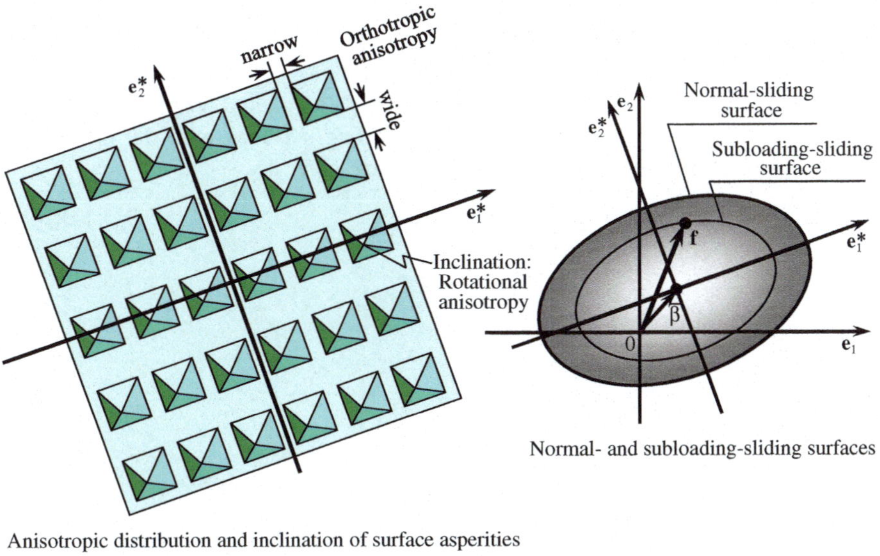

Fig. 18.16 Surface asperity model suggesting the rotational and the orthotropic anisotropy

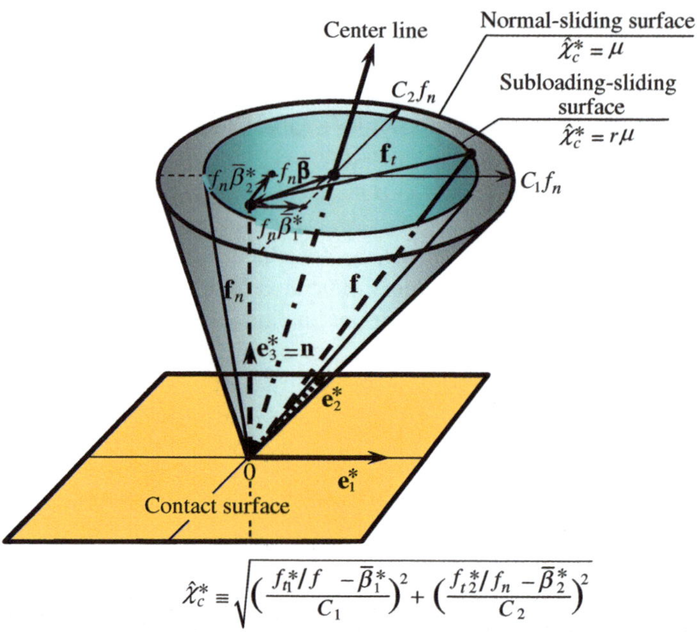

Fig. 18.17 Anisotropic normal- and subloading-sliding surfaces

Equation (18.100) is rewritten by $f_1^* = f_{t1}^*, f_2^* = f_{t2}^*, f_3^* = -f_n^*$ and $\bar{\beta}_1^* = \bar{\beta}_1^*$, $\bar{\beta}_2^* = \bar{\beta}_2^*$, $\bar{\beta}_3^* = 0$ as follows:

$$\left.\begin{array}{l} \mathbf{f} = f_{t1}^*\mathbf{e}_1^* + f_{t2}^*\mathbf{e}_2^* - f_n\mathbf{e}_3^* \\ \bar{\boldsymbol{\beta}} = \bar{\beta}_1{}^*\mathbf{e}_1^* + \bar{\beta}_2{}^*\mathbf{e}_2^* \end{array}\right\} \tag{18.101}$$

Invoking the orthotropic anisotropy proposed by Mroz and Stupkiewicz (1994), the normal-yield sliding and the subloading-sliding surfaces with the orthotropic and the rotational hardenings (see Fig. 18.17) are given by

$$\hat{\chi}_c^* = \mu \tag{18.102}$$

$$\hat{\chi}_c^* = r\mu \tag{18.103}$$

where

$$\hat{\chi}_c^* \equiv \sqrt{\hat{\chi}_{c1}^{*2} + \hat{\chi}_{c2}^{*2}}, \quad \hat{\chi}_{c1}^* \equiv \frac{\hat{\eta}_1^*}{C_1}, \quad \hat{\chi}_{c2}^* \equiv \frac{\hat{\eta}_2^*}{C_2} \tag{18.104}$$

$$\left.\begin{array}{l} \eta_1^* \equiv \dfrac{f_{t1}^*}{f_n}, \quad \eta_2^* \equiv \dfrac{f_{t2}^*}{f_n} \\ \hat{\eta}_1^* \equiv \hat{\eta}_1^* - \bar{\beta}_1^*, \quad \hat{\eta}_2^* \equiv \hat{\eta}_2^* - \bar{\beta}_2^* \end{array}\right\} \tag{18.105}$$

C_1 and C_2 are the material constants designating the orthotropic anisotropy, whereas the $\mathbf{e}_1^*$-direction is chosen for the long axis of ellipsoid in the cross section of sliding-yield surface so that μ designates the friction coefficient for $C_1 = C_2 = 1$ and $\bar{\beta}_{t1}^* = \bar{\beta}_{t2}^* = 0$ leading to the isotropic sliding-yield surface.

The partial derivatives for Eq. (18.104) are given as follows:

$$\left.\begin{aligned}
\frac{\partial \hat{\chi}_{ci}^*}{\partial f_{ti}^*} &= \frac{\partial (f_{ti}^*/f_n - \bar{\beta}_i^*)/C_i}{\partial f_{ti}^*} = \frac{1}{f_n C_i} \text{(no sum)} \\
\frac{\partial \hat{\chi}_{ci}^*}{\partial f_n} &= \frac{\partial (f_{ti}^*/f_n - \bar{\beta}_i^*)/C_i}{\partial f_n} = \frac{-f_{ti}^*}{f_n^2 C_i} = -\frac{\hat{\chi}_{ci}^*}{f_n} \text{(no sum)} \\
\frac{\partial \hat{\chi}_c^*}{\partial \hat{\chi}_{ci}^*} &= \frac{1}{2\hat{\chi}_c^*} 2\hat{\chi}_{ci}^* = \frac{\hat{\chi}_{ci}^*}{\hat{\chi}_c^*} \equiv \hat{\zeta}_{ci}^* \\
\frac{\partial \hat{\chi}_c^*}{\partial f_{ti}^*} &= \frac{\partial \hat{\chi}_c^*}{\partial \hat{\chi}_{ci}^*} \frac{\partial \hat{\chi}_{ci}^*}{\partial f_{ti}^*} = \frac{\hat{\chi}_{ci}^*}{\hat{\chi}_c^*} \frac{1}{f_n C_i} = \frac{1}{f_n} \frac{\hat{\zeta}_{ci}^*}{C_i} \text{(no sum)} \\
\frac{\partial \hat{\chi}_c^*}{\partial f_n} &= \frac{\partial \hat{\chi}_c^*}{\partial \hat{\chi}_{ci}^*} \frac{\partial \hat{\chi}_{ci}^*}{\partial f_n} + \frac{\partial \hat{\chi}_c^*}{\partial \hat{\chi}_{ci}^*} \frac{\partial \hat{\chi}_{ci}^*}{\partial f_n} = -\frac{1}{f_n} (\hat{\zeta}_{ci}^* \chi_{ci}^* + \hat{\zeta}_{ci}^* \chi_{c2}^*)
\end{aligned}\right\} \tag{18.106}$$

Further, it holds from Eqs. (18.23) and (18.106) that

$$\frac{\partial \hat{\chi}_c^*}{\partial \mathbf{f}^*} = \frac{1}{f_n} \left[\frac{\hat{\zeta}_{c1}^*}{C_1} \mathbf{e}_1^* + \frac{\hat{\zeta}_{c2}^*}{C_2} \mathbf{e}_2^* + \left(\hat{\zeta}_{c1}^* \chi_{c1}^* + \hat{\zeta}_{c2}^* \chi_{c2}^* \right) \mathbf{n} \right] \tag{18.107}$$

$$\begin{aligned}
(\mathbf{I} - \mathbf{n} \otimes \mathbf{n}) \frac{\partial \hat{\chi}_c^*}{\partial \mathbf{f}^*} &= (\mathbf{I} - \mathbf{n} \otimes \mathbf{n}) \frac{1}{f_n} \left[\frac{\hat{\zeta}_{c1}^*}{C_1} \mathbf{e}_1^* + \frac{\hat{\zeta}_{c2}^*}{C_2} \mathbf{e}_2^* + (\hat{\zeta}_{c1}^* \chi_{c1}^* + \hat{\zeta}_{c2}^* \chi_{c2}^*) \mathbf{n} \right] \\
&= \frac{1}{f_n} \left(\frac{\hat{\zeta}_{c1}^*}{C_1} \mathbf{e}_1^* + \frac{\hat{\zeta}_{c2}^*}{C_2} \mathbf{e}_2^* \right)
\end{aligned} \tag{18.108}$$

$$\begin{aligned}
\mathbf{C}^e \frac{\partial \hat{\chi}_c^*}{\partial \mathbf{f}^*} &= [\alpha_n \mathbf{e}_3^* \otimes \mathbf{e}_3^* + \alpha_t (\mathbf{e}_1^* \otimes \mathbf{e}_1^* + \mathbf{e}_2^* \otimes \mathbf{e}_2^*)] \\
&\quad \frac{1}{f_n} \left[\frac{\hat{\zeta}_{c1}^*}{C_1} \mathbf{e}_1^* + \frac{\hat{\zeta}_{c2}^*}{C_2} \mathbf{e}_2^* + (\hat{\zeta}_{c1}^* \chi_{c1}^* + \hat{\zeta}_{c2}^* \chi_{c2}^*) \mathbf{n} \right] \\
&= \frac{1}{f_n} \left[\alpha_t \left(\frac{\hat{\zeta}_{c1}^*}{C_1} \mathbf{e}_1^* + \frac{\hat{\zeta}_{c2}^*}{C_2} \mathbf{e}_2^* \right) + \alpha_n (\hat{\zeta}_{c1}^* \chi_{c1}^* + \hat{\zeta}_{c2}^* \chi_{c2}^*) \mathbf{n} \right]
\end{aligned} \tag{18.109}$$

$$\mathbf{t}_v^p = \frac{\frac{\hat{\zeta}_{c1}^*}{C_1}\mathbf{e}_1^* + \frac{\hat{\zeta}_{c2}^*}{C_2}\mathbf{e}_2^*}{\sqrt{\left(\frac{\hat{\zeta}_{c1}^*}{C_1}\right)^2 + \left(\frac{\hat{\zeta}_{c2}^*}{C_2}\right)^2}} \tag{18.110}$$

$$\begin{aligned}\frac{\partial \hat{\chi}_c^*}{\partial \mathbf{f}^*}\mathbf{C}^e \cdot \mathbf{t}_v^p &= \frac{1}{f_n}\left[\alpha_t\left(\frac{\hat{\zeta}_{c1}^*}{C_1}\mathbf{e}_1^* + \frac{\hat{\zeta}_{c2}^*}{C_2}\mathbf{e}_2^*\right) + \alpha_n(\hat{\zeta}_{c1}^*\chi_{c1}^* + \hat{\zeta}_{c2}^*\chi_{c2}^*)\mathbf{n}\right] \cdot \frac{\frac{\hat{\zeta}_{c1}^*}{C_1}\mathbf{e}_1^* + \frac{\hat{\zeta}_{c2}^*}{C_2}\mathbf{e}_2^*}{\sqrt{\left(\frac{\hat{\zeta}_{c1}^*}{C_1}\right)^2 + \left(\frac{\hat{\zeta}_{c2}^*}{C_2}\right)^2}} \\ &= \frac{\alpha_t}{f_n}\sqrt{\left(\frac{\hat{\zeta}_{c1}^*}{C_1}\right)^2 + \left(\frac{\hat{\zeta}_{c2}^*}{C_2}\right)^2}\end{aligned} \tag{18.111}$$

Substituting Eqs. (18.23), (18.101) and (18.107)–(18.111) into Eqs. (18.58) and (18.60), we obtain the sliding velocity versus contact traction rate and its inverse relation as follows:

$$\begin{aligned}\bar{\mathbf{v}} &= \left[\frac{1}{\alpha_n}\mathbf{n}\otimes\mathbf{n} + \frac{1}{\alpha_t}(\mathbf{I}-\mathbf{n}\otimes\mathbf{n})\right]\left(\overset{\circ}{f}{}^*_{t1}\mathbf{e}_1^* + \overset{\circ}{f}{}^*_{t2}\mathbf{e}_2^* - \overset{\circ}{f}_n\mathbf{n}\right) \\ &\quad + \frac{\frac{1}{f_n}\left[\frac{\hat{\zeta}_{c1}^*}{C_1}\mathbf{e}_1^* + \frac{\hat{\zeta}_{c2}^*}{C_2}\mathbf{e}_2^* + \left(\hat{\zeta}_{c1}^*\chi_{c1}^* + \hat{\zeta}_{c2}^*\chi_{c2}^*\right)\mathbf{n}\right] \cdot \left(\overset{\circ}{f}{}^*_{t1}\mathbf{e}_1^* + \overset{\circ}{f}{}^*_{t2}\mathbf{e}_2^* - \overset{\circ}{f}_n\mathbf{n}\right) - m^c}{m^p} \\ &\quad \frac{\frac{\hat{\zeta}_{c1}^*}{C_1}\mathbf{e}_1^* + \frac{\hat{\zeta}_{c2}^*}{C_2}\mathbf{e}_2^*}{\sqrt{\left(\frac{\hat{\zeta}_{c1}^*}{C_1}\right)^2 + \left(\frac{\hat{\zeta}_{c2}^*}{C_2}\right)^2}} \\ &= \frac{1}{\alpha_t}\left(\overset{\circ}{f}{}^*_{t1}\mathbf{e}_1^* + \overset{\circ}{f}{}^*_{t2}\mathbf{e}_2^*\right) - \frac{1}{\alpha_n}\overset{\circ}{f}_n\mathbf{n} \\ &\quad + \frac{\frac{1}{f_n}\left[\frac{\hat{\zeta}_{c1}^*}{C_1}\overset{\circ}{f}{}^*_{t1} + \frac{\hat{\zeta}_{c2}^*}{C_2}\overset{\circ}{f}{}^*_{t2} - \left(\hat{\zeta}_{c1}^*\chi_{c1}^* + \hat{\zeta}_{c2}^*\chi_{c2}^*\right)\overset{\circ}{f}_n\right] - m^c}{m^p}\frac{\frac{\hat{\zeta}_{c1}^*}{C_1}\mathbf{e}_1^* + \frac{\hat{\zeta}_{c2}^*}{C_2}\mathbf{e}_2^*}{\sqrt{\left(\frac{\hat{\zeta}_{c1}^*}{C_1}\right)^2 + \left(\frac{\hat{\zeta}_{c2}^*}{C_2}\right)^2}}\end{aligned} \tag{18.112}$$

$$\overset{\circ}{\mathbf{f}} = \{\alpha_n \mathbf{n}\otimes\mathbf{n} + \alpha_t(\mathbf{e}_1^*\otimes\mathbf{e}_1^* + \mathbf{e}_2^*\otimes\mathbf{e}_2^*)\}\left[\bar{v}_1\mathbf{e}_1^* + \bar{v}_2\mathbf{e}_2^* - \bar{v}_n\mathbf{n}\right.$$

$$-\left\langle\frac{\frac{1}{f_n}\left[\alpha_t\left(\frac{\hat{\zeta}_{c1}^*}{C_1}\mathbf{e}_1^* + \frac{\hat{\zeta}_{c2}^*}{C_2}\right) + \alpha_n(\hat{\zeta}_{c1}^*\chi_{c1}^* + \hat{\zeta}_{c2}^*\chi_{c2}^*)\mathbf{n}\right]\cdot(\bar{v}_1\mathbf{e}_1^* + \bar{v}_2\mathbf{e}_2^* - \bar{v}_n\mathbf{n}) - m^c}{m^p + \frac{\alpha_t}{f_n}\sqrt{\left(\frac{\hat{\zeta}_{c1}^*}{C_1}\right)^2 + \left(\frac{\hat{\zeta}_{c2}^*}{C_2}\right)^2}}\right\rangle$$

$$\left.\frac{\frac{\hat{\zeta}_{c1}^*}{C_1}\mathbf{e}_1^* + \frac{\hat{\zeta}_{c2}^*}{C_2}\mathbf{e}_2^*}{\sqrt{\left(\frac{\hat{\zeta}_{c1}^*}{C_1}\right)^2 + \left(\frac{\hat{\zeta}_{c2}^*}{C_2}\right)^2}}\right] = \alpha_t(\bar{v}_1\mathbf{e}_1^* + \bar{v}_2\mathbf{e}_2^*) - \alpha_n\bar{v}_n\mathbf{n}$$

$$-\alpha_t\left\langle\frac{\frac{1}{f_n}\left[\alpha_t\left(\frac{\hat{\zeta}_{c1}^*}{C_1}\bar{v}_1 + \frac{\hat{\zeta}_{c2}^*}{C_2}\bar{v}_2\right) - \alpha_n\left(\hat{\zeta}_{c1}^*\chi_{c1}^* + \hat{\zeta}_{c2}^*\chi_{c2}^*\right)\bar{v}_n\right] - \bar{m}^c}{m^p + \frac{\alpha_t}{f_n}\sqrt{\left(\frac{\hat{\zeta}_{c1}^*}{C_1}\right)^2 + \left(\frac{\hat{\zeta}_{c2}^*}{C_2}\right)^2}}\right\rangle\frac{\frac{\hat{\zeta}_{c1}^*}{C_1}\mathbf{e}_1^* + \frac{\hat{\zeta}_{c2}^*}{C_2}\mathbf{e}_2^*}{\sqrt{\left(\frac{\hat{\zeta}_{c1}^*}{C_1}\right)^2 + \left(\frac{\hat{\zeta}_{c2}^*}{C_2}\right)^2}} \tag{18.113}$$

The calculation for sliding with the orthotropic anisotropy must be performed in the coordinate system with the principal axes of orthotropic anisotropy, i.e. $(\mathbf{e}_1^*, \mathbf{e}_2^*, \mathbf{n})$.

We examine below the basic response of the present friction model by numerical experiments and comparisons with test data for the linear sliding phenomenon without a normal sliding velocity leading to $\bar{\mathbf{v}}_n = \mathbf{0}$.

The traction rate versus sliding velocity relation is given by substituting $\bar{v}_n = 0$ into Eq. (18.113) as

$$\overset{\circ}{\mathbf{f}} = \alpha_t(\bar{v}_1\mathbf{e}_1^* + \bar{v}_2\mathbf{e}_2^*)$$
$$-\alpha_t\left\langle\frac{\frac{1}{f_n}\alpha_t\left(\frac{\hat{\zeta}_{c1}^*}{C_1}\bar{v}_1 + \frac{\hat{\zeta}_{c2}^*}{C_2}\bar{v}_2\right) - m^c}{m^p + \frac{\alpha_t}{f_n}\sqrt{\left(\frac{\hat{\zeta}_{c1}^*}{C_1}\right)^2 + \left(\frac{\hat{\zeta}_{c2}^*}{C_2}\right)^2}}\right\rangle\frac{\frac{\hat{\zeta}_{c1}^*}{C_1}\mathbf{e}_1^* + \frac{\hat{\zeta}_{c2}^*}{C_2}\mathbf{e}_2^*}{\sqrt{\left(\frac{\hat{\zeta}_{c1}^*}{C_1}\right)^2 + \left(\frac{\hat{\zeta}_{c2}^*}{C_2}\right)^2}} \tag{18.114}$$

where m^p and m^c are given by Eqs. (18.55) and (18.56).

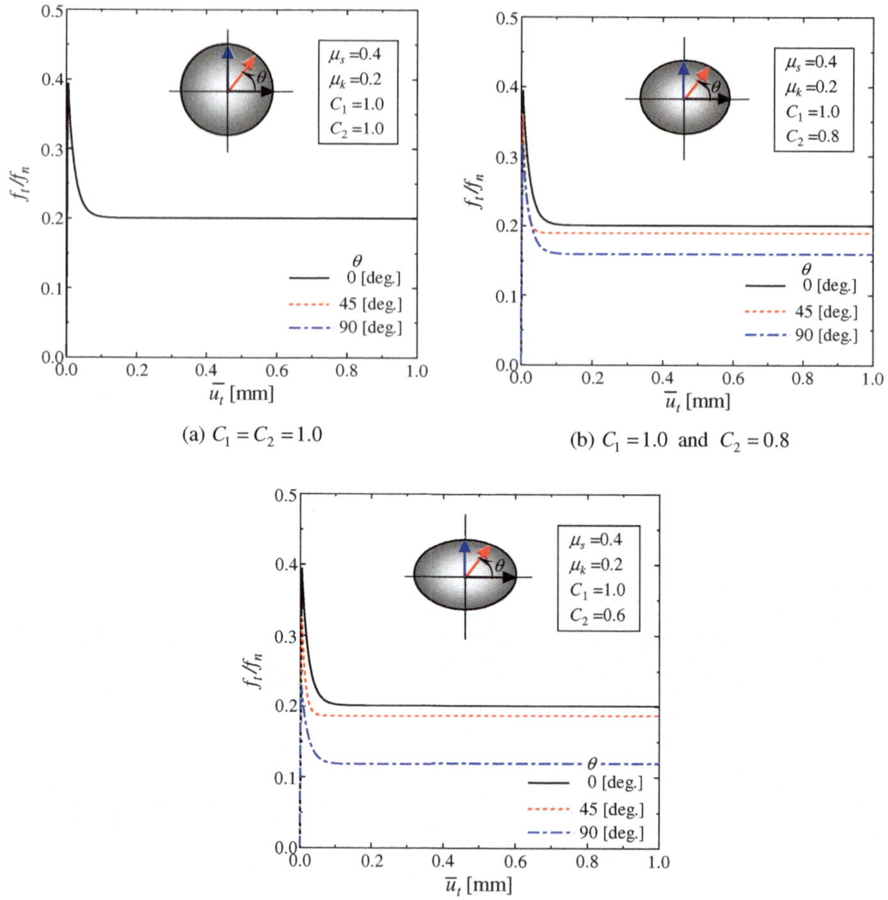

(a) $C_1 = C_2 = 1.0$

(b) $C_1 = 1.0$ and $C_2 = 0.8$

(c) $C_1 = 1.0$ and $C_2 = 0.6$

Fig. 18.18 Influence of rotational anisotropy on relation of contact stress ratio (Ozaki et al. 2012)

In what follows, we demonstrate the basic response of the present anisotropic friction model through numerical experiments for the linear sliding phenomenon without a rigid-body rotation under a constant normal traction. The calculations described in the following have been executed by Dr. S. Ozaki.

The nine material parameters and the initial value are selected as

$$
\begin{aligned}
\mu_0 &= \mu_s = 0.4,\ \mu_k = 0.2 \\
\kappa &= 10\,\text{mm}^{-1},\ \xi = 0.1/\text{s} \\
\tilde{u} &= 1000\,\text{mm}^{-1} \\
\alpha_n &= \alpha_t = 1000\,\text{N/mm}^3
\end{aligned}
$$

under the condition

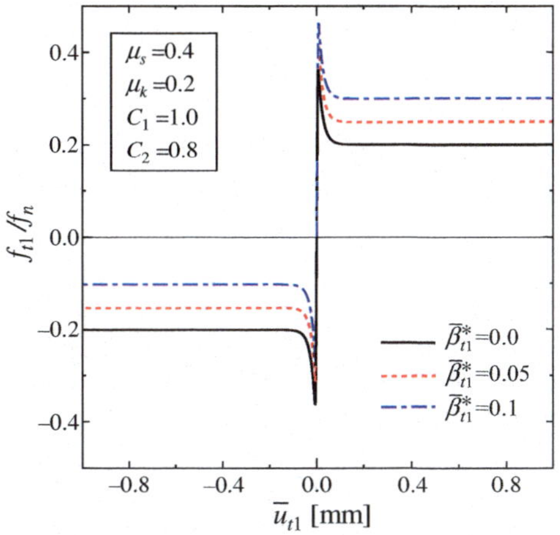

Fig. 18.19 Influence of rotational anisotropy on contact stress (Ozaki and Hashiguchi 2012)

$$f_n = 10\,\mathrm{MPa}, \quad \bar{v}_t = 1.0\,\mathrm{mm/s}$$

The variations in the traction ratio f_t/f_n with the tangential sliding displacement $\bar{u}_t$ is shown in Fig. 18.18. Here, we assume the three sets of the parameters for orthotropy C_1 and C_2 without the rotational anisotropy, i.e., $\bar{\boldsymbol{\beta}} = \mathbf{0}$. Then, the constant sliding velocity of the magnitude $\bar{v}_t = 1.0$ mm/s was given in the directions 0°, 45° and 90° from the base vector $\mathbf{e}_1^*$ of orthotropy. As shown in this figure, the friction behavior varies with the sliding direction because of orthotropic anisotropy.

The influence of the rotational anisotropy, i.e., the parameter $\bar{\boldsymbol{\beta}} = \{\overset{*}{\bar{\beta}_1}\,\overset{*}{\bar{\beta}_2}\,0\}$ on the relation of the traction ratio with the tangential sliding displacement is shown in Fig. 18.19. In this calculation, we set the orthotropic anisotropic parameters as $C_1 = 1.0$ and $C_2 = 0.8$, and set the rotational hardening parameters $\overset{*}{\bar{\beta}_1}$ as 0.0, 0.05, and 0.1 and $\bar{\beta}_2{}^* = 0.0$. Then, the constant sliding velocity $\bar{v}_t = 1.0$ mm/s is given into mutually opposite directions. It is confirmed that the frictional properties for mutually opposite directions of sliding are different from each other. The description of the differences in friction coefficients in opposite directions is important in biomimetic textures and in drive systems of off-the-road vehicles and robots.

Some other verifications of the pertinence of the present model by the comparisons with test data are referred to Ozaki et al. (2012).

18.9 Stick-Slip Phenomenon

When the sliding between solid bodies proceeds in a low velocity, the unstable motion leading to the intermittent vibration phenomenon is induced, which is referred to as the *stick-slip motion*. The unstable motion influences on the performance of machinery and wear, fatigue, durability and acoustic emission systems. It should be avoided in elements such as gears and bearings to produce the smooth movement. In addition, the stick-slip motion is observed in earthquakes in a continual plate sliding type. In what follows, let the stick-slip motion be simulated by the subloading-friction model described in the preceding sections. The stick-slip analysis delineated in this section was executed pertinently by Dr. S. Ozaki (Ozaki and Hashiguchi 2010).

First, let the stick-slip motion be examined qualitatively by the simplest example, i.e., the one-dimensional sliding as shown in Fig. 18.20, while the movement of slider is shown in Fig. 18.20a and the variation of contact tangential stress and the sliding displacement of slider with the time is shown in Fig. 18.20b. The slider is connected to the spring which is pulled in a constant velocity.

(1) The slider stops until the spring force reaches the value causing the static friction as shown in the processes (1-0)–(1-s).
(2) At the moment when the spring force reaches the value causing the static friction, the slider moves until spring force decreases to the value corresponding to the kinetic friction as shown in the process (1-k).
(3) The identical phenomena to the processes (1-m)–(1-k) is repeated as shown in the processes (2-m)–(3-k) and so on.

We now describe mathematically the stick-slip instability, taking account of the acceleration of slider. Denoting the constant velocity at the spring-end induced by the driver by V and the displacement of deriver by U, the spring elongation is given as $(U - \bar{u}_t\ (= \int (V - \bar{v}_t)dt))$. The equation of motion is given by

$$K(U - \bar{u}_t) - f_t S = M\bar{a}_t \tag{18.115}$$

where M is the mass of the slider, K is the spring stiffness, S is the nominal contact area, and $\bar{a}_t$ and $\bar{u}_t$ are the acceleration and the displacement, respectively, of the slider, which are relative values with respect to the fixed base. The tangential contact traction f_t is estimated by Eq. (18.97).

Now, we show the numerical simulation of the test result for the stick-slip behavior measured by Baumberger et al. (1994). The structure of test apparatus is as follows:

Mass of slider: $M = 0.8$ kg, Contact area: $S = 720\ \mathrm{mm}^2$, Spring constant: $K = 58\ \mathrm{N/mm}$, Velocity of deriver of spring: $V = 0.001\ \mathrm{mm/s}$.

The comparison with test data is shown in Fig. 18.21, where the material parameters are chosen as follows:

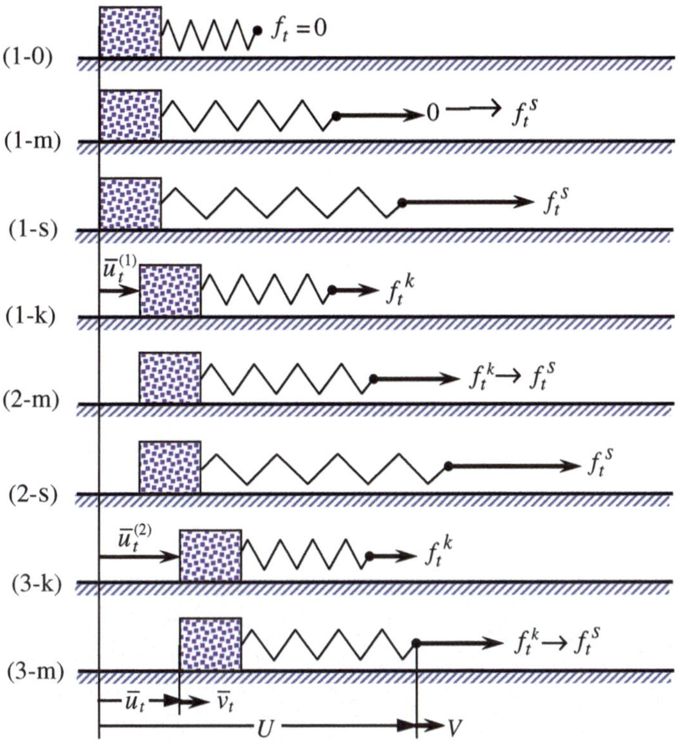

(a) Movement of slider.

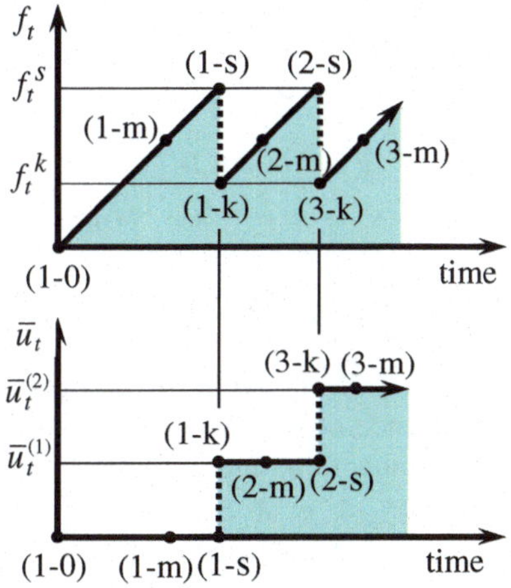

(b) Variations of force and movement of slider.

Fig. 18.20 One-dimensional stick-slip phenomenon

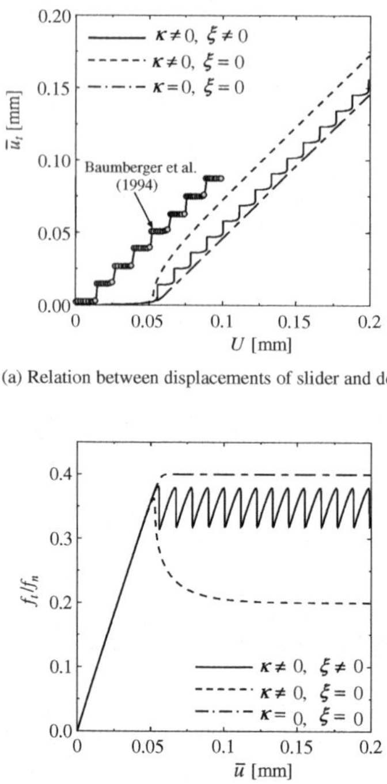

(a) Relation between displacements of slider and deriver

(b) Relation between contact stress ratio versus sliding displacement.

Fig. 18.21 Simulation of stick-slip phenomenon with comparison to test result after Baumberger et al. (1994), (Ozaki and Hashiguchi 2010)

$$\begin{aligned}\mu_0 &= \mu_s = 0.4,\ \mu_k = 0.2\\ \kappa &= 10\ \mathrm{mm}^{-1},\ \xi = 0.1\\ \tilde{u} &= 1000\ \mathrm{mm}^{-1}\\ \alpha_n &= \alpha_t = 1000\ \mathrm{N/mm}^3\end{aligned}$$

The influences of the mass of slider and the spring constant are shown in Figs. 18.22 and 18.23. The larger mass of slider and the weaker spring constant induce the more intense stick-slip behavior as observed in this figure.

Now, examine the influence of deriver velocity on the stick-slip motion. Then, the variations of spring elongation under the increase and decrease of deriver velocity are shown in Fig. 18.24 for $K = 50\ \mathrm{N/mm}$, $M = 1\ \mathrm{kg}$, $S = 1000\ \mathrm{mm}^2$. The material parameters are chosen to be same as the ones for Fig. 18.21. The responses for the cases that the deriver velocity increases from $V = 0.0005\ \mathrm{mm/s}$ and decreases from $V = 0.01\ \mathrm{mm/s}$ linearly are shown in this figure. It is observed that the stick-slip movement converges as the deriver velocity increases and that it commences and amplifies as the velocity decreases.

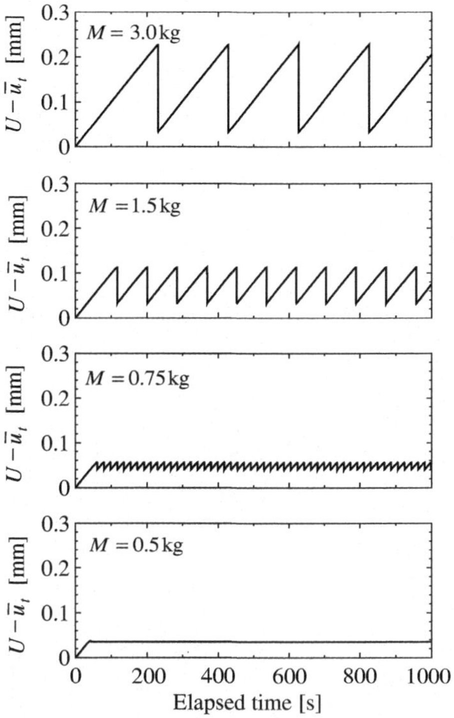

Fig. 18.22 Influence of slider mass on stick-slip behavior (Ozaki and Hashiguchi 2010)

The prediction of earthquake occurrence has been done by the unreliable method by the rate-and-state model (cf. e.g. Dieterich 1978, 1979; Ruina 1980, 1983; Rice and Ruina 1983; Scholz 1998; Rice et al. 2001; Kame 2013) which is quite irrational as described in Sect. 18.1. The reliable prediction of the earthquake would be able to be realized by incorporating the subloading-friction model.

18.10 Generalised Subloading Friction Model: Subloading-Overstress Friction Model

In the subloading friction model, the sliding velocity is postulated to consist of elastic and plastic sliding velocities. It leads to negative rate-sensitivity (i.e., a decrease in friction resistance with increasing sliding velocity) because the adhesion of surface asperities is lost quickly at high sliding velocities and the recovery of adhesion requires time. In contrast, the *positive rate sensitivity* (i.e. an increase in the friction resistance with increasing sliding velocity) is observed in the sliding between lubricated surfaces or between soft solids, e.g. indium, Teflon and various polymers, which is often called the *fluid friction* or the *wet friction*. Its theoretical

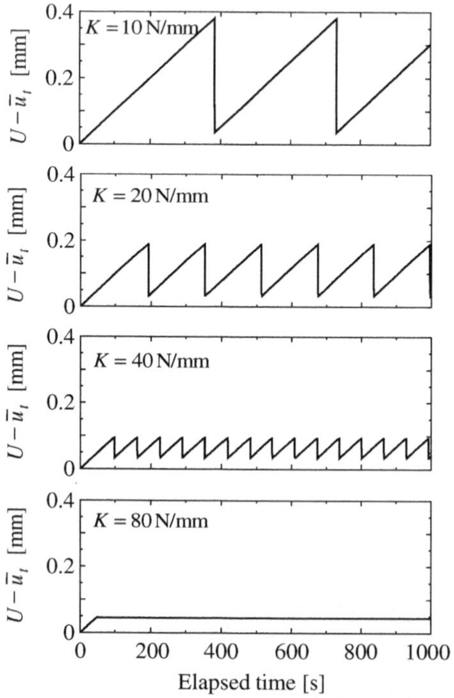

Fig. 18.23 Influence of spring constant on stick-slip behavior (Ozaki and Hashiguchi 2010)

prediction is of importance for mechanical designs of lubricated machinery elements such as gears and bearings in which low friction resistances are desirable and of wheel tires travelling on wet roads in which high friction resistances are required for high breaking performances, and so forth.

In what follows, we consider the generalisation of the subloading friction model to describe not only negative but also positive rate sensitivities (Hashiguchi et al. 2016) by introducing the notion of the overstress which has been adopted in the description of viscoplastic deformation behaviour.

18.10.1 Formulation of Generalised Subloading Friction Model

First, we introduce the viscoplastic sliding velocity $\bar{\mathbf{v}}^{vp}$ instead of the plastic sliding velocity $\bar{\mathbf{v}}^{p}$ in Eq. (18.4); i.e.

$$\bar{\mathbf{v}} = \bar{\mathbf{v}}^{e} + \bar{\mathbf{v}}^{vp}. \tag{18.116}$$

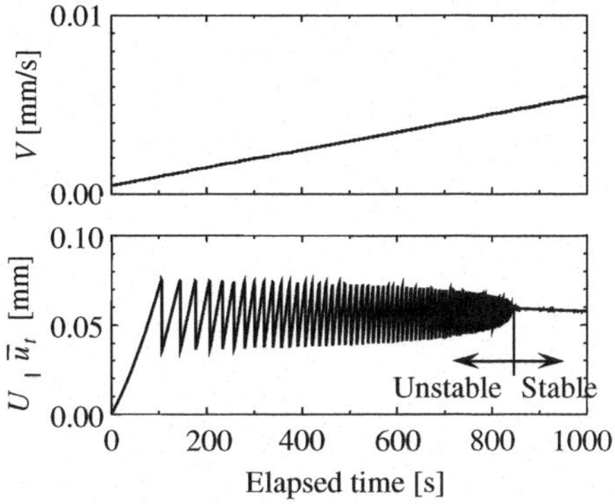

(a) Linear increasing process of deriver velocity

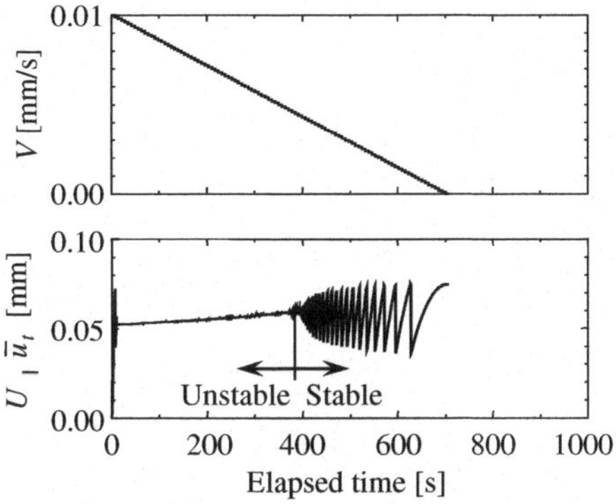

(b) Linear decreasing process of deriver velocity

Fig. 18.24 Variation of spring elongation under linear variation of deriver velocity (Ozaki and Hashiguchi 2010)

Viscoplastic models describing rate-dependent plastic deformation are classified to either the creep or the overstress models. Depending on the ratio of the stress to the yield stress, a viscoplastic strain rate is always induced inappropriately even if the stress is lower than the yield stress in the creep models (e.g., see Norton (1929) for the case without a yield surface and Nakada and Keh (1966), Hutchinson (1976), and Pan and Rice (1983) for the case with a yield surface). In contrast, a viscoplastic strain rate is induced depending on the overstress from the yield surface in the overstress model (Bingham 1922; Hohenemser and Prager 1932; Prager 1961; Perzyna 1963, 1966). In other words, whereas the creep model does not possess a loading criterion for the creep strain rate, the overstress model possesses one for the viscoplastic strain rate. Here, it should be noted that a loading criterion for the viscoplastic strain rate as well as for the plastic strain rate in the ordinary elastoplastic constitutive equation is required for the realistic description of the rate-dependent elastoplastic deformation. The subloading friction model can then be generalised so as to describe both negative and positive rate sensitivities appropriately by incorporating the concept of overstress. First, let the viscoplastic sliding velocity be given as

$$\bar{\mathbf{v}}^{vp} = \frac{1}{\eta_v}\left\langle\frac{f(\mathbf{f})-\mu}{\mu}\right\rangle^n \mathbf{t}_n = \frac{1}{\eta_v}\langle r-1\rangle^n \mathbf{t}_n. \tag{18.117}$$

where η_v and n are material constants, η_v being the viscoplastic coefficient. Hence, by substituting Eqs. (18.22) and (18.117) into Eq. (18.116), we have

$$\bar{\mathbf{v}} = \mathbf{C}^{e-1}\overset{\circ}{\mathbf{f}} + \frac{1}{\eta_v}\langle r-1\rangle^n \mathbf{t}_n, \tag{18.118}$$

$$\overset{\circ}{\mathbf{f}} = \mathbf{C}^e\bar{\mathbf{v}} - \frac{1}{\eta_v}\langle r-1\rangle^n \mathbf{C}^e\mathbf{t}_n, \tag{18.119}$$

$\langle\,\rangle$ denotes the Macaulay bracket; i.e., $s<0:\ \langle s\rangle = 0$ and $s\geq 0:\ \langle s\rangle = s$ (where s is an arbitrary scalar). Note that the loading criterion for the viscoplastic sliding velocity is furnished by introducing the Macaulay brackets. The surface that passes through the current contact stress is similar to the normal friction-yield surface (i.e., similarity in the constant normal contact stress plane for the Coulomb friction-yield surface) and is called the *dynamic loading friction surface*. Renamed the *dynamic loading friction-yield ratio*, r is defined as the ratio of the size of the dynamic loading friction surface to that of the normal friction-yield surface. Here, note that $r\leq 1$ holds in the quasi-static sliding process but $r>1$ holds in the dynamic sliding process. A variation of the internal variables is induced generally by irreversible (inelastic) sliding and hence the plastic sliding velocity $\bar{\mathbf{v}}^p$ in the evolution rules for the internal variables in the elastoplastic sliding equation should be replaced by the viscoplastic sliding velocity $\bar{\mathbf{v}}^{vp}$ in the evolution rules for the internal variables in the elasto-viscoplastic sliding equation. Then, from Eq. (18.38), the evolution rule for the coefficient of friction μ is generalised to

$$\boxed{\dot{\mu} = \underbrace{-\kappa\left(1-\frac{\mu}{\mu_k}\right)||\bar{\mathbf{v}}^{vp}||}_{\text{Negative}} + \underbrace{\xi\left(1-\frac{\mu}{\mu_s}\right)}_{\text{Positive}}} \tag{18.120}$$

Note here that Eqs. (18.118) and (18.119) are rewritten in incremental forms:

$$d\bar{\mathbf{u}} = \mathbf{C}^{e-1}d\mathbf{f} + \frac{1}{\eta_v}\langle r-1\rangle^n \mathbf{t}_n dt, \tag{18.121}$$

$$d\mathbf{f} = \mathbf{C}^e d\bar{\mathbf{u}} - \frac{1}{\eta_v}\langle r-1\rangle^n \mathbf{C}^e\mathbf{t}_n dt, \tag{18.122}$$

where $d\bar{\mathbf{u}}$ is the increment for the sliding displacement; i.e., $d\bar{\mathbf{u}} \equiv \bar{\mathbf{v}}dt$. For a quasi-static deformation, Eqs. (18.118) and (18.119) reduce to

$$\mathbf{0} \cong \mathbf{0} + \frac{1}{\eta_v}\langle r-1\rangle^n \mathbf{t}_n dt, \tag{18.123}$$

$$\mathbf{0} \cong \mathbf{0} - \frac{1}{\eta_v}\langle r-1\rangle^n \mathbf{C}^e \mathbf{t}_n dt, \tag{18.124}$$

which leads to

$$r-1 \to 0, \text{ i.e. } f(\boldsymbol{\sigma}) - \mu \to 0, \tag{18.125}$$

fulfilling the yield condition so that the rate-independent elastoplastic deformation behaviour is described in the stress state lower than the yield stress state; i.e., the sub-yield state. In impact loading, in contrast, Eqs. (18.118) and (18.119) reduce to

$$d\bar{\mathbf{u}} \cong \mathbf{C}^{e-1} d\mathbf{f} + \mathbf{0}, \text{i.e.} \bar{\mathbf{v}} \cong \mathbf{C}^{e-1} \overset{\circ}{\mathbf{f}}, \tag{18.126}$$

$$d\mathbf{f} \cong \mathbf{C}^e d\bar{\mathbf{u}} - \mathbf{0}, \text{i.e.} \overset{\circ}{\mathbf{f}} \cong \mathbf{C}^e \bar{\mathbf{v}}, \tag{18.127}$$

which describe the elastic deformation behaviour with an infinite friction strength such that these equations are inapplicable to sliding phenomenon at high sliding velocities. We then modify these equations as

$$\bar{\mathbf{v}} = \mathbf{C}^{e-1} \overset{\circ}{\mathbf{f}} + \frac{1}{\eta_v} \frac{\langle r-1\rangle^n}{\hat{r}-r} \mathbf{t}_n, \tag{18.128}$$

$$\overset{\circ}{\mathbf{f}} = \mathbf{C}^e \bar{\mathbf{v}} - \frac{1}{\eta_v} \frac{\langle r-1\rangle^n}{\hat{r}-r} \mathbf{C}^e \mathbf{t}_n, \tag{18.129}$$

where $\hat{r}(>1)$ is a material constant. If r approaches $\hat{r}$ (i.e., $r \to \hat{r}$), the viscoplastic sliding velocity becomes infinite so that the limitation is imposed to the dynamic loading friction-yield ratio r. Here, we call $\hat{r}$ the *limit dynamic loading friction-yield ratio*. Equations (18.128) and (18.129) are applicable to sliding behaviours for arbitrary sliding velocity ranging from quasi-static to impact loadings.

The viscoplastic sliding velocity is induced suddenly at the moment that the contact stress reaches the normal friction-yield surface; i.e., when the dynamic-loading friction-yield ratio becomes unity ($r=1$). Therefore, a smooth elastic–viscoplastic transition cannot be described by Eqs. (18.128) and (18.129), which must be further modified as

$$\bar{\mathbf{v}} = \mathbf{C}^{e-1} \overset{\circ}{\mathbf{f}} + \frac{1}{\eta_v} \frac{\langle r-r_s\rangle^n}{\hat{r}-r} \mathbf{t}_n, \tag{18.130}$$

$$\overset{\circ}{\mathbf{f}} = \mathbf{C}^e \bar{\mathbf{v}} - \frac{1}{\eta_v} \frac{\langle r - r_s \rangle^n}{\hat{r} - r} \mathbf{C}^e \mathbf{t}_n, \tag{18.131}$$

where $r_s (0 \leq r_s \leq 1)$ is the *subloading friction-yield ratio* calculated using Eqs. (18.33) and (18.34) by replacing the plastic sliding velocity $\bar{\mathbf{v}}^p$ with the viscoplastic sliding velocity $\bar{\mathbf{v}}^{vp}$; i.e.,

$$\dot{r}_s = \overline{U}(r_s) ||\bar{\mathbf{v}}^{vp}|| \quad \text{for} \quad \bar{\mathbf{v}}^{vp} \neq \mathbf{0}, \tag{18.132}$$

$$r_s = \frac{f(\mathbf{f})}{\mu} \quad \text{for} \quad \bar{\mathbf{v}}^{vp} = \mathbf{0}, \tag{18.133}$$

with

$$\overline{U}(r_s) = \tilde{u} \cot\left(\frac{\pi}{2} r_s\right). \tag{18.134}$$

Thus, the viscoplastic sliding is induced by overstress $f(\mathbf{f}) - r_s \mu$ from the subloading friction surface:

$$f(\mathbf{f}) = r_s \mu, \text{i.e. } r = r_s, \tag{18.135}$$

so that a smooth elastic–viscoplastic transition is described.

Simulations reproducing test data are difficult to perform using Eqs. (18.130) and (18.131) with a viscoplastic term in the power form but, as will be described later, simulations with high accuracy are possible using the equations with the exponential form

$$\boxed{\bar{\mathbf{v}} = \mathbf{C}^{e-1} \overset{\circ}{\mathbf{f}} + \frac{1}{\eta_v} \frac{\langle \mathrm{e}^{n(r - r_s)} - 1 \rangle}{\hat{r} - r} \mathbf{t}_n} \tag{18.136}$$

and

$$\boxed{\overset{\circ}{\mathbf{f}} = \mathbf{C}^e \bar{\mathbf{v}} - \frac{1}{\eta_v} \frac{\langle \mathrm{e}^{n(r - r_s)} - 1 \rangle}{\hat{r} - r} \mathbf{C}^e \mathbf{t}_n} \tag{18.137}$$

which will be used in subsequent sections. The generalized subloading friction model proposed above is referred to as the *subloading overstress friction model.*

The various behaviours derived from the subloading overstress friction model for fluid friction are schematically shown in Fig. 18.25 in which the normalized overstress divided by the normal contact stress f_n is written simply by the term "overstress". The dynamic loading friction-yield ratio r coincides with the subloading friction-yield ratio r_s, i.e. $r = r_s$ in quasi-static sliding and increases above r_s with increasing sliding velocity. However, r does not rise above $\hat{r}$; i.e., $r \leq \hat{r}$, with the equality $r = \hat{r}$ being realised only for impact sliding.

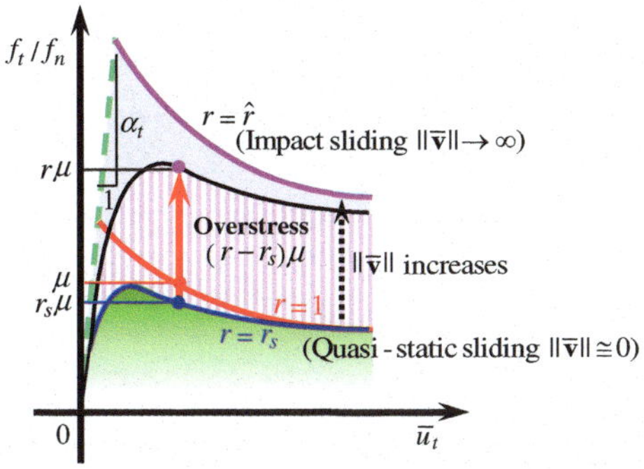

Fig. 18.25 Response of the generalized subloading-friction model

18.10.2 Interpretation of Generalised Subloading Friction Model

Equations (18.120) and (18.137) are described in incremental forms as

$$d\mu = \underbrace{-\kappa\left(1\frac{\mu}{\mu_k}\right)||d\bar{\mathbf{u}}^{vp}||}_{\text{Negative}} + \underbrace{\xi\left(1-\frac{\mu}{\mu_s}\right)dt}_{\text{Positive}} \tag{18.138}$$

$$d\mathbf{f} = \mathbf{C}^e d\bar{\mathbf{u}} - \frac{1}{\eta_v}\frac{\langle e^{n(r-r_s)}-1\rangle}{\hat{r}-r}\mathbf{C}^e\mathbf{t}_n dt, \tag{18.139}$$

where $d\bar{\mathbf{u}}^{vp} \equiv \bar{\mathbf{v}}^{vp}dt$.

We infer the following properties from Eq. (18.138) for the coefficient of friction μ in monotonic sliding at constant sliding velocity (see Fig. 18.26).

(1) In the quasi-static sliding process $(d\mu/dt \cong 0, ||d\bar{\mathbf{u}}^{vp}||/dt \cong 0,\ ||d\bar{\mathbf{u}}||/dt \cong 0, ||d\mathbf{f}||/dt \cong 0)$ in which the terms other than the second terms in the right-hand sides are negligible in Eqs. (18.138) and (18.139), we have $\mu = \mu_s$ from Eq. (18.138) and $r = r_s$ from Eq. (18.139), resulting in $f_t/f_n = r_s\mu_s$ due to Eq. (18.30) with Eq. (18.29). Therefore, the contact stress moves satisfying the subloading surface in quasi-static sliding process, so that the original subloading-friction model is reproduced in that process. In other words, the subloading-overstress model is generalized so as to contain the subloading-friction model formulated for the dry friction in Sect. 2.

(2) In the fast sliding process $(d\mu/dt \to \infty, ||d\bar{\mathbf{u}}^{vp}||/dt \to \infty, ||d\bar{\mathbf{u}}||/dt \to \infty, ||d\mathbf{f}||/dt \to \infty)$ for which the creep part of the second term in the right-hand side is negligible in Eq. (18.138), the friction coefficient μ decreases with the viscoplastic sliding displacement $\bar{u}^{vp} \equiv \int ||\bar{\mathbf{v}}^{vp}||dt$ after reaching a peak.

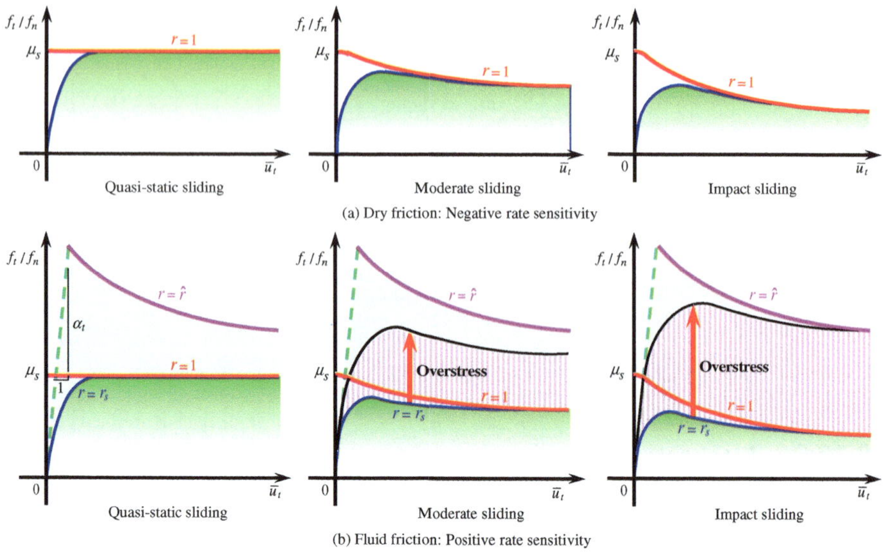

Fig. 18.26 Influence of sliding velocity at constant sliding velocities

Friction resistance depends on the destruction of the adhesion of surface asperities and the shearing of the viscous fluid lying between contact surfaces. The former is dominant in the case of dry friction whereas the latter is dominant in the case of fluid friction since the surfaces are in direct contact with each other in the case of dry friction but are in indirect contact via the viscous medium in the case of fluid friction. We thus infer the following differences between the responses of dry and fluid friction in the monotonic sliding process at a constant sliding velocity (see Fig. 18.27).

(3) The friction resistance in dry friction is obviously larger compared with that in fluid friction, because both the static coefficient of friction μ_s and the kinetic coefficient of friction μ_k in the former are larger compared with those in the latter. In addition, the difference between a peak and a bottom contact stress ratios is larger in the dry friction than in the fluid friction.
(4) The sliding displacement in the transition from static friction to kinetic friction in the case of dry friction is far smaller than that in the case of fluid friction because the sliding displacement required for destruction of the adhesion of surface asperities is far smaller than that required for shear flow in a viscous fluid. Therefore, the contact stress ratio decreases more gradually in the case of fluid friction than in the case of dry friction.
(5) The friction coefficient μ decreases with increasing sliding displacement as described in (1). Then, the dry friction exhibits the negative rate-sensitivity since the contact stress ratio f_t/f_n is given by the friction coefficient itself. On the other hand, the fluid friction exhibits the positive rate-sensitivity since the contact stress ratio is given by the friction coefficient plus the overstress, while

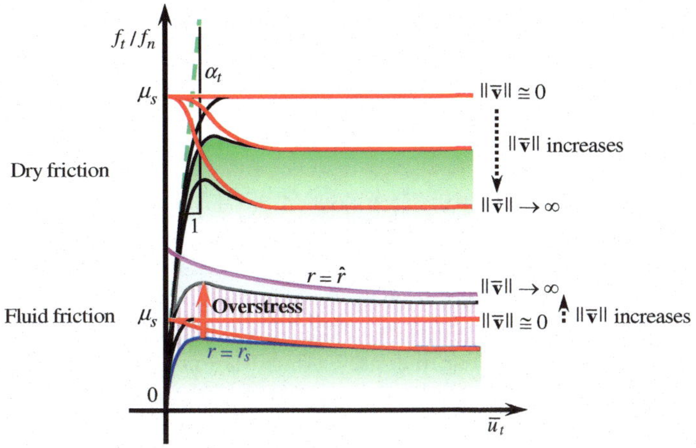

Fig. 18.27 Comparison of responses in dry and fluid frictions at constant sliding velocities

the overstress increases with sliding velocity inducing a higher viscous resistance. The friction resistance is kept constant in the quasi-static sliding process in which the friction coefficient is kept to be the static-friction, and from which the friction resistance decreases and increases in the dry and the fluid friction, respectively, with the increment of sliding velocity.

(6) Consequently, the contact stress ratio $f_t/(\mu f_n)$ approaches unity (i.e., $r = 1$: the normal sliding yield state) in the case of dry friction but becomes greater than unity (i.e., $r > 1$: over the normal sliding-yield state) in fluid friction.

It should be emphasized again that it is not required to use Eqs. (18.58) and/or Eq. (18.60) in the subloading friction model even for calculation of dry friction behaviour. In facts, the dry friction behaviour is generated in the quasi-static sliding process in the subloading overstress friction model. Consequently, we need only to use Eqs. (18.136) and/or (18.137) with Eqs. (18.120) and (18.132)–(18.134) in the subloading overstress friction model for calculations of general sliding behaviour involving both dry and fluid frictions.

Additionally, there exists an intermediate friction between dry and fluid friction where contact surfaces are lubricated slightly, leading to a behaviour in which there is no effect of the sliding velocity on the contact stress ratio under constant sliding velocities. The sliding behaviour ranging from dry to fluid friction covering negative to positive rate sensitivities can be described universally by the generalised subloading friction model. The return-mapping projection in the numerical analysis for the rate-independent elastoplastic deformation (e.g., Simo and Hughes 1998; Hashiguchi 2013b) is regarded to be an application of the overstress concept. Consequently, the subloading overstress friction model is the generalisation of the subloading friction model. In addition, the *Stribeck curve* (Stribeck 1902), which has been used widely to provide a qualitative relationship between frictional resistance f_t and variable $\eta_v \bar{v}_t / f_n$, would also be able to be drawn by changing appropriately the material constants α_n, α_t, μ_s, μ_k, κ, ξ, $\tilde{u}$, η_v, n and by extending the limit dynamic loading friction-yield ratio $\hat{r}$ as

$$\hat{r} = \exp\left(\frac{c_s}{f_n}\right), \tag{18.140}$$

where c_s is a material parameter. It follows from Eq. (18.140) that $\hat{r} \to \infty$ for $f_n \to 0$ and $\hat{r} \to 1$ for $f_n \to \infty$. Therefore, $\hat{r}$ becomes larger for a smaller normal contact stress f_n, noting that the thickness of the fluid layer between contact surfaces increases approaching the fluid friction for a smaller normal contact stress. In contrast, $\hat{r}$ approaches unity so that the dynamic loading friction-yield ratio r can become only slightly larger than the subloading friction-yield ratio r_s for an infinitely large normal contact stress f_n approaching dry friction.

18.10.3 Numerical Experiments

The basic mechanical properties of the subloading overstress friction model formulated in the preceding section are examined below in numerical experiments using the constitutive Eq. (18.136) with Eq. (18.134) and adopting Eq. (18.29) for the Coulomb-type friction-yield stress function $f(\mathbf{f})$.

In numerical simulations, the material constants in the subloading friction model are chosen as follows:

$$\begin{aligned}
&\text{Elastic sliding constant: } \alpha_n = \alpha_t = 1000\,\text{GPa mm}^{-1}\\
&\text{Static friction coefficient: } \mu_s = 0.097,\ \text{Kinetic friction coefficient: } \mu_k = 0.085\\
&\text{Sliding–softening constant: } \kappa = 1.0\,\text{mm}^{-1}\\
&\text{Creep–hardening constant: } \xi = 0.005\,\text{min.}^{-1}\\
&\text{Normal friction–yield ratio evolution constant: } \tilde{u} = 80\,\text{mm}^{-1}
\end{aligned}$$

Furthermore, the three material constants η_v, n and $\hat{r}$ added in the subloading overstress friction model are changed in the following five levels, fixing the other two material constants to be $\eta_v = 1000$, $n = 16$ and $\hat{r} = 1.6$.

$$\begin{aligned}
&\text{Viscoplastic coefficient: } \eta_v = 10, 100, 1000, 10000, 100000\,\text{min.mm}^{-1}\\
&\text{Viscoplastic power coefficient: } n = 4, 8, 16, 32, 64\\
&\text{Limit dynamic loading friction–yield ratio: } \hat{r} = 1.2, 1.6, 2.4, 3.2, 6.4
\end{aligned}$$

The sliding velocity is set to be 0.1, 1, 100 and 1000 mm/min. The calculated results are shown in Fig. 18.28. The contact stress ratio f_t/f_n is larger for higher sliding velocities, exhibiting positive rate sensitivity. Higher contact stress ratios are predicted for larger values of η_v and $\hat{r}$ and for lower values of n. Curves of the

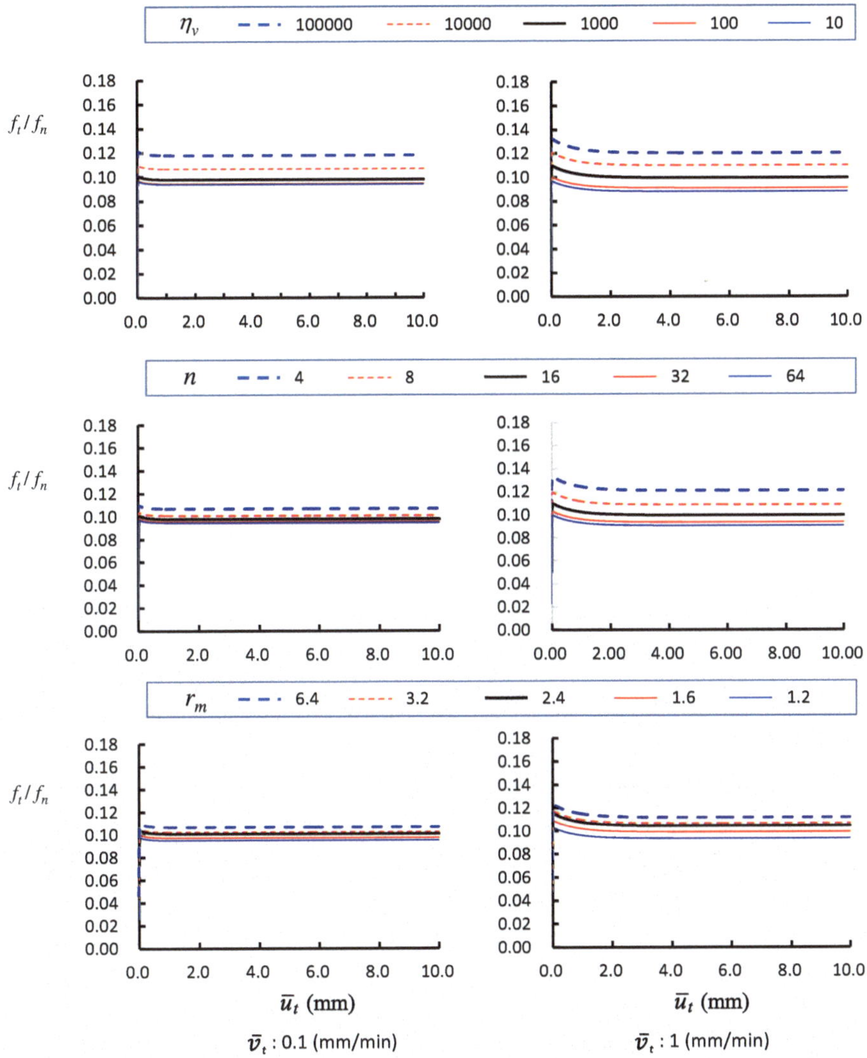

Fig. 18.28 Influence of material parameters in subloading overstress friction model

contact stress ratio versus sliding displacement for various sliding velocities are shown in Fig. 18.29, indicating positive rate sensitivity, where the material constants are chosen as $\eta_v = 1000$, $n = 16$ and $\hat{r} = 1.6$. The effect of the sliding velocity on the peak (maximum) and bottom (minimum) contact stress ratios in the contact stress ratio versus sliding displacement curves, which are read from Fig. 18.29, are plotted in Fig. 18.30. Here, the peak and bottom contact stress ratios increase with sliding velocity in a certain velocity range, whereas they converge to certain values outside this range.

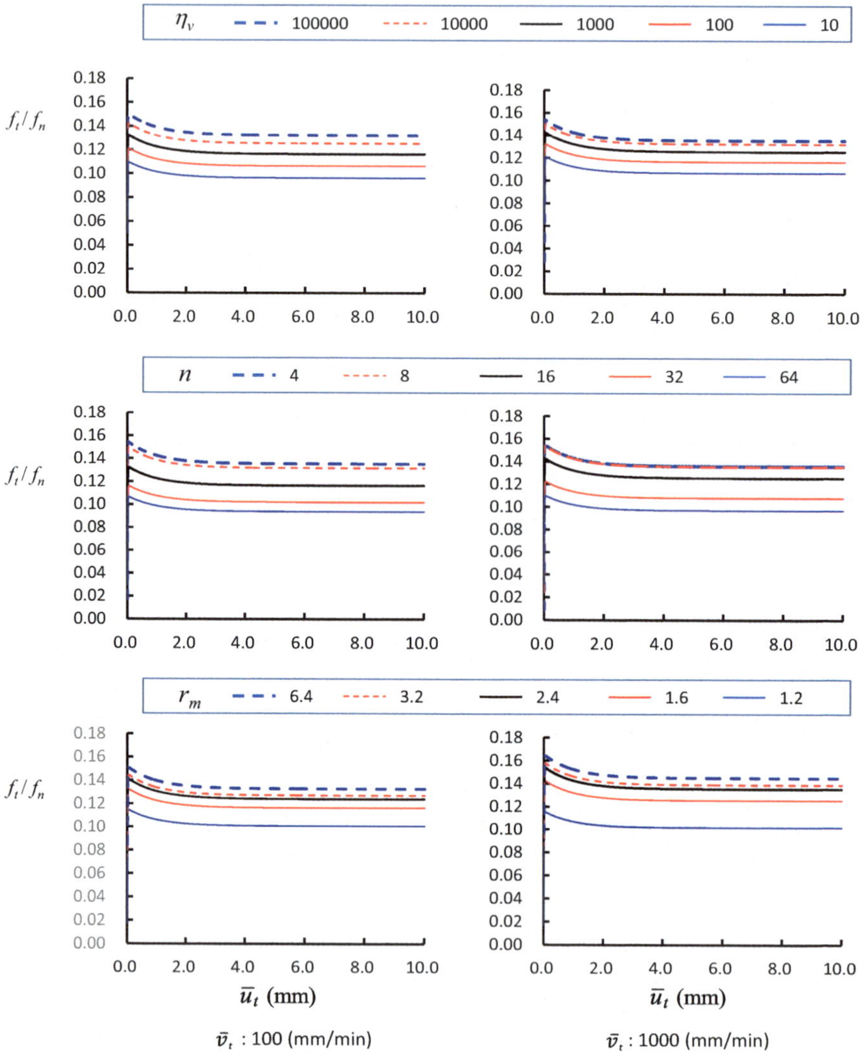

Fig. 18.28 (continued)

18.10.4 Comparison with Test Data

To examine the validity of the generalised subloading overstress model, a lubricated friction test was performed. A schematic diagram of the test apparatus is shown in Fig. 18.31. The test plate, which is placed between the tools, is subjected to the constant normal load (5 kN) and is pulled up at constant velocity. The pulling force is measured by the load cell (maximum load: 100 kN) attached to the upper part of the test plate. The test plate is made of galvannealed steel sheet with the friction

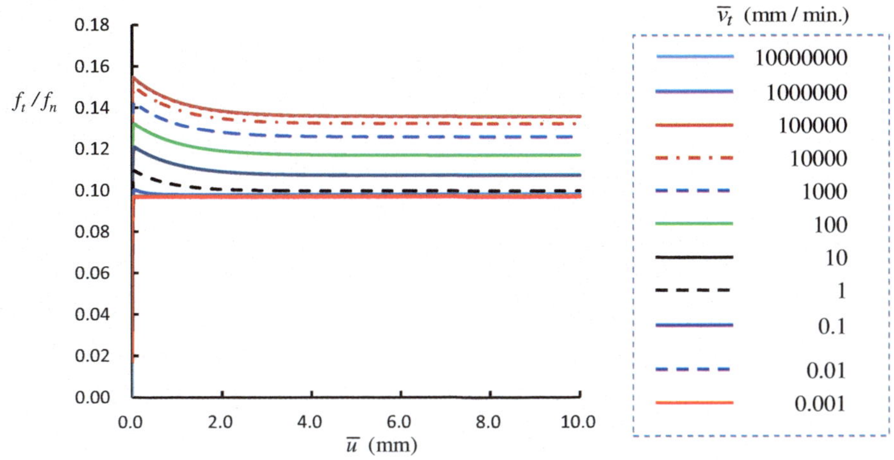

Fig. 18.29 Influence of sliding velocity

area of width 30 mm, height 300 mm and thickness 0.7 mm. The tool steel SKD-11 of width 40 mm, height 30 mm and thickness 20 mm is used for the tools which grasp the test plate. Therefore, the friction contact area for the normal contact stress is 900 mm^2 and therefore the normal contact stress is 5.56 MPa, whereas the friction contact area for the tangential contact stress is 1800 mm^2. The friction surfaces were polished and coated with anti-rust oil prior to the tests. The pulling-up velocity of the test plate is set at five levels: $1, 10, 50, 100, 200\,\mathrm{mm/min}$. The friction test was adopted supposing the press forming of thin sheet metal in the metal forming process.

The measured relationship between the contact stress ratio f_t/f_n and the tangential sliding displacement $\bar{u}_t$ is shown in Fig. 18.32a. The contact stress ratio first

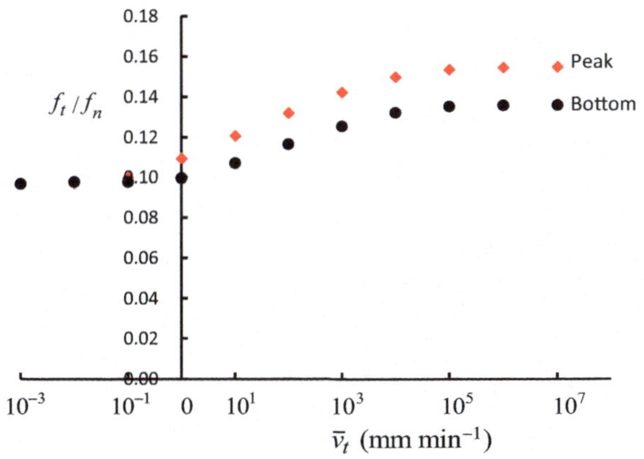

Fig. 18.30 Influence of sliding velocity on contact stress ratio

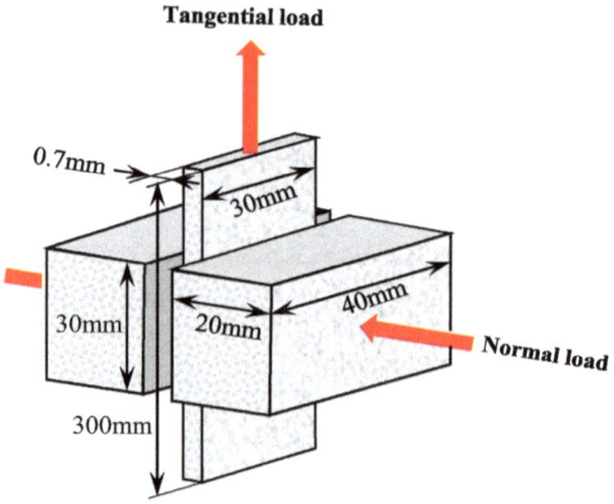

Fig. 18.31 Illustration of friction test apparatus

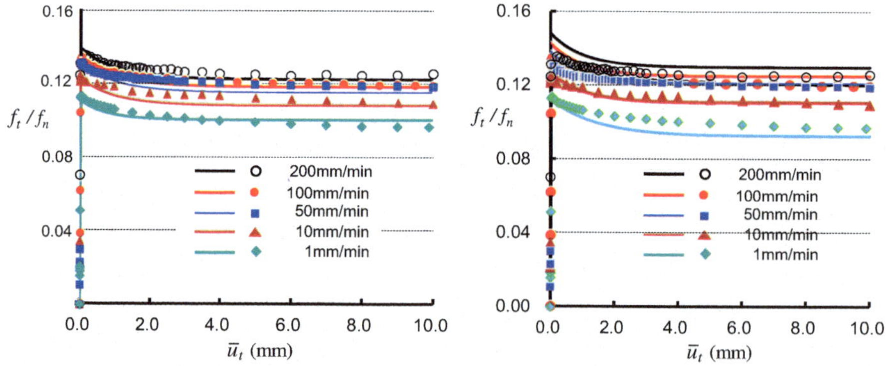

(a) Eq. (63) with exponential function

(b) Eq. (58) with power function

Fig. 18.32 Comparison with test data for various levels of sliding velocity

peaks and then gradually falls to a stationary value, exhibiting a positive rate sensitivity, i.e. larger contact stress ratios at higher sliding velocities.

The simulation of the above-mentioned test result using Eq. (18.136) with Eqs. (18.29) and (18.134) is shown by the solid lines in Fig. 18.32a, using the following values for the material constants.

$$\begin{aligned}
&\alpha_n = \alpha_t = 1000\,\mathrm{GPamm}^{-1}\\
&\mu_s = 0.097,\ \mu_k = 0.085\\
&\kappa = 0.2\,\mathrm{mm}^{-1},\ \xi = 0.009\,\mathrm{min}^{-1}\\
&\tilde{u} = 80\,\mathrm{mm}^{-1}\\
&\eta_v = 950\,\mathrm{min.mm}^{-1},\ n = 16,\ \hat{r} = 1.8
\end{aligned}$$

The simulation of the test result using Eq. (18.130) with the power function is shown by the solid lines in Fig. 18.32b, using the following values for material constants. These material constants are chosen so as to simulate the test result as closely as possible. However, it is impossible even to restrict the range from the highest and the lowest curves to the range in the test curves. Equation (18.136) with the exponential function would be much more appropriate than Eq. (18.130) with the power function for the prediction of real friction behaviour. However, the clarification of the definite physical background should be continued for the future.

$$\begin{aligned}
&\alpha_n = \alpha_t = 1000\,\mathrm{GPamm}^{-1}\\
&\mu_s = 0.097,\ \mu_k = 0.085\\
&\kappa = 1.0\,\mathrm{mm}^{-1},\ \xi = 0.005\,\mathrm{min}^{-1}\\
&\tilde{u} = 80\,\mathrm{mm}^{-1}\\
&\eta_v = 950\,\mathrm{min.mm}^{-1},\ n = 16,\ \hat{r} = 1.8
\end{aligned}$$

Based on Fig. 18.32a, a comparison of the calculated and test results for the influence of the sliding velocity on the peak (maximum) and bottom (minimum) values in the contact stress ratio versus sliding displacement curves is shown in Fig. 18.33, where the close coincidence can be observed.

The subloading-friction model will be relevant to the rate-dependent crystal plasticity as will be described in Chap. 19. Further, instead of the rate-and state model, it will have to be applied also to the prediction of earthquake occurrence in continental-slip type for the steady prevention of earthquake disaster in the near future.

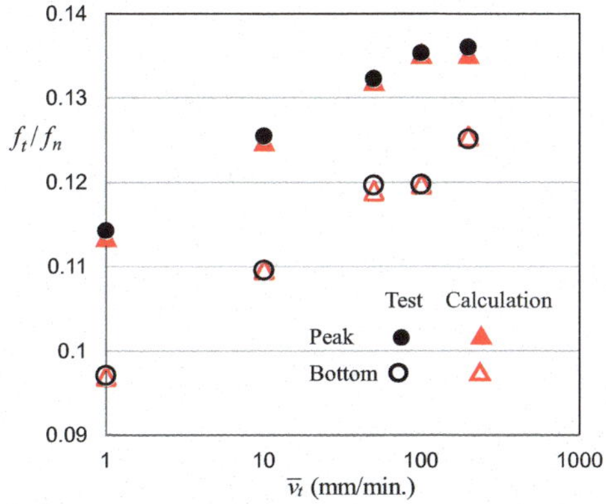

Fig. 18.33 Comparison of calculated and test results for influence of sliding velocity on contact stress ratio

18.11 Dependence of Friction Coefficient on Normal Contact Stress

It can be stated from experiments that the friction coefficient decreases with the increase of contact pressure (cf. Bay and Wanheim, 1976; Dunkin and Kim 1996; Gearing et al., 2001; Stupkiewicz and Mroz 2003). This property cannot be described appropriately by the Coulomb sliding-yield surface in which the tangential contact traction is linearly related to the normal contact traction through the constant friction angle. In what follows, the sliding-yield surface with the nonlinear relation of tangential contact traction and normal contact traction will be introduced and the constitutive relations will be formulated for the description of the decrease of friction coefficient with the increase of normal contact traction.

The subloading-friction model with the decrease of friction coefficient due to the increase of normal contact traction and with the rotational anisotropy will be shown first. Thereafter, it will be extended to incorporate the orthogonal anisotropy in addition to the rotational anisotropy.

(a) Rotational anisotropy

The closed normal sliding-yield and the sliding-subloading surfaces with the rotational hardening can be described as follows:

$$\bar{f} = \bar{F} \tag{18.141}$$

and

$$\bar{f} = r\bar{F} \tag{18.142}$$

respectively, by setting

$$\bar{f} = \bar{f}\,(\mathbf{f},\,\bar{\boldsymbol{\beta}}) = f_n\,\bar{g}(\hat{\chi}) \tag{18.143}$$

in the form analogous to Eq. (11.18), where

$$\hat{\chi} \equiv \frac{||\hat{\boldsymbol{\eta}}||}{\bar{M}} \tag{18.144}$$

$$\hat{\boldsymbol{\eta}} \equiv \bar{\boldsymbol{\eta}} - \bar{\boldsymbol{\beta}},\ \bar{\boldsymbol{\eta}} \equiv \frac{\mathbf{f}_t}{f_n} \tag{18.145}$$

$\bar{\boldsymbol{\beta}}$ is the vector describing the rotation of the sliding-yield surface around the origin of traction space and it does not have the normal component since it describes the anisotropy in the contact (tangential) surface. Hereinafter, let $\bar{\boldsymbol{\beta}}$ be assumed to be a fixed vector on the contact surface leading to $\overset{\circ}{\bar{\boldsymbol{\beta}}}=\mathbf{0}$ for sake of simplicity. Then, it holds that

$$\mathbf{n}\cdot\bar{\boldsymbol{\beta}} = 0 \tag{18.146}$$

$$\dot{\bar{\boldsymbol{\beta}}} = \boldsymbol{\omega}_c\bar{\boldsymbol{\beta}} \tag{18.147}$$

$\bar{M}$ is the material constant denoting the traction ratio f_t/f_n at the maximum point of f_t. Simple examples of the function $\bar{g}(\hat{\chi})$ in the sliding-yield function in Eq. (18.143) are given as follows:

$$\bar{g}(\hat{\chi}) = \exp(\hat{\chi}),\ \bar{g}'(\hat{\chi}) = \exp(\hat{\chi}) \tag{18.148}$$

$$\bar{g}(\hat{\chi}) = 1+\hat{\chi}^2,\ \bar{g}'(\hat{\chi})= 2(\hat{\chi}) \tag{18.149}$$

$$\bar{g}(\hat{\chi})=\exp\hat{\chi}^2/2),\ \bar{g}'(\hat{\chi}) = \hat{\chi}\exp(\hat{\chi}) \tag{18.150}$$

$$\bar{g}(\hat{\chi}) = \frac{1}{1-\hat{\chi}/2},\ \bar{g}'(\hat{\chi}) = \frac{1}{2(1-\hat{\chi}/2)^2} \tag{18.151}$$

All sets of Eq. (18.143) with Eqs. (18.148)–(18.151) exhibit closed surfaces passing through points $f_n = 0$ and $f_n = \bar{F}$ at $f_t = 0$ for $\bar{\boldsymbol{\beta}} = 0$. Equations (18.148) and (18.149) are based on Eq. (11.24) for the original Cam-clay yield surface (Schofield and Wroth 1968) and Eq. (11.26) for the modified Cam-clay yield surfaces (Roscoe and Burland 1968), respectively, for soils described in Sect. 11.2. Equation (18. 150) exhibits the teardrop-shaped surface (Hashiguchi 1972, 1985a; Hashiguchi et al. 2005b) which is reversed from the surface of Eq. (18.148) on the axis of normal contact traction. Equation (18.151) exhibits the parabola-shaped surfaces (Hashiguchi et al. 2005b). The evolution of the function $\bar{F}$ is given analogously to Eq. (18.38) as follows:

$$\dot{\bar{F}} = \underbrace{-\kappa(\frac{\bar{F}}{\bar{F}_k} - 1)||\bar{\mathbf{v}}^p||}_{\text{Negative}} + \underbrace{\xi(1 - \frac{\bar{F}}{\bar{F}_s})}_{\text{Positive}} \tag{18.152}$$

where $\overline{F}_s$ and $\overline{F}_k(\overline{F}_s \geq \overline{F} \geq \overline{F}_k)$ are material constants designating the maximum and minimum values of $\overline{F}$ for static friction and kinetic friction, respectively. $\overline{\mathcal{K}}$ is a material constant specifying the rate of decrease of $\overline{F}$ in the plastic sliding process, and $\overline{\xi}$ is a material constant specifying the rate of recovery of $\overline{F}$ with a time elapse. The normal sliding-yield and the sliding-subloading surface for Eqs. (18.150) and (18.151) with $\overline{\boldsymbol{\beta}} = 0$ are depicted in Figs. 18.34 and 18.35, respectively.

It holds from (18.145) that

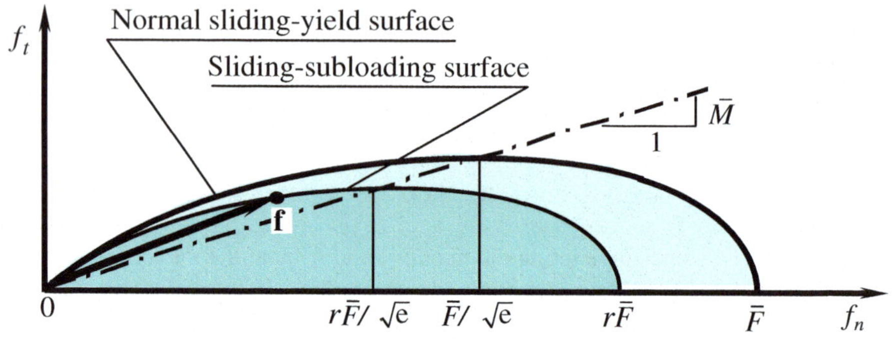

Fig. 18.34 Teardrop-shaped normal sliding-yield and sliding-subloading surfaces

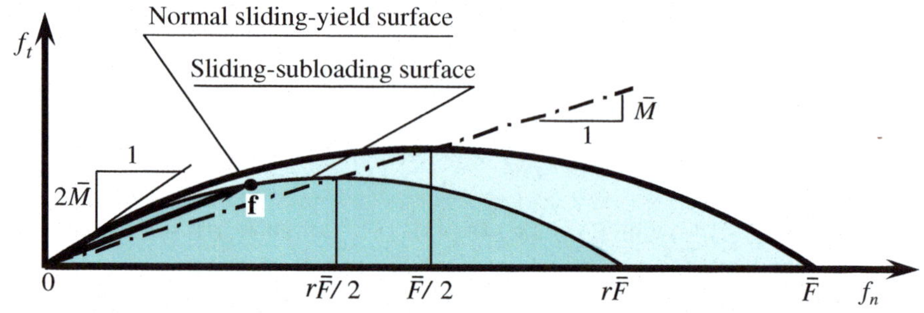

Fig. 18.35 Parabola-shaped normal sliding-yield and sliding-subloading surfaces

$$\frac{\partial \hat{\boldsymbol{\eta}}}{\partial \mathbf{f}} = \frac{\partial \boldsymbol{\eta}}{\partial \mathbf{f}} = \frac{\partial \mathbf{f}_t / f_n}{\partial \mathbf{f}} = \frac{f_n(\mathbf{I} - \mathbf{n} \otimes \mathbf{n}) - \mathbf{f}_t \otimes (-\mathbf{n})}{f_n^2} = \frac{\mathbf{I} - \mathbf{n} \otimes \mathbf{n} + \overline{\boldsymbol{\eta}} \otimes \mathbf{n}}{f_n} \tag{18.153}$$

$$\frac{\partial \|\hat{\boldsymbol{\eta}}\|}{\partial \mathbf{f}} = \frac{\partial \|\hat{\boldsymbol{\eta}}\|}{\partial \hat{\boldsymbol{\eta}}} \frac{\partial \hat{\boldsymbol{\eta}}}{\partial \mathbf{f}} = \hat{\boldsymbol{\tau}} \frac{\mathbf{I} - \mathbf{n} \otimes \mathbf{n} + \overline{\boldsymbol{\eta}} \otimes \mathbf{n}}{f_n} = \frac{1}{f_n}[(\hat{\boldsymbol{\tau}} \cdot \hat{\boldsymbol{\eta}})\mathbf{n} + \hat{\boldsymbol{\tau}}] \tag{18.154}$$

where

$$\hat{\boldsymbol{\tau}} \equiv \frac{\hat{\boldsymbol{\eta}}}{\|\hat{\boldsymbol{\eta}}\|} \ (\mathbf{n} \cdot \hat{\boldsymbol{\tau}} = 0) \tag{18.155}$$

Further, it holds from Eqs. (18.23), (18.46), (18.143), (18.144), (18.153) and (18.154) that

$$\frac{\partial \hat{\chi}}{\partial \mathbf{f}} = \frac{\partial \|\hat{\boldsymbol{\eta}}\| / \bar{M}}{\partial \mathbf{f}} = \frac{1}{\bar{M} f_n}[(\hat{\boldsymbol{\tau}} \cdot \hat{\boldsymbol{\eta}})\mathbf{n} + \hat{\boldsymbol{\tau}}] \tag{18.156}$$

$$\frac{\partial \bar{f}}{\partial \mathbf{f}} = -\bar{g}(\hat{\chi})\bar{\mathbf{n}} + f_n \bar{g}'(\hat{\chi})\frac{1}{\bar{M} f_n}[(\hat{\boldsymbol{\tau}}\cdot\hat{\boldsymbol{\eta}})\bar{\mathbf{n}} + \hat{\boldsymbol{\tau}}] = -\left[\bar{g}(\hat{\chi}) - \frac{\bar{g}'(\hat{\chi})}{\bar{M}}(\hat{\boldsymbol{\tau}}\cdot\hat{\boldsymbol{\eta}})\right]\bar{\mathbf{n}} + \frac{\bar{g}'(\hat{\chi})}{\bar{M}}\hat{\boldsymbol{\tau}} \tag{18.157}$$

$$\left(\frac{\partial f_t}{\partial \mathbf{f}} =\right)(\mathbf{I} - \mathbf{n}\otimes\mathbf{n})\frac{\partial \bar{f}}{\partial \mathbf{f}} = (\mathbf{I} - \mathbf{n}\otimes\mathbf{n})\left\{-\left[\bar{g}(\hat{\chi}) - \frac{\bar{g}'(\hat{\chi})}{\bar{M}}(\hat{\boldsymbol{\tau}}\cdot\hat{\boldsymbol{\eta}})\right]\mathbf{n} + \frac{\bar{g}'(\hat{\chi})}{\bar{M}}\hat{\boldsymbol{\tau}}\right\}$$
$$= \frac{\bar{g}'(\hat{\chi})}{\bar{M}}\hat{\boldsymbol{\tau}} \tag{18.158}$$

$$\mathbf{t}_n = \hat{\boldsymbol{\tau}} \tag{18.159}$$

$$\frac{\partial \bar{f}}{\partial \mathbf{f}}\mathbf{C}^e = \left\{-\left[\bar{g}(\hat{\chi}) - \frac{\bar{g}'(\hat{\chi})}{\bar{M}}(\hat{\boldsymbol{\tau}}\cdot\hat{\boldsymbol{\eta}})\right]\mathbf{n} + \frac{\bar{g}'(\hat{\chi})}{\bar{M}}\hat{\boldsymbol{\tau}}\right\}[\alpha_n\mathbf{n}\otimes\mathbf{n} + \alpha_t(\mathbf{I} - \mathbf{n}\otimes\mathbf{n})]$$
$$- \alpha_n\left[\bar{g}(\hat{\chi}) - \frac{\bar{g}'(\hat{\chi})}{\bar{M}}(\hat{\boldsymbol{\tau}}\cdot\bar{\boldsymbol{\eta}})\right]\mathbf{n} + \alpha_t\frac{\bar{g}'(\hat{\chi})}{\bar{M}}\hat{\boldsymbol{\tau}} \tag{18.160}$$

$$\frac{\partial \bar{f}}{\partial \mathbf{f}}\mathbf{C}^e\cdot\mathbf{t}_n = \left\{-\alpha_n\left[\bar{g}(\hat{\chi}) - \frac{\bar{g}'(\hat{\chi})}{\bar{M}}(\hat{\boldsymbol{\tau}}\cdot\bar{\boldsymbol{\eta}})\right]\mathbf{n} + \alpha_t\frac{\bar{g}'(\hat{\chi})}{\bar{M}}\hat{\boldsymbol{\tau}}\right\}\cdot\hat{\boldsymbol{\tau}} = \alpha_t\frac{\bar{g}'(\hat{\chi})}{\bar{M}} \tag{18.161}$$

The substitution of Eqs. (18.23) and (18.156)–(18.161) into Eqs. (18.58) and (18.60) regarding μ, μ_s, μ_k κ and ξ as $\bar{F}$, $\bar{F}_s$, $\bar{F}_k$ $\bar{\kappa}$ and $\bar{\xi}$, respectively, reads:

$$\bar{\boldsymbol{v}} = [\frac{1}{\alpha_n}\mathbf{n}\otimes\mathbf{n} + \frac{1}{\alpha_t}(\mathbf{I} - \mathbf{n}\otimes\mathbf{n})]\overset{\circ}{\mathbf{f}} + \frac{\{-[\bar{g}(\hat{\chi}) - \frac{\bar{g}'(\hat{\chi})}{\bar{M}}(\hat{\boldsymbol{\tau}}\cdot\boldsymbol{\eta})]\mathbf{n} + \frac{\bar{g}'(\hat{\chi})}{\bar{M}}\hat{\boldsymbol{\tau}}\}\cdot\overset{\circ}{\mathbf{f}} - \bar{m}^c}{\bar{m}^p}\hat{\boldsymbol{\tau}}$$
$$= \left(\frac{1}{\alpha_n} - \frac{1}{\alpha_t}\right)(\mathbf{n}\cdot\overset{\circ}{\mathbf{f}})\mathbf{n} + \frac{1}{\alpha_t}\overset{\circ}{\mathbf{f}} + \frac{-[\bar{g}(\hat{\chi}) - \frac{\bar{g}'(\hat{\chi})}{\bar{M}}(\hat{\boldsymbol{\tau}}\cdot\bar{\boldsymbol{\eta}})](\mathbf{n}\cdot\overset{\circ}{\mathbf{f}}) + \frac{\bar{g}'(\hat{\chi})}{\bar{M}}(\hat{\boldsymbol{\tau}}\cdot\overset{\circ}{\mathbf{f}}) - \bar{m}^c}{\bar{m}^p}\hat{\boldsymbol{\tau}} \tag{18.162}$$

$$\overset{\circ}{\mathbf{f}} = [\alpha_n\mathbf{n}\otimes\mathbf{n} + \alpha_t(\mathbf{I} - \mathbf{n}\otimes\mathbf{n})]$$
$$\left[\bar{\mathbf{v}} - \left\langle\frac{\{-\alpha_n[\bar{g}(\hat{\chi}) - \frac{\bar{g}'(\hat{\chi})}{\bar{M}}(\hat{\boldsymbol{\tau}}\cdot\boldsymbol{\eta})]\mathbf{n} + \alpha_t\frac{\bar{g}'(\hat{\chi})}{\bar{M}}\hat{\boldsymbol{\tau}}\}\cdot\bar{\mathbf{v}} - \bar{m}^c}{\bar{m}^p + \alpha_t\frac{\bar{g}'(\hat{\chi})}{\bar{M}}}\right\rangle\hat{\boldsymbol{\tau}}\right]$$
$$= (\alpha_n - \alpha_t)(\mathbf{n}\cdot\bar{\mathbf{v}})\mathbf{n} + \alpha_t\bar{\mathbf{v}}$$
$$- \alpha_t\left\langle\frac{-\alpha_n[\bar{g}(\hat{\chi}) - \frac{\bar{g}'(\hat{\chi})}{\bar{M}}(\hat{\boldsymbol{\tau}}\cdot\boldsymbol{\eta})](\mathbf{n}\cdot\bar{\mathbf{v}}) + \alpha_t\frac{\bar{g}'(\hat{\chi})}{\bar{M}}(\hat{\boldsymbol{\tau}}\cdot\bar{\mathbf{v}}) - \bar{m}^c}{\bar{m}^p + \alpha_t\frac{\bar{g}'(\hat{\chi})}{\bar{M}}}\right\rangle\hat{\boldsymbol{\tau}} \tag{18.163}$$

where

$$\boxed{\bar{m}^p \equiv -\bar{\kappa}\left(\frac{\bar{F}}{\bar{F}_k} - 1\right)r + \bar{U}(r)\bar{F}} \tag{18.164}$$

$$\boxed{\bar{m}^c \equiv \bar{\xi}\left(1 - \frac{\bar{F}}{\bar{F}_s}\right)r(\geq 0)} \tag{18.165}$$

The numerical calculations by the equations described in this section and the comparisons with test data in various sliding states can be referred to Hashiguchi (2009). The extension to the orthotropic anisotropy, i.e. the subloading-friction model with the dependence of friction coefficient on the normal contact stress and the rotational and the orthotropic anisotropies can be also referred to it.

(b) Rotational and orthogonal anisotropies

The difference of friction coefficients in mutually opposite sliding directions can be described by the aforementioned rotational anisotropy. However, the difference of friction coefficients in the mutually-perpendicular directions cannot be described by the rotational anisotropy. In order to extend so as to describe this difference, let the concept of orthotropy described in Section 18.8 be further incorporated below.

Let Eqs. (18.141) and (18.142) be extended by taking account of the orthotropic anisotropy in addition to the rotational anisotropy as follows:

$$\bar{f}^* = \bar{F} \tag{18.166}$$

and

$$\bar{f}^* = r\bar{F} \tag{18.167}$$

respectively, where

$$\bar{f}^* = f_n g(\hat{\chi}^*_M) \tag{18.168}$$

$$\hat{\chi}^*_M \equiv \sqrt{\hat{\chi}^{*2}_{M1} + \hat{\chi}^{*2}_{M2}},\ \hat{\chi}^*_{M1} \equiv \frac{\hat{\eta}^*_1}{\bar{M}_1},\ \hat{\chi}^*_{M2} \equiv \frac{\hat{\eta}^*_2}{\bar{M}_2} \tag{18.169}$$

$\bar{M}_1$ and $\bar{M}_2$ are the material constants standing for the values of $\bar{M}$ in the maximum and the minimum principal directions of anisotropy, respectively.

The partial derivatives for Eq. (18.168) are given as

$$\left.\begin{aligned}
\frac{\partial \hat{\chi}^*_{M_i}}{\partial f^*_{ti}} &= \frac{\partial (f^*_{ti}/f_n - \overline{\beta}^*_i)/\overline{M}_i}{\partial f^*_{ti}} = \frac{1}{f_n \bar{M}_i} \text{ (no sum)} \\
\frac{\partial \hat{\chi}^*_{Mi}}{\partial f_n} &= \frac{\partial (f^*_{ti}/f_n - \overline{\beta}^*_i)/\overline{M}_i}{\partial f_n} = \frac{-f^*_{ti}}{f_n^2 \bar{M}_i} = -\frac{\chi^*_{Mi}}{f_n} \text{ (no sum)} \\
\frac{\partial \hat{\chi}^*_{M}}{\partial \hat{\chi}^*_{Mi}} &= \frac{1}{2\hat{\chi}^*_M} 2\hat{\chi}^*_{Mi} = \hat{\zeta}^*_{Mi} \\
\frac{\partial \hat{\chi}^*_{M_i}}{\partial f^*_{ti}} &= \frac{\partial \hat{\chi}^*_{M}}{\partial \hat{\chi}^*_{Mi}} \frac{\partial \hat{\chi}^*_{M_i}}{\partial f^*_{ti}} = \frac{\hat{\chi}^*_{Mi}}{\hat{\chi}^*_M} \frac{1}{f_n \bar{M}_i} = \frac{1}{f_n} \frac{\hat{\zeta}^*_{Mi}}{\bar{M}_i} \text{ (no sum)} \\
\frac{\partial \hat{\chi}^*_{M}}{\partial f_n} &= \frac{\partial \hat{\chi}^*_{M}}{\partial \hat{\chi}^*_{M1}} \frac{\partial \hat{\chi}^*_{M1}}{\partial f_n} + \frac{\partial \hat{\chi}^*_{M}}{\partial \hat{\chi}^*_{M2}} \frac{\partial \hat{\chi}^*_{M2}}{\partial f_n} = -\frac{1}{f_n} (\hat{\zeta}^*_{M1} \chi^*_{M1} + \hat{\zeta}^*_{M2} \chi^*_{M2})
\end{aligned}\right\} \quad (18.170)$$

where

$$\hat{\zeta}_{Mi} \equiv \frac{\hat{\chi}^*_{Mi}}{\hat{\chi}^*_M} \quad (18.171)$$

The subscript i takes 1 or 2 and is not summed even when it is repeated in Eq. (18.170).

It holds from Eqs. (18.23) and (18.170) that

$$\frac{\partial \hat{\chi}^*_M}{\partial \mathbf{f}} = \frac{\partial \hat{\chi}^*_M}{\partial f^*_{t1}} \mathbf{e}^*_1 + \frac{\partial \hat{\chi}^*_M}{\partial f^*_{t2}} \mathbf{e}^*_2 - \frac{\partial \hat{\chi}^*_M}{\partial f_n} \mathbf{n} = \frac{1}{f_n} \left[\frac{\hat{\zeta}^*_{M1}}{\bar{M}_1} \mathbf{e}^*_1 + \frac{\hat{\zeta}^*_{M1}}{\bar{M}_2} \mathbf{e}^*_2 + (\hat{\zeta}^*_{M1} \chi^*_{M1} + \hat{\zeta}^*_{M2} \chi^*_{M2}) \mathbf{n} \right] \quad (18.172)$$

$$\begin{aligned}
\frac{\partial \bar{f}^*}{\partial \mathbf{f}} &= \frac{\partial (f_n \bar{g}(\hat{\chi}^*_M))}{\partial \mathbf{f}} = \frac{\partial f_n}{\partial \mathbf{f}} \bar{g}(\hat{\chi}^*_M) + f_n \bar{g}'(\hat{\chi}^*_M) \frac{\partial \hat{\chi}^*_M}{\partial \mathbf{f}} \\
&= -\bar{g}(\hat{\chi}^*_M) \mathbf{n} + f_n \bar{g}'(\hat{\chi}^*_M) \frac{1}{f_n} \left[\frac{\hat{\zeta}^*_{M1}}{\bar{M}_1} \mathbf{e}^*_1 + \frac{\hat{\zeta}^*_{M2}}{\bar{M}_2} \mathbf{e}^*_2 + (\hat{\zeta}^*_{M1} \chi^*_{M1} + \hat{\zeta}^*_{M2} \chi^*_{M2}) \mathbf{n} \right] \\
&= \bar{g}'(\hat{\chi}^*_M) \left(\frac{\hat{\zeta}^*_{M1}}{\bar{M}_1} \mathbf{e}^*_1 + \frac{\hat{\zeta}^*_{M1}}{\bar{M}_2} \mathbf{e}^*_2 \right) - [\bar{g}(\hat{\chi}^*_M) - \bar{g}'(\hat{\chi}^*_M)(\hat{\zeta}^*_{M1} \chi^*_{M1} + \hat{\zeta}^*_{M2} \chi^*_{M2}) \mathbf{n}]
\end{aligned} \quad (18.173)$$

$$
\begin{aligned}
(\mathbf{I}-\mathbf{n}\otimes\mathbf{n})\cdot\frac{\partial\bar{f}^*}{\partial\mathbf{f}} &= (\mathbf{I}-\mathbf{n}\otimes\mathbf{n})\cdot\left\{\bar{g}'(\hat{\chi}_M^*)\left(\frac{\hat{\zeta}_{M1}^*}{\bar{M}_1}\mathbf{e}_1^*+\frac{\hat{\zeta}_{M2}^*}{\bar{M}_2}\mathbf{e}_2^*\right)\right.\\
&\quad\left.-\left[\bar{g}(\hat{\chi}_M^*)-\bar{g}'(\hat{\chi}_M^*)(\hat{\zeta}_{M1}^*\chi_{M1}^*+\hat{\zeta}_{M2}^*\chi_{M2}^*)\right]\mathbf{n}\right\}\\
&= \bar{g}'(\hat{\chi}_M^*)\left(\frac{\hat{\zeta}_{M1}^*}{\bar{M}_1}\mathbf{e}_1^*+\frac{\hat{\zeta}_{M2}^*}{\bar{M}_2}\mathbf{e}_2^*\right)
\end{aligned}
\tag{18.174}
$$

$$
\mathbf{t}_n=\frac{\dfrac{\hat{\zeta}_{M1}^*}{\bar{M}_1}\mathbf{e}_1^*+\dfrac{\hat{\zeta}_{M2}^*}{\bar{M}_2}\mathbf{e}_2^*}{\sqrt{\left(\dfrac{\hat{\zeta}_{M1}^*}{\bar{M}_1}\right)^2+\left(\dfrac{\hat{\zeta}_{M2}^*}{\bar{M}_2}\right)^2}}
\tag{18.175}
$$

$$
\begin{aligned}
\frac{\partial\bar{f}^*}{\partial\mathbf{f}}\cdot\mathbf{C}^e &= \left\{\bar{g}'(\hat{\chi}_M^*)\left(\frac{\hat{\zeta}_{M1}^*}{\bar{M}_1}\mathbf{e}_1^*+\frac{\hat{\zeta}_{M2}^*}{\bar{M}_2}\mathbf{e}_2^*\right)-\left[\bar{g}(\hat{\chi}_M^*)-\bar{g}'(\hat{\chi}_M^*)(\hat{\zeta}_{M1}^*\chi_{M1}^*+\hat{\zeta}_{M2}^*\chi_{M2}^*)\right]\mathbf{n}\right\}\\
&\quad\cdot\left[\alpha_n\mathbf{e}_3^*\otimes\mathbf{e}_3^*+\alpha_t(\mathbf{e}_1^*\otimes\mathbf{e}_1^*+\mathbf{e}_2^*\otimes\mathbf{e}_2^*)\right]\\
&= \alpha_t\bar{g}'(\hat{\chi}_M^*)\left(\frac{\hat{\zeta}_{M1}^*}{\bar{M}_1}\mathbf{e}_1^*+\frac{\hat{\zeta}_{M2}^*}{\bar{M}_2}\mathbf{e}_2^*\right)-\alpha_n\left[\bar{g}(\hat{\chi}_M^*)-\bar{g}'(\hat{\chi}_M^*)(\hat{\zeta}_{M1}^*\chi_{M1}^*+\hat{\zeta}_{M2}^*\chi_{M2}^*)\right]\mathbf{n}
\end{aligned}
\tag{18.176}
$$

$$
\begin{aligned}
\frac{\partial\bar{f}^*}{\partial\mathbf{f}}\cdot\mathbf{C}^e\cdot\mathbf{t}_n &= \left\{\alpha_t g'(\hat{\chi}_M^*)\left(\frac{\hat{\zeta}_{M1}^*}{\bar{M}_1}\mathbf{e}_1^*+\frac{\hat{\zeta}_{M2}^*}{\bar{M}_2}\mathbf{e}_2^*\right)-\alpha_n[g(\hat{\chi}_M^*)\right.\\
&\quad\left.-g'(\hat{\chi}_M^*)(\hat{\zeta}_{M1}^*\chi_{M1}^*+\hat{\zeta}_{M2}^*\chi_{M2}^*)]\mathbf{n}\right\}\frac{\frac{\hat{\zeta}_{M1}^*}{\bar{M}_1}\mathbf{e}_1^*+\frac{\hat{\zeta}_{M2}^*}{\bar{M}_2}\mathbf{e}_2^*}{\sqrt{\left(\frac{\hat{\zeta}_{M1}^*}{\bar{M}_1}\right)^2+\left(\frac{\hat{\zeta}_{M2}^*}{\bar{M}_2}\right)^2}}\\
&= \alpha_t\bar{g}'(\hat{\chi}_M^*)\sqrt{\left(\frac{\hat{\zeta}_{M1}^*}{\bar{M}_1}\right)^2+\left(\frac{\hat{\zeta}_{M2}^*}{\bar{M}_2}\right)^2}
\end{aligned}
\tag{18.177}
$$

Substituting Eqs. (18.23) and (18.173)–(18.177) into Eqs. (18.58) and (18.60) regarding $\mu\,\mu_s$, $\mu_k\,\kappa$ and ξ as $\bar{F}$, $\bar{F}_s$, $\bar{F}_k$, $\bar{\kappa}$ and $\bar{\xi}$, respectively, reads:

$$
\bar{\mathbf{v}} = [\frac{1}{\alpha_n}\mathbf{n}\otimes\mathbf{n} + \frac{1}{\alpha_t}(\mathbf{I}-\mathbf{n}\otimes\mathbf{n})](\mathring{f}_1\mathbf{e}_1^* + \mathring{f}_2\mathbf{e}_2^* - \mathring{f}_n\mathbf{n})
$$

$$
+\frac{\left\{\bar{g}'(\hat{\chi}_M^*)\left(\frac{\hat{\zeta}_{M1}^*}{\bar{M}_1}\mathbf{e}_1^* + \frac{\hat{\zeta}_{M2}^*}{\bar{M}_2}\mathbf{e}_2^*\right) - [\bar{g}(\hat{\chi}_M^*)\bar{g}'(\hat{\chi}_M^*)(\hat{\zeta}_{M1}^*\chi_{M1}^* + \hat{\zeta}_{M2}^*\chi_{M2}^*)]\mathbf{n}\right\} \cdot (\mathring{f}_1\mathbf{e}_1^* + \mathring{f}_2\mathbf{e}_2^* - \mathring{f}_n\mathbf{n}) - \bar{m}^c}{\bar{m}^p}
$$

$$
\frac{\frac{\hat{\zeta}_{M1}^*}{\bar{M}_1}\mathbf{e}_1^* + \frac{\hat{\zeta}_{M2}^*}{\bar{M}_2}\mathbf{e}_2^*}{\sqrt{\left(\frac{\hat{\zeta}_{M1}^*}{\bar{M}_1}\right)^2 + \left(\frac{\hat{\zeta}_{M2}^*}{\bar{M}_2}\right)^2}} = \frac{1}{\alpha_t}(\mathring{f}_1\mathbf{e}_1^* + \mathring{f}_2\mathbf{e}_2^*) - \frac{1}{\alpha_n}\mathring{f}_n\mathbf{n}
$$

$$
+\frac{\bar{g}'(\hat{\chi}_M^*)\left(\frac{\hat{\zeta}_{M1}^*}{\bar{M}_1}\mathring{f}_1 + \frac{\hat{\zeta}_{M2}^*}{\bar{M}_2}\mathring{f}_2\right) + [\bar{g}(\hat{\chi}_M^*) - \bar{g}'(\hat{\chi}_M^*)(\hat{\zeta}_{M1}^*\chi_{M1}^* + \hat{\zeta}_{M2}^*\chi_{M2}^*)]\mathring{f}_n - \bar{m}^c}{\bar{m}^p}
$$

$$
\frac{\frac{\hat{\zeta}_{M1}^*}{\bar{M}_1}\mathbf{e}_1^* + \frac{\hat{\zeta}_{M2}^*}{\bar{M}_2}\mathbf{e}_2^*}{\sqrt{\left(\frac{\hat{\zeta}_{M1}^*}{\bar{M}_1}\right)^2 + \left(\frac{\hat{\zeta}_{M2}^*}{\bar{M}_2}\right)^2}} \tag{18.178}
$$

$$
\mathring{\mathbf{f}} = [\alpha_n\mathbf{n}\otimes\mathbf{n} + \alpha_t(\mathbf{e}_1^*\otimes\mathbf{e}_1^* + \mathbf{e}_2^*\otimes\mathbf{e}_2^*)]\left\{\bar{v}_1\mathbf{e}_1^* + \bar{v}_2\mathbf{e}_2^* - \bar{v}_n\mathbf{n}\right.
$$

$$
-\left\langle\frac{[\alpha_t\bar{g}'(\hat{\chi}_M^*)(\frac{\hat{\zeta}_{M1}^*}{\bar{M}_1}\mathbf{e}_1^* + \frac{\hat{\zeta}_{M2}^*}{\bar{M}_2}\mathbf{e}_2^*) - \alpha_n\{\bar{g}(\hat{\chi}_M^*) - \bar{g}'(\hat{\chi}_M^*)(\hat{\zeta}_{M1}^*\chi_{M1}^* + \hat{\zeta}_{M2}^*\chi_{M2}^*)\}\mathbf{n}]\cdot(\bar{v}_1\mathbf{e}_1^* + \bar{v}_2\mathbf{e}_2^* - \bar{v}_n\mathbf{n}) - \bar{m}^c}{\bar{m}^p + \alpha_t\bar{g}'(\hat{\chi}_M^*)\sqrt{\left(\frac{\hat{\zeta}_{M1}^*}{\bar{M}_1}\right)^2 + \left(\frac{\hat{\zeta}_{M2}^*}{\bar{M}_2}\right)^2}}\right\rangle\left.\frac{\frac{\hat{\zeta}_{M1}^*}{\bar{M}_1}\mathbf{e}_1^* + \frac{\hat{\zeta}_{M2}^*}{\bar{M}_2}\mathbf{e}_2^*}{\sqrt{\left(\frac{\hat{\zeta}_{M1}^*}{\bar{M}_1}\right)^2 + \left(\frac{\hat{\zeta}_{M2}^*}{\bar{M}_2}\right)^2}}\right\}
$$

$$
= \alpha_t(\bar{v}_1\mathbf{e}_1^* + \bar{v}_2\mathbf{e}_2^*) - \alpha_n\bar{v}_n\mathbf{n}
$$

$$
-\alpha_t\left\langle\frac{\alpha_t\bar{g}'(\hat{\chi}_M^*)\left(\frac{\hat{\zeta}_{M1}^*}{\bar{M}_1}\bar{v}_1 + \frac{\hat{\zeta}_{M2}^*}{\bar{M}_2}\bar{v}_2\right) + \alpha_n[\bar{g}(\hat{\chi}_M^*) - \bar{g}'(\hat{\chi}_M^*)(\hat{\zeta}_{M1}^*\chi_{M1}^* + \hat{\zeta}_{M2}^*\chi_{M2}^*)]\bar{v}_n - \bar{m}^c}{\bar{m}^p + \alpha_t\bar{g}'(\hat{\chi}_M^*)\sqrt{\left(\frac{\hat{\zeta}_{M1}^*}{\bar{M}_1}\right)^2 + \left(\frac{\hat{\zeta}_{M2}^*}{\bar{M}_2}\right)^2}}\right\rangle
$$

$$
\frac{\frac{\hat{\zeta}_{M1}^*}{\bar{M}_1}\mathbf{e}_1^* + \frac{\hat{\zeta}_{M2}^*}{\bar{M}_2}\mathbf{e}_2^*}{\sqrt{\left(\frac{\hat{\zeta}_{M1}^*}{\bar{M}_1}\right)^2 + \left(\frac{\hat{\zeta}_{M2}^*}{\bar{M}_2}\right)^2}} \tag{18.179}
$$

The subloading-friction model is inevitable to the rigorous analysis of the crystal plasticity analysis as will be described in Chap. 19. Further, instead of the rate-and state model, it will have to be applied to the prediction of earthquake occurrence in continental-slip type for the steady prevention of earthquake disaster.

References

Anand L (1993) A constitutive model for interface friction. Comput Mech 12:197–213

Baumberger T, Heslot F, Perrin B (1994) Crossover from creep to inertial motion in friction dynamics. Nature 30:544–546

Bay N, Wanheim T (1976) Real area of contact and friction stresses at high pressure sliding contact. Wear 38:201–209

Bingham EC (1922) Fluidity and plasticity. McGraw-Hill, New York

Bowden FP, Tabor D (1958) The friction and lubrication of solids. Clarendon Press, Oxford

Brockley CA, Davis HR (1968) The time-dependence of static friction. J Lubr Tech (ASME) 90:35–41

Bureau L, Baumberger T, Caroli C, Ronsin O (2001) Low-velocity friction between macroscopic solids. CR Acad Sci Paris Ser IV Differ Faces Tribol 2:699–707

Cheng J-H, Kikuchi N (1985) An incremental constitutive relation of uniaxial contact friction for large deformation analysis. J Appl Mech (ASME) 52:639–648

Courtney-Pratt JS, Eisner E (1957) The effect of a tangential force on the contact metallic bodies. Proc Roy Soc A 238:529–550

Curnier A (1984) A theory of friction. Int J Solids Struct 20:637–647

Derjaguin BV, Push VE, Tolstoi DM (1957) A theory of stick-slipping of solids. In: Proceedings of the conference on lubrication and wear, Institute of Mechanical Engineers, London, pp 257–268

Dieterich JH (1978) Time-dependent friction and the mechanism of stick-slip. Pure Appl Geophys 116:790–806

Diteterih JH (1979) Modeling of rock friction 1. Experimental results and constitutive equations. J Geophys Res 84:2161–2168

Dokos SJ (1946) Sliding friction under extreme pressure—I. Trans ASME 68:A148–A156

Drucker DC (1988) Conventional and unconventional plastic response and representation. Appl Mech Rev (ASME). 41:151–167

Ferrero JF, Barrau JJ (1997) Study of dry friction under small displacements and near-zero sliding velocity. Wear 209:322–327

Fredriksson B (1976) Finite element solution of surface nonlinearities in structural mechanics with special emphasis to contact and fracture mechanics problems. Comput Struct 6:281–290

Gearing BP, Moon HS, Anand L (2001) A plasticity model for interface friction: application to sheet metal forming. Int J Plast 17:237–271

Hashiguchi K (1980) Constitutive equations of elastoplastic materials with elastic-plastic transition. J Appl Mech (ASME) 47:266–272

Hashiguchi K (1989) Subloading surface model in unconventional plasticity. Int J Solids Struct 25:917–945

Hashiguchi K (2006) Constitutive model of friction with transition from static- to kinetic-friction-time-dependent subloading-friction model. In: Proceedings of the international symposium plasticity 2006, pp 178–180

Hashiguchi K (2007) Anisotropic constitutive equation of friction with rotational hardening. In: Proceedings of the 13th international symposium on plasticity and its current applications, pp 34–36

Hashiguchi K (2009) Elastoplasticity theory. Lecture note in applied comptutational mechanics, 1st edn. Springer-Verlag, Heidelberg

Hashiguchi K (2013a) General description of elastoplastic deformation/sliding phenomena of solids in high accuracy and numerical efficiency: subloading surface concept. Arch Compt Meth Eng 20:361–417

Hashiguchi K (2013b) Elastoplasticity theory. Lecture note in applied computational mechanics, 2nd edn. Springer-Verlag, Heidelberg

Hashiguchi K, Ozaki S (2007) Constitutive equation of friction with rotational and orthotropic anisotropy. J Appl Mech (JSCE) 10:383–389

Hashiguchi K, Ozaki S (2008a) Constitutive equation for friction with transition from static to kinetic friction and recovery of static friction. Int J Plast 24:2102–2124

Hashiguchi K, Ozaki S (2008b) Anisotropic constitutive equation for friction with transition from static to kinetic friction and vice versa. J Appl Mech (JSCE) 11:271–282

Hashiguchi K, Ozaki S, Okayasu T (2005) Unconventional friction theory based on the subloading surface concept. Int J Solids Struct 42:1705–1727

Hashiguchi K, Ueno M, Kuwayama T, Suzuki N, Yonemura S, Yoshikawa N (2016) Constitutive equation of friction based on the subloading surface concept. Proc R Soc Lond A472:1–24

Hohenemser K, Prager W (1932) Uber die Ansatze der Mechanik isotroper Kontinua. ZAMM 12:216–226

Horowitz F, Ruina A (1989) Slip patterns in a spatially homogeneous fault model. J Geophys Res 94:10279–10298

Howe PG, Benson DP, Puddington IE (1955) London-Van der Waals' attractive forces between glass surface. Can J Chem 33:1375–1383

Hutchinson JW (1976) Bounds and self-consistent estimates for creep of polycrystalline materials. Proc R Soc Lond A 348:101–127

Kame N, Fujita S, Nakatani M, Kusakabe T (2013) Effects of a revised rate- and state-dependent friction law on aftershock triggering model. Tectonophysics 600:187–195

Kato S, Sato N, Matsubayashi T (1972) Some considerations on characteristics of static friction of machine tool sideway. J Lubr Tech (ASME) 94:234–247

Kikuchi N, Oden JT (1988) Contact problem in elasticity: a study of variational inequalities and finite element methods. SIAM, Philadelphia

Michalowski R, Mroz Z (1978) Associated and non-associated sliding rules in contact friction problems. Archiv Mech 30:259–276

Mroz Z, Stupkiewicz S (1994) An anisotropic friction and wear model. Int J Solids Struct 31:1113–1131

Nakada Y, Keh AS (1966) Latent hardening in iron single crystals. Acta Metall 14:961–973

Norton FH (1929) Creep of steel at high temperature. McGraw-Hill, New York

Oden JT, Martines JAC (1986) Models and computational methods for dynamic friction phenomena. Comput Meth Appl Mech Eng 52:527–634

Oden JT, Pires EB (1983a) Algorithms and numerical results for finite element approximations of contact problems with non-classical friction laws. Comput Struct 19:137–147

Oden JT, Pires EB (1983b) Nonlocal and nonlinear friction laws and variational principles for contact problems in elasticity. J Appl Mech (ASME) 50:67–76

Ozaki S, Hashiguchi K (2010) Numerical analysis of stick-slip instability by a rate-dependent elastoplastic formulation for friction. Tribol Int 43:2120–2133

Ozaki S, Hikida K, Hashiguchi K (2012) Elastoplastic formulation for friction with orthotropic anisotropy and rotational hardening. Int J Solids Struct 49:648–657

Pan J, Rice JR (1983) Rate sensitivity of plastic flow and implications for yield surface vertices. J Mech Phys Solids 19:973–987

Perić D, Owen RJ (1992) Computational model for 3-D contact problems with friction based on the penalty method. Int J Numer Meth Eng 35:1289–1309

Perzyna P (1963) The constitutive equations for rate sensitive plastic materials. Quart Appl Math 20:321–332

Perzyna P (1966) Fundamental problems in viscoplasticity. Adv Appl Mech 9:243–377

Prager W (1961) Introduction to mechanics of continua. Ginn & Company, Boston
Rabinowicz E (1951) The nature of the static and kinetic coefficients of friction. J Appl Phys 22:1373–1379
Rabinowicz E (1958) The intrinsic variables affecting the stick-slip process. Proc Phys Soc 71:668–675
Rice JR, Ruina AL (1983) Stability of steady frictional slipping. J Appl Mech (ASME) 50:343–349
Rice JR, Lapusta N, Ranjith K (2001) Rate and state dependent friction and the stability of sliding between elastically deformable solids. J Mech Phys Solids 49:1865–1898
Ruina AL (1980) Friction laws and instabilities: quasistatic analysis of some dry frictional behavior. Ph.D. thesis, Brown University, Providence
Ruina AL (1983) Slip instability and state variable friction laws. J Geophys Res 88:10359–10370
Scholz CH (1998) Rate-and state-variable friction law. Nature 391:37–41
Seguchi Y, Shindo A, Tomita Y, Sunohara M (1974) Sliding rule of friction in plastic forming of metal. Comput Meth. Nonlinear Mech 683–692
Simo JC, Hughes TJR (1998) Computational inelasticity. Springer, New York
Stribeck R (1902) Die Wesentlichen Eigenschaften der Gleit- und Rollenlager. Z Verein Deut Ing (in German) 46:1341–1348
Truesdell C (1955) Hypo-elasticity. J Ration Mech Anal 4:83–133
Wriggers P (2003) Computational contact mechanics. Wiley, Hoboken
Wriggers P, Van Vu T, Stein E (1990) Finite element formulation of large deformation impact-contact problems with friction. Comput Struct 37:319–331
Zhong Z-H (1993) Finite element procedures for contact-impact problems. Oxford University Press, London

Chapter 19
Crystal Plasticity

The crystal plasticity analysis requires the calculation of the slips in numerous slip systems. Therefore, it could not be realized by the concept of the conventional elastoplasticity with the yield surface enclosing a purely-elastic domain, since it requires the yield judgment and the operation to pull-back the resolved shear stress to the critical shear stress. Then, unfortunately the creep crystal-plasticity model proposed by Peirce et al. (1982, 1983) is used widely. It is impertinent such that the creep shear strain rate is always induced even for the state that the resolved shear stress is unloaded from the critical shear stress as known from the defect of the creep model described in Sect. 13.3.

The crystal plasticity analysis can be attained appropriately by introducing the concept of the subloading surface which is endowed with the distinguished advantages: (1) Yield judgment is not required and (2) Automatic controlling function to attract the stress to the yield surface in the plastic loading process is furnished as described in Chaps. 7, 8, and 9. The pertinent formulation for the crystal plasticity analysis based on the subloading concept will be given in this chapter.

19.1 Multiplicative Decomposition of Deformation Gradient Tensor

The intermediate configuration is obtained by excluding the rigid-body rotation in addition to the elastic deformation from the current configuration as was postulated in Chaps. 6 and 12. Therefore, the intermediate configuration is independent of the rigid-body rotation. Consequently, the deformation gradient $\mathbf{F}$ is multiplicatively decomposed into the elastic deformation gradient $\mathbf{F}^e$ composed of the substructure rotation tensor $\mathbf{R}^e$ involving the rigid-body rotation and the right elastic stretch tensor $\mathbf{U}^e$ and the plastic deformation gradient $\mathbf{F}^p$ as follows (see Fig. 19.1):

K. Hashiguchi, *Foundations of Elastoplasticity: Subloading Surface Model*,
DOI 10.1007/978-3-319-48821-9_19

$$d\mathbf{x} = \mathbf{R}^e d\tilde{\mathbf{x}} = \mathbf{R}^e\mathbf{U}^e d\overline{\mathbf{X}} = \mathbf{F}^e d\overline{\mathbf{X}} = \mathbf{F}d\mathbf{X} \tag{19.1}$$

$$\left.\begin{array}{l} d\tilde{\mathbf{x}} = \mathbf{U}^e d\overline{\mathbf{X}} = \mathbf{U}^e\mathbf{F}^p d\mathbf{X} \\ d\overline{\mathbf{X}} = \mathbf{F}^p d\mathbf{X} \end{array}\right\} \tag{19.2}$$

Then, it follows that

$$\boxed{\mathbf{F} = \mathbf{F}^e\mathbf{F}^p = \mathbf{R}^e\mathbf{U}^e\mathbf{F}^p} \tag{19.3}$$

$$\mathbf{F}^e = \mathbf{R}^e\mathbf{U}^e = \mathbf{V}^e\mathbf{R}^e \tag{19.4}$$

$$\mathbf{F}^p = \mathbf{R}^p\mathbf{U}^p = \mathbf{V}^p\mathbf{R}^p \tag{19.5}$$

where $\mathbf{X}$, $\mathbf{x}$ and $\overline{\mathbf{X}}$ are the position vectors of material particles in the initial (reference), the current and the intermediate configuration, respectively. $\tilde{\mathbf{x}}$ is the position vector of material particle in the configuration pulled-back by the substructure rotation $\mathbf{R}^e$ from the current configuration.

The initial and the current configurations are designated by $\mathcal{K}_0$ and $\mathcal{K}$, respectively, and the configurations pulled-back from the current configuration by the substructure rotation $\mathbf{R}^e$ and by the elastic deformation gradient $\mathbf{F}^e$ are designated as $\widetilde{\mathcal{K}}$ and $\overline{\mathcal{K}}$, respectively, as shown in Fig. 19.1.

19.2 Strain Rate and Spin

The following decomposition is obtained from Eq. (19.2).

$$\boldsymbol{l} = \boldsymbol{l}^e + \boldsymbol{l}^p \tag{19.6}$$

where

$$\left.\begin{array}{l} \boldsymbol{l} \equiv \dfrac{\partial \boldsymbol{v}}{\partial \mathbf{x}} = \dot{\mathbf{F}}\mathbf{F}^{-1}, \\ \boldsymbol{l}^e \equiv \dot{\mathbf{F}}^e\mathbf{F}^{e-1}, \; \boldsymbol{l}^p \equiv \mathbf{F}^e\overline{\mathbf{L}}^p\mathbf{F}^{e-1} \\ \overline{\mathbf{L}}^p \equiv \dot{\mathbf{F}}^p\mathbf{F}^{p-1} \end{array}\right\} \tag{19.7}$$

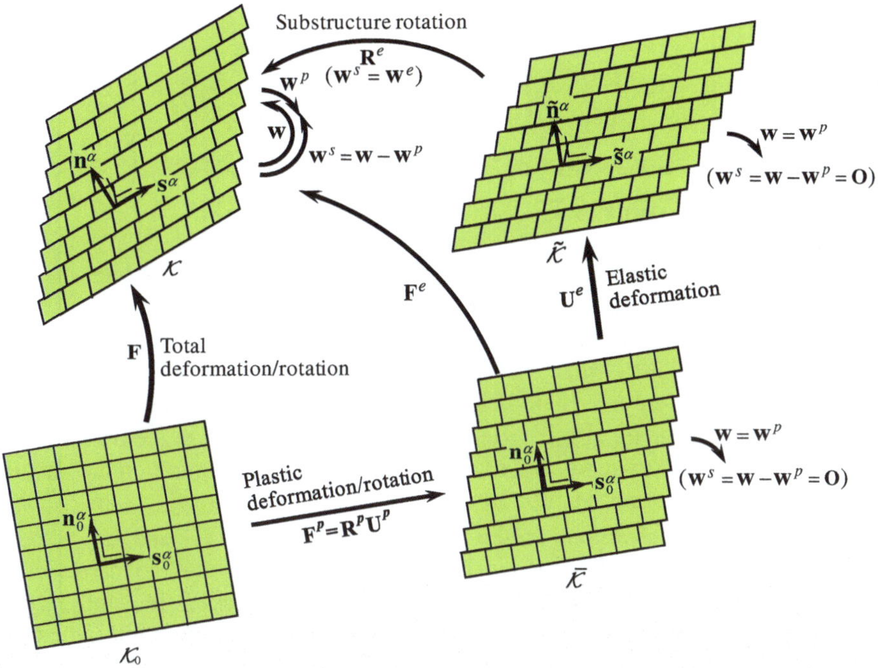

Fig. 19.1 Multiplicative decomposition in crystal plasticity based on isoclinic concept (Mandel 1973, 1974)

Further, the following additive decompositions to the symmetric and the anti-symmetric parts hold.

$$\left.\begin{aligned} \boldsymbol{l} &= \mathbf{d} + \mathbf{w} \\ \boldsymbol{l}^e &= \mathbf{d}^e + \mathbf{w}^e \\ \boldsymbol{l}^p &= \mathbf{d}^p + \mathbf{w}^p \\ \overline{\mathbf{L}}^p &= \overline{\mathbf{D}}^p + \overline{\mathbf{W}}^p \end{aligned}\right\} \tag{19.8}$$

with

$$\left.\begin{aligned} \mathbf{d} &= \mathbf{d}^e + \mathbf{d}^p \\ \mathbf{w} &= \mathbf{w}^e + \mathbf{w}^p \end{aligned}\right\} \tag{19.9}$$

where

$$\left.\begin{array}{ll}\mathbf{d} = \mathrm{sym}[\boldsymbol{l}] = \mathrm{sym}[\dot{\mathbf{F}}\mathbf{F}^{-1}], & \mathbf{w} = \mathrm{ant}[\boldsymbol{l}] = \mathrm{ant}[\dot{\mathbf{F}}\mathbf{F}^{-1}] \\ \mathbf{d}^{e} = \mathrm{sym}[\boldsymbol{l}^{e}] = \mathrm{sym}[\dot{\mathbf{F}}^{e}\mathbf{F}^{e-1}], & \mathbf{w}^{e} = \mathrm{ant}[\boldsymbol{l}^{e}] = \mathrm{ant}[\dot{\mathbf{F}}^{e}\mathbf{F}^{e-1}] \\ \overline{\mathbf{D}}^{p} = \mathrm{sym}\left[\overline{\mathbf{L}}^{p}\right] = \mathrm{sym}[\dot{\mathbf{F}}^{p}\mathbf{F}^{p-1}], & \overline{\mathbf{W}}^{p} = \mathrm{ant}\left[\overline{\mathbf{L}}^{p}\right] = \mathrm{ant}[\dot{\mathbf{F}}^{p}\mathbf{F}^{p-1}] \\ \mathbf{d}^{p} = \mathrm{sym}[\boldsymbol{l}^{p}] = \mathrm{sym}[\mathbf{F}^{e}\overline{\mathbf{L}}^{p}\mathbf{F}^{e-1}], & \mathbf{w}^{p} = \mathrm{ant}[\boldsymbol{l}^{p}] = \mathrm{ant}[\mathbf{F}^{e}\overline{\mathbf{L}}^{p}\mathbf{F}^{e-1}]\end{array}\right\} \quad (19.10)$$

Now, adopt the embedded base $(\mathbf{g}_1, \mathbf{g}^2, \mathbf{g}_3)$ described in Sect. 4.4. Here, the first primary base vector $\mathbf{g}_1$ is chosen parallel to the crystal lattice and denoted by $\mathbf{s}^\alpha$ and the secondary reciprocal base vector $\mathbf{g}^2$ by $\mathbf{n}^\alpha$ in the slip system α, satisfying $\mathbf{s}^\alpha \cdot \mathbf{n}^\alpha = 0$ by Eq. $(4.30)_2$ while they are called the *lattice vectors, director frame, director triad, isoclinic triad*, etc. Limiting to the two-dimensional deformation, $\mathbf{g}_3 (= \mathbf{g}^3)$ is the unit vector $(||\mathbf{g}_3|| = 1)$ and chosen perpendicular to the base vectors $\mathbf{s}^\alpha$ and $\mathbf{n}^\alpha$. Further, the base vectors $\mathbf{s}^\alpha$ and $\mathbf{n}^\alpha$ in the initial configuration are chosen to be the unit vectors and denoted by the symbols $\mathbf{s}_0^\alpha$ and $\mathbf{n}_0^\alpha$ $(||\mathbf{s}_0^\alpha|| = ||\mathbf{n}_0^\alpha|| = 1)$, respectively, which correspond to the reference base vectors $\mathbf{G}_1$ and $\mathbf{G}^2$, respectively, in Sect. 4.4. Reminding the aforementioned assumption that the intermediate configuration is independent of the rigid-body rotation and noting the simple shear deformation along the crystalline lattice under the plastic incompressibility, the base vectors are kept unchanged as $\mathbf{s}_0^\alpha$ and $\mathbf{n}_0^\alpha$ in the process from the initial to the intermediate configurations (see Fig. 19.1). This physical consequence is referred to as the *isoclinic concept* by Mandel (1973, 1974), while "isoclinic" is the Greek word meaning "same (or constant) direction" as was described in Sect. 6.1.

The current primary base vector $\mathbf{s}^\alpha$ and its reciprocal base vector $\mathbf{n}^\alpha$ are related to the initial base vectors $\mathbf{s}_0^\alpha$ and $\mathbf{n}_0^\alpha$ by Eqs. (4.32) and (4.35), replacing $\mathbf{F}$ to $\mathbf{F}^e$, as follows:

$$\left.\begin{array}{l}\mathbf{s}^\alpha = \mathbf{F}^e\mathbf{s}_0^\alpha = \mathbf{s}_0^\alpha\mathbf{F}^{eT},\ \mathbf{s}_0^\alpha = \mathbf{F}^{e-1}\mathbf{s}^\alpha = \mathbf{s}^\alpha\mathbf{F}^{e-T} \\ \mathbf{n}^\alpha = \mathbf{F}^{e-T}\mathbf{n}_0^\alpha = \mathbf{n}_0^\alpha\mathbf{F}^{e-1},\ \mathbf{n}_0^\alpha = \mathbf{F}^{eT}\mathbf{n}^\alpha = \mathbf{n}^\alpha\mathbf{F}^e \\ (\mathbf{s}^\alpha \cdot \mathbf{n}^\alpha = \mathbf{F}^e\mathbf{s}_0^\alpha \cdot \mathbf{F}^{e-T}\mathbf{n}_0^\alpha = \mathbf{s}_0^\alpha \cdot \mathbf{F}^{eT}\mathbf{F}^{e-T}\mathbf{n}_0^\alpha = \mathbf{s}_0^\alpha \cdot \mathbf{n}_0^\alpha = 0)\end{array}\right\} \quad (19.11)$$

noting $\mathbf{T}\mathbf{v} = \mathbf{v}\mathbf{T}^T (T_{ij}v_j = v_jT_{ij})$. The base vectors $\tilde{\mathbf{s}}^\alpha$ and $\tilde{\mathbf{n}}^\alpha$ obtained by excluding the substructure rotation from the current configuration as shown in Fig. 19.1 are related to the initial base vectors $\mathbf{s}_0^\alpha$ and $\mathbf{n}_0^\alpha$ by replacing $\mathbf{F}^e$ to $\mathbf{R}^e$ in Eq. (19.11).

The rates of $\mathbf{s}^\alpha$ and $\mathbf{n}^\alpha$ are given as follows:

$$\left.\begin{array}{l}\dot{\mathbf{s}}^\alpha = \dot{\mathbf{F}}^e\mathbf{s}_0^\alpha = \dot{\mathbf{F}}^e\mathbf{F}^{e-1}\mathbf{s}^\alpha = \boldsymbol{l}^e\mathbf{s}^\alpha = \mathbf{s}^\alpha\boldsymbol{l}^{eT} \\ \dot{\mathbf{n}}^\alpha = \dot{\mathbf{F}}^{e-T}\mathbf{n}_0^\alpha = \dot{\mathbf{F}}^{e-T}\mathbf{F}^{eT}\mathbf{n}^\alpha = -\boldsymbol{l}^{eT}\mathbf{n}^\alpha = -\mathbf{n}^\alpha\boldsymbol{l}^e\end{array}\right\} \quad (19.12)$$

Needless to say, the current base vectors $(\mathbf{s}^\alpha, \mathbf{n}^\alpha)$ and $(\tilde{\mathbf{s}}^\alpha, \tilde{\mathbf{n}}^\alpha)$ are no longer unit vectors.

On the other hand, the simple shear strain γ^α along the slip system α is additively decomposed into the elastic shear strain $\gamma^{e\alpha}$ and the plastic shear strain $\gamma^{p\alpha}$ as follows:

$$\gamma^{\alpha} = \gamma^{e\alpha} + \gamma^{p\alpha}, \quad \dot{\gamma}^{\alpha} = \dot{\gamma}^{e\alpha} + \dot{\gamma}^{p\alpha} \tag{19.13}$$

The difference of plastic displacements, $d\mathbf{u}^{p\alpha}$, in both ends of infinitesimal line element $d\mathbf{X}$, which is induced by plastic shear strain γ^{α} in slip system α, is given by the following equation (see Fig. 19.2).

$$d\mathbf{u}^{p\alpha} = \gamma^{p\alpha}(\mathbf{n}_0^{\alpha} \cdot d\mathbf{X})\mathbf{s}_0^{\alpha} = \gamma^{p\alpha}\mathbf{s}_0^{\alpha} \otimes \mathbf{n}_0^{\alpha} d\mathbf{X} \tag{19.14}$$

Then, $d\mathbf{X}$ changes to

$$d\bar{\mathbf{X}} = d\mathbf{X} + d\mathbf{u}^{p\alpha} = (\mathbf{I} + \gamma^{p\alpha}\mathbf{s}_0^{\alpha} \otimes \mathbf{n}_0^{\alpha})d\mathbf{X} \tag{19.15}$$

in the intermediate configuration.

Noting Eqs. $(19.1)_3$ and (19.15), one has

$$\mathbf{F}^{p\alpha} = \mathbf{I} + \gamma^{p\alpha}\mathbf{s}_0^{\alpha} \otimes \mathbf{n}_0^{\alpha}, \quad \mathbf{F}^{p\alpha-1} = \mathbf{I} - \gamma^{p\alpha}\mathbf{s}_0^{\alpha} \otimes \mathbf{n}_0^{\alpha} \tag{19.16}$$

by virtue of $(\mathbf{I} + \gamma^{p\alpha}\mathbf{s}_0^{\alpha} \otimes \mathbf{n}_0^{\alpha})(\mathbf{I} - \gamma^{p\alpha}\mathbf{s}_0^{\alpha} \otimes \mathbf{n}_0^{\alpha}) = \mathbf{I}$ due to $\mathbf{n}_0^{\alpha} \cdot \mathbf{s}_0^{\alpha} = 0$. Further, noting

$$(\mathbf{I} + \gamma^{p\alpha}\mathbf{s}_0^{\alpha} \otimes \mathbf{n}_0^{\alpha})^{\bullet}(\mathbf{I} - \gamma^{p\alpha}\mathbf{s}_0^{\alpha} \otimes \mathbf{n}_0^{\alpha}) = \dot{\gamma}^{p\alpha}\mathbf{s}_0^{\alpha} \otimes \mathbf{n}_0^{\alpha} - \dot{\gamma}^{p\alpha}\mathbf{s}_0^{\alpha} \otimes \mathbf{n}_0^{\alpha}\gamma^{p\alpha}\mathbf{s}_0^{\alpha} \otimes \mathbf{n}_0^{\alpha} \tag{19.17}$$

it follows that

$$\dot{\mathbf{F}}^{p\alpha}\mathbf{F}^{p\alpha-1} = \dot{\gamma}^{p\alpha}\mathbf{s}_0^{\alpha} \otimes \mathbf{n}_0^{\alpha} \tag{19.18}$$

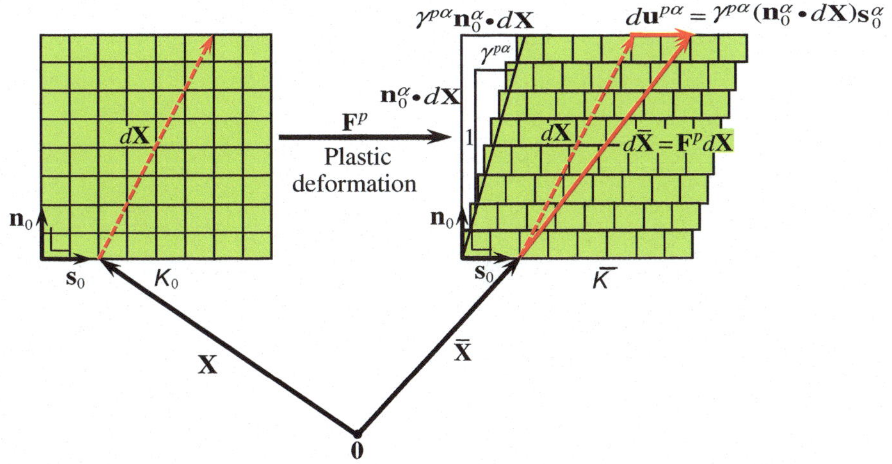

Fig. 19.2 Plastic displacement in end of infinitesimal line element $d\mathbf{X}$, which is induced by plastic shear strain in slip system

The plastic velocity gradient based in the intermediate configuration is given by summing the plastic velocity gradients induced in all the relevant slip systems as follows:

$$\overline{\mathbf{L}}^p = \sum_{\beta=1}^{n} \dot{\mathbf{F}}^{p\beta}\mathbf{F}^{p\beta-1} = \sum_{\beta=1}^{n} \mathbf{s}_0^\beta \otimes \mathbf{n}_0^\beta \dot{\gamma}^{p\beta} \tag{19.19}$$

where n is the number of slip systems. The substitution of Eq. (19.19) into Eq. (19.10) leads to

$$\left.\begin{aligned} \overline{\mathbf{D}}^p &= \sum_{\beta=1}^{n} \overline{\mathbf{P}}^\beta \dot{\gamma}^{p\beta} \\ \overline{\mathbf{W}}^p &= \sum_{\beta=1}^{n} \overline{\mathbf{Q}}^\beta \dot{\gamma}^{p\beta} \end{aligned}\right\} \tag{19.20}$$

where

$$\left.\begin{aligned} \overline{\mathbf{P}}^\alpha &= \mathrm{sym}[\mathbf{s}_0^\alpha \otimes \mathbf{n}_0^\alpha] \\ \overline{\mathbf{Q}}^\alpha &= \mathrm{ant}[\mathbf{s}_0^\alpha \otimes \mathbf{n}_0^\alpha] \end{aligned}\right\} \tag{19.21}$$

Substituting Eq. (19.19) with Eq. (19.11) into Eq. (19.7) reads:

$$\mathbf{l}^p = \sum_{\beta=1}^{n} \mathbf{s}^\beta \otimes \mathbf{n}^\beta \dot{\gamma}^{p\beta} \tag{19.22}$$

Inserting Eq. (19.22) into Eq. (19.10), one has

$$\left.\begin{aligned} \mathbf{d}^p &= \sum_{\beta=1}^{n} \mathbf{p}^\beta \dot{\gamma}^{p\beta} \\ \mathbf{w}^p &= \sum_{\alpha=1}^{n} \mathbf{q}^\beta \dot{\gamma}^{p\beta} \end{aligned}\right\} \tag{19.23}$$

with

$$\left.\begin{aligned} \mathbf{p}^\alpha &= \mathrm{sym}[\mathbf{s}^\alpha \otimes \mathbf{n}^\alpha] \\ \mathbf{q}^\alpha &= \mathrm{ant}[\mathbf{s}^\alpha \otimes \mathbf{n}^\alpha] \end{aligned}\right\} \tag{19.24}$$

Substituting Eq. (19.22) into Eq. (19.6), one has

$$\boldsymbol{l}^e = \boldsymbol{l} - \boldsymbol{l}^p = \boldsymbol{l} - \sum_{\beta=1}^{n} \mathbf{s}^\beta \otimes \mathbf{n}^\beta \dot{\gamma}^{p\beta} = \boldsymbol{l} - \sum_{\beta=1}^{n} (\mathbf{p}^\beta + \mathbf{q}^\beta)\dot{\gamma}^{p\beta} \tag{19.25}$$

$$\left.\begin{aligned} \mathbf{d}^e &= \mathrm{sym}[\boldsymbol{l}^e] = \mathrm{sym}[\dot{\mathbf{F}}^e\mathbf{F}^{e-1}] = \mathbf{d} - \sum_{\beta=1}^{n} \mathbf{p}^\beta \dot{\gamma}^{p\beta} \\ \mathbf{w}^e &= \mathrm{ant}[\boldsymbol{l}^e] = \mathrm{ant}[\dot{\mathbf{F}}^e\mathbf{F}^{e-1}] = \mathbf{w} - \sum_{\beta=1}^{n} \mathbf{q}^\beta \dot{\gamma}^{p\beta} \end{aligned}\right\} \tag{19.26}$$

19.3 Resolved Shear Stress (Rate)

Let the resolved shear stress be defined as follows (Asaro and Rice 1977):

$$\tau^\alpha = \mathbf{s}^\alpha \cdot \boldsymbol{\tau}\mathbf{n}^\alpha = \mathbf{p}^\alpha : \boldsymbol{\tau} \tag{19.27}$$

The time-differentiation of Eq. (19.27) leads to

$$\dot{\tau}^\alpha = \dot{\mathbf{s}}^\alpha \cdot \boldsymbol{\tau}\boldsymbol{n}^\alpha + \mathbf{s}^\alpha \cdot \dot{\boldsymbol{\tau}}\mathbf{n}^\alpha + \mathbf{s}^\alpha \cdot \boldsymbol{\tau}\dot{\mathbf{n}}^\alpha$$

Substituting Eqs. (19.8) and (19.12), the right-hand side of this equation leads:

$$\begin{aligned} &\dot{\mathbf{s}}^\alpha \cdot \boldsymbol{\tau}\mathbf{n}^\alpha + \mathbf{s}^\alpha \cdot \dot{\boldsymbol{\tau}}\mathbf{n}^\alpha + \mathbf{s}^\alpha \cdot \boldsymbol{\tau}\dot{\mathbf{n}}^\alpha \\ &= (\mathbf{d}^e + \mathbf{w}^e)\mathbf{s}^\alpha \cdot \boldsymbol{\tau}\mathbf{n}^\alpha + \mathbf{s}^\alpha \cdot \dot{\boldsymbol{\tau}}\mathbf{n}^\alpha - \mathbf{s}^\alpha \cdot \boldsymbol{\tau}(\mathbf{d}^e + \mathbf{w}^e)^T\mathbf{n}^\alpha \\ &= \mathbf{s}^\alpha \cdot (\mathbf{d}^e + \mathbf{w}^e)^T\boldsymbol{\tau}\mathbf{n}^\alpha + \mathbf{s}^\alpha \cdot \dot{\boldsymbol{\tau}}\mathbf{n}^\alpha - \mathbf{s}^\alpha \cdot \boldsymbol{\tau}(\mathbf{d}^e - \mathbf{w}^e)\mathbf{n}^\alpha \\ &= \mathbf{s}^\alpha \cdot (\dot{\boldsymbol{\tau}} - \mathbf{w}^e\boldsymbol{\tau} + \boldsymbol{\tau}\mathbf{w}^e + \mathbf{d}^e\boldsymbol{\tau} - \boldsymbol{\tau}\mathbf{d}^e)\mathbf{n}^\alpha \end{aligned}$$

leading to

$$\dot{\tau}^\alpha = \mathbf{s}^\alpha \cdot \overset{*}{\boldsymbol{\tau}}\mathbf{n}^\alpha + \mathbf{s}^\alpha \cdot (\mathbf{d}^e\boldsymbol{\tau} - \boldsymbol{\tau}\mathbf{d}^e)\mathbf{n}^\alpha \tag{19.28}$$

where (*) designates the corotational rate based on the spin of crystal lattice subjected to the rigid-body rotation and elastic rotation and $\overset{*}{\boldsymbol{\tau}}$ is given by

$$\overset{*}{\boldsymbol{\tau}} = \mathbf{R}^e(\mathbf{R}^{eT}\boldsymbol{\tau}\mathbf{R}^e)^{\boldsymbol{\cdot}}\mathbf{R}^{eT} = \dot{\boldsymbol{\tau}} - \mathbf{w}^e\boldsymbol{\tau} + \boldsymbol{\tau}\mathbf{w}^e = \mathbf{E} : \mathbf{d}^e \tag{19.29}$$

E being the overall elastic modulus tensor.

Taking account of Eqs. (19.24) and (19.52) and the symmetry of $\boldsymbol{\tau}$ and $\mathbf{d}^e$, one has

$$\begin{aligned}\mathbf{s}^{\alpha}\cdot\overset{*}{\boldsymbol{\tau}}\mathbf{n}^{\alpha} &= s_i^{\alpha}\overset{*}{\tau}_{ij}n_j^{\alpha} = s_i^{\alpha}n_j^{\alpha}\overset{*}{\tau}_{ij} = \mathbf{s}^{\alpha}\otimes\mathbf{n}^{\alpha} : \overset{*}{\boldsymbol{\tau}} = \mathbf{p}^{\alpha} : \overset{*}{\boldsymbol{\tau}}\\ &= \mathbf{p}^{\alpha} : \mathbf{E} : \mathbf{d}^{e} = p_{ij}^{\alpha}E_{ijkl}d_{kl}^{e} = E_{klij}p_{ij}^{\alpha}d_{kl}^{e} = \mathbf{E}\mathbf{p}^{\alpha} : \mathbf{d}^{e}\end{aligned}$$

$$\begin{aligned}\mathbf{s}^{\alpha}\bullet(\mathbf{d}^{e}\boldsymbol{\tau}-\boldsymbol{\tau}\mathbf{d}^{e})\mathbf{n}^{\alpha} &= s_i^{\alpha}d_{ir}^{e}\tau_{rs}n_s^{\alpha} - s_i^{\alpha}\tau_{ir}d_{rs}^{e}n_s^{\alpha}\\ &= \frac{1}{2}[(s_i^{\alpha}n_s^{\alpha}\tau_{sr})d_{ir}^{e} + (\tau_{rs}n_s^{\alpha}s_i^{\alpha})d_{ri}^{e}] - \frac{1}{2}[(n_s^{\alpha}s_i^{\alpha}\tau_{ir})d_{sr}^{e} + (\tau_{ri}s_i^{\alpha}n_s^{\alpha})d_{rs}^{e}]\\ &= \frac{1}{2}\{[(\mathbf{s}^{\alpha}\otimes\mathbf{n}^{\alpha})\boldsymbol{\tau}] : \mathbf{d}^{e} - [(\mathbf{n}^{\alpha}\otimes\mathbf{s}^{\alpha})\boldsymbol{\tau}] : \mathbf{d}^{e}\} - \frac{1}{2}\{[\boldsymbol{\tau}(\mathbf{s}^{\alpha}\otimes\mathbf{n}^{\alpha})] : \mathbf{d}^{e}\\ &\quad - [\boldsymbol{\tau}(\mathbf{n}^{\alpha}\otimes\mathbf{s}^{\alpha})] : \mathbf{d}^{e}\}\\ &= \left[\frac{1}{2}(\mathbf{s}^{\alpha}\otimes\mathbf{n}^{\alpha} - \mathbf{n}^{\alpha}\otimes\mathbf{s}^{\alpha})\boldsymbol{\tau} - \boldsymbol{\tau}\frac{1}{2}(\mathbf{s}^{\alpha}\otimes\mathbf{n}^{\alpha} - \mathbf{n}^{\alpha}\otimes\mathbf{s}^{\alpha}\right] : \mathbf{d}^{e}\\ &= (\mathbf{q}^{\alpha}\boldsymbol{\tau} - \boldsymbol{\tau}\mathbf{q}^{\alpha}) : \mathbf{d}^{e} = \boldsymbol{\beta}^{\alpha} : \mathbf{d}^{e}\end{aligned}$$

Substituting these relations into Eq. (19.28), we have the relation between the resolved shear stress rate versus the global elastic strain rate as follows:

$$\overset{\bullet}{\tau}{}^{\alpha} = \boldsymbol{\Xi}^{\alpha} : \mathbf{d}^{e} \tag{19.30}$$

where

$$\boldsymbol{\Xi}^{\alpha} \equiv \mathbf{E} : \mathbf{p}^{\alpha} + \boldsymbol{\beta}^{\alpha}(\neq \boldsymbol{\Xi}^{\alpha T}) \tag{19.31}$$

Further, substituting Eq. $(19.26)_1$ into Eq. (19.30) reads:

$$\overset{\bullet}{\tau}{}^{\alpha} = \boldsymbol{\Xi}^{\alpha} : (\mathbf{d} - \sum_{\beta=1}^{n}\mathbf{p}^{\beta}\overset{\bullet}{\gamma}{}^{p\beta}) \tag{19.32}$$

19.4 Plastic Shear Strain Rate

The crystal shear yield condition describing the *crystal shear yield region* in the slip system α is given by

$$|\hat{\tau}^{\alpha}| = \tau_y^{\alpha} \tag{19.33}$$

where

$$\hat{\tau}^{\alpha} \equiv \tau^{\alpha} - \chi^{\alpha} \tag{19.34}$$

χ^{α} is the shear kinematic hardening variable and $\tau_y^{\alpha}\,(>0)$ is the shear hardening function, referred to as the *critical shear stress*, in the slip system α.

The associated flow rule to the subloading shear region is adopted for the plastic shear strain rate as follows:

$$\dot{\gamma}^{p\alpha} = \dot{\lambda}^{\alpha}\hat{n}^{\alpha}\ (\dot{\lambda}^{\alpha} \geq 0)(|\dot{\gamma}^{p\alpha}| = \dot{\lambda}^{\alpha}) \tag{19.35}$$

where

$$\hat{n}^{\alpha} \equiv \frac{\partial|\hat{\tau}^{\alpha}|}{\partial\tau^{\alpha}} = \frac{\hat{\tau}^{\alpha}}{|\hat{\tau}^{\alpha}|} = \text{sign}(\hat{\tau}^{\alpha})\ (|\hat{n}^{\alpha}| = 1) \tag{19.36}$$

The material-time derivative of Eq. (19.33) reads:

$$\hat{n}^{\alpha}(\dot{\tau}^{\alpha} - \dot{\chi}^{\alpha}) = \dot{\tau}_y^{\alpha} \tag{19.37}$$

i.e.

$$\dot{\tau}^{\alpha} = \hat{n}^{\alpha}\dot{\tau}_y{}^{\alpha} + \dot{\chi}^{\alpha} \tag{19.38}$$

The rate of critical shear stress is specified by

$$\dot{\tau}_y{}^{\alpha} = \sum_{\beta=1}^{n} h_{\alpha\beta}|\dot{\gamma}^{p\beta}| = \sum_{\beta=1}^{n} h_{\alpha\beta}\dot{\lambda}^{\beta} \tag{19.39}$$

where $h_{\alpha\beta}$ is given by the following matrix which is the function of the plastic shear strain (Peirce et al. 1982).

$$h_{\alpha\beta} = qh(\gamma^p) + (1-q)h(\gamma^p)\delta_{\alpha\beta}(= h_{\beta\alpha}) = \begin{cases} h(\gamma^p) & \text{for} \quad \alpha = \beta \\ qh(\gamma^p) & \text{for} \quad \alpha \neq \beta \end{cases} (1 \leq q \leq 1.4) \tag{19.40}$$

$h(\gamma^p)$ is given by the functional form

$$h(\gamma^p) = h_0 \text{sech}^2\left(\frac{h_0\gamma^p}{\tau_s - \tau_{y0}}\right) \tag{19.41}$$

with

$$\gamma^p \equiv \sum_{\beta=1}^{n} \int_0^t |\dot{\gamma}^{p\beta}|dt \tag{19.42}$$

h_0 and τ_{y0} are the initial values of h and τ_y^{α}, i.e. $h_0 = h(0)$ and $\tau_{y0} = \tau_y^{\alpha}(0)$, respectively, and τ_s is the saturation value of τ_y^{α}, i.e. $\tau_s = \tau_y^{\alpha}(\infty)$.

Assume the following shear nonlinear-kinematic hardening rule.

$$\overset{\bullet}{\chi}{}^{\alpha} = c_{\chi}\left(\overset{\bullet}{\gamma}{}^{p\alpha} - \frac{1}{\zeta_{\chi}\tau_{y}^{\alpha}}|\overset{\bullet}{\gamma}{}^{p\alpha}|\chi^{\alpha}\right) = c_{\chi}\overset{\bullet}{\lambda}{}^{\alpha}\left(\hat{n}^{\alpha} - \frac{1}{\zeta_{\chi}\tau_{y}^{\alpha}}\chi^{\alpha}\right) \tag{19.43}$$

where c_{χ} and ζ_{χ} are the material constants. The latent hardening may be incorporated for the shear kinematic hardening (e.g. Bassani and Wu 1991; Harder 1999; Xu and Jiang 2004).

Substituting Eqs. (19.39) and (19.43) into Eq. (19.38), one has the consistency condition for the subloading crystalline shear region.

$$\overset{\bullet}{\tau}{}^{\alpha} = \hat{n}^{\alpha}\sum_{\beta=1}^{n} h_{\alpha\beta}\overset{\bullet}{\lambda}{}^{\beta} + c_{\chi}\left(\hat{n}^{\alpha} - \frac{1}{\zeta_{\chi}\tau_{y}^{\alpha}}\chi^{\alpha}\right)\overset{\bullet}{\lambda}{}^{\alpha} \tag{19.44}$$

which is rewritten as

$$\overset{\bullet}{\tau}{}^{\alpha} = \sum_{\beta=1}^{n} \bar{h}_{\alpha\beta}\overset{\bullet}{\lambda}{}^{\beta} \tag{19.45}$$

where

$$\bar{h}_{\alpha\beta} \equiv \hat{n}^{\alpha}h_{\alpha\beta} + c_{\chi}\left(\hat{n}^{\alpha} - \frac{1}{\zeta_{\chi}\tau_{y}^{\alpha}}\chi^{\alpha}\right)\delta_{\alpha\beta}(\neq \bar{h}_{\beta\alpha}) \tag{19.46}$$

19.5 Strain Rate Versus Stress Rate Relations

The global constitutive relation is given by Eq. (19.29) i.e.

$$\overset{*}{\boldsymbol{\tau}} = \mathbf{E} : \mathbf{d}^{e} \tag{19.47}$$

On the other hand, the Jaumann rate of the Kirchhoff stress is given as

$$\overset{\circ}{\boldsymbol{\tau}}{}^{w} = \overset{\bullet}{\boldsymbol{\tau}} - \mathbf{w}\boldsymbol{\tau} + \boldsymbol{\tau}\mathbf{w} \tag{19.48}$$

These corotational rates are related as

$$\overset{*}{\boldsymbol{\tau}} = \overset{\circ}{\boldsymbol{\tau}}{}^{w} + \mathbf{w}^{p}\boldsymbol{\tau} - \boldsymbol{\tau}\mathbf{w}^{p} \tag{19.49}$$

noting

$$\dot{\boldsymbol{\tau}} - \mathbf{w}^e\boldsymbol{\tau} + \boldsymbol{\tau}\mathbf{w}^e = \dot{\boldsymbol{\tau}} - (\mathbf{w} - \mathbf{w}^p)\boldsymbol{\tau} + \boldsymbol{\tau}(\mathbf{w} - \mathbf{w}^p). \tag{19.50}$$

The substitution of Eq. (19.23) into Eq. (19.49) leads to

$$\overset{*}{\boldsymbol{\tau}} = \overset{\circ}{\boldsymbol{\tau}}{}^w + \sum_{\beta=1}^{n} \boldsymbol{\beta}^\beta \dot{\gamma}^{p\beta} \tag{19.51}$$

where

$$\boldsymbol{\beta}^\alpha \equiv \mathbf{q}^\alpha\boldsymbol{\tau} - \boldsymbol{\tau}\mathbf{q}^\alpha \ (= -\boldsymbol{\beta}^{\alpha T}) \tag{19.52}$$

Further, substituting Eqs. $(19.9)_1$ and $(19.23)_1$ into Eq. (19.47), one has

$$\overset{*}{\boldsymbol{\tau}} = \mathbf{E} : (\mathbf{d} - \mathbf{d}^p) = \mathbf{E} : \mathbf{d} - \sum_{\beta=1}^{n} \mathbf{E} : \mathbf{p}^\beta \dot{\gamma}^{p\beta} \tag{19.53}$$

The substitution of Eq. (19.53) into Eq. (19.51) reads:

$$\overset{\circ}{\boldsymbol{\tau}}{}^w = \mathbf{E} : \mathbf{d} - \sum_{\beta=1}^{n} \boldsymbol{\Xi}^\beta \dot{\gamma}^{p\beta} \tag{19.54}$$

The strain rate is described by the stress rates from Eq. (19.53) and with Eq. (19.51) as follows:

$$\mathbf{d} = \mathbf{E}^{-1} : \overset{*}{\boldsymbol{\tau}} + \sum_{\beta=1}^{n} \mathbf{p}^\beta \dot{\gamma}^{p\beta} = \mathbf{E}^{-1} : (\overset{\circ}{\boldsymbol{\tau}}{}^w + \sum_{\beta=1}^{n} \boldsymbol{\Xi}^\beta \dot{\gamma}^{p\beta}) \tag{19.55}$$

Equating Eq. (19.32) with Eq. (19.45), it follows that

$$\boldsymbol{\Xi}^\alpha : \left(\mathbf{d} - \sum_{\beta=1}^{n} \mathbf{p}^\beta \hat{n}^\beta \dot{\lambda}^\beta\right) = \sum_{\beta=1}^{n} \bar{h}_{\alpha\beta} \dot{\lambda}^\beta \tag{19.56}$$

which is rewritten as

$$\boldsymbol{\Xi}^\alpha : \mathbf{d} = \sum_{\beta=1}^{n} \overline{M}_{\alpha\beta} \dot{\lambda}^\beta \tag{19.57}$$

where

$$\overline{M}_{\alpha\beta} \equiv \bar{h}_{\alpha\beta} + \boldsymbol{\Xi}^{\alpha} : \mathbf{p}^{\beta}\hat{n}^{\beta} \ (\neq \overline{M}^{T}_{\alpha\beta}) \tag{19.58}$$

The plastic shear strain rate is given from Eq. (19.57) as follows:

$$\dot{\lambda}^{\alpha} = \sum_{\beta=1}^{n} \overline{M}^{-1}_{\alpha\beta} \boldsymbol{\Xi}^{\beta} : \mathbf{d} \ \left(\sum_{\beta=1}^{n} \overline{M}_{\alpha\beta} \dot{\lambda}^{\beta} = \boldsymbol{\Xi}^{\alpha} : \mathbf{d} \right) \tag{19.59}$$

Substituting Eq. (19.59) into Eq. (19.54), it follows that

$$\mathring{\boldsymbol{\tau}}^{w} = \mathbf{E} : \mathbf{d} - \sum_{\alpha=1}^{n} \boldsymbol{\Xi}^{\alpha} \dot{\lambda}^{\alpha} \hat{n}^{\alpha} = \mathbf{E} : \mathbf{d} - \sum_{\alpha=1}^{n} \boldsymbol{\Xi}^{\alpha} \hat{n}^{\alpha} \sum_{\beta=1}^{n} \overline{M}^{-1}_{\alpha\beta} \boldsymbol{\Xi}^{\beta} : \mathbf{d}$$

leading to

$$\mathring{\boldsymbol{\tau}}^{w} = \overline{\mathbf{K}}^{ep} : \mathbf{d} \tag{19.60}$$

where

$$\overline{\mathbf{K}}^{ep} \equiv \mathbf{E} - \sum_{\alpha=1}^{n} \sum_{\beta=1}^{n} \boldsymbol{\Xi}^{\alpha} \otimes \hat{n}^{\alpha} \overline{M}^{-1}_{\alpha\beta} \boldsymbol{\Xi}^{\beta} \ (\neq \overline{\mathbf{K}}^{epT}) \tag{19.61}$$

The loading criterion for the plastic shear strain rate is given by the sign of the plastic multiplier in terms of the shear strain rate as follows:

$$\left.\begin{array}{ll} \dot{\gamma}^{p\alpha} \neq 0 & \text{for } |\hat{\tau}^{\alpha}| = \tau^{\alpha}_{y} \text{ and } \tau^{\alpha}\dot{\hat{\tau}}^{\alpha} > 0 \\ \dot{\gamma}^{p\alpha} = 0 & \text{for others} \end{array}\right\} \tag{19.62}$$

19.6 Uniqueness of Slip Rate Mode

Hill and Rice (1972) proved that the sufficient condition for the uniqueness of the combination of slip rates is the positive-definite of the matrix in Eq. (19.58) in all the slip systems as follows:

Suppose to impose the two strain rates $\mathbf{d}$ and $\bar{\mathbf{d}}$, and designate the slip rates for these strain rates as $\dot{\gamma}^{\alpha}$ and $\dot{\bar{\gamma}}^{\alpha}$, respectively, in the two slip modes, and denote their differences as follows:

$$\Delta\dot{\gamma}^{\alpha} = \dot{\gamma}^{\alpha} - \dot{\bar{\gamma}}^{\alpha} \tag{19.63}$$

$$\Delta\mathbf{d} = \mathbf{d} - \bar{\mathbf{d}} \tag{19.64}$$

The following inequality must be satisfied from Eqs. (19.38) and (19.39), ignoring the kinematic hardening for simplicity.

$$n^{\alpha}\dot{\tau}^{\alpha} \leq \sum_{\beta=1}^{n} (h_{\alpha\beta}\dot{\gamma}^{\beta}n^{\alpha}) \tag{19.65}$$

where

$$n^{\alpha} \equiv \frac{\partial|\tau^{\alpha}|}{\partial\tau^{\alpha}} = \frac{\tau^{\alpha}}{|\tau^{\alpha}|} = \mathrm{sign}(\tau^{\alpha}) \ (|n^{\alpha}| = 1) \tag{19.66}$$

The substitution of Eq. (19.32) into Eq. (19.65) leads to

$$\boldsymbol{\Xi}^{\alpha} : \left(\mathbf{d} - \sum_{\beta=1}^{n} \mathbf{p}^{\beta}\dot{\gamma}^{\beta}\right) n^{\alpha} \leq \sum_{\beta=1}^{n} (h_{\alpha\beta}\dot{\gamma}^{\beta}n^{\alpha}) \tag{19.67}$$

resulting in

$$\left[\boldsymbol{\Xi}^{\alpha} : \left(\mathbf{d} - \sum_{\beta=1}^{n} \mathbf{p}^{\beta}\dot{\gamma}^{\beta}\right) - \sum_{\beta=1}^{n} h_{\alpha\beta}\dot{\gamma}^{\beta}\right] n^{\alpha} \leq 0$$

which is described as

$$\left(\boldsymbol{\Xi}^{\alpha} : \mathbf{d} - \sum_{\beta=1}^{n} M_{\alpha\beta}\dot{\gamma}^{\beta}\right) n^{\alpha} \leq 0 \tag{19.68}$$

where

$$M_{\alpha\beta} \equiv h_{\alpha\beta} + \boldsymbol{\Xi}^{\alpha} : \mathbf{p}^{\beta}n^{\beta} \tag{19.69}$$

First, we assume that the slip system α is active in both modes, i.e. $\dot{\gamma}^{\alpha} \neq 0$, $\dot{\bar{\gamma}}^{\alpha} \neq 0$, so that the following equations hold from Eq. (19.68).

$$\left.\begin{aligned} \left(\boldsymbol{\Xi}^{\alpha} : \mathbf{d} - \sum_{\beta=1}^{n} M_{\alpha\beta}\dot{\gamma}^{\beta}\right) n^{\alpha} = 0 \quad \text{for} \quad \dot{\gamma}^{\alpha} \neq 0 \\ \left(\boldsymbol{\Xi}^{\alpha} : \bar{\mathbf{d}} - \sum_{\beta=1}^{n} M_{\alpha\beta}\dot{\bar{\gamma}}^{\beta}\right) n^{\alpha} = 0 \quad \text{for} \quad \dot{\bar{\gamma}}^{\alpha} \neq 0 \end{aligned}\right\} \tag{19.70}$$

which leads to

$$\boldsymbol{\Xi}^{\alpha}:\Delta\mathbf{d}-\sum_{\beta=1}^{n}M_{\alpha\beta}\Delta\overset{\circ}{\gamma}^{\beta}=0 \tag{19.71}$$

where $\Delta\mathbf{d}\equiv\mathbf{d}-\bar{\mathbf{d}}$ and $\Delta\overset{\circ}{\gamma}^{\beta}\equiv\overset{\circ}{\gamma}^{\beta}-\overset{\circ}{\bar{\gamma}}^{\beta}$. Multiplying $\Delta\overset{\circ}{\gamma}^{\alpha}$ to Eq. (19.71), we have

$$\boldsymbol{\Xi}^{\alpha}:\Delta\mathbf{d}\Delta\overset{\circ}{\gamma}^{\alpha}=\sum_{\beta=1}^{n}M_{\alpha\beta}\Delta\overset{\circ}{\gamma}^{\beta}\Delta\overset{\circ}{\gamma}^{\alpha} \tag{19.72}$$

Next, we assume that the slip system α is active in the first mode but it is inactive in the second mode, i.e. $\overset{\circ}{\gamma}^{\alpha}\neq 0$, $\overset{\circ}{\bar{\gamma}}^{\alpha}=0$, so that

$$\left.\begin{aligned}\left(\boldsymbol{\Xi}^{\alpha}:\mathbf{d}-\sum_{\beta=1}^{n}M_{\alpha\beta}\overset{\circ}{\gamma}^{\beta}\right)n^{\alpha}=0 \quad &\text{for} \quad \overset{\circ}{\gamma}^{\alpha}\neq 0\\ \left(\boldsymbol{\Xi}^{\alpha}:\bar{\mathbf{d}}-\sum_{\beta=1}^{n}M_{\alpha\beta}\overset{\circ}{\bar{\gamma}}^{\beta}\right)n^{\alpha}<0 \quad &\text{for} \quad \overset{\circ}{\bar{\gamma}}^{\alpha}=0\end{aligned}\right\} \tag{19.73}$$

which leads to

$$\left(\boldsymbol{\Xi}^{\alpha}:\Delta\mathbf{d}-\sum_{\beta=1}^{n}M_{\alpha\beta}\Delta\overset{\circ}{\gamma}^{\beta}\right)n^{\alpha}>0 \tag{19.74}$$

Noting

$$n^{\alpha}\Delta\overset{\circ}{\gamma}^{\alpha}=n^{\alpha}\overset{\circ}{\gamma}^{\alpha}>0$$

Equation (19.74) leads to

$$\left(\boldsymbol{\Xi}^{\alpha}:\Delta\mathbf{d}-\sum_{\beta=1}^{n}M_{\alpha\beta}\Delta\overset{\circ}{\gamma}^{\beta}\right)n^{\alpha}n^{\alpha}\Delta\overset{\circ}{\gamma}^{\alpha}>0$$

from which it follows that

$$\boldsymbol{\Xi}^{\alpha}:\Delta\mathbf{d}\Delta\overset{\circ}{\gamma}^{\alpha}>\sum_{\beta=1}^{n}M_{\alpha\beta}\Delta\overset{\circ}{\gamma}^{\beta}\Delta\overset{\circ}{\gamma}^{\alpha} \tag{19.75}$$

Consider the inverse case that the slip system α is inactive in the first mode and active in the second mode, i.e. $\overset{\circ}{\gamma}^{\alpha}=0$, $\overset{\circ}{\bar{\gamma}}^{\alpha}\neq 0$, so that

$$\left.\begin{array}{ll}(\boldsymbol{\Xi}^{\alpha}:\mathbf{d}-\sum_{\beta=1}^{n}M_{\alpha\beta}\overset{\bullet}{\gamma}^{\beta})n^{\alpha}<0 & \text{for} \quad \overset{\bullet}{\gamma}^{\alpha}=0\\ (\boldsymbol{\Xi}^{\alpha}:\mathbf{d}-\sum_{\beta=1}^{n}M_{\alpha\beta}\overset{\bullet}{\gamma}^{\beta})n^{\alpha}=0 & \text{for} \quad \overset{\bullet}{\bar{\gamma}}^{\alpha}\neq 0\end{array}\right\} \tag{19.76}$$

from which it follows that

$$\left(\boldsymbol{\Xi}^{\alpha}:\Delta\mathbf{d}-\sum_{\beta=1}^{n}M_{\alpha\beta}\Delta\overset{\bullet}{\gamma}^{\beta}\right)n^{\alpha}<0$$

for which, noting

$$n^{\alpha}\Delta\overset{\bullet}{\gamma}^{\alpha}=-n^{\alpha}\overset{\bullet}{\bar{\gamma}}^{\alpha}<0$$

it follows that

$$\left(\boldsymbol{\Xi}^{\alpha}:\Delta\mathbf{d}-\sum_{\beta=1}^{n}M_{\alpha\beta}\Delta\overset{\bullet}{\gamma}^{\beta}\right)n^{\alpha}n^{\alpha}\Delta\overset{\bullet}{\gamma}^{\alpha}>0$$

Consequently, the inequality in Eq. (19.75) holds also in this case.

Furthermore, in the case that the slip system α is inactive in both of the first and the second modes, i.e. $\overset{\bullet}{\gamma}^{\alpha}=0, \overset{\bullet}{\bar{\gamma}}^{\alpha}=0$ leading to $\Delta\overset{\bullet}{\gamma}^{\alpha}=0$, so that the equality in Eq. (19.72) holds. Consequently, the following inequality holds in all the above-mentioned four cases.

$$\boldsymbol{\Xi}^{\alpha}:\Delta\mathbf{d}\Delta\overset{\bullet}{\gamma}^{\alpha}\geq\sum_{\beta=1}^{n}M_{\alpha\beta}\Delta\overset{\bullet}{\gamma}^{\beta}\Delta\overset{\bullet}{\gamma}^{\alpha} \tag{19.77}$$

Taking the total sum of Eq. (19.77) for all of slip systems, the following inequality holds.

$$\sum_{\alpha=1}^{n}\boldsymbol{\Xi}^{\alpha}:\Delta\mathbf{d}\Delta\overset{\bullet}{\gamma}^{\alpha}\geq\sum_{\alpha=1}^{n}\sum_{\beta=1}^{n}M_{\alpha\beta}\Delta\overset{\bullet}{\gamma}^{\beta}\Delta\overset{\bullet}{\gamma}^{\alpha} \tag{19.78}$$

$\Delta\mathbf{d}=\mathbf{O}$ holds for given strain rate, and if $M_{\alpha\beta}$ is the positive-definite matrix, the right-hand side in Eq. (19.78) is non-negative, so that the uniqueness of slip mode, i.e. $\Delta\overset{\bullet}{\gamma}^{\alpha}=0$ holds. In other words, $M_{\alpha\beta}$ must be positive-definite in order that the uniqueness of slip mode, i.e. the uniqueness of $\overset{\bullet}{\gamma}^{\alpha}$ holds for given strain rate $\mathbf{d}$. Then, the matrix $M_{\alpha\beta}$ is called the effective slip-systems hardening moduli. The matrix $M_{\alpha\beta}$ is asymmetric, i.e. $M_{\alpha\beta}\neq M_{\beta\alpha}$ and thus it is not the positive-definite matrix, so that the uniqueness of slip mode does not holds in general.

The uniqueness of the matrix $M_{\alpha\beta}$ depends on the hardening coefficient, state of stress and the number and the directions of critical shear stress sensitively. It is not guaranteed and its tendency is remarkable for a higher latent hardening (Hill 1966; Hill and Rice 1972; Havner 1982; Asaro 1983; Franciosi and Zaoli 1991).

19.7 Various Schemes for Calculation of Shear Strain Rates

Big time is required for calculating shear strain rates in numerous slip systems. Then, various schemes for the improvement of the calculation have been proposed to date. Main schemes for the improvement will be explained in this section.

19.7.1 Singular Value Decomposition

It is required to solve Eq. (19.57) in order to calculate shear strain rates in slip systems directly from macroscopic strain rate applied to crystalline. However, the matrix $M_{\alpha\beta}$ is not positive-definite, so that there does not exist a unique solution in general as described in the last section. The *singular value decomposition* is used to calculate the solution with the shortest path (Golub and Van Loan 2013; Press et al. 1988). It has been applied to the crystal plasticity by Anand and Kothari (1996) and used widely by Miehe and Schroder (2001), Knockaert et al. (2000), Yoshida and Kuroda (2012), etc. The singular value decomposition is explained in this subsection.

The general second-order tensor $\mathbf{T}$ in the n-dimensional space which is asymmetric and thus cannot be led to the spectral decomposition in general. On the other hand, designating the eigenvectors of the positive-definite tensors $\mathbf{T}\mathbf{T}^T$ and $\mathbf{T}^T\mathbf{T}$ as $\mathbf{u}_\rho$ and $\mathbf{v}_\rho$ $(\rho = 1, 2, \cdots, n)$, respectively, and their eigenvalues σ_ρ^2 because they are positive, it follows that

$$\left.\begin{aligned} \mathbf{T}\mathbf{T}^T &= \sum_{\rho=1}^{n} \sigma_\rho^2 \mathbf{u}_\rho \otimes \mathbf{u}_\rho \\ \mathbf{T}^T\mathbf{T} &= \sum_{\rho=1}^{n} \sigma_\rho^2 \mathbf{v}_\rho \otimes \mathbf{v}_\rho \end{aligned}\right\} \tag{19.79}$$

i.e.

$$\left.\begin{aligned} \mathbf{T}\mathbf{T}^T\mathbf{u}_\rho &= \sigma_\rho^2 \mathbf{u}_\rho \\ \mathbf{T}^T\mathbf{T}\mathbf{v}_\rho &= \sigma_\rho^2 \mathbf{v}_\rho \end{aligned}\right\} \tag{19.80}$$

while $\mathbf{u}_\rho$, $\mathbf{v}_\rho$ and σ_ρ are not the eigenvectors and eigenvalues except for the symmetric tensor $\mathbf{T}$.

Exploiting the eigen values and vectors defined above, the tensor $\mathbf{T}$ can be led to the following singular value decomposition.

$$\mathbf{T} = \mathbf{U}\boldsymbol{\Sigma}\mathbf{V}^T \tag{19.81}$$

where $\boldsymbol{\Sigma}$ is the diagonalized tensor with the components $\sigma_1, \sigma_2, \cdots, \sigma_n$ and thus it is described as follows:

$$\boldsymbol{\Sigma} = \mathrm{diag}(\sigma_1,\ \sigma_2,\ \cdots,\ \sigma_n) = \begin{bmatrix} \sigma_1 & 0 & \cdots & 0 \\ 0 & \sigma_2 & \cdots & 0 \\ \vdots & \vdots & \ddots & \vdots \\ 0 & 0 & \cdots & \sigma_n \end{bmatrix} \tag{19.82}$$

provided that the order of the magnitudes of these components is $\sigma_1 \geq \sigma_2 \geq \cdots \geq \sigma_n (\geq 0)$. $\mathbf{U}$ and $\mathbf{V}$ are the orthogonal tensors by lining the eigenvectors in column (horizontally).

$$\left.\begin{array}{l} \mathbf{U} = \lfloor \mathbf{u}_1\ \mathbf{u}_2 \cdots \mathbf{u}_n \rfloor = \begin{Bmatrix} \mathbf{u}_1 \\ \mathbf{u}_2 \\ \vdots \\ \mathbf{u}_n \end{Bmatrix} = \begin{bmatrix} u_{11} & u_{12} & \cdots & u_{1n} \\ u_{21} & u_{22} & \cdots & u_{2n} \\ \vdots & \vdots & \ddots & \vdots \\ u_{n1} & u_{n2} & \cdots & u_{nn} \end{bmatrix} \\ \mathbf{V} = \lfloor \mathbf{v}_1\ \mathbf{v}_2 \cdots \mathbf{v}_n \rfloor = \begin{Bmatrix} \mathbf{v}_1 \\ \mathbf{v}_2 \\ \vdots \\ \mathbf{v}_n \end{Bmatrix} = \begin{bmatrix} v_{11} & v_{12} & \cdots & v_{1n} \\ v_{21} & v_{22} & \cdots & v_{2n} \\ \vdots & \vdots & \ddots & \vdots \\ v_{n1} & v_{n2} & \cdots & v_{nn} \end{bmatrix} \end{array}\right\} \tag{19.83}$$

where the components of each vector are lined up in row (vertically), satisfying

$$\mathbf{U}\mathbf{U}^T = \mathbf{U}^T\mathbf{U} = \mathbf{V}\mathbf{V}^T = \mathbf{V}^T\mathbf{V} = \mathbf{I}\,(\|\mathbf{U}\| = \|\mathbf{V}\| = 1) \tag{19.84}$$

noting $(\mathbf{U}\mathbf{U}^T)_{ij} = U_{ir}U_{jr} = (\mathbf{u}_i \cdot \mathbf{e}_r)(\mathbf{u}_j \cdot \mathbf{e}_r) = \mathbf{u}_i \cdot (\mathbf{u}_j \cdot \mathbf{e}_r)\mathbf{e}_r = \mathbf{u}_i \cdot \mathbf{u}_j = \delta_{ij}$. It follows from Eqs. (19.81) and (19.84) that

$$\left.\begin{array}{l} \mathbf{T}\mathbf{T}^T = \mathbf{U}\boldsymbol{\Sigma}\mathbf{V}^T\mathbf{V}\boldsymbol{\Sigma}\mathbf{U}^T = \mathbf{U}\boldsymbol{\Sigma}^2\mathbf{U}^T \\ \mathbf{T}^T\mathbf{T} = \mathbf{V}\boldsymbol{\Sigma}\mathbf{U}^T\mathbf{U}\boldsymbol{\Sigma}\mathbf{V}^T = \mathbf{V}\boldsymbol{\Sigma}^2\mathbf{V}^T \end{array}\right\} \tag{19.85}$$

The pseudo-inverse tensor $\mathbf{T}^{\dagger}$ of $\mathbf{T}$ is defined by

$$\mathbf{T}^{\dagger} \equiv \mathbf{V}\boldsymbol{\Sigma}^{\dagger}\mathbf{U}^{T} \tag{19.86}$$

where

$$\boldsymbol{\Sigma}^{\dagger} = \mathrm{diag}(1/\sigma_1,\ 1/\sigma_2,\ \cdots 1/\sigma_r,\ 0\cdots 0) = \begin{bmatrix} \sigma_1^{-1} & 0\cdots & 0 & 0\cdots & 0 \\ 0 & \sigma_2^{-1}\cdots & 0 & 0\cdots & 0 \\ \vdots & \vdots & \ddots & \vdots\ \vdots & \vdots \\ 0 & 0\cdots & \sigma_r^{-1} & 0\cdots & 0 \\ 0 & 0\cdots & 0 & 0\cdots & 0 \\ \vdots & \vdots & \vdots & \vdots\ \ddots & \vdots \\ 0 & 0\cdots & 0 & 0\cdots & 0 \end{bmatrix} \tag{19.87}$$

provided that we set $1/\sigma_i = 0\ (i = 1, 2, \ldots, r)$ for $\sigma_i = 0\ (i = r+1, 2, \ldots, n)$, obviously fulfilling $\boldsymbol{\Sigma}\boldsymbol{\Sigma}^{\dagger} = \mathbf{I}$. It is confirmed that the following equation holds from Eqs. (19.81), (19.84) and (19.86).

$$\mathbf{T}\mathbf{T}^{\dagger} = \mathbf{U}\boldsymbol{\Sigma}\mathbf{V}^{T}\mathbf{V}\boldsymbol{\Sigma}^{\dagger}\mathbf{U}^{T} = \mathbf{I} \tag{19.88}$$

Consider the following tensor equation with the vectors $\mathbf{x}$ and $\mathbf{c}$.

$$\mathbf{T}\mathbf{x} = \mathbf{c} \tag{19.89}$$

If $\mathbf{T}$ is the singular tensor, there exist numerous solutions for $\mathbf{x}$. The vector $\mathbf{x}$ is expressed noting Eqs. (19.86) and (19.88) as follows:

$$\mathbf{x} = \mathbf{T}^{\dagger}\mathbf{c} = \mathbf{V}\boldsymbol{\Sigma}^{\dagger}\mathbf{U}^{T}\mathbf{c} = \mathbf{V}\mathrm{dia}(1/\sigma_j)\mathbf{U}^{T}\mathbf{c} \tag{19.90}$$

which is called the *singular value decomposition* and calculated from the right to the left. The solution obtained by the singular value decomposition is unique and possesses the shortest path among numerous solutions satisfying the original equation as will be proved as follows.

There exists the zero-dimensional subspace of the vector $\mathbf{x}$ projected to zero vector if $\mathbf{T}$ is the singular tensor. Designating the solution of Eq. (19.90) for an arbitrary tensor $\boldsymbol{\Sigma}^{\dagger}$ possessing non-zero component only in the zero component in Eq. (18.87) by $\mathbf{x}'$, it follows for the vector $\mathbf{x}+\mathbf{x}'$ noting Eqs. (19.86) and (19.90) that

$$\begin{aligned}||\mathbf{x}+\mathbf{x}'|| &= ||\mathbf{T}^{\dagger}\mathbf{c}+\mathbf{x}'|| = ||\mathbf{V}\boldsymbol{\Sigma}^{\dagger}\mathbf{U}^{T}\mathbf{c}+\mathbf{x}'|| \\ &= ||\mathbf{V}(\boldsymbol{\Sigma}^{\dagger}\mathbf{U}^{T}\mathbf{c}+\mathbf{V}^{T}\mathbf{x}')|| = ||\mathbf{V}||\,||\boldsymbol{\Sigma}^{\dagger}\mathbf{U}^{T}\mathbf{c}+\mathbf{V}^{T}\mathbf{x}'|| \\ &= ||\boldsymbol{\Sigma}^{\dagger}\mathbf{U}^{T}\mathbf{c}+\mathbf{V}^{T}\mathbf{x}'||\end{aligned} \tag{19.91}$$

The components in the second term are non-zero in the zero components in the first terms and they are orthogonal and thus independent to each other in the n-dimensional orthogonal coordinate system, so that the minimum of the quantity in Eq. (19.91) holds for $\mathbf{x}' = \mathbf{0}$ leading to the vector $\mathbf{x}$ as the singular value decomposition.

Consider the following tensor based on the above-mentioned singular value decomposition.

$$\overline{\mathbf{T}} = (\mathbf{T} + \varepsilon\mathbf{I}) \tag{19.92}$$

where $\overline{\mathbf{T}}$ is the nonsingular tensor made by adding the infinitesimal perturbation tensor $\varepsilon\mathbf{I}$ to the singular tensor $\mathbf{T}$. Adopting it in Eq. (19.89), we have

$$\mathbf{x} = (\mathbf{T} + \varepsilon\mathbf{I})^{-1}\mathbf{c} \tag{19.93}$$

from which we can obtain the similar solution to that due to the singular value decomposition. This is the simplest solution of the singular equation and called the *diagonal shift method* (Miehe and Schroder 2001).

Number of unknown quantities on shear strain rates is larger than nine given equations on macroscopic strain rate components in the crystal plasticity, so that solution is not determined uniquely. In such case, we may solve by supplementing the lines composed of zero components to the matrix $\mathbf{T}$ and the zero components to the vector $\mathbf{c}$ for the difference of numbers in unknown quantities and given equations. By applying the singular value decomposition to the simultaneous equation made by this method, we can obtain the solution with the shortest path. It corresponds to the minimum shear principle by Taylor (1938).

19.7.2 Regularized Schmid Law

The yield surface in the slip systems is formed by plural intersecting planes in the stress space so that it possesses the sharp corner. Then, the shear strain rates must be calculated in each slip systems. In order to avoid this complicated work, the *regularized Schmid law* has been studied, in which the yield surface with rounded-off corners is formulated and the only one plastic multiplier by applying the associated flow rule to that yield surface is calculated.

The yield condition in slip system is described by

$$\frac{|\tau^{\alpha}|}{\tau_y^{\alpha}} - 1 = 0 \tag{19.94}$$

The slip systems described by Eq. (19.94) give rise to the yield surfaces composed of the planes in number of slip systems, exhibiting the sharp corners at their intersections. A single smooth yield surface can be formulated by smoothing the envelope of the yield surfaces in Eq. (19.94) (Gambin 1991, 2001; Gambin and Barlat 1997; Darrieulat and Piot 1996; Zamiri et al. 2007; Zamiri and Pourboghrat 2010). This is regarded as the invocation of the method to derive the Mises yield surface from the Tresca yield condition by Hosford (1974, 2009). Zamiri et al. (2007) proposed the following simple yield surface.

$$f = \left(\sum_{\alpha=1}^{n} \frac{\tau^{\alpha}}{\tau_y^{\alpha}} \left| \frac{\tau^{\alpha}}{\tau_y^{\alpha}} \right|^{2n-1} - 1 \right)^{1/m} = \left(\sum_{\alpha=1}^{n} \frac{\tau^{\alpha}}{\tau_y^{\alpha}} \left| \frac{\mathbf{p}^{\alpha} : \boldsymbol{\tau}}{\tau_y^{\alpha}} \right|^{2n-1} - 1 \right)^{1/m} = 0 \tag{19.95}$$

noting Eq. (19.27). $m(\geq 1)$ is the material constant for smoothing the corner of yield surface, while the larger m is, the smoother the yield surface is. While Eq. (19.95) is of the power form, in order to avoid the problem on the numerical calculation caused by the power function, Zamiri and Pourboghrat (2010) proposed the yield surface in the logarithm-exponential function.

$$f = \frac{1}{\rho} \ln \left\{ \sum_{\alpha=1}^{n} \exp \left[\rho \left(\left| \frac{\mathbf{p}^{\alpha} : \boldsymbol{\tau}}{\tau_y^{\alpha}} \right| - 1 \right) \right] \right\} = 0 \tag{19.96}$$

where $\rho\,(\geq 1)$ is the material constant for smoothing the corner of yield surface, while the larger ρ is, the smoother the yield surface is.

The plastic strain rate for the above-mentioned yield condition with the associated flow rule is given by

$$\mathbf{d}^p = \dot{\lambda} \frac{\partial f}{\partial \boldsymbol{\tau}} (\dot{\lambda} \geq 0) \tag{19.97}$$

The plastic multiplier $\dot{\lambda}$ is calculated by the formulation based on the consistency condition of the yield condition (Gambin 1997, 2001; Gambin and Barlat 1997) or the return-mapping (Zamiri et al. 2007; Zamiri and Pourboghrat 2010).

The above-mentioned regularized Schmid law reduces the calculation time since only one plastic multiplier for the single yield surface has only to be calculated. However, it would deviate from the primary purpose of the crystal plasticity for deriving the macroscopic behavior from the microscopic physical law, since it replaces the yield conditions in multi slip systems to the global yield surface.

19.7.3 Creep-Type Crystal Plasticity Mode

The crystal elastoplastic constitutive equation described in the last section possesses the difficulties:

(1) The yield judgment whether the resolved shear stresses τ^α reach the critical shear stress τ_y^α is required.
(2) Particular algorithm to pull-back the resolved shear stresses to the critical shear stress must be incorporated.

These procedures must be executed in numerous slip systems and thus the analysis by this constitutive equation is so complicated as actually impossible. Then, the crystal plasticity analyses by the creep model is widely used as will be described below.

Nakada and Keh (1966) first advocated and ten years later Hutchinson (1976) presented the following creep-type rate-dependent equation of crystalline slip rate, which are widely adopted after the review report by Peirce et al. (1982, 1983) for single crystals and Asaro and Needleman (1985) for polycrystal.

$$\dot{\gamma}^{c\alpha} = \dot{\gamma}_0{}^{c\alpha}\left(\frac{\tau^\alpha}{\tau_y^\alpha}\right)\left|\frac{\tau^\alpha}{\tau_y^\alpha}\right|^{(1/m)-1} \tag{19.98}$$

where $\dot{\gamma}_0^{c\alpha}$ is the reference rate of shearing which is taken usually same in all slip systems and m is the material constant, while for $m < 0.02$ the creep slip rate $\dot{\gamma}^{c\alpha}$ is induced abruptly when the magnitude of shear stress, $|\tau^\alpha|$, reaches the shear-yield stress τ_y^α. All the slip systems are active and thus the selection of active ones is not required for Eq. (19.98). However, Eq. (19.98) possesses the following fundamental impertinences.

(1) It depends only on the shear stress because it falls within the framework of the creep-type viscoplasticity. Therefore, the creep crystalline slip rate is determined only by the current shear stress with time and thus an arbitrary deformation rate cannot be given to the material obeying Eq. (19.98) which is independent of the stress rate as far as a large elastic deformation of crystal lattice is not incorporated. In other words, it cannot be applied to rigid-viscoplastic materials. In facts, however, the crystalline slip rate would have to depend not only on the shear stress but also on its rate in a quasi-static deformation at room temperature. On the other hand, the rate-independent equation of plastic slip rate $\dot{\gamma}^{p\alpha}$ must depend on both the shear stress and its rate (magnitude) in the slip system because it falls within the framework of the plasticity.
(2) It belongs to the creep type model without a definite yield stress among the viscoplastic models. Therefore, it predicts always the creep deformation except for the stress-free state, and thus the creep slip rate is induced even when the magnitude of shear stress, $|\tau^\alpha|$, decreases as illustrated in Fig. 19.3 in which a

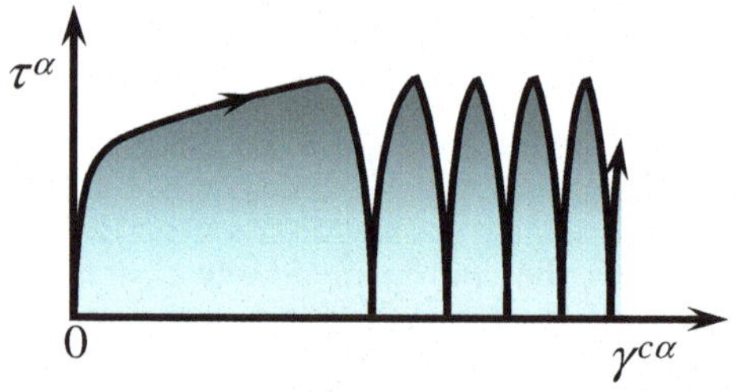

Fig. 19.3 Shear stress versus crystalline slip curves accompanied with excessive mechanical ratcheting predicted by creep model (Peirce et al. 1983)

general trend is intelligibly shown for a moderate value of m. It is unrealistic as known from the fact that any metallic solid has never collapsed under their own weight at room temperature as indicated in p. 201 in Havner (1992). Needless to say, it cannot be applied to the description of cyclic loading behavior since it does not possess a loading criterion and thus it predicts identical deformation behavior in the reloading process and in the unloading process. Therefore, it predicts excessively large mechanical ratcheting as illustrated in Fig. 19.3. In general, the viscoplastic deformation behavior cannot be described pertinently by the creep-type model and instead it can be predicted realistically by the overstress-type model possessing the loading criterion as explained in Sect. 13.3.

As examined above, the creep-type equation of crystalline slip contains fundamental defects. Then, it has been the strong desire over the last half century to find the physically and numerically pertinent rate-independent equation of crystalline slip rate. This fact is declared emotionally as "*The various viscoplastic, finite-strain aggregate calculations reviewed in this section, and similar ones in the literature, are computationally impressive (although which approximate polycrystal model is superior appears to be an open question). However, one hopes that there may soon evolve a theory of rate-dependent crystalline slip in metals that would leave such great structural landmarks as the Eiffel Tower, Empire State Building, and Golden Gate Bridge still standing*" in p. 204 of Havner (1992). The landmark would have been found in the subloading crystal plasticity model described in the next section, while it is not a rate-dependent formulation but is a rate-independent formulation falling within the framework of plasticity based on the subloading surface concept.

All of the afore-mentioned defects in the rate-dependent crystalline slip rate are dissolved by subloading crystal plasticity model. The crucial importance for the incorporation of the subloading concept is manifested most distinctly in the crystal plasticity analysis which is required to calculate slips in numerous number of slip systems, while the yield judgment is not required and the stress is automatically attracted to the yield surface only in the subloading crystal plasticity model.

19.8 Subloading Crystal Plasticity Model

The crystal shear yield condition describing the *crystal shear yield region* in the slip system α is given by

$$|\hat{\tau}^{\alpha}| = \tau_y^{\alpha} \tag{19.99}$$

where

$$\hat{\tau}^{\alpha} \equiv \tau^{\alpha} - \chi^{\alpha} \tag{19.100}$$

χ^{α} is the shear kinematic hardening variable and $\tau_y^{\alpha}\,(>0)$ is the shear hardening function, referred to as the *critical shear stress*, in the slip system α.

Now, incorporate the shear subloading region described by the following relation based on the subloading concept (Hashiguchi 2015).

$$|\hat{\tau}^{\alpha}| = r^{\alpha}\tau_y^{\alpha}, \text{ i.e.} r^{\alpha} \equiv \frac{|\hat{\tau}^{\alpha}|}{\tau_y^{\alpha}} \tag{19.101}$$

where $r^{\alpha}(0 \le r^{\alpha} \le 1)$ is referred to as the *normal-yield shear ratio* which designates always the ratio of $|\tau^{\alpha} - \chi^{\alpha}|$ to the critical shear stress τ_y^{α} not only in the slipping process $(\dot{\gamma}^{p\alpha} \neq 0)$ but also in the non-slipping process $(\dot{\gamma}^{p\alpha} = 0)$.

The associated flow rule to the subloading shear region is adopted for the plastic shear strain rate as follows:

$$\dot{\gamma}^{p\alpha} = \dot{\lambda}^{\alpha}\hat{n}^{\alpha}\ (\dot{\lambda}^{\alpha} \ge 0)\ (|\dot{\gamma}^{p\alpha}| = \dot{\lambda}^{\alpha}) \tag{19.102}$$

The material-time derivative of Eq. (19.101) reads:

$$\hat{n}^{\alpha}(\dot{\tau}^{\alpha} - \dot{\chi}^{\alpha}) = r^{\alpha}\dot{\tau}_y{}^{\alpha} + \dot{r}^{\alpha}\tau_y^{\alpha} \tag{19.103}$$

i.e.

$$\dot{\tau}^{\alpha} = \hat{n}^{\alpha}r^{\alpha}\dot{\tau}_y{}^{\alpha} + \dot{\chi}^{\alpha} + \hat{n}^{\alpha}\dot{r}^{\alpha}\tau_y^{\alpha} \tag{19.104}$$

The rate of critical shear stress is specified by

$$\dot{\tau}_y{}^{\alpha} = \sum_{\beta=1}^{n} h_{\alpha\beta}|\dot{\gamma}^{p\beta}| = \sum_{\beta=1}^{n} h_{\alpha\beta}\dot{\lambda}^{\beta} \tag{19.105}$$

where $h_{\alpha\beta}$ is given by the following matrix which is the function of the plastic shear strain (Peirce et al. 1982).

$$h_{\alpha\beta} = qh(\gamma^p) + (1-q)h(\gamma^p)\delta_{\alpha\beta}(= h_{\beta\alpha}) = \begin{cases} h(\gamma^p) & \text{for } i = j \\ qh(\gamma^p) & \text{for } i \neq j \end{cases} (1 \leq q \leq 1.4) \tag{19.106}$$

$h(\gamma^p)$ is given by the functional form

$$h(\gamma^p) = h_0 \mathrm{sech}^2\left(\frac{h_0 \gamma^p}{\tau_s - \tau_{y0}}\right) \tag{19.107}$$

with

$$\gamma^p \equiv \sum_{\beta=1}^{n} \int_0^t |\dot{\gamma}^{p\beta}| dt \tag{19.108}$$

h_0 and τ_{y0} are the initial values of h and τ_y^α, i.e. $h_0 = h(0)$ and $\tau_{y0} = \tau_y^\alpha(0)$, respectively, and τ_s is the saturation value of τ_y^α, i.e. $\tau_s = \tau_y^\alpha(\infty)$.

Assume the following shear nonlinear-kinematic hardening rule.

$$\dot{\chi}^\alpha = c_\chi\left(\dot{\gamma}^{p\alpha} - \frac{1}{\zeta_\chi \tau_y^\alpha}|\dot{\gamma}^{p\alpha}|\chi^\alpha\right) = c_\chi \dot{\lambda}^\alpha\left(\hat{n}^\alpha - \frac{1}{\zeta_\chi \tau_y^\alpha}\chi^\alpha\right) \tag{19.109}$$

where c_χ and ζ_χ are the material constants. The latent hardening may be incorporated for the shear kinematic hardening (e.g. Bassani and Wu 1991; Harder 1999; Xu and Jiang 2004).

Referring to the Sect. 7.2 on the subloading surface model, let us postulate for the crystalline shear strain rate as follows: *The crystalline plastic shear strain rate is induced when the resolved stress approaches the critical shear stress but only the crystalline elastic shear strain rate is induced when the resolved stress lowers from the critical shear stress*, while the resolved stress rate causes the crystalline elastic shear strain rate inevitably. In other words, **the resolved shear stress approaches the critical shear stress when a crystalline plastic shear strain rate is induced but it lowers from the critical shear stress when only a crystalline elastic shear strain rate occurs**. Here, note that the approaching degree to the resolved shear stress to the critical shear stress is described by the shear normal-yield ratio R^α. Then, let the evolution equation of shear normal-yield ratio r^α be given analogously to Eq. (7.9) for the normal-yield ratio R of the subloading surface model as follows:

$$\dot{r}^\alpha = U(r^\alpha)|\dot{\gamma}^{p\alpha}| = U(r^\alpha)\dot{\lambda}^\alpha \quad \text{for } \dot{\gamma}^{p\alpha} \neq 0 \tag{19.110}$$

where the function $U(r^\alpha)$ fulfills the conditions (see Fig. 19.4):

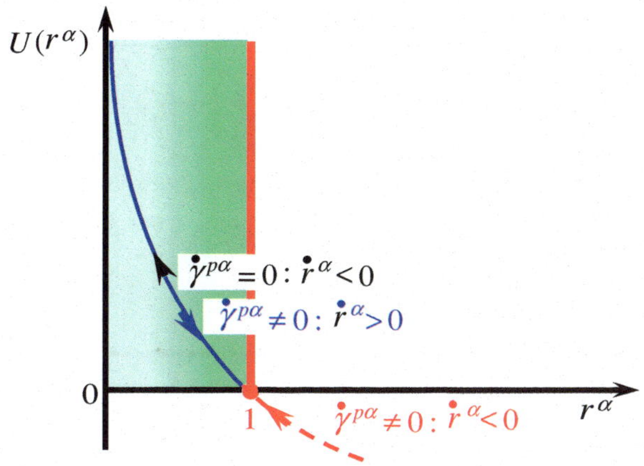

Fig. 19.4 Function $U(r^{\alpha})$ in the evolution rule of shear normal-yield ratio R^{α}

$$U(r^{\alpha})\begin{cases} \to +\infty & \text{for} \quad r^{\alpha}=0 \text{ (elastic state)} \\ >0 & \text{for} \quad 0<r^{\alpha}<1 \text{ (subyield state)} \\ =0 & \text{for} \quad r^{\alpha}=1 \text{ (normal-yield state)} \\ <0 & \text{for} \quad r^{\alpha}>1 \text{ (over normal-yield state)} \end{cases} \tag{19.111}$$

Let the explicit function of $U(r^{\alpha})$ be given by

$$U(r^{\alpha}) = u_c \cot[(\pi/2)r^{\alpha^{n_c}}] \tag{19.112}$$

where u_c and $n_c(\geq 1)$ are the material constants.

The smooth shear stress versus crystalline shear strain curve is depicted and the resolved shear stress is automatically attracted to the critical shear stress because of $\overset{\bullet}{r}^{\alpha}<0$ for $r^{\alpha}>1$ (over shear normal-yield state) in Eq. (19.110) with Eq. $(19.111)_4$ as shown in Fig. 19.5.

Substituting Eqs. (19.105), (19.109) and (19.110) into Eq. (19.103), one has the consistency condition for the subloading crystalline shear region.

$$\overset{\bullet}{\tau}^{\alpha} = \hat{n}^{\alpha} r^{\alpha} \sum_{\beta=1}^{n} h_{\alpha\beta} \overset{\bullet}{\lambda}^{\beta} + c_{\chi}\left(\hat{n}^{\alpha} - \frac{1}{\zeta_{\chi}\tau_y^{\alpha}}\chi^{\alpha}\right)\overset{\bullet}{\lambda}^{\alpha} + \hat{n}^{\alpha} U(r^{\alpha})\tau_y^{\alpha}\overset{\bullet}{\lambda}^{\alpha} \tag{19.113}$$

from which it follows that

$$\overset{\bullet}{\tau}^{\alpha} = \sum_{\beta=1}^{n} \bar{h}_{\alpha\beta}\overset{\bullet}{\lambda}^{\beta} \tag{19.114}$$

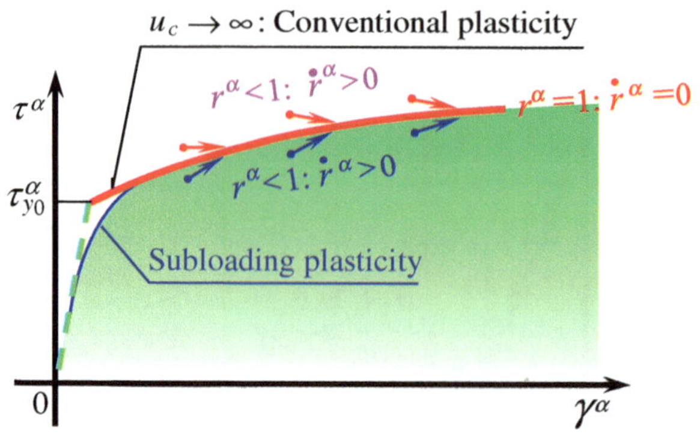

$$\dot{r}^\alpha = U(r^\alpha) |\dot{\gamma}^{p\alpha}| \text{ for } \dot{\gamma}^{p\alpha} \neq 0$$

$$U(r^\alpha) \begin{cases} > 0 \text{ for } r^\alpha < 1 \\ = 0 \text{ for } r^\alpha = 1 \\ < 0 \text{ for } r^\alpha > 1 \end{cases}$$

Fig. 19.5 Resolved shear stress is automatically attracted to critical shear stress in plastic shear process

where

$$\bar{h}_{\alpha\beta} \equiv \hat{n}^\alpha r^\alpha h_{\alpha\beta} + c_\chi \left(\hat{n}^\alpha - \frac{1}{\zeta_\chi \tau_y^\alpha} \chi^\alpha \right) \delta_{\alpha\beta} + \hat{n}^\alpha U(r^\alpha) \tau_y^\alpha \delta_{\alpha\beta} \; (\neq \bar{h}_{\beta\alpha}) \tag{19.115}$$

Equating Eq. (19.32) with (19.114), it follows that

$$\boldsymbol{\Xi}^\alpha : \left(\mathbf{d} - \sum_{\beta=1}^{n} \mathbf{p}^\beta \hat{n}^\beta \dot{\lambda}^\beta \right) = \sum_{\beta=1}^{n} \bar{h}_{\alpha\beta} \dot{\lambda}^\beta \tag{19.116}$$

which is rewritten as

$$\boldsymbol{\Xi}^\alpha : \mathbf{d} = \sum_{\beta=1}^{n} \overline{M}_{\alpha\beta} \dot{\lambda}^\beta \tag{19.117}$$

where

$$\overline{M}_{\alpha\beta} \equiv \bar{h}_{\alpha\beta} + \boldsymbol{\Xi}^\alpha : \mathbf{p}^\beta \hat{n}^\beta \; (\neq \overline{M}_{\beta\alpha}) \tag{19.118}$$

The plastic shear strain rate is given from Eq. (19.117) as follows:

$$\dot{\lambda}^{\alpha} = \sum_{\beta=1}^{n} \overline{M}_{\alpha\beta}^{-1} \boldsymbol{\Xi}^{\beta} : \mathbf{d} \tag{19.119}$$

Substituting Eq. (19.119) into Eq. (19.54), it follows that

$$\overset{\circ}{\boldsymbol{\tau}}{}^{w} = \mathbf{E} : \mathbf{d} - \sum_{\alpha=1}^{n} \boldsymbol{\Xi}^{\alpha} \dot{\lambda}^{\alpha} \hat{n}^{\alpha} = \mathbf{E} : \mathbf{d} - \sum_{\alpha=1}^{n} \boldsymbol{\Xi}^{\alpha} \hat{n}^{\alpha} \sum_{\beta=1}^{n} \overline{M}_{\alpha\beta}^{-1} \boldsymbol{\Xi}^{\beta} : \mathbf{d}$$

leading to

$$\overset{\circ}{\boldsymbol{\tau}}{}^{w} = \overline{\mathbf{K}}^{ep} : \mathbf{d} \tag{19.120}$$

where

$$\overline{\mathbf{K}}^{ep} \equiv \mathbf{E} - \sum_{\alpha=1}^{n} \sum_{\beta=1}^{n} \boldsymbol{\Xi}^{\alpha} \otimes \hat{n}^{\alpha} \overline{M}_{\alpha\beta}^{-1} \boldsymbol{\Xi}^{\beta} \ (\neq \overline{\mathbf{K}}^{epT}) \tag{19.121}$$

The loading criterion for the plastic shear strain rate is given by the sign of the plastic multiplier in terms of the shear strain rate as follows:

$$\left. \begin{array}{ll} \dot{\gamma}^{p\alpha} \neq 0 & \text{for} \quad \dot{\lambda}^{\alpha} > 0 \\ \dot{\gamma}^{p\alpha} = 0 & \text{for} \quad \dot{\lambda}^{\alpha} \leq 0 \end{array} \right\} \tag{19.122}$$

The deformation analysis by the forward-Euler calculation method is performed as follows:

(1) First calculate the plastic multipliers $\dot{\lambda}^{\alpha}$ by solving Eq. (19.117) for the input of the strain rate **d**.
(2) If $\dot{\lambda}^{\alpha}$ is positive, calculate the plastic shear strain rate $\dot{\gamma}^{p\alpha} = n^{\alpha}\dot{\lambda}^{\alpha}$, the critical shear stress rate $\dot{\tau}_{y}^{\alpha}$ by Eq. (19.105) and the kinematic hardening rate $\dot{\chi}^{\alpha}$ by Eq. (19.109), the rate of normal-yield shear ratio $\dot{r}^{\alpha}$ by Eq. (19.110), the resolved shear stress rate $\dot{\tau}^{\alpha}$ by Eq. (19.32), the Jaumann rate of the Kirchhoff stress $\overset{\circ}{\boldsymbol{\tau}}{}^{w}$ by Eq. (19.54) and the rates of the base vectors $\dot{\mathbf{s}}^{\alpha}$, $\dot{\mathbf{n}}^{\alpha}$ of slip system by substituting Eq. (19.25) into Eq. (19.12). Then, update all these variables.
(3) If $\dot{\lambda}^{\alpha}$ is negative, set $\dot{\tau}_{y}^{\alpha} = 0$, $\dot{\chi}^{\alpha} = 0$, and calculate $\dot{\tau}^{\alpha}$ by Eq. (19.32), $\overset{\circ}{\boldsymbol{\tau}}{}^{w}$ by Eq. (19.54) and $\dot{\mathbf{s}}^{\alpha}$, $\dot{\mathbf{n}}^{\alpha}$ by substituting Eq. (19.25) into Eq. (19.12) under setting $\dot{\gamma}^{p\alpha} = 0$. Then, update all these variables. Thereafter, update r^{α} by Eq. (19.101).
(4) Move to the calculation for the next incremental step in which the updated values obtained in the above-mentioned processes are substituted.

19.9 Subloading-Overstress-Crystal Plasticity Model

Based on the extension of the overstress model by the subloading surface model in Eq. (13.29) or (13.30), let the viscoplastic slip rate $\overset{\bullet}{\gamma}^{vp\alpha}$ be given by

$$\overset{\bullet}{\gamma}^{vp\alpha}=\frac{1}{\mu_{cs}}\frac{\langle r^{\alpha}-r_s^{\alpha}\rangle^n}{r_{cm}-r^{\alpha}}n^{\alpha} \tag{19.123}$$

or

$$\overset{\bullet}{\gamma}^{vp\alpha}=\frac{1}{\mu_{cs}}\frac{\langle \exp[n(r^{\alpha}-r_s^{\alpha})]-1\rangle}{r_{cm}-r^{\alpha}}n^{\alpha} \tag{19.124}$$

where μ_{cs} is the material constant standing for the *crystalline viscous coefficient*, $r_{cm}(\gg 1)$ is the material constant, called the *limit dynamic-loading ratio*, specifying the maximum value of r^{α} renamed as the *dynamic-loading ratio*, and $n(\gg 1)$ is the material constant.

The evolution rule of the *subloading ratio* $r_s^{\alpha}(0\leq r_s^{\alpha}\leq 1)$ is given by

$$\left.\begin{array}{lll}\overset{\bullet}{r}_s{}^{\alpha}=U(r_s^{\alpha})|\overset{\bullet}{\gamma}^{vp\alpha}| & \text{for} & r^{\alpha}>r_s^{\alpha}\;(\overset{\bullet}{\gamma}^{vp\alpha}\neq 0)\\ r_s^{\alpha}=r^{\alpha}=|\tau^{\alpha}-\chi^{\alpha}|/\tau_y^{\alpha} & \text{for} & \text{others}\;(\overset{\bullet}{\gamma}^{vp\alpha}=0)\end{array}\right\} \tag{19.125}$$

following Eq. (19.110), where the function $U(r_s^{\alpha})$ is given as

$$U(r_s^{\alpha})=u_c\cot\left[(\pi/2)r_s^{\alpha n_c}\right] \tag{19.126}$$

The smooth resolved shear stress versus crystalline plastic shear strain curve is described always as shown in Fig. 19.6.

The evolution rule of the internal variables are given by replacing the plastic shear strain rate in the elastoplastic sliding equation to the viscoplastic shear strain rate. Then, the rate of critical shear stress is given from Eq. (19.105) as

$$\overset{\bullet}{\tau}_y{}^{\alpha}=\sum_{\beta=1}^{n}h_{\alpha\beta}|\overset{\bullet}{\gamma}^{vp\beta}| \tag{19.127}$$

$$h_{\alpha\beta}=qh(\gamma^{vp})+(1-q)h(\gamma^{vp})\delta_{\alpha\beta}=(h_{\beta\alpha})\;(1\leq q\leq 1.4) \tag{19.128}$$

$$h(\gamma^{vp})=h_0\mathrm{sech}^2\left(\frac{h_0\gamma^{vp}}{\tau_s-\tau_{y0}}\right),\;\gamma^{vp}\equiv\sum_{\beta=1}^{n}\int_0^t|\overset{\bullet}{\gamma}^{vp\beta}|dt \tag{19.129}$$

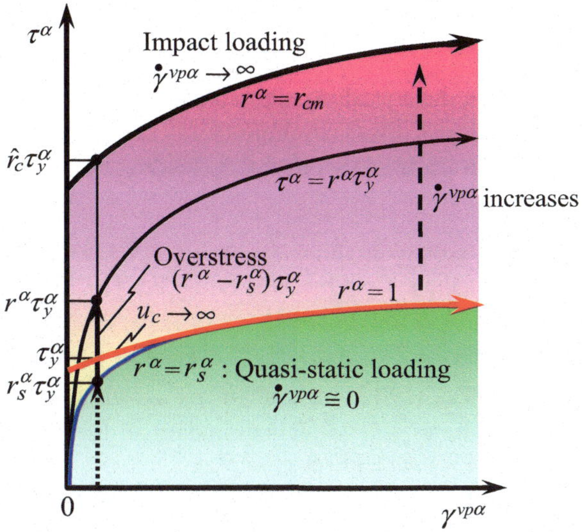

Fig. 19.6 Resolved shear stress versus crystalline plastic shear strain curve predicted by subloading-overstress-crystal plasticity model

and the rate of shear kinematic hardening variable is given from Eq. (19.109) as

$$\overset{\bullet}{\chi}^\alpha = c_\chi\left(\overset{\bullet}{\gamma}^{vp\alpha} - \frac{1}{\zeta_\chi \tau_y^\alpha}|\overset{\bullet}{\gamma}^{vp\alpha}|\chi^\alpha\right) \tag{19.130}$$

$\overset{\circ}{\boldsymbol{\tau}}^w$ and $\overset{\bullet}{\tau}^\alpha$ are given following Eqs. (19.54) and (19.32) by

$$\overset{\circ}{\boldsymbol{\tau}}^w = \mathbf{E}:\mathbf{d} - \sum_{\beta=1}^{n} \boldsymbol{\Xi}^\beta \overset{\bullet}{\gamma}^{vp\beta} \tag{19.131}$$

$$\overset{\bullet}{\tau}^\alpha = \boldsymbol{\Xi}^\alpha : \left(\mathbf{d} - \sum_{\beta=1}^{n} \mathbf{p}^\beta \overset{\bullet}{\gamma}^{vp\beta}\right) \tag{19.132}$$

The elastic velocity gradient is given following Eq. (19.25) by

$$\boldsymbol{l}^* = \boldsymbol{l} - \boldsymbol{l}^{vp} = \boldsymbol{l} - \sum_{\beta=1}^{n} (\mathbf{p}^\beta + \mathbf{q}^\beta)\overset{\bullet}{\gamma}^{vp\beta} \tag{19.133}$$

The deformation analysis by the forward-Euler calculation method is performed as follows:

(1) For $r^{\alpha} > r_s^{\alpha}$, calculate the crystalline slip rate $\dot{\gamma}^{vp\alpha}$ by Eq. (19.123) or (19.124), the rate of critical shear stress $\dot{\tau}_y^{\alpha}$ by Eq. (19.127), the rate of shear kinematic hardening variable $\dot{\chi}^{\alpha}$ by Eq. (19.130), the rate of resolved shear stress $\dot{\boldsymbol{\tau}}^{\alpha}$ by Eq. (19.132), the rate of subloading shear ratio $\dot{r}_s^{\alpha}$ by Eq. $(19.125)_1$, the Jaumann rate of the Kirchhoff stress $\overset{\circ}{\boldsymbol{\tau}}^{w}$ by Eq. (19.131) and the rates of the base vectors $\dot{\mathbf{s}}^{\alpha}$, $\dot{\mathbf{n}}^{\alpha}$ of slip system by substituting Eq. (19.133) into Eq. (19.12). Thereafter, calculate r^{α} by Eq. (19.101).
(2) For the other leading to $\dot{\gamma}^{vp\alpha} = 0$, set $\dot{\tau}_y^{\alpha} = 0$, $\dot{\chi}^{\alpha} = 0$, and calculate $\dot{\tau}^{\alpha}$ by Eq. (19.132), $\overset{\circ}{\boldsymbol{\tau}}^{w}$ by Eq. (19.131) and $\dot{\mathbf{s}}^{\alpha}$, $\dot{\mathbf{n}}^{\alpha}$ by substituting Eq. (19.133) into Eq. (19.12) under setting $\dot{\gamma}^{vp\alpha} = 0$. Then, update all these variables. Further, calculate r^{α} by Eq. (19.101) and set $r_s^{\alpha} = r^{\alpha}$.
(3) Move to the calculation for the next incremental step in which the updated values obtained in the above-mentioned processes are substituted.

19.10 Extension to Description of Cyclic Loading Behavior

The crystal plasticity model formulated in the preceding sections is based on the initial subloading surface model in which the similarity-center of the shear normal-yield and the shear subloading regions, i.e. the shear elastic-core is fixed at the shear kinematic hardening variable point. Therefore, unrealistically large shear strain accumulation is predicted, while open hysteresis loops are depicted as illustratively shown in Fig. 19.7. As described in Sect. 9.1. It will be extended to describe cyclic loading behavior by letting the similarity-center of the shear normal-yield and the shear subloading regions move with a plastic shear strain as shown in Fig. 19.7. The extended shear crystalline model formulated by Hashiguchi (2015) will be described in this section.

The subloading shear region is given instead of Eq. (19.101) as follows:

$$|\bar{\tau}^{\alpha}| = r^{\alpha}\tau_y^{\alpha} \tag{19.134}$$

where

$$\bar{\tau}^{\alpha} \equiv \tau^{\alpha} - \bar{\chi}^{\alpha} \tag{19.135}$$

$\bar{\chi}^{\alpha}$ stands for the conjugate (similar) point in the shear subloading shear region to the point χ^{α} in the normal-yield shear region. By letting c^{α} denote the similarity-center of the shear normal-yield and the shear subloading regions, which

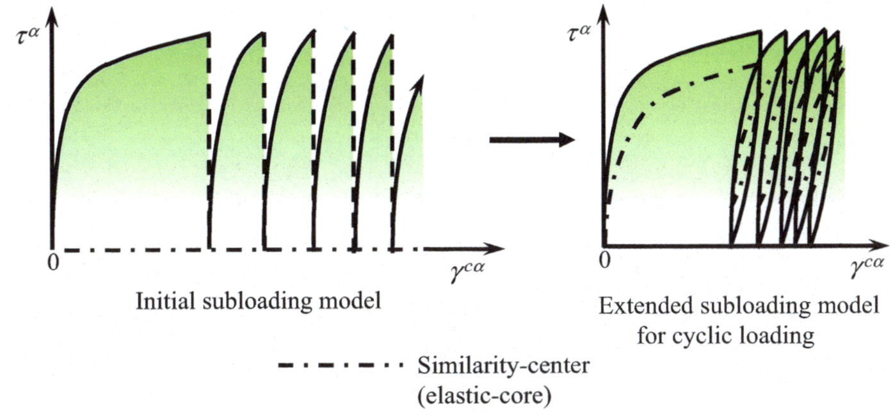

Fig. 19.7 Modification of subloading model to describe cyclic loading behavior

is called *shear elastic-core* since the most elastic shear behavior is induced when the resolved shear stress lies on it fulfilling $r^\alpha = 0$, the following relations hold by virtue of the similarity of the shear subloading region to the shear normal-yield region (see Fig. 19.8).

$$\left.\begin{array}{l} \bar{c}^\alpha \equiv c^\alpha - \bar{\chi}^\alpha, \hat{c}^\alpha \equiv c^\alpha - \chi^\alpha \\ \bar{c}^\alpha = r^\alpha \hat{c}^\alpha \end{array}\right\} \tag{19.136}$$

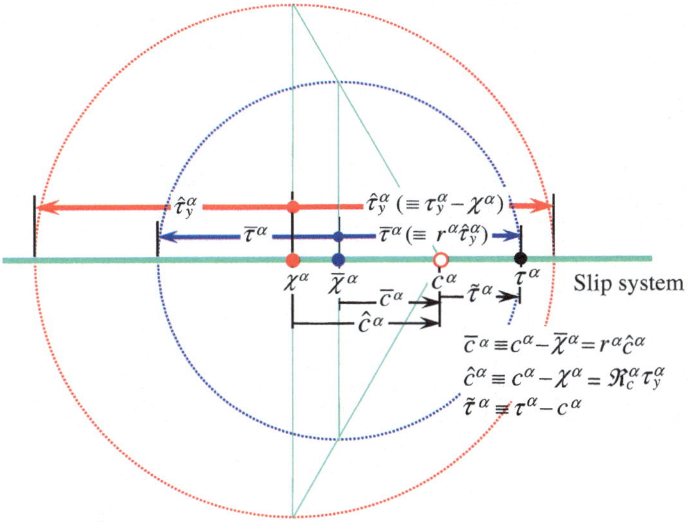

Fig. 19.8 Resolved and critical shear stresses, shear back-stress and shear elastic-core in slip system

It follows from Eqs. (19.135) and (19.136) that

$$\bar{\chi}^{\alpha} = c^{\alpha} - r^{\alpha}\hat{c}^{\alpha} \tag{19.137}$$

$$\bar{\tau}^{\alpha} = \tilde{\tau}^{\alpha} + r^{\alpha}\hat{c}^{\alpha} \tag{19.138}$$

where

$$\tilde{\tau}^{\alpha} \equiv \tau^{\alpha} - c^{\alpha} \tag{19.139}$$

The associated flow rule for the extended shear subloading region is adopted for the plastic shear strain rate as follows:

$$\dot{\gamma}^{p\alpha} = \dot{\lambda}^{\alpha}\bar{n}^{\alpha} \ (\dot{\lambda}^{\alpha} \geq 0) \ (|\dot{\gamma}^{p\alpha}| = \dot{\lambda}^{\alpha}) \tag{19.140}$$

where

$$\bar{n}^{\alpha} \equiv \frac{\partial|\bar{\tau}^{\alpha}|}{\partial\tau^{\alpha}} = \frac{\bar{\tau}^{\alpha}}{|\bar{\tau}^{\alpha}|} = \mathrm{sign}(\bar{\tau}^{\alpha}) \ (|\bar{n}^{\alpha}| = 1) \tag{19.141}$$

The shear kinematic hardening rule is given from Eq. (19.109) with Eq. (19.140) as

$$\dot{\chi}^{\alpha} = c_{\chi}\left(\dot{\gamma}^{p\alpha} - \frac{1}{\zeta_{\chi}\tau_y^{\alpha}}|\dot{\gamma}^{p\alpha}|\chi^{\alpha}\right) = c_{\chi}\dot{\lambda}^{\alpha}\left(\bar{n}^{\alpha} - \frac{1}{\zeta_{\chi}\tau_y^{\alpha}}\chi^{\alpha}\right) \tag{19.142}$$

Now, let the following shear *elastic-core region* be introduced, which always passes through the *shear elastic-core* c^{α} and maintains the similarity to the shear normal-yield surface with respect to the shear kinematic-hardening variable χ^{α}.

$$\hat{c}^{\alpha} = \Re_c^{\alpha}\tau_y^{\alpha}, \ \text{i.e.} \Re_c^{\alpha} = \hat{c}^{\alpha}/\tau_y^{\alpha} \tag{19.143}$$

where $\Re_c^{\alpha}$ designates the ratio of the size of the shear elastic-core region to the normal-yield region (see Fig. 19.8) so that let it be called the *shear elastic-core yield ratio*. Then, let it be postulated that the shear elastic-core can never reach the shear normal-yield region designating the fully-plastic shear state so that the shear elastic-core does not go over the following *limit shear elastic-core region.*

$$\hat{c}^{\alpha} = \xi_c\tau_y^{\alpha} \tag{19.144}$$

where ξ_c (<1) is material parameter and the following inequality must be satisfied.

$$\hat{c}^{\alpha} \leq \xi_c\tau_y^{\alpha}, \ \text{i.e.} \Re_c^{\alpha} \leq \xi_c \tag{19.145}$$

The evolution rule of the shear elastic-core is given analogously to Eq. (19.15) as follows:

$$\overset{\circ}{c}^{\alpha} = c_c\left(\dot{\gamma}^{p\alpha} - \frac{\Re_c^{\alpha}}{\xi_c}|\dot{\gamma}^{p\alpha}|\hat{n}_c^{\alpha}\right) = c_c\dot{\lambda}^{\alpha}\left(\bar{n}^{\alpha} - \frac{\Re_c^{\alpha}}{\xi_c}\hat{n}_c^{\alpha}\right) \tag{19.146}$$

where c_c is the material constant and

$$\hat{n}_c \equiv \frac{\hat{c}^{\alpha}}{|\hat{c}^{\alpha}|}\ (|\hat{n}_c| = 1) \tag{19.147}$$

The material-time derivative of Eq. (19.134) reads:

$$\bar{n}^{\alpha}(\overset{\circ}{\tau}^{\alpha} - \overset{\circ}{\bar{\chi}}^{\alpha}) = r^{\alpha}\overset{\circ}{\tau}_y{}^{\alpha} + \overset{\circ}{r}^{\alpha}\tau_y^{\alpha} \tag{19.148}$$

where $\overset{\circ}{\bar{\chi}}^{\alpha}$ is described from Eq. (19.137) as

$$\overset{\circ}{\bar{\chi}}^{\alpha} = r^{\alpha}\overset{\circ}{\chi}^{\alpha} + (1 - r^{\alpha})\overset{\circ}{c}^{\alpha} - \overset{\circ}{r}^{\alpha}\hat{c}^{\alpha} \tag{19.149}$$

Substituting Eq. (19.149) into Eq. (19.148), one has

$$\bar{n}^{\alpha}\left\{\overset{\circ}{\tau}^{\alpha} - \left[r^{\alpha}\overset{\circ}{\chi}^{\alpha} + (1 - r^{\alpha})\overset{\circ}{c}^{\alpha} - \overset{\circ}{r}^{\alpha}\hat{c}^{\alpha}\right]\right\} = r^{\alpha}\overset{\circ}{\tau}_y^{\alpha} + \overset{\circ}{r}^{\alpha}\tau_y^{\alpha} \tag{19.150}$$

which is rewritten as

$$\bar{n}^{\alpha}\overset{\circ}{\tau}^{\alpha} = \bar{n}^{\alpha}r^{\alpha}\overset{\circ}{\chi}^{\alpha} + \bar{n}^{\alpha}(1 - r^{\alpha})\overset{\circ}{c}^{\alpha} - \bar{n}^{\alpha}\overset{\circ}{r}^{\alpha}\hat{c}^{\alpha} + r^{\alpha}\overset{\circ}{\tau}_y^{\alpha} + \overset{\circ}{r}^{\alpha}\tau_y^{\alpha} \tag{19.151}$$

i.e.

$$\overset{\circ}{\tau}^{\alpha} = \bar{n}^{\alpha}r^{\alpha}\overset{\circ}{\tau}_y^{\alpha} + r^{\alpha}\overset{\circ}{\chi}^{\alpha} + (1 - r^{\alpha})\overset{\circ}{c}^{\alpha} + (\bar{n}^{\alpha}\tau_y^{\alpha} - \hat{c}^{\alpha})\overset{\circ}{r}^{\alpha} \tag{19.152}$$

Here, noting the relation

$$\begin{aligned}\bar{n}^{\alpha}\tau_y^{\alpha} - \hat{c}^{\alpha} &= \frac{\bar{\tau}^{\alpha}}{|\bar{\tau}^{\alpha}|}\frac{|\bar{\tau}^{\alpha}|}{r^{\alpha}} - \frac{\bar{c}^{\alpha}}{r^{\alpha}} = \frac{\bar{\tau}^{\alpha} - \bar{c}^{\alpha}}{r^{\alpha}} \\ &= \frac{(\tau^{\alpha} - \bar{\chi}^{\alpha}) - (c^{\alpha} - \bar{\chi}^{\alpha})}{r^{\alpha}} = \frac{\tilde{\tau}^{\alpha}}{r^{\alpha}}\end{aligned} \tag{19.153}$$

Equation (19.152) is rewritten as

$$\overset{\circ}{\tau}^{\alpha} = \bar{n}^{\alpha}r^{\alpha}\overset{\circ}{\tau}_y^{\alpha} + r^{\alpha}\overset{\circ}{\chi}^{\alpha} + (1 - r^{\alpha})\overset{\circ}{c}^{\alpha} + \frac{\overset{\circ}{r}^{\alpha}}{r^{\alpha}}\tilde{\tau}^{\alpha} \tag{19.154}$$

Substituting Eqs. (19.105), (19.110), (19.142) and (19.146) into Eq. (19.154), it follows that

$$\begin{aligned}\overset{\bullet}{\tau}^{\alpha} &= \bar{n}^{\alpha} r^{\alpha} \sum_{\beta=1}^{n} h_{\alpha\beta} \overset{\bullet}{\lambda}^{\beta} + r^{\alpha} c_{\chi} \overset{\bullet}{\lambda}^{\alpha} \left(\bar{n}^{\alpha} - \frac{1}{\zeta_c \tau_y^{\alpha}} \chi^{\alpha} \right) \\ &+ (1 - r^{\alpha}) c_c \overset{\bullet}{\lambda}^{\alpha} (\bar{n}^{\alpha} - \frac{\mathfrak{R}_c^{\alpha}}{\xi_c} \hat{n}_c^{\alpha}) + \frac{U(r^{\alpha})}{r^{\alpha}} \tilde{\tau}^{\alpha} \overset{\bullet}{\lambda}^{\alpha}\end{aligned} \tag{19.155}$$

which is rewritten as

$$\overset{\bullet}{\tau}^{\alpha} - \sum_{\beta=1}^{n} \bar{h}_{\alpha\beta} \overset{\bullet}{\lambda}^{\beta} = 0 \tag{19.156}$$

where

$$\begin{aligned}\bar{h}_{\alpha\beta} &\equiv \bar{n}^{\alpha} r^{\alpha} h_{\alpha\beta} + r^{\alpha} c_{\chi} \left(\bar{n}^{\alpha} - \frac{1}{\zeta_{\chi} \tau_y^{\alpha}} \chi^{\alpha} \right) \delta_{\alpha\beta} \\ &+ (1 - r^{\alpha}) c_c \left(\bar{n}^{\alpha} - \frac{\mathfrak{R}_c^{\alpha}}{\xi_c} \hat{n}_c^{\alpha} \right) \delta_{\alpha\beta} + \frac{U(r^{\alpha})}{r^{\alpha}} \tilde{\tau}^{\alpha} \delta_{\alpha\beta}\end{aligned} \tag{19.157}$$

The overall constitutive equation itself described in Sect. 19.8 holds only with the replacement of $\bar{M}_{\alpha\beta}$ to the following equation.

$$\bar{M}_{\alpha\beta} \equiv \bar{h}_{\alpha\beta} + \boldsymbol{\Xi}^{\alpha} : \mathbf{p}^{\beta} \bar{n}^{\beta} (\neq \overline{M}_{\beta\alpha}) \tag{19.158}$$

References

Anand L, Kothari M (1996) A computational procedure for rate-independent crystal plasticity. J Mech Phys Solids 44:525–558

Asaro RJ (1983) Micromechanics of crystals and polycrystals. Adv Appl Mech 23

Asaro RJ, Needleman A (1985) Texture development and strain hardening in rate dependent polycrystals. Acta Metall 33:923–953

Asaro RJ, Rice JR (1977) Strain localization in ductile single crystals. J Mech Phys Solids 25:309–338

Bassani JL, Wu TY (1991) Latent hardening in single crystals II: theory analytical characterization and predictions. Proc Royal Soc London A 435:21–41

Darrieulat M, Piot D (1996) A method of generalized analytical yield surfaces of crystalline materials. Int J Plast 12:575–610

Franciosi P, Zaoui A (1991) Crystal hardening and the issue of uniqueness. Int J Plast 7:295–311

Gambin W (1991) Refined analysis of elastic-plastic crystals. Int J Solids Struct 29:2013–2021

Gambin W (2001) Plasticity and textures. Kluwer Academic Publishers, Dordrecht

Gambin W, Barlat F (1997) Modeling of deformation texture development based on rate independent crystal plasticity. Int J Plast 13:75–85

Golub GH, Van Loan CF (2013) In: Matrix computations, 4th edn. John Hopkins University Press. Baltimore, Maryland
Harder J (1999) A crystallographic model for the study of local deformation processes in polycrystals. Int J Plast 15:605–624
Hashiguchi K (2015) Crystal plasticity based on extended subloading surface model. In: Proceedings of 2nd science meeting of Kyushu Branch of Society of Material Science, Japan, B17
Havner KS (1982) The theory of finite plastic deformation of crystalline solids. In Mechanics of solids—Rodney Hill 60th anniversary volume, Pergamon, pp 265–302
Havner KS (1992) Finite plastic deformation of crystalline solids. Cambridge University Press, Cambridge
Hill R (1966) Generalized constitutive relations for incremental deformation of metal crystals. J Mech Phys Solids 14:95–102
Hill R, Rice JR (1972) Constitutive analysis of elastic-plastic crystals at arbitrary strain. J Mech Phys Solids 20:401–413
Hosford WF (1974) A generalized isotropic yield criterion. J Appl Mech (ASME) 41:607–609
Hosford WF (2009) Mechanical behavior of solids. Cambridge University Press, Cambridge
Hutchinson JW (1976) Bounds and self-consistent estimates for creep of polycrystalline materials. Proc Roy Soc London A 348:101–127
Knockaert R, Chastel Y, Massoni (2000) Rate-independent crystalline plasticity, application to FCC materials. Int J Plast 16:179–198
Mandel J (1973) Equations constitutives directeurs dans les milieux plastiques at viscoplastiques. Int J Solids Struct 9:725–740
Mandel J (1974) Director vectors and constitutive equations for plastic and viscoplastic media. In: Sawczuk A (ed) Problems of plasticity, proceedings of international symposium foundation of plasticity. Noordhoff Int. Publ., Leyden, Netherland, pp 135–141
Miehe C, Schroder J (2001) A comparative study of stress update algorithms for rate-independent and rate-dependent crystal plasticity. Int J Numer Meth Eng 50:273–298
Nakada Y, Keh AS (1966) Latent hardening in iron single crystals. Acta Metall 14:961–973
Peirce D, Asaro JR, Needleman A (1982) Overview 21: an analysis of nonuniform and localized deformation in ductile single crystals. Act Metall 30:1087–1119
Peirce D, Asaro JR, Needleman A (1983) Overview 32: material rate dependence and localized deformation in crystal solids. Act Metall 31:1951–1976
Press WH, Teukolsky SA, Vetterling WT, Flannery BP (1988) Numerical recipies. In: The art of scientific computing. Cambridge University Press, New York
Taylor GI (1938) Plastic strain in metals. J Inst Metals 62:307–324
Xu B, Jiang Y (2004) A cyclic plasticity model for single crystals. Int J Plast 20:2161–2178
Yoshida K, Kuroda M (2012) Comparison of bifurcation and imperfection analyses of localized necking in rate-independent polycrystalline sheets. Int J Solids Struct 49:2073–2084
Zamiri A, Pourbogharat F (2010) A novel yield function for single crystal based on combined constraints optimization. Int J Plast 26:731–746
Zamiri A, Pourbogharat F, Barlat F (2007) An effective computational algorithm for rate-independent crystal plasticity based on a single crystal yield surface with an application to tube hydroforming. Int J Plast 23:1126–1147

Chapter 20
Implicit Stress Integration: Return-Mapping and Consistent Tangent Modulus Tensor

Constitutive equations of irreversible deformation, e.g. elastoplastic, viscoelastic and viscoplastic deformations are described in rate forms in which the stress rate and the strain rate are related to each other through the tangent modulus. Therefore, numerical calculations are executed in their incremental forms by the input of load (stress) increment or displacement (deformation) increment, while the time increment is also input in rate-(or time-)dependent constitutive equations, e.g. viscoelastic and viscoplastic ones. The algorithm to pull-back the stress to the yield surface is required for the numerical calculation by elastoplastic constitutive equations other than the subloading surface model possessing the automatic-controlling function to attract the stress to the yield surface. Then, numerical calculation by the *explicit* (or *forward Euler*) *method* results in the deviation from the exact solution or fails without convergence in the case that the tangent modulus is not constant, if an incremental step is taken large. This is caused by the fact that the tangent modulus after increment differs from that at the initiation of increment, so that the stress versus strain relation calculated using the tangent moduli at initiations of each increment deviates gradually from an exact solution. Therefore, numerical calculations by quite small increments have to be executed in the forward-Euler method.

In order to overcome the above-mentioned limitation in numerical calculations by the forward-Euler method, the *return-mapping method* which is the *implicit* (or *backward Euler*) *method* improving drastically numerical calculations by elastoplastic constitutive equations other than the subloading surface model has been developed by Simo and Taylor (1986) and the other workers, extending and generalizing the *classical radial return algorithm* (Yamada et al. 1968; Wilkins 1964), the *midpoint return map* (Rice and Tracy 1973), the *extended radial return method* (Krieg and Key 1976), the *elastic-trial (predictor)/plastic corrector method* (cf. Krieg and Krieg 1977), etc. It has taken the place of the traditional calculation method using directly elastoplastic tangent modulus in the rate form of constitutive equation (cf. e.g. Argyris 1965; Zienkiewicz 1977; Hinton and Owen 1980).

K. Hashiguchi, *Foundations of Elastoplasticity: Subloading Surface Model*,
DOI 10.1007/978-3-319-48821-9_20

Further, the consistent (algorithmic) tangent modulus has been formulated, which enables us to calculate the tangent modulus accurately in the implicit method.

The return mapping algorithm and the consistent tangent modulus tensor will be deliberated in this chapter. Their distinctive advantages can be exerted by the applications to the finite strain elastoplasticity based on the multiplicative decomposition (Hashiguchi and Yamakawa 2012), because the constitutive equation in the rate form becomes rather complicated as shown in Sect. 12.7. The incorporation of return mapping algorithm and the consistent tangent modulus tensor would improve also the efficiency of calculation by the subloading surface model in the case of the loading process near proportional loading. The infinitesimal strain hyperelastic-based plasticity will be used for the explanation because the fundamental notions and procedures can be captured concisely by the infinitesimal elastoplasticity.

20.1 Hyperelastic Constitutive Equation

In the return mapping, a trial stress is calculated first by inputting a strain increment on the premise that a purely elastic deformation is induced, and thereafter a stress relaxation is induced by the generation of plastic strain rate until the yield condition (the subloading surface equation in the subloading surface model) is satisfied. During the stress relaxation process, the total strain must be fixed in order to fulfill exactly the compatibility condition between surrounding finite elements and the plastic strain rate is calculated exactly based on the overstress from the yield or the subloading surface. In this process, the stress must be evaluated exactly by the elastic strain which is calculated by subtracting the plastic strain from the fixed strain. Therefore, the hyperelastic constitutive equation describing the exact relation between the stress and the elastic strain must be incorporated in the return-mapping scheme. Here, note that the capability of describing the finite deformation and rotation in the hypoelastic-based plasticity is spoiled by replacing it to the infinitesimal hyperelastic-based plasticity.

Let the infinitesimal strain in Eq. (2.50) be additively decomposed into the elastic and the plastic parts as

$$\boldsymbol{\varepsilon} = \boldsymbol{\varepsilon}^e + \boldsymbol{\varepsilon}^p \tag{20.1}$$

Further, consider the state that the deformation and rotation are infinitesimal fulfilling

$$\overset{\circ}{\boldsymbol{\sigma}} = \dot{\boldsymbol{\sigma}}, \quad \mathbf{d} = \dot{\boldsymbol{\varepsilon}}, \quad \mathbf{d}^e = \dot{\boldsymbol{\varepsilon}}^e, \quad \mathbf{d}^p = \dot{\boldsymbol{\varepsilon}}^p \tag{20.2}$$

Hereinafter, the hyperelastic constitutive equation is given based on Eq. (5.24) by the following equation, where $\psi(\boldsymbol{\varepsilon}^e)$ is the strain energy function.

$$\boxed{\boldsymbol{\sigma} = \frac{\partial \psi(\boldsymbol{\varepsilon}^e)}{\partial \boldsymbol{\varepsilon}^e} = \frac{\partial \psi(\boldsymbol{\varepsilon} - \boldsymbol{\varepsilon}^p)}{\partial (\boldsymbol{\varepsilon} - \boldsymbol{\varepsilon}^p)}} \tag{20.3}$$

For the particular strain energy function

$$\psi(\boldsymbol{\varepsilon}^e) = \frac{1}{2} L (\mathrm{tr}\boldsymbol{\varepsilon}^e)^2 + G \mathrm{tr}\boldsymbol{\varepsilon}^{e2} \tag{20.4}$$

based on Eq. (5.25), the stress is given by the linear relation to the elastic strain $\boldsymbol{\varepsilon}^e$.

$$\boldsymbol{\sigma} = L(\mathrm{tr}\boldsymbol{\varepsilon}^e)\mathbf{I} + 2G\boldsymbol{\varepsilon}^e \tag{20.5}$$

or

$$\boldsymbol{\sigma} = \left(L + \frac{2}{3}G\right)(\mathrm{tr}\boldsymbol{\varepsilon}^e)\mathbf{I} + 2G\boldsymbol{\varepsilon}^{e'} \tag{20.6}$$

which is rewritten as

$$\boldsymbol{\sigma} = K(\mathrm{tr}\boldsymbol{\varepsilon}^e)\mathbf{I} + 2G\boldsymbol{\varepsilon}^{e'} \tag{20.7}$$

and further

$$\boxed{\boldsymbol{\sigma} = \mathbf{E} : \boldsymbol{\varepsilon}^e = \mathbf{E} : (\boldsymbol{\varepsilon} - \boldsymbol{\varepsilon}^p)} \tag{20.8}$$

based on Eqs. (5.33) and (20.1), where the elastic modulus tensor $\mathbf{E}$ is shown in Eqs. (5.35) and (5.42).

The rate form of Eq. (20.8) is given as

$$\dot{\boldsymbol{\sigma}} = \mathbf{E} : \dot{\boldsymbol{\varepsilon}}^e = \mathbf{E} : (\dot{\boldsymbol{\varepsilon}} - \dot{\boldsymbol{\varepsilon}}^p) \tag{20.9}$$

for the linear elastic material with $\mathbf{E} = \mathrm{const}$. Equation (20.9) is rewritten as

$$\dot{\boldsymbol{\sigma}} = \dot{\boldsymbol{\sigma}}^e + \dot{\boldsymbol{\sigma}}^p \tag{20.10}$$

noting Eq. (6.47) with Eq. (6.48), where

$$\dot{\boldsymbol{\sigma}}^e \equiv \mathbf{E} : \dot{\boldsymbol{\varepsilon}}, \quad \dot{\boldsymbol{\sigma}}^p \equiv -\mathbf{E} : \dot{\boldsymbol{\varepsilon}}^p \tag{20.11}$$

as shown in Fig. 20.1. Here, $\dot{\boldsymbol{\sigma}}^e$ is the elastic stress rate and the plastic relaxation stress rate $\dot{\boldsymbol{\sigma}}^p$ is described by the plastic flow rule as follows:

$$\boxed{\dot{\boldsymbol{\sigma}}^p = -\mathbf{E} : \dot{\lambda}\mathbf{n}} \tag{20.12}$$

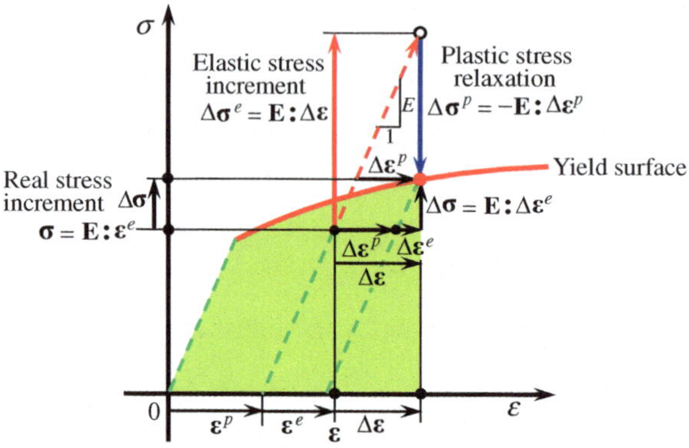

Fig. 20.1 Elastic stress increment and plastic relaxation stress increment

The linear hyperelasticity is adopted below in explanations for the sake of simplicity, although nonlinear hyperelasticity can be adopted for the return-mapping with the consistent tangent modulus.

20.2 Return Mapping

Suppose the state that the stress $\boldsymbol{\sigma}_{\mathrm{n}}$, the plastic strain $\boldsymbol{\varepsilon}_{\mathrm{n}}^{p}$ and the internal variables in the n-th step are already known by performing the calculations in the incremental steps of n times. At the beginning of calculation in the step n + 1, one calculates first the elastoplastic deformation under a given boundary condition by use of the global tangent stiffness matrix obtained in the previous step n. Then, calculate the strain increment $\Delta\boldsymbol{\varepsilon}$ from the deformation in each numerical integration point. Further, calculate the trial stress by inputting the strain increment $\Delta\boldsymbol{\varepsilon}$, postulating that only the elastic deformation is induced. This process and the trial stress calculated in this process are called the *elastic trial* (or *predictor*) *step* and the *elastic trial stress*, respectively. Designating the trial stress and the elastic strain calculated in this step by $\boldsymbol{\sigma}_{\mathrm{n+1}}^{\mathrm{trial}}$ and $\boldsymbol{\varepsilon}_{\mathrm{n+1}}^{e\mathrm{trial}}$, respectively, they are related from Eq. (20.8) by the hyperelastic relation as follows:

$$\boldsymbol{\sigma}_{\mathrm{n+1}}^{\mathrm{trial}} = \mathbf{E} : \boldsymbol{\varepsilon}_{\mathrm{n+1}}^{e\mathrm{trial}} = \mathbf{E} : (\boldsymbol{\varepsilon}_{\mathrm{n+1}} - \boldsymbol{\varepsilon}_{\mathrm{n}}^{p}) \tag{20.13}$$

where

$$\boldsymbol{\varepsilon}_{\mathrm{n+1}} = \boldsymbol{\varepsilon}_{\mathrm{n}} + \Delta\boldsymbol{\varepsilon}_{\mathrm{n+1}} = \boldsymbol{\varepsilon}_{\mathrm{n}}^{p} + \boldsymbol{\varepsilon}_{\mathrm{n+1}}^{e\mathrm{trial}} = \boldsymbol{\varepsilon}_{\mathrm{n+1}}^{e} + \boldsymbol{\varepsilon}_{\mathrm{n+1}}^{p} = \mathrm{const.} \tag{20.14}$$

$$\boldsymbol{\varepsilon}_{\mathrm{n+1}}^{e\mathrm{trial}} = \boldsymbol{\varepsilon}_{\mathrm{n}}^{e} + \Delta\boldsymbol{\varepsilon}_{\mathrm{n+1}} = \boldsymbol{\varepsilon}_{\mathrm{n+1}} - \boldsymbol{\varepsilon}_{\mathrm{n}}^{p} \tag{20.15}$$

$$\boldsymbol{\varepsilon}^{p\text{trial}}_{\text{n}+1} = \boldsymbol{\varepsilon}^{p}_{\text{n}}, \quad H^{\text{trial}}_{\text{n}+1} = H_{\text{n}}, \quad \boldsymbol{\alpha}^{\text{trial}}_{\text{n}+1} = \boldsymbol{\alpha}_{\text{n}} \tag{20.16}$$

where $\boldsymbol{\varepsilon}_{\text{n}+1}$ is the fixed strain throughout each elastic trial and plastic corrector process until the calculation step $\text{n}+1$ is finished.

The following loading criterion is imposed just after the elastic trial step in the conventional elastoplastic model.

$$\left.\begin{array}{ll} f(\hat{\boldsymbol{\sigma}}^{\text{trial}}_{\text{n}+1}) - F(H_n) \leq 0 & : \Delta\boldsymbol{\varepsilon}^{p}_{\text{n}+1} = \mathbf{O}, \quad \boldsymbol{\sigma}^{\text{Final}}_{\text{n}+1} = \boldsymbol{\sigma}^{\text{trial}}_{\text{n}+1} \\ \text{Otherwise} & : \Delta\boldsymbol{\varepsilon}^{p}_{\text{n}+1} \neq \mathbf{O}, \quad \boldsymbol{\sigma}^{\text{Final}}_{\text{n}+1} \neq \boldsymbol{\sigma}^{\text{trial}}_{\text{n}+1} \end{array}\right\} \tag{20.17}$$

where

$$\hat{\boldsymbol{\sigma}}^{\text{trial}}_{\text{n}+1} \equiv \boldsymbol{\sigma}^{\text{trial}}_{\text{n}+1} - \boldsymbol{\alpha}_{\text{n}} \tag{20.18}$$

and $\boldsymbol{\sigma}^{\text{Final}}_{\text{n}+1}$ designates the finally obtained correct solution.

The plastic relaxation is obviously required if the subloading surface expands in the elastic trial step. Therefore, it can be stated that only the eastic deformation is induced if the subloading surface contracts after the elastic trial step in the monotonic loading process. On the other hand, it should be noted that the contraction of the subloading surface in the elastic trial step does not necessarily mean that the plastic relaxation not required after the elastic trial step, since large increments is usually input in the return-mapping calculation. In fact, the plastic relaxation is required if the subloading surface once contracts and then expands in the elastic trial step as shown in Fig. 20.2a for the initial subloading surface model and in Fig. 20.2b for the extended subloading surface model. Here, the plastic relaxation must be induced in a general loading process if the stress increment in the elastic trial step, i.e. $\boldsymbol{\sigma}^{\text{trial}}_{\text{n}+1} - \boldsymbol{\sigma}_{\text{n}}$ makes a sharp angle with the outward-normal $\hat{\mathbf{n}}^{\text{trial}}_{\text{n}+1}$ of the subloading surface formed after the elastic trial step. Then, the following loading criterion must be imposed just after the elastic trial step in the initial subloading model ($\mathbf{c} = \boldsymbol{\alpha}$) (Hashiguchi 2016).

$$\boxed{\begin{array}{ll} f(\hat{\boldsymbol{\sigma}}^{\text{trial}}_{\text{n}+1}) - R_e F(H_{\text{n}}) \leq 0 \text{ or } \hat{\mathbf{n}}^{\text{trial}}_{\text{n}+1} : (\boldsymbol{\sigma}^{\text{trial}}_{\text{n}+1} - \boldsymbol{\sigma}_{\text{n}}) \leq 0 & : \Delta\boldsymbol{\varepsilon}^{p}_{\text{n}+1} = \mathbf{O}, \quad \boldsymbol{\sigma}^{\text{Final}}_{\text{n}+1} = \boldsymbol{\sigma}^{\text{trial}}_{\text{n}+1} \\ \text{Otherwise} & : \Delta\boldsymbol{\varepsilon}^{p}_{\text{n}+1} \neq \mathbf{O}, \quad \boldsymbol{\sigma}^{\text{Final}}_{\text{n}+1} \neq \boldsymbol{\sigma}^{\text{trial}}_{\text{n}+1} \end{array}} \tag{20.19}$$

where

$$\hat{\mathbf{n}}^{\text{trial}}_{\text{n}+1} \equiv \frac{\partial f(\hat{\boldsymbol{\sigma}}^{\text{trial}}_{\text{n}+1})}{\partial \boldsymbol{\sigma}^{\text{trial}}_{\text{n}+1}} \bigg/ \left\| \frac{\partial f(\hat{\boldsymbol{\sigma}}^{\text{trial}}_{\text{n}+1})}{\partial \boldsymbol{\sigma}^{\text{trial}}_{\text{n}+1}} \right\|, \quad \hat{\boldsymbol{\sigma}}^{\text{trial}}_{\text{n}+1} \equiv \boldsymbol{\sigma}^{\text{trial}}_{\text{n}+1} - \boldsymbol{\alpha}_{\text{n}} \tag{20.20}$$

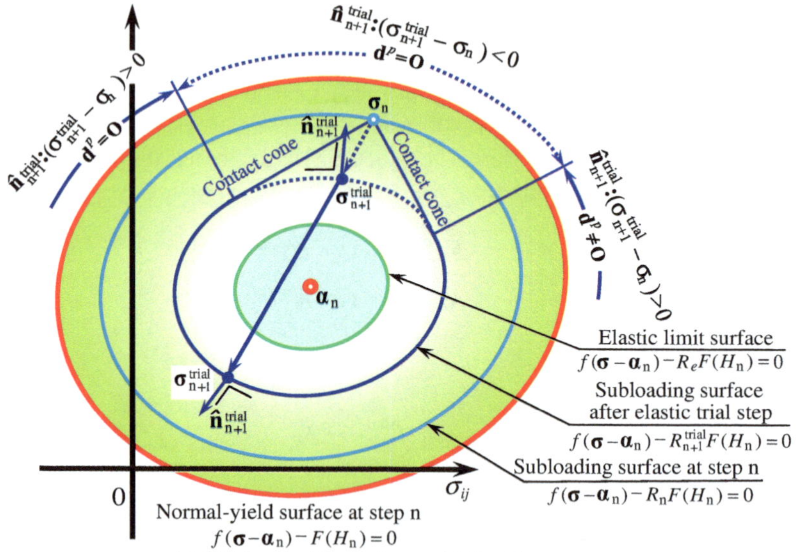

(a) Initial subloading surface model ($\mathbf{c}=\boldsymbol{\alpha}$)

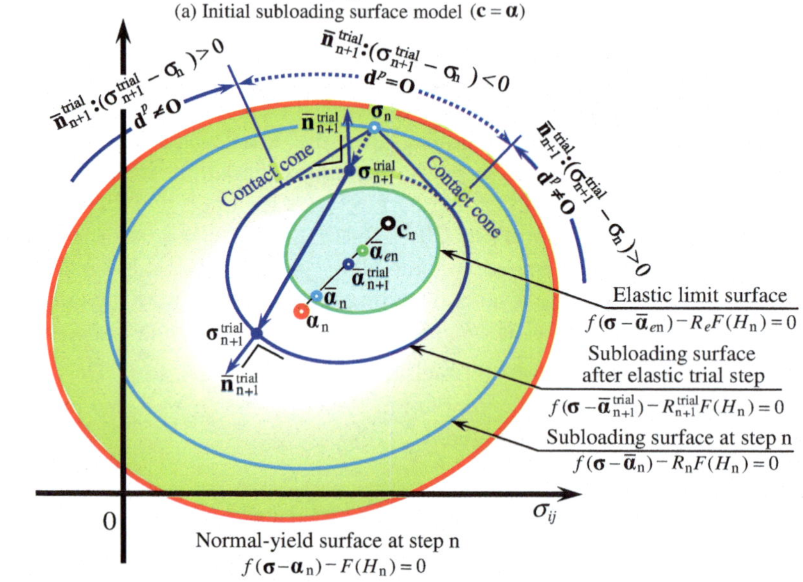

(b) Extended subloading surface model ($\mathbf{c}\neq\boldsymbol{\alpha}$)

Fig. 20.2 Loading criterion when the elastic trial stress is directed inwards the subloading surface at the step n in return-mapping projection for subloading surface model

Analogously, the following loading criterion must be imposed in the extended subloading model ($\mathbf{c} \neq \boldsymbol{\alpha}$) (Hashiguchi 2016).

$f(\boldsymbol{\sigma}^{\text{trial}}_{\text{n}+1} - \bar{\boldsymbol{\alpha}}_{e\text{n}}) - R_e F(H_{\text{n}}) \leq 0$ or $\bar{\mathbf{n}}^{\text{trial}}_{\text{n}+1} : (\boldsymbol{\sigma}^{\text{trial}}_{\text{n}+1} - \boldsymbol{\sigma}_{\text{n}}) \leq 0$	$: \Delta\boldsymbol{\varepsilon}^p_{\text{n}+1} = \mathbf{O}, \quad \boldsymbol{\sigma}^{\text{Final}}_{\text{n}+1} = \boldsymbol{\sigma}^{\text{trial}}_{\text{n}+1}$
Otherwise	$: \Delta\boldsymbol{\varepsilon}^p_{\text{n}+1} \neq \mathbf{O}, \quad \boldsymbol{\sigma}^{\text{Final}}_{\text{n}+1} \neq \boldsymbol{\sigma}^{\text{trial}}_{\text{n}+1}$

(20.21)

where the variables are given, noting Eqs. (9.4)–(9.7) as follows:

$$\bar{\boldsymbol{\alpha}}_{e\text{n}} = \mathbf{c}_{\text{n}} - R_e \hat{\mathbf{c}}_{\text{n}} \tag{20.22}$$

$$\bar{\boldsymbol{\alpha}}_{\text{n}} = \mathbf{c}_{\text{n}} - R_{\text{n}} \hat{\mathbf{c}}_{\text{n}} \tag{20.23}$$

$$\bar{\boldsymbol{\alpha}}^{\text{trial}}_{\text{n}+1} = \mathbf{c}_{\text{n}} - R^{\text{trial}}_{\text{n}+1} \hat{\mathbf{c}}_{\text{n}} \tag{20.24}$$

$$\hat{\mathbf{c}}_{\text{n}} \equiv \mathbf{c}_{\text{n}} - \boldsymbol{\alpha}_{\text{n}} \tag{20.25}$$

$$\bar{\boldsymbol{\sigma}}^{\text{trial}}_{\text{n}+1} \equiv \boldsymbol{\sigma}^{\text{trial}}_{\text{n}+1} - \bar{\boldsymbol{\alpha}}^{\text{trial}}_{\text{n}+1} \tag{20.26}$$

$$\bar{\mathbf{n}}^{\text{trial}}_{\text{n}+1} \equiv \frac{\partial f(\bar{\boldsymbol{\sigma}}^{\text{trial}}_{\text{n}+1})}{\partial \boldsymbol{\sigma}^{\text{trial}}_{\text{n}+1}} \Bigg/ \left\| \frac{\partial f(\bar{\boldsymbol{\sigma}}^{\text{trial}}_{\text{n}+1})}{\partial \boldsymbol{\sigma}^{\text{trial}}_{\text{n}+1}} \right\| \tag{20.27}$$

Here, $R^{\text{trial}}_{\text{n}+1}$ is calculated from

$$f(\bar{\boldsymbol{\sigma}}^{\text{trial}}_{\text{n}+1}) = R^{\text{trial}}_{\text{n}+1} F(H_{\text{n}}) \tag{20.28}$$

which is explicitly given for the Mises yield function, noting Eqs. (10.30) and (10.32), as follows:

$$\sqrt{3/2} || \tilde{\boldsymbol{\sigma}}^{\text{trial}\prime}_{\text{n}+1} + R^{\text{trial}}_{\text{n}+1} \hat{\mathbf{c}}'_{n} || = R^{\text{trial}}_{\text{n}+1} F(H_{\text{n}}) \tag{20.29}$$

leading to

$$R^{\text{trial}}_{\text{n}+1} = \frac{\tilde{\boldsymbol{\sigma}}^{\text{trial}\prime}_{\text{n}+1} : \hat{\mathbf{c}}'_{\text{n}} + \sqrt{(\tilde{\boldsymbol{\sigma}}^{\text{trial}\prime}_{\text{n}+1} : \hat{\mathbf{c}}'_{\text{n}})^2 + [(2/3)(F(H_{\text{n}}))^2 - ||\hat{\mathbf{c}}'_{\text{n}}||^2] \, ||\tilde{\boldsymbol{\sigma}}^{\text{trial}\prime}_{\text{n}+1}||^2}}{(2/3)(F(H_{\text{n}}))^2 - ||\hat{\mathbf{c}}'_{\text{n}}||^2} \tag{20.30}$$

where

$$\tilde{\boldsymbol{\sigma}}^{\text{trial}}_{\text{n}+1} \equiv \boldsymbol{\sigma}^{\text{trial}}_{\text{n}+1} - \mathbf{c}_{\text{n}} \tag{20.31}$$

Further, the loading criterion for the multiplicative finite strain theory described in Chap. 12 is given by the replacements

$$\bar{\mathbf{n}} \rightarrow \underline{\overline{\mathbf{N}}},\ \Delta\boldsymbol{\varepsilon}^{p} \rightarrow \Delta\bar{\lambda}\underline{\overline{\mathbf{N}}} \tag{20.32}$$

in addition to Eq. (12.51) in the loading criterion of Eq. (20.21) for the extended subloading surface model.

In the second cases of Eqs. (20.17), (20.19) and (20.21) in which it is judged such the the plastic strain rate should be induced, the stress must be reduced through the plastic relaxation process by generating the plastic strain increment which is offset by the decrease of elastic strain increment in order to fix the strain until the yield surface or the subloading surface equation will be satisfied. That is, the calculation is repeated until the yield condition

$$f(\hat{\boldsymbol{\sigma}}_{n+1}) - F(H_{n+1}) = 0 \tag{20.33}$$

in the conventional elastoplasticity and the subloading surface equation

$$\left.\begin{array}{ll} f(\hat{\boldsymbol{\sigma}}_{n+1}) - R_{n+1}F(H_{n+1}) = 0 & \text{for initial subloading surface model} \\ f(\bar{\boldsymbol{\sigma}}_{n+1}) - R_{n+1}F(H_{n+1}) = 0 & \text{for extended subloading surface model} \end{array}\right\} \tag{20.34}$$

in the subloading surface model is satisfied within a prescribed tolerance (see Fig. 20.3), while R_{n+1} must be calculated by the evolution rule in Eq. (7.9) or Eq. (7.19) in the analytical integration. This process is called the *plastic corrector step*. In what follows, the plastic corrector step will be explained.

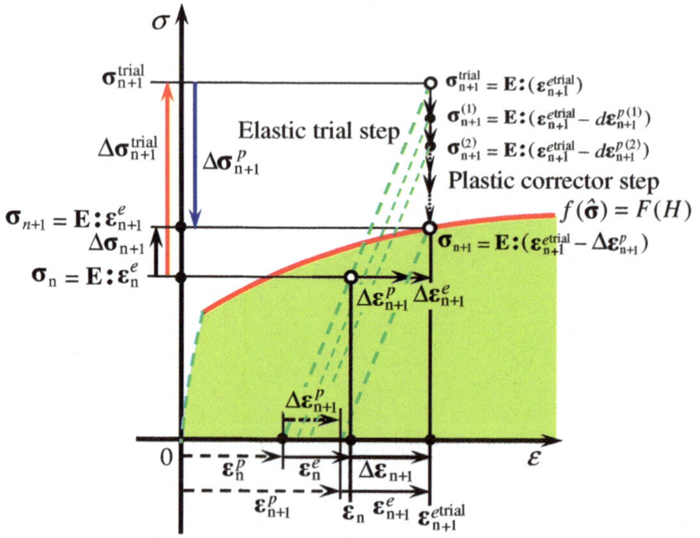

Fig. 20.3 Basic concept of return-mapping projection

Designating the stress and the elastic strain in the plastic corrector step by $\boldsymbol{\sigma}_{n+1}$ and $\boldsymbol{\varepsilon}^{e}_{n+1}$, respectively, and the plastic strain increment induced in this step as $\Delta\boldsymbol{\varepsilon}^{p}_{n+1}$, the stress $\boldsymbol{\sigma}_{n+1}$ is calculated from Eq. (20.8) by the hyperelastic constitutive equation

$$\boldsymbol{\sigma}_{n+1} = \mathbf{E} : \boldsymbol{\varepsilon}^{e}_{n+1} = \mathbf{E} : [\boldsymbol{\varepsilon}_{n+1} - (\boldsymbol{\varepsilon}^{p}_{n} + \Delta\boldsymbol{\varepsilon}^{p}_{n+1})] = \mathbf{E} : (\boldsymbol{\varepsilon}^{etrial}_{n+1} - \Delta\boldsymbol{\varepsilon}^{p}_{n+1}) \quad (20.35)$$

The stress must be calculated exactly by the hyperelastic constitutive Eq. (20.35) and the plastic strain rate must be calculated exactly based on the overstress from the yield surface. However, the solution obtained after the corrector step does not usually fulfill the equilibrium equation in the nodal points inside the finite element body and on its boundary surface. Therefore, the calculation has to be repeated again by returning all the state variables at the end of the last step n and inputting the strain increment modified to cancel the residual forces until the yield condition or the subloading surface equation in each element and the equilibrium equation in the whole body will be fulfilled within a prescribed tolerance.

Substituting

$$\Delta\boldsymbol{\varepsilon}^{p}_{n+1} = \boldsymbol{\varepsilon}^{p}_{n+1} - \boldsymbol{\varepsilon}^{p}_{n} = \Delta\lambda_{n+1}\mathbf{n}_{n+1} \quad (20.36)$$

into Eq. (20.35), one has

$$\boldsymbol{\sigma}_{n+1} = \mathbf{E} : (\boldsymbol{\varepsilon}_{n+1} - \Delta\boldsymbol{\varepsilon}^{p}_{n+1}) = \mathbf{E} : (\boldsymbol{\varepsilon}_{n+1} - \Delta\lambda_{n+1}\mathbf{n}_{n+1}) \quad (20.37)$$

Further, the stress increment is decomposed from Eq. (20.10) as follows:

$$\Delta\boldsymbol{\sigma}_{n+1} = \Delta\boldsymbol{\sigma}^{e}_{n+1} + \Delta\boldsymbol{\sigma}^{p}_{n+1} \quad (20.38)$$

where

$$\Delta\boldsymbol{\sigma}^{e}_{n+1} \equiv \mathbf{E} : \Delta\boldsymbol{\varepsilon}_{n+1}, \quad \Delta\boldsymbol{\sigma}^{p}_{n+1} \equiv -\mathbf{E} : \Delta\boldsymbol{\varepsilon}^{p}_{n+1} \quad (20.39)$$

The *closest point projection* as the fully implicit method and the *cutting plane projection* as the incomplete implicit method have been proposed for the calculation in the corrector step as will be described in the subsequent sections.

20.3 Closest Point Projection

The stress is pulled-back to the closest point on the yield surface so that the line connecting the current stress and the closest point is normal to the yield surface in the closest point projection. The closest point projection (Ortiz and Popov 1985; Simo and Ortiz 1985) is the complete implicit method. The geometrical interpretation of the closest point projection is illustrated in Fig. 20.4. Here, the plastic strain increment is induced towards the normal of the forthcoming dynamic-loading

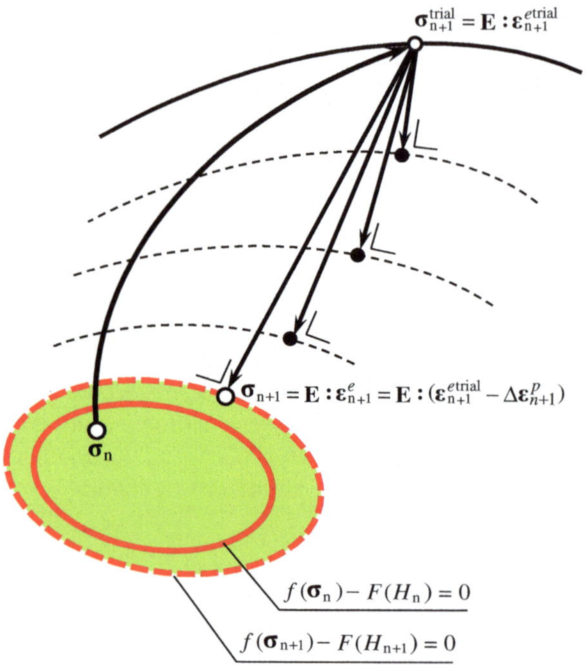

Fig. 20.4 Closest-point projection

surface in the plastic corrector step because of the exact *implicit* (or *backward Euler*) *projection*.

The simultaneous equation consisting of the yield condition, the stress and internal variables is solved numerically in this projection. Therefore, analytical partial derivatives by all variables must be derived as the preparation task which is cumbersome and difficult for sophisticated constitutive equation. On the other hand, the plastic strain rate can be obtained by calculating only the positive plastic multiplier involved in the plastic flow rule without the use of the explicit equation of plastic strain rate expressed by the plastic modulus which is complicated in sophisticated plastic constitutive equation.

There exist several calculation methods in the closest pint projection as will be described below.

20.3.1 General Solution by Simultaneous Equation

This is the most general method to solve the simultaneous equation describing the yield condition and evolution rules of internal variables. In what follows, it will be explained based on the paper (Anjiki et al. 2016) in which the finite strain elastoplasticity with the subloading surface model formulated by Prof. Yamakawa

(Hashiguchi and Yamakawa 2012) is simplified to the infinitesimal elastoplasticity and extended to the extended subloading surface model.

All the equation which must be satisfied are shown below

Plastic flow rule:

$$\mathbf{Y}_\sigma \equiv \boldsymbol{\varepsilon}^e_{n+1} + \Delta\boldsymbol{\varepsilon}^p_{n+1} - \boldsymbol{\varepsilon}^{etrial}_{n+1} = \mathbf{E}^{-1} : \boldsymbol{\sigma}_{n+1} + \Delta\lambda_{n+1}\hat{\mathbf{n}}_{n+1} - \boldsymbol{\varepsilon}^{etrial}_{n+1} = \mathbf{O} \quad (20.40)$$

Isotropic hardening rule:

$$Y_H \equiv H_{n+1} - H_n - \sqrt{2/3}\Delta\lambda_{n+1} = 0 \quad (20.41)$$

Kinematic hardening rule:

$$\mathbf{Y}_\alpha \equiv \boldsymbol{\alpha}_{n+1} - \boldsymbol{\alpha}_n - \left(c_k \hat{\mathbf{n}}_{n+1} - \frac{1}{\sqrt{3/2}\zeta F_{n+1}} \boldsymbol{\alpha}_{n+1} \right) \Delta\lambda_{n+1} = \mathbf{O} \quad (20.42)$$

Subloading surface:

$$Y_s \equiv f(\hat{\boldsymbol{\sigma}}_{n+1}) - R_{n+1}F_{n+1} = \sqrt{3/2}\|\hat{\boldsymbol{\sigma}}_{n+1}\| - R_{n+1}F_{n+1} = 0 \quad (20.43)$$

noting Eqs. (7.6), (10.1) and (10.8), where

$$\boldsymbol{\sigma}_{n+1} = \mathbf{E} : (\boldsymbol{\varepsilon}_{n+1} - \boldsymbol{\varepsilon}^p_{n+1}) \quad (20.44)$$

$$\hat{\mathbf{n}}_{n+1} = \frac{\partial f(\hat{\boldsymbol{\sigma}}_{n+1})}{\partial \hat{\boldsymbol{\sigma}}_{n+1}} \Big/ \left\| \frac{\partial f(\hat{\boldsymbol{\sigma}}_{n+1})}{\partial \hat{\boldsymbol{\sigma}}_{n+1}} \right\| \quad (20.45)$$

$$F_{n+1} = F_0\{1 + h_1[1 - \exp(-h_2 H_{n+1})]\} \quad (20.46)$$

$$R_{n+1} = \frac{2}{\pi}(1 - R_e)\cos^{-1}\left[\cos\left(\frac{\langle R_n - R_e\rangle}{1 - R_e}\frac{\pi}{2}\right)\exp\left(-u\frac{\varepsilon^p_{n+1} - \varepsilon^p_n}{1 - R_e}\frac{\pi}{2}\right)\right] + R_e \quad (20.47)$$

There are fourteen unknowns of the components of $\boldsymbol{\varepsilon}^p_{n+1}$, $\boldsymbol{\alpha}_{n+1}$, H_{n+1} and $\Delta\lambda_{n+1}$, while we do not need to regard the other variables contained in Eqs. (20.44)–(20.47) as unknown variables since they can be calculated from them.

Equations (20.40)–(20.43) is represented in the simultaneous equation

$$\boxed{\mathbf{Y}(\mathbf{X}) = \mathbf{0}} \quad (20.48)$$

where

$$\mathbf{Y} \equiv \begin{Bmatrix} \mathbf{Y}_\sigma \\ \mathbf{Y}_\alpha \\ Y_H \\ Y_s \end{Bmatrix}, \quad \mathbf{X} \equiv \begin{Bmatrix} \boldsymbol{\sigma}_{\mathrm{n}+1} \\ \boldsymbol{\alpha}_{\mathrm{n}+1} \\ H_{\mathrm{n}+1} \\ \Delta\lambda_{\mathrm{n}+1} \end{Bmatrix} \tag{20.49}$$

In order to solve Eq. (20.48) numerically, linearizing it by means of Taylor expansion and taking the first order term, we have

$$\mathbf{Y}(\mathbf{X}^{(\mathrm{k}+1)}) \cong \mathbf{Y}(\mathbf{X}^{(\mathrm{k})}) + \mathbf{J}(\mathbf{X}^{(\mathrm{k})}) \cdot d\mathbf{X} = \mathbf{0} \tag{20.50}$$

where

$$\mathbf{J} = \begin{bmatrix} \overbrace{\dfrac{\partial \mathbf{Y}_\sigma}{\partial \boldsymbol{\sigma}_{\mathrm{n+1}}}}^{6} & \overbrace{\dfrac{\partial \mathbf{Y}_\sigma}{\partial \boldsymbol{\alpha}_{\mathrm{n+1}}}}^{6} & \overbrace{\dfrac{\partial \mathbf{Y}_\sigma}{\partial H_{\mathrm{n+1}}}}^{1} & \overbrace{\dfrac{\partial \mathbf{Y}_\sigma}{\partial (\Delta\lambda_{\mathrm{n+1}})}}^{1} \\ \dfrac{\partial \mathbf{Y}_\alpha}{\partial \boldsymbol{\sigma}_{\mathrm{n+1}}} & \dfrac{\partial \mathbf{Y}_\alpha}{\partial \boldsymbol{\alpha}_{\mathrm{n+1}}} & \dfrac{\partial \mathbf{Y}_\alpha}{\partial H_{\mathrm{n+1}}} & \dfrac{\partial \mathbf{Y}_\alpha}{\partial (\Delta\lambda_{\mathrm{n+1}})} \\ \dfrac{\partial Y_H}{\partial \boldsymbol{\sigma}_{\mathrm{n+1}}} & \dfrac{\partial Y_H}{\partial \boldsymbol{\alpha}_{\mathrm{n+1}}} & \dfrac{\partial Y_H}{\partial H_{\mathrm{n+1}}} & \dfrac{\partial Y_H}{\partial (\Delta\lambda_{\mathrm{n+1}})} \\ \dfrac{\partial Y_s}{\partial \boldsymbol{\sigma}_{\mathrm{n+1}}} & \dfrac{\partial Y_s}{\partial \boldsymbol{\alpha}_{\mathrm{n+1}}} & \dfrac{\partial Y_s}{\partial H_{\mathrm{n+1}}} & \dfrac{\partial Y_s}{\partial (\Delta\lambda_{\mathrm{n+1}})} \end{bmatrix} \begin{matrix} \}6 \\ \}6 \\ \}1 \\ \}1 \end{matrix} \tag{20.51}$$

which is the square matrix of 14 × 14 since $\mathbf{Y}_\sigma$ and $\mathbf{Y}_\alpha$ are 6 rows by 1 column, and $\boldsymbol{\varepsilon}^p_{\mathrm{n}+1}$ $\boldsymbol{\alpha}_{\mathrm{n}+1}$ are 1 row and 6 columns.

Equation (20.50) is the simultaneous equation for the unknown $d\mathbf{X}$, and thus, solving Eq. (20.51), it is updated by

$$d\mathbf{X} = -[\mathbf{J}(\mathbf{X}^{(\mathrm{k})})]^{-1}\mathbf{Y}(\mathbf{X}^{(\mathrm{k})}) \rightarrow \mathbf{X}^{(\mathrm{k}+1)} = \mathbf{X}^{(\mathrm{k})} + d\mathbf{X} = \mathbf{X}^{(\mathrm{k})} - [\mathbf{J}(\mathbf{X}^{(\mathrm{k})})]^{-1}\mathbf{Y}(\mathbf{X}^{(\mathrm{k})}) \tag{20.52}$$

where (k) designates the number of the repeated calculations.

The elements in Eq. (20.51) are given below, where the subscript n + 1 added to the unknown variables are omitted hereafter for the sake of simplicity.

$$\frac{\partial \mathbf{Y}_\sigma}{\partial \boldsymbol{\sigma}} = \frac{\partial(\mathbf{E}^{-1} : \boldsymbol{\sigma} + \Delta\lambda\hat{\mathbf{n}} - \boldsymbol{\varepsilon}^{e\mathrm{trial}})}{\partial \boldsymbol{\sigma}} = \mathbf{E}^{-1} : \frac{\partial \boldsymbol{\sigma}}{\partial \boldsymbol{\sigma}} + \Delta\lambda \frac{\partial \hat{\mathbf{n}}}{\partial \hat{\boldsymbol{\sigma}}} : \frac{\partial \hat{\boldsymbol{\sigma}}}{\partial \boldsymbol{\sigma}} = (\mathbf{E}^{-1} + \Delta\lambda \frac{\partial \hat{\mathbf{n}}}{\partial \hat{\boldsymbol{\sigma}}}) : \boldsymbol{\mathcal{S}} \tag{20.53}$$

$$\frac{\partial \mathbf{Y}_{\sigma}}{\partial \boldsymbol{\alpha}} = \frac{\partial(\mathbf{E}^{-1} : \boldsymbol{\sigma} + \Delta\lambda\hat{\mathbf{n}} - \boldsymbol{\varepsilon}^{\text{etrial}})}{\partial \boldsymbol{\alpha}} = \Delta\lambda \frac{\partial \hat{\mathbf{n}}}{\partial \hat{\boldsymbol{\sigma}}} : \frac{\partial \hat{\boldsymbol{\sigma}}}{\partial \boldsymbol{\alpha}} = -\Delta\lambda \frac{\partial \hat{\mathbf{n}}}{\partial \hat{\boldsymbol{\sigma}}} : \boldsymbol{\mathcal{S}} \tag{20.54}$$

$$\frac{\partial \mathbf{Y}_{\sigma}}{\partial H} = \frac{\partial(\mathbf{E}^{-1} : \boldsymbol{\sigma} + \Delta\lambda\hat{\mathbf{n}} - \boldsymbol{\varepsilon}^{\text{etrial}})}{\partial H} = \mathbf{O} \tag{20.55}$$

$$\frac{\partial \mathbf{Y}_{\sigma}}{\partial(\Delta\lambda)} = \frac{\partial(\mathbf{E}^{-1} : \boldsymbol{\sigma} + \Delta\lambda\hat{\mathbf{n}} - \boldsymbol{\varepsilon}^{\text{etrial}})}{\partial(\Delta\lambda)} = \hat{\mathbf{n}} \tag{20.56}$$

$$\frac{\partial \mathbf{Y}_{\alpha}}{\partial \boldsymbol{\sigma}} = \frac{\partial\left[\boldsymbol{\alpha} - \boldsymbol{\alpha}_n - \left(c_k\hat{\mathbf{n}} - \frac{1}{\sqrt{3/2}\zeta F}\boldsymbol{\alpha}\right)\Delta\lambda\right]}{\partial \boldsymbol{\sigma}} = -c_k\Delta\lambda \frac{\partial \hat{\mathbf{n}}}{\partial \hat{\boldsymbol{\sigma}}} : \boldsymbol{\mathcal{S}} \tag{20.57}$$

$$\frac{\partial \mathbf{Y}_{\alpha}}{\partial \boldsymbol{\alpha}} = \frac{\partial\left[\boldsymbol{\alpha} - \boldsymbol{\alpha}_n - \left(c_k\hat{\mathbf{n}} - \frac{1}{\sqrt{3/2}\zeta F}\boldsymbol{\alpha}\right)\Delta\lambda\right]}{\partial \boldsymbol{\alpha}} = \frac{\partial \boldsymbol{\alpha}}{\partial \boldsymbol{\alpha}} - \left(c_k \frac{\partial \hat{\mathbf{n}}}{\partial \hat{\boldsymbol{\sigma}}} : \frac{\partial \hat{\boldsymbol{\sigma}}}{\partial \boldsymbol{\alpha}} - \frac{1}{\sqrt{3/2}\zeta F}\frac{\partial \boldsymbol{\alpha}}{\partial \boldsymbol{\alpha}}\right)\Delta\lambda$$

$$= \left[\left(1 + \frac{1}{\sqrt{3/2}\zeta F}\Delta\lambda\right)\boldsymbol{\mathcal{I}} + c_k\Delta\lambda\frac{\partial \hat{\mathbf{n}}}{\partial \hat{\boldsymbol{\sigma}}}\right] : \boldsymbol{\mathcal{S}} \tag{20.58}$$

$$\frac{\partial \mathbf{Y}_{\alpha}}{\partial H} = \frac{\partial\left[\boldsymbol{\alpha} - \boldsymbol{\alpha}_n - \left(c_k\hat{\mathbf{n}} - \frac{1}{\sqrt{3/2}\zeta F}\boldsymbol{\alpha}\right)\Delta\lambda\right]}{\partial H} = \mathbf{O} \tag{20.59}$$

$$\frac{\partial \mathbf{Y}_{\alpha}}{\partial(\Delta\lambda)} = \frac{\partial\left[\boldsymbol{\alpha} - \boldsymbol{\alpha}_n - \left(c_k\hat{\mathbf{n}} - \frac{1}{\sqrt{3/2}\zeta F}\boldsymbol{\alpha}\right)\Delta\lambda\right]}{\partial(\Delta\lambda)} = -\left(c_k\hat{\mathbf{n}} - \frac{1}{\sqrt{3/2}\zeta F}\boldsymbol{\alpha}\right) \tag{20.60}$$

$$\frac{\partial Y_H}{\partial \boldsymbol{\sigma}} = \frac{\partial(H - H_{\text{n}} - \sqrt{2/3}\Delta\lambda)}{\partial \boldsymbol{\sigma}} = \mathbf{O} \tag{20.61}$$

$$\frac{\partial Y_H}{\partial \boldsymbol{\alpha}} = \frac{\partial(H - H_{\text{n}} - \sqrt{2/3}\Delta\lambda)}{\partial \boldsymbol{\alpha}} = \mathbf{O} \tag{20.62}$$

$$\frac{\partial Y_H}{\partial H} = \frac{\partial(H - H_{\text{n}} - \sqrt{2/3}\Delta\lambda)}{\partial H} = 1 \tag{20.63}$$

$$\frac{\partial Y_H}{\partial \Delta\lambda} = \frac{\partial(H - H_{\text{n}} - \sqrt{2/3}\Delta\lambda)}{\partial \Delta\lambda} = -\sqrt{2/3} \tag{20.64}$$

$$\frac{\partial Y_s}{\partial \boldsymbol{\sigma}} = \frac{\partial(\sqrt{3/2}||\hat{\boldsymbol{\sigma}}'|| - RF)}{\partial \boldsymbol{\sigma}} = \sqrt{\frac{3}{2}}\frac{\partial||\hat{\boldsymbol{\sigma}}'||}{\partial \hat{\boldsymbol{\sigma}}} : \frac{\partial \hat{\boldsymbol{\sigma}}}{\partial \boldsymbol{\sigma}} = \sqrt{\frac{3}{2}}\frac{\hat{\boldsymbol{\sigma}}'}{||\hat{\boldsymbol{\sigma}}'||} : \boldsymbol{\mathcal{S}} = \sqrt{\frac{3}{2}}\hat{\mathbf{n}} : \boldsymbol{\mathcal{S}} \tag{20.65}$$

$$\frac{\partial Y_s}{\partial \boldsymbol{\alpha}} = \frac{\partial(\sqrt{3/2}||\hat{\boldsymbol{\sigma}}'|| - RF)}{\partial \hat{\boldsymbol{\sigma}}} : \frac{\partial \hat{\boldsymbol{\sigma}}}{\partial \boldsymbol{\alpha}} = \sqrt{\frac{3}{2}} \frac{\partial ||\hat{\boldsymbol{\sigma}}'||}{\partial \hat{\boldsymbol{\sigma}}} : \frac{\partial \hat{\boldsymbol{\sigma}}}{\partial \boldsymbol{\alpha}} = -\sqrt{\frac{3}{2}} \hat{\mathbf{n}} : \mathcal{S} \tag{20.66}$$

$$\frac{\partial Y_s}{\partial H} = \frac{\partial(\sqrt{3/2}||\hat{\boldsymbol{\sigma}}'|| - RF)}{\partial H} = -RF' \tag{20.67}$$

$$\frac{\partial Y_s}{\partial \Delta\lambda} = \frac{\partial(\sqrt{3/2}||\hat{\boldsymbol{\sigma}}'|| - RF)}{\partial \Delta\lambda} = -F \frac{\partial R}{\partial \Delta\lambda} \tag{20.68}$$

$$\frac{\partial R}{\partial \Delta\lambda} = u \frac{\cos\left(\dfrac{\pi}{2} \dfrac{\langle R_{\mathrm{n}} - R_e \rangle}{1 - R_e}\right)}{\sin\left(\dfrac{\pi}{2} \dfrac{\langle R - R_e \rangle}{1 - R_e}\right)} \exp\left(-u \frac{\pi}{2} \frac{\Delta\lambda}{1 - R_e}\right) \tag{20.69}$$

noting $R_n = R_0$ and $\Delta\lambda = \varepsilon^p - \varepsilon_0^p$ in Eq. (7.19). Here, $\partial \hat{\mathbf{n}} / \partial \hat{\boldsymbol{\sigma}}$ is given as follows:

$$\frac{\partial \hat{\mathbf{n}}}{\partial \hat{\boldsymbol{\sigma}}} = \frac{\mathcal{I}'||\hat{\boldsymbol{\sigma}}'|| - \hat{\boldsymbol{\sigma}}' \otimes \frac{\hat{\boldsymbol{\sigma}}'}{||\hat{\boldsymbol{\sigma}}'||}}{||\hat{\boldsymbol{\sigma}}'||^2} = \frac{1}{||\hat{\boldsymbol{\sigma}}'||}\left(\mathcal{I}' - \frac{\hat{\boldsymbol{\sigma}}'}{||\hat{\boldsymbol{\sigma}}'||} \otimes \frac{\hat{\boldsymbol{\sigma}}'}{||\hat{\boldsymbol{\sigma}}'||}\right)$$

$$\frac{\partial \hat{n}_{ij}}{\partial \hat{\sigma}_{kl}} = \frac{1}{||\hat{\boldsymbol{\sigma}}'||}\left(\mathcal{I}'_{ijkl} - \frac{\hat{\sigma}'_{ij}}{||\hat{\boldsymbol{\sigma}}'||} \frac{\hat{\sigma}'_{kl}}{||\hat{\boldsymbol{\sigma}}'||}\right) \tag{20.70}$$

The iteration calculation by Eqs. (20.50) and (20.52) is continued until the residual becomes sufficiently small: that is, it is judged that the convergence of solution is attained when the following equation in terms of the residual norm is fulfilled.

$$\frac{||\boldsymbol{\sigma}_{\mathrm{n}+1}|| - ||\boldsymbol{\sigma}_{\mathrm{n}}||}{||\boldsymbol{\sigma}_{\mathrm{n}}||} < \mathrm{TOL} \quad \text{or} \quad \frac{||\boldsymbol{\sigma}_{\mathrm{n}+1}|| - ||\boldsymbol{\sigma}_{\mathrm{n}}||}{F} < \mathrm{TOL} \tag{20.71}$$

The above-mentioned return-mapping formulation for the back stress is based on the hypoelasticity. A more efficient return-mapping can be executed by calculating the back stress $\boldsymbol{\alpha}$ exploiting Eq. (6.109) in which the stored part $\boldsymbol{\varepsilon}_{ks}^{p\prime}$ is obtained by subtracting the dissipative part $\boldsymbol{\varepsilon}_{kd}^{p\prime}$ from the deviatoric plastic strain $\boldsymbol{\varepsilon}^{p\prime}$, while the elastic-core can be calculated efficiently exploiting Eq. (9.23) (or Eq. (9.25) for the simplest case).

20.3.2 Single Equation for Plastic Multiplier

The return mapping with a high generality was explained in the last section, in which simultaneous equation for yield condition and evolution rules are solved for plural unknown variables. The closest point projection for the particular simple case in which relevant equations are reduced into a single equation containing only one unknown variable of the plastic multiplier will be described in this subsection.

(1) **Single equation with Hessian matrix**

The return-mapping formulation in this type is explained for the isotropic hardening material.

The residual of the plastic strain rate in the k-th iteration is given by

$$\mathbf{R}_{n+1}^{(k)} = -(\boldsymbol{\varepsilon}_{n+1}^{p(k)} - \boldsymbol{\varepsilon}_{n}^{p}) + \Delta\lambda_{n+1}^{(k)}\mathbf{n}_{n+1}^{(k)} \tag{20.72}$$

The calculation is proceeded until the residual is reduced less than a prescribed tolerance. $\Delta\lambda_{n+1}^{(k)}$ designates the accumulation of plastic multiplier induced during the n + 1 step. The following relation holds in the calculation in each time, noting Eq. (20.72).

$$\begin{aligned}\mathbf{R}_{n+1}^{(k+1)} &= \mathbf{R}_{n+1}^{(k)} + d\mathbf{R}_{n+1}^{(k)} \\ &= \mathbf{R}_{n+1}^{(k)} + d[-(\boldsymbol{\varepsilon}_{n+1}^{p(k)} - \boldsymbol{\varepsilon}_{n}^{p}) + \Delta\lambda_{n+1}^{(k)}\mathbf{n}_{n+1}^{(k)}] \\ &= \mathbf{R}_{n+1}^{(k)} - d\boldsymbol{\varepsilon}_{n+1}^{p(k)} + d(\Delta\lambda_{n+1}^{(k)}\mathbf{n}_{n+1}^{(k)}) \\ &= \mathbf{R}_{n+1}^{(k)} - d\boldsymbol{\varepsilon}_{n+1}^{p(k)} + d\Delta\lambda_{n+1}^{(k)}\mathbf{n}_{n+1}^{(k)} + \Delta\lambda_{n+1}^{(k)}\frac{\partial\mathbf{n}_{n+1}^{(k)}}{\partial\boldsymbol{\sigma}_{n+1}^{(k)}} : d\boldsymbol{\sigma}_{n+1}^{(k)}\end{aligned} \tag{20.73}$$

Here, note that $\Delta(\ )$ and $d(\ )$ designate the increment during the step n + 1 and in the time k, respectively. Because of

$$\boldsymbol{\sigma}_{n+1}^{(k)} = \mathbf{E} : \boldsymbol{\varepsilon}_{n+1}^{e(k)} = \mathbf{E} : (\boldsymbol{\varepsilon}_{n+1} - \boldsymbol{\varepsilon}_{n+1}^{p(k)})$$

it follows that

$$d\boldsymbol{\varepsilon}_{n+1}^{p(k)} = -\mathbf{E}^{-1} : d\boldsymbol{\sigma}_{n+1}^{(k)} \tag{20.74}$$

Substituting Eq. (20.74) into Eq. (20.73), we have

$$\mathbf{R}_{n+1}^{(k+1)} = \mathbf{R}_{n+1}^{(k)} + \left(\mathbf{E}^{-1} + \Delta\lambda_{n+1}^{(k)}\frac{\partial\mathbf{n}_{n+1}^{(k)}}{\partial\boldsymbol{\sigma}_{n+1}^{(k)}}\right) : d\boldsymbol{\sigma}_{n+1}^{(k)} + d\Delta\lambda_{n+1}^{(k)}\mathbf{n}_{n+1}^{(k)}$$

i.e.

$$\mathbf{R}_{n+1}^{(k+1)} = \mathbf{R}_{n+1}^{(k)} + \boldsymbol{\Xi}_{n+1}^{(k-1)} : d\boldsymbol{\sigma}_{n+1}^{(k)} + d\Delta\lambda_{n+1}^{(k)}\mathbf{n}_{n+1}^{(k)} \tag{20.75}$$

where

$$\boxed{\boldsymbol{\Xi}_{\mathrm{n+1}}^{(\mathrm{k})} \equiv \left(\mathbf{E}^{-1} + \Delta\lambda_{\mathrm{n+1}}^{(\mathrm{k})} \frac{\partial \mathbf{n}_{\mathrm{n+1}}^{(\mathrm{k})}}{\partial \boldsymbol{\sigma}_{\mathrm{n+1}}^{(\mathrm{k})}}\right)^{-1}} \tag{20.76}$$

is referred to as the *Hessian matrix* which contains the second-order partial derivative of yield function. Here, the inverse operation is applied in Eq. (20.76) in order that the Hessian matrix has the dimension of stress like the elastic modulus tensor.

Setting the quantity in Eq. (20.75) to be zero, one has

$$\mathbf{R}_{\mathrm{n+1}}^{(\mathrm{k})} + \boldsymbol{\Xi}_{\mathrm{n+1}}^{(\mathrm{k-1})} : d\boldsymbol{\sigma}_{\mathrm{n+1}}^{(\mathrm{k})} + d\Delta\lambda_{\mathrm{n+1}}^{(\mathrm{k})} \mathbf{n}_{\mathrm{n+1}}^{(\mathrm{k})} = \mathbf{O} \tag{20.77}$$

from which it follows that

$$\mathbf{n}_{\mathrm{n+1}}^{(\mathrm{k})} : \boldsymbol{\Xi}_{\mathrm{n+1}}^{(\mathrm{k})} : \mathbf{R}_{\mathrm{n+1}}^{(\mathrm{k})} + \mathbf{n}_{\mathrm{n+1}}^{(\mathrm{k})} : d\boldsymbol{\sigma}_{\mathrm{n+1}}^{(\mathrm{k})} + d\Delta\lambda_{\mathrm{n+1}}^{(\mathrm{k})} \mathbf{n}_{\mathrm{n+1}}^{(\mathrm{k})} : \boldsymbol{\Xi}_{\mathrm{n+1}}^{(\mathrm{k})} : \mathbf{n}_{\mathrm{n+1}}^{(\mathrm{k})} = 0 \tag{20.78}$$

Now, introduce the following function for the yield condition in Eq. (6.30).

$$g(\boldsymbol{\sigma}, F(H)) \equiv f(\boldsymbol{\sigma}) - F(H) \tag{20.79}$$

Applying the Taylor expansion to Eq. (20.79) and taking the first and the second terms, one has

$$\begin{aligned} g(\boldsymbol{\sigma} + d\boldsymbol{\sigma}, F(H) + dF(H)) &= g(\boldsymbol{\sigma}, F(H)) + \frac{\partial g(\boldsymbol{\sigma}, F(H))}{\partial \boldsymbol{\sigma}} : d\boldsymbol{\sigma} + \frac{\partial g(\boldsymbol{\sigma}, F(H))}{\partial F(H)} dF(H) \\ &= f(\boldsymbol{\sigma}) - F(H) + \frac{\partial f(\boldsymbol{\sigma})}{\partial \boldsymbol{\sigma}} : d\boldsymbol{\sigma} - dF(H) \end{aligned} \tag{20.80}$$

If the yield condition is satisfied already in the elastoplastic process, i.e. $f(\boldsymbol{\sigma}) - F(H) = 0$, it follows from the transformation of Eqs. (6.35)–(6.40) that

$$\frac{\partial f(\boldsymbol{\sigma})}{\partial \boldsymbol{\sigma}} : d\boldsymbol{\sigma} - dF(H) = \left\| \frac{\partial f(\boldsymbol{\sigma})}{\partial \boldsymbol{\sigma}} \right\| (\mathbf{n} : d\boldsymbol{\sigma} - d\Delta\lambda M^p) \tag{20.81}$$

Substituting Eq. (20.81) into Eq. (20.80) and setting it to be zero, one has

$$f(\boldsymbol{\sigma}) - F(H) + \left\| \frac{\partial f(\boldsymbol{\sigma})}{\partial \boldsymbol{\sigma}} \right\| (\mathbf{n} : d\boldsymbol{\sigma} - d\Delta\lambda M^p) = 0 \tag{20.82}$$

from which one has

$$\mathbf{n} : d\boldsymbol{\sigma} = -\frac{f(\hat{\boldsymbol{\sigma}}) - F(H)}{\left\| \frac{\partial f(\hat{\boldsymbol{\sigma}})}{\partial \boldsymbol{\sigma}} \right\|} + d\Delta\lambda M^p \tag{20.83}$$

The substitution of Eq. (20.83) into Eq. (20.78) leads to

$$\mathbf{n}_{n+1}^{(k)} : \boldsymbol{\Xi}_{n+1}^{(k)} : \mathbf{R}_{n+1}^{(k)} - \frac{f(\boldsymbol{\sigma}_{n+1}^{(k)}) - F(H_{n+1}^{(k)})}{\left\| \frac{\partial f(\boldsymbol{\sigma}_{n+1}^{(k)})}{\partial \boldsymbol{\sigma}_{n+1}^{(k)}} \right\|} + d\Delta\lambda_{n+1}^{(k)} M_{n+1}^{p(k)} + d\Delta\lambda_{n+1}^{(k)} \mathbf{n}_{n+1}^{(k)} : \boldsymbol{\Xi}_{n+1}^{(k)} : \mathbf{n}_{n+1}^{(k)} = 0$$

noting that $d\boldsymbol{\sigma}_{n+1}^{(k)}$ is the stress increment in the k-time, from which $d\Delta\lambda_{n+1}^{(k)}$ is given as follows:

$$\boxed{d\Delta\lambda_{n+1}^{(k)} = \frac{(f(\hat{\boldsymbol{\sigma}}_{n+1}^{(k)}) - F(H_{n+1}^{(k)}))/\left\| \frac{\partial f(\hat{\boldsymbol{\sigma}}_{n+1}^{(k)})}{\partial \boldsymbol{\sigma}_{n+1}^{(k)}} \right\| - \mathbf{n}_{n+1}^{(k)} : \boldsymbol{\Xi}_{n+1}^{(k)} : \mathbf{R}_{n+1}^{(k)}}{M_{n+1}^{p(k)} + \mathbf{n}_{n+1}^{(k)} : \boldsymbol{\Xi}_{n+1}^{(k)} : \mathbf{n}_{n+1}^{(k)}}} \tag{20.84}$$

Substituting Eqs. (20.72) and (20.76) into Eq. (20.84), one has

$$d\Delta\lambda_{n+1}^{(k)} = \frac{1}{M_{n+1}^{p(k)} + \mathbf{n}_{n+1}^{(k)} : \left(\mathbf{E}^{-1} + \Delta\lambda_{n+1}^{(k)} \frac{\partial \mathbf{n}_{n+1}^{(k)}}{\partial \boldsymbol{\sigma}_{n+1}^{(k)}} \right)^{-1} : \mathbf{n}_{n+1}^{(k)}} \left[(f(\hat{\boldsymbol{\sigma}}_{n+1}^{(k)}) - F(H_{n+1}^{(k)}))/\left\| \frac{\partial f(\hat{\boldsymbol{\sigma}}_{n+1}^{(k)})}{\partial \boldsymbol{\sigma}_{n+1}^{(k)}} \right\| - \mathbf{n}_{n+1}^{(k)} : \left(\mathbf{E}^{-1} + \Delta\lambda_{n+1}^{(k)} \frac{\partial \mathbf{n}_{n+1}^{(k)}}{\partial \boldsymbol{\sigma}_{n+1}^{(k)}} \right)^{-1} : [-(\boldsymbol{\varepsilon}_{n+1}^{p(k)} - \boldsymbol{\varepsilon}_n^p) + \Delta\lambda_{n+1}^{(k)} \mathbf{n}_{n+1}^{(k)}] \right] \tag{20.85}$$

which can be calculated by the known variables.

The stress in the k time calculation is updated from Eq. (20.77) as

$$d\boldsymbol{\sigma}_{n+1}^{(k)} = -\boldsymbol{\Xi}_{n+1}^{(k)} : (\mathbf{R}_{n+1}^{(k)} + d\Delta\lambda_{n+1}^{(k)} \mathbf{n}_{n+1}^{(k)}) \tag{20.86}$$

by which the plastic strain increment is given by

$$d\boldsymbol{\varepsilon}_{n+1}^{p(k)} = -\mathbf{E}^{-1} : d\boldsymbol{\sigma}_{n+1}^{(k)} = \mathbf{E}^{-1} : \boldsymbol{\Xi}_{n+1}^{(k)} : (\mathbf{R}_{n+1}^{(k)} + d\Delta\lambda_{n+1}^{(k)} \mathbf{n}_{n+1}^{(k)}) \tag{20.87}$$

noting Eq. (20.74). Then, the plastic strain, the stress and the isotropic hardening variable are updated as follows:

$$\boldsymbol{\varepsilon}^{p(\mathrm{k}+1)}_{\mathrm{n}+1} = \boldsymbol{\varepsilon}^{p(\mathrm{k})}_{\mathrm{n}+1} + \mathbf{E}^{-1} : \boldsymbol{\Xi}^{(\mathrm{k})}_{\mathrm{n}+1} : (\mathbf{R}^{(\mathrm{k})}_{\mathrm{n}+1} + d\Delta\lambda^{(\mathrm{k})}_{\mathrm{n}+1}\mathbf{n}^{(\mathrm{k})}_{\mathrm{n}+1}) \tag{20.88}$$

$$\boldsymbol{\sigma}^{(\mathrm{k}+1)}_{\mathrm{n}+1} = \mathbf{E} : (\boldsymbol{\varepsilon}_{\mathrm{n}+1} - \boldsymbol{\varepsilon}^{p(\mathrm{k}+1)}_{\mathrm{n}+1}) \tag{20.89}$$

$$H^{(\mathrm{k}+1)}_{\mathrm{n}+1} = H^{(\mathrm{k})}_{\mathrm{n}+1} + \sqrt{2/3}||d\Delta\lambda^{(\mathrm{k})}_{\mathrm{n}+1}|| \tag{20.90}$$

These iteration calculation has to be repeated until $\mathbf{R}^{(\mathrm{k})}_{\mathrm{n}+1}$ will converge within a given tolerance.

(2) **Isotropic Mises material without Hessian matrix**

The particularly simple case was considered in the foregoing, in which equations for yield condition and evolution rules for internal variables can be reduced into a single equation. However, it also requires the second-order partial derivative of yield function since it contains the Hessian matrix, which requires usually the cumbersome mathematical manipulation. Here, a simpler example which is not concerned with the Hessian Matrix will be described for von Mises model only with the isotropic hardening, referring to de Souza Neto et al. (2008).

First, the following relations hold for the isotropic Mises yield condition by the input of deviatoric elastic strain increment, i.e. $\mathrm{tr}\Delta\boldsymbol{\varepsilon} = 0$ in the trial step.

$$\boldsymbol{\varepsilon}'^{\,e\mathrm{trial}}_{\mathrm{n}+1} - \boldsymbol{\varepsilon}'^{\,e}_{\mathrm{n}+1} = \Delta\boldsymbol{\varepsilon}'^{\,p}_{\mathrm{n}+1} = \Delta\lambda_{\mathrm{n}+1}\frac{\boldsymbol{\sigma}'_{\mathrm{n}+1}}{||\boldsymbol{\sigma}'_{\mathrm{n}+1}||} \tag{20.91}$$

$$H_{n+1} = H_n + \sqrt{\frac{2}{3}}\Delta\lambda_{\mathrm{n}+1} \tag{20.92}$$

$$\sigma^{eq\mathrm{trial}}_{\mathrm{n}+1} \equiv \sqrt{\frac{3}{2}}||\boldsymbol{\sigma}'^{\mathrm{trial}}_{\mathrm{n}+1}|| = \sqrt{6}G||\boldsymbol{\varepsilon}^{e'\mathrm{trial}}_{\mathrm{n}+1}|| \tag{20.93}$$

$$\boldsymbol{\sigma}_{\mathrm{n}+1} = \sigma_{m_{\mathrm{n}+1}}\mathbf{I} + \boldsymbol{\sigma}'_{\mathrm{n}+1} \tag{20.94}$$

with

$$\left.\begin{aligned} \sigma_{m_{\mathrm{n}+1}} &= \sigma^{\mathrm{trial}}_{m_{\mathrm{n}+1}} \\ \boldsymbol{\sigma}'_{\mathrm{n}+1} &= \mathbf{E} : \boldsymbol{\varepsilon}'^{\,e}_{\mathrm{n}+1} = \mathbf{E} : (\boldsymbol{\varepsilon}'^{\,e\mathrm{trial}}_{\mathrm{n}+1} - \Delta\boldsymbol{\varepsilon}'^{\,p}_{\mathrm{n}+1}) = \boldsymbol{\sigma}'^{\,\mathrm{trial}}_{\mathrm{n}+1} - \mathbf{E} : \Delta\lambda_{\mathrm{n}+1}\frac{\boldsymbol{\sigma}'_{\mathrm{n}+1}}{||\boldsymbol{\sigma}'_{\mathrm{n}+1}||} \\ &= \boldsymbol{\sigma}'^{\,\mathrm{trial}}_{\mathrm{n}+1} - 2G\Delta\lambda_{\mathrm{n}+1}\frac{\boldsymbol{\sigma}'_{\mathrm{n}+1}}{||\boldsymbol{\sigma}'_{\mathrm{n}+1}||} \end{aligned}\right\} \tag{20.95}$$

where σ^{eq} is the equivalent stress defined in Eq. (6.55). It follows from Eq. $(20.95)_2$ that

$$\left(1 + \frac{2G\lambda_{n+1}}{\|\boldsymbol{\sigma}'_{n+1}\|}\right)\boldsymbol{\sigma}'_{n+1} = \boldsymbol{\sigma}'^{\,trial}_{n+1} \tag{20.96}$$

from which we have

$$\frac{\boldsymbol{\sigma}'_{n+1}}{\|\boldsymbol{\sigma}'_{n+1}\|} = \frac{\boldsymbol{\sigma}'^{\,trial}_{n+1}}{\|\boldsymbol{\sigma}'^{\,trial}_{n+1}\|} \tag{20.97}$$

and

$$\boldsymbol{\sigma}'_{n+1} = \boldsymbol{\sigma}'^{\,trial}_{n+1} - 2G\Delta\lambda_{n+1}\frac{\boldsymbol{\sigma}'^{\,trial}_{n+1}}{\|\boldsymbol{\sigma}'^{\,trial}_{n+1}\|} = \left(1 - \frac{2G\Delta\lambda_{n+1}}{\|\boldsymbol{\sigma}'^{\,trial}_{n+1}\|}\right)\boldsymbol{\sigma}'^{\,trial}_{n+1} \tag{20.98}$$

$$\sigma^{eq}_{n+1} = \sqrt{\frac{3}{2}}\|\boldsymbol{\sigma}'_{n+1}\| = \sqrt{\frac{3}{2}}\left(1 - \frac{2G\Delta\lambda_{n+1}}{\|\boldsymbol{\sigma}'^{\,trail}_{n+1}\|}\right)\|\boldsymbol{\sigma}'^{trail}_{n+1}\| = \sigma^{eqtrail}_{n+1} - \sqrt{6}G\Delta\lambda_{n+1} \tag{20.99}$$

Substituting Eqs. (20.92) and (20.99) into the yield condition

$$g(\sigma^{eq}_{n+1}, F(H_{n+1})) = \sigma^{eq}_{n+1} - F(H_{n+1}) = 0 \tag{20.100}$$

we have

$$g_{n+1} = \sigma^{eq\,trial}_{n+1} - \sqrt{6}G\Delta\lambda_{n+1} - F(H_n + \sqrt{2/3}\Delta\,\lambda_{n+1}) = 0 \tag{20.101}$$

The further substitution of Eq. (20.93) into Eq. (20.101) leads to

$$g_{n+1} = \sqrt{6}G\|\boldsymbol{\varepsilon}'^{e}_{n} + \Delta\boldsymbol{\varepsilon}'\| - \sqrt{6}G\Delta\lambda_{n+1} - F(H_n + \sqrt{2/3}\Delta\lambda_{n+1}) = 0 \tag{20.102}$$

Equation (20.101) or (20.102) is the nonlinear equation of $\Delta\lambda_{n+1}$, so that it has to be solved by numerical method such as Newton-Raphson method described later.

Substituting Eqs. $(20.95)_1$ and (20.98) into Eq. (20.94), the stress and the strain are updated by the following equations.

$$\boldsymbol{\sigma}_{n+1} = \sigma^{trial}_{m_{n+1}}\mathbf{I} + \left(1 - \frac{2G\Delta\lambda_{n+1}}{\|\boldsymbol{\sigma}'^{trial}_{n+1}\|}\right)\boldsymbol{\sigma}'^{trial}_{n+1} = \boldsymbol{\sigma}^{trial}_{n+1} - 2G\Delta\lambda_{n+1}\frac{\boldsymbol{\sigma}'^{trial}_{n+1}}{\|\boldsymbol{\sigma}'^{trial}_{n+1}\|} \tag{20.103}$$

$$\boldsymbol{\varepsilon}^{e}_{n+1} = \frac{1}{3}(\mathrm{tr}\boldsymbol{\varepsilon}^{etrial}_{n+1})\mathbf{I} + \frac{1}{2G}\boldsymbol{\sigma}'_{n+1} \tag{20.104}$$

$$\boldsymbol{\varepsilon}^{p}_{n+1} = \boldsymbol{\varepsilon}^{p}_{n} + \Delta\lambda_{n+1}\frac{\boldsymbol{\sigma}'_{n+1}}{||\boldsymbol{\sigma}'_{n+1}||} \tag{20.105}$$

(**Remark**) Newton-Raphson method for Eq. (20.102):

$$g_{n+1} = g_{n} + \frac{\partial g_{n}}{\partial \Delta\lambda_{n+1}} d\Delta\lambda_{n+1} = 0 \rightarrow d\Delta\lambda_{n+1} = -\frac{g_{n}}{\dfrac{\partial g_{n}}{\partial \Delta\lambda_{n+1}}} \tag{20.106}$$

where it follows from Eq. (20.101) that

$$\frac{\partial g_{n}}{\partial \Delta\lambda_{n+1}} = -\sqrt{6}G - \frac{dF_{n}}{dH_{n}}\frac{dH_{n}}{d\lambda_{n+1}} = -\sqrt{6}G - \sqrt{2/3}F'_{n} \tag{20.107}$$

Then, $\Delta\lambda$ can be updated by

$$\Delta\lambda_{n+1} = \Delta\lambda_{n} + d\Delta\lambda_{n} = \Delta\lambda_{n} + \frac{g_{n}}{-\dfrac{\partial g_{n}}{\partial \Delta\lambda_{n+1}}} = \Delta\lambda_{n} + \frac{g_{n}}{\sqrt{6}G + \sqrt{2/3}F'_{n}} \tag{20.108}$$

The update-calculation is performed until Eq. (20.102) converges within a given tolerance.

If the kinematic hardening is incorporated, however, Eq. (20.96) does not hold and thus the second-order partial derivative is necessary requiring the calculation with the treatment of Hessian matrix.

20.4 Cutting Plane Projection

The closest point projection described in the last section requires the preparation task of the calculation of the second-order partial derivative of yield function, although it is the fully implicit method with a high efficiency. Therefore, it would be difficult to be applied to materials with yield function and/or plastic potential function(s) which is not simple. The incomplete implicit method, called the *cutting plane projection*, was proposed, which does not require the second-order partial derivative of yield function (Simo and Ortiz 1985; Ortiz and Simo 1986). The plastic strain rate is induced in the normal direction to the yield surface on which the current stress lies in the cutting plane projection, while it is induced in normal direction to the yield surface on which the implicit stress will settle down in the closet point projection in Sect. 20.3.

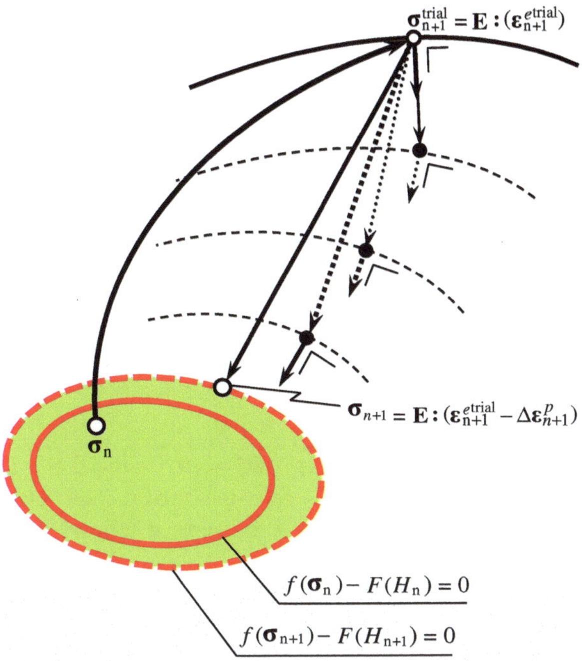

Fig. 20.5 Cutting-plane projection

The geometrical interpretation of the cutting plane projection is illustrated in Fig. 20.5. Here, the plastic relaxation stress rate $\dot{\boldsymbol{\sigma}}^p$ is induced to the inward-normal of the current dynamic-loading surface because of the explicit projection. The cutting plane projection for the extended subloading surface model is described below (Hashiguchi et al. 2014; Suzuki et al. 2014).

Introduce the following equation based on the subloading surface in Eq. (9.2):

$$g(\bar{\boldsymbol{\sigma}}, R, F) \equiv f(\bar{\boldsymbol{\sigma}}) - RF \tag{20.109}$$

Applying the Taylor expansion to Eq. (20.109) and taking the first and the second terms, one has

$$\begin{aligned} & g(\bar{\boldsymbol{\sigma}} + d\bar{\boldsymbol{\sigma}}, R + dR, F + dF) \\ & = g(\bar{\boldsymbol{\sigma}}, R, F) + \frac{\partial g(\bar{\boldsymbol{\sigma}}, R, F)}{\partial \bar{\boldsymbol{\sigma}}} : d\bar{\boldsymbol{\sigma}} + \frac{\partial g(\bar{\boldsymbol{\sigma}}, R, F)}{\partial R} dR + \frac{\partial g(\bar{\boldsymbol{\sigma}}, R, F)}{\partial F} dF \\ & = f(\bar{\boldsymbol{\sigma}}) - RF + \frac{\partial f(\bar{\boldsymbol{\sigma}})}{\partial \bar{\boldsymbol{\sigma}}} : d\boldsymbol{\sigma} - \frac{\partial f(\bar{\boldsymbol{\sigma}})}{\partial \bar{\boldsymbol{\sigma}}} : d\bar{\boldsymbol{\alpha}} - dRF - RdF \end{aligned} \tag{20.110}$$

If the subloading surface is satisfied already in the elastoplastic process, i.e. $f(\bar{\boldsymbol{\sigma}}) - RF = 0$, Eq. (20.110) leads to

$$\frac{\partial f(\bar{\boldsymbol{\sigma}})}{\partial \boldsymbol{\sigma}} : d\boldsymbol{\sigma} - \frac{\partial f(\bar{\boldsymbol{\sigma}})}{\partial \boldsymbol{\sigma}} : d\bar{\boldsymbol{\alpha}} - dRF - RdF = \left\| \frac{\partial f(\bar{\boldsymbol{\sigma}})}{\partial \bar{\boldsymbol{\sigma}}} \right\| (\bar{\mathbf{n}} : d\boldsymbol{\sigma} - d\bar{\lambda}\overline{M}^p) \quad (20.111)$$

by virtue of Eq. (9.8). The equation obtained by putting Eq. (20.111) to be zero is no more than the incremental form, i.e. the consistency condition of the subloading surface. Equation (20.111) becomes

$$\frac{\partial f(\bar{\boldsymbol{\sigma}})}{\partial \boldsymbol{\sigma}} : d\boldsymbol{\sigma} - \frac{\partial f(\bar{\boldsymbol{\sigma}})}{\partial \boldsymbol{\sigma}} : d\bar{\boldsymbol{\alpha}} - dRF - RdF = -\left\| \frac{\partial f(\bar{\boldsymbol{\sigma}})}{\partial \bar{\boldsymbol{\sigma}}} \right\| d\bar{\lambda}(\bar{\mathbf{n}} : \mathbf{E} : \bar{\mathbf{n}} + \overline{M}^p) \quad (20.112)$$

by substituting Eq. (20.12) regarding $d\boldsymbol{\sigma} = d\boldsymbol{\sigma}^p$ in the plastic corrector step. Further, substituting Eq. (20.112) into the incremental term in the Taylor expansion in Eq. (20.110) and putting it to be zero, it follows that

$$f(\bar{\boldsymbol{\sigma}}) - RF - \left\| \frac{\partial f(\bar{\boldsymbol{\sigma}})}{\partial \bar{\boldsymbol{\sigma}}} \right\| d\bar{\lambda}(\bar{\mathbf{n}} : \mathbf{E} : \bar{\mathbf{n}} + \overline{M}^p) = 0 \quad (20.113)$$

from which $d\bar{\lambda}$ is obtained as follows:

$$d\bar{\lambda} = \frac{f(\bar{\boldsymbol{\sigma}}) - RF}{\left\| \frac{\partial f(\bar{\boldsymbol{\sigma}})}{\partial \bar{\boldsymbol{\sigma}}} \right\| (\overline{M}^p + \bar{\mathbf{n}} : \mathbf{E} : \bar{\mathbf{n}})} \quad (20.114)$$

Equation (20.114) shows the plastic strain increment which has to be input when the excessive stress $f(\bar{\boldsymbol{\sigma}}) - RF$ remains, so that the stress is reduced onto the subloading surface.

Taking account of Eq. (20.114), the plastic strain increment in the k + 1 time is calculated by the stress and the internal variables in the k time as follows:

$$\boxed{d\Delta\bar{\lambda}_{n+1}^{(k+1)} = \frac{f(\bar{\boldsymbol{\sigma}}_{n+1}^{(k)}) - R_{n+1}^{(k)} F_{n+1}^{(k)}}{\left\| \frac{\partial f(\bar{\boldsymbol{\sigma}}_{n+1}^{(k)})}{\partial \bar{\boldsymbol{\sigma}}_{n+1}^{(k)}} \right\| (\overline{M}_{n+1}^{p(k)} + \bar{\mathbf{n}}_{n+1}^{(k)} : \mathbf{E} : \bar{\mathbf{n}}_{n+1}^{(k)})}} \quad (20.115)$$

$$d\Delta\boldsymbol{\varepsilon}_{n+1}^{p(k)} = d\Delta\bar{\lambda}_{n+1}^{(k+1)} \bar{\mathbf{n}}_{n+1}^{(k)} \quad (20.116)$$

The subloading surface and the plastic multiplier are given for Mises yield condition as follows:

$$\sqrt{\frac{3}{2}}||\bar{\boldsymbol{\sigma}}'^{(k+1)}_{n+1}|| - R^{(k+1)}_{n+1}F(H^{(k+1)}_{n+1}) = 0 \tag{20.117}$$

$$d\Delta\bar{\lambda}^{(k+1)}_{n+1} = \frac{||\bar{\boldsymbol{\sigma}}'^{(k+1)}_{n+1}|| - \sqrt{2/3}R^{(k)}_{n+1}F^{(k)}_{n+1}}{\overline{M}^{p(k)}_{n+1} + 2G} \tag{20.118}$$

$$\bar{\mathbf{n}}^{(k)}_{n+1} = \frac{\bar{\boldsymbol{\sigma}}'^{(k)}_{n+1}}{||\bar{\boldsymbol{\sigma}}'^{(k)}_{n+1}||} \tag{20.119}$$

The plastic modulus $\overline{M}^{p(k)}_{n+1}$ is given from Eq. (9.39) as

$$\begin{aligned}\bar{M}^{p(k)}_{n+1} \equiv \bar{\mathbf{n}}^{(k)}_{n+1} : \Bigg[&\sqrt{\frac{2}{3}}\frac{F'^{(k)}_{n+1}}{F^{(k)}_{n+1}} f^{(k)}_{Hn_{n+1}} \bar{\boldsymbol{\sigma}}^{(k)}_{n+1} + R^{(k)}_{n+1}\mathbf{f}^{(k)}_{kn_{n+1}} + \frac{U^{(k)}_{n+1}}{R^{(k)}_{n+1}}\tilde{\boldsymbol{\sigma}}^{(k)}_{n+1} \\ &+ c(1 - R^{(k)}_{n+1})\left(\bar{\mathbf{n}}^{(k)}_{n+1} - \frac{\Re^{(k)}_{c_{n+}}}{\xi}\hat{\mathbf{n}}^{(k)}_{c_{n+1}}\right)\Bigg]\end{aligned} \tag{20.120}$$

$$F^{(k)}_{n+1} = F_0\{1 + h_1[1 - \exp(-h_2 H^{(k)}_{n+1})]\} \tag{20.121}$$

$$F'^{(k)}_{n+1} = F_0\, h_1\, h_2 \exp(-h_2 H^{(k)}_{n+1}) \tag{20.122}$$

$$f^{(k)}_{Hn\,n+1} = \sqrt{\frac{3}{2}} \tag{20.123}$$

$$\mathbf{f}^{(k)}_{kn\,n+1} = c_k\left(\bar{\mathbf{n}}^{(k)}_{n+1} - \frac{1}{\sqrt{3/2}\zeta F^{(k)}_{n+1}}\boldsymbol{\alpha}^{(k)}_{n+1}\right) \tag{20.124}$$

$$\tilde{\boldsymbol{\sigma}}^{(k)}_{n+1} = \boldsymbol{\sigma}^{(k)}_{n+1} - \mathbf{c}^{(k)}_{n+1}, \quad \hat{\mathbf{c}}^{(k)}_{n+1} = \mathbf{c}^{(k)}_{n+1} - \boldsymbol{\alpha}^{(k)}_{n+1} \tag{20.125}$$

$$\bar{\boldsymbol{\alpha}}^{(k)}_{n+1} = \mathbf{c}^{(k)}_{n+1} - R^{(k)}_{n+1}\hat{\mathbf{c}}^{(k)}_{n+1}, \quad \bar{\boldsymbol{\sigma}}^{(k)}_{n+1} = \boldsymbol{\sigma}^{(k)}_{n+1} - \bar{\boldsymbol{\alpha}}^{(k)}_{n+1} \tag{20.126}$$

$$u^{(k)}_{n+1} = \bar{u}\exp(u_s \Re^{(k)}_{c_{n+1}}\mathbf{n}^{(k)}_{c_{n+1}} : \mathbf{n}^{(k)}_{n+1}) \tag{20.127}$$

$$U^{(k)}_{n+1} = u^{(k)}_{n+1}\cot\left(\frac{\pi}{2}\frac{\langle R^{(k)}_{n+1} - R_e\rangle}{1 - R_e}\right) \tag{20.128}$$

$$\Re_{c_{n+1}}^{(k)} = \frac{\sqrt{3/2}\|\hat{\mathbf{c}}_{n+1}^{(k)}\|}{F(H_{n+1}^{(k)})} \tag{20.129}$$

$$\hat{\mathbf{n}}_{c_{n+1}}^{(k)} = \frac{\hat{\mathbf{c}}_{n+1}^{(k)}}{\|\hat{\mathbf{c}}_{n+1}^{(k)}\|} \tag{20.130}$$

The variables are updated as follows:

$$\boldsymbol{\varepsilon}_{n+1}^{p(k+1)} = \boldsymbol{\varepsilon}_{n+1}^{p(k)} + d\Delta\boldsymbol{\varepsilon}_{n+1}^{p(k+1)} \tag{20.131}$$

$$\boldsymbol{\sigma}_{n+1}^{(k+1)} = \mathbf{E} : (\boldsymbol{\varepsilon}_{n+1} - \boldsymbol{\varepsilon}_{n+1}^{p(k+1)}) \tag{20.132}$$

$$H_{n+1}^{(k+1)} = H_{n+1}^{(k)} + \sqrt{\frac{2}{3}}\|d\Delta\boldsymbol{\varepsilon}_{n+1}^{p(k+1)}\| \tag{20.133}$$

$$F_{n+1}^{(k+1)} = F_0\{1 + h_1[1 - \exp(-h_2 H_{n+1}^{(k+1)})]\} \tag{20.134}$$

$$F_{n+1}^{\prime(k+1)} = F_0\, h_1\, h_2 \exp(-h_2 H_{n+1}^{(k+1)}) \tag{20.135}$$

$$\boldsymbol{\alpha}_{n+1}^{(k+1)} = \boldsymbol{\alpha}_{n+1}^{(k)} + \mathbf{f}_{kn\,n+1}^{(k)}\|d\Delta\boldsymbol{\varepsilon}_{n+1}^{p(k+1)}\| \tag{20.136}$$

$$\mathbf{c}_{n+1}^{(k+1)} = \mathbf{c}_{n+1}^{(k)} + c\left(\bar{\mathbf{n}}_{n+1}^{(k)} - \frac{\Re_{c_{n+1}}^{(k)}}{\xi}\hat{\mathbf{n}}_{c_{n+1}}^{(k)}\right)\|d\Delta\boldsymbol{\varepsilon}_{n+1}^{p(k+1)}\| \tag{20.137}$$

$$R_{n+1}^{(k+1)} = R_{n+1}^{(k)} + U_{n+1}^{(k)} d\Delta\bar{\lambda}_{n+1}^{(k+1)} \tag{20.138}$$

$$R_{n+1}^{(k)} = \frac{\tilde{\boldsymbol{\sigma}}_{n+1}^{\prime(k)} : \hat{\mathbf{c}}_{n+1}^{\prime(k)} + \sqrt{(\tilde{\boldsymbol{\sigma}}_{n+1}^{\prime(k)} : \hat{\mathbf{c}}_{n+1}^{\prime(k)})^2 + (\frac{2}{3}F_{n+1}^{(k)2} - \|\hat{\mathbf{c}}_{n+1}^{\prime(k)}\|^2)\|\tilde{\boldsymbol{\sigma}}_{n+1}^{\prime(k)}\|^2}}{\frac{2}{3}F_{n+1}^{(k)2} - \|\hat{\mathbf{c}}_{n+1}^{\prime(k)}\|^2} \quad \text{for } d\Delta\bar{\lambda}_{n+1}^{(k+1)} = 0 \tag{20.139}$$

where the stress is updated by Eq. (20.89).

20.5 Consistent Tangent Modulus Tensor

The stress update calculation can be performed in a high efficiency by the return-mapping scheme described in the preceding sections.

1. However, the stress state obtained by the return-mapping scheme would not fulfill the equilibrium equation on the stress (rate) boundary and the equilibrium

equation inside the body in the finite elements in general because the strain is fixed in the plastic corrector step.

2. Therefore, the implicit tangent modulus tensor, called the *consistent* (or *algorithmic*) *tangent modulus tensor*, is calculated at the end of each plastic corrector step. It can be regarded as an algorithmic or a time-discrete counterpart of the continuum tangent modulus tensor, which is essential for efficient and numerically stable solution of nonlinear elastoplastic problems by an iterative algorithm with quadratic convergence (Simo and Taylor 1985),
3. Then the additional strain increment is calculated in order to exclude the residual force by using the global stiffness matrix based on the consistent tangent modulus tensors,
4. The return-mapping is performed again from the final state of the previous step n by inputting the strain increment supplemented the additional strain increment, where all the state variables, i.e. the stress and internal variables at the end of the previous step n are used at the beginning of the return-mapping,
5. The procedures described in 1–4 are repeated until the convergence less than a prescribed tolerance for the yield condition in each element and the equilibrium condition in the whole body.

This procedure is required for the nodal forces and displacement increments, using the global tangent stiffness modulus tensor based on the consistent tangent modulus tensor in FEM analysis. The analytical and numerical calculations of the consistent tangent modulus will be explained in this section.

20.5.1 Analytical Method

Firstly, analytical derivation of the consistent tangent modulus is described in this subsection, which has been formulated by Hughes and Pister (1978), Simo and Taylor (1985), etc.

(1) **Isotropic hardening material**

The analytical calculation of the consistent tangent modulus is explained below for the isotropic-kinematic hardening material assumed in 20.3.2 (1).

The differentiation of Eq. (20.37) with the substitution of Eq. (20.36) reads:

$$\begin{aligned} d\boldsymbol{\sigma}_{\mathrm{n}+1} &= \mathbf{E} : d\boldsymbol{\varepsilon}^{e}_{\mathrm{n}+1} = \mathbf{E} : (d\boldsymbol{\varepsilon}_{n+1} - d\Delta\boldsymbol{\varepsilon}^{p}_{\mathrm{n}+1}) \\ &= \mathbf{E} : \left(d\boldsymbol{\varepsilon}_{n+1} - \Delta\lambda_{\mathrm{n}+1} \frac{\partial \mathbf{n}_{\mathrm{n}+1}}{\partial \boldsymbol{\sigma}_{\mathrm{n}+1}} : d\boldsymbol{\sigma}_{\mathrm{n}+1} - d\Delta\lambda_{\mathrm{n}+1} \mathbf{n}_{\mathrm{n}+1} \right) \end{aligned} \tag{20.140}$$

Applying $\mathbf{E}^{-1}$ to Eq. (20.140), we have

$$\mathbf{E}^{-1} : d\boldsymbol{\sigma}_{n+1} + \Delta\lambda_{n+1} \frac{\partial \mathbf{n}_{n+1}}{\partial \boldsymbol{\sigma}_{n+1}} : d\boldsymbol{\sigma}_{n+1} = d\boldsymbol{\varepsilon}_{n+1} - d\Delta\lambda_{n+1} \mathbf{n}_{n+1} \tag{20.141}$$

i.e.

$$\left(\mathbf{E}^{-1} + \Delta\lambda_{n+1} \frac{\partial \mathbf{n}_{n+1}}{\partial \boldsymbol{\sigma}_{n+1}} \right) : d\boldsymbol{\sigma}_{n+1} = d\boldsymbol{\varepsilon}_{n+1} - d\Delta\lambda_{n+1} \mathbf{n}_{n+1} \tag{20.142}$$

which leads to

$$d\boldsymbol{\sigma}_{n+1} = \boldsymbol{\Xi}_{n+1} : (d\boldsymbol{\varepsilon}_{n+1} - d\Delta\lambda_{n+1} \mathbf{n}_{n+1}) \tag{20.143}$$

where $\boldsymbol{\Xi}$ is the Hessian matrix defined in Eq. (20.76) and given in this case as follows:

$$\boldsymbol{\Xi}_{n+1} = \left(\mathbf{E}^{-1} + \Delta\lambda_{n+1} \frac{\partial \mathbf{n}_{n+1}}{\partial \boldsymbol{\sigma}_{n+1}} \right)^{-1} \tag{20.144}$$

Hereafter, assuming the isotropic-kinematic hardening material, the consistency condition is given from Eq. (6.40) by

$$\mathbf{n}_{n+1} : d\boldsymbol{\sigma}_{n+1} - d\Delta\lambda_{n+1} M^p_{n+1} = 0 \tag{20.145}$$

Substituting Eq. (20.143) into Eq. (20.145), we have

$$\mathbf{n}_{n+1} : \boldsymbol{\Xi}_{n+1} : (d\boldsymbol{\varepsilon}_{n+1} - d\Delta\lambda_{n+1} \mathbf{n}_{n+1}) - d\Delta\lambda_{n+1} M^p_{n+1} = 0 \tag{20.146}$$

leading to

$$d\Delta\lambda_{n+1} = \frac{\mathbf{n}_{n+1} : \boldsymbol{\Xi}_{n+1} : d\boldsymbol{\varepsilon}_{n+1}}{M^p_{n+1} + \mathbf{n}_{n+1} : \boldsymbol{\Xi}_{n+1} : \mathbf{n}_{n+1}} \tag{20.147}$$

The substitution of Eq. (20.147) into Eq. (20.143) reads:

$$d\boldsymbol{\sigma}_{n+1} = \left(\boldsymbol{\Xi}_{n+1} - \frac{\boldsymbol{\Xi}_{n+1} : \mathbf{n}_{n+1} \otimes \mathbf{n}_{n+1} : \boldsymbol{\Xi}_{n+1}}{M^p_{n+1} + \mathbf{n}_{n+1} : \boldsymbol{\Xi}_{n+1} : \mathbf{n}_{n+1}} \right) : d\boldsymbol{\varepsilon}_{n+1} \tag{20.148}$$

from which the consistent tangent modulus tensor $\mathbf{K}^{ep,\mathrm{algo}}_{n+1}$ is given by

$$\boxed{\mathbf{K}^{ep,\mathrm{algo}}_{n+1} \equiv \frac{\partial \boldsymbol{\sigma}_{n+1}}{\partial \boldsymbol{\varepsilon}_{n+1}} = \boldsymbol{\Xi}_{n+1} - \frac{\boldsymbol{\Xi}_{n+1} : \mathbf{n}_{n+1} \otimes \mathbf{n}_{n+1} : \boldsymbol{\Xi}_{n+1}}{M^p_{n+1} + \mathbf{n}_{n+1} : \boldsymbol{\Xi}_{n+1} : \mathbf{n}_{n+1}}} \tag{20.149}$$

(2) **Isotropic Mises material without Hessian matrix**

The derivation of the simple consistent tangent modulus tensor without the Hessian matrix is shown for the isotropic Mises metal, referring to de Souza Neto et al. (2008).

It follows from Eq. (20.103) that

$$\boldsymbol{\sigma}_{n+1} = \mathbf{E} : \boldsymbol{\varepsilon}^{etrial}_{n+1} - 2G\Delta\lambda_{n+1}\frac{\mathbf{E} : \boldsymbol{\varepsilon}^{e\prime\,trial}_{n+1}}{\|\boldsymbol{\sigma}^{\prime\,trial}_{n+1}\|} = \left(\mathbf{E} - \frac{4G^2\Delta\lambda_{n+1}}{\|\boldsymbol{\sigma}^{\prime trial}_{n+1}\|}\mathcal{I}'\right) : \boldsymbol{\varepsilon}^{etrial}_{n+1} \tag{20.150}$$

On the other hand, one has

$$\begin{aligned}\sigma^{eqtrial}_{n+1} &= \sqrt{3/2}\|\boldsymbol{\sigma}^{\prime\,trial}_{n+1}\| = \sqrt{3/2}\|\mathbf{E} : \boldsymbol{\varepsilon}^{e\prime\,trial}_{n+1}\| = 2G\sqrt{3/2}\|\boldsymbol{\varepsilon}^{e\prime\,trial}_{n+1}\| \\ &= 2G\sqrt{3/2}\|\mathcal{I}'\boldsymbol{\varepsilon}^{etrial}_{n+1}\|\end{aligned} \tag{20.151}$$

The consistent tangent modulus is given by

$$\begin{aligned}\frac{\partial\boldsymbol{\sigma}_{n+1}(\boldsymbol{\varepsilon}^{e}_{n+1})}{\partial\boldsymbol{\varepsilon}_{n+1}} &= \frac{\partial\boldsymbol{\sigma}_{n+1}(\boldsymbol{\varepsilon}_{n+1} - \boldsymbol{\varepsilon}^{p}_{n+1})}{\partial\boldsymbol{\varepsilon}^{etrial}_{n+1}} : \frac{\partial\boldsymbol{\varepsilon}^{etrial}_{n+1}}{\partial\boldsymbol{\varepsilon}_{n+1}} \\ &= \frac{\partial\boldsymbol{\sigma}_{n+1}(\boldsymbol{\varepsilon}^{etrial}_{n+1} + \boldsymbol{\varepsilon}^{p}_{n} - \boldsymbol{\varepsilon}^{p}_{n+1})}{\partial\boldsymbol{\varepsilon}^{etrial}_{n+1}} : \frac{\partial\boldsymbol{\varepsilon}^{etrial}_{n+1}}{\partial\boldsymbol{\varepsilon}_{n+1}} \\ &= \frac{\partial\boldsymbol{\sigma}_{n+1}(\boldsymbol{\varepsilon}^{etrial}_{n+1} - \Delta\lambda_{n+1}\boldsymbol{\sigma}^{\prime\,trial}_{n+1}/\|\boldsymbol{\sigma}^{\prime\,trial}_{n+1}\|)}{\partial\boldsymbol{\varepsilon}^{etrial}_{n+1}} : \frac{\partial(\boldsymbol{\varepsilon}_{n+1} - \boldsymbol{\varepsilon}^{p}_{n})}{\partial\boldsymbol{\varepsilon}_{n+1}} \\ &= \frac{\partial\boldsymbol{\sigma}_{n+1}(\boldsymbol{\varepsilon}^{etrial}_{n+1} - \Delta\lambda_{n+1}\boldsymbol{\sigma}^{\prime\,trial}_{n+1}/\|\boldsymbol{\sigma}^{\prime\,trial}_{n+1}\|)}{\partial\boldsymbol{\varepsilon}^{etrial}_{n+1}}\end{aligned} \tag{20.152}$$

Therefore, the consistent tangent modulus can be obtained by the partial derivative of $\boldsymbol{\sigma}_{n+1}$ by $\boldsymbol{\varepsilon}^{etrial}_{n+1}$ instead of $\boldsymbol{\varepsilon}_{n+1}$.

The partial derivative of Eq. (20.150) by $\boldsymbol{\varepsilon}^{etrial}_{n+1}$ leads to the following equation, noting that $\Delta\lambda_{n+1}$ also depends on $\boldsymbol{\varepsilon}^{etrial}_{n+1}$.

$$\begin{aligned}\frac{\partial\boldsymbol{\sigma}_{n+1}}{\partial\boldsymbol{\varepsilon}^{etrial}_{n+1}} &= \mathbf{E} - \frac{4G^2\Delta\lambda_{n+1}}{\|\boldsymbol{\sigma}^{\prime\,trial}_{n+1}\|}\mathcal{I}' - \frac{4G^2}{\|\boldsymbol{\sigma}^{\prime\,trial}_{n+1}\|}\boldsymbol{\varepsilon}^{e\prime\,trial}_{n+1} \otimes \frac{\partial\Delta\lambda_{n+1}}{\partial\boldsymbol{\varepsilon}^{etrial}_{n+1}} \\ &\quad + \frac{4G^2\Delta\lambda_{n+1}}{\|\boldsymbol{\sigma}^{\prime\,trial}_{n+1}\|^2}\boldsymbol{\varepsilon}^{e\prime\,trial}_{n+1} \otimes \frac{\partial\|\boldsymbol{\sigma}^{\prime\,trial}_{n+1}\|}{\partial\boldsymbol{\varepsilon}^{etrial}_{n+1}}\end{aligned} \tag{20.153}$$

where the following relation is used.

$$
\left[\frac{\partial}{\partial \boldsymbol{\varepsilon}^{etrial}_{n+1}}\left(\frac{4G^2\Delta\lambda_{n+1}}{||\boldsymbol{\sigma}'^{trial}_{n+1}||}\mathcal{I}':\boldsymbol{\varepsilon}^{etrial}_{n+1}\right)\right]_{ijkl}=\frac{\partial}{\partial \varepsilon^{etrial}_{n+1,kl}}\left(\frac{4G^2\Delta\lambda_{n+1}}{||\boldsymbol{\sigma}'^{trial}_{n+1}||}\mathcal{I}'_{ijrs}\varepsilon^{etrial}_{n+1,rs}\right)
$$

$$
=\frac{4G^2}{||\boldsymbol{\sigma}'^{trial}_{n+1}||}\frac{\partial\Delta\lambda_{n+1}}{\partial \varepsilon^{etrial}_{n+1,kl}}\mathcal{I}'_{ijrs}\varepsilon^{etrial}_{n+1,rs}-\frac{4G^2\Delta\lambda_{n+1}}{||\boldsymbol{\sigma}'^{trial}_{n+1}||^2}\frac{\partial||\boldsymbol{\sigma}'^{trial}_{n+1}||}{\partial \varepsilon^{etrial}_{n+1,kl}}\mathcal{I}'_{ijrs}\varepsilon^{etrial}_{n+1,rs}+\frac{4G^2\Delta\lambda_{n+1}}{||\boldsymbol{\sigma}'^{trial}_{n+1}||}\mathcal{I}'_{ijrs}\frac{\partial \varepsilon^{etrial}_{n+1,rs}}{\partial \varepsilon^{etrial}_{n+1,kl}}
$$

$$
=\frac{4G^2}{||\boldsymbol{\sigma}'^{trial}_{n+1}||}\varepsilon'^{etrial}_{n+1,ij}\frac{\partial\Delta\lambda_{n+1}}{\partial \varepsilon^{etrial}_{n+1,kl}}-\frac{4G^2\Delta\lambda_{n+1}}{||\boldsymbol{\sigma}'^{trial}_{n+1}||^2}\varepsilon'^{etrial}_{n+1,ij}\frac{\partial||\boldsymbol{\sigma}'^{trial}_{n+1}||}{\partial \varepsilon^{etrial}_{n+1,kl}}+\frac{4G^2\Delta\lambda_{n+1}}{||\boldsymbol{\sigma}'^{trial}_{n+1}||}\mathcal{I}'_{ijkl}
$$

Besides, the following relations hold for the quantities included in Eq. (20.153).

$$
\frac{\boldsymbol{\varepsilon}'^{etrial}_{n+1}}{||\boldsymbol{\sigma}'^{trial}_{n+1}||}=\frac{1}{2G}\frac{\boldsymbol{\sigma}'^{trial}_{n+1}}{||\boldsymbol{\sigma}'^{trial}_{n+1}||} \tag{20.154}
$$

$$
\frac{\partial||\boldsymbol{\sigma}'^{trial}_{n+1}||}{\partial \boldsymbol{\varepsilon}^{etrial}_{n+1}}=\frac{\partial||\mathbf{E}:\boldsymbol{\varepsilon}'^{etrial}_{n+1}||}{\partial \boldsymbol{\varepsilon}^{etrial}_{n+1}}=2G\frac{\partial||\boldsymbol{\varepsilon}'^{etrial}_{n+1}||}{\partial \boldsymbol{\varepsilon}^{etrial}_{n+1}}=2G\frac{\boldsymbol{\varepsilon}'^{etrial}_{n+1}}{||\boldsymbol{\varepsilon}'^{etrial}_{n+1}||}=2G\frac{\boldsymbol{\sigma}'^{trial}_{n+1}}{||\boldsymbol{\sigma}'^{trial}_{n+1}||} \tag{20.155}
$$

Further, noting the following equation obtained by taking the partial derivative of Eq. (20.101)

$$
\frac{\partial \sigma^{eqtrial}_{n+1}}{\partial \boldsymbol{\varepsilon}^{etrial}_{n+1}}-\sqrt{6}G\frac{\partial\Delta\lambda_{n+1}}{\partial \boldsymbol{\varepsilon}^{etrial}_{n+1}}-\sqrt{2/3}\frac{dF(H_{n+1})}{dH_{n+1}}\frac{\partial\Delta\lambda_{n+1}}{\partial \boldsymbol{\varepsilon}^{etrial}_{n+1}}=\mathbf{O}
$$

the partial derivative of $\Delta\lambda_{n+1}$ is given by

$$
\begin{aligned}
\frac{\partial\Delta\lambda_{n+1}}{\partial \boldsymbol{\varepsilon}^{etrial}_{n+1}}&=\frac{1}{\sqrt{6}G+\sqrt{2/3}\frac{dF(H_{n+1})}{dH_{n+1}}}\frac{\partial \sigma^{eqtrial}_{n+1}}{\partial \boldsymbol{\varepsilon}^{etrial}_{n+1}}=\frac{1}{\sqrt{6}G+\sqrt{2/3}\frac{dF(H_{n+1})}{dH_{n+1}}}\frac{\partial(\sqrt{3/2}||\mathbf{E}:\boldsymbol{\varepsilon}'^{etrial}_{n+1}||)}{\partial \boldsymbol{\varepsilon}^{etrial}_{n+1}}\\
&=\frac{2G\sqrt{3/2}}{\sqrt{6}G+\sqrt{2/3}\frac{dF(H_{n+1})}{dH_{n+1}}}\frac{\boldsymbol{\sigma}'^{trial}_{n+1}}{||\boldsymbol{\sigma}'^{trial}_{n+1}||}
\end{aligned} \tag{20.156}
$$

Substituting Eqs. (20.154), (20.155) and (20.156) into Eq. (20.153), we have

$$
\begin{aligned}
\mathbf{K}^{ep,algo}_{n+1}&=\frac{\partial\boldsymbol{\sigma}_{n+1}}{\partial\boldsymbol{\varepsilon}_{n+1}}=\frac{\partial\boldsymbol{\sigma}_{n+1}}{\partial\boldsymbol{\varepsilon}^{etrial}_{n+1}}=\mathbf{E}-\frac{4G^2\Delta\lambda_{n+1}}{||\boldsymbol{\sigma}'^{trial}_{n+1}||}\mathcal{I}'-\frac{4G^2}{2G}\frac{\boldsymbol{\sigma}'^{trial}_{n+1}}{||\boldsymbol{\sigma}'^{trial}_{n+1}||}\otimes\frac{2G\sqrt{3/2}}{\sqrt{6}G+\sqrt{2/3}\frac{dF(H_{n+1})}{dH_{n+1}}}\frac{\boldsymbol{\sigma}'^{trial}_{n+1}}{||\boldsymbol{\sigma}'^{trial}_{n+1}||}\\
&\quad+\frac{4G^2\Delta\lambda_{n+1}}{||\boldsymbol{\sigma}'^{trial}_{n+1}||}\frac{1}{2G}\frac{\boldsymbol{\sigma}'^{trial}_{n+1}}{||\boldsymbol{\sigma}'^{trial}_{n+1}||}\otimes 2G\frac{\boldsymbol{\sigma}'^{trial}_{n+1}}{||\boldsymbol{\sigma}'^{trial}_{n+1}||}
\end{aligned}
$$

Further, taking account of Eq. (5.35) into this equation, the consistent tangent modulus $\mathbf{K}^{ep,\mathrm{algo}}_{n+1}$ is given as

$$\boxed{\mathbf{K}^{ep,\mathrm{algo}}_{n+1} = K\mathcal{I} + \left(2G - \frac{4G^2\Delta\lambda_{n+1}}{\|\boldsymbol{\sigma}'^{\mathrm{trial}}_{n+1}\|}\right)\mathcal{I}' + \left(\frac{4G^2\Delta\lambda_{n+1}}{\|\boldsymbol{\sigma}'^{\mathrm{trial}}_{n+1}\|} - \frac{4\sqrt{3/2}G^2}{\sqrt{6}G + \sqrt{2/3}\frac{dF(H_{n+1})}{dH_{n+1}}}\right)\frac{\boldsymbol{\sigma}'^{\mathrm{trial}}_{n+1}}{\|\boldsymbol{\sigma}'^{\mathrm{trial}}_{n+1}\|} \otimes \frac{\boldsymbol{\sigma}'^{\mathrm{trial}}_{n+1}}{\|\boldsymbol{\sigma}'^{\mathrm{trial}}_{n+1}\|}} \tag{20.157}$$

20.5.2 Numerical Method

The analytical method for the calculation of the consistent tangent modulus requires the mathematical preparation task including the second-order partial-derivative of the yield function in general as described in the last subsection, so that it leads to the difficulty in case of sophisticated constitutive equations. The numerical methods for the calculation of the consistent tangent modulus are described in the following.

(a) **Inverse matrix method**

The inverse matrix method is explained below (de Sauza Neto et al. 2008).

If $\boldsymbol{\varepsilon}^{e\mathrm{trial}}_{n+1}$ is not constant but changes with a plastic deformation, the simultaneous equation in Eq. (20.50) becomes as follows:

$$\begin{bmatrix} \frac{\partial \mathbf{Y}_\sigma}{\partial \boldsymbol{\sigma}_{n+1}} & \frac{\partial \mathbf{Y}_\sigma}{\partial \boldsymbol{\alpha}_{n+1}} & \frac{\partial \mathbf{Y}_\sigma}{\partial H_{n+1}} & \frac{\partial \mathbf{Y}_\sigma}{\partial (\Delta\lambda_{n+1})} \\ \frac{\partial \mathbf{Y}_\alpha}{\partial \boldsymbol{\sigma}_{n+1}} & \frac{\partial \mathbf{Y}_\alpha}{\partial \boldsymbol{\alpha}_{n+1}} & \frac{\partial \mathbf{Y}_\alpha}{\partial H_{n+1}} & \frac{\partial \mathbf{Y}_\alpha}{\partial (\Delta\lambda_{n+1})} \\ \frac{\partial Y_H}{\partial \boldsymbol{\sigma}_{n+1}} & \frac{\partial Y_H}{\partial \boldsymbol{\alpha}_{n+1}} & \frac{\partial Y_H}{\partial H_{n+1}} & \frac{\partial Y_H}{\partial (\Delta\lambda_{n+1})} \\ \frac{\partial Y_s}{\partial \boldsymbol{\sigma}_{n+1}} & \frac{\partial Y_s}{\partial \boldsymbol{\alpha}_{n+1}} & \frac{\partial Y_s}{\partial H_{n+1}} & \frac{\partial Y_s}{\partial (\Delta\lambda_{n+1})} \end{bmatrix} \begin{Bmatrix} d\boldsymbol{\sigma}_{n+1} \\ d\boldsymbol{\alpha}_{n+1} \\ dH_{n+1} \\ d(\Delta\lambda_{n+1}) \end{Bmatrix} = [\mathbf{J}] \begin{Bmatrix} d\boldsymbol{\sigma}_{n+1} \\ d\boldsymbol{\alpha}_{n+1} \\ dH_{n+1} \\ d(\Delta\lambda_{n+1}) \end{Bmatrix} = \begin{Bmatrix} d\boldsymbol{\varepsilon}^{e\mathrm{trial}}_{n+1} \\ \mathbf{O} \\ 0 \\ 0 \end{Bmatrix} \tag{20.158}$$

Let the inverse relation of Eq. (20.158) be described as follows:

$$\begin{bmatrix} \mathbb{D}_{11} & \mathbb{D}_{12} & \mathbf{D}_{13} & \mathbf{D}_{14} \\ \mathbb{D}_{21} & \mathbb{D}_{22} & \mathbf{D}_{23} & \mathbf{D}_{24} \\ \mathbf{D}_{31} & \mathbf{D}_{32} & D_{33} & D_{34} \\ \mathbf{D}_{41} & \mathbf{D}_{42} & D_{43} & D_{44} \end{bmatrix} \begin{Bmatrix} d\boldsymbol{\varepsilon}^{e\mathrm{trial}}_{n+1} \\ \mathbf{O} \\ 0 \\ 0 \end{Bmatrix} = \begin{Bmatrix} d\boldsymbol{\sigma}_{n+1} \\ d\boldsymbol{\alpha}_{n+1} \\ dH_{n+1} \\ d(\Delta\lambda_{n+1}) \end{Bmatrix} \tag{20.159}$$

where the fourth-order tensors, the second-order tensors and the scalars are denoted by $\mathbb{D}_{ij}, \mathbf{D}_{ij}, D_{ij}$, respectively.

$$\mathbf{d}\boldsymbol{\sigma}_{n+1} = \mathbb{D}_{11} : \mathbf{d}\boldsymbol{\varepsilon}^{etrial}_{n+1} \tag{20.160}$$

leading to

$$\mathbf{K}^{ep,\text{algo}}_{n+1} = \frac{\mathbf{d}\boldsymbol{\sigma}_{n+1}}{\mathbf{d}\boldsymbol{\varepsilon}^{etrial}_{n+1}} = \mathbb{D}_{11} \tag{20.161}$$

Therefore, the consistent tangent modulus tensor is obtained by calculating the inverse of the matrix $\mathbf{J}$.

(b) **Perturbation method**

The perturbation method was developed by Miehe (1996), and it has been applied to the finite strain theory by Eidel and Gruttmann (2003) and Menzel and Steinmann (2003), Menzel et al. (2005), Hashiguchi and Yamakawa (2012) and to geomaterials by Perez-Foguet et al. (2000a, b, 2001).

Suppose that the stress $\boldsymbol{\sigma}_{n+1}$ in the n + 1 step was calculated already by the return-mapping method under the input of the strain increment $\Delta\boldsymbol{\varepsilon}_{n+1}$. Let it be denoted as $\boldsymbol{\sigma}_{n+1}(\boldsymbol{\varepsilon}_{n+1})(=\boldsymbol{\sigma}_{n+1}(\boldsymbol{\varepsilon}_n + \Delta\boldsymbol{\varepsilon}_{n+1}))$. Further, perform again the return-mapping calculation to obtain a new stress $\boldsymbol{\sigma}_{n+1}(\boldsymbol{\varepsilon}_n + \Delta\boldsymbol{\varepsilon}_{n+1} + \varepsilon\mathbf{e}_i \otimes \mathbf{e}_j)$ in the n + 1 step by the input of the strain increment $\Delta\boldsymbol{\varepsilon}_{n+1} + \varepsilon\mathbf{e}_i \otimes \mathbf{e}_j$, where $\varepsilon\mathbf{e}_i \otimes \mathbf{e}_j$ $(\varepsilon \ll ||\Delta\boldsymbol{\varepsilon}_{n+1}||)$ is the perturbation strain increment, which possess same infinitesimal components ε in all six directions (see Fig. 20.6). Here, the calculation has to be started by returning all the internal state variables to the state at the end of the step n. The consistent tangent modulus is given by

$$\boxed{\mathbf{K}^{ep,\text{algo}}_{n+1} = \frac{\partial(\boldsymbol{\sigma}_{n+1}(\boldsymbol{\varepsilon}_n + \Delta\boldsymbol{\varepsilon}_{n+1} + \varepsilon\mathbf{e}_r \otimes \mathbf{e}_s) - \boldsymbol{\sigma}_{n+1}(\boldsymbol{\varepsilon}_n + \Delta\boldsymbol{\varepsilon}_{n+1}))}{\partial(\varepsilon\mathbf{e}_r \otimes \mathbf{e}_s)} \quad \text{(no sum)}}$$
$$(\mathbf{K}^{ep,\text{algo}}_{n+1})_{ijkl} = \frac{\sigma_{n+1_{ij}}(\boldsymbol{\varepsilon}_n + \Delta\boldsymbol{\varepsilon}_{n+1} + \varepsilon\mathbf{e}_k \otimes \mathbf{e}_l) - \sigma_{n+1_{ij}}(\boldsymbol{\varepsilon}_n + \Delta\boldsymbol{\varepsilon}_{n+1})}{\varepsilon} \tag{20.162}$$

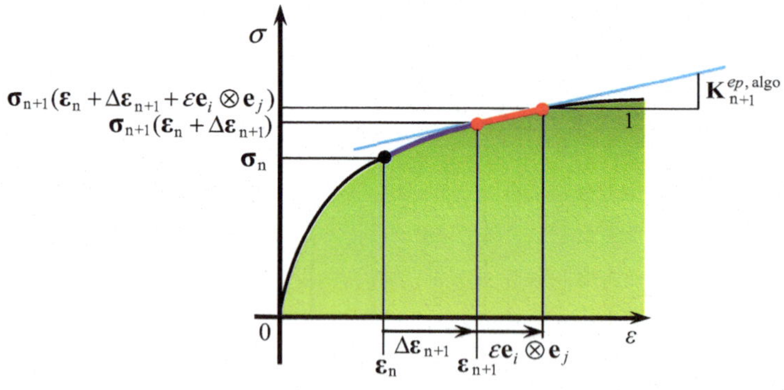

Fig. 20.6 Calculation of consistent tangent modulus by perturbation method

The $6 \times 6 = 36$ components of the fourth-order tensor $(\mathbf{K}_{\mathrm{n}+1}^{ep,\mathrm{algo}})_{ijkl}$ are determined through the six time calculations for $(k, l) = (1,1), (2,2), (3,3), (1,2), (2,3), (3,1)$ setting only each component of perturbation strain increment to be non-zero in Eq. (20.162). The common perturbed strain component is roughly taken as $\varepsilon = 1.0 \times 10^{-6}$–$1.0 \times 10^{-8}$.

20.6 Procedure for FEM Analysis

The return-mapping and the consistent tangent modulus tensor were deliberated in the preceding sections. Their application procedure to the finite element analysis is briefly summarized below.

1. To begin the calculation in the step n + 1 after obtaining the correct solution in the step n which fulfills the boundary condition and the yield condition, we calculate the deformation of the whole material body under the given boundary condition, exploiting the global elastoplastic stiffness tangent matrix in the step n, fixing internal variables.
2. The stress is calculated by the hyperelastic constitutive equation for the strain increments in each finite element based on the deformation calculated in the process 1. It is called the elastic trial step. The stress is judged to be correct and thus the calculation in the step n + 1 is finished if it does not go over the yield surface or the subloading surface at the end of the step. On the other hand, the calculation is transferred to the following plastic corrector step if it goes over them.
3. The stresses are reduced to the yield surface by the plastic corrector step within a prescribed tolerance, fixing the strain in each finite element. Here, the stress is always calculated by the hyperelastic equation by input of the fixed strain minus the plastic strain increment accumulated during the n+1 step. However, the nodal forces calculated from these stresses in each element do not satisfy the stress boundary condition on the surface of body and/or the null force condition in the nodal points inside the body since they were calculated fixing the strains in each element. Then, imbalance nodal force vector $\mathbb{R}_{\mathrm{n}+1}^{(\mathrm{k})}$ are induced. Therefore, we must perform again the calculations in the elastic trial and plastic corrector procedures by input of the corrected strain increment such that imbalance nodal force vector will decrease less than a prescribed tolerance. The corrected strain increments will be found by the updated calculations based on the following processes 4 and 5.
4. We calculate the consistent tangent modulus tensors by the methods described in Sect. 20.5. In the numerical method, we calculate the another stress increments through the elastic trial and plastic corrector procedures by inputting the strain increments which are given by the latest strain increments plus the perturbed infinitesimal strain increment. Then, we calculate the consistent tangent modulus tensors by the analytical or numerical methods described on Sect. 20.5.

5. We calculate the global tangent stiffness matrix $\mathbb{K}^{(k)}_{n+1}$ from the consistent tangent modulus tensors obtained for each element. Then, we calculate the corrector $\delta\mathbf{u}$ of nodal displacement increment vector by solving the equilibrium equation $\mathbb{K}^{(\kappa)}_{n+1}\delta\mathbf{u} = -\mathbb{R}^{(\kappa)}_{n+1}$, from which the corrected nodal displacement incremental vector $\Delta\mathbf{u}^{(k+1)}_{n+1} = \Delta\mathbf{u}^{(k)}_{n+1} + \delta\mathbf{u}$ is calculated. Further, the corrected strain increments which should be given in the next iteration calculation are calculated by $\Delta\boldsymbol{\varepsilon}^{(k+1)}_{n+1} = \mathrm{sym}[\partial\Delta\mathbf{u}^{(k+1)}_{n+1}/\partial\mathbf{x}^{(k+1)}_{n+1}]$.
6. Calculations in the elastic trial and the plastic corrector steps are performed again in each element by inputting the corrected strain increments $\Delta\boldsymbol{\varepsilon}^{(k+1)}_{n+1}$. The calculation in the step n + 1 is finished if the residual norm $\left\|\mathbb{R}^{(k+1)}_{n+1}\right\|$ is reduced less than a prescribed tolerance by repeating these calculations.

Note that the last values of internal variables must not be used but those at the end of the step n must be used always at the beginning of elastic trial processes, because the formers do not fulfill the equilibrium but the latter fulfills it. That is, we have to return values of internal variables to those at the end of the step n in every time.

20.7 Viscoplastic Material: Overstress Model

The return-mapping and the consistent tangent modulus tensor can be formulated for the overstress model, in which the stress exists outside the yield surface in general. Then, the values of stress and state variables are calculated by solving the finite difference equations of the stress rate versus strain rate relation and the evolution equations described by the values before and after the increment (Simo and Hughes 1998; de Souza Neto et al. 2008). In what follows, the equations required for the return-mapping for the conventional overstress model described in Chap. 13 are first formulated for the material with the equivalent viscoplastic strain hardening.

One has the relations

$$\boldsymbol{\varepsilon}_{n+1} = \boldsymbol{\varepsilon}_n + \Delta\boldsymbol{\varepsilon}_{n+1} = \boldsymbol{\varepsilon}^{vp}_n + \boldsymbol{\varepsilon}^{etrial}_{n+1} = \boldsymbol{\varepsilon}^{e}_{n+1} + \boldsymbol{\varepsilon}^{vp}_{n+1} = \text{const.} \tag{20.163}$$

$$\boldsymbol{\varepsilon}^{vptrial}_{n+1} = \boldsymbol{\varepsilon}^{vp}_n, \quad H^{trial}_{n+1} = H_n \tag{20.164}$$

and

$$\left.\begin{array}{ll} f(\boldsymbol{\sigma}^{trial}_{n+1}) - F(H_n) \leq 0 & : \Delta\boldsymbol{\varepsilon}^{vp}_{n+1} = \mathbf{O},\ \boldsymbol{\sigma}^{Final}_{n+1} = \boldsymbol{\sigma}^{trial}_{n+1} \\ \text{Otherwise} & : \Delta\boldsymbol{\varepsilon}^{vp}_{n+1} \neq \mathbf{O},\ \boldsymbol{\sigma}^{Final}_{n+1} \neq \boldsymbol{\sigma}^{trial}_{n+1} \end{array}\right\} \tag{20.165}$$

Let the viscoplastic corrector step be explained for the second case in Eq. (20.165). The viscoplastic strain rate and the isotropic hardening function of metals are described in the differential form from Eqs. (10.6) and (13.6) as follows:

$$d\boldsymbol{\varepsilon}^{vp}_{n+1} = \frac{1}{\bar{\mu}}\left\langle \frac{f(\boldsymbol{\sigma}_{n+1})}{F(H_{n+1})} - 1 \right\rangle^{n} \mathbf{n}_{n+1}(\boldsymbol{\sigma}_{n+1})\Delta t \tag{20.166}$$

$$H_{n+1} = H_n + \sqrt{\frac{2}{3}}||\Delta\boldsymbol{\varepsilon}^{vp}_{n+1}|| = H_n + \sqrt{\frac{2}{3}}\frac{1}{\bar{\mu}}\left\langle \frac{f(\boldsymbol{\sigma}_{n+1})}{F(H_{n+1})} - 1 \right\rangle^{n} \Delta t \tag{20.167}$$

Noting

$$\Delta\boldsymbol{\varepsilon}^{vp}_{n+1} = \boldsymbol{\varepsilon}^{etrial}_{n+1} - \boldsymbol{\varepsilon}^{e}_{n+1} = \boldsymbol{\varepsilon}^{e}_{n} + \Delta\boldsymbol{\varepsilon}_{n+1} - \boldsymbol{\varepsilon}^{e}_{n+1} \tag{20.168}$$

$$\boldsymbol{\sigma}_{n+1} = \mathbf{E} : \boldsymbol{\varepsilon}^{e}_{n+1} \tag{20.169}$$

Equations (20.166) and (20.167) lead to

$$\boldsymbol{\varepsilon}^{e}_{n} + \Delta\boldsymbol{\varepsilon}_{n+1} - \boldsymbol{\varepsilon}^{e}_{n+1} - \frac{1}{\bar{\mu}}\left\langle \frac{f(\mathbf{E} : \boldsymbol{\varepsilon}^{e}_{n+1})}{F(H_{n+1})} - 1 \right\rangle^{n} \mathbf{n}_{n+1}(\mathbf{E} : \boldsymbol{\varepsilon}^{e}_{n+1})\Delta t = \mathbf{O} \tag{20.170}$$

$$H_{n+1} - H_n - \sqrt{\frac{2}{3}}\frac{1}{\bar{\mu}}\left\langle \frac{f(\mathbf{E} : \boldsymbol{\varepsilon}^{e}_{n+1})}{F(H_{n+1})} - 1 \right\rangle^{n} \Delta t = 0 \tag{20.171}$$

One can obtain the unique equation which contains only the unknown variable $\boldsymbol{\varepsilon}^{e}_{n+1}$ by substituting Eq. (20.171) into Eq. (20.170), whilst $\Delta\boldsymbol{\varepsilon}$ and Δt are the input variables. One has only to continue the calculation until the simultaneous equation and the boundary condition will be fulfilled within a given tolerance. However, it is the nonlinear equation of $\boldsymbol{\varepsilon}^{e}_{n+1}$ and thus the cumbersome preparation tasks is required, which includes the second-order partial derivative of the yield function $f(\boldsymbol{\sigma})$.

Next, let the return-mapping be extended for the subloading-overstress model described in Sect. 12.5 for which one has:

$$\left.\begin{array}{rl} f(\boldsymbol{\sigma}^{trial}_{n+1}) - R_{s_n}F(H_n) \leq 0 & : \Delta\boldsymbol{\varepsilon}^{vp}_{n+1} = \mathbf{O},\ \boldsymbol{\sigma}^{Final}_{n+1} = \boldsymbol{\sigma}^{trial}_{n+1} \\ \text{Otherwise} & : \Delta\boldsymbol{\varepsilon}^{vp}_{n+1} \neq \mathbf{O},\ \boldsymbol{\sigma}^{Final}_{n+1} \neq \boldsymbol{\sigma}^{trial}_{n+1} \end{array}\right\} \tag{20.172}$$

The viscoplastic corrector step for the second case in Eq. (20.172) is shown below.

It follows from Eqs. (6.56), (13.21), (13.28) and (13.29) that

$$d\boldsymbol{\varepsilon}^{vp}_{n+1} = \frac{1}{\bar{\mu}}\frac{\langle R_{n+1} - R_{s_{n+1}}\rangle^{n}}{R_m - R_{n+1}} \mathbf{n}_{n+1}(\boldsymbol{\sigma}_{n+1})\Delta t \tag{20.173}$$

$$R_{n+1} = \frac{f(\boldsymbol{\sigma}_{n+1})}{F(H_{n+1})} \tag{20.174}$$

$$H_{n+1} = H_n + \sqrt{\frac{2}{3}}\|\boldsymbol{\varepsilon}^{vp}_{n+1}\Delta t\| = H_n + \sqrt{\frac{2}{3}}\frac{1}{\bar{\mu}}\frac{\langle R_{n+1} - R_{s_{n+1}}\rangle^n}{R_m - R_{n+1}}\Delta t \tag{20.175}$$

$$R_{s_{n+1}} = \frac{2}{\pi}(1 - R_e)\cos^{-1}\left[\cos\left(\frac{\pi}{2}\frac{R_{s_n} - R_e}{1 - R_e}\right)\exp\left(-\frac{\pi}{2}u\frac{u^{vp}_{n+1} - u^{vp}_n}{1 - R_e}\right)\right] + R_e \tag{20.176}$$

Noting Eqs. (20.168) and (20.169), Eqs. (20.173)–(20.176) lead to

$$\boldsymbol{\varepsilon}^e_n + \Delta\boldsymbol{\varepsilon}_{n+1} - \boldsymbol{\varepsilon}^e_{n+1} - \frac{1}{\bar{\mu}}\frac{\langle R_{n+1} - R_{s_{n+1}}\rangle^n}{R_m - R_{n+1}}\mathbf{n}_{n+1}(\mathbf{E} : \boldsymbol{\varepsilon}^e_{n+1})\Delta t = \mathbf{O} \tag{20.177}$$

$$H_{n+1} - H_n - \sqrt{\frac{2}{3}}\frac{1}{\bar{\mu}}\frac{\langle R_{n+1} - R_{s_{n+1}}\rangle^n}{R_m - R_{n+1}}\Delta t = 0 \tag{20.178}$$

$$R_{n+1} = \frac{f(\mathbf{E} : \boldsymbol{\varepsilon}^e_{n+1})}{F(H_{n+1})} \tag{20.179}$$

$$R_{s_{n+1}} = \frac{2}{\pi}(1 - R_e)\cos^{-1}\left[\cos\left(\frac{\pi}{2}\frac{R_{s_n} - R_e}{1 - R_e}\right)\exp\left(\frac{\pi}{2}u\frac{\|\boldsymbol{\varepsilon}_{n+1} - \boldsymbol{\varepsilon}^e_{n+1}\| - \|\boldsymbol{\varepsilon}_{n+1} - \boldsymbol{\varepsilon}^e_n\|}{1 - R_e}\right)\right] + R_e \tag{20.180}$$

One can obtain the unique equation which contains only the unknown variable $\boldsymbol{\varepsilon}^e_{n+1}$ by substituting Eqs. (20.178)–(20.180) into Eq. (20.177), whilst $\Delta\boldsymbol{\varepsilon}$ and Δt are the input variables. One has only to continue the calculation until the simultaneous equation and the boundary condition are fulfilled within a given tolerance. However, it is the nonlinear equation of $\boldsymbol{\varepsilon}^e_{n+1}$ and thus the cumbersome preparation tasks is required, which includes the second-order partial derivative of the yield function $f(\boldsymbol{\sigma})$.

The return-mapping for the viscoplasticity is formulated above by the closest-point projection but it cannot be formulated by the cutting-plane projection since there is no base state such as the yield surface or the subloading surface to which the stress has to be returned.

The consistent tangent modulus tensor can be obtained by the identical way to the numerical method described in 20.5.

20.8 Subloading-Friction Model

In this section the return-mapping will be formulated for the subloading-friction model described in Chap. 18. Both of the closest-point and the cutting-plane projections can be formulated as will be described in the following.

The time-integration of the sliding-velocity in Eq. (18.4) leads to the additive decomposition of the sliding displacement $\bar{\mathbf{u}}$ into the elastic sliding displacement $\bar{\mathbf{u}}^e$ and the plastic sliding displacement $\bar{\mathbf{u}}^p$, i.e.

$$\bar{\mathbf{u}} = \bar{\mathbf{u}}^e + \bar{\mathbf{u}}^p \tag{20.181}$$

Further, assume that the contact traction $\mathbf{f}$ is given by the hyperelastic equation as

$$\mathbf{f} = \frac{\partial \psi(\bar{\mathbf{u}}^e)}{\partial \bar{\mathbf{u}}^e} \tag{20.182}$$

Here, adopt the following simplest potential function

$$\psi(\bar{\mathbf{u}}^e) = \frac{1}{2}\bar{\mathbf{u}}^e \cdot \mathbf{C}^e \bar{\mathbf{u}}^e \tag{20.183}$$

leading to

$$\mathbf{f} = \frac{\partial \psi(\bar{\mathbf{u}}^e)}{\partial \bar{\mathbf{u}}^e} = \mathbf{C}^e \bar{\mathbf{u}}^e \tag{20.184}$$

The rate form of Eq. (20.184) conforms to Eq. (18.22).

The contact traction rate is given from Eq. (18.22) or from Eq. (20.184) as

$$\overset{\bullet}{\mathbf{f}} = \mathbf{C}^e(\bar{\mathbf{v}} - \bar{\mathbf{v}}^p) = \overset{\bullet}{\mathbf{f}}{}^e + \overset{\bullet}{\mathbf{f}}{}^p \tag{20.185}$$

where

$$\overset{\bullet}{\mathbf{f}}{}^e \equiv \mathbf{C}^e \bar{\mathbf{v}}, \ \overset{\bullet}{\mathbf{f}}{}^p \equiv -\mathbf{C}^e \bar{\mathbf{v}}^p \tag{20.186}$$

The plastic relaxation contact traction rate $\overset{\bullet}{\mathbf{f}}{}^p$ is given by substituting Eq. (18.49) into Eq. $(20.186)_2$ as follows:

$$\overset{\bullet}{\mathbf{f}}{}^p = -\mathbf{C}^e \dot{\bar{\lambda}}\, \mathbf{t}_n \tag{20.187}$$

Now, suppose that the contact traction $\mathbf{f}_{\mathrm{n}}$ in the step n is calculated already by giving n-time inputs of sliding increments. Then, for the n + 1 step calculation, first calculate the elastic trial contact traction $\mathbf{f}_{\mathrm{n+1}}^{\mathrm{elastic}}$, supposing that the input displacement increment $\Delta\bar{\mathbf{u}}$ is induced as the elastic sliding process as follows:

$$\mathbf{f}^{\text{elastic}}_{\text{n}+1} = \mathbf{C}^e(\bar{\mathbf{u}}^e_{\text{n}} + \Delta\bar{\mathbf{u}}_{\text{n}+1}) = \mathbf{f}_{\text{n}} + \mathbf{C}^e\Delta\bar{\mathbf{u}}_{\text{n}+1} \tag{20.188}$$

Here, if $\mathbf{f}^{\text{elastic}}_{\text{n}+1}$ lies inside the subloading-sliding surface in the previous step, i.e. if $f(\mathbf{f}^{\text{elastic}}_{\text{n}+1})$ is smaller than $r_{\text{n}}\mu_{\text{n}}$, it is judged that the calculation as the elastic sliding process was correct so that $\mathbf{f}^{\text{elastic}}_{\text{n}+1}$ is determined as the correct traction $\mathbf{f}^{\text{Final}}_{\text{n}+1}$ in the step $\text{n}+1$. However, if $\mathbf{f}^{\text{elastic}}_{\text{n}+1}$ lies outside the subloading-sliding surface in the previous step, it is judged that the calculation as the elastic sliding process was incorrect so that a plastic relaxation has to be induced until the subloading surface will be satisfied. Then, it holds that

$$\left.\begin{array}{rl} f(\mathbf{f}^{\text{elastic}}_{\text{n}+1}) \leq r_{\text{n}}\mu_{\text{n}} & : \ \bar{\mathbf{v}}^p_{\text{n}+1} = \mathbf{0},\ \mathbf{f}^{\text{Final}}_{\text{n}+1} = \mathbf{f}^{\text{elastic}}_{\text{n}+1} \\ \text{Otherwise} & : \ \bar{\mathbf{v}}^{p(1)}_{\text{n}+1} \neq \mathbf{0},\ \mathbf{f}^{\text{Final}}_{\text{n}+1} \neq \mathbf{f}^{\text{elastic}}_{\text{n}+1} \end{array}\right\} \tag{20.189}$$

The plastic sliding corrector step in the second case in Eq. (20.189) will be formulated in the following.

20.8.1 Closest-Point Projection

The residual of the plastic strain rate in the k-time in the calculation step $\text{n}+1$ is given by

$$\mathbf{R}^{(\text{k})}_{\text{n}+1} = -(\bar{\mathbf{u}}^{p(\text{k})}_{\text{n}+1} - \bar{\mathbf{u}}^p_{\text{n}}) + \Delta\bar{\lambda}^{(\text{k})}_{\text{n}+1}\mathbf{t}^{(\text{k})}_{n_{\text{n}+1}} \tag{20.190}$$

The calculation is proceeded until the residual is reduced less than a prescribed tolerance. $\Delta\bar{\lambda}^{(\text{k})}_{\text{n}+1}$ designates the accumulation of plastic multiplier induced in the $\text{n}+1$ step. The residual of the plastic sliding displacement in each time is updated from Eq. (20.190) as follows:

$$\begin{aligned} \mathbf{R}^{(\text{k}+1)}_{\text{n}+1} &= \mathbf{R}^{(\text{k})}_{\text{n}+1} + d\mathbf{R}^{(\text{k})}_{\text{n}+1} \\ &= \mathbf{R}^{(\text{k})}_{\text{n}+1} + d\left[-(\bar{\mathbf{u}}^{p(\text{k})}_{\text{n}+1} - \bar{\mathbf{u}}^p_{\text{n}}) + \Delta\bar{\lambda}^{(\text{k})}_{\text{n}+1}\mathbf{t}^{(\text{k})}_{n_{\text{n}+1}}\right] \\ &= \mathbf{R}^{(\text{k})}_{\text{n}+1} - d\bar{\mathbf{u}}^{p(\text{k})}_{\text{n}+1} + d(\Delta\bar{\lambda}^{(\text{k})}_{\text{n}+1}\mathbf{t}_{n^{(\text{k})}_{\text{n}+1}}) \\ &= \mathbf{R}^{(\text{k})}_{\text{n}+1} - d\bar{\mathbf{u}}^{p(\text{k})}_{\text{n}+1} + d\Delta\bar{\lambda}^{(\text{k})}_{\text{n}+1}\mathbf{t}_{n^{(\text{k})}_{\text{n}+1}} + \Delta\bar{\lambda}^{(\text{k})}_{\text{n}+1}\frac{\partial\mathbf{t}_{n^{(\text{k})}_{\text{n}+1}}}{\partial\mathbf{f}^{(\text{k})}_{\text{n}+1}} : d\mathbf{f}^{(\text{k})}_{\text{n}+1} \end{aligned} \tag{20.191}$$

Here, one has

$$\mathbf{f}^{(\mathrm{k})}_{\mathrm{n}+1} = \mathbf{C}^e \bar{\mathbf{u}}^{e(\mathrm{k})}_{\mathrm{n}+1} = \mathbf{C}^e (\bar{\mathbf{u}}_{\mathrm{n}+1} - \bar{\mathbf{u}}^{p(\mathrm{k})}_{\mathrm{n}+1})$$

from Eq. (18.27), which leads to

$$d\bar{\mathbf{u}}^{p(\mathrm{k})}_{\mathrm{n}+1} = -\mathbf{C}^{e-1} d\mathbf{f}^{(\mathrm{k})}_{\mathrm{n}+1} \tag{20.192}$$

noting $\bar{\mathbf{u}}_{\mathrm{n}+1} = \mathrm{const.}$ Substituting Eq. (20.192) into Eq. (20.191), we have

$$\mathbf{R}^{(\mathrm{k}+1)}_{\mathrm{n}+1} = \mathbf{R}^{(\mathrm{k})}_{\mathrm{n}+1} + \left(\mathbf{C}^{e-1} + \Delta\bar{\lambda}^{(\mathrm{k})}_{\mathrm{n}+1} \frac{\partial \mathbf{t}^{(\mathrm{k})}_{n\mathrm{n}+1}}{\partial \mathbf{f}^{(\mathrm{k})}_{\mathrm{n}+1}} \right) d\mathbf{f}^{(\mathrm{k})}_{\mathrm{n}+1} + d\Delta\bar{\lambda}^{(\mathrm{k})}_{\mathrm{n}+1} \mathbf{t}^{(\mathrm{k})}_{n\mathrm{n}+1}$$

i.e.

$$\mathbf{R}^{(\mathrm{k}+1)}_{\mathrm{n}+1} = \mathbf{R}^{(\mathrm{k})}_{\mathrm{n}+1} + \bar{\mathbf{\Xi}}^{(\mathrm{k}-1)}_{\mathrm{n}+1} d\mathbf{f}^{(\mathrm{k})}_{\mathrm{n}+1} + d\Delta\bar{\lambda}^{(\mathrm{k})}_{\mathrm{n}+1} \mathbf{t}^{(\mathrm{k})}_{n\mathrm{n}+1} \tag{20.193}$$

where

$$\boxed{\bar{\mathbf{\Xi}}^{(\mathrm{k})}_{\mathrm{n}+1} \equiv \left(\mathbf{C}^{e-1} + \Delta\bar{\lambda}^{(\mathrm{k})}_{\mathrm{n}+1} \frac{\partial \mathbf{t}^{(\mathrm{k})}_{n_{\mathrm{n}+1}}}{\partial \mathbf{f}^{(\mathrm{k})}_{\mathrm{n}+1}} \right)^{-1}}$$

is the Hessian matrix which contains the second-order partial derivative of yield function.

Setting the residual in Eq. (20.193) to be zero, one has

$$\mathbf{R}^{(\mathrm{k})}_{\mathrm{n}+1} + \bar{\mathbf{\Xi}}^{(\mathrm{k}-1)}_{\mathrm{n}+1} d\mathbf{f}^{(\mathrm{k})}_{\mathrm{n}+1} + d\Delta\bar{\lambda}^{(\mathrm{k})}_{\mathrm{n}+1} \mathbf{t}^{(\mathrm{k})}_{n_{\mathrm{n}+1}} = \mathbf{0} \tag{20.195}$$

from which it follows that

$$\frac{\partial f(\mathbf{f}^{(\mathrm{k})}_{\mathrm{n}+1})}{\partial \mathbf{f}^{(\mathrm{k})}_{\mathrm{n}+1}} \bullet \bar{\mathbf{\Xi}}^{(\mathrm{k})}_{\mathrm{n}+1} \mathbf{R}^{(\mathrm{k})}_{\mathrm{n}+1} + \frac{\partial f(\mathbf{f}^{(\mathrm{k})}_{\mathrm{n}+1})}{\partial \mathbf{f}^{(\mathrm{k})}_{\mathrm{n}+1}} d\mathbf{f}^{(\mathrm{k})}_{\mathrm{n}+1} + d\Delta\lambda^{(\mathrm{k})}_{\mathrm{n}+1} \frac{\partial f(\mathbf{f}^{(\mathrm{k})}_{\mathrm{n}+1})}{\partial \mathbf{f}^{(\mathrm{k})}_{\mathrm{n}+1}} \bullet \bar{\mathbf{\Xi}}^{(\mathrm{k})}_{\mathrm{n}+1} \mathbf{t}^{(\mathrm{k})}_{n_{\mathrm{n}+1}} = 0 \tag{20.196}$$

Now, introduce the following function based on the subloading-sliding surface equation in Eq. (18.30).

$$g(\mathbf{f}, r, \mu) \equiv f(\mathbf{f}) - r\mu \tag{20.197}$$

Applying the Taylor expansion to this equation and taking only the first derivative, one has

$$\begin{aligned} g(\mathbf{f}+d\mathbf{f}, r+dr, \mu+d\mu) &= g(\mathbf{f}, r, \mu) \\ &+ \frac{\partial g(\mathbf{f}, r, \mu)}{\partial \mathbf{f}} d\mathbf{f} + \frac{\partial g(\mathbf{f}, r, \mu)}{\partial r} dr + \frac{\partial g(\mathbf{f}, r, \mu)}{\partial \mu} d\mu \\ &= f(\mathbf{f}) - r\mu + \frac{\partial f(\mathbf{f})}{\partial \mathbf{f}} \bullet d\mathbf{f} - \mu dr - r d\mu \end{aligned} \tag{20.198}$$

The quantity consisting of the last three incremental terms in the last equation of Eq. (20.198) is no more than the quantity in incremental form of the consistency condition of the subloading-sliding surface. It is expressed by the following equation, noting that Eq. (18.44) is transformed to Eq. (18.54).

$$\frac{\partial f(\mathbf{f})}{\partial \mathbf{f}} \bullet d\mathbf{f} - \mu dr - r d\mu = \frac{\partial f(\mathbf{f})}{\partial \mathbf{f}} \bullet d\mathbf{f} - m^p d\bar{\lambda} - m^c dt \tag{20.199}$$

Substituting Eq. (20.199) into Eq. (20.198) and setting Eq. (20.198) to be zero, one has

$$f(\mathbf{f}) - r\mu + \frac{\partial f(\mathbf{f})}{\partial \mathbf{f}} \bullet d\mathbf{f} - m^p d\bar{\lambda} - m^c dt = 0 \tag{20.200}$$

from which one has

$$\frac{\partial f(\mathbf{f})}{\partial \mathbf{f}} \bullet d\mathbf{f} = -(f(\mathbf{f}) - r\mu) + m^p d\bar{\lambda} + m^c dt \tag{20.201}$$

Substituting Eq. (20.201) into Eq. (20.196), it follows that

$$\begin{aligned} &\frac{\partial f(\mathbf{f}_{n+1}^{(k)})}{\partial \mathbf{f}_{n+1}^{(k)}} \bullet \boldsymbol{\Xi}_{n+1}^{(k)} \mathbf{R}_{n+1}^{(k)} - (f(\mathbf{f}_{n+1}^{(k)}) - r_{n+1}^{(k)} \mu_{n+1}^{(k)}) + m_{n+1}^{p\,(k)} d\Delta\bar{\lambda}_{n+1}^{(k)} + m_{n+1}^{c(k)} dt \\ &+ d\Delta\bar{\lambda}_{n+1}^{(k)} \frac{\partial f(\mathbf{f}_{n+1}^{(k)})}{\partial \mathbf{f}_{n+1}^{(k)}} \bullet \boldsymbol{\Xi}_{n+1}^{(k)} \mathbf{t}_{n_{n+1}}^{(k)} = 0 \end{aligned} \tag{20.202}$$

from which one has

$$d\Delta\bar{\lambda}_{n+1}^{(k)} = \frac{(f(\mathbf{f}_{n+1}^{(k)}) - r_{n+1}^{(k)} \mu_{n+1}^{(k)}) - m^c dt - \dfrac{\partial f(\mathbf{f}_{n+1}^{(k)})}{\partial \mathbf{f}_{n+1}^{(k)}} \bullet \boldsymbol{\Xi}_{n+1}^{(k)} \mathbf{R}_{n+1}^{(k)}}{\bar{m}_{n+1}^{p\,(k)} + \dfrac{\partial f(\mathbf{f}_{n+1}^{(k)})}{\partial \mathbf{f}_{n+1}^{(k)}} \bullet \boldsymbol{\Xi}_{n+1}^{(k)} \mathbf{t}_{n_{n+1}}^{(k)}} \tag{20.203}$$

where $m_{n+1}^{p(k)}$ and $m_{n+1}^{c(k)}$ are given from Eqs. (18.55) and (18.56) by the following equation.

$$m_{n+1}^{p\,(k)} = -\kappa\left(\frac{\mu_{n+1}^{(k)}}{\mu_k} - 1\right) r_{n+1}^{(k)} + \overline{U}(r_{n+1}^{(k)})\mu_{n+1}^{(k)} \tag{20.204}$$

$$m_{v n+1}^{c\,(k)} = \xi\left(1 - \frac{\mu_{n+1}^{(k)}}{\mu_s}\right) r_{n+1}^{(k)} \tag{20.205}$$

The contact traction increment in k time calculation is given from Eq. (20.195) as

$$d\mathbf{f}_{n+1}^{(k)} = -\bar{\mathbf{\Xi}}_{n+1}^{(k)} : (\mathbf{R}_{n+1}^{(k)} + d\Delta\bar{\lambda}_{n+1}^{(k)}\mathbf{t}_{n_{n+1}}^{(k)}) \tag{20.206}$$

by which the plastic sliding increment is given by

$$d\bar{\mathbf{u}}_{n+1}^{p\,(k)} = -\mathbf{C}^{e-1} : d\mathbf{f}_{n+1}^{(k)} = \mathbf{C}^{e-1} : \bar{\mathbf{\Xi}}_{n+1}^{(k)} : (\mathbf{R}_{n+1}^{(k)} + d\Delta\bar{\lambda}_{n+1}^{(k)}\mathbf{t}_{n_{n+1}}^{(k)}) \tag{20.207}$$

Then, the plastic sliding displacement and the contact traction are updated as follows:

$$\bar{\mathbf{u}}_{n+1}^{p(k+1)} = \bar{\mathbf{u}}_{n+1}^{p\,(k)} + \mathbf{C}^{e-1} : \bar{\mathbf{\Xi}}_{n+1}^{(k)} : (\mathbf{R}_{n+1}^{(k)} + d\Delta\bar{\lambda}_{n+1}^{(k)}\mathbf{t}_{n_{n+1}}^{(k)}) \tag{20.208}$$

$$\mathbf{f}_{n+1}^{(k+1)} = \mathbf{E} : (\bar{\mathbf{u}}_{n+1} - \bar{\mathbf{u}}_{n+1}^{p(k+1)}) \tag{20.209}$$

μ is updated by Eq. (18.38) as

$$\mu_{n+1}^{(k+1)} = \mu_{n+1}^{(k)} - \kappa\left(\frac{\mu_{n+1}^{(k)}}{\mu_k} - 1\right)||d\bar{\mathbf{u}}_{n+1}^{p\,(k)}|| + \xi\left(1 - \frac{\mu_{n+1}^{(k)}}{\mu_s}\right)dt_{n+1}^{(k)} \tag{20.210}$$

which can be analytically integrated for $\xi = 0$ as follows:

$$\mu_{n+1}^{(k+1)} = (\mu_n - \mu_k)\exp\left[-\frac{\kappa}{\mu_k}(\bar{u}_{n+1}^{p(k+1)} - \bar{u}_n^p)\right] + \mu_k \tag{20.211}$$

based on Eq. (18.42). Further, r is updated by the following analytical integration based on Eq. (18.37).

$$r_{n+1}^{(k+1)} = \frac{2}{\pi}\cos^{-1}\left\{\cos(\frac{\pi}{2}r_n)\exp\left[-\frac{\pi}{2}\tilde{u}(\bar{u}_{n+1}^{p(k+1)} - \bar{u}_n^p)\right]\right\} \tag{20.212}$$

Equation (20.203) can be calculated by the known variables.

These iteration calculation has to be repeated until $\mathbf{R}_{n+1}^{(k)}$ will converge within a given tolerance.

20.8.2 Cutting-Plane Projection

Let the relaxation calculation in the k + 1 times be performed after relaxation calculations of k times obtaining $\mathbf{f}_{n+1}^{(k)}, \mu_{n+1}^{(k)}$ and $r_{n+1}^{(k)}$. Denoting the plastic sliding increment expected in this calculation as $\bar{\mathbf{v}}_{n+1}^{p(k+1)}dt$, the traction increment $d\mathbf{f}_{n+1}^{(k+1)}$, called the *plastic predictor*, is given from Eq. (20.187) by

$$d\mathbf{f}_{n+1}^{(k+1)} = d\mathbf{f}_{n+1}^{p(k+1)} = -\mathbf{C}^e d\bar{\lambda}_{n+1}^{(k+1)}\mathbf{t}_{n_{n+1}}^{(k)} \tag{20.213}$$

Repeating these calculations, it is expected that the following subloading sliding surface equation is fulfilled.

$$f(\mathbf{f}_{n+1}^{(k+1)}) - r_{n+1}^{(k+1)}\mu_{n+1}^{(k+1)} = 0 \tag{20.214}$$

In what follows, we formulate $\bar{\lambda}_{n+1}^{(k+1)}dt$ which has to be input into Eq. (20.213) in the iteration calculation.

By inserting Eq. (20.187) into the term of contact traction increment in the right-hand side of Eq. (20.199), one has

$$\frac{\partial f(\mathbf{f})}{\partial \mathbf{f}} \bullet d\mathbf{f} - \mu dr - r d\mu = -\frac{\partial f(\mathbf{f})}{\partial \mathbf{f}} \bullet \mathbf{C}^e d\bar{\lambda}\,\mathbf{t}_n - m^p d\bar{\lambda} - m^c dt \tag{20.215}$$

Further, substituting Eq. (20.215) into the Taylor expansion in Eq. (20.198), one has

$$f(\mathbf{f}) - r\mu - \frac{\partial f(\mathbf{f})}{\partial \mathbf{f}} \bullet \mathbf{C}^e d\bar{\lambda}\,\mathbf{t}_n - m^p d\bar{\lambda} - m^c dt \tag{20.216}$$

By putting the quantity in Eq. (20.216) to be zero, $d\bar{\lambda}$ is obtained as follows:

$$d\bar{\lambda} = \frac{f(\mathbf{f}) - r\mu - m^c dt}{m^p + \dfrac{\partial f(\mathbf{f})}{\partial \mathbf{f}} \bullet \mathbf{C}^e \mathbf{t}_n}, \quad \bar{\mathbf{v}}^p dt = \left\langle \frac{f(\mathbf{f}) - r\mu - m^c dt}{m^p + \dfrac{\partial f(\mathbf{f})}{\partial \mathbf{f}} \bullet \mathbf{C}^e \mathbf{t}_n} \right\rangle \mathbf{t}_n \tag{20.217}$$

Equation $(20.217)_1$ shows the magnitude of plastic sliding velocity which has to be input in order that the contact traction returns to the subloading sliding surface when the quantity $f(\mathbf{f}) - r\mu - m^c dt$ remains.

The plastic sliding increment used for updating the contact stress and the internal variables in the k + 1 calculation is given from Eq. (20.217) by

$$\bar{\mathbf{v}}^{p\,(\mathrm{k}+1)}_{\mathrm{n}+1}dt = \left\langle \frac{f(\mathbf{f}^{(\mathrm{k})}_{\mathrm{n}+1}) - r^{(\mathrm{k})}_{\mathrm{n}+1}\mu^{(\mathrm{k})}_{\mathrm{n}+1} - m^{c\,(\mathrm{k})}_{v\,\mathrm{n}+1}dt}{m^{p\,(\mathrm{k})}_{\mathrm{n}+1} + \dfrac{\partial f(\mathbf{f}^{(\mathrm{k})}_{\mathrm{n}+1})}{\partial \mathbf{f}^{(\mathrm{k})}_{\mathrm{n}+1}} \bullet \mathbf{C}^{e}\mathbf{t}^{(\mathrm{k})}_{n_{\mathrm{n}+1}}} \right\rangle \mathbf{t}^{(\mathrm{k})}_{n_{\mathrm{n}+1}} \tag{20.218}$$

The solution for the step n + 1 will be determined by performing the iteration calculation for updating **f**, μ and r until the sliding-subloading surface equation (20.214) is fulfilled in the range less than a required tolerance.

20.9 Objective Time-Integration Algorithm of Rate Formulation

The return-mapping method described in the preceding sections are concerned with calculation for deformation without a material rotation. Efficient time-integration method has been developed for calculation of deformation under a material rotation for elastoplastic constitutive equations with constant hypoelastic moduli by executing the procedure of three processes, i.e. the pull-back operation, the time-integration and the push-forward operation reflecting the physical meaning of the convected rate described in Sect. 4.4 (Flanagan and Taylor 1987; Hughes and Winget 1980; Pinsky et al. 1983; Simo and Hughes 1998; de Souza Neto et al. 2008). Needless to say, it is useful also for time-integration of simple hypoelastic constitutive equation with constant hypoelastic moduli. The explicit procedures in this method will be described in this section.

(1) **Pull-back operation**

The original constitutive relation described in the current configuration, i.e. *spatial description*, is firstly pulled back to the reference configuration, i.e. *convected description* which is independent of rigid-body rotation, leading to the rotation-free (insensitive) tensor. To this end, all the variables, i.e. the stress (rate), the back stress (rate) and the strain rate must be pulled back from the current to the reference configurations. Here, let the four types of the pull-back operation in Eq. (4.44) be described collectively as

$$\widehat{\mathbf{t}} = \mathbf{\Lambda}^{T}\mathbf{t}\mathbf{\Lambda} \tag{20.219}$$

The Cotter-Rivlin, the Oldroyd and the Green-Naghdi rates in Eqs. (4.66), (4.61) and (4.68) are related to $\mathbf{\Lambda} = \mathbf{F}, \mathbf{F}^{-T}$ and $\mathbf{R}$, respectively. The tensor $\mathbf{t}$ in the rotation-free configuration is denoted as $\widehat{\mathbf{t}}$. On the other hand, the Jaumann rate is required to satisfy

$$\dot{\mathbf{\Lambda}} = \mathbf{w}\mathbf{\Lambda} \tag{20.220}$$

leading to

$$\dot{\mathbf{\Lambda}}_{\mathrm{n}+1} = \mathbf{w}_{\mathrm{n}+1/2}\mathbf{\Lambda}_{\mathrm{n}+1} \tag{20.221}$$

noting $\mathring{\mathbf{t}}^{w} = \mathbf{\Lambda}(\mathbf{\Lambda}^{T}\mathbf{t}\mathbf{\Lambda})^{\boldsymbol{\cdot}}\mathbf{\Lambda}^{T} = \dot{\mathbf{t}} + \mathbf{t}\dot{\mathbf{\Lambda}}\mathbf{\Lambda}^{T} + \mathbf{\Lambda}\dot{\mathbf{\Lambda}}^{T}\mathbf{t}$ with $\mathbf{w} = \dot{\mathbf{\Lambda}}\mathbf{\Lambda}^{T} = -\mathbf{\Lambda}\dot{\mathbf{\Lambda}}^{T}$ leading to $\dot{\mathbf{\Lambda}} = \mathbf{w}\mathbf{\Lambda}^{-T} = \mathbf{w}\mathbf{\Lambda}(\mathbf{\Lambda}^{-1} = \mathbf{\Lambda}^{T})$. Equation (20.220) is fulfilled by setting

$$\boxed{\mathbf{\Lambda}_{\mathrm{n}+1} = \exp[\mathbf{w}_{\mathrm{n}+1/2}\Delta t]\mathbf{\Lambda}_{\mathrm{n}}} \tag{20.222}$$

where $\Delta t \equiv t_{\mathrm{n}+1} - t_{\mathrm{n}}$, because of

$$\frac{\partial \mathbf{\Lambda}_{\mathrm{n}+1}}{\partial t_{\mathrm{n}+1}} = \frac{\partial(\exp[\mathbf{w}_{\mathrm{n}+1/2}\Delta t]\mathbf{\Lambda}_{\mathrm{n}})}{\partial t_{\mathrm{n}+1}} = \mathbf{w}_{\mathrm{n}+1/2}\exp[\mathbf{w}_{\mathrm{n}+1/2}\Delta t]\mathbf{\Lambda}_{\mathrm{n}} = \mathbf{w}_{\mathrm{n}+1/2}\mathbf{\Lambda}_{\mathrm{n}+1} \tag{20.223}$$

noting t_{n} and $\mathbf{\Lambda}_{\mathrm{n}}$ are constant and

$$\frac{\partial \exp(\mathbf{w}t)}{\partial t} = \mathbf{w}\exp(\mathbf{w}t) \tag{20.224}$$

noting

$$\begin{aligned}\exp(\mathbf{w}t) &= \sum_{n=0}^{\infty}\frac{1}{n!}(\mathbf{w}t)^{n} = \mathbf{I} + (\mathbf{w}t) + \frac{1}{2!}(\mathbf{w}t)^{2} + \frac{1}{3!}(\mathbf{w}t)^{3} + \cdots \\ &= \sum_{n=0}^{\infty}\frac{1}{n!}(w_{P}t)^{n}\mathbf{e}_{P}\otimes\mathbf{e}_{P} = (1 + w_{P}t + \frac{1}{2!}(w_{P}t)^{2} + \frac{1}{3!}(w_{P}t)^{3} + \cdots)\mathbf{e}_{P}\otimes\mathbf{e}_{P}\end{aligned} \tag{20.225}$$

based on Eq. (1.285), and

$$\begin{aligned}\frac{\partial \exp(\mathbf{w}t)}{\partial t} &= w_{P}\sum_{n=0}^{\infty}\frac{1}{n!}(w_{P}t)^{n}\mathbf{e}_{P}\otimes\mathbf{e}_{P} \\ &= w_{P}\left[1 + w_{P}t + \frac{1}{2!}(w_{P}t)^{2} + \frac{1}{3!}(w_{P}t)^{3} + \cdots\right]\mathbf{e}_{P}\otimes\mathbf{e}_{P}\end{aligned} \tag{20.226}$$

due to $\partial[(w_{P}t)^{n}/n!]/\partial t = w_{P}[(w_{P}t)^{n-1}/(n-1)!]$.

(2) Time-integration process

The stress in the elastic trial process is given by

$$\widehat{\boldsymbol{\sigma}}^{\text{trial}}_{\text{n}+1} = \widehat{\boldsymbol{\sigma}}_{\text{n}} + \Delta t\mathbf{E} : \widehat{\mathbf{d}}_{\text{n}+1/2} \tag{20.227}$$

where

$$\widehat{\boldsymbol{\sigma}}_{\text{n}+1} = \boldsymbol{\Lambda}^T_{\text{n}+1}\boldsymbol{\sigma}_{\text{n}+1}\boldsymbol{\Lambda}_{\text{n}+1},\ \boldsymbol{\sigma}_{\text{n}+1} = \boldsymbol{\Lambda}_{\text{n}+1}\widehat{\boldsymbol{\sigma}}_{\text{n}+1}\boldsymbol{\Lambda}^T_{\text{n}+1} \tag{20.228}$$

$$\widehat{\mathbf{d}}_{\text{n}+1/2} = \boldsymbol{\Lambda}^T_{\text{n}+1/2}\mathbf{d}_{\text{n}+1/2}\boldsymbol{\Lambda}_{\text{n}+1/2},\ \mathbf{d}_{\text{n}+1/2} = \boldsymbol{\Lambda}_{\text{n}+1/2}\widehat{\mathbf{d}}_{\text{n}+1/2}\boldsymbol{\Lambda}^T_{\text{n}+1/2} \tag{20.229}$$

The substitution of Eq. (20.229) into Eq. (20.227) reads:

$$\begin{aligned}\boldsymbol{\sigma}^{\text{trial}}_{\text{n}+1} &= \boldsymbol{\Lambda}_R\boldsymbol{\sigma}_{\text{n}}\boldsymbol{\Lambda}^T_R + \Delta t\mathbf{E} : \boldsymbol{\Lambda}_r\mathbf{d}_{\text{n}+1/2}\boldsymbol{\Lambda}^T_r \\ &= \boldsymbol{\Lambda}_R\boldsymbol{\sigma}_{\text{n}}\boldsymbol{\Lambda}^T_R + \Delta t\mathbf{E} : \boldsymbol{\Lambda}_r\text{sym}[\nabla_{\text{n}+1/2}\Delta\mathbf{u}_{\text{n}+1/2}]\boldsymbol{\Lambda}^T_r\end{aligned} \tag{20.230}$$

where

$$\boldsymbol{\Lambda}_R \equiv \boldsymbol{\Lambda}_{\text{n}+1}\boldsymbol{\Lambda}^T_{\text{n}} = \exp(\Delta t\mathbf{w}_{\text{n}+1/2}) = \exp(\text{ant}[\nabla_{\text{n}+1/2}\Delta\mathbf{u}_{\text{n}+1/2}]) \tag{20.231}$$

$$\boldsymbol{\Lambda}_r \equiv (\boldsymbol{\Lambda}_R)^{1/2} = \boldsymbol{\Lambda}_{\text{n}+1}\boldsymbol{\Lambda}^T_{\text{n}+1/2} = \exp[(\Delta t/2)\mathbf{w}_{\text{n}+1/2}] = \exp(\text{ant}[\nabla_{\text{n}+1/2}\Delta\mathbf{u}_{\text{n}+1/2}]/2) \tag{20.232}$$

noting

$$\begin{aligned}\boldsymbol{\sigma}^{\text{trial}}_{\text{n}+1} &= \boldsymbol{\Lambda}_{\text{n}+1}(\widehat{\boldsymbol{\sigma}}_{\text{n}} + \Delta t\mathbf{E} : \widehat{\mathbf{d}}_{\text{n}+1/2})\boldsymbol{\Lambda}^T_{\text{n}+1} \\ &= \boldsymbol{\Lambda}_{\text{n}+1}(\boldsymbol{\Lambda}^T_{\text{n}}\boldsymbol{\sigma}_{\text{n}}\boldsymbol{\Lambda}_{\text{n}} + \Delta t\mathbf{E} : \boldsymbol{\Lambda}^T_{\text{n}+1/2}\mathbf{d}_{\text{n}+1/2}\boldsymbol{\Lambda}_{\text{n}+1/2})\boldsymbol{\Lambda}^T_{\text{n}+1} \\ &= \boldsymbol{\Lambda}_{\text{n}+1}\boldsymbol{\Lambda}^T_{\text{n}}\boldsymbol{\sigma}_{\text{n}}(\boldsymbol{\Lambda}_{\text{n}+1}\boldsymbol{\Lambda}^T_{\text{n}})^T + \Delta t\mathbf{E} : \boldsymbol{\Lambda}_{\text{n}+1}\boldsymbol{\Lambda}^T_{\text{n}+1/2}\mathbf{d}_{\text{n}+1/2}(\boldsymbol{\Lambda}_{\text{n}+1}\boldsymbol{\Lambda}^T_{\text{n}+1/2})^T\end{aligned}$$

Equation (20.231) is given by exploiting Eq. (20.222).

Taking only the first and second terms in the Taylor expansion of Eq. (20.225), one has the approximations for $\boldsymbol{\Lambda}_r$ and $\boldsymbol{\Lambda}_R$ in Eq. (20.232) (Hughes and Winget 1980):

$$\boldsymbol{\Lambda}_r \cong \begin{cases}\mathbf{I} + (1/2)\Delta t\mathbf{w}_{\text{n}+1/2} \\ \{\exp[-(1/2)\Delta t\mathbf{w}_{\text{n}+1/2}]\}^{-1} \cong [\mathbf{I} - (1/2)\Delta t\mathbf{w}_{\text{n}+1/2}]^{-1}\end{cases} \tag{20.233}$$

$$\boldsymbol{\Lambda}^T_r \cong \begin{cases}\mathbf{I} - (1/2)\Delta t\mathbf{w}_{\text{n}+1/2} \\ [\mathbf{I} + (1/2)\Delta t\mathbf{w}_{\text{n}+1/2}]^{-1}\end{cases} \tag{20.234}$$

$$\mathbf{\Lambda}_R \cong \begin{cases} (\mathbf{I}-(1/2)\Delta t\mathbf{w}_{n+1/2})^{-1}(\mathbf{I}+(1/2)\Delta t\mathbf{w}_{n+1/2}) \\ (\mathbf{I}+(1/2)\Delta t\mathbf{w}_{n+1/2})(\mathbf{I}-(1/2)\Delta t\mathbf{w}_{n+1/2})^{-1} \end{cases} \tag{20.235}$$

$$\mathbf{\Lambda}_R^T \cong \begin{cases} (\mathbf{I}-(1/2)\Delta t\mathbf{w}_{n+1/2})(\mathbf{I}+(1/2)\Delta t\mathbf{w}_{n+1/2})^{-1} \\ (\mathbf{I}+(1/2)\Delta t\mathbf{w}_{n+1/2})^{-1}(\mathbf{I}-(1/2)\Delta t\mathbf{w}_{n+1/2}) \end{cases} \tag{20.236}$$

where the orthogonality in the approximations of $\mathbf{\Lambda}_R$ and $\mathbf{\Lambda}_r$ in Eqs. (20.233) and (20.235) can be confirmed readily.

Then, the return-mapping calculation is executed in the rotation-free configuration, i.e. convected description.

(3) **Push-forward process**

Finally, the stress is described based on Eq. (20.219) by the push-forward from the convected to the spatial description as follows:

$$\boldsymbol{\sigma}_{n+1} = \mathbf{\Lambda}_{n+1}\widehat{\boldsymbol{\sigma}}_{n+1}\mathbf{\Lambda}_{n+1}^T, \quad \mathbf{d}_{n+1} = \mathbf{\Lambda}_{n+1}\widehat{\boldsymbol{d}}_{n+1}\mathbf{\Lambda}_{n+1}^T \tag{20.237}$$

The return-mapping is described thoroughly in this chapter. However, it should be kept in mind that there exists the limitation in the validity of the return-mapping. For instance, small incremental steps must be input for a curved loading path because of the loading path-dependence of irreversible deformation and for a material rotation in the return-mapping. In addition, the application of the return-mapping to the calculation for the nonhardening region in metals described in Sect. 10.2 would be difficult because of the simultaneous return-mappings of the back stress to the non-hardening surface and the stress to the yield (or subloading) surface. In facts, the formulations of the non-hardening region are abandoned in the FEM program with the return-mapping by the Ohno-Wang model (Ohno et al. 2013) and the Yoshida-Uemori model (Ghaei and Green 2010).

20.10 Forward-Euler and Return-Mapping Methods for Stress Integration

Stress integration is performed by the forward (explicit) Euler method or the backward (implicit) Euler method due to the return-mapping projection. The features of these methods are delineated in this section.

Forward-Euler method: A repetitive calculation is not performed to obtain a stress which satisfies the yield condition. Here, note the following features.

(1) The elastoplastic modulus tensor and the evolution rules of internal variables are used in calculations.
(2) Quite small loading increments must be input in the plastic constitutive equation (conventional and cyclic kinematic hardening models) assuming the yield surface enclosing purely elastic domain in order that the stress does not deviate

from the yield surface. However, larger loading increments can be input in the subloading surface model possessing the automatic controlling function to pull-back the stress to the yield surface.

(3) Various minute deformation behaviors, e.g. the tangential-inelastic strain rate and the cyclic stagnation of the isotropic hardening can be easily incorporated within the framework of the hypoelastic-based plasticity.
(4) The spin of material is easily considered based on the corotational rates of stress and internal variables.

Return-mapping: It has been developed as the method to pull-back the stress to the yield surface in constitutive equations assuming the yield surface enclosing a purely-elastic domain, i.e. the conventional and the cyclic kinematic hardening models. Here, note the following facts.

(1) Repetitive calculation due to the elastic prediction and the plastic correction is required in an incremental step.
(2) Large loading increments can be input in proportional loading path. However, small loading increments must be input and thus calculation time is extended in calculations under non-linear loading path and cyclic loading process.
(3) The elastoplastic modulus tensor is not used but instead the yield condition, the flow rue and evolution rules of internal variables are used in the calculation.
(4) Material properties which can be taken into account are limited as it is difficult to incorporate the tangential-inelastic strain rate required to describe non-proportional loading, the isotropic hardening stagnation of metals required to describe cyclic loading behavior, etc.

Consequently, the forward-Euler method is more appropriate to the deformation analysis of materials with various minute material properties along a complex loading path by adopting the subloading surface model. In contrast, the return-mapping method is pertinent to the deformation analysis of materials with simple material properties along near proportional loading path.

The description in this chapter is highly supported by the discussions with Prof. Y. Yamakawa and his advices on numerical analysis.

The computer programming is expressed by the matrix form. Here, the second-order symmetric tensors possesses six independent values so that the relation between them is expressed by using 6×6 matrix, which is called the Voigt expression. Some important notes on the Voigt expression are given in **Appendix I**.

References

Amorosi A, Boldini D, Germano V (2007) Implicit integration of a mixed isotropic–kinematic hardening plasticity model for structured clays. Int J Numer Anal Meth Geomech 32:1173–1203

Anjiki T, Oka M, Hashiguchi K (2016) Elastoplastic analysis by complete implicit stress-update algorithm based on the extended subloading surface model. Trans Japan Soc Mech Eng. doi:10.1299/transjsme.16-00029 (in Japanese)

Argyris JH (1965) Elasto-plastic matrix analysis of three dimensional continua. J. Roy Aeronaut Soc 69:231–262

Borja RI, Tamagnini C (1998) Cam-clay plasticity, Part III: extension of the infinitesimal model to include finite strains. Comp Meth Appl Mech Eng 155:73–95

Callari C, Auricchio F, Sacco E (1998) A finite-strain Cam-clay model in the framework of multiplicative elasto-plasticity. Int J Plast 14:1155–1187

de Souza Neto EA, Perić D, Owen DJR (2008) Computational methods for plasticity. Wiley, Chichester, UK

Eidel B, Gruttmann F (2003) Elastoplastic orthotropy at finite strains: multiplicative formulation and numerical implementation. Compt Mater Sci 28:732–742

Flanagan DP, Taylor LM (1987) An accurate numerical algorithm for stress integration with finite rotations. Comput Meth Appl Mech Eng 62:305–320

Ghaei A, Green DE (2010) Numerical implementation of Yoshida-Uemori two-surface plasticity model using a fully implicit integration scheme. Compt Mater Sci 48:195–205

Hashiguchi K (2016) Loading criterion in return-mapping for subloading surface model. Proc Comput Mech Div JSME, 03-6

Hashiguchi K, Yamakawa Y (2012) Introduction to finite strain theory for continuum elasto-plasticity, vol Wiley Series in computational mechanics. Wiley, Chichester, UK

Hashiguchi K, Suzuki N, Ueno M (2014) Elastoplastic deformation analysis by return-mapping and consistent tangent modulus tensor based on subloading surface model, (1st Report, Formulation of return-mapping). Trans Japan Soc Mech Eng. doi:10.1299/transjsme.2014smm0083 (in Japanese)

Hinton E, Owen DRJ (1980) Finite elements in plasticity: theory and practice. Pineridge Press, Swansea, UK

Houlsby GT (1985) The use of a variable shear modulus in elastic-plastic models for clays. Comput Geotech 1:3–13

Houlsby GT, Amorosi A, Rojas E (2005) Elastic moduli of soils dependent on pressure: a hyperelastic formulation. Geotechnique 55(5):383–392

Hughes TJR, Pister KS (1978) Consistent linearization in mechanics of solids and structures. Comput Struct 9:391–397

Hughes TJR, Winget J (1980) Finite rotation effects in numerical integration of rate consistent equations arising in large-deformation analysis. Int J Numer Meth Eng 15:1862–1867

Krieg RD, Key SW (1976) Implementation of a time dependent plasticity theory into structural computer programs. In: Strickin JA, Saczlski KJ (eds) Constitutive equations in viscoplasticity: computational and engineering aspects. AMD-20, ASME

Krieg RD, Krieg DB (1977) Accuracies of numerical solution methods for the elastic-perfectly plastic models. J. Pressure Vessel Tech (ASME) 99:510–515

Menzel A, Steinmann P (2003) On the spatial formulation of anisotropic multiplicative elasto-plasticity. Compt Meth Appl Mech Eng 192:3431–3470

Menzel A, Ekh M, Runesson K, Steinmann P (2005) A framework for multiplicative elastoplasticity with kinematic hardening coupled to anisotropic damage. Int J Plast 21:397–434

Miehe C (1996) Numerical computation of algorithmic (consistent) tangent moduli in large-strain computational in elasticity. Comput Meth Appl Mech Eng 134:223–240

Niemunis A, Cudny M (1998) On hyperelasticity for clays. Comput Geotech 23:221–236

Ohno N, Tsuda M, Kamei T (2013) Elastoplastic implicit integration algorithm applicable to both plane stress and three-dimensional stress states. Finite Elem Anal Design 66:1–11

Ortiz M, Popov EP (1985) Accuracy and stability of integration algorithms for elastoplastic constitutive relations. Int J Numer Meth Eng 21:1561–1576

Ortiz M, Simo JC (1986) An analysis of a new class of integration algorithms for elastoplastic constitutive relations. Int J Numer Meth Eng 23:353–366

Pérez-Foguet A, Rodréguez-Ferran A, Huerta A (2000a) Numerical differentiation for non-trivial consistent tangent matrices: an application to the MRS-Lade model. Int J Numer Meth Eng 48:159–184

Pérez-Foguet A, Rodríguez-Ferran A, Huerta A (2000b) Numerical differentiation for local and global tangent operators in computational plasticity. Compt Meth Appl Mech Eng 189:277–296

Pérez-Foguet A, Rodríguez-Ferran A, Huerta A (2001) Consistent tangent matrices for substepping schemes. Compt Meth Appl Mech Eng 190:4627–4647

Pinsky PM, Ortiz M, Pister KS (1983) Numerical integration of rate constitutive equations in finite deformation analysis. Comput Meth Appl Mech Eng 193:5223–5256

Rice JR, Tracey DM (1973) Computational fracture mechanics. In: Feves SJ (ed) Proceedings of symposium numerical method of structure mechanics. Academic Press, New York, p 585

Simo JC, Hughes TJR (1998) Computational inelasticity. Springer, New York

Simo JC, Ortiz M (1985) A unified approach to finite deformation elastoplasticity based on the use of hyperelastic constitutive equations. Compt Meth Appl Mech Eng 49:221–245

Simo JC, Taylor RL (1985) Consistent tangent operators for rate-independent elastoplasticity. Comput Meth Appl Mech Eng 48:101–118

Simo JC, Taylor RL (1986) A return mapping algorithm for plane stress elastoplasticity. Int J Numer Meth Eng 22:649–670

Suzuki N, Hashiguchi K, Ueno M (2014): Elastoplastic deformation analysis by return-mapping and consistent tangent modulus tensor based on subloading surface model (2st Report, Deformation analyses of machine elements). Trans Japan Soc Mech Eng doi:10.1299/transjsme.2014smm0356 (in Japanese)

Tamagnini C, Castellanza R, Nova R (2002) A generalized backward Euler algorithm for the numerical integration of an isotropic hardening elastoplastic model for mechanical and chemical degradation of bonded geomaterials. Int J Numer Anal Meth Geomech 26:963–1004

Wilkins ML (1964) Calculation of elastoplastic flow. In: Alder B et al (eds) Methods of computational physics 3. Academic Press, New York

Yamada Y, Yoshimura N, Sakurai T (1968) Plastic stress-strain matrix and its application for the solution of elastic-plastic problems by finite element method. Int J Mech Sci 10:343–354

Yamakawa Y, Hashiguchi K, Ikeda K (2010a) Implicit stress-update algorithm for isotropic Cam-clay model based on the subloading surface concept at finite strains. Int J Plast 26:634–658

Yamakawa Y, Yamaguchi Y, Hashiguchi K, Ikeda K (2010b) Formulation and implicit stress-update algorithm of the extended subloading surface Cam-clay model with kinematic hardening for finite strains. J Appl Mech (ASCE) 13:411–412 (in Japanese)

Zienkiewicz OC (1977) The finite element method, 3rd edn. McGraw-Hill, London

Chapter 21
On Formulations from Thermodynamic View-Point

Thermodynamic laws must be satisfied in all natural phenomena, while, needless to say, an elastoplastic constitutive equation is not also an exception. However, the thermodynamic explanation is hardly described in literatures on elastoplastic constitute equations, especially hypoelastic-based plastic constitutive equation. On the other hand, the formulations from the aspect of thermodynamics are described in many literatures on constitutive equations of damage phenomena (cf. e.g. Lemaitre and Chaboche 1990; Lematire 1992; Lemaitre and Desmoral 2005; de Sauza Neto et al. 2008; Murakmi 2012) and transformation phenomena (cf. e.g. Bartel et al. 2011; Mahnken 2012) and the finite strain theory based on the multiplicative decomposition of the deformation gradient (cf. e.g. Menzel et al. 2005; Vladimirov et al. 2008, 2010; Wallin et al. 2003; Wallin and Ristinmaa 2005), which are based on the hyperelasiticy. What is the difference of these situations? Is the thermodynamics, especially the second law of thermodynamics inevitably necessary or useful for the formulations of plastic deformation of solids? We cannot find a literature which explains directly on these basic questions. The answer to the question to whether a useful information for formulations of constitutive equations on irreversible deformation or of plastic flow rule is derived based on the thermodynamics will be provided without an ambiguity in this chapter, while it was criticized in Hashiguchi (2001b). The explanations in Sects. 21.1–21.6 are referred widely to Fung (1965) and Lubarda (2002).

The readers may skip over this chapter, since the thermodynamical approaches to the formulation of plastic constitutive equation would be not be physically meaningful as will be described below.

K. Hashiguchi, *Foundations of Elastoplasticity: Subloading Surface Model*,
DOI 10.1007/978-3-319-48821-9_21

21.1 First Law of Thermodynamics: Conservation Law of Energy

The total energy possessed in a material is given by the sum of the kinetic energy K and the internal energy U:

$$K \equiv \int_v \frac{1}{2}\rho \mathbf{v} \cdot \mathbf{v} dv \tag{21.1}$$

$$U \equiv \int_v \rho u dv \tag{21.2}$$

where u is the internal energy per unit mass. On the other hand, the changing rate of the total energy in a material is given by the sum of the power $\dot{W}$ done by the traction vector $\mathbf{f}$ and the body force vector $\mathbf{b}$ and the rates of heat flux and generation $\dot{Q}$, i.e.

$$\dot{W} = \int_s \mathbf{f} \cdot \mathbf{v} ds + \int_v \mathbf{b} \cdot \mathbf{v} dv = \int_s \boldsymbol{\sigma}\mathbf{n} \cdot \mathbf{v} ds + \int_v \rho \mathbf{b} \cdot \mathbf{v} dv \tag{21.3}$$

$$\dot{Q} = -\int_s \mathbf{q} \cdot \mathbf{n} ds + \int_v \rho r dv = \int_v (-\nabla \cdot \mathbf{q} + \rho r) dv \tag{21.4}$$

where $\mathbf{q}$ is the heat flux vector so that the heat $\mathbf{q}dt$ flows out during the time-increment dt from the unit area of a material surface. The heat flowing out from the infinitesimal area ds is given by $\mathbf{q} \cdot \mathbf{n} ds$ where $\mathbf{n}$ is the unit outward-normal vector of the material surface. The scalar r is the rate of heat input per unit mass from heat source. The kinetic energy K and the internal energy U are the state quantities but $\int \dot{W} dt$ and $\int \dot{Q} dt$ are not state quantities which change the total energy consisting of K and U. Then, it follows by virtue of the energy conservation law that

$$\dot{K} + \dot{U} = \dot{W} + \dot{Q} \ (dK + dU = \delta W + \delta Q) \tag{21.5}$$

where the incremental symbols $d(\,)$ and $\delta(\,)$ are used for the state variables and the non-state variables, respectively.

Noting the relation

$$\left(\int_v \frac{1}{2}\rho\mathbf{v}\cdot\mathbf{v}dv\right)^{\bullet} = \int_v \left[\left(\frac{1}{2}\rho\mathbf{v}\cdot\mathbf{v}\right)^{\bullet} + \left(\frac{1}{2}\rho\mathbf{v}\cdot\mathbf{v}\right)\nabla\cdot\mathbf{v}\right]dv$$
$$= \int_v \left[\frac{1}{2}\dot{\rho}\,\mathbf{v}\cdot\mathbf{v} + \rho\,\dot{\mathbf{v}}\cdot\mathbf{v} + \left(\frac{1}{2}\rho\mathbf{v}\cdot\mathbf{v}\right)\nabla\cdot\mathbf{v}\right]dv$$
$$= \int_v \left[\rho\,\dot{\mathbf{v}}\cdot\mathbf{v} + \left(\frac{1}{2}\mathbf{v}\cdot\mathbf{v}\right)(\dot{\rho} + \rho\nabla\cdot\mathbf{v})\right]dv = \int_v \rho\,\dot{\mathbf{v}}\cdot\mathbf{v}dv$$

obtained using the Reynolds' transportation theorem in Eq. (2.160) and the continuity equation in Eq. (3.26), the rate of kinetic energy is given by

$$\dot{K} = \int_v \rho\,\dot{\mathbf{v}}\cdot\mathbf{v}dv \tag{21.6}$$

Noting the relation

$$\left(\int_v \rho u dv\right)^{\bullet} = \int_v \rho\,\dot{u}dv$$

based on Eq. (3.27), the rate of internal energy is given by

$$\dot{U} = \int_v \rho\,\dot{u}dv \tag{21.7}$$

Further, by the relation

$$\int_s \boldsymbol{\sigma}\mathbf{n}\cdot\mathbf{v}ds = \int_s \sigma_{ij}n_j v_i ds = \int_v \frac{\partial\sigma_{ij}v_j}{\partial x_i}dv$$
$$= \int_v \left(\frac{\partial\sigma_{ij}}{\partial x_i}v_j + \sigma_{ij}\frac{\partial v_j}{\partial x_i}\right)dv = \int_v (\nabla\boldsymbol{\sigma}\cdot\mathbf{v} + \boldsymbol{\sigma}:\mathbf{d})dv$$

the work rate in Eq. (21.3) is rewritten as

$$\dot{W} = \int_v \boldsymbol{\sigma}:\mathbf{d}dv + \int_v (\nabla\boldsymbol{\sigma} + \rho\mathbf{b})\cdot\mathbf{v}dv \tag{21.8}$$

The substitution of Eqs. (21.4), (21.6), (21.7) and (21.8) into Eq. (21.5) leads to

$$\rho\,\dot{\mathbf{v}}\cdot\mathbf{v}+\rho\,\dot{u}=\boldsymbol{\sigma}:\mathbf{d}+(\nabla\boldsymbol{\sigma}+\rho\mathbf{b})\cdot\mathbf{v}-\nabla\cdot\mathbf{q}+\rho r$$

Taking account of the equilibrium equation in Eq. (3.31) to this relation, one has

$$\rho\,\dot{u}=\boldsymbol{\sigma}:\mathbf{d}-\nabla\cdot\mathbf{q}+\rho r \tag{21.9}$$

or

$$\dot{u}=\frac{1}{\rho}\boldsymbol{\sigma}:\mathbf{d}-\frac{1}{\rho}\nabla\cdot\mathbf{q}+r \tag{21.10}$$

which is the first law of thermodynamics and called the Fourier-Kirchhoff-Neumann's (or simply Neumann's) heat energy equation.

21.2 Energy Equation in Reference Configuration

The power per unit mass in the current configuration is transformed to that in the reference configuration, noting the relation $\rho\det\mathbf{F}=\rho_0$ and $\mathbf{A}:\mathbf{BC}=A_{ij}(B_{ir}C_{rj})=(A_{ij}C_{rj})B_{ir}=\mathbf{AC}^T:\mathbf{B}$ for arbitrary second-order tensors $\mathbf{A},\mathbf{B},\mathbf{C}$ and Eq. (3.19) as follows:

$$\begin{aligned}\frac{1}{\rho}\boldsymbol{\sigma}:\mathbf{d}&=\frac{1}{\rho}\boldsymbol{\sigma}:\frac{1}{2}\left[\frac{\partial\dot{\mathbf{x}}}{\partial\mathbf{x}}+\left(\frac{\partial\dot{\mathbf{x}}}{\partial\mathbf{x}}\right)^T\right]=\frac{1}{\rho}\boldsymbol{\sigma}:\frac{\partial\dot{\mathbf{x}}}{\partial\mathbf{x}}\\&=\frac{1}{\rho}\boldsymbol{\sigma}:\frac{\partial\dot{\mathbf{x}}}{\partial\mathbf{X}}\frac{\partial\mathbf{X}}{\partial\mathbf{x}}=\frac{\det\mathbf{F}}{\rho_0}\boldsymbol{\sigma}:\left(\frac{\partial\mathbf{x}}{\partial\mathbf{X}}\right)^{\bullet}\frac{\partial\mathbf{X}}{\partial\mathbf{x}}=\frac{\det\mathbf{F}}{\rho_0}\boldsymbol{\sigma}:\dot{\mathbf{F}}\mathbf{F}^{-1}\\&=\frac{1}{\rho_0}(\det\mathbf{F})\boldsymbol{\sigma}\mathbf{F}^{-T}:\dot{\mathbf{F}}=\frac{1}{\rho_0}\boldsymbol{\Pi}:\dot{\mathbf{F}}=\frac{1}{\rho_0}\mathbf{T}^{(n)}:\dot{\mathbf{E}}^{(n)}\end{aligned} \tag{21.11}$$

denoting the work-conjugate pair of the general stress and the strain by $\mathbf{T}^{(n)}$ and $\mathbf{E}^{(n)}$, respectively.

One has also

$$\nabla=\nabla_0\mathbf{F}^{-1}=\mathbf{F}^{-T}\nabla_0 \tag{21.12}$$

$$(\partial/\partial x_i=(\partial/\partial X_r)(\partial X_r/\partial x_i)=(\partial X_r/\partial x_i)(\partial/\partial X_r))$$

Further, noting the relation

$$\mathbf{q}\bullet\mathbf{n}ds = \mathbf{q}_0\bullet\mathbf{n}_0 ds_0 = \mathbf{q}_0\bullet\mathbf{F}^T\mathbf{n}ds/\det\mathbf{F} = \mathbf{F}\mathbf{q}_0\bullet\mathbf{n}ds/\det\mathbf{F}$$

which is derived by substituting $\mathbf{q}\bullet\mathbf{n}ds = \mathbf{q}_0\bullet\mathbf{n}_0 ds_0$ into $\mathbf{n}_0 ds_0 = \mathbf{F}^T\mathbf{n}ds/\det\mathbf{F}$ due to Eq. (2.147), one has

$$\mathbf{q} = \frac{1}{\det\mathbf{F}}\mathbf{F}\mathbf{q}_0 = \frac{\rho}{\rho_0}\mathbf{F}\mathbf{q}_0 \tag{21.13}$$

It follows from Eqs. (21.12) and (21.13) that

$$\frac{1}{\rho}\nabla\bullet\mathbf{q} = \frac{1}{\rho_0}(\mathbf{F}^{-T}\nabla_0)\bullet(\mathbf{F}\mathbf{q}_0) = \frac{1}{\rho_0}\nabla_0\bullet\mathbf{q}_0 \tag{21.14}$$

Equation (21.9) is transformed by Eqs. (21.11) and (21.13) to the energy equation in the reference configuration.

$$\rho_0\,\dot{u} = \mathbf{T}^{(n)}:\overset{\bullet}{\mathbf{E}}{}^{(n)} - \nabla_0\bullet\mathbf{q}_0 - \rho_0 r \tag{21.15}$$

or

$$\dot{u} = \frac{1}{\rho_0}\mathbf{T}^{(n)}:\overset{\bullet}{\mathbf{E}}{}^{(n)} - \frac{1}{\rho_0}\nabla_0\bullet\mathbf{q}_0 + r \tag{21.16}$$

21.3 Second Law of Thermodynamics: Clausius-Duhem Inequality

The first law of thermodynamics is relevant to the conservation of energy but irrelevant to the directions of variations of work and heat. The second law imposes the restriction to the direction of the thermodynamic process which is generated actually. The thermodynamic process is generated naturally to the direction in which the entropy describing the degree of randomness of material state increases. The entropy is generated by the irreversible mechanical phenomena, e.g. the heat conduction, the heat diffusion and the purely-plastic slip which does not induce the variation of internal structure in addition to the heat flux and the heat supply by the heat generation in the heat source. Here, the rate of entropy caused only by the input/output of heat is given by Eq. (21.4) as follows (Truesdell and Noll 1965; Malvern 1969):

$$\int_v \left(-\nabla \cdot \frac{\mathbf{q}}{\theta} + \rho \frac{r}{\theta}\right) dv = \int_v \left(-\frac{1}{\rho}\nabla \cdot \frac{\mathbf{q}}{\theta} + \frac{r}{\theta}\right) \rho dv \tag{21.17}$$

where $\theta\,(> 0)$ is the absolute temperature. The quantities in Eq. (21.17) is divided by absolute temperature since the degree of randomness is low for lower temperature (randomness is zero in the zero absolute temperature) so that the rate of variation in randomness, i.e. the rate of variation in entropy is higher in a lower temperature. The quantity defined by Eq. (21.17) is the *entropy supply*.

The entropy variation in the system based on the heat flux and generation in the reversible process is given by Eq. (21.17) but it is larger than that given by Eq. (21.17) in the general irreversible process. Then, the general rate of entropy is given by the following inequality, designating the entropy per unit mass by η.

$$\int_v \frac{d\eta}{dt} dv \geq \int_v \left(-\frac{1}{\rho}\nabla \cdot \frac{\mathbf{q}}{\theta} + \frac{r}{\theta}\right) \rho dv \tag{21.18}$$

Here, the rate of entropy production in the whole system satisfies the following equation, designating the rate of entropy production induced by the irreversible deformation inducing micromechanical changes per unit mass as γ.

$$\int_v \frac{d\eta}{dt} dv = \int_v \left(-\frac{1}{\rho}\nabla \cdot \frac{\mathbf{q}}{\theta} + \frac{r}{\theta} + \gamma\right) \rho dv \tag{21.19}$$

where the rate of entropy production γ is zero in the reversible process but positive in the irreversible process so that the following inequality holds, i.e.

$$\gamma \geq 0 \tag{21.20}$$

The following relation holds in each material point, i.e. locally.

$$\dot{\eta} = \overset{\bullet}{\eta}{}^{e} + \overset{\bullet}{\eta}{}^{i} = -\frac{1}{\rho}\nabla \cdot \left(\frac{\mathbf{q}}{\theta}\right) + \frac{r}{\theta} + \gamma \tag{21.21}$$

$$\dot{\eta} = \overset{\bullet}{\eta}{}^{e} + \overset{\bullet}{\eta}{}^{i} = -\frac{1}{\rho_0}\nabla_0 \cdot \left(\frac{\mathbf{q}_0}{\theta}\right) + \frac{r}{\theta} + \gamma \tag{21.22}$$

where

$$\overset{\bullet}{\eta}{}^{e} = \begin{cases} -\dfrac{1}{\rho}\nabla \cdot \left(\dfrac{\mathbf{q}}{\theta}\right) + \dfrac{r}{\theta} \\ -\dfrac{1}{\rho_0}\nabla_0 \cdot \left(\dfrac{\mathbf{q}_0}{\theta}\right) + \dfrac{r}{\theta} + \gamma \end{cases} \tag{21.23}$$

$$\dot{\eta}^{i} = \gamma\,(\geq 0) \tag{21.24}$$

Here, noting the relations

$$\nabla \cdot \left(\frac{\mathbf{q}}{\theta}\right) = \frac{1}{\theta}\nabla \cdot \mathbf{q} - \frac{1}{\theta^2}\mathbf{q} \cdot \nabla\theta \tag{21.25}$$

$$\nabla_0 \cdot \left(\frac{\mathbf{q}_0}{\theta}\right) = \frac{1}{\theta}\nabla_0 \cdot \mathbf{q}_0 - \frac{1}{\theta^2}\mathbf{q}_0 \cdot \nabla_0\theta \tag{21.26}$$

Eqs. (21.21) and (21.22) are described as follows:

$$\dot{\eta} = -\frac{1}{\rho\theta}\nabla \cdot \mathbf{q} + \frac{1}{\rho\theta^2}\mathbf{q} \cdot \nabla\theta + \frac{r}{\theta} + \gamma \tag{21.27}$$

$$\dot{\eta} = -\frac{1}{\rho_0\theta}\nabla_0 \cdot \mathbf{q}_0 + \frac{1}{\rho_0\theta^2}\mathbf{q}_0 \cdot \nabla_0\theta + \frac{r}{\theta} + \gamma \tag{21.28}$$

Substituting Eqs. (21.10) and (21.16) into these equations, one has

$$\theta\,\dot{\eta} = -\frac{1}{\rho}\boldsymbol{\sigma} : \mathbf{d} + \dot{u} + \frac{1}{\rho\theta}\mathbf{q} \cdot \nabla\theta + \theta\gamma \tag{21.29}$$

$$\theta\,\dot{\eta} = -\frac{1}{\rho_0}\mathbf{T}^{(n)} : \dot{\mathbf{E}}^{(n)} + \dot{u} + \frac{1}{\rho_0\theta}\mathbf{q}_0 \cdot \nabla_0\theta + \theta\gamma \tag{21.30}$$

On the other hand, inserting Eqs. (21.21) and (21.22) into the inequality (21.20), it follows that

$$\dot{\eta} + \frac{1}{\rho}\nabla \cdot \left(\frac{\mathbf{q}}{\theta}\right) - \frac{r}{\theta} \geq 0 \tag{21.31}$$

$$\dot{\eta} + \frac{1}{\rho}\nabla_0 \cdot \left(\frac{\mathbf{q}_0}{\theta}\right) - \frac{r}{\theta} \geq 0 \tag{21.32}$$

where the equalities hold only in reversible process. These inequalities can be described as

$$\dot{\eta} \geq -\frac{1}{\rho}\nabla \cdot \left(\frac{\mathbf{q}}{\theta}\right) + \frac{r}{\theta} \tag{21.33}$$

$$\dot{\eta} \geq -\frac{1}{\rho}\nabla_0 \cdot \left(\frac{\mathbf{q}_0}{\theta}\right) + \frac{r}{\theta} \tag{21.34}$$

which insist that the rate of entropy is larger than the sum of the heat flux and the heat generation divided by the absolute temperature. These inequalities are referred to as the second law of thermodynamics or *Clausius-Duhem inequality* or the *entropy inequality*. Here, Eqs. (21.31) and (21.32) are described by (21.26) as follows:

$$\theta \dot{\eta} + \frac{1}{\rho} \mathbf{\nabla} \cdot \mathbf{q} - \frac{1}{\rho \theta} \mathbf{q} \cdot \mathbf{\nabla} \theta - r \geq 0 \tag{21.35}$$

$$\theta \dot{\eta} + \frac{1}{\rho_0} \mathbf{\nabla}_0 \cdot \mathbf{q}_0 - \frac{1}{\rho_0 \theta} \mathbf{q}_0 \cdot \mathbf{\nabla}_0 \theta - r \geq 0 \tag{21.36}$$

which are simplified for the process without the heat flux and heat generation as follows:

$$\dot{\eta} \geq 0 \tag{21.37}$$

Consequently, a natural phenomenon proceeds in the direction in which the entropy increases. Besides, the thermodynamic equilibrium state is the stable state that the variation does not occurs spontaneously without an action of external agency and thus the maximal state of entropy. In other words, the entropy increases when the state of system varies and the variation of entropy stops when it becomes maximum compared with those in the neighborhood systems. Equation (21.37) is called the Gibbs' condition for thermodynamic equilibrium.

21.4 Generalized Force and Displacement: Onsagar's Principle

Prior to entering to the main issue, consider the heat conduction phenomenon as an actual example of entropy production (Fung 1965).

Considering the slender bar with a section of area da surrounded by the adiabatic wall, choosing the x-axis to the longitudinal direction (see Fig. 21.1) and designating the heat flux per a unit area in the x-direction as q, the increment of heat in the part with an infinitesimal length dx is given as

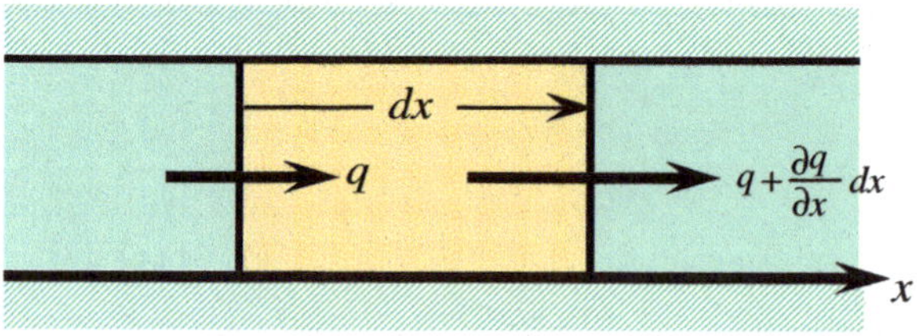

Fig. 21.1 One-dimensional heat conduction

$$dQ = \left[q dt - \left(q + \frac{\partial q}{\partial x} dx \right) dt \right] da = -\frac{\partial q}{\partial x} dadxdt \tag{21.38}$$

for which the rate of entropy is given noting $\dot{\eta} = \dot{Q}/[\rho(dadx)\theta]$ by

$$\dot{\eta}\, \rho dadx = \frac{\dot{Q}}{\theta} = -\frac{1}{\theta}\frac{\partial q}{\partial x} dadx \tag{21.39}$$

Taking account of

$$\frac{\partial}{\partial x}\left(\frac{q}{\theta}\right) = \frac{1}{\theta}\frac{\partial q}{\partial x} - \frac{q}{\theta^2}\frac{\partial \theta}{\partial x}$$

in Eq. (21.39), it follows that

$$\dot{\eta} = -\frac{1}{\rho}\frac{\partial}{\partial x}\left(\frac{q}{\theta}\right) - \frac{1}{\rho}\frac{q}{\theta^2}\frac{\partial \theta}{\partial x} \tag{21.40}$$

The first term in right-hand in this equation is the reversible part shown in Eq. (21.23). In contrast, the second term in the right-hand side is the irreversible part due to the heat conduction, i.e. entropy production, which can be extended to the three-dimensional state as follows:

$$\dot{\eta}^{i} = \gamma = -\frac{1}{\rho}\frac{\nabla \theta}{\theta} \bullet \frac{\mathbf{q}}{\theta} \tag{21.41}$$

which shows the scalar product of the generalized force $-(\nabla\theta)/\theta$ and its related generalized flux $\mathbf{q}/\theta$.

In what follows, we consider the generalized irreversible phenomenon, referring to the above-mentioned heat conduction phenomenon.

Designating the generalized force causing the energy dissipation by f_i and the generalized displacement causing the entropy production by ξ_i, the rate of entropy production is described by the following equation:

$$\theta\gamma = f_i \dot{\xi}_i \tag{21.42}$$

The relations $f_i = -(1/\rho)(\nabla_i\theta)/\theta$ and $\dot{\xi}_i = q_i$ hold from Eq. (21.41) in the above-mentioned example of heat conduction.

Onsager (1931) assumed the following equation, noting that the rate of the generalized displacement, i.e. the flux $\dot{\xi}_i$ depends linearly on the thermodynamic generalized force f_i.

$$\dot{\xi}_i = \Lambda_{ij} f_j \tag{21.43}$$

where the coefficient Λ_{ij} is called the phenomenological coefficient. For instance, consider the one-dimensional heat diffusion as follows (Onsagar 1931; Fung 1965):

$$\left.\begin{aligned} \dot{\xi}_1 &= \Lambda_{11} f_1 + \Lambda_{12} f_2 \\ \dot{\xi}_2 &= \Lambda_{21} f_1 + \Lambda_{22} f_2 \end{aligned}\right\} \tag{21.44}$$

where $\dot{\xi}_1$ and $\dot{\xi}_2$ are the heat flux and the mass flux, respectively, and f_1 and f_2 are the temperature gradient and the concentration gradient of the particular ingredient, respectively. Here, Λ_{11} is the thermal conductivity in the x_1-direction, Λ_{22} the mass diffusion coefficient in the x_2-direction, and Λ_{12} and Λ_{21} are the influence coefficients describing the interfere of the two irreversible phenomena for the heat conduction and the heat diffusion, respectively. Λ_{12} is relevant to the *Dufour effect* describing the generation of heat flux and Λ_{21} to the *Storet effect* describing the generation of the mass diffusion when the temperature gradient exists. Therefore, The upper equation in Eq. (21.44) is extension of the following *Fourier heat conduction law* to include the Dufour effect.

$$q_i = -k \nabla_i \theta \tag{21.45}$$

where k is the thermal conductivity. In contrast, the lower equation in Eq. (21.44) is the extension of the following Fick's mass diffusion to include the Soret effect.

$$p_i = -j \nabla_i \phi \tag{21.46}$$

where $\mathbf{p}$ is the mass flux, j the diffusion coefficient, ϕ the mass density.

Substituting Eq. (21.43) into Eq. (21.42), one has

$$\theta \gamma = \Lambda_{ij} f_i f_j = \Lambda_{ji} f_i f_j \tag{21.47}$$

by which the symmetry

$$\Lambda_{ij} = \Lambda_{ji} \tag{21.48}$$

holds. This is referred to as the *Onsager reciprocal relation* (cf. Ziegler 1983; Germain et al. 1983).

21.5 Thermodynamic Potential

The extensive function which possesses the all the information for thermodynamic property and is described by the independent generalized displacements is referred to the thermodynamic potential. It is also called the complete thermodynamic function. One can define various thermodynamic functions with dimension of energy, depending on the choice of extensive variables. Further, by comparing the rate of the thermodynamic potential with the total differential equation, the conjugate pairs of the generalized forces and the generalized displacements included in the thermodynamic potential are derived as will be described below, referring to Lubarda (2002).

Substituting Eq. (21.42) into Eqs. (21.29) and (21.30), we have the following equations.

$$\dot{u} = \frac{1}{\rho}\boldsymbol{\sigma} : \mathbf{d} + \theta\,\dot{\eta} - f_j\dot{\xi}_j - \frac{1}{\rho\theta}\mathbf{q}\cdot\nabla\theta \tag{21.49}$$

$$\dot{u} = \frac{1}{\rho_0}\mathbf{T}^{(n)} : \dot{\mathbf{E}}^{(n)} + \theta\,\dot{\eta} - f_j\dot{\xi}_j - \frac{1}{\rho_0\theta}\mathbf{q}_0\cdot\nabla_0\theta \tag{21.50}$$

from which it is known that the rate of the internal energy is given by the sum of (1) the energy due to the power done to the material, (2) the reversible rate of entropy which is given by subtracting the irreversible rate of entropy from the rate of entropy and (3) the input rate of heat.

Regarding $\mathbf{E}^{(n)}$, η, ξ_i as the independent variables, the internal energy u is described by

$$u = u(\mathbf{E}^{(n)}, \eta, \xi_i) \tag{21.51}$$

The total differentiation of Eq. (21.51) leads to

$$\dot{u} = \frac{\partial u}{\partial \mathbf{E}^{(n)}} : \dot{\mathbf{E}}^{(n)} + \frac{\partial u}{\partial \eta}\dot{\eta} + \frac{\partial u}{\partial \xi_i}\dot{\xi}_i \tag{21.52}$$

Comparing Eq. (21.52) with Eq. (21.50), the following relation are obtained for the case that the gradient of temperature is negligible.

$$\mathbf{T}^{(n)} = \rho_0\frac{\partial u}{\partial \mathbf{E}^{(n)}}, \quad \theta = \frac{\partial u}{\partial \eta}, \quad f_i = -\frac{\partial u}{\partial \xi_i} \tag{21.53}$$

Then, u can be regarded to be the thermodynamic potential of $\mathbf{T}^{(n)}$, θ, f_i.

Inserting Eqs. (21.29) and (21.30) into the inequality Eq. (21.20), the Clausius-Duhem inequality is rewritten as

$$\frac{1}{\rho}\boldsymbol{\sigma}:\mathbf{d}-\dot{u}-\frac{1}{\rho\theta}\mathbf{q}\cdot\nabla\theta+\theta\dot{\eta}\geq 0 \tag{21.54}$$

$$\frac{1}{\rho_0}\mathbf{T}^{(n)}:\dot{\mathbf{E}}^{(n)}-\dot{u}-\frac{1}{\rho_0\theta}\mathbf{q}_0\cdot\nabla_0\theta+\theta\dot{\eta}\geq 0 \tag{21.55}$$

which is simplified in the state that the work is not done and the heat flux is negligible, leading to the isothermal process, as follows:

$$\dot{u}\leq 0 \tag{21.56}$$

Therefore, the phenomenon proceeds in the direction to which the internal energy decreases. The equilibrium state is defined as the minimal state of the internal energy and thus it is regarded as the thermodynamically stable state that the state does not change naturally.

Introduce the *Helmholtz free energy function* (strain energy function) ψ which is related to the internal energy u by the following equation.

$$\psi(\mathbf{E}^{(n)},\theta,\xi_i)=u-\theta\eta(\mathbf{E}^{(n)},\eta,\xi_i) \tag{21.57}$$

where

$$\theta=\frac{\partial u}{\partial\eta},\quad \eta=-\frac{\partial\psi}{\partial\theta} \tag{21.58}$$

hold and thus the function ψ is the Legendre transformation u with respect to θ,η.

(Note) Legendre transformation:

Noting Fig. 21.2, consider the straight line which has the inclination p and intersects with the curve which is convex to the downward and is described by the function

$$y=f(x) \tag{21.59}$$

The straight line which possesses the minimum value of the interception with the y-axis is the one contacting with the curve. Here, designate the minimum value of the interception with the y-axis by $\min\limits_{x}[f(x)-px]$ which is defined as the Legendre transformation, i.e.

$$f^*(p)=\min_{x}[f(x)-px] \tag{21.60}$$

where $\min\limits_{x}[\,]$ designates the minimum value of the value inside the bracket $[\,]$ when x changes. Using this notation, the tangent line is described by

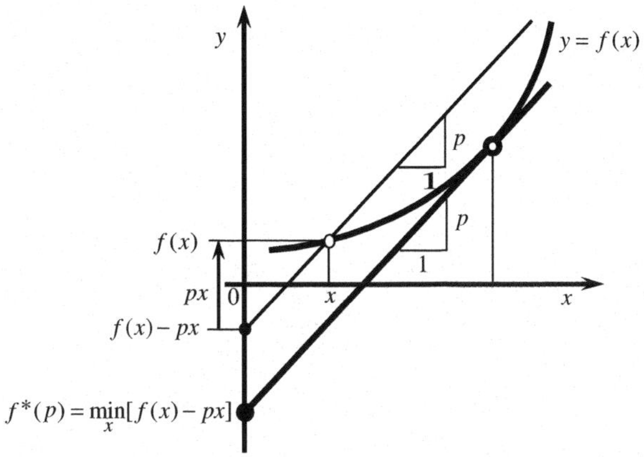

Fig. 21.2 Legendre transformation

$$y = px + f^*(p) \tag{21.61}$$

It follows from Eqs. (21.59) and (21.61) that

$$f(x) = px + f^*(p), \quad \text{i.e.}\, f^*(p) = f(x) - px \tag{21.62}$$

from which one has

$$p = \frac{\partial f}{\partial x}, \quad x = -\frac{\partial f^*}{\partial p} \tag{21.63}$$

Equation (21.62) is the *Legendre transformation* transforming the function $f(x)$ to $f^*(p)$ in which the argument x is replaced to $p = \partial f / \partial x$ which is the inclination of the tangent line of the curve $y = f(x)$ as described above. Besides, the replacements $x \to \theta, f \to \psi; p \to -\eta, f^* \to u$ leads Eq. (21.62) for the general Legendre transformation to Eq. (21.57) for the relation of Helmholtz energy function and the internal energy.

It follows from Eq. (21.57) that

$$\dot{\psi} = \dot{u} - \dot{\theta}\eta - \theta\dot{\eta}, \quad \dot{u} = \dot{\psi} + \theta\dot{\eta} + \dot{\theta}\eta, \quad \theta\dot{\eta} = -\dot{\psi} + \dot{u} - \dot{\theta}\eta \tag{21.64}$$

The substitutions of Eqs. (21.49) and (21.50) into Eq. $(21.64)_1$ leads to

$$\dot{\psi} = \frac{1}{\rho}\boldsymbol{\sigma} : \mathbf{d} - \eta\dot{\theta} - f_j\dot{\xi}_j - \frac{1}{\rho\theta}\mathbf{q} \cdot \nabla\theta \tag{21.65}$$

$$\dot{\psi} = \frac{1}{\rho_0}\mathbf{T}^{(n)} : \dot{\mathbf{E}}^{(n)} - \eta\dot{\theta} - f_j\dot{\xi}_j - \frac{1}{\rho_0}\mathbf{q}_0 \cdot \nabla_0\theta \tag{21.66}$$

Then, $\dot{\psi}$ designates the power per unit density which is done in the reversible process ($\dot{\xi}_j = 0$) at the adiabatic state ($\mathbf{q} = \mathbf{0}$) under the constant temperature ($\dot{\theta} = 0$).

Regarding $\mathbf{E}^{(n)}$, θ , ξ_i to be the independent variables, the Helmholtz free energy is described by

$$\psi = \psi(\mathbf{E}^{(n)}, \theta, \xi_j) \tag{21.67}$$

The total differentiation of Eq. (21.67) leads to

$$\dot{\psi} = \frac{\partial \psi}{\partial \mathbf{E}^{(n)}} : \dot{\mathbf{E}}^{(n)} + \frac{\partial \psi}{\partial \theta}\dot{\theta} + \frac{\partial \psi}{\partial \xi_i}\dot{\xi}_i \tag{21.68}$$

Comparing Eq. (21.68) with Eq. (21.66) and ignoring the scalar product of the heat flux and temperature gradient ($\mathbf{q} \cdot \nabla\theta = \mathbf{q}_0 \cdot \nabla_0\theta = 0$), one has

$$\mathbf{T}^{(n)} = \rho_0 \frac{\partial \psi}{\partial \mathbf{E}^{(n)}}, \quad \eta = -\frac{\partial \psi}{\partial \theta}, \quad f_i = -\frac{\partial \psi}{\partial \xi_i} \tag{21.69}$$

Substituting Eqs. (21.10) and (21.16) into Eq. $(21.64)_3$, one has

$$\theta\dot{\eta} = \frac{1}{\rho}\boldsymbol{\sigma} : \mathbf{d} - \dot{\psi} - \frac{1}{\rho}\nabla \cdot \mathbf{q} + r - \dot{\theta}\eta \tag{21.70}$$

$$\theta\dot{\eta} = \frac{1}{\rho_0}\mathbf{T}^{(n)} : \dot{\mathbf{E}}^{(n)} - \dot{\psi} - \frac{1}{\rho_0}\nabla_0 \cdot \mathbf{q}_0 + r - \dot{\theta}\eta \tag{21.71}$$

The substitutions of these equations into the Clausius-Duhem inequalities in Eqs. (21.35) and (21.36) are described as follows:

$$\frac{1}{\rho}\boldsymbol{\sigma} : \mathbf{d} - \dot{\psi} - \frac{1}{\rho\theta}\mathbf{q} \cdot \nabla\theta - \dot{\theta}\eta \geq 0 \tag{21.72}$$

$$\frac{1}{\rho_0}\mathbf{T}^{(n)} : \dot{\mathbf{E}}^{(n)} - \dot{\psi} - \frac{1}{\rho_0\theta}\mathbf{q}_0 \cdot \nabla_0 - \dot{\theta}\eta \geq 0 \tag{21.73}$$

which are simplified to the following inequality in the isothermal state that a work is not done and there does not exist a heat flow.

$$\dot{\psi} \leq 0 \tag{21.74}$$

Therefore, in such state, the process proceeds towards the direction to which the Helmholtz free decreases. The thermodynamic equilibrium state means the minimum state of the Helmholtz free energy.

Further, introduce the Legendre transformation of the Helmholtz energy ψ, i.e. the *Gibbs' free energy function* (complementary strain energy function) ϕ defined as the complementary energy to ψ.

$$\psi(\mathbf{E}^{(n)}, \theta, \xi_i) + \phi(\mathbf{T}^{(n)}, \theta, \xi_i) = \frac{1}{\rho_0}\mathbf{T}^{(n)} : \mathbf{E}^{(n)} \tag{21.75}$$

which is obtained from the Legendre transformation equation (21.62) by the replacement of

$$x \to \mathbf{E}^{(n)}, f \to \psi; p \to \mathbf{T}^{n}, f^* \to -\phi$$

It follows from Eq. (21.75) that

$$\overset{\bullet}{\psi}(\mathbf{E}^{(n)}, \theta, \xi_i) + \overset{\bullet}{\phi}(\mathbf{T}, \theta, \xi_i) = \frac{1}{\rho_0}\mathbf{T}^{(n)} : \overset{\bullet}{\mathbf{E}}^{(n)} + \frac{1}{\rho_0}\overset{\bullet}{\mathbf{T}}^{(n)} : \mathbf{E}^{(n)} \tag{21.76}$$

Substituting Eq. (21.66) into Eq. (21.76), one has

$$\begin{aligned}
&\overset{\bullet}{\phi}(\mathbf{T}^{(n)}, \theta, \xi_i) \\
&= \frac{1}{\rho_0}\overset{\bullet}{\mathbf{T}}^{(n)} : \mathbf{E}^{(n)} + \frac{1}{\rho_0}\mathbf{T}^{(n)} : \overset{\bullet}{\mathbf{E}}^{(n)} - \frac{1}{\rho_0}\mathbf{T}^{(n)} : \overset{\bullet}{\mathbf{E}}^{(n)} + \eta\,\overset{\bullet}{\theta} + \frac{1}{\rho_0}\mathbf{q}_0 \bullet \nabla_0\theta + f_j\overset{\bullet}{\xi}_j
\end{aligned}$$

which leads to

$$\overset{\bullet}{\phi} = \frac{1}{\rho_0}\mathbf{E}^{(n)} : \overset{\bullet}{\mathbf{T}}^{(n)} + \eta\,\overset{\bullet}{\theta} + f_j\overset{\bullet}{\xi}_j + \frac{1}{\rho_0}\mathbf{q}_0 \bullet \nabla_0\theta \tag{21.77}$$

On the other hand, the Gibbs free energy ϕ is described for the independent variables $\mathbf{T}^{(n)}$, θ, ξ_i as

$$\phi = \phi(\mathbf{T}^{(n)}, \theta, \xi_i) \tag{21.78}$$

the time-differentiation of which leads to

$$\overset{\bullet}{\phi} = \frac{\partial\phi}{\partial\mathbf{T}^{(n)}} : \overset{\bullet}{\mathbf{T}}^{(n)} + \frac{\partial\phi}{\partial\theta}\overset{\bullet}{\theta} + \frac{\partial\phi}{\partial\xi_i}\overset{\bullet}{\xi}_i \tag{21.79}$$

Comparing Eq. (21.77) with Eq. (21.79), it follows that

$$\mathbf{E}^{(n)} = \rho_0\frac{\partial\phi}{\partial\mathbf{T}^{(n)}}, \quad \eta = \frac{\partial\phi}{\partial\theta}, \quad f_i = \frac{\partial\phi}{\partial\xi_i} \tag{21.80}$$

in the state that the temperature gradient is negligible. Therefore, the Gibbs free energy ϕ is regarded as the thermodynamic potential for $\mathbf{E}^{(n)}$, η and f_i and possesses the role of the complementary strain energy.

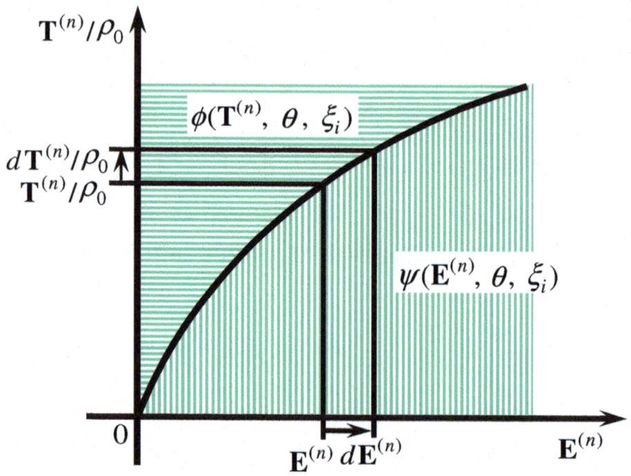

Fig. 21.3 Legendre transformation of Helmholtz free energy ψ and the Gibbs free energy ϕ

The relation of the Helmholtz free energy ψ and the Gibbs free energy ϕ is schematically illustrated in Fig. 21.3 by Eqs. (21.66), (21.75) and (21.77).

ψ is measure-invariant for a given geometrical change but ϕ is not measure-invariant since $\mathbf{T}^{(n)}:\mathbf{E}^{(n)}$ depends on the chosen stress measure $\mathbf{T}^{(n)}$ and the strain $\mathbf{E}^{(n)}$. However, $\mathbf{T}^{(n)}:d\mathbf{E}^{(n)}$ is measure-invariant so that $\mathbf{E}^{(n)}$ and $\mathbf{T}^{(n)}$ are work-conjugate. The substitution of Eq. (21.73) into Clausius-Duhem inequality in Eq. (21.76) leads to

$$-\frac{1}{\rho_0}\mathbf{E}^{(n)}:\dot{\mathbf{T}}^{(n)}+\dot{\phi}-\frac{1}{\rho_0}\mathbf{q}_0\cdot\nabla_0\theta-\dot{\theta}\,\eta\geq 0 \tag{21.81}$$

Equation (21.81) is simplified in the isothermal process in which a work is not done and a heat does not flows as follows:

$$\dot{\phi}\geq 0 \tag{21.82}$$

Therefore, the process proceeds to the direction in which the Gibbs energy increases. In addition, The thermodynamic equilibrium state satisfies the maximal state of Gibbs energy since it is the state in which a variation does not occur naturally.

Finally, the enthalpy function (heat function) h is defined as the Legendre transformation to the internal energy.

$$h(\mathbf{T}^{(n)},\eta)=\frac{1}{\rho_0}\mathbf{T}^{(n)}:\mathbf{E}^{(n)}-u(\mathbf{E}^{(n)},\eta)=\phi(\mathbf{T}^{(n)},\theta)-\theta\eta \tag{21.83}$$

which is obtained from the Legendre transformation equation (21.62) by the replacement of

$$x\rightarrow\mathbf{T}^{(n)},f\rightarrow h;p\rightarrow\ :\mathbf{E}^{(n)},f^*\rightarrow -u$$

or

$$x\rightarrow\eta,f\rightarrow h;p\rightarrow -\theta,f^*\rightarrow\phi$$

.

Substituting Eq. (21.50) or (21.77) into Eq. (21.83), one has

$$\dot{h}=\frac{1}{\rho_0}(\mathbf{T}^{(n)}:\mathbf{E}^{(n)})^{\bullet}-\dot{u}=\dot{\phi}-\eta\,\dot{\theta}-\theta\,\dot{\eta} \tag{21.84}$$

which is obtained by the time-differentiation of (21.83), the rate of the enthalpy is given by

$$\dot{h}=\frac{1}{\rho_0}\mathbf{E}^{(n)}:\dot{\mathbf{T}}^{(n)}-\theta\,\dot{\eta}+f_j\dot{\xi}_j+\frac{1}{\rho_0\theta}\mathbf{q}_0\bullet\nabla_0\theta \tag{21.85}$$

which means that the enthalpy is a part of internal energy released as heat under the constant state of stress $\mathbf{T}^{(n)}$. The enthalpy h is described by the independent variables $\mathbf{T}^{(n)}$, η, ξ_i as follows:

$$h=h(\mathbf{T}^{(n)},\eta,\xi_i) \tag{21.86}$$

The time-differentiation of Eq. (21.86) leads to

$$\dot{h}=\frac{\partial h}{\partial\mathbf{T}^{(n)}}:\dot{\mathbf{T}}^{(n)}+\frac{\partial h}{\partial\eta}\dot{\eta}+\frac{\partial h}{\partial\xi_i}\dot{\xi}_i \tag{21.87}$$

Further, comparing Eqs. (21.85) and (21.87), one has

$$\mathbf{E}^{(n)}=\rho_0\frac{\partial h}{\partial\mathbf{T}^{(n)}},\quad\theta=-\frac{\partial h}{\partial\eta},\quad f_j=\frac{\partial h}{\partial\xi_i} \tag{21.88}$$

Therefore, the enthalpy $h^{(n)}$ is the thermodynamic potential to $\mathbf{E}^{(n)}$, θ and f_i and plays the role of the complementary strain energy.

The substitution of Eq. (21.84) into Eq. (21.55), the Clausius-Duhem inequality is expressed as follows:

$$-\frac{1}{\rho_0}\mathbf{E}^{(n)}:\dot{\mathbf{T}}^{(n)}+\dot{h}^{(n)}-\frac{1}{\rho_0\theta}\mathbf{q}_0\bullet\nabla_0\theta+\theta\,\dot{\eta}\geq 0 \tag{21.89}$$

which is simplified in the isothermal state without a work and heat and entropy fluxes as follows:

$$\overset{\bullet}{h}^{(n)} \geq 0 \tag{21.90}$$

Therefore, the process proceeds to the direction in which the enthalpy increases in this state, and the thermodynamic equilibrium state satisfies the maximal state of enthalpy.

It follows from Eqs. (21.53), (21.58), (21.69), (21.80) and (21.88) that

$$\left.\begin{array}{l} \dfrac{\partial u}{\partial \eta} = \theta = -\dfrac{\partial h}{\partial \eta}, \rho_0 \dfrac{\partial u}{\partial \mathbf{E}^{(n)}} = \mathbf{T}^{(n)} = \rho_0 \dfrac{\partial \psi}{\partial \mathbf{E}^{(n)}} \\ -\dfrac{\partial \psi}{\partial \theta} = \eta = \dfrac{\partial \phi}{\partial \theta}, \rho_0 \dfrac{\partial \phi}{\partial \mathbf{T}^{(n)}} = \mathbf{E}^{(n)} = \rho_0 \dfrac{\partial h}{\partial \mathbf{T}^{(n)}} \end{array}\right\} \tag{21.91}$$

Further, taking the cross-differentiation of Eqs. (21.53), (21.69), (21.80) and (21.88), the *Maxwell relations* are obtained as follows:

$$\left.\begin{array}{l} \dfrac{\partial \mathbf{T}^{(n)}(\mathbf{E}^{(n)}, \eta, \xi)}{\partial \eta} = \rho_0 \dfrac{\partial \theta(\mathbf{E}^{(n)}, \eta, \xi)}{\partial \mathbf{E}^{(n)}}, \dfrac{\partial \mathbf{E}^{(n)}(\mathbf{T}^{(n)}, \theta, \xi)}{\partial \theta} = \rho_0 \dfrac{\partial \hat{\eta}(\mathbf{T}^{(n)}, \theta, \xi)}{\partial \mathbf{T}^{(n)}} \\ \dfrac{\partial \mathbf{T}^{(n)}(\mathbf{E}^{(n)}, \theta, \xi)}{\partial \theta} = -\rho_0 \dfrac{\partial \bar{\eta}(\mathbf{E}^{(n)}, \theta, \xi)}{\partial \mathbf{E}^{(n)}}, \dfrac{\partial \mathbf{E}^{(n)}(\mathbf{T}^{(n)}, \eta, \xi)}{\partial \eta} = -\rho_0 \dfrac{\partial \theta(\mathbf{T}^{(n)}, \eta, \xi)}{\partial \mathbf{T}^{(n)}} \end{array}\right\} \tag{21.92}$$

and

$$\frac{\partial \mathbf{T}^{(n)}(\mathbf{E}^{(n)}, \theta, \xi)}{\partial \xi_i} = -\rho_0 \frac{\partial \bar{f}_i(\mathbf{E}^{(n)}, \theta, \xi)}{\partial \mathbf{E}^{(n)}}, \quad \frac{\partial \mathbf{E}^{(n)}(\mathbf{T}^{(n)}, \theta, \xi)}{\partial \xi_i} = \rho_0 \frac{\partial \hat{f}_i(\mathbf{T}^{(n)}, \theta, \xi)}{\partial \mathbf{T}^{(n)}} \tag{21.93}$$

where the hats ($^{-}$) and ($^{\wedge}$) are added for the independent variables $\mathbf{E}^{(n)}$ and $\mathbf{T}^{(n)}$, respectively.

The fourth-order tensor

$$\mathbb{E}^{(n)} = \frac{\partial \mathbf{T}^{(n)}}{\partial \mathbf{E}^{(n)}} = \frac{\partial^2 (\rho_0 \psi)}{\partial \mathbf{E}^{(n)} \otimes \partial \mathbf{E}^{(n)}} \tag{21.94}$$

$$\mathbb{M}^{(n)} = \frac{\partial \mathbf{E}^{(n)}}{\partial \mathbf{T}^{(n)}} = \frac{\partial^2 (\rho_0 \psi)}{\partial \mathbf{T}^{(n)} \otimes \partial \mathbf{T}^{(n)}} \tag{21.95}$$

are the isothermal elastic modulus and the compliance (inverse tensor) for the conjugate pair of strain and stress ($\mathbf{E}^{(n)}$, $\mathbf{T}^{(n)}$) satisfying the relation

$$\mathbb{E}^{(n)} : \mathbb{M}^{(n)} = \frac{\partial \mathbf{T}^{(n)}}{\partial \mathbf{E}^{(n)}} : \frac{\partial \mathbf{E}^{(n)}}{\partial \mathbf{T}^{(n)}} = \mathcal{I} \tag{21.96}$$

As known from Eqs. (21.94) and (21.95), $\mathbb{E}^{(n)}$ and $\mathbb{M}^{(n)}$ possess the major symmetries

$$\mathbb{E}^{(n)}_{ijkl} = \mathbb{E}^{(n)}_{klij},\ \mathbb{M}^{(n)}_{ijkl} = \mathbb{M}^{(n)}_{klij} \tag{21.97}$$

and they possess also the minor symmetries because of the symmetries of $\mathbf{E}^{(n)}$ and $\mathbf{T}^{(n)}$.

The above-mentioned thermodynamic potential function are shown in Table 21.1 for the reversible process $(\theta\gamma = f_j \overset{\bullet}{\xi}_j = 0)$ without the gradient of temperature $(\nabla_0\theta = \mathbf{0})$ based on Eqs. (21.50), (21.66), (21.77) and (21.85). While there are various definitions of signs of the thermodynamic functions, they are chosen here as their time-derivatives show the influences of the rates of entropy and temperature to the power and the compliance power for the internal energy and the Helmholtz free energy and for the Gibbs free energy and the enthalpy, respectively.

21.6 Specific Heat and Latent Heat

The specific heat is the quantity of heat required to rise up a unit temperature. Denoting the specific heat in the process under the constant strain (no power) as $C_{E^{(n)}}$, i.e. $C_{E^{(n)}} = -\partial q/\partial\theta|_{E^{(n)}}$ where q is the heat required to rise up unit temperature for unit mass, the rate of internal energy is identical to the inflow of heat and thus $du = -dq$ holds. Then, it follows that

$$C_{E^{(n)}}d\theta = -\frac{\partial q}{\partial\theta}d\theta = \frac{\partial u}{\partial\theta}d\theta$$

On the other hand, it follows under the constant stress from Eqs. (21.4), (21.16) and (21.84) that

$$dh = \frac{1}{\rho_0}d(\mathbf{T}^{(n)} : \mathbf{E}^{(n)}) - du = \frac{1}{\rho_0}\mathbf{T}^{(n)} : d\mathbf{E}^{(n)} - du = dq$$

Denoting the specific heat in this case as $C_{T^{(n)}}$, i.e. $C_{T^{(n)}} = -\partial q/\partial\theta|_{T^{(n)}}$, one has

$$C_{T^{(n)}}d\theta = -\frac{\partial q}{\partial\theta}d\theta = -\frac{\partial h}{\partial\theta}d\theta$$

and thus it follows that

$$\left.\begin{aligned} C_{E^{(n)}} &= \frac{\partial u}{\partial\theta} \\ C_{T^{(n)}} &= -\frac{\partial h}{\partial\theta} \end{aligned}\right\} \tag{21.98}$$

Table 21.1 Thermodynamic potential functions

Potential function	Legendre transformation	Hyperelastic relations	Gibbs relations
Internal energy $u(\mathbf{E}^{(n)},\eta)$	$u(\mathbf{E}^{(n)},\eta)=\psi(\mathbf{E}^{(n)},\theta)+\theta\eta$	$\mathbf{T}^{(n)}=\rho_0\dfrac{\partial u}{\partial \mathbf{E}^{(n)}}$ $\theta=\dfrac{\partial u}{\partial \eta}$	$\dot{u}=\dfrac{1}{\rho_0}\mathbf{T}^{(n)}:\dot{\mathbf{E}}^{(n)}+\theta\,\dot{\eta}$
Helmholtz free energy $\psi(\mathbf{E}^{(n)},\theta)$	$\psi(\mathbf{E}^{(n)},\theta)=u(\mathbf{E}^{(n)},\eta)-\theta\eta$	$\mathbf{T}^{(n)}=\rho_0\dfrac{\partial \psi}{\partial \mathbf{E}^{(n)}}$ $\eta=-\dfrac{\partial \psi}{\partial \theta}$	$\dot{\psi}=\dfrac{1}{\rho_0}\mathbf{T}^{(n)}:\dot{\mathbf{E}}^{(n)}-\eta\,\dot{\theta}$
Gibbs free energy $\phi(\mathbf{T}^{(n)},\ \theta)$	$\phi(\mathbf{T}^{(n)},\ \theta)=-\psi(\mathbf{E}^{(n)},\theta)+\dfrac{1}{\rho_0}\mathbf{T}^{(n)}:\mathbf{E}^{(n)}$	$\mathbf{E}^{(n)}=\rho_0\dfrac{\partial \phi}{\partial \mathbf{T}^{(n)}}$ $\eta=\dfrac{\partial \phi}{\partial \theta}$	$\dot{\phi}=\dfrac{1}{\rho_0}\mathbf{E}^{(n)}:\dot{\mathbf{T}}+\eta\,\dot{\theta}$
Entalpy $h^{(n)}(\mathbf{T}^{(n)},\ \eta)$	$h(\mathbf{T}^{(n)},\ \eta)=\phi(\mathbf{T}^{(n)},\ \theta)-\theta\eta$	$\mathbf{E}^{(n)}=\rho_0\dfrac{\partial h}{\partial \mathbf{T}^{(n)}}$ $\theta=-\dfrac{\partial h}{\partial \eta}$	$\dot{h}=\dfrac{1}{\rho_0}\mathbf{E}^{(n)}:\dot{\mathbf{T}}^{(n)}-\theta\,\dot{\eta}$

$C_{E^{(n)}}$ and $C_{T^{(n)}}$ are noting but the specific heat at constant volume and the specific heat at constant pressure, respectively.

Taking account of Eqs. $(21.58)_1$ and $(21.88)_2$ in Eq. (21.98), $C_{E^{(n)}}$ and $C_{T^{(n)}}$ are described as follows:

$$\left.\begin{aligned} C_{E^{(n)}} &= \frac{\partial \bar{u}}{\partial \theta} = \frac{\partial \bar{u}}{\partial \bar{\eta}}\frac{\partial \bar{\eta}}{\partial \theta} = \theta\frac{\partial \bar{\eta}}{\partial \theta} \\ C_{T^{(n)}} &= \frac{\partial \hat{h}}{\partial \theta} = -\frac{\partial \hat{h}}{\partial \hat{\eta}}\frac{\partial \hat{\eta}}{\partial \theta} = \theta\frac{\partial \hat{\eta}}{\partial \theta} \end{aligned}\right\} \tag{21.99}$$

where

$$\left.\begin{aligned} \bar{u} &= u(\mathbf{E}^{(n)},\theta), \bar{\eta} = \eta(\mathbf{E}^{(n)},\theta) \\ \hat{u} &= u(\mathbf{T}^{(n)},\theta), \hat{\eta} = \eta(\mathbf{T}^{(n)},\theta) \end{aligned}\right\} \tag{21.100}$$

It is noted from Eq. $(21.80)_2$ that the first partial derivative of Gibbs free energy by a temperature, $\partial\phi/\partial\theta$, is the entropy and from Eq. (21.99) that the second partial derivative $\partial^2\phi/\partial\theta^2$ is the specific heat divided by the temperature.

The heat required to increase the temperature $d\theta$ and to give the strain increment $d\mathbf{E}^{(n)}$ or the stress increment $d\mathbf{T}^{(n)}$ per unit volume is $-dq$ (flow in). Then, it follows that

$$-dq = \begin{cases} C_{E^{(n)}}d\theta + \ell_{E^{(n)}} : d\mathbf{E}^{(n)} \\ C_{T^{(n)}}d\theta + \ell_{T^{(n)}} : d\mathbf{T}^{(n)} \end{cases} \tag{21.101}$$

where $\ell_{E^{(n)}}$ and $\ell_{T^{(n)}}$ are the latent heats for the strain increment $d\mathbf{E}^{(n)}$ and the stress increment $d\mathbf{T}^{(n)}$, respectively. Therefore, the temperature variation induced by the strain increment $d\mathbf{E}^{(n)}$ and the stress increment $d\mathbf{T}^{(n)}$ are described as follows:

$$d\theta = -\frac{1}{C_{E^{(n)}}}\ell_{E^{(n)}} : d\mathbf{E}^{(n)}, \quad d\theta = -\frac{1}{C_{T^{(n)}}}\ell_{T^{(n)}} : d\mathbf{T}^{(n)} \tag{21.102}$$

The variation of entropy under the reversible condition is given from Eq. (21.101) as follows:

$$d\eta = \frac{-dq}{\theta} = \begin{cases} \dfrac{C_{E^{(n)}}}{\theta}d\theta + \dfrac{\ell_{E^{(n)}}}{\theta} : d\mathbf{E}^{(n)} \\ \dfrac{C_{T^{(n)}}}{\theta}d\theta + \dfrac{\ell_{T^{(n)}}}{\theta} : d\mathbf{T}^{(n)} \end{cases} \tag{21.103}$$

On the other hand, the entropy is described in the cases for which the pair of temperature θ and the strain $\mathbf{E}^{(n)}$ and the pair of the temperature θ and the stress $\mathbf{T}^{(n)}$ are chosen as the arguments as follows:

$$d\eta = \begin{cases} \dfrac{\partial \bar{\eta}}{\partial \theta} d\theta + \dfrac{\partial \bar{\eta}}{\partial \mathbf{E}^{(n)}} : d\mathbf{E}^{(n)} \\ \dfrac{\partial \hat{\eta}}{\partial \theta} d\theta + \dfrac{\partial \hat{\eta}}{\partial \mathbf{T}^{(n)}} : d\mathbf{T}^{(n)} \end{cases} \tag{21.104}$$

Comparing Eqs. (21.103) and (21.104), one has

$$\ell_{E^{(n)}} = \theta \frac{\partial \bar{\eta}}{\partial \mathbf{E}^{(n)}}, \quad \ell_{T^{(n)}} = \theta \frac{\partial \hat{\eta}}{\partial \mathbf{T}^{(n)}} \tag{21.105}$$

The mutual relations in Eq. (21.92) yield

$$\rho_0 \frac{\partial \bar{\eta}}{\partial \mathbf{E}^{(n)}} = -\frac{\partial \mathbf{T}^{(n)}}{\partial \theta}, \quad \rho_0 \frac{\partial \hat{\eta}}{\partial \mathbf{T}^{(n)}} = \frac{\partial \mathbf{E}^{(n)}}{\partial \theta} \tag{21.106}$$

and thus the latent heats are described as follows:

$$\ell_{E^{(n)}} = -\frac{1}{\rho_0} \theta \frac{\partial \mathbf{T}^{(n)}}{\partial \theta}, \quad \ell_{T^{(n)}} = \frac{1}{\rho_0} \theta \frac{\partial \mathbf{E}^{(n)}}{\partial \theta} \tag{21.107}$$

where the minus sign of $\ell_{E^{(n)}}$ means that the compression stress is induced by the increase of temperature under the constant strain.

The following relation is derived from Eq. (21.103).

$$C_{E^{(n)}} + \ell_{E^{(n)}} : \frac{\partial \mathbf{E}^{(n)}}{\partial \theta} = C_{T^{(n)}} + \ell_{T^{(n)}} : \frac{\partial \mathbf{T}^{(n)}}{\partial \theta} \tag{21.108}$$

Substituting Eq. (21.107) into Eq. (21.108), we have

$$C_{E^{(n)}} + \frac{\rho_0}{\theta} \ell_{E^{(n)}} : \ell_{T^{(n)}} = C_{T^{(n)}} - \frac{\rho_0}{\theta} \ell_{T^{(n)}} : \ell_{E^{(n)}} \tag{21.109}$$

Further, substituting Eq. (21.105) into

$$\frac{\partial \hat{\eta}}{\partial \mathbf{T}^{(n)}} = \frac{\partial \bar{\eta}}{\partial \mathbf{E}^{(n)}} : \frac{\partial \mathbf{E}^{(n)}}{\partial \mathbf{T}^{(n)}} = \frac{\partial \bar{\eta}}{\partial \mathbf{E}^{(n)}} : \mathbb{M}^{(n)} \tag{21.110}$$

it follows that

$$\ell_{T^{(n)}} = \mathbb{M}^{(n)} : \ell_{E^{(n)}} \tag{21.111}$$

Further, substituting Eq. (21.111) into Eq. (21.109), one has

$$C_{T^{(n)}} - C_{E^{(n)}} = 2\frac{\rho_0}{\theta}\ell_{E^{(n)}} : \mathbb{M}^{(n)} : \ell_{E^{(n)}} \tag{21.112}$$

where the elastic compliance $\mathbb{M}^{(n)}$ is the positive-definite tensor so that one has

$$C_{T^{(n)}} - C_{E^{(n)}} \geq 0 \tag{21.113}$$

21.7 Derivation of Various Flow Rules from Second Law of Thermodynamics

The derivations of the associated flow rule are described based on the Drucker's and the Ilyushin's postulates in Sect. 6.6. On the other hand, the flow rules of the plastic strain rate and the conjugate dissipative rates for internal variables are adopted widely by the thermodynamic formalism in the damage phenomena (cf. e.g. Lemaitre and Chaboche 1990; Lematire 1992; Lemaitre and Desmoral 2005; de Sauza Neto et al. 2008; Murakmi 2012) and transformation phenomena (cf. e.g. Bartel et al. 2011; Mahnken 2012) and the finite strain theory based on the multiplicative decomposition of the deformation gradient (cf. e.g. Menzel et al. 2005; Vladimirov 2008, 2010; Wallin et al. 2003; Wallin and Ristinmaa 2005). They are explained and their pertinences will be discussed in this section.

21.7.1 Formulation by Infinitesimal Strain Theory

To capture the basic issue in brief, we first adopt the infinitesimal strain and consider the simple case that the variation of density is negligible, i.e. $\dot{\rho} = 0$.

(a) **Derivation based on the second law of thermodynamics**

Assuming the isothermal and adiabatic process ($\dot{\theta} = 0, \mathbf{q} = \mathbf{0}$) and replacing the strain rate $\mathbf{d}$ to the material-time derivative of infinitesimal strain $\dot{\boldsymbol{\varepsilon}}$, Eq. (21.72) is described as follows:

$$\frac{1}{\rho}\boldsymbol{\sigma} : \dot{\boldsymbol{\varepsilon}} - \dot{\psi} \geq 0 \tag{21.114}$$

Thus, the Clausius-Duhem inequality means the positiveness of the dissipative power, noting that $\dot{\psi}$ is the storage power as described in Sect. 21.5.

As described in Sect. 6.9, the stress and back stress should be formulated by the hyperelastic relations based on the elastic strain $\boldsymbol{\varepsilon}^e$ and the storage part of the plastic strain $\boldsymbol{\varepsilon}^p_{ks}$. Then, the Helmholtz free energy function ψ is given by the addition of the terms of the functions of $\boldsymbol{\varepsilon}^e$ and $\boldsymbol{\varepsilon}^p_{ks}$ and the isotropic hardening variable H as follows:

$$\psi = \psi^e(\boldsymbol{\varepsilon}^e) + \psi^k(\boldsymbol{\varepsilon}^p_{ks}) + \psi^F(H) \tag{21.115}$$

Then, the stress $\boldsymbol{\sigma}$, the back stress $\boldsymbol{\alpha}$ and the isotropic hardening function F are given by

$$\boldsymbol{\sigma} = \frac{\partial \psi^e(\boldsymbol{\varepsilon}^e)}{\partial \boldsymbol{\varepsilon}^e}, \quad \boldsymbol{\alpha} = \frac{\partial \psi^k(\boldsymbol{\varepsilon}^p_{ks})}{\partial \boldsymbol{\varepsilon}^p_{ks}}, \quad F = \frac{\partial \psi^F(H)}{\partial H} \tag{21.116}$$

The $\boldsymbol{\sigma}$, $\boldsymbol{\alpha}$ and F are the thermodynamic forces conjugate to the generalized thermodynamic displacements $\boldsymbol{\varepsilon}^e$, $\boldsymbol{\varepsilon}^p_{ks}$ and H, respectively.

The time-derivative of the Helmholtz free energy is given by

$$\dot{\psi} = \frac{\partial \psi^e(\boldsymbol{\varepsilon}^e)}{\partial \boldsymbol{\varepsilon}^e} : \dot{\boldsymbol{\varepsilon}}^e + \frac{\partial \psi^k(\boldsymbol{\varepsilon}^p_{ks})}{\partial \boldsymbol{\varepsilon}_{ks}{}^p} : \dot{\boldsymbol{\varepsilon}}^p_{ks} + \frac{\partial \psi^F(H)}{\partial H} \dot{H} \tag{21.117}$$

Substituting Eq. (21.117) with Eq. (21.116), the inequality (21.114) is expressed as follows:

$$\begin{aligned}
&\frac{1}{\rho}\boldsymbol{\sigma} : \dot{\boldsymbol{\varepsilon}} - \left(\frac{\partial \psi^e(\boldsymbol{\varepsilon}^e)}{\partial \boldsymbol{\varepsilon}^e} : \dot{\boldsymbol{\varepsilon}}^e + \frac{\partial \psi^k(\boldsymbol{\varepsilon}^p_{ks})}{\partial \boldsymbol{\varepsilon}^p_{ks}} : \dot{\boldsymbol{\varepsilon}}^p_{ks} + \frac{\partial \psi^F(H)}{\partial H} \dot{H} \right) \\
&= \boldsymbol{\sigma} : (\dot{\boldsymbol{\varepsilon}}^e + \dot{\boldsymbol{\varepsilon}}^p) - \frac{\partial \psi^e(\boldsymbol{\varepsilon}^e)}{\partial \boldsymbol{\varepsilon}^e} : \dot{\boldsymbol{\varepsilon}}^e - \frac{\partial \psi^k(\boldsymbol{\varepsilon}^p_s)}{\partial \boldsymbol{\varepsilon}^p_{ks}} : (\dot{\boldsymbol{\varepsilon}}^p - \dot{\boldsymbol{\varepsilon}}_{kd}{}^p) - \frac{\partial \psi^F(H)}{\partial H} \dot{H} \\
&= \boldsymbol{\sigma} : (\dot{\boldsymbol{\varepsilon}}^e + \dot{\boldsymbol{\varepsilon}}^p) - \boldsymbol{\sigma} : \dot{\boldsymbol{\varepsilon}}^e - \frac{1}{\rho}\boldsymbol{\alpha} : (\dot{\boldsymbol{\varepsilon}}^p - \dot{\boldsymbol{\varepsilon}}_{kd}{}^p) - F\dot{H} \\
&= (\boldsymbol{\sigma} - \boldsymbol{\alpha}) : \dot{\boldsymbol{\varepsilon}}^p + \boldsymbol{\alpha} : \dot{\boldsymbol{\varepsilon}}^p_{kd} - F\dot{H} \geq 0
\end{aligned}$$

i.e.

$$\hat{\boldsymbol{\sigma}} : \dot{\boldsymbol{\varepsilon}}^p + \boldsymbol{\alpha} : \dot{\boldsymbol{\varepsilon}}^p_{kd} - F\dot{H} \geq 0 \tag{21.118}$$

where the minus sign is added because $F\dot{H}$ is regarded to be the storage energy as the isotropic hardening.

Now, we introduce the homogeneous positive functions $g^p(\hat{\boldsymbol{\sigma}})$ and $g^k(\boldsymbol{\alpha})$ of $\hat{\boldsymbol{\sigma}}$ and $\boldsymbol{\alpha}$, respectively, and function $g^F(F)$ of isotropic hardening function F in degree-one, and let the irreversible deformation measures $\dot{\boldsymbol{\varepsilon}}^p$, $\dot{\boldsymbol{\varepsilon}}^p_{kd}$ and $\dot{H}$ be given by

$$\overset{\bullet}{\boldsymbol{\varepsilon}}^{p} = \overset{\bullet}{\lambda}\frac{\partial g^{p}(\hat{\boldsymbol{\sigma}})}{\partial \boldsymbol{\sigma}}, \quad \overset{\bullet}{\boldsymbol{\varepsilon}}^{p}_{kd} = \overset{\bullet}{\lambda}\frac{\partial g^{k}(\boldsymbol{\alpha})}{\partial \boldsymbol{\alpha}}, \quad \dot{H} = -\overset{\bullet}{\lambda}\frac{\partial g^{F}(F)}{\partial F} \tag{21.119}$$

Equation (21.119)$_3$ is assumed *a priori* in some literatures (e.g. cf. Ottossen and Rstinmaa 2005). The inequality in Eq. (21.118) is satisfied by substituting Eq. (21.119) as follows:

$$\overset{\bullet}{\lambda}\left(\frac{1}{\rho}g^{p}(\hat{\boldsymbol{\sigma}}) + \frac{1}{\rho}g^{k}(\boldsymbol{\alpha}) + g^{F}(H)\right) \geq 0 \tag{21.120}$$

Further, if we adopt $f(\hat{\boldsymbol{\sigma}})$ for $g^{p}(\hat{\boldsymbol{\sigma}})$, we obtains the associated flow rule for the plastic strain rate $\overset{\bullet}{\boldsymbol{\varepsilon}}^{p}$. However, there is no inevitability in adopting the potential function for the irreversible deformation measures so that the above-mentioned discussion is regarded as the logic in so-called *predetermined harmony*.

If we adopt the simplest function

$$\psi^{k}(\boldsymbol{\varepsilon}^{p}_{s}) = \frac{1}{2}c\boldsymbol{\varepsilon}^{p}_{ks}:\boldsymbol{\varepsilon}^{p}_{ks} \tag{21.121}$$

for in the Helmholtz free energy of the back stress, we have

$$\boldsymbol{\alpha} = c\boldsymbol{\varepsilon}^{p}_{ks} = c(\boldsymbol{\varepsilon}^{p} - \boldsymbol{\varepsilon}^{p}_{kd}) \tag{21.122}$$

$$\overset{\bullet}{\boldsymbol{\alpha}} = c(\overset{\bullet}{\boldsymbol{\varepsilon}}^{p} - \overset{\bullet}{\boldsymbol{\varepsilon}}^{p}_{kd}) \tag{21.123}$$

Further, if we adopt the explicit function

$$g^{k}(\boldsymbol{\alpha}) = \frac{1}{2b}\boldsymbol{\alpha}:\boldsymbol{\alpha} \tag{21.124}$$

for $g^{\alpha}(\boldsymbol{\alpha})$,it follows from Eq. (21.119)$_2$ that

$$\overset{\bullet}{\boldsymbol{\varepsilon}}^{p}_{kd} = \frac{1}{b}\boldsymbol{\alpha}\overset{\bullet}{\lambda} \tag{21.125}$$

The substitution of Eq. (21.125) into Eq. (21.123) leads to the following equation of the back stress.

$$\overset{\bullet}{\boldsymbol{\alpha}} = c(\overset{\bullet}{\boldsymbol{\varepsilon}}^{p} - \frac{1}{b}\boldsymbol{\alpha}\overset{\bullet}{\lambda}) = c\left(\frac{\partial g^{p}(\hat{\boldsymbol{\sigma}})}{\partial \boldsymbol{\sigma}} - \frac{1}{b}\boldsymbol{\alpha}\right)\overset{\bullet}{\lambda} \tag{21.126}$$

which is no more than the nonlinear kinematic hardening rule of Armstrong-Frederick (1966) if the yield function is adopted for $g^{p}(\hat{\boldsymbol{\sigma}})$. Further, if we assume

$$\psi^F(H) = F_0\left[(1+h_1)H + \frac{h_1}{h_2}\exp(-h_2 H)\right] \tag{21.127}$$

it follows that

$$F(H) = \frac{\partial \psi^F(H)}{\partial H} = F_0[1 + h_1\{1 - \exp(-h_2 H)\}] \tag{21.128}$$

Furthermore, if we adopt the potential function of isotropic hardening

$$g^F(F) = -\sqrt{\frac{2}{3}}F \tag{21.129}$$

in Eq. $(21.119)_3$, we obtain the following equivalent plastic strain of metals.

$$\dot{H} = \sqrt{\frac{2}{3}}\dot{\lambda} \tag{21.130}$$

(b) **Formulation based on Maximum dissipation**

The formulation of flow rules have been often formulated based on the Maximum dissipation (cf. e.g. Lubliner 1984)

Let the dissipative power in the right-hand side of the inequality in Eq. (21.118) be denoted as $\mathfrak{D}$, i.e.

$$\mathfrak{D} \equiv \hat{\boldsymbol{\sigma}} : \dot{\boldsymbol{\varepsilon}}^p + \boldsymbol{\alpha} : \dot{\boldsymbol{\varepsilon}}^p_{kd} - F\dot{H} \tag{21.131}$$

Then, introduce the *principle of maximum plastic work* (Hill 1948, 1950) insisting that the dissipative power is maximum under the yield condition:

$$f(\hat{\boldsymbol{\sigma}}) - F(H) = 0 \tag{21.132}$$

Now, consider to make the *Lagrange function* with the incidental (collateral) condition under the fulfillment of the yield condition

$$\mathfrak{L} \equiv \mathfrak{D} - \dot{\lambda}[f(\hat{\boldsymbol{\sigma}}) - F(H)] = \frac{1}{\rho}\hat{\boldsymbol{\sigma}} : \dot{\boldsymbol{\varepsilon}}^p + \boldsymbol{\alpha} : \dot{\boldsymbol{\varepsilon}}^p_{kd} - F\dot{H} - \dot{\lambda}[f(\hat{\boldsymbol{\sigma}}) - F(H)] = 0 \tag{21.133}$$

maximum in the variations of the thermodynamic forces $\boldsymbol{\sigma}, \boldsymbol{\alpha}$ and F. Then, it must hold that

$$\begin{aligned}
\delta\mathfrak{L} &= \delta\hat{\boldsymbol{\sigma}}:\dot{\boldsymbol{\varepsilon}}^p + \delta\boldsymbol{\alpha}:\dot{\boldsymbol{\varepsilon}}^p_{kd} - \delta F\dot{H} - \dot{\lambda}[\delta f(\hat{\boldsymbol{\sigma}}) - \delta F(H)] - \delta\dot{\lambda}[f(\hat{\boldsymbol{\sigma}}) - F(H)] \\
&= \dot{\boldsymbol{\varepsilon}}^p:\delta\boldsymbol{\sigma} - \dot{\boldsymbol{\varepsilon}}^p:\delta\boldsymbol{\alpha} + \dot{\boldsymbol{\varepsilon}}^p_{kd}:\delta\boldsymbol{\alpha} - \delta F\dot{H} - \dot{\lambda}\frac{\partial f(\hat{\boldsymbol{\sigma}})}{\partial\boldsymbol{\sigma}}:\delta\boldsymbol{\sigma} + \dot{\lambda}\frac{\partial f(\hat{\boldsymbol{\sigma}})}{\partial\boldsymbol{\sigma}}:\delta\boldsymbol{\alpha} + \dot{\lambda}\,\delta F(H) \\
&= \left(\dot{\boldsymbol{\varepsilon}}^p - \dot{\lambda}\frac{\partial f(\hat{\boldsymbol{\sigma}})}{\partial\boldsymbol{\sigma}}\right):\delta\boldsymbol{\sigma} + \left(\dot{\boldsymbol{\varepsilon}}^p_{kd} + \dot{\lambda}\frac{\partial f(\hat{\boldsymbol{\sigma}})}{\partial\boldsymbol{\sigma}}\right):\delta\boldsymbol{\alpha} - (\dot{H} - \dot{\lambda})\delta F(H) = 0
\end{aligned}$$

from which we have

$$\dot{\boldsymbol{\varepsilon}}^p = \dot{\lambda}\frac{\partial f(\hat{\boldsymbol{\sigma}})}{\partial\boldsymbol{\sigma}}, \quad \dot{\boldsymbol{\varepsilon}}^p_{kd} = -\dot{\lambda}\frac{\partial f(\hat{\boldsymbol{\sigma}})}{\partial\boldsymbol{\sigma}} = -\dot{\boldsymbol{\varepsilon}}^p, \quad \dot{H} = \dot{\lambda} \tag{21.134}$$

Therefore, the plastic strain rate and its dissipative part are imposed the strong restriction as they are given only by the associated flow rule. In particular, the dissipative rate of kinematic hardening becomes the inverse direction of the plastic strain rate and the rate of isotropic hardening is limited to the plastic multiplier so that the generality is remarkably lost.

We should not impose the restriction to material property only by the thermodynamic formalism. It should be noted that the thermodynamics is to be the universal law which must be satisfied by any phenomena in the natural world.

21.7.2 Multiplicative Finite Strain Theory

Assume that the Helmholtz free energy function ψ is given in the additive form of the storage variables $\overline{\mathbf{C}}^e$, $\widehat{\mathbf{C}}^p_{ks}$ and H, i.e.

$$\psi = \psi^e(\overline{\mathbf{C}}^e) + \psi^k(\widehat{\mathbf{C}}^p_s) + \psi^F(H) \tag{21.135}$$

and that the stress $\overline{\mathbf{S}}^e$, the back stress $\widehat{\mathbf{S}}^k$ and the isotropic hardening variable F are given by

$$\overline{\mathbf{S}} = 2\frac{\partial\psi^e(\overline{\mathbf{C}}^e)}{\partial\overline{\mathbf{C}}^e}, \quad \widehat{\mathbf{S}}_k = 2\frac{\partial\psi^k(\widehat{\mathbf{C}}^p_{ks})}{\partial\widehat{\mathbf{C}}^p_{ks}}, \quad F = \frac{\partial\psi^F(H)}{\partial H} \tag{21.136}$$

The back stress is expressed in the intermediate configuration as follows:

$$\overline{\mathbf{S}}_k = {}^{p}_{ks}\overrightarrow{\overline{\mathbf{S}}}^{GG} = 2\mathbf{F}^p_{ks}\frac{\partial\psi^k(\widehat{\mathbf{C}}^p_{ks})}{\partial\widehat{\mathbf{C}}^p_{ks}}\mathbf{F}^{p\;T}_{ks} \tag{21.137}$$

Substituting Eqs. (21.136) and (21.137), the Mandel stress and the back stress in the intermediate Configuration are given as follows:

$$\overline{\mathbf{M}} \equiv \overline{\mathbf{C}}^e\overline{\mathbf{S}} = 2\overline{\mathbf{C}}^e\frac{\partial\psi^e(\overline{\mathbf{C}}^e)}{\partial\overline{\mathbf{C}}^e} \tag{21.138}$$

$$\overline{\mathbf{M}}_k = \overline{\mathbf{C}}^p_{ks}\overline{\mathbf{S}}_k = \overline{\mathbf{S}}_k = \mathbf{F}^p_{ks}\widehat{\mathbf{S}}_k\mathbf{F}^{p\;T}_{ks} = 2\mathbf{F}^p_{ks}\frac{\partial\psi^k(\widehat{\mathbf{C}}^p_s)}{\partial\widehat{\mathbf{C}}^p_{ks}}\mathbf{F}^{p\;T}_{ks} \tag{21.139}$$

The Clausius-Dehem inequality in Eq. (21.114) is expressed in the reference configuration as follows:

$$\mathbf{S}:\frac{1}{2}\dot{\mathbf{C}} - \dot{\psi} \le 0 \tag{21.140}$$

The first term in Eq. (21.140) is based on

$$\begin{aligned}\boldsymbol{\tau}:\mathbf{d} &= \mathbf{F}\mathbf{S}\mathbf{F}^T:(\boldsymbol{l}^T+\boldsymbol{l})/2 = \mathbf{S}:\mathbf{F}^T(\boldsymbol{l}^T+\boldsymbol{l})\mathbf{F}/2\\ &= \mathbf{S}:(\mathbf{F}^T\mathbf{F}^{-T}\dot{\mathbf{F}}^T\mathbf{F}+\mathbf{F}^T\dot{\mathbf{F}}\mathbf{F}^{-1}\mathbf{F})/2 = \mathbf{S}:(\dot{\mathbf{F}}^T\mathbf{F}+\mathbf{F}^T\dot{\mathbf{F}})/2\\ &= \mathbf{S}:\dot{\mathbf{C}}/2\end{aligned}$$

noting the tensor formulation ($\mathbf{A},\mathbf{B},\mathbf{S},\mathbf{T}:$ second-order tensor)

$$\mathbf{ASB}:\mathbf{T} = A_{ir}S_{rs}B_{sj}T_{ij} = S_{rs}A_{ir}T_{ij}B_{sj} = \mathbf{S}:\mathbf{A}^T\mathbf{T}\mathbf{B}^T \tag{21.141}$$

Substituting Eq. (21.135) into Eq. (21.140), we have

$$\mathbf{S}:\frac{1}{2}\dot{\mathbf{C}} - \left(\frac{\partial\psi^e(\overline{\mathbf{C}}^e)}{\partial\overline{\mathbf{C}}^e}:\dot{\overline{\mathbf{C}}}^e + \frac{\partial\psi^k(\widehat{\mathbf{C}}_{ks}{}^p)}{\partial\widehat{\mathbf{C}}_{ks}{}^p}:\dot{\widehat{\mathbf{C}}}^p_{ks} + \frac{\partial\psi^F(H)}{\partial H}\dot{H}\right) \ge 0 \tag{21.142}$$

i.e.

$$\mathbf{S}:\frac{1}{2}\dot{\mathbf{C}} - \left(\frac{1}{2}\overline{\mathbf{S}}:\dot{\overline{\mathbf{C}}}^e + \frac{1}{2}\widehat{\mathbf{S}}_k:\dot{\widehat{\mathbf{C}}}^p_{ks} + F\dot{H}\right) \ge 0 \tag{21.143}$$

Here, noting Eqs. $(12.6)_2$, we have

$$
\begin{aligned}
\dot{\overline{\mathbf{C}}}^e &= (\mathbf{F}^{p-T}\mathbf{C}\mathbf{F}^{p-1})^{\bullet} \\
&= \mathbf{F}^{p-T}\,\dot{\mathbf{C}}\,\mathbf{F}^{p-1} + \dot{\mathbf{F}}^{p-T}\mathbf{C}\mathbf{F}^{p-1} + \mathbf{F}^{p-T}\mathbf{C}\dot{\mathbf{F}}^{p-1} \\
&= \mathbf{F}^{p-T}\,\dot{\mathbf{C}}\,\mathbf{F}^{p-1} + \dot{\mathbf{F}}^{p-T}\mathbf{F}^{pT}\mathbf{F}^{p-T}\mathbf{C}\mathbf{F}^{p-1} + \mathbf{F}^{p-T}\mathbf{C}\mathbf{F}^{p-1}\mathbf{F}^{p}\dot{\mathbf{F}}^{p-1} \\
&= \mathbf{F}^{p-T}\,\dot{\mathbf{C}}\,\mathbf{F}^{p-1} - \mathbf{F}^{p-T}\dot{\mathbf{F}}^{pT}\overline{\mathbf{C}}^e - \overline{\mathbf{C}}^e\dot{\mathbf{F}}^{p}\mathbf{F}^{p-1} \\
&= \mathbf{F}^{p-T}\,\dot{\mathbf{C}}\,\mathbf{F}^{p-1} - (\overline{\mathbf{L}}^{pT}\overline{\mathbf{C}}^e + \overline{\mathbf{C}}^e\overline{\mathbf{L}}^p) = \mathbf{F}^{p-T}\,\dot{\mathbf{C}}\,\mathbf{F}^{p-1} - \mathrm{sym}[\overline{\mathbf{C}}^e\overline{\mathbf{L}}^p]
\end{aligned}
\tag{21.144}
$$

and

$$
\begin{aligned}
\frac{1}{2}\overline{\mathbf{S}}:\dot{\overline{\mathbf{C}}}^e &= \frac{1}{2}\overline{\mathbf{S}}:[\mathbf{F}^{p-T}\,\dot{\mathbf{C}}\,\mathbf{F}^{p-1} - (\overline{\mathbf{L}}^{pT}\overline{\mathbf{C}}^e + \overline{\mathbf{C}}^e\overline{\mathbf{L}}^p)] \\
&= \mathbf{F}^{p-1}\overline{\mathbf{S}}F^{p-T}:\frac{1}{2}\dot{\mathbf{C}} - \frac{1}{2}\overline{\mathbf{C}}^e\overline{\mathbf{S}}:(\overline{\mathbf{L}}^{pT} + \overline{\mathbf{L}}^p) = \mathbf{S}:\frac{1}{2}\dot{\mathbf{C}} - \overline{\mathbf{M}}:\overline{\mathbf{D}}^p
\end{aligned}
\tag{21.145}
$$

Analogously, noting Eqs. $(12.6)_4$ and (12.19), we have

$$
\begin{aligned}
\dot{\widehat{\mathbf{C}}}^p_{ks} &= (\mathbf{F}^{p-T}_{kd}\mathbf{C}^p\mathbf{F}^{p-1}_{kd})^{\bullet} \\
&= \mathbf{F}^{p-T}_{kd}(\dot{\mathbf{F}}^{pT}\mathbf{F}^p + \mathbf{F}^{pT}\dot{\mathbf{F}}^p)\mathbf{F}^{p-1}_{kd} + \dot{\mathbf{F}}^{p-T}_{kd}\mathbf{C}^p\mathbf{F}^{p-1}_{kd} + \mathbf{F}^{p-T}_{kd}\mathbf{C}^p\dot{\mathbf{F}}^{p-1}_{kd} \\
&= \mathbf{F}^{p-T}_{kd}(\mathbf{F}^{pT}\mathbf{F}^{p-T}\dot{\mathbf{F}}^{pT}\mathbf{F}^p + \mathbf{F}^{pT}\dot{\mathbf{F}}^p\mathbf{F}^{p-1}\mathbf{F}^p)\mathbf{F}^{p-1}_{kd} + \dot{\mathbf{F}}^{p-T}_{kd}\mathbf{C}^p\mathbf{F}^{p-1}_{kd} + \mathbf{F}^{p-T}_{kd}\mathbf{C}^p\dot{\mathbf{F}}^{p-1}_{kd} \\
&= \mathbf{F}^{pT}_{ks}(\mathbf{F}^{p-T}\dot{\mathbf{F}}^{pT} + \dot{\mathbf{F}}^p\mathbf{F}^{p-1})\mathbf{F}^p_{ks} + \dot{\mathbf{F}}^{p-T}_{kd}\mathbf{F}^{pT}_{kd}\mathbf{F}^{p-T}_{kd}\mathbf{C}^p\mathbf{F}^{p-1}_{kd} + \mathbf{F}^{p-T}_{kd}\mathbf{C}^p\mathbf{F}^{p-1}_{kd}\mathbf{F}^p_{kd}\dot{\mathbf{F}}^{p-1}_{kd} \\
&= \mathbf{F}^{pT}_{ks}(\overline{\mathbf{L}}^{pT} + \overline{\mathbf{L}}^p)\mathbf{F}^p_{ks} - \mathbf{F}^{p-T}_{kd}\dot{\mathbf{F}}^{pT}_{kd}\mathbf{F}^{p-T}_{kd}\mathbf{C}^p\mathbf{F}^{p-1}_{kd} - \mathbf{F}^{p-T}_{kd}\mathbf{C}^p\mathbf{F}^{p-1}_{kd}\dot{\mathbf{F}}^p_{kd}\mathbf{F}^{p-1}_{kd} \\
&= 2\mathbf{F}^{pT}_{ks}\overline{\mathbf{D}}^p\mathbf{F}^p_{ks} - (\widehat{\mathbf{L}}^{pT}_{kd}\widehat{\mathbf{C}}^p_{ks} + \widehat{\mathbf{C}}^p_{ks}\widehat{\mathbf{L}}^p_{kd}) = 2\mathbf{F}^{pT}_{ks}\overline{\mathbf{D}}^p\mathbf{F}^p_{ks} - 2\mathrm{sym}[\widehat{\mathbf{C}}^p_{ks}\widehat{\mathbf{L}}^p_{kd})]
\end{aligned}
\tag{21.146}
$$

It follows from Eq. (21.146) that

$$
\begin{aligned}
\frac{1}{2}\widehat{\mathbf{S}}_k:\dot{\widehat{\mathbf{C}}}^p_{ks} &= \frac{1}{2}\widehat{\mathbf{S}}_k:[2\mathbf{F}^{pT}_{ks}\overline{\mathbf{D}}^p\mathbf{F}^p_{ks} - (\widehat{\mathbf{L}}^{pT}_{kd}\widehat{\mathbf{C}}^p_{ks} + \widehat{\mathbf{C}}^p_{ks}\widehat{\mathbf{L}}^p_{kd})] \\
&= \widehat{\mathbf{S}}_k:[\mathbf{F}^{pT}_{ks}\overline{\mathbf{D}}^p\mathbf{F}^p_{ks} - \frac{1}{2}(\widehat{\mathbf{L}}^{pT}_{kd}\widehat{\mathbf{C}}^p_{ks} + \widehat{\mathbf{C}}^p_{ks}\widehat{\mathbf{L}}^p_{kd})] \\
&= \mathbf{F}^p_{ks}\widehat{\mathbf{S}}_k\mathbf{F}^{pT}_{ks}:\overline{\mathbf{D}}^p - \frac{1}{2}\widehat{\mathbf{C}}^p_{ks}\widehat{\mathbf{S}}_k:(\widehat{\mathbf{L}}^{pT}_{kd} + \widehat{\mathbf{L}}^p_{kd}) = \overline{\mathbf{M}}_k:\overline{\mathbf{D}}^p - \widehat{\mathbf{M}}_k:\widehat{\mathbf{D}}^p_{kd}
\end{aligned}
\tag{21.147}
$$

Substituting Eqs. (21.145) and (21.147), the inequality in Eq. (21.143) is expressed as follows:

$$\mathbf{S}:\frac{1}{2}\dot{\mathbf{C}}-(\mathbf{S}:\frac{1}{2}\dot{\mathbf{C}}-\overline{\mathbf{M}}:\overline{\mathbf{D}}^p+\overline{\mathbf{M}}_k:\overline{\mathbf{D}}^p-\widehat{\mathbf{M}}_k:\widehat{\mathbf{D}}^p_{kd}+F\dot{H})\geq 0$$

i.e.

$$\hat{\overline{\mathbf{M}}}:\overline{\mathbf{D}}^p+\widehat{\mathbf{M}}_k:\widehat{\mathbf{D}}^p_{kd}-F\dot{H}\geq 0 \tag{21.148}$$

where

$$\hat{\overline{\mathbf{M}}}\equiv\overline{\mathbf{M}}-\overline{\mathbf{M}}_k \tag{21.149}$$

Introducing the positive homogeneous functions $g^p(\hat{\overline{\mathbf{M}}})$ and $g^k(\widehat{\mathbf{M}}_k)$ of $\hat{\overline{\mathbf{M}}}$ and $\widehat{\mathbf{M}}_k$ in degree-one and the isotropic hardening function $g^F(F)$ of F and formulating the rates of irreversible deformation variables $\overline{\mathbf{D}}^p$, $\widehat{\mathbf{D}}^p_{kd}$ and the rate of isotropic hardening variable by these potential functions, i.e.

$$\overline{\mathbf{D}}^p=\dot{\lambda}\frac{\partial g^p(\hat{\overline{\mathbf{M}}})}{\partial\overline{\mathbf{M}}},\quad \widehat{\mathbf{D}}^p_{kd}=\dot{\lambda}\frac{\partial g^k(\widehat{\mathbf{M}}_k)}{\partial\widehat{\mathbf{M}}_k},\quad \dot{H}=-\dot{\lambda}\frac{\partial g^F(F)}{\partial F} \tag{21.150}$$

the inequality in Eq. (21.148) is satisfied as follows:

$$\dot{\lambda}(g^p(\hat{\overline{\mathbf{M}}})+g^k(\widehat{\mathbf{M}}_k)+g^F(F))\geq 0 \tag{21.151}$$

Here, it should be noted in Eq. (21.150) that $\overline{\mathbf{D}}^p$ and $\widehat{\mathbf{D}}^p_{kd}$ are the dissipative quantities but $\dot{H}$ is the storage quantity.

Assuming the following equation for explicit function of $g^k(\widehat{\mathbf{M}}_k)$

$$g^k(\widehat{\mathbf{M}}_k)=\frac{1}{2b}\widehat{\mathbf{M}}_k:\widehat{\mathbf{M}}_k \tag{21.152}$$

where b is the material parameter, it follows that

$$\widehat{\mathbf{D}}^p_{kd}=\frac{1}{2b}\widehat{\mathbf{M}}_k\dot{\lambda} \tag{21.153}$$

Substituting Eqs. (21.146) and (21.153) into Eq. $(21.136)_2$, we have the nonlinear-kinematic hardening rule of Armstrong-Frederick (1966). Adopting the yield function $f(\hat{\overline{\mathbf{M}}})$ for $g^p(\hat{\overline{\mathbf{M}}})$, we obtain the associated flow rule for the plastic strain rate as formulated by Valdimirov et al. (2008, 2010).

As known from Eq. (12.32) and (12.38), $\overline{\mathbf{M}}$ and $\widehat{\mathbf{M}}_k$ are not symmetric tensors and thus the first and the second equations in Eq. (21.150) do not hold except for the case that the potential functions ψ^e and ψ^k are the functions only of the storage variables $\overline{\mathbf{C}}^e$ and $\widehat{\mathbf{C}}^p_{ks}$ leading to the isotropy because of the symmetries of $\overline{\mathbf{D}}^p$ and $\widehat{\mathbf{D}}^p_{kd}$.

In what follows, consider the formulation of the plastic flow rule by principle of maximum dissipation.

The dissipation power in the left-hand side of the inequality in Eq. (21.148) is described as

$$\mathfrak{D} \equiv \hat{\overline{\mathbf{M}}}{:}\overline{\mathbf{D}}^p + \widehat{\mathbf{M}}_k{:}\widehat{\mathbf{D}}^p_{kd} - F\dot{H} \tag{21.154}$$

The Lagrange function with the incidental (a collateral) condition of the yield condition

$$f(\hat{\overline{\mathbf{M}}}) - F(H) = 0 \tag{21.155}$$

is given as

$$\mathfrak{L} \equiv \mathfrak{D} - \dot{\lambda}[f(\hat{\overline{\mathbf{M}}}) - F(H)] = \hat{\overline{\mathbf{M}}}:\overline{\mathbf{D}}^p + \widehat{\mathbf{M}}_k:\widehat{\mathbf{D}}^p_{kd} - F\dot{H} - \dot{\lambda}[f(\hat{\overline{\mathbf{M}}}) - F(H)] = 0 \tag{21.156}$$

Now, consider to maximize $\mathfrak{L}$ for the variation of $\overline{\mathbf{M}}$, $\widehat{\mathbf{M}}_k$ and F, i.e.

$$\begin{aligned}
\delta\mathfrak{L} &= \delta\hat{\overline{\mathbf{M}}}:\overline{\mathbf{D}}^p + \delta\widehat{\mathbf{M}}_k:\widehat{\mathbf{D}}^p_{kd} - \delta F\dot{H} - \dot{\lambda}[\delta f(\hat{\overline{\mathbf{M}}}) - \delta F(H)] - \delta\dot{\lambda}[f(\hat{\overline{\mathbf{M}}}) - F(H)] \\
&= \overline{\mathbf{D}}^p:\delta\hat{\overline{\mathbf{M}}} + \widehat{\mathbf{D}}^p_{kd}:\delta\widehat{\mathbf{M}}_k - \delta F\dot{H} \\
&\quad - \dot{\lambda}\frac{\partial f(\hat{\overline{\mathbf{M}}})}{\partial\overline{\mathbf{M}}}:\delta\overline{\mathbf{M}} + \dot{\lambda}\frac{\partial f(\hat{\overline{\mathbf{M}}})}{\partial\overline{\mathbf{M}}}:\delta\widehat{\mathbf{M}}_k + \dot{\lambda}\,\delta F(H) \\
&= \left(\overline{\mathbf{D}}^p - \dot{\lambda}\frac{\partial f(\hat{\overline{\mathbf{M}}})}{\partial\overline{\mathbf{M}}}\right):\delta\overline{\mathbf{M}} + \left(\widehat{\mathbf{D}}^p_{kd} + \dot{\lambda}\frac{\partial f(\hat{\overline{\mathbf{M}}})}{\partial\overline{\mathbf{M}}}\right):\delta\widehat{\mathbf{M}}_k - (\dot{H} - \dot{\lambda})\delta F(H) = 0
\end{aligned}$$

which is satisfied if

$$\overline{\mathbf{D}}^p = \dot{\lambda}\frac{\partial f(\hat{\overline{\mathbf{M}}})}{\partial\overline{\mathbf{M}}},\ \widehat{\mathbf{D}}^p_{kd} = -\dot{\lambda}\frac{\partial f(\hat{\overline{\mathbf{M}}})}{\partial\overline{\mathbf{M}}} = -\overline{\mathbf{D}}^p,\ \dot{H} = \dot{\lambda} \tag{21.157}$$

so that the associated flow rule holds for the plastic strain rate.

21.8 Impertinence in Formulation of Plastic Flow Rule by Second Law of Thermodynamics

As described in the beginning of this chapter, the hyperelasticity based on the elastic potential is adopted in the formulations of constitutive equations for the damage, the phase-transformation phenomena and the finite strain theory based on the multiplicative decomposition of deformation gradient, since the large elastic deformation is concerned in them. The derivations of the flow rules for the plastic strain rate and the thermodynamic dissipative strain rates conjugate to internal variables in the potential types from the second law of thermodynamics, i.e. the Clausius-Duhem inequality and further the derivation of the associated flow rule from the principle of maximum dissipation are widely written in many literatures (cf. e.g. Lemaitre and Chaboche 1990; Lemaitre 1992; Haupt 2000; Holzapfel 2000; Wallin et al. 2003; Dettmer and Reese 2004; Houlsby et al. 2006; Wallin and Ristinmaa 2005; Voyiadjis and Kattan 2006: de Souza Neto et al. 2008; Vladimirov et al. 2008, 2010; Murakami 2012; Belytschko et al. 2014).

However, it would be reckless to expect that a valuable information is provided for the explicit formulation (on such a question which is better the associate flow rule and the potential flow rule or the non-potential flow rule) of the irreversible constitutive equation of solids exhibiting complex behavior contrastive to ideal gas behavior from the thermodynamic second law, i.e. the Clausius-Duhem inequality or the *principle of increase of entropy* which is to be the universal law governing all the phenomena in the natural world. Further, the principle of maximum dissipation is merely to be a postulate. The formulations of plastic constitutive relation by use of them are to be the easygoing pre-established harmonies so as to lead to the potential flow rule and further the associated flow rule.

Only the stress for the elastic strain and internal variables for storage parts of plastic strain are required the formulations by the potential functions, which are based on the first laws of thermodynamics. In contrast, the potential theory does not hold physically in general, although only the potential theory has been formulated by quoting the second law of thermodynamics. Note that a rule with the generality has to be derived from the universal law: the second law of thermodynamics. Then, the derivation of the potential theory would be misuse of the second law of thermodynamics. In facts, any flow rule which is derived independent of the second law of thermodynamics does not violate the second law of thermodynamics. Moreover, it would be impossible to derive a flow rule which can be proved to violate the second law of thermodynamics.

In order to avoid the misunderstanding, one should not say that the plastic potential flow rule is derived from the thermodynamics but one should say merely that the plastic potential flow rule does not violate the thermodynamic laws. It is desirable to formulate constitutive equations clarifying the definite physical backgrounds, without falling into the thermodynamic formalism.

Besides, it would be harmful for the education of mechanics to describe as if the plastic flow rules are derived based on the second law of thermodynamics or as if the second law of thermodynamics is inevitable for the formulation of plastic constitutive relations.

References

Armstrong PJ, Frederick CO (1966) A mathematical representation of the multiaxial Bauschinger effect. CEGB report RD/B/N 731 (or in materials at high temperature) 24:1–26

Bartel T, Menzel A, Svendsen B (2011) Thermodynamic and relaxation-based modeling of the interaction between martensitic phase transformations and plasticity. J Mech Phys Solids 59:1004–1019

Belytschko T, Liu WK, Moran B (2014) Nonlinear finite elements for continua and structures, 2nd edn. Wiley, New York

de Souza Neto EA, Perić D, Owen DJR (2008) Computational methods for plasticity. Wiley, Chichester

Dettmer W, Reese S (2004) On the theoretical and numerical modelling of Armstrong-Frederic kinematic hardening in the finite strain regime. Compt Meth Appl Mech Eng 193:87–116

Fung YC (1965) Foundations of solid mechanics. Prentice-Hall, Englewood Cliffs

Germain P, Nguyen QS, Suquet P (1983) Continuum thermodynamics. J Appl Mech 50:1010–1020

Hashiguchi K (2001) On the thermomechanical approach to the formulation of plastic constitutive equations. Soils Found 41(4):89–94

Haupt P (2000) Continuum mechanics and theory of materials. Springer, Wien

Hill R (1948) A variational principle of maximum plastic work in classical plasticity. Quart J Mech Appl Math 1:18–28

Hill R (1950) The mathematical theory of plasticity. Oxford University Press, London

Holzapfel GA (2000) Nonlinear solid mechanics: a continuum approach for engineering. Wiley, New York

Houlsby GT, Puzrin AM (2006) Principles of hyperelasticity: an approach to plasticity theory based on thermodynamic principles. Springer, Heidelberg

Lemaitre JA (1992) A course on damage mechanics. Springer, Heidelberg

Lemaitre JA, Chaboche J-L (1990) Mechanics of solid materials. Cambridge University Press, Cambridge

Lemaitre JA, Desmoral R (2005) Engineering damage mechanics. Springer, Heidelberg

Lubarda VA (2002) Elastoplasticity Theory. CRC Press, Boca Ranton, Florida

Lubliner J (1984) A maximum-dissipation principle in generalized plasticity. Acta Mech 52:225–237

Malvern LE (1969) Introduction to the mechanics of a continuous medium. Prentice-Hall, Englewood Cliffs

Menzel A, Ekh M, Runesson K, Steinmann P (2005) A framework for multiplicative elastoplasticity with kinematic hardening coupled to anisotropic damage. Int J Plast 21:397–434

Murakami S (2012) Continuum damage mechanics: a continuum mechanics approach to the analysis of damage and fracture. Springer, Dordrecht

Onsager L (1931) Reciprocal relations in irreversible processes. I and II. Phys Rev 37 and 38:405–426 and 2265–2279

Ottosen NS, Ristinmaa M (2005) The mechanics of constitutive modeling. Elsevier, Amsterdam

Truesdell C, Noll W (1965) In: Flugge S (ed) The nonlinear field theories of mechanics, encyclopedia of physics, vol III/3. Springer, Berlin

Vladimirov IN, Pietryga MP, Reese S (2008) On the modeling of nonlinear kinematic hardening at finite strains with application to springback -comparison of time integration algorithm. Int J Numer Meth Eng 75:1–28

Vladimirov IN, Pietryga MP, Reese S (2010) Anisotroipc finite elastoplasticity with nonlinear kinematic and isotropic hardening and application to shear metal forming. Int J Plast 26:659–687

Voyiadjis GZ, Kattan PI (2006) Damage mechanics. In: Mechanical engineering, 2nd edn. CRC Press, New York

Wallin M, Ristinmaa M (2005) Deformation gradient based kinematic hardening model. Int J Plast 21:2025–2050

Wallin M, Ristinmaa M, Ottesen NS (2003) Kinematic hardening in large strain plasticity. Eur J Mech A/Solids 22:341–356

Ziegler H (1983) An introduction to thermomechanics, 2nd edn. North-Holland, Amsterdam

Eratum to: Foundations of Elastoplasticity: Subloading Surface Model

Eratum to: K. Hashiguchi, *Foundations of Elastoplasticity: Subloading Surface Model*, DOI 10.1007/978-3-319-48821-9

An error in the production process unfortunately led to publication of this book prematurely, before incorporation of the final corrections. The version supplied here has been corrected.

The original version of this book was revised. The erratum to this book is available at 10.1007/978-3-319-48821-9

K. Hashiguchi, *Foundations of Elastoplasticity: Subloading Surface Model*,
DOI 10.1007/978-3-319-48821-9_22

Appendix A
Projection of Area

Consider the projection of the area having the unit normal vector **n** onto the surface having the normal vector **m** in Fig. A.1.

Now, suppose the plane (□abcd in Fig. A.1) which contains the unit normal vectors **m** and **n**. Then, consider the line $\overline{\mathrm{ef}}$ obtained by cutting the area having the unit normal vector **n** by this plane. Further, divide the area having the unit normal vector **n** to the narrow bands perpendicular to this line and their projections onto the surface having the normal vector **m**. The lengths of projected bands are same as the those of the original bands but the projected widths $d\bar{b}$ are obtained by multiplying the scalar product of the unit normal vectors, i.e. $\mathbf{m}\cdot\mathbf{n}$ to the original widths db. Eventually, the projected area $d\bar{a}$ is related to the original area da as follows:

$$d\bar{a} = \mathbf{m}\cdot\mathbf{n}\,da \tag{A.1}$$

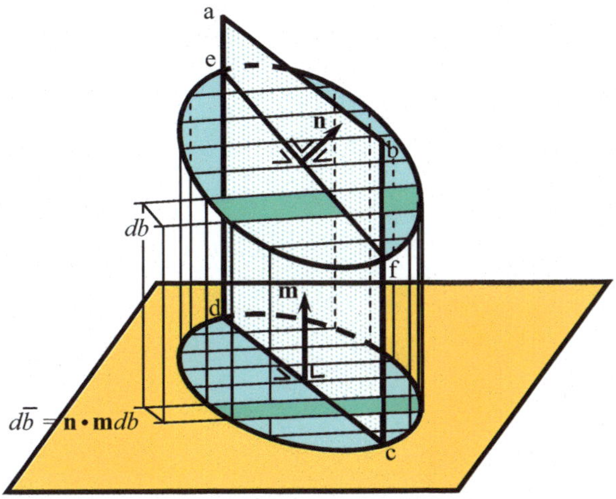

Fig. A.1 Projection of area

K. Hashiguchi, *Foundations of Elastoplasticity: Subloading Surface Model*,
DOI 10.1007/978-3-319-48821-9

Appendix B
Covariant and Contravariant Base Vectors and Components

Consider the general curvilinear coordinate system $(\theta^1, \theta^2, \theta^3)$ with the primary base $\{\mathbf{a}_i\}$ and the locally defined coordinate system $(\theta_1, \theta_2, \theta_3)$ with the reciprocal base $\{\mathbf{a}^i\}$. The infinitesimal line-element $d\mathbf{x}$ is described from Eq. (1.47) in these coordinate systems as follows:

$$d\mathbf{x} = \begin{cases} \dfrac{\partial \mathbf{x}}{\partial \theta^i} d\theta^i = d\theta^i \mathbf{a}_i (= (d\mathbf{x} \cdot \mathbf{a}^i)\mathbf{a}_i) \\ \dfrac{\partial \mathbf{x}}{\partial \theta_i} d\theta_i = d\theta_i \mathbf{a}^i (= (d\mathbf{x} \cdot \mathbf{a}_i)\mathbf{a}^i) \end{cases} \tag{A.2}$$

from which we have

$$d\theta^i = d\mathbf{x} \cdot \mathbf{a}^i, \quad d\theta_i = d\mathbf{x} \cdot \mathbf{a}_i \tag{A.3}$$

$$\mathbf{a}_i = \frac{\partial \mathbf{x}}{\partial \theta^i}, \quad \mathbf{a}^i = \frac{\partial \mathbf{x}}{\partial \theta_i} \tag{A.4}$$

satisfying Eq. (1.46), i.e.

$$\mathbf{a}_i \cdot \mathbf{a}^j = \frac{\partial \mathbf{x}}{\partial \theta^i} \cdot \frac{\partial \mathbf{x}}{\partial \theta^j} = \frac{\partial \theta^r \mathbf{a}_r}{\partial \theta^i} \cdot \frac{\partial \theta_s \mathbf{a}^s}{\partial \theta_j} = \delta^{ir} \delta_{js} \delta_r^s = \delta_j^i \tag{A.5}$$

Now, consider the another curvilinear coordinate system $(\theta^{*1}, \theta^{*2}, \theta^{*3})$ with the base $\{\mathbf{a}_i^*\}$ and the locally defined coordinate system $(\theta_1^*, \theta_2^*, \theta_3^*)$ with the reciprocal base $\{\mathbf{a}^{*i}\}$. The following coordinate transformation rules hold for the base vectors by the chain rule of differentiation.

$$\mathbf{a}_i^* = \frac{\partial \mathbf{x}}{\partial \theta^{*i}} = \frac{\partial \mathbf{x}}{\partial \theta^j} \frac{\partial \theta^j}{\partial \theta^{*i}} = \frac{\partial \theta^j}{\partial \theta^{*i}} \mathbf{a}_j, \quad \mathbf{a}_i = \frac{\partial \theta^{*j}}{\partial \theta^i} \mathbf{a}_j^* \left(= \frac{\partial \theta^{*j}}{\partial \theta^i} \frac{\partial \theta^r}{\partial \theta^{*j}} \mathbf{a}_r = \delta_i^r \mathbf{a}_r \right) \tag{A.6}$$

$$\mathbf{a}^{*i} = \frac{\partial \theta^{*i}}{\partial \theta^j} \mathbf{a}^j, \quad \mathbf{a}^i = \frac{\partial \theta^i}{\partial \theta^{*j}} \mathbf{a}^{*j} \left(= \frac{\partial \theta^i}{\partial \theta^{*j}} \frac{\partial \theta^{*j}}{\partial \theta^r} \mathbf{a}^r = \delta_r^i \mathbf{a}^r \right) \tag{A.7}$$

K. Hashiguchi, *Foundations of Elastoplasticity: Subloading Surface Model*,
DOI 10.1007/978-3-319-48821-9

The vector $\mathbf{v}$ is described by

$$\mathbf{v} = \begin{cases} v^i\mathbf{a}_i = v^i \dfrac{\partial \theta^{*j}}{\partial \theta^i}\mathbf{a}_j^* = v^j \dfrac{\partial \theta^{*i}}{\partial \theta^j}\mathbf{a}_i^* = v^{*i}\mathbf{a}_i^* \\ v_i\mathbf{a}^i = v_i \dfrac{\partial \theta^{i}}{\partial \theta^{*j}}\mathbf{a}^{*j} = v_j \dfrac{\partial \theta^{j}}{\partial \theta^{*i}}\mathbf{a}^{*i} = v_i^*\mathbf{a}^{*i} \end{cases} \tag{A.8}$$

from which one has the transformation rules:

$$\left.\begin{aligned} v^{*i} &= \frac{\partial \theta^{*i}}{\partial \theta^j} v^j, \quad v^i = \frac{\partial \theta^i}{\partial \theta^{*j}} v^{*j} \\ v_i^* &= \frac{\partial \theta^j}{\partial \theta^{*i}} v_j, \quad v_i = \frac{\partial \theta^{*j}}{\partial \theta^i} v_j^* \end{aligned}\right\} \tag{A.9}$$

In the analogous way, one has the following transformation rules for the second-order tensor.

$$\left.\begin{aligned} t^{*ij} &= \frac{\partial \theta^{*i}}{\partial \theta^r}\frac{\partial \theta^{*j}}{\partial \theta^s} t^{rs}, \quad & t^{ij} &= \frac{\partial \theta^i}{\partial \theta^{*r}}\frac{\partial \theta^j}{\partial \theta^{*s}} t^{*rs} \\ t^{*i}_{\cdot j} &= \frac{\partial \theta^{*i}}{\partial \theta^r}\frac{\partial \theta^s}{\partial \theta^{*j}} t^{r}_{\cdot s}, \quad & t^{i}_{\cdot j} &= \frac{\partial \theta^i}{\partial \theta^{*r}}\frac{\partial \theta^{*s}}{\partial \theta^j} t^{*r}_{\cdot s} \\ t_i^{*\cdot j} &= \frac{\partial \theta^r}{\partial \theta^{*i}}\frac{\partial \theta^{*j}}{\partial \theta^s} T_r^{\cdot s}, \quad & t_i^{\cdot j} &= \frac{\partial \theta^{*r}}{\partial \theta^i}\frac{\partial \theta^j}{\partial \theta^{*s}} t_r^{*\cdot s} \\ t_{ij}^* &= \frac{\partial \theta^r}{\partial \theta^{*i}}\frac{\partial \theta^s}{\partial \theta^{*j}} t_{rs}, \quad & t_{ij} &= \frac{\partial \theta^{*r}}{\partial \theta^i}\frac{\partial \theta^{*s}}{\partial \theta^j} t_{rs}^* \end{aligned}\right\} \tag{A.10}$$

It can be recognized from Eqs. (A.6), (A.7), (A.9) and (A.10) that the transformation rules of the base vectors and the components of vector and tensor are classified into the type with the subscript and the other type with the superscript: The subscript $(\ )_i$ and the superscript $(\ )^i$ are put in the base vector or component when the partial-derivative operator $\partial \theta^i$ is placed in the denominator and the numerator, respectively, in their transformation rules. The former and the latter are referred to as the "*covariant*" and the "*contravariant*", respectively, base vector, component and description of vector and tensor. These notations are originated from the following mathematical background.

The vector whose components obey the following transformation rule which is automatically derived by the *chain rule of differentiation* (e.g. the gradient of scalar potential function $\partial f/\partial \theta^{*i} = (\partial f/\partial \theta^j)\partial \theta^j/\partial \theta^{*i}$) is classified to the covariant vector.

$$v^{*i} = \frac{\partial \theta^j}{\partial \theta^{*i}} v^j$$

In contrast, the vector whose components obey the following transformation rule is classified to the contravariant vector (e.g. the velocity vector $d\theta^{*i}/dt = (\partial \theta^{*i}/\partial \theta^j)d\theta^j/dt$).

$$v^{*i} = \frac{\partial \theta^{*i}}{\partial \theta^j} v^j$$

Tensors are classified to covariant and contravariant tensors analogously. This classification based on the coordinate transformation is also applied to both of the base vectors and the coordinate components as shown in Eqs. (A.6), (A.7), (A.9) and (A.10). Again note that the covariant one and the contravariant one are combined inversely since the scale transformations are inverse in the base vector and its component in general.

Appendix C
Logarithmic Spin

We adopt the Eulerian-logarithmic strain tensor $\boldsymbol{e}^{(0)}$ in Eq. (2.68) and its corotational rate $\overset{\circ}{\boldsymbol{e}}{}^{(0)\mathrm{E}}$ with the Eulerian spin $\boldsymbol{\Omega}^{E}$. Further, consider to add a spin $\mathbf{x}$ to $\boldsymbol{\Omega}^{E}$ so that the corotational rate $\overset{\circ}{\boldsymbol{e}}{}^{(0)\mathrm{Log}}$ with the spin $\boldsymbol{\Omega}^{E}+\mathbf{x}$ coincides to the strain rate $\mathbf{d}$, i.e.

$$\begin{aligned}\mathbf{d} &= \overset{\circ}{\boldsymbol{e}}{}^{(0)\mathrm{Log}} \equiv \dot{\boldsymbol{e}}^{(0)} + \mathbf{e}^{(0)}\boldsymbol{\Omega}^{\mathrm{Log}} - \boldsymbol{\Omega}^{\mathrm{Log}}\mathbf{e}^{(0)} \\ &= \dot{\boldsymbol{e}}^{(0)} + \boldsymbol{e}^{(0)}(\boldsymbol{\Omega}^{E}+\mathbf{x}) - (\boldsymbol{\Omega}^{E}+\mathbf{x})\boldsymbol{e}^{(0)} \\ &\left(= \dot{\boldsymbol{e}}^{(0)} + \boldsymbol{e}^{(0)}\boldsymbol{\Omega}^{E} - \boldsymbol{\Omega}^{E}\boldsymbol{e}^{(0)} + (\boldsymbol{e}^{(0)}\mathbf{x} - \mathbf{x}\boldsymbol{e}^{(0)}) = \overset{\circ}{\boldsymbol{e}}{}^{(0)E} + (\boldsymbol{e}^{(0)}\mathbf{x} - \mathbf{x}\boldsymbol{e}^{(0)})\right)\end{aligned} \tag{A.11}$$

with the logarithmic spin

$$\boldsymbol{\Omega}^{\mathrm{Log}} = \boldsymbol{\Omega}^{E} + \mathbf{x} \tag{A.12}$$

The relation

$$d_{\alpha\alpha} = (\overset{\circ}{\boldsymbol{e}}{}^{(0)E})_{\alpha\alpha} = (\overset{\circ}{\boldsymbol{e}}{}^{(0)})_{\alpha\alpha} = \frac{\dot{\lambda}_{\alpha}}{\lambda_{\alpha}} \ (\text{no sum}) \tag{A.13}$$

holds for the diagonal components, noting Eqs. (2.72) and (2.143), and

$$\begin{aligned}d_{\alpha\beta} &= \left(-\left(\mathbf{x}\mathbf{e}^{(0)} - \mathbf{e}^{(0)}\mathbf{x}\right)\right)_{\alpha\beta} = -x_{\alpha\sigma}e_{\sigma\beta} + e_{\alpha\sigma}x_{\sigma\beta} \\ &= -x_{\alpha\beta}\ln\lambda_{\beta} + \ln\lambda_{\alpha}x_{\alpha\beta} = x_{\alpha\beta}\ln(\lambda_{\alpha}/\lambda_{\beta}) \quad \text{for} \quad \alpha \neq \beta\end{aligned}$$

K. Hashiguchi, *Foundations of Elastoplasticity: Subloading Surface Model*,
DOI 10.1007/978-3-319-48821-9

for the diagonal components, from which it follows that

$$x_{\alpha\beta} = \frac{d_{\alpha\beta}}{\ln(\lambda_\alpha/\lambda_\beta)} \quad \text{for } \alpha \neq \beta \tag{A.14}$$

Taking account of Eqs. (A.12) and (2.109) for the Eulerian spin into Eq. (A.12), we have the logarithmic spin

$$\begin{aligned}
\boldsymbol{\Omega}^{\mathrm{Log}} &= \sum_\alpha \sum_\beta \Omega^{\mathrm{Log}}_{\alpha\beta} \mathbf{n}^{(\alpha)} \otimes \mathbf{n}^{(\beta)} \\
&= \sum_\alpha \sum_\beta \left\{ w_{\alpha\beta} + \left[\frac{1+(\lambda_\alpha/\lambda_\beta)^2}{1-(\lambda_\alpha/\lambda_\beta)^2} + \frac{1}{\ln(\lambda_\alpha/\lambda_\beta)} \right] d_{\alpha\beta} \right\} \mathbf{n}^{(\alpha)} \otimes \mathbf{n}^{(\beta)} \\
&= \mathbf{w} + \sum_\alpha \sum_\beta \left[\frac{1+(\lambda_\alpha/\lambda_\beta)^2}{1-(\lambda_\alpha/\lambda_\beta)^2} + \frac{1}{\ln(\lambda_\alpha/\lambda_\beta)} \right] \mathbf{P}_\alpha \mathbf{d} \mathbf{P}_\beta
\end{aligned} \tag{A.15}$$

where $\mathbf{P}_\sigma$ is defined by the Eulerian triad $\{\mathbf{n}^{(\sigma)}\}$ as

$$\mathbf{P}_\sigma = \mathbf{n}^{(\sigma)} \otimes \mathbf{n}^{(\sigma)} \text{ (no sum)} \tag{A.16}$$

noting

$$\begin{aligned}
\mathbf{P}_\alpha \mathbf{d} \mathbf{P}_\beta &= \mathbf{n}^{(\alpha)} \otimes \mathbf{n}^{(\alpha)} d_{\tau\gamma} \mathbf{n}^{(\tau)} \otimes \mathbf{n}^{(\gamma)} \mathbf{n}^{(\beta)} \otimes \mathbf{n}^{(\beta)} = \mathbf{n}^{(\alpha)} \otimes \delta_{\alpha\tau} d_{\tau\gamma} \delta_{\lambda\beta} \mathbf{n}^{(\beta)} \\
&= d_{\alpha\beta} \mathbf{n}^{(\alpha)} \otimes \mathbf{n}^{(\beta)}
\end{aligned}$$

The logarithmic spin is applied to the stress and internal variables in the hypoelastic-based plastic constitutive equation. The corotational rates of the Cauchy stress is described as follows:

$$\overset{\circ}{\boldsymbol{\sigma}}{}^{\mathrm{Log}} = \mathbf{E} : \mathbf{d}^e = \mathbf{E} : (\mathbf{d} - \mathbf{d}^p) \tag{A.17}$$

In what follows, it will be verified that the hypoelastic elastic equation with the corotational rates due to the logarithmic spin leads to the hyperelastic equation.

The hypoelastic equation based on the logarithmic corotational rates in the Hooke's type is described by Eq. (5.54) as

$$\overset{\circ}{\boldsymbol{\sigma}}{}^{\mathrm{Log}} = \mathbf{E} : \mathbf{d} = (K - \frac{2}{3}G)(\mathrm{tr}\,\mathbf{d})\mathbf{I} + 2G\mathbf{d} \tag{A.18}$$

Here, noting Eq. (2.60) the substitution of

$$(\ln \mathbf{V})^{\circ \mathrm{Log}} = \mathbf{d} \tag{A.19}$$

leads Eq. (A.18) to

$$\overset{\circ}{\boldsymbol{\sigma}}{}^{\mathrm{Log}} = \left(K - \frac{2}{3}G\right)\left[\mathrm{tr}(\ln\mathbf{V})^{\circ \mathrm{Log}}\right]\mathbf{I} + 2G(\ln\mathbf{V})^{\circ \mathrm{Log}} \tag{A.20}$$

If the elastic parameters K and G are constants, Eq. (A.20) is rewritten as

$$\overset{\circ}{\boldsymbol{\sigma}}{}^{\mathrm{Log}} = \left[\left(K - \frac{2}{3}G\right)[\mathrm{tr}(\ln \mathbf{V})]\mathbf{I} + 2G(\ln \mathbf{V})\right]^{\circ \mathrm{Log}} \tag{A.21}$$

which results in

$$\boldsymbol{\sigma} = \left(K - \frac{2}{3}G\right)[\mathrm{tr}(\ln\mathbf{V})]\mathbf{I} + 2\mu(\ln\mathbf{V}) \tag{A.22}$$

Eq. (A.22) is to be the hyperelastic equation.

The hypoelastic equation with the corotational rate based on the continuum spin causes the oscillation of stress in the simple shear deformation as was shown by Dienes (1979). Further, the hypoelastic-based plasticity with the kinematic hardening and the corotational rate based on the continuum spin causes the oscillation of stress in the simple shear deformation as was shown by Nagtegaal and de Jong (1982). The oscillation is excluded by use of the logarithmic corotational rates as shown by the numerical experiments (Xiao et al. 1997; Brepols et al. 2014; Shutov and Ihlemann 2014). However, it should be noted that the logarithmic corotational rates possesses the following limitations.

1. The hyperelastic equation is obtained only from the Hooke's type hypoelastic equation with the constant elastic parameters. Therefore, it is applied to metals but inapplicable to pressure-dependent materials, e.g. soils, rocks and concretes the elastic parameters of which depend on the pressure.
2. The fact that the rotation of the substructure in material is not determined by the geometrical change of the external appearance of material is ignored. The physically meaningful spin is the spin of substructure in material which is suppressed by the plastic dissipative deformation from the rigid-body rotation so that the plastic spin must be incorporated in addition to the spin based on the geometrical change of the outside appearance as described in Sect. 6.2 and Chap. 16.

Appendix D
Euler's Theorem for Homogeneous Function

The homogeneous function of degree n is defined to fulfill the relation

$$f(ax_1, \, ax_2, \cdots, \, ax_m) = a^n f(x_1, \, x_2, \cdots, \, x_m) \qquad (A.23)$$

for the variables $x_1, x_2, \cdots, x_m$ (m: number of variables), letting a denote an arbitrary scalar constant. Then, consider the homogeneous function given by the polynomial expression:

$$
\begin{aligned}
f(x_1, \, x_2, \, \cdots, \, x_m) &= \sum_{i=1}^{s} c_i x_1^{n_1^i} x_2^{n_2^i} \cdots x_m^{n_m^i} \\
&= c_1 x_1^{n_1^1} x_2^{n_2^1} \cdots x_m^{n_m^1} + c_2 x_1^{n_1^2} x_2^{n_2^2} \cdots x_m^{n_m^2} + \cdots + c_s x_1^{n_1^s} x_2^{n_2^s} \cdots x_m^{n_m^s}
\end{aligned}
\qquad (A.24)
$$

where s is the number of terms of polynomial expression and c_i are constants, provided to fulfill

$$\sum_{j=1}^{m} n_j^i = n \quad \text{for each } i \qquad (A.25)$$

Eq. (A.24) leads to

$$
\begin{aligned}
\sum_{j=1}^{m} \frac{\partial f(x_1, \, x_2, \, \cdots, \, x_m)}{\partial x_j} x_j &= \sum_{i=1}^{s} c_i \sum_{j=1}^{m} n_j^i x_1^{n_1^i} x_2^{n_2^i} \cdots x_j^{n_j^i - 1} \cdots x_m^{n_m^i} x_j \\
&= n \sum_{i=1}^{s} c_i x_1^{n_1^i} x_2^{n_2^i} \cdots x_m^{n_m^i}
\end{aligned}
$$

© Springer International Publishing AG 2017

K. Hashiguchi, *Foundations of Elastoplasticity: Subloading Surface Model*,

DOI 10.1007/978-3-319-48821-9

Then, it holds that

$$\sum_{j=1}^{m} \frac{\partial f(x_1, x_2, \cdots, x_m)}{\partial x_j} x_j = n f(x_1, x_2, \cdots, x_m) \tag{A.26}$$

which is called the *Euler's theorem for homogeneous function*.
For the simple example ($m = 3, n = 4, s = 3$):

$$f(x, y, z) = \alpha x^4 + \beta x^3 y + \gamma x^2 yz$$

Eq. (A.26) is confirmed as follows:

$$\frac{\partial f}{\partial x} x + \frac{\partial f}{\partial y} y + \frac{\partial f}{\partial z} z = (4\alpha x^3 + 3\beta x^2 y + 2\gamma xyz)x + (\beta x^3 + \gamma x^2 z)y + \gamma x^2 y \cdot z = 4f$$

Eq. (A.26) yields Eq. (6.32) for the yield function ($n = 1$).

Appendix E
Outward-Normal Vector of Surface

The quantity $(\partial f(\mathbf{t})/\partial \mathbf{t})\mathbf{:}d\mathbf{t}$ is regarded as the scalar product of the vectors $\partial f(\mathbf{t})/\partial \mathbf{t}$ and $d\mathbf{t}$ in the nine-dimensional space $(t_{11}, t_{22}, t_{33}, \cdots, t_{31}, t_{13})$. Here, it holds that

$$\frac{\partial f(\mathbf{t})}{\partial \mathbf{t}} : d\mathbf{t} \begin{cases} > 0 : d\mathbf{t} \text{ is directed outward-normal to surface} \\ = 0 : d\mathbf{t} \text{ is directed tangential to surface} \\ < 0 : d\mathbf{t} \text{ is directed inward-normal to surface} \end{cases} \tag{A.27}$$

Therefore, $\partial f(\mathbf{t})/\partial \mathbf{t}$ is interpreted to be the vector designating the outward-normal of the surface described by $f(\mathbf{t}) = \text{const}$. Therefore, $\partial f(\boldsymbol{\sigma})/\partial \boldsymbol{\sigma}$ designates the outward-normal of the yield surface $f(\boldsymbol{\sigma}) = F$.

K. Hashiguchi, *Foundations of Elastoplasticity: Subloading Surface Model*,
DOI 10.1007/978-3-319-48821-9

Appendix F
Relationships of Material Constants in $\ln v - \ln p$ and $e - \ln p$ Linear Relations

The following relation holds from Eqs. (11.3) and (11.14) for $p_e = 0$, provided that Eq. (2.142) holds for elastic volumetric strain, i.e. $\varepsilon_v^e = \ln(1+\varepsilon_v^e)$.

$$-\widetilde{\kappa}\ln\frac{p}{p_0} = \ln\left(1 - \frac{\kappa}{1+e_0}\ln\frac{p}{p_0}\right) \tag{A.28}$$

from which one has

$$\widetilde{\kappa} = \frac{\ln\left(1 - \frac{\kappa}{1+e_0}\ln\frac{p}{p_0}\right)}{-\ln\frac{p}{p_0}} \tag{A.29}$$

It follows from Eq. (A.29) for infinitesimal deformation under $p \cong p_0$ that

$$\lim_{p\to p_0}\widetilde{\kappa} = \lim_{p\to p_0}\frac{\ln\left(1 - \frac{\kappa}{1+e_0}\ln\frac{p}{p_0}\right)}{-\ln\frac{p}{p_0}} = \lim_{p\to p_0}\frac{\dfrac{-\frac{\kappa}{1+e_0}\frac{1}{p}}{1-\frac{\kappa}{1+e_0}\ln\frac{p}{p_0}}}{-\frac{1}{p}} = \frac{\kappa}{1+e_0} \tag{A.30}$$

resulting in

$$\widetilde{\kappa} \cong \frac{\kappa}{1+e_0} \tag{A.31}$$

Further, substituting Eqs. $(11.3)_2$ and $(11.14)_2$ into $\varepsilon_v^p = \ln(1+\varepsilon_v^p)$ due to Eq. (2.142), it follows that

$$-(\widetilde{\lambda} - \widetilde{\kappa})\ln\frac{p_y}{p_{y0}} = \ln\left(1 - \frac{\lambda-\kappa}{1+e_0}\ln\frac{p_y}{p_{y0}}\right)$$

i.e.

K. Hashiguchi, *Foundations of Elastoplasticity: Subloading Surface Model*,
DOI 10.1007/978-3-319-48821-9

$$-(\tilde{\lambda}-\tilde{\kappa}) = \lim_{p_y \to p_{y0}} \frac{\ln\left(1-\dfrac{\lambda-\kappa}{1+e_0}\ln\dfrac{p_y}{p_{y0}}\right)}{\ln\dfrac{p_y}{p_{y0}}} = \lim_{p_y \to p_{y0}} \frac{\dfrac{-\dfrac{\lambda-\kappa}{1+e_0}\dfrac{1}{p_y}}{1-\dfrac{\lambda-\kappa}{1+e_0}\ln\dfrac{p_y}{p_{y0}}}}{\dfrac{1}{p_y}} = -\frac{\lambda-\kappa}{1+e_0}$$

from which, noting Eq. (A.31), one has

$$\lim_{p_y \to p_{y0}} \tilde{\lambda}-\tilde{\kappa} \cong \frac{\lambda-\kappa}{1+e_0} \tag{A.32}$$

Based on Eqs. (A.30) and (A.32), $\tilde{\kappa}$ and $\tilde{\lambda}$ may be given by

$$\tilde{\kappa} = \frac{\kappa}{1+e_0}, \quad \tilde{\lambda} = \frac{\lambda}{1+e_0} \tag{A.33}$$

which can be calculated from a plenty of data on λ and κ accumulated in the past. Then, the analysis would be improved over the finite deformation by using Eq. (11.4) with Eq. (A.33) instead of Eq. (11.15) or (11.17).

Needless to say, one has to determine the material parameters $\tilde{\lambda}$ and $\tilde{\kappa}$ directly from test data for soils without the data of λ and κ or for the case that an accurate formulation is required. Here, note that the curve fitting of $\ln v - \ln p$ linear relation to test data is easier than the fitting of the $e - \ln p$ linear relation to test data because real soil behavior is far near to the former than the latter.

Appendix G
Derivation of Eq. (11.22)

Differentiation of Eq. (11.19) under the condition $f(\boldsymbol{\sigma})$ = const. leads to

$$g(\eta_m)dp + pg'(\eta_m)\left(\frac{\partial \eta_m}{\partial p}dp + \frac{\partial \eta_m}{\partial \|\boldsymbol{\sigma}'\|}d\|\boldsymbol{\sigma}'\|\right)$$
$$= g(\eta_m)dp + pg'(\eta_m)\left(-\frac{\|\boldsymbol{\sigma}'\|}{p^2 M}dp + \frac{1}{pM}d\|\boldsymbol{\sigma}'\|\right) = 0$$

from which it follows that

$$\frac{d\|\boldsymbol{\sigma}'\|}{dp} = \frac{-g(\eta_m) + \dfrac{\|\boldsymbol{\sigma}'\|}{pM}g'(\eta_m)}{\dfrac{1}{M}g'(\eta_m)} = M\left(-\frac{g(\eta_m)}{g'(\eta_m)} + \eta_m\right) \tag{A.34}$$

Taking account of $d\|\boldsymbol{\sigma}'\|/dp = 0$ at $\eta_m = 1$ in Eq. (A.34), one has Eq. (11.22).

K. Hashiguchi, *Foundations of Elastoplasticity: Subloading Surface Model*,
DOI 10.1007/978-3-319-48821-9

Appendix H
Convexity of Two-Dimensional Curve

When the curve is described by the polar coordinates (r, θ) as shown in Fig. A.2, the following relation holds.

$$\tan\alpha = \frac{rd\theta}{dr} \tag{A.35}$$

where α is the angle measured from the radius vector to the tangent line in the anti-clockwise direction. Eq. (A.35) is rewritten as

$$\cot\alpha = \frac{r'}{r}, \tag{A.36}$$

where $(\ )'$ designates the first order differentiation with respect to θ.

The equation of the tangent line at (r, θ) of the curve $r = r(\theta)$ is described by the following equation by using the current coordinates (R, Θ) on the tangent line.

$$R\cos\{\Theta - [\theta - (\pi/2 - \alpha)]\} = r\cos(\pi/2 - \alpha)$$

which is rewritten as

$$-R\sin(\Theta - \theta - \alpha) = r\sin\alpha \rightarrow \frac{1}{R\sin(\Theta - \theta - \alpha)} = -\frac{1}{r\sin\alpha}$$
$$\rightarrow \frac{1}{R} = \frac{1}{r}\cos(\Theta - \theta) - \frac{1}{r}\cot\alpha\sin(\Theta - \theta)$$

Substituting Eq. (A.36) to this equation and noting $(1/r)' = -r'/r^2$, one has the relation

$$\frac{1}{R} = \frac{1}{r}\cos(\Theta - \theta) + \left(\frac{1}{r}\right)'\sin(\Theta - \theta), \tag{A.37}$$

Equation (A.37) is rewritten by applying the Taylor expansion to $\cos\vartheta$ and $\sin\vartheta$ as

K. Hashiguchi, *Foundations of Elastoplasticity: Subloading Surface Model*,
DOI 10.1007/978-3-319-48821-9

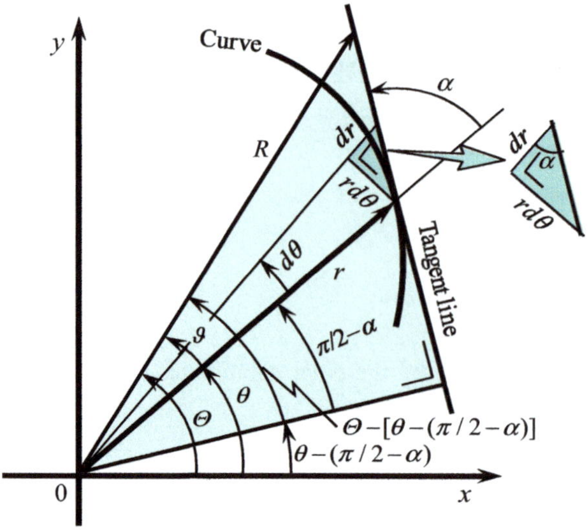

Fig. A.2 Curve in the polar coordinate (r, q)

$$\frac{1}{R(\Theta)} = \frac{1}{r(\theta)}\cos\vartheta + \left(\frac{1}{r(\theta)}\right)'\sin\vartheta = \frac{1}{r(\theta)}\left(1 - \frac{1}{2}\vartheta^2 + \cdots\right) + \left(\frac{1}{r(\theta)}\right)'\left(\vartheta - \frac{1}{6}\vartheta^3 + \cdots\right) \tag{A.38}$$

where $\vartheta \equiv \Theta - \theta$. On the other hand, the radius $r(\Theta)$ ($\Theta = \theta + \vartheta$) of the curve is described by the Taylor expansion as follows:

$$\frac{1}{r(\Theta)} = \frac{1}{r(\theta)} + \left(\frac{1}{r(\theta)}\right)'\vartheta + \frac{1}{2}\left(\frac{1}{r(\theta)}\right)''\vartheta^2 + \cdots \tag{A.39}$$

Equations (A.38) and (A.39) lead to

$$\frac{1}{r(\Theta)} - \frac{1}{R(\Theta)} = \frac{1}{2}\left[\frac{1}{r(\theta)} + \left(\frac{1}{r(\theta)}\right)''\right]\vartheta^2 + \cdots. \tag{A.40}$$

In order that the curve is convex ($r(\Theta) \le R(\Theta)$), the following inequality, i.e. *convexity condition* must hold from Eq. (A.40).

$$\frac{1}{r} + \left(\frac{1}{r}\right)'' > 0 \tag{A.41}$$

Appendix I
Matrix Representation of Tensor Relations

The representations of the second-order symmetric tensor **t** and the fourth-order symmetric tensor **T** by displaying their component is called the *Voigt representations*.

$$\mathbf{t} = \{t_{ij}\} = \{t_{11}\, t_{22}\, t_{33}\, t_{23}\, t_{32}\, t_{31}\, t_{13}\, t_{12}\, t_{21}\}^T = \{t_1\, t_2\, t_3\, t_4\, t_5\, t_6\}^T \tag{A.42}$$

$$\mathbf{T} = [T_{ijkl}] = \begin{bmatrix} T_{1111} & T_{1122} & T_{1133} & T_{1123} & T_{1132} & T_{1131} & T_{1113} & T_{1112} & T_{1121} \\ T_{2211} & T_{2222} & T_{2233} & T_{2223} & T_{2232} & T_{2231} & T_{2213} & T_{2212} & T_{2221} \\ T_{3311} & T_{3322} & T_{3333} & T_{3323} & T_{3332} & T_{3331} & T_{3313} & T_{3312} & T_{3321} \\ T_{2311} & T_{2322} & T_{2333} & T_{2323} & T_{2332} & T_{2231} & T_{2313} & T_{2312} & T_{2321} \\ T_{3211} & T_{3222} & T_{3233} & T_{3223} & T_{3232} & T_{3231} & T_{3213} & T_{3212} & T_{3221} \\ T_{3111} & T_{3122} & T_{3133} & T_{3123} & T_{3132} & T_{3131} & T_{3113} & T_{3112} & T_{3121} \\ T_{1311} & T_{1322} & T_{1333} & T_{1323} & T_{1332} & T_{1331} & T_{1313} & T_{1312} & T_{1321} \\ T_{1211} & T_{1222} & T_{1233} & T_{1223} & T_{1232} & T_{1231} & T_{1213} & T_{1212} & T_{1221} \\ T_{2111} & T_{2122} & T_{2133} & T_{2123} & T_{2132} & T_{2131} & T_{2113} & T_{2112} & T_{2121} \end{bmatrix}$$

$$= \begin{bmatrix} T_{11} & T_{12} & T_{13} & T_{14} & T_{15} & T_{16} \\ T_{21} & T_{22} & T_{23} & T_{24} & T_{25} & T_{26} \\ T_{31} & T_{32} & T_{33} & T_{34} & T_{35} & T_{36} \\ T_{41} & T_{42} & T_{43} & T_{44} & T_{45} & T_{46} \\ T_{51} & T_{52} & T_{53} & T_{54} & T_{55} & T_{56} \\ T_{61} & T_{62} & T_{63} & T_{64} & T_{65} & T_{66} \end{bmatrix} \tag{A.43}$$

K. Hashiguchi, *Foundations of Elastoplasticity: Subloading Surface Model*,
DOI 10.1007/978-3-319-48821-9

where the subscript numbers are replaced as $11 \rightarrow 1, 22 \rightarrow 2, 33 \rightarrow 3$, $23, 32 \rightarrow 4$, $31, 13 \rightarrow 5, 12, 21 \rightarrow 6$ $31, 13 \rightarrow 5, 12, 21 \rightarrow 6$

If $\mathbf{T}$ satisfies the major symmetry $T_{ijkl} = T_{klij}$, i.e. $T_{rs} = T_{sr}$, it can be represented as

$$\mathbf{T} = \begin{bmatrix} T_{11} & T_{12} & T_{13} & T_{14} & T_{15} & T_{16} \\ & T_{22} & T_{23} & T_{24} & T_{25} & T_{26} \\ & & T_{33} & T_{34} & T_{35} & T_{36} \\ & & & T_{44} & T_{45} & T_{46} \\ & Sym. & & & T_{55} & T_{56} \\ & & & & & T_{66} \end{bmatrix} \tag{A.44}$$

The fourth-order tracing identity tensor $\mathcal{T}$ in Eq. (1.143) possesses both of the minor and the major symmetries and thus it is represented as follows:

$$\mathcal{T} = [\delta_{ij}\delta_{kl}] = \begin{bmatrix} \delta_{11} & \delta_{11} & \delta_{11} & \delta_{22} & \delta_{11} & \delta_{33} & \delta_{11} & \delta_{23} & \delta_{11} & \delta_{31} & \delta_{11} & \delta_{12} \\ & & \delta_{22} & \delta_{22} & \delta_{22} & \delta_{33} & \delta_{22} & \delta_{23} & \delta_{22} & \delta_{31} & \delta_{22} & \delta_{12} \\ & & & & \delta_{33} & \delta_{33} & \delta_{33} & \delta_{23} & \delta_{33} & \delta_{31} & \delta_{33} & \delta_{12} \\ & & & & & & \delta_{12} & \delta_{12} & \delta_{12} & \delta_{31} & \delta_{12} & \delta_{12} \\ & & Sym. & & & & & & \delta_{31} & \delta_{31} & \delta_{31} & \delta_{12} \\ & & & & & & & & & & \delta_{12} & \delta_{12} \end{bmatrix}$$

$$= \begin{bmatrix} 1 & 1 & 1 & 0 & 0 & 0 \\ 1 & 1 & 1 & 0 & 0 & 0 \\ 1 & 1 & 1 & 0 & 0 & 0 \\ 0 & 0 & 0 & 0 & 0 & 0 \\ 0 & 0 & 0 & 0 & 0 & 0 \\ 0 & 0 & 0 & 0 & 0 & 0 \end{bmatrix} \tag{A.45}$$

The fourth-order tracing identity tensor $\mathcal{I}$ in Eq. (1.144) possesses both of the minor and the major symmetries and thus it is represented as follows:

$$
\mathcal{I} = [\delta_{ik}\delta_{jl}] = \begin{bmatrix}
\delta_{11}\ \delta_{11} & \delta_{12}\ \delta_{12} & \delta_{13}\ \delta_{13} & \delta_{12}\ \delta_{13} & \delta_{13}\ \delta_{11} & \delta_{11}\ \delta_{12} \\
 & \delta_{22}\ \delta_{22} & \delta_{23}\ \delta_{23} & \delta_{22}\ \delta_{23} & \delta_{23}\ \delta_{21} & \delta_{21}\ \delta_{22} \\
 & & \delta_{33}\ \delta_{33} & \delta_{32}\ \delta_{33} & \delta_{33}\ \delta_{31} & \delta_{31}\ \delta_{32} \\
 & & & \delta_{22}\ \delta_{33} & \delta_{23}\ \delta_{31} & \delta_{21}\ \delta_{32} \\
 & & & & \delta_{11}\ \delta_{22} & \delta_{12}\ \delta_{23} \\
 & & & & & \delta_{11}\ \delta_{22}
\end{bmatrix}
$$

$$
= \begin{bmatrix}
1 & 0 & 0 & 0 & 0 & 0 \\
0 & 1 & 0 & 0 & 0 & 0 \\
0 & 0 & 1 & 0 & 0 & 0 \\
0 & 0 & 0 & 1 & 0 & 0 \\
0 & 0 & 0 & 0 & 1 & 0 \\
0 & 0 & 0 & 0 & 0 & 1
\end{bmatrix}
\tag{A.46}
$$

The following matrix must be adopted to transform between the infinitesimal scientific strain $\boldsymbol{\varepsilon}$ and the infinitesimal engineering strain $\widetilde{\boldsymbol{\varepsilon}}$, i.e. $\widetilde{\boldsymbol{\varepsilon}} = \widetilde{\boldsymbol{I}} : \boldsymbol{\varepsilon}$ and $\boldsymbol{\varepsilon} = \boldsymbol{I} : \widetilde{\boldsymbol{\varepsilon}}$.

$$
\widetilde{\mathcal{I}} = \begin{bmatrix}
1 & 0 & 0 & 0 & 0 & 0 \\
0 & 1 & 0 & 0 & 0 & 0 \\
0 & 0 & 1 & 0 & 0 & 0 \\
0 & 0 & 0 & 2 & 0 & 0 \\
0 & 0 & 0 & 0 & 2 & 0 \\
0 & 0 & 0 & 0 & 0 & 2
\end{bmatrix}, \quad
\underset{\sim}{\mathcal{I}} = \begin{bmatrix}
1 & 0 & 0 & 0 & 0 & 0 \\
0 & 1 & 0 & 0 & 0 & 0 \\
0 & 0 & 1 & 0 & 0 & 0 \\
0 & 0 & 0 & 1/2 & 0 & 0 \\
0 & 0 & 0 & 0 & 1/2 & 0 \\
0 & 0 & 0 & 0 & 0 & 1/2
\end{bmatrix}
\tag{A.47}
$$

The *deviatoric projection tensor* $\mathcal{I}'$ in Eq. (1.146) possesses both of the minor and the major symmetries and thus it is represented as follows:

$$\mathcal{I}' = \mathcal{I} - \frac{1}{3}\mathcal{T} = \begin{bmatrix} 1 & 0 & 0 & 0 & 0 & 0 \\ 0 & 1 & 0 & 0 & 0 & 0 \\ 0 & 0 & 1 & 0 & 0 & 0 \\ 0 & 0 & 0 & 1 & 0 & 0 \\ 0 & 0 & 0 & 0 & 1 & 0 \\ 0 & 0 & 0 & 0 & 0 & 1 \end{bmatrix} - \frac{1}{3}\begin{bmatrix} 1 & 1 & 1 & 0 & 0 & 0 \\ 1 & 1 & 1 & 0 & 0 & 0 \\ 1 & 1 & 1 & 0 & 0 & 0 \\ 0 & 0 & 0 & 0 & 0 & 0 \\ 0 & 0 & 0 & 0 & 0 & 0 \\ 0 & 0 & 0 & 0 & 0 & 0 \end{bmatrix}$$

$$= \begin{bmatrix} 2/3 & -1/3 & -1/3 & 0 & 0 & 0 \\ -1/3 & 2/3 & -1/3 & 0 & 0 & 0 \\ -1/3 & -1/3 & 2/3 & 0 & 0 & 0 \\ 0 & 0 & 0 & 1 & 0 & 0 \\ 0 & 0 & 0 & 0 & 1 & 0 \\ 0 & 0 & 0 & 0 & 0 & 1 \end{bmatrix} \tag{A.48}$$

Here, note that the following matrix must be adopted to transform the engineering strain to the scientific strain, i.e. $\boldsymbol{\varepsilon}' = \widetilde{\mathcal{I}}' : \widetilde{\boldsymbol{\varepsilon}}$.

$$\widetilde{\mathcal{I}}' = [\widetilde{\mathcal{I}}] - \frac{1}{3}[\mathcal{T}] = \begin{bmatrix} 1 & 0 & 0 & 0 & 0 & 0 \\ 0 & 1 & 0 & 0 & 0 & 0 \\ 0 & 0 & 1 & 0 & 0 & 0 \\ 0 & 0 & 0 & 1/2 & 0 & 0 \\ 0 & 0 & 0 & 0 & 1/2 & 0 \\ 0 & 0 & 0 & 0 & 0 & 1/2 \end{bmatrix} - \frac{1}{3}\begin{bmatrix} 1 & 1 & 1 & 0 & 0 & 0 \\ 1 & 1 & 1 & 0 & 0 & 0 \\ 1 & 1 & 1 & 0 & 0 & 0 \\ 0 & 0 & 0 & 0 & 0 & 0 \\ 0 & 0 & 0 & 0 & 0 & 0 \\ 0 & 0 & 0 & 0 & 0 & 0 \end{bmatrix}$$

$$= \begin{bmatrix} 2/3 & -1/3 & -1/3 & 0 & 0 & 0 \\ -1/3 & 2/3 & -1/3 & 0 & 0 & 0 \\ -1/3 & -1/3 & 2/3 & 0 & 0 & 0 \\ 0 & 0 & 0 & 1/2 & 0 & 0 \\ 0 & 0 & 0 & 0 & 1/2 & 0 \\ 0 & 0 & 0 & 0 & 0 & 1/2 \end{bmatrix}$$

The stress-strain relation is represented as follows:

$$\begin{Bmatrix} \sigma_1 \\ \sigma_2 \\ \sigma_3 \\ \sigma_4 \\ \sigma_5 \\ \sigma_6 \end{Bmatrix} = \begin{bmatrix} E_{11} & E_{12} & E_{13} & E_{14} & E_{15} & E_{16} \\ & E_{22} & E_{23} & E_{24} & E_{25} & E_{26} \\ & & E_{33} & E_{34} & E_{35} & E_{36} \\ & & & E_{44} & E_{45} & E_{46} \\ & Sym. & & & E_{55} & E_{56} \\ & & & & & E_{66} \end{bmatrix} \begin{Bmatrix} \varepsilon_1 \\ \varepsilon_2 \\ \varepsilon_3 \\ 2\varepsilon_4 \\ 2\varepsilon_5 \\ 2\varepsilon_6 \end{Bmatrix} \tag{A.49}$$

which is represented in terms of the engineering strain $\widetilde{\varepsilon}_i$ as follows:

$$\begin{Bmatrix} \sigma_1 \\ \sigma_2 \\ \sigma_3 \\ \sigma_4 \\ \sigma_5 \\ \sigma_6 \end{Bmatrix} = \begin{bmatrix} E_{11} & E_{12} & E_{13} & E_{14} & E_{15} & E_{16} \\ & E_{22} & E_{23} & E_{24} & E_{25} & E_{26} \\ & & E_{33} & E_{34} & E_{35} & E_{36} \\ & & & E_{44} & E_{45} & E_{46} \\ & Sym. & & & E_{55} & E_{56} \\ & & & & & E_{66} \end{bmatrix} \begin{Bmatrix} \widetilde{\varepsilon}_1 \\ \widetilde{\varepsilon}_1 \\ \varepsilon_3 \\ \widetilde{\varepsilon}_4 \\ \widetilde{\varepsilon}_5 \\ \widetilde{\varepsilon}_6 \end{Bmatrix} \tag{A.50}$$

$\mathbf{E} : \bar{\mathbf{n}}$ is calculated as follows:

$$\begin{aligned}(\mathbf{E} : \bar{\mathbf{n}})_{ij} &= E_{ij11}\bar{n}_{11} + E_{ij22}\bar{n}_{22} + E_{ij33}\bar{n}_{33} + E_{ij23}\bar{n}_{23} \\ &\quad + E_{ij32}\bar{n}_{32} + E_{ij31}\bar{n}_{31} + E_{ij13}\bar{n}_{13} + E_{ij12}\bar{n}_{12} + E_{ij21}\bar{n}_{21}\end{aligned}$$

which is represented using $\overline{N}_{ij} = \bar{n}_{ij}(i = j), \overline{N}_{ij} = 2\bar{n}_{ij}(i \neq j)$ as follows:

$$(\mathbf{E} : \bar{\mathbf{n}})_{ij} = E_{ij11}\overline{N}_{11} + E_{ij22}\overline{N}_{22} + E_{ij33}\overline{N}_{33} + E_{ij23}\overline{N}_{23} + E_{ij31}\overline{N}_{31} + E_{ij12}\overline{N}_{12} \tag{A.51}$$

which is further described in the Voigt representation as

$$(\mathbf{E} : \bar{\mathbf{n}})_i = E_{i1}\overline{N}_1 + E_{i2}\overline{N}_2 + E_{i3}\overline{N}_{33} + E_{i4}\overline{N}_4 + E_{i5}\overline{N}_5 + E_{i6}\overline{N}_6 = \sum_{j=1}^{6} E_{ij}\overline{N}_j \tag{A.52}$$

The scalar product $\bar{\mathbf{n}} : \mathbf{E} : \bar{\mathbf{n}}$ is represented as

$$\begin{aligned}\bar{\mathbf{n}} : \mathbf{E} : \bar{\mathbf{n}} &= E_{ijkl}\bar{n}_{kl}\bar{n}_{ij} = (\mathbf{E} : \bar{\mathbf{n}})_{ij}\bar{n}_{ij} \\ &= \sum_{i,j=1}^{3} [(E_{ij11}\bar{n}_{11} + E_{ij22}\bar{n}_{22} + E_{ij33}\bar{n}_{33} \\ &\quad + E_{ij23}\bar{n}_{23} + E_{ij32}\bar{n}_{32} + E_{ij31}\bar{n}_{31} + E_{ij13}\bar{n}_{13} + E_{ij12}\bar{n}_{12} + E_{ij21}\bar{n}_{21})\bar{n}_{ij}] \\ &= \sum_{i,j=1}^{3} [E_{ij11}\bar{n}_{11} + E_{ij22}\bar{n}_{22} + E_{ij33}\bar{n}_{33} \\ &\quad + 2(E_{ij23}\bar{n}_{23} + E_{ij31}\bar{n}_{31} + E_{ij12}\bar{n}_{12})\bar{n}_{ij}] \\ &= \sum_{i,j=1}^{3} [(E_{ij11}\overline{N}_{11} + E_{ij22}\overline{N}_{22} + E_{ij33}\overline{N}_{33} + E_{ij23}\overline{N}_{23} + E_{ij31}\overline{N}_{31} + E_{ij12}\overline{N}_{12})\bar{n}_{ij}]\end{aligned}$$

$$
\begin{aligned}
&= (E_{1111}\overline{N}_{11} + E_{1122}\overline{N}_{22} + E_{1133}\overline{N}_{33} + E_{1123}\overline{N}_{23} + E_{1131}\overline{N}_{31} + E_{1112}\overline{N}_{12})\bar{n}_{11} \\
&\quad + (E_{2211}\overline{N}_{11} + E_{2222}\overline{N}_{22} + E_{2233}\overline{N}_{33} + E_{2223}\overline{N}_{23} + E_{2231}\overline{N}_{31} + E_{2212}\overline{N}_{12})\bar{n}_{22} \\
&\quad + (E_{3311}\overline{N}_{11} + E_{3322}\overline{N}_{22} + E_{3333}\overline{N}_{33} + E_{3323}\overline{N}_{23} + E_{3331}\overline{N}_{31} + E_{3312}\overline{N}_{12})\bar{n}_{33} \\
&\quad + 2(E_{2311}\overline{N}_{11} + E_{2322}\overline{N}_{22} + E_{2333}\overline{N}_{33} + E_{2323}\overline{N}_{23} + E_{2331}\overline{N}_{31} + E_{2312}\overline{N}_{12})\bar{n}_{23} \\
&\quad + 2(E_{3111}\overline{N}_{11} + E_{3122}\overline{N}_{22} + E_{3133}\overline{N}_{33} + E_{3123}\overline{N}_{23} + E_{3131}\overline{N}_{31} + E_{3112}\overline{N}_{12})\bar{n}_{31} \\
&\quad + 2(E_{1211}\overline{N}_{11} + E_{1222}\overline{N}_{22} + E_{1233}\overline{N}_{33} + E_{1223}\overline{N}_{23} + E_{1231}\overline{N}_{31} + E_{1212}\overline{N}_{12})\bar{n}_{12} \\
&= (E_{1111}\overline{N}_{11} + E_{1122}\overline{N}_{22} + E_{1133}\overline{N}_{33} + E_{1123}\overline{N}_{23} + E_{1131}\overline{N}_{31} + E_{1112}\overline{N}_{12})\overline{N}_{11} \\
&\quad + (E_{2211}\overline{N}_{11} + E_{2222}\overline{N}_{22} + E_{2233}\overline{N}_{33} + E_{2223}\overline{N}_{23} + E_{2231}\overline{N}_{31} + E_{2212}\overline{N}_{12})\overline{N}_{22} \\
&\quad + (E_{3311}\overline{N}_{11} + E_{3322}\overline{N}_{22} + E_{3333}\overline{N}_{33} + E_{3323}\overline{N}_{23} + E_{3331}\overline{N}_{31} + E_{3312}\overline{N}_{12})\overline{N}_{33} \\
&\quad + (E_{2311}\overline{N}_{11} + E_{2322}\overline{N}_{22} + E_{2333}\overline{N}_{33} + E_{2323}\overline{N}_{23} + E_{2331}\overline{N}_{31} + E_{2312}\overline{N}_{12})\overline{N}_{23} \\
&\quad + (E_{3111}\overline{N}_{11} + E_{3122}\overline{N}_{22} + E_{3133}\overline{N}_{33} + E_{3123}\overline{N}_{23} + E_{3131}\overline{N}_{31} + E_{3112}\overline{N}_{12})\overline{N}_{31} \\
&\quad + (E_{1211}\overline{N}_{11} + E_{1222}\overline{N}_{22} + E_{1233}\overline{N}_{33} + E_{1223}\overline{N}_{23} + E_{1231}\overline{N}_{31} + E_{1212}\overline{N}_{12})\overline{N}_{12}
\end{aligned}
$$

which is represented in the Voigt representation as follows:

$$
\bar{\mathbf{n}} : \mathbf{E} : \bar{\mathbf{n}} = \overline{EN_1}\overline{N}_1 + \overline{EN_2}\overline{N}_2 + \overline{EN_3}\overline{N}_3 + \overline{EN_4}\overline{N}_4 + \overline{EN_5}\overline{N}_5 + \overline{EN_6}\overline{N}_6 = \sum_{i=1}^{6} \overline{EN_i}\overline{N}_i \tag{A.53}
$$

Analogously, the normal part of the strain rate to the subloading surface is given by

$$
\begin{aligned}
(\mathbf{d}_n)_{ij} &= (\bar{\mathbf{n}} \otimes \bar{\mathbf{n}} : \mathbf{d})_{ij} = \bar{n}_{ij}\bar{n}_{rs}d_{rs} \\
&= \bar{n}_{ij}[\bar{n}_{11}d_{11} + \bar{n}_{22}d_{22} + \bar{n}_{33}d_{33} + 2(\bar{n}_{23}d_{23} + \bar{n}_{31}d_{31} + \bar{n}_{12}d_{12})]
\end{aligned}
$$

which is described in terms of the engineering strain rate as

$$
(\mathbf{d}_n)_{ij} = \bar{n}_{ij}(\bar{n}_{11}\tilde{d}_{11} + \bar{n}_{22}\tilde{d}_{22} + \bar{n}_{33}\tilde{d}_{33} + \bar{n}_{23}\tilde{d}_{23} + \bar{n}_{31}\tilde{d}_{31} + \bar{n}_{12}\tilde{d}_{12})
$$

which is further represented in the Voigt form as

$$
(\mathbf{d}_n)_i = (\bar{\mathbf{n}} \otimes \bar{\mathbf{n}} : \mathbf{d})_i = \bar{n}_i(\bar{n}_1\tilde{d}_1 + \bar{n}_2\tilde{d}_2 + \bar{n}_3\tilde{d}_3 + \bar{n}_4\tilde{d}_4 + \bar{n}_5\tilde{d}_5 + \bar{n}_6\tilde{d}_6) = \bar{n}_i \sum_{r=1}^{6} \bar{n}_r\tilde{d}_r \tag{A.54}
$$

Appendix J
Computer Programs of Hashiguchi (Subloading Surface) Models

The computer programs for analyses of the elastoplastic deformation of metals based on the subloading surface model and of the friction phenomena based on the subloading-friction model are provided below so that the readers will understand the formulation of the models clearly by pursuing each line in the programs. In addition, the readers can perform readily the FE analyses of the boundary-value problems involving the frictional boundary by installing these programs into the FE program (through user-subroutine in case of commercial software). Besides, they are installed as the standard uploaded (ready-made) programs in the commercial FEM software Marc of MSC Software Ltd. after the 2017 version by the leading support of Dr. Motoharu TATEISHI (MSC Software, Ltd.). The following programs were composed by the author (Technical Adviser of MSC Software Ltd.; Emeritus Prof., Kyushu Univ.), Masami UENO (Emeritus Prof., Univ. Ryukyus), Toshiyuki OZAKI (Kyushu Electric Eng. Consult. Inc.) and Shingo OZAKI (Yokohama National Univ.).

The infinitesimal engineering strain (increment) is used in the program as described in Appendix I and the material-time derivatives are used for simplicity, which will be extended readily in terms of the strain rate and the Jaumann rates of the stress and the internal variables.

(a) Computer program of subloading surface model for elastoplastic deformation of metals

The three versions: i) Isotropic and kinematic hardening version in Chapt. 7, ii) Version eliminating the isotropic hardening stagnation and the tangential-inelastic strain rate and iii) Full version in Chapt. 9 and 10 taken account of all behaviors are provided in the following.

The conventional elastoplasticity model with the isotropic and the kinematic hardenings is drastically improved by the version i) so as to need no the yield judgment and to be furnished with the automatic controlling function to attract the stress to the yield surface in the plastic deformation process, although it is quite simple.

K. Hashiguchi, *Foundations of Elastoplasticity: Subloading Surface Model*,
DOI 10.1007/978-3-319-48821-9

(a) Computer program of subloading surface model for elastoplastic deformation of metals

i) Isotropic and kinematic hardening version

composed by Koichi Hashiguchi and Masami Ueno.

Sub-program Code	Variables/Equations
Subroutine calc_exsub()	
implicit none	
--	
Declaration of variables	
--	
!**	
!* *	
!* Calculations of variables *	
!* *	
!**	
!	
CLU = 1.0d0	
!	
! Stress	
SIGM = (SIG(1) + SIG(2) + SIG(3)) / 3.0d0	σ_m
do i = 1, 6	
SD_M(i) = SIG(i) - SIGM * I_M(i)	$\boldsymbol{\sigma}'$
end do	
SD = 0.0d0	
do i = 1, 3	
SD = SD + SD_M(i) ** 2.0	$\sigma_1'^2+\sigma_2'^2+\sigma_3'^2$
end do	
do i = 4, 6	
SD = SD + 2.0d0 * SD_M(i) ** 2.0	$\sigma_1'^2+\sigma_2'^2+\sigma_3'^2$ $+2(\sigma_4'^2+\sigma_5'^2+\sigma_6'^2)$
end do	
SD = dsqrt(SD)	$\lVert\boldsymbol{\sigma}'\rVert$
' Back-stress	
ALM = (AL (1) + AL (2) + AL (3)) / 3.0d0	α_m (=0)
do i = 1, 6	
ALD_M(i) = AL (i) - ALM * I_M(i)	$\boldsymbol{\alpha}'$ (=$\boldsymbol{\alpha}$)
end do	
ALD = 0.0d0	

`do i = 1, 3`	
`    ALD = ALD + ALD_M(i) **2.0`	
`end do`	
`do i = 4, 6`	$\alpha_1'^2+\alpha_2'^2+\alpha_3'^2$
`    ALD = ALD + 2.0d0 * ALD_M(i) **2.0`	$+2(\alpha_4'^2+\alpha_5'^2+\alpha_6'^2)$
`end do`	
`ALD = dsqrt(ALD)`	$\Vert\boldsymbol{\alpha}'\Vert$
`! Stress(SIG) - Back-stress(ALM)`	
`do i = 1, 6`	
`    SA(i) = SIG(i) - AL(i)`	$\hat{\boldsymbol{\sigma}}\equiv\boldsymbol{\sigma}-\boldsymbol{\alpha}$: Eq. (6.86)
`end do`	
`SAM = (SA(1) + SA(2) + SA(3)) / 3.0d0`	$\hat{\sigma}_m$
`do i = 1, 6`	
`    SAD_M(i) = SA (i) - SAM * I_M(i)`	$\hat{\boldsymbol{\sigma}}'$
`end do`	
`SAD = 0.0d0`	
`do i =1, 3`	
`    SAD = SAD + SAD_M(i) **2.0`	
`end do`	
`do i=4, 6`	$\hat{\sigma}_1'^2+\hat{\sigma}_2'^2+\hat{\sigma}_3'^2$
`    SAD = SAD + 2.0d0 * SAD_M(i) **2.0`	$+2(\hat{\sigma}_4'^2+\hat{\sigma}_5'^2+\hat{\sigma}_6'^2)$
`end do`	
`SAD = dsqrt(SAD)`	$\Vert\hat{\boldsymbol{\sigma}}'\Vert$
`! Normal-yield ratio`	
`R1 =dsqrt(3.0d0 / 2.0d0) * SAD/ F`	$R=\dfrac{\sqrt{\frac{3}{2}}\Vert\hat{\boldsymbol{\sigma}}'\Vert}{F}$: Eq. (7.9)
`if (R1 < 1.0d0) then`	with Eq. (6.55)
`    CLU = 0.0d0`	
`end if`	
`!*********************************`	
`!*                               *`	
`!*    Elastic coefficient matrix *`	
`!*                               *`	
`!*********************************`	
`SG = EL / (2.0d0 * (1.0d0 + POI))`	$G=E/[2(1+\nu)]$
`KV = EL *SG / (3.0d0 * (3.0d0 * SG - EL))`	$K=EG/[3(3G-E)]$

EML1 = KV + (4.0d0 / 3.0d0) * SG	$K+(4/3)G$
EML2 = KV - (2.0d0 / 3.0d0) * SG:	$K-(2/3)G$
EML3 = SG	G
do i = 1, 6	
do j = 1, 6	
EL_M(i, j) = 0.0d0	
end do	
end do	
EL_M(1, 1) = EML1	$E_{11}=K+(4/3)G$
EL_M(2, 2) = EML1	$E_{22}=K+(4/3)G$
EL_M(3, 3) = EML1	$E_{33}=K+(4/3)G$
EL_M(4, 4) = EML3	$E_{44}=G$
EL_M(5, 5) = EML3	$E_{55}=G$
EL_M(6, 6) = EML3	$E_{66}=G$
EL_M(1, 2) = EML2	$E_{12}=K-(2/3)G$
EL_M(2, 3) = EML2	$E_{23}=K-(2/3)G$
EL_M(3, 1) = EML2	$E_{31}=K-(2/3)G$
EL_M(2, 1) = EML2	$E_{21}=K-(2/3)G$
EL_M(3, 2) = EML2	$E_{32}=K-(2/3)G$
EL_M(1, 3) = EML2	$E_{13}=K-(2/3)G$
IEML1 = 1.0d0 / (9.0d0 * KV) + 1.0d0 / (3.0d0 * SG)	$1/(9K)+1/(3G)$
IEML2 = 1.0d0 / (9.0d0 * KV) - 1.0d0 / (6.0d0 * SG)	$1/(9K)-1/(6G)$
IEML3 = 1.0d0 / SG	$1/G$
do i = 1, 6	
do j = 1, 6	
IEL_M(i, j) = 0.0d0	
end do	
end do	
IEL_M(1, 1) = IEML1	$E^{-1}_{11}=1/(9K)+1/(3G)$
IEL_M(2, 2) = IEML1	$E^{-1}_{22}=1/(9K)+1/(3G)$
IEL_M(3, 3) = IEML1	$E^{-1}_{33}=1/(9K)+1/(3G)$
IEL_M(4, 4) = IEML3	$E^{-1}_{44}=1/G$
IEL_M(5, 5) = IEML3	$E^{-1}_{55}=1/G$
IEL_M(6, 6) = IEML3	$E^{-1}_{66}=1/G$
IEL_M(1, 2) = IEML2	$E^{-1}_{12}=1/(9K)-1/(6G)$
IEL_M(2, 3) = IEML2	$E^{-1}_{23}=1/(9K)-1/(6G)$

`IEL_M(3, 1) = IEML2`	$E_{31}^{-1} = 1/(9K) - 1/(6G)$
`IEL_M(2, 1) = IEML2`	$E_{21}^{-1} = 1/(9K) - 1/(6G)$
`IEL_M(3, 2) = IEML2`	$E_{32}^{-1} = 1/(9K) - 1/(6G)$
`IEL_M(1, 3) = IEML2`	$E_{13}^{-1} = 1/(9K) - 1/(6G)$
`!******************** ************************`	
`! Normalized outward-normal of subloading surface`	

`do i = 1, 3`
`N(i) = SAD_M(i) / SAD`
`end do`
	$\hat{N}_i = \hat{n}_i = \frac{\hat{\sigma}'_i}{\lVert\hat{\boldsymbol{\sigma}}'\rVert}$ $(i=1\sim3)$

`do i = 4, 6`
`N(i) = SAD_M(i) / SAD*2.0`
`end do`
	$\hat{N}_i = 2\hat{n}_i = 2\frac{\hat{\sigma}'_i}{\lVert\hat{\boldsymbol{\sigma}}'\rVert}$ $(i=4\sim6)$

`UR = UA / (dtan(PAI / 2.0d0) * R1))`	$U(R) = u\cot[(\pi/2)R]$: Eq. (7.12)

`! E:n`
`do i = 1, 6`
`EN(i) = 0.0d0`
`end do`	

`do i = 1, 6`
`do j = 1, 6`
`EN(i) = EN(i) + EL_M(i,j) * N(j)`
`end do`
`end do`
	$E_{ij}\hat{N}_j$ $(\mathbf{E}:\hat{\mathbf{n}})$

`! n:E:n`
`trNEN = 0.0d0`
`do i = 1, 6`
`trNEN = trNEN + (N(i) * EN(i))`
`end do`
	$\hat{\mathbf{n}}:\mathbf{E}:\hat{\mathbf{n}} = \hat{N}_i E_{ij} \hat{N}_j$

`! Plastic modulus with isotropic-hardening stagnation`	
`do i = 1, 6` `MP1(i) = (Fdh / F) * dsqrt(2.0d0 / 3.0d0) * SAD_M(i)`	$\sqrt{\frac{2}{3}}\frac{F'}{F}\hat{\boldsymbol{\sigma}}$
`MP2(i) = R1 * Ck * (SAD_M(i) / SAD - &` `1.0d0 / (dsqrt(3.0d0 / 2.0d0) * Zeta * F) * AL(i))`	$c_k R\left(\hat{\mathbf{n}} - \frac{1}{\sqrt{3/2}\zeta F}\boldsymbol{\alpha}\right)$
`MP3(i) = (UR / R1) * SIG(i)`	$\frac{U}{R}\hat{\boldsymbol{\sigma}}$
`MP_N(i) = MP1(i) + MP2(i) + MP3(i)` `end do`	

```
      MP = 0.0d0
      do I = 1,6
            MP = MP + N(i) * MP_N(i)
      end do


20    continue


      do i = 1, 6
         do j = 1, 6
               EPM_M(i,j) = EL_M(i,j) - (CLU * EN(i) * &
                     EN(j) / (MP + trNEN))
               Bm_m(i,j) = EPM_M(i,j)
         end do
      end do

!  Solve Simulteneous Equation (Gauss method) *************

      do i = 1,6
         Dq_m(i) = Ds_m(i)
         Dz_m(i) = De_m(i)
      end do

      if (C_m(1) + C_m(2) + C_m(3) + C_m(4) + C_m(5) +&
            C_m(6) < 6) Then

         do i = 1,6
            if (C_m(i) == 1) Then
               do j = 1,6
                  if (C_m(j) == 0) Then
                     Dq_m(j) = Dq_m(j) - Bm_m(j, i) * Dz_m(i)
                  end if
               end do
               do j = 1,6
                  Bm_m(i, j) = 0.0d0
                  Bm_m(j, i) = 0.0d0
               end do
                  Bm_m(i, i) = 1.0d0
                  Dq_m(i) = 0.0d0
            end if
         end do

         do l = 1,5
            if (C_m(l) == 0) Then
                  Pt = Bm_m(l, l)
                  do j = 1,6
                        Bm_m(l, j) = Bm_m(l, j) / Pt
                  end do
                  Dq_m(l) = Dq_m(l) / Pt
                  do i = l + 1,6
```

$$\bar{M}^p \equiv \hat{\mathbf{n}}:\left[\sqrt{\frac{2}{3}}\frac{F'}{F}\hat{\boldsymbol{\sigma}} + c_k R\left(\hat{\mathbf{n}} - \frac{1}{\sqrt{3/2}\zeta F}\boldsymbol{\alpha}\right) + \frac{U}{R}\boldsymbol{\sigma}\right]$$
: Eq. (10.12)

Eq. (7.34):

$$\mathbf{K}^{ep} = \mathbf{E} - \frac{\mathbf{E}:\hat{\mathbf{n}}\otimes\hat{\mathbf{n}}:\mathbf{E}}{\bar{M}^p + \hat{\mathbf{n}}:\mathbf{E}:\hat{\mathbf{n}}}$$

<pre>   Pt = Bm_m(i, l)
                do j = 1,6
                   Bm_m(i, j) = Bm_m(i, j) - Pt * Bm_m(l, j)
                end do
                Dq_m(i) = Dq_m(i) - Pt * Dq_m(l)
             end do
          end if
    end do
 	
    do l = 1,6
          if (C_m(l) == 0) Then
             u = l
          end if
    end do
 	
    Dz_m(u) = Dq_m(u) / Bm_m(u, u)
 	
    do l = u - 1,1,-1
          if (C_m(l) == 0) Then
             Pt = Dq_m(l)
             do j = l + 1,6
                Pt = Pt - Bm_m(l, j) * Dz_m(j)
             end do
             Dz_m(l) = Pt
          end if
    end do
 	
  end if
 	
  do i = 1,6
     if (C_m(i) == 0) Then
        Dq_m(i) = Ds_m(i)
     else
        do j = 1,6
           Dq_m(i) = Dq_m(i) + EPM_M (i, j) * Dz_m(j)
        end do
     end if
  end do</pre>	

<pre>!
! Stress and strain increments
  do i = 1, 6
     if (C_m(i) == 0) then
        DW_M(i) = Ds_m(i)
        DX_M(i) = Dz_m(i)
     end if
     if (C_m(i) == 1) then
        DW_M(i) = Dq_m(i)
        DX_M(i) = De_m(i)
     end if
  end do</pre>
	$d\boldsymbol{\sigma}$
	
	$d\boldsymbol{\varepsilon}$ (Eng. strain increment)

<pre> DWM = (DW_M(1) + DW_M(2) + DW_M(3)) / 3.0d0</pre>	$d\sigma_m$

! Deviatoric stress increment do i = 1, 6 DWD_M(i) = DW_M(i) - (DWM * I_M(i)) end do	$d\boldsymbol{\sigma}'$
! trN_dSigma trNED trNdS = 0.0d0 trNED = 0.0d0 do I = 1,6 trNdS = trNdS + (N(i) * DW_M(i)) trNED = trNED + (EN(i) * DX_M(i)) end do	$\hat{\mathbf{n}}:d\boldsymbol{\sigma} = \hat{N}_i d\sigma_i$ $\hat{\mathbf{n}}:\mathbf{E}:d\boldsymbol{\varepsilon} = \hat{N}_i E_{ij} d\varepsilon_j$

! Plastic multiplier
LM = trNdS / Mp
LLM = trNED / (Mp + trNEN)
 	
if (CLU == 0.0d0) then
LM = 0.0d0
LLM = 0.0d0
GoTo 10
end if
 	
if (LLM < 0.0d0) then
LM = 0.0d0
CLU = 0.0d0
LLM = 0.0d0
GoTo 20
end if
	Eq. (9.38):
	$d\bar{\lambda} = \dfrac{\hat{\mathbf{n}}:d\boldsymbol{\sigma}}{\bar{M}^p}$
	Eq. (9.41):
	$d\bar{\Lambda} = \dfrac{\hat{\mathbf{n}}:\mathbf{E}:d\boldsymbol{\varepsilon}}{\bar{M}^p + \hat{\mathbf{n}}:\mathbf{E}:\hat{\mathbf{n}}}$

!!!!!!!!!!!!!!!!!!!!!!!!!!!!!!!!!!!!!!! ! Strain increment ! !!!!!!!!!!!!!!!!!!!!!!!!!!!!!!!!!!!!!!!	
! Plastic strain increment 10 continue do i = 1, 6 DP_M(i) = CLU * LM * N(i) end do	Plastic engineering strain increment: $d\boldsymbol{\varepsilon}^p = d\bar{\Lambda}\hat{\mathbf{N}}$
DPX = 0.0d0 do i = 1,3 DPX = DPX + DP_M(i)**2.0 end do do i = 4,6 DPX = DPX + (DP_M(i)**2.0) / 2.0d0 end do	$d\varepsilon_1^{p2} + d\varepsilon_2^{p2} + d\varepsilon_3^{p2}$ $d\varepsilon_1^{p2} + d\varepsilon_2^{p2} + d\varepsilon_3^{p2}$ $+(d\varepsilon_4^{p2} + d\varepsilon_4^{p2} + d\varepsilon_4^{p2})/2$

DPX = dsqrt(DPX)	$\lVert d\boldsymbol{\varepsilon}^p \rVert$
DR = UR * DPX	$U\lVert d\boldsymbol{\varepsilon}^p \rVert$
! Elastic engineering strain increment	
do i = 1,6	
Dex_M(i) = 0.0d0	
do j = 1,3	
Dex_M(i) = Dex_M(i) + (IEL_M(i, j) * & DW_M(j))	$d\varepsilon_i^e = E_{ij} d\sigma_j$ (j=1~3)
end do	
do j = 4,6	
Dex_M(i) = Dex_M(i) + 2.0d0 * (IEL_M(i, j) * & DW_M(j))	$d\varepsilon_i^e = 2E_{ij} d\sigma_j$ (j=4~6)
end do	
end do	
! Total strain increment	
do i = 1, 6	
D_M(i) = Dex_M(i) + DP_M(i)	$d\boldsymbol{\varepsilon} = d\boldsymbol{\varepsilon}^e + d\boldsymbol{\varepsilon}^p$
end do	
! Stress and strain	
do i = 1, 6	
SIG(i) = SIG(i) + DW_M(i)	$\boldsymbol{\sigma}$
E_M(i) = E_M(i) + DX_M(i)	$\boldsymbol{\varepsilon}$
Ee_M(i) = Ee_M(i) + Dex_M(i)	$\boldsymbol{\varepsilon}^e$
Ep_M(i) = Ep_M(i) + DP_M(i)	$\boldsymbol{\varepsilon}^p$
end do	
! Increments of back stress and elastic-core	
do i = 1, 6	Eq. (10.8):
Da_M(i) = Ck * (SAD_M(i) / SAD - 1.0d0 / & (dsqrt(3.0d0 / 2.0d0) * Zeta * F) * AL (i)) * DPX	$d\boldsymbol{\alpha} = c_k\left(\hat{\mathbf{n}} - \frac{1}{\sqrt{3/2}\zeta F}\boldsymbol{\alpha}\right)d\bar{\lambda}$
AL(i) = AL(i) + Da_M(i)	$\boldsymbol{\alpha}$
end do	
	Eq. (10.6):
Hdh = dsqrt(2.0d0 / 3.0d0) * DPX	$dH = \sqrt{\frac{2}{3}}\, d\bar{\lambda}$
! Isotropic hardening	
H = H + Hdh	H

`Fdh = F0 * H1 * H2 * exp(-H2 * H)`	Eq. (10.5): $F' = F_0 h_1 h_2 \exp(-h_2 H)$
`F = F0 * (1.0d0 + H1 * (1.0d0 - exp(-H2 * H)))`	Eq. (10.5): $F(H) = F_0[1 + h_1\{1 - \exp(-h_2 H)\}]$

```
return
end Subroutine calc_exsub




Subroutine Input_Parameter(m, nn)
implicit none
!++++++++++++++++++++++++++
!
!   Input material constants
!
!++++++++++++++++++++++++++
!
      integer*8 m,nn

      open(100,file='para.dat',access='sequential',err=999,&
                    status='old',form='formatted')
      go to 9000
  999 continue
      write(*,'(a10,d12.5)')'open error in para.dat'
      stop
 9000 continue
!
!----------------------------------------------
      read(100,*) F0
      read(100,*) EL
      read(100,*) POI
      read(100,*) H1
      read(100,*) H2
      read(100,*) Ck
      read(100,*) Zeta
      read(100,*) UA
      close (100)
!
      if(m.eq.1.and.nn.eq.1) then
       write(6,'(a10,d12.5)') 'F0          =', F0
       write(6,'(a10,d12.5)') 'EL          =', EL
       write(6,'(a10,d12.5)') 'POI         =', POI
       write(6,'(a10,d12.5)') 'H1          =', H1
       write(6,'(a10,d12.5)') 'H2          =', H2
       write(6,'(a10,d12.5)') 'Ck          =', Ck
       write(6,'(a10,d12.5)') 'Zeta        =', Zeta
       write(6,'(a10,d12.5)') 'UA          =', UA
      end if

return
end subroutine Input_Parameter
```

```
Subroutine Inicon()
implicit none
!-------------------------------------------------
!   Set initial conditions
!
!-------------------------------------------------
        PAI=3.14159265358979
        RAD=PAI / 180.0d0
!
        do i = 1, 6
           Ds_m(i) = 0.0d0
           De_m(i) = 0.0d0
           SIG (i) = 0.0d0
           AL (i) = 0.0d0
           C_m(i) = 0
           E_M(i) = 0.0d0
           Ep_M(i) = 0.0d0
           Ee_M(i) = 0.0d0
           Ep_M(i) = 0.0d0
           D_M(i) = 0.0d0
           Dex_M(i) = 0.0d0
           DP_M(i) = 0.0d0
        end do
!
        SIG (1) = 0.0000001
        AL (1) = 0.0000000001
        AL (2) = - AL (1) / 2.0d0
        AL (3) = - A L(1) / 2.0d0
!
        DPX = 0.0d0
        H = 0.0d0
        Hdh = 0.0d0
        F = F0
        Fdh = F0 * H1 * H2

        I_M(1) = 1.0d0
```

I_M(2) = 1.0d0 I_M(3) = 1.0d0 I_M(4) = 0.0d0 I_M(5) = 0.0d0 I_M(6) = 0.0d0 return end subroutine Inicon	

ii) Simplified version without tangential-inelastic strain rate and isotropic hardening stagnation

composed by Koichi Hashiguchi and Masami Ueno.

Sub-program Code	Variables/Equations
Subroutine calc_exsub()	
implicit none	

Declaration of variables	

!***	
!* *	
!* Calculations of variables *	
!* *	
!***	
!	
CLU = 1.0d0	
!	
! Stress	
SIGM = (SIG(1) + SIG(2) + SIG(3)) / 3.0d0	σ_m
do i = 1, 6	
SD_M(i) = SIG(i) - SIGM * I_M(i)	$\boldsymbol{\sigma}'$
end do	
SD= 0.0d0	
do i = 1, 3	
SD = SD + SD_M(i) ** 2.0	
end do	
do i = 4, 6	$\sigma_1'^2+\sigma_2'^2+\sigma_3'^2$
SD = SD + 2.0d0 * SD_M(i) ** 2.0	$+2(\sigma_4'^2+\sigma_5'^2+\sigma_6'^2)$
end do	
SD = dsqrt(SD)	$\|\boldsymbol{\sigma}'\|$
' Back-stress	
ALM = (AL (1) + AL (2) + AL (3)) / 3.0d0	α_m (=0)
do i = 1, 6	
ALD_M(i) = AL (i) - ALM * I_M(i)	$\boldsymbol{\alpha}'$ (=$\boldsymbol{\alpha}$)
end do	

`ALD = 0.0d0`	
`do i = 1, 3`	
`    ALD = ALD + ALD_M(i) **2.0`	
`end do`	
`do i = 4, 6`	
`    ALD = ALD + 2.0d0 * ALD_M(i) **2.0`	$\alpha_1'^2+\alpha_2'^2+\alpha_3'^2$ $+2(\alpha_4'^2+\alpha_5'^2+\alpha_6'^2)$
`end do`	
`ALD = dsqrt(ALD)`	$\|\boldsymbol{\alpha}'\|$
`! Stress(SIG) - Back-stress(ALM)`	
`do i = 1, 6`	
`    SA(i) = SIG(i) - AL(i)`	$\hat{\boldsymbol{\sigma}} \equiv \boldsymbol{\sigma} - \boldsymbol{\alpha}$: Eq. (6.86)
`end do`	
`SAM = (SA(1) + SA(2) + SA(3)) / 3.0d0`	$\hat{\sigma}_m$
`do i = 1, 6`	
`    SAD_M(i) = SA (i) - SAM * I_M(i)`	$\hat{\boldsymbol{\sigma}}'$
`end do`	
`SAD = 0.0d0`	
`do i = 1, 3`	
`    SAD = SAD + SAD_M(i) **2.0`	
`end do`	
`do i = 4, 6`	
`    SAD = SAD + 2.0d0 * SAD_M(i) **2.0`	$\hat{\sigma}_1'^2+\hat{\sigma}_2'^2+\hat{\sigma}_3'^2$ $+2(\hat{\sigma}_4'^2+\hat{\sigma}_5'^2+\hat{\sigma}_6'^2)$
`end do`	
`SAD = dsqrt(SAD)`	$\|\hat{\boldsymbol{\sigma}}'\|$
`' Elastic core`	
`CSM = (CS(1) + CS(2) + CS(3)) / 3.0d0`	c_m
`do i = 1, 6`	
`    CSD_M(i) = CS(i) - CSM * I_M(i)`	$\mathbf{c}'$
`end do`	
`CSD = 0.0d0`	
`do i =1, 3`	
`    CSD = CSD + CSD_M(i) **2.0`	
`end do`	
`do i = 4, 6`	
`    CSD = CSD + 2.0d0 * CSD_M(i) **2.0`	$c_1'^2+c_2'^2+c_3'^2$ $+2(c_4'^2+c_5'^2+c_6'^2)$
`end do`	

`CSD = dsqrt(CSD)`	$\lVert \mathbf{c}' \rVert$
`! Stress(SIG) - Similarity center(CS)`	
`do i = 1, 6`	
`SC(i) = SIG(i) - CS(i)`	$\tilde{\boldsymbol{\sigma}} \equiv \boldsymbol{\sigma} - \mathbf{c}$: Eq. (9.7)
`end do`	
`SCM = (SC(1) + SC(2) + SC(3)) / 3.0d0`	$\tilde{c}_m$
`do i = 1, 6`	
`SCD_M(i) = SC(i) - SCM * I_M(i)`	$\tilde{\boldsymbol{\sigma}}'$
`end do`	
`SCD = 0.0d0`	
`do i = 1, 3`	
`SCD = SCD + SCD_M(i) **2.0`	
`end do`	
`do i = 4, 6`	
`SCD = SCD + 2.0d0 * SCD_M(i) **2.0`	$\tilde{\sigma}_1'^2 + \tilde{\sigma}_2'^2 + \tilde{\sigma}_3'^2 + 2(\tilde{\sigma}_4'^2 + \tilde{\sigma}_5'^2 + \tilde{\sigma}_6'^2)$
`end do`	
`SCD = dsqrt(SCD)`	$\lVert \tilde{\boldsymbol{\sigma}}' \rVert$
`! Similarity center(CS) - Back-stress(AL)`	
`do i = 1, 6`	
`CA(i) = CS(i) - AL(i)`	$\hat{\mathbf{c}} \equiv \mathbf{c} - \boldsymbol{\alpha}$: Eq. (9.4)
`end do`	
`CAM = (CA(1) + CA(2) + CA(3)) / 3.0d0`	$\hat{c}_m$
`do i = 1, 6`	
`CAD_M(i) = CA(i) - CAM * I_M(i)`	$\hat{\mathbf{c}}'$
`end do`	
`CAD = 0.0d0`	
`do i = 1, 3`	
`CAD = CAD + CAD_M(i) **2.0`	
`end do`	
`do i = 4, 6`	
`CAD = CAD + 2.0d0 * CAD_M(i) **2.0`	$\hat{c}_1'^2 + \hat{c}_2'^2 + \hat{c}_3'^2 + 2(\hat{c}_4'^2 + \hat{c}_5'^2 + \hat{c}_6'^2)$
`end do`	
`CAD = dsqrt(CAD)`	$\lVert \hat{\mathbf{c}}' \rVert$
`! Normal-yield ratio`	
`RA = 0.0d0`	
`do i = 1, 3`	

RA = RA + SCD_M(i) * CAD_M(i)	
end do	
do i = 4, 6	$\tilde{\boldsymbol{\sigma}}':\hat{\mathbf{c}}' = \tilde{\sigma}_1'\hat{c}_1' + \tilde{\sigma}_2'\hat{c}_2' + \tilde{\sigma}_3'\hat{c}_3' + 2(\tilde{\sigma}_4'\hat{c}_4' + \tilde{\sigma}_5'\hat{c}_5' + \tilde{\sigma}_6'\hat{c}_6')$
RA = RA + 2.0d0 * SCD_M(i) * CAD_M(i)	
end do	
RC = (2.0d0 / 3.0d0) * (F**2.0) - (CAD**2.0)	$(2/3)F^2 - \lVert\hat{\mathbf{c}}'\rVert^2$
RB = dsqrt(RA**2.0) + (RC*(SCD**2.0))	$\sqrt{(\tilde{\boldsymbol{\sigma}}':\hat{\mathbf{c}}')^2 + [(2/3)F^2 - \lVert\hat{\mathbf{c}}'\rVert^2]\lVert\tilde{\boldsymbol{\sigma}}'\rVert^2)}$
R1 = (RA + RB) / RC	Eq. (10.32): $R = \dfrac{\tilde{\boldsymbol{\sigma}}':\hat{\mathbf{c}}' + \sqrt{(\tilde{\boldsymbol{\sigma}}':\hat{\mathbf{c}}')^2 + (\frac{2}{3}F^2 - \lVert\hat{\mathbf{c}}'\rVert^2)\lVert\tilde{\boldsymbol{\sigma}}'\rVert^2}}{\frac{2}{3}F^2 - \lVert\hat{\mathbf{c}}'\rVert^2}$
if (R1 < 0.0d0) then	
CLU = 0.0d0	
end if	
! Conjugate point of back-stress on subloading surface	
do i = 1, 6	
ALB(i) = CS(i) - (R1 * CA(i))	$\bar{\boldsymbol{\alpha}} = \mathbf{c} - R\hat{\mathbf{c}}$: Eq. (9.5)
end do	
ALBM = (ALB(1) + ALB(2) + ALB(3)) / 3.0d0	$\bar{\alpha}_m$
do i = 1, 6	
ALBD_M(i) = ALB(i) - ALBM * I_M(i)	$\bar{\boldsymbol{\alpha}}'$
end do	
ALBD = 0.0d0	
do i = 1, 3	
ALBD = ALBD + ALBD_M(i) **2.0	
end do	
do i =4, 6	$\bar{\alpha}_1^{2\prime} + \bar{\alpha}_2^{2\prime} + \bar{\alpha}_3^{2\prime} + 2(\bar{\alpha}_4^{2\prime} + \bar{\alpha}_5^{2\prime} + \bar{\alpha}_6^{2\prime})$
ALBD = ALBD + 2.0d0 * ALBD_M(i) **2.0	
end do	
ALBD = dsqrt(ALBD)	$\lVert\bar{\boldsymbol{\alpha}}'\rVert$
! Stress(SIG) - Conjugate point of back-stress on subloading surface	
do i = 1, 6	
SALB(i) = SIG(i) - ALB(i)	$\bar{\boldsymbol{\sigma}} = \boldsymbol{\sigma} - \bar{\boldsymbol{\alpha}}$: Eq. (9.3)
end do	
SALBM = (SALB(1) + SALB(2) + SALB(3)) / 3.0d0	$\bar{\sigma}_m$

do i = 1, 6	
SALBD_M(i) = SALB(i) - SALBM * I_M(i)	$\bar{\boldsymbol{\sigma}}'$
end do	
SALBD = 0.0d0	
do i = 1, 3	
SALBD = SALBD + SALBD_M(i) **2.0	
end do	
do i = 4, 6	$\bar{\sigma}_1^{2\prime} + \bar{\sigma}_2^{2\prime} + \bar{\sigma}_3^{2\prime}$
SALBD = SALBD + 2.0d0 * SALBD_M(i) **2.0	$+2(\bar{\sigma}_4'^2 + \bar{\sigma}_5'^2 + \bar{\sigma}_6'^2)$
end do	
SALBD = dsqrt(SALBD)	$\lVert\bar{\boldsymbol{\sigma}}'\rVert$
!********************************	
!* *	
!* Elastic coefficient matrix *	
!* *	
!********************************	
SG = EL / (2.0d0 * (1.0d0 + POI))	$G = E/[2(1+\nu)]$
KV = EL *SG / (3.0d0 * (3.0d0 * SG - EL))	$K = EG/[3(3G-E)]$
EML1 = KV + (4.0d0 / 3.0d0) * SG	$K+(4/3)G$
EML2 = KV - (2.0d0 / 3.0d0) * SG:	$K-(2/3)G$
EML3 = SG	G
do i = 1, 6	
do j = 1, 6	
EL_M(i, j) = 0.0d0	
end do	
end do	
EL_M(1, 1) = EML1	$E_{11} = K+(4/3)G$
EL_M(2, 2) = EML1	$E_{22} = K+(4/3)G$
EL_M(3, 3) = EML1	$E_{33} = K+(4/3)G$
EL_M(4, 4) = EML3	$E_{44} = G$
EL_M(5, 5) = EML3	$E_{55} = G$
EL_M(6, 6) = EML3	$E_{66} = G$
EL_M(1, 2) = EML2	$E_{12} = K-(2/3)G$
EL_M(2, 3) = EML2	$E_{23} = K-(2/3)G$
EL_M(3, 1) = EML2	$E_{31} = K-(2/3)G$
EL_M(2, 1) = EML2	$E_{21} = K-(2/3)G$

EL_M(3, 2) = EML2	$E_{32} = K-(2/3)G$
EL_M(1, 3) = EML2	$E_{13} = K-(2/3)G$
IEML1 = 1.0d0 / (9.0d0 * KV) + 1.0d0 / (3.0d0 * SG)	$1/(9K)+1/(3G)$
IEML2 = 1.0d0 / (9.0d0 * KV) – 1.0d0 / (6.0d0 * SG)	$1/(9K)-1/(6G)$
IEML3 = 1.0d0 / SG	$1/G$
do i = 1, 6	
do j = 1, 6	
IEL_M(i, j) = 0.0d0	
end do	
end do	
IEL_M(1, 1) = IEML1	$E_{11}^{-1} = 1/(9K)+1/(3G)$
IEL_M(2, 2) = IEML1	$E_{22}^{-1} = 1/(9K)+1/(3G)$
IEL_M(3, 3) = IEML1	$E_{33}^{-1} = 1/(9K)+1/(3G)$
IEL_M(4, 4) = IEML3	$E_{44}^{-1} = 1/G$
IEL_M(5, 5) = IEML3	$E_{55}^{-1} = 1/G$
IEL_M(6, 6) = IEML3	$E_{66}^{-1} = 1/G$
IEL_M(1, 2) = IEML2	$E_{12}^{-1} = 1/(9K)-1/(6G)$
IEL_M(2, 3) = IEML2	$E_{23}^{-1} = 1/(9K)-1/(6G)$
IEL_M(3, 1) = IEML2	$E_{31}^{-1} = 1/(9K)-1/(6G)$
IEL_M(2, 1) = IEML2	$E_{21}^{-1} = 1/(9K)-1/(6G)$
IEL_M(3, 2) = IEML2	$E_{32}^{-1} = 1/(9K)-1/(6G)$
IEL_M(1, 3) = IEML2	$E_{13}^{-1} = 1/(9K)-1/(6G)$
!******************** ************************	
! Normalized outward-normal of subloading surface	
do i = 1, 3 N(i) = SALBD_M(i) / SALBD end do	$\bar{N}_i = \bar{n}_i = \dfrac{\bar{\sigma}'_i}{\lVert\bar{\boldsymbol{\sigma}}'\rVert}$ $(i{=}1{\sim}3)$
do i = 4, 6 N(i) = SALBD_M(i) / SALBD * 2.0d0 end do	$\bar{N}_i = 2\bar{n}_i = 2\dfrac{\bar{\sigma}'_i}{\lVert\bar{\boldsymbol{\sigma}}'\rVert}$ $(i{=}4{\sim}6)$
do i = 1, 6 NN(i) = SALBD_M(i) / SALBD end do	$\bar{\mathbf{n}} = \dfrac{\bar{\boldsymbol{\sigma}}'}{\lVert\bar{\boldsymbol{\sigma}}'\rVert}$: Eq. (10.2)
! Normalized outward-normal of elastic-core surface	
do i = 1, 3	

```
          NS(i) = CAD_M(i) / CAD
     end do
     do i = 4, 6
          NS(i) = CAD_M(i) / CAD * 2.0d0
     end do
```

$\hat{n}_{ci}=\frac{\hat{c}'_i}{\|\hat{\mathbf{c}}'\|}$ (i=1~3)

$\hat{n}_{ci}=2\frac{\hat{c}'_i}{\|\hat{\mathbf{c}}'\|}$ (i=4~6)

```
     do i = 1, 6
          NsN(i) = CAD_M(i) / CAD
     end do
```

$\hat{\mathbf{n}}_c=\frac{\hat{\mathbf{c}}'}{\|\hat{\mathbf{c}}'\|}$: Eq. (10.4)

```
! Scalar product of outward-normal of elastic-core and subloading
surface
     Ssigma = 0.0d0
     do i = 1, 3
          Ssigma = Ssigma + (NS(i) * N(i))
     end do
     do i=4, 6
          Ssigma = Ssigma + ((NS(i) * N(i)) * 2.0d0)
     end do
```

$C_\sigma \equiv \hat{\mathbf{n}}_c:\bar{\mathbf{n}}=\hat{n}_{ci}\bar{N}_i$ (i=1~3) $+2\hat{n}_{ci}\bar{N}_i$ (i=4~6) : Eq. (9.46)

```
! Elastic-core yield ratio
     RSC = dsqrt(2.0d0 / 3.0d0) * CAD / F
```

$\mathcal{R}_c=\sqrt{\frac{3}{2}}\frac{\|\hat{\mathbf{c}}'\|}{F}$: Eq. (10.4)

```
     if ((R – Re) > 0.0d0 ) then
        UR = UA * exp(UB * RSC * Ssigma) / &
            (dtan(PAI / 2.0d0) * ((R – Re) / (1.0d0 – Re))**NNu))
     else
        UR = 1.0d0 * (10.0d0 ** 23.0)
     end if
```

Eqs. (7.20) and (9.47): $U(R)=\bar{u}\exp(u_c\mathcal{R}_cC_\sigma)\cot\left[\frac{\pi}{2}\left\langle\frac{R-R_e}{1-R_e}\right\rangle^n\right]$

```
! E:n
     do i =1, 6
          EN(i) = 0.0d0
     end do
     do i = 1, 6
          do j = 1, 6
               EN(i) = EN(i) + EL_M(i,j) * N(j)
          end do
     end do
```

$E_{ij}\bar{N}_j$ $(\mathbf{E}:\bar{\mathbf{n}})$

```
! n:E:n
     trNEN = 0.0d0
     do i = 1, 6
          trNEN = trNEN + (N(i) * EN(i))
     end do
```

$\bar{\mathbf{n}}:\mathbf{E}:\bar{\mathbf{n}}=\bar{N}_iE_{ij}\bar{N}_j$

```
! Plastic modulus with isotropic-hardening stagnation

      do i = 1, 6
         MP1(i) = (Fdh / F) * dsqrt(2.0d0 / 3.0d0) * SALB(i)
```

$$\sqrt{\frac{2}{3}}\frac{F'}{F}\bar{\boldsymbol{\sigma}}$$

```
         MP2(i) = R1 * Ck * (SALBD_M(i) / SALBD - &
                  1.0d0 / (dsqrt(3.0d0 / 2.0d0) * Zeta * F) * AL(i))
```

$$c_k R\left(\bar{\mathbf{n}} - \frac{1}{\sqrt{3/2}\,\zeta F}\boldsymbol{\alpha}\right)$$

```
         MP3(i) = (UR / R1) * SC(i) + Cs*(1.0d0 - R1) * &
                  (SALBD_M(i) / SALBD - &
                  RSC / Ks * CAD_M(i) / CAD)
```

$$\frac{U}{R}\tilde{\boldsymbol{\sigma}} + c(1-R)\left(\bar{\mathbf{n}} - \frac{\Re_c}{\xi}\hat{\mathbf{n}}_c\right)$$

```
         MP_N(i) = MP1(i) + MP2(i) + MP3(i)
      end do

      MP = 0.0d0
      do i = 1,6
          MP = MP + (N(i) * MP_N(i))
      end do
```

$$\bar{M}^p \equiv \bar{\mathbf{n}}:\left[\sqrt{\frac{2}{3}}\frac{F'}{F}\bar{\boldsymbol{\sigma}} + c_k R\left(\bar{\mathbf{n}} - \frac{1}{\sqrt{3/2}\,\zeta F}\boldsymbol{\alpha}\right) + \frac{U}{R}\tilde{\boldsymbol{\sigma}} + c(1-R)\left(\bar{\mathbf{n}} - \frac{\Re_c}{\xi}\hat{\mathbf{n}}_c\right)\right]$$

```
20    continue

      do i = 1, 6
        do j = 1, 6
            EPM_M(i,j) = EL_M(i,j) - (CLU * EN(i) * &
                     EN(j) / (MP + trNEN))

            Bm_m(i,j) = EPM_M(i,j)
        end do
      end do
```

Eq. (9.42):

$$\mathbf{K}^{ep} = \mathbf{E} - \frac{\mathbf{E}:\bar{\mathbf{n}}\otimes\bar{\mathbf{n}}:\mathbf{E}}{\bar{M}^p + \bar{\mathbf{n}}:\mathbf{E}:\bar{\mathbf{n}}}$$

```
!   Solve Simulteneous Equation (Gauss method) *************

      do i = 1,6
        Dq_m(i) = Ds_m(i)
        Dz_m(i) = De_m(i)
      end do

      if (C_m(1) + C_m(2) + C_m(3) + C_m(4) + C_m(5) +&
               C_m(6) < 6) Then

        do i = 1,6
           if (C_m(i) == 1) Then
             do j = 1,6
               if (C_m(j) == 0) Then
                  Dq_m(j) = Dq_m(j) - Bm_m(j, i) * Dz_m(i)
               end if
             end do
             do j = 1,6
               Bm_m(i, j) = 0.0d0
               Bm_m(j, i) = 0.0d0
```

```
                end do
                    Bm_m(i, i) = 1.0d0
                    Dq_m(i) = 0.0d0
              end if
        end do

        do l = 1,5
              if (C_m(l) == 0) Then
                    Pt = Bm_m(l, l)
                    do j = 1,6
                          Bm_m(l, j) = Bm_m(l, j) / Pt
                    end do
                    Dq_m(l) = Dq_m(l) / Pt
                    do i = l + 1,6
                       Pt = Bm_m(i, l)
                       do j = 1,6
                          Bm_m(i, j) = Bm_m(i, j) - Pt * Bm_m(l, j)
                       end do
                       Dq_m(i) = Dq_m(i) - Pt * Dq_m(l)
                    end do
              end if
        end do

        do l = 1,6
              if (C_m(l) == 0) Then
                    u = l
              end if
        end do

        Dz_m(u) = Dq_m(u) / Bm_m(u, u)

        do l = u - 1, 1, -1
              if (C_m(l) == 0) Then
                 Pt = Dq_m(l)
                 do j = l + 1,6
                    Pt = Pt - Bm_m(l, j) * Dz_m(j)
                 end do
                 Dz_m(l) = Pt
              end if
        end do

      end if

      do i = 1,6
         if (C_m(i) == 0) Then
              Dq_m(i) = Ds_m(i)
         else
              do j = 1,6
                 Dq_m(i) = Dq_m(i) + EPM_M (i, j) * Dz_m(j)
              end do
         end if
      end do
!
! Stress and strain increments
      do i = 1, 6
```

if (C_m(i) == 0) then
DW_M(i) = Ds_m(i)
DX_M(i) = Dz_m(i)
end if
	$d\boldsymbol{\sigma}$

if (C_m(i) == 1) then
DW_M(i) = Dq_m(i)
DX_M(i) = De_m(i)
end if
end do
	$d\boldsymbol{\varepsilon}$ (Eng. strain increment)

DWM = (DW_M(1) + DW_M(2) + DW_M(3)) / 3.0d0	$d\sigma_m$
! Deviatoric stress increment do i = 1, 6 DWD_M(i) = DW_M(i) - (DWM * I_M(i)) end do	$d\boldsymbol{\sigma}'$
! trN_dSigma trNED trNdS = 0.0d0 trNED = 0.0d0 do i = 1,6 trNdS = trNdS + (N(i) * DW_M(i)) trNED = trNED + (EN(i) * DX_M(i)) end do	$\bar{\mathbf{n}}:d\boldsymbol{\sigma}=\bar{N}_i d\sigma_i$ $\bar{\mathbf{n}}:\mathbf{E}:d\boldsymbol{\varepsilon}=\bar{N}_i E_{ij} d\varepsilon_j$
! Plastic multiplier LM = trNdS / Mp LLM = trNED / (Mp + trNEN)	$d\bar{\lambda}=\frac{\bar{\mathbf{n}}:d\boldsymbol{\sigma}}{\bar{M}^p}$: Eq. (9.38) $d\bar{\Lambda}=\frac{\bar{\mathbf{n}}:\mathbf{E}:d\boldsymbol{\varepsilon}}{\bar{M}^p+\bar{\mathbf{n}}:\mathbf{E}:\bar{\mathbf{n}}}$: Eq. (9.41)

if (CLU== 0.0d0) then
LM = 0.0d0
LLM = 0.0d0
GoTo 10
end if	

if (LLM < 0.0d0) then
LM = 0.0d0
CLU = 0.0d0
LLM = 0.0d0
GoTo 20
end if	

!!! ! Strain increment ! !!!	
! Plastic strain increment	

10 continue do i = 1, 6 DP_M(i) = CLU * LM * N(i) end do	Plastic engineering strain increment: $d\boldsymbol{\varepsilon}^p = d\bar{\Lambda}\bar{\mathbf{N}}$
DPX = 0.0d0 do i = 1,3 DPX = DPX + DP_M(i)**2.0 end do do i = 4,6 DPX = DPX + (DP_M(i)**2.0) / 2.0d0 end do DPX = dsqrt(DPX) DR = UR * DPX	$d\varepsilon_1^{p2} + d\varepsilon_2^{p2} + d\varepsilon_3^{p2}$ $d\varepsilon_1^{p2} + d\varepsilon_2^{p2} + d\varepsilon_3^{p2}$ $+(d\varepsilon_4^{p2} + d\varepsilon_4^{p2} + d\varepsilon_4^{p2})/2$ $\|d\boldsymbol{\varepsilon}^p\|$
! Elastic engineering strain increment do i = 1,6 Dex_M(i) = 0.0d0 do j = 1,3 Dex_M(i) = Dex_M(i) + (IEL_M(i, j) * & DW_M(j)) end do do j = 4,6 Dex_M(i) = Dex_M(i) + 2.0d0*(IEL_M(i, j) * & DW_M(j)) end do end do	$d\varepsilon_i^e = E_{ij} d\sigma_j$ (j=1~3) $d\varepsilon_i^e = 2E_{ij} d\sigma_j$ (j=4~6)
! Total strain increment do i = 1, 6 D_M(i) = Dex_M(i) + DP_M(i) end do	$d\boldsymbol{\varepsilon} = d\boldsymbol{\varepsilon}^e + d\boldsymbol{\varepsilon}^p$
! Stress and strain do i = 1, 6 SIG(i) = SIG(i) + DW_M(i) E_M(i) = E_M(i) + DX_M(i) Ee_M(i) = Ee_M(i) + Dex_M(i) Ep_M(i) = Ep_M(i) + DP_M(i) end do	$\boldsymbol{\sigma}$ $\boldsymbol{\varepsilon}$ $\boldsymbol{\varepsilon}^e$ $\boldsymbol{\varepsilon}^p$
! Increments of back stress and elastic-core do i = 1, 6	

Da_M(i) = Ck * (SALBD_M(i) / SALBD - 1.0d0 / & (dsqrt(3.0d0 / 2.0d0) *Zeta * F) * AL (i)) * DPX Dsim_M(i) = Cs * ((SALBD_M(i) / SALBD) - & (RSC / Ks * CAD_M(i) / CAD)) * DPX end do	Eq. (10.8): $d\boldsymbol{\alpha} = c_k\left(\bar{\mathbf{n}} - \frac{1}{\sqrt{3/2}\zeta F}\boldsymbol{\alpha}\right)d\bar{\lambda}$ Eq. (9.15): $d\mathbf{c} = c\left(\bar{\mathbf{n}} - \frac{\mathfrak{R}_c}{\xi}\hat{\mathbf{n}}_c\right)d\bar{\lambda}$
! Increments of back stress and elastic-core do i = 1, 6 AL(i) = AL(i) + Da_M(i) CS(i) = CS(i) + Dsim_M(i) end do	 $\boldsymbol{\alpha}$ $\mathbf{c}$
Hdh=dsqrt(2.0d0 / 3.0d0) * DPX	Eq. (10.6): $dH = \sqrt{\frac{2}{3}}d\bar{\lambda}$
! Isotropic hardening H = H + Hdh Fdh = F0 * H1 * H2 * exp(-H2 * H) F = F0 * (1.0d0 + H1 * (1.0d0 - exp(-H2 * H)))	 H Eq. (10.5): $F' = F_0 h_1 h_2 \exp(-h_2 H)$ Eq. (10.5): $F(H) = F_0[1 + h_1\{1 - \exp(-h_2 H)\}]$
return end Subroutine calc_exsub	
Subroutine Input_Parameter(m, nn) implicit none !++++++++++++++++++++++++++ ! ! Input material constants ! !++++++++++++++++++++++++++ ! integer*8 m, nn open(100,file = 'para.dat', access='sequential', err=999,& status='old', form='formatted') go to 9000 999 continue write(*,'(a10,d12.5)') 'open error in para.dat' stop 9000 continue ! !--- read(100,*) F0 read(100,*) EL read(100,*) POI read(100,*) H1	

```
        read(100,*) H2
        read(100,*) Ck
        read(100,*) Zeta
        read(100,*) Cs
        read(100,*) UA
        read(100,*) UB
        read(100,*) Ks
        read(100,*) Re
        read(100,*) NNu
        close (100)
!
        if(m.eq.1.and.nn.eq.1) then
          write(6,'(a10,d12.5)') 'F0          =', F0
          write(6,'(a10,d12.5)') 'EL          =', EL
          write(6,'(a10,d12.5)') 'POI         =', POI
          write(6,'(a10,d12.5)') 'H1          =', H1
          write(6,'(a10,d12.5)') 'H2          =', H2
          write(6,'(a10,d12.5)') 'Ck          =', Ck
          write(6,'(a10,d12.5)') 'Zeta        =', Zeta
          write(6,'(a10,d12.5,)') 'Cs         =', Cs
          write(6,'(a10,d12.5)') 'UA          =', UA
          write(6,'(a10,d12.5)') 'UB          =', UB
          write(6,'(a10,d12.5)') 'Ks          =', Ks
          write(6,'(a10,d12.5)') 'Re          =', Re
          write(6,'(a10,d12.5)') 'NNu         =', NNu
        end if

return
end subroutine Input_Parameter



Subroutine Inicon()
implicit none
!-----------------------------------------------
!
!   Set initial conditions
!
!-----------------------------------------------
        PAI=3.14159265358979
        RAD=PAI/180.0d0
!
        do i = 1, 6
          Ds_m(i) = 0.0d0
          De_m(i) = 0.0d0
          SIG (i) = 0.0d0
          AL (i) = 0.0d0
```

```
            C_m(i) = 0
            E_M(i) = 0.0d0
            CS (i) = 0.0d0
            Ep_M(i) = 0.0d0
            Ee_M(i) = 0.0d0
            Ep_M(i) = 0.0d0
            D_M(i) = 0.0d0
            Dex_M(i) = 0.0d0
            DP_M(i) = 0.0d0
        end do
!
        SIG (1) = 0.0000001
        CS (1) = 0.00000001
        AL (1) = 0.0000000001
        AL (2) = - AL (1) / 2.0d0
        AL (3) = - AL(1) / 2.0d0
!
        CS(2) = - CS(1) / 2.0d0
        CS(3) = - CS(1) / 2.0d0
        DPX = 0.0d0
        H = 0.0d0
        Hdh = 0.0d0
        F = F0
        Fdh = F0 * H1 * H2
!
        I_M(1) = 1.0d0
        I_M(2) = 1.0d0
        I_M(3) = 1.0d0
        I_M(4) = 0.0d0
        I_M(5) = 0.0d0
        I_M(6) = 0.0d0

return
end subroutine Inicon
```

iii) Full version

composed by Koichi Hashiguchi, Masami Ueno and Toshiyuki Ozaki.

Sub-program Code	Variables/Equations
Subroutine calc_exsub() implicit none --- *Declaration of variables* --- !*** !* * !* Calculations of variables * !* * !*** ! CLU = 1.0d0 ! ! Stress	
SIGM = (SIG(1) + SIG(2) + SIG(3)) / 3.0d0	σ_m
do i = 1, 6 SD_M(i) = SIG(i) - SIGM * I_M(i)	$\boldsymbol{\sigma}'$

end do
    SD= 0.0d0
    do i = 1, 3
      SD = SD + SD_M(i) ** 2.0
    end do	

do i = 4, 6
      SD = SD + 2.0d0 * SD_M(i) ** 2.0
    end do
	$\sigma_1'^2+\sigma_2'^2+\sigma_3'^2$
	$+2(\sigma_4'^2+\sigma_5'^2+\sigma_6'^2)$

SD = dsqrt(SD)	$\\|\boldsymbol{\sigma}'\\|$
' Back-stress	
ALM = (AL (1) + AL (2) + AL (3)) / 3.0d0	α_m (=0)
do i = 1, 6 ALD_M(i) = AL (i) - ALM * I_M(i)	$\boldsymbol{\alpha}'$ (= $\boldsymbol{\alpha}$)
end do ALD = 0.0d0	

`do i = 1, 3`	
`ALD = ALD + ALD_M(i) **2.0`	
`end do`	
`do i = 4, 6`	$\alpha_1'^2+\alpha_2'^2+\alpha_3'^2$
`ALD = ALD + 2.0d0 * ALD_M(i) **2.0`	$+2(\alpha_4'^2+\alpha_5'^2+\alpha_6'^2)$
`end do`	
`ALD = dsqrt(ALD)`	$\lVert\boldsymbol{\alpha}'\rVert$
`! Stress(SIG) - Back-stress(ALM)`	
`do i = 1, 6`	
`SA(i) = SIG(i) - AL(i)`	$\hat{\boldsymbol{\sigma}}\equiv\boldsymbol{\sigma}-\boldsymbol{\alpha}$: Eq. (6.86)
`end do`	
`SAM = (SA(1) + SA(2) + SA(3)) / 3.0d0`	$\hat{\sigma}_m$
`do i = 1, 6`	
`SAD_M(i) = SA (i) - SAM * I_M(i)`	$\hat{\boldsymbol{\sigma}}'$
`end do`	
`SAD = 0.0d0`	
`do i = 1, 3`	
`SAD = SAD + SAD_M(i) **2.0`	
`end do`	
`do i = 4, 6`	$\hat{\sigma}_1'^2+\hat{\sigma}_2'^2+\hat{\sigma}_3'^2$
`SAD = SAD + 2.0d0 * SAD_M(i) **2.0`	$+2(\hat{\sigma}_4'^2+\hat{\sigma}_5'^2+\hat{\sigma}_6'^2)$
`end do`	
`SAD = dsqrt(SAD)`	$\lVert\hat{\boldsymbol{\sigma}}'\rVert$
`' Elastic core`	
`CSM = (CS(1) + CS(2) + CS(3)) / 3.0d0`	c_m
`do i = 1, 6`	
`CSD_M(i) = CS(i) - CSM * I_M(i)`	$\mathbf{c}'$
`end do`	
`CSD = 0.0d0`	
`do i = 1, 3`	
`CSD = CSD + CSD_M(i) **2.0`	
`end do`	
`do i = 4, 6`	$c_1'^2+c_2'^2+c_3'^2$
`CSD = CSD + 2.0d0 * CSD_M(i) **2.0`	$+2(c_4'^2+c_5'^2+c_6'^2)$
`end do`	
`CSD = dsqrt(CSD)`	$\lVert\mathbf{c}'\rVert$

! Stress(SIG) - Similarity center(CS)	
do i = 1, 6	
SC(i) = SIG(i) - CS(i)	$\tilde{\boldsymbol{\sigma}} \equiv \boldsymbol{\sigma} - \mathbf{c}$: Eq. (9.7)
end do	
SCM = (SC(1) + SC(2) + SC(3)) / 3.0d0	$\tilde{c}_m$
do i = 1, 6	
SCD_M(i) = SC(i) - SCM * I_M(i)	$\tilde{\boldsymbol{\sigma}}'$
end do	
SCD = 0.0d0	
do i = 1, 3	
SCD = SCD + SCD_M(i) **2.0	
end do	
do i = 4, 6	
SCD = SCD + 2.0d0 * SCD_M(i) **2.0	$\tilde{\sigma}_1'^2+\tilde{\sigma}_2'^2+\tilde{\sigma}_3'^2$ $+2(\tilde{\sigma}_4'^2+\tilde{\sigma}_5'^2+\tilde{\sigma}_6'^2)$
end do	
SCD = dsqrt(SCD)	$\lVert\tilde{\boldsymbol{\sigma}}'\rVert$
! Similarity center(CS) - Back-stress(AL)	
do i = 1, 6	
CA(i) = CS(i) - AL(i)	$\hat{\mathbf{c}} \equiv \mathbf{c} - \boldsymbol{\alpha}$: Eq. (9.7)
end do	
CAM = (CA(1) + CA(2) + CA(3)) / 3.0d0	$\hat{c}_m$
do i = 1, 6	
CAD_M(i) = CA(i) - CAM * I_M(i)	$\hat{\mathbf{c}}'$
end do	
CAD = 0.0d0	
do i = 1, 3	
CAD = CAD + CAD_M(i) **2.0	
end do	
do i = 4, 6	
CAD = CAD + 2.0d0 * CAD_M(i) **2.0	$\hat{c}_1'^2+\hat{c}_2'^2+\hat{c}_3'^2$ $+2(\hat{c}_4'^2+\hat{c}_5'^2+\hat{c}_6'^2)$
end do	
CAD = dsqrt(CAD)	$\lVert\hat{\mathbf{c}}'\rVert$
! Normal-yield ratio	
RA = 0.0d0	
do i = 1, 3	
RA = RA + SCD_M(i) * CAD_M(i)	

end do	
do i = 4, 6	$\tilde{\boldsymbol{\sigma}}':\hat{\mathbf{c}}' = \tilde{\sigma}'_1\hat{c}'_1+\tilde{\sigma}'_2\hat{c}'_2+\tilde{\sigma}'_3\hat{c}'_3 + 2(\tilde{\sigma}'_4\hat{c}'_4+\tilde{\sigma}'_5\hat{c}'_5+\tilde{\sigma}'_6\hat{c}'_6)$
RA = RA + 2.0d0 * SCD_M(i) * CAD_M(i)	
end do	
RC = (2.0d0 / 3.0d0) * (F**2.0) - (CAD**2.0)	$(2/3)F^2 - \lVert\hat{\mathbf{c}}'\rVert^2$
RB = dsqrt(RA**2.0) + (RC * (SCD**2.0))	$\sqrt{(\tilde{\boldsymbol{\sigma}}':\hat{\mathbf{c}}')^2 + [(2/3)F^2 - \lVert\hat{\mathbf{c}}'\rVert^2]\lVert\tilde{\boldsymbol{\sigma}}'\rVert^2}$
R1 = (RA + RB) / RC	Eq. (10.32): $R = \dfrac{\tilde{\boldsymbol{\sigma}}':\hat{\mathbf{c}}' + \sqrt{(\tilde{\boldsymbol{\sigma}}':\hat{\mathbf{c}}')^2 + \left(\frac{2}{3}F^2 - \lVert\hat{\mathbf{c}}'\rVert^2\right)\lVert\tilde{\boldsymbol{\sigma}}'\rVert^2}}{\frac{2}{3}F^2 - \lVert\hat{\mathbf{c}}'\rVert^2}$
if (R1 < 0.0d0) then	
CLU = 0.0d0	
end if	
! Conjugate point of back-stress on subloading surface	
do i = 1, 6	
ALB(i) = CS(i) - (R1 * CA(i))	$\bar{\boldsymbol{\alpha}} = \mathbf{c} - R\hat{\mathbf{c}}$: Eq. (9.5)
end do	
ALBM = (ALB(1) + ALB(2) + ALB(3)) / 3.0d0	$\bar{\alpha}_m$
do i = 1, 6	
ALBD_M(i) = ALB(i) - ALBM * I_M(i)	$\bar{\boldsymbol{\alpha}}'$
end do	
ALBD = 0.0d0	
do i = 1, 3	
ALBD = ALBD + ALBD_M(i) **2.0	
end do	
do i = 4, 6	$\bar{\alpha}_1^{2\prime} + \bar{\alpha}_2^{2\prime} + \bar{\alpha}_3^{2\prime} + 2(\bar{\alpha}_4^{2\prime} + \bar{\alpha}_5^{2\prime} + \bar{\alpha}_6^{2\prime})$
ALBD = ALBD + 2.0d0 * ALBD_M(i) **2.0	
end do_	
ALBD = dsqrt(ALBD)	$\lVert\bar{\boldsymbol{\alpha}}'\rVert$
! Stress(SIG) - Conjugate point of back-stress on subloading surface	
do i = 1, 6	
SALB(i) = SIG(i) - ALB(i)	$\bar{\boldsymbol{\sigma}} = \boldsymbol{\sigma} - \bar{\boldsymbol{\alpha}}$: Eq. (9.3)
end do	
SALBM = (SALB(1) + SALB(2) + SALB(3)) / 3.0d0	$\bar{\sigma}_m$
do i = 1, 6	
SALBD_M(i) = SALB(i) - SALBM * I_M(i)	$\bar{\boldsymbol{\sigma}}'$

end do
SALBD = 0.0d0
do i = 1, 3
    SALBD = SALBD + SALBD_M(i) **2.0
end do
do i = 4, 6
    SALBD = SALBD + 2.0d0 * SALBD_M(i) **2.0
end do
SALBD = dsqrt(SALBD)
	$\bar{\sigma}_1^{2\prime}+\bar{\sigma}_2^{2\prime}+\bar{\sigma}_3^{2\prime}$
	$+2(\bar{\sigma}_4^{\prime 2}+\bar{\sigma}_5^{\prime 2}+\bar{\sigma}_6^{\prime 2})$
	$\lVert\bar{\boldsymbol{\sigma}}'\rVert$

!******************************** !* * !* Elastic coefficient matrix * !* * !********************************	
SG = EL / (2.0d0 * (1.0d0 + POI)) KV = EL *SG / (3.0d0 * (3.0d0 * SG - EL)) EML1 = KV + (4.0d0 / 3.0d0) * SG EML2 = KV - (2.0d0 / 3.0d0) * SG: EML3 = SG	$G=E/[2(1+\nu)]$ $K=EG/[3(3G-E)]$ $K+(4/3)G$ $K-(2/3)G$ G

do i = 1, 6
    do j = 1, 6
        EL_M(i, j) = 0.0d0
    end do
end do	

EL_M(1, 1) = EML1	$E_{11}=K+(4/3)G$
EL_M(2, 2) = EML1	$E_{22}=K+(4/3)G$
EL_M(3, 3) = EML1	$E_{33}=K+(4/3)G$
EL_M(4, 4) = EML3	$E_{44}=G$
EL_M(5, 5) = EML3	$E_{55}=G$
EL_M(6, 6) = EML3	$E_{66}=G$
EL_M(1, 2) = EML2	$E_{12}=K-(2/3)G$
EL_M(2, 3) = EML2	$E_{23}=K-(2/3)G$
EL_M(3, 1) = EML2	$E_{31}=K-(2/3)G$
EL_M(2, 1) = EML2	$E_{21}=K-(2/3)G$
EL_M(3, 2) = EML2	$E_{32}=K-(2/3)G$
EL_M(1, 3) = EML2	$E_{13}=K-(2/3)G$

IEML1 = 1.0d0 / (9.0d0 * KV) + 1.0d0 / (3.0d0*SG)	$1/(9K)+1/(3G)$
IEML2 = 1.0d0/(9.0d0*KV) – 1.0d0/(6.0d0*SG)	$1/(9K)-1/(6G)$
IEML3 = 1.0d0 / SG	$1/G$
do i = 1, 6	
do j = 1, 6	
IEL_M(i, j) = 0.0d0	
end do	
end do	
IEL_M(1, 1) = IEML1	$E_{11}^{-1}=1/(9K)+1/(3G)$
IEL_M(2, 2) = IEML1	$E_{22}^{-1}=1/(9K)+1/(3G)$
IEL_M(3, 3) = IEML1	$E_{33}^{-1}=1/(9K)+1/(3G)$
IEL_M(4, 4) = IEML3	$E_{44}^{-1}=1/G$
IEL_M(5, 5) = IEML3	$E_{55}^{-1}=1/G$
IEL_M(6, 6) = IEML3	$E_{66}^{-1}=1/G$
IEL_M(1, 2) = IEML2	$E_{12}^{-1}=1/(9K)-1/(6G)$
IEL_M(2, 3) = IEML2	$E_{23}^{-1}=1/(9K)-1/(6G)$
IEL_M(3, 1) = IEML2	$E_{31}^{-1}=1/(9K)-1/(6G)$
IEL_M(2, 1) = IEML2	$E_{21}^{-1}=1/(9K)-1/(6G)$
IEL_M(3, 2) = IEML2	$E_{32}^{-1}=1/(9K)-1/(6G)$
IEL_M(1, 3) = IEML2	$E_{13}^{-1}=1/(9K)-1/(6G)$
!******************** ************************	
! Normalized outward-normal of subloading surface	
do i = 1, 3 N(i) = SALBD_M(i) / SALBD end do	$\bar{N}_i=\bar{n}_i=\frac{\bar{\sigma}'_i}{\lVert\bar{\boldsymbol{\sigma}}'\rVert}$ (i=1~3)
do i = 4, 6 N(i) = SALBD_M(i) / SALBD * 2.0d0 end do	$\bar{N}_i=2\bar{n}_i=2\frac{\bar{\sigma}'_i}{\lVert\bar{\boldsymbol{\sigma}}'\rVert}$ (i=4~6)
do i = 1, 6 NN(i) = SALBD_M(i) / SALBD end do	$\bar{\mathbf{n}}=\frac{\bar{\boldsymbol{\sigma}}'}{\lVert\bar{\boldsymbol{\sigma}}'\rVert}$: Eq. (10.2)
! Normalized outward-normal of elastic-core surface	
do i = 1, 3 NS(i) = CAD_M(i) / CAD end do	$\hat{n}_{ci}=\frac{\hat{c}'_i}{\lVert\hat{\mathbf{c}}'\rVert}$ (i=1~3)
do i = 4, 6	

`NS(i) = CAD_M(i) / CAD * 2.0d0` `end do`	$\hat{n}_{ci}=2\frac{\hat{c}'_i}{\|\hat{\mathbf{c}}'\|}$ $(i$=4~6)

`do i = 1, 6`
`NsN(i) = CAD_M(i) / CAD`
`end do`
	$\hat{\mathbf{n}}_c=\frac{\hat{\mathbf{c}}'}{\|\hat{\mathbf{c}}'\|}$: Eq. (10.4)

`! Scalar product of outward-normal of elastic-core and subloading surface`
`Ssigma = 0.0d0`
`do i = 1, 3`
`Ssigma = Ssigma + (NS(i) * N(i))`
`end do`
`do i = 4, 6`
`Ssigma = Ssigma + ((NS(i) * N(i)) * 2.0d0)`
`end do`
	$C_\sigma\equiv\hat{\mathbf{n}}_c:\bar{\mathbf{n}}=\hat{n}_{ci}\bar{N}_i\,(i$=1~3)
	$+2\hat{n}_{ci}\bar{N}_i\,(i$=4~6)
	: Eq. (9.46)

`! Elastic-core yield ratio` `RSC = dsqrt(2.0d0 / 3.0d0) * CAD / F`	$\mathfrak{R}_c=\sqrt{\frac{3}{2}}\frac{\|\hat{\mathbf{c}}'\|}{F}$: Eq. (10.4)

`if ((R – Re) > 0.0d0 ) then`
`UR = UA * exp(UB * RSC * Ssigma) / &`
`(dtan(PAI / 2.0d0) * ((R – Re) / (1.0d0 – Re))**NNu))`
`else`
`UR = 1.0d0 * (10.0d0 ** 23.0)`
`end if`
	Eqs. (7.20) and (9.47):
	$U(R)=\bar{u}\exp(u_c\mathfrak{R}_cC_\sigma)$
	$\cot\left[\frac{\pi}{2}\left\langle\frac{R-R_e}{1-R_e}\right\rangle^n\right]$

`! E:n`
`do i = 1, 6`
`EN(i) = 0.0d0`
`end do`
`do i = 1, 6`
`do j = 1, 6`
`EN(i) = EN(i) + EL_M(i,j) * N(j)`
`end do`
`end do`
	$E_{ij}\bar{N}_j$ $(\mathbf{E}:\bar{\mathbf{n}})$

`! n:E:n`
`trNEN = 0.0d0`
`do i = 1, 6`
`trNEN = trNEN + (N(i) * EN(i))`
`end do`
	$\bar{\mathbf{n}}:\mathbf{E}:\bar{\mathbf{n}}=\bar{N}_iE_{ij}\bar{N}_j$

`!!!!!!!!!!!!!!!!!!!!!!!!!!!!!!!!!!!!!!!!!!!!!!!!!!!`
`! !`
`! Stagnation of isotropic hardening !`
`! (based on back-stress) !`	

! ! !!!	
do i = 1, 6 ALP_HARD_M(i) = AL(i) - ALHARD_M(i) end do	$\tilde{\boldsymbol{\alpha}} \equiv \boldsymbol{\alpha} - \boldsymbol{\rho}$: Eq. (10.14)
ALP_HARD=0.0d0 do i = 1, 3 ALP_HARD = ALP_HARD + & (ALP_HARD_M(i)**2.0) end do do i = 4, 6 ALP_HARD = ALP_HARD + & 2.0d0*(ALP_HARD_M(i)**2.0) end do	$\tilde{\alpha}_1^2+\tilde{\alpha}_2^2+\tilde{\alpha}_3^2$ $+2(\tilde{\alpha}_4^2+\tilde{\alpha}_5^2+\tilde{\alpha}_6^2)$
ALP_HARD = dsqrt(ALP_HARD)	$\Vert\tilde{\boldsymbol{\alpha}}\Vert$
R_HARD = dsqrt(2.0d0 / 3.0d0) * ALP_HARD / K_HARD	$\tilde{R} = \sqrt{\frac{3}{2}}\frac{\Vert\tilde{\boldsymbol{\alpha}}\Vert}{\tilde{K}}$: Eqs. (10.15), (10.16)

if (ALP_HARD == 0.0d0) then
do i = 1, 6
NHARD_M(i) = 0.0d0
end do
else
do i = 1, 6
NHARD_M(i) = ALP_HARD_M(i) / &
ALP_HARD
end do
end if
	$\tilde{\mathbf{n}} = \frac{\tilde{\boldsymbol{\alpha}}}{\Vert\tilde{\boldsymbol{\alpha}}\Vert}$: Eq. (10.24)

! Plastic modulus with isotropic-hardening stagnation	
do i = 1, 6 Fkn_M(i) = Ck * (SALBD_M(i) / SALBD - & 1.0d0 / (dsqrt(3.0d0 / 2.0d0) * Zeta * F) * AL(i)) end do	Eq. (10.9) $\bar{\mathbf{f}}_{kn} = c_k\left(\bar{\mathbf{n}} - \frac{1}{\sqrt{3/2}\zeta F}\boldsymbol{\alpha}\right)$
Fkn = 0.0d0 do i = 1, 3 Fkn = Fkn + (Fkn_M(i)**2.0) end do do i = 4, 6 Fkn = Mp2 + 2.0d0 * (Fkn_M(i)**2.0) end do	$\bar{\mathbf{f}}_{kn} : \bar{\mathbf{f}}_{kn}$
Fkn = dsqrt(Fkn)	$\Vert\bar{\mathbf{f}}_{kn}\Vert$

```
trNhardFkn = 0.0d0
do i = 1, 3
   trNhardFkn = trNhardFkn + (NHARD_M(i) * &
                     Fkn_M(i) / Fkn)
end do
do i = 4, 6
   trNhardFkn = trNhardFkn + 2.0d0 * (NHARD_M(i) * &
                     Fkn_M(i) / Fkn)
end do
```

$\tilde{\mathbf{n}}:\frac{\bar{\mathbf{f}}_{kn}}{\|\bar{\mathbf{f}}_{kn}\|}$

```
if (trNhardFkn < 0.0d0) then
    trNhardFkn = 0.0d0
end if

do i = 1, 6
    MP1(i) = (Fdh / F) * dsqrt(2.0d0 / 3.0d0) * &
              (RHARD**HARD_MYU) * &
              trNhardFkn * SALB(i)
    MP2(i) = R2 *Fkn_M(i)
    MP3(i) = (UR / R1) * SC(i) + Cs * (1.0d0-R1) * &
              (SALBD_M(i) / SALBD - &
              RSC / Ks * CAD_M(i) / CAD)
    MP_N(i) = MP1(i) + MP2(i) + MP3(i)
end do
```

$\sqrt{\frac{2}{3}}\frac{F'}{F}\tilde{R}^{\upsilon}\langle\tilde{\mathbf{n}}:\frac{\bar{\mathbf{f}}_{kn}}{\|\bar{\mathbf{f}}_{kn}\|}\rangle\bar{\boldsymbol{\sigma}}$

$R\bar{\mathbf{f}}_{kn}$

$\frac{U}{R}\tilde{\boldsymbol{\sigma}}+c(1-R)\left(\bar{\mathbf{n}}-\frac{\mathfrak{R}_c}{\xi}\hat{\mathbf{n}}_c\right)$

Eq. (10.29)

$$\bar{M}^p \equiv \bar{\mathbf{n}}:\Big[\sqrt{\frac{2}{3}}\frac{F'}{F}\tilde{R}^{\upsilon}\langle\tilde{\mathbf{n}}:\frac{\bar{\mathbf{f}}_{kn}}{\|\bar{\mathbf{f}}_{kn}\|}\rangle\bar{\boldsymbol{\sigma}} + R\bar{\mathbf{f}}_{kn} + \frac{U}{R}\tilde{\boldsymbol{\sigma}}+c(1-R)\left(\bar{\mathbf{n}}-\frac{\mathfrak{R}_c}{\xi}\hat{\mathbf{n}}_c\right)\Big]$$

```
MP = 0.0d0
do i = 1,6
     MP = MP + (N(i) * MP_N(i))
end do


! Tangential-inelastic strain increment
     TR = Cbar*R1**NNt

     TCOM = TR/ (1.0d0 + TR)
```

Eq. (7.62): $T(R)=\tilde{c}R^{\tilde{n}}$

Eq. (9.55): $\frac{T(R)}{1+T(R)}$

```
do j = 1, 6
    do i = 1, 6
         TIBP_M(i,j) = II_M(i,j) - &
              (I_M(i) * I_M(j) / 3.0d0) - (NN(i) * NN(j))
    end do
end do
```

Eq. (9.51):

$\bar{\mathfrak{T}}'_t \equiv \mathcal{I}'-\bar{\mathbf{n}}'\otimes\bar{\mathbf{n}}'$
$= \mathcal{I}-(1/3)\mathbf{I}\otimes\mathbf{I}-\bar{\mathbf{n}}'\otimes\bar{\mathbf{n}}'$

```
20      continue

        do i = 1, 6
          do j = 1, 6
                EPM_M(i,j) = EL_M(i,j) - (CLU * EN(i) * &
                        EN(j) / (MP + trNEN)) -    &
                        (2.0d0 * SG * TCOM) * TIBP_M(i,j)
                Bm_m(i,j) = EPM_M(i,j)
          end do
        end do

!   Solve Simulteneous Equation (Gauss method) *************

        do i = 1,6
          Dq_m(i) = Ds_m(i)
          Dz_m(i) = De_m(i)
        end do

        if (C_m(1) + C_m(2) + C_m(3) + C_m(4) + C_m(5) +&
                C_m(6) < 6) Then

          do i = 1,6
              if (C_m(i) == 1) Then
                  do j = 1,6
                      if (C_m(j) == 0) Then
                          Dq_m(j) = Dq_m(j) - Bm_m(j, i) * Dz_m(i)
                      end if
                  end do
                  do j = 1,6
                      Bm_m(i, j) = 0.0d0
                      Bm_m(j, i) = 0.0d0
                  end do
                      Bm_m(i, i) = 1.0d0
                      Dq_m(i) = 0.0d0
              end if
          end do

          do l = 1,5
                if (C_m(l) == 0) Then
                    Pt = Bm_m(l, l)
                    do j = 1,6
                        Bm_m(l, j) = Bm_m(l, j) / Pt
                    end do
                    Dq_m(l) = Dq_m(l) / Pt
                    do i = l + 1,6
                        Pt = Bm_m(i, l)
                        do j = 1,6
                            Bm_m(i, j) = Bm_m(i, j) - Pt * Bm_m(l, j)
                        end do
                        Dq_m(i) = Dq_m(i) - Pt * Dq_m(l)
                    end do
                end if
          end do
```

Eq. (9.55):

$$\mathbf{K}^{ep} = \mathbf{E} - \frac{\mathbf{E}:\bar{\mathbf{n}}\otimes\bar{\mathbf{n}}:\mathbf{E}}{\bar{M}^{p} + \bar{\mathbf{n}}:\mathbf{E}:\bar{\mathbf{n}}} - 2G\frac{T(R)}{1+T(R)}\bar{\bar{\mathbf{T}}}'_t$$

<pre>do l = 1,6
    if (C_m(l) == 0) Then
        u = l
    end if
end do
 	
Dz_m(u) = Dq_m(u) / Bm_m(u, u)
 	
do l = u - 1, 1, -1
    if (C_m(l) == 0) Then
        Pt = Dq_m(l)
        do j = l + 1,6
            Pt = Pt - Bm_m(l, j) * Dz_m(j)
        end do
        Dz_m(l) = Pt
    end if
end do
 	
end if
 	
do i = 1,6
    if (C_m(i) == 0) Then
        Dq_m(i) = Ds_m(i)
    else
        do j = 1,6
            Dq_m(i) = Dq_m(i) + EPM_M (i, j) * Dz_m(j)
        end do
    end if
end do
!
! Stress and strain increments
do i = 1, 6
    if (C_m(i) == 0) then
        DW_M(i) = Ds_m(i)
        DX_M(i) = Dz_m(i)
    end if
    if (C_m(i) == 1) then
        DW_M(i) = Dq_m(i)
        DX_M(i) = De_m(i)
    end if
end do
 	
DWM = (DW_M(1) + DW_M(2) + DW_M(3)) / 3.0d0
 	
! Deviatoric stress increment
do i = 1, 6
    DWD_M(i) = DW_M(i) - (DWM * I_M(i))
end do
 	
! trN_dSigma   trNED
trNdS = 0.0d0
trNED = 0.0d0</pre>
	$d\boldsymbol{\sigma}$
	
	$d\boldsymbol{\varepsilon}$ (Eng. strain increment)
	
	$d\sigma_m$
	
	$d\boldsymbol{\sigma}'$

```
        do i = 1,6
             trNdS = trNdS + (N(i) * DW_M(i))
             trNED = trNED + (EN(i) * DX_M(i))
        end do
```

$\bar{\mathbf{n}}:d\boldsymbol{\sigma}=\bar{N}_i d\sigma_i$

$\bar{\mathbf{n}}:\mathbf{E}:d\boldsymbol{\varepsilon}=\bar{N}_i E_{ij} d\varepsilon_j$

```
! Plastic multiplier
        LM = trNdS / Mp
        LLM = trNED / (Mp + trNEN)
```

Eq. (9.38): $d\bar{\lambda}=\dfrac{\bar{\mathbf{n}}:d\boldsymbol{\sigma}}{\bar{M}^p}$

Eq. (9.41): $d\bar{\Lambda}=\dfrac{\bar{\mathbf{n}}:\mathbf{E}:d\boldsymbol{\varepsilon}}{\bar{M}^p+\bar{\mathbf{n}}:\mathbf{E}:\bar{\mathbf{n}}}$

```
        if (CLU== 0.0d0) then
            LM = 0.0d0
            LLM = 0.0d0
            GoTo 10
        end if

        if (LLM < 0.0d0) then
            LM = 0.0d0
            CLU = 0.0d0
            LLM = 0.0d0
            GoTo 20
        end if

!!!!!!!!!!!!!!!!!!!!!!!!!!!!!!!!!!!!!!!
!       Strain increment             !
!!!!!!!!!!!!!!!!!!!!!!!!!!!!!!!!!!!!!!!

! Plastic strain increment

10   continue
        do i = 1, 6
            DP_M(i) = CLU * LM * N(i)
        end do
```

Plastic engineering strain increment: $d\boldsymbol{\varepsilon}^p=d\bar{\Lambda}\bar{\mathbf{N}}$

```
        DPX=0.0d0
        do i = 1,3
            DPX = DPX + DP_M(i)**2.0
        end do
        do i = 4,6
            DPX = DPX + (DP_M(i)**2.0) / 2.0d0
        end do
        DPX = dsqrt(DPX)
```

$d\varepsilon_1^{p2}+d\varepsilon_2^{p2}+d\varepsilon_3^{p2}+(d\varepsilon_4^{p2}+d\varepsilon_4^{p2}+d\varepsilon_4^{p2})/2$

$\|d\boldsymbol{\varepsilon}^p\|$

```
        DR = UR * DPX
```

$U\|d\boldsymbol{\varepsilon}^p\|$

```
! Elastic engineering strain increment
     do i = 1,6
        Dex_M(i) = 0.0d0
```

do j = 1,3	
Dex_M(i) = Dex_M(i) + (IEL_M(i, j) * & DW_M(j))	$d\varepsilon_i^e = E_{ij}d\sigma_j$ (j=1~3)
end do	
do j = 4,6	
Dex_M(i) = Dex_M(i) + 2.0d0 * (IEL_M(i, j) * & DW_M(j))	$d\varepsilon_i^e = 2E_{ij}d\sigma_j$ (j=4~6)
end do	
end do	
! Total strain increment	
do i = 1, 6	
D_M(i) = Dex_M(i) + DP_M(i)	$d\boldsymbol{\varepsilon} = d\boldsymbol{\varepsilon}^e + d\boldsymbol{\varepsilon}^p$
end do	
! Stress and strain	
do i = 1, 6	
SIG(i) = SIG(i) + DW_M(i)	$\boldsymbol{\sigma}$
E_M(i) = E_M(i) + DX_M(i)	$\boldsymbol{\varepsilon}$
Ee_M(i) = Ee_M(i) + Dex_M(i)	$\boldsymbol{\varepsilon}^e$
Ep_M(i) = Ep_M(i) + DP_M(i)	$\boldsymbol{\varepsilon}^p$
end do	
! Increments of back stress and elastic-core	
do i = 1, 6	Eq. (10.8):
Da_M(i) = Ck * (SALBD_M(i) / SALBD - 1.0d0 / & (dsqrt(3.0d0 / 2.0d0) *Zeta * F) * AL (i)) * DPX	$d\boldsymbol{\alpha} = c_k\left(\bar{\mathbf{n}} - \frac{1}{\sqrt{3/2}\zeta F}\boldsymbol{\alpha}\right)d\bar{\lambda}$
Dsim_M(i) = Cs * ((SALBD_M(i) / SALBD) - & (RSC / Ks * CAD_M(i) / CAD)) * DPX	Eq. (9.15): $d\mathbf{c} = c\left(\bar{\mathbf{n}} - \frac{\mathcal{R}_c}{\xi}\hat{\mathbf{n}}_c\right)d\bar{\lambda}$
end do	
! Increments of back stress and elastic-core	
do i = 1, 6	
AL(i) = AL(i) + Da_M(i)	$\boldsymbol{\alpha}$
CS(i) = CS(i) + Dsim_M(i)	$\mathbf{c}$
end do	
! Stagnation of isotropic hardening	
trNhardDa = 0.0d0	
do i = 1, 3	
trNhardDa = trNhardDa + & (NHARD_M(i) * Da_M(i))	
end do	
do i = 4, 6	

<pre>   trNhardDa = trNhardDa + &
                              2.0d0 * (NHARD_M(i) * Da_M(i))
        end do</pre>
	$\tilde{\mathbf{n}}:d\boldsymbol{\alpha}=\tilde{n}_i d\alpha_i\ (i=1\sim 3)$
	$+2\tilde{n}_i d\alpha_i\ (i=4\sim 6)$

<pre>   if (trNhardN < 0.0d0) then
            trNhardN = 0.0d0
        end if
 	
        if (trNhardDa < 0.0d0) then
            trNhardDa = 0.0d0
        end if</pre>	

<pre>   Khard_dot = Chard * (R_HARD ** Hard_Zita) * &
                              * trNhardDa
        K_HARD = K_HARD + Khard_dot</pre>
	Eq. (10.21):
	$d\tilde{K}=C\tilde{R}^{\varsigma}\langle\tilde{\mathbf{n}}:d\boldsymbol{\alpha}\rangle$
	$\tilde{K}$

<pre>! cap(Dot)
        do i = 1, 6
            DAlhard_M(i) = (1.0d0 - Chard) * &
                        (R_HARD ** Hard_Zita) * &
                        trNhardDa * NHARD_M(i)
        end do</pre>
	Eq. (10.22):
	$d\boldsymbol{\rho}=(1-C)\tilde{R}^{\varsigma}\langle\tilde{\mathbf{n}}:d\boldsymbol{\alpha}\rangle\tilde{\mathbf{n}}$

<pre>! cap
        do i = 1, 6
            ALHARD_M(i) = ALHARD_M(i) + &
                        DAlhard_M(i)
        end do</pre>
	$\boldsymbol{\rho}$

<pre> Hdh = dsqrt(2.0d0 / 3.0d0) * (R_HARD** HARD_MYU) * & DPX * trNhardFkn</pre>	Eq. (10.27): $dH=\sqrt{\frac{2}{3}}\tilde{R}^{\upsilon}\langle\tilde{\mathbf{n}}:\frac{\bar{\mathbf{f}}_{kn}}{\lVert\bar{\mathbf{f}}_{kn}\rVert}\rangle d\bar{\lambda}$

<pre>! Isotropic hardening
        H = H + Hdh
        Fdh = F0 * H1 * H2 * exp(-H2 * H)
        F = F0 * (1.0d0 + H1 * (1.0d0 - exp(-H2 * H)))</pre>
	Eq. (10.5):
	$F'=F_0 h_1 h_2 \exp(-h_2 H)$
	Eq. (10.5):
	$F(H)=F_0\{1+h_1[1-\exp(-h_2 H)]\}$

<pre>return
end Subroutine calc_exsub
 	
Subroutine Input_Parameter(m, nn)
implicit none
!++++++++++++++++++++++++++
!
!   Input material constants
!
!++++++++++++++++++++++++++
!
        integer*8 m,nn</pre>	

```
        open(100,file='para.dat', access='sequential', err=999, &
                      status='old', form='formatted')
        go to 9000
   999 continue
        write(*,'(a10,d12.5)') 'open error in para.dat'
        stop
  9000 continue
!
!---------------------------------------------
        read(100,*) F0
        read(100,*) EL
        read(100,*) POI
        read(100,*) H1
        read(100,*) H2
        read(100,*) Ct
        read(100,*) Ck
        read(100,*) Cbar
        read(100,*) Zeta
        read(100,*) Cs
        read(100,*) Chard
        read(100,*) Hard_Zita
        read(100,*) HARD_MYU
        read(100,*) UA
        read(100,*) UB
        read(100,*) Ks
        read(100,*) Re
        read(100,*) NNu
        read(100,*) NNt
        close (100)
!
        if(m.eq.1.and.nn.eq.1) then
          write(6,'(a10,d12.5)') 'F0          =', F0
          write(6,'(a10,d12.5)') 'EL          =', EL
          write(6,'(a10,d12.5)') 'POI         =', POI
          write(6,'(a10,d12.5)') 'H1          =', H1
          write(6,'(a10,d12.5)') 'H2          =', H2
          write(6,'(a10,d12.5)') 'Ct          =', Ct
          write(6,'(a10,d12.5)') 'Ck          =', Ck
          write(6,'(a10,d12.5)') 'Zeta        =', Zeta
          write(6,'(a10,d12.5,)') 'Cs         =', Cs
          write(6,'(a10,d12.5,)') 'Cbar       =', Cbar
          write(6,'(a10,d12.5)') 'Chard       =', Chard
          write(6,'(a10,d12.5)') 'Hard_Zita   =', Hard_Zita
          write(6,'(a10,d12.5)') 'HARD_MYU =', HARD_MYU
          write(6,'(a10,d12.5)') 'UA          =', UA
          write(6,'(a10,d12.5)') 'UB          =', UB
          write(6,'(a10,d12.5)') 'Ks          =', Ks
          write(6,'(a10,d12.5)') 'Re          =', Re
          write(6,'(a10,d12.5)') 'NNu         =', NNu
          write(6,'(a10,d12.5)') 'NNt         =', NNt
        end if

return
end subroutine Input_Parameter
```

```
Subroutine Inicon()
implicit none
!-----------------------------------------------
!
!  Set initial conditions
!
!-----------------------------------------------
        PAI = 3.14159265358979
        RAD = PAI / 180.0d0
!
        do i = 1, 6
          Ds_m(i) = 0.0d0
          De_m(i) = 0.0d0
          SIG (i) = 0.0d0
          AL (i) = 0.0d0
          C_m(i) = 0
          E_M(i) = 0.0d0
          CS (i) = 0.0d0
          Ep_M(i) = 0.0d0
          Ee_M(i) = 0.0d0
          Ep_M(i) = 0.0d0
          D_M(i) = 0.0d0
          Dex_M(i) = 0.0d0
          DP_M(i) = 0.0d0
          ALHARD_M(i) = 0.0d0
        end do
!
        SIG (1) = 0.0000001
        CS (1) = 0.000000001
        AL (1) = 0.0000000001
        AL (2) = -AL (1) / 2.0d0
        AL (3) = -A (1) / 2.0d0
!
        CS (2) = -CS (1) / 2.0d0
```

```
        CS (3) = -CS (1) / 2.0d0
        DPX = 0.0d0
        H = 0.0d0
        Hdh = 0.0d0
        F = F0
        Fdh = F0 * H1 * H2
        ALP_HARD = 0.0d0
!
        do i = 1, 3
          ALP_HARD _M(i) = AL (i) - ALHAD_M(i)
          ALP_HARD = ALP_HARD + ALP_HARD_M(i)**2.0
        end do
        do i = 4, 6
          ALP_HARD _M(i) = AL (i) - ALHARD_M(i)
          ALP_HARD = ALP_HARD + &
                        2.0d0*ALP_HARD_M(i)**2.0
        end do
        ALP_HARD = dsqrt(ALP_HARD)
        Khard = (dsqrt (3.0d0 / 2.0d0) * ALP_HARD) + &
                  0.00000000001
!
        I_M(1) = 1.0d0
        I_M(2) = 1.0d0
        I_M(3) = 1.0d0
        I_M(4) = 0.0d0
        I_M(5) = 0.0d0
        I_M(6) = 0.0d0
        do i = 1, 6
          do j = 1, 6
                  II_M(i, j) = 0.0d0
          end do
        end do
        II_M(1, 1) = 1.0d0
        II_M(2, 2) = 1.0d0
        II_M(3, 3) = 1.0d0
        II_M(4, 4) = 0.5d0
        II_M(5, 5) = 0.5do
```

II_M(6, 6) = 0.5d0 return end subroutine Inicon	

(b) Computer program of subloading-friction model

composed by Koichi Hashiguchi, Masami Ueno and Shingo Ozaki

Sub-program code	Variables/Equations
subroutine Calc_friction ()	
implicit none	
***************************	
Declaration of variables	
***************************	
Clu = 1.0d0	
PAI = 3.14159265358979	
! Normal and tangential contact stresses	Eq. (18.13):
fn = -(Nb_v(1) * F_v(1) + Nb_v(2) * F_v(2) +&	$f_n = -\mathbf{f} \bullet \mathbf{n}$
Nb_v(3) * F_v(3))	
do i = 1, 3	Eq. (18.12):
Fn_v(i) = -fn * Nb_v(i)	$\mathbf{f}_n \ (= \mathbf{f} \bullet \mathbf{n}) = -f_n \mathbf{n}$
Ft_v(i) = F_v(i) - Fn_v(i)	$\mathbf{f}_t = \mathbf{f} - \mathbf{f}_n$
end do	
if (Ft_v(1) == 0.0d0) then	
Ft_v(1) = 0.0000001	
end if	
ft = dsqrt(Ft_v(1) **2.0 + Ft_v(2) **2.0 +&	$f_t = \sqrt{(\mathbf{f}_t)_1^2 + (\mathbf{f}_t)_2^2 + (\mathbf{f}_t)_3^2}$
Ft_v(3) ** 2.0)	
	Eq. (18.36):
! Create U(r) mbp and mbc	$\bar{U}(r) = \breve{u}\cot\left(\frac{\pi}{2}r\right)$
Ur = ub / (dtan((PAI / 2.0d0) * r))	Eq. (18.55):
mbp = - kapa * (Myu / Myuk – 1.0d0) * r +&	$m^p \equiv -\kappa\left(\frac{\mu}{\mu_k} - 1\right) r + \bar{U}(r)\mu$
Ur * Myu	
mbc = ksi * (1.0d0 - Myu / Myus) * r	Eq. (18.56):
	$m^c \equiv \xi\left(1 - \frac{\mu}{\mu_s}\right) r$

! Frictional elasto-plastic constitutive matrix ----------	Eq. (18.13):
Tft_v(1) = Ft_v(1) / ft	$t_{f1} = (\mathbf{f}_t)_1 / \lVert\mathbf{f}_t\rVert$
Tft_v(2) = Ft_v(2) / ft	$t_{f2} = (\mathbf{f}_t)_2 / \lVert\mathbf{f}_t\rVert$
Tft_v(3) = dsqrt(Ft_v(1)**2.0 +Ft_v(2)**2.0)/ ft	$t_{f3} = \sqrt{t_{f1}^2 + t_{f2}^2} = \sqrt{(\mathbf{f}_t)_1^2 + (\mathbf{f}_t)_2^2} / \lVert\mathbf{f}_t\rVert \ (=1)$
! Elastic modulus do i = 1, 3 do j = 1, 3 if (i== j) then	
Ce_m(i,j)=Alft else	Eq. (18.23): $C^e_{ij} = \alpha_t \ (i=j)\,(i,j=1,2)$

Ce_m(i, j) = 0.0d0
end if
end do
end do
	$C^e_{ij} = 0 \,(i \neq j)$

! Alfa*Tft_v(i)	
Al_ft_v(1) = Alft * Tft_v(1)	$\alpha_t t_{f1} = \alpha_t (\mathbf{f}_t)_1 / \lVert\mathbf{f}_t\rVert$
Al_ft_v(2) = Alft * Tft_v(2)	$\alpha_t t_{f2} = \alpha_t (\mathbf{f}_t)_2 / \lVert\mathbf{f}_t\rVert$
Al_ft_v(3) = Alfn * dsqrt(Ft_v(1) **2.0 + & Ft_v(2) **2.0) / fn	α_n / f_n
! Variables in elasto-plastic modulus	
Mbp_Al_fn = mbp + Alft / fn	Eq. (18.85): $m^p + \alpha_t / f_n$
Al_fn = Alft / fn	α_t / f_n
Al_fn_Mbp_Al_fn = Al_fn / Mbp_Al_fn	$(\alpha_t / f_n)/(m^p + \alpha_t / f_n)$
Al_Mbc_Mbp_Al_fn = mbc * Alft * dt / & Mbp_Al_fn	$\alpha_t m^c dt / (m^p + \alpha_t / f_n)$
20 continue	
! Elastoplastic modulus do i = 1, 3 do j = 1, 3 Cep_m(i, j) = 0.0d0 Bm_m(i, j)= 0.0d0	

```
            end do
        end do


    do i = 1, 3
        Cep_T_v(i) = Al_Mbc_Mbp_Al_fn * &
                    Tft_v(i)
        do j = 1, 3
            Cep_m(i, j) = Ce_m(i, j) - Clu * &
                        Al_fn_Mbp_Al_fn * &
                        Tft_v(i) * Al_ft_v(j)
            Bm_m(i, j) = Cep_m(i, j)
        end do
    end do

! Solve constitutive equation  ----------------------
    do i = 1, 3
        Dq_v(i) = Df_v(i) - Cep_T_v(i)
        Dz_v(i) = DUt_v(i)
    end do

    if (C_m(1) + C_m(2) + C_m(3) < 3) Then
        do = 1, 3
        if (C_m(i) == 1) Then
            do j = 1, 3
            if (C_m(j) == 0) Then
                Dq_v(j) = Dq_v(j) - Bm_m(j, i) * Dz_v(i)
            end if
            end do

            do j = 1, 3
            Bm_m(i, j) = 0.0d0
            Bm_m(j, i) = 0.0d0
            end do
            Bm_m(i, i) = 1.0d0
            Dq_v(i) = 0.0d0
        end if
        end do

! Gauss method
    do l = 1, 2
        if (C_m(l) == 0) Then
        Pt = Bm_m(l, l)
        do j = l, 3
            Bm_m(l, j) = Bm_m(l, j) / Pt
```

Eq. (18.85):
$$[\alpha_t m^c dt / (m^p + \alpha_t / f_n)] t_{f_i}$$

Eq. (18.76):
$$\mathbf{K}^{ep} = \mathbf{C}^e - \frac{\alpha_t \mathbf{t}_f \otimes (\alpha_t / f_n) \mathbf{t}_f}{m^p + \alpha_t / f_n}$$

Creep term m^c in Eq. (18.85) is irrelevant to sliding increment so that it is subtracted from the contact stress increment before solving simultaneous equation:

$$df_i$$
$$-[\alpha_t m^c dt / (m^p + \alpha_t / f_n)] t_{f_i}$$

$$d\bar{u}_i$$

```
                end do
                Dq_v(l) = Dq_v(l) / Pt
                do i = l + 1, 3
                    Pt = Bm_m(i, l)
                    do j = l, 3
                        Bm_m(i, j) = Bm_m(i, j) - Pt * Bm_m(l, j)
                    end do
                    Dq_v(i) = Dq_v(i) - Pt * Dq_v(l)
                end do
            end if
        end do

        do l = 1, 3
            if (C_m(l) == 0) Then
                ii = l
            end if
        end do

        Dz_v(ii) = Dq_v(ii) / Bm_m(ii, ii)

        do l = ii – 1, 1, -1
            if (C_m(l) == 0) Then
                Pt = Dq_v(l)
                do j = l + 1, 3
                    Pt = Pt - Bm_m(l, j) * Dz_v(j)
                end do
                Dz_v(l) = Pt
            end if
        end do
    end if

    do i = 1, 3
        if (C_m(i) == 0) Then
            Dq_v(i) = Df_v(i)
        else
            Dq_v(i) = Cep_T_v(i)
            do j = 1, 3
                Dq_v(i) = Dq_v(i) + Cep_m(i, j) * Dz_v(j)
            end do
        end if
    end do

! Contact force increment and sliding displacement increment
    do i = 1, 3
        Dw_v(i) = Dq_v(i)                                     
        Dx_v(i) = Dz_v(i)                                     
    end do
! Plastic multiplier
    LM = ((Tft_v(1) * Alft * Dz_v(1) + Tft_v(2) *  &
        Alft * Dz_v(2) + (ft / fn) * Alfn *&
```

df_i

$d\bar{u}_i$

Eq. (18.74):

$$d\bar{\Lambda} = \{[\alpha_t t_{f1} d\bar{u}_1 + \alpha_t t_{f2} d\bar{u}_2 + \alpha_n (f_t / f_n) d\bar{u}_3] / f_n - m^c dt\} / (m^p + \alpha_t / f_n)$$

Dz_v(3)) / (fn) - mbc * dt) / (mbp + Alft / fn)
if (Clu == 0.0d0) then
go to 20
end if
if (LM < 0.0d0) then
Clu = 0.0d0
go to 10
end if
10 continue	

! Plastic sliding displacement increment and its accumulation do i = 1, 3 Dpx_v(i) = 0.0d0 Dpx_v(i) = Clu * LM * Tft_v(i) end do	Eq. (18.49): $d\bar{\mathbf{u}}^p = d\bar{\lambda}\mathbf{t}_n$ $(d\bar{u}_i^p = d\bar{\lambda}t_{fi})$
Dpx = dsqrt(Dpx_v(1) **2.0 + & Dpx_v(2) **2.0 + Dpx_v(3) **2.0)	$d\bar{u}^p$ $= \sqrt{(d\bar{\mathbf{u}}^p)_1^2 + (d\bar{\mathbf{u}}^p)_2^2 + (d\bar{\mathbf{u}}^p)_3^2}$
TDpx = TDpx + Dpx ubp = TDpx	$\bar{u}^p$
! Revolution of normal sliding-yield ratio and friction coefficient dr = Ur * Dpx	Eq. (18.33): $dr = \bar{U}(r)\lVert d\bar{\mathbf{u}}^p \rVert$
r = r + dr	r
DMyu = kapa * (1.0d0 – Myu / Myuk) * Dpx +& ksi * (1.0d0 -Myu / Myus) * dt	Eq. (18.38): $d\mu$ $= \kappa\left(1-\frac{\mu}{\mu_k}\right)d\bar{u}^p + \xi\left(1-\frac{\mu}{\mu_s}\right)dt$
Myu = Myu + DMyu	μ
! Sliding displacement increment and its accumulation do i = 1, 3 DUn_v(i) = (Dx_v(1) * Nb_v(1) + & Dx_v(2) * Nb_v(2) +& Dx_v(3) * Nb_v(3)) * Nb_v(i)	Eq. (18.2): $d\bar{\mathbf{u}}_n = (d\bar{\mathbf{u}} \bullet \mathbf{n})\mathbf{n}$
DUt_v(i) = Dx_v(i) - DUn_v(i)	Eq. (18.2): $d\bar{\mathbf{u}}_t = d\bar{\mathbf{u}} - d\bar{\mathbf{u}}_n$
DUpn_v(i) = (Dpx_v(1) * Nb_v(1) + & Dpx_v(2) * Nb_v(2) +& Dpx_v(3) * Nb_v(3)) * Nb_v(i)	Eq. (18.8) $d\bar{\mathbf{u}}_n^p = (d\bar{\mathbf{u}}^p \bullet \mathbf{n})\mathbf{n}$

```
            DUpt_v(i) = Dpx_v(i) - DUpn_v(i)
        end do

        do i = 1, 3
            F_v(i) = F_v(i) + Dw_v(i)
            U_v(i) = U_v(i) + Dx_v(i)
            Un_v(i) = Un_v(i) + DUn_v(i)
            Ut_v(i) = Ut_v(i) + DUt_v(i)
            Up_v(i) = Up_v(i) + Dpx_v(i)
            Upn_v(i) = Upn_v(i) + DUpn_v(i)
            Upt_v(i) = Upt_v(i) + DUpt_v(i)
        end do

return
end subroutine Calc_friction



Subroutine Input_Parameter(m, nn)
implicit none
!
!++++++++++++++++++++++++
!
! input material constants
!
!++++++++++++++++++++++++
!
        integer*8 m,nn

        open(10,file='para.dat',access='sequential',err=999,&
                        status='old',form='formatted')
        go to 9000
   999 continue
        write(*,'(a10,d12.5)')'open error in para.dat'
        stop
  9000 continue

! Material constants
-----------------------------------------------------------------------
!       Alfn    Alft                            elastcity
!       kapa    ksi     Myuk    Myus            friction coefficient
!       m       n                               sliding velocity
!       ub                                      U(rs)
!
!
! Calculation conditions ---------------------------------------------
```

Eq. (18.8): $d\bar{\mathbf{u}}_t^p = d\bar{\mathbf{u}}^p - d\bar{\mathbf{u}}_n^p$

$\mathbf{f}$

$\bar{\mathbf{u}}$

$\bar{\mathbf{u}}_n$

$\bar{\mathbf{u}}_t$

$\bar{\mathbf{u}}^p$

$\bar{\mathbf{u}}_n^p$

$\bar{\mathbf{u}}_t^p$

```
!       Cm1     Cm2         0  free  1  fixed(sliding velocity)
!       Vt1     Vt2         sliding velocity
!       ft1     ft2         sliding force
!       dt                  time increment
!       Umax                maximum sliding displacement
!
!--------------------------------------------- ---------------------------------
!
        read(10,*) Alfn
        read(10,*) Alft
        read(10,*) kapa
        read(10,*) ksi
        read(10,*) Myuk
        read(10,*) Myus
        read(10,*) ub
        read(10,*) m
        read(10,*) n
        read(10,*) Cm1
        read(10,*) Cm2
        read(10,*) Vt1
        read(10,*) Vt2
        read(10,*) ft1
        read(10,*) ft2
        read(10,*) dem2
        read(10,*) dt
        read(10,*) Umax
        Close (10)
!
        if(m.eq.1.and.nn.eq.1) then
          write(6,'(a10,d12.5)') 'Alftn          =', Alftn
          write(6,'(a10,d12.5)') ' Alft          =', Alft
          write(6,'(a10,d12.5)') ' kapa          =', kappa
          write(6,'(a10,d12.5)') ' ksi           =', ksi
          write(6,'(a10,d12.5)') ' Myuk          =', Myuk
          write(6,'(a10,d12.5)') ' ub            =', ub
          write(6,'(a10,d12.5)') 'm              =',m
          write(6,'(a10,d12.5)') 'n              =',n
          write(6,'(a10,d12.5)') ' Cm1           =', Cm1
          write(6,'(a10,d12.5)') ' Cm2           =', Cm2
          write(6,'(a10,d12.5,)') 'Vt1           =',Vt1
          write(6,'(a10,d12.5)') 'Vt2            =',Vt2
          write(6,'(a10,d12.5)') 'ft1            =',ft1
          write(6,'(a10,d12.5)') 'ft2            =',ft2
          write(6,'(a10,d12.5)') 'dem2           =', dem2
          write(6,'(a10,d12.5)') 'dt             =',dt
          write(6,'(a10,d12.5)') 'Umax           =',Umax
        endif

        return
        end subroutine Input_Parameter



Subroutine Inicon
```

```
implicit none

      do i = 1, 3
          U_v(i) = 0.0d0
          DUn_v(i) = 0.0d0
          DUt_v(i) = 0.0d0
          DUpn_v(3) = 0.0d0
          DUpt_v(3) = 0.0d0
          Un_v(i) = 0.0d0
          Up_v(i) = 0.0d0
          Ut_v(i) = 0.0d0
          Upn_v(i) = 0.0d0
          Upt_v(i) = 0.0d0
          F_v(i) = 0.0d0
          Fn_v(i) = 0.0d0
          Ft_v(i) = 0.0d0
          Df_v(i) = 0.0d0
          Tfn_v(i) = 0.0d0
          Tft_v(i) = 0.0d0
          Al_ft_v(i)=0.0d0
          Cep_T_v(i)=0.0d0
          Dpx_v(i)=0.0d0
          Nb_v(i) = 0.0d0
      end do

      Myu = Myus
      Myuo = Myu
      F_v(3) = fn3
      Fn_v(3) = fn3
      fno = dsqrt(F_v(1) ** 2.0 + F_v(2) ** 2.0 + F_v(3) ** 2.0)
      fto = 0.0d0

      Tb_v(1) = 1.0d0
      Tb_v(2) = 1.0d0
      Tb_v(3) = 0.0d0

      ubpo = 0.0d0
      Dpx = 0.0d0
      TDpx = 0.0d0
      LM = 0.0d0
      r = 0.000000001

      do i = 1, 3
         do j = 1, 3
            I_m(i, j) = 0.0d0
         end do
      end do

      I_m(1,1) = 1.0d0
      I_m(2,2) = 1.0d0
      I_m(3,3) = 1.0d0

      C_m(1)=Cm1
      C_m(2)=Cm2
      C_m(3)=0
```

```
        dx1 = Vt1 * dt
        dx2 = Vt2 * dt
        dx3 = 0.0d0

        Vt_v(1) = Vt1
        Vt_v(2) = Vt2
        Vt_v(3) = 0.0d0

        Nstep = Int(Umax / dx1)

        Nb_v(1) = 0.0d0
        Nb_v(2) = 0.0d0
        Nb_v(3) = 1.0d0

        DUt_v(1) = dx1
        DUt_v(2) = dx2
        Df_v(3) = 0.0d0

return
end subroutine Inicon
```

References

Brepols T, Vladimirov IN, Reese S (2014) Numerical comparison of isotropic hypo- and hyperelastic-based plasticity models with application to industrial forming processes. Int J Plast 63:18–48

Dienes JK (1979) On the analysis of rotation and stress rate in deforming bodies. Acta Mech 32:217–232

Nagtegaal JC, De Jong JE (1982) Some aspects of non-isotropic workhardening in finite strain plasticity. In: Lee EH, Mallett RL (eds) Plasticity of metals and finite strain: theory, experiment and computation. Division of Applied Mechanics, Stanford University and Department of Engineering, Rensselaer Polytechnic Institute, pp 65–102

Shutov AV, Ihlemann J (2014) Analysis of some basic approaches to finite strain elasto-plasticity in view of reference change. Int J Plast 63:193–197

Xiao H, Bruhns OT, Meyers A (1997) Logarithmic strain, logarithmic spin and logarithmic rate. Acta Mech 124:89–105

All Referred Bibliography

Monographs on solid mechanics and tensor analysis

Aris R (1962) Vectors, tensors, and the basic equations of fluid mechanics. Dover Publ. Inc., New York
Asaro RJ, Lubarda V (2006) Mechanics of solids and materials. Cambridge University Press, Cambridge
Bazant ZP, Cedolin L (1991) Stability of structures: elastic, inelastic, fracture and damage theories. Oxford University Press, New York
Becker E, Burger W (1975) Kontinuumsmechanik. B. G. Teubner, Stuttgart
Belytschko T, Liu WK, Moran B (2014) Nonlinear finite elements for continua and structures, 2nd edn. Wiley, New York
Bensson J, Cailletaud G, Chaboche J-L, Forest S, Blétry M (2001) Non-linear mechanics of materials. Springer, Heidelberg
Bertram A (2008) Elasticity and plasticity of large deformations. Springer, Berlin
Besseling JF, Van der Giessen E (1994) Mathematica modelling of inelastic deformation. Chapman & Hall, London
Bingham EC (1922) Fluidity and plasticity. McGraw-Hill, New York
Biot MA (1956) Mechanics of incremental deformations. Wiley, New York
Bonet J, Wood RD (1997) Nonlinear continuum mechanics for finite element analysis. Cambridge University Press, Cambridge
Borja RI (2013) Plasticity: modeling & computation. Springer, Heidelberg
Bowden FP, Tabor D (1958) The friction and lubrication of solids. Clarendon Press, Oxford, UK
Bowen RM, Wang C-C (2008) Introduction to vectors & tensors, 2nd edn. Dover Publ. Inc., New York
Chadwick P (1999) Continuum mechanics: concise theory and problems. Dover Publ. Inc., New York
Chakrabarty J (1987) Theory of plasticity. McGraw-Hill, New York
Chakrabarty J (2000) Applied plasticity. Springer, New York
Cristescu N (1967) Dynamic plasticity. North-Holland, Amsterdam
Chung TJ (2007) General continuum mechanics. Cambridge University Press, Cambridge, UK
De Borst R, Crisfield MA, Remmers JJC, Verhoosed CV (2012) Nonlinear finite element analysis of solids and structures, 2nd edn. Wiley, Chichester
Desai C, Siriwardane HJ (1984) Constitutive laws for engineering materials with emphasis on geomaterials. Prentice-Hall Inc., New York
de Souza Neto EA, Perić D, Owen DJR (2008) Computational methods for plasticity. Wiley, Chichester, UK

K. Hashiguchi, *Foundations of Elastoplasticity: Subloading Surface Model*,
DOI 10.1007/978-3-319-48821-9

Dimitrienko YI (2010) Tensor analysis and nonlinear tensor functions. Kluwer Academic Publishing, Dordrecht

Dimitrienko YI (2011) Nonlinear continuum mechanics and large inelastic deformations. Springer, Dordrecht

Doghri I (2000) Mechanics of deformable solids. Springer, Berlin

Duvaut G, Lions JL (1972) Les Inequations en Mechanique et en Physique. Dunod, Paris (Duvaut G, Lions, JL (1976) Inequalities in Mechanics and Physics. Springer, New York)

Dvorkin EN, Goldschmit MB (2006) Nonlinear continua. Springer, Berlin

Ellyin F (1997) Fracture damage, crack growth and life prediction. Chapman & Hall, London

Ericksen JL (1991) Introduction to the thermodynamics of solids. Chapman and Hall, London

Eringen AC (1962) Nonlinear theory of continuous media. McGraw-Hill, New York

Eringen AC (1967) Mechanics of continua. Wiley, New York

Fischer-Cripps AC (2000) Introduction to contact mechanics. Springer, New York

Flugge W (1972) Tensor analysis and continuum mechanics. Springer, Berlin

Fung YC (1969) A first course in continuum mechanics. Prentice-Hall, Englewood Cliffs

Gambin W (2001) Plasticity and textures. Kluwer Academic Publ., Dordrecht

Golub GH, Loan CFV (2013) Matrix computations, 4th edn. John Hopkins University Press, Baltimore

Gurtin ME (1981) An introduction to continuum mechanics. Academic Press, New York

Gurtin ME, Fried E, Anand L (2010) The mechanics and thermodynamics of continua. Cambridge University Press, New York

Han W, Reddy BD (1999) Plasticity—mathematical theory and numerical analysis. Springer, New York

Hashiguchi K (2009) Elastoplasticity theory, 1st edn. In: Lecture notes in applied and computational mechanics. Springer, Heidelberg

Hashiguchi K (2013) Elastoplasticity theory, 2nd edn. In: Lecture notes in applied and computational mechanics. Springer, Heidelberg

Hashiguchi K, Yamakawa Y (2012) Introduction to finite strain theory for continuum elasto-plasticity. In: Wiley Series in Computational Mechanics. Wiley, UK

Haupt P (2000) Continuum mechanics and theory of materials. Springer, Wien

Havner KS (1992) Finite plastic deformation of crystalline solids. Cambridge University Press, Cambridge

Hecker SS (1972) Experimental investigation of corners in yield surface. Acta Mech 13:69–86

Hill R (1950) The mathematical theory of plasticity. Oxford University Press, London

Hinton E, Owen DRJ (1980) Finite elements in plasticity: theory and practice. Pineridge Press, Swansea, UK

Hisada T (1992) Tensor analysis for nonlinear finite element method. Maruzen Publ., Inc., Tokyo (in Japanese)

Hisada T, Noguchi H (1995) Foundations and applications in nonlinear finite element method. Maruzen Publishing, Tokyo (in Japanese)

Holzapfel GA (2000) Nonlinear solid mechanics: a continuum approach for engineering. Wiley, Ltd

Hosford WF (2009) Mechanical behavior of solids. Cambridge University Press

Houlsby GT, Puzrin AM (2006) Principles of hyperelasticity; an approach to plasticity theory based on thermodynamic principles. Springer, Heidelberg

Hunter SC (1983) Mechanics of continuous media. Ellis Horwood, Chichester

Ibrahimbegovic A (2009) Nonlinear solid mechanics: theoretical formulations and finite element solution methods. Springer, Dordrecht

Ilyushin AA (1963) Plasticity—foundation of the general mathematical theory. Izdatielistbo Akademii Nauk CCCR (Publisher of the Russian Academy of Sciences), Moscow

Itskov M (2010) Tensor algebra and tensor analysis for engineers with application to continuum mechanics, 2nd edn. Springer, Dordrecht

Jaunzemis W (1967) Continuum mechanics. The Macmillan, New York

Jeffreys H (1931) Cartesian tensors. Cambridge University Press, London
Kachanov LM (1971) Foundation of theory of plasticity. North-Holland, Amsterdam
Khan AS, Huang S (1995) Continuum theory of plasticity. Wiley, New York
Kitagawa H (1987) Theory of elasticity and plasticity. Shokabo Publ. Ltd., Tokyo (in Japanese)
Kyoya T (2008) Note on continuum mechanics, Japan. Assoc. for Nonlinear CAE, Tokyo (in Japanese)
Leigh DC (1968) Nonlinear continuum mechanics: an introduction to the continuum physics and mechanical theory of the nonlinear mechanical behavior of materials. McGraw-Hill, New York
Lemaitre JA (1992) A course on damage mechanics. Springer, Heidelberg
Lemaitre JA, Chaboche J-L (1990) Mechanics of solid materials. Cambridge University Press, Cambridge
Lemaitre JA, Desmoral R (2005) Engineering damage mechanics. Springer, Heidelberg
Lubarda VA (2002) Elastoplasticity theory. CRC Press, Boca Ranton
Lubliner J (1990) Plasticity theory. Macmillan, New York
Malvern, LE (1969) Introduction to the mechanics of a continuous medium. Prentice-Hall, Englewood Cliffs, New Jersey
Marsden JE, Hughes TJR (1983) Mathematical foundation of elasticity. Prentice-Hall, Englewood Cliffs, New Jersey
Martin JB (1975) Plasticity: foundation and general results. MIT Press, Cambridge
Maugin GA (1992) The thermomechanics of plasticity and fracture. Cambridge University Press, Cambridge
Murakami S (2012) Continuum damage mechanics: a continuum mechanics approach to the analysis of damage and fracture. Springer, Dordrecht
Nadai A (1963) Theory of flow and fracture of solids. McGraw-Hill, New York
Negahban M (2012) The mechanical and thermodynamical theory of plasticity. CRC Press, Boca Raton, Florida
Nemat-Nasser S (2004) Plasticity: a treatise on finite deformation of heterogeneous inelastic materials. Cambridge University Press, New York, NY
Norton FH (1929) Creep of steel at high temperature. McGraw-Hill, New York, NY
Oden JT (2011) An introduction to mathematical modeling: a course in mechanics, wiley series in continuum mechanics. Wiley, New York
Odqvist FKG, Hult JAH (1962) Kriechfestigkeit Metallischer Werkstoffe. Springer, Berlin
Odqvist FKG (1966) Mathematical theory of creep and creep rupture. Oxford University Press, London
Ogden RW (1984) Non-linear elastic deformations. Ellis-Horwood, Chichester, UK
Ottosen NS, Ristinmaa M (2005) The mechanics of constitutive modeling. Elsevier, Amsterdam
Prager W (1961) Introduction to mechanics of continua. Ginn & Comp., Boston, MA
Press WH, Teukolsky SA, Vetterling WT, Flannery BP (1992) Numerical recipies, the art of scientific computing. Cambridge University Press, New York
Rabotnov YN (1969) Creep problems in structural members. North-Holland, Amsterdam
Saada AS, Bianchini G (1989) Proceedings of international workshop on constitutive equations for granular non-cohesive soils. Balkema, Amsterdam
Sawczuk A (ed) (1974) Foundations of plasticity and problems of plasticity. In: Proceedings of international symposium on foundation of plasticity. Noordhoff Int. Publ., Leyden
Sawczuk A, Bianchi G (eds) (1985) Plasticity today: modelling, methods and applications. Elsevier Applied Science Publishers, London
Schofield AN, Wroth CP (1968) Critical state soil mechanics. McGraw-Hill, London
Sedov LI (1966) Foundation of the non-linear mechanics of continua. Pergamon, Oxford
Simo JC (1998) Numerical analysis and simulation of plasticity. In: Ciarlet PG, Lions JL (eds) Handbook of numerical analysis, vol 6. Part 3 (Numerical Methods for Solids). Elsevier, Amsterdam
Simo JC, Hughes TJR (1998) Computational inelasticity. Springer, New York

Skrzypek JJ, Hetnarski RB (1993) Plasticity and creep. Theory, example and problems. CRC Press, London
Sokolnikoff IS (1964) Tensor analysis: theory and applications to geometry and mechanics of continua. Wiley, New York
Spencer AJM (1980) Continuum mechanics. Longman, London
Taylor DW (1948) Fundamentals of soil mechanics. Wiley, Chichester, UK
Truesdell C (1977) A first course in rational continuum mechanics, vol 1, general concepts. Academic Press, New York
Truesdell C, Noll W (1965) The nonlinear field theories of mechanics. In: Flugge S (ed) Encyclopedia of Physics, vol III/3. Springer, Berlin
Truesdell C, Toupin R (1960) The classical field theories. In: Flugge S (ed) Encyclopedia of Physics, vol III/1. Springer, Berlin
Vardoulakis I, Sulem J (1995) Bifurcation analysis in geomechanics. Blackie Academic & Profess, London
Voyiadjis GZ, Kattan PI (2005) Damage mechanic (Mechanical Engineering). CRC Press, New York
Wriggers P (2003) Computational contact mechanics. Wiley, Hoboken, NJ
Wu H-C (2004) Continuum mechanics and plasticity. Chapman & Hall/CRC, New York
Ziegler H (1983) An introduction to thermomechanics, 2nd edn. North-Holland, Amsterdam
Zienkiewicz OC (1977) The finite element method, 3rd edn. McGraw-Hill, London
Zyczkowski M (1981) Combined loading in the theory of plasticity. PWN-Polish Scientific Publishers, Warsaw

Research Articles

Aifantis EC (1984) On the microstructural origin of certain inelastic models. J Eng Mater Tech 106:326–330
Alonso EE, Gens A, Josa A (1990) A constitutive model for partially saturated soils. Geotechnique 40:405–430
Amorosi A, Boldini D, Germano V (2007) Implicit integration of a mixed isotropic–kinematic hardening plasticity model for structured clays. Int J Numer Anal Methods Geomech 32:1173–1203
Anand L (1993) A constitutive model for interface friction. Comput Mech 12:197–213
Anand L, Kothari M (1996) A computational procedure for rate-independent crystal plasticity. J Mech Phys Solids 44:525–558
Anjiki T, Oka M, Hashiguchi K (2016) Elastoplastic analysis by complete implicit stress-update algorithm based on the extended subloading surface model. Trans Japan Soc Mech Eng. doi:10.1299/transjsme.16-00029 (in Japanese)
Argyris JH (1965) Elasto-plastic matrix analysis of three dimensional continua. J Roy Aeronaut Soc 69:231–262
Argyris JH, Faust G, Szimma J, Warnke EP, William KJ (1973) Recent developments in the finite element analysis of PCRV. In: Proceedings of 2nd international conference SMIRT. Berlin
Armstrong PJ, Frederick CO (1966) A mathematical representation of the multiaxial Bauschinger effect. CEGB Report RD/B/N 731 (or in Mater High Temp 24:1–26 (2007))
Asaoka A, Nakano M, Noda T (1997) Soil-water coupled behaviour of heavily overconsolidated clay near/at critical state. Soil Found 37(1):13–28
Asaro RJ (1983) Micromechanics of crystals and polycrystals. Adv Appl Mech 23:1–115
Asaro RJ, Needleman A (1985) Texture development and strain hardening in rate dependent polycrystals. Acta Metall 33:923–953
Asaro RJ, Rice JR (1977) Strain localization in ductile single crystals. J Mech Phys Solids 25:309–338

Barlat F, Yoon JW, Cazacu O (2007) On linear transformations of stress tensors for the description of plastic anisotropy. Int J Plast 23:876–896

Bartel T, Menzel A, Svendsen B (2011) Thermodynamic and relaxation-based modeling of the interaction between martensitic phase transformations and plasticity. J Mech Phys Solids 59:1004–1019

Bassani JL, Wu TY (1991) Latent hardening in single crystals II: theory analytical characterization and predictions. Proc Royal Soc London A 435:21–41

Batdorf SB, Budiansky B (1949) A mathematical theory of plasticity based on the concept of slip. NACA Tech Note 1871:1–31

Baumberger T, Heslot F, Perrin B (1994) Crossover from creep to inertial motion in friction dynamics. Nature 30:544–546

Bay N, Wanheim T (1976) Real area of contact and friction stresses at high pressure sliding contact. Wear 38:201–209

Betton J (1986) Application of tensor functions to the formulation of constitutive equations involving damage and initial anisotropy. Eng Fract Mech 25:573–584

Bishop AW, Webb DL, Lewin PI (1965) Undisturbed samples of London clay from the Ashford Common shaft: strength-effective stress relationships. Geotechnique 15:1–31

Bland DR (1957) The associated flow rule of plasticity. J Mech Phys Solids 6:71–78

Borja RI (2004) Cam–Clay plasticity. Part V: a mathematical framework for three-phase deformation and strain localization analyses of partially saturated porous media. Comput Methods Appl Mech Eng 193:5301–5338

Borja RI, Sama KM, Sanz PF (2003) On the numerical integration of three-invariant elastoplastic constitutive models. Comput Methods Appl Mech Eng 192:1227–1258

Borja RI, Tamagnini C (1998) Cam–Clay plasticity. Part III: extension of the infinitesimal model to include finite strains. Comp Meth Appl Mech Eng 155:73–95

Brepols T, Vladimirov IN, Reese S (2014) Numerical comparison of isotropic hypo- and hyperelastic-based plasticity models with application to industrial forming processes. Int J Plast 63:18–48

Brockley CA, Davis HR (1968) The time-dependence of static friction. J Lubr Tech 90:35–41

Budiansky B (1959) A reassessment of deformation theories of plasticity. J Appl Mech 20:259–264

Bureau L, Baumberger T, Caroli C, Ronsin O (2001) Low-velocity friction between macroscopic solids. C R Acad Sci (Paris, Series IV, Different faces of Tribology) 2:699–707

Burland JB (1965) The yielding and dilatation of clay, correspondence. Geotechnique 15:211–214

Butterfield R (1979) A natural compression law for soils (an advance on e-log p′). Geotechnique 29:469–480

Callari C, Auricchio F, Sacco E (1998) A finite-strain Cam–Clay model in the framework of multiplicative elasto-plasticity. Int J Plast 14:1155–1187

Carlson DE, Hoger A (1986) The derivative of a tensor-valued function of a tensor. Quart Appl Math 406:409–423

Castro G (1969) Liquefaction of sands. PhD Thesis, Harvard Soil Mechanics Series 81

Chaboche JL (1982) The concept of effective stress applied to elasticity and viscoplasticity in the presence of anisotropic damage. In: Boehler JP (eds) Mechanical behavior of anisotropic solids. Matrinus Nijhoff Publ., Hague

Chaboche JL (1988) Continuum damage mechanics. In: Part I: general concept; Part II: damage growth, crack initiation, and crack growth. J Appl Mech 55:59–72

Chaboche JL (1989) Constitutive equations for cyclic plasticity and cyclic viscoplasticity. Int J Plast 5:247–302

Chaboche JL (1991) On some modifications of kinematic hardening to improve the description of ratcheting effects. Int J Plast 7:661–678

Chaboche JL (2008) A review of some plasticity and viscoplasticity constitutive theories. Int J Plast 24:1642–1693

Chaboche JL, Dang-Van K, Cordier G (1979) Modelization of the strain memory effect on the cyclic hardening of 316 stainless steel. In: Transactions of the 5th international conference on SMiRT. Berlin, Division L, Paper No L 11/3

Chaboche JL, Rousselier G (1983) On the plastic and viscoplastic constitutive equations, Parts I and II. J Pressure Vessel Tech 165:153–164

Cheng J-H, Kikuchi N (1985) An incremental constitutive relation of uniaxial contact friction for large deformation analysis. J Appl Mech 52:639–648

Christoffersen J, Hutchinson JW (1979) A class of phenomenological corner theories of plasticity. J Mech Phys Solids 27:465–487

Chowdhury EQ, Nakai T, Tawada M, Yamada S (1999) A model for clay using modified stress under various loading conditions with the application of subloading concept. Soils Found 39(6):103–116

Chu CC, Needleman A (1980) Void nucleation effects in biaxially stretched sheets. J Eng Technol ASME 102:249–256

Collins IF, Hilder T (2002) A theoretical framework for constructing elastic/plastic constitutive models of triaxial tests. Int J Nemer Anal Methods Geomech 26:1313–1347

Coombs WM, Crouch RS (2011) Algorithmic issues for three-invariant hyperplastic critical state models. Comput Methods Appl Mech Eng 200:2297–2318

Coombs WM, Crouch RS, Augarde CE (2013) A unique Critical State two-surface hyperplasticity model for fine-grained particulate media. J Mech Phys Solids 61:175–189

Cordebois JP, Sidoroff F (1982a) Damage induced elastic anisotropy. In: Boehler JP (ed) Mechanical behavior of anisotropic solids. Martinuus Nijhoff Publ., pp 761–774

Cordebois JP, Sidoroff F (1982b) Endommagement anisotrope en elasticite et plasticite. J Mech Theory Appl: 45–60

Cosserat E, Cosserat F (1909) Theorie des corps deformables. In: Davaux E (trans) Chwolson OD (ed) Traite de Physique, 2nd edn. Paris, pp 953–1173

Cotter BA, Rivlin RS (1955) Tensors associated with time-dependent stresses. Quart Appl Math 13:177–182

Courtney-Pratt JS, Eisner E (1957) The effect of a tangential force on the contact metallic bodies. Proc Roy Soc A 238:529–550

Cundall P, Board M (1988) A microcomputer program for modeling large-strain plasticity problems. In: Prepare for the 6th international conference numerical methods in geomechanics. Innsbruck, Austria, pp 2101–2108

Curnier A (1984) A theory of friction. Int J Solids Struct 20:637–647

Dafalias YF (1983) Corotational rates for kinematic hardening at large plastic deformations. J Appl Mech 50:561–565

Dafalias YF (1984) The plastic spin concept and a simple illustration of its role in finite plastic transformation. Mech Mater 3:223–233

Dafalias YF (1985a) The plastic spin. J Appl Mech 52:865–871

Dafalias YF (1985b) A missing link in the macroscopic constitutive formulation of large plastic deformations. In: Sawczuk A, Bianchi G (eds) Plasticity today. International symposium on recent trends and results in plasticity. Elsevier, pp 135–151

Dafalias YF (1986) Bounding surface plasticity. I: mathematical foundation and hypoplasticity. J Eng Mech 112:966–987

Dafalias YF (1998) Plastic spin: necessity or redundancy? Int J Plast 14:909–931

Dafalias YF (2011) Finite elastic-plastic deformations: beyond the plastic spin. Theor Appl Mech 38:321–345

Dafalias YF, Herrmann LR (1980) A bounding surface soil plasticity model. In: Proceedings, international symposium of soils under cyclic and transient loadings. University of Swansea, pp 335–345

Dafalias YF, Popov EP (1975) A model of nonlinearly hardening materials for complex loading. Acta Mech 23:173–192

Dafalias YF, Popov EP (1976) Plastic internal variables formalism of cyclic plasticity. J Appl Mech 43:645–651
Dafalias YF, Popov EP (1977) Cyclic loading for materials with a vanishing elastic domain. Nucl Eng Design 41:293–302
Darrieulat M, Piot D (1996) A method of generalized analytical yield surfaces of crystalline materials. Int J Plast 12:575–610
Dashner PA (1986) Invariance considerations in large strain elasto-plasticity. J Appl Mech 53:55–60
de Borst R, Groen AE (1999) Towards efficient and robust elements for 3-D plasticity. Comput Struct 70:23–34
del Peiro G (1979) Some properties of the set of fourth-order tensors, with application to elasticity. J Elast 9:245–261
Derjaguin BV, Push VE, Tolstoi DM (1957) A theory of stick-slipping of solids. In: Proceedings of the conference lubrication and wear (Inst Mech Eng). London, pp 257–268
Denis S, Gautier E, Simon A, Beck G (1985) Stress-phase transformation interactions—basic principles, modelling, and calculation of internal stresses. Mater Sci Technol 1:805–814
Delobelle P, Robinet P, Bocher L (1995) Experimental study and phenomenological modelization of ratchet under uniaxial and biaxial loading on austenitic stainless steel. Int J Plast 11:295–330
Denis S, Gautier E, Simon A, Beck G (1985) Stress-phase transformation interactions–basic principles, modelling, and calculation of internal stresses. Mater Sci Technol 1:805–814
Dettmer W, Reese S (2004) On the theoretical and numerical modelling of Armstrong–Frederic kinematic hardening in the finite strain regime. Compt Meth Appl Mech Eng 193:87–116
Dienes JK (1979) On the analysis of rotation and stress rate in deforming bodies. Acta Mech 32:217–232
Dieterich JH (1978) Time-dependent friction and the mechanism of stick-slip. Pure Appl Geophys 116:790–806
Diteterih JH (1979) Modeling of rock friction 1: experimental results and constitutive equations. J Geophys Res 84:2161–2168
Dokos SJ (1946) Sliding friction under extreme pressure–I. Trans ASME 68:A148–A156
Drucker DC (1951) A more fundamental approach to plastic stress-strain relations. In: Proceedings of the 1st U.S. national congress of applied mechanics (ASME), vol 1. pp 487–491
Drucker DC (1988) Conventional and unconventional plastic response and representation. Appl Mech Rev 41:151–167
Drucker DC, Gibson RE, Henkel DJ (1957) Soil mechanics and workhardening theories of plasticity. Trans Am Soc Civil Eng 122:338–346
Drucker DC, Prager W (1952) Soil mechanics and plastic analysis or limit design. Quart Appl Math 10:157–165
Dunkin JE, Kim DE (1996) Measurement of static friction coefficient between flat surfaces. Wear 193:186–192
Duszek MK, Perzyna P (1991) On combined isotropic and kinematic hardening effects in plastic flow process. Int J Plast 9:351–363
Eidel B, Gruttmann F (2003) Elastoplastic orthotropy at finite strains: multiplicative formulation and numerical implementation. Compt Mater Sci 28:732–742
Ellyin F (1989) An anisotropic hardening rule for elastoplastic solids based on experimental observations. J Appl Mech 56:499–507
Ellyin F, Xia Z (1989) A rate-independent constitutive model for transient non-proportional loading. J Mech Phys Solids 37:71–91
Fardshisheh F, Onat ET (1974) Representations of elastoplastic behavior by means of state variables. In: Sawczuk A (ed) Problems of plasticity. pp 89–115
Ferrero JF, Barrau JJ (1997) Study of dry friction under small displacements and near-zero sliding velocity. Wear 209:322–327
Flanagan DP, Taylor LM (1987) An accurate numerical algorithm for stress integration with finite rotations. Comput Meth Appl Mech Eng 62:305–320

Franciosi P, Zaoui A (1991) Crystal hardening and the issue of uniqueness. Int J Plast 7:295–311
Fredriksson B (1976) Finite element solution of surface nonlinearities in structural mechanics with special emphasis to contact and fracture mechanics problems. Comput Struct 6:281–290
Fukutake K, Ohtsuki M, Sato M (1990) Analysis of saturated dense sand-structure system and comparison with results from shaking table test. Earthq Eng Struct Dyn 19:977–992
Gambin W, Barlat F (1997) Modeling of deformation texture development based on rate independent crystal plasticity. Int J Plast 13:75–85
Gearing BP, Moon HS, Anand L (2001) A plasticity model for interface friction: application to sheet metal forming. Int J Plast 17:237–271
Germain P, Nguyen QS, Suquet P (1983) Continuum thermodynamics. J Appl Mech 50:1010–1020
Ghaei A, Green DE (2010) Numerical implementation of Yoshida–Uemori two-surface plasticity model using a fully implicit integration scheme. Compt Mater Sci 48:195–205
Golub GH (2013) Matrix computations, 4th edn. The Johns Hopkins Univ Press Baltimore, USA
Gotoh M (1985) A class of plastic constitutive equations with vertex effect. Int J Solids Struct 21:1101–1163
Goya M, Ito K (1991) An expression of elastic-plastic constitutive laws incorporating vertex formulation and kinematic hardening. J Appl Mech 58:617–622
Green AE, Naghdi PM (1965) A general theory of an elastic-plastic continuum. Arch Ration Mech Anal 18:251–281
Gudehus G (1973) Elastoplastische stoffgleichungen fur trockenen sand. Ing Arch 42:151–169 (in German)
Gudehus G (1979) A comparison of some constitutive laws for soils under radially symmetric loading and unloading. In: Wittke W (ed) Proceedings of 3rd international conference on numerical methods in geomechanics. Aachen, Balkema (Rotterdam), pp 1309–1323
Gurson AL (1977) Continuum theory of ductile rupture by void nucleation and growth: part I—yield criteria and flow rules for porous media. J Eng Mater Technol 99:2–15
Han C-S, Chung K, Wagoner RH, Oh S-I (2003) A multiplicative finite elasto-plastic formulation with anisotropic yield functions. Int J Plast 19:197–211
Harder J (1999) A crystallographic model for the study of local deformation processes in polycrystals. Int J Plast 15:605–624
Harrysson M, Ristinmaa M (2007) Description of evolving anisotropy at large strains. Mech Mater 39:267–282
Hashiguchi K (1972) On a yielding of frictional materials—a hardening law. In: Proceedings of the 27th annual meeting, JSCE. pp 105–108 (in Japanese)
Hashiguchi K (1974) Isotropic hardening theory of granular media. In: Proceedings of Japan society of civil engineering, vol 227. pp 45–60 (in Japanese)
Hashiguchi K (1977) An expression of anisotropy in a plastic constitutive equation of soils. In: Murayama S, Schofield AN (eds) Constitutive equations of soils (Proc 9th Int Conf Soil Mech Found Eng, Spec Ses. 9). Tokyo, JSSMFE, pp 302–305
Hashiguchi K (1978) Plastic constitutive equations of granular materials. In: Cowin SC, Satake M (eds) Proceedings US–Japan seminar on continuum-mechanical and statistical approaches in the mechanics of granular materials. Sendai, pp 321–329
Hashiguchi K (1980) Constitutive equations of elastoplastic materials with elastic-plastic transition. J Appl Mech 47:266–272
Hashiguchi K (1981) Constitutive equations of elastoplastic materials with anisotropic hardening and elastic-plastic transition. J Appl Mech 48:297–301
Hashiguchi K (1985a) Macrometric approaches -static- intrinsically time-independent. In: Constitutive laws of soils (Proc Discuss Ses 1A, 11th Int Conf Soil Mech Found Eng). San Francisco, pp 25–65
Hashiguchi K (1985b) Subloading surface model of plasticity. In: Constitutive laws of soils (Proc Discuss Ses 1A, 11th Int Conf Soil Mech Found Eng). San Francisco. pp 127–130

Hashiguchi K (1986) Elastoplastic constitutive model with a subloading surface. In: Proceedings of international conference on computational mechanics. pp IV65–70
Hashiguchi K (1988) A mathematical modification of two surface model formulation in plasticity. Int J Solids Struct 24:987–1001
Hashiguchi K (1989) Subloading surface model in unconventional plasticity. Int J Solids Struct 25:917–945
Hashiguchi K (1993a) Fundamental requirements and formulation of elastoplastic constitutive equations with tangential plasticity. Int J Plast 9:525–549
Hashiguchi K (1993b) Mechanical requirements and structures of cyclic plasticity models. Int J Plast 9:721–748
Hashiguchi K (1995) On the linear relations of *V-lnp* and *lnv-lnp* for isotropic consolidation of soils. Int J Numer Anal Methods Geomech 19:367–376
Hashiguchi K (1997) The extended flow rule in plasticity. Int J Plast 13:37–58
Hashiguchi K (1998) The tangential plasticity. Met Mater 4:652–656
Hashiguchi K (2000) Fundamentals in constitutive equation: continuity and smoothness conditions and loading criterion. Soils Found 40(3):155–161
Hashiguchi K (2001a) Description of inherent/induced anisotropy of soils: rotational hardening rule with objectivity. Soils Found 41(6):139–145
Hashiguchi K (2001b) On the thermomechanical approach to the formulation of plastic constitutive equations. Soils Found 41(4):89–94
Hashiguchi K (2002) A proposal of the simplest convex-conical surface for soils. Soils Found 42(3):107–113
Hashiguchi K (2005) Subloading surface model with tangential relaxation. In: Proc Int Symp Plast 05: 259–261
Hashiguchi K (2006) Constitutive model of friction with transition from static- to kinetic-friction (Time dependent subloading-friction model). In: Proceedings of international symposium on plasticity. pp 178–180
Hashiguchi K (2007a) General corotational rate tensor and replacement of material-time derivative to corotational derivative of yield function. Comput Model Eng Sci 17:55–62
Hashiguchi K (2007b) Anisotropic constitutive equation of friction with rotational hardening. In: Proceedings of the 13th international symposium on plasticity & its current applications. pp 34–36
Hashiguchi K (2007c) Extended overstress model for general rate of deformation including impact load. In: Proceedings of the 13th international symposium on plasticity & its current applications. pp 37–39
Hashiguchi K (2007d) Yield condition of soils with tensile yield strength and rotational hardening. Proc Int Conf Compt Exp Eng Sci 07:1441–1446
Hashiguchi K (2008) Verification of compatibility of isotropic consolidation characteristics of soils to multiplicative decomposition of deformation gradient. Soils Found 48:597–602
Hashiguchi K (2011) General interpretations and tensor symbols for pull-back, push-forward and convected derivative. Proc JSME 24th Comp Mech Conf: 669–671
Hashiguchi K (2013a) General description of elastoplastic deformation/sliding phenomena of solids in high accuracy and numerical efficiency: subloading surface concept. Arch Compt Meth Eng 20:361–417
Hashiguchi K (2015a) Subloading-damage constitutive equation. Proc Compt Eng Conf Jpn 20:D-2-4
Hashiguchi K (2015b) Crystal plasticity based on extended subloading surface model. In: Proceeding of the 2nd scientific meeting of Kyushu branch of society of material science. Japan, B17
Hashiguchi K (2015c) Cyclic stagnation of isotropic hardening in metals. In: Proceeding of the 2nd scientific meeting of Kyushu branch of society of material science. Japan, B18
Hashiguchi K (2016a) Exact formulation of subloading surface model: unified constitutive law for irreversible mechanical phenomena in solids. Arch Compt Meth Eng 23:417–447

Hashiguchi K (2016b) Multiplicative finite strain theory based on subloading surface model. Proc Comput Eng Conf: B-8-3

Hashiguchi K (2016c) Loading criterion in return-mapping for subloading surface model. Proc Comput Mech Div: 03–6

Hashiguchi K (2016d) Exact multiplicative finite strain theory based on subloading surface model. Proc Mater Mech Div: GS-26

Hashiguchi K, Chen Z-P (1998) Elastoplastic constitutive equations of soils with the subloading surface and the rotational hardening. Int J Numer Anal Meth Geomech 22:197–227

Hashiguchi K, Mase T (2007) Extended yield condition of soils with tensile strength and rotational hardening. Int J Plast 23:1939–1956

Hashiguchi K, Mase T (2011) Physical interpretation and quantitative prediction of cyclic mobility by the subloading surface model. Jpn Geotech J 6:225–241 (in Japanese)

Hashiguchi K, Okamura K (2014) Subloading-phase transformation model. Proc 27th JSME Comput Mech Div Conf: pp.OS17–1707

Hashiguchi K, Okayasu T, Saitoh K (2005a) Rate-dependent inelastic constitutive equation: the extension of elastoplasticity. Int J Plast 21:463–491

Hashiguchi K, Ozaki S (2007) Constitutive equation of friction with rotational and orthotropic anisotropy. J Appl Mech 10:383–389

Hashiguchi K, Ozaki S (2008a) Constitutive equation for friction with transition from static to kinetic friction and recovery of static friction. Int J Plast 24:2102–2124

Hashiguchi K, Ozaki S (2008b) Anisotropic constitutive equation for friction with transition from static to kinetic friction and vice versa. J Appl Mech 11:271–282

Hashiguchi K, Ozaki S, Okayasu T (2005) Unconventional friction theory based on the subloading surface concept. Int J Solids Struct 42:1705–1727

Hashiguchi K, Protasov A (2004) Localized necking analysis by the subloading surface model with tangential-strain rate and anisotropy. Int J Plast 20:1909–1930

Hashiguchi K, Saitoh K, Okayasu T, Tsutsumi S (2002) Evaluation of typical conventional and unconventional plasticity models for prediction of softening behavior of soils. Geotechnique 52:561–573

Hashiguchi K, Suzuki N, Ueno M (2014) Elastoplastic deformation analysis by return-mapping and consistent tangent modulus tensor based on subloading surface model, (1st Report, Formulation of return-mapping). Trans Japan Soc Mech Eng (in Japanese). doi:10.1299/transjsme.2014smm0083

Hashiguchi K, Tsutsumi S (2001) Elastoplastic constitutive equation with tangential stress rate effect. Int J Plast 17:117–145

Hashiguchi K, Tsutsumi S (2003) Shear band formation analysis in soils by the subloading surface model with tangential stress rate effect. Int J Plast 19:1651–1677

Hashiguchi K, Tsutsumi S (2006) Gradient plasticity with the tangential subloading surface model and the prediction of shear band thickness of granular materials. Int J Plast 22:767–797

Hashiguchi K, Ueno M (1977) Elastoplastic constitutive laws of granular materials, constitutive equations of soils. In: Proceedings of the 9th international conference on soil mechanics and foundation engineering. Specialty session 9, Tokyo, JSSMFE, pp 73–82

Hashiguchi K, Ueno M (2016) Elastoplastic constitutive equation of metals under cyclic loading. Int J Eng Sci 111:86–112

Hashiguchi K, Ueno M, Kuwayama T, Suzuki N, Yonemura S, Yoshikawa N (2016) Constitutive equation of friction based on the subloading-surface concept. Proc Royal Soc London 472:20160212. doi:10.1098/rspa.2016.0212

Hashiguchi K, Ueno M, Ozaki T (2012) Elastoplastic model of metals with smooth elastic-plastic transition. Acta Mech 223:985–1013

Hashiguchi K, Yoshimaru T (1995) A generalized formulation of the concept of nonhardening region. Int J Plast 11:347–365

Hassan S, Kyriakides S (1992) Ratcheting in cyclic plasticity. Part I: uniaxial behavior. J Appl Mech 8:91–116

Hassan T, Taleb T, Krishna S (2008) Influence of non-proportional loading on ratcheting responses and simulations by two recent cyclic plasticity models. Int J Plast 24:1863–1889

Haupt P (1985) On the concept of an intermediate configuration and its application to a representation of viscoelastic-plastic material behavior. Int J Plast 1:303–316

Havner KS (1982) The theory of finite plastic deformation of crystalline solids. In: Hopkins HG, Sewell MJ (eds) Mechanics of solids—Rodney Hill 60th anniversary volume, Pergamon, pp 265–302

Hecker SS (1972) Experimental investigation of corners in yield surface. Acta Mech 13:69–86

Hencky H (1924) Zur Theorie plastischer Deformationen und der hierdurch im Material herforgerufenen Nachspannungen. Z A M M 4:323–334

Higuchi R, Okamura K (2016) Prediction of residual stress change due to cyclic loading—validation of advantage of sub-loading surface model

Hill R (1948a) Theory of yielding and plastic flow of anisotropic metal. Proceedings of royal society, London, A193, 281–297

Hill R (1948b) A variational principle of maximum plastic work in classical plasticity. Quart J Mech Appl Math 1:18–28

Hill R (1959) Some basic principles in the mechanics of solids without a natural time. J Mech Phys Solids 7:225–229

Hill R (1966) Generalized constitutive relations for incremental deformation of metal crystals. J Mech Phys Solids 14:95–102

Hill R (1967) On the classical constitutive relations for elastic/plastic solids. Recent Prog Appl Mech 241–249

Hill R (1968) On the constitutive inequalities for simple materials—1. J Mech Phys Solids 16:229–242

Hill R (1978) Aspects of invariance in solid mechanics. Adv Appl Mech 18:1–75

Hill R (1983) On the intrinsic eigenstates in plasticity with generalized variables. Math Proc Cambridge Phil Soc 93:177–189

Hill R (1990) Constitutive modeling of orthotropic plasticity in sheet metals. J Mech Phys Solids 38:241–249

Hill R, Rice JR (1972) Constitutive analysis of elastic-plastic crystals at arbitrary strain. J Mech Phys Solids 20:401–413

Hoger A, Carlson DE (1984) On the derivative of the square root of a tensor and Guo's theorem. J Elast 14:329–336

Hohenemser K, Prager W (1932) Uber die Ansatze der Mechanik isotroper Kontinua. Z A M M 12:216–226

Holsapple KA (1973) A finite elastic-plastic theory and invariance requirements. Acta Mech 17:277–290

Horowitz F, Ruina A (1989) Slip patterns in a spatially homogeneous fault model. J Geophys Res 94:10279–10298

Hosford WF (1974) A generalized isotropic yield criterion. J Appl Mech 41:607–609

Houlsby GT (1985) The use of a variable shear modulus in elastic-plastic models for clays. Comput Geotech 1:3–13

Houlsby GT, Amorosi A, Rojas E (2005) Elastic moduli of soils dependent on pressure: a hyperelastic formulation. Geotechnique 55(5):383–392

Howe PG, Benson DP, Puddington IE (1955) London-Van der Waals' attractive forces between glass surface. Can J Chem 33:1375–1383

Hughes TJR, Pister KS (1978) Consistent linearization in mechanics of solids and structures. Comput Struct 9:391–397

Hughes TJR, Shakib F (1986) Pseudo-corner theory: a simple enhancement of J_2-flow theory for applications involving non-proportional loading. Eng Comput 3:116–120

Hughes TJR, Winget J (1980) Finite rotation effects in numerical integration of rate consistent equations arising in large-deformation analysis. Int J Numer Meth Eng 15:1862–1867

Hutchinson JW (1976) Bounds and self-consistent estimates for creep of polycrystalline materials. Proc Roy Soc London A 348:101–127

Iai S, Ohtsuki O (2005) Yield and cyclic behaviour of a strain space multiple mechanism model for granular materials. Int J Numer Anal Meth Goemech 29:417–442

Ikegami K (1979) Experimental plasticity on the anisotropy of metals. Proc Euromech Colloquium 115:201–242

Ilyushin AA (1961) On the postulate of plasticity. Appl Math Meek 25:746–752 (Translation of O postulate plastichnosti. Prikladnaya Mathematika i Mekkanika 25:503–507)

Inoue T, Raniecki B (1978) Determination of thermal-hardening stress in steels by use of thermoplasticity theory. J Mech Phys Solids 26:187–212

Inoue T, Watanabe Y, Okamura K, Narazaki M, Shichino H, Ju D-Y, Kanamori H, Ichitani K (2007) Metallo-thermo-mechanical simulation of carburized quenching process by several codes—a benchmark project-. Key Eng Mater 340–341, 1061–1066

Ishihara K, Tatsuoka F, Yasuda S (1975) Undrained deformation and liquefaction of sand under cyclic stresses. Soils Found 15:29–44

Itasca Consulting Group (2006) FLAC3D, "Fast lagrangian analysis of continua in 3 dimensions". Minneapolis, Minnesota, USA

Ito K (1979) New flow rule for elastic-plastic solids based on KBW model with a view to lowering the buckling stress of plates and shells. Tech Report Tohoku Univ 44:199–232

Iwan WD (1967) On a class of models for the yielding behavior of continuous and composite systems. J Appl Mech 34:612–617

Jaumann G (1911) Geschlossenes System physicalisher und chemischer Differentialgesetze, Sitzber. Akad Wiss Wien (IIa) 120:385–530

Jiang Y, Zhang J (2008) Benchmark experiments and characteristic cyclic plasticity deformation. Int J Plast 24:1481–1515

Johnson GR, Cook WH (1983) A constitutive model and data for metals subjected to large strain, high strain rates and high temperatures. In: Proceedings of 7th international symposium on ballistics. The Hague, pp 541–547

Kachanov LM (1958) On rupture time under condition of creep. Izvestia Akademi Nauk SSSR. Otd Tekhn Nauk 8:26–31 (in Russian)

Kame N, Fujita S, Nakatani M, Kusakabe T (2013) Effects of a revised rate- and state-dependent friction law on aftershock triggering model. Tectonophys 600:187–195

Kato S, Sato N, Matsubayashi T (1972) Some considerations on characteristics of static friction of machine tool sideway. J Lubr Tech 94:234–247

Khojastehpour M, Hashiguchi K (2004a) The plane strain bifurcation analysis of soils by the tangential-subloading surface model. Int J Solids Struct 41:5541–5563

Khojastehpour M, Hashiguchi K (2004b) Axisymmetric bifurcation analysis in soils by the tangential-subloading surface model. J Mech Phys Solids 52:2235–2262

Khojastehpour M, Murakami Y, Hashiguchi K (2006) Antisymmetric bifurcation in a circular cylinder with tangential plasticity. Mech Mater 38:1061–1071

Kikuchi N, Oden JT (1988) Contact problem in elasticity: a study of variational inequalities and finite element methods. SIAM, Philadelphia

Kintzel O, Bazar Y (2006) Fourth-order tensors—tensor differentiation with applications to continuum mechanics, part I: classical tensor analysis. Z A M M 86:291–311

Kiyota T, Kozeki J, Sato T, Kuwano S (2009a) Aging effects on small strain shear moduli and liquefaction properties of in-situ frozen and reconstituted sandy soils. Soils Found 49:259–274

Kiyota T, Kozeki J, Sato T, Tsutsumi Y (2009b) Effects of sample disturbance on small strain characteristics and liquefaction properties of Holocene and pleistocene sandy soils. Soils Found 49:509–523

Knockaert R, Chastel Y, Massoni YCE (2000) Rate-independent crystalline plasticity, application to FCC materials. Int J Plast 16:179–198

Kobayashi M, Ohno N (2002) Implementation of cyclic plasticity models based on a general form of kinematic hardening. Int J Numer Meth Eng 53:2217–2238

Kohgo Y, Nakano M, Miyazaki T (1993) Verification of the generalized elastoplastic model for unsaturated soils. Soil Found 33(4):64–73

Koiter WT (1953) Stress-strain relations, uniqueness and variational theories for elastic-plastic materials with a singular yield surface. Quart Appl Math 11:350–354

Kolymbas D, Wu W (1993) Introduction to plasticity. In: Modern approaches to plasticity. Elsevier, pp 213–224

Kratochvil J (1971) Finite-strain theory of crystalline elastic-inelastic materials. J Appl Phys 42:1104–1108

Krieg RD (1975) A practical two surface plasticity theory. J Appl Mech 42:641–646

Krieg RD, Key SW (1976) Implementation of a time dependent plasticity theory into structural computer programs. In: Strickin JA, Saczlski KJ (eds) Constitutive equations in viscoplasticity: computational and engineering aspects. AMD-20, ASME

Krieg RD, Krieg DB (1977) Accuracies of numerical solution methods for the elastic-perfectly plastic models. J Pressure Vessel Tech 99:510–515

Kroner E (1960) Allgemeine kontinuumstheoreie der versetzungen und eigenspannnungen. Arch Ration Mech Anal 4:273–334

Kuroda M (1997) Interpretation of the behavior of metals under large plastic shear deformations: a macroscopic approach. Int J Plast 13:359–383

Ladevéze P, Lemaitre JA (1984) Damage effective stress in quasi unilateral conditions. In: 16th international congress for applied mechanics. Lyngby, Denmark

Lee EH (1969) Elastic-plastic deformation at finite strain. J Appl Mech 36:1–6

Lee J-Y, Barlat F, Lee M-G (2015) Constitutive and friction modeling for accurate springback analysis of advanced high strength steel sheets. Int J Past 71:113–135

Lee EH, Liu DT (1967) Finite-strain elastic-plastic theory with application to plane-wave analysis. J Appl Phys 38:19–27

Lemaitre JA (1971) Evaluation of dissipation and damage in metals subjected to dynamic loading. In: Proceedings of international congress for mechanics behavior of materials 1 (ICM 1). Kyoto

Lemaitre JA (2001) Handbook of materials behavior models. Academic Press, San Diego

Lemaitre JA, Dosmorat R, Sauzay M (2000) Anisotropic damage law of evolution. Eur J Mech A/Solids 19:182–208

Lion A (2000) Constitutive modeling in finite thermoviscoplasticity: a physical approach based on nonlinear rheological models. Int J Plast 16:469–494

Loret B (1983) On the effects of plastic rotation in the finite deformation of anisotropic elastoplastic materials. Mech Mater 2:287–304

Lubarda VA (2004) Constitutive theories based on the multiplicative decomposition of deformation gradient: thermoplasticity, elastoplasticity, and biomechanics. Appl Mech Rev 57:95–108

Lubarda VA, Lee EH (1981) A correct definition of elastic and plastic deformation and its computational significance. J Appl Mech 48:35–40

Lubliner J (1984) A maximum-dissipation principle in generalized plasticity. Acta Mech 52:225–237

Mahnken R, Wolff M, Schneidt A, Bohm M (2012) Multi-phase transformations at large strains—thermodynamic framework and simulation. Int J Plast 39:1–26

Mandel J (1964) Contribution theorique a l'eude de l'ecrouissage et des lois de l'ecoulement plastique. Proc 11th Int Congr Appl Mech: 502–509

Mandel J (1965) Generalisation de la theorie de plasticite de W.T. Koiter. Int J Solids Struct 1:273–295

Mandel J (1971) Plastidite classique et viscoplasticite. In: Course & Lectures, No. 97, Int Center Mech Sci. Udine, Springer, Heidelberg

Mandel J (1973) Equations constitutives directeurs dans les milieux plastiques at viscoplastiques. Int J Solids Struct 9:725–740

Mandel J (1974) Director vectors and constitutive equations for plastic and viscoplastic media. In: Sawczuk A (ed) Problems of plasticity (Proc Int Symp Foundation of Plasticity). Noordhoff Int Publ., Leyden, pp 135–141

Mase T, Hashiguchi K (2009) Numerical analysis of footing settlement problem by subloading surface model. Soils Found 49:207–220

Masing G (1926) Eigenspannungen und Verfestigung beim Messing. Proc 2nd Int Congr Appl Mech Zurich: 332–335

Matsuoka H, Nakai T (1974) Stress-deformation and strength characteristics of soil under three different principal stress. In: Proceedings of the Japan society of civil engineers, vol 232. pp 59–70

Matsuoka H, Yao YP, Sun DA (1999) The Cam–Clay model revised by SMP criterion. Soils Found 39(1):81–95

McDowell DL (1985) An experimental study of the structure of constitutive equations for nonproportional cyclic plasticity. J Eng Mater Tech 107:307–315

McDowell DL (1989) Evaluation of intersection conditions for two-surface plasticity theory. Int J Plast 5:29–50

Mengoni M, Ponthot JP (2015) A generic anisotropic continuum damage model integration scheme adaptable to both ductile damage and biological damage-like situations. Int J Plast 66:46–70

Menzel A, Steinmann P (2003) On the spatial formulation of anisotropic multiplicative elasto-plasticity. Compt Meth Appl Mech Eng 192:3431–3470

Menzel A, Ekh M, Runesson K, Steinmann P (2005) A framework for multiplicative elastoplasticity with kinematic hardening coupled to anisotropic damage. Int J Plast 21:397–434

Michalowski R, Mroz Z (1978) Associated and non-associated sliding rules in contact friction problems. Archiv Mech 30:259–276

Miehe C (1996) Numerical computation of algorithmic (consistent) tangent moduli in large-strain computational inelasticity. Comput Methods Appl Mech Eng 134:223–240

Miehe C, Schroder J (2001) A comparative study of stress update algorithms for rate-independent and rate-dependent crystal plasticity. Int J Numer Meth Eng 50:273–298

Mindlin RD (1963) Influence of couple-stresses on stress concentrations. Exp Mech 3:1–7

Mooney M (1940) A theory of large elastic deformation. J Appl Phys 11(9):582–592

Mroz Z (1966) On forms of constitutive laws for elastic-plastic solids. Arch Mech Stos 18:3–35

Mroz Z (1967) On the description of anisotropic workhardening. J Mech Phys Solids 15:163–175

Mroz Z (1976) A non-linear hardening model and its application to cyclic plasticity. Acta Mech 25:51–61

Mroz Z, Norris VA, Zienkiewicz OC (1981) An anisotropic, critical state model for soils subject to cyclic loading. Geotechnique 31:451–469

Mroz Z, Stupkiewicz S (1994) An anisotropic friction and wear model. Int J Solids Struct 31:1113–1131

Muhlhaus HB, Vardoulakis I (1987) The thickness of shear bands in granular materials. Geotechnique 37:271–283

Murakami S (1988) Mechanical modelling of material damage. J Appl Mech 55:280–286

Murakami S, Ohno N (1981) A continuum theory of creep and creep damage. In: Proceedings of 3rd IUTAM symposium on creep in structures. pp 422–444

Nagtegaal JC, De Jong JE (1982) Some aspects of non-isotropic workhardening in finite strain plasticity. In: Lee EH, Mallett RL (eds) Plasticity of metals and finite strain: theory, experiment and computation. Division of Applied Mechanics, Stanford University and Department of Mechanical Engineering, Rensselaer Polytechnic Institute, pp 65–102

Nakada Y, Keh AS (1966) Latent hardening in iron single crystals. Acta Metall 14:961–973

Nakai T, Hinokio M (2004) A simple elastoplastic model for normally and over consolidated soils with unified material parameters. Soils Found 44(2):53–70

Nakai T, Mihara Y (1984) A new mechanical quantity for soils and its application to elastoplastic constitutive models. Soils Found 24(2):82–941

Needleman A, Rice JR (1978) Limits to ductility set by plastic flow localization. In: Koistinen DP, Wang N-M (eds) Mechanics of sheet metal forming. Plenum press, New York, pp 237–265

Needleman A, Tvergaard V (1985) Material strain-rate sensitivity in round tensile bar. In: Proceedings of international symposium on plastic instability. Pressure de l'cole nationale des Ponts et Shausseses, Paris, pp 251–262

Niemunis A, Cudny M (1998) On hyperelasticity for clays. Comput Geotech 23:221–236

Nova R (1977) On the hardening of soils. Arch Mech Stos 29:445–458

Oden JT, Martines JAC (1986) Models and computational methods for dynamic friction phenomena. Comput Meth Appl Mech Eng 52:527–634

Oden JT, Pires EB (1983a) Algorithms and numerical results for finite element approximations of contact problems with non-classical friction laws. Comput Struct 19:137–147

Oden JT, Pires EB (1983b) Nonlocal and nonlinear friction laws and variational principles for contact problems in elasticity. J Appl Mech 50:67–76

Ogden RW (1982) Elastic deformations of rubberlike solids. In: Hopkins HG, Sewell MJ (eds) Mechanics of solids: Rodney Hill 60th anniversary volume. Pergamon, Oxford, UK, pp 499–537

Ohno N (1982) A constitutive model of cyclic plasticity with a non-hardening strain region. J Appl Mech 49:721–727

Ohno N, Abdel-Karim M (2000) Uniaxial ratchetting of 316FR steel at room temperature—part II: constitutive modeling and simulation. J Eng Mater Tech 122:35–41

Ohno N, Kachi Y (1986) A constitutive model of cyclic plasticity for nonlinearly hardening materials. J Appl Mech 53:395–403

Ohno N, Tsuda M, Kamei T (2013) Elastoplastic implicit integration algorithm applicable to both plane stress and three-dimensional stress states. Finite Elem Anal Des 66:1–11

Ohno N, Wang JD (1993) Kinematic hardening rules with critical state of dynamic recovery, part I: Formulation and basic features for ratcheting behavior. Part II: application to experiments of ratcheting behavior. Int J Plast 9:375–403

Oka F, Yashima A, Taguchi A, Yamashita S (1999) A cyclic elasto-plastic constitutive model for sand considering a plastic-strain dependence of the shear modulus. Geotechnique 49:661–680

Okahara M, Takagi S, Mori H, Koike S, Tatsuda M, Tatsuoka F, Morimoto H (1989) Largescale plane strain bearing capacity tests of shallow foundation on sand (part 1). In: Proceedings of 24th annual meeting Japan. Society geotechnical engineering, pp 1239–1242 (in Japanese)

Okamura K (2006a) Reviews and perspective on hardening simulation. In: Proceedings of the Japan institute of metals and materials and iron and steel institute, Japan, Kyushu-branch, pp 1–12 (in Japanese)

Okamura K (2006b) Actuarity and scope on simulation of heat treatment: I: material properties and database. J Soc Mater Sci Japan 55:529–535 (in Japanese)

Okamura K (2015) Simulation of heat treatment process. Netu-shori (Heat Treatment) 55:86–92 (in Japanese)

Okamura K, Kawashima H (1988a) Finite element analysis of thermal stress in heat treatment. Netu-shori (Heat Treatment) 28:141–148 (in Japanese)

Okamura K, Kawashima H (1988b) Analysis of residual deformation of a gear during quenching. In: Proceedings of the 32nd Japan congress on materials research. pp 323–329 (in Japanese)

Okamura K, Yamamoto K, Fukumoto M (2005) Material properties for quenching simulation and assessment on computational results. Proc 3rd Asian Conf Heat Treat Mater: 353–355

Oldroyd JG (1950) On the formulation of rheological equations of state. Proc Roy Soc London A200: 523–541

Onsager L (1931) Reciprocal relations in irreversible processes. I. Phys Rev 37:405–426; and Reciprocal relations in irreversible processes. II. Phys Rev 38:2265–2279

Ortiz M, Popov EP (1985) Accuracy and stability of integration algorithms for elastoplastic constitutive relations. Int J Numer Meth Eng 21:1561–1576

Ortiz M, Simo JC (1986) An analysis of a new class of integration algorithms for elastoplastic constitutive relations. Int J Numer Meth Eng 23:353–366
Ozaki S, Hashiguchi K (2010) Numerical analysis of stick-slip instability by a rate-dependent elastoplastic formulation for friction. Tribol Int 43:2120–2133
Ozaki S, Hikida K, Hashiguchi K (2012) Elastoplastic formulation for friction with orthotropic anisotropy and rotational hardening. Int J Solids Struct 49:648–657
Pan J, Rice JR (1983) Rate sensitivity of plastic flow and implications for yield surface vertices. J Mech Phys Solids 19:973–987
Peirce D, Asaro JR, Needleman A (1982) Overview 21: an analysis of nonuniform and localized deformation in ductile single crystals. Act Metall 30:1087–1119
Peirce D, Asaro JR, Needleman A (1983) Overview 32: material rate dependence and localized deformation in crystal solids. Act Metall 31:1951–1976
Pérez-Foguet A, Rodréguez-Ferran A, Huerta A (2000a) Numerical differentiation for non-trivial consistent tangent matrices: an application to the MRS-Lade model. Int J Numer Meth Eng 48:159–184
Pérez-Foguet A, Rodríguez-Ferran A, Huerta A (2000b) Numerical differentiation for local and global tangent operators in computational plasticity. Compt Meth Appl Mech Eng 189:277–296
Pérez-Foguet A, Rodríguez-Ferran A, Huerta A (2001) Consistent tangent matrices for substepping schemes. Compt Meth Appl Mech Eng 190:4627–4647
Perić D, Owen RJ (1992) Computational model for 3-D contact problems with friction based on the penalty method. Int J Numer Meth Eng 35:1289–1309
Perzyna P (1963) The constitutive equations for rate sensitive plastic materials. Quart Appl Math 20:321–332
Perzyna P (1966) Fundamental problems in viscoplasticity. Adv Appl Mech 9:243–377
Pietruszczak St, Mroz Z (1983) On hardening anisotropy of K0-consolidated clays. Int J Numer Anal Methods Geomech 7:19–38
Pietruszczak ST, Niu X (1993) On the description of localized deformation. Int J Numer Anal Methods Geomech 17:791–805
Pinsky PM, Ortiz M, Pister KS (1983) Numerical integration of rate constitutive equations in finite deformation analysis. Comput Meth Appl Mech Eng 193:5223–5256
Prager W (1945) Strain hardening under combined stress. J Appl Phys 16:837–840
Prager W (1949) Recent development in the mathematical theory of plasticity. J Appl Mech 20:235–241
Prager W (1956) A new methods of analyzing stresses and strains in work hardening plastic solids. J Appl Mech 23:493–496
Prager W (1961) Linearization in visco-plasticity. Ing Archiv 15:152–157
Rabinowicz E (1951) The nature of the static and kinetic coefficients of friction. J Appl Phys 22:1373–1379
Rabinowicz E (1958) The intrinsic variables affecting the stick-slip process. Proc Phys Soc 71:668–675
Rice JR, Lapusta N, Ranjith K (2001) Rate and state dependent friction and the stability of sliding between elastically deformable solids. J Mech Phys Solids 49:1865–1898
Rice JR, Ruina AL (1983) Stability of steady frictional slipping. J Appl Mech 50:343–349
Rice JR, Tracey DM (1973) Computational fracture mechanics. In: Feves SJ (ed) Proceedings of symposium numerical methods in structural mechanics. Urbana, Illinois, Academic Press, New York, p 585
Rivlin RS (1948) Large elastic deformations of isotropic materials. IV. Further developments of the general theory. Phil Trans Royal Soci London Ser A Math Phys Sci 241(835):379–397
Rohde J, Jeppsson A (2000) Literature review of heat treatment simulations with respect to phase transformation, residual stresses and distortion. Scand J Metall 29:47–62
Roscoe KH, Burland JB (1968) On the generalized stress-strain behaviour of 'wet' clay. In: Engineering plasticity. Cambridge University Press, Cambridge, pp 535–608

Rudnicki JW, Rice JR (1975) Conditions for the localization of deformation in pressure-sensitive dilatant materials. J Mech Phys Solids 23:371–394
Ruina AL (1980) Friction Laws and instabilities: quasistatic analysis of some dry frictional behavior. PhD Thesis, Brown University, Providence
Ruina AL (1983) Slip instability and state variable friction laws. J Geophys Res 88:10359–10370
Satake M (1972) A proposal of new yield criterion for soils. Proc Jpn Soc Civil Eng 189:79–88 (in Japanese)
Scholz CH (1998) Rate-and state-variable friction law. Nature 391:37–41
Seguchi Y, Shindo A, Tomita Y, Sunohara M (1974) Sliding rule of friction in plastic forming of metal. Compt Meth Nonlinear Mech: 683–692
Sekiguchi H, Ohta H (1977) Induced anisotropy and its time dependence in clays. Constitutive equations of soils (Proc spec session 9, 9th ICSFME). Tokyo, pp 229–238
Seth BR (1964) Generalized strain measure with applications to physical problems. In: Second-order effects inelasticity, plasticity, and fluid dynamics. Pergamon, Oxford
Sewell MJ (1973) A yield-surface comer lowers the buckling stress of an elastic-plastic plate under compression. J Mech Phys Solids 21:19–45
Sewell M J (1974) A plastic flow at a yield vertex. J Mech Phys Solids 22:469–490
Sheng D, Sloan SW, Yu HS (2000) Aspects of finite element implementation of critical state models. Comput Mech 26:185–196
Shield RT, Ziegler H (1958) On Prager's hardening rule. Z ang Math Mech 9:260–276
Shutov AV, Ihlemann J (2014) Analysis of some basic approaches to finite strain elasto-plasticity in view of reference change. Int J Plast 63:193–197
Siddiquee MSA, Tanaka T, Tatsuoka F, Tani K, Morimoto T (1999) Numerical simulation of bearing capacity characteristics of strip footing on sand. Soils Found 39(4):93–109
Simo JC (1987) A J_2-flow theory exhibiting a corner-like effect and suitable for large-scale computation. Comput Meth Appl Mech Eng 62:169–194
Simo JC, Kennedy JG, Govindjee S (1988) Non-smooth multisurface plasticity and viscoplasticity—loading unloading conditions and numerical algorithms. Int J Numer Meth Eng 26:2161–2185
Simo JC, Meschke G (1993) A new class of algorithms for classical plasticity extended to finite strains. Appl Geomater Comput Mech 11:253–278
Simo JC, Ortiz M (1985) A unified approach to finite deformation elastoplasticity based on the use of hyperelastic constitutive equations. Compt Meth Appl Mech Eng 49:221–245
Simo JC, Taylor RL (1985) Consistent tangent operators for rate-independent elastoplasticity. Comput Meth Appl Mech Eng 48:101–118
Simo JC, Taylor RL (1986) A return mapping algorithm for plane stress elastoplasticity. Int J Numer Meth Eng 22:649–670
Skempton AW, Brown JD (1961) A landslide in boulder clay at Selset, Yorkshire. Geotechnique 11:280–293
Sloan SW, Randolph MF (1982) Numerical prediction of collapse loads using finite element methods. Int J Numer Anal Methods Geomech 6:47–76
Stallebrass SE, Taylor RN (1997) The development and evaluation of a constitutive model for the prediction of ground movements in overconsolidated clay. Geotechnique 47:235–253
Stark TD, Ebeling RM, Vettel JJ (1994) Hyperbolic stress-strain parameters for silts. J Geotech Eng 120:420–441
Steinmann P, Larsson R, Runesson K (1997) On the localization properties of multiplicative hyperelasto-pastic continua with strong discontinuities. Int J Solids Struct 8:969–990
Stribeck R (1902) Die Wesentlichen Eigenschaften der Gleit- und Rollenlager. Z Verein Deut Ing 46:1341–1348 (in German)
Stupkiewicz S, Mroz Z (2003) Phenomenological model of real contact area evolution with account for bulk plastic deformation in metal forming. Int Plast 19:323–344
Sun L, Wagoner RH (2011) Complex unloading behavior: nature of the deformation and its consistent constitutive representation. Int J Past 27:1126–1144

Suzuki N, Hashiguchi K, Ueno M (2014) Elastoplastic deformation analysis by return-mapping and consistent tangent modulus tensor based on subloading surface model (2nd Report, Deformation analyses of machine elements). Trans Jpn Soc Mech Eng. doi:10.1299/transjsme.2014smm0356 (in Japanese)

Tamagnini C, Castellanza R, Nova R (2002) A generalized backward Euler algorithm for the numerical integration of an isotropic hardening elastoplastic model for mechanical and chemical degradation of bonded geomaterials. Int J Numer Anal Meth Geomech 26:963–1004

Tanahashi T (1985) Mechanics of continua. Rikoh Tosho Publ, Tokyo (in Japanese)

Tanaka T, Kawamoto O (1988) Three dimensional finite element collapse analysis for foundations and slopes using dynamic relaxation. Proc Numer Meth Geomech Innsbruck: 1213–1218

Tanaka T, Sakai T (1993) Progressive failure effect of trap-door problems with granular materials. Soils Found 33(1):11–22

Tani K (1986) Mechanism of bearing capacity of shallow foundation. Master thesis, University of Tokyo (in Japanese)

Tatsuoka F, Ikuhara O, Fukushima S, Kawamura T (1984) On the relation of bearing capacity of shallow footing on model sand ground and element test strength. In: Proceedings of symposium on assessment of deformation and failure. Strength of sandy soils and sand grounds, Japan (Soc Geotech Eng). pp 141–148 (in Japanese)

Taylor GI (1938) Plastic strain in metals. J Inst Metals 62:307–324

Topolnicki M (1990) An elasto-plastic suboading surface model for clay with isotropic and kinematic mixed hardening parameters. Soils Found 30(2):103–113

Truesdell C (1955) Hypo-elasticity. J Ration Mech Anal 4:83–133

Tsutsumi S, Hashiguchi K (2005) General non-proportional loading behavior of soils. Int J Plast 21:1941–1969

Tsutsumi S, Toyosada M, Hashiguchi K (2006) Extended subloading surface model incorporating elastic limit concept. Proc Plast 06:217–219

Tvergaard V (1982) On localization in ductile materials containing spherical voids. Int J Fract 18:237–252

Tvergaard V, Needleman A (1984) Analysis of the cup-cone fracture in a round tensile bar. Acta Metall 32:157–169

Van der Giessen E (1989) Micromechanical and thermodynamic aspects of the plastic spin. Int J Plast 7:365–386

Vermeer PA (1982) A simple shear band analysis using compliances. In: Proceedings of the IUTAM symposium. Deformation and failure of granular materials. Balkema, Amsterdam, pp 493–499

Vladimirov IN, Pietryga MP, Reese S (2008) On the modeling of nonlinear kinematic hardening at finite strains with application to springback comparison of time integration algorithm. Int J Numer Meth Eng 75:1–28

Vladimirov IN, Pietryga MP, Reese S (2010) Anisotroipc finite elastoplasticity with nonlinear kinematic and isotropic hardening and application to shear metal forming. Int J Plast 26:659–687

von Mises R (1923) Mechanik der plastischen Formanderung von Kristallen. Z angew Math Mech 8:161–185

Wallin M, Ristinmaa M (2005) Deformation gradient based kinematic hardening model. Int J Plast 21:2025–2050

Wallin M, Ristinmaa M, Ottesen NS (2003) Kinematic hardening in large strain plasticity. Europ J Mech A/Solids 22:341–356

Wang CCA (1970) A new representation theorem for isotropic functions: an answer to Professor G. F. Smith's criticism of my paper on representations for isotropic functions. Arch Ratl Mech Anal 36:166–223

Wang Z-L, Dafalias YF, Shen C-K (1990) Bounding surface hypoplasticity model for sand. J Eng Mech 116:983–1001

Wang Z-Q, Dui G-S (2008) Two-point constitutive equations and integration algorithms for isotropic-hardening rate-independence elastoplastic materials in large deformation. Int J Numer Methods Eng 75:1435–1456

Wagoner RH, Lim H, Lee M-G (2013) Advanced isuuess in springback. Int J Plast 45:3–20

Weber G, Anand L (1990) Finite deformation constitutive equations and a integration procedure for isotropic, hyperelastic-viscoplastic solids. Comput Mech Appl Mech Eng 79:173–202

Wesley LD (1990) Influence of structure and composition on residual soils. J Geotech Eng 116:589–603

Wilde P (1977) Two invariants depending models of granular media. Arch Mech Stos 29:799–809

Wilkins ML (1964) Calculation of elastoplastic flow. In: Alder B et al (eds) Methods of computational physics, vol 3. Academic Press, New York

White PS (1975) Elastic-plastic solids as simple materials. Quart J Mech Appl Math 28:483–496

Wongsaroj J, Soga K, Mair RJ (2007) Modeling of long-term ground response to tunneling under St James's Park, London. Geotechnique 57:75–90

Wriggers P, Vu Van T, Stein E (1990) Finite element formulation of large deformation impact-contact problems with friction. Comput Struct 37:319–331

Xia Z, Ellyin F (1994) Biaxial ratcheting under strain or stress-controlled axial cycling with constant hoop stress. J Appl Mech 61:422–428

Xiao H (1995) Unified explicit basis-free expressions for time rate and conjugate stress of an arbitrary Hill's strain. Int Solids Struct 32:3327–3340

Xiao H, Bruhns OT, Meyers A (1997) Logarithmic strain. Logarithmic spin and logarithmic rate. Acta Mech 124:89–105

Xiao H, Bruhns OT, Meyers A (1999) Existence and uniqueness of the integrable-exactly hypoelastic equation $\overset{\circ}{\boldsymbol{\tau}}{}^{*} = \lambda(\mathrm{tr}\, D)I + 2\mu D$ and its significance to finite inelasticity. Acta Mech 138:31–50

Xu B, Jiang Y (2004) A cyclic plasticity model for single crystals. Int J Plast 20:2161–2178

Yamada S, Takamori T, Sato K (2010) Effects on reliquefaction resistance produced by changes in anisotropy during liquefaction. Soils Found 50:9–25

Yamada Y, Yoshimura N, Sakurai T (1968) Plastic stress-strain matrix and its application for the solution of elastic-plastic problems by finite element method. Int J Mech Sci 10:343–354

Yamakawa Y, Hashiguchi K (2011) Elastoplasticity theory: numerical methods for finite elastoplastic constitutive equation, part 3. In: Science of machine, vol 63. Yokendo, Tokyo, pp 251–254 (in Japanese)

Yamakawa Y, Hashiguchi K, Ikeda K (2010a) Implicit stress-update algorithm for isotropic Cam–Clay model based on the subloading surface concept at finite strains. Int J Plast 26:634–658

Yamakawa Y, Yamaguchi Y, Hashiguchi K, Ikeda K (2010b) Formulation and implicit stress-update algorithm of the extended subloading surface Cam–Clay model with kinematic hardening for finite strains. J Appl Mech 13:411–412 (in Japanese)

Yamamoto Y (1998) Evaluation of seismic behavior of clay and sand grounds. PhD Thesis, Yamaguchi University

Yoshida F, Hamasaki H, Uemori T (2015) Modeling of anisotropic hardening of sheet metals including description of the Bauschinger effect. Int J Plast 75:170–188

Yoshida K, Kuroda M (2012) Comparison of bifurcation and imperfection analyses of localized necking in rate-independent polycrystalline sheets. Int J Solids Struct 49:2073–2084

Yoshida F, Uemori T (2002a) Elastic-plastic behavior of steel sheets under in-plane cyclic tension-compression at large strain. Int J Plast 18:633–659

Yoshida F, Uemori T (2002b) A model of large-strain cyclic plasticity describing the Bauschinger effect and workhardening stagnation. Int J Plast 18:661–686

Yoshida F, Uemori T (2003) A model of large-strain cyclic plasticity and its application to springback simulation. Int J Mech Sci 45:1687–1702

Zamiri A, Pourbogharat F (2010) A novel yield function for single crystal based on combined constraints optimization. Int J Plast 26:731–746

Zamiri A, Pourbogharat F, Barlat F (2007) An effective computational algorithm for rate-independent crystal plasticity based on a single crystal yield surface with an application to tube hydroforming. Int J Plast 23:1126–1147

Zaremba S (1903) Su une forme perfectionnee de la theorie de la relaxation. Bull Int Acad Sci 594–614 (in French)

Zbib HM, Aifantis EC (1988) On the concept of relative and plastic spins and its implications to large deformation theories. Part I: hypoelasticity and vertex-type plasticity. Acta Mech 75:15–33

Zhang F, Ye B, Noda T, Nakano M, Nakai K (2007) Explanation of cyclic mobility of soils: approach by stress-induced anisotropy. Soils Found 47:635–648

Ziegler H (1959) A modification of Prager's hardening rule. Quart Appl Phys 17:55–60

Index

K. Hashiguchi, *Foundations of Elastoplasticity: Subloading Surface Model*,
DOI 10.1007/978-3-319-48821-9

T

U

V

Printed by Printforce, the Netherlands